Space Antenna Handbook

Space Antenna Handbook

Editors

William A. Imbriale
Jet Propulsion Laboratory, California Institute of Technology, USA

Steven (Shichang) Gao
Surrey Space Centre, University of Surrey, UK

Luigi Boccia
University of Calabria, Italy

WILEY

A John Wiley & Sons, Ltd., Publication

This edition first published 2012

Registered office
John Wiley & Sons Ltd, The Atrium, Southern Gate, Chichester, West Sussex, PO19 8SQ, United Kingdom

For details of our global editorial offices, for customer services and for information about how to apply for permission to reuse the copyright material in this book please see our website at www.wiley.com.

Library of Congress Cataloging-in-Publication Data

Space antenna handbook/editors, William A. Imbriale, Steven Gao, Luigi Boccia.
p. cm.
Includes bibliographical references and index.
ISBN 978-1-119-99319-3 (cloth)
1. Antennas (Electronics) 2. Space vehicles–Radio antennas. 3. Artificial satellites in telecommunication. I. Imbriale, W. A. (William A.) II. Gao, Steven. III. Boccia, Luigi.
TK7871.6.S63 2012
629.47′43—dc23

2011046036

A catalogue record for this book is available from the British Library.

ISBN: 978-1-119-99319-3

Set in 10/12 pt, Times by Thomson Digital, Noida, India

Contents

Preface

Due to the special environment of space and the launch vehicle dynamics to get there, spacecraft antenna requirements and design are quite different from those of terrestrial antennas. There are few books focusing on the special needs of space antennas. One such work is *Spaceborne Antennas for Planetary Exploration* (John Wiley & Sons, Inc., 2006), which covers the work from JPL/NASA only. Thus, there is a need for a comprehensive book which presents an up-to-date development of space antennas from leading engineers around the world.

This book addresses a broad range of topics on antennas for space applications. The aim of the book is two-fold. First, it introduces the reader to the fundamental methodologies of space antenna design, modeling, and analysis along with the state of the art and future technological developments. Each of the topics is specialized and contextualized to the space sector. Furthermore, case studies are provided in chapters to demonstrate how to design and implement antennas in practical scenarios. Next, the book presents a detailed review of antenna designs for some popular applications such as satellite communications, spaceborne synthetic aperture radar (SAR), global navigation satellite system (GNSS) receivers, radio astronomy, small satellites, and deep space applications.

Since the book covers such a wide range of topics, from basic principles to technologies and practical case studies, it is suitable for a wide range of audiences including beginners, students, researchers, and experienced engineers. The technical terms in the text assume that the reader is familiar with basic engineering and mathematical concepts as well as material typically found in a senior-level course in electromagnetics.

The book is divided into three sections: antenna development; space antenna technology; and space antennas for specific applications. The first section covers antenna basics and modeling as well as the specialized needs for space antennas with respect to the environment and materials including the mechanical and thermal considerations required for space antennas. There is also a chapter on system architecture depicting the critical role antennas play in the overall spacecraft design. The second section describes in detail technologies associated with mesh reflector antennas, array antennas, and printed reflectarray antennas. It provides a historical perspective as well as highlighting the emerging technologies. The third section covers the specific applications of satellite communications, spaceborne SAR, GNSS receivers, radio astronomy, small satellites, and deep space. The concluding chapter provides a broad outlook into the future development of space antennas. Thus the reader is provided with a comprehensive and logical path from the basics of space antenna development to the specific aspects related to individual applications.

William A. Imbriale, Steven (Shichang) Gao and Luigi Boccia
February 2012

Acknowledgments

The editors would like to express their thankfulness to all chapters' authors whose valuable contribution had brought the book project to reality. The support from Richard Davies, Anna Smart, Susan Barclay, and Tiina Ruonamaa from Wiley was also extremely important and it is most appreciated. The editors are also grateful to Abhishan Sharma, Neville Hankins and to all typesetters for their careful work.

William Imbriale would like to thank his wife Carol for her patience and understanding during the editing of the book. Steven Gao would like to express his deep appreciation to his wife, Jun, and his daughter, Karen, for their support. Luigi Boccia would like to manifest his gratitude and admiration to his wife, MariAntonietta, his son, Antonio Francesco, and his daughter, Caterina Dilia, for their encouragement and loving care. He also would like to honorably mention his parents for their guidance and strength.

William A. Imbriale, Steven (Shichang) Gao and Luigi Boccia
February 2012

Acronyms

ABC	Absorbing Boundary Conditions
AFR	Array-Fed Reflector
AIT	Assembly, Integration and Test
AIT-AIV	Assembly, Integration and Test–Assembly, Integration and Validation
AMSU-A/B	Advanced Microwave Sounding Unit version A/B
BFN	Beam-Forming Network
CALISTO	Cryogenic Aperture Large Infrared Space Telescope Observatory
CEM	Computational Electromagnetics
CFIE	Combined Fields Integral Equation
CFRP	Carbon Fibre Reinforced Plastic
CFRS	Carbon Fibre Reinforced Silicon
CMB	Cosmic Microwave Background
COBE	Cosmic Background Explorer Satellite
CTE	Coefficient of Thermal Expansion
CW	Continuous Wave
DARS	Digital Audio Radio Service
DB	Distribution Board
DBF	Digital Beam Forming
DGR	Dual-Gridded Reflector
DK	Dielectric Constant
DLGA	Descent Low Gain Antenna
DMB	Digital Multimedia Broadcasting
DMR	Differential Microwave Radiometer
DOA	Direction Of Arrival
DOS	Denial Of Service
DRA	Direct Radiating Arrays
DTH	Direct To Home
DUHF	Descent Stage UHF
DUT	Device Under Test
DVB	Digital Video Broadcasting
EBG	Electronic Band Gap
EFIE	Electric Field Integral Equation
EGNOS	European Geostationary Navigation Overlay Service
EGSE	Electrical Ground Support Equipment

EHP	Equalisation Heat Pipes
EIRP	Equivalent Isotropic Radiated Power
EM	Electro Magnetic
EMC	Electro Magnetic Compatibility
EOS	Earth Observing System
ESD	Electro Static Discharge
FAFR	Focal Array Fed Reflector
FCS	Feeding Control System
FDTD	Finite Difference Time Domain
FEM	Finite Element Method
FF	Far Field
FMM	Fast Multipole Method
FOV	Field Of View
FSS	Frequency-Selective Surface
G/T	Gain over Temperature
GBAS	Ground-based augmentation systems
GD	Group Delay
GEO	Geostationary Earth Orbit
GFRP	Glass Fibre Reinforced Plastic
GLONASS	GLObal NAvigation Satellite System
GNSS	Global Navigation Satellite Systems
GO	Geometrical Optics
GPS	Global Positioning System
HFI	High Frequency Instrument
HGA	High Gain Antenna
HMPA	Hybrid Matrix Power Amplification
HP	Heat Pipes
HPA	High-Power Amplifier
HPBW	Half-Power Beamwidth
HTCC	High Temperature Cofired Ceramics
HW	Hard Ware
I/F	Inter Face
ICU	Interface Control Unit
IM	Inter Modulation
INET	Input NETwork
InSAR	Interferometric Synthetic Aperture Radar
ITU	International Telecommunication Union
LAAS	Local Area Augmentation System
LDA	Large Deployable Antenna
LEO	Low Earth Orbit
LFI	Low Frequency Instrument
LGA	Low Gain Antenna
LHCP	Left Hand Circular Polarisation
LHP	Loop Heat Pipes
LNA	Low Noise Amplifier
LO	Local Oscillator
LTCC	Low Temperature Cofired Ceramics

MCCM	Muti Chip Complex Module
MCM	Multi Chip Module
MEMS	micro-electromechanical systems
MEO	Medium Earth Orbit
MFIE	Magnetic Field Integral Equation
MFPB	Multiple feed per beam
MGA	Medium Gain Antenna
MHS	Microwave Humidity Sounder
MLFMA	Multilevel Fast Multipole Method
MLI	Multi Layer Insulator
MLS	Microwave Limb Sounder
MMIC	Monolithic Microwave Integrated Circuit
MOM	Method Of Moments
MPA	Medium Power Amplifier
MPA	Multi-Port Amplifier (chapter 18)
MSA	Mechanically Steered Antenna
MSAS	Multi-functional Satellite Augmentation System
MSG	Meteosat Second Generation
MSS	Mobile Satellite System
MSU	Microwave Sounding Unit
MTG	Meteosat Third Generation
MTI	Multiple Target Indicator
NF	Noise figure
OBFN	Optical Beam Forming Networks
OBP	On-Board digital Processor
OMT	Orthomode Transducer
ONET	Output NETwork
OSR	Optical Solar Reflector
PAA	Phase Array Antenna
PAE	Power-Added Efficiency
PCB	Printed Circuit Board
PEC	Perfectly Electrically Conducting
PEC	Patch Excited Cup (chapter 15)
PFD	Power Flux Density
PIM	Passive Inter Modulation
PIMP	Passive Inter Modulation Products
PLGA	Parachute Cone Low Gain Antenna
PML	Perfectly Matched Layers
PNF	Planar Near Field
PO	Physical Optics
PRN	Pseudo Random Noise
PSU	Power Supply Unit
PTD	Physical Theory of Diffraction
PUHF	Parachute Cone UHF
QHA	Quadrifilar Helix Antenna
QoS	Quality of Service
RF	Radio Frequency

RFCB	Radio Frequency and Control Board
RHCP	Right Hand Circular Polarisation
RHGA	Rover High Gain Antenna
RLGA	Rover Low Gain Antenna
RWG	Rao–Wilton–Glission
SAP	Shorted Annular Patch
SAR	Synthetic Aperture Radar
SBAS	Satellite-based augmentation systems
SBR	Shooting and Bouncing Ray
S/C	Space Craft
SEU	Single-Event Upset
SEL	Single-Event Latch-up
SEB	Single-Event Burnout
SFB	Single Feed per Beam
SFPB	Single Feed Per Beam
SGH	Standard Gain Horn
SiGe	silicon–germanium
SIW	substrate integrated waveguide
SLI	Single layer Insulator
SLL	Side Lobe Level
SNF	Spherical Near Field
SSPA	Solid State Power Amplifier
SSSL	Suspended Substrate Strip Line
SW	Software
T/R	Transmit/Receive
THP	Transport Heat Pipes
TLGA	Tilted Low Gain Antenna
TPS	Thermal Protection System
TT&C	Telemetry, Tracking and Command
TWT	Travelling Wave Tubes
TWTA	Travelling Wave Tube Amplifier
UHF	Ultra High Frequency
UMTS	Universal Mobile Telecommunication System
UV	Ultra Violet
UTD	Uniform Theory of Diffraction
VHF	Very High Frequency
VLBI	Very Long Baseline Interferometry
VSAT	Very Small-Aperture Terminal
VSWR	Voltage Standing Wave Ratio
WAAS	Wide Area Augmentation System
WMAP	Wilkinson Microwave Anisotropy Probe
XP	Crosspolar
XPD	Cross-Polar Discrimination
XPI	Crosspolar Isolation

Contributors

Eduardo Alonso, EADS CASA Espacio, Spain

David Álvarez, EADS CASA Espacio, Spain

Eric Amyotte, MDA, Canada

Silvia Arenas, EADS CASA Espacio, Spain

Luigi Boccia, University of Calabria, Italy

Olav Breinbjerg, Technical University of Denmark, Denmark

Paula R. Brown, Jet Propulsion Laboratory, California Institute of Technology, USA

Miguel Bustamante, EADS CASA Espacio, Spain

Jennifer Campuzano, EADS CASA Espacio, Spain

Pasquale Capece, Thales Alenia Space Italia, Rome, Italy

Francisco Casares, EADS CASA Espacio, Spain

Chi-Chih Chen, The Ohio State University, USA

Jacqueline C. Chen, Jet Propulsion Laboratory, California Institute of Technology, USA

Keith Clark, Surrey Satellite Technology Ltd, UK

Luis E. Cuesta, EADS CASA Espacio, Spain

Tie Jun Cui, School of Information Science and Engineering, Southeast University, Nanjing, People's Republic of China

L. Salghetti Drioli, European Space Research and Technology Centre (ESTEC) – European Space Agency (ESA), The Netherlands

Jose A. Encinar, Technical University of Madrid, Spain

Mohammad Fakharzadeh, Center for Intelligent Antenna and Radio Systems (CIARS), Department of Electrical and Computer Engineering, University of Waterloo, Canada

Paolo Focardi, Jet Propulsion Laboratory, California Institute of Technology, USA

Luis F. de la Fuente, EADS CASA Espacio, Spain

Steven (Shichang) Gao, Surrey Space Centre, University of Surrey, UK

Quiterio Garcia, EADS CASA Espacio, Spain

Vicente García, EADS CASA Espacio, Spain

Paul F. Goldsmith, Jet Propulsion Laboratory, California Institute of Technology, USA

Richard E. Hodges, Jet Propulsion Laboratory, California Institute of Technology, USA

William A. Imbriale, Jet Propulsion Laboratory, California Institute of Technology, USA

Jerzy Lemanczyk, European Space Research and Technology Centre (ESTEC) – European Space Agency (ESA), The Netherlands

Cyril Mangenot, European Space Research and Technology Centre (ESTEC) – European Space Agency (ESA), The Netherlands

Moazam Maqsood, Surrey Space Centre, University of Surrey, UK

Luís Martins Camelo, MDA, Canada

Kevin Maynard, Surrey Satellite Technology Ltd, UK

Fernando Monjas, EADS CASA Espacio, Spain

Antonio Montesano, EADS CASA Espacio, Spain

Margarita Naranjo, EADS CASA Espacio, Spain

Xue Wei Ping, School of Information Science and Engineering, Southeast University, Nanjing, People's Republic of China

Yahya Rahmat-Samii, University of California at Los Angeles (UCLA), USA

Heiko Ritter, European Space Research and Technology Centre (ESTEC) – European Space Agency (ESA), The Netherlands

Antoine G. Roederer, Delft University of Technology – IRCTR, The Netherlands

Safieddin Safavi-Naeini, Center for Intelligent Antenna and Radio Systems (CIARS), Department of Electrical and Computer Engineering, University of Waterloo, Canada

J. Santiago-Prowald, European Space Research and Technology Centre (ESTEC) – European Space Agency (ESA), The Netherlands

José Luis Serrano, EADS CASA Espacio, Spain

Hans Juergen Steiner, Astrium GmbH, Germany

Michael A. Thorburn, Space Systems/Loral, USA

Andrea Torre, Thales Alenia Space Italia, Rome, Italy

Ana Trastoy, EADS CASA Espacio, Spain

Jiadong Xu, Northwestern Polytechnical University, Xi'an, People's Republic of China

Wen Ming Yu, School of Information Science and Engineering, Southeast University, Nanjing, People's Republic of China

Jan Zackrisson, RUAG Aerospace, Sweden

Jian Feng Zhang, School of Information Science and Engineering, Southeast University, Nanjing, People's Republic of China

Xiao Yang Zhou, School of Information Science and Engineering, Southeast University, Nanjing, People's Republic of China

1

Antenna Basics

Luigi Boccia[1] and Olav Breinbjerg[2]

[1] *University of Calabria, Italy*
[2] *Technical University of Denmark, Denmark*

1.1 Introduction

Antennas radiate and receive electromagnetic waves by converting guided waves supported by a guiding structure into radiating waves propagating in free space and vice versa. This function has to be accomplished by fulfilling specific requirements which affect the antenna design in different ways. In general, a number of antennas are installed in a satellite and their requirements vary depending on the application and on the mission. They can be roughly classified into three types: antennas for telemetry, tracking and control (TT&C), high-capacity antennas, and antennas for space instruments or for other specific applications. Several examples of the latter class are reported in the third section of this book.

This chapter provides an overview of the basic antenna parameters and antenna types, and it presents other basic concepts related to the space environment which will introduce the reader to the development of antennas for space applications. Although many basic definitions are presented, the chapter is not intended to provide a comprehensive background to antennas. For this reason, the reader should refer to the extensive literature available on the subject, some of which we list as references.

The chapter is organized as follows. In the first part, the main antenna parameters will be given in accordance with the IEEE Standard Definition of Terms for Antennas [1] and with the IEEE Standard Test Procedures for Antennas [2] which will be adopted throughout the book. In the second part of the chapter, basic antenna types commonly employed in spaceborne applications will be presented. In the third part of the chapter, antenna development will be related to the space environment by introducing fundamental concepts such as multipaction and outgassing.

Space Antenna Handbook, First Edition. Edited by William A. Imbriale, Steven (Shichang) Gao and Luigi Boccia.

1.2 Antenna Performance Parameters

Numerous parameters exist for characterizing the performance of antennas and in the following subsections the most significant of these are reviewed. The relevance of these antenna parameters will be seen in Chapter 3 where they are combined into the Friis transmission formula which links the available power of the transmitter to the received power of the receiver in a radio communication system.

1.2.1 Reflection Coefficient and Voltage Standing Wave Ratio

For a multi-port antenna as shown in Figure 1.1, the scattering parameters, S_{ij}, relate the equivalent voltage of the outgoing wave at port i, V_i^-, to the equivalent voltage of the incoming wave at port j, V_j^+, that is, $V_i^- = S_{ij}V_j^+$ [3]. The reflection coefficient at the i' port is

$$\Gamma_i \equiv V_i^-/V_i^+ = S_{ii} + \sum_{j \neq i} S_{ij}V_j^+/V_i^+ \tag{1.1}$$

For a single-port antenna, or for a multi-port antenna with all other ports matched (thus $V_j^+ = 0$ for $j \neq i$), the reflection coefficient Γ_i equals the scattering coefficient S_{ii} and, if the antenna is passive, the magnitude of the reflection coefficient is then less than or equal to 1. Note that the reflection coefficient is defined in terms of equivalent voltage which requires the existence of a well-defined mode in the port of the antenna. Furthermore, the voltage is defined at a specific position – the reference plane – in the antenna port, and the reflection coefficient is thus referenced to that position.

The voltage standing wave ratio (VSWR) is the ratio of the maximum and minimum voltages on the transmission line connected to the antenna, and it follows directly from the reflection coefficient Γ as

$$\text{VSWR} = \frac{1+|\Gamma|}{1-|\Gamma|} \tag{1.2}$$

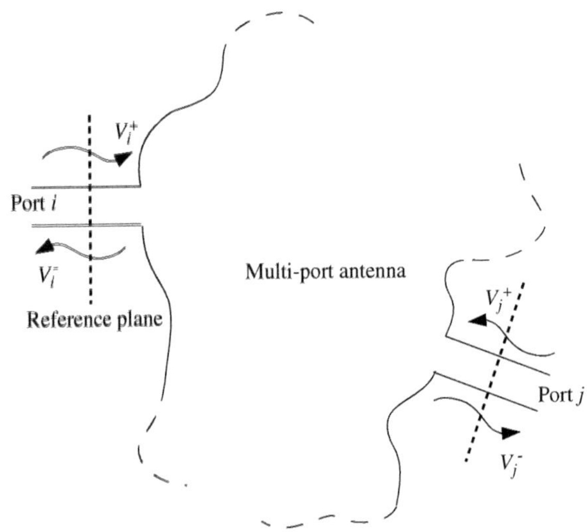

Figure 1.1 *Arbitrary multi-port antenna.*

The scattering parameters are the main representation of antenna behavior with respect to the circuit to which the antenna is connected. This is particularly true for passive antennas while more complex parameters are required for active antennas.

1.2.2 Antenna Impedance

The input impedance of an antenna Z_A is the ratio of the voltage V and current I at the port of the antenna when the antenna is isolated in free space; that is, without the presence of other antennas or scattering structures. Thus, this is sometimes referred to as the isolated input impedance. Since voltage and current are not practical quantities at radio frequencies (RFs), the input impedance is usually determined from the reflection coefficient Γ and the characteristic impedance Z_C of the transmission line connected to the port of the antenna; that is,

$$Z_A \equiv \frac{V}{I} = Z_C \frac{1+\Gamma}{1-\Gamma} \tag{1.3}$$

For a linear multi-port antenna the voltage at the ith port can be related to the currents at all ports as

$$V_i = Z_{ii} I_i + \sum_{j \neq i} Z_{ij} I_j \tag{1.4}$$

where Z_{ii} is the self-impedance of the ith port and Z_{ij} is the mutual impedance between the ith and jth ports. The input impedance of the ith port is then

$$Z_{A,i} \equiv V_i / I_i = Z_{ii} + \sum_{j \neq i} Z_{ij} I_j / I_i \tag{1.5}$$

which is seen to depend on the excitations (currents) of the other ports and therefore differs from the isolated input impedance. Thus, the input impedance of a port in a multi-port system is sometimes referred to as the active input impedance. Even the self-impedance, which is seen from above to equal the active input impedance when all other ports are open-circuited (zero current), is generally different from the isolated input impedance since the open-circuited ports may still act as scattering structures. For an antenna array, see Section 1.4, with identical antenna elements and thus identical isolated input impedances, the active input impedances may differ due to the mutual coupling. Furthermore, if the excitation of the ports is changed, for example, to scan the main beam in a phased array, the active input impedance of an individual port can vary drastically and become very poorly matched to the transmission line characteristic impedance.

If the scattering parameters are arranged in a scattering matrix $\overline{\overline{S}}$ and the self- and mutual impedances in an impedance matrix $\overline{\overline{Z}}$, the relationship between these, for a multi-port antenna with the common characteristic impedance of the transmission lines on the ports Z_C, can be expressed as ($\overline{\overline{U}}$ is the unit matrix)

$$\overline{\overline{Z}} = Z_C(\overline{\overline{U}} + \overline{\overline{S}}) \cdot (\overline{\overline{U}} - \overline{\overline{S}})^{-1} \tag{1.6}$$

$$\overline{\overline{S}} = (\overline{\overline{Z}} + Z_C\overline{\overline{U}})^{-1} \cdot (\overline{\overline{Z}} - Z_C\overline{\overline{U}}) \cdot \tag{1.7}$$

1.2.3 Radiation Pattern and Coverage

The radiation pattern is a 'mathematical function or graphical representation of the radiation properties of the antenna as a function of space coordinates' [1]. In the most common case, antenna radiation patterns are determined in the far-field region [4]. This region is 'where the angular field distribution is essentially independent of the distance from a specified point in the antenna region' [1]. Typically, the far-field region is identified by those distances greater than $2D^2/\lambda$, D being the maximum overall dimension of the antenna and λ the free-space wavelength. In the far-field region of any antenna the radiated field takes a particularly simple form. For time-harmonic fields, and using phasor notation with the suppressed time factor $\exp(j\omega t)$ with ω the angular frequency and t time, the far-field can be expressed as

$$\lim_{r \to \infty} \boldsymbol{E}(\boldsymbol{r}) = \boldsymbol{P}(\boldsymbol{a}_r)\frac{e^{-jkr}}{r} \tag{1.8}$$

Thus, the radiated electric field $\boldsymbol{E}$ at the position of the position vector $\boldsymbol{r}$ can be expressed as the product of a pattern function $\boldsymbol{P}$ that depends only on the direction $\boldsymbol{a}_r$ of the position vector and the term $\exp(-jkr)/r$ that depends only on the length r of the position vector. Furthermore, the pattern function $\boldsymbol{P}$ has only transverse components w.r.t. $\boldsymbol{a}_r$; that is, $\boldsymbol{P} \cdot \boldsymbol{a}_r = 0$. The position vector $\boldsymbol{r}$ is referenced to the origin of the antenna coordinate system. Note that the pattern function $\boldsymbol{P}$ defines all radiation properties that are particular for the antenna.

The parameter represented by the radiation pattern is typically a normalized magnitude of the pattern function or one of its components, the directivity or partial directivity, or the gain or partial gain – but it may be the phase of a polarization-phase vector component, the axial ratio, or the tilt angle as well; these parameters are reviewed in the following subsections. The graphical representation may be two or three dimensional with the transmission/reception direction typically expressed by the polar θ and azimuthal ϕ coordinates of the antenna coordinate system for a full-sphere pattern or the projected coordinates $u = \sin\theta\cos\phi$ and $v = \sin\theta\sin\phi$ for a hemispherical pattern.

An antenna can be defined as directional when it can 'radiate or receive electromagnetic waves more effectively in some directions than in others' [1]. In order to discriminate between directional and non-directional antennas, the half-wave dipole is normally taken as reference while the antenna directivity is generally compared to the ideal isotropic radiator [5]. Normally, the portion of the radiation pattern of a directive antenna where the radiation intensity is maximum is defined as the main lobe. Side, minor, back and grating lobes can also be identified. The first three types are related to the direction and to the intensity of radiation while the last one can be present only in an antenna array environment.

1.2.3.1 Half-Power Beamwidth

The half-power beamwidth (HPBW) is identified in a cut of a radiation pattern as the angle between the two directions in which the radiation intensity is half of its maximum value (see Figure 1.2a). HPBW characterizes the behavior of the antenna in its main lobe but it does not take into account the amount of power radiated out of the main beam. For this reason, parameters are normally used to more accurately evaluate the antenna's directional performance.

1.2.3.2 Coverage

The coverage C of an antenna is the range of transmission/reception directions over which one or more antenna parameters meet certain specifications. In most cases, the coverage C refers to the directivity or gain, or the

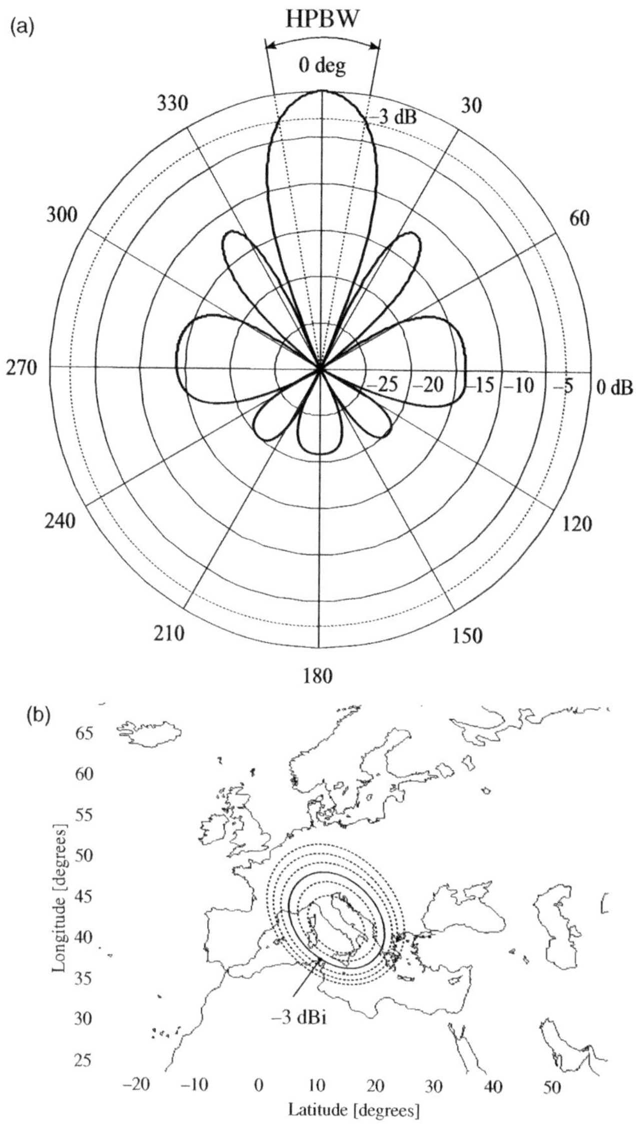

Figure 1.2 Radiation pattern: (a) half-power beam width (HPBW); (b) footprint example (1 dBi per circle).

co-polarized partial directivity or gain, and is thus the range over which the relevant parameter is larger than a specified minimum value; this could be 3 dB below the maximum value. When the antenna points towards the Earth, it is convenient to express the coverage in terms of Earth footprint, which is the projection of the satellite antenna pattern onto the Earth's surface (see Figure 1.2b). The footprint is that portion of the Earth's surface where the antenna points with a given gain. For some applications the footprint corresponds simply to a circle in a (θ, ϕ) coordinate system, calling for a pencil-beam antenna, while for other applications the coverage is the projected shape of a country, calling for a shaped-beam antenna. Clearly, both the footprint and the coverage C can be determined from the radiation pattern and thus the pattern function $\boldsymbol{P}$.

1.2.4 Polarization

The polarization of an antenna in a given direction is the polarization of the plane wave transmitted (or received) by the antenna in the far field. Polarization is classified as linear when the electric field in a given direction is always directed along a line. Pure linear polarization is an ideal case as all antennas generate both a co-polarization field, that is, the polarization the antenna is intended to radiate, and a cross-polarization field, that is, in the case of linearly polarized fields, the component of the electric field orthogonal to the desired polarization. For this reason, the electric field vector normally describes an ellipse and the polarization is classified as elliptical. If the axes of the ellipse are equal, then the polarization is referred to as circular. It is worth noticing that the polarization of an antenna is normally defined by taking into account the radiating wave. Satellite–Earth communication links typically adopt circularly polarized (CP) signals. Indeed, the use of linear polarization would lead to high polarization mismatches arising from alignment issues or from the Faraday rotation effect of the ionosphere [6–8].

Antenna polarization can be described in terms of the polarization-phase vector, $\boldsymbol{p}$, that is, a unit vector that represents the polarization as well as the phase of the radiated field of an antenna. The simple distance dependence of the phase due to the term $\exp(-jkr)/r$ is not included in the polarization-phase vector, and from the pattern function $\boldsymbol{P}$ the polarization-phase vector is thus defined as

$$\boldsymbol{p} \equiv \boldsymbol{P}/|\boldsymbol{P}| \quad \text{with} \quad |\boldsymbol{P}| = \sqrt{\boldsymbol{P}\cdot\boldsymbol{P}^*} \tag{1.9}$$

The pattern function $\boldsymbol{P}$ can be decomposed w.r.t. two orthogonal polarization unit vectors referred to as the co- and cross-polarization unit vectors; that is,

$$\boldsymbol{P} = \boldsymbol{P}\cdot\hat{\boldsymbol{a}}_{co}^{*}\hat{\boldsymbol{a}}_{co} + \boldsymbol{P}\cdot\hat{\boldsymbol{a}}_{cross}^{*}\hat{\boldsymbol{a}}_{cross} \equiv P_{co}\hat{\boldsymbol{a}}_{co} + P_{cross}\hat{\boldsymbol{a}}_{cross} \tag{1.10}$$

where the polarization unit vectors, $\hat{\boldsymbol{a}}_{co}$ and $\hat{\boldsymbol{a}}_{cross}$, are typically the linearly polarized θ and ϕ unit vectors of the standard spherical coordinate system, the linearly polarized unit vectors according to Ludwig's third definition [9], or the circularly polarized unit vectors defined from either of these linearly polarized unit vectors. Obviously, the polarization-phase vector can be decomposed in the same manner; that is, $\boldsymbol{p} \equiv p_{co}\hat{\boldsymbol{a}}_{co} + p_{cross}\hat{\boldsymbol{a}}_{cross}$.

The polarization-phase vector can also be represented in terms of the polarization ellipse with its axial ratio, tilt angle, and sense of rotation. In each direction of observation $\hat{\boldsymbol{a}}_r$, a local right-hand orthogonal, rectangular xyz coordinate system is defined with the unit vectors $\hat{\boldsymbol{a}}_x$ and $\hat{\boldsymbol{a}}_y$ transverse to, and the unit vector $\hat{\boldsymbol{a}}_z$ parallel to, the direction of observation $\hat{\boldsymbol{a}}_r$. The polarization-phase vector is now decomposed as $\boldsymbol{p} \equiv p_x\hat{\boldsymbol{a}}_x + p_y\hat{\boldsymbol{a}}_y$. The axial ratio (AR) can then be expressed as

$$\text{AR} = \sqrt{\frac{1+|\boldsymbol{p}\cdot\boldsymbol{p}|}{1-|\boldsymbol{p}\cdot\boldsymbol{p}|}} \tag{1.11}$$

while the tilt angle (TA) w.r.t. the direction $\hat{\boldsymbol{a}}_x$ is

$$TA = \arctan\left(\text{Re}\left(p_y exp\left(-\frac{j}{2}\arg(\boldsymbol{p}\cdot\boldsymbol{p})\right)\right)\Big/\text{Re}\left(p_x exp\left(-\frac{j}{2}\arg(\boldsymbol{p}\cdot\boldsymbol{p})\right)\right)\right) \tag{1.12}$$

and the sense of rotation (SOR) is

$$\text{SOR} = \begin{cases} \text{right-hand if } 0 < \arg(p_x) - \arg(p_y) < \pi \\ \text{left-hand if } \pi < \arg(p_x) - \arg(p_y) < 2\pi \end{cases} \tag{1.13}$$

Alternatively, the AR can be determined from the magnitude of the right- and left-hand circularly polarized components of the electric field, $\boldsymbol{E}_{RHCP}$ and $\boldsymbol{E}_{LHCP}$ respectively. The expression for AR can thus be cast in the form

$$\text{AR} = \left| \frac{|\boldsymbol{P}_{RHCP}| + |\boldsymbol{P}_{LHCP}|}{|\boldsymbol{P}_{RHCP}| - |\boldsymbol{P}_{LHCP}|} \right| \tag{1.14}$$

Differences between the polarization of the incident wave and the receiving antenna are normally referred to as polarization mismatch. In general, they can cause strong link losses which can be taken into account by using different figures of merit. One possibility is to use co-polarization and cross-polarization field patterns. Alternatively, the polarization efficiency e_p could be defined as [1]

$$e_p = |\hat{\boldsymbol{p}}_i \cdot \hat{\boldsymbol{p}}_a|^2 \tag{1.15}$$

where $\hat{\boldsymbol{p}}_i$ and $\hat{\boldsymbol{p}}_a$ are the polarization vectors of the incident wave and of the receiving antenna respectively. If the polarization of the incident wave and that of the receiving antenna are the same, the inner product defined in Equation 1.15 is equal to 1.

1.2.5 Directivity

The directivity D of an antenna is the ratio of the far-field radiation intensity in a given direction to the average radiation intensity over the radiation sphere. The radiation intensity in a given direction, U, is the radiated power per solid angle Ω and thus $U = |\boldsymbol{P}|^2/2\eta_0$ with η_0 being the free-space intrinsic impedance and $|\boldsymbol{P}|^2$ the power radiated in a given direction. Thus the directivity D in a given direction can be expressed as

$$D \equiv \frac{4\pi U}{\boldsymbol{P}_{rad}} = \frac{2\pi|\boldsymbol{P}|^2}{\eta_0 \boldsymbol{P}_{rad}} = \frac{4\pi|\boldsymbol{P}|^2}{\int_{4\pi} |\boldsymbol{P}|^2 \, d\Omega} \tag{1.16}$$

P_{rad} is the total radiated power, which can be calculated by integrating the power radiated in a given direction over the entire radiating sphere. When the direction is not specified, the maximum directivity is usually taken.

Antenna directivity can be discriminated in terms of polarization by defining partial directivities. The partial directivities, D_{co} and D_{cross}, in a given direction for the co- and cross-polarized components can then be expressed as

$$D_{co} = \frac{4\pi|P_{co}|^2}{\int_{4\pi} |P|^2 \, d\Omega} \quad \text{and} \quad D_{cross} = \frac{4\pi|P_{cross}|^2}{\int_{4\pi} |P|^2 \, d\Omega} \tag{1.17}$$

1.2.6 Gain and Realized Gain

The gain G of an antenna in a given direction is the ratio of the radiation intensity to the average radiation intensity over the radiation sphere if all accepted power is radiated isotropically. In mathematical form, this can be written as

$$G \equiv \frac{U}{P_{acc}/4\pi} = \frac{2\pi|\boldsymbol{P}|^2}{\eta_0 \boldsymbol{P}_{acc}} \tag{1.18}$$

where P_{acc} is the power accepted in input by the antenna. The antenna gain can be related to the directivity by taking into account the antenna radiation efficiency, e_{cd} which can be defined as the ratio of the radiated power to the accepted power. Using the definitions of directivity and gain, it can thus be expressed as

$$e_{cd} \equiv \frac{G}{D} \tag{1.19}$$

For a lossless antenna where all accepted power is also radiated, $P_{rad} = P_{acc}$ and $e_{cd} = 1$, the gain G equals the directivity D. However, for most practical antennas the loss is non-negligible, $P_{rad} \neq P_{acc}$ and $e_{cd} < 1$, and it is important to distinguish between gain and directivity. When the direction of radiation is not stated, the direction of maximum radiation is normally presumed. In analogy to partial directivity, partial gain can be defined to discriminate the antenna gain w.r.t. the polarization of the radiated fields.

According to IEEE standards, the definition of antenna gain does not include reflection losses and polarization mismatches. The realized gain $G_{realized}$ of an antenna is the ratio of the radiation intensity to the average radiation intensity over the radiation sphere if all incident power is radiated; it thus includes the effect of the impedance mismatch at the antenna terminals and can be expressed as

$$G_{realized} \equiv \frac{U}{P_{in}/4\pi} = G(1-|\Gamma|^2) = e_0 D \tag{1.20}$$

where $e_0 = e_{cd}(1-|\Gamma|^2)$ is the overall antenna efficiency. The relevance of realized gain is clear from the Friis transmission formula that comprises the product of the gain and the impedance mismatch factor for both the transmitter and the receiver (see Chapter 3 for further details). However, since the loss and the mismatch are two completely different mechanisms, it is still important to use gain and reflection coefficients separately and to distinguish between gain and realized gain.

1.2.7 Equivalent Isotropically Radiated Power

Equivalent isotropically radiated power (EIRP) in a given direction is defined as 'the gain of a transmitting antenna multiplied by the net power accepted by the antenna from the connected transmitter' [1]. EIRP can be written as

$$\text{EIRP} = P_T G_T \tag{1.21}$$

where P_T is the net power accepted by the antenna from the transmitter and G_T is the gain of the transmitting antenna. In order to include transmitter output power, P_{Tx}, and interconnection losses between transmitter and antenna, L_c, Equation 1.21 can be changed to

$$\text{EIRP} = \frac{P_{Tx} G_T}{L_c} \tag{1.22}$$

The EIRP definition is important because it allows calculation of absolute power and field strength values and it makes possible comparisons between different emitters regardless of the type of configuration.

1.2.8 Effective Area

The effective area A_{eff} of a receiving antenna is the ratio of the available power at the terminals of the antenna to the power density of a polarization-matched incident plane wave. The effective area A_{eff} can be measured itself, but in most situations it is found from its reciprocity-based relation to the gain G as

$$A_{eff} = \frac{\lambda^2}{4\pi}G \tag{1.23}$$

1.2.9 Phase Center

In IEEE standards, the phase center is defined as 'the location of a point associated with an antenna such that, if it is taken as the center of a sphere whose radius extends into the far-field, the phase of a given field component over the surface of the radiation sphere is essentially constant'. As the size of real antennas is not null, the phase center depends on the observation direction. In general, the phase center is calculated experimentally by measuring the phase pattern at different cut planes [10].

For some applications, knowing the location of the phase center is very important. For example, in a reflector antenna the phase center of the feed needs to be located at the focal point of the paraboloid. Another example where the phase center location is critical is the global navigation satellite system (GNSS) [11]. Indeed, one of the parameters which defines the accuracy of high-precision GNSSs is the invariance of the phase center which should be highly stable in order to minimize positioning errors.

1.2.10 Bandwidth

The bandwidth (BW) of an antenna is the range of frequencies over which one or more antenna parameters meet certain specifications. In most cases, BW refers to the reflection coefficient Γ and is thus the range over which Γ is less than a specified maximum allowable value, $\Gamma_{\max}$, with the tacit assumption that other parameters remain within their specifications too. BW depends strongly on the value of $\Gamma_{\max}$ and it is important that this be stated explicitly. With f_u and f_l denoting the upper and lower limits of the frequency range, respectively, the fractional bandwidth (FBW) is

$$\text{FBW} = \frac{f_u - f_l}{f_c} \quad \text{with} \quad f_c = \frac{f_u + f_l}{2} \tag{1.24}$$

with the condition that the center frequency f_c coincides with the frequency of operation.

When multiple antenna parameters have to be considered, BW is given by the minimum range of frequencies over which specifications are satisfied. Typically, link budget calculations pose stringent requirements on the antenna gain and coverage, on the polarization efficiency, and on the reflection coefficient over the system bandwidth.

1.2.11 Antenna Noise Temperature

The antenna noise temperature T_A of a receiving antenna is the temperature (in kelvin) that, through the formula $P_n = KT_A\text{BW}$, gives the noise power P_n at the terminals of the antenna; K is Boltzmann's constant and BW the bandwidth [12]. In terms of the background noise temperature $T_B(\Omega)$ over the radiation sphere of the antenna, expressing the noise from the sky, the satellite structure and the Earth, and the antenna physical temperature T, the antenna noise temperature T_A in the radio frequency range can be expressed as

$$T_A = \frac{\eta_{rad}}{4\pi} \int_{4\pi} T_B(\Omega) D(\Omega)\, d\Omega + (1-\eta_{rad})T \tag{1.25}$$

While all previous antenna parameters relate only to the antenna itself, and any influence of the surroundings on these is considered parasitical, the antenna noise temperature relates equally to the antenna and its surroundings and it is particular also for the latter. Equation 1.25 shows that the antenna noise temperature can be calculated from the directivity D, the radiation efficiency η_{rad} and the background temperature T_B.

1.3 Basic Antenna Elements

Depending on the electrical and physical requirements, spacecraft antenna design can be based on different classes of radiators. In this section a basic overview of some of the most common antenna types is provided and includes references to relevant chapters of this book.

1.3.1 Wire Antennas

The dipole antenna is the most representative type of wire radiator. In the most common case, it consists of a linear wire with a feed point at the center as shown in Figure 1.3a. The radiation properties depend on the current distribution along its main axis, this current being mainly related to the dipole length. Some radiation occurs in all directions with the exception of the dipole axis itself. Due to the rotational symmetry of the dipole around its main axis (z-axis in Figure 1.3a), the radiation pattern is symmetrical over the azimuthal ϕ coordinate. In Figure 1.3b, c the radiation pattern of an ideal half-wavelength dipole is shown. Its maximum directivity is 2.15 dB and its HPBW is equal to 78°. The behavior of a dipole antenna changes when the dipole interacts with the spacecraft. For this reason, the ideal pattern shown on Figure 1.3b, c is valid only for isolated dipoles and it does not take into account the interactions with the spacecraft as shown in the examples of Chapter 2.

The monopole antenna is formed by replacing one-half of a dipole with an infinite ground plane perpendicular to the dipole axis as shown in Figure 1.3d. Using image theory [13], the fields above the ground plane can be found by substituting the ground plane with image currents forming the missing half of the dipole. The radiating behaviour of these two wire antennas is similar, but the monopole radiation below the ground plane is ideally null. For this reason, the directivity of a monopole antenna of length l is twice that of the equivalent dipole antenna of double length, $2l$.

Since the early spacecraft missions (see Chapter 7), wire antennas have been widely used in space exploration. Due to their omnidirectional radiation properties, dipole and monopole antennas are generally used to send or receive telemetry and command signals during launch, when the spacecraft attitude is out of control or in other circumstances when the high-directivity antennas cannot be employed.

1.3.2 Horn Antennas

Another important type of antenna which has found wide application in space missions is the horn antenna. In general, horn antennas are employed in satellite missions to produce wide-beam coverage such as Earth

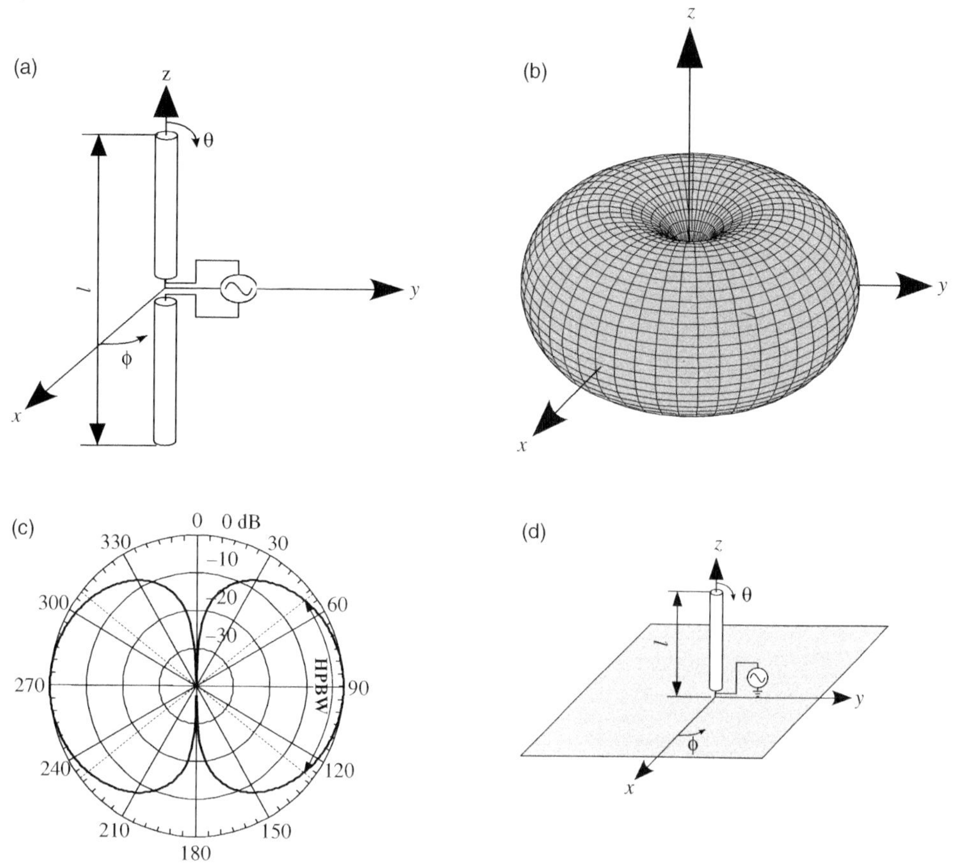

Figure 1.3 *Dipole and monopole antennas: (a) dipole antenna geometry; (b) 3D normalized amplitude radiation pattern (dB) of a dipole antenna; (c) elevation plane normalized amplitude pattern (dB); (d) monopole antenna geometry.*

coverage or to feed reflector antennas. Horn antennas are designed to provide a smooth transition between the feeding waveguide and a wider aperture which serves to focus the main lobe. Horn antennas belong to the category of aperture antennas, their radiation characteristics being determined by the field distribution across the aperture. The most common type of horn antenna is the pyramidal horn shown in Figure 1.4a. The horn provides a transition of length d between a feeding section of rectangular waveguide of height a and width b and a radiating aperture of height A and width B. In the most common case, the waveguide is excited by a single TE_{10} mode. In this case the dominant horn polarization would be linear with the main electric field component directed along the z-axis. Horn polarization can also be circular or dual linear depending on the modes excited in the waveguide section. Knowing the waveguide dimensions and the gain specifications, the pyramidal horn geometry can be defined through simple analytical formulas derived from the hypothesis of an aperture terminating in an infinite flange [14]. In general, the finiteness of the terminating flange can lead to inaccuracies which can be overcome through full-wave analysis. As a general rule, for a given horn length, d, as the aperture width, B, increases, the gain increases until it reaches a maximum after which it starts to decrease.

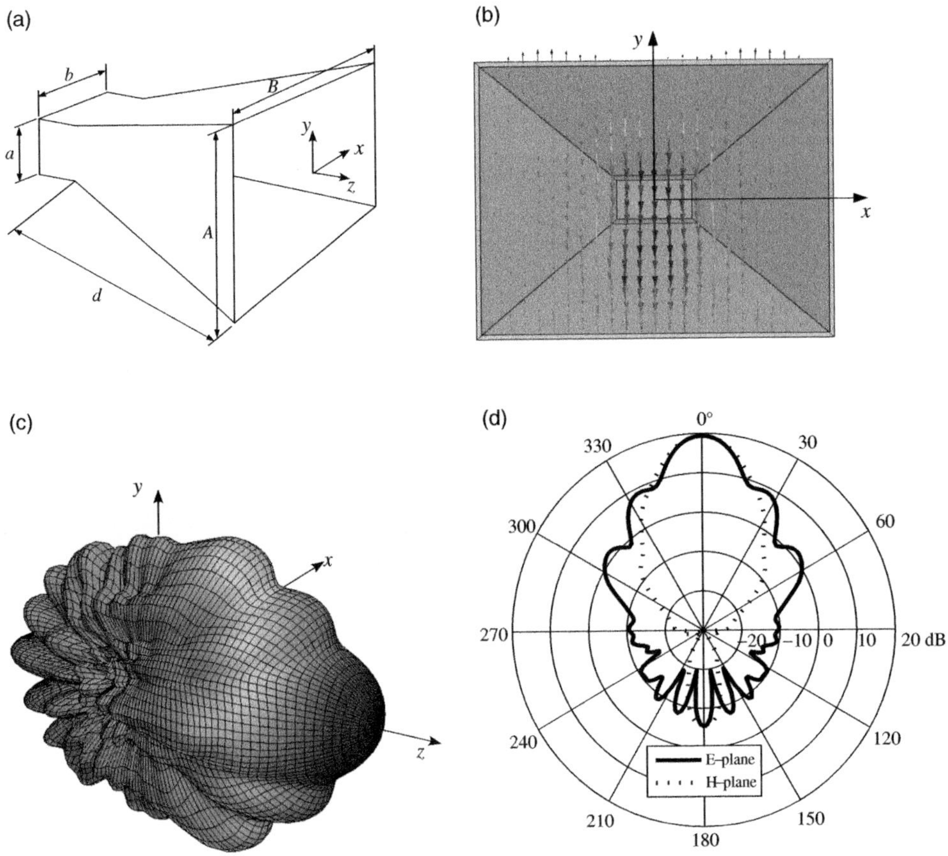

Figure 1.4 Pyramidal horn antenna: (a) geometry; (b) electric field distribution on the antenna aperture; (c) typical 3D normalized amplitude radiation pattern (dB) of a horn antenna; (d) typical gain pattern on the E- and H-plane of a vertically polarized pyramidal horn.

Figure 1.4 shows the field patterns of a pyramidal horn simulated through full-wave software [15]. Results were obtained from a rectangular horn antenna with $A = 120$ mm, $B = 90$ mm and $d = 120$ mm at a frequency of 10 GHz. The horn is fed through a standard section of WR102 waveguide excited in its fundamental mode. As can be observed in Figure 1.4b, the electric field vector on the antenna aperture is polarized along the y-axis. The y–z plane is thus referred to as the *E-plane* as it contains the $\boldsymbol{E}$-field vector and the maximum direction of maximum radiation. Similarly, the x–z plane is referred to as the *H-plane*. The dominant polarization is linear (vertical) polarization. For the proposed example, the gain is around 19 dB at10 GHz while the HPBW is equal to 19° and 20° in the H- and E-plane, respectively. The asymmetry of the beam amplitude in the two main planes is a common problem of pyramidal horn antennas. Another limitation is related to the diffraction arising from the horn flanges and, in particular, from those that are perpendicular to the electric field vector. In general, such diffraction produces back radiation and sidelobes which are indeed more evident in the E-plane.

Another important type of horn antenna is the conical horn whose geometry is shown in Figure 1.5. The conical horn aperture is circular and, in the most typical configuration, is fed by a section of circular waveguide which is typically excited by a TE_{11} mode. The behavior of a conical horn is similar to that of a pyramidal horn. The directivity can be expressed as [16]

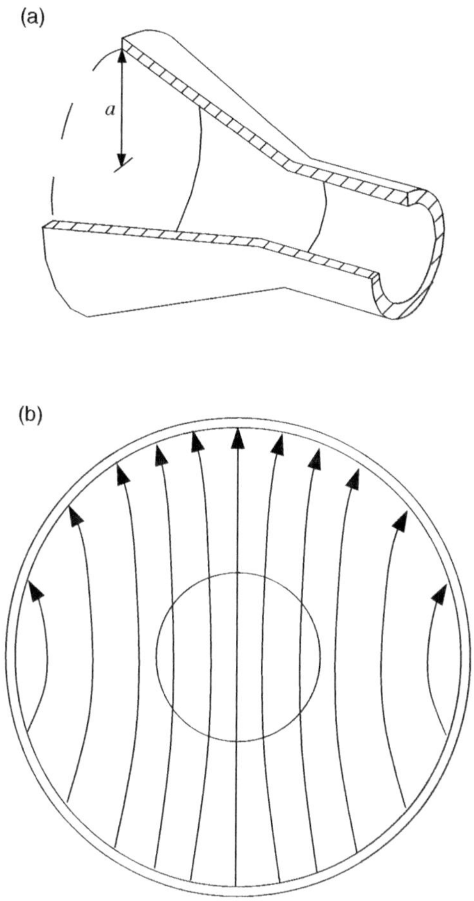

Figure 1.5 Conical horn antenna: (a) geometry; (b) electric field distribution of a smooth-walled conical horn aperture.

$$D_c(\text{dB}) = 10\log_{10}\left[\varepsilon_{ap}\frac{4\pi}{\lambda^2}(\pi a^2)\right] \tag{1.26}$$

where a is the aperture radius and ε_{ap} is the aperture efficiency. Although the conical horn is geometrically symmetric, its pattern is asymmetric and it suffers from similar limitations to that of pyramidal horns. In particular, the conical horn can present high cross-polar levels, which can be easily explained by looking at the transverse electric field distribution on the antenna aperture as shown in Figure 1.5b for vertical polarization. As can be observed, components of the electric field are also present along the y-axis. In the far field, such components would give rise to an electric field horizontally polarized with peaks of intensity at $\pm 45°$. Poor polarization performance can be a severe limitation both in radio astronomy applications and in satellite communication systems as reported in Section 12.4.

The lack of symmetry in pyramidal and conical horn antennas can cause severe limitations in terms of efficiency, increasing losses when global coverage is required and generating spillover losses when horns are used as reflector feeds. A common way to improve the field distribution across the horn aperture is to

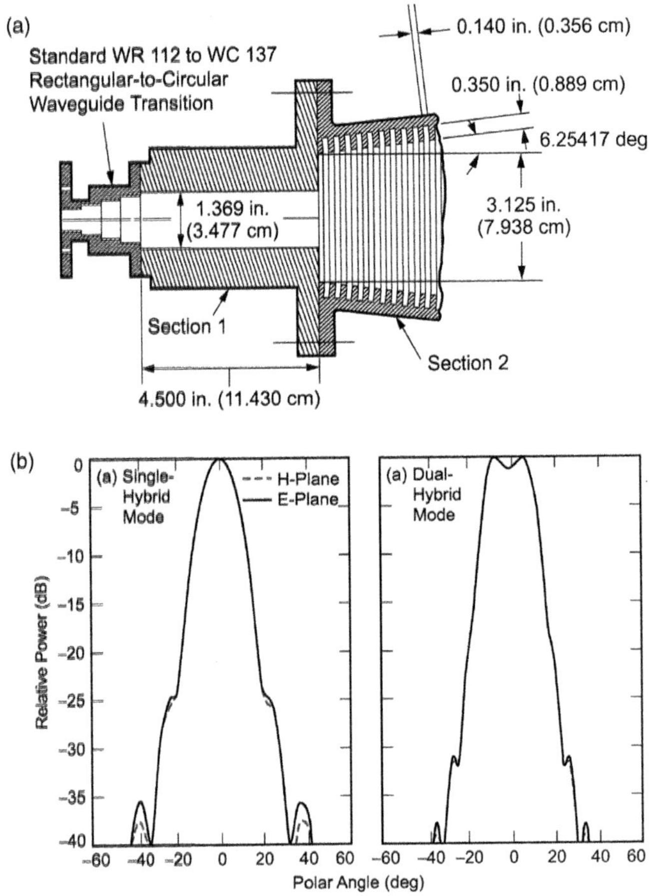

Figure 1.6 *Dual-hybrid-mode feed horn: (a) geometry; (b) normalized radiation patterns [22].*

employ grooved walls [17]. Corrugations perpendicular to the walls are designed to provide a capacitive reactance which inhibits surface wave propagation thus avoiding spurious diffraction from the edges. For pyramidal horns, the corrugations are usually placed only on the E-plane walls as edge currents on the H-plane walls are negligible. However, most corrugated horns are conical horns, this type of antenna being easier to fabricate. An example of a conical corrugated horn is shown in Figure 1.6a. As the groove response is polarization independent, the fundamental mode of a corrugated horn is the hybrid mode HE_{11} that can be associated to a combination of a TE_{11} and TM_{11} modes in a smooth circular waveguide. In general, the two modes are optimally phased to yield a highly symmetric field distribution across the aperture which, in turn, generates a symmetric radiation pattern ideally with very low sidelobes [18]. The performance of this type of radiator can be further optimized by using a Gaussian profiled conical horn [19]. In this case, the radius increases longitudinally following the expansion law of a Gaussian beam. As a result, the field distribution at the horn mouth is almost perfectly Gaussian, thus generating a far-field pattern ideally without sidelobes.

Another technique which can be employed to improve the horn pattern is to use a multimode approach. In this case, higher order modes can be deliberately excited with a specific phase and amplitude relationship, improving the horn radiation performance [20]. When even more demanding performance is required, multi-hybrid-mode corrugated horn antennas can be designed as shown in [21] and in [22] for Deep Space Network antennas.

1.3.3 Reflectors

Reflector antennas are by far the most common antenna element for applications requiring high gain and directivity. This class of antennas has been widely employed in space missions since the early days of space exploration (see Section 7.2). Over the years, their concepts evolved both mechanically and functionally to meet technical requirements of increasing complexity. In this section only a basic review of this type of antenna is provided, the interested reader being directed to following chapters and to the referenced literature [16,22–25] for further study.

1.3.3.1 Main Reflector Parameters

Although reflector antennas can be made in different types, shapes and configurations, they all essentially consist of a passive reflecting surface illuminated by a smaller primary feed. Reflector antenna performance is influenced by several parameters, as follows.

Spillover and Aperture Illumination Efficiency Reflector efficiency is highly influenced by the feed radiation characteristics. In particular, an ideal reflector should be uniformly illuminated and all power should be focused on the reflecting surface. The portion of the feed power that does not reach the reflector is referred to as spillover loss while the ability to uniformly feed the parabola is referred to as illumination efficiency. Since primary feeds have a tapered radiation pattern, a compromise between spillover losses and illumination efficiency must be considered to maximize the aperture gain.

Aperture Blockage Feed and mechanical support structures located in front of the aperture, partially block field radiation in the far field. This phenomenon is referred to as aperture blockage and its main effect is to reduce the on-axis gain and to increase the sidelobe amplitude level. The reduction of efficiency due to aperture blockage varies depending on the feed configuration and aperture size.

Axial and Lateral Defocusing Axial and lateral defocusing are the errors generated by displaced feed positions along the reflector axis and orthogonally to the reflector axis respectively. Axial displacements generate a broader beamwidth while lateral defocusing causes beam squints [26,27].

Reflector Surface Deviation Deviations from the curvature surface cause a distortion of the reflector antenna radiation pattern [28]. The effect of surface deviation can be significantly high in deployable reflector antennas as outlined in Chapter 5.

Feed Feed selection and design have a major role in the correct and efficient operation of a reflector system. In general, the feed type depends on the system requirements in terms of frequency band, radiation characteristics and efficiency. Although simpler antenna types can be used, the best performance is usually achieved through horn antennas with Gaussian beam characteristics [29].

1.3.3.2 Basic Reflector Types

Some of the most common reflector systems are shown in Figure 1.7. The simplest form of reflector antenna is the parabolic reflector shown in Figure 1.7a. This configuration benefits from the geometrical properties of the

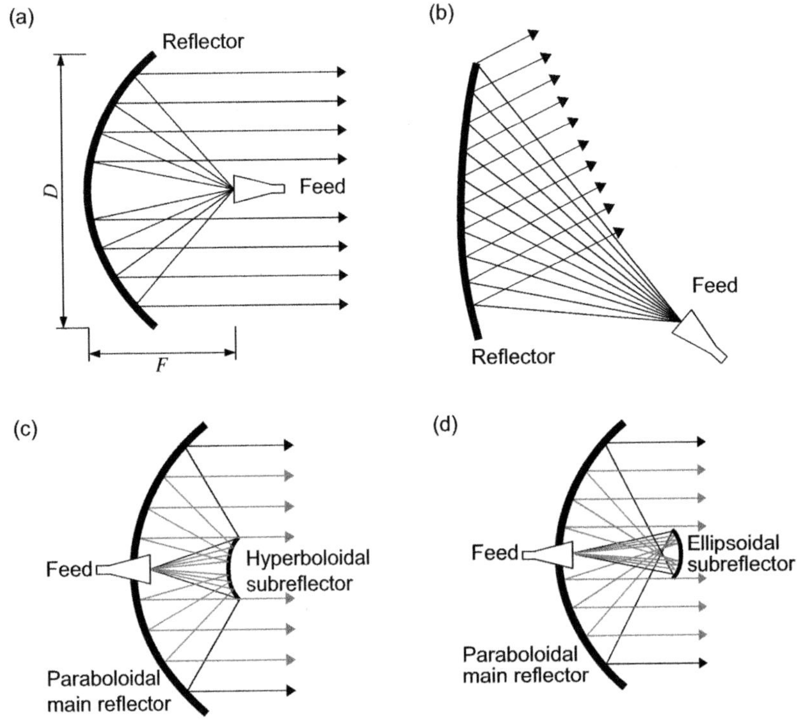

Figure 1.7 *Reflector antenna configurations: (a) on-focus parabolic reflector; (b) off-axis reflector; (c) Cassegrain reflector; (d) Gregorian reflector.*

parabola since spherical waves radiated by a source placed at the focal point are transformed into plane waves directed along the aperture rotation axis. This type of reflector generates a pencil beam whose characteristics are mainly controlled by the aperture diameter, D, the focal length, F, the reflecting surface curvature, F/D, and the pattern and size of the feed antenna. The electrical performance of this elementary reflector system is limited by the effect of aperture blockage [30]. As a possible solution to this problem, configurations employing an offset feed and a sectioned parabolic reflector [31] can be considered as illustrated in Figure 1.7b. In this case, the blockage effect of the feed is negligible and the direction of maximum radiation can be controlled by optimally shaping the reflector surface. The absence of feed blockage can be particularly important for those applications where multiple-feed systems are needed. Compared to the axisymmetric configuration, the main drawbacks of this type of reflector system are related to the large cross-polar fields for linear polarization [32]. Depolarization effects are due to reflector curvature and they can be reduced by selecting a relatively large F/D ratio [33]. However, when it is not possible to increase the reflector curvature, polarization rotation can be cancelled by using a polarization grid [34] or by optimally designing the primary feed [35]. When offset reflector antennas are illuminated by a circularly polarized primary feed, high cross-polar fields generate angular displacements of the main beam [32,36]. Beam squinting can be counteracted by using reflectors with large curvatures or by employing compensation techniques at feed level [37,38].

For larger apertures, a more compact feed arrangement can be realized by employing smaller subreflectors. Classical axisymmetric geometries for the Cassegrain and Gregorian reflector types are shown in Figure 1.7c and d respectively. In both systems, the primary feed is located on the rear of the main paraboloidal reflector. In the Cassegrain arrangement the subreflector is a section of a hyperboloid located within the focus of the main reflector, while in the Gregorian configuration the subreflector is an ellipsoid located outside the focus of the main reflector. Both systems have similar electrical features but Cassegrain designs are more commonly used in satellite applications.

Shaped Reflectors Dual reflectors have higher efficiency and reduced sidelobes with respect to the on-focus fed parabolic reflector [39]. In particular, it has been demonstrated [40] that aperture efficiency can be improved by controlling the shape of the main and sub-reflector surfaces to improve aperture energy distribution. Varying the shape of the reflector surface has a direct impact on the illumination function which can be controlled in both amplitude and phase, thus reducing both spillover losses and illumination efficiency.

Cross-polarization Reduction Offset dual-reflector antennas can be designed to have very limited cross-polar components. In particular, the optogeometrical condition for eliminating cross-polarization [41] depends on whether the subreflector surface is concave or convex, on the eccentricity and on the angles of the axes of the main reflector surface and subreflector surface, and on the axis of the primary radiation.

Contoured- or Multiple-Beam Reflectors Contoured- or multiple-beam configurations can be obtained through specific offset dual-reflector arrangements. In the most conventional approach, contoured-beam patterns can be achieved using a multi-feed dual-reflector system [42]. In this case, the desired coverage contour is achieved by superposing overlapping spots generated by different feeds whose fields are then combined through a beam-forming network. This approach is also used when multiple beams have to be generated from a single antenna. In this latter case, individual beam-forming networks for each beam have to be implemented. Digital beam forming can also be employed for implementing beam scanning capabilities [43].

Alternatively, it is also possible to generate contoured beams by using a single feed and by shaping the reflector surface [44]. Shaped reflectors are the most common design approach for single-beam applications in satellite applications due to lower weight and lower spillover losses w.r.t. a single-feed design [45].

Deployable Reflector Antennas Reflector antennas have evolved significantly over the years, boosted by space-related research. In particular, significant improvements have been achieved in terms of aperture size through the employment of deployable structures which can be larger than 20 m, as described in Chapter 8. Space-related research continues to lead the technological development of reflector antennas as is evident from the list of future configurations reported in Section 18.4.

1.3.4 Helical Antennas

Helical antennas are widely used in satellite communication systems mainly because of their circular polarization and wide-band features. In its simplest form, a helical antenna consists of a conducting wire wound in the form of a helix as shown in Figure 1.8a. Generally, this type of antenna is fed through a coaxial transition and includes a ground plane. The radiation characteristics of this antenna and its input impedance depend on the helix diameter, d, on the wire diameter, t, on the pitch, p, and on the number of turns, N.

The helix antenna has different modes of radiation. In normal mode (or broadside mode) the helix length is short compared to the wavelength and its behavior is similar to a short dipole [16]. This type of antenna radiates in directions normal to its axis (Figure 1.8b) and can be designed to operate in linear polarization or circular polarization. In this configuration, the helix behavior is highly sensitive to the antenna dimensions.

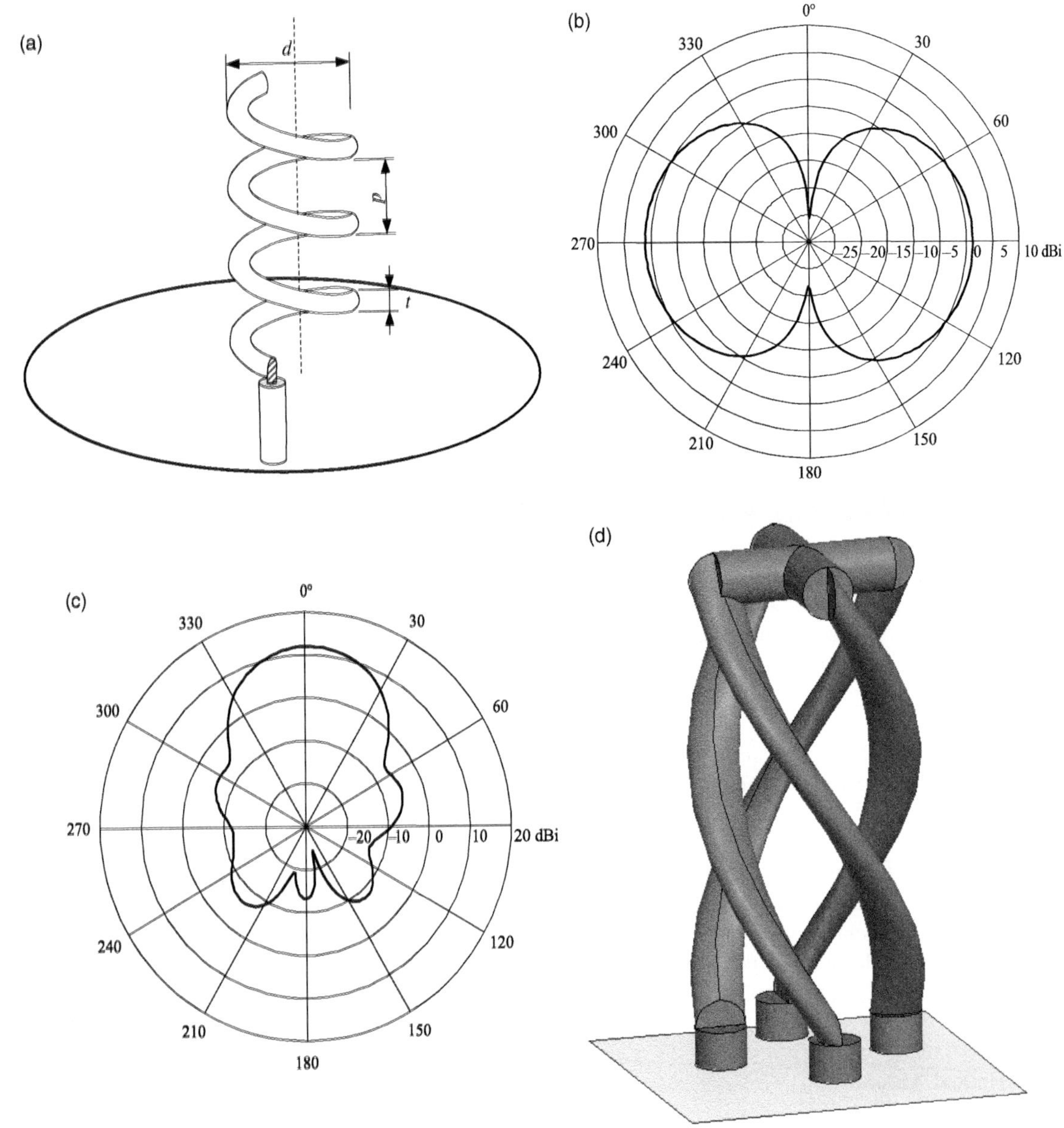

Figure 1.8 Helical antenna: (a) monofilar helical antenna geometry; (b) typical radiation pattern of a helical antenna operating in normal mode; (c) typical co-polar gain pattern of a helical antenna operating in axial mode; (d) short-circuited quadrifilar helical antenna (QHA).

In axial mode (or end-fire mode) the helical antenna has a main lobe directed along its axis, as shown in Figure 1.8c. This operating mode is achieved when both the helix diameter, d, and the pitch, p, are large fractions of the wavelength [46]. Helical antennas operating in axial mode are circularly polarized and they are normally installed on a ground plane. However, when the diameter of the ground plane of a conventional helical antenna is less than the diameter of the helix, the helix radiates with its main beam in the backfire direction when the pitch angle is small [47].

The helix radiation characteristics can be controlled by changing the geometrical parameters of the antenna or by varying the number of wires [48–50]. For example, quadrifilar helical antennas (QHAs) (Figure 1.8d) are widely used for TT&C [51]. QHAs consist of four helical wires equally spaced and circumferentially located 90° apart from each other and sequentially fed with 90° of phase shift.

1.3.5 Printed Antennas

In the past few decades, microstrip antennas [52] have been one of the most commonly used antennas for space applications and, in all likelihood, will play a key role also in the coming years. In its most classical configuration, a microstrip radiator consists of a metallic patch element printed on a thin insulating dielectric layer placed above a ground plane. Figure 1.9 shows the two most popular microstrip antenna configurations: the rectangular patch antenna and the circular patch antenna. Since their first introduction [53,54], printed antennas have become a very popular research topic gaining the attention of both the industrial and the academic communities. Thousands of papers have been published on this subject, introducing many improvements and contributing to a rapid evolution of the early concept and widespread diffusion in many applications.

1.3.5.1 Features and Limitations

The diffusion of microstrip radiators is mainly due to their unique features, which are outlined below. Microstrip antennas are very low profile, of light weight and can be conformal to the mounting surface. These characteristics can be extremely important in several military, commercial or space applications where

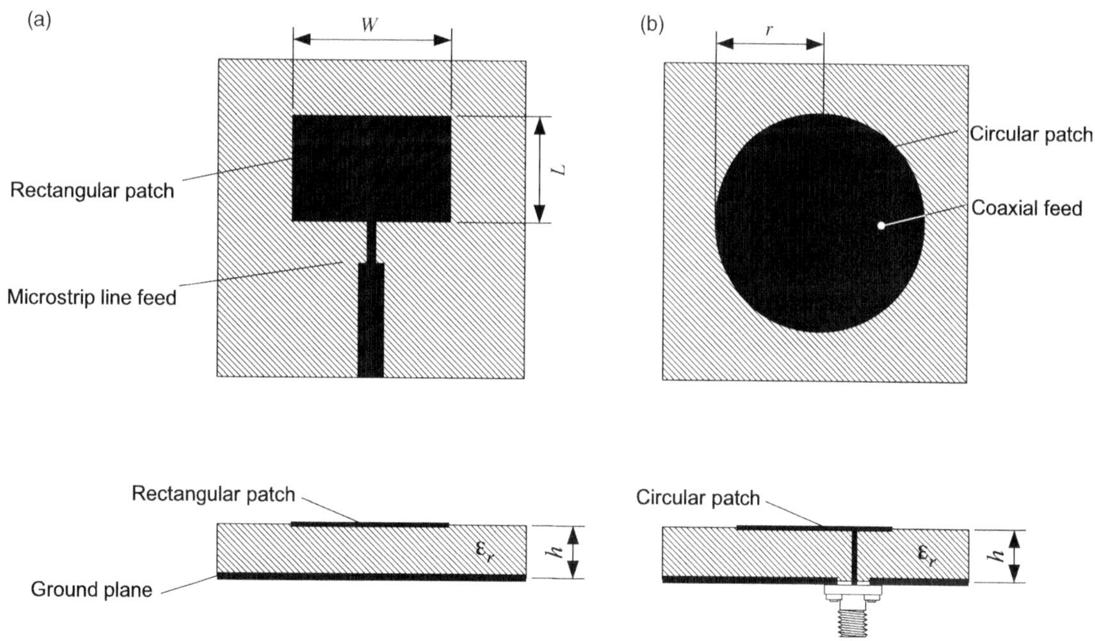

Figure 1.9 *Basic types of microstrip antennas: (a) rectangular patch; (b) circular patch.*

physical constrains are of prime concern. Depending on the type of materials, on the configuration and on the required fabrication process, microstrip antennas can also be low cost when compared to other types of antenna elements. Microstrip technology is naturally flexible, making possible the design of antennas of different shapes and configurations using single or multilayer arrangements and covering multiple bands. Furthermore, integration of printed antennas in microwave integrated circuits (MICs) is straightforward and high degrees of integration levels can be reached.

The main operational limitations of microstrip antennas are due to their narrow bandwidth. Indeed, a classical microstrip antenna would normally have a bandwidth of a few percent. Moreover, when compared to other radiators (e.g., horns, reflectors), the efficiency of microstrip antennas is much lower and the gain of a single patch is usually around 5–7 dBi. Another major disadvantage of printed radiators is related to their low power handling capability. This limitation is due to the small distance between the radiating patch and the ground plane. Depending on the substrate material characteristics and thickness, and on the thickness of metal layers, a microstrip radiator can be designed to handle hundreds of watts [55]. However, due to the multipacting breakdown effect [56], microstrip power handling in space is significantly reduced with respect to the expected value of the Earth's atmosphere. This aspect will be reconsidered in Section 1.5.1.

1.3.5.2 Basic Characteristics

In this subsection a rectangular patch antenna is taken as reference to discuss the basic radiation characteristics of microstrip antennas. A rectangular patch antenna consists of a rectangular patch of width W and length L printed on a substrate having relative dielectric permittivity ε_r of thickness h as shown in Figure 1.9. Generally, dielectric thickness is a fraction of wavelength ($0.003 \leq \lambda_0 \leq 0.05$ where λ_0 is the free-space wavelength) [16] while metal layers are tens of microns thick. The relative dielectric constant depends on the type of dielectric material. It mainly influences the resonant patch length L, the bandwidth and the patch efficiency.

A microstrip antenna designed to operate in its fundamental mode can be related to a half-wavelength resonator with two radiating edges. As can be observed in Figure 1.10, the electric field distribution at the patch radiating borders can be associated to that of two slots. This equivalence is the basis of the so-called transmission line model [57–59] that is the most intuitive way to represent a rectangular patch antenna. Yet this model does not capture many important physical phenomena which take place on a rectangular patch antenna. One of the effects which is not included in the transmission line model is the far-field radiation of the so-called non-radiating edges that are the patch borders orthogonal to the feed line axis (Figure 1.9a). The electric field associated to these borders for the fundamental mode is shown in Figure 1.10. It can be demonstrated that their

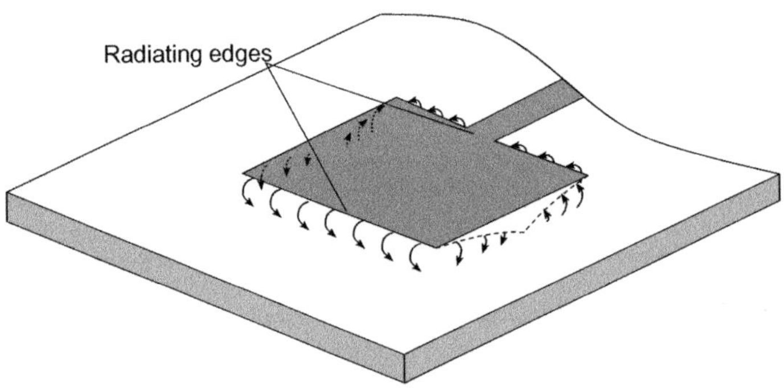

Figure 1.10 *Electric field distribution at the edges of a rectangular patch antenna excited in its fundamental mode.*

contribution to the radiation pattern on the H- and E-plane is virtually null [16]. A more accurate analytical representation can be obtained by treating the antenna region as a cavity bounded by electric conductors (patch and ground plane) and by magnetic walls along the perimeter of the patch. Although the cavity model provides a more realistic depiction of microstrip antenna behavior for different radiator shapes, it is normally used only for a first rough approximation of the antenna geometry or to understand design principles and physical insights. Indeed, the most common design approach is based on one of the commercially available simulators which make use of the full-wave techniques discussed in Chapter 2 of this handbook.

The patch configurations shown in Figures 1.9 and 1.10 radiate a linearly polarized field. In general, the polarization purity of a microstrip radiator is poor, as discussed in detail in Chapter 14. Patch configurations with improved linear or circular polarization performance will be presented later on in this subsection.

Typical radiation patterns of a rectangular patch are shown in Figure 1.11. In general terms, microstrip radiators are wide beam antennas. Their radiation performance is directly related to the equivalent magnetic current densities at the patch borders. For a given resonant frequency and dielectric material, the patch length, L, cannot be modified. Directivity can be indeed slightly changed by controlling the patch width, W. Typical gain values for a standard single-patch radiator are usually in the range from 5 to 7 dBi. The antenna gain and, consequently, the efficiency are strongly influenced by the characteristics of the dielectric material and by metal losses. Another type of loss in a microstrip antenna is related to surface wave excitation. Surface waves are generated at the discontinuity between the substrate and the dielectric above the antenna (e.g., air or free space). Surface wave power propagates at the dielectric interface causing efficiency reduction, spurious radiation and diffraction from the ground plane border, and mutual coupling in array scenarios [60].

1.3.5.3 Feeding Techniques

The electromagnetic behavior of a microstrip antenna is strongly influenced by the feed techniques. Illustrations of the most common feeding methods are shown in Figure 1.12. Feeding techniques based on coaxial probes are implemented by soldering the outer connector of a coaxial cable to the ground plane and elongating the inner conductor to fit flush against the patch. This technique is usually implemented when the antenna has to be attached to a standard 50 Ω coaxial probe. However, it is possible to use the same coaxial configuration also in multilayer microstrip circuits. The connector should be located on the patch E-plane axis and the position has to be selected to match the coaxial feed characteristic impedance. When the height of the dielectric is too high, the metal pin penetrating inside the substrate provides an inductive reactance which shrinks the bandwidth and makes this configuration unsuitable for thick structures. In general, pin inductance can be compensated by adding a capacitive load [61]. The vertical currents excited by the coaxial probe generate spurious radiation which is indeed evident by looking at the asymmetries present in the E-plane co-polar pattern of Figure 1.11a.

Another common technique for feeding microstrip antennas is to use a simple microstrip transmission line feed as shown in Figures 1.9a and 1.10. In this case, a microstrip transmission line is connected to the radiating border of a patch. In order to match the characteristic impedance of the microstrip line with the patch input impedance two approaches can be adopted: using an impedance transformer (e.g., quarter-wavelength transformer) or inserting the feed line inside the patch. Both radiating element and feed line are printed on the same layer. Although this configuration is simple to fabricate, the leakage radiation from the feed line can significantly deteriorate the radiation pattern. A similar phenomenon takes place also when proximity feed arrangements are used. In this configuration (Figure 1.12c), the feeding microstrip line is printed on an additional metal layer underneath the patch radiator. Another common feeding scheme is the aperture-coupled technique (Figure 1.12d) proposed in [62]. A microstrip line printed back to back with the patch radiator is coupled to the antenna by means of a slot on the ground plane. Slot coupling provides better bandwidth,

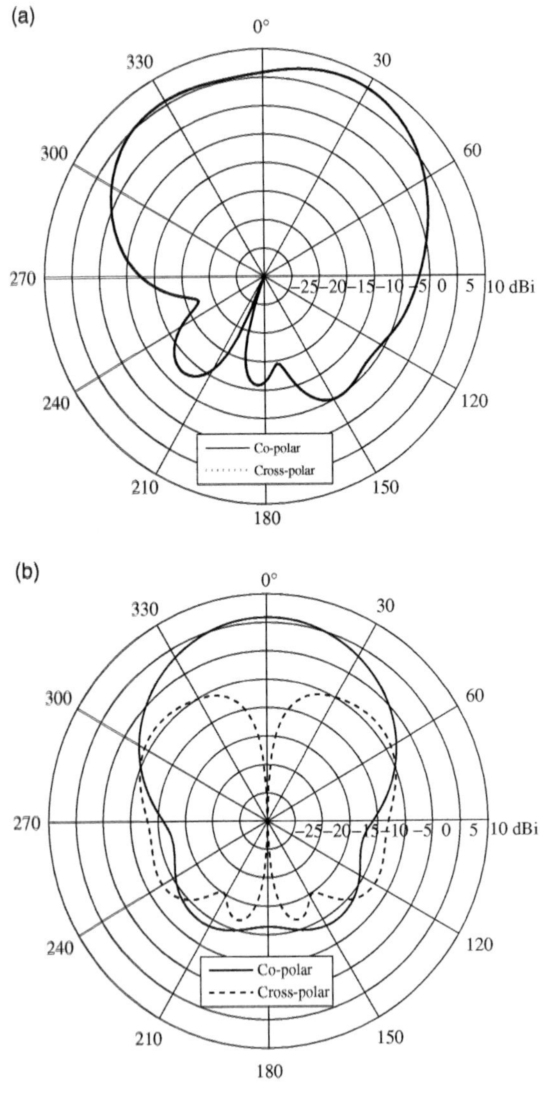

Figure 1.11 *Typical radiation pattern of a rectangular patch antenna with coaxial feed: (a) E-plane and (b) H-plane co-polar and cross-polar gain.*

minimizes spurious radiation from the microstrip lines and avoids vertical elements and soldering. The main limitation of this solution is related to possible unwanted radiation from the slot. Proximity coupling, aperture coupling and all other non-contact feeding techniques provide better performance in terms of passive intermodulation distortion [4] (see Section 1.5.2 for details).

1.3.5.4 Materials and Fabrication Processes

The selection of the dielectric material is of key importance in the design of microstrip antennas, affecting mechanical, thermal and electrical performance. The dielectric substrate in a microstrip antenna mainly serves

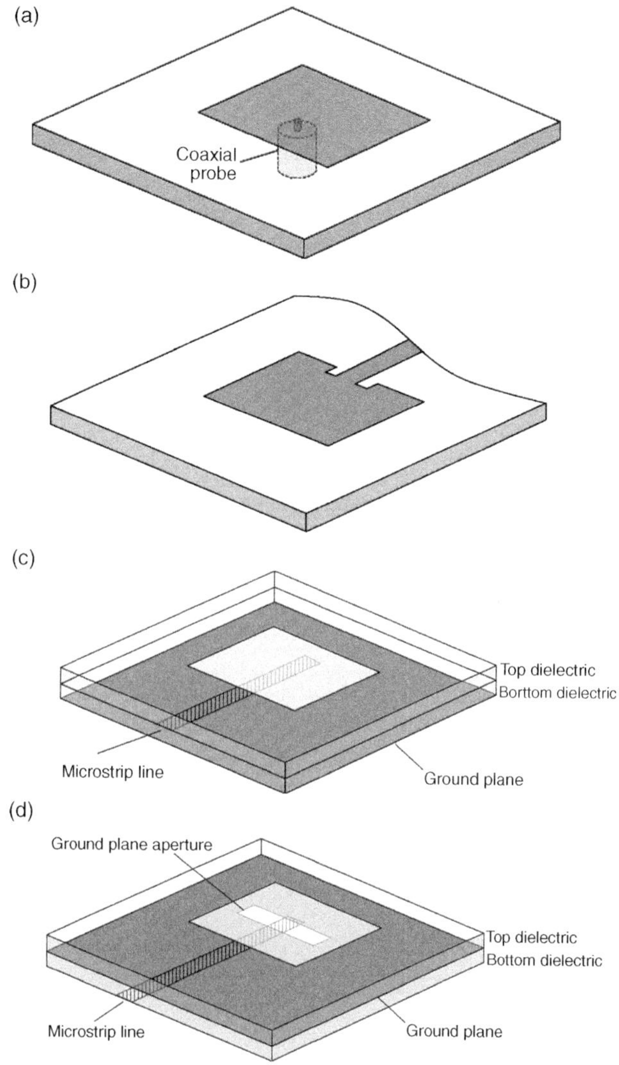

Figure 1.12 *Common microstrip antenna feeding techniques: (a) coaxial probe; (b) microstrip line; (c) proximity coupling; (d) aperture coupling.*

as mechanical support for the patch providing uniform spacing and mechanical stability. Lower values of relative dielectric constant (normally between 1 and 2) can be obtained by using polystyrene foam or honeycomb structures. Dielectrics based on fiberglass reinforced Teflon, also known as PTFE (polytetrafluoroethylene), typically provide a relative dielectric constant between 2 and 4. Higher values can be obtained through materials based on ceramic, quartz or alumina. However, these materials should be carefully employed as they provide a reduction in the patch size at the expense of radiation efficiency and mechanical stability. Another important selection driver for microstrip antenna substrates are dielectric losses. Acceptable dielectric losses are usually related to the application requirements and to the antenna architecture. In general, low tangent loss results in higher dielectric cost.

For a satellite microstrip antenna, the thermal behavior of the substrate and the temperature dependence of its main parameters are of primary importance. Indeed, a microstrip antenna mounted in a spacecraft operates under large ranges of thermal variations. At Earth-like distances from the Sun, temperatures of 273 ± 100 K can be expected, whereas larger ranges can be expected for interplanetary missions. For example, variations in the relative dielectric constant have a direct effect on the antenna operating frequency. For this reason, antenna bandwidth is usually evaluated through a sensitivity study including temperature effects. Material behavior in a space environment is discussed in Chapter 4 while basic effects such as multipaction and outgassing are described in Section 1.5. Aspects related to thermal conductivity, heat dissipation and mechanical stability are discussed in detail in Chapter 5.

In the last few years, the complexity of microstrip antennas has constantly increased. In particular, multilayer configurations with a high integration level and many vertical transitions are becoming evermore popular. In general, the development of multilayer assemblies can be particularly difficult as layers with different materials are usually employed for the circuit elements and for the antennas. Indeed, dielectrics of low dielectric constant are preferred for radiating structures whereas materials with high permittivity are usually employed for microwave circuits. This difference usually results in different coefficients of thermal expansion which can generate mechanical deformations of the multilayer structure. For this reason, new types of materials and fabrication processes are attracting the attention of many researchers. In particular, interesting results [63] have been obtained with liquid crystal polymers (LCPs). LCPs are low dielectric-constant, low loss-tangent [64] materials with very good package hermeticity and a low cost [65]. LCP has gained attention especially as a potential high-performance microwave substrate and packaging material for multilayer arrays and for highly integrated circuits [66].

Another interesting solution for multilayer antennas is the usage of low-temperature cofired ceramic (LTCC). This technology allows the implementation of flexible multilayer configurations with a high integration density and many vertical transitions. LTCC multilayer circuits are produced by firing in a single laminate multiple tape layers where conductive, dielectric and/or resistive pastes are selectively deposited to form transmission lines, resistors, inductors, and so on [67]. Although LTCC is mainly used for microwave integrated circuits, several interesting results have been obtained also for antenna elements [68–72]. LTCC antenna examples will be discussed in Chapter 11 along with examples of other emerging technologies for on-chip and in-package antenna integration.

1.3.5.5 Microstrip Antenna Configurations

It is very difficult to enumerate all the possible microstrip antenna configurations as new designs are proposed in every issue of specialized journals. In this subsection we discuss only a basic subset of possible configurations mainly related to dual or circular polarization operation and to multiband or wide-band applications.

Dual and Circular Polarization Operation In dual-polarized microstrip antennas two orthogonal modes have to be excited. Excitation can be actuated by two orthogonal feeds as shown in Figure 1.13a. Each feed is designed to excite a single mode and to be isolated as much as possible from the orthogonal mode. Ideally, orthogonal location of the feeds accomplishes high isolation between the two ports because each feed is located in the area where fields of the orthogonal modes are virtually null. However, in practice feed coupling is one of the major challenges in the design of dual-polarized antennas. Dual-feed arrangements provide narrow-band performance and are well suited only for thin substrates. An improved frequency response can be achieved by using two oppositely located feeds for each mode as shown in Figure 1.13b. The use of two feeds with 180° of phase difference reinforces the polarization mode, helps cancel unwanted feed radiation and

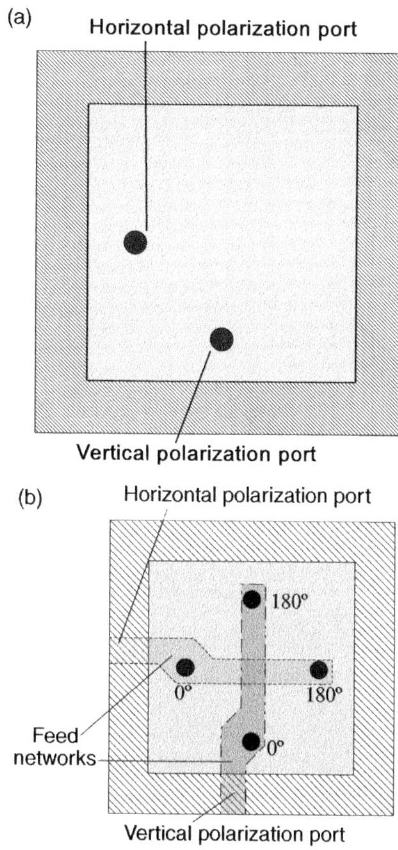

Figure 1.13 Dual-polarization generation in a square microstrip antenna: (a) dual feed; (b) four-feed configuration.

suppresses higher order modes of thick substrates [73]. Examples of dual-polarized antennas for SAR applications are reported in Chapter 13.

A microstrip antenna is circularly polarized when it radiates two orthogonally polarized electric fields with $\pm 90^\circ$ of phase difference between them. CP excitation is thus achieved by exciting two orthogonal modes in a patch radiator. All different circular polarization techniques for microstrip antennas can be grouped into two main classes: perturbative and multi-feed. Figure 1.14 shows examples of both techniques. To the first class of architectures belong single-feed microstrip antennas where perturbation of the patch shape is used to excite orthogonal modes with $\pm 90^\circ$ of phase difference. Typical single-feed CP configurations include square patches with truncated diagonal corners, circular patches with notches, elliptical patches or rectangular patches [16]. Although simple to fabricate, this configuration provides narrow-band CP performance.

In the second case, circular polarization is enforced by exciting a patch through multiple feeds orthogonally located and with an appropriate phase difference. Such a technique usually provides higher polarization purity, suppresses higher order modes and provides wider bandwidth. The main drawback is related to mutual coupling between multiple feeds and to the feeding network size and complexity. Examples of CP GNSS

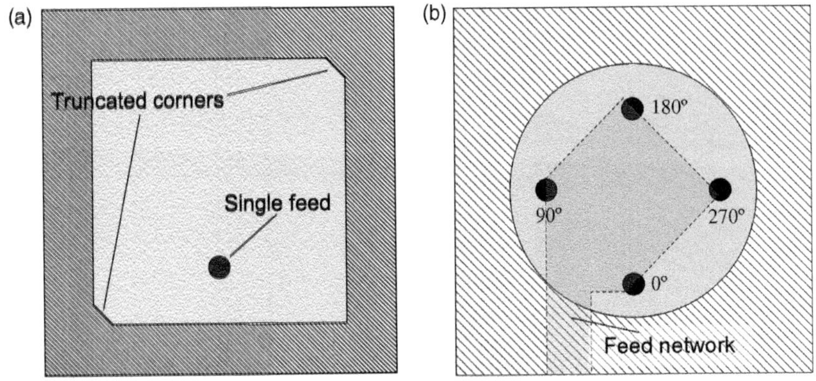

Figure 1.14 Circular polarization generation in microstrip antennas: (a) perturbative technique example, square patches with truncated corners; (b) multi-feed example: four- feed LHCP Rx antenna.

microstrip antennas are discussed in Chapter 14 while examples of polarization enhancement techniques at array level are described in Chapter 9.

Bandwidth Enhancement Techniques Bandwidth enhancement of microstrip antennas is usually required for many practical applications. Bandwidth can be increased by lowering the Q factor of the microstrip antenna. This can be achieved by using thick substrates or by using materials with low relative dielectric constant. In both cases, insurgence of higher order modes should be carefully considered. Another common bandwidth enhancement technique is to use radiators with multiple contiguous resonances. This method can be implemented by using parasitic stacked patches [74] or through reactive loading by means of shaped slots, notches, cuts, pins or posts. Wide-band can be achieved also at feed level by designing broadband matching networks [75] or by means of reactive feeds such as L-shaped probes [76–78].

1.4 Arrays

Antenna arrays are a set of antennas arranged to provide highly directive patterns. They yield an increment in the aperture area which can be controlled geometrically and electrically by optimally setting the location and the excitation of the array elements. The geometry of an arbitrary array of N elements is represented in Figure 1.15. In its most general form, the radiation pattern of an arbitrary array of N antennas can be written as

$$E_{total}(\theta, \phi) = \sum_{n=1}^{N} A_n \, F_n(\theta_n, \phi_n) e^{-j(k_0|r_n| + \beta_n)} \tag{1.27}$$

where:

- $E_{total}(\theta, \phi)$ is the total far field radiated by the array in the (θ, ϕ) direction
- A_n is the amplitude factor of the nth array element

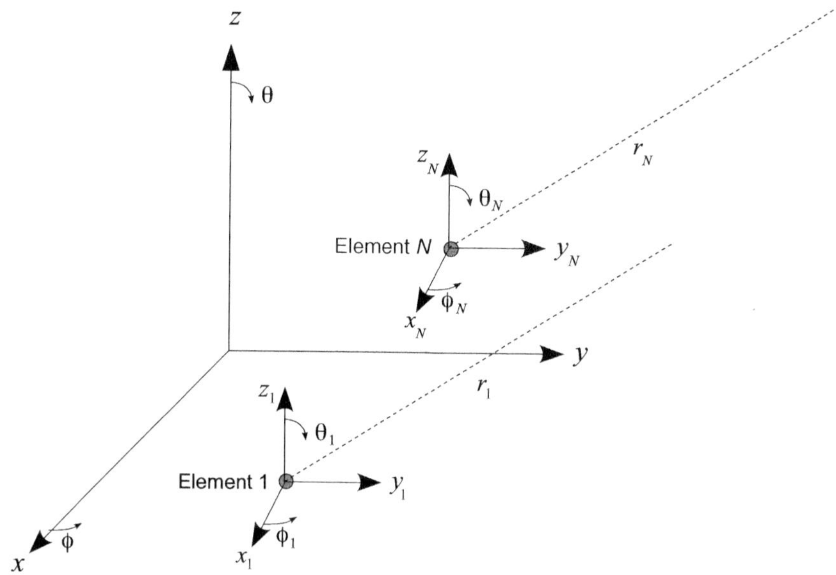

Figure 1.15 *Far-field geometry of an arbitrary array of N elements.*

- $F_n(\theta_n, \phi_n)$ is the pattern function of the nth array element defined w.r.t. the local coordinate system
- k_0 is the free-space propagation constant
- $|r_n|$ is the distance from the observation point of the nth array element
- β_n is the phase of the nth array element.

More simple and compact expressions of the array pattern can be obtained when identical antennas are employed and regular geometries are adopted. In these cases, the total pattern can be decomposed into two contributions: the element pattern, $E_{element}(\theta, \phi)$, and the so-called array factor, $AF(\theta, \phi)$:

$$E_{total}(\theta, \phi) = E_{element}(\theta, \phi)AF(\theta, \phi) \tag{1.28}$$

The array factor is a function of the array geometry, the inter-element distance, the element excitation in amplitude and phase, the number of elements and the frequency. The simplest case is that of the so-called uniform array: a 'linear array of identically oriented and equally spaced radiating elements having equal current amplitudes and equal phase increments between excitation currents' [1]. For a uniform linear array of N elements arranged along the z-axis, with inter-element distance d and progressive phase shift β, the array factor can be written as

$$AF(\theta, \phi) = \frac{\sin(N\psi/2)}{\sin(\psi/2)} \tag{1.29}$$

where $\psi = k_0 d\cos\theta + \beta$. When $\psi = 0$ the array factor has a maximum which corresponds to the main array lobe. Other array factor maxima are found when $\psi/2 = \pm r\pi$ with $r = 1, 2, \ldots$. Lobes corresponding to these maxima are referred to as *grating lobes* and they have the same array factor amplitude of the main lobe. As a consequence, directions of maximum amplitude of the array factor can be written as

$$\theta_{\max} = \arccos[(\lambda/2\pi d)(-\beta \pm 2r\pi)], \quad r = 0, 1, 2, \ldots \tag{1.30}$$

Both the main lobe ($r = 0$) and grating lobe directions depend on the array spacing, d, and on sequential phase shift, β. However, grating lobes appear only when the argument of the cosine in Equation 1.30 is less than in module 1. For example, for a uniform linear array with equiphase distribution ($\beta = 0$) the main beam direction is $\theta_{\max,r=0} = 90°$, while grating lobes do not appear for inter-element spacings of less than a wavelength. For $d = \lambda$ grating lobes are in the visible region at $\theta_{\max,r=1} = 0°$ and $180°$. In general, grating lobes are an unwanted effect that has to be carefully avoided, especially in phased-array scenarios.

In general, the geometry of the array and its excitations should be defined through a synthesis procedure where the starting point is a set of given requirements specified on the array radiation pattern [79–81]. The target of the synthesis process is to find an array geometry and excitation distribution to suitably approximate the desired pattern.

1.4.1 Array Configurations

In this subsection a basic review of the most common array types is presented.

1.4.1.1 Direct Radiating Arrays (DRA)s

The simplest case is that of an array of regular geometry whose elements are excited through a beam-forming network (BFN). BFNs distribute and/or collect the power of individual elements to a single port. Power distribution is accomplished so that each element receives the signal with a desired amplitude and phase. In their most typical form, BFNs are implemented through binary power combiners/dividers as shown in Figure 1.16a. Usually, DRA efficiency is limited by BFN losses which become evident in arrays with a large number of elements. In some cases, it is possible to obtain good radiation performance by selectively removing some of the DRA elements. This configuration is referred to as a thinned array [82–85]. In other cases, a synthesis technique can be used to design a so-called sparse array whose elements are uniformly excited and located in a non-regular grid [82,83,86–88].

1.4.1.2 Phased Arrays

In many applications, the main antenna beam has to be moved dynamically to point in different directions. Although mechanical scanning is possible under certain conditions, a common solution is represented by phased arrays. In phased-array antennas beams are formed by shifting the phase of the signal emitted from each radiating element. An example of a corporate-fed linear phased array is shown in Figure 1.16b. In this case, the direction of maximum radiation, $\theta_{\max}$ in Equation 1.30, is controlled by changing the phase of each array element, φ_i, so that different progressive phase shifts, β, can be achieved. A common problem in large phased arrays with wide-band elements is referred to as scan blindness. This phenomenon is generated by inter-element mutual coupling. Indeed, the active impedance of a given radiator in the array changes as a function of the amplitude and phase distribution across the entire array. As a result, it can happen that when the array is scanned, at certain angles the module of the reflection coefficient of the array rapidly increases to 1 and the

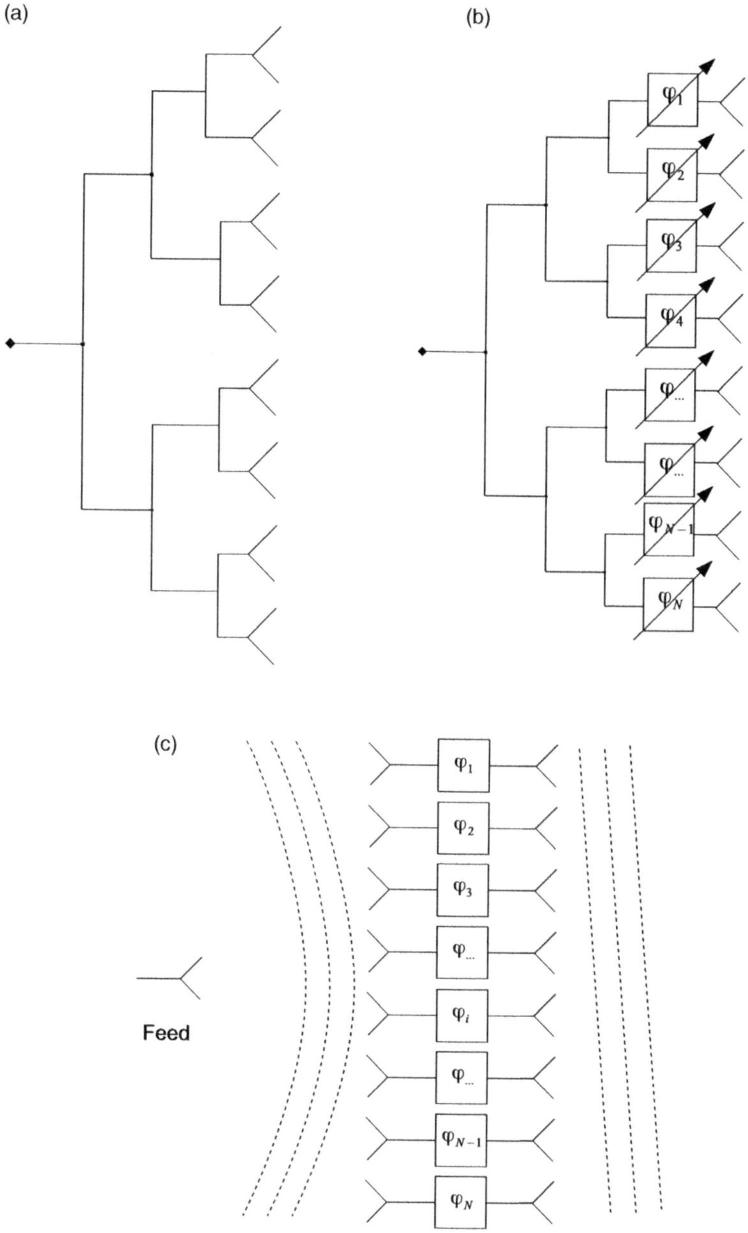

Figure 1.16 *Array configurations: (a) direct radiating array (DRA); (b) phased array; (c) transmit array.*

array pattern forms a null [89,90]. Therefore, scan blindness limits the scan range and lowers the antenna efficiency.

This book discusses several examples of phased arrays for space applications. Some of them are presented in Chapter 9 'Microstrip Array Technologies for Space Applications', in Chapter 10 'Printed Reflectarray Antennas for Space Applications' and in Chapter 13 'Antennas for Spaceborne Synthetic Aperture Radar'.

1.4.1.3 Reflectarrays

Reflectarrays [91] are flat reflectors illuminated by an external feed. The reflecting surface is realized through a spatially fed antenna array as shown in Figure 10.1 for the case of a printed reflectarray. In other terms, reflectarrays are discrete flat reflectors where the reflected field is controlled by each element of the array which should be designed to reradiate the incident field at a proper phase. A detailed review of reflectarray antennas for satellite applications is given in Chapter 10 of this handbook.

1.4.1.4 Transmit Arrays

Transmit array antennas (Figure 1.16c), also called lens antennas, are planar discrete lenses that operate a phase-front transformation by converting an incident spherical wavefront into an outgoing plane wave propagating in a specified direction. Transmit array antennas are an attractive solution for achieving high gain at millimeter wave frequencies where the free-space feeding improves the radiation efficiency by eliminating the losses that occur with corporate feed networks. With respect to reflectarrays, they are inherently less affected by surface errors if employed in deployable configurations.

1.5 Basic Effects of Antennas in the Space Environment

Satellites must survive for the duration of their mission with the required stability and survivability. For this reason, mechanical and thermal behavior as well as material characteristics are a major issue for all satellite components. From this point of view, since antennas are located in the external satellite body, they are highly exposed to radiation and to thermal variations. In this section we will provide a basic description of the most important phenomena involving antennas in space. An in-depth description of specific space environmental threats to space antenna constituent materials is presented in Chapter 4 while the mechanical and thermal behavior of satellite antennas is provided in Chapter 5. Moreover, Chapter 6 presents a detailed description of the most important tests required for space antenna assessment before flight.

1.5.1 Multipaction

Multipaction [92], or multiple impaction, is a resonance type of discharge that occurs between two electrodes with RF fields, usually in a vacuum or low-pressure condition. In a high-vacuum environment, an electron may have a free path that is larger than the gap between the electrodes. When the electron, accelerated by the electric field, collides with the electrode it might cause the emission of secondary electrons from the material surface. If the impact energy, the frequency and the distance between the two conductors are favorable, a resonant multiplication of the number of electrons takes place resulting in operational impairments and potential physical damage. The generation of multipaction depends upon several constraints [93]:

1. the vacuum condition (*usually lower than* 10^{-3}*torr*);
2. the applied RF voltage (*depends on the material and on the angle of incidence of the primary electrons*);
3. the electrode geometry and the operating frequency (*the gap size should correspond to a multiple half cycle of the applied RF voltage to satisfy the condition of electron resonance*);
4. the material surface (*material contaminations or impurities enhance the possible occurrence of multipaction*).

In RF space systems, multipaction limits the power handling capability of RF systems and can cause loss and distortion of the RF signal (increase in noise figure or bit error rate) as well as damage to RF components or subsystems due to excess RF power being reflected back or dissipated by them [94].

Multipactor prevention represents a very significant problem in the design and implementation of high-power antenna feed networks for communications satellites. In general, multipaction can be avoided by optimally modifying the device geometries, by laminating electrodes with particular materials or by reducing the RF power levels. For instance, in a microstrip antenna, the multipaction effect can be estimated by assuming that the two electrodes are the patch and its ground plane. High power handling cannot be achieved when thin dielectrics are used. Design rules or RF design tools such as the Multipactor Calculator [95] developed by the European Space Agency (ESA) can be used to achieve design margins that preclude the onset of multipaction.

1.5.2 Passive Inter-Modulation

Passive inter-modulation (PIM) is comparable to the phenomenon taking place in active devices due to their inherent nonlinearity when two or more RF carriers are mixed in a passive system and form unwanted signals. Nonlinearity in a passive device such as an antenna can be caused by several physical effects which can be grouped into two categories: contact nonlinearity and material nonlinearity. Causes of contact nonlinearity include the formation of a junction capacitance due to thin oxide layers between conductors, the presence of contaminating particles or mechanical imperfections on the surfaces, the tunnel/Schottky effect, contact resistance caused by two dissimilar metals or loose metal-to-metal contacts (some metals, like stainless steel, are more susceptible than others, like Al or Ti alloys) [93]. Material nonlinearities are generally due to hysteresis effect in ferromagnetic materials or to insufficient thickness of plated metal, causing RF heating. Contact sources can be excited by relatively low energy. For this reason, the accuracy of the manufacturing and assembly process as well as materials employed in a satellite RF system have to be accurately validated in an attempt to detect and eliminate any possible PIM source. A material selection guide for PIM reduction is proposed in Section 4.3.3.

The PIM interference caused by antenna nonlinearity has a serious impact on the performance of high-power multi-frequency communication systems, especially when the antenna is shared by the transmitter and the receiver at the same time [96]. Typical satellite communication systems are designed to avoid PIM orders below VII or XI. However, passive inter-modulation evaluation is not entirely amenable to theoretical design and experimental assessment is required in most cases.

1.5.3 Outgassing

Outgassing is generated by pockets of gas trapped or absorbed in materials or components during manufacture. In a vacuum environment, trapped gas evolves causing a material to lose volatile mass particles acting as contaminants to other surfaces and harming the satellite.

All materials used for space flight applications should satisfy the outgassing requirements [97–99] recognized by NASA, ESA and other space agencies. Contamination requirements are usually expressed in terms of total mass loss (TML), collected volatile condensable materials (CVCM)s and recovered mass loss (RML). Typical acceptance criteria are TML $< 1.0\%$, CVCM $< 0.10\%$ and RML $< 1.0\%$. Screening test data of materials that have successfully passed the NASA outgassing standard is listed in the database ‘Outgassing Data for Selecting Spacecraft Materials’ [100]. Materials included in this database are approved for use in a space environment unless more stringent constraints are required by a specific application.

References

1. IEEE (1983) *IEEE Standard Definitions of Terms for Antennas.*
2. IEEE (1979) *IEEE Standard Test Procedures for Antennas.*
3. Pozar, D.M. (1997) *Microwave Engineering*, 2nd edn, John Wiley & Sons, Inc.
4. Balanis, C. (2008) *Modern Antenna Handbook*, John Wiley & Sons, Inc.
5. Balanis, C.A. (1996) *Antenna Theory: Analysis and Design*, 2nd edn, John Wiley & Sons, Inc.
6. Davies, K. (1965) *Ionospheric Radio Propagation*, US Department of Commerce.
7. Sorensen, E. (1961) Magneto-ionic Faraday rotation of the radio signals on 40 MC from satellite 1957α (Sputnik I). *IRE Transactions on Antennas and Propagation*, **9**(3), 241–247.
8. Brookner, E., Hall, W.M. and Westlake, R.H. (1985) Faraday loss for L-band radar and communications systems. *IEEE Transactions on Aerospace and Electronic Systems*, **21**(4), 459–469.
9. Ludwig, A. (1973) The definition of cross polarization. *IEEE Transactions on Antennas and Propagation*, **21**(1), 116–119.
10. Teichman, M. (1970) Precision phase center measurements of horn antennas. *IEEE Transactions on Antennas and Propagation*, **18**(5), 689–690.
11. Best, S.R. (2004) A 7-turn multi-step quadrifilar helix antenna providing high phase center stability and low angle multipath rejection for GPS applications. IEEE Antennas and Propagation Society International Symposium, 2004, vol. 3, pp. 2899–2902.
12. Dijk, J., Jeuken, M., and Maanders, E.J. (1968) Antenna noise temperature. *Proceedings of the Institution of Electrical Engineers*, **115**(10), 1403–1410.
13. Balanis, C.A. (1989) *Advanced Engineering Electromagnetics*, Solution Manual, John Wiley & Sons, Inc.
14. Stutzman, W.L. and Thiele, G.A. (1997) *Antenna Theory and Design*, 2nd edn, John Wiley & Sons, Inc.
15. Ansys Corporation, Ansys HFSS, version 13, Canonsburg, PA.
16. Balanis, C.A. (2005) *Antenna Theory: Analysis and Design*, 3rd edn, Wiley-Interscience.
17. Simmons, A.J. and Kay, A.F. (1966) The scalar feed – a high performance feed for large paraboloid reflectors. Design and Construction of Large Steerable Aerials, IEE Conference Publication, vol. 21.
18. Clarricoats, P.J.B. (1984) *Corrugated Horns for Microwave Antennas*, The Institution of Engineering and Technology.
19. Teniente, J., Goni, D., Gonzalo, R. and del-Rio, C. (2002) Choked Gaussian antenna: extremely low sidelobe compact antenna design. *IEEE Antennas and Wireless Propagation Letters*, **1**(1), 200–202.
20. Ludwig, A. (1966) Radiation pattern synthesis for circular aperture horn antennas. *IEEE Transactions on Antennas and Propagation*, **14**(4), 434–440.
21. Granet, C., Bird, T.S. and James, G.L. (2000) Compact multimode horn with low sidelobes for global earth coverage. *IEEE Transactions on Antennas and Propagation*, **48**(7), 1125–1133.
22. Imbriale, D.W.A. (2003) *Large Antennas of the Deep Space Network*, 1st edn, Wiley-Interscience.
23. Love, A.W. and Love, A.W. (1978) *Reflector Antennas*, Illustrated edn, IEEE Press.
24. Imbriale, W.A. (2008) Reflector antennas, in *Modern Antenna Handbook*, 1st edn, Wiley-Interscience.
25. Rahmat-Samii, Y. (1988) Reflector antennas, in *Antenna Handbook: Antenna Theory*, 1st edn, vol. 2, Van Nostrand Reinhold.
26. Imbriale, W., Ingerson, P. and Wong, W. (1974) Large lateral feed displacements in a parabolic reflector. *IEEE Transactions on Antennas and Propagation*, **22**(6), 742–745.
27. Ruze, J. (1965) Lateral-feed displacement in a paraboloid. *IEEE Transactions on Antennas and Propagation*, **13**(5), 660–665.
28. Ruze, J. (1966) Antenna tolerance theory—a review. *Proceedings of the IEEE*, **54**(4), 633–640.
29. McEwan, N.J. and Goldsmith, P.F. (1989) Gaussian beam techniques for illuminating reflector antennas. *IEEE Transactions on Antennas and Propagation*, **37**(3), 297–304.
30. Brain, D.J. and Rudge, A.W. (1984) Efficient satellite antennas. *Electronics and Power*, **30**(1), 51–56.
31. Rudge, A.W. and Adatia, N.A. (1978) Offset-parabolic-reflector antennas: a review. *Proceedings of the IEEE*, **66**(12), 1592–1618.

32. Chu, T.-S. and Turrin, R. (1973) Depolarization properties of offset reflector antennas. *IEEE Transactions on Antennas and Propagation*, **21**(3), 339–345.
33. Strutzman, W. and Terada, M. (1993) Design of offset-parabolic-reflector antennas for low cross-pol and low sidelobes. *IEEE Antennas and Propagation Magazine*, **35**(6), 46–49.
34. Chu, T.S. (1977) Cancellation of polarization rotation in an offset paraboloid by a polarization grid. *Bell System Technical Journal*, **56**, 977–986.
35. Lier, E. and Skyttemyr, S.A. (1994) A shaped single reflector offset antenna with low cross-polarization fed by a lens horn. *IEEE Transactions on Antennas and Propagation*, **42**(4), 478–483.
36. Fiebig, D., Wohlleben, R., Prata, A., and Rusch, W.V. (1991) Beam squint in axially symmetric reflector antennas with laterally displaced feeds. *IEEE Transactions on Antennas and Propagation*, **39**(6), 774–779.
37. Eilhardt, K., Wohlleben, R. and Fiebig, D. (1994) Compensation of the beam squint in axially symmetric, large dual reflector antennas with large-ranging laterally displaced feeds. *IEEE Transactions on Antennas and Propagation*, **42** (10), 1430–1435.
38. Prasad, K.M. and Shafai, L. (1988) Improving the symmetry of radiation patterns for offset reflectors illuminated by matched feeds. *IEEE Transactions on Antennas and Propagation*, **36**(1), 141–144.
39. Hannan, P. (1961) Microwave antennas derived from the Cassegrain telescope. *IRE Transactions on Antennas and Propagation*, **9**(2), 140–153.
40. Mittra, R. and Galindo-Israel, V. (1980) Shaped dual reflector synthesis. *IEEE Antennas and Propagation Society Newsletter*, **22**(4), 4–9.
41. Tanaka, H. and Mizusawa, M. (1975) Elimination of cross polarization in offset dual-reflector antennas. *Electronics and Communications in Japan*, **58**, 71–78.
42. Raab, A. and Farrell, K. (1978) A shaped beam multifeed 14/12GHz antenna for ANIK-B. IEEE Antennas and Propagation Society International Symposium, vol. 16, pp. 416–419.
43. Huber, S., Younis, M., Patyuchenko, A. and Krieger, G. (2010) Digital beam forming concepts with application to spaceborne reflector SAR systems. 11th International Radar Symposium (IRS), pp. 1–4.
44. Ramanujam, P., Lopez, L.F., Shin, C. and Chwalek, T.J. (1993) A shaped reflector design for the DirecTv direct broadcast satellite for the United States. IEEE Antennas and Propagation Society International Symposium, AP-S. Digest, vol. 2. pp. 788–791.
45. Ramanujam, P. and Law, P.H. (1999) Shaped reflector and multi-feed paraboloid – a comparison. IEEE Antennas and Propagation Society International Symposium, vol. 2, pp. 1136–1139.
46. King, H. and Wong, J. (1980) Characteristics of 1 to 8 wavelength uniform helical antennas. *IEEE Transactions on Antennas and Propagation*, **28**(2), 291–296.
47. Nakano, H., Yamauchi, J. and Mimaki, H. (1988) Backfire radiation from a monofilar helix with a small ground plane. *IEEE Transactions on Antennas and Propagation*, **36**, 1359–1364.
48. Jordan, R.L. (1980) The Seasat – q synthetic aperture radar system. *IEEE Journal of Oceanic Engineering*, **5**(2), 154–164.
49. Cahill, R., Cartmell, I., van Dooren, G. *et al.* (1998) Performance of shaped beam quadrifilar antennas on the METOP spacecraft. IEE Proceedings – Microwaves, Antennas and Propagation, 145 (1), 19–24.
50. Lier, E. and Melcher, R. (2009) A modular and lightweight multibeam active phased receiving array for satellite applications: design and ground testing. *IEEE Antennas and Propagation Magazine*, **51**(1), 80–90.
51. Kilgus, C. (1969) Resonant quadrifilar helix. *IEEE Transactions on Antennas and Propagation*, **17**(3), 349–351.
52. James, J.R. and IE Engineers (1989) *Handbook of Microstrip Antennas*, IET.
53. Deschamp, G.A. (1953) Microstrip microwave antennas. Paper presented at the Antenna Applications Symposium.
54. Gutton, H. and Baissinot, G. (1955) Flat aerial for ultra high frequencies, French Patent 703113.
55. Pozar, D.M. and Schaubert, D. (1995) *Microstrip Antennas: The Analysis and Design of Microstrip Antennas and Arrays*, John Wiley & Sons, Inc.
56. Woo, R. (1968) Multipacting breakdown in coaxial transmission lines. *Proceedings of the IEEE*, **56**(4), 776–777.
57. Munson, R. (1974) Conformal microstrip antennas and microstrip phased arrays. *IEEE Transactions on Antennas and Propagation*, **22**(1), 74–78.

58. Derneryd, A. (1978) A theoretical investigation of the rectangular microstrip antenna element. *IEEE Transactions on Antennas and Propagation*, **26**(4), 532–535.
59. Pues, H. and van de Capelle, A. (1984) Accurate transmission-line model for the rectangular microstrip antenna. *IEE Proceedings H (Microwaves, Optics and Antennas)*, **131**(6), 334–340.
60. Jackson, D.R., Williams, J.T., Bhattacharyya, A.K. *et al.* (1993) Microstrip patch designs that do not excite surface waves. *IEEE Transactions on Antennas and Propagation*, **41**(8), 1026–1037.
61. Hall, P.S. (1987) Probe compensation in thick microstrip patches. *Electronics Letters*, **23**(11), 606.
62. Pozar, D.M. (1985) Microstrip antenna aperture-coupled to a microstripline. *Electronics Letters*, **21**(2), 49–50.
63. DeJean, G., Bairavasubramanian, R., Thompson, D. *et al.* (2005) Liquid crystal polymer (LCP): a new organic material for the development of multilayer dual-frequency/dual-polarization flexible antenna arrays. *Antennas and Wireless Propagation Letters*, **4**(1), 22–25.
64. Jayaraj, K., Noll, T.E. and Singh, D. (1995) RF characterization of a low cost multichip packaging technology for monolithic microwave and millimeter wave integrated circuits. Proceedings of the URSI International Symposium on Signals, Systems, and Electronics, ISSSE'95, pp. 443–446.
65. Farrell, B. and St Lawrence, M. (2002) The processing of liquid crystalline polymer printed circuits. Proceedings of the 52nd Electronic Components and Technology Conference, pp. 667–671.
66. Brownlee, K., Raj, P.M., Bhattacharya, S.K. *et al.* (2002) Evaluation of liquid crystal polymers for high performance SOP application. Proceedings of the 52nd Electronic Components and Technology Conference, pp. 676–680.
67. Imanaka, Y. (2010) *Multilayered Low Temperature Cofired Ceramics (LTCC) Technology*, Softcover reprint of hardcover 1st edn, 2005, Springer.
68. Wi, S.-H., Zhang, Y.P., Kim, H. *et al.* (2011) Integration of antenna and feeding network for compact UWB transceiver package. *IEEE Transactions on Components, Packaging and Manufacturing Technology*, **1**(1), 111–118.
69. Seki, T., Honma, N., Nishikawa, K. and Tsunekawa, K. (2005) A 60-GHz multilayer parasitic microstrip array antenna on LTCC substrate for system-on-package. *IEEE Microwave and Wireless Components Letters*, **15**(5), 339–341.
70. Wi, S.-H., Kim, J.-S., Kang, N.-K. *et al.* (2007) Package-level integrated LTCC antenna for RF package application. *IEEE Transactions on Advanced Packaging*, **30**(1), 132–141.
71. Chen, S.C., Liu, G.C., Chen, X.Y. *et al.* (2010) Compact dual-band GPS microstrip antenna using multilayer LTCC substrate. *IEEE Antennas and Wireless Propagation Letters*, **9**, 421–423.
72. Sanadgol, B., Holzwarth, S., Milano, A. and Popovich, R. (2010) 60GHz substrate integrated waveguide fed steerable LTCC antenna array. Proceedings of the Fourth European Conference on Antennas and Propagation (EuCAP), pp. 1–4.
73. Chiba, T., Suzuki, Y. and Miyano, N. (1982) Suppression of higher modes and cross polarized component for microstrip antennas. Antennas and Propagation Society International Symposium, vol. 20, pp. 285–288.
74. Croq, F. and Pozar, D.M. (1991) Millimeter-wave design of wide-band aperture-coupled stacked microstrip antennas. *IEEE Transactions on Antennas and Propagation*, **39**(12), 1770–1776.
75. Pues, H.F. and Van de Capelle, A.R. (1989) An impedance-matching technique for increasing the bandwidth of microstrip antennas. *IEEE Transactions on Antennas and Propagation*, **37**(11), 1345–1354.
76. Park, J., Na, H.-G., and Baik, S.-H. (2004) Design of a modified L-probe fed microstrip patch antenna. *IEEE Antennas and Wireless Propagation Letters*, **3**(1), 117–119.
77. Guo, Y.-X., Chia, M.Y., Chen, Z.N. and Luk, K.-M. (2004) Wide-band L-probe fed circular patch antenna for conical-pattern radiation. *IEEE Transactions on Antennas and Propagation*, **52**(4), 1115–1116.
78. Mak, C.L., Luk, K.M., Lee, K.F. and Chow, Y.L. (2000) Experimental study of a microstrip patch antenna with an L-shaped probe. *IEEE Transactions on Antennas and Propagation*, **48**(5), 777–783.
79. Schelkunoff, S. (1943) A mathematical theory of linear arrays. *Bell System Technical Journal*, **22**(1), 80–107.
80. Dolph, C.L. (1946) A current distribution for broadside arrays which optimizes the relationship between beam width and side-lobe level. *Proceedings of the IRE*, **34**(6), 335–348.
81. Bucci, O.M., D'Elia, G., Mazzarella, G. and Panariello, G. (1994) Antenna pattern synthesis: a new general approach. *Proceedings of the IEEE*, **82**(3), 358–371.

82. Guiraud, C., Cailloce, Y. and Caille, G. (2007) Reducing direct radiating array complexity by thinning and splitting into non-regular sub-arrays. Paper presented at the 29th ESA Antenna Workshop on Multiple Beams and Reconfigurable Antennas, pp. 211–214.
83. Toso, G., Mangenot, C. and Roederer, A.G. (2007) Sparse and thinned arrays for multiple beam satellite applications. Paper presented at the 29th ESA Antenna Workshop on Multiple Beams and Reconfigurable Antennas, pp. 207–210.
84. Leeper, D.G. (1999) Isophoric arrays–massively thinned phased arrays with well-controlled sidelobes. *IEEE Transactions on Antennas and Propagation*, **47**(12), 1825–1835.
85. Haupt, R.L. (1994) Thinned arrays using genetic algorithms. *IEEE Transactions on Antennas and Propagation*, **42**(7), 993–999.
86. Leahy, R.M. and Jeffs, B.D. (1991) On the design of maximally sparse beamforming arrays. *IEEE Transactions on Antennas and Propagation*, **39**(8), 1178–1187.
87. Goodman, N.A. and Stiles, J.M. (2003) Resolution and synthetic aperture characterization of sparse radar arrays. *IEEE Transactions on Aerospace and Electronic Systems*, **39**(3), 921–935.
88. Chen, K., Yun, X., He, Z. and Han, C. (2007) Synthesis of sparse planar arrays using modified real genetic algorithm. *IEEE Transactions on Antennas and Propagation*, **55**(4), 1067–1073.
89. Pozar, D. and Schaubert, D. (1984) Scan blindness in infinite phased arrays of printed dipoles. *IEEE Transactions on Antennas and Propagation*, **32**(6), 602–610.
90. Milligan, T.A. (2005) *Modern Antenna Design*, John Wiley & Sons, Ltd.
91. Huang, J. and Encinar, J.A. (2008) *Reflectarray Antennas*, John Wiley & Sons, Inc.
92. Vaughan, J.R. (1988) Multipactor. *IEEE Transactions on Electron Devices*, **35**(7), 1172–1180.
93. Kudsia, C., Cameron, R. and Tang, W.-C. (1992) Innovations in microwave filters and multiplexing networks for communications satellite systems. *IEEE Transactions on Microwave Theory and Techniques*, **40**(6), 1133–1149.
94. Yu, M. (2007) Power-handling capability for RF filters. *IEEE Microwave Magazine*, **8**(5), 88–97.
95. European Space Agency (2007) Multipactor Calculator, http://multipactor.esa.int/.
96. Lui, P.L. (1990) Passive intermodulation interference in communication systems. *Electronics & Communication Engineering Journal*, **2**(3), 109–118.
97. ASTM (2009) ASTM E1559. *Standard Test Method for Contamination Outgassing Characteristics of Spacecraft Materials*.
98. The European Cooperation for Space Standardization (2000) ECSS-Q70-02. *Thermal vacuum outgassing test for the screening of space materials*.
99. ASTM (2007) ASTM E595. *Standard Test Method for Total Mass Loss and Collected Volatile Condensable Materials from Outgassing in a Vacuum Environment*.
100. NASA (2011) Outgassing Data for Selecting Spacecraft Materials, http://outgassing.nasa.gov.

2

Space Antenna Modeling

Jian Feng Zhang, Xue Wei Ping, Wen Ming Yu, Xiao Yang Zhou, and Tie Jun Cui

School of Information Science and Engineering, Southeast University, Nanjing, China

2.1 Introduction

Since World War II, numerous techniques for solving scientific and engineering problems have been developed based on numerical approaches with the aid of electronic computers [1]. In 1952, computer-aided simulation methods were first used on the weather forecast and gained great success. In the past few decades, various numerical methods have been developed and used in different areas, producing many new branches such as computational electromagnetics (CEM), computational fluid dynamics (CFD), computational physics, computational chemistry, and so on. CEM is a new subject that combines numerical methods with electronic computers to solve problems in electromagnetics.

In space applications, communications between various systems are mainly achieved through electromagnetic waves generated by different types of antennas, such as the helix antenna for telemetry and tracking control, the antenna array for the data downlink and navigation, and so on. Radiation patterns of such antennas should be well designed to meet practical requirements. Theoretically, this can be determined by solving Maxwell's equations. However, for most practical antennas, the closed-form solutions of Maxwell's equations are too complex to obtain by pencil and paper. Hence numerical simulation methods based on CEM have found their way into antenna design and analysis. Besides determining the field distributions and radiation patterns, CEM can also provide crucial information and an understanding of electromagnetic operations in the space antenna system, which is difficult or even impossible to achieve by experimental methods or analytical calculations. As space antenna systems become more complicated, numerical simulation methods will be increasingly important, with a concurrent need to reduce the cost and time from design to manufacture. It is therefore very important for modern space antenna engineers to have a good understanding of CEM.

Space Antenna Handbook, First Edition. Edited by William A. Imbriale, Steven (Shichang) Gao and Luigi Boccia.

2.1.1 Maxwell's Equations

The goal of this chapter is how to solve Maxwell's equations efficiently derived from real problems. We start by tracing a little of the history of the development of the electromagnetic theory and CEM. Maxwell's equations were completed in 1864 by James Clerk Maxwell [2]. This is one of the greatest discoveries up to the present, and unifies traditional electromagnetism and the theory of light. Today, Maxwell's equations appear in a neat form in the time domain:

$$\nabla \times \mathbf{E} = -\frac{\partial \mathbf{B}}{\partial t} \tag{2.1}$$

$$\nabla \times \mathbf{H} = \frac{\partial \mathbf{D}}{\partial t} + \mathbf{J} \tag{2.2}$$

$$\nabla \cdot \mathbf{B} = 0 \tag{2.3}$$

$$\nabla \cdot \mathbf{D} = \rho \tag{2.4}$$

However, the original equations that Maxwell derived in 1864 were component forms of (2.1)–(2.4) in the Cartesian coordinate system. Through the efforts of Oliver Heaviside and Heinrich Hertz, concise expressions for (2.1)–(2.4) were obtained. Hence, in these early times the equations were called HH (Heaviside and Hertz) equations. Later, as the suggestion of Albert Einstein, they were named Maxwell's equations. Besides these four equations, another fundamental equation, known as the equation of continuity for electric current, is given by

$$\nabla \cdot \mathbf{J} = -\frac{\partial \rho}{\partial t} \tag{2.5}$$

2.1.2 CEM

Many numerical methods have been developed in CEM, including high-frequency methods, full-wave simulation methods, mode matching methods, and so on. Each of these methods has its advantages and disadvantages, and is optimal for a portion of electromagnetic problems. The advice given by J. C. Rautio, the founder of Sonnet Software, at the beginning of 2003 [3], 'No single EM tool can solve all problems; an informed designer must select the appropriate tool for the appropriate problem', is also valid for the digital design of space antennas as their profiles are becoming more and more complicated. A satellite usually has tens of antennas to accomplish different applications, such as communication, navigation, and detection. The purpose of this section is to give the space antenna designer a brief introduction to the most popular algorithms in CEM, so the optimal analysis technique can be easily chosen for the given structure.

2.1.2.1 High-Frequency Methods

High-frequency methods mainly include physical optics (PO), geometrical optics (GO) [4], the uniform theory of diffraction (UTD) [5], the physical theory of diffraction (PTD) [6], and the shooting and bouncing ray (SBR) method [7], among others. These methods require both mathematical and physical approximations in dealing with Maxwell's equations. In general, the accuracy of these methods increases asymptotically with frequency. For suitable problems, they are very powerful, but the intrinsic approximations of the physics limit their usage for general problems. When the frequency is low, or the structure is inhomogeneous, the accuracy cannot be guaranteed; even the mesh density and digital precision are greatly increased.

2.1.2.2 *Full-Wave Methods*

Full-wave methods mainly include the method of moments (MOM) [8–15], the finite element method (FEM) [16–22], and the finite difference time-domain (FDTD) method [23–29]. Theoretically, Maxwell's equations can be accurately solved by such methods, since no physical approximations are made. However, actually, the precision of these methods is strongly related to mesh discretization. Although full-wave methods are valid at any frequency, they are not suitable for extremely electrically large problems as the computational overhead increases rapidly with the electrical size of the structure. To distinguish them from the asymptotic high-frequency methods, the full-wave methods are also called low-frequency methods. In space antenna design, the full-wave methods are more suitable than the high-frequency methods because of their higher accuracy. Hence, only the popular full-wave methods are discussed in detail in this chapter.

Method of Moments MOM is a versatile numerical method for solving integral equations and was pioneered by R. F. Harrington [8]. One of the main advantages of MOM lies in the variational nature of its solution, which implies that even if the unknown field is modeled to first-order accuracy, the solution is accurate to second order. MOM involves a good amount of preprocessing of Maxwell's equations because it makes use of Green's function. The use of Green's function not only allows the solution of open problems in radiation, scattering, planar circuits, and antennas in an efficient manner, but also makes MOM free from numerical dispersion. For perfectly conducting and partially homogeneous objects, only the interfaces between different regions need to be meshed (surface meshing). Hence the size of the coefficient matrix is much smaller than in other methods, which in turn makes it more efficient. For the FEM or FDTD method, however, in order to make the fields fully deteriorated, besides the whole region occupied by objects, the open region near the objects should also be meshed (volume meshing). The disadvantage of MOM is that it requires prior knowledge of Green's function for the problem. Additionally, compared to differential equation-based methods, its application to inhomogeneous dielectric objects is much more difficult. Furthermore, MOM will generate a dense matrix, which is more expensive to solve, especially for extremely electrically large problems.

FEM FEM is a numerical method for solving boundary value problems characterized by a partial differential equation and a set of boundary conditions. It was first proposed by structural analysts and its mathematical base was provided by R. Courant in 1943 [16]. The application of FEM in electromagnetics was first reported in 1968 [17]. With the developments over 50 years, it has now become a standard numerical tool in many disciplines, for example, in solid and structural mechanics, fluid dynamics, acoustics, thermal conduction, and electromagnetics.

The main advantage of FEM is that only a highly sparse linear system is required to be solved, thus the memory requirement scales linearly with the number of degrees of freedom. Further, the natural treatment of complex media makes FEM a very popular and powerful method in dealing with serious inhomogeneous structures. However, no method is flawless, and FEM also has its disadvantages. First, solving the linear system is the main obstacle that affects the efficiency, because the coefficient matrix is usually ill-conditioned, which makes iterative solvers converge very slowly. Another disadvantage of FEM is the difficulty in dealing with open problems. For antenna radiation and scattering problems, artificial absorbing boundary conditions (ABC) s, such as the perfectly matched layers [18,19], should be applied around the object to simulate the radiation condition. This increases the complexity and instability. Although it has some disadvantages, FEM is very mature and some books systematically provide a clear, conceptual picture, such as those in [20–22].

FDTD Method The FDTD method is another popular differential equation-based computational algorithm for electromagnetic problems. FDTD was first proposed in 1966 by K. S. Yee in [23], where it was used to solve electromagnetic scattering from perfectly electric conducting (PEC) cylinders. However, not much attention was paid to this method due to the limitation of computer techniques. In the late 1980s, with the development of computers, the method was revived and developed further. Up to now, it has been used for a variety of applications in time-varying electromagnetics, such as scattering, antenna design, electromagnetic compatibility analysis, and so on.

The FDTD method is simple to program, highly efficient and easily adopted. Different from MOM and FEM, it can provide information on the model in the time domain, which is very useful in many cases. Fourier transforms are performed to obtain frequency-domain information if needed. Compared to FEM, the FDTD method is massively parallelizable. Furthermore, it is also a matrix-free technique. Therefore, the computational complexity of FDTD scales only linearly with the number of degrees of freedom of the problem. Another advantage is that no matrix has to be stored, which reduces memory requirements and makes it possible to solve problems with a large number of unknowns. The major drawback of this method is that the errors introduced by staircase approximations generated by uniform hexahedral meshing cannot be eliminated. In recent years, the application of conformal meshes has alleviated this drawback to some extent [28]. The FDTD method is now considered to be a basic tool in CEM, and detailed instructions and important extensions are given in many papers and books [25–29].

2.2 Methods of Antenna Modeling

2.2.1 Basic Theory

All numerical methods are developed to solve a boundary value problem characterized by the governing equations and a set of boundary conditions. In electromagnetics, Maxwell's equations can describe any electromagnetic behavior, which therefore forms the foundation of CEM. These equations can be written in both differential and integral form. The differential form in the time domain is usually stated as (2.1)–(2.4). When dealing with time-varying electromagnetic phenomena, the frequency-domain Maxwell's equations are usually used

$$\nabla \times \mathbf{E} = i\omega \mathbf{B} \tag{2.6}$$

$$\nabla \times \mathbf{H} = -i\omega \mathbf{D} + \mathbf{J} \tag{2.7}$$

$$\nabla \cdot \mathbf{B} = 0 \tag{2.8}$$

$$\nabla \cdot \mathbf{D} = \rho \tag{2.9}$$

$$\nabla \cdot \mathbf{J} = i\omega\rho \tag{2.10}$$

in which the time dependence $e^{-i\omega t}$ is assumed and suppressed throughout this chapter. For simple media, the field quantities satisfy the following constitutive relations:

$$\mathbf{D} = \varepsilon\mathbf{E}, \quad \mathbf{B} = \mu\mathbf{H}, \quad \mathbf{J} = \sigma\mathbf{E} \tag{2.11}$$

where ε, μ, and σ denote the permittivity, permeability, and conductivity of the medium respectively. The goal of CEM is to develop methods to solve Maxwell's equations efficiently.

2.2.2 Method of Moments

2.2.2.1 *Surface Integral Equations*

First, as shown in Figure 2.1, a PEC object with boundary S embedded in an unbounded homogeneous medium characterized by (ε, μ) is considered. For this case, there will be an induced surface current $\mathbf{J}(\boldsymbol{r})$ on S under the illumination of the incident field. From electromagnetic theory, we obtain the following electric field integral equation (EFIE):

$$i\omega\mu\hat{t}\cdot\int_S \bar{\mathbf{G}}_{\mathrm{e}}(\boldsymbol{r},\boldsymbol{r}')\cdot\mathbf{J}(\boldsymbol{r}')\,d\boldsymbol{r}' = -\hat{t}\cdot\mathbf{E}^{\mathrm{inc}}(\boldsymbol{r}), \quad \boldsymbol{r}\in S \tag{2.12}$$

in which

$$\bar{\mathbf{G}}_{\mathrm{e}}(\boldsymbol{r},\boldsymbol{r}') = \left[\bar{\mathbf{I}} + \frac{\nabla\nabla}{k^2}\right] g(\boldsymbol{r},\boldsymbol{r}') \tag{2.13}$$

is the electric dyadic Green's function and

$$g(\boldsymbol{r},\boldsymbol{r}') = \frac{e^{ik|r-r'|}}{4\pi|\boldsymbol{r}-\boldsymbol{r}'|} \tag{2.14}$$

is the scalar Green's function of the background, and $k = \omega\sqrt{\mu\varepsilon}$ is the wavenumber. Similarly, the magnetic field integral equation (MFIE)

$$\mathbf{J}(\boldsymbol{r}) - \hat{n}\times\int_S \bar{\mathbf{G}}_{\mathrm{m}}(\boldsymbol{r},\boldsymbol{r}')\cdot\mathbf{J}(\boldsymbol{r}')\,d\boldsymbol{r}' = \hat{n}\times\mathbf{H}^{\mathrm{inc}}(\boldsymbol{r}) \tag{2.15}$$

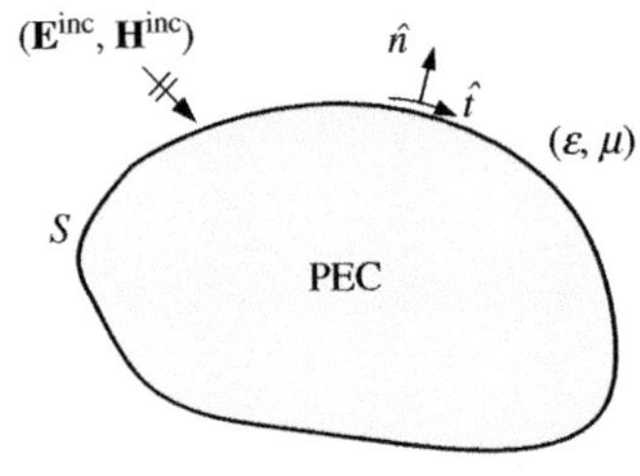

Figure 2.1 *A PEC object under the illumination of electromagnetic waves.*

can be obtained, where

$$\bar{\mathbf{G}}_{\mathrm{m}}(\boldsymbol{r},\boldsymbol{r}') = \nabla \times [g(\boldsymbol{r},\boldsymbol{r}')\bar{\mathbf{I}}] \tag{2.16}$$

is the magnetic dyadic Green's function. In (2.15), there is a singularity when $\boldsymbol{r} \to \boldsymbol{r}'$. As in [14], the singularity can be extracted, and (2.15) can be rewritten as

$$\frac{1}{2}\mathbf{J}(\boldsymbol{r}) - \hat{n} \times \mathrm{PV}\int_S \nabla \times [\mathbf{J}(\boldsymbol{r}')g(\boldsymbol{r},\boldsymbol{r}')]\,d\boldsymbol{r}' = \hat{n} \times \mathbf{H}^{\mathrm{inc}}(\boldsymbol{r}), \quad \boldsymbol{r} \in S \tag{2.17}$$

where PV stands for the principal value of the integral.

One should note that the EFIE is valid for both open and closed surfaces but MFIE is valid only for the closed one. Another issue that should be addressed is the internal resonance problem. When the working frequency coincides with the resonant frequencies, the EFIE and MFIE will both break down [9]. In order to solve this, a combined field integral equation (CFIE) is proposed, which combines the EFIE and MFIE together, that is,

$$\mathrm{CFIE} = \alpha\mathrm{EFIE} + \eta(1-\alpha)\mathrm{MFIE} \tag{2.18}$$

where α is a parameter ranging from zero to one and η is the intrinsic impedance.

2.2.2.2 *Basic Principle of MOM*

MOM, as systematically developed by Harrington [8], has been made much more mature in the past few decades. Generally, we consider the operator equation

$$\bar{\mathbf{L}} \cdot \mathbf{f} = \mathbf{g} \tag{2.19}$$

where $\bar{\mathbf{L}}$ is a linear operator, $\mathbf{g}$ is the source function which is known, and $\mathbf{f}$ is the function to be determined. Usually, $\mathbf{f}$ and $\mathbf{g}$ are in different spaces, $\mathcal{F}$ and $\mathcal{G}$.

To obtain the numerical solution of $\mathbf{f}$, we expand the unknown function using a set of basis functions $\{\mathbf{f}_n\}$ as

$$\mathbf{f}(\boldsymbol{r}) = \sum_{n=1}^{N} a_n \mathbf{f}_n(\boldsymbol{r}) \tag{2.20}$$

in which a_n is the unknown coefficient to be determined. Substituting (2.20) into (2.19) yields

$$\sum_{n=1}^{N} a_n \bar{\mathbf{L}} \cdot \mathbf{f}_n(\boldsymbol{r}) = \mathbf{g}(\boldsymbol{r}) \tag{2.21}$$

which is in the space $\mathcal{G}$. In order to determine the coefficient a_n, we make projections of (2.21) on N testing functions, $\mathbf{w}_1, \mathbf{w}_2, \ldots, \mathbf{w}_N$, to obtain

$$\sum_{n=1}^{N} a_n \langle \mathbf{w}_m, \bar{\mathbf{L}} \cdot \mathbf{f}_n(\boldsymbol{r}) \rangle = \langle \mathbf{w}_m, \mathbf{g}(\boldsymbol{r}) \rangle \tag{2.22}$$

by defining the inner product as

$$\langle \mathbf{w}, \mathbf{f} \rangle = \int \mathbf{w}^*(\mathbf{r}) \cdot \mathbf{f}(\mathbf{r})\, d\mathbf{r} \tag{2.23}$$

Thus, (2.19) can be transformed into the matrix equation

$$\bar{\mathbf{Z}} \cdot \mathbf{a} = \mathbf{b} \tag{2.24}$$

where an element of matrix $\bar{\mathbf{Z}}$ is given by

$$Z_{mn} = \langle \mathbf{w}_m, \bar{\mathbf{L}} \cdot \mathbf{f}_n(\mathbf{r}) \rangle \tag{2.25}$$

and the source term is given by

$$b_m = \langle \mathbf{w}_m, \mathbf{g}(\mathbf{r}) \rangle \tag{2.26}$$

In MOM, the most important thing is the choice of basis and testing (or weighting) functions [8]. There are two types of basis functions: the entire-domain basis function and the sub-domain basis function. It is very difficult to find appropriate entire-domain basis functions for complicated two-dimensional (2D) and three-dimensional (3D) problems. Hence, sub-domain basis functions are commonly used. If the basis and testing functions are in the same space, that is, $\mathcal{F}_N = \mathcal{G}_N$, then we can take $\mathbf{w}_n = \mathbf{f}_n$. This is the well-known Galerkin method, the most popular testing procedure.

2.2.2.3 RWG Basis Function

For an arbitrary PEC object, the current will flow on the surface and hence we need to solve a surface integral equation. For the discretization of an arbitrary PEC surface, a triangular mesh is the simplest and most efficient way.

For the triangular mesh, Rao, Wilton, and Glission proposed a rooftop basis function [15]

$$\mathbf{f}_n(\mathbf{r}) = \begin{cases} \dfrac{l_n}{2A_n^+}\boldsymbol{\rho}_n^+, & \mathbf{r} \in T_n^+ \\ \dfrac{l_n}{2A_n^-}\boldsymbol{\rho}_n^-, & \mathbf{r} \in T_n^- \\ 0, & \text{otherwise} \end{cases} \tag{2.27}$$

which is defined on a triangular pair T_n^+ and T_n^-, as shown in Figure 2.2a. Obviously, this is a sub-domain basis, which is usually called the Rao–Wilton–Glission (RWG) basis function. The RWG basis is associated with interior edges, that is, non-boundary edges of triangular patches. The RWG basis function has a very clear physical significance. From its definition, we can clearly see that:

1. The electric current flows along the boundary of the triangular pair T_n^+ and T_n^-, and hence no line charges exist along the outer boundary.

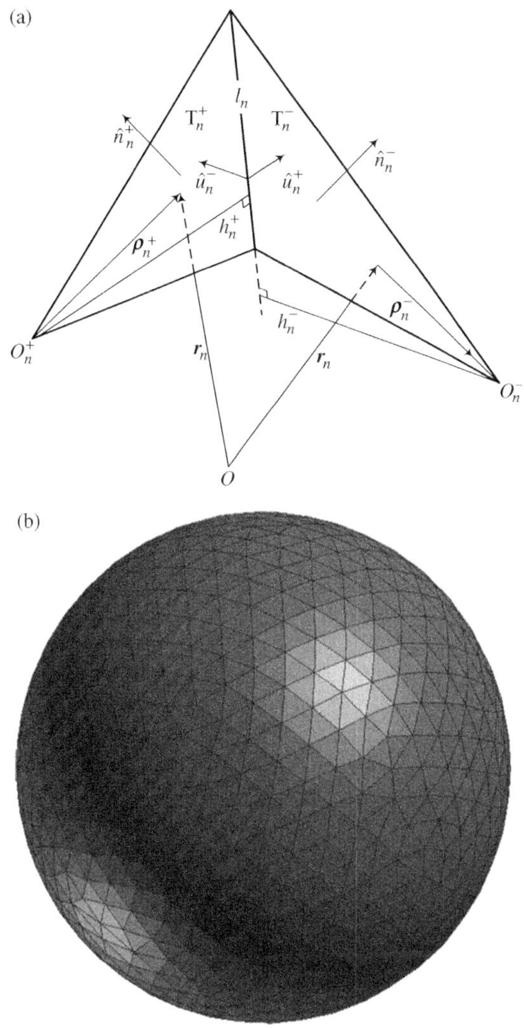

Figure 2.2 RWG basis function and mesh discretization in MOM. (a) A triangular pair associated with the interior edge l_n. (b) A spherical surface discretized by triangular meshes.

2. On the interior edge l_n ($\mathbf{r} \in l_n$), we have

$$\mathbf{f}_n^+ \cdot \hat{u}_n^+ = \frac{l_n}{2A_n^+}\boldsymbol{\rho}_n^+ \cdot \hat{u}_n^+ = \frac{l_n h_n^+}{2A_n^+} = 1$$
$$\mathbf{f}_n^- \cdot \hat{u}_n^- = \frac{l_n}{2A_n^-}\boldsymbol{\rho}_n^- \cdot \hat{u}_n^- = \frac{l_n h_n^-}{2A_n^-} = -1$$

where $\hat{u}_n^{\pm}$ is the normal unit vector of the edge l_n within T_n^+ and T_n^-. Therefore, the normal component of electric current across the interior edge l_n is continuous. As a consequence, there is no line charge accumulated on this edge.

3. The surface charges on T_n^+ and T_n^- are both constants, which form a pulse doublet. Clearly, the total surface charge on the triangular pair is zero.

From above analysis, the RWG basis function does not introduce any net charge, which is physically correct.

2.2.2.4 MOM Implementation Using the RWG Basis Function

Now we consider a MOM implementation of a PEC object using the RWG basis function. For the PEC object embedded in a simple medium characterized by (ε, μ), EFIE can be written as

$$i\omega\hat{t}\cdot\mathbf{A}(\boldsymbol{r}) - \hat{t}\cdot\nabla\Phi(\boldsymbol{r}) = -\hat{t}\cdot\mathbf{E}^{\text{inc}}(\boldsymbol{r}) \tag{2.28}$$

where

$$\mathbf{A}(\boldsymbol{r}) = \mu\int_S g(\boldsymbol{r},\boldsymbol{r}')\mathbf{J}(\boldsymbol{r}')\,ds' \tag{2.29}$$

$$\Phi(\boldsymbol{r}) = \frac{1}{i\omega\varepsilon}\int_S g(\boldsymbol{r},\boldsymbol{r}')\nabla'\cdot\mathbf{J}(\boldsymbol{r}')\,ds' \tag{2.30}$$

Now we test (2.28) through the Galerkin procedure. Since $\mathbf{w}_m(\boldsymbol{r}) = \mathbf{f}_m(\boldsymbol{r})$ is along the tangential direction $\hat{t}$, we have

$$i\omega\langle\mathbf{f}_m, \mathbf{A}\rangle - \langle\mathbf{f}_m, \nabla\Phi\rangle = -\langle\mathbf{f}_m, \mathbf{E}^{\text{inc}}\rangle \tag{2.31}$$

which can be written in matrix form as

$$\bar{\mathbf{Z}}^{\text{e}}\cdot\mathbf{I} = \mathbf{V}^{\text{e}} \tag{2.32}$$

where the impedance matrix element is given by

$$Z^{\text{e}}_{mn} = i\omega\mu\int_{S_m} ds\int_{S_n}\mathbf{f}_m(\boldsymbol{r})\cdot[g(\boldsymbol{r},\boldsymbol{r}')\mathbf{f}_n(\boldsymbol{r}')]\,ds' + \frac{1}{i\omega\varepsilon}\int_{S_m}\nabla\cdot\mathbf{f}_m(\boldsymbol{r})ds\int_{S_n} g(\boldsymbol{r},\boldsymbol{r}')\nabla'\cdot\mathbf{f}_n(\boldsymbol{r}')\,ds' \tag{2.33}$$

and the excitation vector term is given by

$$V^{\text{e}}_m = -\int_{S_m}\mathbf{f}_m(\boldsymbol{r})\cdot\mathbf{E}^{\text{inc}}(\boldsymbol{r})\,ds \tag{2.34}$$

Similarly, the MOM implementation of the MFIE is expressed as

$$\bar{\mathbf{Z}}^{\text{m}}\cdot\mathbf{I} = \mathbf{V}^{\text{m}} \tag{2.35}$$

in which

$$Z_{mn}^{\mathrm{m}} = \frac{1}{2}\int_{S_m} \mathbf{f}_m(\boldsymbol{r}) \cdot \mathbf{f}_n(\boldsymbol{r}')\, ds - \mathrm{PV}\int_{S_m} ds \int_{S_n} \mathbf{f}_m(\boldsymbol{r}) \cdot \hat{n}(\boldsymbol{r}) \times \nabla \times [g(\boldsymbol{r}, \boldsymbol{r}')\mathbf{f}_n(\boldsymbol{r}')]\, ds' \tag{2.36}$$

and

$$V_m^{\mathrm{m}} = \int_{S_m} \mathbf{f}_m(\boldsymbol{r}) \cdot \left[\hat{n}(\boldsymbol{r}) \times \mathbf{H}^{\mathrm{inc}}(\boldsymbol{r})\right] ds \tag{2.37}$$

Combining above equations, we obtain the MOM implementation of CFIE.

2.2.3 FEM

The principle of FEM is to replace the entire computational domain by a number of small sub-domains in which the unknown function is represented by simple interpolation functions with unknown coefficients. A finite element analysis of the boundary value problems mainly includes the following basic steps:

1. Determining the suitable governing equations and boundary conditions.
2. Generation of suitable FEM meshes.
3. Selecting the interpolation and weighting functions and transforming the governing equations into matrix equations by the Ritz variational method or the Galerkin method.
4. Solving the linear systems.
5. Postprocessing.

To deal with time-harmonic electromagnetic problems with FEM, the governing equations should be predetermined. Usually, FEM deals with the problem directly in terms of the electric or magnetic field. The differential equation for **E** can be obtained by eliminating **H** from (2.6) and (2.7)

$$\nabla \times \left(\mu_r^{-1} \nabla \times \mathbf{E}\right) - k_0^2 \varepsilon_r \mathbf{E} = ik_0 Z_0 \mathbf{J} \tag{2.38}$$

Similarly, the wave equation for **H** can be obtained:

$$\nabla \times \left(\varepsilon_r^{-1} \nabla \times \mathbf{H}\right) - k_0^2 \mu_r \mathbf{H} = \nabla \times \varepsilon_r^{-1} \mathbf{J} \tag{2.39}$$

In most cases, (2.38) is preferred. The boundary conditions most widely used are

$$\hat{n} \times \mathbf{E} = 0, \quad \text{on } \Gamma_1 \tag{2.40}$$

$$\hat{n} \times \nabla \times \mathbf{E} = 0, \quad \text{on } \Gamma_2 \tag{2.41}$$

where Γ_1 is the boundary of the PEC target, and Γ_2 is the boundary of the perfectly magnetic conducting (PMC) target.

Traditionally, two classical methods can be adopted to solve (2.38). One is the Galerkin method and the other is the Ritz variational method. Here the former is briefly introduced. The Galerkin method is a weighted residual method which seeks the solution by weighting the residual of the differential equation. Defining $\tilde{\mathbf{E}}$ to be the approximation of the electric field $\mathbf{E}$, this can be expanded as

$$\tilde{\mathbf{E}} = \sum_{i=1}^{N} a_i \mathbf{N}_i \tag{2.42}$$

where $\mathbf{N}_i$ is called the interpolation function. Substitution of $\tilde{\mathbf{E}}$ for $\mathbf{E}$ in (2.38) would result in a non-zero residual:

$$\mathbf{r} = \frac{1}{\mu_r}\nabla \times \nabla \times \tilde{\mathbf{E}} - k_0^2 \varepsilon_r \tilde{\mathbf{E}} - ik_0 Z_0 \mathbf{J} \tag{2.43}$$

The best value of $\tilde{\mathbf{E}}$ will be the one that reduces the residual $\mathbf{r}$ to the least value at all points of the whole domain. Thus, the weighted residual method enforces the condition

$$\mathbf{R}_i = \int_V \mathbf{W}_i \cdot \left(\frac{1}{\mu_r}\nabla \times \nabla \times \tilde{\mathbf{E}} - k_0^2 \varepsilon_r \tilde{\mathbf{E}}\right) dV - ik_0 Z_0 \int_V \mathbf{W}_i \cdot \mathbf{J}\, dV = 0 \tag{2.44}$$

where $\mathbf{R}_i$ denotes the weighted residual integral and $\mathbf{W}_i$ is the chosen weighted function. For the Galerkin method, we have $\mathbf{W}_i = \mathbf{N}_i$. Then (2.44) becomes

$$\int_V \mathbf{N}_i \cdot \left(\frac{1}{\mu_r}\nabla \times \nabla \times \tilde{\mathbf{E}} - k_0^2 \varepsilon_r \tilde{\mathbf{E}}\right) dV = ik_0 Z_0 \int_V \mathbf{N}_i \cdot \mathbf{J}\, dV \tag{2.45}$$

This equation can be transformed into a linear system, and hence the unknown coefficients a_i can be obtained by various matrix solvers.

An important step in the Galerkin method is the selection of the interpolation function defined over the entire domain. For most 3D electromagnetic problems, this is very difficult, even impossible. To overcome this obstacle, the computational domain is meshed by using sub-domain elements. In each sub-domain, the electric field can be represented with simple interpolation functions. Thus, the Galerkin method can be applied to each sub-domain, which is the basic idea of FEM. In FEM simulation, different types of mesh elements can be used, such as tetrahedral, hexahedral, triangular prism, and even curvilinear elements. Among all these types, the tetrahedral element is the simplest and best suited for arbitrary volume domains as illustrated in Figure 2.3a.

When the domain is meshed, the interpolation function should be determined. For tetrahedral elements, the Whitney functions [21] are most widely adopted, which are of the following form:

$$\mathbf{N}_i = (\zeta_{i_1} \nabla \zeta_{i_2} - \zeta_{i_2} \nabla \zeta_{i_1}) l_i \tag{2.46}$$

where i_1 and i_2 denote the local numbers of the two nodes associated with edge i, and ζ_i is the volume coordinate. For any point P in the tetrahedron, denoting V_i as the volume defined by point P and three other nodes of the tetrahedron except node i, and V^e as the volume of the tetrahedron, ζ_i is defined by

$$\zeta_i = V_i / V^e \tag{2.47}$$

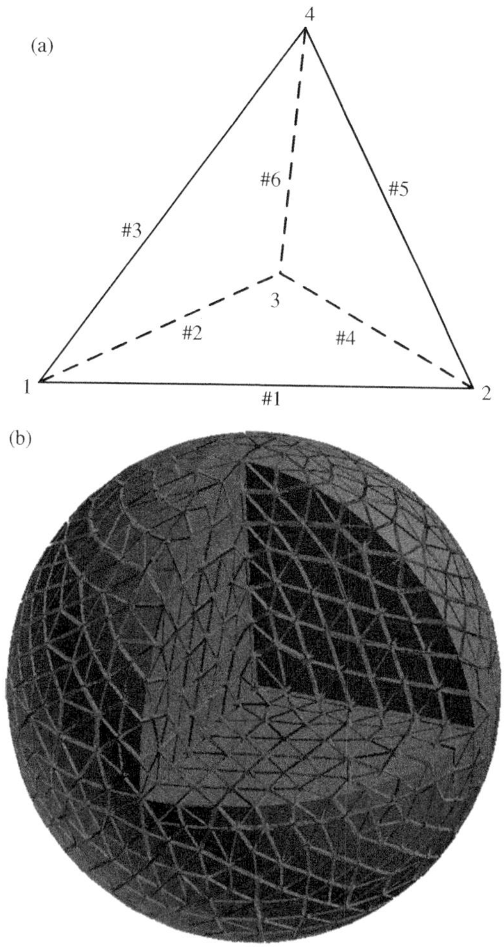

Figure 2.3 *Illustration of discretization in FEM. (a) Linear tetrahedral element. (b) A solid sphere discretized by tetrahedral (one-eighth of the sphere has been cut away to reveal interior view).*

Let $\hat{t}_i$ be the unit vector pointing from node i_1 to node i_2. Then we have

$$\hat{t}_i \cdot \nabla\zeta_{i_1} = -1/l_i \quad \text{and} \quad \hat{t}_i \cdot \nabla\zeta_{i_2} = 1/l_i \tag{2.48}$$

Therefore

$$\hat{t}_i \cdot \mathbf{N}_i = \left(\zeta_{i_1} + \zeta_{i_2}\right) = 1 \tag{2.49}$$

which means that the tangential component of $\mathbf{N}_i$ along edge i is a constant. Moreover, $\mathbf{N}_i$ has no tangential component along the other five edges. Thus, the coefficient corresponding to $\mathbf{N}_i$ represents the tangential component of the electric field on edge i.

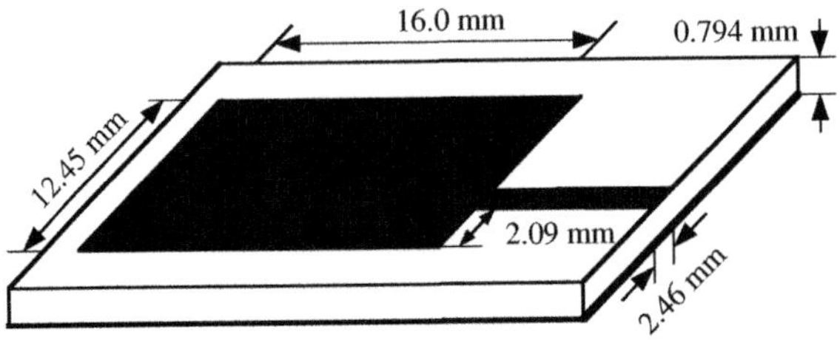

Figure 2.4 *Line-fed microstrip patch antenna.*

By applying the Galerkin method to each element, a local matrix equation can be obtained which has the same form as (2.45) except V is replaced by V^e. Putting all local matrix equations together and eliminating unknown edges on the boundary, a linear system is obtained, which can be written as

$$\bar{\mathbf{A}}\mathbf{x} = \mathbf{b} \tag{2.50}$$

The stiff matrix $\bar{\mathbf{A}}$ is a symmetric, highly sparse, and ill-conditioned matrix. For most problems, the average number of non-zero elements per row in $\bar{\mathbf{A}}$ is less than 20. Thus, sparse matrix storage formats are often used to store $\bar{\mathbf{A}}$. Usually, the solution of (2.50) is the most time-consuming step in the whole FEM simulation. Thus, an efficient solver is of vital importance. One powerful directive solver is the multi-frontal solver. Another kind of solver widely used is the preconditioned Krylov subspace iterative method, such as ICCG, ICGMRES, and so on.

In the simulation of antenna radiation, the domain of interest is open to the surrounding free space, that is, the computational domain is, in theory, infinite. In order to make it solvable, the infinite space must be truncated. One method is to combine FEM with MOM to come up with a hybrid modeling scheme. Another is to construct an appropriate boundary condition placed at some distance from the antenna. Such a boundary condition is called as an absorbing boundary condition (ABC). ABC can be implemented using a variety of techniques, such as the perfectly matched layer (PML) [19].

As an example, a line-fed microstrip patch antenna is investigated by using FEM. The configuration of the patch antenna is shown in Figure 2.4. As illustrated in Figure 2.5, PML is used to truncate the simulation region. The distance between PML and the antenna is about $\frac{1}{4}\lambda$, and the thickness of PML is also $\frac{1}{4}\lambda$. Figure 2.6 shows a comparison of the return loss of the patch antenna computed by the FEM and FDTD method. There are a few discrepancies between the two curves, because the structure cannot be accurately represented by the regular FDTD lattice.

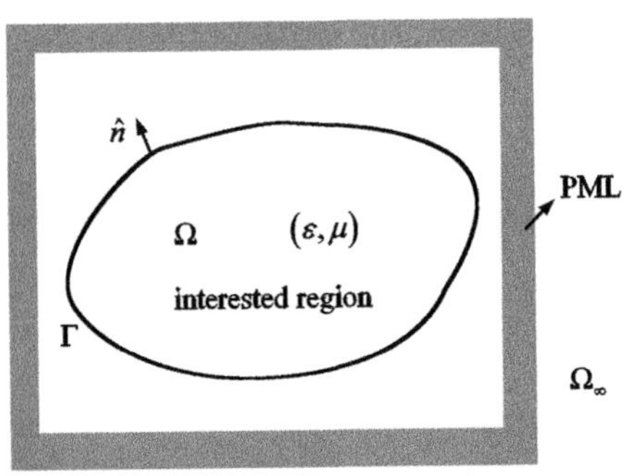

Figure 2.5 *Truncation of simulation domain by PML.*

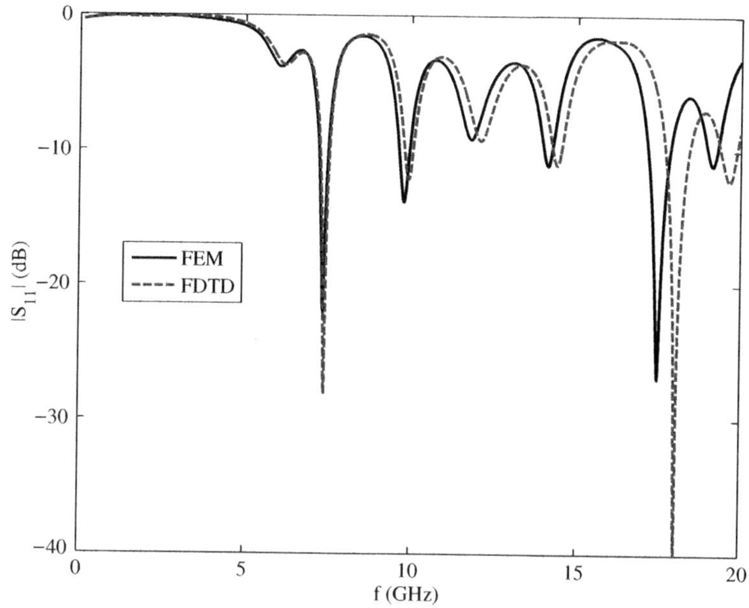

Figure 2.6 Return loss of the patch antenna computed by the FEM and FDTD method.

2.2.4 FDTD Method

The FDTD method seeks approximate solutions to differential equations using the finite difference, whose procedure can be summarized as follows:

1. Divide the continuous computational region into a grid of nodes.
2. Approximate derivatives in the given partial differential equation by finite differences.
3. Solve the finite difference equations for the value at the next time step at each node subject to boundary conditions, which is called leapfrogging, specifically in the context of the FDTD method.

The FDTD method directly discretizes the time-domain Maxwell's equations into difference equations. Maxwell's time-domain equations, (2.1) and (2.2), can be cast into more general and symmetric forms by including the virtual magnetic current, and are

$$\nabla \times \mathbf{E} = -\frac{\partial(\mu\mathbf{H})}{\partial t} - \sigma_{\mathrm{M}}\mathbf{H} \tag{2.51}$$

$$\nabla \times \mathbf{H} = \frac{\partial(\varepsilon\mathbf{E})}{\partial t} + \sigma_{\mathrm{E}}\mathbf{E} \tag{2.52}$$

Suppose μ and ε do not vary with time. Then (2.51) and (2.52) can be expanded into six equations:

$$\frac{\partial H_x}{\partial t} = \frac{1}{\mu}\left(\frac{\partial E_y}{\partial z} - \frac{\partial E_z}{\partial y} - \sigma_{\mathrm{M}} H_x\right) \tag{2.53}$$

$$\frac{\partial H_y}{\partial t} = \frac{1}{\mu}\left(\frac{\partial E_z}{\partial x} - \frac{\partial E_x}{\partial z} - \sigma_{\mathrm{M}} H_y\right) \tag{2.54}$$

$$\frac{\partial H_z}{\partial t} = \frac{1}{\mu}\left(\frac{\partial E_x}{\partial y} - \frac{\partial E_y}{\partial x} - \sigma_{\mathrm{M}} H_z\right) \tag{2.55}$$

$$\frac{\partial E_x}{\partial t} = \frac{1}{\varepsilon}\left(\frac{\partial H_z}{\partial y} - \frac{\partial H_y}{\partial z} - \sigma_{\mathrm{E}} E_x\right) \tag{2.56}$$

$$\frac{\partial E_y}{\partial t} = \frac{1}{\varepsilon}\left(\frac{\partial H_x}{\partial z} - \frac{\partial H_z}{\partial x} - \sigma_{\mathrm{E}} E_y\right) \tag{2.57}$$

$$\frac{\partial E_z}{\partial t} = \frac{1}{\varepsilon}\left(\frac{\partial H_y}{\partial x} - \frac{\partial H_x}{\partial y} - \sigma_{\mathrm{E}} E_z\right) \tag{2.58}$$

Each field component is a function of time t and spatial (x, y, z). In order to transform (2.53)–(2.58) into difference equations, the time domain should be uniformly discretized and the spatial domain should be meshed with Yee cells as illustrated in Figure 2.7a. In the Yee cell, the six field components are distributed on the surface. Each electric field component is located at the center of the edge, while the magnetic field components are located on the center of the surface.

When the mesh is determined, a function $u(x, y, z, t)$ is defined in the discrete coordinate space as

$$u(x, y, z, t) = u(i\Delta x, j\Delta y, k\Delta z, n\Delta t) = u^n(i, j, k) \tag{2.59}$$

Here, Δx, Δy, and Δz denote the mesh length along the x, y, and z directions, and Δt is the time step. To discretize Maxwell's equations, three difference forms are commonly considered: forward, backward, and central difference. For a function $u(x)$, a reasonable approximation for the first-order derivative would be

$$\frac{du}{dx} \approx \frac{u(x+\Delta x) - u(x)}{\Delta x}, \quad \text{forward difference} \tag{2.60}$$

$$\frac{du}{dx} \approx \frac{u(x) - u(x-\Delta x)}{\Delta x}, \quad \text{backward difference} \tag{2.61}$$

$$\frac{du}{dx} \approx \frac{u(x+\Delta x/2) - u(x-\Delta x/2)}{\Delta x}, \quad \text{central difference} \tag{2.62}$$

According to the Taylor expansion, the truncation error of central difference is $O(\Delta x^2)$, which is the least among the three methods. Thus, central difference is often used in the FDTD method. Taking the central

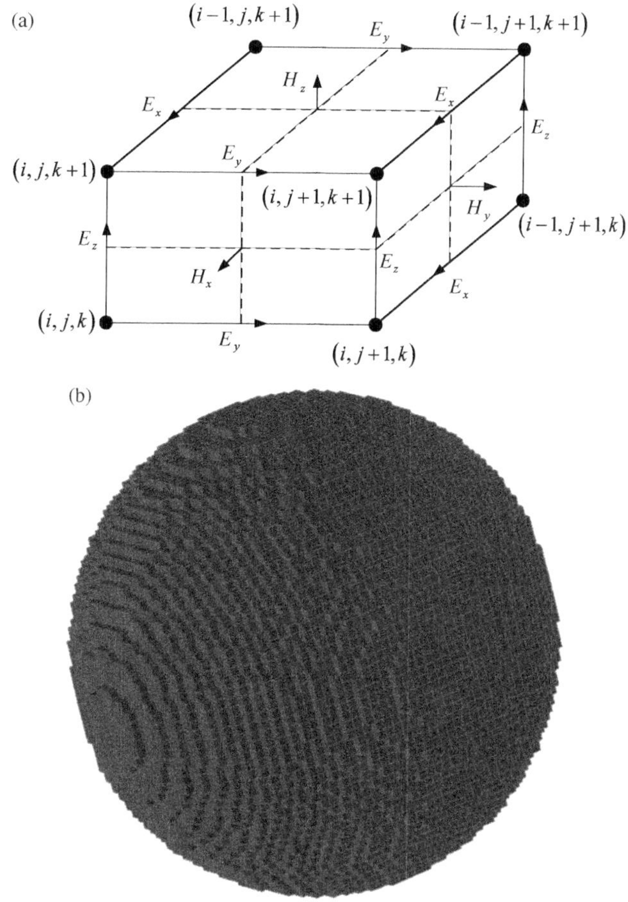

Figure 2.7 Yee cell and mesh discretization in the FDTD method. (a) Field distribution in Yee cell. (b) A solid sphere meshed by hexahedra.

difference approximations for both the time and spatial derivatives in (2.56) gives

$$\frac{E_x^{n+1}\left(i+\frac{1}{2},j,k\right)-E_x^{n}\left(i+\frac{1}{2},j,k\right)}{\Delta t}=\frac{1}{\varepsilon\left(i+\frac{1}{2},j,k\right)}$$
$$\times\left[\frac{H_z^{n+\frac{1}{2}}\left(i+\frac{1}{2},j+\frac{1}{2},k\right)-H_z^{n+\frac{1}{2}}\left(i+\frac{1}{2},j-\frac{1}{2},k\right)}{\Delta y}-\frac{H_y^{n+\frac{1}{2}}\left(i+\frac{1}{2},j,k+\frac{1}{2}\right)-H_y^{n+\frac{1}{2}}\left(i+\frac{1}{2},j,k-\frac{1}{2}\right)}{\Delta z}-\sigma_E E_x^{n+\frac{1}{2}}\left(i+\frac{1}{2},j,k\right)\right] \tag{2.63}$$

For the convenience of programming, $E_x^{n+\frac{1}{2}}(i+\frac{1}{2},j,k)$ is usually approximated by

$$E_x^{n+\frac{1}{2}}\left(i+\frac{1}{2},j,k\right)=\frac{E_x^{n+1}(i+\frac{1}{2},j,k)+E_x^{n}(i+\frac{1}{2},j,k)}{2} \tag{2.64}$$

Thus (2.64) can be transformed to

$$E_x^{n+1}\left(i+\tfrac{1}{2},j,k\right)=k_{\mathrm{E}}^{\mathrm{E}}E_x^{n}\left(i+\tfrac{1}{2},j,k\right)$$
$$+k_{\mathrm{H}}^{\mathrm{E}}\left[\frac{H_z^{n+\frac{1}{2}}\left(i+\frac{1}{2},j+\frac{1}{2},k\right)-H_z^{n+\frac{1}{2}}\left(i+\frac{1}{2},j-\frac{1}{2},k\right)}{\Delta y}-\frac{H_y^{n+\frac{1}{2}}\left(i+\frac{1}{2},j,k+\frac{1}{2}\right)-H_y^{n+\frac{1}{2}}\left(i+\frac{1}{2},j,k-\frac{1}{2}\right)}{\Delta z}\right] \tag{2.65}$$

Here,

$$k_{\mathrm{E}}^{\mathrm{E}}=\frac{1-\dfrac{\sigma_{\mathrm{E}}(i+\frac{1}{2},j,k)\Delta t}{2\varepsilon(i+\frac{1}{2},j,k)}}{1+\dfrac{\sigma_{\mathrm{E}}(i+\frac{1}{2},j,k)\Delta t}{2\varepsilon(i+\frac{1}{2},j,k)}} \tag{2.66}$$

$$k_{\mathrm{H}}^{\mathrm{E}}=\frac{\Delta t}{\varepsilon(i+\frac{1}{2},j,k)}\cdot\frac{1}{1+\dfrac{\sigma_{\mathrm{E}}(i+\frac{1}{2},j,k)\Delta t}{2\varepsilon(i+\frac{1}{2},j,k)}} \tag{2.67}$$

The other five difference functions can be similarly obtained [26]. From the above difference equations, the new electric component at any cell is only related to its value on the previous time step and the surrounding magnetic components on the previous time step. Therefore, at any given time step, computation of a field component may proceed one point at a time.

In the FDTD method, Δx, Δy, Δz, and Δt must be selected to meet the Courant–Fredrichs–Lewy (CFL) stability criterion in order to guarantee the convergence of the algorithm [24]

$$\Delta t\leq\frac{1}{c\sqrt{\frac{1}{(\Delta x)^2}+\frac{1}{(\Delta y)^2}+\frac{1}{(\Delta z)^2}}} \tag{2.68}$$

which indicates that the maximum value of the time step is determined by the minimum grid cell size. Another problem related to the selection of the mesh size in the FDTD method is numerical dispersion. Compared to FEM, numerical dispersion is much more serious. Dispersion means that the propagation velocity of the wave depends on frequency, but in a non-dispersive medium the group velocity of electromagnetic waves is not related to frequency. However, as the FDTD method is only an approximation of Maxwell's equations, the dispersion phenomenon occurs even in a non-dispersive medium, and is called numerical dispersion. In qualitative terms, numerical dispersion can deteriorate the pulse waveform and cause numerical anisotropy, cumulative phase error, and non-physical refraction. Therefore, numerical dispersion is an undesirable non-physical effect inherently present in the FDTD method, which is an important factor that affects the accuracy of the method. To decrease numerical dispersion, the space step must be carefully selected. Usually, the meshing condition $\Delta\leq 0.1\lambda$ needs be satisfied, where Δ is the maximum space step and λ is the minimum wavelength of the problem. Additionally, unlike MOM and FEM, the mesh discretization error in the FDTD method is much

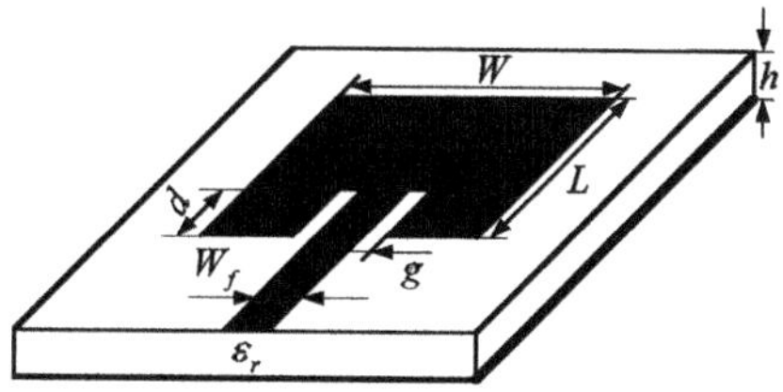

Figure 2.8 A microstrip antenna with an inset feed from its radiating edge.

more serious, as depicted in Figure 2.7b, since the curved surface cannot be accurately modeled by hexahedra as staircase approximations. Such error can be alleviated to some extent by conformal techniques [28].

As in FEM, artificial boundary conditions should be added in the FDTD method for open problems. The boundary truncation techniques in FEM, such as PML and the Mur absorbing boundary condition, can also be applied in the FDTD method by transforming the formulas from the frequency domain into the time domain.

As an example, a microstrip patch antenna with an inset feed is simulated with the FDTD method. The structure of the antenna is shown in Figure 2.8. Parameters in the figure are $L = 16.3$ mm, $W = 25.15$ mm, $W_f = 2.286$ mm, $d = 3.668$ mm, $g = 0.762$ mm and $h = 0.787$ mm.

Figure 2.9 plots the FDTD simulation results for the complex input impedance of the antenna. The phase reference is at the inset feeding point. Figure 2.10 plots both the simulation and measurement results of the input return loss. The simulation results are in good agreement with the measured results considering the uncertainties in both thickness and dielectric constant of the substrate.

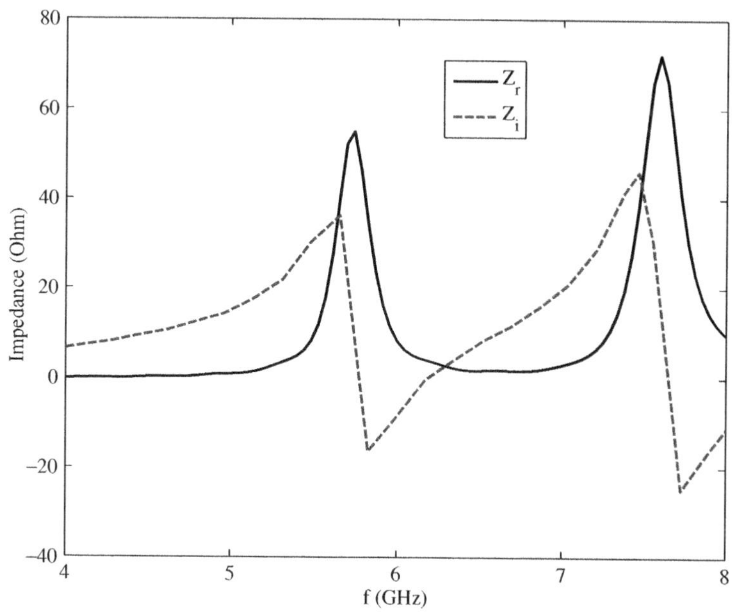

Figure 2.9 Simulation results for the input impendence of the inset-fed patch antenna.

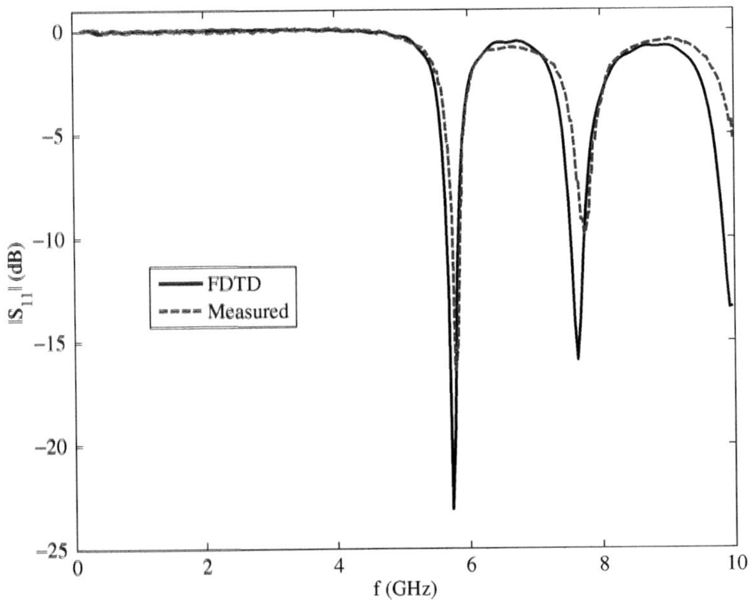

Figure 2.10 *Simulated and measured input return loss of the inset-fed patch antenna.*

2.3 Fast Algorithms for Large Space Antenna Modeling

2.3.1 Introduction

Although the ability of computers is greatly enhanced nowadays, there are still many difficulties for traditional numerical methods in solving complex electromagnetic problems. As a result, many fast algorithms have been developed to improve the efficiency of the MOM, FEM, and FDTD methods over the past few decades. One viable way of raising efficiency is to reduce the solution time and memory needed by the linear system formed by numerical methods. The multilevel fast multipole algorithm (MLFMA), which can greatly accelerate the product of matrix and vector, is the most famous method for reducing the computational complexity of MOM. Another method to reduce the memory requirement is the adoption of a high-order basis, which can greatly reduce the unknowns in the linear system generated by FEM and MOM. Besides, to reduce the iterations of the linear system, efficient iterative solvers and preconditioners have also been extensively investigated. Sometimes, hybrid methods, such as the combination of MOM (or FEM) with high-frequency methods, is another way usually adopted to improve the solvability. In the following, MLFMA and high-order basis for FEM are introduced.

2.3.2 MLFMA

The Fast multipole method (FMM) was first proposed by Greengard and Rokhlin [30] in 1987 for particle simulations, a static problem associated with the Laplace equation. In this method, it is unnecessary to store a matrix, and hence the computational complexity (including memory requirements and computational time) is of $O(N)$.

FMM was extended to dynamic problems in 1990 by Rokhlin to solve the Helmholtz equation for 2D problems [31]. Later, some other work on FMM was reported during 1990–1994 [32–35]. In these methods, the computational complexity is of $O(N^{1.5})$. Significant progress on FMM was made in 1994 with the

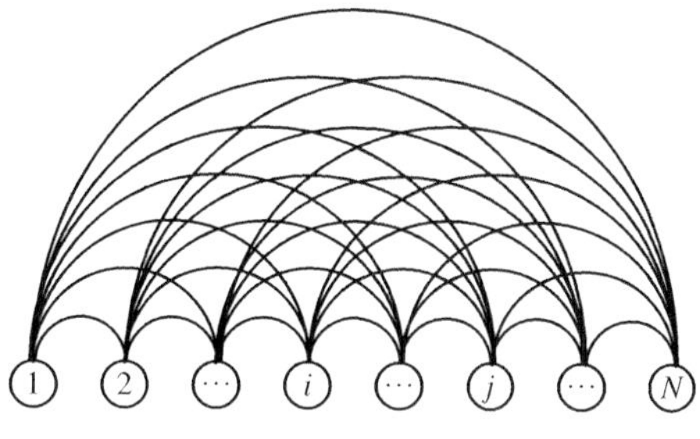

Figure 2.11 *A one-level network where* N^2 links are required.

introduction of a multilevel version by Song and Chew [36] and other groups [37], called the multilevel fast multipole algorithm (MLFMA), and whose computational complexity is reduced to O(Nlog$N$).

By MOM, the integral equation can be transformed into a linear system, that is,

$$\bar{\mathbf{Z}} \cdot \mathbf{I} = \mathbf{V} \tag{2.69}$$

As we know, direct interactions of all current elements, which correspond to the N unknowns, must be considered in the conventional MOM, producing $O(N^2)$ computational complexity.

The cooperative behavior of the above current elements is similar to a telephone network consisting of N users where each user is connected directly to all other users. Hence, N^2 telephone lines are required to set up such a telephone network, as shown in Figure 2.11.

In order to reduce the number of telephone lines to connect all N users, a hub system is introduced, as shown in Figure 2.12. First, all users are divided into M groups according to their locations. The users in a single group share the same hub. Then, M^2 links are used to connect all the hubs together. In this way, the number of telephone lines is greatly reduced.

Applying the above idea to the scattering problem, and translating the idea into mathematical language, the matrix element in (2.69) can be expressed as

$$Z_{ij} = \mathbf{V}_{im}^{\mathrm{T}} \cdot \bar{\mathbf{T}}_{mn} \cdot \mathbf{V}_{nj} \tag{2.70}$$

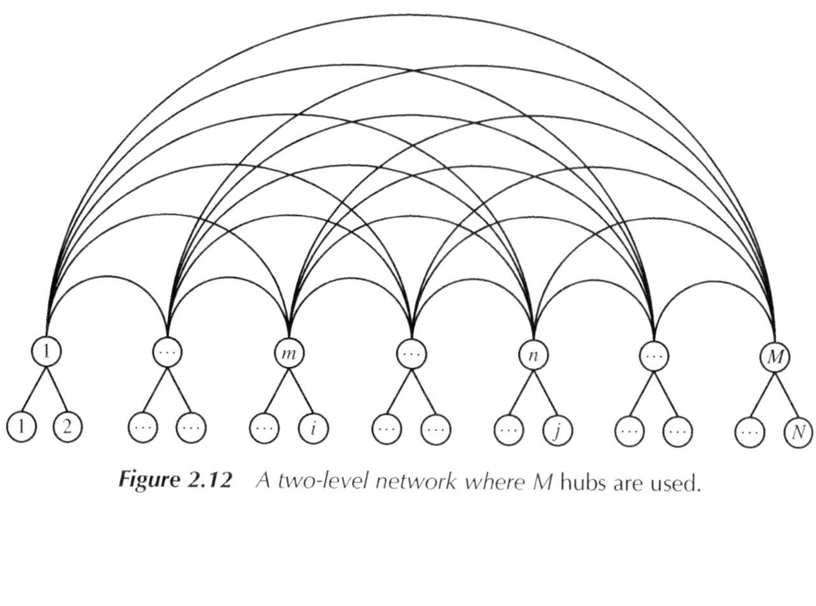

Figure 2.12 *A two-level network where M* hubs are used.

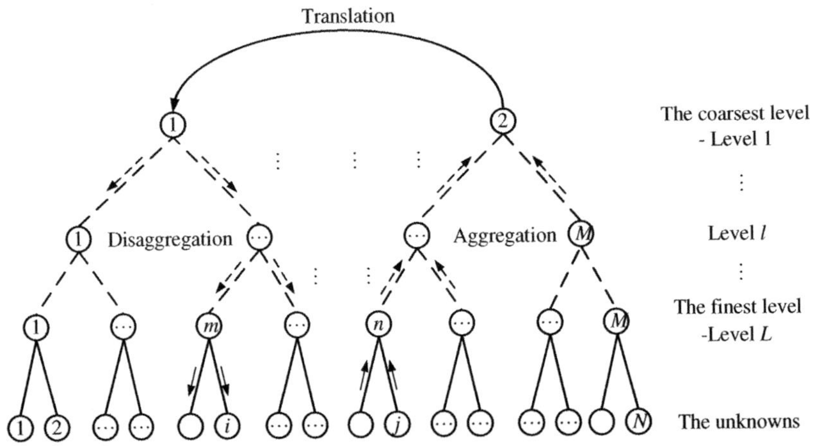

Figure 2.13 A multilevel algorithm where the aggregation, translation, and disaggregation are involved.

The matrix element Z_{ij} is a conduit to transmit field information from the current cell j to the current cell i in the one-level strategy. After grouping and introducing 'hubs,' the centers of groups, the field information from cell j is first transmitted to hub n using the vector $\mathbf{V}_{nj}$, then the information on hub n is translated to hub m using the translator matrix $\bar{\mathbf{T}}_{mn}$, and finally the information on hub m is transmitted to cell i using the vector $\mathbf{V}_{im}^{\mathrm{T}}$.

In the above procedure, a scalar number Z_{ij} has been expressed as a product of a vector, matrix, and vector. It seems that simple information transmission has been replaced by a complicated one. Actually, the translator $\bar{\mathbf{T}}_{mn}$ is reused for all cells within groups m and n. If $\bar{\mathbf{T}}_{mn}$ is a dense matrix, the complexity is the same order as that in the direct solver. If $\bar{\mathbf{T}}_{mn}$ is diagonal, however, information transmission can be more efficient. Fortunately, $\bar{\mathbf{T}}_{mn}$ can be diagonalized by using either method proposed by Rokhlin [38] or Chew [39]. As a consequence, the computational complexity of FMM drops from $O(N^2)$ to $O(N^{1.5})$. However, the two-level algorithm becomes less efficient when the problem under consideration is very large since the group number M will also be large in this case. In order to solve large-scale problems more efficiently, a natural extension of the two-level FMM to a multilevel one was developed, called MLFMA [36,37]. A simple nesting of the two-level algorithm to the multilevel one will not yield an $O(N\log N)$ algorithm. Interpolators and anterpolators must be inserted between different levels to achieve an $N \log N$ algorithm [40].

The basic principle of MLFMA is illustrated in Figure 2.13. All of the unknowns are first aggregated to groups in the finest level, level L. The groups are then aggregated to the upper level, and this process is continued until to the coarsest level, level 1. At this coarsest level, field information is exchanged. Then it is disaggregated from the coarsest level to the lower level, and this process is continued back to the finest level. In the following, the 3D version of MLFMA is illustrated in detail.

2.3.2.1 Addition Theorem for 3D Problems

In order to realize the hub connection idea introduced above, that is, to evaluate the direct interaction by a multilevel procedure, the addition theorem arising from coordinate system translation theory must be used. In electromagnetics, such a theorem corresponds to the expansion of Green's function. In the 3D case, it can be expanded as

$$\frac{e^{ik|r_1+r_2|}}{|\mathbf{r}_1+\mathbf{r}_2|} = ik\sum_{l=0}^{\infty}(-1)^l(2l+1)j_l(kr_1)h_l^{(1)}(kr_2)P_l(\hat{r}_1\cdot\hat{r}_2) \tag{2.71}$$

where $r_1 < r_2, \boldsymbol{r}_1 = \hat{r}_1 r_1$, and $\boldsymbol{r}_2 = \hat{r}_2 r_2$; $j_l(kr_1)$ is the spherical Bessel function of the first kind, $h_l^{(1)}(kr_2)$ is the spherical Hankel function of the first kind, and $P_l(\hat{r}_1 \cdot \hat{r}_2)$ is the Legendre polynomial. The above equation describes the spherical harmonic expansion of a spherical wave function. When $l \to \infty$, the spherical Bessel function $j_l(kr_1)$ approaches zero and the spherical Hankel function $h_l^{(1)}(kr_2)$ approaches infinity. However, it is much faster for the Bessel function to approach zero than for the Hankel function to approach infinity. Hence, the series (2.71) can be truncated as

$$\frac{e^{ik|\boldsymbol{r}_1+\boldsymbol{r}_2|}}{|\boldsymbol{r}_1+\boldsymbol{r}_2|} = ik\sum_{l=0}^{L}(-1)^l(2l+1)j_l(kr_1)h_l^{(1)}(kr_2)P_l(\hat{r}_1\cdot\hat{r}_2) \tag{2.72}$$

Based on Rokhlin's analysis [38], the truncation number L satisfies

$$L = kD + \beta(kD)^{1/3} \tag{2.73}$$

where D is the group diameter and β is a constant which depends on the accuracy. Using the elementary identity

$$4\pi i^l j_l(kr_1)P_l(\hat{r}_1\cdot\hat{r}_2) = \int e^{ik\hat{k}\cdot\boldsymbol{r}_1}P_l\left(\hat{k}\cdot\hat{r}_2\right)d^2\hat{k} \tag{2.74}$$

where the integral is defined on a unit sphere, the Ewald sphere, the spherical wave function (2.72) can be further expanded as the summation of plane-wave functions

$$\frac{e^{ik|\boldsymbol{r}_1+\boldsymbol{r}_2|}}{|\boldsymbol{r}_1+\boldsymbol{r}_2|} = \frac{ik}{4\pi}\int e^{ik\hat{k}\cdot\boldsymbol{r}_1}\sum_{l=0}^{L}i^l(2l+1)h_l^{(1)}(kr_2)P_l\left(\hat{k}\cdot\hat{r}_2\right)d^2\hat{k} \tag{2.75}$$

where $\hat{k} = \hat{x}\sin\theta\cos\phi + \hat{y}\sin\theta\sin\phi + \hat{z}\cos\theta$ and $d^2\hat{k} = \sin\theta\, d\theta\, d\phi$. Introducing

$$\mathrm{T}_L\left(\hat{k}\cdot\hat{r}_2\right) = \sum_{l=0}^{L}i^l(2l+1)h_l^{(1)}(kr_2)P_l\left(\hat{k}\cdot\hat{r}_2\right) \tag{2.76}$$

(2.75) can be written as

$$\frac{e^{ik|\boldsymbol{r}_1+\boldsymbol{r}_2|}}{|\boldsymbol{r}_1+\boldsymbol{r}_2|} = \frac{ik}{4\pi}\int e^{ik\hat{k}\cdot\boldsymbol{r}_1}\mathrm{T}_L\left(\hat{k}\cdot\hat{r}_2\right)d^2\hat{k} \tag{2.77}$$

Now we consider two points i and j in the computational domain. Suppose that i is the observation point and j is the source point. Observation point i is in group m and source point j is in group n, as shown in Figure 2.14. From Figure 2.14, the spatial vector from the source point to the observation point can be written as

$$\boldsymbol{r}_{ij} = \boldsymbol{r}_i - \boldsymbol{r}_j = \boldsymbol{r}_{im} + \boldsymbol{r}_{mn} + \boldsymbol{r}_{nj} \tag{2.78}$$

If groups m and n are not overlapping or closely adjacent, then

$$\left|\boldsymbol{r}_{im} + \boldsymbol{r}_{nj}\right| < r_{mn} \tag{2.79}$$

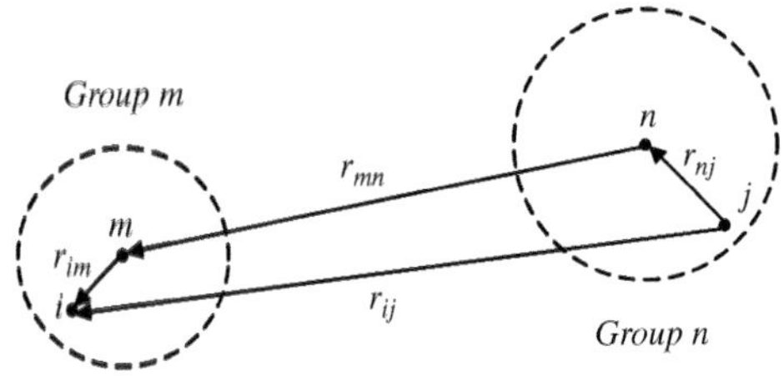

Figure 2.14 The observation point in group m and the source point in group n.

Therefore, (2.77) gives

$$\frac{e^{ik|\boldsymbol{r}_i-\boldsymbol{r}_j|}}{|\boldsymbol{r}_i-\boldsymbol{r}_j|} = \frac{ik}{4\pi}\int e^{ik\cdot(\boldsymbol{r}_{im}+\boldsymbol{r}_{nj})}\mathrm{T}_{mn}\left(\hat{k}\cdot\hat{r}_{mn}\right)d^2\hat{k} \tag{2.80}$$

where

$$\mathrm{T}_{mn}\left(\hat{k}\cdot\hat{r}_{mn}\right) = \sum_{l=0}^{L} i^l(2l+1)h_l^{(1)}(kr_{mn})P_l\left(\hat{k}\cdot\hat{r}_{mn}\right) \tag{2.81}$$

is the translator, and $e^{ik\cdot\boldsymbol{r}_{nj}}$ and $e^{ik\cdot\boldsymbol{r}_{im}}$ are radiation and receiving pattern functions [41].

2.3.2.2 *Multipole Expansions of Dyadic Green's Functions*

From (2.81), the translator T_{mn} is only related to group centers m and n. Therefore, the nabla operator for the field point, ∇, can be written as

$$\nabla = ik\hat{k} \tag{2.82}$$

Hence, the electric field dyadic Green's function (2.13) can be expressed as

$$\bar{\mathbf{G}}_\mathrm{e}(\boldsymbol{r}_i,\boldsymbol{r}_j) = \frac{ik}{(4\pi)^2}\int\left(\bar{\mathbf{I}}-\hat{k}\hat{k}\right)e^{ik\cdot(\boldsymbol{r}_{im}+\boldsymbol{r}_{nj})}\mathrm{T}_{mn}\left(\hat{k}\cdot\hat{r}_{mn}\right)d^2\hat{k} \tag{2.83}$$

Similarly, the magnetic field dyadic Green's function (2.16) can be expressed as

$$\bar{\mathbf{G}}_\mathrm{m}(\boldsymbol{r}_i,\boldsymbol{r}_j) = -\left(\frac{k}{4\pi}\right)^2\int\left(\hat{k}\times\bar{\mathbf{I}}\right)e^{ik\cdot(\boldsymbol{r}_{im}+\boldsymbol{r}_{nj})}\mathrm{T}_{mn}\left(\hat{k}\cdot\hat{r}_{mn}\right)d^2\hat{k} \tag{2.84}$$

2.3.2.3 *Multipole Expansions of Impedance Matrix*

EFIE Substituting (2.83) into (2.33), we have

$$\mathbf{Z}_{ij}^\mathrm{e} = -\eta\left(\frac{k}{4\pi}\right)^2\int\mathbf{V}_{fim}^\mathrm{e}\left(\hat{k}\right)\cdot\mathrm{T}_{mn}\left(\hat{k}\cdot\hat{r}_{mn}\right)\cdot\mathbf{V}_{snj}^\mathrm{e}\left(\hat{k}\right)d^2\hat{k} \tag{2.85}$$

where

$$\mathbf{V}^{\mathrm{e}}_{fim}\left(\hat{k}\right) = \int_S \left(\bar{\mathbf{I}} - \hat{k}\hat{k}\right) \cdot \mathbf{f}_i(\boldsymbol{r}_{im}) e^{i\boldsymbol{k}\cdot\boldsymbol{r}_{im}}\, ds \tag{2.86}$$

$$\mathbf{V}^{\mathrm{e}}_{snj}\left(\hat{k}\right) = \int_S \left(\bar{\mathbf{I}} - \hat{k}\hat{k}\right) \cdot \mathbf{f}_j(\boldsymbol{r}_{nj}) e^{i\boldsymbol{k}\cdot\boldsymbol{r}_{nj}}\, ds' \tag{2.87}$$

are receiving and radiation patterns for EFIE.

MFIE Similarly, substituting (2.84) into (2.36), and noting that the first term on the right-hand side will disappear when $m \neq n$, we have

$$Z^m_{ij} = \left(\frac{k}{4\pi}\right)^2 \int \mathbf{V}^{\mathrm{m}}_{fim}\left(\hat{k}\right) \cdot \mathrm{T}_{mn}\left(\hat{k}\cdot\hat{r}_{mn}\right) \cdot \mathbf{V}^{\mathrm{m}}_{snj}\left(\hat{k}\right) d^2\hat{k} \tag{2.88}$$

where

$$\mathbf{V}^{\mathrm{m}}_{fim}\left(\hat{k}\right) = -\hat{k} \times \int_S e^{i\boldsymbol{k}\cdot\boldsymbol{r}_{im}} \mathbf{f}_i(\boldsymbol{r}_{im}) \times \hat{n}\, ds \tag{2.89}$$

$$\mathbf{V}^{\mathrm{m}}_{snj}\left(\hat{k}\right) = \int_S e^{i\boldsymbol{k}\cdot\boldsymbol{r}_{nj}} \mathbf{f}_j\left(\boldsymbol{r}_{nj}\right) ds' \tag{2.90}$$

are receiving and radiation patterns for MFIE. From (2.89), the receiving pattern has only θ and ϕ components with respect to $\hat{k}$. In the radiation pattern, we have

$$\mathbf{f}_j\left(\boldsymbol{r}_{nj}\right) = \bar{\mathbf{I}} \cdot \mathbf{f}_j\left(\boldsymbol{r}_{nj}\right) = \left(\hat{r}\hat{r} + \hat{\theta}\hat{\theta} + \hat{\phi}\hat{\phi}\right) \cdot \mathbf{f}_j\left(\boldsymbol{r}_{nj}\right) \tag{2.91}$$

Hence, only θ and ϕ components are required in the radiation pattern, which is equivalent to $\mathbf{V}^{\mathrm{e}}_{snj}\left(\hat{k}\right)$. As a consequence, (2.88) can be rewritten as

$$Z^m_{ij} = \frac{k^2}{(4\pi)^2} \int_S \mathbf{V}^{\mathrm{m}}_{fim}\left(\hat{k}\right) \cdot \mathrm{T}_{mn}\left(\hat{k}\cdot\hat{r}_{mn}\right) \cdot \mathbf{V}^{\mathrm{e}}_{snj}\left(\hat{k}\right) d^2\hat{k} \tag{2.92}$$

CFIE The ith equation of (2.18) is

$$\sum_{j=1}^{N} Z_{ij} I_j = V_i \tag{2.93}$$

where

$$Z_{ij} = \alpha Z_{ij}^{e} + \eta(1-\alpha)Z_{ij}^{m} \tag{2.94}$$

$$V_i = \alpha V_i^{e} + \eta(1-\alpha)V_i^{m} \tag{2.95}$$

As a consequence, the multipole expansion of the coefficient matrix element is

$$Z_{ij} = \eta\left(\frac{k}{4\pi}\right)^2 \int_S \mathbf{V}_{fim}\left(\hat{k}\right)\cdot \mathbf{T}_{mn}\left(\hat{k}\cdot\hat{r}_{mn}\right)\cdot \mathbf{V}_{snj}\left(\hat{k}\right)\, d^2\hat{k} \tag{2.96}$$

where

$$\mathbf{V}_{fim}\left(\hat{k}\right) = \alpha \mathbf{V}_{fim}^{e}\left(\hat{k}\right) + (1-\alpha)\mathbf{V}_{fim}^{m}\left(\hat{k}\right) \tag{2.97}$$

$$\mathbf{V}_{snj}\left(\hat{k}\right) = \mathbf{V}_{snj}^{e}\left(\hat{k}\right) \tag{2.98}$$

are the receiving and radiation patterns for CFIE.

2.3.2.4 *Factorization of the Green's Function*

Consider the case of multilevel reference points, as shown in Figure 2.15, where $m_2, m_3, \ldots, m_L$ are multilevel reference points to observation point i, and $n_2, n_3, \ldots, n_L$ are multilevel reference points to source point j. Then we have

$$\boldsymbol{r}_{ij} = \boldsymbol{r}_{im} + \boldsymbol{r}_{m_2n_2} + \boldsymbol{r}_{nj} \tag{2.99}$$

where

$$\boldsymbol{r}_{im} = \boldsymbol{r}_{im_L} + \boldsymbol{r}_{m_Lm_{L-1}} + \cdots + \boldsymbol{r}_{m_3m_2} \tag{2.100}$$

$$\boldsymbol{r}_{nj} = \boldsymbol{r}_{n_2n_3} + \cdots + \boldsymbol{r}_{n_{L-1}n_L} + \boldsymbol{r}_{n_Lj} \tag{2.101}$$

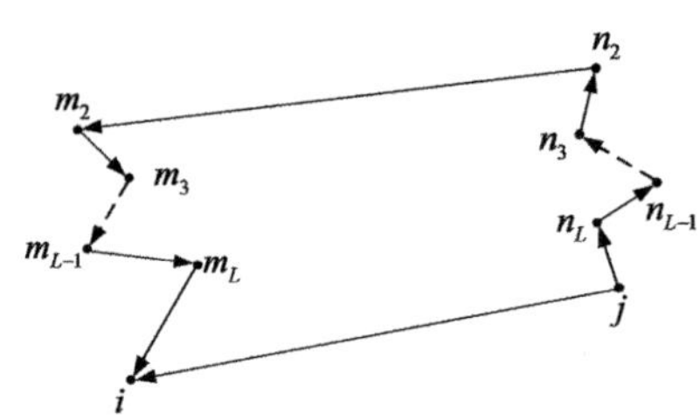

Figure 2.15 *The observation and source points with multilevel reference points.*

When $r_{m_2n_2} > |\boldsymbol{r}_{im} + \boldsymbol{r}_{nj}|$, the addition theorem (2.80) gives

$$\frac{e^{ik|\boldsymbol{r}_i-\boldsymbol{r}_j|}}{|\boldsymbol{r}_i-\boldsymbol{r}_j|} = \frac{ik}{4\pi}\int_S e^{i\boldsymbol{k}\cdot(\boldsymbol{r}_{im}+\boldsymbol{r}_{nj})}\mathrm{T}_{m_2n_2}\left(\hat{k}\cdot\hat{r}_{m_2n_2}\right) d^2\hat{k} \tag{2.102}$$

where $\mathrm{T}_{m_2n_2}\left(\hat{k}\cdot\hat{r}_{m_2n_2}\right)$ is the translator between the top-level reference points. Thus, the scalar Green's function can be factorized at multilevel as

$$\frac{e^{ik|\boldsymbol{r}_i-\boldsymbol{r}_j|}}{|\boldsymbol{r}_i-\boldsymbol{r}_j|} = \frac{ik}{4\pi}\int_S e^{i\boldsymbol{k}\cdot\boldsymbol{r}_{im_L}}e^{i\boldsymbol{k}\cdot\boldsymbol{r}_{m_Lm_{L-1}}}\ldots e^{i\boldsymbol{k}\cdot\boldsymbol{r}_{m_3m_2}}\mathrm{T}_{m_2n_2}\left(\hat{k}\cdot\hat{r}_{m_2n_2}\right)e^{i\boldsymbol{k}\cdot\boldsymbol{r}_{n_2n_3}}\ldots e^{i\boldsymbol{k}\cdot\boldsymbol{r}_{n_{L-1}n_L}}e^{i\boldsymbol{k}\cdot\boldsymbol{r}_{n_Lj}}\, d^2\hat{k} \tag{2.103}$$

2.3.2.5 Octree Structure Construction

In MLFMA, the PEC object under consideration is first discretized using small elements by MOM. Each element is about 0.1λ to 0.15λ in size. Then we set up an octree structure and consider the interactions between the *i*th and *j*th current elements under the tree structure. The construction of the octree is described as follows:

- Choose the smallest box that contains the PEC scatterer, which is regarded as the zeroth-level box. Divide it into eight sub-boxes, which are regarded as first-level boxes. Divide each of the first-level boxes into eight sub-sub-boxes, which are regarded as second-level boxes. This procedure is repeated up to the finest level L, as shown in Figure 2.16. At the finest level, each small box contains at most a few current elements.
- Because there is only one zeroth-level box, and the eight first-level boxes are near neighbors to each other where the addition theorem is invalid, the coarsest level (or the top level) has to be chosen as two or higher.
- When element i is the near neighbor of element j, the field interactions have to be computed by the traditional method using the unfactorized Green's function.
- The far-element calculation is divided into (i) the aggregation process, (ii) the translation process, and (iii) the disaggregation process. The aggregation process is to compute the radiation patterns (outgoing fields) of the sources at different levels starting from the finest level to the coarsest level in the inverted tree.

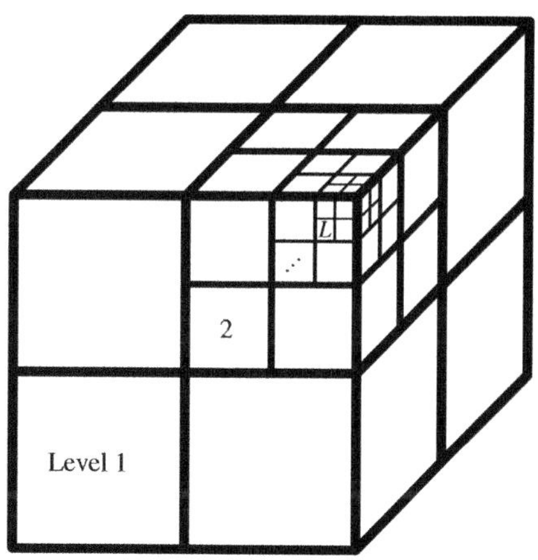

Figure 2.16 Octree structure for 3D problems.

The translator converts the outgoing waves into incoming waves, and, next, the disaggregation process converts these incoming waves from the coarsest level into incoming waves at the finest level, and, finally, into received fields at the desired field points.

2.3.3 Hierarchical Basis for FEM

The adoption of high-order basis functions is an efficient way to reduce the memory requirements and computation time of MOM and FEM, which have received much attention in recent years. For MOM, the combination of high-order basis with MLFMA is not too satisfactory. Thus the application of high-order basis in FEM is mainly discussed here. High-order basis can be classified into two categories, the interpolatory [42] and the hierarchical [43], of which the hierarchical basis is considered more promising as it permits the application of different orders within different computational domains, and p-refinement is permitted. A set of basis functions is referred to as hierarchical if the basis functions forming the nth-order space are a subset of the basis functions forming the $(n+1)$th-order space. For a large class of electromagnetic problems, the hierarchical basis is attractive for FEM discretization. In this section, the application of hierarchical basis in FEM is mainly discussed.

For a tetrahedron, the lowest order basis is the Whitney basis, which is also called the zeroth-order curl-conforming ($\mathbf{H}_0$(curl)) basis. Many researchers have outlined the hierarchical $\mathbf{H}_1$(curl) basis, including Webb [43] and Andersen [44], among others. The approaches outlined by Webb are promising not only because they are hierarchical, but also because they are split into two groups – pure gradient basis functions and solenoid-like vector basis functions. The Webb bases are of the following forms:

$$\begin{aligned} \mathbf{w}_i &= (\zeta_{i1}\nabla\zeta_{i2}-\zeta_{i2}\nabla\zeta_{i1})l_i, && \text{(one per edge)} \\ \mathbf{g}_i &= \zeta_{i1}\nabla\zeta_{i2}+\zeta_{i2}\nabla\zeta_{i1}, && \text{(one per edge)} \\ \mathbf{f}_j &= \zeta_{j3}(\zeta_{j1}\nabla\zeta_{j2}-\zeta_{j2}\nabla\zeta_{j1}), && \text{(two per edge)} \end{aligned} \tag{2.104}$$

where i_1 and i_2 are the local numbers of the two nodes associated with edge i, and j_1, j_2, and j_3 are the local numbers of the nodes associated with face j. The six basis functions $\mathbf{w}_i$ are the Whitney bases, which span the zeroth-order curl-conforming space. The gradient functions $\mathbf{g}_i$ and the second-order face-type non-gradient total variation (TV) subspace $\mathbf{f}_j$ are added to make the basis of the $\mathbf{H}_1$(curl) space complete. It can be seen that the above set of basis functions is hierarchical:

$$\begin{aligned} \mathbf{H}_0(\text{curl}) &= \text{span}\{\mathbf{w}_i\} \\ \mathbf{H}_1(\text{curl}) &= \mathbf{H}_0(\text{curl}) \oplus \text{span}\{\mathbf{g}_i\} \oplus \text{span}\{\mathbf{f}_j\} \end{aligned} \tag{2.105}$$

In total, there are 20 bases. The distribution of the hierarchical bases in a tetrahedral element is shown in Figure 2.17.

In order to test the ability of $\mathbf{H}_1$(curl) in improving the precision of FEM, the smallest eigenvalue of a rectangular cavity is calculated with $\mathbf{H}_0$(curl), $\mathbf{H}_1$(curl), and $\mathbf{H}_2$(curl) respectively. The errors between the numerical results and the analytical solutions are plotted in Figure 2.18, which shows that the result given by the high-order FEM is more accurate. A comprehensive test showed that, to obtain the same precision, the memory needed by using $\mathbf{H}_1$(curl) basis is only about a half of that by using $\mathbf{H}_0$(curl) basis, and the unknowns by using $\mathbf{H}_1$(curl) basis is about one-fifth of that by using $\mathbf{H}_0$(curl) basis. Thus $\mathbf{H}_1$(curl) basis is very attractive for improving the efficiency of FEM.

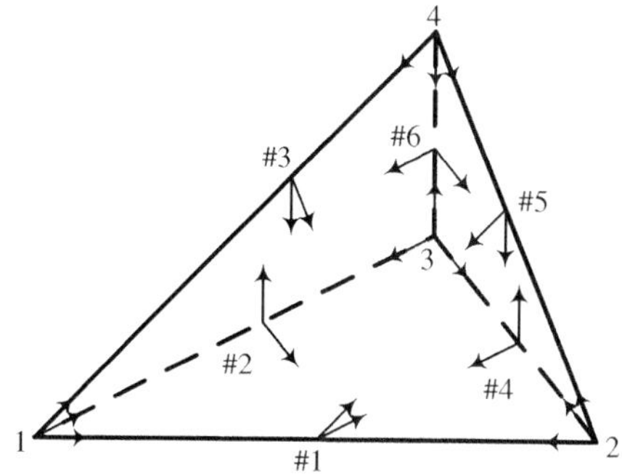

Figure 2.17 *Distribution of basis functions in a tetrahedron.*

The highly sparse linear system generated by FEM is usually solved by iterative solvers. One kind of powerful iterative algorithm is the Krylov subspace iterative methods [45] which show great superiority in solving large, sparse linear systems, such as the bi-conjugate gradient (BiCG) and the generalized minimal residual (GMRES) methods, and so on. One attractive characteristic of these kinds of methods is that they solve the linear system only in terms of matrix–vector multiplication. However, the convergence rate of the Krylov subspace iterative methods is closely related to the condition number of the coefficient matrix. When used to solve a TV-FEM equation resulting from the discretization of 3D electrically large bodies, this kind of method may converge very slowly. Even if the powerful incomplete Cholesky preconditioner is used [46], the operating cost will still be extremely high.

$\mathbf{H}_1$(curl) basis has the property of wavelet basis. Therefore, it permits the construction of fast solvers, which is very attractive for FEM. As mentioned above, the coefficient matrix generated by the Whitney

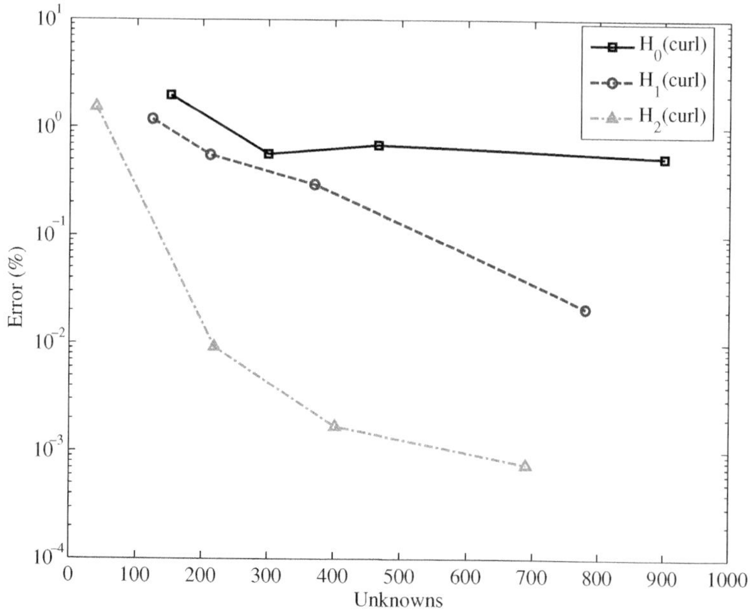

Figure 2.18 *Error versus unknowns for the eigenvalue of the rectangular cavity.*

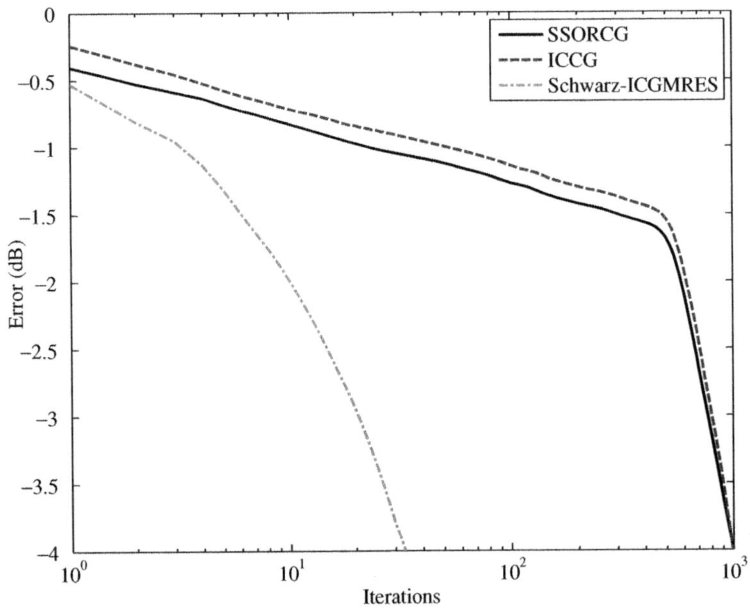

Figure 2.19 Convergence curves of three different iterative methods (SSORCG, Symmetric Successive Over-Relaxation Conjugate Gradient).

basis is ill-conditioned. This is mainly caused by the spurious DC modes contained in the null space of the curl operator and oversampling of the low-frequency physical modes. Therefore, iterative solvers, including the most powerful ICCG solver, converge very slowly. With the help of a hierarchical basis, more efficient solvers can be constructed. Here a kind of efficient solver is introduced. By numbering the unknowns from a low-order group to a high-order group, the linear system can be written in block form as

$$\begin{bmatrix} \bar{\mathbf{A}}_{11} & \bar{\mathbf{A}}_{12} \\ \bar{\mathbf{A}}_{21} & \bar{\mathbf{A}}_{22} \end{bmatrix} \begin{bmatrix} \mathbf{x}_1 \\ \mathbf{x}_2 \end{bmatrix} = \begin{bmatrix} \mathbf{b}_1 \\ \mathbf{b}_2 \end{bmatrix} \tag{2.106}$$

Here $\bar{\mathbf{A}}_{11}$ is the matrix in the $\mathbf{H}_0$(curl) space, which occupies only a small fraction of the stiff matrix. The unknown vector $\mathbf{x}_1$ is formed by the coefficients of $\mathbf{w}_{ij}$ basis functions, which contain most of the low-frequency parts of the solution, while $\mathbf{x}_2$ contains most of the rapidly oscillating modes. As a result, the

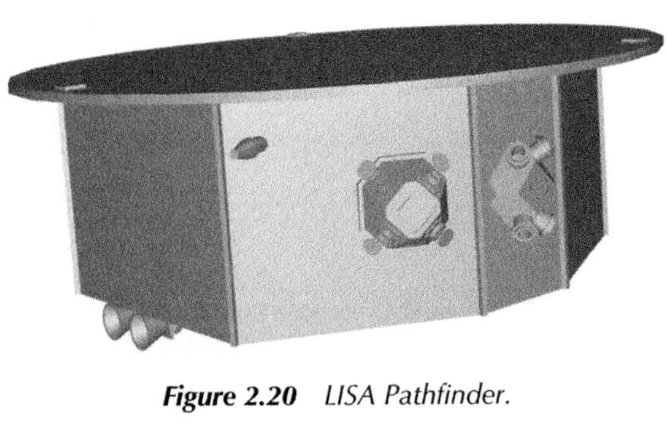

Figure 2.20 LISA Pathfinder.

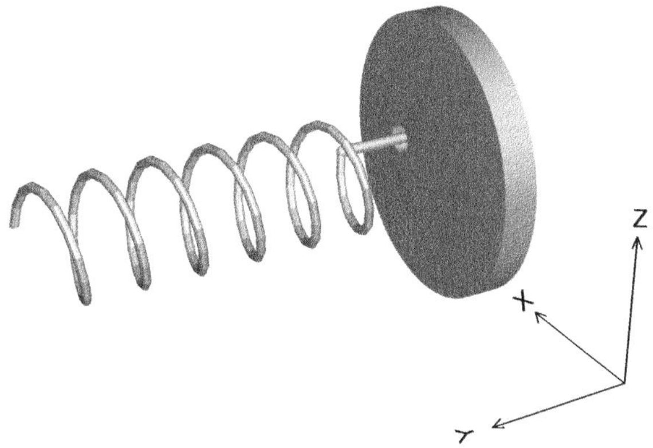

Figure 2.21 *Helix antenna.*

Schwarz method can be used to solve (2.106). Through Schur decomposition, the stiff matrix $\bar{\mathbf{A}}$ can be written as

$$\begin{bmatrix} \bar{\mathbf{A}}_{11} & \bar{\mathbf{A}}_{12} \\ \bar{\mathbf{A}}_{21} & \bar{\mathbf{A}}_{22} \end{bmatrix} = \begin{bmatrix} \bar{\mathbf{I}} & \mathbf{0} \\ \bar{\mathbf{A}}_{21}\bar{\mathbf{A}}_{11}^{-1} & \bar{\mathbf{I}} \end{bmatrix} \begin{bmatrix} \bar{\mathbf{A}}_{11} & \mathbf{0} \\ \mathbf{0} & \bar{\mathbf{A}}_{22} - \bar{\mathbf{A}}_{21}\bar{\mathbf{A}}_{11}^{-1}\bar{\mathbf{A}}_{12} \end{bmatrix} \begin{bmatrix} \bar{\mathbf{I}} & \bar{\mathbf{A}}_{11}^{-1}\bar{\mathbf{A}}_{12} \\ \mathbf{0} & \bar{\mathbf{I}} \end{bmatrix} \tag{2.107}$$

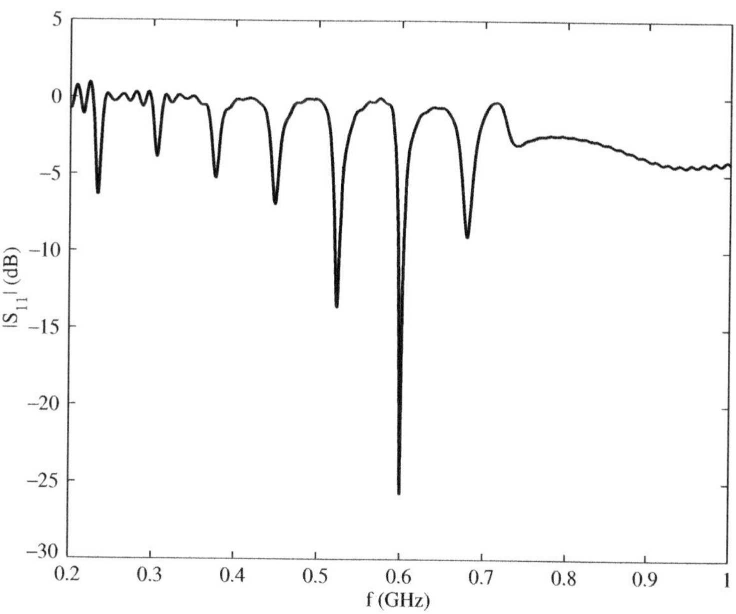

Figure 2.22 *The* S_{11} parameter of the helix antenna.

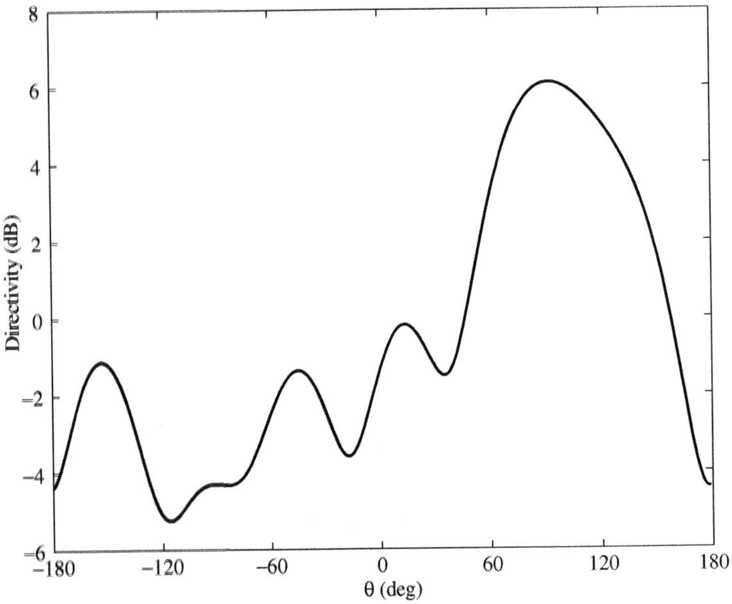

Figure 2.23 The directivity of the helix antenna.

If we define

$$\begin{bmatrix}\mathbf{y}_1\\ \mathbf{y}_2\end{bmatrix}=\begin{bmatrix}\bar{\mathbf{I}} & \bar{\mathbf{A}}_{11}^{-1}\bar{\mathbf{A}}_{12}\\ \mathbf{0} & \bar{\mathbf{I}}\end{bmatrix}\begin{bmatrix}\mathbf{x}_1\\ \mathbf{x}_2\end{bmatrix},\quad \begin{bmatrix}\mathbf{b}_1\\ \mathbf{b}_2\end{bmatrix}=\begin{bmatrix}\bar{\mathbf{I}} & 0\\ \bar{\mathbf{A}}_{21}\bar{\mathbf{A}}_{11}^{-1} & \bar{\mathbf{I}}\end{bmatrix}\begin{bmatrix}\tilde{\mathbf{b}}_1\\ \tilde{\mathbf{b}}_2\end{bmatrix} \tag{2.108}$$

then the system (2.106) is transformed into two equations

$$\bar{\mathbf{A}}_{11}\mathbf{y}_1=\tilde{\mathbf{b}}_1 \tag{2.109}$$

$$\left(\bar{\mathbf{A}}_{22}-\bar{\mathbf{A}}_{21}\bar{\mathbf{A}}_{11}^{-1}\bar{\mathbf{A}}_{12}\right)\mathbf{y}_2=\tilde{\mathbf{b}}_2 \tag{2.110}$$

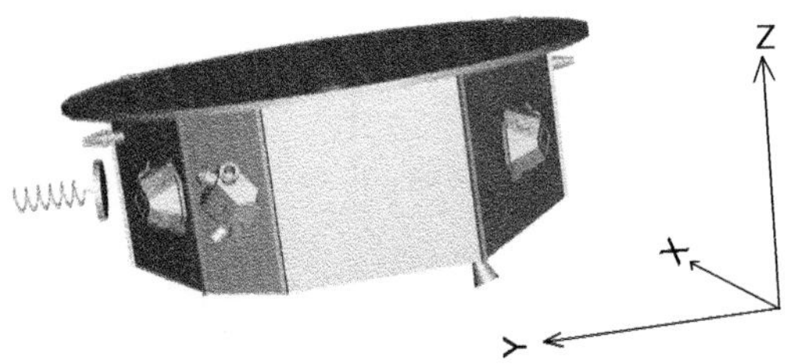

Figure 2.24 LISA Pathfinder with one helix antenna.

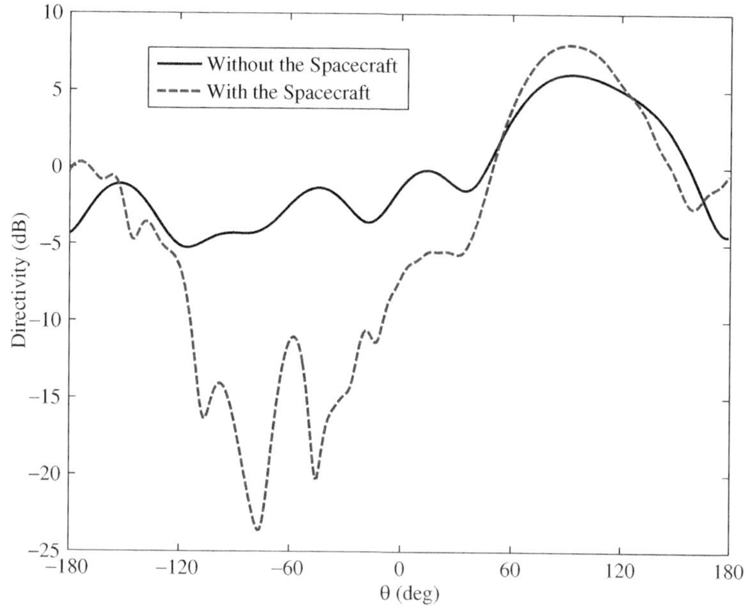

Figure 2.25 *Comparison of the directivity with/without the spacecraft.*

As $\bar{\mathbf{A}}_{11}$ occupies only a much smaller fraction of the stiff matrix $\bar{\mathbf{A}}$, (2.109) can be solved efficiently with a suitable direct solver, while (2.110) can be solved efficiently with the powerful ICGMRES solver. Using such a scheme, (2.106) can also be solved very efficiently. Figure 2.19 shows a comparison of convergence curves of the different iterative methods for the patch antenna, which showed the great advantage of the Schwarz method compared to conventional iterative methods.

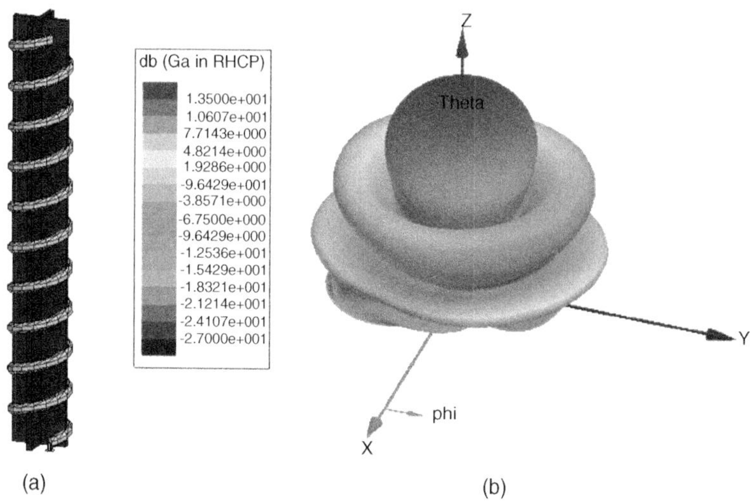

Figure 2.26 *Helix antenna element design: (a) helix antenna; (b) the 3D radiation pattern.*

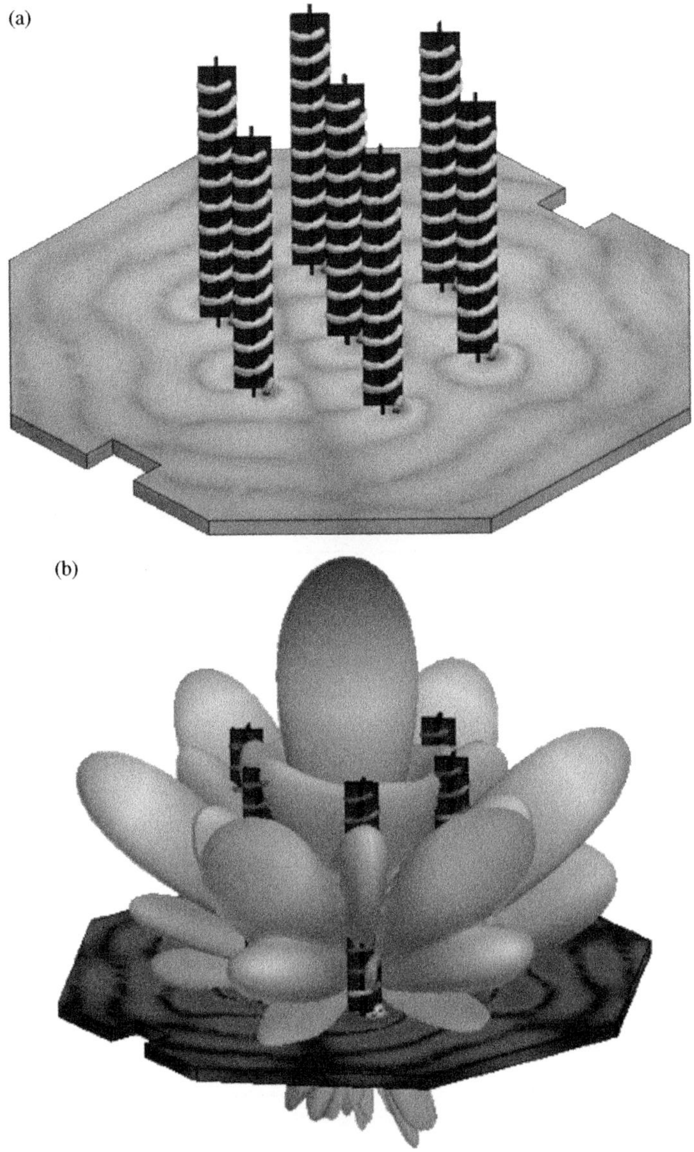

Figure 2.27 Antenna array design. (a) Seven-element array with finite ground plane. (b) The 3D radiation pattern with all elements excited.

2.4 Case Studies: Effects of the Satellite Body on the Radiation Patterns of Antennas

In this section, the effects of the satellite body on the radiation patterns of space antennas are studied. The emphasis is on illustrating the irreplaceable status of CEM in analyzing the influence of the platform on antenna performance. Since the accuracy and efficiency of the numerical methods have been demonstrated in many books, such as [21,26,41], details are omitted here.

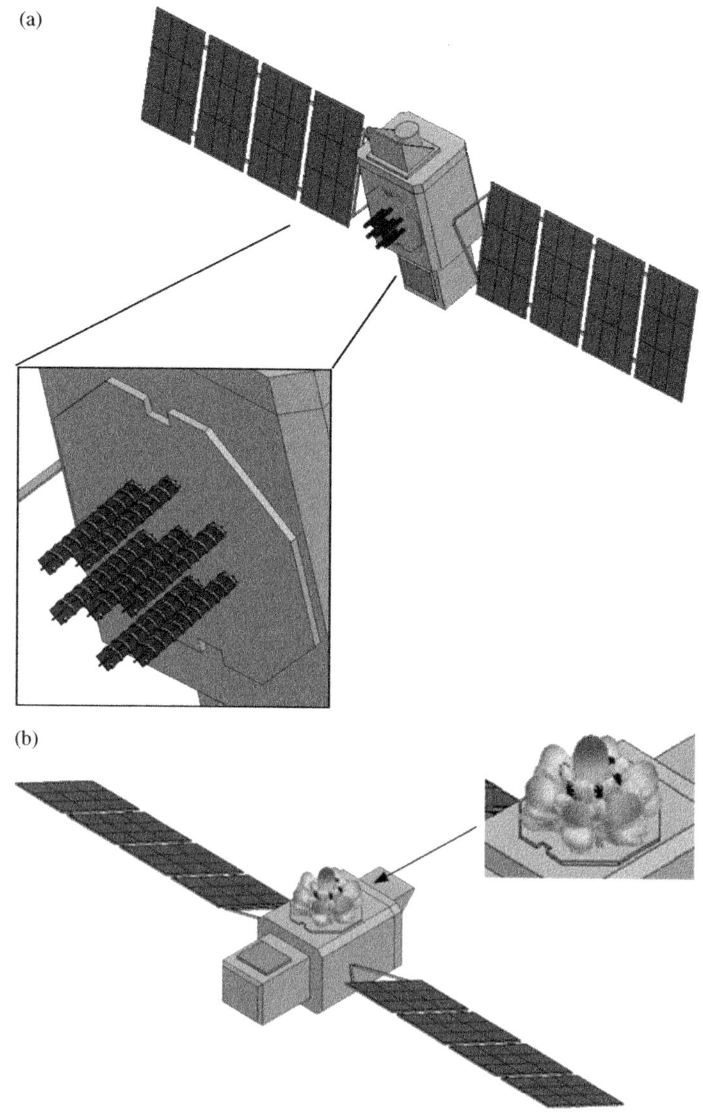

Figure 2.28 Influence of the spacecraft on the antenna array. (a) Spacecraft with the seven-element array. (b) The 3D radiation pattern of the array when equipped on the spacecraft. (c) Directivity of the seven-element array with/without the spacecraft. (d) The field distribution on two perpendicular planes.

The Laser Interferometer Space Antenna (LISA) Pathfinder, as shown in Figure 2.20, the second of the European Space Agency's Small Missions for Advanced Research in Technology (SMART), is a dedicated technology demonstrator for the joint ESA/NASA LISA mission [47]. The scientific objective of the LISA Pathfinder mission consists of the first in-flight test of low-frequency gravitational wave detection metrology, which will tell us more about the way space and time are interconnected. This technology has been developed to mitigate part of the risk associated with LISA, since it needs to achieve many things simultaneously. All the

(c)

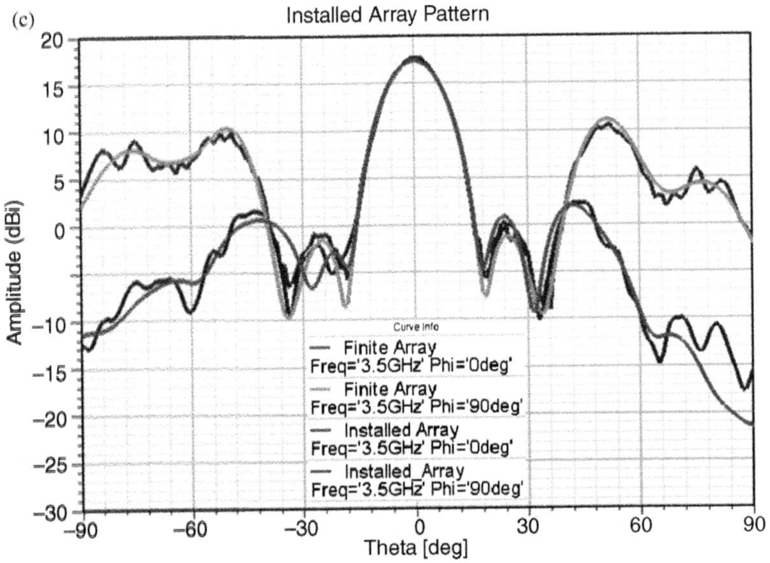

(d)

Figure 2.28 (Continued)

technologies are essential not just for LISA; they also lie at the heart of any future space-based test of Einstein's theory of general relativity.

In the following, the effects of the spacecraft on a helix antenna often used in telemetry and tracking control subsystems are studied by full-wave simulation methods. First, the radiation pattern of a helix antenna alone as shown in Figure 2.21 is calculated. The radius and pitch of the helix is 0.2λ and 0.1λ respectively. From Figure 2.22, we can clearly see that the resonance frequency is about 600 MHz and the bandwidth is narrow, which is the nature of the traditional helix antenna. Figure 2.23 shows the directivity of the antenna, in which the maximum directivity is achieved at the direction of the axis of the antenna.

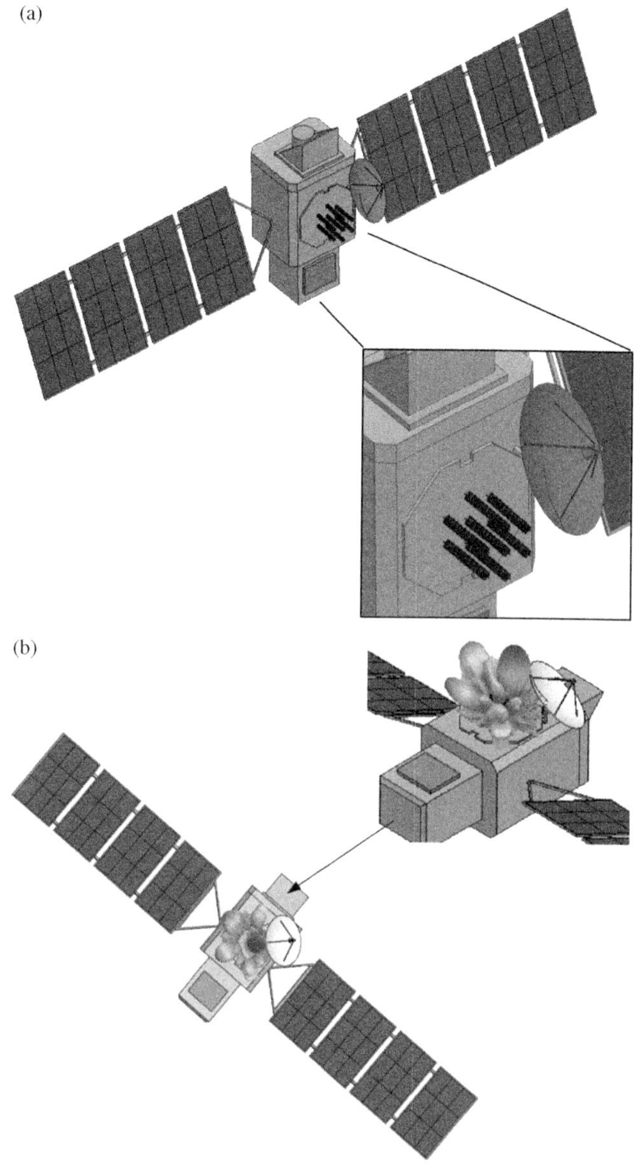

Figure 2.29 Effects of the spacecraft and the reflector antenna on the seven-element array. (a) Spacecraft with the reflector antenna and seven-element array. (b) The 3D radiation pattern of the array when equipped on the spacecraft and the reflector antenna present. (c) Directivity of the seven-element array with/without the spacecraft and the reflector antenna. (d) The field distribution on the whole structure.

From electromagnetic theory, one can qualitatively predict that the radiation pattern will change once the antenna has been equipped on the spacecraft; however, more useful quantitative analyses are very difficult to perform without the help of CEM. To illustrate this, the one-antenna case shown in Figure 2.24 is studied, in which the helix antenna is the same as that in Figure 2.21. The directivity is given in Figure 2.25. Clearly, the radiation pattern of the helix antenna alone is dramatically modified by the LISA platform.

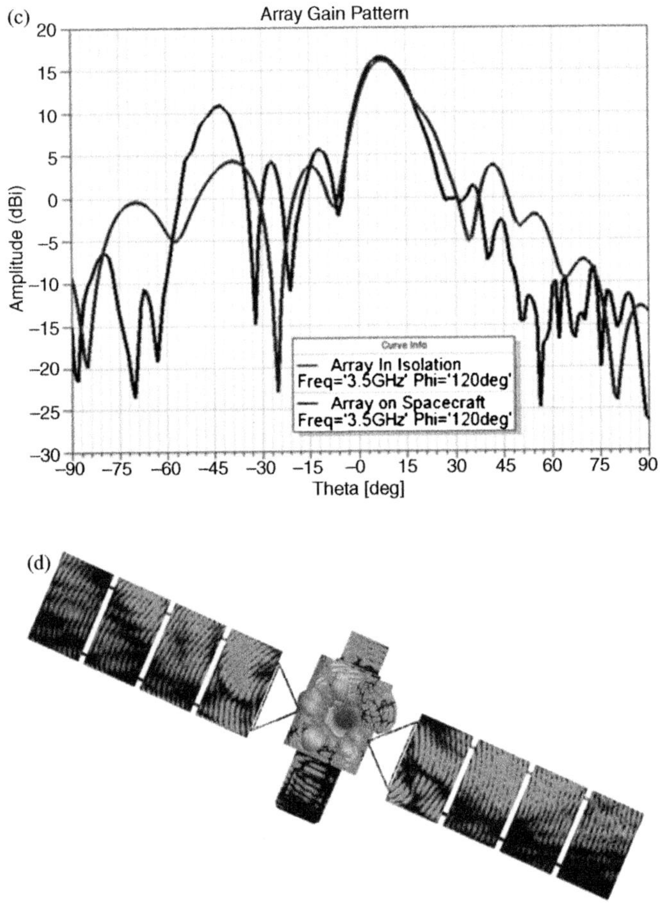

Figure 2.29 *(Continued)*

In the following, more complicated cases are studied by FEM, which have been provided by the Ansoft Company. In the above case, only a single antenna was considered. In practice, the antenna array is much more useful and widely used to achieve higher directivity. As shown in Figure 2.26a, a helix antenna with 10 turns is simulated, whose radius, pitch spacing, and wire radius are 0.16λ, 0.25λ, and 0.03λ respectively. The 3D radiation pattern is presented in Figure 2.26b.

Second, a seven-element array with a finite ground plane as illustrated in Figure 2.27a is studied. The 3D far-field pattern is given in Figure 2.27b, which is much better than that in the single-element case.

In order to study the influence of the spacecraft on the designed performance of the antenna array, a spacecraft with the seven-element array is considered, as in Figure 2.28a. Figure 2.28b is the 3D far-field pattern in this case. From Figure 2.28c, we can clearly see that the radiation pattern changes very little after the array has been installed on the spacecraft, which meets the design goals. The field distribution on two perpendicular planes of the structure is plotted in Figure 2.28d.

As mentioned above, in order to accomplish different applications, various antennas will be installed on one satellite. Reflector antennas are often used for communications. To investigate the effects of a reflector antenna on the performance of the seven-element array, the structure in Figure 2.29a was studied. From Figure 2.29b and c, it is clear that the radiation pattern of the seven-element array has changed when the reflector antenna is present, especially for the sidelobe which has been increased. We emphasize that such results are obtained under a small scan angle of 10°, but other detrimental effects may be present for lager scan angles. With the benefit of the full-wave method, as shown in Figure 2.29d, the physical insight can be deeply understood by the field distribution picture.

2.5 Summary

This chapter reviews various numerical methods applied broadly in space antenna simulations, in which three popular full-wave simulation methods (MOM, FEM, and FDTD) are specifically depicted. First, the basic theory of such methods and their applications in antenna simulations are briefly given. Then the related fast algorithms are discussed. As illustrated, each method has its own merits and demerits in antenna simulation. MOM can deal with open problems naturally and needs only surface discretization, but can be difficult in dealing with serious inhomogeneous materials. FEM is a versatile technique in dealing with complex structures and inhomogeneous materials, but it requires considerable computation time and memory. The FDTD method has good computational complexity, being matrix free, and can give abundant time information on the physical model, but its accuracy is affected by the staircase approximation and numerical dispersion. For some complex cases, the adoption of one single method is inefficient, or even impossible to solve some problems. In such circumstances, hybrid methods are preferred, such as a combination of MOM or FEM with high-frequency methods, a combination of FEM, BEM, and MLFMA, and so on. Particularly, in [48] the hybrid FEBI-MLFMA-UTD method is introduced for electrically large antenna modeling, which combines four methods (FEM, BEM, MLFMA, and UTD) to obtain the best efficiency. A good algorithm should have the following properties: short CPU time, compact memory, and good precision. The choice of an optimal method must achieve a good compromise between these requirements.

Acknowledgments

We would like to thank Ansoft, USA, for providing some figures (Figures 2.26–2.29) in the case studies.

References

1. Ceruzzi, P.E. (2003) *A History of Modern Computing*, The MIT Press, Cambridge, MA.
2. Maxwell, J.C. (1891) *A Treatise on Electricity and Magnetism*, 3rd edn, Clarendon Press, Oxford.
3. Rautio, J.C. (2003) Planar electromagnetic analysis. *IEEE Microwave Magazine*, **4**(1), 35–41.
4. Kline, M. and Kay, I.W. (1965) *Electromagnetic Theory and Geometrical Optics*, Wiley-Interscience, New York.
5. Kouyoumjian, R.G. and Pathak, P.H. (1974) A uniform geometrical theory of diffraction for an edge in a perfectly conducting surface. *Proceedings of the IEEE*, **62**(11), 1448–1461.
6. Ufimtsev, P.Y. (2007) *Fundamentals of the Physical Theory of Diffraction*, John Wiley & Sons, Inc., Hoboken, NJ.
7. Ling, H., Chou, R.C., Lee, S.-W. *et al.* (1989) Shooting and bouncing rays: calculating the RCS of an arbitrarily shaped cavity. *IEEE Transactions on Antennas and Propagation*, **37**(2), 194–205.
8. Harrington, R.F. (1968) *Field Computation by Moment Methods*, Macmillan, New York.
9. Mautz, J.R. and Harrington, R.F. (1978) H-field, E-field, and combined-field solutions for conducting bodies of revolution. *Archiv fuer Elektronik und Uebertragungstechnik*, **32**, 157–164.
10. Mautz, J.R. and Harrington, R.F. (1979) Electromagnetic scattering from a homogeneous material body of revolution. *Archiv fuer Elektronik und Uebertragungstechnik*, **33**, 71–80.

11. Wu, T.K. and Tsai, L.L. (1977) Scattering from arbitrarily-shaped lossy dielectric bodies of revolution. *Radio Science*, **12**, 709–718.
12. Medgyesi-Mitschang, L.N. and Putnam, J.M. (1984) Electromagnetic scattering from axially inhomogeneous bodies of revolution. *IEEE Transactions on Antennas and Propagation*, **32**(8), 797–806.
13. Huddleston, P.L., Medgyesi-Mitschang, L.N., and Putnam, J.M. (1986) Combined field integral equation formulation for scattering by dielectrically coated conducting bodies. *IEEE Transactions on Antennas and Propagation*, **34**(4), 510–520.
14. Poggio, A.J. and Miller, E.K. (1973) Integral equation solutions of three-dimensional scattering problems, in *Computer Techniques for Electromagnetics*, Pergamon, Oxford, pp. 159–264.
15. Rao, S.M., Wilton, D.R., and Glisson, A.W. (1982) Electromagnetic scattering by surfaces of arbitrary shape. *IEEE Transactions on Antennas and Propagation*, **30**(3), 409–418.
16. Courant, R. (1943) Variational methods for the solution of problems of equilibrium and vibrations. *Bulletin of the American Mathematical Society*, **49**, 1–23.
17. Alett, P.L., Baharani, A.K., and Zienkiewicz, O.C. (1968) Application of finite elements to the solution of Helmholtz's equation. *IEE Proceedings*, **115**, 1762–1766.
18. Sacks, Z.J., Kingsland, D.M., Lee, R. *et al.* (1995) A perfectly matched anisotropic absorber for use as an absorbing boundary condition. *IEEE Transactions on Antennas and Propagation*, **43**(12), 1460–1463.
19. Berenger, J.P. (1994) A perfectly matched layer for the absorption of electromagnetic waves. *Journal of Computational Physics*, **114**, 185–200.
20. Silvester, P.P. and Ferrari, R.L. (1990) *Finite Elements for Electrical Engineers*, 2nd edn, Cambridge University Press, Cambridge.
21. Jin, J.M. (1993) *The Finite Element Method in Electromagnetics*, John Wiely & Sons, Inc., New York.
22. Volakis, J.L., Arindam, C., and Kempel, L.C. (1998) *Finite Element Method for Electromagnetics: Antennas, Microwave Circuits, and Scattering Applications*, Wiley-IEEE, New York.
23. Yee, K.S. (1966) Numerical solution of initial boundary value problems involving Maxwell's equations in isotropic media. *IEEE Transactions on Antennas and Propagation*, **14**(3), 302–307.
24. Courant, R., Friedrichs, K., and Lewy, H. (1967) On the partial difference equations of mathematical physics. *IBM Journal of Research and Development*, **11**(2), 215–234.
25. Kunz, K.S. and Luebbers, R.J. (1993) *The Finite Difference Time-domain Method for Electromagnetics*, CRC Press, Boca Raton, FL.
26. Taflove, A. (1995) *Computational Electrodynamics: The Finite-Difference Time-Domain Method*, Artech House, Norwood, MA.
27. Yang, P. and Liou, K.N. (1996) Finite-difference time-domain method for light scattering by small ice crystals in three-dimensional space. *Journal of the Optical Society of America A*, **13**(10), 2072–2085.
28. Dey, S. and Mittra, R. (1997) A locally conformal finite-difference time-domain (FDTD) algorithm for modeling three-dimensional perfectly conducting objects. *Microwave and Guided Wave Letters, IEEE*, **7**(9), 273–275.
29. Namiki, T. (2000) 3-D ADI-FDTD method-unconditionally stable time-domain algorithm for solving full vector Maxwell's equations. *IEEE Transactions on Microwave Theory and Techniques*, **48**(10), 1743–1748.
30. Greengard, L. and Rokhlin, V. (1987) A fast algorithm for particle simulations. *Journal of Computational Physics*, **73**, 325–348.
31. Rokhlin, V. (1990) Rapid solution of integral equations of scattering theory in two dimensions. *Journal of Computational Physics*, **86**(2), 414–439.
32. Engheta, N., Murphy, W.D., Rokhlin, V. *et al.* (1992) The fast multipole method (FMM) for electromagnetic scattering problems. *IEEE Transactions on Antennas and Propagation*, **40**(6), 634–641.
33. Hamilton, L.R., Stalzer, M.A., Turley, R.S. *et al.* (1993) Scattering computation using the fast multipole method. IEEE Antennas and Propagation Society International Symposium, 2 June, pp. 852–855.
34. Coifman, R., Rokhlin, V., and Wandzura, S. (1993) The fast multiple method for the wave equation: a pedestrian prescription. *IEEE Antennas and Propagation Magazine*, **35**(3), 7–12.
35. Lu, C.C. and Chew, W.C. (1993) Fast algorithm for solving hybrid integral equations. *IEE Proceedings – Microwaves, Antennas and Propagation*, **140**, 455–460.

36. Lu, C.C. and Chew, W.C. (1994) A multilevel algorithm for solving boundary-value scattering. *Microwave and Optical Technology Letters*, **7**(10), 466–470.
37. Dembart, B. and Yip, E. (1995) A 3D fast multipole method for electromagnetics with multiple levels. 11th Annual Review of Progress in Applied Computational Electromagnetics, vol. 1, pp. 621–628.
38. Rokhlin, V. (1993) Diagonal forms of translation operators for the Helmholtz equation in three dimensions. *Applied and Computational Harmonic Analysis*, **1**(1), 82–93.
39. Chew, W.C., Koc, S., Song, J.M. *et al.* (1997) A succinct way to diagonalize the translation matrix in three dimensions. *Microwave and Optical Technology Letters*, **15**(3), 144–147.
40. Song, J.M. and Chew, W.C. (2001) Interpolation of translation matrix in MLFMA. *Microwave and Optical Technology Letters*, **30**(2), 109–114.
41. Chew, W.C., Jin, J.M., Michielssen, E. *et al.* (2001) *Fast and Efficient Algorithms in Computational Electromagnetics*, Artech House, Norwood, MA.
42. Graglia, R.D., Wilton, D.R., and Peterson, A.F. (1997) Higher order interpolatory vector bases for computational electromagnetics. *IEEE Transactions on Antennas and Propagation*, **45**(3), 329–342.
43. Webb, J.P. (1999) Hierarchal vector basis functions of arbitrary order for triangular and tetrahedral finite elements. *IEEE Transactions on Antennas and Propagation*, **47**(8), 1244–1253.
44. Andersen, L.S. and Volakis, J.L. (1998) Hierarchical tangential vector finite elements for tetrahedra. *Microwave and Guided Wave Letters, IEEE*, **8**(3), 127–129.
45. Kelley, C.T. (1995) *Iterative Methods for Linear and Nonlinear Equations*, Society for Industrial and Applied Mathematics, Philadelphia.
46. Benzi, M. (2002) Preconditioning techniques for large linear systems: a survey. *Journal of Computational Physics*, **182**(2), 418–477.
47. McNamara, P. and Racca, G. (2009) Introduction to LISA Pathfinder, LISA-LPF-RP-0002, ESA.
48. Tzoulis, A. and Eibert, T.F. (2007) Antenna modeling with the hybrid finite element – boundary integral – multilevel fast multipole – uniform geometrical theory of diffraction method. 2nd International ITG Conference on Antennas, March, pp. 91–95.

3

System Architectures of Satellite Communication, Radar, Navigation and Remote Sensing

Michael A. Thorburn

Space Systems/Loral, USA

3.1 Introduction

To fully appreciate the satellite antenna, one must have a basic understanding of the overall system of which it is a part. In this chapter the elements of satellite systems are described with an eye toward those that directly impact the satellite antenna functional or performance requirements. In the first sections of the chapter an overview is provided of the classes of satellite systems used for communication, radar, navigation and remote sensing. The key features of these systems that are related to the antenna are described and the primary performance parameters of the system that have an impact on the antenna requirements are identified. The next sections of the chapter provide the derivations of the equations that govern the performance of the satellite system as they relate to antenna requirements. The satellite link equation is introduced and the major components of the satellite are identified that correspond to these parameters. Additionally the effects of the atmosphere on the link are quantified. The chapter concludes with a discussion of the major subsystems of a satellite and an elementary discussion of the orbital considerations.

3.2 Elements of Satellite System Architecture

The key elements of a satellite system architecture are (1) the mission and its functional requirements, (2) the architecture of the payload system, (3) the partition of the functionality of the system to the space segment and the ground segment, (4) the basic orbital considerations of the satellite, and (5) the bus subsystem requirements to meet the mission objectives and control the orbit.

Space Antenna Handbook, First Edition. Edited by William A. Imbriale, Steven (Shichang) Gao and Luigi Boccia.

In the following sections, the first two of these key elements are considered. They are the elements that most intimately tie together the antenna functional and performance requirements. A discussion of the remaining elements will follow later in the chapter.

3.3 Satellite Missions

Throughout this handbook the antennas of satellite communication, radar, navigation and remote sensing missions are discussed. In this chapter, each mission is described briefly and the corresponding mission performance parameters that are of greatest importance to the definition of the antennas are identified.

3.4 Communications Satellites

Communications satellites are commonplace in modern society. Applications include television broadcast, mobile telephone networks and data transmission. The convention is to subdivide the satellites by their communications mission.

Each mission is characterized by the service that it provides, the number of users it can support and the geographic area over which the service is provided. Additional features include the connectivity that it offers to its users, the spectrum that it is allocated, the power levels that are used, the scheme by which multiple users can share the satellite resource and the ability of the satellite to be reconfigured. The characteristics of greatest importance to the antenna definition are the spectrum allocation, the power levels and the coverage areas. Additionally, the antenna design is impacted significantly by the frequency and polarization reuse used to provide the desired capacity.

The major mission categories are Fixed Satellite Services (FSS), Broadcast Satellite Services (BSS), Digital Audio Radio Services (DARS), Direct to Home Internet Services (DTH) and Mobile Communications Services (Mobile). Sometimes BSS systems are referred to as Direct Broadcast Service (DBS) systems.

3.4.1 Fixed Satellite Services

FSS systems provide a communications connection between users separated by great distances. As such these systems are referred to as point-to-point systems. The connection is usually in the form of a two-way communication circuit (see Figure 3.1). For example, FSS systems provide telephone services to shipping fleets, video services for new agencies, or credit card verification services for individual gasoline service stations.

FSS systems operate at C-band and at Ku-band with a considerable range of power levels and coverage areas that vary from small geographic spots to continental or hemispherical areas.

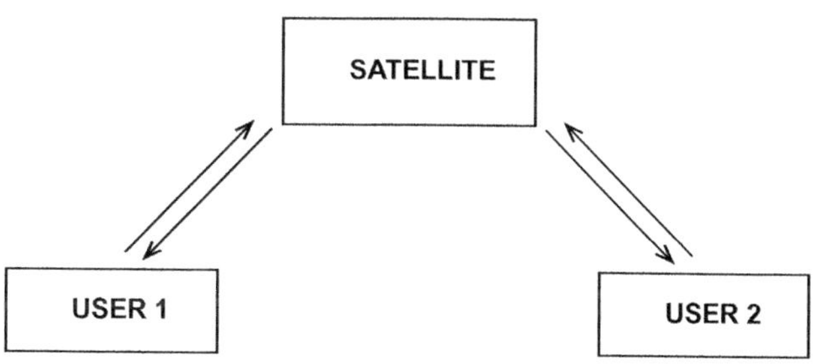

Figure 3.1 *Two-way point-to-point satellite communication.*

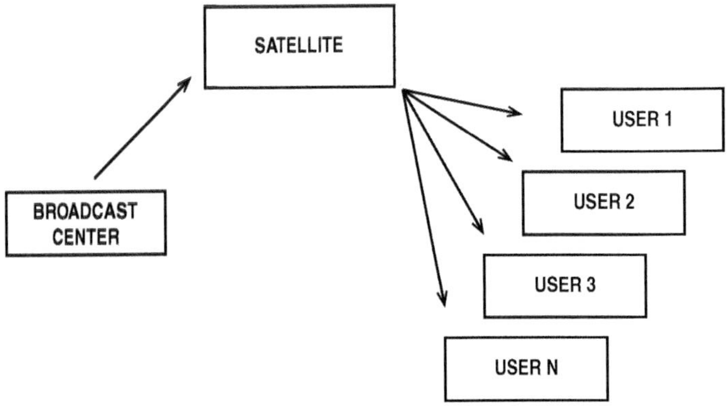

Figure 3.2 One-way point to many-point satellite communication.

3.4.2 Broadcast Satellite Services (Direct Broadcast Services)

BSS (or DBS) systems are used to provide television and radio services to many users over large geographic areas throughout the world by connecting them to broadcast centers. As such the architectures of the BSS system are one point to many points. That is, one broadcast center uplinks signals to the satellite, which in turn transmits them to many users in one or more distinct coverage areas (see Figure 3.2). Examples of BSS systems include satellite television systems.

BSS systems operate at C-band, Ku-band and K-band. One example of these services is C-band broadcast television, which has been available for decades but requires individual subscribers to have a 3 m satellite dish in the backyard. Ku-band BSS systems also include television subscription services but these require much smaller user systems, typically with dishes of about 0.75 m diameter.

3.4.3 Digital Audio Radio Services

DARS systems are relatively new. Like BSS systems, they are point to many-point architectures. These systems are used to provide subscriber-based radio services to properly equipped automobiles and portable radios. The systems operate at lower frequencies and with significant satellite power to accommodate the portability or mobility of the user systems.

3.4.4 Direct to Home Broadband Services

Like DARS systems, the DTH systems are relatively new. They are growing significantly as a business segment and in particular over the past decade significant investments have been made in broadband DTH satellite Internet services. The architectures are one point to many points but distinguish themselves from BSS systems by being two-way connections (Figure 3.3). Additionally, the modern DTH system achieves the capacity required by the business plan through an aggressive reuse of frequency and polarization. This does not increase the complexity of the communications system significantly but does increase the size of the system directly. The frequency reuse also puts increasing performance demands on the antenna subsystem as the sidelobe performance becomes the limiting factor to overall system capacity.

3.4.5 Mobile Communications Services

Mobile systems are designed to provide user-to-user connections, chiefly for telephone or data transmission. They are point-to-point systems. The feature that distinguishes mobile from FSS systems is the number of

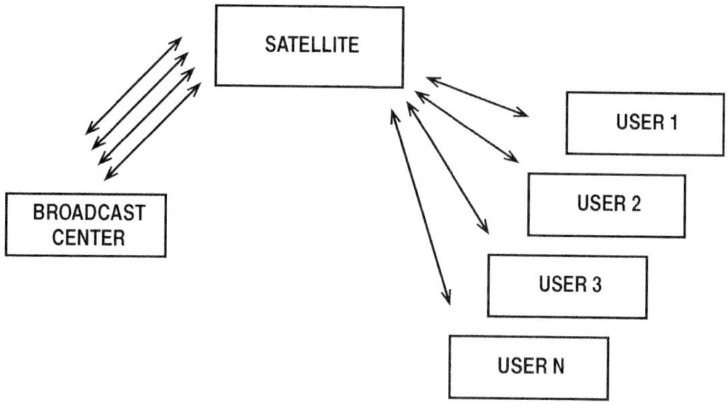

Figure 3.3 *Two-way point to many-point satellite communication.*

users and the fact that many users share a common satellite communications circuit. The multiple access scheme employed by the system dictates the requirements to support these users. The reader can find considerable discussion of multiple access techniques in the references [1–10].

3.5 Radar Satellites

Radar satellites are a subset of the larger group of remote sensing satellites. There are many examples of radar systems used in meteorology and the earth sciences. Additionally radar satellites can be part of an aircraft tracking system. What makes radar satellites interesting is the microwave electronics composing the radar system. As a result, although the performance parameters are different, the analytical processes involved in the engineering are very similar to those used for communications satellites (see Figure 3.4).

3.6 Navigational Satellites

Obviously the most well-known navigational satellite system is the global positioning system (GPS). GPS is a constellation of satellites orbiting the Earth and transmitting signals to users. The microwave electronics is very similar to that in the communication system satellites but additional requirements are imposed on stable clocks.

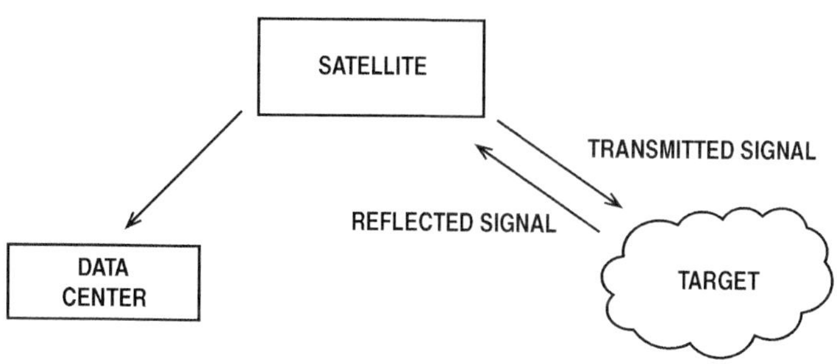

Figure 3.4 *Radar satellite configuration.*

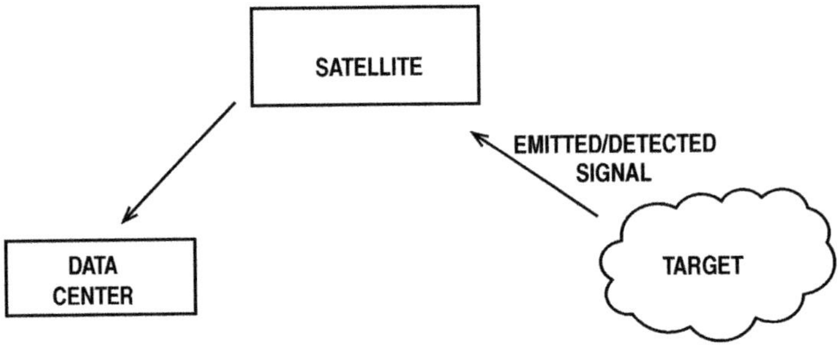

Figure 3.5 Passive remote sensing satellite configuration.

3.7 Remote Sensing Satellites

Satellites used for meteorology, earth or space science, or remote observations constitute the family of remote sensing satellites. As described above, radar satellites are a subset of these. Continuing with that definition, the remainder of this class of satellites falls into the category of those doing passive remote sensing (see Figure 3.5). The chief characteristic of this class of satellites is the remote sensing receiver.

3.8 Architecture of Satellite Command and Control

Every satellite, whether used for communications, navigation, or remote sensing, has a command and control system (see Figure 3.6) that results in antenna requirements. The satellite must be able to receive commands to enable control of its various subsystems and must be able to send telemetry to inform the command and control station of its configuration status and health. This capability must be present through all of the phases of a satellite's life including launch, orbit raising, on-orbit maneuvers, on-station operations, contingency operations and eventually, at the end of the satellite's life, de-orbit. The architecture of this system is illustrated below.

3.9 The Communications Payload Transponder

There are two features that help categorize the family of communications payload systems. One is whether the system is regenerative or non-regenerative; that is, whether the satellite communications payload

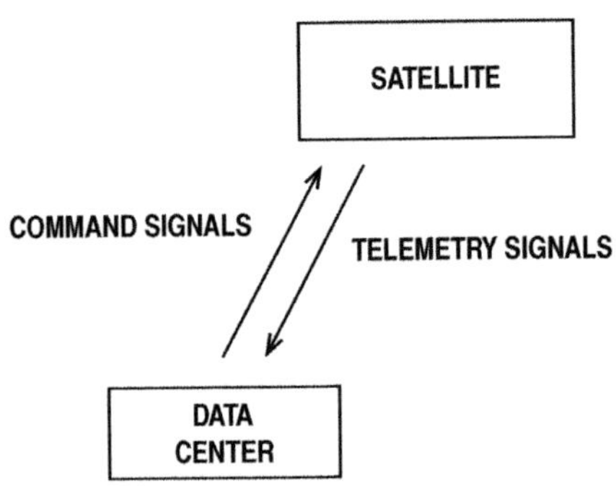

Figure 3.6 Satellite command and control.

demodulates, processes and then remodulates the signals or whether the system simply converts the waveform for retransmission without any demodulation or processing. Those systems in the first category are referred to as processed payloads. Those in the second category are referred to as repeaters. The second feature differentiates the repeaters into two groups. A repeater can be either an analog repeater where the amplification, filtering and routing of the signals are done entirely by analog components such as mechanical switches, analog filters, low-noise amplifiers and power amplifiers, or alternatively, if the signal is digitized somewhere along its path, referred to as a digital channelizer as some portion of the routing, filtering and amplification is done digitally.

3.9.1 Bent-Pipe Transponders

By far, the configuration of most communications satellite payloads is that of non-regenerative microwave repeaters (or equivalently bent-pipe frequency-translation transponders). The simplicity and DC power efficiency of these systems is what makes them the dominant configuration. With these systems, the uplink signals are amplified, their frequencies are translated, the circuits are established, the signals are filtered and then amplified again for retransmission to the ground via the downlink. No processing of the signal, other than the frequency translation, is done on board the satellite. The satellite communications payload serves to establish the link completing the communications circuit and preserve the integrity of the signal by diminishing the signal-to-noise ratio as little as is practical. As such, the communications payload is thought of as a circuit-switched system. Each circuit is called a transponder. Each transponder is a series of components that provides the communications channel between the uplink signal received at the uplink antenna and the downlink signal transmitted by the downlink antenna.

3.9.2 Digital Transponders

The reconfiguration of the bent-pipe transponder is carried out by commanding switches (mostly mechanical switches) to a different state and possibly commanding some active components, such as microwave oscillators, to turn on and others to turn off. These commands take time to execute and there are reliability impacts associated with changing the state of any of these components. Furthermore, the number of possible states for the communications payload is established by the number of independent switches, oscillators, filters and other components within the payload. As digital components establish more flight heritage and power efficiency, there are some applications that require a high degree of reconfigurability and the introduction of a digital channelizer which can replace the switch matrix; some channel filters and some variable gain preamplifiers is a good trade.

3.9.3 Regenerative Repeater

If the requirements of the mission either result in limitations due to available bandwidth or need some measure of security, a fully regenerative repeater may be warranted. Regardless of whether the communications payload is analog, digital or regenerative, it will have many transponders. The set of transponders, together with the antennas on the satellite, make up the communications payload.

3.10 Satellite Functional Requirements

There are many functional requirements of a satellite. For the purposes of this chapter the discussion will be limited to those of greatest importance to the antenna definition. For each of the satellite mission types the key feature of the architecture that impacts the antennas is the links to the satellite control center, data center or users. Therefore, characterizing the communications system is the key requirement. The next sections of this chapter will focus on this and thereafter a few comments regarding the satellite bus will be made.

3.10.1 Key Performance Concepts: Coverage, Frequency Allocations

A fundamental performance parameter for a satellite that is flowed directly to the antenna subsystem is coverage. The antenna pattern determines on which portion of the Earth are ground stations able to communicate with the satellite. In many applications, the antenna pattern is a simple spot beam and the coverage area specifies the location where the spot beam must be pointed and the accuracy to which it must be pointed. In other applications, the desired coverage area may be highly shaped and hence the antenna pattern must be formed by a phased array or shaped reflector. In either case, the coverage area is that portion of the Earth over which communications at the required levels of performance must be achieved.

Another basic element of the definition of the satellite system is the regulatory apportionment of frequency bands for use in satellite communications. The International Telecommunication Union (ITU) and the Federal Communications Commission (FCC) manage frequency allocations for satellites. These organizations license a satellite to operate at a set of specific frequency bands at a specific orbital location. A summary of some of the allocated frequency bands was is shown in Table 3.1.

3.10.2 Architecture of the Communications Payload

Broadly interpreted, the purpose of the communications payload is to deliver bits of information from one geographic location to another with sufficient quality to meet the objectives of the overarching communications system. The measures of the effectiveness of the communications payload are in overall link distortion (line quality) and payload capacity (or throughput).

The key functional requirements of the communications payload system are frequency plan, coverage area and service type, as well as quality of service, capacity, configurability and reliability. An example illustrating some of these requirements is given in Table 3.2 where three categories of satellite systems found in the USA are described in terms of their frequency band and the service they provide.

Table 3.1 *Frequency bands.*

Frequency range (GHz)	Band designation
0.1–0.3	VHF
0.3–1.0	UHF
1.0–2.0	L
2.0–4.0	S
4.0–8.0	C
8.0–12.0	X
12.0–18.0	Ku
18.0–27.0	K
27.0–40.0	Ka
40.0–75	V
75–110	W
110–300	mm
300–3000	µm

Table 3.2 Typical FSS and BSS parameters.

	High power	Medium power	Low power
Band	Ku	Ku	C
Downlink (GHz)	12.2–12.7	11.7–12.2	3.7–4.2
Uplink (GHz)	17.3–17.8	14.0–14.5	5.925–6.425
Service	BSS	FSS	FSS
Primary use	DBS	Point to point	Point to point
Additional use	Point to point	DBS	DBS
Satellite EIRP range (dBW)	51–60	40–48	33–37

3.10.3 Satellite Communications System Performance Requirements

Antennas are always a critical component of the satellite communications system. In remote sensing applications, antennas may be a critical component of the sensor subsystem as well. Most of the examples in this chapter will emphasize the connection of the antenna and the communications system.

Beginning with the next sections of this chapter, the microwave free-space communications link and the principal components of the satellite which establish the performance of the link will be presented. This is a natural starting place and one that illustrates the importance of the antenna subsystem to the satellite system and its role in the satellite microwave front end (or receiver) and the satellite microwave transmitter.

The complexity of the link will unfold one term after another in the following sections. The impact of losses in the satellite RF system and also in the propagation path will be introduced. The development of the link carrier-to-noise ratio will be introduced and the impact of interference will be presented.

3.11 The Satellite Link Equation

To understand the satellite link equation, one must understand the fundamental components of each microwave system contained in the link (see Figure 3.7), the basic principles of how they work, and the effect of the medium through which the signal propagates.

For every free-space microwave link the fundamental components are:

1. the microwave transmitter block
2. the transmit antenna
3. the microwave receiver block

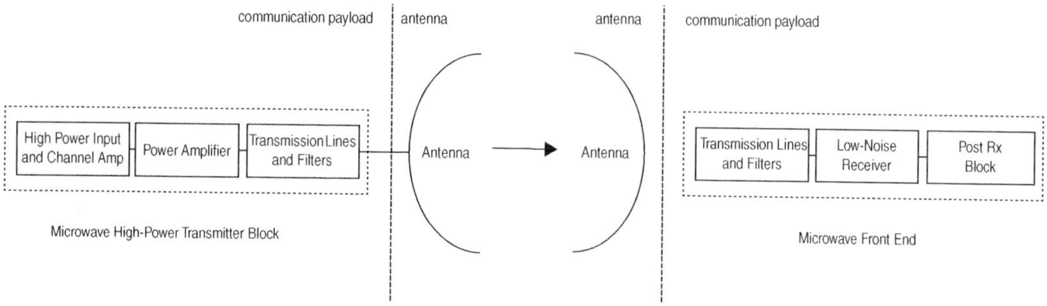

Figure 3.7 Microwave link.

4. the receive antennas
5. the physical separation
6. the propagation medium.

Each component of the link is characterized by a few key parameters. The microwave transmitter block is characterized by its output power and the linearity of the transfer function relating its input signal to its output signal. The transmit antenna and receive antennas are both characterized by their spatial gain, their effective aperture and their ohmic losses. The microwave receiver block is characterized by its gain and by its noise figure or equivalent noise temperature. The propagation medium is characterized by attenuation of the signal propagating through it and by the additional noise it adds to the equation. In the next section, each of these components is addressed.

3.12 The Microwave Transmitter Block

The typical satellite microwave transmitter block (see Figure 3.8) contains a group of components at its input which work together to provide a signal of the proper amplitude level and frequency to a power amplifier and this is followed by one or more filters and a transmission line (most typically a waveguide).

The key parameters which capture the performance of this block are the maximum power level as measured at the output of the block and the nonlinearity of the transfer function. The power level provides a measure of the signal strength and the linearity provides a measure of the signal distortion. In this configuration, the linearity of the block is dominated by the linearity of the power amplifier.

Every power amplifier has a characteristic 'saturated power' which is the maximum amount of power, $P_{out,sat}$, that the power amplifier will provide over the range of possible input power levels. There is usually one unique input power level which corresponds to the saturated output power level and that input power level is referred to as $P_{in,sat}$.

Over a range of input power levels the microwave transmitter block has a nearly linear transfer function. In this range the input power and the output power are proportional and the proportionality constant is the gain of the power amplifier. This is referred to as the linear region of the transmitter block. Eventually, as the input power increases, the corresponding increase in the output power diminishes. This is characterized by writing the output power as a power series of the input power (see Figure 3.9):

$$P_{out} = \sum_k g_k P_{in}^k \tag{3.1}$$

The 'fundamental' term is the first term of this series:

$$P_{out,1} = g_1 P_{in} \tag{3.2}$$

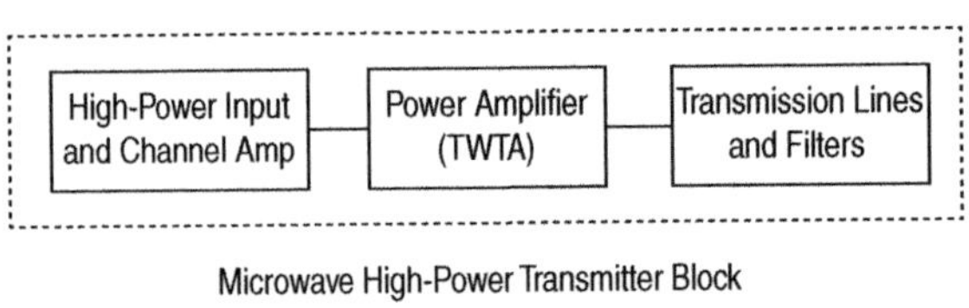

Figure 3.8 Microwave high-power transmitter block.

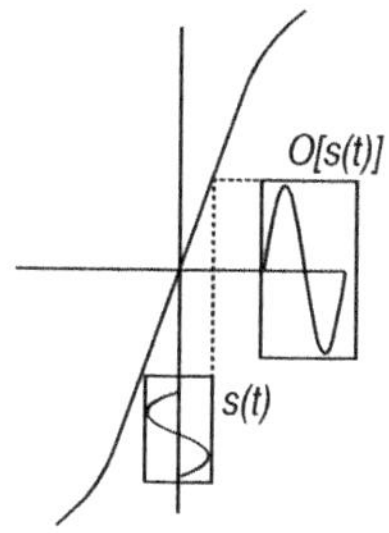

Figure 3.9 *Transfer function.*

The 'third-order product' is the third term of the series:

$$P_{out,3} = g_3 P_{in}^3 \tag{3.3}$$

When written in terms of decibels, these two expressions become:

$$P_{out,1}(\text{dBm}) = G_1(\text{dB}) + P_{in}(\text{dBm}) \tag{3.4}$$

and

$$P_{out,3}(\text{dBm}) = G_3(\text{dB}) + 3P_{in}(\text{dBm}) \tag{3.5}$$

When plotted on a dB/dB graph these lines intersect at what is defined to be the third-order intercept point (see Figure 3.10). This point gives a usual measure of the nonlinearity in the power amplifier through the quasi-linear region of operation.

3.12.1 Intercept Point

With this approach, the linearity of the transmitter block is characterized by the intercept point. In the linear region the power associated with the fundamental is much higher than that associated with the third-order product. A transmitter block which has a larger range of input power levels for which the third-order product is much lower than the fundamental is said to be more linear and its intercept point will be much higher. This

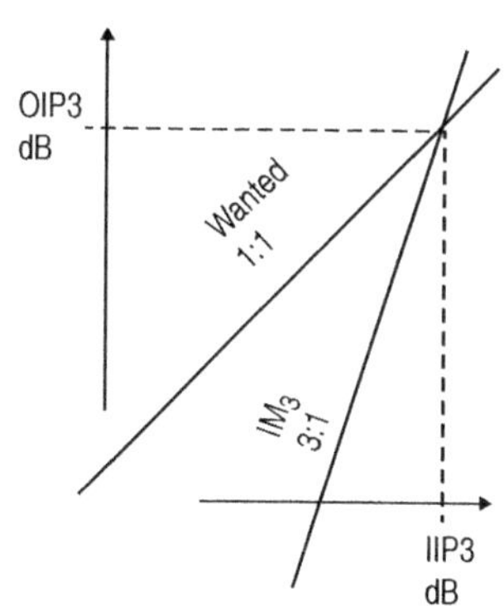

Figure 3.10 *Intercept point.*

concept leads to a different, yet equivalent term which is commonly used to characterize the nonlinearity of a component: the ratio of the carrier power to the power in the third-order product. This is referred to as C3IM or C/3IM and is defined as a function of either input or output power:

$$\mathrm{C/3IM(dB)} = P_{out,1}(\mathrm{dBm}) - P_{out,3}(\mathrm{dBm}) \tag{3.6}$$

It can be shown to equal

$$\mathrm{C/3IM(dB)} = 2\left(P_{out,IP3}(\mathrm{dBm}) - P_{out,1}(\mathrm{dBm})\right) \tag{3.7}$$

The intercept point of the transmitter block is important and it can be found from the corresponding intercept points of each of the components within the block. If one considers a cascade (or line-up) of two components, then the intercept of the cascade is found from the equation

$$\frac{1}{P_{in,ip3}} = \left(\frac{1}{P_{in,ip3,1}} + \frac{g_1}{P_{in,ip3,2}}\right) \tag{3.8}$$

and applying this iteratively gives the corresponding result for the basic satellite transmitting block:

$$\frac{1}{P_{in,ip3}} = \left(\frac{1}{P_{in,ip3,1}} + \frac{g_1}{P_{in,ip3,2}} + \frac{g_1 g_2}{P_{in,ip3,3}}\right) \tag{3.9}$$

This is often more convenient when referenced to the output power and so

$$\frac{1}{P_{out,ip3}} = \left(\frac{1}{g_2 g_3 P_{out,ip3,1}} + \frac{1}{g_3 P_{out,ip3,2}} + \frac{1}{P_{out,ip3,3}}\right) \tag{3.10}$$

While this is the worst case result, for the general case not all the intermodulation products add in phase with one another. Assuming the relative phase is random leads to the result

$$\left(\frac{1}{P_{out,ip3}}\right)^2 = \left(\frac{1}{g_2 g_3 P_{out,ip3,1}}\right)^2 + \left(\frac{1}{g_3 P_{out,ip3,2}}\right)^2 + \left(\frac{1}{P_{out,ip3,3}}\right)^2 \tag{3.11}$$

3.12.2 Output Backoff

Since the C/3IM increases at a lower operating power level, the level at which the transmitter will be operated becomes an operational parameter. The common measure of output power level references the saturated level of the transponder and is referred to as output backoff (OBO). The transmitted power at this OBO level is denoted

$$P_{out,OBO} = \frac{P_{out,1,sat}}{\mathrm{OBO}} \tag{3.12}$$

Only the power in the fundamental term of the power series is considered as this represents the useful power.

3.12.3 The Transmit Antenna and EIRP

The microwave high-power transmitter block provides a signal to the transmit antenna. For the link equation, the primary parameter of interest on the transmit antenna is the antenna gain. The effective isotropic radiated power (EIRP) is the metric of interest:

$$\text{EIRP} = P_{out} * G \tag{3.13}$$

EIRP is a function of angular position. When a value is associated with EIRP, it is usually either peak EIRP (maximum EIRP) or the EIRP at the edge of the coverage. This usually references the saturated output power of the transmitter and is called saturated EIRP:

$$\text{EIRP}_{sat,peak} = P_{out,sat} * G_{peak} \tag{3.14}$$

In the analysis of a link, one must consider the EIRP provided to any point in the coverage area of the antenna and so the edge-of-coverage EIRP is defined:

$$\text{EIRP}_{sat,eoc} = P_{out,sat} * G_{eoc} \tag{3.15}$$

where the gain at the edge of coverage or the minimum gain within the coverage replaces the peak gain.

The operational level of EIRP is defined as the EIRP guaranteed over the coverage area at the operational OBO level and is denoted

$$\text{EIRP}_{obo,eoc} = P_{out,obo} * G_{eoc} \tag{3.16}$$

Care has to be taken to define the interface between the transmitter and the antenna and make sure that all interconnecting waveguide is accounted for on one side of that interface or the other (Figure 3.11). The convention is that the output power and antenna gain are both defined for this interface. One can see that the EIRP improves and the C/3IM degrades as the transmitter is driven to saturation.

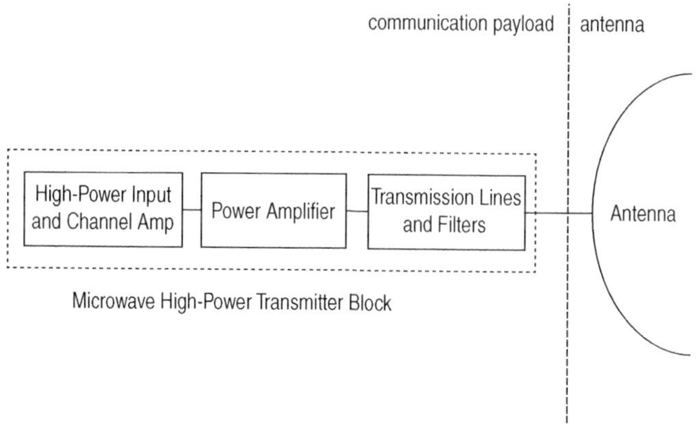

Figure 3.11 High-power transmitter and antenna.

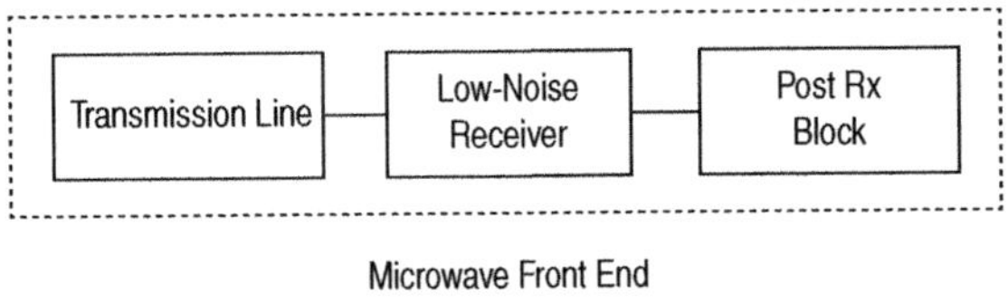

Figure 3.12 *Microwave front end.*

3.13 Rx Front-End Block

The typical satellite microwave receiver block contains a transmission line (waveguide, in most applications), a low-noise amplifier or receiver and a post-receiver block (see Figure 3.12) which routes and filters the signals in preparation for retransmission.

3.13.1 Noise Figure and Noise Temperature

The key performance parameter for the microwave front end is the noise figure. Its relationship to the system noise temperature, the minimum detectable signal and the link signal-to-noise ratio will be covered in subsequent sections.

The noise figure of a component is defined in terms of the ratio of the input signal-to-noise ratio and the output signal-to-noise ratio (Figure 3.13).

Here F is the noise figure, G is the gain and T_0 is the reference temperature of the component under consideration. The noise figure is defined by the expression

$$F = \frac{(S/N)_{input}}{(S/N)_{output}} \tag{3.17}$$

when

$$N_{input} = kT_0B \tag{3.18}$$

and since

$$S_{output} = g * S_{input} \tag{3.19}$$

we have

$$N_{output} = kT_0BFg \tag{3.20}$$

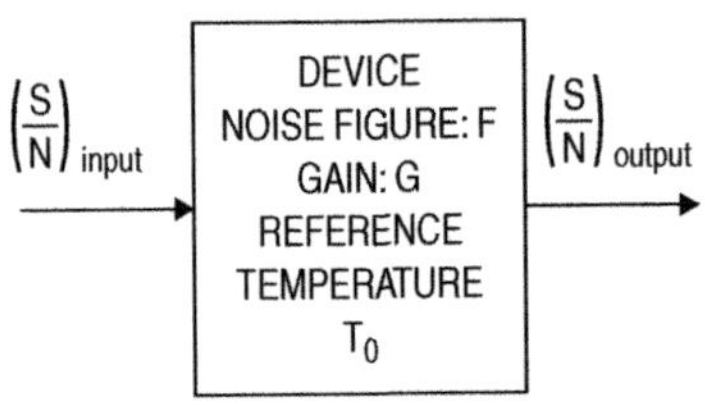

Figure 3.13 *Two-port device.*

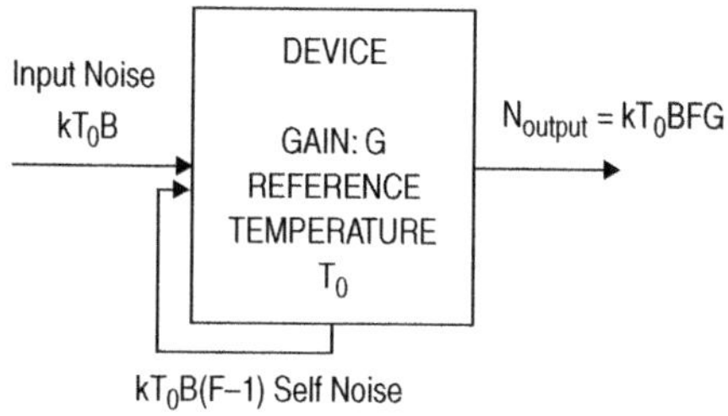

Figure 3.14 *Noise model for two-port device.*

It is convenient to think about the noise generated by the component referenced to its own input and a self-generated term, added together (Figure 3.14),

$$N_{output} = kT_0Bg + (F-1)kT_0Bg \tag{3.21}$$

With this understanding of the noise figure of a device, a cascade of components can be considered. The total equivalent noise figure of a cascade of two components (Figure 3.15) can be written as

$$F_{Total} = F_1 + \frac{F_2-1}{g_1} \tag{3.22}$$

It can be shown that for a two-port passive component the noise figure F is equal to the loss of the component L.

For satellite communications systems, the front end often can be represented by the cascade or line-up (Figure 3.16).

It is convenient to introduce the concept of equivalent noise temperature where

$$T_{equivalent} = T_{reference}(F-1) \tag{3.23}$$

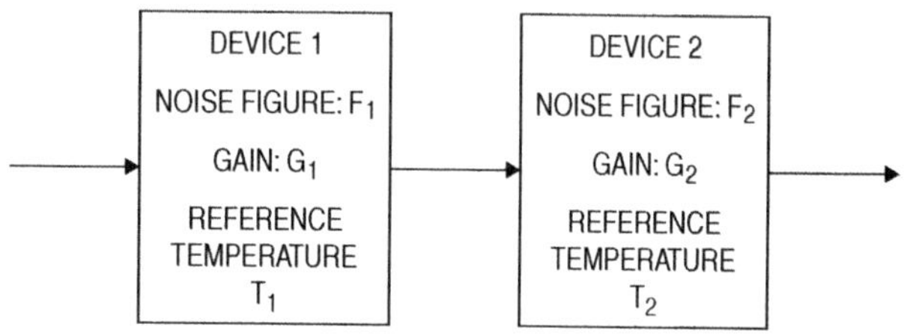

Figure 3.15 *Cascaded two-port devices.*

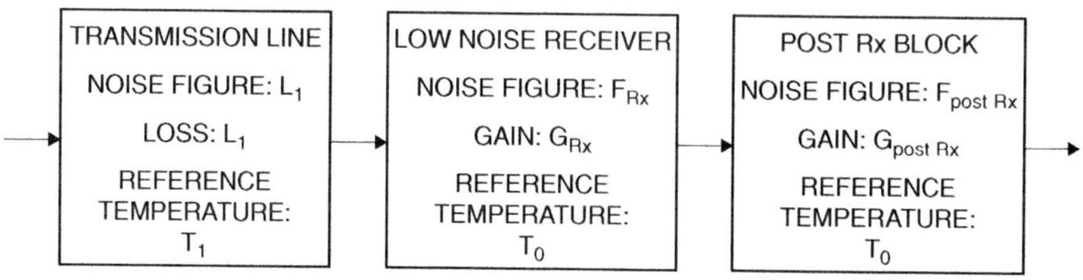

Figure 3.16 *Cascade of multiple two-port devices.*

giving the result

$$T_{Front_End} = T_1(L_1 - 1) + \frac{T_0(F_{Rx} - 1)}{1/L_1} + \frac{T_0(F_{Post_{RX}} - 1)}{g_{Rx}/L_1} \tag{3.24}$$

3.14 Received Power in the Communications System's RF Link

In satellite communication systems, the most critical parameters are related to the amount of power transmitted at one end of the link and the amount of power received at the other end of the link. The basic relationship between these two parameters is described by the equation

$$P_r = \frac{P_t G_t}{4\pi R^2} A_{effective} \tag{3.25}$$

where G_t is the gain of the transmit antenna in the link, P_t is the transmitted power, R is the spatial separation of the transmitter and receiver, and $A_{effective}$ is the effective aperture of the receive antenna system as measured at the antenna interface to the communication system's receiver system.

The effective aperture can be written in terms of the gain of the receive system and the frequency of the carrier through the expression

$$A_{effective} = \frac{\lambda^2}{4\pi} G_r \tag{3.26}$$

Combining these expressions gives the result

$$P_r = P_t G_t \left(\frac{\lambda}{4\pi R}\right)^2 G_r \tag{3.27}$$

Using the definition for the EIRP and defining the space loss as that loss due to the spatial separation of the transmit and receive systems

$$L_{space} = \left(\frac{4\pi R}{\lambda}\right)^2 \tag{3.28}$$

results in an expression for the received power in the link which is written as

$$P_r = \frac{\text{EIRP} * G_r}{L_{space}} \tag{3.29}$$

In the simple derivation above, the angular dependence of the gain of the antennas is ignored as are a number of physical parameters that are important to the communications system. These include the polarization losses of the link and the absorption losses within the link.

It goes without saying that the gain or directivity is a function of spatial coordinates. For satellite performance evaluation, it is typical to reference (u,v) coordinates and for ground station performance it is typical to reference elevation and azimuth angles (θ,ϕ).

3.14.1 The Angular Dependencies of the Uplink and Downlink

The uplink equation describes the power received by the satellite as a function of the EIRP of the transmitting ground station, the gain of the satellite's receive antenna and the losses in the path between the ground station and the satellite. Starting with our general expression, we recognize that the received power depends on the angular characteristics of both the earth station transmit and satellite receive antennas. Using the standard definitions for the coordinate systems, we have

$$P_r^{satellite} = \frac{P_t^{ground\_station} * G_t^{ground\_station}(\theta, \phi) * G_r^{satellite}(u, v)}{L_{space}\left(\vec{x}_{satellite}, \vec{x}_{ground\_station}\right)} \tag{3.30}$$

Similarly the downlink equation describes the power received by the ground station as a function of the EIRP of the transmitting satellite, the gain of the ground station's receive antenna and the losses in the path between the ground station and the satellite. Therefore our general expression reduces to

$$P_r^{ground\_station} = \frac{P_t^{satellite} * G_t^{satellite}(u, v) * G_r^{ground\_station}(\theta, \phi)}{L_{space}\left(\vec{x}_{satellite}, \vec{x}_{ground\_station}\right)} \tag{3.31}$$

3.15 Additional Losses in the Satellite and Antenna

As mentioned in Chapter 1, there are a number of other losses that one needs to consider when evaluating the overall link performance. In particular, a careful accounting of ohmic losses as well as mismatch terms such as antenna/payload interface losses and reflections (return losses). Antenna polarization impurities and the polarization distortion through propagation must be accounted for. There are other terms, resulting from limitations associated with manufacturing or integration tolerances that result in a degradation to the directivity that also must be included. Among these mis-directivity terms are antenna misalignments, reflector surface tolerances and thermal distortion.

3.15.1 Additional Losses due to Propagation Effects and the Atmosphere

Atmospheric losses are modeled in a fashion similar to those of any passive block in the communications chain. The losses associated with the atmosphere are grouped into those that are due to the absorption of electromagnetic energy due to the gases in the atmosphere and those that are due to rain.

The total attenuation is written as

$$A_k(\mathrm{dB}) = \alpha_k \ell_{effective,k} \tag{3.32}$$

where α_k is the specific attenuation which, because the atmosphere is made up of nitrogen and oxygen and smaller amounts of other gases and can also contain a significant amount of water vapor, is a function of frequency and polarization of the electromagnetic field and has units of webers/length. The effective path length through the layer is strictly a geometric function of the thickness of the absorptive layer and the angle that the direction of propagation of the electromagnetic wave takes with respect to the layer normal, and can be written as

$$\ell_{effective,k} = \frac{h_k - h_0}{\sin(\theta_{elevation})} \tag{3.33}$$

where h_k is the effective height of the atmosphere and h_0 is the height of the ground antenna in the link.

Interpolation formulas from empirical data are used to calculate the losses associated with the atmosphere or rain. Atmospheric absorption losses vary with frequency. Of particular interest is the water vapor resonance absorption at 22.3 GHz and the oxygen molecular absorption at 60 GHz. This is shown in the graphs in Figure 3.17.

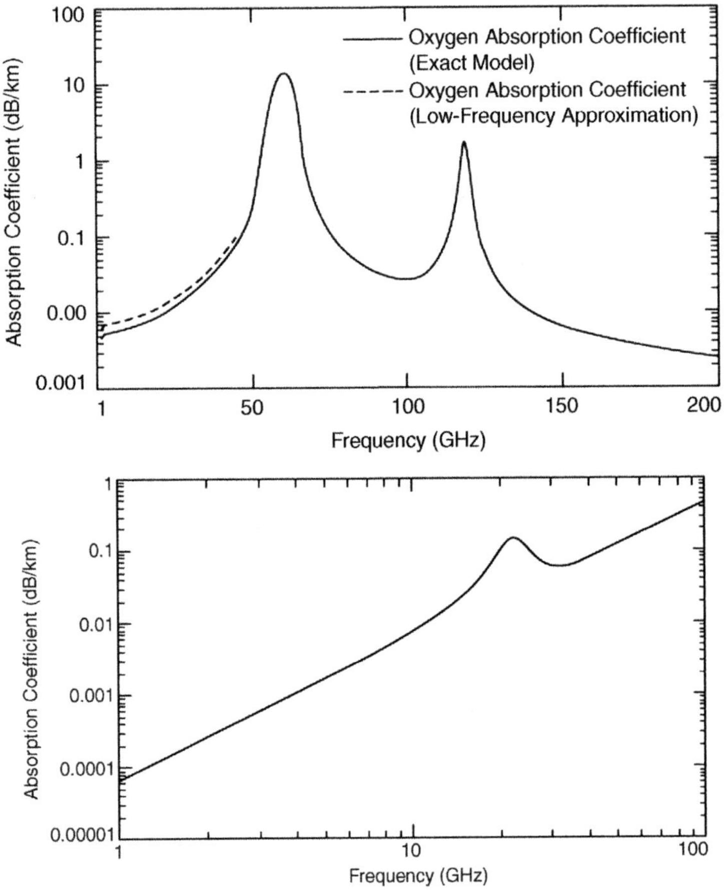

Figure 3.17 *Oxygen absorption (top) and water vapor absorption (bottom).*

Ippolito's model [2] for the rain loss uses a statistically determined reduction factor (Table 3.3). The specific attenuation of rain is modeled by

$$\alpha_k = a_k r_p^{b_k} \tag{3.34}$$

where a_k and b_k are functions of frequency and polarization and r_p is the rain rate in mm/h (Table 3.4).

Table 3.3 *Reduction factors.*

P	r_p
0.001%	$\frac{10}{10+L_G}$
0.01%	$\frac{90}{90+4L_G}$
0.1%	$\frac{180}{90+4L_G}$
1%	1

Table 3.4 Specific attenuation coefficients.

Frequency (GHz)	a_h	a_v	b_h	b_v
1	0.000 0387	0.000 0352	0.912	0.88
4	0.000 65	0.000 591	1.121	1.075
8	0.004 54	0.003 95	1.327	1.31
12	0.018 8	0.016 8	1.217	1.2
20	0.075 1	0.069 1	1.099	1.065
25	0.124	0.113	1.061	1.03
30	0.187	0.167	1.021	1

3.15.2 Ionospheric Effects – Scintillation and Polarization Rotation

In addition to the absorption of microwave signals from water vapor and other gases in the atmosphere, ionospheric scintillations, a variation in the amplitude, phase, polarization or angle of arrival of radio waves, are caused by irregularities in the ionosphere which change with time. The most significant effect is fading of the signal, which can be severe in magnitude and last several minutes.

3.16 Thermal Noise and the Antenna Noise Temperature

3.16.1 The Interface between the Antenna and the Communications System

The figure of merit of a receive system is a combination of antenna and communications system parameters. The reference plane defining the interface between the antenna and the communications system must be specified. Once this plane is specified, the corresponding apportionment of the system noise temperature between the input noise, the contribution of the antenna and the contribution of the communications system can be determined.

Given the diagram in Figure 3.18 for the interface between the antenna and the communications system above, the expression for the system noise temperature is

$$T_{system} = T_A + \frac{(L_{antenna} - 1)}{L_{antenna}} T_{physical}^{antenna} + T_{Front\_End} \quad (3.35)$$

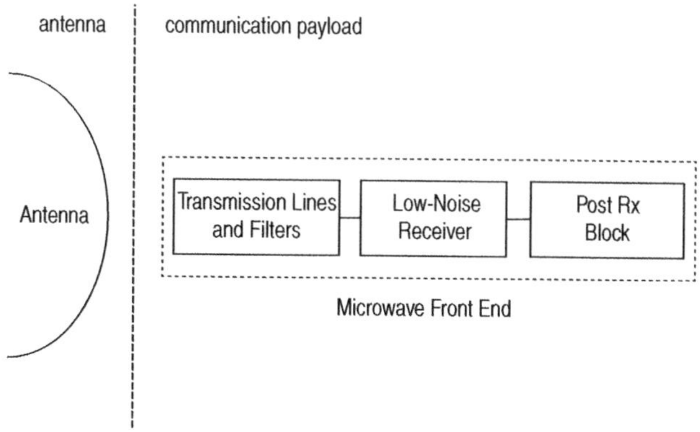

Figure 3.18 Receive antenna and receiver front end.

where

$$T_A(\theta_0,\phi_0)=\frac{1}{4\pi}\iint T_b(\theta,\phi)G(\theta-\theta_0,\phi-\phi_0)\sin(\theta)\,d\theta\,d\phi \tag{3.36}$$

and where the physical temperature of the antenna components such as the feed are explicitly identified as well as the equivalent temperature of the receiver front end. In these equations G is the gain of the receive antenna as measured at the antenna input and T_b is the brightness temperature which is empirically determined and tabularized by the ITU and other organizations.

3.16.2 The Uplink Signal to Noise

When defining the satellite antenna performance, the uplink signal-to-noise performance is of keen interest.

The uplink noise in a transponder referenced to the input of a satellite communications system can be written as

$$N^{uplink}_{transponder}=kT^{satellite}_{system}B_{transponder} \tag{3.37}$$

with its dependence on the equivalent system noise temperature and the transponder bandwidth identified specifically.

Combining this with the equation for the uplink power in a transponder channel gives

$$\frac{P^{satellite}_{r}}{N^{uplink}_{transponder}}=\mathrm{EIRP}_{earth\_station}(\theta,\phi)*\frac{G^{satellite}_{r}(u,v)}{T^{satellite}_{system}}*\frac{1}{kB_{transponder}}*\frac{1}{L_{path}} \tag{3.38}$$

As such it is explicit that the key satellite parameter for the uplink is the G/T figure of merit of the satellite system.

3.17 The SNR Equation and Minimum Detectable Signal

The minimum detectable signal (MDS) is an important parameter characterizing the sensitivity of the front end of a transponder. The performance of every demodulator and detector combination can be characterized by the signal-to-noise ratio (SNR) required to achieve a given bit error rate (BER). Any particular mission has an acceptable BER and hence has a threshold SNR required for satisfactory performance. Given the level of noise input to the system, the combination of this data results in the minimum signal level that will produce the acceptable BER (Figure 3.19).

The governing equation is

$$\mathrm{MDS}=\left(\frac{S}{N}\right)_{threshold}N_{output} \tag{3.39}$$

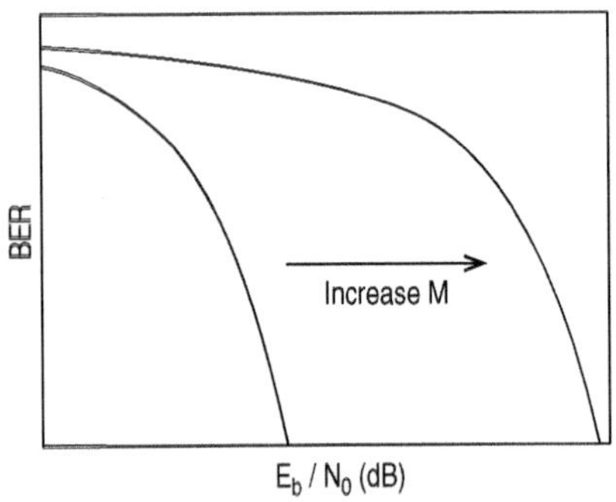

Figure 3.19 Bit error rate as function of signal to noise.

If the input noise is written in terms of the system noise temperature this equation takes the form

$$\text{MDS} = \left(\frac{S}{N}\right)_{threshold} \left(N_{input} + (F-1)kT_0B\right) \tag{3.40}$$

3.18 Power Flux Density, Saturation Flux Density and Dynamic Range

The link equation gives the relationship between uplink power flux density (PFD) and the ground station EIRP:

$$\text{PFD}_{satellite} = \frac{\text{EIRP}_{ground\_station}}{4\pi R^2} \tag{3.41}$$

This can be generalized somewhat by taking into account the other losses discussed earlier in the chapter:

$$\text{PFD}_{satellite} = \frac{\text{EIRP}_{ground\_station}}{L_{Total}} * \frac{4\pi}{\lambda^2} \tag{3.42}$$

The received power is the product of the PFD and the effective aperture of the receive antenna

$$P_{received} = \text{PFD}_{satellite} * A_{effective} \tag{3.43}$$

The saturated flux density (SFD) is defined to be that minimum PFD which drives the payload transponder into saturation. Since this is usually a function of variable gain state, there is usually a range of SFD. It is an important parameter in characterizing the performance of a satellite payload transponder. The saturated power level of the transponder is easy to obtain by measurement, and establishing the minimum PFD level required to drive the transponder into saturation for each of its given gain states is very useful.

The dynamic range of a transponder is established by the MDS at the lower end and the linearity at the higher end (Figure 3.20).

3.18.1 Important Relationship between PFD and Gain State of the Satellite Transponder

It is important to note that in the case of low PFD, or minimum SFD, the gain of the satellite is maximized and the resulting noise figure for the transponder is minimized. This configuration provides the most sensitive

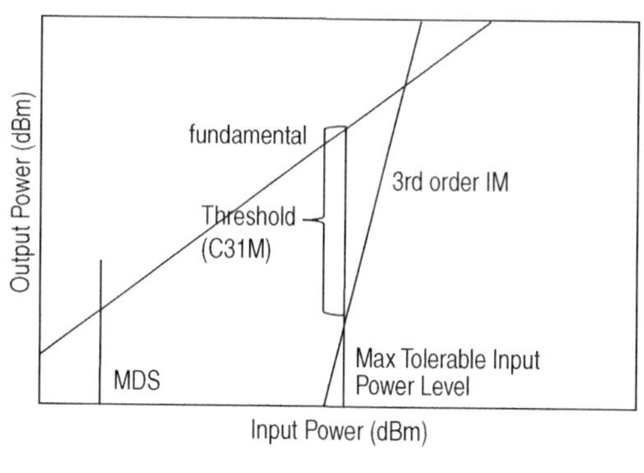

Figure 3.20 Nonlinear power in/out curves.

transponder front end and hence the MDS for the satellite. In this configuration, the linearity of the transponder is usually not an issue as the signal power levels at all of the intermediate active components in the transponder chain are comparatively low.

When the uplink PFD is high, or approaches the maximum rated SFD, the gain of the satellite transponder is minimized so that the active components are not driven into the nonlinear realm. A consequence of the reduction of the gain of the transponder is that its noise figure and hence the corresponding MDS is increased.

Understanding these limitations a priori will in general lead to better designs, where the MDS (and correspondingly the system noise figure) is minimized when needed and the nonlinearities are minimized when needed. Overspecifying the MDS or linearity requirements to be applied to the entire range of PFDs may well lead to a suboptimal design throughout the entire dynamic range.

3.19 Full-Duplex Operation and Passive Intermodulation

Passive intermodulation (PIM) products appear as noise to the microwave front end. Consequently, the presence of PIM energy effectively desensitizes the receiver. This is most conveniently analyzed by considering the impact to the effective system temperature.

$$P_{IM3} = \text{PIM power} \tag{3.44}$$

At communication system input, the total noise power density is

$$N_{0,input} = kT_{sys} + \frac{P_{PIM}}{B} \tag{3.45}$$

A new modified system noise temperature can be defined as

$$T'_{sys} = T_{sys} + \frac{P_{PIM}}{kB} \tag{3.46}$$

and so

$$\frac{T'_{sys}}{T_{sys}} = 1 + \frac{P_{PIM}}{kBT_{sys}} \tag{3.47}$$

A common figure of merit is that the equivalent PIM noise contribution should not degrade the system noise temperature by more than 0.1 dB, which amounts to the PIM noise being about 2% of the system thermal noise.

3.20 Gain and Gain Variation

When planning the communications system, it is important to understand how the link power levels change as a function of time. Clouds may drift in or out of a beam. The satellite pointing varies with time. The satellite antennas may distort due to thermal effects, thereby changing the gain. The output power of the satellite TWTA or the noise figure of the satellite microwave front end may change over time due to thermal effects.

The measure of performance for the satellite is EIRP variation and *G*/*T* variation. In either case, the terms are deconstructed into an antenna performance term and a repeater performance term.

For EIRP variation we see

$$\Delta\text{EIRP} = (\Delta P_{out})G + P_{out}(\Delta G) \tag{3.48}$$

and for *G*/*T* variation we have

$$\Delta(G/T) = \frac{\Delta G}{T} - \frac{G(\Delta T)}{T^2} \tag{3.49}$$

It can be shown that in both cases it is the relative variation of *G* that is important. Thus

$$\frac{\Delta\text{EIRP}}{\text{EIRP}} = \frac{(\Delta P_{out})}{P_{out}} + \frac{(\Delta G)}{G} \tag{3.50}$$

and

$$\frac{\Delta(G/T)}{G/T} = \frac{\Delta G}{G} - \frac{(\Delta T)}{T} \tag{3.51}$$

The time variation of the gain is predominantly a function of thermal effects and pointing error

$$\frac{\Delta G}{G} = \frac{\sqrt{\varepsilon_{pointing}^2 + \varepsilon_{thermal}^2}}{G} \tag{3.52}$$

Commonly the error is cast as a ratio of the gain plus variance to the gain itself

$$\frac{G+\Delta G}{G} = 1 + \frac{\sqrt{\varepsilon_{pointing}^2 + \varepsilon_{thermal}^2}}{G} \tag{3.53}$$

From this the requirements for pointing and thermal deformation of the antennas are derived.

3.21 Pointing Error

As described in the previous section, the pointing error is a major component of the gain variation of the coverage area. It is a major consideration when evaluating the performance of an antenna within a communications system. In fact, the general approach is to define the minimum achievable gain including the pointing error. The pointing error is a function of a number of items and is statistical in nature. The contributors are not necessarily correlated. It is common to assume that the pointing errors in elevation and in cross-elevation are Gaussian distributed; then the resulting RF pointing error is Rayleigh distributed.

Some of the contributors to pointing error are deterministic and occur only occasionally. One example of this is thruster firing for purposes of station keeping. The station-keeping maneuver induces a pointing error but it only occurs a small fraction of the time. This results in a natural time-weighted calculation of pointing error when evaluating the link performance.

The pointing error may have a yaw component. The result of this depends on the coverage area. Yaw errors matter the most for coverage areas furthest away from the axis of the yaw error.

Autotrack systems are commonplace to reduce antenna pointing errors. These systems provide RF feedback on the pointing. The performance of the autotrack depends on the autotrack feed, the combining network behind the feed and the resulting error voltage signals. The residual pointing error also needs to take into account the physical limitations of the antenna positioning system. There may be a commanded-position error, a servo control loop error, or an error in the pointing error estimate.

In the next sections the satellite bus will be described. The satellite pointing systems involve satellite bus systems that include both the attitude control and the station-keeping systems. Furthermore, the structural frequency of the antennas and subreflectors in on-orbit position are also factors. The mounting of the antenna on the overall spacecraft structure is a factor too. Reducing pointing error is complex and the satellite designer or satellite antenna designer needs to consider even small contributions from the deployed structural resonance frequencies, the deployment repeatability, the mechanism rotation rate and even the size of a single step.

Major pointing disturbances may result from the solar array thermal response to eclipse transition or from the solar array interaction with thrusters in north/south station keeping. While there are many mechanism-related errors, some pointing errors are a result of simple manufacturing or integration tolerance issues such as antenna alignment tolerances.

3.22 Remaining Elements of Satellite System Architecture

The satellite system architecture is typically partitioned into a space segment and a ground segment (Figure 3.21).

The space segment is composed of the satellite and the command/telemetry facility that includes those systems on the ground that support the flight operations of the satellite.

The ground segment contains satellite ground stations, including both user terminals and ground communications infrastructure.

3.23 Orbits and Orbital Considerations

The mission of the satellite will dictate the orbit in which the satellite will operate. There are essentially four different types of orbits. Two major types will be discussed here to illustrate the differences.

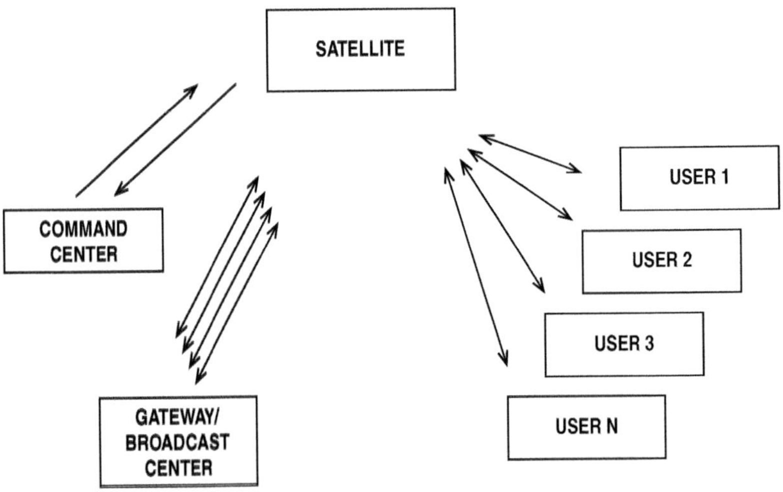

Figure 3.21 Space and ground segments.

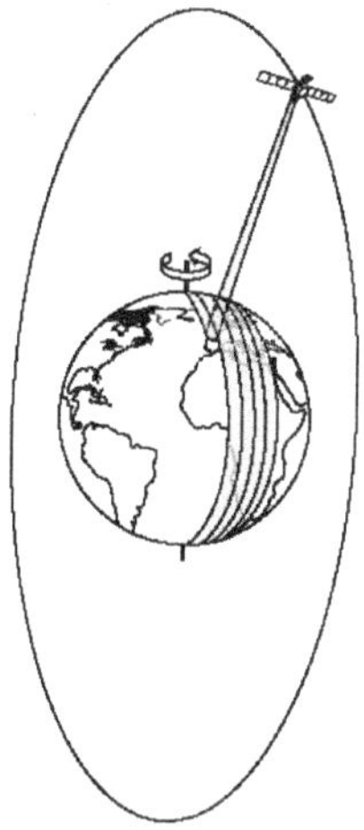

Figure 3.22 Example: polar orbit.

Low Earth orbits are characterized by an orbit altitude of approximately 400 km and are often used for remote sensing satellites because the reduction in the distance to the target inherently allows for better resolution. Polar orbits (Figure 3.22) take advantage of the fact that the Earth rotates about its polar axis to enable mapping of the surface of the Earth completely. Geostationary orbits (Figure 3.23), with an orbit altitude of approximately 36 000 km above the equator, are useful because the satellite can appear to be fixed from the point of view of users on the Earth. Highly elliptical orbits allow for the satellite to be within the field of view of the northern hemisphere for most of its orbit (Figure 3.24).

Orbits are a natural result of Newton's law of universal gravitation whereby the gravitational force of attraction between two bodies varies as the product of their masses and inversely as the square of the distance r between them, and is directed along a line connecting their centers. The result of this is captured in Kepler's laws, which state that the path followed by a satellite around the Earth will be an ellipse and that, for equal time periods, a satellite will sweep out equal areas in its orbital plane, focused at the barycenter or center of mass (Figure 3.25).

Kepler's third law states that the square of the periodic time of orbit is proportional to the cube of the mean distance between two bodies. Following these basic principles, the mechanics of the orbits under consideration for the satellite mission can be determined.

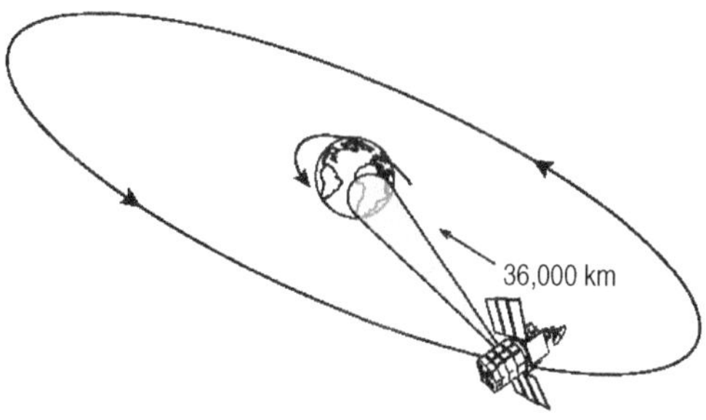

Figure 3.23 Example: geostationary orbit.

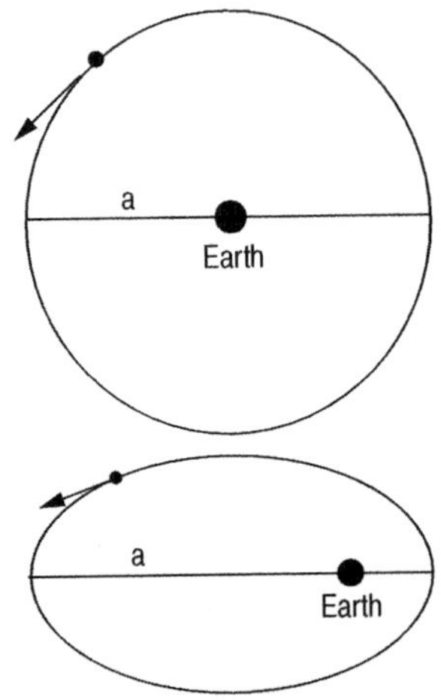

Figure 3.24 Example: circular and elliptical orbits.

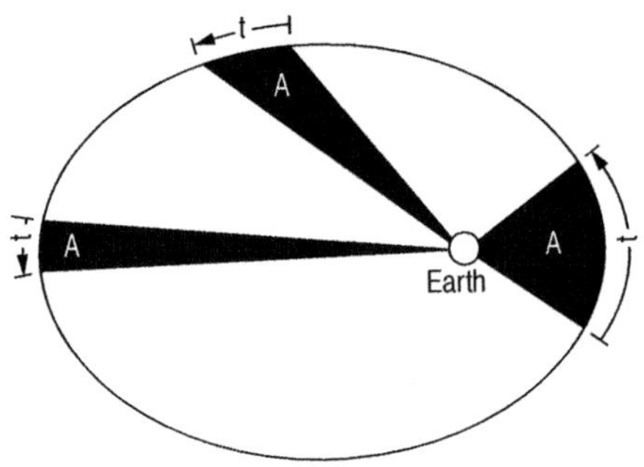

Figure 3.25 Example: Kepler's third law.

3.24 Spacecraft Introduction

Those parts of the satellite that are not part of the payload are said to be part of the satellite bus. The bus is essentially those components which make the satellite a spacecraft and it includes all of those subsystems that are required for the satellite to function and carry out its mission but which are not part of the payload.

The antenna subsystem may be classified either as part of the payload or as part of the bus. For example, when considering communications satellites, antennas are obviously an integral part of the payload. The antenna performance is a critical parameter in the design of the communications link. This is in contrast to, for example, a remote sensing satellite where the antenna is still a key part of the communications system but the communications system is considered to be part of the satellite bus. For a remote sensing satellite the sensor and related components compose the payload.

In this section, the key components of the satellite bus are described briefly. The communications system was discussed separately. The overarching theme of the chapter, to concentrate on those elements that are of greatest importance to the antenna designer, will continue here.

3.25 Spacecraft Budgets (Mass, Power, Thermal)

Along with the mission performance parameters, the most critical characteristics of any satellite are captured in three overarching satellite metrics: the satellite mass, the satellite DC power consumption, and the satellite thermal dissipation capability. These parameters establish the temporal period over which the satellite will be able to perform its mission, the launch vehicle that will be required to place the satellite in its proper orbit and, to a large extent, the cost of the satellite.

3.25.1 Satellite Mass

The mass of the satellite impacts the launch requirements directly. There are two different mass metrics. The first is the dry mass, which characterizes the mass of all of the components on the satellite, whether part of the bus or part of the payload. The second metric is the wet mass, which includes the dry mass plus all of the fuel on board the satellite which is required for launch, orbit placement and orbit station keeping throughout the life of the satellite.

A launch vehicle will be rated to lift a satellite of a given mass to a given position in orbit around the Earth. Satellite operators usually request that satellites are designed so that they could be launched on a set of different launch vehicles to insure the mission of the satellite against either the availability or the cost of a given launch vehicle at the desired time. The satellite designer then has to keep the overall mass from exceeding the capabilities of the least capable launch vehicle selected to be included in the set of possible launch vehicles. This upper bound on the mass of the satellite directly impacts the amount of hardware that can be included to support the satellite mission. Therefore detailed mass estimates are kept throughout the design and build of the satellite. Initially some margin is applied to the estimate to account for uncertainty but as the design matures and parts are selected and eventually measured, this margin is retired. The mass budget is a key element of any satellite design. Great effort is expended to make it as accurate as possible and hence enable the most capable satellite possible given the launch vehicle limitations.

3.25.2 Satellite Power

The satellite power is often directly proportional to the overall capability of the satellite to perform its mission. For communications satellites, it is generally true that the satellite mission architect runs out of available power before running out of available bandwidth. Henceforth, when a satellite payload is defined, the power required to drive that payload is determined in a fashion similar to that for the mass budget. Each item that consumes DC power is identified and tabulated. Most of these items are part of the payload but some are components of the satellite bus. With the catalog of components and rated component power consumption at hand, the aggregate power required and the requisite *power generation* capability (solar arrays) and *power storage* capability (batteries) are defined. Like the mass estimates, some margin is required early in the design phase but the overall budget is managed closely and margin is retired as the design matures. Of course the components of the power subsystem themselves have mass and so typically some iteration is required to maximize the overall satellite capability to achieve its mission and the balance of mass and power required.

3.25.3 Satellite Thermal Dissipation

Every component on the satellite that consumes DC power must dissipate some of it in heat. The satellite must be designed to transport this heat to the exterior of the satellite and then radiate it into space to prevent a thermal-runaway condition. Therefore, just as important as satellite mass budgets and satellite power budgets is the satellite thermal dissipation budget. In very general terms, satellite thermal dissipation can be traded

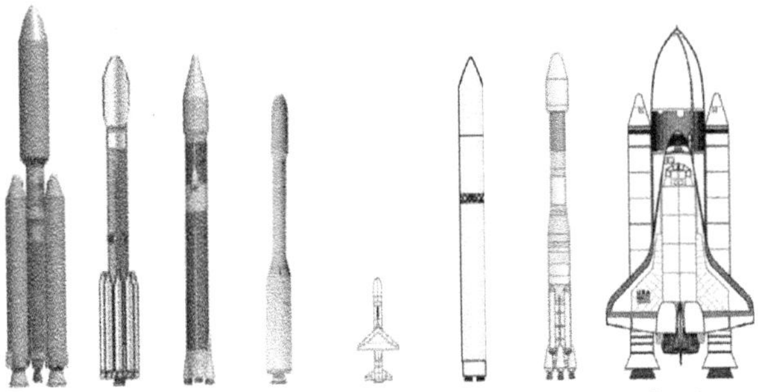

Figure 3.26 *Launch vehicles.*

against satellite mass. (More dissipation requires more radiators which requires more mass!) As a consequence, along with satellite mass and satellite power, the satellite thermal dissipation rounds out the three most important system-level satellite parameters. Substantial design effort is made to ensure there is sufficient margin in each of the three parameters and that the amount of margin is balanced in such a way as to maximize both performance of the payload and the life of the satellite.

3.26 Orbital Mission Life and Launch Vehicle Considerations

The owners and operators of a satellite are most concerned with the satellite accomplishing its mission in a financially rational way. This overarching requirement translates to a small handful of key requirements. The satellite must maximize its on-orbit mission life and it must fit within the parameters for the desired launch vehicle.

There are only a small number of different launch vehicles and they come in a small countable set of options. The need to fit into the fairing of the launch vehicle constrains the size of the satellite and also the size of the antennas mounted on it. The most common launch vehicles include the Russian Proton and the French Ariane rockets (Figure 3.26).

In general, satellites are launched by expendable launch vehicles. The main technical feature determining the suitability of the launch vehicle is its ability to deliver a given amount of mass to the desired orbit. Most commonly, this is referred to as either delivered mass to GEO or delivered mass to LEO.

The reader can review the references [1–10] for more detailed information about launch vehicles. For the purposes of this section, it is important to understand that the larger the satellite, the larger the required launch vehicle. Launch vehicles come in a small set of discrete sizes and hence satellites tend to come in discrete families of size as compared to a continuum of sizes from small to large.

The relationship between the size of the launch vehicle and that of the satellite on board is driven by economics. Satellite service providers tend to pack as big a satellite as possible on a given launch vehicle to amortize the cost of the launch over the mission.

3.27 Environment Management (Thermal, Radiation)

Satellites are expected to operate in an extremely severe environment. Some effects are easily overcome. For example, when a satellite is in a stable orbit, the gravitational force is counterbalanced by the centripetal force, thus creating a 'zero-gravity' environment. In such an environment, any liquid fuel system will require a

Table 3.5 Thermal environment for major subsystems.

Battery	0–20 °C
Power regulator	0–40 °C
Payload transponder	0–50 °C
Earth sensor	−5 to 50 °C
Antenna elements	−100 to 150 °C
Solar arrays	−100 to 85 °C

bladder to operate properly. Others pose a more serious threat. For example, the continuous reduction of atmosphere that the satellite encounters as it is progresses through launch may result in critical pressure levels that would support corona events (typically catastrophic voltage breakdown) within the communications system.

One significant environment issue is thermal management. The thermal characteristics of the satellite in the vacuum of space with thermal radiation as the only mechanism for shedding unwanted heat pose many technological challenges in terms of establishing thermal conduction paths and selecting materials which are robust to significant thermal gradients. A typical range of thermal specifications for satellite components on orbit is shown in Table 3.5.

The thermal subsystem is designed to provide a stable temperature for all of the elements on the satellite. Passive thermal designs include widespread use of thermal blanketing to protect the satellite from radiation from the Sun and, for LEO satellites, the radiation from the Earth's albedo. Heaters are used to provide a low-temperature bound for some communications and bus hardware.

Additionally the solar arrays and solid state electronics systems are jeopardized by cosmic rays, protons from solar flares and tiny meteorites. Without the benefit of the Earth's atmosphere to absorb unwanted radiation, the satellite relies almost exclusively on radiation shielding to provide protection of the solid state devices in the communications subsystem or satellite bus. Essentially this becomes a tradeoff between the life span of the electronics on board, the need for redundant units to be used in the event of a failure of a primary unit and mass.

The radiation environment is usually captured as an equivalent field strength.

3.28 Spacecraft Structure (Acoustic/Dynamic)

Fundamentally, the objectives of the satellite bus components are to provide collectively a stable platform for the satellite payload. The spacecraft structure must carry a number of subsystems. The structure needs to be stable thermally as well as dynamically.

3.29 Satellite Positioning (Station Keeping)

Once on orbit, the satellite must be kept in its proper orbit station. To achieve this goal every satellite bus has the station-keeping subsystem. The satellite bus must maintain its orbital location in the presence of the non-uniform gravitational field of the Earth and the gravitational reaction to the position of the Sun and the Moon.

The ellipticity of the Earth causes satellites to drift slowly toward 75°E or 105°W. This drift in longitude necessitates east/west station-keeping maneuvers. Typically these are accomplished through the use of pulse jets approximately once every two weeks. For frequency coordination, satellites must be kept within a box defined by $\pm 0.1^\circ$ of designated longitude at C-band and to within 0.05° at Ku-band.

The pull of the Sun and the Moon cause the satellite to drift in latitude about 0.85° each year. The result is a north/south station-keeping maneuver. These maneuvers are more fuel intensive. As with longitudinal drift, the satellite must be kept within a box defined by ±0.1° of designated latitude (or to within 0.05° in Ku-band).

Lastly the attitude variation allowed for the satellite is about ±0.1% of nominal height.

These allowances define a box in which the satellite needs to stay. For C-band, it is approximately 150 km × 150 km × 72 km and for Ku-band it is approximately 75 km × 75 km × 72 km (150 km = 0.1°; 75 km = 0.05° from geostationary orbit).

The orbital corrections, when required, are typically carried out by command from a satellite control team operating the telemetry, command and tracking earth stations. Placing a satellite in inclined orbit may allow N/S station keeping to be dispensed with, at the expense of ground station complexity.

3.30 Satellite Positioning (Attitude Control)

The satellite attitude defines its orientation in space with respect to the Earth. The attitude and control system keeps the satellite pointing in the correct direction. For the antenna engineer, the pointing of the satellite is obviously an extremely important parameter and hence the satellite attitude control subsystem an extremely important subsystem.

The satellite typically has an earth sensor, which uses horizontal detectors to determine the location of the rim of the Earth. The system is used to keep the antenna pointed, keep sensors pointed and keep solar arrays pointed. This system must operate continuously as there are many disturbance torques in space that alter the satellite attitude. These include the gravity field variations of the Earth, the gravity tug by the Sun and the Moon, and the solar radiation itself. Alternatively, the satellite may make use of a star tracker or an *in situ* RF uplink to provide more accurate pointing feedback (Figure 3.27).

Unlike station-keeping processes involving operators on the ground sending commands to fire thrusters or take some equivalent action, the attitude and control processes are commonly controlled by satellite. Torques are commonly controlled passively and rely on gravity gradients or, for a class of spin-stabilized satellites, the spin vector. More complex systems are active and involve using one or more momentum wheels (spinning flywheels).

3.31 Power Subsystem

Satellite power subsystems are designed to generate the power required for satellite operation and to store the power so that the satellite service is not interrupted when the power generation is diminished. The power subsystem provides electrical energy to run the satellite and communications payload. Key components of the subsystem are the solar cells, power conditioners and batteries.

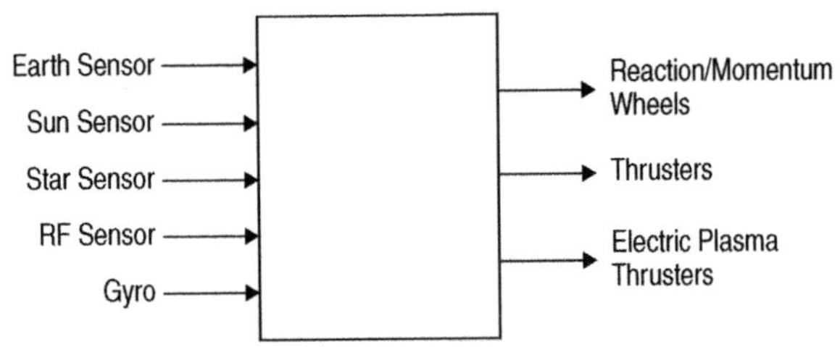

Figure 3.27 *Attitude and control system.*

The primary electrical power for a satellite is generated by solar cells. For example, a typical communications satellite may need to provide DC power up to ~20 kW. Most of this is consumed by the satellite's payload transmitter. The direct efficiency of the solar cells is typically in the range of 20–25% at the beginning of the satellite's lifetime (beginning of life, BOL). The cells degrade over time and exposure to the environment and so at the end of the nominal GEO satellite life of 15 years (end of life, EOL) the cells are only 5–10% efficient.

The energy storage elements (batteries) are sized to meet the power needs of the satellite for those conditions when the solar array cannot be effectively utilized. These include both the power required through launch and the power required when the Earth blocks the satellite from the Sun in what is called an eclipse.

Earth will eclipse satellites two seasons a year, during the spring and autumnal equinoxes. The periods of eclipse begin to occur 23 days before the autumnal (or vernal) equinox and last until 23 days after the autumnal (or vernal) equinox. The eclipses themselves last up to 72 minutes. Battery power must then be sufficient to provide ~20 kW DC power for that period, hence defining the overall requirement.

3.32 Tracking, Telemetry, Command and Monitoring

The centerpiece of every satellite is a subsystem called telemetry, tracking and command (TT&C) or sometimes telemetry, command and ranging (TCR). This is composed of a relatively simple communications system along with the computer which drives and monitors all of the satellite bus functions.

The TT&C system has three purposes, essential to the satellite. The first is to collect telemetry, which is the data collected from many sensors on board the satellite that is relayed to the ground so that the state of the satellite and its components can be monitored. The second is issue commands whereby the satellite relays specific control and operations information from the ground to the satellite components. And the third is tracking or ranging, which is essentially a way to allow for the determination of current orbit position and movement. For obvious reasons, these systems are required to be highly reliable and are usually protected with encryption technology. Fundamentally, the TT&C system allows an earth station to control the subsystems or the spacecraft and monitor their function.

References

1. Roddy, D. (2006) *Satellite Communications*, 4th edn, McGraw-Hill, New York.
2. Ippolito, L. Jr. (2008) *Satellite Communications Systems Engineering – Atmospheric Effects, Satellite Link Design and System Performance*, John Wiley & Sons, Ltd, Chichester.
3. Martin, D. (2000) *Communication Satellites*, 4th edn, The Aerospace Press, El Segundo, CA.
4. Larson, W. and Wertz, J. (1992) *Space Mission Analysis and Design*, 2nd edn, Microcosm, Torrance, CA.
5. Pratt, T. and Bostian, C. (1986) *Satellite Communications*, John Wiley & Sons, Inc., New York.
6. International Telecommunications Union (2002) *Handbook on Satellite Communications*, 3rd edn, Wiley-Interscience, Geneva.
7. Elbert, B. (1997) *The Satellite Communication Applications Handbook*, Artech House, Norwood, MA.
8. Morgan, W. and Gordon, G. (1989) *Communications Satellite Handbook*, John Wiley & Sons, Inc., New York.
9. Gagliardi, R. (1984) *Satellite Communications*, Wadsworth, Monterey, CA.
10. Maral, G. and Bousquet, M. (1998) *Satellite Communications Systems – Systems, Techniques and Technology*, 3rd edn, John Wiley & Sons, Ltd, Chichester.

4

Space Environment and Materials

J. Santiago-Prowald[1] **and L. Salghetti Drioli**[2]

[1]*ESA-ESTEC Mechanical Department, The Netherlands*
[2]*ESA-ESTEC Electrical Department, The Netherlands*

4.1 Introduction

The aim of this chapter is to describe the natural and induced space environments to which space antennas are exposed and to provide a discussion on materials and manufacturing processes suitable for space antennas, in view of the demanding performance and environmental requirements. Furthermore, the chapter provides some insight into the measures and verification methods necessary to mitigate the risks of the harsh space environment. Emphasis is put on the tests required for material screening, characterization and qualification for use in space hardware.

Within the families of metallic, polymeric, ceramic and composite materials, the ones that are currently more extensively used in space antennas are polymer composites. Hence, their properties, processing and characterization deserve dedicated attention. Because of length limitations and the wide scope of the subjects, it is not possible to provide a complete reference text on the entirety of the space environment and materials. Instead, an introductory overview of the related disciplines is provided, as well as some specific material data valid for preliminary selection and design.

4.2 The Space Environment of Antennas

Space antennas are always exterior spacecraft appendages and as such are subject to many environmental threats which can cause degradation of the constituent materials. Threats are represented by photon radiation, charged particle radiation, temperature effects and thermal cycling, impacts of micrometeoroids and debris, contamination, and low Earth orbit atomic oxygen. Moreover, each mission is characterized by its own unique set of environmental variables such as mission duration, orbital parameters and consequent view angle of the

Space Antenna Handbook, First Edition. Edited by William A. Imbriale, Steven (Shichang) Gao and Luigi Boccia.

spacecraft antennas to the Sun and solar events. All these factors need to be taken into account in space antenna design and material selection.

As well as the threats represented by the above-mentioned phenomena, it is extremely important to consider structural and thermal properties as design drivers. In fact, reflector antenna structures are designed for stiffness, stability and strength. Thermal control designs rely on thermal insulations with the required solar absorptance and IR emittance. Degradation of these thermo-optical properties can cause an undesirable change in temperature of the antenna components.

One of the major difficulties in this field is to understand the degradation of materials due to the harsh space environment. To this end, analyses of retrieved flight hardware or dedicated experiments based on space exposure have been used in the past. However, this kind of investigation is rare, while the most common means to understand material degradation remains ground laboratory studies. Also this second approach presents disadvantages as the representativeness of the intended mission environment achieved by the ground experiment environment is rather limited. Due to the difficulties in exactly simulating the space effects, complex calibrations and cautious interpretation of the results are required. A combination of space exposures, ground laboratory studies and computational modelling is most useful for assuring the durability of space antenna materials.

This chapter includes descriptions of specific space environmental threats to space antenna constituent materials. A significant emphasis is placed on the effects of Earth orbit environments, because the majority of space missions have been flown in Earth orbits which have provided a significant amount of data on materials effects.

4.2.1 The Radiation Environment

This subsection will present the major contributors to the radiation in the space environment [1], namely trapped particle belts, solar particle events and cosmic rays.

4.2.1.1 Trapped Particle Belts

The radiation belts consist principally of electrons with an energy up to a few MeV and protons with an energy up to several hundred MeV. These are trapped in the Earth's magnetic field; their motions in the field consist of a gyration about field lines, a bouncing motion between the magnetic mirrors found near the Earth's poles, and a drift motion around the Earth (Figure 4.1).

Radiation belts and solar event particles cause radiation damage to electronic components, solar cells and certain materials. Unshielded locations (e.g. space antenna materials) are exposed to very high doses because the lowest energy particles are absorbed in the first few millimetres of the material. This can degrade thermal control materials, radomes, and so on [3].

The International Space Station (ISS), Space Shuttle, EnviSat and other low-altitude missions fly in the inner edge of the radiation belt. This region is dominated by the 'South Atlantic Anomaly (SAA)', an area of enhanced radiation caused by the offset and tilt of the geomagnetic axis with respect to the Earth's rotational axis (Figure 4.2). This brings part of the radiation belt to lower altitudes.

Besides the SAA, the polar horns also play a role in radiation analysis at low altitudes (e.g. ISS-type orbit). Polar horns are parts of the outer radiation belts, which are close to the Earth. As can be seen in the simulation below, increased radiation flux due to the polar horns can be expected between 60° and 90° latitude. The SAA is clearly visible in the South Atlantic region around 30° and up to 50° latitude (Figure 4.3).

At low Earth orbit (around 400–500 km) altitudes, there are important interactions between the trapped radiation belts and the atmosphere, giving rise to a strong asymmetry in fluxes from the east and west. This is important for attitude-stabilized spacecraft such as the ISS [1].

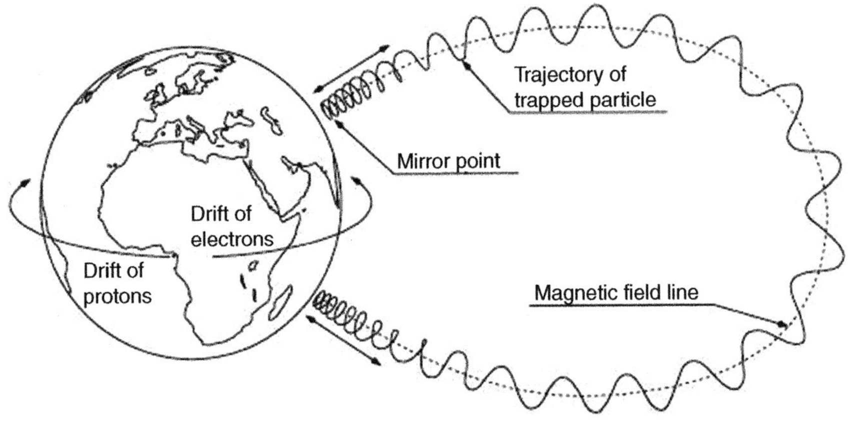

Figure 4.1 Basic motion of trapped particles in Earth's magnetic field [2].

During solar events, large fluxes of energetic protons are produced which reach the Earth. The August 1972 event produced a peak flux in excess of 1×10^6 protons/cm^2/s above 10 MeV energy [1]. Such events are unpredictable in their time of occurrence, magnitude, duration or composition. The Earth's magnetic field shields a region of near-Earth space from these particles (geomagnetic shielding) but they easily reach polar regions and high altitudes such as the geostationary orbit.

4.2.1.2 Cosmic Rays

Cosmic rays originate outside the solar system. Fluxes of these particles are low but, because they include heavy, energetic ('HZE') ions of elements such as iron, they cause intense ionization as they pass through matter. It is difficult to shield against these types of ions, and therefore they constitute a significant hazard. They give rise to single-event upset (SEU), single-event latch-up (SEL) and single-event burnout (SEB) phenomena in large-scale integrated electronic components, as well as interference.

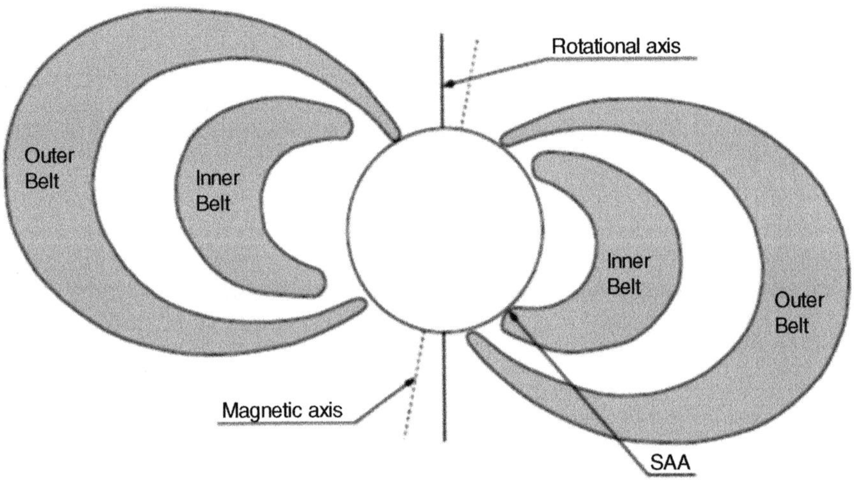

Figure 4.2 Earth radiation belts with the South Atlantic Anomaly indicated [2].

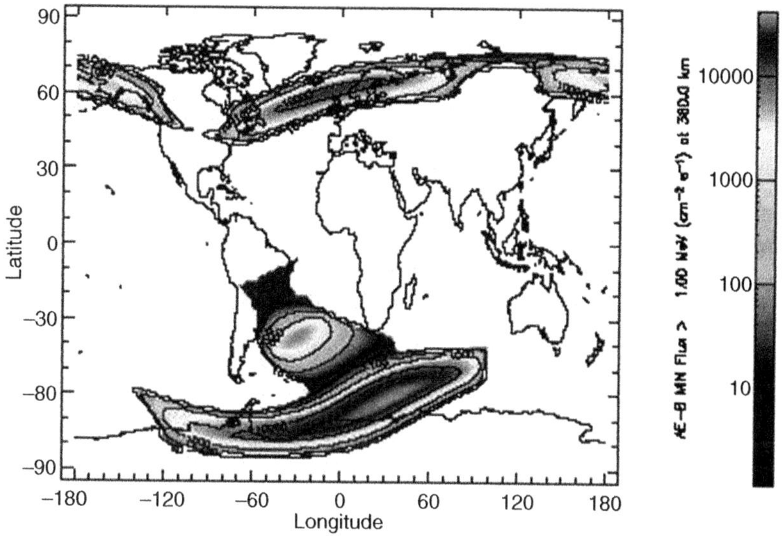

Figure 4.3 Trapped electron flux at 380 km altitude; polar horns and SAA are evident [2].

4.2.2 The Plasma Environment

'Low-energy' charged particles constitute the plasma environment. A body in a plasma is subject to a flux of electrons and ions. Although by definition a plasma contains equal number densities of positive and negative charges, the electrons are much more mobile and the flux of electrons to an 'uncharged' surface normally exceeds the flux of ions. The net effect, seen in laboratory plasmas and often in space, is that surfaces charge negatively with respect to the local plasma. The surface continues to attempt to acquire the amount of charge necessary to establish a current balance: to reduce the flux of electrons and enhance the flux of ions, until equilibrium is reached at the floating potential. In space, under solar UV and energetic particle irradiation, the current balance also includes secondary and photoemitted electrons. As a consequence, the charging of spacecraft surfaces is a function of the energy distribution of the plasma (e.g. temperature), the area of the sunlit surfaces and the properties of the materials on the spacecraft surfaces. In particular, plasma environments at high altitudes and in polar orbits result in electrostatic charging of spacecraft surfaces (including antennas) and at low altitude cause leakage of power from exposed solar arrays and electromagnetic perturbations.

Material selection to reduce the electrostatic charging in high-altitude plasma is based on the following properties:

- dielectric thickness;
- dielectric constant;
- dielectric resistivity (generally not a constant in space but dependent on illumination, temperature, radiation and field);
- surface resistivity;
- secondary electron emission yield as a function of incident electron or ion energy;
- photoemission current (from solar illumination).

At low altitudes the plasma is normally cold and dense. This means that the plasma effectively screens any spacecraft-generated electric field and that surface charging to a high-potential level is normally not possible.

In addition the plasma interacts with electromagnetic waves. The dispersive properties and inhomogeneity of the ionospheric plasma can be a concern for several space-based systems, for example telecommunication devices, global positioning systems and RADAR.

4.2.3 The Neutral Environment

4.2.3.1 *Atomic Oxygen*

When a spacecraft is due to fly in the low Earth orbit (LEO) environment, it has to confront atomic oxygen. In fact, at those altitudes (e.g. between 180 and 650 km), the solar radiation has sufficient energy to break the diatomic oxygen bond and the probability of reassociation or ozone (O_3) formation is very weak [4–6].

The density of atomic oxygen depends during the day on the solar heating and in particular, due to the relative motion of the atmosphere with respect to the Earth, the peak of this density occurs around 3 p.m. Moreover, the production rate of atomic oxygen is strongly linked to solar activity and its UV radiation variation. Depending on the orbit and the spacecraft velocity vector, the incidence angle of the atomic oxygen flux can vary significantly during the orbital cycle. All the above-mentioned effects make the prediction of the flux of atomic oxygen almost impossible and enforce some mitigation actions that must be considered in the design of surfaces exposed to this kind of environment.

In fact, atomic oxygen strongly interacts with materials. Clear evidence that was observed with the advent of the Space Shuttle missions is the so-called glow phenomenon that occurs when atomic oxygen and other LEO atmospheric species impact spacecraft surfaces, causing the creation of short-lived excited state species that emit visible radiation near the surfaces of the spacecraft [7–9].

In general, atomic oxygen interacts with polymers, carbon and many metals to form oxygen bonds with atoms on the exposed surface. Depending on the chemical reaction in place, atomic oxygen can result in the erosion of materials, surface oxidation or gradual conversion of silicones to high-modulus silica. This latter process is, however, limited to the outer atomic layers. The silicate layer resists further oxidation and hence silicones are considered stable to the atomic oxygen attack.

Over the years, thin-polymer erosion by atomic oxygen in LEO has presented a challenging spacecraft performance and durability problem, and to counteract this risk three mitigation actions have mainly been put in place [10–14].

The first method to reduce the effect of atomic oxygen on thin polymers and increase their durability is based on the use of coatings, generally made of SiO_2, Al_2O_3, indium tin oxide, Ge, Si, Al and Au. Up to 100 nm of sputtering of vapour-deposited protective coatings is generally used.

The second possible mitigation approach is based on modification of the surface of polymers to make them more durable to atomic oxygen. This can be achieved by either implanting metal atoms into the surface of the polymer, or modifying the chemical composition of the surface to incorporate silicon atoms into the surface and near the surface. The areal density of metal atoms that can be placed into the polymer surface has a direct impact on the reduction of the atomic oxygen erosion yield.

The other possible way of reducing the negative effects of atomic oxygen is to use alternative polymers that contain metal atoms which develop a protective coating on exposure to atomic oxygen.

4.2.3.2 *The Contamination*

Differently from the effects analysed up till now, the contamination is mostly caused by spacecraft self-generated degradation. In fact, sources of contamination can include thruster propellants and burn residue, outgassing of spacecraft materials, vented gases from spacecraft systems, fluids released from the spacecraft by dumping or leakage. On the other hand, micrometeoroids, orbital debris and space environment interactions with materials can also produce contaminants, such as volatile products of atomic oxygen reactions and UV induced or radiation-induced chain scission products in polymer materials. Moreover, contaminants could further be modified by space environment effects such as atomic oxygen, UV and radiation interactions.

Build-up of molecular or particulate spacecraft contamination can cause degradation in transmittance, reflectance, solar absorptance, and thermal emittance of surfaces. Some particularly detrimental cases of space environment interactions with spacecraft contamination include atomic oxygen oxidation of outgassed silicones to produce a non-eroding silica-based layer, and UV or ionizing radiation interactions with contaminants to produce a contaminant film. Almost all spacecraft have silicones on board in the form of adhesives, potting compounds and lubricants. Although most LEO spacecraft designers make efforts to use only silicones that are vacuum stripped to eliminate or reduce the amount of volatile short-chain content, silicone fragments are often released in the vacuum environment in LEO, with the process being further enhanced with atomic oxygen and radiation-induced bond breaking. The resulting silicone fragments can deposit on surfaces that are exposed to atomic oxygen.

4.2.4 Space Environment for Typical Spacecraft Orbits

In general, depending on the mission orbit of the spacecraft, the designer has to cope with several different environmental constraints. In particular, the environment for space missions in LEO is characterized by dense neutral gas (atomic oxygen), ionospheric plasma, solar UV radiation and orbital debris. In medium Earth orbit (MEO) spacecraft have to cope with solar UV radiation, trapped radiation belts and the plasma sphere. For years MEO orbits have been, for the most part, unattractive due to the belief that the Van Allen radiation belts create a radiation environment unsuitable for satellites. JPL performed extensive analyses of this radiation and the effects of shielding. The altitude of the GPS constellation was selected at 20 200 km because, at this altitude, the radiation environment was considered the highest acceptable limit. Europe is now developing its satellite navigation system (Galileo) based on an orbit at 23 000 km. This very harsh environment is putting additional constraints on material selection for the constellation of spacecraft. In particular, one of the two validation elements of the project (GIOVE-A and GIOVE-B) is flying an environmental monitor unit to provide more data on space weather at that particular altitude.

The best known and less critical environment is represented by the geosynchronous orbit (GEO) where high-energy plasma sheets, solar UV radiation and a combination of solar flares and cosmic rays are present.

4.2.5 Thermal Environment

Antennas flying on board spacecraft are subjected to significant temperature variations. The number of thermal cycles and temperature extremes depend on the orbit, the spacecraft and the antenna materials. In particular, the duration of the eclipse periods is a function of the orbit plane angle to the Sun (the so-called beta angle), the orbit altitude and the season. For LEO, for example, the eclipse fraction can range from 0 to 40%. In the case of GEO, due to the inclination of the Earth, eclipses can only occur around the vernal and autumn equinoxes, and the maximum duration is 72 minutes. On the other hand, the influence of Earth's IR and albedo are very relevant for LEO, but almost negligible for GEO. Consequently, the temperatures experienced by the materials during thermal cycling depend upon the orbit parameters, the thermo-optical properties (solar absorptivity, IR emissivity, transmissivity), Sun incidence, radiative and conductive couplings with the surrounding parts of the spacecraft, and the influence of equipment or components that produce heat.

On-orbit extreme temperatures and thermal cycling pose a threat to material durability for various reasons. First, the main consideration is the thermal qualification range under which material integrity and properties can be guaranteed. In particular, for inhomogeneous materials in intimate contact, such as composites or surface coatings, a mismatch in the coefficients of thermal expansion (CTEs) may lead to cracking or debonding when the material experiences significant temperature excursions. Second, all properties of materials can be temperature dependent. For instance, during the course of on-orbit thermal cycling, a metal or polymer may experience decreased strength, making it more vulnerable to damage. Because radiation damage is generally more concentrated at the exposed surface and diminishes through the thickness, one can consider that a radiation-damaged polymer no longer has homogeneous properties throughout its thickness, making it vulnerable to the effects of a mismatch in CTE. Coatings such as Z-93, a zinc oxide pigment/potassium silicate binder white paint applied to aluminium substrates, have been observed to microcrack after thermal cycling [15,16]. This is an expected result of the CTE mismatch between the coating and the substrate. However, with proper substrate surface preparations, delamination and spalling are prevented in such a way that, despite a 'mud-tiled' appearance due to microcracking, the painted surfaces exhibit durability to orbital thermal cycling.

Due to the thermal environment that antennas will encounter during the mission, the design of these units has to take into account all these aspects and materials have to be selected accordingly. In particular, due to exposure to the Sun's direct radiation, deep space sink and the absence of convection in the space vacuum, extreme temperatures can be reached, typically in the range between −190 and 160 °C. This chapter will only address material selection for preventing issues related to the thermal environment, while Chapter 5 will address design aspects and analysis of thermoelastic distortions and consequent antenna mis-pointing due to material thermal deformations.

Antennas must be protected by thermal hardware to limit the temperature range and gradients, and to control the heat exchange with the platform, as well as the thermoelastic distortions. Passive thermal control devices consist of multilayer insulation (MLI), sunshields, coatings and paints. The choice of outer-layer material is driven by thermo-optical properties, glint prevention, electrical grounding, atomic oxygen and protection from micrometeoroid impacts. Usually, aluminized Kapton, black Kapton, white paint or Betacloth are employed. Materials having low solar absorptance (α_s) are used for radiators and reflective surfaces, designed to minimize solar radiation absorption. Thermal balance can be maintained over the spacecraft's lifetime only if the material maintains its properties, i.e. its beginning-of-life (or BOL) α_s value, or equivalently, its α_s/ε ratio. Unfortunately, optical properties can be seriously affected by the radiation and thermal cycling environment. As an example, the white paint coating commonly used for the thermal protection of antennas (feed chains and reflectors) in GEO orbits can be severely degraded in MEO orbits in a relatively short time [16]. *In situ* observations of thermal control coating degradation on the GPS Block I satellites are shown in Figure 4.4.

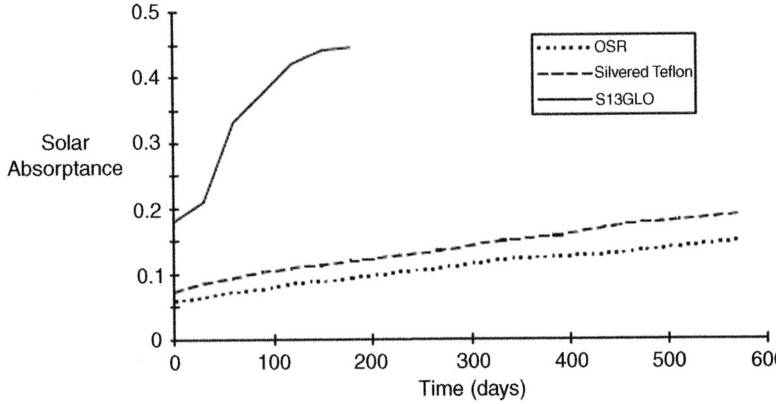

Figure 4.4 *Degradation in thermal control materials seen on the GPS Block I spacecraft (OSR, Optical Solar Reflector) [18]. Reproduced by permission of NASA and Dr Alan Tribble.*

Much of this degradation is associated with photochemical deposition of contamination. Depending on the orbit, there are a variety of other mechanisms that contribute to the degradation of surface materials, such as the solar UV, atomic oxygen, radiation, and the impact of micrometeoroids and debris.

Other passive thermal protection is represented by either radomes/sunshields or coatings. For example, reflectors and array antenna radiating surfaces are protected by sunshields generally made of Kapton–germanium foil. Sunshields are thin sheets of RF transparent material introducing the lowest RF performance degradation (losses and depolarization), while providing the required thermo-optical properties and electrical grounding. However, these materials are quite sensitive to humidity and have to be handled with extreme care and in a controlled atmosphere. This imposes a particular philosophy for testing antenna units mounting this kind of thermal hardware, in order not to deteriorate the sunshield on the ground. This issue is generally resolved by delivering a sunshield for the test programme and a flight sunshield to be installed shortly before launch.

Helix antennas, due to their three-dimensional geometry, are often covered with rigid radomes, generally glass-fibre-reinforced polymers, coated with white paint. In fact it would be much more difficult to design sunshields conformal with the shape of the antenna.

For very specific applications, such as antennas on launchers, the thermal hardware has to protect the antenna unit against the heat fluxes due to aerodynamic heating effects (aero-thermal loads). In these particular cases special thermal protection is used. For example, NORCOAT is a white-coloured material developed by Astrium based on silicone rubber with fillers wrapped on a silica fabric. The material shows excellent thermo-optical properties, is an ablative thermal protection for low to medium aerodynamic fluxes, but is RF transparent. All telemetry antennas mounted on Ariane 4 and 5 launchers and the ones that will be installed on Vega launchers are protected with NORCOAT.

4.2.6 Launch Environment

The launch campaign normally represents the most demanding combination of mechanical and thermal loads experienced by space hardware from the point of view of survival and potential degradation of mission performance. Launch loads evolve during the different phases of the launch, are transient in nature and of very high levels. They result from the rocket propulsion system characteristics, especially at ignition and liftoff, and the dynamic response of the launcher during the atmospheric flight, ignition and separation events of rocket stages, fairing separation, exposure to the aero-thermal flow and direct Sun incidences at the higher altitudes of the launch trajectory. Remarkably, a very characteristic environmental load for large appendages of light weight and stiff architecture, such as antennas and solar generators, is the acoustic load. The acoustic field is of very high intensity, often reaching 148 dB (qualification sound pressure level) in the frequency range covered by the octaves of 31.5–2000 Hz. It is a sizing load for the strength of these appendages. In general, the environmental loads can be found as a preliminary specification in the launcher manuals, although detailed predictions are only available after dedicated analyses.

Concerning mechanical loads, the main differences between missions of the same launcher are due to the mutual dynamic coupling of rocket and spacecraft, which is predicted through the coupled loads dynamic analysis (CLA). The CLA, performed by the launcher authority by integrating the finite elements models of launcher and spacecraft and using representative forcing functions, provides the time history and frequency spectra predictions of the launch loads at the interfaces, as well as relevant interior parts. Therefore, a very complex and relevant task for the spacecraft authority is to design and develop all spacecraft subsystems according to the required qualification or flight acceptance loads, taking into account the need to preview the environmental loading conditions, even before the CLA is available. Very often, derogations of the qualification loads are requested in order to protect sensitive components from overloading during on-ground verification. This can only be approved, in principle, after detailed verification of the margins against the final CLA.

Concerning thermal loads, a thermal coupled load analysis also has to be performed in order to conclude on the margins during launch and early orbit phases. The situation is in most cases not as critical as for mechanical loads, due to the lower variability of thermal conditions and better predictability at BOL. The exceptions are often related to special features of the spacecraft, such as the presence of cryogenic units or large appendages that could be detrimentally exposed to aero-thermal and particles fluxes or long eclipse periods. A typical sizing thermal load for antennas concerning hot cases, often discovered during late stages of development, is the combination of aero-thermal fluxes and direct Sun incidence while still attached to the launcher. Free-molecular heating can reach values of the order of 1000 W/m^2. Other typical cases of concern are transfer orbits such as geosynchronous transfer orbit (GTO), which could result in very long eclipses and demanding cold cases for telecommunications spacecraft. Dedicated measures are required for mitigating these thermal load cases.

Due to the complexity of the topic, a detailed description of the coupled loads analyses is beyond the scope of this book. In the following subsections a pragmatic approach is adopted. Instead of a lengthy and probably incomplete launch environment discussion, descriptions of typical conditions during qualification tests of antennas are provided.

4.2.6.1 *Mechanical and Thermal Loads for Antenna Qualification*

The typical mechanical requirements and the corresponding qualification tests are described hereafter for a general case, although significant differences in test sequences and levels can be found for particular cases:

- *Stiffness.* Static stiffness of the structure should be such that stresses and deformations during loading or operation do not result in damage, permanent deformation or functional degradation. However, a dynamic stiffness measurement is typically required. This is carried out by dedicated modal tests, although often replaced by a modal survey test on the shaker, taking advantage of other tests. A 100 Hz minimum value of the antenna fundamental frequency is desirable in order to minimize coupling with spacecraft dynamics and reducing testing, especially sinusoidal vibration. When not possible, a 50–60 Hz minimum fundamental eigenfrequency can be acceptable if the lateral modes of the spacecraft are sufficiently separated. Exceptions to these rules may require dedicated CLAs. The verification of this requirement is performed by dedicated modal testing or during sinusoidal vibration testing.
- *Quasi-static loads.* Structural strength is commonly verified by imposing either static accelerations at the centre of mass or equivalent forces at the interfaces of the element. This is performed by static load tests. Alternatively, a shaker can be employed by applying sinusoidal vibration at frequencies well below the fundamental eigenfrequency of the element tested, if the design loads can be reached. The structural design load, however, consists of a combination of axes, covering the real maximum expected flight loads. The quasi-static load specification should cover the low-frequency loads (below antenna fundamental modes) resulting from flight events, and a quasi-static rectification component of the high-frequency random vibration and acoustic loads transmitted through the structures or impinging on the antenna. These contributions are conservatively combined linearly, due to the difficulty of predicting the phase correlation of dynamic loads. The quasi-static component is an output of the spacecraft CLA. This analysis normally does not include the rectification effects of random loads. Typical values of antenna overall quasi-static design loads are ± 20 to $\pm 35g$ on each spacecraft axis individually, and combinations of ± 20 to $\pm 30g$ perpendicularly to the mounting plane plus ± 15 to $\pm 25g$ in plane. The precise values depend on the mass, the stiffness, the exposed surface and the location and mounting conditions of the antenna on the spacecraft during launch.
- *Sinusoidal vibration.* The coupled launcher/spacecraft low-frequency dynamic response during launch, typically below 100 Hz, creates loads at the antenna interfaces which can excite the antenna eigenmodes.

This environmental load case is verified by shaker tests sweeping the 5–100 Hz band with acceleration amplitudes of the order of 5 to 20*g*, typically. The particular value depends on the antenna mass and coupling with the spacecraft structure and it has to be confirmed by CLA. Due to the need of early equipment test specifications, the environmental conditions must be predicted by similarity or preliminary analysis of the whole spacecraft, making judicious assumptions on flight loads and structural parameters. During sinusoidal testing it is normal practice to notch down the load input at resonance frequencies if the inputs can create artificially high loads not covered by the static sizing loads. This is mainly due to the fact that the shaker table has a higher impedance than the actual mounting conditions on the spacecraft. Therefore, notching based on interface forces down to the quasi-static levels, the so-called primary notching, is considered normally as acceptable. Other notching inputs based on strength limits require authorization by the spacecraft authority.

- *Random vibration and acoustic.* The decision on whether to perform random vibration or acoustic tests depends largely on the architecture of the antenna. Large and lightweight reflectors are sensitive to acoustic pressure, while smaller size units are more sensitive to random vibrations transmitted through their interfaces. However, the source of both phenomena is the same, namely the excitation of the spacecraft structures due to the highly intense acoustic field under the launcher fairing. As a general approximate rule, the acoustic eigenresponse is predominant for surfaces larger than 1 m wave field blocking size and with an areal mass lower than 3 kg/m^2. More details are provided in the next section and in Chapter 5. Random vibration testing, on the other hand, is one of the mandatory tests for electronic units. The aim is to verify the survival of the interfaces and electrical components to high-frequency vibrations. Typical random vibration qualification levels are of the order of 5 to 10*g* RMS (20–2000 Hz) in the out-of-plane direction, depending mainly on the mass of the unit, the mounting conditions of units and the location on the spacecraft. Another important consideration for the case of electrical connections is the order in which thermal and vibration tests are conducted, since often a failure of electrical connections cannot be easily detected after thermal cycling.
- *Shock testing.* This is required for verification of the mechanical integrity after the release of stowed structures, as well as the transient loads transmitted through the spacecraft structure resulting from the release of other subsystems (e.g. solar generators), separation events (spacecraft and fairing separation) and rocket stage burn-off and ignition. The tests should reproduce as far as possible the predicted shock response spectra (SRS), although at early phases, simplified design loads can be specified in terms of half-sine pulses. The particular value depends on the spacecraft and launcher architectures, but also the type of release devices and their installation preload, which are design aspects.
- *Thermoelastic and hygroelastic distortions.* These are specific tests for high accuracy and stability antennas. The reflecting surfaces of reflectors should be kept within the necessary shape with respect to the gain loss budget. This often translates into thermal stability requirements better than 200 microns RMS for Ku-band reflectors of about 2.4 m projected aperture. Even more stringent distortions are necessary if sidelobe levels, cross-polar and other requirements are taken into account. The way to verify this on the ground is by thermal distortion tests in thermal cycling chambers, ideally under vacuum. Thermal cycling under vacuum also allows verification of the hygroelastic performances. The measurement methods are related to the required accuracy. For example, video/photogrammetry can provide accuracies in the range of 10–20 microns per point. This may not be sufficient for applications above Ka-band, in which case laser interferometers, laser RADAR or theodolites may be necessary. Since typically the qualification or acceptance temperatures cannot be reached in combination with these distortion measurement instruments, the tests can be performed with smaller temperature ranges, and then be employed for tuning the finite element models for in-orbit predictions.
- *Thermal cycling, thermal vacuum and thermal balance.* These are the thermal tests required for qualification to the expected space environment. In principle, the thermal cycling test submits the

hardware to hot and cold extremes in a number of cycles at ambient pressure. Cycling serves for qualifying materials and manufacturing processes, assemblies of parts, modules, actuators and mechanisms. Thermal cycling is therefore an environmental screening test. Thermal vacuum tests, on the other hand, reproduce the flight environment more realistically and serve mainly for performance verification by functional testing of units. In the case of antennas, due to the large proportion of polymer composite materials and complex interfaces, thermal cycling in vacuum is the preferred thermal test. The number of cycles is often a discussion point, although typically eight cycles in vacuum are required for antenna qualification. Thermal balance tests are usually performed at subsystem or spacecraft level, but often also on units when needed. These tests are necessary for verifying the thermal control hardware, its operation and for the correlation of thermal model parameters.

Other tests are typically performed for verification of operational life, release and deployment functions, and pointing accuracy. These are tests related mostly to the operation of mechanisms.

4.2.6.2 Acoustic Load and Testing

Launcher liftoff and transonic flight are the most demanding phases for equipment mounted on the lightweight structural panels of spacecraft with respect to the response to high-frequency vibrations. These vibrations result from the excitation of large structures of the spacecraft by the acoustic loads generated at the base of the launcher: rocket engine sound radiation, turbulent mixing of jets in the atmosphere and reflections of sound on the ground and launch pads. The acoustic loads on the payloads inside the fairing are random and broadband in nature, and are assumed to be ergodic and stationary for analysis and testing purposes.

Sound pressure levels often reach 148 dB for qualification. The qualification and acceptance for the flight of spacecraft and their equipment includes acoustic testing, specified customarily as reverberating chamber tests. However, launcher authorities describe the acoustic field as diffuse, although flight loads do not strictly obey this characterization. Such verification procedures have been accepted historically on the grounds that launch loads are covered by safety factors. However, this is the main source of discrepancies between test results and analytical predictions, especially at low frequencies, where test chambers fail to generate a homogeneous and diffuse pressure field with a white noise frequency response. This can be seen in Figure 4.5, where the fine-resolution

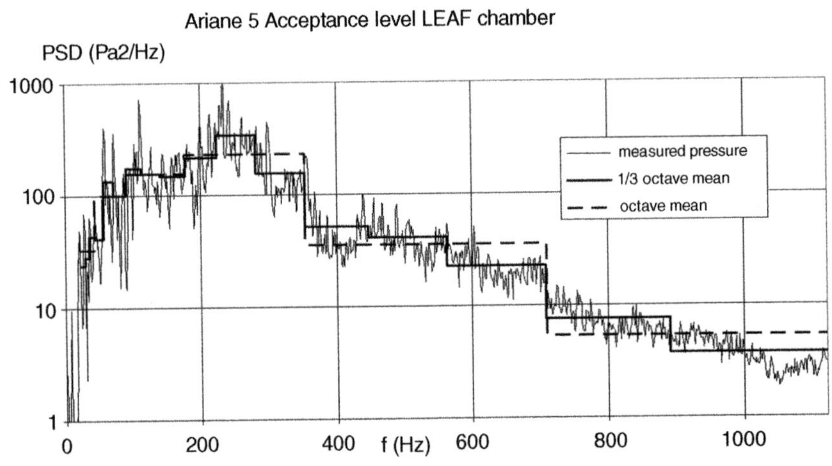

Figure 4.5 *Fine-resolution pressure PSD compared with octave and one-third octave averages. Measurement data from ESA's Large European Acoustic Facility.*

pressure PSD measurement is compared with the classical octave and third-octave specifications. Even if the octave and one-third octave curves are within the test tolerances and fulfil the specifications of the tests, it can be seen that the modal behaviour of an acoustic chamber is a reality, especially below the chamber cutoff.

The use of reverberating chambers is, however, recommended, due to their controllability and quality of pressure field diffusion, compared with other facilities. Other test methods exist, for example direct field acoustic excitation, although larger dispersions of pressure are obtained.

4.3 Selection of Materials in Relation to Their Electromagnetic Properties

4.3.1 RF Transparent Materials and Their Use

As already discussed in Section 4.2.5, in several cases antennas on board a spacecraft need to be protected by radomes or sunshields. In general, the function of these protective structures is mainly thermal as sunshields need to limit the absorption of the solar fluxes by the antenna materials. Regarding RF behaviour, the material constituting the sunshield has to be transparent allowing the electromagnetic waves to radiate out of the antenna. Therefore, in the selection of the RF properties of the sunshield material, losses and depolarization are as important as its thermo-optical properties. Germanium-coated polyimide films are well suited for antenna cover applications due to the coating's unique combination of RF transparency and ability to bleed off static charges. The polyimide substrates typically used are Kapton, non-conductive Black Kapton, or conductive Black Kapton. The germanium coating has a thickness in the range of 1000–1750 Å. When more structural radomes are needed, materials mostly used are glass reinforced fibre plastic (GFRP) painted with special coatings or, as already mentioned in Section 4.2.5, the NORCOAT.

4.3.2 RF Conducting Materials and Their Use

The primary function of an antenna is to transmit and/or receive electromagnetic waves and therefore to a large extent it has to be made of RF conducting materials. Waveguide feeds are primarily aluminium assemblies with special surface treatment to make the aluminium durable to the harsh environmental conditions. Alodine 1200 for example is a chemical used to produce a protective coating on aluminium which ranges in colour from light iridescent gold to tan. The coating produced minimizes corrosion and provides an improved bond for paint. Alodine 1200 is used in an immersion, spray and brush-on application. When waveguide losses are considered critical a silver coating instead of an Alodine one is preferred. On the other hand, with respect to power handling, the silver plating has lower margins with respect to those achievable with Alodine, making the final selection of the RF conductive coating a challenge.

When it comes to reflector antennas, the choice is almost fixed: polymer composite materials. In particular, reflectors are sandwich constructions consisting of three elements: two thin, high-stiffness facing (skins) and a low-density core. The three elements act as a continuous structure due to an adhesive bonding.

In several cases, two surface skins are made of CFRP (Carbon Fibre Reinforced Plastic), consisting of several layers of unidirectional fibre or woven fabric, while for the honeycomb supporting structure typically aluminium, carbon, Kevlar or Nomex are used. From an RF standpoint, the active surface of the sandwich structure of the reflector needs to be characterized in terms of reflectivity, ohmic losses, depolarization, and so on. Reflectivity properties in particular depend strongly on material and process selection, field polarization versus fibre orientation, field angle of incidence, frequency, as well as manufacturing issues and metallization (use of vapour-deposited aluminium (VDA), Kapton aluminium sheets, copper strips).

4.3.3 Material Selection Golden Rules for PIM Control

On several occasions, the thermal blankets described in Section 4.2.5 were responsible for intermodulation problems on board satellites. In fact, loose contacts between metals act as diodes (nonlinear junctions) generating intermodulation frequencies when illuminated by high-power signals. The power associated with these undesired frequencies could be very harmful for the sensitive receivers on board. Satellite design will always have to consider these issues from the very early stages of development in order to implement a proper mitigation plan to avoid any problems during flight.

Passive RF components should be built following good-quality standards to prevent passive intermodulation (PIM) interference. The design should take into account multicarrier operation and intermodulation products should be minimized.

In general for waveguide components a number of 'PIM prevention rules' have been derived. These rules are summarized in the following and are strongly recommended in the development of all electrical components on board spacecraft. It is well known that screws are possible sources of PIM, especially when exposed to thermal cycling. To reduce the risk introduced by this, the use of waveguide components without tuning screws is highly recommended. This is now possible due to the availability of very accurate electromagnetic models that allow the realization of components with performances perfectly aligned with predictions, thus avoiding the need for tuning screws. Other possible sources of PIM are flanges between different components or even within the same component. To avoid this, single-block fabrication of the largest set of components possible is recommended to further reduce the risk. Wherever it is essential to have an interface and therefore a flange, high-pressure flanges are used to guarantee strong contact between metal parts. It is also very important to avoid dissimilar metal contact at interface flanges. If a reference surface finish (say Alodine) is selected and some components (diplexers, polarizers) need an internal silver plating for reducing the insertion loss, then it is necessary to master the manufacturing and surface treatment of the flanges in order to have the component silver plated internally but with accessing flanges in Alodine. In this particular case, the junction between the Alodine and the silver plating is performed internally along a plane waveguide section where no connection cut is present, thus again reducing the PIM generation risk to a minimum.

4.4 Space Materials and Manufacturing Processes

Materials for the manufacture of space antennas include metals and their alloys, organic polymers, ceramics and composites. Due to their adaptable properties and increasing maturity, polymer matrix composites are becoming the most relevant material family. In fact, the properties of composites are 'created' at the same time as the part is produced. In all cases, the demanding functional requirements, the harsh space environment and the life cycle characteristics narrow down the selection range of materials. Furthermore, also as important as the materials themselves are the accompanying manufacturing, conditioning and finishing processes. The material nature, the associated processes and the environmental conditions determine the properties. This section describes some of the main materials, the manufacturing and conditioning processes, and the characterization tests needed to determine their behaviour and ultimately the antenna performance.

4.4.1 Metals and Their Alloys

Metals and their alloys are still the most numerous and readily available source of materials. Their properties (see Table 4.1) depend strongly on the particular alloy composition and product form (plate, sheet, rod, extrusion, etc.), as well as the thermal and mechanical treatments applied during production of the raw material and the history of the part. Due to mass constraints, lightweight alloys must be selected for bulky parts.

Table 4.1 *Design values for typical metal alloys. For strength data, B-basis: at least 90% of the population of values is expected to exceed the property value, with a confidence of 95%; S-basis is equivalent to a specification minimum value. Sources: [19] and data sheets*

Material	E(GPa)	$\rho(10^3\ \mathrm{kg/m^3})$	ν	F_{ty}(MPa)	F_{tu}(MPa)	CTE(ppm/°C)
Clad 2024-T81 sheet	72.4 (primary)	2.77	0.33	372 (S)	427 (S)	22.5
6061-T6 sheet	68.3	2.71	0.33	255 (B)	296 (B)	22.7
7075-T73 sheet	71.0	2.80	0.33	386 (S)	462 (S)	22.0
AlBeMet 162 extruded bar annealed	192.0	2.07	0.17	320	430	14.0
Magnesium alloy AZ31B-H24 sheet	45.0	1.77	0.35	200 (S)	269 (S)	26.0
Ti–6Al–4V annealed	116.5	4.43	0.31	1034 (S)	965.3 (S)	8.8
Stainless steel A286	201.0	7.95	0.31	660 (S)	960 (S)	16.2
Stainless steel A2-70	190.0	8.0	0.29	450 (S)	700 (S)	
Stainless steel 15-5PH H1075 bar	197.0	7.85	0.27	862 (S)	1000 (S)	12.0
INVAR36 annealed	141.0	8.14	0.29	276	492	1.3–2.7
Imphy INVAR M93	142.0	8.125	0.26	280	470	1.1–2.5
Copper–beryllium C17200 TH1 strip	128.0	8.25	0.27	1034 (S)	1210 (S)	16.5
Kovar	160.0	8.1		270		5.0–6.1

4.4.1.1 Aluminium Alloys

The main advantages of aluminium alloys are low density, high stiffness and strength, high electrical and thermal conductivities, good corrosion resistance, fracture toughness, machinability and availability. They are designated by their composition and thermal/mechanical treatment [19]. The most common aluminium alloys used for space hardware are the wrought alloys, namely 7075 in several tempers (T73, T651, T7351), 6061 (T6, T651), 5056-H39, 6082-T6 and 2024 (T3, T81).

The 2024 is an Al–Cu alloy, widely available in a variety of forms and tempers. It is employed for instance as a foil in the manufacture of honeycomb cores (2024-T3 and T81). In this case, prior to bonding, the foil has to be cleaned and treated with chemical conversion coatings for corrosion protection. Some of the forms and treatments of 2024 are susceptible to stress-corrosion cracking (SCC).

The 5000 series alloys contain magnesium as the principal alloying element and are especially suited for operation at low temperatures. At high temperatures they may show susceptibility to SCC. The 5056 (4.5–5.6% Mg) alloy is used for honeycomb core foils (5056-H39).

The 6000 aluminium alloys contain magnesium and silicon as their principal alloying elements. In particular the 6061 and 6082 are among the most commonly used (especially in T6 and T651 tempers) for electronic unit housings. The 6063-T6 extruded alloy is employed for waveguides and heat pipes.

The 7000 alloys contain zinc as the principal alloying element, as well as magnesium and copper. They are among the aluminium alloys of highest strength and resistance to corrosion. Due to its outstanding properties and variety of product forms, the 7075 alloy with temper T73 or T7351 is the workhorse alloy for space structures. However, its T6 temper may result in SCC in some product forms, which should be avoided.

4.4.1.2 Titanium Alloys

Titanium alloys offer excellent mechanical and thermal properties, especially strength-to-density ratio, corrosion resistance and high-temperature behaviour. In particular Ti–6Al–4V is an alpha–beta alloy with good strength at cryogenic as well as elevated temperatures. It is weldable and normally used in annealed or solution-treated and aged (STA) condition. In addition to its higher strength and stiffness than aluminium, its CTE is close to one-third and the high-temperature behaviour of its properties, including electrical conductivity, is more stable. Titanium alloys are used for highly loaded brackets and inserts, screws and washers, and, in special situations, also for radio frequency chain (RFC) components. Precaution has to be exercised with some metal coatings, such as cadmium and silver, which may produce brittle behaviour at high temperatures. Titanium parts have been successfully manufactured and qualified for space employing additive layer manufacturing (ALM) as an interesting alternative to machining methods for complex geometries, which allows better exploitation of the specific properties of titanium.

4.4.1.3 Steel and Iron Alloys

Steel and other iron alloys are normally employed for high strength-parts, that is fasteners, inserts and brackets. Due to their high density and magnetic characteristics, these alloys should be used in small quantities and as far as possible from reflective/radiating elements. For instance, A-286 is an Fe–Cr–Ni precipitation-hardened alloy with excellent strength at high temperatures, used in sheet assemblies and fasteners. If high temperatures are not present during the life of the part, an austenitic stainless steel such as A2-70 or A2-80 can be employed. Among the high-strength stainless steels is the martensitic 15-5PH. This alloy is supplied in either annealed or overaged condition and is heat treated after fabrication.

Within the family of Fe–Ni alloys, INVAR offers an exceptionally low thermal expansion coefficient. It is employed in high-accuracy/stability dies and moulds for reinforced polymer curing, but also in flight parts that require high thermal stability, for example interface brackets with pointing stability responsibility, inserts, quasi-isostatic supports and other stable components. This alloy preserves and even improves its properties at cryogenic temperatures, but not at high temperatures. The austenitic face-centred cubic crystal microstructure of INVAR makes it paramagnetic and ductile, but sensitive to creep at high temperatures. Fe–Ni–Co alloys such as Kovar also exhibit low expansion coefficients and are used in connectors and electronic and optical component mounts.

4.4.1.4 Beryllium and Beryllium–Aluminium Alloys

Beryllium and its alloys, normally processed by powder metallurgy, offer significant advantages compared with aluminium in terms of mass and overall mechanical and thermal properties (except for fracture toughness). An example is the AlBeMet 162 composite metal (Brush&Wellman), with 62% Be and 37% A. It is formed by hot pressing gas-atomized pre-alloyed powder, resulting in a fine and uniform microstructure. It is available in two forms, highly isostatic pressed (HIP) or extruded. However, the particles resulting from machining beryllium and its derivatives are toxic, which has limited their use in addition to the high cost of the raw material and the limited supply. Beryllium has been considered for the manufacture of waveguides and feed elements, lightweight heat sinks, radiators, motor and mechanism parts, and integrally machined stiffened plates. In the case of RF chain components, silver plating may be required to improve PIM behaviour.

4.4.1.5 Magnesium Alloys

Magnesium alloys provide lower density and higher specific stiffness compared with aluminium. One of the interesting properties of magnesium alloys is their intrinsic vibration damping behaviour. Their lower absolute stiffness modulus and higher CTE have to be taken into account when mechanical and thermal distortions are relevant. The main difficulty is the chemical activity, which can result in oxidation and explosive behaviour during machining. The high galvanic potential leads to corrosion when in contact with dissimilar materials, such as aluminium. The surface can be protected by chemical conversion, anodic oxidation, plating or coating with resins.

4.4.1.6 Copper and Copper–Beryllium

Copper alloys are known for their electrical and thermal conductivities. Copper is ductile and hard and can be readily alloyed. However, its mechanical properties degrade quickly above 70 °C. Cu–Be is the highest strength copper-base alloy, with good corrosion resistance and better behaviour in high-temperature environments. It is used for antenna parts and mechanical springs, due to its high elastic limit, high fatigue strength and good electrical conductivity. As well as for other beryllium alloys, dust particles of this alloy require careful handling due to health risks.

4.4.1.7 Molybdenum and Wolfram

Molybdenum and wolfram (tungsten) alloys are used in special cases for antenna components. In spite of their high density, these metals exhibit exceptional mechanical and thermal behaviour, especially when drawn in thin wires and due to their ductility. The electrical conductivity and PIM products are greatly improved after gold plating. These gilded wires are employed, for instance, in the knitted meshes of large deployable antennas. The excellent conductivity of these alloys, low CTE and high-temperature properties make them suitable for high-power applications, such as in high-energy laser optics.

4.4.2 Polymer Matrix Composites

At the present time polymer matrix composites are the most extensively used family of materials for the manufacture of space antennas, taking over from aluminium alloys in large-size components. The main reasons are the favourable stiffness-to-mass ratio, the improved thermoelastic stability, the possibility of tailoring the stiffness and strength properties, and the functionalization of the composite material to the specific application. This latter point is quite important for antennas, since the close coupling of mechanical, thermal and electrical properties can determine the selection of materials. Composite materials can be designed to be reflective or transparent to certain RF bands, whether polarization sensitive or not, thermoelastically stable, thermally conductive or isolating, mechanically stiff or strong in certain directions. For instance, pitch carbon (graphite) fibres can be up to 10 times more conductive than aluminium in the fibre direction. These combinations of physical behaviours are not always possible with metal alloys or ceramic matrix composites, although the driving factors for their success have been the significant mass reduction in general and, for antennas, the excellent thermoelastic stability over a wide range of temperatures.

The most common polymer composite types used for antennas, as well as the majority of space structures, are based on long-fibre unidirectional laminas or woven fabrics of carbon fibres in epoxy or cyanate ester resins. Some of the fibres and resin types are described in more detail hereafter (see also Table 4.2).

4.4.2.1 Glass and Quartz Fibres

Glass fibres, especially E-glass, are the most common fibres in commercial-grade composites for ground applications due to their low cost, high strength, corrosion resistance and non-conductive behaviour. In the context of antennas, glass fibres are used as reinforcement in RF transparent components, printed circuit plates and thermally isolating parts. In the case of demanding dielectric constant and tangent loss requirements, quartz fibres, made of ultrapure silica, provide the best performances within the glass fibre family but at higher cost. In addition, quartz fibres have the lowest CTE among glass fibres, although lower mechanical stiffness and strength. A remarkable feature of glass fibres, due to their amorphous microstructure, is their isotropic nature, which is not the case for organic and carbon fibres.

Table 4.2 *Fibre properties. Sources: [20,21] and suppliers' data sheets*

Fibres	E (GPa)	ρ (10^3 kg/m^3)	F_{tu} (MPa)	Diameter (μm)	CTE (ppm/°C)	Thermal cond. (W/m °C)
Toray T300	230.0	1.76	3530	7	−0.5	10
Toho Tenax HTA40	238.0	1.76	3950	7	−0.1	
Toray M55J	540.0	1.91	4020	5	−1.1	156
NGF YSH-50A	520.0	2.1			−1.4	140
NGF YSH-70A	720.0	2.15	3600		−1.5	250
Mitsubishi K13C2U	900.0	2.2	3800	9		620
DuPont Kevlar K49	135.0	1.45	3000	12		
DuPont Nomex	17.0	1.38	640	15		
Glass-E	73.5	2.54	3500	5–20	4.9–6.5	
Quartz (99.999% SiO_2)	69.0	2.15	3400		0.54	

4.4.2.2 *Aramid Fibres*

Aramid fibres, such as Kevlar (DuPont), combine excellent dielectric behaviour, low negative CTE and significantly better mechanical properties than glass and quartz fibres along the fibre axis. All of these are very important properties for antennas. The aromatic polyamide nature of Kevlar (para-polyaramid) results in highly oriented molecular chains, but in the transverse direction the chains are linked by hydrogen bonds resulting in low transverse strength. Aramid fibres in general have good tensile strength and stiffness, low density and excellent toughness, but produce composites with poor longitudinal compression, transverse tension and interlaminar shear strength (ISS). Other important drawbacks are the moisture absorption and, concerning the space environment, sensitivity to UV radiation and high-temperature limits. In spite of these drawbacks, Kevlar is extensively used in RF transparent surfaces and honeycomb cores as an alternative to glass and quartz fibres, with superior tensile stiffness, strength and CTE. Kevlar is commonly used in the form of unidirectional (UD) and fabric prepregs and in braided ropes. A lower grade aramid is Nomex (DuPont), commonly impregnated in phenolic resin as paper for honeycomb cores. Other aromatic organic fibres include Twaron (AKZO para-aramid), very similar to Kevlar 49, and PBO 'Zylon' (Toyobo). The latter has exceptional mechanical properties but exhibits very high sensitivity to UV radiation. Vectran (Celanese) is an aromatic polyester with very low creep, and therefore interesting for use in braided cords that are required to stay in tension for a long time, such as in deployable guyed truss structures.

4.4.2.3 *Carbon Fibres*

Carbon and graphite fibres are electrically and thermally conductive, as opposed to aramid or glass fibres, and provide a wide range of mechanical and thermal properties to polymer composites. In addition, they are resistant to radiation and corrosion and exhibit the best balance of mechanical properties and a low negative CTE. They are classified according to their elastic properties as high-strength ($E < 300$ GPa), high-modulus ($E < 500$ GPa) and ultrahigh-modulus fibres ($E > 500$ GPa). Their variety of properties is due to the fibre precursor (polyacrylonitrile (PAN), filament, petroleum pitch or coal tar pitch) and the temperatures during their production processes. PAN-based fibres offer the best balance of properties and pitch fibres provide the highest thermal and electrical conductivities and elastic modulus at the expense of lower strength and higher cost. The processing of PAN carbon fibres consists of heating up to 1500 °C in continuous furnaces, for the carbonization of the PAN filaments. Further thermal treatment results in graphitization. In the case of pitch fibres, the process starts by melting and spinning the petroleum or coal tar pitch, followed by processes similar to those for PAN fibres, although at higher temperatures. In both cases the final surface treatment and sizing processes are very important for improving the adherence of fibres to the polymer matrices and protecting them against abrasion. Carbon fibres are typically provided in untwisted tows of 1000, 3000, 6000, 12 000 or more fibres. Twisted yarns are also common. These tows and yarns are employed in several product forms such as unidirectional prepregs, tapes and fabrics. A large number of manufacturers and types of PAN fibres especially are available. However, high-performance PAN and pitch carbon fibres, necessary for the high performance required by antennas, are produced in smaller amounts and are increasingly in demand by all sectors of industry, including the leisure and sports articles, automotive, wind turbine and aerospace industries.

4.4.2.4 *Epoxy Resins*

Epoxies have been the most widely used thermoset resin as the matrix material and adhesive in aerospace structures for decades. The excellent adhesion and strength over wide temperature ranges are the reasons for their extensive use. A large choice of commercial epoxy resins and adhesives is available. As a matrix material they are applied for impregnating fibre tows, UD laminas, fabric laminas, injection moulding, resin transfer moulding (RTM) and many other composite fabrication processes and products.

The processing parameters and the required performance of the end product drive the selection, mainly:

- Viscosity as a function of temperature during the curing, gel time and pot life are the main parameters to consider in relation to the manufacturing processes.
- The glass transition temperature (T_g) of the cured resin is controlled first by the curing and post-curing temperatures, but also cure duration and the particular resin chemistry, because of the cross-link density generated. A high T_g is associated with high stiffness and strength, but also brittleness. Toughened resins have typically lower T_g.
- Epoxy resins have a tendency to absorb water, which has also an effect in reducing T_g. The main consequence is hygroscopic distortions produced by moisture content changes due to thermal cycling and vacuum exposure. In modern composites, the hygroscopic distortions resulting from moisture release in thermal vacuum cycling are comparable with the thermoelastic ones.
- The high value of the thermal expansion coefficient and the brittleness, especially of the high curing-temperature epoxies, might result in microcracking, especially when combined with high-modulus fibres.

4.4.2.5 *Cyanate Ester Resins*

These resins present several advantages compared with epoxies, in particular for antenna structures. In addition to a typically higher T_g, this resin family exhibits low moisture uptake, lower outgassing and better microcracking behaviour than most high-T_g epoxies. Some cyanate ester resins do not require elevated curing temperatures to achieve high T_g, which is an interesting feature for manufacturing high-accuracy reflectors or mirrors. The dielectric constant and loss tangent are better than for most thermoset resins due to the absence of internal polarity (Table 4.3). On the other hand, cyanate esters have lower adherence than epoxies and hence are not replacing them as adhesives. In order to improve adherence to fibres, these are seized with epoxy even if used in a cyanate ester matrix.

4.4.2.6 *Bismaleimide and Polyimide Resins*

Other thermoset resins are used especially in high-temperature applications. These are bismaleimide and polyimide resins. In both cases, their use is limited to niche applications due to processing difficulty, the products of the chemical reactions, the high-temperature processes involved and the potential brittleness and corrosion problems of the manufactured parts. However, these composites represent an alternative to the more expensive ceramic matrix composites.

Table 4.3 Resin properties. Sources: [20,21] and suppliers' data sheets

Resin system	E (GPa)	ρ (10^3 kg/m^3)	F_{tu} (MPa)	CTE (ppm/°C)	T_g (°C)	Dielectric constant	Loss tangent
Hexcel M18 epoxy	3.5	1.16	81.1		198		
Hexcel 8552 epoxy	4.67			65.0			
Tencate RS3 cyanate ester	2.96	1.19	80.0	43.2	250	2.67 (2–18 GHz)	0.005 (2–18 GHz)
ACG LTM123 cyanate ester		1.19		53.0	210	2.77 (10–14 GHz)	0.005 (10–14 GHz)
Bryte EX-1515 cyanate ester		1.17		61.0	177	2.8 (10 GHz)	0.004 (10 GHz)
UBE PETI 365E polyimide for prepreg	2.8	1.33	128.0	51.0	357		

4.4.2.7 *Manufacturing Processes*

The manufacture of fibre-reinforced polymer parts involves specific precursor products and fabrication processes [20–22]:

- *Prepreg layup, oven and autoclave curing.* Prepregs are composite product forms of an uncured resin impregnating the fibres with a closely controlled mass fraction. There are prepregs of unidirectional tows and tapes, as well as woven fabrics, made from a large variety of fibres and resins. Unidirectional and woven prepreg tapes are used for hand layup and oven or autoclave curing. Autoclaves, as opposed to ovens or hot-press curing, allow the combination of high temperatures with vacuum bagging and homogeneous pressure in a controlled manufacturing process, standardized and suited for high-quality parts. This translates into excellent compaction of the laminates and bonds, lower voids and porosity contents, and higher fibre volume fractions. Furthermore, autoclaves are necessary for complex operations such as co-curing of assembled structures (one-shot operation involving skin curing and bonding to core in the case of a sandwich), co-bonding (one-shot curing and bonding of joining elements) and bake-out, in addition to the resin curing or post-curing. Manual layup and autoclave curing is the standard process combination for high-accuracy antenna reflectors and other spacecraft structures.
- *Filament winding.* Filament winding employs rovings, tows or twisted yarns, either prepreg or directly impregnated in a resin bath prior to winding. A wet fibre band is routed on a rotating mandrel, for example in an helical pattern, producing balanced angles $\pm\theta$ interleaved in each ply. The best mechanical properties are obtained with untwisted and well-collimated tows. This technique is used only for revolution symmetric objects, such as cylinders and cones. The winding angles can be tailored to the application, with preference for fibres orientations close to the part axis. Special methods allow fibres to be aligned with the part axis. Filament-wound parts are normally cured in ovens rather than autoclaves, which can be acceptable when the tension during winding is sufficient to guarantee compaction. High volume fractions are achieved normally ($V_f = 0.7$), but higher porosity and voids are common when compared with manual layup and autoclave curing. Typical applications for antennas are the production of rods and thin beams for subreflector assemblies and supporting strut trusses. Diameters of rods can be as small as 15 mm. Smaller diameters can be achieved by pultrusion. In addition to the tube itself, the integration or bonding of the end fittings requires specialized engineering procedures, due to the differential thermal expansion of the joined parts and the complex loading conditions. In the case of space antennas, the thermal cycling of this joint is a design driver.

- *Fibre placement (FP).* Due to a sophisticated machine head and numerical control (NC) programming, fibre tow pregs are deposited with low tension and rolled on complex mould surfaces, allowing for the automated layup of non-revolution parts, even with concave shapes. The objective is to increase the production accuracy and speed of large and complex-shaped components. Mechanical properties of fibre-placed parts are comparable with hand-laid parts, in spite of the higher ply angle accuracy achievable. Application of this technique to antennas is not common, although new types of sandwich composite skins are enabled with tow placement control, potentially replacing open-weave fabrics.
- *Resin transfer moulding (RTM).* Resin transfer is a variant of liquid moulding, in which a low-viscosity resin is injected by pressure into a die containing a dry fibre preform. The method is used for the production of complex parts in one piece with tight dimensional tolerances and finishes in several surfaces. Fibre volume fractions of 0.5 to 0.6 are typical. In these techniques, resin rheology and chemistry in relation to the porosity and size of the preform drive the process. The production cost is higher than manual layup due to the preforms, the tooling and tuning of the injection. Resin film infusion (RFI) and vacuum-assisted RTM (VARTM) are the variants mostly employed in the aerospace industry. Vacuum assistance allows void contents to be reduced if good sealing of the mould is implemented. The curing operation does not require an autoclave in this case and typically the injection pressure is kept high until gellation, in order to minimize void nucleation and growth. Parts that can be manufactured with RTM for antennas are mainly CFRP brackets and large inserts.

In all these processes one of the enemies to beat is the formation of voids and porosity within the cured part. Voids can be entrapped air resulting from defective lamination or injection, as well as volatiles locked in the matrix during curing. Air bubbles can be controlled with proper resin condition and tackiness of the prepregs during lamination, as well as suitable tooling and debulking operations with the assistance of vacuum. Meanwhile, intralaminar voids and pores build up and grow when the vapour pressure of volatiles, mainly water and solvents in the resin, becomes larger than the pressure in the liquid resin during curing. The vapour pressure of volatiles increases rapidly with the curing temperature. This is one of the main reasons for applying a vacuum to the laminated part prior to resin gellation and the high hydrostatic pressures in an autoclave or resin injection systems. Environmental control of prepregs and resins during their life is therefore very important, especially during layup, in order to control the quality of the end product.

Another important effect that requires control during polymer composite manufacture is the build-up of residual stresses, which is extremely important for the production of accurate parts. These are a consequence of the high curing temperatures and the thermal expansion difference between fibres and matrix, as well as the tools and moulds. The higher the curing temperatures and the stiffer the fibres, the more the residual stresses in the manufactured part. The consequences of these stresses are mainly warping of thin laminates, spring-in of angular parts and microcracking of the resin. Methods of mitigating these problems include: use of symmetric and balanced laminates with good ply alignment accuracy; when possible the careful selection of the heat-up profile of the curing cycle, taking into account the exothermic character of thermoset resins' cross-linking reactions; selection of the mould material matching the cured part's CTE; and careful implementation of all the tooling required.

4.4.3 Ceramics and Ceramic Matrix Composites

This vast family of high-performance materials, including carbon–carbon (C–C) and silicon carbide (SiC) and their variants, is generally based on complex and lengthy fabrication processes, leading to high cost and long lead times. These difficulties inhibit their regular use in commercially driven scenarios, but in cases where environmental and performance requirements are especially demanding they are among the few feasible options. In particular ceramic matrix composites (CMC)s deserve special attention due to their improved

Table 4.4 *Ceramic and CMC properties. Sources: [24,25] and providers' data sheets*

Material	E (GPa)	ρ (10^3 kg/m^3)	ν	K_{IC} (MN m$^{-3/2}$)	F_{tu} (MPa)	CTE (ppm/°C)
SiC (Boostec) Alpha SiC	420	3.10–3.21	0.16	3.5	450 (3-point bending, Weibull 10)	4.0 (20–500 °C) 2.0 (70–293 K)
C/SiC (Astrium/IABG)	238	2.70			210	2.0
HB CeSiC (ECM)	350	2.96	0.18	3.7	266 (4-point bending, Weibull 13)	2.3

fracture toughness. They are the best choice, and sometimes the only one, when cryogenic and/or high-temperature operation with dimensional stability, stiffness and light weight are the selection criteria. The purpose of fibre reinforcement in a ceramic matrix is to provide fail-safe behaviour in the case of crack propagation in the matrix. Generally the strength of a CMC is far above the stress level where crack initiation occurs. Thus, the presence and formation of cracks in operation must always be taken as a given, which does not necessarily mean mechanical failure.

The shape accuracy over wide temperature ranges is crucial for space optics, where these materials are often traded off alongside optical glass and special CFRP developments. Some of the basic materials are described in Table 4.4. Important characteristics to evaluate when selecting these materials are shrinkage and shape control during manufacture, porosity, fracture toughness in relation to the particular failure mode and its statistics, the homogeneity and non-isotropic nature of the properties and the repeatability of the production batches [23–25].

A remarkable example with extremely demanding dimensional accuracy and stability requirements in harsh environments is ESA's Herschel 3.5 m telescope main mirror, manufactured in SiC-brazed petals. Even if these technologies are more suited to space optics than antennas, some missions may benefit from the high-temperature stability of ceramic composites, as in hot planetary environments. This is the case for ESA's Bepi-Colombo mission to Mercury, where CMCs have been deeply studied and traded off in several critical components of the high-gain antenna, such as the main reflector, waveguides and subreflector supporting struts.

Silicon carbide (SiC) is the most extended high-performance structural ceramic material for space applications. It is available in several forms and production methods. The chemical and crystalline constitutions of SiC are extremely stable compared with most metals, from cryogenic to high temperatures (above 1600 °C for beta crystals and above 2000 °C for alpha crystals). The production methods include hot press, hot isostatic press (HIP) and sintering of SiC powder for the production of complex parts. Chemical vapour deposition (CVD) is applied for coatings and surface finishing, achieving fine optical scatter. Different microstructures, compositions and levels of shrinkage are obtained in each production method. The blanks can be ground and joined by brazing.

In particular SiC-100 (Boostec) is obtained by sintering a green body obtained by isostatic pressing SiC powder at about 1400 bar. The green body can be machined by NC to near-net shapes and lightened to walls 1 mm thick. The sintering takes place at about 2100 °C without additional pressure, resulting in a typical shrinkage of about 15%. After grinding, typical residual defects are a few microns peak to valley (PTV) and a roughness R_a of the order of 0.3 microns.

C/SiC is the most employed CMC in space applications, mainly in optical mirrors and benches. It can be obtained from several processes, such as infiltration of a C–C green body with Si at high temperatures. The main variants of C/SiC are due to the use of long or short fibre preforms and the infiltration methods. Silicon infiltration can be performed by chemical vapour infiltration (CVI) or liquid phase infiltration (LPI).

For example, during LPI liquid Si reacts with the carbon matrix or within a slurry, as well as the surface of the fibres, converting into an SiC matrix surrounding C fibres or their remains. In large parts, where gravity can have an influence on the Si concentration, C cakes can be employed to absorb excess Si. The blanks are readily machinable and joined by glue at the green body stage, prior to infiltration. Machining or grinding of the finished C/SiC is possible, although more delicate.

The particular variant called Cesic (ECM) is produced from a felt of short, chopped, randomly oriented C fibres in a phenolic resin for the production of C–C blanks. The green body is then infiltrated by LPI under vacuum at temperatures above 1600 °C. The result is a composite of SiC with controlled inclusions of C and Si phases. Cesic can be machined with diamond tools and spark erosion.

4.5 Characterization of Mechanical and Thermal Behaviour

Emphasis is put in this section on polymer composite materials due to their increasing relevance and use in space antennas, as well as to the relative complexity of their characterization. In fact, the properties of composite materials are inherent to the production of the part; that is, the properties are created during manufacture. They do not result just from the chemical nature of the fibres and resins. The screening tests described in the following are valid for all space materials, however.

4.5.1 Thermal Vacuum Environment and Outgassing Screening

The materials constituting the antennas experience temperatures that range typically between −190 and 160 °C for GEO orbits; for example, deep space and planetary missions can easily exceed that range. See Chapter 5 for more details. These temperatures are reached in a cyclic fashion due to the orbital motions, pointing of the spacecraft and antennas, shadows and reflections from spacecraft parts, and the activation of operational modes. Especially, long eclipse periods or deep space pointing configurations can lead to very low temperatures, as well as steep thermal transients and gradients. This is combined with the effects of vacuum, residual atmospheric gases such as atomic oxygen, UV radiation and charged particles. In addition, outgassing from the material itself has to be controlled to avoid contamination of sensitive optical and thermo-optical coatings, and malfunction of equipment.

For the early screening of materials and processes, it is normal practice to verify their integrity and properties after thermal cycling in vacuum. In ESA standards, for example, this is covered by ECSS-Q-ST-70-04A [26]. A typical material screening test may include 100 cycles of ±100 °C, unless the qualification temperatures and life cycles with the required margins are specified. A vacuum level of 10^{-5} Pa is requested. The extent of physical and mechanical properties to be examined before and after cycling depends on the intended application. In the case of composite materials, micrographs and mechanical testing are required. Contamination control evaluation (performed with witness samples), UV absorption, IR spectrometry analysis of contaminant deposits on CaF_2 or ZnSe windows and precision gravimetry.

Outgassing behaviour is another typical screening test to be performed on materials for space. In ESA standards this is covered by ECSS-Q-ST-70-02C [27]. The sample is submitted for 24 h to 125 °C and a vacuum level below 10^{-3} Pa, accompanied by a collector plate. The following values are determined:

- Collected volatile condensable material (CVCM): quantity of outgassed matter from a test specimen that condenses on a collector maintained at a specific temperature for a specific time.
- Total mass loss (TML): total mass loss of material outgassed from a specimen that is maintained at a specific constant temperature and operating pressure for a specified time.

- Water vapour regained (WVR): mass of water vapour regained by the specimen after the final reconditioning phase of the test.
- Recovered mass loss (RML): total mass loss of the specimen itself without the absorbed water (RML = TML − WVR).

As a minimum, the outgassing screening parameters must be TML < 1.0%, RML < 1.0%, CVCM < 0.10%. Under certain conditions, TML > 1.0% can be accepted. For materials used in the fabrication of optical devices, or in their vicinity, the acceptance limits can be more stringent. The concepts and test methods mentioned are in line with the ones developed by NASA and ASTM standards.

4.5.2 Fundamental Characterization Tests of Polymers and Composites

Polymers and polymer composites require development and characterization testing in order to tune the manufacturing processes, determine the suitability of a certain matrix resin or adhesive to a certain product, and for quality control. A battery of physico-chemical tests is therefore necessary. A brief description of the standard parameters and tests performed on thermoset resins is described in the following. For more details the reader is referred to [20–22].

Rheology tests are intended for the evaluation of resin or adhesive suitability for the foreseen manufacturing process and determination of the heating rates during cure. The viscosity is measured with a rheometer as a function of time for given heating rates. The viscosity of a mixed resin changes with time after starting the cross-linking reactions. First the viscosity drops due to initial thinning of the uncured components when heated, then gellation builds up as cross-linking develops, raising the viscosity drastically. Gellation occurs when the viscosity reaches 1000 poise, typically. The particular viscosity values, gel time and heating rates have to be adapted to the process needs, depending, for example, on whether the resin is injected in a preform or used as adhesive between the core and skins of a sandwich. Rheological phenomena are also closely linked to physico-chemical compatibility in relation to capillary forces (wetting and surface tension) and the effects of gravity forces during curing.

Chemical analytical techniques, typically high-performance liquid chromatography (HPLC), IR spectroscopy or mass spectroscopy, are employed for the verification of the chemical composition of resins and volatiles.

Thermal analytical testing methods are used for the determination of curing temperatures, degree of cure, reaction heats, rates of cure and thermal stability. The following test methods are the most common and standardized ones:

- Differential scanning calorimetry (DSC) is one of the most useful analytical techniques for polymers. It provides heat rates as a function of time (isothermal DSC) or temperature (dynamic DSC) during the curing of a resin sample. From a plot of dynamic DSC it is possible to determine graphically: T_i (onset of polymerization temperature), T_{exo} (cure or exothermic temperature), T_m (minor or intermediate exothermic temperature), T_f (final cure temperature). By further inspection of the curve it is possible to determine the reaction heat and the degree of curing. In addition, DSC of cured resin or laminate samples is a common method for measuring the glass transition temperature T_g, which is obtained by observing discontinuities in the heat capacity of the sample. DSC is sometimes a non-conclusive method for T_g determination due to the difficult interpretation of the curves.
- Thermo-mechanical analysis (TMA) consists of the measurement of the thermal expansion of a sample of cured resin or composite while it is heated. The CTE (slope of the thermal strain curve) experiences discontinuities at transition temperatures such as the glass transition. Typically CTE increases above T_g for thermosets. Although this technique provides a measure of CTE it is not intended for its accurate determination.

- Dynamic mechanical analysis (DMA) is one of the most accurate techniques for the determination of T_g, among other properties. It consists of measuring the complex elastic or shear modulus ($E^* = E' + jE''$, $G^* = G' + jG''$) under sinusoidal excitation as a function of temperature or frequency. The real part, the storage modulus, falls drastically above the glass transition. Therefore a measure of T_g is provided by the onset of storage modulus decay or, more pragmatically, the crossing of the tangent lines approximating the two regimes. The loss tangent ($\tan \delta = E''/E'$) maximum also provides an estimate of the Tg upper bounds. Note that the glass transition depends on the moisture content of the sample. Therefore both dry and wet T_g measurements are considered.
- Thermo-gravimetric analysis (TGA) consists of the accurate mass loss measurement as a function of temperature. The instrument allows determination of the amount of moisture desorption or volatile emission and the decomposition temperature. When it is combined with mass or IR spectrometry, the chemical nature of the gases released can be analysed.

4.5.2.1 *Prepreg and Laminate Physical Properties*

Typical physical properties that are measured on prepregs using standard laboratory techniques are: areal mass, resin content (weight fraction), water and volatile contents. In addition, tack of the uncured prepreg, which can be measured by flat-wise tension tests, has to be in the right range to allow the layup operation and reduce entrapped air.

After curing, the physical parameters typically determined are fibre volume fraction, ply thickness and areal weight. Void content and microcrack density are strongly indicative of composite manufacturing quality. Porosity and voids in particular can be assessed with specific ultrasonic inspection. However, microsections and microscope visual inspection are still the best tools for the determination of voids and microcrack density and their distribution.

4.5.2.2 *Batch Testing and Qualification Testing*

Batch testing should be performed by the supplier in order to certify the characteristics of the produced materials. However, due to variability, potential degradation and for quality control reasons, it is normal practice for the user to verify the batch properties with incoming inspections. For instance, typical composite prepreg batch testing includes:

- prepreg areal mass;
- resin chemistry;
- fibre/resin content;
- cure properties (curing temperature and pressure, heat rates, dwell time, gel time);
- shelf life and shop life;
- volatile content and composition;
- ply tack.

Qualification testing may include all the tests performed for incoming inspections, as well as the mechanical, thermal and electrical performances of the composite cured by the user in unidirectional configurations and the design laminate configuration. The parameters measured depend on the application, including typically:

- ply thickness;
- tensile and compressive strength in the longitudinal and transverse directions;
- tensile and compressive stiffness moduli in the longitudinal and transverse directions;

- shear modulus and strength in plane;
- interlaminar shear strength (ILSS), a measure of the adhesion between plies;
- laminate glass transition (T_g) and decomposition temperature;
- CTE and the coefficient of moisture expansion (CME);
- moisture uptake;
- void contents and microcrack density;
- thermal conductivity;
- thermo-optical properties (BOL and EOL);
- electrical conductivity (DC);
- radioelectric reflection or transmission losses and depolarization;
- adhesive bonding to other materials (lap shear, flat-wise tensile, peel).

4.5.3 Characterization of Mechanical Properties

The main mechanical properties that are required for mechanical design verification and materials qualification are, as a minimum:

- tensile and compressive stiffness modulus;
- tensile and compressive strength (yield and ultimate values);
- shear modulus and strength;
- Poisson ratio;
- ILSS in the case of composite materials or laminated shells;
- core crushing (specific for core materials);
- fracture toughness;
- fatigue crack propagation.

When considering joints or assembled structures like sandwich plate configurations or interfaces between parts, the following tests are typically conducted:

- joint strength (lap-shear and peel strength for bonded joints, pull-out for bolted joints, joint material bearing);
- short bending (three-point) for shear load transfer characterization;
- long bending (three- or four-point) for bending load transfer characterization;
- flat-wise tensile (transverse pull-out capability of bondings such as core to skin).

These lists are by no means exhaustive but can be used as a starting point for specific mechanical characterization test campaigns. The particular conditions of the materials have to be carefully considered. The tests must be conducted under representative conditions (cycled or aged) and, if temperature sensitivity is present, at the expected temperatures. In addition, a sufficient number of samples need to be tested for drawing relevant statistics of each property measured. The industry standard for composites is typically B-value statistics, although often A values are necessary.

Stiffness modulus and strength are measured by means of pull-out machines, allowing the simultaneous measurement of force and elongation at a determined rate. The characteristics of the elongation curve provide the nature of the material, the linearity of its behaviour and the strength and strain limits (yield or ultimate). By measuring the transverse strain in the same test, the Poisson ratio can be determined. The slope of the stress–strain curve provides the stiffness moduli. The test should be performed on all material axes, in order to determine the anisotropy and the material principal axes. Compressive tests are also necessary, due to the fact that the same behaviour may not necessarily be present as for tensile stresses. It is also possible to use the pull-out machine to perform compressive and other tests, such as joint strength, shear tests, bending, and so on.

In summary, the mechanical characterization of materials and structural joints represents an intense and costly activity, due to the large number of parameters to be measured and the large number of samples to be produced and tested.

4.5.4 Thermal and Thermoelastic Characterization

Thermo-optical properties and their degradation in the space environment are the most important thermal properties to be characterized for surfaces directly exposed to radiative heat exchange. In general, the materials typically employed, including coatings and paints, are sufficiently characterized and qualified. However, measurements are often necessary. Solar absorptivity is commonly determined based on the indirect measurement of reflectivity ($\alpha_s = 1-\rho-\tau$, with the transmittivity usually negligible). Reflectivity is typically measured in an integrating sphere reflectometer within the wavelength range from 0.25 to 2.5 μm. The nature of the reflection, specular or diffuse, can be differentiated. Specularity can have an influence on aspects such as Sun concentration due to reflection on certain spacecraft surfaces. The effects of specularity have to be understood in advance and may require mitigation measures. As for the IR hemispherical emissivity, it can be measured directly by means of a dedicated thermal balance test on samples. The normal emissivity, as well as the angular dependence of solar absorptivity, can be readily measured with specific equipment.

Other thermal properties that require measurement are thermal conductivity (or conductance) and heat capacity. Conductivity measurements seem to experience large uncertainty due to their inherent difficulty. Relevant factors in the uncertainty are the level of vacuum and condition of the material sample, as well as the contact resistance of setup surfaces. Typical measurement techniques are the guarded hot-plate method and the laser flash method.

The measurement of CTE, defined as the secant or tangent slope (average or local) of the thermal strain curve, is a rather challenging task considering the high stability of the materials employed. The classical push-rod dilatometers (coupled with linear variable differential transformer (LVDT) or capacitive sensors) reach their accuracy limit in the CTE range where precisely the high-stability materials are expected to operate, that is CTE $\approx 1.0 \times 10^{-6}$ m/m °C. In specific cases, as important as the CTE value is the difference between materials, which increases the accuracy requirements even more. In optical applications it is common to require CTE measurement uncertainties of the order of 1.0×10^{-8} m/m °C. In these cases, only interferometric instruments under vacuum can provide the accuracy. Another constraint on the test is the need to cover very wide temperature ranges, including cryogenic operation.

Acknowledgements

The authors wish to thank E. Daly and L. Pambaguian for fruitful discussions on the space environment, radiation effects and ceramic materials, as well as the editors for their patience and dedication.

References

1. Daly, E.J. (1998) The evaluation of space radiation environments for ESA projects. *ESA Journal*, **12**(2).
2. ESA, Space environments and effects, http://www.esa.int/TEC/Space_Environment (accessed 29 November 2011).
3. ESA (1993) The Radiation Design Handbook, ESA PSS-01-609. ESTEC, Noordwijk.

4. Dever, J., Banks, B., de Groh, K. and Miller, S. (2005) Degradation of spacecraft materials, in *Handbook of Environmental Degradation of Materials*, William Andrew, Norwich, NY., pp. 465–501.
5. Dickerson, R.E., Gray, H.B. and Haight, G.P. (1979) *Chemical Principles*, 3rd edn, Benjamin Cummings, Menlo Park, CA, p. 457.
6. US Standard Atmosphere (1976) NASA Technical Memo TMX-74335.
7. Mende, S.B., Swenson, G.R. and Clifton, K.S. (1984) Space plasma physics: atmospheric emissions photometric imaging experiment. *Science*, **225**, 191–193.
8. Banks, B.A., de Groh, K.K. and Miller, S.K. (2005) Low Earth orbital atomic oxygen interactions with spacecraft materials. *Materials Research Society Symposium Proceedings*, **851**.
9. Caledonia, G.E. (1989) Laboratory simulations of energetic atom interactions occuring in low Earth orbit, in *Rarefied Gas Dynamics: Space Related Studies*, vol. **116** (eds E.O. Munts, D.P. Weaver and D.H. Campbell), Progress in Astronautics and Aeronautics Series, AIAA, Menlo Park, CA, pp. 129–142.
10. Banks, B.A., Mirtich, M.J., Rutledge, S.K. and Swec, D.M. (1984) Sputtered coatings for protection of spacecraft polymers. 11th International Conference on Metallurgical Coatings (AVS), San Diego, CA, April 9–13, NASA TM-83706.
11. Dever, J.A., Rutledge, S.K., Hambourger, P.D. *et al.* (1996) Indium tin oxide-magnesium fluoride co-deposited films for spacecraft applications. International. Conference on Metallurgical Coatings (AVS), San Diego, CA, April 24–26, NASA TM-1988-208499.
12. Banks, B.A., Snyder, A., Miller, S.K. and Demko, R. (2002) Issues and consequences of atomic oxygen undercutting of protected polymers in low earth orbit. 6th International Conference on Protection of Materials and Structures from Space Environment, Toronto, Canada, May 1–3, NASA TM-2002-211577.
13. Rutledge, S.K. and Mihelcic, J.A. (1990) The effect of atomic oxygen on altered and coated Kapton surfaces for spacecraft applications in low Earth orbit, in *Materials Degradation in Low Earth Orbit* (eds V. Srinivasan and B. Banks), TMS, Warrendale, PA, pp. 35–48.
14. Goode, D.C., Williams, A.W., Wood, N.J. and Binzakaria, A. (1994) Photothermal imaging of gold and vermiculite coated Kapton exposed to atomic oxygen. ESA Proceedings of the 6th International Symposium on Materials in a Space Environment, SEE N95-27568 09-23, pp. 201–206.
15. Daneman, S.A., Babel, H.W. and Hasegawa, M.M. (1994) Selection rationale, application, optical properties, and life verification of Z-93 for the space station, Report 94H0632, McDonnell Douglas Aerospace.
16. Dittberner, G.J., Gerber, A. Jr., Tralli, D.M. and Bajpai, S.N. (2005) Medium Earth orbit (MEO) as a venue for future NOAA satellite systems. Proceedings of the EUMETSAT Meteorological Satellite Conference, Dubrovnik.
17. Pence, W.R. and Grant, T.J. (1981) Alphas measurement of thermal control coatings on Navstar global positioning system spacecraft. AIAA 16th Thermophysics Conference.
18. Tribble, A.C., Boyadjian, B., Davis, J., Haffner, J. and McCullough, E. (1996) Contamination control engineering design guidelines for the aerospace community, NASA Contractor Report 4740.
19. FAA (2003) MIL-HDBK-5J: Metallic Materials and Elements for Aerospace Vehicle Structures, DoD Handbook.
20. ESA (1994) Structural Materials Handbook, vol. 1: Polymer Composites, ESA PSS-03-203. ESTEC, Noordwijk.
21. Campbell, F.C. (2004) *Manufacturing Processes of Advanced Composites*, Elsevier, Oxford.
22. ESA (1990) Guidelines for Carbon and other Advanced Fibre Prepreg Procurement Specifications, ESA PSS 03-207. ESTEC, Noordwijk.
23. FAA (2002) MIL-HDBK-17-5. Composite Materials Handbook, vol. 5: Ceramic Matrix Composites, DoD Handbook.
24. Paquin, R. (2000) Mechanical and thermal properties of optical and structural materials, SC219 Short Course Notes, SPIE International Symposium on Optical Science and Technology.
25. Harnisch, B., Kunkel, B., Deyerler, M., Bauereisen, S. and Papenburg, U. (1998) Ultra-lightweight C/SiC mirrors and structures. *ESA Bulletin*, **95**, August.
26. ECSS (1999) *Thermal cycling test for the screening of space materials and processes*, ECSS-Q-ST-70-04A.
27. ECSS (2010) *Thermal vacuum outgassing test for the screening of space material*, ECSS-Q-ST-70-02C.

5

Mechanical and Thermal Design of Space Antennas

J. Santiago-Prowald and Heiko Ritter

ESA–ESTEC Mechanical Department, The Netherlands

5.1 Introduction: The Mechanical–Thermal–Electrical Triangle

Antennas belong to the category of external spacecraft appendages most severely exposed to thermal, radiation and mechanical environments, both during launch and in orbit. In addition, the requirements for RF performance, dimensional accuracy and stability, mass and stiffness drive their design and verification. The only solution to these demands is the use of the most advanced materials, manufacturing technologies and engineering methods. The latter imply a strong coordination of the disciplines involved, that is electrical, thermal and mechanical engineering. Often the same part or component has to fulfil a combination of functions, which means it has to respond to design and functional requirements originated by different disciplines. This is schematically sketched in Figure 5.1 taking into account the system engineering process.

In summary, the antenna requirements derived from user, mission and system requirements, combined with environmental and verification requirements, translate into RF, mechanical and thermal requirements. In this process, system-level analyses are necessary for the derivation of functions, environments and life cycle definition. In the case of moving parts due to deployment, latching/locking, pointing or steering functions, additional requirements result from the operation of the required mechanisms.

Following the requirements definition and the resulting concept design, the next step is the definition of a verification matrix. The aim is to define a suitable and affordable combination of analysis, testing and inspection methods intended to demonstrate the fulfilment of the requirements. Nevertheless, the verification process often results in a demanding on-ground and in-orbit test plan and prototype philosophy. Due to the characteristics of the space environment and the necessary levels of reliability, safety and conservatism, the testing conditions are typically more demanding than the real operational and non-operational conditions during the life of the hardware. The consequence is that the antenna product is designed to survive and

Space Antenna Handbook, First Edition. Edited by William A. Imbriale, Steven (Shichang) Gao and Luigi Boccia.

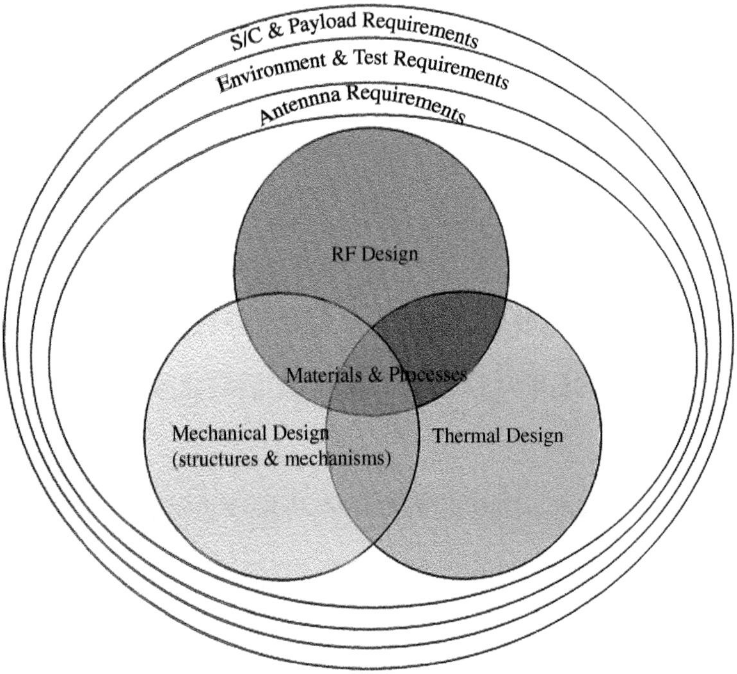

Figure 5.1 *The multidisciplinary scenario of spacecraft (S/C) antenna design with the mechanical–thermal–electrical triangle at the core.*

demonstrate the functionality during the test campaign and hence is over-engineered. This has to be carefully taken into account during the early design phases, combining all the disciplines involved.

5.1.1 Antenna Product

The antenna subsystem can be described as an industrial product by the units in its product tree. In the cases of both side-wall and Earth-deck reflector antennas, a reflector assembly may include:

- A main reflector (including thermal and RF coatings).
- A subreflector in the case of multiple optics (including thermal and RF coatings).
- Supporting elements (struts, tower structures, interface parts).
- Alignment, grounding and bonding devices.

The radio-frequency chain (RFC) comprises:

- Feeder assembly and waveguides (with coatings and thermal control devices).
- RFC support structures (towers, feed mount, waveguide mount, thermal control devices).

Mechanisms if present are typically:

- Antenna deployment and pointing/trimming mechanism.
- Steering, scanning or shaping mechanisms.

- Antenna hold-down and release mechanisms.
- Latching/locking mechanisms.

Specific thermal control hardware for the antennas includes:

- Multilayer insulation (MLI) and sunshields
- Thermal paints or coatings.
- Thermal radiators.
- Heat pipes and/or conductive links (e.g. thermal straps).
- Decoupling elements (brackets or thermal washers).
- Heaters in the case of active control.

In general, the thermal hardware is an integral part of the reflector assembly, RFC and mechanisms, although the overall thermal control and its verification require subsystem treatment, typically coupled to the spacecraft thermal control.

The case of phased array antennas is different with respect to an industrial implementation, especially in the case of active antennas. Array antennas benefit from space heritage mainly in the field of synthetic aperture radar. The basic unit with functional entity is the sub-array (SA): that is, the smallest set of radiating elements with electrical connectivity mounted on a common support. Feed networks, ground planes and insulating layers can be integrated in the SA together with the radiators themselves. This results in a strong coupling of RF, thermal and mechanical functions, making this type of unit a case-by-case and multidisciplinary development. Especially challenging from the thermo-mechanical point of view is the design of thermal links for heat rejection, as well as the thermo-optical characteristics of the external surfaces. Remarkably, the structural links and interfaces of radiating elements and circuits, considering the thermal expansion characteristics of the materials employed, can turn into key design aspects often showing up only in the most complex phases of a test campaign, hence creating unwanted delays and cost overruns.

5.1.2 Configuration, Materials and Processes

The global geometry of the antenna is sized by electrical requirements (e.g. antenna gain), which has a major impact on mechanical and thermal design. The subsequent configuration aspects are defined by a combination of electrical, thermal and mechanical requirements in order to determine the feasibility of the design. For instance, the properties of the materials and finishes of external surfaces determine the electrical performance, such as reflection losses and depolarization, but at the same time also govern the response to the thermal environment and thereby the thermal balance of the antenna. Another example is the selection of adhesives for dual-gridded reflectors, able to cope with a wide temperature range and to survive the mechanical stresses, while exhibiting low losses and the necessary dielectric constant within the operational temperatures. This is again a consequence of the strong links within the mechanical–thermal–electrical triangle.

Material properties should provide the best possible compromise of electrical, mechanical and thermal performance during the whole life cycle. All materials of space antennas must be space qualified and sufficiently characterized for the space environment according to established standards (see Chapter 4 for more details). Typical materials for antennas are:

- Metal alloys of aluminium, titanium, copper and steel, with specification of their source and shape, thermal treatments, surface finish and corrosion protection.
- Polymers (thermoset resins, mostly epoxy and cyanate ester, and thermoplastics).
- Fibre-reinforced polymer composites (based on carbon, glass, aramid, quartz fibres).

- Ceramic matrix composites.
- Coatings.

In all cases, the qualification status of materials and processes is connected to the specific application, to the thermal and mechanical environment and to the expected life cycle. The selection of materials, manufacturing processes, verification of requirements and configuration control are specific to the space industrial implementation. For instance, the ECSS standards (www.ecss.nl) provide the framework of these activities with respect to their programme implementation and preliminary description.

5.1.3 Review of Requirements and Their Verification

The detailed specification of requirements and their verification approach is one of the most critical tasks of antenna engineering. The list provided in Table 5.1 is only a simplified example for a reflector assembly that may include mechanisms or their interfaces, showing the intensity of the requirements definition task. It is not intended to be exhaustive, but provides an initial template applicable to most cases.

Verification of the above requirements should be accomplished by one or more of the following verification methods: test, analysis, review of design or inspection. In order to achieve qualification of the design and acceptance for flight, a suitable model philosophy should be implemented (see Chapter 6 for more details on model philosophy). The qualification of space antennas is accomplished at several levels: materials and manufacturing processes (see Chapter 4 for more details), parts, equipment and assemblies, or subsystem level. The details of these steps depend largely on the industrial implementation and the requirements of users, the spacecraft prime contractor and the launcher authority. Nevertheless, the final objective of qualification is the formal demonstration that the design implementation, manufacturing and integration have resulted in hardware and software conforming to the specification requirements.

5.2 Design of Antenna Structures

The antenna structure is designed to meet RF performance requirements, with careful consideration of the predicted mechanical and thermal environments and material properties. In the initial design loop, mechanical and thermal requirements have to be included, due to the strong physical coupling and multiple functions of the parts. In this chapter, the focus is on antenna reflector structures due to their generality. Phased-array antenna structures can be regarded as a particular case in what concerns mechanical and thermal design, with probably the exceptions of thermal control strategy, some specific materials and manufacturing technologies.

Special attention will be given in this chapter to the sandwich plate architecture with composite face skins, since this architecture is the most efficient one for both the mechanical and electrical requirements.

5.2.1 Typical Design Solutions for Reflectors

Taking into account the type of antenna, structural architectures can be classified as follows:

- *Thick sandwich shell reflectors.* This configuration is typically used for apertures $D < 1.5\,\text{m}$. It is a structurally simple construction. It is basically the simplest and most robust structure for reflectors in spite of the need to embed inserts for load transfer at interfaces (Figure 5.2).
- *Thin sandwich shell with backing structure.* In order to cover larger diameters with reasonable mass and stiffness, a thin shell is supported by a backing structure either integrated in the rear skin or joined by

Table 5.1 *Typical reflector assembly requirements.*

RF design requirements	Frequency and polarization plan Reflection/transmission losses Reflective surface profile Passive intermodulation products Shape and pointing accuracy (after manufacturing) Shape and pointing stability in operational environment Depolarization Power flux density distribution
Mechanical design requirements	Reference frames, coordinate systems Launch envelope (spacecraft/launcher allowable volume) Location and stiffness of interfaces Deployed geometry Mass and inertia (structural and total) Stiffness requirements (stowed and deployed) Deployment kinematics, duration and operations Alignment tolerances On-ground handling and loads Stowed configuration design loads: • Acceleration limit loads • Interface forces and moments • Shock loads • Random vibration and acoustic loads • Sinusoidal vibration loads • Rapid depressurization loads (venting) Release and deployment loads On-orbit loads (stowed and deployed): • Thermoelastic loads • AOCS manoeuvres, station keeping accelerations
Thermal design requirements	Stowed interfaces and heat fluxes, incl. aero-thermal during launch Release/deployment interfaces and heat fluxes Thermal environment during deployment Thermal environment deployed - operating cases Thermal environment deployed - non-operating cases Thermal limits of materials Definition of thermal control hardware, surface optical properties and coatings Definition of heater profile and control policy Temperature telemetry
Design requirements for mechanisms and pyrotechnic devices	Actuators torque/force and margins Lubrication of contact surfaces Interface loads Stiffness Pointing accuracy, range and speed Release shock emission Firing/non-firing conditions of release devices Handling and safety Deployment/pointing/steering sensing, control and telemetry Deployment reliability

(continued)

Table 5.1 (Continued)

Electrical design requirements	EMC Electrostatic discharge Electrical bonding of metallic/non-metallic parts (DC) Grounding and isolation Harness and connectors
Space environment	Electromagnetic radiation (Sun spectrum) Planetary IR radiation and albedo Particles (electrons and protons) Magnetic field Gravity gradient Micrometeoroids and debris Moisture release, outgassing Spacecraft reaction control thruster (RCT) plumes, Sun reflection, shadow, entrapment
Materials and processes	Materials and processes characterization and qualification Coatings qualification Dissimilar metals Thermal vacuum cycling and outgassing screening tests Stress corrosion cracking characteristics Thermoelastic and hygroelastic stability
Design verification and testing Environmental tests	Mass and physical properties measurement Stiffness or modal test Quasi-static acceleration test Sinusoidal vibration test Random vibration or acoustic test Thermal vacuum cycling test Functional/RF tests PIM test EMC/ESD tests Test approach for mechanisms: • Qualification/acceptance tests • Functional and life tests
Ground support equipment	
Handling, transportation and storage	
Quality assurance provisions and reliability	
Quality assurance plan	
Safety requirements	
Fracture control plan (for manned vehicles, reduced plan for unmanned vehicles)	

bonded cleats (Figure 5.3). The two extreme cases represent either a weak or strong interaction of shell and backing structure. A weakly supported thin shell requires a very high inherent stability. The design of the joints deserves special attention, as well as the thermal control of shell and structure and their relative distortions. The acoustic load case is often the sizing one in terms of strength.

- *Monolithic stiffened shells.* This architecture is obtained by replacing the sandwich with a single, thick stable skin, reinforced with a grid of integrated stiffeners made of the same or similar material (and process). This architecture (Figure 5.4), inherited from telescope mirror technology, is usually employed for high-frequency applications. In spite of the apparent design simplicity, the micromechanical

(a)

(b)

(c)

Figure 5.2 *Examples of thick-shell reflectors and antennas. (a) SIRAL antenna subsystem of CryoSat (courtesy of RUAG Space AB, Sweden, under ESA contract). (b) S-X-band Venus Express dichroic high-gain antenna (HGA) (courtesy of EADS-CASA, under ESA contract). (c) Full carbon Ka-band Gregorian antenna (STANT) during vibration testing (courtesy of HPS-GmbH, under ESA contract).*

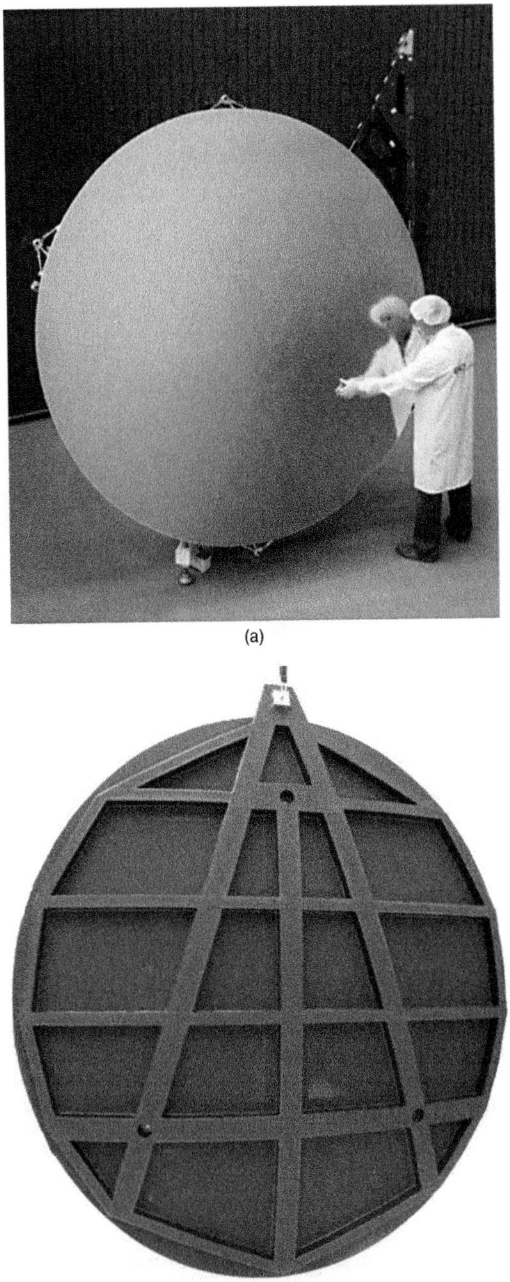

(a)

(b)

Figure 5.3 Examples of thin shell reflectors with backing structures. (a) Ka-Sat 2.6 m Ka-band reflector (courtesy of EADS-CASA Espacio). (b) STAAR 2.4 m Ka-band reflector EM (courtesy of RUAG Space AB, Sweden, under ESA ARTES contract).

Figure 5.4 *Example of monolithic stiffened shell reflector. Herschel telescope 3.5 m M1 rear view. Courtesy of EADS Astrium, under ESA contract.*

and thermal behaviours are very complex, especially when employing composite materials. The overall performances might not be optimal for large diameters unless mass and stiffness constraints are relaxed.

- *Ultralight shells.* This is a particular case of a thin sandwich shell with a backing structure (Figure 5.5). The distinction is made by the use of open fabrics for the manufacture of the reflector shell, thus reducing the acoustic loads significantly and allowing for mass reduction. In the particular type in which the sandwich is replaced by a single skin with elastic folding capability, projected apertures up to 6 m can be achieved. Passive intermodulation requirements are to be considered carefully in this type of reflector.

Figure 5.5 *Example of ultralight reflector: ASAS 3.8 m shaped reflector. Courtesy of Astrium-ST, under ESA contract.*

- *Dual-gridded reflectors.* Due to the strong requirements on transmission losses, depolarization and dielectric constants, material selection is very limited and drives the design. In the case of composite materials, organic or quartz fibres are necessary. The thermal performance of materials often represents a limiting factor. Another distinctive technological aspect is the design and integration of the polarization grid, which is embedded or 'printed' on the transparent sandwich structure (Figure 5.6). As an alternative, stand-alone full carbon conductive grids are proposed.
- *Reflectarrays.* Although not strictly of the same category as the above, these represent a limiting case of mechanical design in between reflectors and phased-array antennas (Figure 5.7). The technologies employed in reflectarrays for manufacturing the sandwich are related to some of the dual-gridded materials and processes. The major difference with respect to classical reflectors is the translation of shape accuracy requirements into plate flatness tolerances and patch phase stability requirements.

A common characteristic of all the architectures is the extensive use of the sandwich structure concept. Therefore, the main focus of this chapter is to provide the design descriptions and understanding of all the

(a)

(b)

Figure 5.6 *Examples of dual-gridded reflectors. (a) The 2.1 m shaped gridded reflector for C- and Ku-bands during thermoelastic distortion testing (courtesy of ASTRIUM-ST). (b) The 1.2 m full carbon dual-gridded reflector demonstrator for the Ku/Ka-band during shape measurement (courtesy of HPS-GmbH, under ESA contract).*

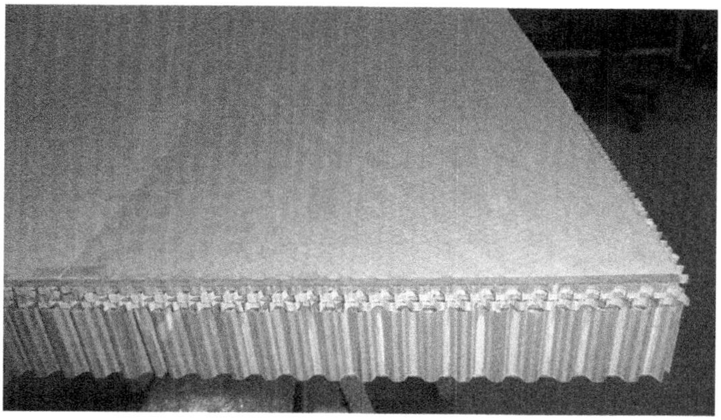

Figure 5.7 Example of reflectarray. Development model of reflectarray, close-up of plate during manufacture. Courtesy of UPM and EADS-CASA, under ESA contract.

implications of this structural configuration in relation to the disciplines involved in space antenna design. The sandwich structure is constructed by separating the external face skins of a plate by means of a lightweight core, thus providing a far higher moment of inertia with negligible mass impact. It therefore represents a well-known and efficient structural solution for increasing bending stiffness and overall strength, while preserving or reducing mass. It allows for efficient use of the material needed to bear mechanical loads, providing the surface shape accuracy and stability, as well as the electrical conductivity required by antennas.

5.2.2 Structural Description of the Sandwich Plate Architecture

The defining characteristics of the sandwich structure are as follows:

- The skins are responsible for carrying in-plane membrane forces and the tension–compression stresses generated by bending.
- The core has to provide global and local stability, bear out-of-plane shear loads, survive local crushing and accommodate inserts for interfaces.
- The adhesive joining of skins to core has to bear shear loads, and survive tensile flat-wise and peeling stresses. In addition, adhesives and potting compounds are needed to join inserts and stabilize loaded core sections.
- The thermoelastic distortions should be minimized by the selection of stable materials and appropriate manufacturing processes, especially when polymer matrix composites are employed. Nevertheless, the differential expansions of the structure elements and the boundary conditions play a dominant role. An important characteristic to control is the symmetry of the skins.

5.2.3 Thermal Description of the Sandwich Plate Architecture

As a consequence of the mechanical configuration, the following aspects are relevant to the thermal behaviour of a sandwich structure:

- The skins with their exterior coating govern the thermal balance of the antenna, since they control the radiative link to the exterior. Their solar absorptivity and IR emissivity determine the dominant heat fluxes. In addition, the skins constitute the conductive heat path in the plane and across the plane towards the core.

- The core transmits heat across the plate and is therefore responsible for the transverse thermal gradients. The internal radiative links among cell walls and skins within the core are usually negligible, but should be taken into account for accurate predictions.
- Normally, the adhesive joining skins to core is not considered in thermal models, in spite of showing resistance to heat conduction. Under proper manufacturing conditions, the heat flux between core and skins should not be interrupted by the adhesive.
- The RF losses of the skin materials and printed or embedded circuits can generate heat dissipation in those applications with high RF power. Then, the in-plane conductive path is relevant for a uniform temperature distribution.

5.2.4 Electrical Description of the Sandwich Plate Architecture in Relation to Thermo-mechanical Design

The main aspects to consider for the RF performances are as follows:

- The reflective face of antenna reflectors is subject to tight manufacturing tolerances and stability requirements, closely related to the operational frequency. It should be noted that first-order plate theory and finite elements method (FEM) analysis normally describe the neutral surface very accurately, but not necessarily the external layers. This must be considered in general, but especially for high-frequency applications.
- The electrical conductivity of the face skin in the reflective side determines the reflection losses. Very often the skin depth ($\delta_s \propto 1/\sqrt{\pi f \mu \sigma}$) is used as a reference design parameter.
- Depolarization has to be characterized at the material level, especially with composite materials, due to the non-isotropic character of the outer layers and details of the surface microstructure.
- Passive intermodulation products can be excited by certain types of composite materials, such as some fabrics. Special care has to be taken in general with coatings, chemical interactions and corrosion during manufacture and testing.
- Core print-through, intracell buckling, skin wrinkling and other local elastic instabilities need to be kept under control in relation to the shortest RF wavelength of interest.
- Transmission losses and the dielectric behaviour of all materials involved in the sandwich plate are a major concern in the transmission mode. In this case, the requirements imposed on the materials limit their selection extremely.
- Thermo-optical ageing, thermoelastic distortions and other distortions such as moisture desorption and gravity release are the main degradation sources of antenna gain, pointing and sidelobe isolation during the mission lifetime. It is relevant to understand their mechanisms and the interactions between the thermal, mechanical and electrical performances.

5.3 Structural Modelling and Analysis

Antenna structures, in general, belong to the category of flat plates or doubly curved shallow shells. As a first approximation, especially bearing in mind numerical FEM analysis, the flat plate theory provides a very useful and didactic starting point. The following subsections address the structural analysis of plates and the implications and specific aspects concerning antennas.

5.3.1 First-Order Plate Theory

The concepts of stress and strain in isotropic or orthotropic solids and their analysis follow the classical work in the theory of elasticity. The text hereafter is self-explanatory, although for more details the reader is referred to the specialized literature [1].

Plates are planar structures with a finite thickness. In this subsection only homogeneous isotropic plates are considered and hence only two material engineering constants are employed: Young's stiffness modulus E and the Poisson ratio ν. The dominant external loads are normal to the plane, which result in shear forces across the thickness and bending moments. In-plane forces can be considered to be linearly superposed to first order; that is, the membrane stresses can be added to the shear and bending stresses produced by out-of-plane loads. Given an element of the plate, internal in-plane forces, shear forces and moments per unit side length are defined as the integral of stresses across the thickness and their first moments, as can be observed in Figure 5.8.

As a first-order approximation, Kirchhoff's hypotheses are commonly accepted (Figure 5.9):

1. The normal displacement is only a function of plane coordinates $w(x, y)$, $\partial w/\partial z = 0$ and the thickness remains constant during the deformation process.
2. Every straight line normal to the middle plane remains straight and normal to the deformed middle surface.

This field of displacements can then be written as

$$\begin{aligned} u &= u_0 - z\frac{\partial w}{\partial x} \\ v &= v_0 - z\frac{\partial w}{\partial y} \\ w &= w(x, y) \end{aligned} \tag{5.1}$$

where u_0 and v_0 are a result of in-plane membrane loads and w of out-of-plane loads. This leads to the strain tensor

$$\begin{aligned} \varepsilon_x &\equiv \frac{\partial u}{\partial x} = \frac{\partial u_0}{\partial x} - z\frac{\partial^2 w}{\partial x^2} \\ \varepsilon_y &\equiv \frac{\partial v}{\partial y} = \frac{\partial v_0}{\partial y} - z\frac{\partial^2 w}{\partial y^2} \\ 2\varepsilon_{xy} &\equiv \frac{\partial u}{\partial y} + \frac{\partial v}{\partial x} = \frac{\partial u_0}{\partial y} + \frac{\partial v_0}{\partial x} - z\frac{\partial^2 w}{\partial x \partial y} \end{aligned} \tag{5.2}$$

which can be expressed in vector form as

$$\varepsilon = \begin{Bmatrix} \varepsilon_x \\ \varepsilon_y \\ 2\varepsilon_{xy} \end{Bmatrix} = \begin{Bmatrix} \varepsilon_{x0} \\ \varepsilon_{y0} \\ 2\varepsilon_{xy0} \end{Bmatrix} + z \begin{Bmatrix} K_x \\ K_y \\ K_{xy} \end{Bmatrix} \quad \text{or} \quad \varepsilon = \varepsilon_0 + z\mathbf{K} \tag{5.3}$$

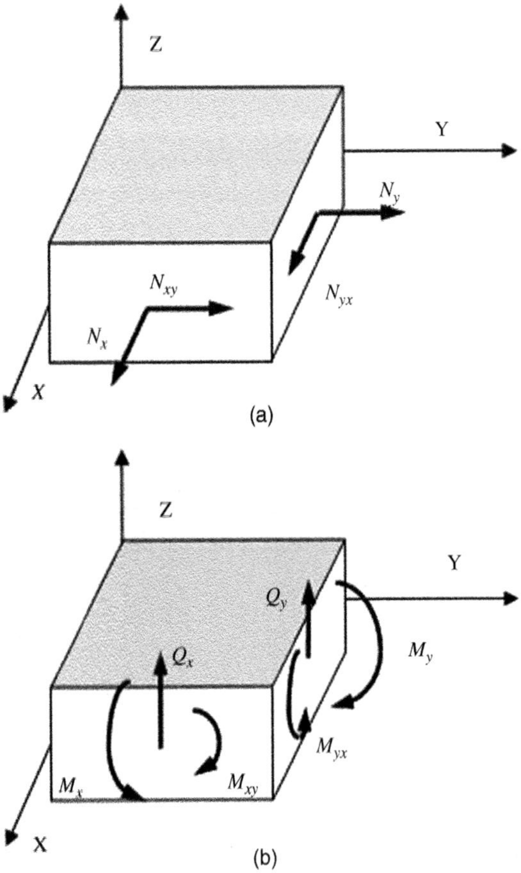

Figure 5.8 *Membrane forces, shear forces and moment definitions.*

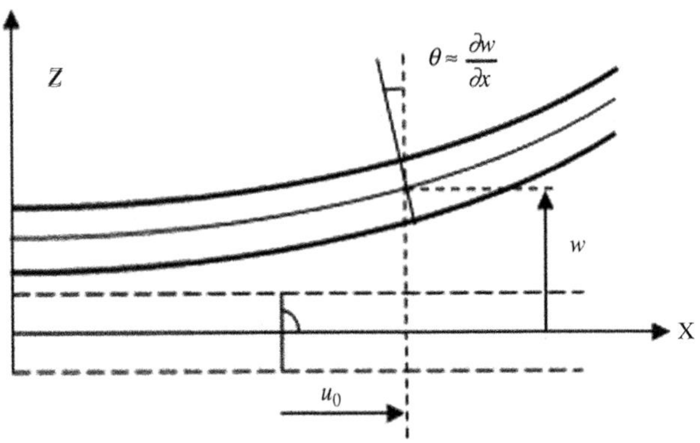

Figure 5.9 *Displacement field of a Kirchhoff plate in the XZ plane.*

The stress tensor (assuming plane tension state) is therefore

$$\sigma_x \equiv \frac{E}{1-\nu^2}\left(\varepsilon_x + \nu\varepsilon_y\right) = \sigma_{x0} - \frac{E}{1-\nu^2} z \left(\frac{\partial^2 w}{\partial x^2} + \nu \frac{\partial^2 w}{\partial y^2}\right)$$

$$\sigma_y \equiv \frac{E}{1-\nu^2}\left(\varepsilon_y + \nu\varepsilon_x\right) = \sigma_{y0} - \frac{E}{1-\nu^2} z \left(\frac{\partial^2 w}{\partial y^2} + \nu \frac{\partial^2 w}{\partial x^2}\right) \quad \text{or} \quad \boldsymbol{\sigma} = \mathbf{C}\varepsilon \tag{5.4}$$

$$\sigma_{xy} \equiv \frac{E}{1+\nu}\varepsilon_{xy} = \sigma_{xy0} - \frac{E}{1+\nu} z \frac{\partial^2 w}{\partial x \partial y}$$

In terms of the curvature vector

$$\boldsymbol{\sigma} = \begin{Bmatrix} \sigma_x \\ \sigma_y \\ \sigma_{xy} \end{Bmatrix} = \begin{Bmatrix} \sigma_{x0} \\ \sigma_{y0} \\ \sigma_{xy0} \end{Bmatrix} + \frac{Ez}{1-\nu^2} \begin{bmatrix} 1 & \nu & 0 \\ \nu & 1 & 0 \\ 0 & 0 & (1-\nu)/2 \end{bmatrix} \begin{Bmatrix} K_x \\ K_y \\ K_{xy} \end{Bmatrix} \quad \text{or} \quad \boldsymbol{\sigma} = \boldsymbol{\sigma}_0 + z\mathbf{CK} \tag{5.5}$$

Inserting these stresses into the moment definitions, their linear relation with curvature and therefore the bending stiffness coefficient is obtained as

$$\mathbf{M} = \frac{Eh^3}{12(1-\nu^2)} \begin{bmatrix} 1 & \nu & 0 \\ \nu & 1 & 0 \\ 0 & 0 & (1-\nu)/2 \end{bmatrix} \begin{Bmatrix} K_x \\ K_y \\ K_{xy} \end{Bmatrix} \quad \text{or} \quad \mathbf{M} = \mathbf{DK} \tag{5.6}$$

Finally, it is straightforward to obtain from these expressions the linear distribution of strains and stresses across the plate thickness (Figure 5.10):

$$\begin{aligned} \sigma_x &= \sigma_{x0} + \frac{12M_x}{h^3} z \\ \sigma_y &= \sigma_{y0} + \frac{12M_y}{h^3} z \\ \sigma_{xy} &= \sigma_{xy0} + \frac{12M_{xy}}{h^3} z \end{aligned} \tag{5.7}$$

$$\begin{aligned} \varepsilon_x &= \varepsilon_{x0} + zK_x \\ \varepsilon_y &= \varepsilon_{y0} + zK_y \\ \varepsilon_{xy} &= \varepsilon_{xy0} + zK_{xy} \end{aligned}$$

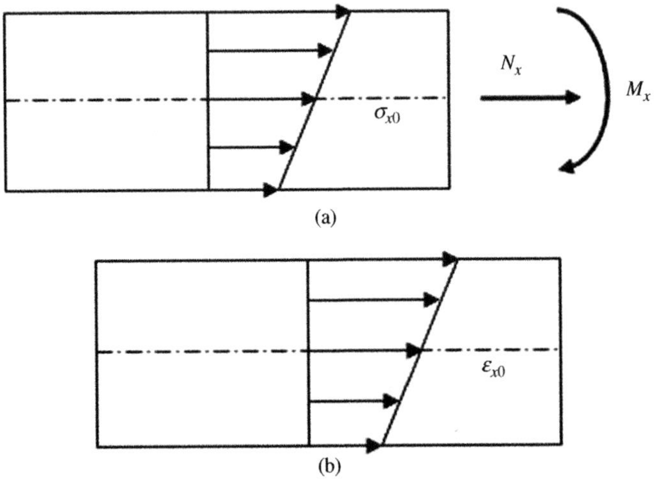

Figure 5.10 *Stresses and strains in first-order homogeneous Kirchhoff plate.*

5.3.2 Higher Order Plate Theories

Higher order theories retain more terms in the approximations of strains and stresses in order to improve the accuracy of the representation of the displacement field. This, however, often results in largely increased analytical complexity without necessarily improving the accuracy of predictions.

On the other hand, an important improvement is provided by the modelling of shear strains and thickness effects. The Reissner–Mindlin [2] theory allows the release of the Kirchhoff condition of normality on the straight lines and includes a shear correction factor. This formulation is very useful when working with thick plates and is also available in commercial finite element software. However, it should be used with caution due to the loss of accuracy in the thin plate limit. It should be kept in mind that most of the antenna plate structures can be considered as thin plates or shells. The exceptions are commonly treated with solid elements which offer a more consistent formulation than plate elements.

5.3.3 Classical Laminated Plate Theory

In the case of thin multilayer plates or shells, it is still assumed customarily that Kirchhoff's hypotheses apply and there is a perfect adherence between layers. Therefore, the linear variation of strains across the thickness holds, while stresses are a multilinear function due to the different material properties of each layer, as shown in Figure 5.11.

By using the constitutive law of each lamina and the definition of forces and moments, the final formulation of the stress/force status of the plate can be written as follows, where the membrane/bending duality is kept:

$$\begin{Bmatrix} \mathbf{N} \\ \mathbf{M} \end{Bmatrix} = \begin{bmatrix} \mathbf{A} & \mathbf{B} \\ \mathbf{B} & \mathbf{D} \end{bmatrix} \begin{Bmatrix} \varepsilon_0 \\ \mathbf{K} \end{Bmatrix} \tag{5.8}$$

with

$$\mathbf{A} = \int_{-h/2}^{h/2} C\,dz = \sum_{k=1}^{n} C_k[z_k - z_{k-1}] \quad \text{(membrane or in-plane stiffness matrix)}$$

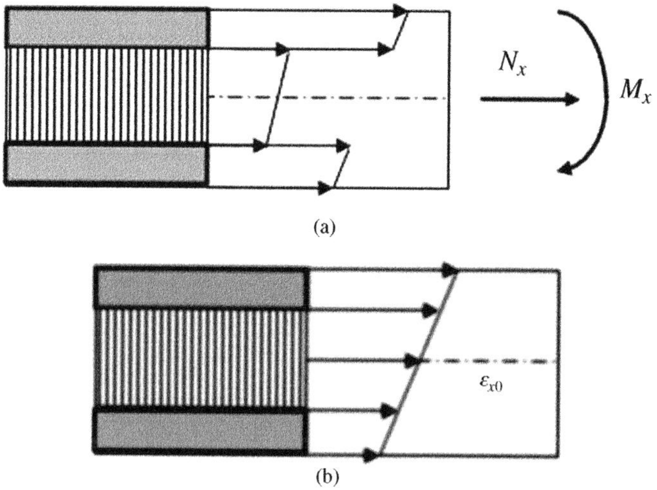

Figure 5.11 Stresses and strains in laminated plate.

$$\mathbf{B} = \int_{-h/2}^{h/2} \mathbf{C} z\, dz = \sum_{k=1}^{n} \frac{C_k}{2}\left[z_k^2 - z_{k-1}^2\right] \quad \text{(bending-membrane coupling matrix)}$$

$$\mathbf{D} = \int_{-h/2}^{h/2} \mathbf{C} z^2\, dz = \sum_{k=1}^{n} \frac{C_k}{3}\left[z_k^3 - z_{k-1}^3\right] \quad \text{(bendingstiffness matrix)}$$

This is the so-called **ABD** matrix notation, customarily employed in finite element software.

5.3.4 Homogeneous Isotropic Plate Versus Symmetric Sandwich Plate

Let us consider an aluminium uniform plate ($E = 70$ GPa, $\nu = 0.33$) with $h = 2.5$ mm thickness and compare its bending stiffness behaviour with a symmetric sandwich plate made of two face skins of aluminium, $t_s = h/2 = 1.25$ mm thick, and a core material of negligible mass and $h_c = 25$ mm. As defined, both plates have the same total mass.

The particular case of the homogeneous isotropic plate yields the well-known relations

$$\mathbf{C} = \frac{E}{1-\nu^2}\begin{bmatrix} 1 & \nu & 0 \\ \nu & 1 & 0 \\ 0 & 0 & (1-\nu)/2 \end{bmatrix}, \quad \mathbf{A} = \frac{Eh}{1-\nu^2}\begin{bmatrix} 1 & \nu & 0 \\ \nu & 1 & 0 \\ 0 & 0 & (1-\nu)/2 \end{bmatrix}, \quad \mathbf{B} = 0$$
$$\mathbf{D} = \frac{Eh^3}{12(1-\nu^2)}\begin{bmatrix} 1 & \nu & 0 \\ \nu & 1 & 0 \\ 0 & 0 & (1-\nu)/2 \end{bmatrix} \tag{5.9}$$

This yields the value of the bending stiffness coefficient

$$\frac{Eh^3}{12(1-\nu^2)} = 102.3\ \mathrm{N\ m}$$

The case of the sandwich plate results in

$$\mathbf{C} = \frac{E_s}{1-\nu_s^2}\begin{bmatrix} 1 & \nu_s & 0 \\ \nu_s & 1 & 0 \\ 0 & 0 & (1-\nu_s)/2 \end{bmatrix}$$

for each of the skins, and for the sandwich plate

$$\mathbf{A} = \frac{2E_s t_s}{1-\nu_s^2}\begin{bmatrix} 1 & \nu_s & 0 \\ \nu_s & 1 & 0 \\ 0 & 0 & (1-\nu_s)/2 \end{bmatrix}, \quad \mathbf{B} = 0$$

$$\mathbf{D} = \frac{2E_s\left[\left(\frac{h_c}{2}+t_s\right)^3 - \left(\frac{h_c}{2}\right)^3\right]}{3(1-\nu_s^2)}\begin{bmatrix} 1 & \nu_s & 0 \\ \nu_s & 1 & 0 \\ 0 & 0 & (1-\nu_s)/2 \end{bmatrix} \tag{5.10}$$

Noting that $t_s \ll h_c$, the bending stiffness coefficient is given by

$$\frac{E_s h_c^2 t_s}{2(1-\nu_s^2)} = 1227.4\ \mathrm{N\ m}$$

5.3.5 Skins Made of Composite Material

Instead of metal skins, the most efficient and common solution for reflectors is the use of carbon fibre reinforced polymers (CFRPs). The electrical conductivity of the carbon fibres and their volume fraction V_f in the mixture have to be such that the reflection losses are kept below 0.1 dB typically for any incidence angle compared with an aluminium plate. Due to the fact that the fibres introduce a non-isotropic behaviour, this should be understood and determined for the RF, mechanical and thermal properties. The fundamental micromechanical characteristics are briefly described in this subsection.

Considering a laminate made of several plies of unidirectional CFRP, it is necessary to describe first the characteristics of a single ply and then the full laminate. Fibres are aligned in a ply pre-impregnated in a matrix of isotropic material with a typical ply thickness of 60–120 μm. The resulting ply material is anisotropic and follows the 'rule of mixtures' for the density, axial stiffness and Poisson ratio [3,4]:

$$\begin{aligned} \rho &= \rho_f V_f + \rho_m V_m \\ E_1 &= E_{Lf} V_f + E_m V_m \\ \nu_{12} &= \nu_f V_f + \nu_m V_m \\ E_2 &\approx E_{Tf} E_m \left(E_{Tf} V_m + E_m V_f\right)^{-1}, \quad G_{12} \approx G_f G_m \left(G_f V_m + G_m V_f\right)^{-1} \end{aligned} \tag{5.11}$$

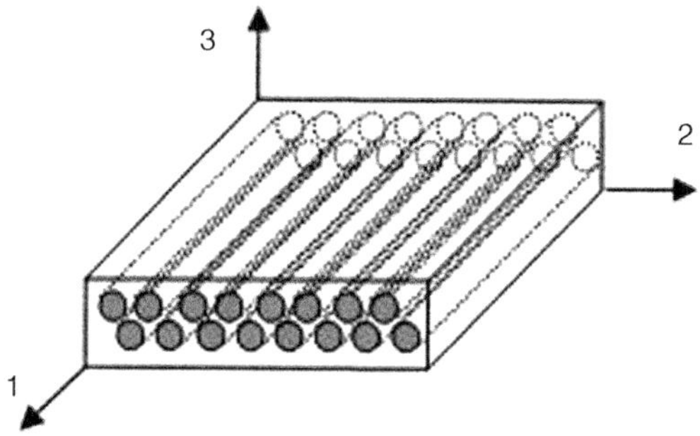

Figure 5.12 *Fibres in unidirectional composite ply.*

In Figure 5.12 the directions 1 and 2 are in plane, longitudinal and transverse to the fibres respectively. The fibre properties are normally different in the longitudinal and transverse directions and hence shear and transverse stiffness follow a more complex behaviour than the rule of mixtures. The approximate expressions in (5.11) for E_2 and the in-plane shear modulus G_{12} are the ones proposed by Jones [3,4]. Alternative semi-empirical expressions for E_2 and G_{12} have been developed by Halpin and Tsai or Puck and others [3] with better accuracy in particular cases, but not in general for any fibre–resin combination (Table 5.2). Note that for unidirectional prepregs of high-modulus fibres and typical fibre volume fractions (V_f between 0.5 and 0.7) E_1 is dominated by the fibre and significantly larger than E_2, as expected. Out-of-plane properties may be derived by assuming that the ply is transversely isotropic: $E_3 = E_2$, $G_{13} = G_{12}$, $\nu_{13} = \nu_{12}$.

By laminating several plies of unidirectional prepreg with different orientations the anisotropy in the plane can be reduced, while still producing a thin skin of composite material with reduced weight and significantly higher stiffness compared with the same thickness of aluminium. The formulation provided above for the **ABD** matrices holds also for laminated plates, with the only precaution of transforming the material stiffness tensor $\mathbf{C}_k$ of each lamina from the material reference system into the plate reference. In the

Table 5.2 *Fibre, matrix and laminate properties. Data sources: (1) data sheet, (2) PSS=03–203 [3], (3) HPS-GmbH measured data, (4) Kueh and Pellegrino [5].*

Material		E_L (GPa)	E_T (GPa)	G_{12} (GPa)	ν_{12}	CTE_L (m/m K)
Fibres	Toray T300 (1) (4)	230.0	23.1	9.0	0.2	−0.5 E-6
	Toray M55J (1) (2)	540.0	6.0	18.	0.36	−1.1 E-6
	NGF YSH 50A (1) (4)	520.0	40.0	14.3	0.2	−1.4 E-6
	DuPont Kevlar K49 (1) (2)	124.0	7.0	12.0	0.36	-4.0×10^{-6}
	Glass E (2)	73.0	73.0	30.0	0.22	3.0×10^{-6}
Resin systems	Hexcel M18 epoxy (1)	3.5	1.3	0.38		
	Hexcel 8552 epoxy (1) (4)	4.67	1.70	0.37	65×10^{-6}	
	Tencate RS3 cyanate ester (1)	2.96			43.2×10^{-6}	
	UBE PETI-365E polyimide (1)	2.8			51.0	
Laminas	YSH70-100 fabric/LTM123, $V_f = 0.5$ (3)	159.8	160.0	3.6	0.02	-0.15×10^{-6}
	Mitsubishi K63B12 UD/epoxy, $V_f = 0.6$ (1)	530.0	4.8			-1.2×10^{-6}

particular case of symmetrical laminates the **B** matrix is again zero, hence there is no membrane-bending coupling.

Especially relevant is the case of quasi-isotropic laminates, which exhibit RF, thermal and mechanical behaviour almost independently of the direction in the plane. This can be achieved for instance with a laminate defined by the orientations (0, 45, 90, −45). In this case, by using, for example, high-modulus fibres in an epoxy matrix, the skin thickness and mass are less than half the stiffness equivalent to aluminium. In addition, the thermal behaviour is also close to isotropic in the plane and the thermoelastic distortions are well controlled by the fibre coefficient of thermal expansion, which can be of the order of 1×10^{-6}/°C. As for the RF reflection properties, due to the small thickness of each lamina, the overall reflected field is almost independent of the incidence angle for frequencies typically up to the Ku-band. This is due to the small lamina thickness and the transparency of the resin. However, especially in the case of circular polarization, it is often necessary to use fabric instead of unidirectional plies for the higher frequency bands.

5.3.6 Honeycomb Core Characteristics

The core material has to provide stability to the skins and keep them in place. Of all the possibilities, including open and closed cell foams, the honeycomb configuration is still the most common core type for sandwich construction due to the light weight, good mechanical properties and maturity of the manufacturing processes. The technology nowadays exhibits a high level of standardization and cost control from which space structures benefit extensively. A first classification of honeycomb cores is given by the cell shape and bonding of the ribbons, as shown in Figure 5.13.

The manufacturing processes allow control of the density and the properties in all directions. Specific values of cell diameter and wall thickness are then fixed for every material. Typical materials are aluminium alloy foil (Table 5.3) and composites based on glass, Kevlar, Nomex, carbon and other fibres.

Specific core materials and processes are, however, required for special cases like ultrastable antenna reflectors and telescope mirrors. Thermally stable honeycomb cores are made of low thermal expansion materials like carbon or Kevlar fibres. Normally pitch carbon fibres are used in order to improve thermal conductivity and stability. Figure 5.14 shows a carbon core with conformability capabilities.

Specific processes for the aluminium cores used for space are the corrosion protection such as chrome conversion and the need to perforate the metal foil in order to allow venting of air enclosed in the cells. A particularity of the core sheets, namely the anticlastic curvature effect (a honeycomb core sheet tends to preserve the mean curvature when submitted to uniaxial bending), is one of the reasons for splitting the core of doubly curved surfaces into joined portions. Another reason for splicing the core is a reduction in the anisotropy of its in-plane properties by changing the orientation of the portions in distributed directions. This can be relevant when manufacturing dimensionally stable shells. The joining of the core parts among them is performed with an adhesive paste or by carefully designing the cutting and placement of the portions in order to properly distribute the shear stiffness through the seams. Other manufacturing methods include the thermal pre-shaping of the core when made of plastic material, or machining in general, and the reticulation of the adhesive prior to bonding.

5.3.7 Failure Modes of Sandwich Plates

Stresses in the different elements of the sandwich structure can lead to several failure modes (Figure 5.15). Bending moments and membrane forces create in-plane tension or compression of the skins, out-of plane shear forces result in shear stresses in the core and adhesive, normal distributed compressive forces result in

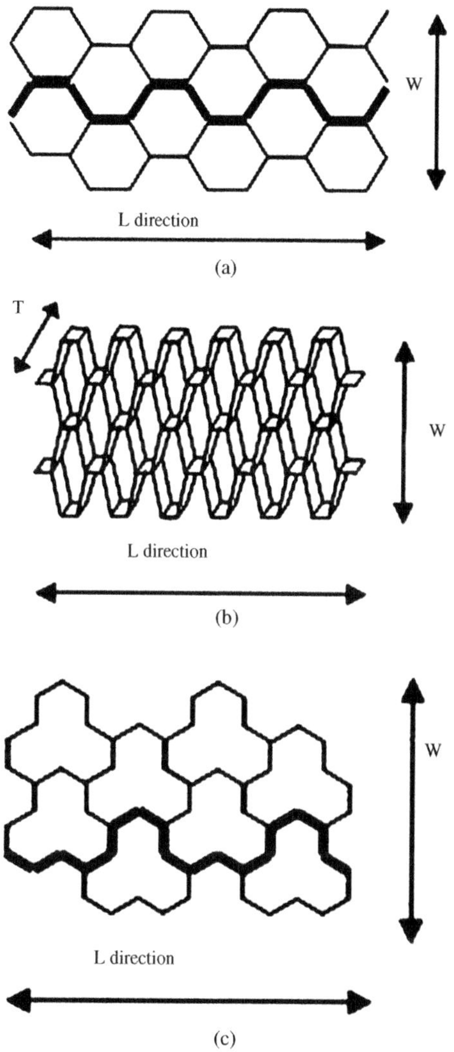

Figure 5.13 *Honeycomb types.*

compression of the core. In addition to skin rupture and global elastic instability (buckling), other failure modes specific to the sandwich structure need to be analysed and designed for.

The following expressions quantify the stresses in each failure mode:

Skin failure	$\sigma_s = \pm \frac{M}{h_c t_s} \pm \frac{N}{2t_s}$
Core shear	$\tau_c = \frac{Q}{h_c}$
Core crushing	$\sigma_c = \frac{F}{A}$

Table 5.3 *Mechanical properties of aluminium hexagonal honeycomb cores [3].*

Type	NIDA designation	d_c (mm)	ρ_c (kg/m^3)	E_c (MPa)	G_L (MPa)	G_T (MPa)	τ_L (MPa)	τ_T (MPa)
1/4-5056-.002p	6–58	6.4	69	3.21	462	186	2.24	1.31
1/4-5056-.0015p	6–48	6.4	54	2.17	345	152	1.59	0.90
1/4-5056-.0007p	6–20	6.4	26	0.55	138	83	0.54	0.26
3/8-5056-.0007p	9–20	9.6	16	0.24	103	62	0.31	0.17

In addition, stability limits are established for the major local stability failure modes [3,6,7]:

Dimpling (intracellular buckling)	$\sigma_{cr,d} = 2.0\frac{E_s}{1-\nu_s^2}\left(\frac{t_s}{d_c}\right)^2$
Face wrinkling	$\sigma_{cr,w} = 0.33\sqrt{\frac{E_s E_c t_s}{(1-\nu_s^2)h_c}}$
Shear crimping	$\sigma_{cr,c} = 0.5G_c\frac{h_c}{t_s}$

These local instability expressions are valid for sandwich plates made of isotropic skins and honeycomb cores under uniaxial loading. Interaction correction formulae are available for biaxial and combined shear loading. In addition to these, global elastic stability should be demonstrated, although it rarely represents a designing load case for antennas.

5.3.8 Mass Optimization of Sandwich Plate Architecture for Antennas

The mass optimization for a given bending stiffness, skin strength or dimpling stability leads to different ratios of core height to skin thickness [7]. The objective function is the mass per unit area $w = \rho_c h_c + 2\rho_s t_s + 2\rho_a t_a$, where the adhesive layers are kept constant and therefore do not play a role in the optimization analysis. Minimization of the mass keeping the bending stiffness constant results in a ratio of core height to skin thickness given by $\rho_c h_c = 4\rho_s t_s$. Reciprocally, the same result can be obtained by maximizing the bending stiffness for a given mass. Note that the density of the core is typically 20 to 100 times lower than that of the skin and hence this result means that the core should be as thick as possible. This again confirms the efficiency of sandwich structures versus monolithic plates when bending stiffness is the driving requirement. Typical core thicknesses of antenna shells range from 5 to 25 mm. In Table 5.4 the optimal ratio of core height to skin thickness is given for several constraints.

Figure 5.14 Carbon fibre honeycomb core. Courtesy of HPS and Invent GmbH, under ESA contract.

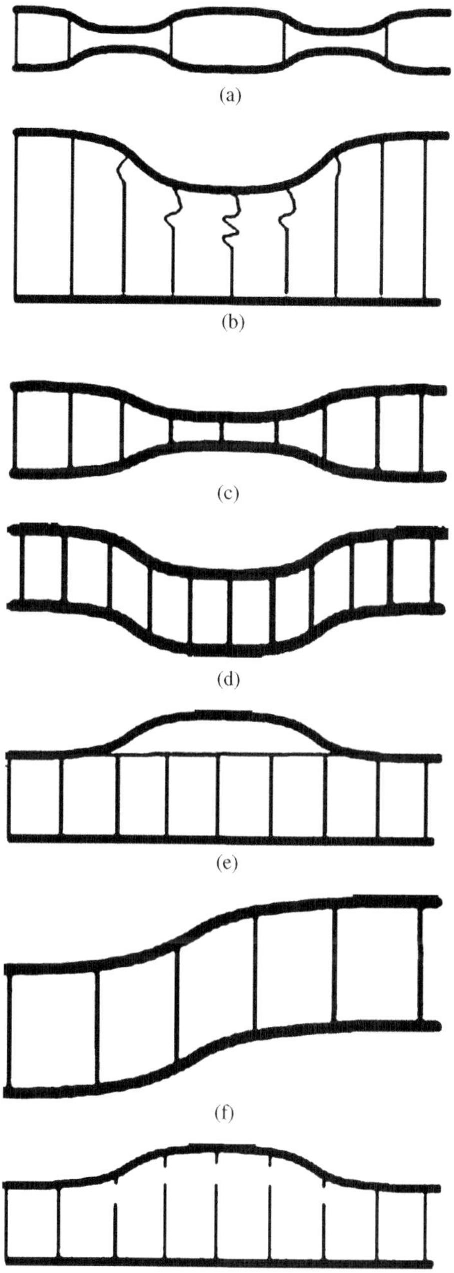

Figure 5.15 *Local failure modes [6]. (a) Intracellular buckling (face dimpling). (b) Core crushing. (c/d) Symmetric/antisymmetric face wrinkling. (e) Flat-wise tensile rupture of bond. (f) Shear crimping. (g) Tensile rupture of core proper.*

Table 5.4 *Core-to-skin ratio for minimum mass with given mechanical performance. After [7].*

Constraint	Formulation	Result of mass optimization
Bending stiffness	$D = \dfrac{E_s h_c^2 t_s}{2(1-\nu_s^2)}$	$\dfrac{h_c}{t_s} = 4\dfrac{\rho_s}{\rho_c}$
Skin strength (under limit bending load)	$M = \sigma_{cr,s} h_c t_s$	$\dfrac{h_c}{t_s} = 2\dfrac{\rho_s}{\rho_c}$
Dimpling (intracell buckling)	$\sigma_{cr,d} \propto \left(\dfrac{t_s}{d_c}\right)^2$	$\dfrac{h_c}{t_s} = \dfrac{2}{3}\dfrac{\rho_s}{\rho_c}$
Face wrinkling	$\sigma_{cr,w} \propto \left(\dfrac{t_s}{h_c}\right)^{1/2}$	$\dfrac{h_c}{t_s} = \dfrac{4}{3}\dfrac{\rho_s}{\rho_c}$

The particular case of shells for space antennas means that dynamic stiffness must be considered as one of the most critical requirements to be met. Due to the large size of some antennas, this can result in rather thick cores. For this reason, the specific stiffness of the skin material must be as high as possible. Above 2 m diameter, one of the best configurations is therefore a thin shell with a backing structure that takes the role of stiffening and transferring the loads through the interfaces. The delicate task then becomes the careful design of the connecting cleats, in order to stiffen the whole structure and transfer loads as necessary, without destroying the thermoelastic stability of the shell. In this configuration, the strength and local stability design criteria become more relevant. From Table 5.4, it can be seen that sparse values for relative core-to-skin thickness provide optimal strength or stability performances. It is then necessary to bring other parameters into the optimal design process, such as the core cell internal diameter d_c, the stiffness and the strength of skins and core, and the distribution and type of connections between shell and backing structure. In particular, the dimpling case can be relevant for ultrastable antennas due to the thermoelastic stresses in orbit.

5.3.9 Finite Element Analysis

The FEM, currently the industry standard for structural analysis, is a numerical solution technique based on the discretization of the domain into small elements, the interpolation of the unknown variables by shape functions within the elements and the integration of the governing equations in discrete form by weighted residuals or variational procedures [8]. It can be applied to linear and nonlinear differential equations, to various material models and used to couple structural analysis with multidisciplinary problems involving fluid, thermal, acoustic, optical and electromagnetic phenomena.

The structural element types most commonly used for antenna structures are bars, beams, plates (triangular and quadrilateral), 3D solids (tetrahedra and hexahedra), rigid elements (bars and constraints) and discrete elements (springs and masses). The formulation of each element and handling of the degrees of freedom (DOF) is particular to each software tool. These details are beyond the scope of this text, but the interested reader is referred to the software user manuals and FEM literature. In Table 5.5 several types of elements are described for some of the most popular software implementations of structural FEM.

The full mathematical model consists of the mesh generated from CAD files by the discretization of beams, thin shell surfaces and solid bodies into elements, plus the material and property definitions of each element, the coordinate systems (global and local), the boundary conditions and constraints, the external loads and the detailed definition of the solution sequence. It is not uncommon for finite element (FE) models of an antenna structure to contain several hundred thousand degrees of freedom, as is the case for the Ka-Sat reflector model shown in Figure 5.16.

Table 5.5 Types of elements in finite element codes (not exhaustive).

		NASTRAN	ABAQUS	ANSYS
Line elements (bars, beams)	Bar/beam (axial force and torsion, transverse shear and bending)	CBAR	B31	BEAM4
		CBEAM	B32	BEAM24
		CROD		BEAM44
Surface elements (membranes, plates, shells)	Triangular	CTRIA3	S3(R)	SHELL
		CTRIA6		
	Quadrilateral	CQUAD4	S4(R)(5)	SHELL91
		CQUAD8	S8(R)(5)	SHELL93
			S9R5	SHELL99
				SHELL150
Solid elements	Tetrahedron	CTETRA	C3D4	SOLID92
			C3D10	SOLID148
	Hexahedron	CHEXA	C3D8	SOLID45
			C3D20	SOLID46
				SOLID64
				SOLID95
Rigid elements	Rigid bar	RBAR, RBE2, RBE3, MPC	RB3D2, MPC	
Discrete elements	Spring (1 DOF linear or rotational stiffness connection)	CELAS2	SPRING1	COMBIN14
			SPRING2	
			SPRINGA	
	Mass (1 DOF point mass or inertia)	CONMASS	MASS	MASS21
			ROTARYI	

FE mesh generation guidelines depend strongly on the purpose of the analysis and the complexity of the design. Concerning the size of the models and the types of elements to use, typical guidelines are as follows:

- As a general rule, the number and complexity of elements have to be kept to the strict minimum for sufficient resolution and mesh size insensitivity. In a wider sense, it is advisable to generate simplified models that allow the capture of global behaviour with less CPU time and the possibility to run multiple-parameter sensitivity analyses. For local small-scale phenomena, such as stress concentration in interfaces, an efficient strategy is to develop separate decoupled models or employ substructures, condensation and matrix models.
- Static analyses often require higher resolution than low- to mid-frequency dynamic response analyses, although no separate FE models are customarily generated in the industrial environment for cost reasons. This translates into higher CPU time and memory needs for the dynamic analyses, compromising in some cases the capabilities of computers and software.
- The 3D solid elements introduce an explosion of DOFs compared with 2D plate elements. Whenever possible, 3D elements should not be used to analyse structures such as thin plates and shells. The exception may be the detailed thermoelastic distortions for ultrastable applications or cases in which accuracy in the

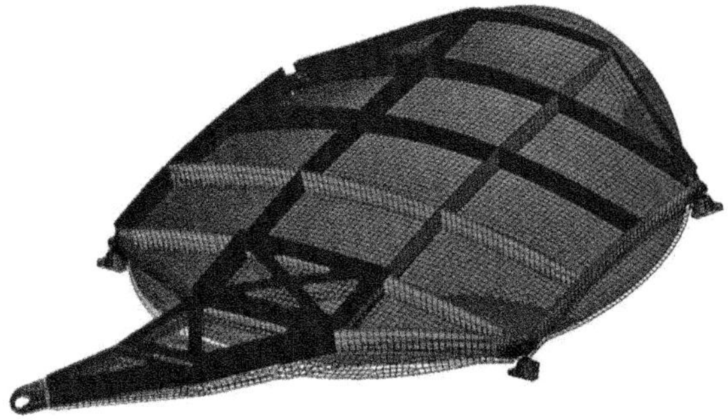

Figure 5.16 *Ka-Sat 2.6 m antenna reflector FE model. First eigenmode shape superposed on FEM mesh. Courtesy of EADS-CASA.*

transverse direction is required. This is the case, for example, for the thermoelastic FE model of the SICR reflector shown in Figure 5.17 (see also Figure 5.18).

- For plates, quadrilateral elements offer more accuracy than triangular elements and reproduce the stiffness better. The use of triangular elements should be minimized strictly to transitions and filling of irregular geometries, but if possible eliminated.
- For acoustic analyses, the maximum element size on wetted surfaces should be less than a quarter of the shortest wavelength of the input load. Due to the relevance of the low-frequency response, detailed modelling of core materials and skin anisotropy is not necessary. It suffices to properly consider the equivalent shell bending behaviour. Membrane, shell–membrane coupling and shear strain energy are not considered mandatory for the strength analysis.

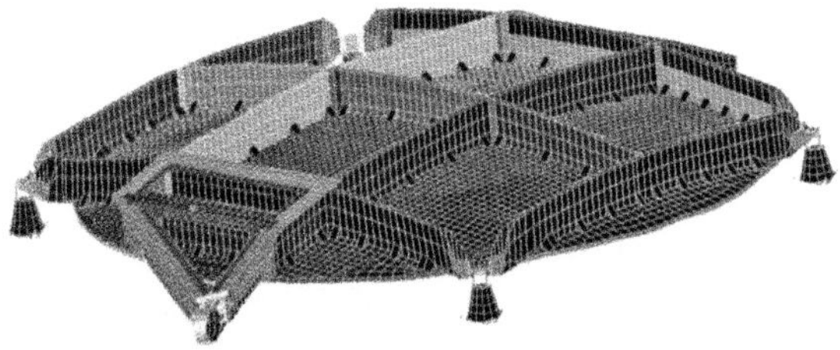

Figure 5.17 *Thermo-elastic FE model of SICR 2.4 m shaped Ka-band reflector with 3D core elements. Courtesy of EADS-CASA under ESA ARTES contract.*

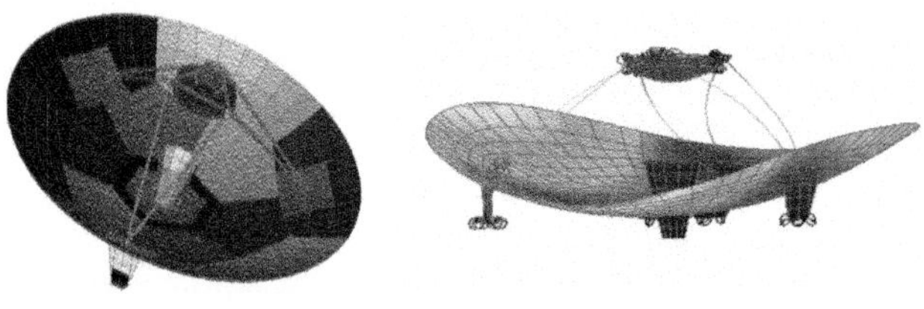

Figure 5.18 *FE model of the Mars Express HGA and first mode shape plot. Courtesy of EADS-CASA under ESA contract.*

FE model quality checks have been standardized in order to guarantee the accuracy and convergence of the analyses. The following sequence of checks is from ECSS-E-ST-32-03C [9]:

- Model geometry (free nodes, coincidences, free edges).
- Model topology (warping of shell elements, normals consistency, aspect ratio).
- Convergence of results to mesh elements size.
- Rigid body motion mass matrix and strain energy check.
- Free–free eigenfrequencies ratio to first elastic mode $< 10^{-4}$.
- Unit static load check versus constraint reactions.
- Residual work (ε) ratio shall be less than 10^{-8}.
- Zero-stress check under thermal loads, dummy isotropic material properties and isostatic boundary conditions. Typically for aluminium and $\Delta T = 100\,\text{K}$, a maximum residual stress of 0.1 MPa shall appear.

From a programmatic point of view, major difficulties for the structural analyses arise when specifying interface loads and boundary conditions of the antenna subsystem, which is an output of system-level analyses involving spacecraft/payload and launcher coupling. These specifications should be frozen if possible at the antenna preliminary design review (PDR), although often suffer updates later on. Prior to tests, analytical verification, based on the FE analysis, consists typically of the following sequence:

- Eigenfrequencies and modes for several boundary conditions (free–free, stowed, deployed).
- Quasi-static launch loads.
- Low-frequency (100 Hz) sinusoidal vibration with notching profile if applicable.
- Random vibration (2000 Hz).
- Acoustic load case.
- Thermoelastic stresses (stowed and deployed) and distortions (deployed).
- Transient loads (deployment release and end shocks, launcher- and spacecraft-induced shocks).

These analyses are normally linear, but also can be nonlinear and involve pre-tension in parts of the system, coupled thermal–structural loads and coupled fluid–structure interactions. Shock analysis is rarely performed due to the large uncertainties in parameters that control high-frequency wave propagation. Qualification of the shock load environment by heritage or test is the normal approach.

5.3.10 Acoustic Loads on Antennas

The acoustic load case deserves special attention since it is the sizing one in relation to structural strength for most antennas of large dimensions. Antenna structures experience intense vibration levels during launch in response to the acoustic field inside the launcher fairing. This acoustic field originates from the rocket engine noise and emissions, the turbulent mixing of jets in the atmosphere and the reflections of sound on the ground and launch pads. It is a random and broadband load, typically assumed to behave as a stationary diffuse field with a white noise spectrum within each octave. The main physical parameters playing a role are the dimensions of the body blocking the acoustic waves, the spatial correlation between the acoustic waves and the natural mode shapes, the mass, structural damping and stiffness of the structure, and the boundary conditions [10,11].

Antenna structures are especially sensitive to the acoustic load case due to their shape and dimensions, as well as their stiffness and low mass and damping. In other words, their configuration makes them excellent acoustic receivers and hence their dynamics is highly excited in acoustic wave fields. Plates of more than 1 m are able significantly to block the pressure field by scattering. As a consequence vibrations are excited and the wetted surfaces become acoustic radiators in their natural modes of vibration, modifying in turn the impinging wave field. This coupling effect, together with the wide-band random nature, the formation of acoustic cavities and the nonlinearity of the damping, are the main reasons for the difficulty in accurately predicting the vibro-acoustic response. In addition, due to the fact that the fundamental natural frequencies of the plate or shell in launch configuration are typically in the frequency range of 30–100 Hz, the stress levels experienced by the materials are significant and very often close to the strength limit. In relation to the frequency range of excitation, which exceeds 10 kHz for most launchers, the highest pressure load appears in the low-frequency bands (31, 63, 125, 250, 500 Hz octaves). Experience shows that the accumulated RMS acceleration, and especially the force response, is covered typically by more than 95% within those octaves. This allows use of deterministic tools for the analysis of pressure distribution, as well as acceleration, force and stress/strain responses.

The formulation of the vibro-acoustic problem for a structure immersed in an infinite fluid domain involves a coupled solution of the equations for fluid variables (only fluid pressure p for an ideal inviscid potential flow) and structural variables (stress tensor σ or displacement vector $\mathbf{w}$). The system of equations to be solved is normally expressed as [12]

$c^2\nabla^2 p - \ddot{p} = -\rho_0 c^2 \dot{q}$	Wave equation in the time domain with distributed sources
$\left.\begin{array}{l}\nabla\cdot\sigma - \rho_s\ddot{\mathbf{w}} = 0 \\ \sigma_{ij} = C_{ijkl}\dfrac{\partial w_k}{\partial x_l}\end{array}\right\}$	Elastic solid momentum and constitutive equations
$\left.\begin{array}{l}\dfrac{\partial p}{\partial n} + \rho_0\ddot{w} = 0 \\ \sigma_{ij}n_j + pn_i = 0\end{array}\right\}$	Solid/fluid interface continuity
$\lim_{r\to\infty} r\left(c\dfrac{\partial p}{\partial r} + \dot{p}\right) = 0$	Sommerfeld radiation condition (non-reflection at infinity)

Implicit assumptions in this formulation are the ideal, linear and isotropic behaviour of waves in air, as well as the continuity towards the surface of the solid. The structural dynamics could be considered nonlinear if necessary, which is not the case for typical space antenna structures. This system of differential equations can be solved numerically with the FEM applied to both domains, the fluid and the structure discretized in FEs. The same formulation can be used for interior problems.

Furthermore, the wave equation is customarily transformed into the frequency domain giving the Helmholtz equation $\nabla^2 P + k^2 P = -4\pi F(\mathbf{r})$, where $k = 2\pi/\lambda$ is the wavenumber in air, with Dirichlet ($P = P_0$) or Neumann ($\partial_n P = N_0$) boundary conditions. The solution can be expressed in terms of Green's function $G(\mathbf{r}|\mathbf{r_0})$, representing the field at $\mathbf{r}$ due to a unit source at $\mathbf{r}_0$. By applying Green's theorems, the problem can be written in the form of a surface integral, leading by discretization and interpolation to the boundary element method (BEM) for the fluid-domain surfaces, as an alternative to the FEM applied to the whole fluid-domain volume. This finds special interest when solving exterior problems, as is the case for most antennas. For more details on the numerical solutions see [12,13]. Commercial software tools can be classified as:

non-coupled solution (rigid scattering);
coupled solution FEM (fluid)/FEM (structure);
coupled solution BEM (fluid)/FEM (structure).

The numerical integration schemes are built either in the frequency or time domains. The most efficient fluid–structure coupling methods exploit the structural eigenmodes for projecting the fluid unknowns onto the surface and thus reducing the size of the problem. Time-domain solutions can be used for transient analysis, for example blast waves.

In cases where the structural modal density becomes too high for deterministic methods, statistical energy analysis and hybrid methods [14] provide approximate solutions in the high- and intermediate-frequency ranges, at the expense of spatial resolution and accuracy of the stress and force predictions. These are governed by the low-frequency response, where modal density typically does not allow the drawing of statistics and detailed local stress predictions require a fine resolution, which is not possible with statistical tools.

As an engineering method, the criticality of an antenna structure can be assessed with a small number of merit figures: the cutoff frequency of the blocked pressure, the fundamental eigenfrequencies of the structure and the joint acceptance of the mode shape to the pressure distribution. It is possible to provide a simplified and sufficiently conservative representation of the actual pressure distribution on the surface of a plate in a diffuse field [11]. To a first approximation, a fully coupled fluid–structure interaction analysis is not necessary. In particular, the acoustic radiation of the plate can often be neglected while retaining other effects such as the air mass loading and the rigidly scattered pressure distribution on the surfaces. This simplification only affects some of the frequency response peaks by slightly overestimating amplitudes while correctly capturing the frequency-dependent force input.

The scattering of plane waves by rigid plates can be obtained with Rayleigh's simplified integral or by the exact solution with Green's formulation. The circular and square plates immersed in plane-wave fields have been extensively studied [15,16]. In the limiting case of short wavelengths compared with the plate size ($ka \gg 1$, with k the wavenumber and a the plate radius), $p/p_\infty = 2$ on the illuminated side (perfect reflection) and $p/p_\infty = 0$ on the shadow side (perfect shadow). This crude constant pressure description, also known as Kirchhoff's approximation, only gives an indication of the pressure blocking of large, rigid flat plates. The approximation using Rayleigh's integral (distribution of point sources on the surface behaving as if located within a rigid infinite baffle) [17]

$$p(|\mathbf{r}-\mathbf{r_0}|, t) = \frac{\rho}{2\pi}\int_S \frac{j\omega v_n e^{j(\omega t - k|\mathbf{r}-\mathbf{r_0}|)}}{|\mathbf{r}-\mathbf{r_0}|}\, dS \tag{5.12}$$

provides solutions for plane waves of any incidence with a fair description of rigid scattering phenomena. In particular, for a plane wave impinging normally onto a circular rigid plate of radius a, the pressure at the centre is

$$p/p_\infty = \sqrt{(2-\cos ka)^2 + \sin^2 ka}$$

The solution of Rayleigh's integral for general incidence angle and shape of the plate is provided in Wiener's classic paper, together with experimental results. All solutions depend on the similarity parameter $ka = 2\pi fa/c$, which allows for generalization of the solution to any plate size. In addition, it can be seen that for $ka \ll 1$, the plate does not block the incoming wave and hence is transparent to it. Concerning the pressure distribution along a radius, it is maximum at the centre within the frequencies of interest, and drops to the boundary condition $p/p_\infty = 1$ at the edge. The shape of the curve changes as ka increases, approaching the constant pressure limiting value. The obvious analogies with optics can therefore be drawn.

Pressure distribution curves in a diffuse field can be synthesized by superposition of a uniform distribution of plane waves in the 4π steradian space. In Figure 5.19, the pressure curve at the centre of the circular plate for several incident angles of plane waves and their combination into a diffuse field (simulated with 26 plane waves uniformly distributed) is provided as a function of the non-dimensional frequency parameter ka.

The mechanical load input on a thin plate is in fact the pressure jump across the faces. For the diffuse field it has been observed by numerical analysis and measurement on typical antenna plate structures that the pressure jump is close to duplicating the pressure power spectral density (PSD) level on each side of the plate, above a certain frequency. In the low frequencies, the phase difference of the pressures tends to zero, as well as the pressure jump. Figure 5.20 shows the pressure jump on a thin, circular rigid plate at the centre for a unit white noise diffuse field. As can be seen, the interaction occurs mainly at frequencies higher than the so-called cutoff frequency, f_{co}. No significant pressure jump builds up below f_{co} and hence the dynamic excitation of the structure there will be low. Above f_{co}, the distribution of the load input evolves from zero at

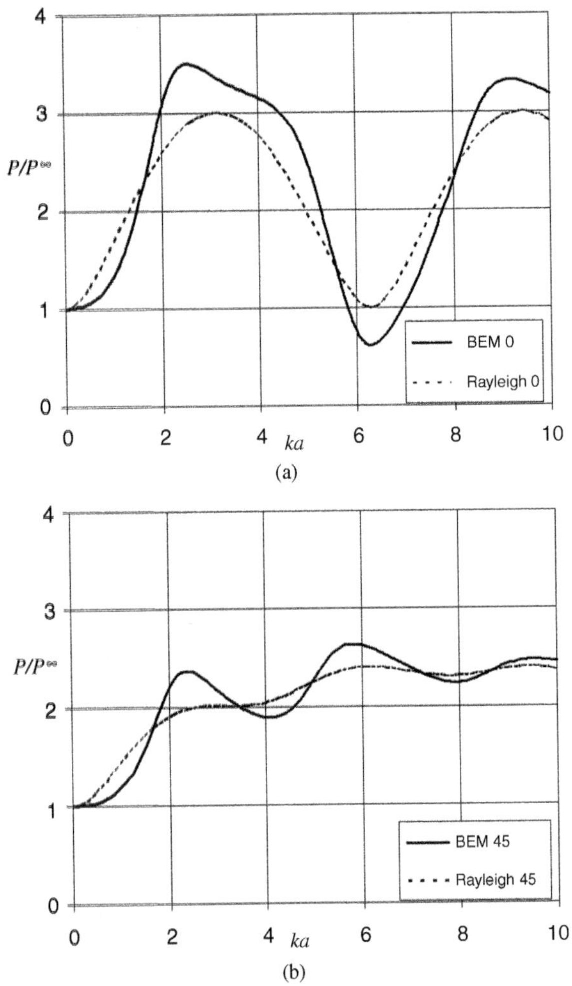

Figure 5.19 *Pressure ratio at the centre of a rigid circular plate of radius a computed with BEM (RAYON-3D) and the Rayleigh integral for several incident angles with respect to the normal and the diffuse field. Harmonic unit input considered in all cases.*

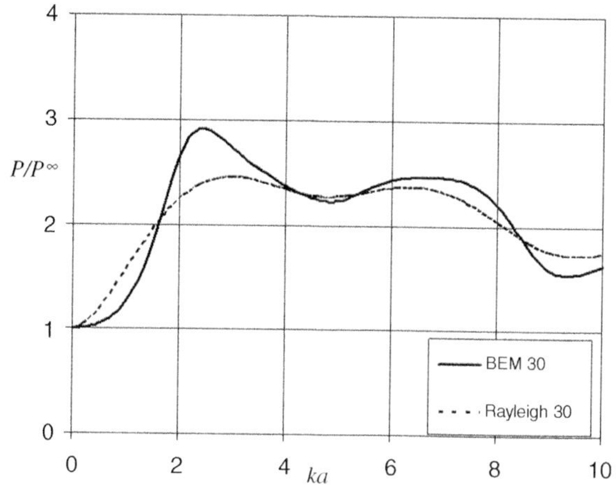

(c)

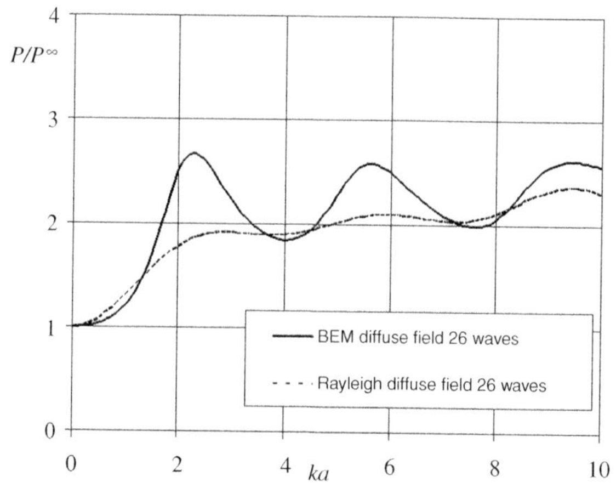

(d)

Figure 5.19 *Continued.*

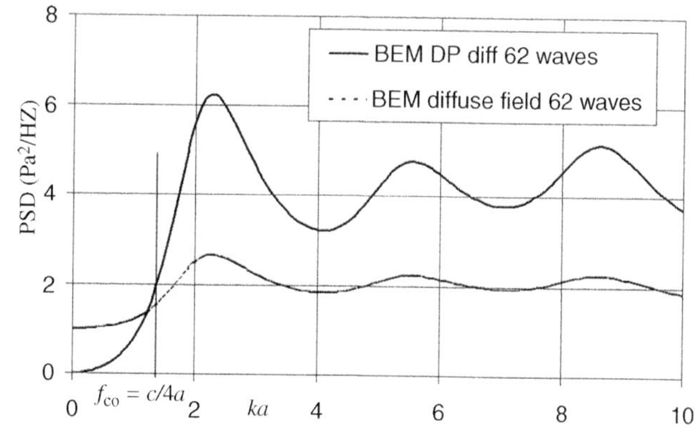

Figure 5.20 *Pressure and pressure jump PSD for 1 Pa^2/Hz white noise input at the centre of a thin and rigid circular plate of radius a. BEM (RAYON-3D) computation.*

the edges to a pressure jump about four times the value of the acoustic diffuse far-field input. The same conclusions can be deduced for square and rectangular plates, taking as reference dimension L the longest of the sides. It is concluded that the cutoff frequency that establishes the threshold of pressure loading of the panel is defined in terms of the largest dimension of the structure receiving the waves, L, and is well approximated by $f_{co} = c/2L$.

This behaviour of the pressure on the surface of plates has been determined numerically as well as experimentally [10,18], although surface pressure measurements are very scarce. Pressure relaxation due to the elasticity of the plates and the transmission of acoustic waves through the thickness can often be observed. Nevertheless, it can be accepted that the rigidly blocked pressure gives a conservative estimate. For instance, in Figure 5.21, the rigidly blocked pressure at the centre of a circular plate does not differ significantly from the experimental curve and the computation with BEM including acoustic radiation. In fact, the radiation results in an overall reduction of the blocked pressure. The measurement of pressure, taken in this case with a

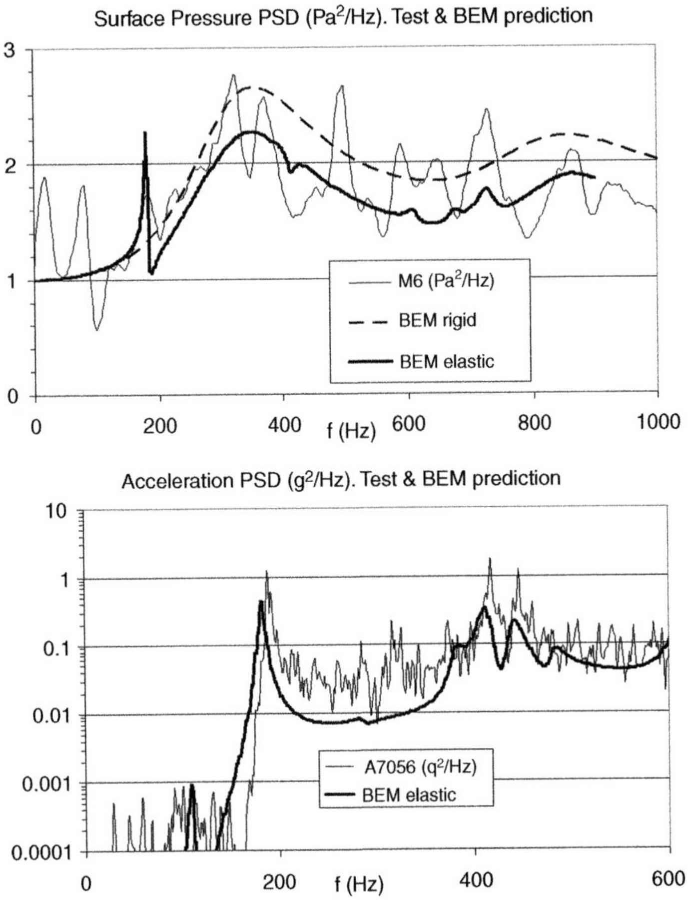

Figure 5.21 *Experimental and numerical results (RAYON-3D) at the centre of a flat circular CFRP plate of 700 mm diameter. Unit white noise input. Tests performed in the 200 m^3 IA-CSIC reverberating chamber. Test data courtesy of EADS-CASA under ESA contract.*

microphone placed 50 mm from a CFRP plate of 700 mm diameter, shows large peaks with respect to the smooth curve of the BEM computations, which are due to the modal behaviour of the reverberating chamber and the plate structure. The chamber cavity modes are clearly visible below 100 Hz, where the diffusion of the field is not guaranteed. Meanwhile, the fundamental structural mode of the plate, near 200 Hz in this case, has a strong radiation efficiency and hence leads to a relaxation of the surface pressure. This is well captured by the BEM computation. The amplitude of the acceleration response seems slightly underestimated by the numerical prediction. This is a typical situation, a consequence of an overestimation of the structural damping and the inertia of the instrumentation. The overall acceleration response is, however, well determined by the BEM elastic computation, typically not an easy task given the variability of the reverberating field input (1–2 dB depending on the octave band) and the structural parameters. The results shown in the figure serve to illustrate the usefulness of the methodology described in the previous paragraphs, based on using the rigid scattering as a first approximation to the acoustic load input of antenna structures.

The dynamic response of a structural normal mode can be highly excited if the eigenfrequency falls within the frequency range of high pressure loading and the mode shape is such that it correlates with the pressure distribution on the surface. This latter effect is quantified by the joint acceptance functions (JAFs) of the modes to the acoustic waves, defined as

$$j_{nn}^2(f) = \int_{A_1}\int_{A_2} \phi_n(y_1)\phi_n(y_2)\rho(y_1, y_2, f)\, dS_1\, dS_2 \tag{5.13}$$

where $\rho(|y_1 - y_2|, f) = \sin(k|y_1 - y_2|)/(k|y_1 - y_2|)$ is the diffuse field cross-correlation function, which depends on the distance between the two surface points and the frequency. These functions can be computed numerically for every structure and mode. Normally only autocorrelations are significant. In the case of typical structures such as beams, cylinders and plates, tabulated data are available [19]. As a general rule, mode shapes that displace large effective volumes of air show a high JAF value over a wide frequency band and thus respond actively in a diffuse field. This is illustrated in Figure 5.22 for the case of the 2.6 m Ka-Sat reflector, showing the pressure RMS value computed over the external surfaces with BEM (VA-One RAYON solver). Hence, the so-called drum mode can be expected to show a high response.

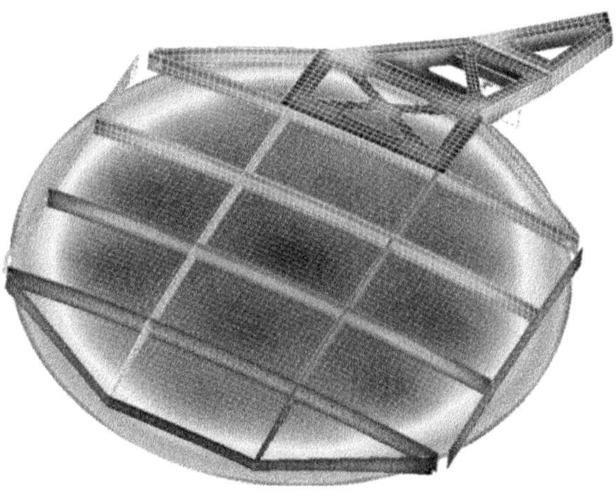

Figure 5.22 *Ka-Sat 2.6 m antenna reflector. Acoustic RMS pressure computed with BEM VA-One. Courtesy of EADS-CASA under ESA ARTES contract.*

5.4 Thermal and Thermoelastic Analysis

5.4.1 The Thermal Environment of Space Antennas

Antenna reflectors, together with solar arrays and deployable booms, are the spacecraft appendages most exposed to the space environment, with the added complexity of surface stability and pointing requirements. In order to illustrate the thermal design and analysis of typical antenna structures, the physics of the problem is treated in this subsection through a simplified case: the flat plate in a non-perturbed space environment (no planetary or spacecraft interaction modelled). Let us consider the symmetric sandwich plate of Section 5.3 made of two aluminium face skins, $t_s = 1.25$ mm thick, and a 6–20 core of $h_c = 25$ mm. Both sides are painted with standard space-qualified white paint, for example SG-121FD. The case of normal Sun incidence onto one face skin is considered, while the other is facing deep space. The steady-state equilibrium temperatures are calculated by solving the system of thermal balance equations of the two skins:

$$\left.\begin{aligned} P_1 &= \dot{Q}_1^{rad} + \dot{Q}_c \\ \dot{Q}_c &= \dot{Q}_2^{rad} \end{aligned}\right\} \tag{5.14}$$

where:

$P_1 = G_s\alpha_s A$	Input solar flux to the front skin
$\dot{Q}_1^{rad} = \epsilon_1\sigma(T_1{}^4 - T_0{}^4)A$	Radiation emitted by the front skin (hemispherically)
$\dot{Q}_2^{rad} = \epsilon_2\sigma(T_2{}^4 - T_0{}^4)A$	Radiation emitted by the back skin (hemispherically)
$\dot{Q}_c = C_s(T_1 - T_2)$	Conductive heat flux from the front to the back skin

This is in fact a two-node thermal model (Figure 5.23). For node 1 (the front skin), the impinging solar radiation must be balanced by the heat radiated by the plate to the front hemispherical space, plus the heat transported inwards by the core. The whole plate is assumed to have a uniform in-plane temperature distribution. Similarly, the back skin receives heat through the core and also radiates into the other half space. This results in the system of nonlinear equations

$$\begin{aligned} G_s\alpha_s &= \varepsilon_1\sigma\left(T_1^4 - T_0^4\right) + \frac{C_s}{A}(T_1 - T_2) \\ \varepsilon_2\sigma\left(T_2^4 - T_0^4\right) &= \frac{C_s}{A}(T_1 - T_2) \end{aligned} \tag{5.15}$$

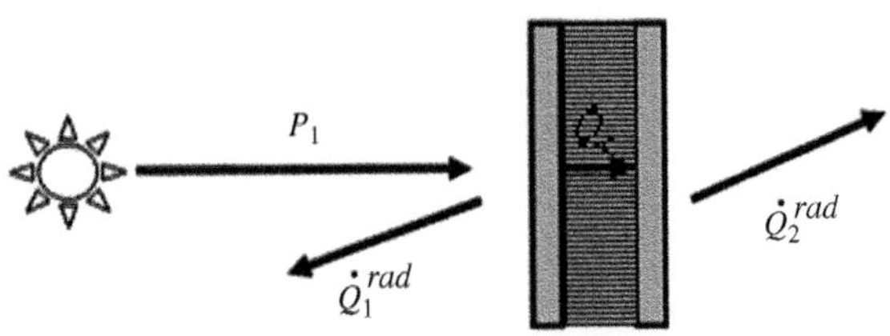

Figure 5.23 *Two-node thermal model of flat sandwich plate with normal Sun incidence.*

The parameters considered for the white paint are: solar absorptivity (BOL) $\alpha_s = 0.2$, infrared emissivity $\varepsilon = 0.8$. The Stefan–Boltzmann constant is $\sigma = 5.67 \times 10^{-8}$ W/(m^2 K^4). This set of equations fully describes the thermal environment of the plate.

5.4.2 Transverse Thermal Conductance Model of the Sandwich Plate

The 1D Fourier's law

$$\dot{Q}_c = -kA\frac{dT}{dx}$$

leads to the finite difference formulation of the conductive heat coupling between two thermal nodes

$$\dot{Q}_{c,1-2} = -\frac{kA}{L}(T_2 - T_1)$$

The thermal conductance, $C_L = kA/L$ (W/K), is the factor governing conductive coupling in this case. The flux of thermal energy can be controlled by the foil material conductivity k, the heat path cross-sectional area A or the distance L.

Hence, the transverse conductance of a honeycomb core is defined by the heat path along the honeycomb walls in the normal direction to the plate (Figure 5.24).

For a hexagonal cell honeycomb, the effective cross-sectional area ratio is $1.5 \times t_c \times N$, with N the number of ribbons per unit plate side length, which is $N = 1/(w_c/2)$. The factor 1.5 expresses the compression of the ribbon, that is its fully developed length. This leads to the transverse conductance of the hexagonal honeycomb core:

$$\frac{C_{Z,core}}{A} = \frac{3kt_c}{w_c h_c} \tag{5.16}$$

where A is the total panel area, k is the thermal conductivity of the foil material (e.g. 117 W/(m K) for A5056), t_c the thickness of the foil (typically 10–100 μm) and h_c the thickness of the core. For the complete sandwich, the overall conductance has to take into account the combination of skins, adhesive and core. For this purpose an electrical analogy is applied, representing the heat transfer from skin to skin by resistors in series (Figure 5.25):

$$\frac{1}{C_{Z,sandwich}} = \frac{2}{C_{Z,skin}} + \frac{2}{C_{adhesive}} + \frac{1}{C_{Z,core}} \tag{5.17}$$

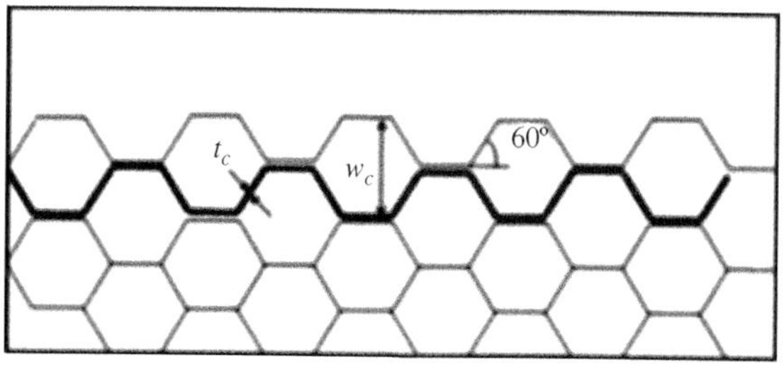

Figure 5.24 *Honeycomb core parameters for conduction model.*

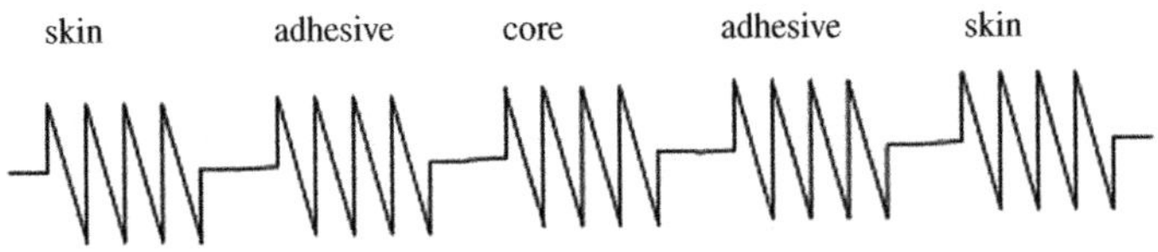

Figure 5.25 Thermal conductance model of sandwich plate.

In general, the dominant term is the resistance of the core since the contact fillet of the adhesive generally provides a good heat path when proper bonding is achieved, as can be seen in Figure 5.26.

More detailed core thermal models are available, such as the semi-empiric Swann and Pittman model and its variants, which include radiation exchange within the cells due to the temperature difference of the cell faces [20]. In practice, it is often possible to linearize the radiation effects inside the sandwich cells and employ an effective conductance.

5.4.3 Thermal Balance of the Flat Sandwich Plate

It is of great interest to study the effect of the core thickness at the equilibrium temperatures. This is shown in Figure 5.27, where variation of the thickness is reflected as variation of the core conductance by the relation $C_s \propto 1/h_c$.

The two limits of the behaviour shown in the figure serve as initial guesses for the numerical solution of the nonlinear system of equations. The limit of high conductance is equivalent to collapsing the two thermal nodes into one, which results in a single equation that can be solved directly: $G_s\alpha_s = 2\varepsilon_1\sigma(T_1^4 - T_0^4) \Rightarrow T_1 = 236\,\text{K}$. Incidentally, this equation also serves to estimate the heat fluxes in the nominal situation (by applying the equilibrium of node 2). Therefore $\dot{Q}_c = \dot{Q}_2^{rad} \approx 140\,\text{W}$, and $T_1 - T_2 \approx G_s\alpha_s/(2C_s/A)$, which yields $T_1 - T_2 \approx 7.8\,\text{K}$ for $C_s/A = 18\,\text{W/(m}^2\,\text{K)}$, the 25 mm thick core conductance. It can be seen that these estimates are not far from the exact solution of the thermal balance equations: $T_1 = 239.3\,\text{K}$, $T_2 = 232.0\,\text{K}$, $T_1 - T_2 = 7.3\,\text{K}$.

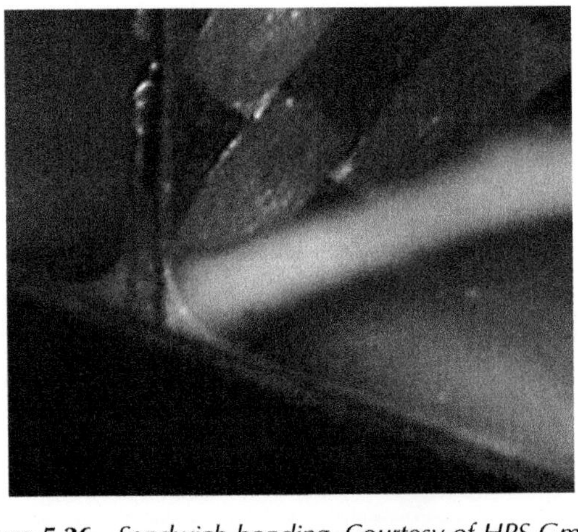

Figure 5.26 Sandwich bonding. Courtesy of HPS GmbH.

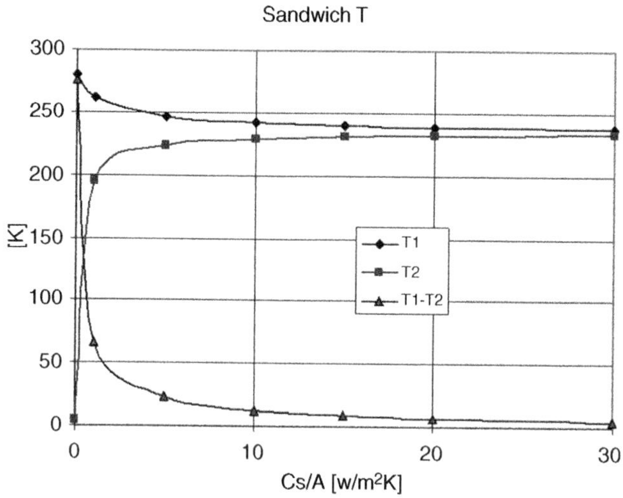

Figure 5.27 *Equilibrium temperatures for the skins as a function of core conductance.*

The limit of negligible conductance also gives interesting information as an estimate of a hot case: $G_s\alpha_s = \varepsilon_1\sigma(T_1^4 - T_0^4) \Rightarrow T_1 = 280\ \text{K}$. In this limit, the back skin would be in equilibrium with deep space, therefore $T_2 = T_0 = 4\ \text{K}$, which is obvious and does not give much new information. These temperatures for the hot case are not yet the maximum expected values, since the radiation influx model should include planet albedo and IR, spacecraft heat fluxes and Sun reflections, heat dissipation due to RF losses and feed chain or other units if mounted on the plate.

The reality of the antenna's thermal environment is far more complex than the simple isolated flat plate, due to the curvatures of reflectors and interactions with other surfaces of the antenna and spacecraft. Inclined Sun incidences create lateral illumination, as well as shadows and reflections, producing significant thermal gradients in all directions. Therefore, it is of utmost importance to model accurately all thermal cases and map thermal model nodes into structural FE nodes, with the required spatial resolution for capturing gradients in all directions. Furthermore, the time response is also relevant. This refers to the seasonal variations, but especially eclipses and shadows from spacecraft elements, which can create thermal transients of typically more than 100 K in a matter of minutes. It should be noted that, due to the low mass and low heat capacity of the materials employed, thermal transients have an immediate effect on temperatures. Proper thermal insulation is typically required for reducing the effects of transients and gradients.

5.4.4 Thermal Distortions of a Flat Plate in Space

Considering the temperatures derived in the previous analysis of the isolated flat plate, it is possible to perform a simple estimate of the thermal distortions generated by the transverse thermal gradient. The sandwich plate, when isostatically supported, becomes uniformly curved into a spherical surface due to the expansion of the hot face skin in relation to the cold one (Figure 5.28).

Hence, at first sight the out-of-plane distortion depends on the core height and the thermal expansion coefficient of the skin material. However, as shown, the transverse thermal gradient also depends on the core height, among other parameters. It is possible to assess the distortion as a function of h_c, the thermal and thermo-optical parameters of the materials. Using the approximations of the thermal gradient and the

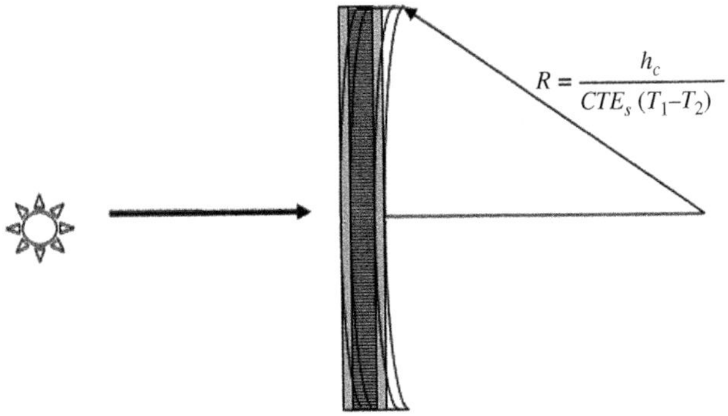

Figure 5.28 *Distortion of flat sandwich plate due to transverse thermal gradient.*

dependence of the core conductance on the core parameters, a rather simple result is obtained, namely the independence of the distortion on the core height to a first approximation:

$$R \approx \frac{6kt_c}{\mathrm{CTE}_s G_s \alpha_s w_c} \tag{5.18}$$

For aluminium skins (CTE $= 23 \times 10^{-6}$/K) and the 6–20 core, this results in $R \approx 140$ m. A more accurate analysis is shown in Figure 5.29, which employs the exact temperatures calculated in the previous paragraphs. Nevertheless, the variation of distortions with core height is still very smooth and does not reveal any instability. On the contrary, the value assessed above for R is reached asymptotically for the high-conductance core.

This first-order approximation is, however, not that accurate for the distortions. The combination of second-order effects is such that the first-order balance is not reached in most cases. The reason is that thin shells are very sensitive to imperfections and perturbations and, if proper boundary conditions are not implemented, the distortions are not negligible. Thermal gradients in the plane belong to the most important perturbations to this idealized picture.

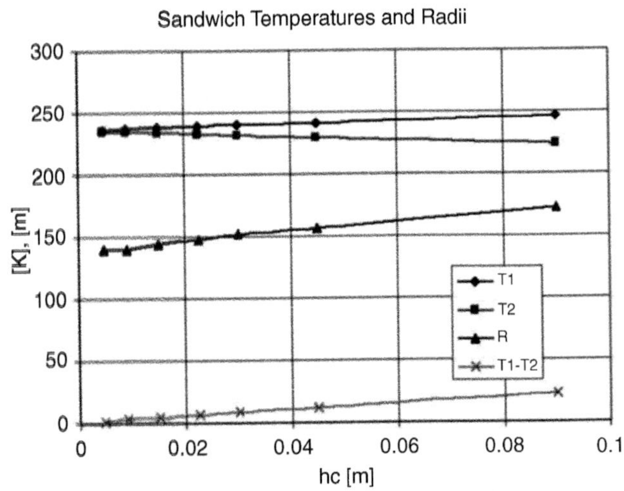

Figure 5.29 *Temperatures and curvature as a function of thickness.*

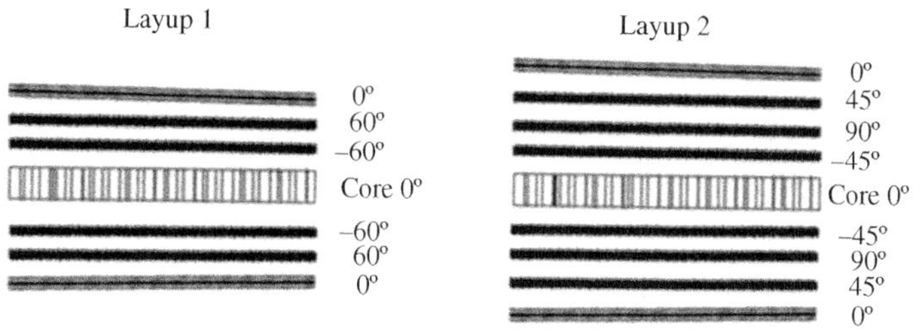

Figure 5.30 *Lamination of CFRP parabolic reflector.*

5.4.5 Thermoelastic Stability of an Offset Parabolic Reflector

As a further step, let us consider a typical offset parabolic reflector with $F/D = 1$, made of carbon fibre composite skins and an aluminium honeycomb core 5 mm thick. The boundary conditions represent the connection to an antenna deployment and pointing mechanism. The material properties for the CFRP and core (Figure 5.30) are given below.

CFRP lamina properties (in material axis):

Density	$\rho = 1800\,\text{kg/m}^3$	
Young's moduli	$E_1 = 270\,\text{GPa}$	$E_2 = 5.9\,\text{GPa}$
Poisson's ratio	$\nu_{12} = 0.30$	
In-plane shear modulus	$G_{12} = 4.1\,\text{GPa}$	
Coefficients of thermal expansion	$\alpha_1 = -0.96 \times 10^{-6}/°\text{C}$	$\alpha_2 = 38 \times 10^{-6}/°\text{C}$

Aluminium honeycomb core properties (in material axis):

Density	$\rho = 26\,\text{kg/m}^3$	
Young's moduli	$E_1 = 183.4\,\text{MPa}$	$E_2 = 155\,\text{MPa}$
Poisson's ratio	$\nu_{12} = 0.33$	
In-plane shear modulus	$G_{12} = 82.7\,\text{MPa}$	
Out-of-plane shear modulus	$G_{13} = 173.9\,\text{MPa}$	$G_{23} = 82.74\,\text{MPa}$
Coefficients of thermal expansion	$\alpha = 23 \times 10^{-6}/°\text{C}$	

The two considered in this study laminates result in a symmetric and balanced sandwich with a good approximation to the quasi-isotropic behaviour. However, each skin of laminate 1 is not symmetric. Two thermal loads are analysed with an FEM code, in order to assess the relative behaviour of the two different laminate configurations (Figure 5.31). The uniform case corresponds to 40 °C and the gradient case corresponds to a transverse temperature difference of 20 °C.

The influence of the core thickness on the peak distortion is shown in Figure 5.32. It can be seen that there is a minimum core thickness for which the thermal stability is well behaved. Lower values seem to indicate uncontrolled distortions in the gradient case, reflecting to some extent the complex interactions discussed in previous paragraphs.

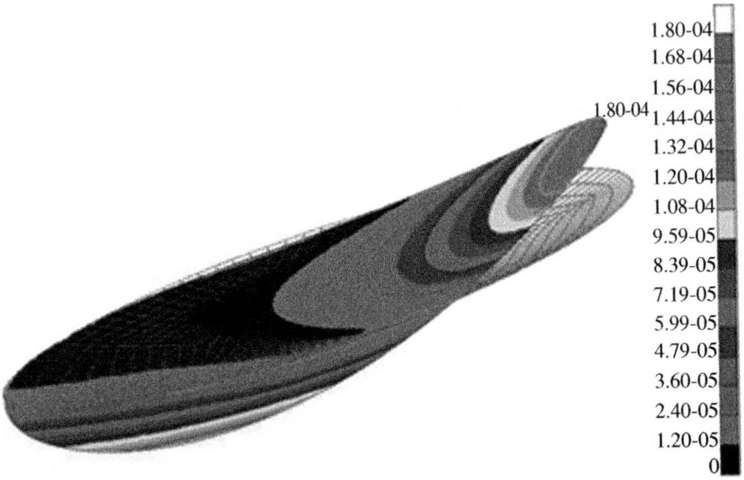

Figure 5.31 Distortions of 5 mm core CFRP skins (layup2) parabolic reflector with temperature transverse gradient case. Displacements in m.

The large number of parameters, the multiple configurations, mission profiles and the demanding stability requirements produce a very rich panoply for the thermal and thermoelastic design and analysis of space antennas.

5.4.6 Thermal Analysis Tools

A geometrical mathematical model (GMM) is typically established for the thermal analyses using commercial software like ESARAD, Thermica, Thermal Desktop or TMG (among others). The geometrical model is then used to calculate view factors, radiative couplings and orbital heating rates (solar, planet IR and albedo). These are then imported into the thermal mathematical model (TMM), which is typically established in a thermal network solver like ESATAN, Sinda, Samcef Thermal or TMG (among others).

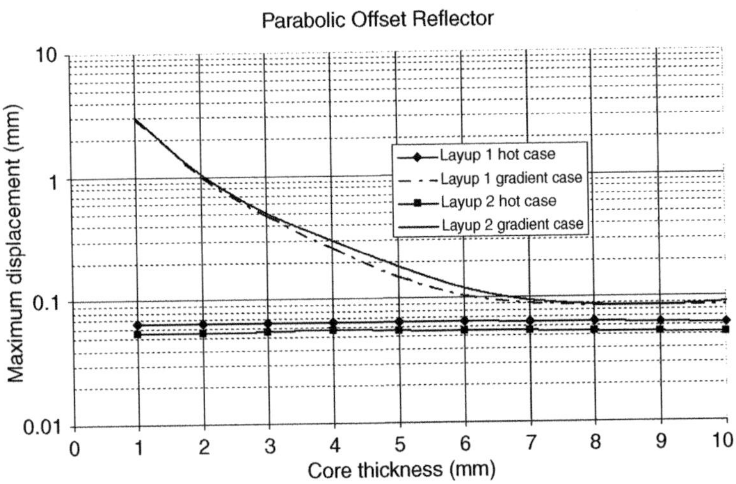

Figure 5.32 Sensitivity of peak thermal distortion to core thickness for given uniform temperature and thermal gradient cases.

5.4.7 Thermal Analysis Cases

A number of worst thermal cases are typically defined to analyse the thermal behaviour of the subsystem or component. These analyses are always transient, taking into account orbital conditions, seasonal variations and spacecraft configuration. For reflector antennas such cases typically include the following:

1. Cold cases (typically minimum Sun incidence or eclipse, low albedo, minimum dissipation).
2. Hot cases (typically maximum Sun incidence with Sun perpendicular to the dish, maximum dissipation or during maximum aero-thermal flux).
3. Maximum temperature gradient cases, determining the pointing accuracy (lateral Sun incidence, shadow or reflection).

5.4.8 Thermal Model Uncertainty and Margins

Uncertainties need to be considered on the calculated extreme temperatures (cold and hot case). Typically in the early project phases (up to phase B) an uncertainty of 10 K is applied in both directions. Once the thermal models are more evolved or correlated by a thermal balance test, this uncertainty can typically be reduced. The values to be applied have then to be derived by a dedicated sensitivity analysis during which the input parameters which are subject of uncertainty are varied within the relevant ranges. Such parameters are:

- Environmental parameters (solar incidence, planet IR radiation and albedo, sink temperature)
- Physical parameters (thermo-optical properties, thermal conductivities, specific heat capacity, contact resistances)
- Modelling parameters (view factors)

The relevant set of sensitivity analyses has to be defined case by case.

The 'calculated temperature range' extended by the model uncertainties is then called the 'predicted temperature range'. On top of this, typically an acceptance margin of 5 K is applied, leading to the 'acceptance temperature range'. Finally an additional qualification margin of typically 5 K is applied, leading to the 'qualification temperature range'.

5.5 Thermal Control Strategies

5.5.1 Requirements and Principal Design Choices

The primary task of antenna thermal control is to keep the temperatures of the antenna components within the temperature limits mostly dictated by the chosen materials, and to properly distribute thermal loads for minimizing thermoelastic stresses and distortions such that the stringent pointing, gain, sidelobe level, cross-polarization, C/I and overall RF requirements can be met. The chosen means of the thermal control of reflector antennas are typically fully passive and consist of multilayer insulation (MLI), thermal paints and/or coatings to adapt the thermo-optical properties of external and internal surfaces, and the choice of the structural materials with typically low CTEs and their differences.

A specific requirement for the antenna thermal design is that sunshields mounted in front of the reflecting or radiating elements need to be RF transparent. This excludes the use of metallized foils in these areas and needs to be taken into consideration in the choice of any surface finish or coating applied to reflectors or inside the horns or RF radiators.

Reflectors are often equipped with a white paint or coating in order to minimize heat absorption and to reduce temperature gradients due to non-homogeneous illumination or shadows. The rear side is often covered by MLI to avoid temperature gradients due to partial illumination or reflections from spacecraft surfaces. However, it has to be noted that, while the front side of the dish is exposed to the Sun, the mounting of MLI on the back will lead to higher temperature levels on the dish due to the missing possibility of heat rejection from the back. For several missions with high solar irradiation it might therefore become necessary not to use a full MLI on the rear side of the dish. On the other hand, for missions designed for a cold environment far away from the Sun, black coatings might be chosen to increase the amount of heat absorbed by the reflector assembly.

Typically, the antenna pointing mechanism is also insulated from the external environment using MLI. A dedicated radiator surface can be foreseen to reject the heat dissipated by the motors. Heaters (controlled by, for example, thermistors) might have to be installed to compensate for extensively rejected heat during cold cases (e.g. during eclipses). In special cases, particularly for missions exposed to a very hot external environment, heat pipes might be required to remove excessive heat from the mechanisms and keep them within their operational temperature limits.

The case of phased-array antennas is specific in the sense that every development requires a dedicated thermal control design. Thermal control of highly dissipating electronic units and RF components requires extensive use of heat pipes for heat transport, equalizing heat pipes, loop heat pipes, radiator surfaces and heaters, in addition to MLIs and sunshields. Massive support plates are sometimes used to serve as cold plates with high thermal inertia. The operating logic of active thermal control can become rather complex in relation to the operational, non-operational and survival modes.

5.5.2 Thermal Control Components

In the following, the essential thermal control components are briefly described, with emphasis on the antenna application. Further information can be found in the specialized thermal control literature [21,22].

5.5.2.1 *MLI*

MLI is composed of a number of thin, lightweight reflective screens typically coated on one or both sides with vapour-deposited aluminium or gold in order to maximize the reflection of thermal radiation in the IR bandwidth ($\rho > 0.97$). Depending on the expected temperatures, typically employed sheet materials are Mylar and Kapton, but also metal foils of aluminium or titanium alloys are used. In order to minimize conductive heat transfer the reflective screens might be crinkled or embossed/dimpled, leading to just point contact between the foils. Alternatively, the foils are separated by thin spacers made of, for example, Dacron or glass fibre fabrics or nets. The external space-exposed layer is typically coated in order to adapt its thermo-optical properties to the expected environment. The whole stack is then held together by either stitches or non-metallic sewing threads and attached to the spacecraft components by non-metallic stand-offs or Velcro. Electrical grounding typically needs to be foreseen to avoid excessive electrostatic charging.

The insulation performance of MLI blankets is often characterized by an effective emissivity, ε_{eff}, defined as

$$\varepsilon_{eff} = q/(\sigma T^4 - \sigma T_{inf}^4) \qquad (5.19)$$

where q is the heat leaking through the blanket (in W/m^2), T_{inf} is the external sink temperature and T the resulting inner side temperature. The insulation performance is usually verified by dedicated tests. However, since the performance is strongly dependent on the manufacturing quality and can significantly vary depending on the local configuration, adequate margins need to be applied in the thermal design analysis on the performance derived by test.

5.5.2.2 *Thermal Paints, Surface Coatings and Finishes*

Thermal paints and surface coatings are typically applied to tailor the thermo-optical surface properties such that the effect of the external environment on the respective components or subsystem is minimized.

On reflector dishes often white paints or white surface coatings are applied. They provide a low absorptivity of solar irradiation while having a high IR emissivity.

Black paints or coatings provide a high absorption of solar irradiation and a high IR emissivity. They are typically applied in areas without direct solar incidence in order to level out the temperatures by radiation. However, they might also be applied on reflectors operating in specifically cold environments, for example on missions to the outer Solar System.

Metallic surface finishes (like aluminium or germanium) typically provide a very low IR emissivity, but have a high ratio of solar absorptivity to IR emissivity. However, with vapour deposition techniques very thin coatings can be achieved such that the IR emissivity of the substrate becomes more dominant, allowing the overall IR emissivity to be tailored.

Table 5.6 provides an indication of the typical ranges of thermo-optical properties for some surface finishes.

Paints and coatings generally undergo ageing. In particular the solar absorptivity of white coatings often gradually increases under UV radiation and from particles (protons and electrons but also atomic oxygen in low Earth orbits). These effects are typically accelerated by elevated temperatures. In addition, the adhesion to the substrate might be degraded under extended exposure to the temperature and radiation environment. Extensive characterization and qualification testing are therefore inevitable, in particular for new materials or applications in environments not previously experienced.

Furthermore, the electrical conductivity of the paint may be required for limiting electrostatic charging. In this case galvanic potential differences, skin depth and passive intermodulation (PIM) requirements should be taken into account. On the other hand, non-conductive paints are a source of additional RF reflection losses of the order of 0.05–0.1 dB in the Ku-band. In all cases, thermal paints increase thermoelastic distortions, especially at low temperatures, due to the large CTE and increased stiffness.

5.5.2.3 *Heat Pipes*

Heat pipes are sometimes integrated in specific cases to remove the heat dissipated within the pointing mechanisms (motor and rotary joint dissipation) and electronic units. This is particularly the case when the specific geometrical design and/or the orbit orientations do not allow selection of a dedicated surface which can be used as thermal radiator in the near vicinity of the heat source.

Heat pipes allow relatively high amounts of heat with rather small temperature gradients to be transported. They are based on tubes made of aluminium with capillary wicks. A two-phase medium (typically ammonia) is used to transport the heat. The medium fills the wicks in liquid form. When heat is applied to one end of the pipe (the evaporator section), the medium will evaporate. The differential vapour pressure will push that vapour

Table 5.6 *Typical range of thermo-optical properties of surface finishes (indicative values).*

Surface finish	Solar absorptivity α_s (BOL)	IR emissivity ϵ	α_s/ϵ
White paints	0.15–0.25	0.8–0.9	$\sim$ 0.2–0.3
Black paints	$>$ 0.9	0.8–0.92	$>$ 1
Optical solar reflectors	0.05–0.16	0.7–0.8	$<$ 0.2
Vapour-deposited aluminium (VDA)	0.08–0.17	$\sim$ 0.04 (for thick coating)	$\sim$ 2–4
Vapour-deposited germanium	$\sim$ 0.5	$\sim$ 0.1 (for thick coating)	$\sim$ 5
Black Kapton with Ge coating	0.5–0.6	0.82	0.6–0.73
Bare CFRP	$\sim$ 0.92	0.74–0.8	1.1–1.2

away from the heated area towards the opposite end of the pipe (the condenser section). There the heat is extracted, which causes the medium to condense and to move back into the wick. However, this transport will now have caused a liquid pressure gradient in the opposite direction which forces the liquid to move through the wick back to the evaporator section. This process will be continuous as long as heat is applied to the evaporator end and extracted at the condenser end.

5.5.3 Thermal Design Examples

The following design examples are based on missions developed in recent years by the European Space Agency (ESA).

Example 1: SIRAL antenna subsystem on CryoSat Earth Explorer [23]

A specific example is the thermal design of the SIRAL instrument antennas on the CryoSat Earth Explorer satellite. CryoSat is a radar altimeter mission developed within ESA's Living Planet Programme with the aim of determining the thickness variations of the Earth's continental ice sheets and marine sea ice cover. Interferometry is performed using a second antenna to resolve steeper terrain slopes. After a launch failure in 2005, CryoSat-2 was developed and successfully launched in 2010. CryoSat-2 operates in a highly inclined low Earth orbit with its main instrument pointing to nadir.

The altimeter/interferometer SIRAL is composed of a Ku-band (13.7 GHz) antenna consisting of two centre-fed Cassegrain reflector systems, see Figure 5.2. The two identical antennas each have a main and subreflector assembly linked with struts, and a chain with feed assembly and waveguide.

From a thermal control point of view, the main design driver for the antenna system is the stability of the interferometric phase. Variations in the RF path length difference between both antennas will cause phase errors. Differences in the thermoelastic deformations of both antennas therefore need to be kept small compared with the wavelength. Extremely high thermal stability and high structural stiffness are therefore required.

As a consequence, the reflectors, struts and bench are made of ultrahigh-modulus CFRP skins and CFRP honeycomb. All feed elements are made in Invar providing high thermal stability due to its extremely low CTE. In order to decouple the reflectors from the external environment and to avoid Sun trapping, single-layer insulation (SLI) tents cover the antenna front sides (Figure 5.33). The SLI is made of black Kapton foil with a germanium coating on both sides, where the external side is optimized such that the ratio of solar absorptivity and IR emissivity is minimized. The active sides of the reflectors are metallized with a 1 μm VDA coating in order to minimize the heat absorbed by radiation from the SLI tent. For the same reason, the strut parts between the main reflector and the SLI tent are covered with VDA-coated Kapton tape. The reflector's rear sides and the base plate are left as uncoated CFRP providing high IR emissivity and thereby allowing heat exchange

Figure 5.33 SIRAL antenna subsystem of CryoSat under thermal tent. Courtesy of RUAG Space AB, Sweden, under ESA contract.

between both antennas to level out the temperatures. White paint is applied to the waveguides to equalize temperatures when only one waveguide is dissipating.

The described antenna thermal design is primarily aimed at minimizing the temperature differences between both antennas which occur particularly in orbits with the Sun incident on the side of one antenna. At the same time, the thermal design also makes temperature changes slower and more insensitive to changes in the external environment, and thereby minimizes temperature gradients over the reflector thickness which would otherwise lead to thermoelastic deformations.

Example 2: BepiColombo HGA in Mercury environment [24]

BepiColombo was selected by ESA as a cornerstone mission of the Horizon 2000 Programme and is under development in cooperation with ISAS/JAXA of Japan. Since Mercury is the innermost planet of the Solar System, the European MPO spacecraft will be exposed to extreme conditions with up to 14.500 W/m^2 of direct solar illumination at Mercury's perihelion. In addition, while passing over the planet's subsolar point, planetary IR of up to 5000 W/m^2 and albedo of up to 1200 W/m^2 need to be taken into account.

These extreme thermal conditions represent one of the main drivers of the design of the three antennas (HGA, MGA, LGA). Despite the application of white coatings, temperatures above 450 °C are expected, hence excluding the use of many classical space materials like carbon fibre epoxy composites and aluminium. Titanium has therefore been chosen for the symmetric Gregorian dual-reflector assembly (1.1 m diameter), the RF chain and parts of the pointing mechanism, providing a high limit temperature and a low CTE.

Several dedicated white thermal coatings have been developed, aiming at low solar absorption while providing high emission for thermal IR. In addition to the high operating temperatures and the required low RF losses, the main difficulty with the coating is its degradation under the very high UV radiation load combined with electron and proton radiation, leading to increased solar absorption.

A dedicated thermal shield made of titanium had to be designed to protect the feed horn from the accumulated heat due to reflections. Because of the high temperatures, a gold coating was chosen for the feed to reduce RF loss.

The antenna pointing mechanism, the waveguides, the antenna support structure and the release mechanism will all be insulated from the external environment by means of a specially developed high-temperature MLI. This HT-MLI contains several VDA-coated titanium and aluminium screens on the hot side, while VDA-coated Upilex-S layers are used in the colder areas. Special glass fibre spacers are used to separate the reflective screens. On its external side the HT-MLI contains a ceramic tissue (Nextel) which provides the required low solar absorption and high IR emission coefficient and was chosen because of its relative stability with respect to ageing under UV radiation at high temperatures.

For the pointing mechanism, special high-temperature hardware capable of operating up to 260 °C had to be selected. However, since the actuators dissipate more than 20 W, dedicated means for rejecting this heat had to be found. Classical radiator surfaces were found not to be feasible since they would lead to too high a heat loss during cold cases. It was therefore decided to transfer the heat generated by the mechanisms via axially grooved ammonia heat pipes towards the spacecraft's main radiator. The heat pipes are attached to the azimuth stage. In addition, a dedicated conductive link had to be introduced to link the elevation stage conductively.

Acknowledgements

The authors wish to thank the valuable contributions of ESA and industry colleagues during the preparation of the manuscript and prior supporting activities, among them I. Ngan, G. Gil, J. L. Riobóo, E. Checa, L. Scialino, M. Hamdi, V. Sohtell, V. Daniel, V. Gomez-Molinero, E. Ozores, O. Castro, J. J. Llorente, K. van't Klooster, L. Datashvili and S. Pellegrino. The authors also wish to thank the editors for their dedication.

References

1. Timoshenko, S. and Woinowsky-Krieger, S. (1959) *Theory of Plates and Shells*, 2nd edn, McGraw-Hill, New York.
2. Mindlin, R. (1951) Influence of rotary inertia and shear on flexural motions of isotropic, elastic plates. *Journal of Applied Mechanics*, **18**, 31–38.
3. ECSS-E-HB-32-20 (2011) Space Engineering – Structural Materials Handbook, European Cooperation for Space Standardization, ESA-ESTEC Requirements & Standards Division, Noordwijk.
4. Jones, R.M. (1975) *Mechanics of Composite Materials*, Hemisphere, New York.
5. Kueh, A. and Pellegrino, S. (2007) Triaxial Weave Fabric Composites, ESA Contractor Report WO-SMH-01.
6. Sullins, R.T., Smith, G.W., and Spier, E.E. (1969) Manual for Structural Stability Analysis of Sandwich Plates and Shells, NASA CR1457.
7. Allen, H.G. (1969) *Analysis and Design of Structural Sandwich Panels*, Pergamon Press, Oxford.
8. Zienkiewicz, O.C. and Cheung, Y.K. (1967) *The Finite Element Method in Structural and Continuum Mechanics*, McGraw-Hill, London.
9. ECSS (July 2008) *Structural Finite Element Models*, ECSS-E-ST-32-03C.
10. Riobóo, J.L., Santiago-Prowald, J. and Garcia-Prieto, R. (2001) Qualitative vibroacoustic response prediction of antenna-like structures during launch into orbit. Proceedings of the European Conference on Spacecraft Structures, Materials and Mechanical Testing, 2000, European Space Agency, ESA SP-468.
11. Santiago-Prowald, J. and Rodrigues, G. (2009) Qualification of spacecraft equipment: early prediction of vibroacoustic environment. *Journal of Spacecraft and Rockets*, **46**(6), 1309–1317.
12. Hamdi, M.A. (1988) Méthodes de discrétisation par éléments finis et éléments finis de frontière, in *Rayonnement Acoustique de Structures: Vibroacoustique, Interaction Fluide-Structure* (ed. C. Lesueur), Editiones Eyrolles, Paris.
13. Santiago-Powald, J., Ngan, I. and Henriksen, T. (2007) Vibro-acoustic analysis of spacecraft structures: tools and methodologies. Proceedings of NAFEMS Workshop on Modelling Vibro-Acoustics and Shock, NAFEMS and Coventry University.
14. Cotoni, V., Shorter, P. and Langley, R. (2007) Numerical and experimental validation of a hybrid finite element-statistical energy analysis method. *Journal of the Acoustical Society of America*, **122**(1), 259–270.
15. Leitner, A. (1949) Diffraction of sound by a circular disk. *Journal of the Acoustical Society of America*, **21**(4), 331–334.
16. Wiener, F. (1949) The diffraction of sound by rigid disks and rigid square plates. *Journal of the Acoustical Society of America*, **21**(4), 334–347.
17. Fahy, F. (1985) *Sound and Structural Vibration, Radiation, Transmission and Response*, Academic Press, London.
18. Santiago-Prowald, J., Ngan, I. and Rodrigues, G. (2009) Vibroacoustic and random vibration benchmarks. European Conference on Spacecraft Structures, Materials and Mechanical Testing, Toulouse, 15–17 September.
19. ESA (1996) Structural Acoustics Design Manual, ESA-PSS-03-204, ESTEC, Noordwijk.
20. Daryabeigi, K. (2001) Heat transfer in adhesively bonded honeycomb core panels. 35th AIAA Thermophysics Conference, Paper 2001-2825.
21. Karam, R. (1998) *Satellite Thermal Control for System Engineers*, vol. **181**, AIAA, Cambridge, MA.
22. Gilmore, D. (2002) *Spacecraft Thermal Control Handbook*, 2nd edn, vol. **1**, The Aerospace Press and AIAA, El Segundo, CA.
23. Honnen, K., Rauscher, U. and Woxlin, K. (2003) Thermal Design of CryoSat, the first ESA Earth Explorer Opportunity Mission, SAE 2003-01-2467.
24. Noschese, P., Milano,M., Zampolini,E. *et al.* (2010) BepiColombo mission to Mercury: design status of the high temperature high gain antenna, ESA 32nd Antenna Workshop, Noordwijk, The Netherlands.

6

Testing of Antennas for Space

Jerzy Lemanczyk[1], Hans Juergen Steiner[2], and Quiterio Garcia[3]

[1]*ESTEC–ESA, The Netherlands*
[2]*Astrium GmbH, Germany*
[3]*EADS CASA Espacio, Spain*

6.1 Introduction

For space antenna applications, the antenna engineer will not only be required to design an antenna to rigorous performance requirements, but also be faced with other very important issues. These include:

- Fitting the antenna into a confined 3D envelope which is sometimes smaller than the antenna itself. Hence the term *deployable* antennas and the need to test antenna deployment from its stowed launch configuration.
- It is essential that the antenna and its deployment mechanisms, if present, survive the heavy launch loads to perform in orbit.
- In-orbit performance can be compromised through thermal effects caused by large temperature variations or extreme cold for deep space antennas where RF performance testing is mainly done at ambient temperatures.
- Material selection and thermal shields and blankets are needed to maintain thermal stability. Low Earth orbit satellites have the added potential problem caused by the presence of atmospheric oxygen in the orbital path.

Therefore, in the context of space antenna applications, many disciplines are required and need to be put to the test before launch.

On satellites and on the antennas they employ, mechanical and material issues can become crucially important design drivers in addition to the required electrical performance. In this chapter, all testing with respect to antenna design and verification will be addressed, covering acceptance, performance and environmental aspects to meet launch survival and in-orbit performance requirements.

Space Antenna Handbook, First Edition. Edited by William A. Imbriale, Steven (Shichang) Gao and Luigi Boccia.

The structure of the chapter goes from unit model definitions and design verification to testing requirements for a realized design. To illustrate the various aspects, the chapter will present the development of a dual-polarized aperture-coupled patch antenna with integrated feed network used in the SMOS (The European Space Agency's Soil Moisture and Ocean Salinity mission launched in November 2009) synthetic aperture radiometric interferometer, which required very special design, performance and verification issues to be solved.

The chapter has three main sections covering antenna testing as a development and verification tool. The first section after this introduction will include definitions for the antenna model philosophy and the type of testing to be performed, either as a design verification or to verify the integrity of the electrical, mechanical and thermal models. The reason for this approach is to detect, as soon as possible, any deficiencies in the design which can be present in terms of RF design and selected materials, mechanical design and thermal design, before extensive testing on advanced models begins. The process starts with the task of verifying, through the correlation of measured data with the model predictions, the theoretical design models through to final performance verification before delivery to the launch site for integration on the launch vehicle.

The next section will present the different types of test facilities that would normally be used in the development of a space-qualified antenna. This will cover RF testing facilities for electrical performance and include near-field antenna measurement systems as well as far-field systems and compact antenna test ranges, vibration, shock and thermal vacuum facilities. The selection of techniques in the different disciplines will be a function of the performance requirements, launch loads and the in-orbit environment.

The deployment must be verified prior to launch and as deployment occurs in zero-*g* conditions, a zero-*g* deployment rig (MGSE) must be defined for this because the antenna's mechanical and structural properties are designed for a very low-gravity environment in order to minimize mass and related launch costs. However, this can place constraints on RF testing in particular.

Finally, to illustrate the topics raised in the preceding sections, the complete design, development and verification sequence used for the SMOS antenna elements (69 in total plus three spares) will be presented. This includes the selection of the test facilities to be used as a function of the design and in-orbit performance requirements.

6.2 Testing as a Development and Verification Tool

6.2.1 Engineering for Test

At a very early stage in a space project, all the defined requirements are summarized in a compliance table to indicate a convincing concept on how to demonstrate the entire performance after realization. All requirements are related to different levels of ranging of significance. The performance requirements relevant to the space mission should be tested at all levels of realization. In cases where recurring hardware is used, or if the design adaptation from a previous project is very small or not impacting performance, a reduced effort for testing is common. On the other hand, the level of confidence for very high reliability of the space components requires at least a performance check after each step of integration, handling or transportation.

6.2.1.1 Dedicated RF Interfaces, Test Couplers

During subsystem testing, the interface connector of the antennas will be available for testing the typical RF performance, such as radiation characteristics, gain, EIRP (Equivalent isotropically Radiated Power), group delay and PIM (Phase Intermodulation) products. After integration into the spacecraft, the antenna ports are connected as a last step to the transponder or associated spacecraft electronics. At this stage of antenna integration, the relevant satellite system functions can be measured with the antenna in the loop, for example

G/T, amplitude–frequency response (AFR), saturated flux density (SFD), system PIM. To be able to execute these tests, additional interfaces are necessary to provide access to the feeding section of the antenna. During the design phase, a suitable coupler needs to be implemented in the interconnecting link between the payload and the antenna. A typical candidate for waveguide interconnections is the so-called cross-coupler with a coupling factor between −20 and −30 dB. The coupler has to be designed under the constraint of avoiding any influence on the radiated signal. This means it must have very low insertion loss and excellent return loss performance. The test coupler must be calibrated very accurately as a part of the S/C antenna and must fulfil all qualification and acceptance levels because the antenna itself will be considered flight hardware. The calibrated test coupler provides the opportunity to verify the required acceptance test results and to revalidate the antenna performance up until final delivery, shortly before integration of the satellite onto the launcher.

6.2.1.2 Optical Interfaces

An additional important aspect during the design phase, which will have a strong bearing on testability, is the optical interfaces (mirrors and/or mirror cubes) mounted on all components of an antenna which have a clear impact on the accuracy of radiation direction, the so-called boresight of the antenna. As an example, deployable reflector antennas consist of a large feeding section with single or multiple feeds which together do not form a single-unit structure. The feed system and the reflector structure must be integrated during component and subsystem tests on a mock-up of the spacecraft or a representative structure whose sole purpose is to locate the feed and reflector correctly with respect to each other (Figure 6.1).

By means of optical targets like small mirrors or retro-reflectors, all the optical targets attached to each subassembly must be measured by using theodolites or laser trackers to locate them all with six degrees of freedom with respect to a reference optical mirror cube defining the entire antenna assembly. During the RF testing, the optical equipment generates the correlation between the antenna components and the coordinate system of the RF test facility, which can also be defined using optical interfaces. By applying triangulation, the optical boresight can be correlated to the RF boresight. For this process, the transformation S/W from the laser tracker suppliers can usually be used. This optical data will be used during integration of the antenna onto the spacecraft structure to ensure that the antenna boresight fits the attitude and orbit control system of the fully integrated and operational satellite.

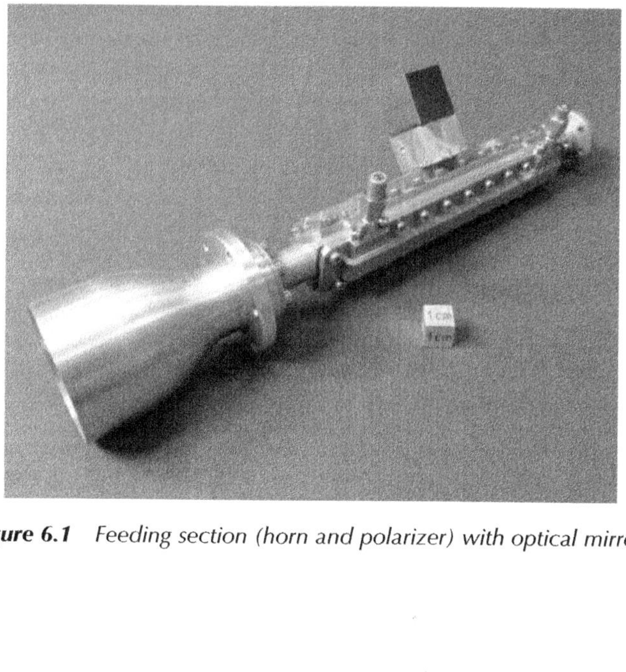

Figure 6.1 Feeding section (horn and polarizer) with optical mirror.

6.2.2 Model Philosophy and Definitions

The elements, assemblies and subassemblies that form part of the space flight equipment are defined in the model philosophy of the space flight project. Depending on previous know-how and on the background and maturity of the equipment design, they can be categorized as follows:

- Category A: Recurring and fully qualified equipment.
- Category B: Previously qualified equipment with limited changes.
- Category C: Modified equipment.
- Category D: New equipment.

Depending on this categorization, the philosophy of what models to build and test for the specific application falls into one of the approaches indicated in Table 6.1, which affects not only the test campaign that the unit is submitted to, but also the manufacturing or build standard to be employed for a specific equipment. It is worth mentioning that the model philosophy is both project and equipment dependent, thus implying that it can be tailored for a specific unit of equipment within a space flight project.

The following guidelines are intended to help defining the models involved in the verification process and to select the associated model philosophy. Examples of typical models and philosophies applied to different types of projects are presented together with suggestions for their effective application. The concept of the hardware matrix and associated data is also introduced.

6.2.2.1 Model Descriptions

Various types of models can be employed according to the verification requirements applicable in each project [1], and a short description of the major physical models is presented in this section. Table 6.2 provides a schematic summary of the models with related objectives, representativeness and applicability.

Mock-up (MU): MUs are used in support of the design definition for overall architecture analyses, configuration design and assessment, interface control and definition, human factor assessment, operational procedure evaluation, layout optimization.

Breadboard (BB): A BB unit is electrically representative of the equipment or part of it. It is used to validate a new or critical feature of the design. Formally (i.e. in the ECSS documentation), BBs are also called development models (DM)s. They are applicable to every type of product (e.g. electronic box, mechanisms, structural parts and thermal equipment) and may be subjected to functional and environmental testing.

Integration model (IM): An IM is used for functional and interface tests and for failure mode investigations. The integration models (sometimes also called electrical models) are functionally representative of the end items in terms of electronics and software. Commercial parts are utilized, but they are typically procured from the same manufacturer of the high-reliability parts to be used in the flight model.

Table 6.1 *Guidelines for qualification approach vs equipment category (see text below for explanation of abbreviations).*

Equipment category	Baseline approach	Alternate approach
A	FM	—
B	PFM	EQM + PFM or QM + FM
C	EQM + PFM	QM + FM
D	EQM + PFM	QM + FM

Table 6.2 Model use and objectives.

Model	Objectives	Representativeness	Applicability	Remarks
Mock-up (MU)	• Interface (I/F) layout optimization/ assessment • Integration procedure validation • Accommodation checks	• Geometrical configuration • Layouts • Interfaces	• System/element levels	• According to their representativeness MUs are classified as: – Low fidelity – High fidelity (to be maintained under configuration control)
Breadboard (BB) or development model (DM)	• Confirmation of design feasibility	• Total conformity with functional electrical and S/W required in agreement with verification objectives (size, shape and I/Fs could not be representative)	• All levels	• Development testing
Integration model (IM)	• Functional development • S/W development • Procedure validation	• Functional representativeness • Commercial parts • Simulators of missing parts	• All levels	• Development testing • It could be considered something in between an MU and an EM • Sometimes also called an electrical model
Suitcase	• Simulation of functional and RF performances	• Flight design • Commercial parts • Functional representativeness	• Equipment level • System level	• Qualification testing
Structural model (SM)	• Qualification structural design • Validation of structural mathematical model	• Flight standard with respect to structural parameters • Equipment structural dummies	• Subsystem (SS) level (structure) • Sometimes it could be considered system level if it involves other SS or is merged with the system test flow	• Qualification testing

(continued)

Table 6.2 (*Continued*)

Model	Objectives	Representativeness	Applicability	Remarks
	• Validation of structural mathematical model			
Thermal model (TM)	• Qualification of thermal design • Validation of thermal mathematical model	• Flight standard with respect to thermal parameters • Equipment thermal dummies	• SS level (thermal control) • Sometime it could be considered system level if it involves other SS or is merged with the system test flow	• Qualification testing
Structural–thermal model (STM)	• SM and TM objectives	• SM and TM representativeness • Equipment thermo-structural dummies	• System level	• Qualification testing
Engineering model (EM)	• Functional qualification failure survival demonstration and parameter drift checking	• Flight representative in form-fit function • Flight design without redundancies and hi-rel parts	• All levels	• Partial functional qualification testing
Engineering qualification model (EQM)	• Functional qualification of design and I/Fs • EMC	• Full flight design • MIL-grade parts procured from the same manufacturer of hi-rel parts	• All levels	• Functional qualification testing
Qualification model (QM)	• Design qualification	• Full flight design and flight standard	• Equipment level • SS level	• Qualification testing
Flight model (FM)	• Flight use	• Full flight design and flight standard	• All levels	• Acceptance testing
Protoflight model (PFM)	• Flight use design qualification	• Full flight design and flight standard	• All levels	• Protoflight qualification testing
Flight spare (FS)	• Spare for flight use	• Full flight design and flight standard	• Equipment level	• Acceptance testing

Structural model (SM): The SM is fully representative of the flight model for structural aspects. It is used for qualification of structural design, for mathematical model correlation and for validation of test facilities/ ground support equipment (GSE) and related procedures. It usually includes structural dummies of the equipment and representative mechanical parts of other subsystems (e.g. mechanisms, solar panels).

Thermal model (TM): The TM is fully representative of the thermal properties of the item. It is used for qualification of the thermal design and for the correlation of mathematical models. It includes a representative structure with thermal dummies of the equipment and representative thermal parts of other subsystems.

Structural–thermal model (STM): The STM combines the objectives of the structural model and thermal model.

Engineering model (EM): An EM is fully representative of the flight model in form, fit and function. A lower standard may be used for electrical parts, materials and processes. An EM is acceptable for use with flight hardware and must be equipped with gold-plated connectors, pins and sockets. An EM is subject to configuration control and its as-built status should be reported.

Engineering qualification model (EQM): An EQM is fully representative of the flight model (FM) in form, fit and function. It must be built to full flight standard in accordance with the product assurance and configuration requirements except for the electrical, electronic and electromechanical (EEE) parts: a lower standard may be used for the EEE parts, but they must be of the same manufacture, type and package as the FM; only the screening and testing may be different from the FM. An EQM is acceptable for use with flight hardware and must be equipped with gold-plated connectors, pins and sockets. An EQM is subject to a full equipment-level qualification test sequence.

Qualification model (QM): A QM is built to full flight standard in accordance with the product assurance and configuration requirements. Flight standards are used for the EEE parts: they must be of the same manufacture, type and package as for the FM; the screening and testing will be at the same level as the FM. A QM is subject to a full equipment-level qualification test sequence and environment, including qualification levels, margins and durations. This includes life test when it is required to be demonstrated. A QM is not acceptable for flight, unless agreed.

Protoflight model (PFM): A PFM is built to full flight standard in accordance with the product assurance and configuration requirements. Flight standard EEE parts are used. A PFM is subject to the protoflight qualification and acceptance test sequence and environment, including qualification levels and acceptance duration and margins as agreed with the customer. After a successful test sequence, a PFM is acceptable for flight, unless agreed.

Flight model (FM): An FM is built to full flight standard in accordance with the product assurance and configuration requirements. Flight standard EEE parts are used. An FM is subject to the acceptance test sequence, including acceptance levels and acceptance duration and margins as agreed with the customer. An FM is acceptable for flight.

Flight spare (FS): The FS is the spare end item for flight. It is subjected to formal acceptance testing. It can be built from refurbished qualification items with the condition of not being identified as a single point failure in the corresponding failure mode effects and criticality analysis (FMECA).

Suitcase: The suitcase is used to test the links with the ground segment or other external infrastructures. It is designed to simulate functional performance in terms of both data handling (e.g. tele-command and telemetry as formats, bit rates, packet type) and RF.

6.2.2.2 Model Philosophies

Several types of model philosophies, linked to the model descriptions above, may be employed according to verification requirements: prototype, protoflight and hybrid model philosophies.

The *prototype model approach* is the most conservative and is used in projects for which all possible measures are taken to achieve minimum risk as it occurs in manned projects. The prototype approach makes extensive use of the above-defined models to cover verification necessities, which minimizes project risks and allows parallel activities to be performed on different models and to complete the qualification activities prior to acceptance. The obvious disadvantage is the high cost.

Generally, the qualification testing can be conducted on the QM, according to the project requirements and objectives. However, the FM equipment is subjected to complete acceptance testing with specific qualification levels and duration.

The *protoflight model approach* is applied to projects in which no critical technology is employed and qualified hardware is extensively available. It is also used in case a compromise is permitted to reduce cost, accepting a moderate level of risk. The approach is based on a single model to be flown after it has been subjected to a protoflight qualification and acceptance test campaign. The advantage of this approach is its lower cost and the obvious disadvantage is that the testing activities run in series on the same model, which affects the development schedule. With this approach the qualification tests are performed on the same model to be flown, normally with qualification levels and reduced (acceptance) durations.

Finally, the *hybrid model approach* philosophy is a compromise between prototype and protoflight options. The hybrid approach always assumes a protoflight model is flown after a protoflight test campaign whose scope is reduced with respect to that of the pure protoflight approach. This model philosophy is used in projects where a new design or critical impact areas are subject to qualification during the verification programme. Specific qualification tests in the critical areas are carried out on dedicated models. The advantages and disadvantages of this approach are intermediate between those of the prototype and the protoflight approaches in terms of risks, costs and schedule. This philosophy represents a good compromise. In fact, it permits one to perform some parallel activities, use QM and EQM (if foreseen in the model philosophy of the project) as integration spares, comply with the lead times of high-reliability components or even consider the possible use of commercial off-the-shelf (COTS) components. With respect to testing, a combination of the prototype and protoflight rules is generally considered, with the possibility to perform qualification testing on dedicated models (e.g. STM, QM or EQM). In critical areas, acceptance testing should be performed only on the PFM.

6.2.2.3 Model Testing During Equipment Build Phase

The qualification and acceptance test campaigns usually required for antennas and associated RF equipments include the tests shown on Table 6.3. As indicated above, this list may be tailored for each project to cover specific areas that are critical for the project mission, depending on the category and maturity of each unit. These test agreements need the approval of all the parties involved.

Figure 6.2 shows a test flow sequence covering the qualification campaign of one generic RF unit. Specific testing to cover particular aspects of each unit should be included in the detailed test plan for each unit. The flow covers most of the tests usually required for RF and antenna units.

6.2.2.4 Development Testing

The objective of development testing is to support the design feasibility and to assist the evolution of the design. The testing requirements are dependent upon the maturity of the design and upon the operational requirements of the specific project and, whenever possible, they should be conducted over a wide range of operating conditions that exceed the project design limits to identify marginal design features and to confirm the following:

- performance margins;
- manufacturability;
- testability;

Table 6.3 Typical EQM and FM testing on electronics unit and antennas.

Test	Electronic box		Antennas		Notes
	EQM/ PFM	FM	EQM/ PFM	FM	
Mechanical					
Mass/dimensions/interfaces	×	×	×	×	
Centre of gravity (CoG)/ moment of inertia (MOI)	×	×	×	×	
Sinusoidal vibration	×		×		Sinusoidal survey only during acceptance
Random vibration	×	×	×	×	Random or acoustic, depending on the type of unit
Acoustic noise	×	×			
Shock	×		×		Test only EQM not PFM/FM
Depressurization/corona discharge	×		×		
Thermal					
Thermal cycling	×	×	×		
Thermal vacuum	×	×	×		Thermal cycling is deemed acceptable for a recurring unit qualified in thermal vacuum
EMC					
Bonding, isolation	×	×	×		
Conducted emission	×	×			
Conducted susceptibility	×	×			
Radiated emission	×	×			
Radiated susceptibility	×	×			
ESD	×		×		
Functional/performance					
Functional tests	×	×	×	×	
Hardware (HW)/software (SW) tests	×	×			For equipment with embedded SW, tests representing the HW/SW qualification of the equipment
RF tests	×	×	×	×	Specific RF testing for telemetry, tracking and control units
Deployment tests			×	×	

- maintainability;
- reliability;
- life expectancy;
- failure modes; and
- compatibility with safety requirements.

Development tests can be conducted on mock-ups, development (breadboards) and integration models. Adequate records of test configuration, test results and other pertinent data should be maintained so that this information is available to supplement other portions of the verification programme and during the functioning of qualification and flight models, when failure occurs.

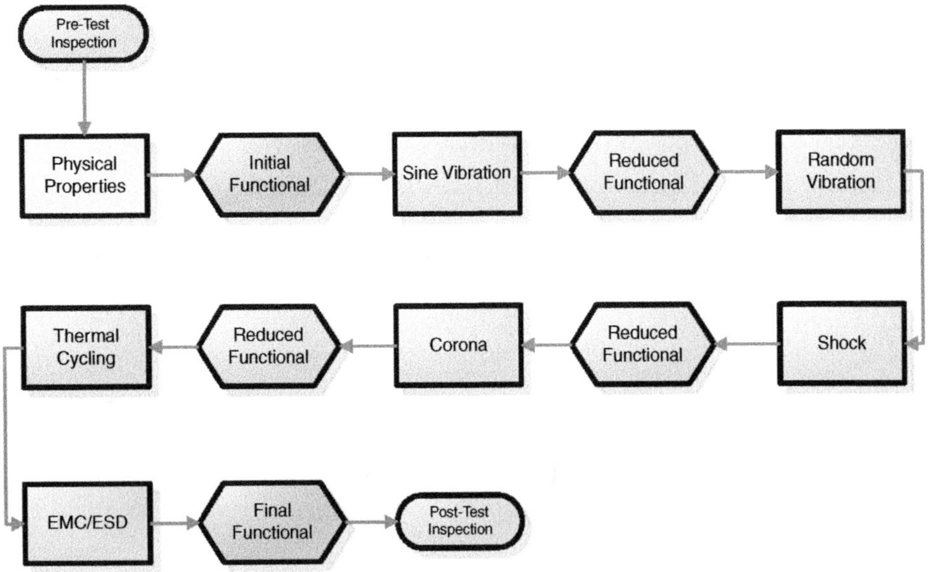

Figure 6.2 *Typical qualification flow chart.*

6.2.2.5 *Qualification Testing*

The objective of qualification testing is to demonstrate that the items perform satisfactorily in the intended environments with sufficient margins, guaranteeing that the design implementation and manufacturing methods conform to the specification requirements. For this purpose, the qualification test levels should exceed the maximum predicted levels by a safety factor, or margin, which assures that, even with the worst combination of test tolerances, the flight levels will not exceed the qualification test levels. This qualification margin is intended to avoid qualification test levels and durations that are less severe than those expected in flight, as well as providing test levels which cover minor differences between qualification and flight units.

Qualification testing is conducted on dedicated QMs that are produced with the same characteristics as the flight item or using lower quality parts instead of space-qualified high-reliability ones, without affecting the test validity (EQM) (Figure 6.3).

In cases where the environmental testing of a complete element is not possible, qualification of an element can be achieved by a combination of tests on various assembly levels, supported by analysis and similarity assessment, review of design, and inspection. In this case, the test requirements of the subassembly are derived from the upper level qualification requirements.

6.2.2.6 *Acceptance Testing*

The purpose of acceptance testing is to demonstrate conformance to specification and to act as quality control screens to detect manufacturing defects, workmanship errors, the start of failures and other performance anomalies which are not readily detectable by normal inspection techniques. Acceptance tests are conducted on all the flight products (including spares).

The acceptance test levels cover the maximum flight levels encountered during the service life (except for equipment thermal tests where a margin should be used to cover the uncertainties of the mathematical model), with durations lower than those of qualification testing that do not degrade the equipment. Acceptance tests are formal tests conducted to demonstrate the adequacy and readiness of an item for delivery and subsequent usage

Figure 6.3 *Gaia PAA EQM radiating prism equipped with sensor accelerometers in the vibration table.*

and are conducted on flight models under environmental conditions no more severe than those expected during the mission.

6.2.2.7 Protoflight Testing

The protoflight test levels and durations are a combination of qualification and acceptance testing in which they are used to qualify margins with acceptance.

6.2.2.8 EGSE and MGSE

All space projects require ground support equipment (GSE) for proper handling and testing of space segment elements. A significant part of this is the electrical ground support equipment (EGSE). Details of the GSE specifications are included in the applicable ECSS standard [2]. In short, the EGSE has to allow monitoring of all telemetry functions of the space flight equipment, commanding through the appropriate electronic interface units, and testing and evaluation of the functional performance of the device under test.

The major signal interfaces to be verified for any electronic unit are telemetry data and test point measurement data. Tele-commanding through the EGSE provides the capability to send data to the test object for control and data loading purposes both from user terminals and from other software. Tele-commanding through the EGSE also has the objective of control, validation and verification.

Verification or test procedure execution through the EGSE provides the capability to automate the tests required to guarantee the functional performance of the unit under test (Figure 6.4). Usually, it is built in a specific language that permits the execution of scripts to command the unit through the EGSE and the specific test equipment (e.g. vector network analyser, power meter, anechoic chamber test equipment, signal analyser, etc.).

The EGSE has to be submitted to a test and validation campaign to guarantee specification compliance and to verify that it cannot damage the equipment it is designed to control and monitor. This requirement applies to the functional test campaign, to the EMC test campaign and to any other environmental test that involves functional testing (as in the case of thermal vacuum - TVAC - operational testing).

A similar set of requirements has to be considered for the mechanical ground support equipment (MGSE) that has to be designed to permit testing or assembly of the flight hardware unit. Typically, the MGSE is used

Figure 6.4 State-of-the-art lightweight zero-g device for large carbon fibre reflectors.

for the equipment assembly, as shown in Figure 6.5, vibration testing, testing in an anechoic chamber for EMC or for radiation pattern testing, antenna deployment testing, and so on.

6.2.3 Electrical Model Correlation

It is very important in the early design and development stage to establish mathematical models which will then be correlated with the antenna testing. The goal here is to be able to reduce the amount of actual testing required with model refinement, thus reducing development cost.

The correlated model will provide the designers with the possibility to obtain reliable predictions for antenna behaviour in circumstances where testing is not envisaged or not even possible.

The typical components of an antenna are feed systems, reflectors and beam-forming networks. The feed system itself can consist of various different subcomponents such as a polarizer, orthomode transducer (OMT), transitions, couplers, including test couplers, diplexers, triplexers, horns, and so on.

As described in Chapter 1, typical design parameters for antennas are gain (edge of coverage), peak gain, beamwidth, sidelobe level, cross-polar level, isolation, return loss, and so forth.

Today most of these parameters can be predicted in a very accurate manner using powerful computers and appropriate software. For some types of components, there are exact methods to compute the performance. For

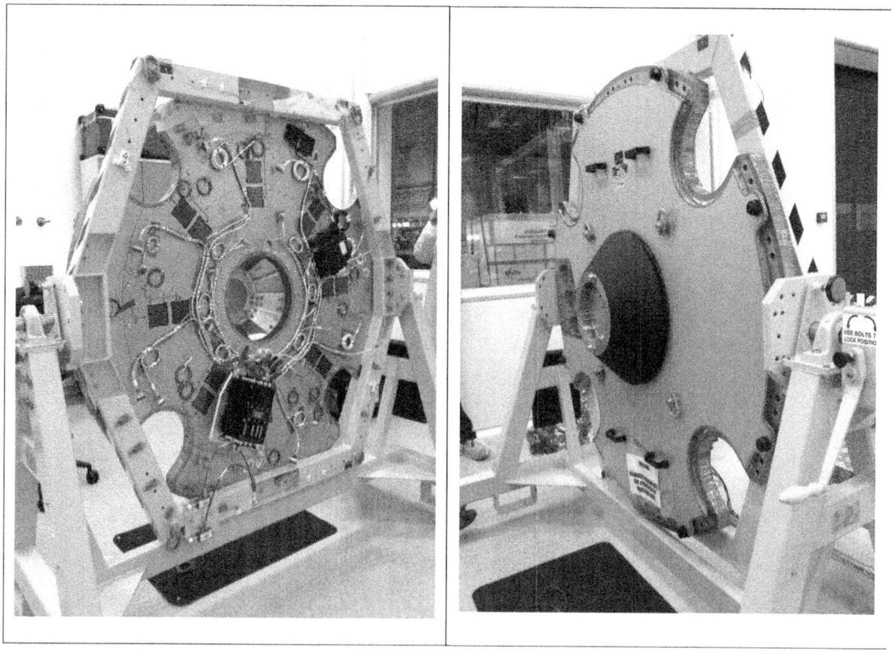

Figure 6.5 *Tilting trolley MGSE in Gaia PAA.*

example, rotational symmetric horns can be analysed by using mode matching for the inner structure and the method of moments for the outer contour. This approach offers an exact solution in principle, although in reality there remains a negligible truncation error. The performance of other waveguide components can be predicted by using computational techniques like the finite element method (FEM) or the finite difference time-domain method (FDTM) with high accuracy.

Radiation performance via reflectors or radiating structures can be theoretically predicted exactly by the method of moments. Due to their sometimes large size with respect to wavelength, approximation methods like physical optics (PO) or the physical theory of diffraction (PTD) need to be applied. These methods are also very accurate when the size of the object to be analysed is large with respect to wavelength. In some cases, hybrid methods (e.g. method of moments plus the geometrical theory of diffraction (GTD)) might be the best choice.

6.2.3.1 Electrical Modelling Uncertainties

Even if the analysis is exact for the nominal designed geometry, because of manufacturing tolerances the performance will be different from the predicted one. Another important aspect is that most of the single antenna components might be modelled and analysed in quite a good manner and the performance prediction for each of these components could be very good. However, when combining individual components into a feed chain assembly or beam-forming network, the interaction between the single components can result in unpredicted RF behaviour.

A further point of uncertainty is the assumption of perfect excitation with the fundamental mode. In reality, when assembling the components, higher modes can be excited by an imperfect interface as a result of manufacturing and/or assembly tolerances. This can have a negative effect on the performance of the entire feed chain. In particular, this is true for multilayer structures. Normally, to reduce the effect of misalignment between different parts, dowel pins are employed. However, the dowel holes can only be manufactured to a finite accuracy. Naturally, these effects have more impact on components at higher frequencies (e.g. Ka-band).

Modelling errors, as well as computation errors, need to be reflected in an uncertainty budget; manufacturing errors should be taken into account as well. The required amount of uncertainty needs to be analysed by sensitivity analysis. Manufacturing errors not only appear for the feed chain, but also for the reflector, beamforming network and array element location accuracy. Therefore, another typical contribution to the uncertainty budget is the manufacturing error of the reflector if a reflector system is being designed, or the array geometry including planarity requirements. The performance of the entire antenna needs to be designed to have sufficient margins to cover these uncertainties.

During the manufacturing and verification phase of the hardware, the uncertainty budget can be adjusted to the actual performance of the subcomponents. Ideally, the measured deviation of each component considered in the uncertainty budget is covered by the foreseen contributions.

6.2.3.2 *Validity of Model Correlation*

The level of confidence with respect to the performance prediction depends mainly on two aspects:

1. Qualification status of the hardware.
2. Complexity of the designed hardware.

For category A or category B equipment which has been previously manufactured and tested and which is again used in the same configuration, the prediction should be quite accurate. Also for category C equipment, the prediction might be quite reliable, but this depends on the degree of modification because the behaviour of the antenna with respect to a reliable prediction of performance can change.

Attention must also be paid when using recurring components but assembled in a different way. For example, a well-known OMT and a horn with known performance can perform unexpectedly when assembled together in a new feed chain type. Indeed, although the performance of both components is known, the interaction between them can lead to surprises.

For instance, a very simple type of antenna is the linearly polarized global horn antenna consisting of a horn (smooth wall or corrugated) and a square/rectangular-to-circular transition. Due to the fact that horns of this type can be analysed in an almost exact manner and the transition is quite a simple and well-predictable component, the performance of the entire antenna can be predicted to great accuracy. The model correlation can easily be performed after measurement of the manufactured and integrated components. If the accommodation of the global horn antenna on the spacecraft was simulated in an RF representative model to predict possible scattering, the measurement results could be used to serve as the source instead of the ideal feed. Here, the scattering effects can be evaluated with the actual feed pattern. A final measurement to predict the in-orbit performance needs to be performed when the global horn antenna is installed on the satellite along with all the other antennas and equipments.

Regarding much more complex antennas, such as a reflector antenna fed by a feed array manufactured in a multilayer structure, the performance prediction has a higher degree of uncertainty and the process of model correlation and design verification is more complex. An example is given in the next subsection. The performance of an antenna installed on a satellite is mainly influenced by two effects. One effect is the RF scattering on the satellite structure and the other is the interaction with other antennas.

6.2.3.3 *Correlation Example*

As an example, consider an offset parabolic reflector antenna operating in C-band. As the feed system, a multifeed array with 145 horns is used, providing eight different circularly polarized beams.

Figure 6.6 Upper layer of the BFN.

A typical test sequence to verify the design and to get a more precise prediction of the final performance is:

1. Initial beam-forming network (BFN) measurements on the integrated BFN.
2. Intermediate testing of the integrated feed array.
3. Intermediate testing at antenna level.
4. Final testing at spacecraft level.

Each of these tests enables a comparison of the prediction at each stage. This is very important during the manufacture and integration of such a complex antenna. Therefore, it becomes possible to detect any malfunction of the device under test (DUT) at an early phase of the manufacturing process, integration or at least at an early stage in the final test sequence. It is essential getting this information at the earliest stage irrespective of whether the malfunction is correlated to a design error or a manufacturing failure.

Initial BFN Measurement Figure 6.6 shows the upper layer of a transmit feed array. The actual excitation coefficients are measured at all BFN polarizer interface points using a network analyser in a swept mode over the complete frequency bands.

The measured BFN excitation coefficients are used to predict the primary field pattern of the feed array. Therefore, the horn pattern needs to be analysed corresponding to its location in the array (mutual coupling must not be neglected!). The resulting patterns are afterwards converted into spherical wave expansion coefficients which are used as the source for calculation of the field distribution over the reflector in order to achieve the secondary far-field. The far-field performance is the fail/pass criterion for the hardware at this stage.

Depending on the heritage and on the sensitivity analysis, but also on the feasibility, the design has to foresee a tuning possibility at this stage which can be applied if necessary.

Intermediate Testing of the Integrated Feed Array After having integrated the entire feed array, the next level of model correlation is to measure the RF patterns of the array. These patterns are then used to predict the secondary pattern via the reflector. This is accomplished by replacing the ideal feed in the RF model with the measured feed pattern. A typical way to perform this replacement is to employ a spherical wave expansion. Thus, the influence of near-field effects is ensured.

There is also the option to measure the array pattern under extreme environmental conditions with respect to temperature. This gives a good indication of the performance variation of the entire antenna system in orbit under different environmental conditions. Figure 6.7 shows the setup for a feed array in a thermal simulation chamber installed in a cylindrical near-field range.

Figure 6.7 Setup with feed array in the climate box at top of the azimuth turntable in the cylindrical near-field range.

Another interesting analysis at this stage is to incorporate real reflector data into the computational model. In the case where the reflector is already manufactured, the measurement surface data can be included in the analysis instead of an assumed ideal reflector surface. This computation can also serve as the acceptance criterion for the use of the manufactured reflector. If the reflector is not already manufactured, one way to get an indication of the final performance as quickly as possible is to use the mould data of the reflector.

It is worth mentioning that, for very large reflector apertures, such as with large deployable antennas, this can be the only way to determine the final antenna performance as a terrestrial measurement may not be practical or even possible.

Intermediate Testing at Antenna Level For measurement of the antenna pattern, the manufactured hardware needs to be installed on a mock-up of the spacecraft structure. A proper mutual alignment of the antenna components is important as well as the correct alignment of the antenna in the measurement facility. Figure 6.8 shows an example of a multi-feed array illuminating a reflector installed in a compensated compact test range. Important parts of the spacecraft are simulated by the mock-up.

Final Testing at Spacecraft Level To achieve the most accurate prediction for the in-orbit performance, testing the antenna installed on the satellite should be performed. This measurement then includes all scattering effects with respect to the satellite structure and the interaction with the surrounding equipment and other antennas.

Example Conclusion Figure 6.9 summarizes the level of confidence with respect to the final performance of an antenna during the design and manufacturing process. At the level of design, all the uncertainties are present. With ongoing manufacturing and testing, the uncertainties decrease and the actual performance becomes known to a high confidence [3].

Figure 6.8 Transmit antenna including mock-up installed in the CCR75/60 of EADS Astrium.

6.2.4 Thermal Testing and Model Correlation

6.2.4.1 Thermal Design

Components of a Space Antenna System The main components of a space antenna system are the feed or feed chain, transmit (Tx) subsystem and a receive (Rx) subsystem, which can be an assembly of several components such as horns, polarizers, transitions, diplexers, OMTs, and so on, as well as cables, waveguides or any other electrical and RF interface. For mounting the antenna system to the spacecraft and also for alignment purposes a support structure is generally necessary.

These components have to be designed so that they fulfil the operational and non-operational thermal requirements over the full life/mission cycle of the spacecraft, in line with the interface requirements specified at the spacecraft level.

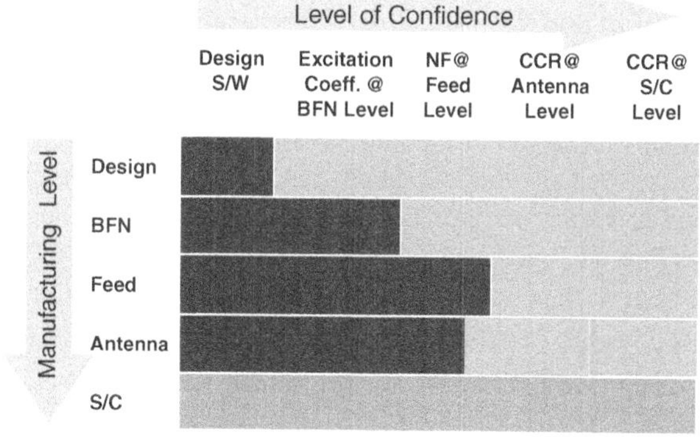

Figure 6.9 Level of confidence during antenna design and manufacturing (NF, Near Field; CCR, Compensated Compact Range; S/C, Spacecraft).

Table 6.4 *Example: Component temperature limits.*

Item		Material	Temperature (°C) Non-operating survival	Operating Qualification	Operating Acceptance
Tx feed assembly:					
	Horn body	Aluminium 3.2315	−180/ + 160	−160/ + 160	−160/ + 160
	Main bracket	Aluminium 3.2315	−180/ + 160	−160/ + 160	−160/ + 160
	Rear bracket	Aluminium 3.3214 T6	−180/ + 160	−160/ + 160	−160/ + 160
	Transition	Aluminium 3.3214	−180/ + 160	−160/ + 160	−160/ + 160
	Septum polarizer	Aluminium 3.4364 T7364	−180/ + 160	−160/ + 160	−160/ + 160
	HF window	Kapton	−269/ + 360	−269/ + 360	−209/ + 360
	Absorber (in load)	Eccosorb MF	−180/ + 180	−180/ + 180	−180/ + 180
	Surface treatment	Yellow chromating	−180/ + 200	−180/ + 200	−180/ + 200
	Surface treatment	Silver plating	−150/ + 160	−150/ + 160	−150/ + 160
	Adhesive	Hysol EA 931 NA	−175/ + 204	−175/ + 177	−175/ + 177
Rx feed assembly:					
	Horn body	Aluminium 3.2315	−180/ + 160	−160/ + 160	−160/ + 160
	Main bracket	Aluminium 3.2315	−180/ + 160	−160/ + 160	−l60/ + 160
	Polarizer	Aluminium 3.3214 T6	−180/ + 160	−160/ + 160	−160/ + 160
	Struts	Rexolite 1422	−190/ + 200	−190/ + 125	−190/ + 125
	Feed-through	Teflon PTFE	−200/ + 260	−200/ + 260	−200/ + 260
	Surface treatment	Yellow chromating	−180/ + 200	−180/ + 200	−180/ + 200
	Surface treatment	Silver plating	−150/ + 160	−150/ + 160	−150/ + 160
	Adhesive	Hysol EA 934 NA	−175/ + 204	−175/ + 177	−175/ + 177
	Adhesive	Stycast 1266 A + B	−180/ + 121	−180/ − 121	−180/ + 121
	Adhesive (screw locking)	Solithane C113-300	−170/ − 120	−170/ + 120	−170/ + 120
	Output port	CuBe	−120/ + 105	−120/ + 105	−120/ + 105
	Load	CuBe	−125/ + 125	−125/ + 125	−125/ + 125
Thermal	Sunshield	Ge-coated Black Kapton	−180/ + 180	−180/ + 180	−180/ + 180
Hardware	Stand-off	Vespel	−200/ + 300	−200/ + 300	−200/ + 300
	White paint	HINCOM/NS43G	−180/ + 180	−180/ + 180	−180/ + 180

Thermal and Interface Constraints and Requirements In general, the components of an antenna system are realized using aluminium or titanium. Thanks to the properties of these materials, they can cope with a wide temperature range. Limiting factors are often the surface treatments, glues or special mechanisms such as the hold-down and release mechanism (HRM) of a deployable reflector.

Therefore, it is common practice to create a table containing the temperature limits (minimum and maximum) for the operational and non-operational modes for all parts and components. Further, a table with the specified or required interface and boundary conditions should be generated. These two tables (Table 6.4 and Table 6.5) are then the basic values for the definition of the thermal design.

In some cases, the interface heat fluxes between the subsystem and the spacecraft are restricted to a certain value or the interface is specified as adiabatic for the design and analysis.

Thermal Design Elements Depending on the allowed temperature range, external loads such as solar radiation or free molecular heating, internal electrical or RF dissipation, the preferred design solution is usually a passive thermal design with no active thermal control elements. If a component has to be kept at a

Table 6.5 *Example: S/C interface temperatures*

	On station		Transfer & Deployment	
	Hot	Cold	Hot	Cold
+X/−X wall	55	−10	40	−20
+Y/−Y wall	55	−10	30	−30
Extensions to +Y/−Y wall	55	−10	30	−30
+Z face	55	−10	30	−20

Standard geo-orbit S/C pointing: +X = east, +Y = south, +Z = earth.

specific temperature by heaters or within a defined temperature range, which must be controlled by the spacecraft, this is called active thermal control. The main thermal control elements and their purpose are given in Table 6.6.

An example of a C-band Tx and Rx feed chain is shown in Figure 6.10.

6.2.4.2 *Thermal Modelling and Analysis*

To define the thermal design with the necessary thermal control elements, a thermal model must be developed for the antenna system. This model should consist of all the main parts of the antenna, but also of all the elements of the spacecraft that have a major influence on the environmental conditions for the antenna.

Mainly there are two different methodologies for the generation of a thermal model:

1. The node model with a non-geometric node solver like ESATAN. The parts and components of the antenna system will be subdivided in discrete lumped mass nodes and the capacitances of the nodes and the conductances between the nodes are the input data for the solver and build the thermal mathematical model (TMM). The radiative couplings, and external loads coming from the Sun's radiation, Earth or albedo radiation, and so on, depend on the geometric configuration. As a consequence, a geometrical mathematical model (GMM) has to be created with an additional software package such as ESARAD, for

Table 6.6 *Main thermal control elements.*

Passive:	
Surface coating and paints	Adjust IR and/or solar loads
Multilayer insulation (MLI)	Reduce thermal fluxes/heat losses
Sunshield (Ge-coated Kapton foil)	Reduce absorbed solar flux, avoid cavity effects
Radiator (with optical solar reflector (OSR) or second surface mirror (SSM)	Component temperature control by increased radiation and reduction of absorbed solar flux
Thermal straps	Increase of conductive link
Heat pipes	Increase heat conduction/distribution between hot and cold zones (e.g. on radiators)
Thermal washers	Thermal decoupling (using materials like Vespel, Diveritt, titanium), limitation of interface fluxes
Active:	
Electrical heaters	Keep component at a defined temperature or within a temperature range controlled by thermostats or thermistors

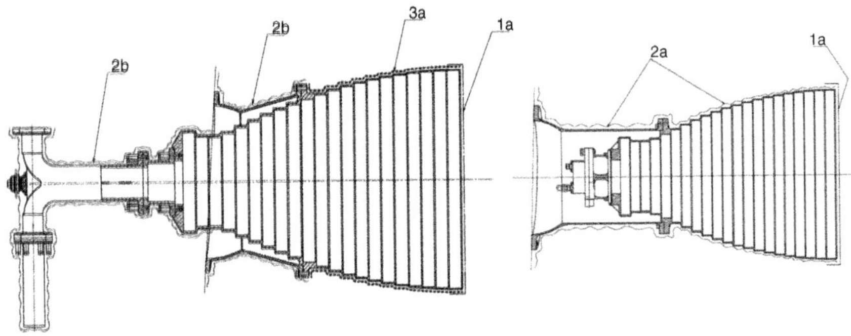

TX Feed Chain RX Feed Chain

No.	Thermal Control Element	Purpose
1	**Sun Shield**	reduce heat fluxes during sun incidence
1a	∘ over horn aperture	reduce heat losses during no sun incidence
2	**Feed MLI**	reduce heat fluxes during sun incidence
2a	∘ RX feed	reduce heat losses during no sun incidence
	– horn front part	
	– main bracket	
2b	∘ TX feed	
	– horn rear part	
	– transition	
	– polarizer	
	– load	
	– feed flange	
	– main bracket	
3	**White Paint**	increase radiative coupling to space
3a	∘ front part of TX horn (radiator)	reduce absorbed solar irradiation
4	**Uncoated Areas (Yellow Chromated)**	no coating allowed due to electrical reasons
4a	∘ RX feed	no coating necessary
	– all parts of feed	
	– main bracket	
4b	∘ TX feed	
	– horn rear part	
	– main and rear bracket	

Figure 6.10 *Example of a C-band Tx and Rx feed chain.*

calculation of the radiation heat exchange and external loads. Internal loads such as electrical and RF dissipation losses come from the RF performance analysis and dissipation loss measurements.

2. The FE model, where the geometry can be used as provided by a CAD system and can be meshed with the FE modelling tools of the software. Fully 3D, 2D and axisymmetric models can be generated to achieve high resolution. Node capacitances and inter-node conductances are calculated and managed internally. The radiation heat exchange can be calculated internally, but often, depending on the resolution, simplifications or groupings of surface elements are necessary to limit the computational time for the analysis.

Well-known thermal software analysis tools for these two model types are:

- Node model tools and solver:
  - ESATAN
  - SINDA
- FE model software:
  - IDEAS/TMG
  - ANSYS.

And for the construction of the radiation geometry and calculation of the radiative heat exchange, the following are commonly used:

- ESARAD
- THERMICA
- NEVADA.

Both types of models allow steady-state and transient simulations. Depending on the mission scenario, different orbital cases in operational or non-operational modes at beginning of life (BOL) or end of life (EOL) condition and varying orbital and seasonal conditions should be simulated. Typical cases include:

- Launch mode and heating under fairing;
- Free molecular heating (FHM) in the upper atmosphere;
- Transfer orbit;
- General or geosynchronous orbit;
- Emergency Sun reacquisition (ESR).

The layout of the thermal design must be chosen so that all temperatures (minimum and maximum, for operational and non-operational modes) are within the material and qualification limits of the parts and components. Therefore the calculated temperatures must be compared with the values from the temperature limit table by adding, depending on the margin philosophy, a model uncertainty and an additional acceptance and qualification margin to the simulation results.

Depending on the design logic process and on the availability of material property data, various sensitivity analyses have to be performed for parameters with a high uncertainty to check the influence on temperatures and heat fluxes. If a high sensitivity is detected, measurements of the properties for the used material, if not already available, should be performed to reduce the uncertainties.

Further reduction of the model uncertainties and validation of the thermal model are achieved by correlation of the test data with measured temperatures, heat fluxes and power values. Generally, measured temperatures from thermally stabilized conditions are used for the correlation, comparing it with the results from a steady-state analysis. But even transient calculations and measurements can be used to detect the lack of a model. The model adaptations should be applied within a justifiable physical range in order to model parts where uncertainties coming from idealization and simplification exist. Further contact values and heat transfer coefficients, MLI losses and, to a lesser extent, surface optical properties where only theoretical or manufacturer-supplied values are used can be included. Automated processes exist for the model correlation by optimizing selected parameters in a predefined range to fit best with a given temperature distribution. However, very often it is better to perform this process manually, step by step, to see the influence of each parameter. Indeed, the optimizer gives only a final set of the varied parameters without indicating separately the effect of individual variations.

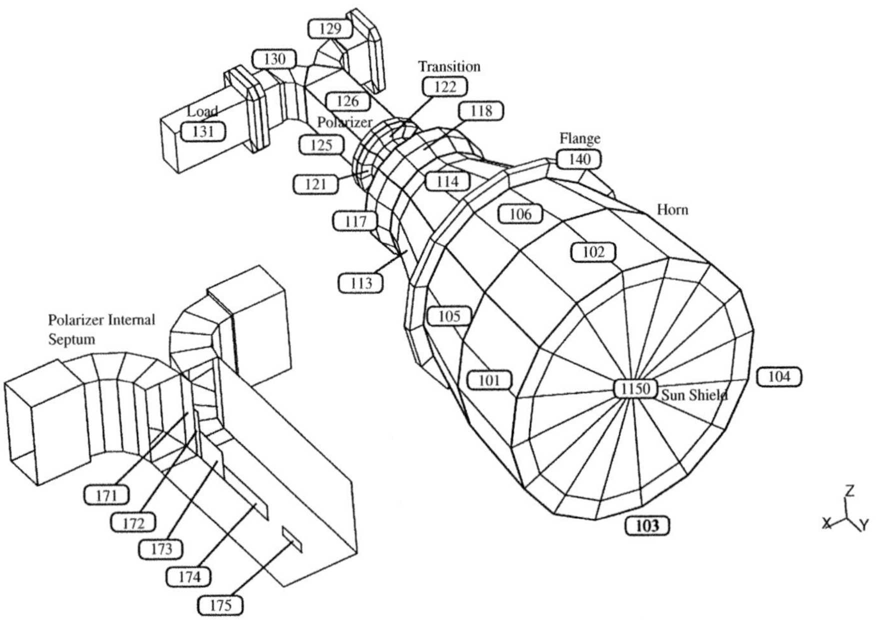

Figure 6.11 C-band Tx feed chain, nodal subdivision for ESATAN analysis.

At subsystem level, detailed thermal models (GMM/TMM) for the thermal analysis during the design process are used. For the overall analysis at system level, it is necessary to extract condensed/reduced thermal models (RGMM/RTMM) which must be representative of, and therefore correlated with, the detailed models. It is possible to automate the condensation process to a certain extent; for example, radiative couplings can be simply calculated by summation of the values for the nodes from the detailed model combined in a condensed node. But if the number of radiation surfaces is to be reduced, it is necessary to create a new geometrical model as well as approximate the geometry with simplified surfaces. Here, it is important that the direction of the surface normal and the average area are still representative if compared with the detailed geometry. For the node temperatures, mass or heat capacity weighted averages of the detailed model have to be compared with the node temperatures of the reduced model. If the adaptation is performed with the maximum and minimum temperatures occurring in the detailed model, the resulting calculation of the heat fluxes in the reduced model will be incorrect.

It is sometimes necessary to verify the thermal design or the thermal control subsystem (TCS) before a final, complete antenna subsystem unit is available, though a thermal representative model is needed. In its simplest form, this can be a mass dummy, but often this mock-up needs to fulfil the requirements for interface temperatures and heat fluxes as well and a design must be defined in such a way that this thermal representative test model (TRTM) behaves in a similar manner to the real unit to allow verification of the TCS or part of it, or another control element at the group or complete system level.

An example of the C-band Tx feed chain nodal subdivision for ESATAN analysis can be seen in Figure 6.11.

6.2.4.3 *Thermal Testing*

There are different types of thermal testing to be performed depending on the scope of the test and qualification status of the antenna unit.

Thermal Balance Test This kind of test is mainly performed under vacuum conditions with a flight representative configuration. For special applications, testing at ambient pressure is also possible. The aim is to reach steady-state conditions with defined temperature stability at a cold or hot temperature level:

- To verify the thermal mathematical model for cold and hot conditions.
- To prove that all components work and endure the extreme temperatures (high and low level).

Solar Simulation Test If it is necessary to simulate the influence of solar fluxes for a unit or configuration, a solar simulation test can be performed. This type of test, which is not very common for antenna applications, is normally performed under vacuum conditions by using lamps with an approximated solar spectrum. However, facilities also exist using the real solar spectrum by focusing the external sunlight with mirrors onto the test specimen. Short tests with flashes (e.g. xenon) are also performed under ambient conditions.

Thermal Cycling Test Thermal cycling can be performed in vacuum conditions or placing the test object in an inert gas like nitrogen. A cold and hot temperature level is defined, extracted from the minimum and maximum allowable temperature. A certain number of cycles have then to be performed. This kind of test can be performed on the part or component level for qualification or with a complete subsystem assembly for qualification or acceptance testing. For assembly configurations the temperature levels have to be restricted by the lowest maximum temperature and the highest minimum temperature from the component limits, otherwise partial heater control has to be used to handle the different temperature ranges.

Functional Testing under Thermal Conditions The functional testing for a unit under thermal conditions can be performed either in ambient pressure conditions or in vacuum depending on the scope of the functional check and the influence of the environmental conditions on function and performance. If, for example, corona effects can appear, a certain vacuum is necessary to avoid such effects and testing at ambient pressure is not possible.

The following documentation usually has to be prepared for the test and evaluation:

- Before test:
  - Test specification: scope of test and details for configuration and test phases.
  - Thermistor/thermocouple/heater location.
  - Definition of sensors and heater locations and types.
  - Test prediction.
  - Analysis of the test configuration and test phases.
  - Test procedure (step by step).
  - Definition of the activities and actions before, during and after test.
- After test:
  - Test evaluation report.
  - Summary of the test results.
  - Model correlation.
  - Thermal model adaptation based on test results.

An example of application is shown in Figures 6.12 and 6.13.

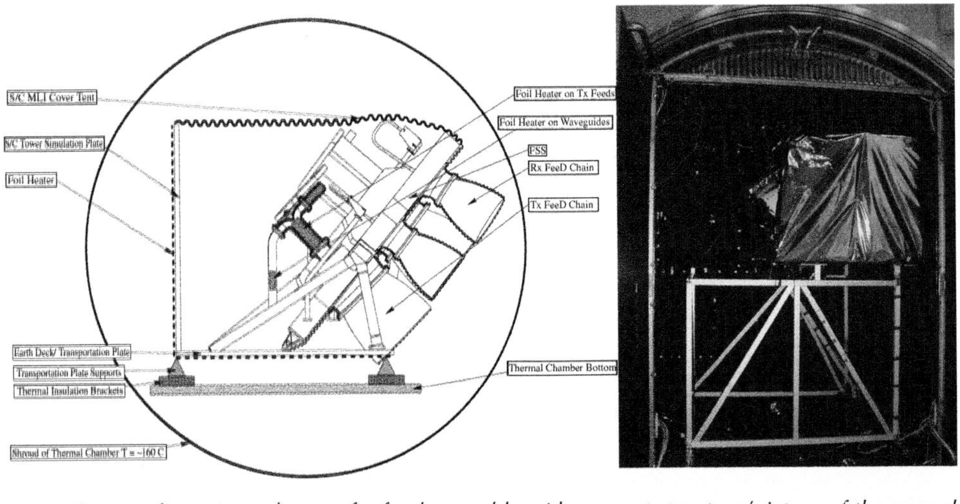

Figure 6.12 Test configuration scheme of a feed assembly with support structure/picture of the assembly on a support frame in the vacuum chamber.

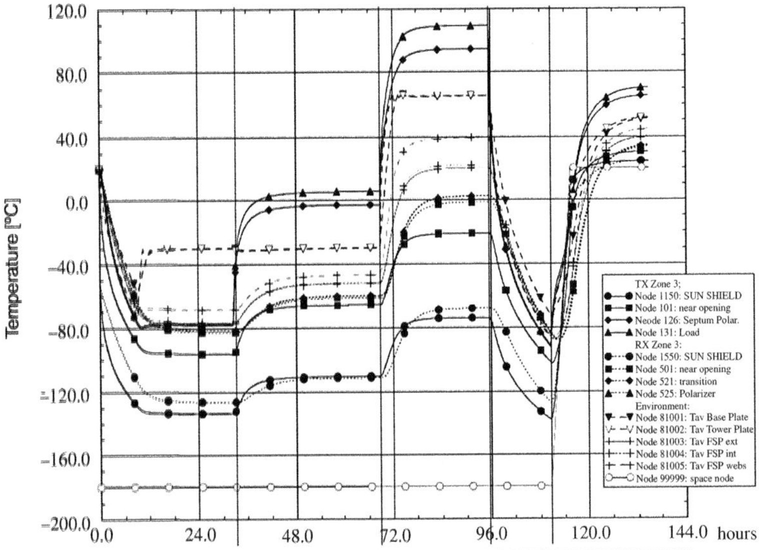

Test Phase No. :		1	2	3	4	5	6
Phase:		Pump Down	Stabilisation	Cold Level	Hot Level	Transient	Recovery
Condition:		trans.	Balance	Balance	Balance	transient	transient
Heater Power:	Tx Feeds	0 W	0 W	20.3 W / 20.3 W	40.5 W / 40.5 W	0 W	13.5 W / 13.5 W
	Rx Feeds	0W	0W	0W	0W	0 W	0 W
	Waveguides	0W	0W	3.3 W/ 2.1 W	6.6 W/ 4.2 W	0 W	2.2 W / 1.4 W
	Base Plate	0W	123.4W	113.7 W	439.3 W	0 W	123.4 W
	Tower Plate	0W	110.4W	96.6 W	380.9 W	0 W	110.4 W
Interface Temp.:	Base Plate	−65 C	−30 C	−30 C	+65 C	−80 C	+20 C
	Tower Plate	−65 C	−30 C	−30C	−65 C	−80 C	+20 C
Shroud Temperature:		amb. →	−180 C	−180 C	−180 C	−180 C	+20 C
Vacuum:		amb. →	1.E−5 mbar	1.E−5 mbar	1.E−5 mbar	1.E−5 mbar	ambient
Duration:							

Figure 6.13 Test phases and conditions: environment, temperature level, heater power and estimated test duration.

6.3 Antenna Testing Facilities

6.3.1 Far-Field Antenna Test Ranges

The most direct way to obtain the radiation pattern of an antenna is to mount it on a tower which has rotation capabilities and to have a measuring antenna far enough away that the phase taper of the measuring antenna's phase pattern across the test object is less than a quarter of the wavelength at the measurement frequency.

This is the standard definition of the far field and the required separation distance is twice the square of the antenna diameter divided by its wavelength. It does introduce errors as the taper means that the test antenna is not illuminated by a plane wave. The most immediate effects on a typical antenna would be tapering of the main beam and smearing out of the first pattern null. As the subject of far-field antenna pattern measurements is well established, the reader is referred to the literature [4].

Far-field antenna test ranges are capable of measuring all the performance parameters for antennas as outlined in Chapter 1. However, the accuracy requirements for space antennas many times cannot be met on a far-field range. This is especially true for high gain antennas where a very large separation distance is required. To mitigate the short distance of the fully enclosed far-field range, an open outdoor range is used where the separation distance between probe and test object can be hundreds of metres. In all cases of such far-field antenna test range geometries, special care is taken to minimise the effect of the specular reflection coming from the ground. At the same time it is worth noting that the classic Reflection Range antenna test range actually does the opposite and is used primarily for gain determination of low gain antennas. Here, the range has a conducting ground plane as an integral part of the facility and the antenna under test receives two known signals, the direct wave and the image wave. With this knowledge, the correct gain can be determined. These ranges are seldom used in space applications except for the measurement of very low frequency antennas in the VHF bands.

Low gain antennas can be suitably measured in a fully enclosed far field range. It is worthwhile to note that these antennas tend to be TTC antennas and they need to be characterised over almost the entire 4π sphere in amplitude and phase. The conventional far field range is an amplitude only facility. If phase measurements are available, then the same constraints apply here as for any spherical near field measurement. The phase information is needed if the radiation pattern is to be used in a scattering analysis of the antenna attached to a spacecraft. It needs to be mentioned also that if measurements of a low gain antenna mounted on a spacecraft are to be carried out, the far field distance being based on the size of the satellite and not just the antenna; in effect, the satellite or satellite mockup become part of the antenna.

It is essential for space applications, especially at the later stages when flight hardware is being produced, that the antenna not be exposed to the elements and is kept in a clean and controlled environment. In a fully enclosed antenna range, the environment can be controlled and one needs only assess possible contamination from microwave absorber particle fallout. If the fallout is too great, the test object will need to be cocooned.

An outdoor range would require a flight antenna in all circumstances to be cocooned to protect it from the elements, so one must be patient and await favourable weather conditions before tests can take place. A far-field test range candidate which combines the large separation distance of a conventional outdoor far-field range and the environment of an anechoic chamber is called a semi-open range. Here the test object is enclosed in an anechoic room with one wall removed or RF transparent. This allows sensitive flight hardware to be protected from the elements in a controlled environment.

6.3.2 Compact Antenna Test Ranges

6.3.2.1 Background and Introduction

The transportation capabilities of satellite launchers have been improving and are even continuing to rise. The size of payload systems and communication antenna platforms follow this trend and are also driven by cost saving and risk reduction decisions. Currently, antenna platforms with dimensions up to 12 m in width for several reflector and horn antennas working in different frequency bands are state of the art. The dimensions of such large platforms and also large satellite buses, for example the European Eurostar 3000+ and 4000, were mainly enabled by the high load capacity of existing space rockets. The Atlas V, Ariane 5 and Proton can be mentioned as examples in this context. Actually, communication satellites such as the American INTELSAT-X, the European ASTRA-1K and INMARSAT-4 as well as the Russian EXPRESS-AM1 are among such a class of modern and extremely large communication satellites [5].

In the case of near-field test facilities, large scanner systems were developed for detecting the near field in planar, cylindrical and spherical cases (Section 6.3.3). In the field of compact ranges, the compensated compact range type with a double-reflector arrangement is well suited for testing communication satellite antennas and payload applications. The goal of high-performance requirements for the quiet zone and measurement accuracy is given in Section 6.3.2.2. A tradeoff, Section 6.3.2.3, will give an overview of the performance for different types of compact ranges. The principal design and quiet zone performance data of an optimized compact range system are given in Section 6.3.2.4. Section 6.3.2.5 will show the typical performance of a state-of-the-art test facility. Finally, Section 6.3.2.6 will demonstrate the special features of a compensated compact range as a test bed for antennas in the loop, for example during payload testing and end-to-end testing of fully integrated satellites.

6.3.2.2 *Facility Requirements*

The design of state-of-the-art as well as future test facilities for antenna testing is driven by the requirements of satellite design engineers and programme managers in order to guarantee the electromagnetic performance of the systems. Typical requirements are as follows:

1. Frequency range: 0.8 up to 200, 500 GHz and beyond for Earth observation missions.
2. Quiet zone size: ≥ 8 m.
3. Silent volume dimension: ≥ 14 m.
4. Co-polar amplitude performance: $< \pm 0.3$ dB in the quiet zone (the volume where the amplitude and phase fronts satisfy the specified flatness).
5. Co-polar phase performance: $< \pm 3^\circ$ in the quiet zone.
6. Cross-polar performance: < -42 dB in the quiet zone.

The required size of the quiet zone and silent volume, that is the maximum width of the plane-wave dimension for scanned Rx and Tx quiet zones, is mainly determined by the dimensions of the telecommunication satellite itself and its very large antenna farms or deployable antennas which have to be tested.

The co-polar amplitude and phase performance requirements are derived from the desired constant amplitude and phase illumination within all beams and frequencies on Earth. The cross-polar performance requirement is determined by the desired polarization diversity for the telecommunication satellite channels.

6.3.2.3 *Facility Tradeoff*

Physically, the basic element of a compact range system is a collimating system for electromagnetic waves. The collimating system is realizable as a transmitting or reflecting unit to transform a spherical waveform into a plane wave. The plane-wave zone is known as the quiet zone (QZ) area of a compact range (Figure 6.14 and Figure 6.15). Technically, metallic reflectors or lenses are used for the collimating process. Due to the frequency dependence of lens systems, the reflector types have many advantages compared with transmitting systems, for example shape of surface, diffraction reduction at the edges.

A special candidate of wave transformer of the transmitting type is the hologram [6]. The superposition of all the interference patterns generated by the hologram provides a plane-wave field in the test zone, see Figure 6.16.

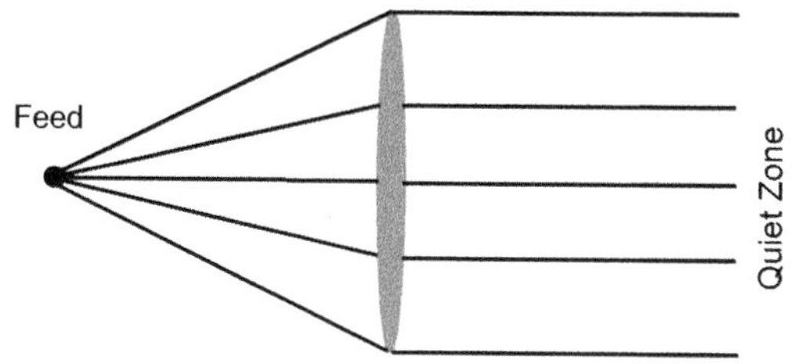

Figure 6.14 Collimating system by transforming lens.

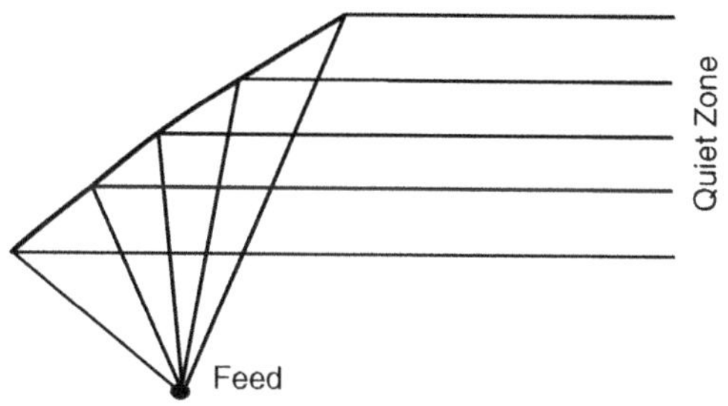

Figure 6.15 Collimating system by transforming reflector.

The narrow-band performance of holograms (typically 2%), single linear polarization operation, edge diffraction and high transmission losses are offset by their relatively low cost, ease of manufacture and portability, allowing the range to be brought to the test object. In addition, the accuracy requirement for hologram etching is much lower than for the surface accuracy requirement for a reflector system, a clear advantage at submillimetre wavelength frequencies.

Due to the above limitations, the hologram has no practical relevance in technical applications of space antenna testing.

The main test parameters and their technical consequences for a reflector compact range (Figure 6.17 and Figure 6.18) are given in the following list [7,8]:

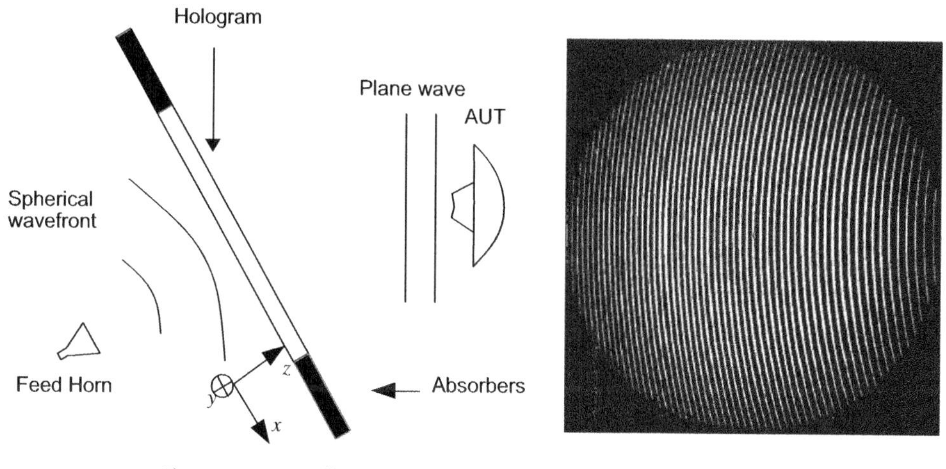

Figure 6.16 Collimating system by holographic transformation.

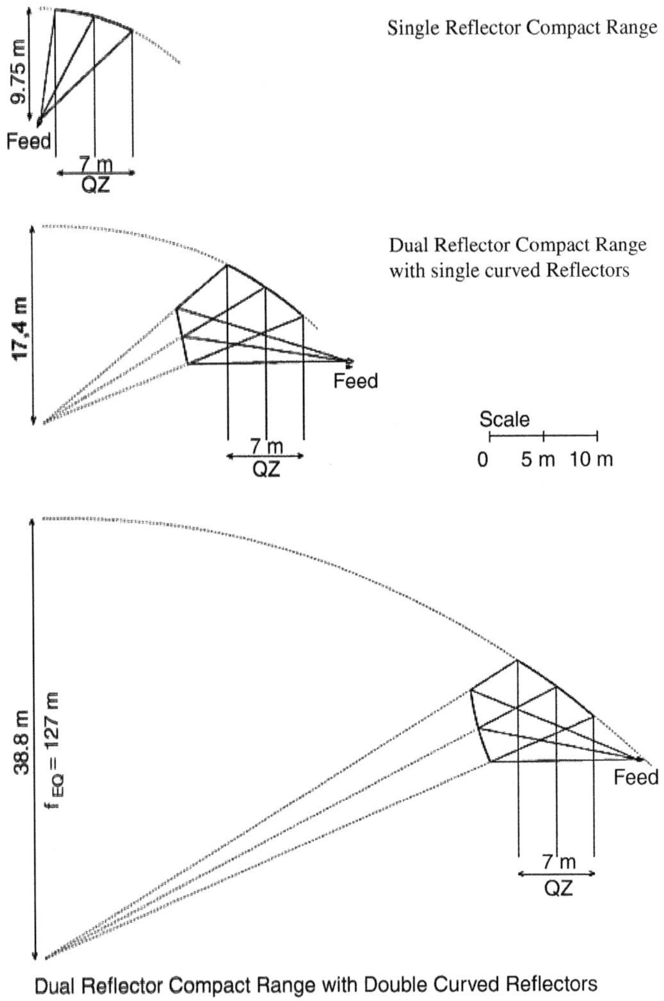

Figure 6.17 *Comparison of different reflector concepts for compact range systems.*

■ Broadband operation	⇨ Reflector compact range needed
■ Large AUT dimensions	⇨ Selection of suitable range type ⇨ Large reflector dimension ⇨ Large test chamber
■ High measurement accuracy	⇨ Selection of suitable compact range type ⇨ Special RF instrumentation
■ High cross-polar purity	⇨ Selection of suitable compact range type ⇨ Compensated compact range
■ Satellite system parameters (EIRP, *G/T*, AFR, PIM)	⇨ Selection of suitable compact range type ⇨ Tailored RF instrumentation
■ Measurement of radar cross-section (RCS)	⇨ Selection of suitable compact range type ⇨ Tailored RF instrumentation ⇨ Gating techniques (soft, hard)

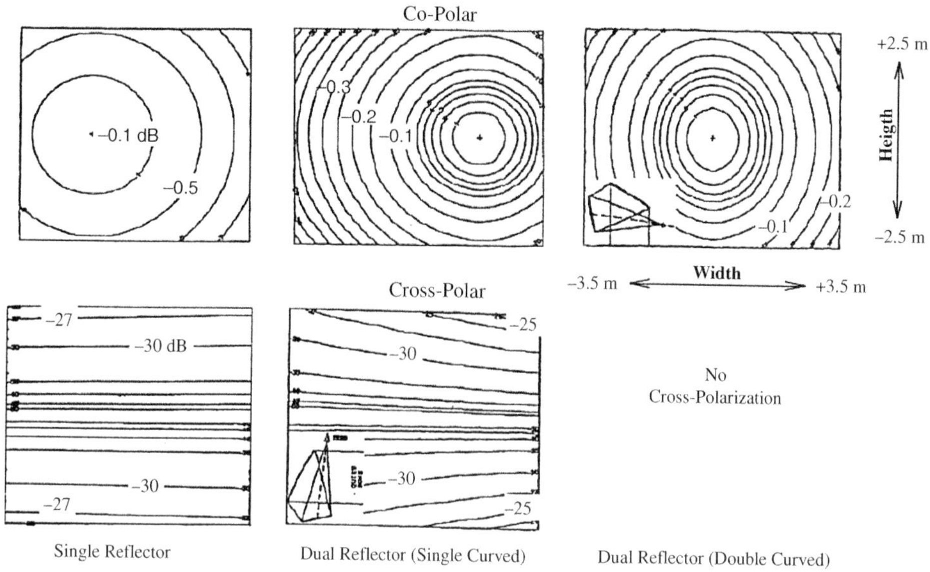

Figure 6.18 *Co- and cross-polar fields in quiet zone for different types of compact ranges.*

6.3.2.4 Facility Design

The selection of the best suited compact range reflector concept is based mainly on the performance requirements of the application [9]:

- Co- and cross-polar measurement accuracy.
- Cross-polar purity.
- Scanning capability.
- Room efficiency.
- Frequency range.
- Type of device under test.
- Test configuration.

Due to the different requirements for telecom, Earth observation, science and navigation antenna applications, high broadband operation is required from the test system. This leads to the main design goal for the system design to be frequency independent. Space antenna requirements, such as accurate measurement of sidelobe level down to −70 dB, cross-polar purity better than −45 dB and excellent gain determination, require a compensated design (Figure 6.19) of the field distribution and very low phase and amplitude ripple as well as low taper function.

6.3.2.5 Facility Performance Validation

To validate the RF performance of the compact ranges there are three different methods:

Plane-wave probing for determination of the quiet zone field planarity: The measurements are performed with reference standard gain horns in five different frequency bands from 1 to 100 GHz in the centre as well

Figure 6.19 A compensated compact range of the Cassegrain type.

as two scanned quiet zone(s). The related data evaluation comprises the amplitude and phase of the co-polar field and the amplitude of the cross-polar field, see Figure 6.20.

Antenna measurements for antenna pattern and gain comparison: The pattern and gain of two different calibrated reference antennas such as the two reference antennas from EADS Astrium GmbH are usually measured. The results can be compared with measurement data from previous test campaigns in different compact and near-field test facilities. Examples of this validation has been performed up to 500 GHz [10].

Payload parameter testing: The most important payload parameters EIRP, IPFD (Input Power Flux Density) and *G/T* are measured for a payload reference unit. The results are compared with calculated and previous measured values. To validate the required payload parameters the use of a specially designed payload test unit guarantees very high accuracy and repeatability [11]. An example is shown in Figure 6.21.

6.3.2.6 Features and Advantages for Space Application

Payload Testing Payload tests at the spacecraft level are mainly based on the following parameters [11]:

- Radiation pattern, coverage, gain
- Beam efficiency
- Optical boresight determination
- EIRP
- Saturated flux density
- Amplitude frequency response
- *G/T*, group delay
- PIM.

Figure 6.20 Mounted scanner for plane-wave probing.

The well-known Compensated Compact Range (CCR) with a nominal test zone of $5.5 \times 5.0 \times 6.0$ m and its instrumentation allows the characterization of the above-listed test parameters. A general problem here is the increasing size of the satellite, approximately 8 m, and more the fact that often the antennas are mounted on the opposite sidewalls of the spacecraft. To overcome this problem, the CCR has a special feature to generate multiple quiet zones to be centred around the respective satellite antennas, see Figure 6.22. The multiple quiet zones are generated by lateral movement of the range feed outside the focal point, resulting in a semi-plane wave. The high quality of the scanned plane waves is due to the long focal length of the compact range reflector layout. Most of the payload parameters above can be measured directly or have to be calculated from a set of measurement values.

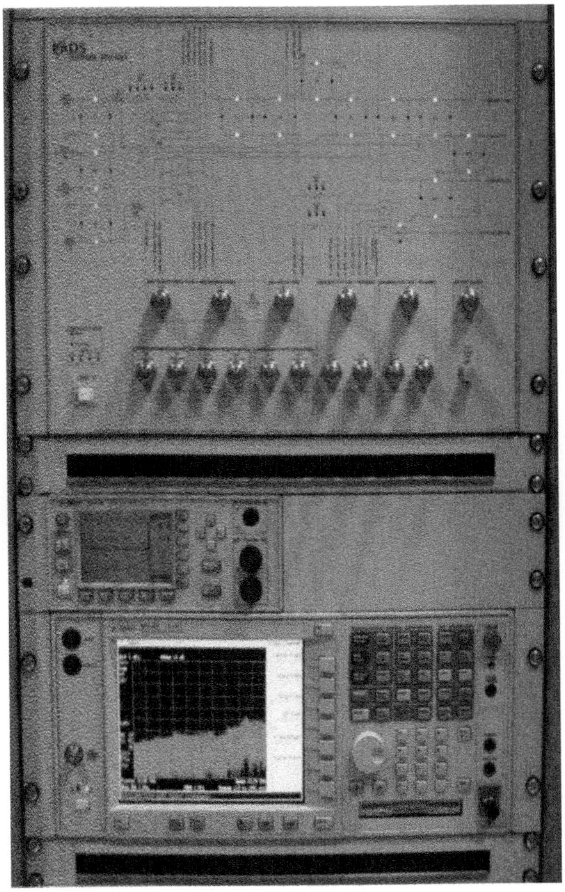

Figure 6.21 Computer-controlled payload test setup.

The multiple plane-wave capability introduces the opportunity to create two or more independent quiet zones in different positions, operating at different frequencies and polarizations. A typical CCR test setup for payload testing is shown in Figure 6.23.

With this test facility and test configuration, payload antenna and transponder tests can be performed in a very short time frame and to high measurement accuracy [12]. Its flexible use and its outstanding performance in frequency range and measurement accuracy led to the decision of more than 15 spacecraft manufactures worldwide to install a CCR for space antenna and payload testing.

Optical Antenna Alignment, Boresight Determination The RF boresight of an antenna after integration on a satellite has to be identical to the nominal boresight (design direction) with tolerances of the order of 0.01–0.02°. To meet this requirement the following tasks have to be undertaken:

1. Internal antenna alignment (i.e. feed, subreflector, main reflector with respect to each other and to the S/C).
2. Determination of the nominal boresight with respect to the facility system.
3. Determination of RF boresight with respect to the facility system.

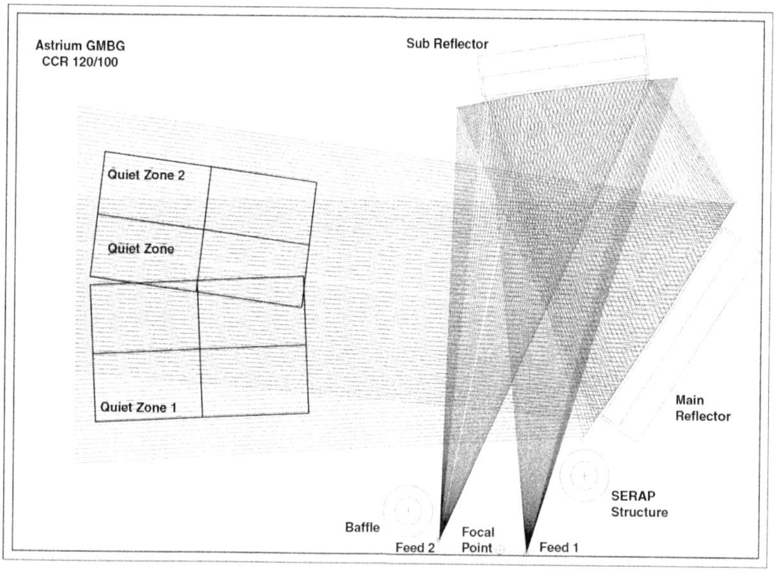

Figure 6.22 Virtually enlarged test zone by generating multiple QZs in CCR (SERAP, Serrated Absorber Protection).

The internal antenna alignment has been pretested and optimized in subsystem tests. The adjustment parameters have been determined and frozen in references like tooling balls or mirror cubes. At system-level antenna integration, the adjustment values should be realized by shimming the feed and reflector or fitting the reflector deployment angle. In some cases not all the axes are adjustable and the resulting boresight errors have to be compensated by feed displacement or reflector tilting. Finally the resulting boresight has to be verified by

Figure 6.23 Payload testing with antenna farm in CCR.

RF tests. The nominal boresight of the antenna is defined by the orientation of vectors with respect to the S/C system and materialized by two sides of a mirror cube. This allows these directions to be picked up by theodolites and to compare them with reference directions of the RF measurement facility. Finally the nominal boresight and – if needed – the nominal polarization direction of the antenna are known with respect to the references of the facility in coordinates of the antenna positioner. The boresight of the RF measurement facility (CCR) depends on the range feed location. For a range feed located at the focal point of the CCR, the RF boresight will be identical to the geometrical boresight of the facility, when it has been measured and is known with respect to references (mirror cube, theodolite, and so on). The accuracy of this process, including calibration errors, is of the order of 0.01°. For a range feed in a scanned location, the RF boresight of the facility will be tilted. The angle of arrival (AOA) of a constant phase front defines the RF boresight and can be predicted by commercial software (e.g. GRASP). A calibrated feed positioner is needed, which allows location of the range feed precisely at a desired position and a table with the predicted AOAs to be drawn up. An accuracy of the feed positioner of 1 mm results in a pointing error of approximately 0.003°. The accuracy of this process, including feed positioner calibration errors and reference accuracy, is of the order of 0.015°. A spacecraft with two antennas on the left and right sides can be measured with two range feeds in the scanned location. The RF pattern of each antenna can be calibrated knowing the nominal boresight defined by the S/C reference and the scanned facility boresight of each range feed. This allows verification of the RF pattern measurements at the system level and calibrated end-to-end tests [13].

6.3.3 Near-Field Measurements and Facilities

6.3.3.1 What is a Near-Field Measurement?

From electromagnetic theory it is known that the radiation pattern of an antenna changes as a function of the separation distance between the test antenna and the measuring antenna. From the previous section, the concept of the Raleigh distance is presented and is the accepted border between the far field (often referred to as the Fraunhofer region) and the subject of this subsection, the near field (also referred to as the Fresnel region). In fact, there is no absolute border and the classic far-field definition is only one of practicality, only an asymptotic result, the true far-field, which is the where an antenna will be illuminated by a single plane wave.

Therefore, all measurement facilities for antennas attempt in their specific way to produce a plane-wave illumination of the test object. Near-field measurements accomplish this by exploiting the Fourier relationship between measurements made at a finite distance from the antenna and the true far field. This relationship is based on measuring the radiation pattern of an antenna in both amplitude and phase for transformation to the far field. In order to effect a near-field to far-field transformation, it is necessary to describe the radiation pattern of an antenna as a finite expansion of orthogonal wave modes. Therefore, no matter which scheme for near-field measurement is employed, eventually it will be necessary to determine mathematically a set of planar, cylindrical or spherical wave functions and their coefficients which can then be used to determine the far field. These wave functions and the determination of their coefficients are applied in their respective geometries. There are therefore three basic types of near-field antenna test facilities: planar near field, cylindrical near field and spherical near field.

6.3.3.2 Probe Correction

Common to all three methods is the subject of probe correction. The probe correction accounts for the radiation properties of the measuring antenna and its effect on the sampled data. For the basic geometries, the theoretical basis for the interaction of the probe and the AUT was derived in the 1960s and 1970s. Kerns' classic monograph from the National Bureau of Standards on plane-wave near-field antenna–antenna interaction came first [14], and the cylindrical solution was completed by Leach and Paris [15] of the Georgia Institute of Technology in their classic 1973 publication based on the earlier work of Brown and Jull [16]. The complete probe-corrected spherical solution based on the Lorentz reciprocity theorem was first presented by F. Jensen in his PhD thesis [17] to the Technical University of Denmark in 1970.

If one names the antenna transmitting to or receiving from the AUT as the probe, it may be necessary, in order to achieve higher accuracy, to correct for the probe's own polarization characteristics and radiation pattern. If no polarization correction is employed, the effects on the measured co-polar pattern will be minimal while the cross-polar pattern will be more influential. This is true for even a perfectly linear polarized probe if the polarization axis in not aligned to the measurement frame coordinate system.

Probe pattern correction makes use of the probe's radiation pattern to account for the fact that if a directive probe is used, in the near field, the AUT subtends an angle of view seen from the probe. If this angle of view is well within the main beam pattern of the probe, then no correction is necessary. However, the probe pattern will cause attenuation of the received signal from directions of the antenna viewed by a part of the probe pattern attenuated w.r.t. to the boresight direction. In such a situation, this attenuation, in the case of a high-gain antenna, causes a narrowing of the main beam with a consequent 'optimistic' directivity value. To carry out probe pattern correction, the radiation pattern of the probe must be known in amplitude and phase.

One could conclude that use of a directive probe only complicates the transformation process as well as the probe calibration requirement (recall that the radiation pattern of the probe must be known), with little value added, and that the same result could be attained with the use of a low-gain probe such as an open-ended waveguide. Two very important points, however, can make probe correction very useful:

1. In planar near-field ranges, even the use of an open-ended waveguide will cause attenuation as the ratio of the antenna size in wavelengths to the antenna/probe separation distance increases, necessitating probe pattern correction.
2. In spherical near-field measurements, the directivity of the probe can be used to suppress unwanted facility stray signals such as room reflections.

6.3.3.3 Planar Near Field

The most common implementation of a planar near-field antenna test range is the vertical XY scanning device as shown in Figure 6.24. The measuring antenna or probe is moved in the vertical and horizontal directions, scanning in one direction. Note that the probe must be kept away from the reactive fields of the antenna. The higher the directivity of an antenna, the less influence these reactive fields will have on the result. At the same time, the probe must be kept as close to the antenna as possible to reduce the physical dimensions of the scan plane. Besides the reactive field, there is also the problem of multiple reflections between the aperture of the AUT and the near-field probe. The closer the near-field probe to the AUT, the more severe the probe–AUT interaction.

From the outset, it should be said that planar near-field measurements give their best results when used to measure highly directive antennas. The size of the required plane depends on the physical dimensions of the antenna, the separation distance between the probe and test antenna aperture, and the required extent for knowledge of the radiation pattern in angular degrees from the boresight. For example, if we wish to measure a reflector antenna at 10 GHz which has a diameter of 1 m, with a probe separation distance of 0.25 m, and require valid information out to 30°, then, using the empirically developed Figure 6.25, the scan plane will need to be at least 1.3×1.3 m.

Figure 6.24 *Planar near-field test range.*

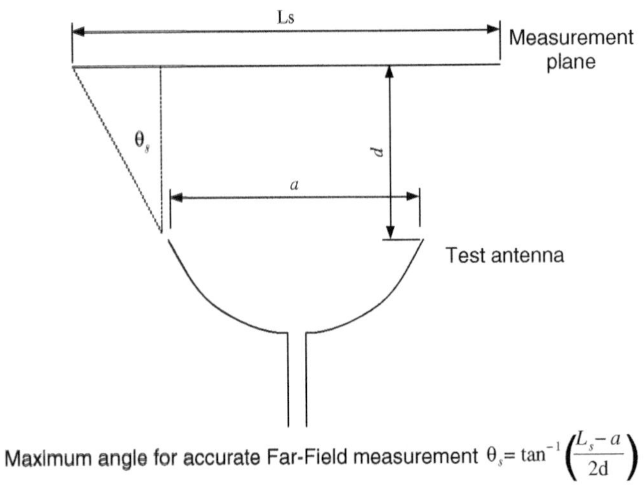

Figure 6.25 *Scan plane dimension definition for reflector antennas. Courtesy of NIST.*

Planar near-field measurements produce results in the far-field plane parallel to the sampling plane. There is a very useful rule of thumb with respect to the adequacy of the size of the plane being scanned on the accuracy of the far field in the far-field plane over the entire half space. That is, provided one's planar scanner is large enough, sample scans can be made to determine at which point in the near-field scan the relative field strength falls to a level, in the order of magnitude, 40 dB below the maximum levels measured near the centre of the scan. This information can then be used to determine the scan plane dimensions. Knowing that most of the radiated energy of the AUT has been sampled in amplitude and phase, then the far-field result in the half space should be valid to the accuracy of one's published error budget.

The sampling rule for planar near-field measurements is that the maximum distance between sample points should be no greater than half a wavelength. This, however, can be relaxed with judicious sampling based on ensuring that the aliasing products generated by the undersampling lie outside the angular area of interest.

The plane of the near-field measurement is usually rectangular. This means that the sampling occurs along lines in X or in Y with steps between the sample lines. Along the sampling line, the probe can be either stepped or in continuous movement depending on the scope of the measurement:

1. Number of frequencies to be measured and whether or not there is a need for collocation of the measurements for all frequencies.
2. Averaging/integration factor to be used for each frequency at the point of measurement.

Continuous probe movement along the scan has the advantage of being able to reduce considerably the time required to sample the entire plane. However, the scan speed will be limited by the speed of the measurement range receiver system at the desired signal-to-noise ratio and the distance between sample points. Ideally, a measurement point is just that, a point. However, measurements have a finite execution time and averaging or integration can be used to improve the signal-to-noise ratio of the acquired data, thus causing a smearing of the measurement point.

In addition, if measurement at each point consists of several frequencies, the faster the probe movement, the larger the actual distance between the first frequency and the last frequency.

The smearing of data points can also be aggravated if, in order to reduce measurement time, one decides to scan in both directions. If a scan is executed, one can choose to return the probe to the starting point and execute a step in the other coordinate and then initiate the next scan. This is commonly known as retrace scanning.

There is also bidirectional scanning where, instead of retracing to the starting point of the probe, the scan is initiated in the opposite direction. The degree to which one can tolerate the systematic errors being introduced is a function of the required measurement accuracy.

Finally, a few words on the near-field probe w.r.t. the measurement acquisition time are necessary. For general near-field measurements, it is generally required to measure both polarizations. If the probe has only one polarization, then it will be required to execute two complete planar acquisitions thereby doubling the measurement duration. And the longer the time required for a single acquisition, the greater the risk that system drift can cause an error to be introduced. This can, however, be greatly reduced with the use of tie scans. The tie scan is a single scan which goes across the acquired scan and is then used for data normalization. One property of XY planar near-field scanning is that if the AUT is known to be highly linearly polarized, and if only the co-polar field is of interest, it is only necessary to acquire the co-polar field.

Planar Near-Field Geometries In the previous subsection, the basic geometry was the vertical XY plane scanner. There are other methods and geometries to acquire a planar near-field measurement and far-field result and these are summarized here.

Horizontal Planar Scanner This is identical to the vertical scanner in operation, the only difference being that the scan plane, situated above the AUT, is horizontal.

The advantage of this scheme is that it allows the AUT to be 'rolled in' and placed beneath the scanner. There is no need for an antenna tower to fix the AUT to. An antenna placed on such a tower must be sufficiently robust to avoid distortion due to gravity; this is less of a problem for a horizontal scanner and it is easier to implement supports beneath the antenna.

It is easier, however, in the vertical scanner to change the measurement distance between the AUT and the scan plane by just moving the antenna tower closer or further away from the scan plane. The horizontal scan plane is also structurally more complicated. Access to the probe is not as immediate as with a vertical scanner.

There is also the aspect of antenna 'safety' and cleanliness, which, in this geometry, can become an issue. If the scanner is situated in a radio anechoic chamber, then the particle fallout from the absorbing material, and dust in general, is potentially more of a problem as it will settle in or on the aperture of the AUT.

Plane Polar Scanner Part of the complication of the horizontal planar scanner is its mechanical complexity. One method developed at JPL, USA, and exploited for the NASA Galileo mission, was to limit the horizontal scan to a one dimension linear scan. The planar data set is then completed by rotating the antenna under test beneath the linear scanner [18].

This necessitates the use of a rotating table on which the antenna under test is to be mounted. While such equipment is readily available, the advantage of just rolling in your antenna to be measured, as in the case of the horizontal XY scanner of the previous section, is lost.

The sampling grid will also not be the normal equally spaced samples in X and Y but will be like in the Figure 6.26. The JPL development relied on carrying out the near field to far field transformation by first describing the measured field as a series of Jacobi-Bessel functions. Another method which will allow one to exploit the simpler, and more direct FFT, is to effect a dense enough sample pattern to then interpolate the data onto a regular XY grid.

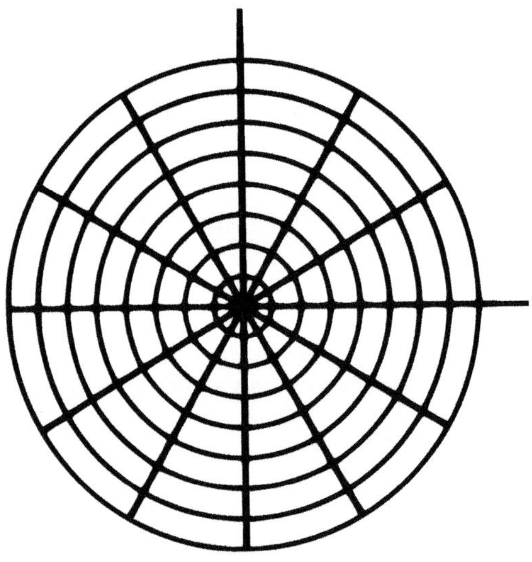

Figure 6.26 *The plane-polar sampling grid.*

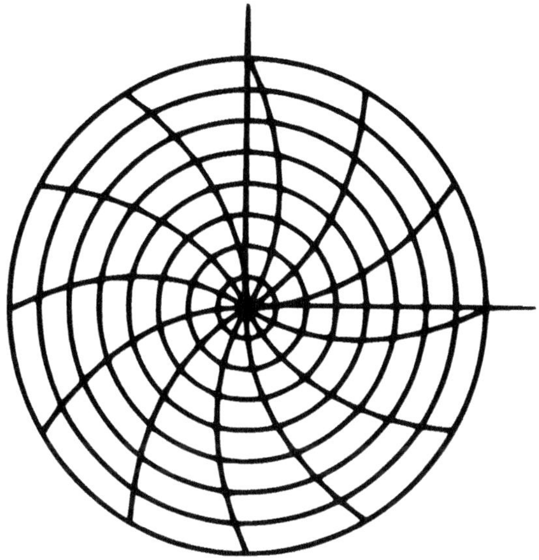

Figure 6.27 *The bipolar sampling grid.*

If the probe does not have a rotating stage itself, then the sampled field will no longer have the same polarisation orientation over the plane. Precise and accurate knowledge of the probe's polarisation characteristics, however, will allow one to reconstruct the field at all sample points to align with the antenna under test's coordinate system.

The Bipolar Planar Scanner Combing the advantages of the XY Horizontal scanner and the simplicity of the Plane Polar scanner leads to what is called the Bipolar scanner. In this method, the antenna under test is again 'rolled in' and placed beneath the scanner which consists of a rotating gantry arm with a linear translation of the probe along the arm.

Here, the sampling grid (Figure 6.27) does not lend itself to a direct transformation algorithm and the process of sufficient sampling to interpolate onto a regular XY grid is to be preferred. It should be noted that if the measuring probe does not have a rotation stage, the samples polarisation components in amplitude in phase will be rotated from point to point. As in the case of the Plane Polar Scanner, how probe polarisation is treated, is not to be forgotten.

6.3.3.4 Cylindrical Near Field

The cylindrical near-field scanner (Figure 6.28) samples data on a circular cylinder. This has several advantages over its planar counterpart. First of all, the scanning itself is simplified in that there is a linear scanner stage for the probe antenna and a rotating table, or azimuth positioner, on which the AUT is mounted; this is the same as the plane polar near-field scanner except for the fact that now the linear translation stage is parallel to the rotation axis. It also provides greater coverage of the antenna's radiation pattern in that now it is possible to measure field strengths beyond the azimuth $\pm 90^{\circ}$ of the planar scanner, at least in one dimension. The same considerations and restrictions in the other dimension, however, apply to the planar sampled field.

The near field to far field transformation is based on the use of cylindrical waveguide functions, the cylindrical wave expansion, based on Bessel and Hankel functions. The original treatise was already done by

Figure 6.28 The cylindrical near-field scanner at EADS Astrium, Ottobrunn. The vertical linear stage is on the left and the antenna under test is mounted on the azimuth positioner. Courtesy of EADS Astrium.

Brown and Jull in 1961 [19] where the sample field is described in cylindrical wave functions was deconvolved with the probe using the Lorentz Reciprocity Theorem; the probe field in this case was based on a plane wave expansions of the probe. The original work by Brown and Jull was then expanded and completed by Leach and Paris [20].

Thus the increased knowledge obtained on the antenna and the simplification of the mechanical implementation are paid for with a more complex processing to obtain the far field.

6.3.3.5 Spherical Near Field

Much of the ground breaking theoretical background for a near field to far field transformation algorithm was done in the 1960s at the Technical University of Denmark and was first published in the PhD thesis of F. Jensen [17]. However the general spherical wave expansions of the antenna under test and the probe could not be deconvolved because of the non-orthogonality of the spherical wave functions themselves. Work on a solution continued at both TUD and NBS; NBS was first to publish the solution which is based on constraining the probe antenna to having a rotationally symmetric aperture. In other words, the probe antenna because of its rotational symmetry, only has two azimuthal indices in the spherical wave function set. This allowed for the reduction of the probe and antenna under test field to be expressed as two series of orthogonal spherical wave functions [21]. The doctoral thesis of Flemming Holm Larsen made the last step by reworking the spherical transmission formula in terms the of scattering matrix theory [22] as done by Kerns [14] and is the basis for the commercially available spherical near field to far field transformation employing full probe correction, i.e. pattern and polarisation.

There are now many implementation schemes for spherical near field antenna measurement ranging from a fixed probe to a fixed antenna under test with a scanning probe. The original spherical near field range at TUD employed a fixed probe with the antenna under test mounted on a model head which allowed for roll rotation. This system, mounted on an azimuth positioner, allows for the acquisition of E-theta and E-phi over the entire sphere where the roll axis of the system corresponds to the Z-axis of the conventional spherical coordinate system; note that there will be some blockage due to the mast of the positioning system partially blocking the back radiation of the antenna under test.

The data is gathered on a spherical surface and it is essential that all axes of rotation intersect, otherwise there is no centre of a sphere and therefore, the phase measurements will be invalid. There are essentially two set ups with a fixed probe where the antenna rotates:

1. Roll over Azimuth: This is the case where the antenna is pointed along the Z axis of the standard spherical coordinate system as shown in (Figure 6.29) The antenna is then said to be polar pointing in the measurement.
2. Elevation over Azimuth: This is the case where the antenna is pointing along the X axis of the standard spherical coordinate system. In this case, the antenna is said to be equatorial pointing in the measurement.

Each system has its advantages and disadvantages. For instance, polar pointing antennas are rotated in the roll axis over a full 360 degrees. As the antenna is designed for a zero-g operation, it can deform during rotation and invalidate the measurement result and even possible damage the structure itself. In other situations, for

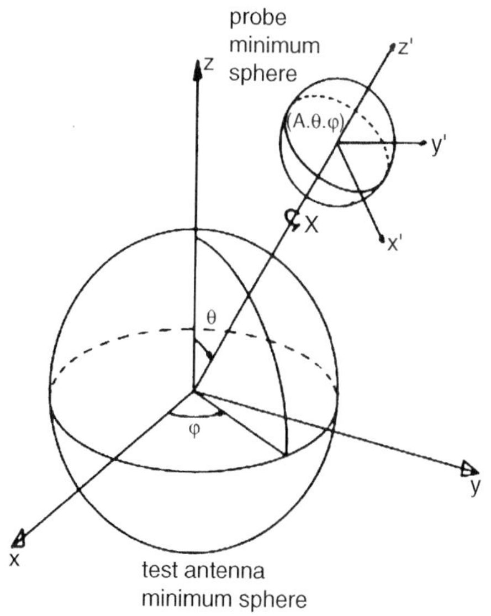

Figure 6.29 *The spherical near field test coordinate system.*

Figure 6.30 The Satimo probe array for spherical near-field antenna measurements.

example with a full spacecraft or spacecraft mock-up, the roll rotation is very impractical, especially with respect to the MGSE needed to mount the device to the test positioner, which has its limitations in this configuration. The more complex and large the test object, the more one wishes to use an Elevation over Azimuth system. It should be said that there is no one ideal solution. It is always a trade off between requirements and practicality and must be assessed in each individual case as a function of performance and cost. For a complete treatise on spherical near field antenna measurements the reader is referred to the definitive work by J.E Hansen [23].

One solution employed in the past to avoid roll or even elevation rotation of the test object, is to have a probe rotating on an articulated boom. Here, the probe itself is rotated about the test object, see Fig. 6.29 for one axis of the spherical coordinate system while the azimuth rotation is as usual.

As with all the other geometries of near field antenna measurements, the data acquisition is time consuming. This is the reason why near field systems are not generally used for final system level performance checks even though they can perform them.

To address the data acquisition time issue, there is available a technique using probe arrays which can be scanned electronically. There is any number of variations on how these arrays can implemented, including planar and cylindrical near field measurements. The system was developed and commercialised by Satimo, France. An example of the most common configuration is shown in Figure 6.30. While these systems are very fast, they do have limited bandwidth and the distance from the test object to the centre of rotation is fixed, as well as the data point spacing does not have the same flexibility of a conventional scanning system.

6.3.4 Environmental Test Facilities and Mechanical Testing

Modern high-tech products, especially in the space flight industries, are required to operate under extreme conditions which can vary as a function of the Earth orbit or deep space application. It is easy to understand that every component for the spacecraft must operate correctly not only in the manufacturer's integration halls, but also during the mission which comes after extreme launch loads and stress. For example, as there is no gravity and no atmosphere, these conditions must be simulated on Earth.

The absence of the atmosphere, for example, leads to a completely different heat exchange between the satellite's components and can, in the worst case, cause damage and mission failure if these conditions are not taken into account during the design phase. All effects acting on a spacecraft from launch, orbit transfer and final orbit are referred to as the *environmental conditions* to which the spacecraft is exposed.

Due to the exposed position of space antennas on the spacecraft, it is extremely important to ensure that the design withstands all these environmental conditions before launch, as the spacecraft is no longer accessible for repair purposes once in orbit. For this reason spacecraft manufacturers provide modern *test laboratories* equipped with a large variety of different facilities for simulating these environmental conditions and for the operational testing of partly or fully integrated systems.

State-of-the-art environmental test facilities are integrated in large test halls meeting the requirements of US Federal Standard 209B, class 8 (formerly 100.000) concerning cleanliness. The facilities ensure the safe and efficient testing and handling of specimens. In addition, the test laboratory provides facilities, tools and measurement devices for further measurements with respect to the:

- centre of gravity
- moment of inertia
- structural balancing (static and dynamic)
- mass determination
- spin testing.

It should be possible to perform complete functional performance test and intermediate check sequences at one location in order to avoid to the extent possible the risks associated with transportation of the spacecraft as well as being able to fulfil the usually tight schedule of modern spacecraft programmes. To perform environmental testing, the real effects acting on an antenna system must usually be simulated by the following tests:

1. Climatic tests, which involve combining different temperatures and humidity, possibly in addition to pressure to simulate atmospheric effects adapted to selected places within the Earth's atmosphere. An example for a nitrogen-operated thermal test facility can be seen in Figure 6.31.

Figure 6.31 Climatic test chamber with nitrogen cooling system.

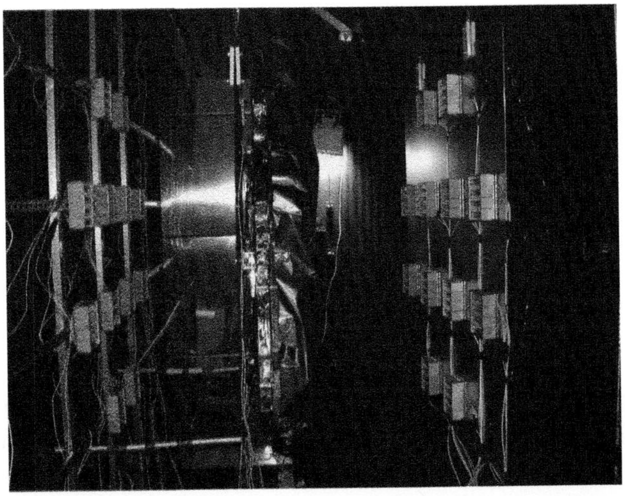

Figure 6.32 *Thermal vacuum chamber with solar simulator by IR radiator array.*

2. Vacuum or thermal vacuum tests which simulate conditions in space without an atmosphere, considering heat transfer by radiation.
3. Tests on the design and functional performance of the thermal subsystem and the complete spacecraft system are undertaken in thermal testing facilities (also called heat balance, space simulation or thermal vacuum facilities). Under simulated thermal space environmental conditions, verification of the thermal model calculation and qualification and acceptance of the functional performance of the thermal control subsystem and the spacecraft system have to be achieved. The essential space conditions which are simulated in thermal testing facilities are vacuum conditions, radiating heat sink and solar radiated input. A thermal vacuum test chamber with halogen radiators is shown in Figure 6.32.
4. Vibration tests simulate all kinds of mechanical stresses on an antenna system which might be applied during real missions due to the enormous oscillations of the engine or launcher, mechanical shocks due to handling, firing propulsion engines, or snap-into-place events. A typical test setup for the vibration test of a top-deck reflector antenna is shown in Figures 6.33 and 6.34. To generate the required test level for a typical telecommunication antenna, electrical power of approximately 500 kW is needed. The maximum possible shock vector will be of the order of 500 kN. To avoid unacceptable coupling with other sensitive test systems an independent seismic block is needed for the shaker operation.
5. Special test setups may be required, in some cases, for the combined environmental simulation of vibration and temperature, for instance.
6. Acoustic noise tests are performed to qualify the space antennas, especially large deployable antennas, against the acoustic environment encountered during the launch of the spacecraft, in particular at liftoff and in the transonic regime. Re-entry vehicles are subject to a combined environment of thermal loading and fluctuating pressure fields. This thermo-acoustic condition can be simulated in a special *progressive wave tube* which may be extended by a flame heating system. By means of a certain number of microphones, the noise field can be controlled and monitored. One of the largest acoustic noise test facilities in Europe, shown in Figure 6.35, is installed at IABG in Munich, Germany. This facility is also able to perform system tests for the complete, installed, largest telecommunication satellites.

Figure 6.33 *Vibration test setup of top-deck antenna with horizontal test axis; moving coil shaker system (left side) and lateral moving slip table (with specimen).*

Figure 6.34 *Vibration test setup of top-deck antenna for vertical test axis; moving coil shaker system of 500 kN with mounted test antenna at adapter cube.*

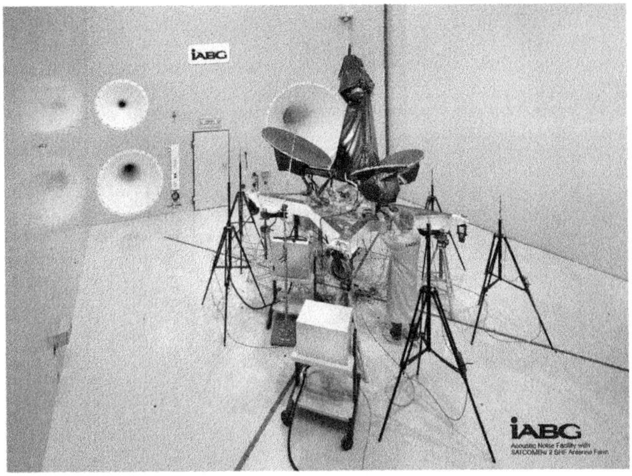

Figure 6.35 Acoustic noise test chamber with top-deck antennas for telecom satellite. In the back wall are the outlets of three sonic horns for different frequency bands.

6.3.5 PIM Testing

6.3.5.1 PIM Testing on the Component Level

All critical antenna items with respect to the entire antenna performance are the possible occurrence of passive intermodulation products (PIMP)s in the case of multiple-carrier high-power operation. Of special interest are products which fall into the satellite receiver frequency band. Potential antenna–PIM sources are mainly within the feed chain but also reflector structures like carbon fibre materials, gridded and large-mesh reflectors are typical candidates for PIM sources.

The setup (Figure 6.36) consists of two high-power carrier signals generated by two synthesizers in conjunction with two high-power amplifiers (HPAs). Two carrier signals are combined in a triplexer to feed the Tx port of the DUT. The Rx port of the triplexer is connected via a bandpass filter, a low-loss elliptical waveguide and a low-noise amplifier (LNA) to a spectrum analyser outside the chamber.

The tests should usually be performed in an anechoic chamber (Figure 6.37) with standard EMC shielding to avoid any reflections or other disturbances from walls or from outside the chamber. The PIM health check should be performed under ambient environmental conditions.

Additional thermal monitoring of the absorber wall within the facility might be implemented to avoid dangerous thermal overloading of the absorber material. IR cameras are well suited for thermal monitoring of the radiated field.

6.3.5.2 PIM Testing on the Antenna System Level

The principle of the test setup for the PIM (performance test) of satellite system testing is shown in Figure 6.38. For this test two uplink carriers are radiated into different satellite receive channels. The magnitude of the uplink carriers F1 and F2 is chosen so that the satellite transmits each carrier F1* and F2* in the downlink with equal $EIRP_{sat}$. Generally any PIM signals, if at all, will be generated by the passive components (e.g. waveguide flanges, O-MUX, etc.) within the satellite transmit signal path. The frequency difference between the uplink carriers F1 and F2 is chosen with respect to the satellite frequency plan in such a way that a worst

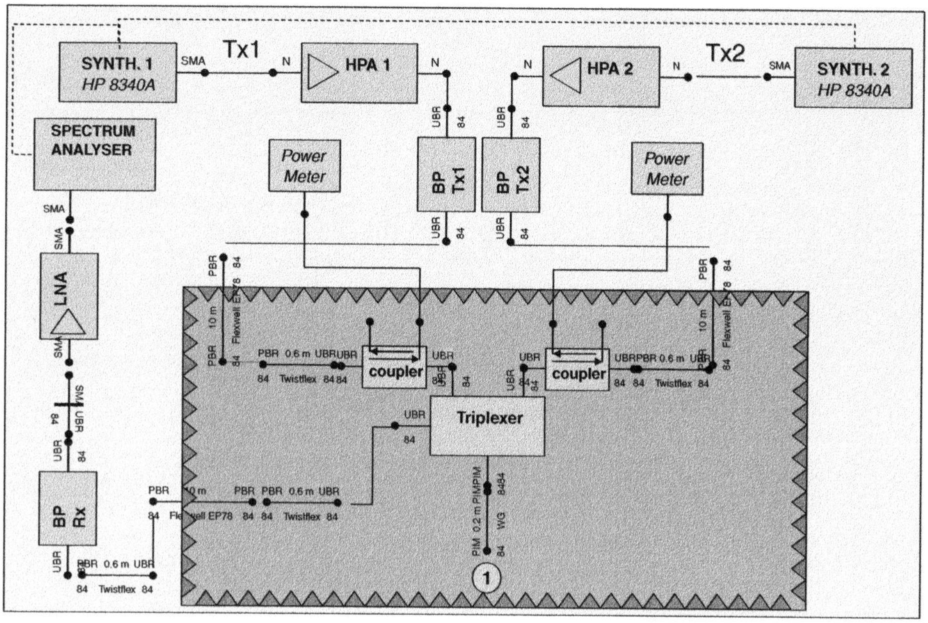

Figure 6.36 Generic PIM test setup for third-order PIM products.

case situation is produced. This situation is so that a possible product of PIM and the frequency

$$\text{PIM}-\text{frequency} = \pm M^*(\text{F1}^*) \pm N^*(\text{F2}^*)$$

(with M, N the harmonic order, $M + N$ an odd number) is generated by the nominal satellite transmit signals F1* and F2*, and falls into the satellite receive frequency band. This PIM product then acts like an uplink signal

Figure 6.37 PIM setup with wooden MGSE and anechoic environment.

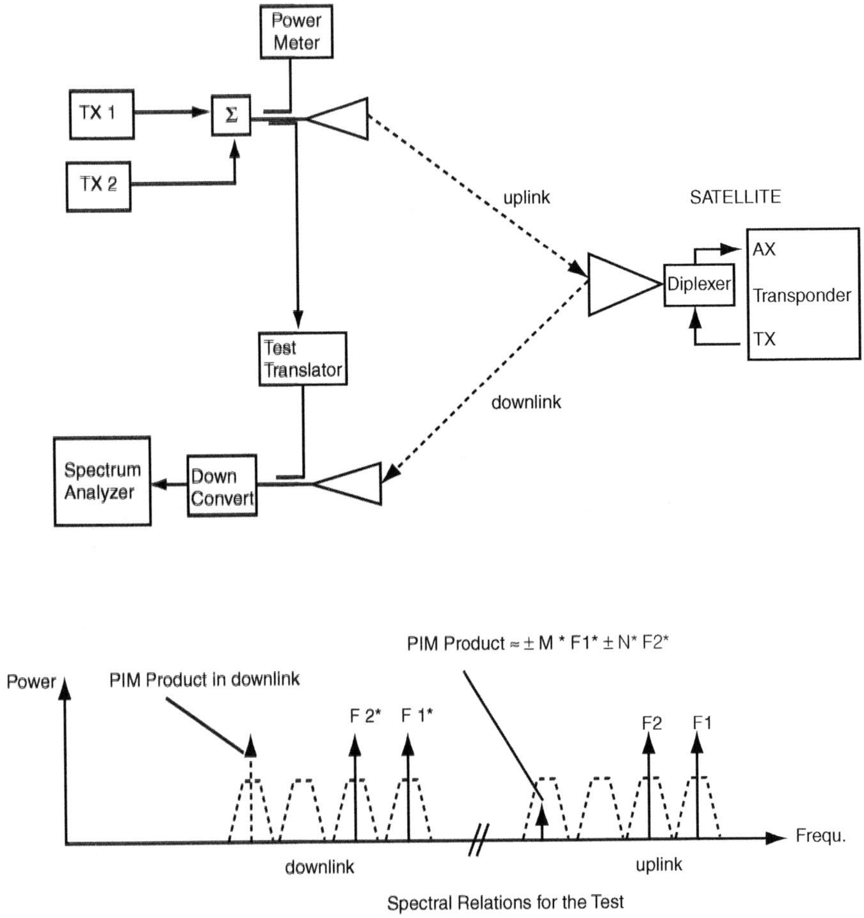

Figure 6.38 *PIM measurement principle setup.*

to the satellite. The transponder frequency is converted and transmitted in the downlink, where it can be detected [11].

6.4 Case Study: SMOS

The SMOS (Soil Moisture and Ocean Salinity) mission is the second of the ESA's Earth Explorer Opportunity Missions, within the ESA's Living Planet Programme. SMOS has been the result of cooperation among three European institutions dedicated to the promotion of the space industry and science, namely ESA, the CNES (the French National Centre for Space Studies) and the CDTI (the Spanish Centre for Industrial Technological Development).

Microwave radiometry at L-band is considered the most adequate technique and the appropriate frequency band to measure soil moisture and ocean salinity. However, the main drawback of L-band radiometry is that to get the minimum desired ground resolution requires the use of a very large antenna, and therefore an instrument such as the one required in this mission created a significant technological challenge.

Figure 6.39 SMOS in orbit. Courtesy of ESA.

In order to overcome the technological problems associated with real antenna apertures, aperture synthesis was employed, which had been successfully demonstrated a few years earlier with the ESTAR (Electronically Steered Thinned Array Radiometer) instrument, which was the first one-dimensional synthetic aperture radiometer.

The concept of synthetic aperture interferometry radiometry was developed in the 1950s to obtain high-resolution radio images of celestial bodies. In [24] it was proposed to use it for Earth observation as a way of increasing the angular resolution of individual antennas.

Initial studies of 2D aperture synthesis for Earth remote sensing were carried out until 1996 [25]. The conclusion was that, to meet the scientific requirements for passive imaging from space of soil moisture and ocean salinity at L-band ($\lambda = 21$ cm), large antenna arrays using aperture synthesis principles offer clear advantages over mechanically or electronically steered antennas and push-broom radiometers. A T-shaped design suitable for space platform accommodation was merged with the design used for the Very Large Array (VLA) Radio Telescope in New Mexico, namely a Y-shape with an exponential element distribution. The MIRAS (Microwave Imaging Radiometer with Aperture Synthesis) instrument has a Y-shaped configuration with equally spaced antenna elements, which has important advantages for ground resolution and grating lobes and is suitable for spaceborne applications [25,26].

The SMOS mission (Figure 6.39) comprised a satellite in a low Sun synchronous orbit, with an altitude of 755 km and a revisit time of three days, that passively measures the electromagnetic noise generated by the Earth at L-band (1.4 GHz) with a spatial resolution of 50 km, a radiometric sensitivity of between 2 and 4.5 K (depending on the observed area) and a radiometric accuracy of less than 1 K. The satellite was launched in November 2009.

6.4.1 The SMOS MIRAS Instrument

Built by a consortium of over 20 European companies led by EADS-CASA Espacio, the MIRAS instrument is the only instrument carried on board the SMOS mission (Figure 6.40). The main subsystems of the payload [27] are the:

- thermo-mechanical architecture
- one-hub structure

Figure 6.40 *SMOS MIRAS during integration at EADS Astrium. Courtesy of EADS Astrium.*

- three-arm structure
- three deployment mechanisms
- hold-down mechanisms for the three arms
- thermal control hardware (MLI, radiators, thermistors, thermocouples and heaters).

For the electrical architecture, there are:

- 66 LICEF units (antennas + receiver) including optical modulators;
- 1 correlator and control unit (CCU) with on-board SW, digital correlator and optical demodulator;
- 1 calibration system (CAS);
- 12 control and monitoring node units (CMNs);
- 3 noise injection radiometers (NIRs);
- 1 optical harness to implement the link to carry the IQ signals (the optical hardware distributed through the CMNs, LICEFs, CCU and harness is called MOHA (MIRAS Optical HArness)).
- 1 pyro-unit to trigger the arms' deployment by releasing the hold-down mechanisms;
- 1 X-band assembly to download the scientific data to ground.

With its arms extended, the MIRAS instrument has a wingspan of 8 m and a weight of 360 kg.

The 66 LICEF (Light Cost-Effective Front-end) units and the 3 NIRs are distributed along the three arms and the central hub structure. Each LICEF contains a dual-polarization antenna and a switched filtered receiver at L-band (1404–1423 MHz) that selects the active polarization (see Figure 6.41). Each LICEF amplifies, filters and downconverts to IF the noise signal captured by the antenna elements. A 1-bit analogue-to-digital converter (ADC) converts the signal that is then optically sampled in the MOHA unit included in the LICEF enclosure (Figure 6.42) and transmitted to the CCU via a fibre-optic harness link where it is downconverted and correlated [28]. The MOHA subsystem also includes an adequate number of optical splitters to perform the optical signal distribution along the MIRAS arms and central hub. SMOS is the first European space mission to use on-board fibre-optic communications extensively. The NIRs are composed of two identical LICEF receivers plus one LICEF antenna, which permits the simultaneous detection of both antenna polarizations.

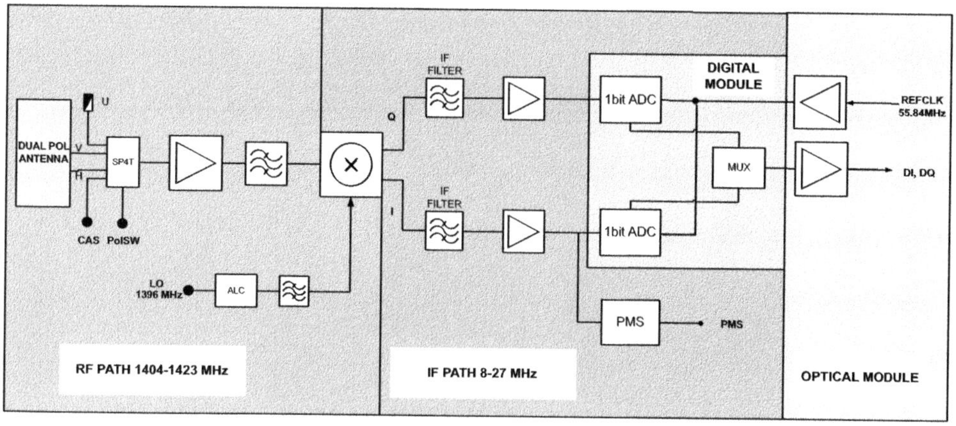

Figure 6.41 *LICEF receiver block diagram.*

The NIRs [29,30], in conjunction with the internal CAS, are necessary in MIRAS to calibrate the instrument [31].

The switch at the receiver input selects one of the two antenna outputs (H or V), uncorrelated noise from a resistor at ambient temperature (U) or correlated noise from the noise distribution network (CAS).

Each of the LICEFs, LICEF-NIRs and the associated calibration network are controlled and monitored by the CMN units distributed in the payload. There is one CMN placed in each of the 12 segments that comprise the arms' structure of the SMOS PLM (Figure 6.43). The CMN acts as a remote terminal of the instrument control function of the CCU, allowing the latter to reach the elements that are located far from it.

The CMN establishes the measurement configuration of the LICEF units when commanded by the CCU, sets the configuration of the calibration network and turns on/off the solid state switch to modulate the heat injection to the segment structure. Further, the CMN allows two modes of operation associated with the LICEF and LICEF-NIR settings: observation mode and calibration mode.

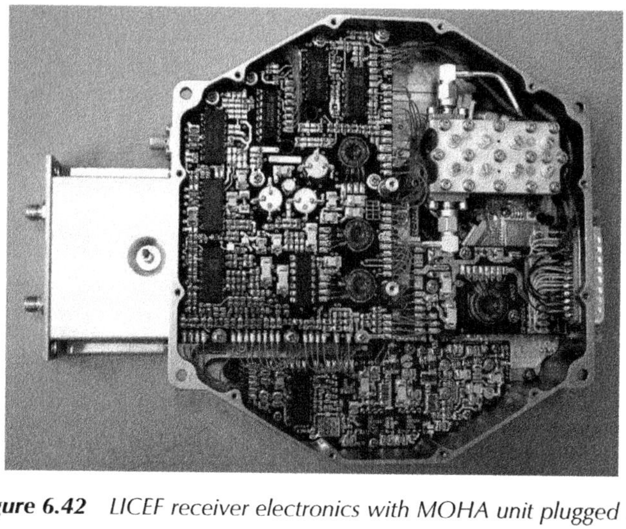

Figure 6.42 *LICEF receiver electronics with MOHA unit plugged in.*

Figure 6.43 Integration of units in one arm segment (LICEFs are along the axis, CMN is the black unit above the LICEFs, optical harness (at bottom) distributed along the arm, and one noise source of the CAS can be identified in the bottom right of the photo).

The local oscillator that reaches each of the LICEF units is also generated in the CMN (25th harmonic of the reference clock signal from the CCU) and distributed to the arm segment units through coaxial cables.

As for the mechanical aspects, the central hub and the arms are made of carbon fibre, which gives a high degree of rigidity to withstand the loads sustained by the MIRAS instrument during launch, with minimum weight. The deployment mechanisms, described in [32,33], were designed to provide a synchronized deployment among the three arms, with pure mechanical control to avoid the need for power consumption or servo-control electronic modules and forcing a slow deployment to minimize perturbations to the satellite's attitude and reduce moments at the end of the movement. The operation is based on a spring-driven deployment that includes a speed regulator per arm to render negligible the end arm shock on the spacecraft. Extensive testing was performed at element level as well as at subsystem level in order to demonstrate the proper behaviour of the deployment mechanisms. Successful testing at satellite level (including vibration, thermal cycling, acoustic and zero-*g* deployment tests) confirmed the adequacy of the design to the assigned function.

Thermal control is defined to maintain all the payload equipment within the applicable temperature range. It is based on a mainly passive design supported by heater systems. The passive part of the thermal hardware consists of a flexible second surface mirror (FSSM) coating for the thermal radiators, MLI blankets covering the deployment mechanism brackets, Ge-coated Black Kapton foil covering the LICEF antennas, and thermal doublers to spread the heat within the antenna receiver array (Figure 6.44). The heater part includes electrical resistance heaters to control the equipment temperature in the instrument measurement/calibration modes powered through the CMNs and controlled through the CCU resident algorithm, which is fed back with temperature readings from sensors installed on LICEF and NIR receivers. A second group of heaters is controlled by the spacecraft platform in the safe hold and standby modes to keep equipment temperatures above non-operational values [34].

The thermal control system ensures that the temperature between any two LICEFs is less than 6 °C. Further, to guarantee that temperature variation effects on the performance of MIRAS are minimal, the LICEFs are calibrated frequently to align their performance. Calibration takes place by injecting correlated noise from the CAS to all the LICEFs and correlating the signals from each LICEF as if the instrument were in observation mode.

Figure 6.44 *SMOS payload during AIT activities at the Plestezk launch site. MIRAS arms and PROTEUS solar panels stowed. Protective Ge Kapton foils cover the payload.*

6.4.2 SMOS Model Philosophy

The model philosophy in SMOS was intended to allow the full qualification of each unit and subsystem as well as the qualification of the overall SMOS payload.

The overall SMOS model philosophy consisted of a sequence of three models, STM, EM and PFM, and each of the instrument subsystems (from the antenna elements to the CCU and software) were the subject of low-level model philosophies that included the development of breadboards, EMs, EQMs and FMs or PFMs that were submitted to their respective qualification and acceptance procedures. In all, recalling the definitions included in Section 6.2, the SMOS model philosophy falls within the hybrid model approach, in which qualification is performed at subassembly level, and a set of tests is conducted at system level, which is denoted as the PFM system model.

6.4.2.1 MIRAS STM

The STM of SMOS consisted of a structure and thermal hardware conforming to the flight standard design and configuration, equipped with a set of QM and mass dummy elements, namely:

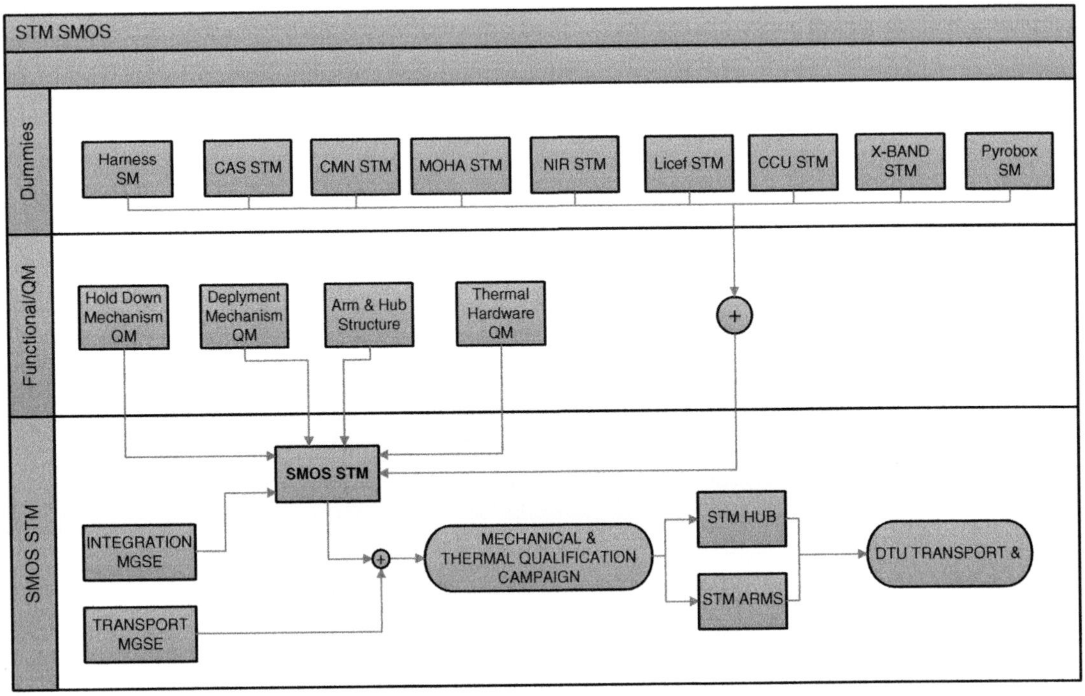

Figure 6.45 SMOS STM model flow.

- QM for the hub structure.
- Flight standard build structure for three arms.
- QM for the deployment and hold-down mechanisms for one arm.
- SM for the deployment and hold-down mechanisms for two arms.
- Structural and thermal dummies for electronic units (masses, dimensions, centre of gravity, thermal dissipation and harness interface characteristics).
- Mass representative electrical harness, optical harness and RF cabling.

The block diagram of the STM model is shown in Figure 6.45.

It is worth mentioning that the STM, apart from being used for the qualification of several mechanical and thermal aspects of the payload, was also used to refine, in advance, all the MGSEs that would be used in the PFM, such as the integration stands for the assembly of the arms, fit check with the PFM transport dolly tool and PFM transport container, and fine tweaking of the deployment zero-*g* test setup.

The qualification campaign on the STM SMOS included mass measurement, sinusoidal vibration test, acoustic test, pyroshock, and deployment and thermal balance tests. They were performed in the ESA facilities at ESTEC, Noordwijk. The objective of the vibration test was to validate the MIRAS structural design (Figure 6.46) for the launch and to confirm the mechanical environments specified for each of the subsystems. The TBT (Thermal Balance Test) performed on the STM was used for validating the mathematical model hypothesis and for tuning the thermal control subsystem parameters.

Once the STM campaign was completed, the arm segments and the central hub were reused to perform a dry run of the test setup in the anechoic test range facilities located at the DTU-ESA Spherical Near-Field Antenna Test Facility in Denmark, the facility selected to characterize the radiation patterns of the LICEF units. It

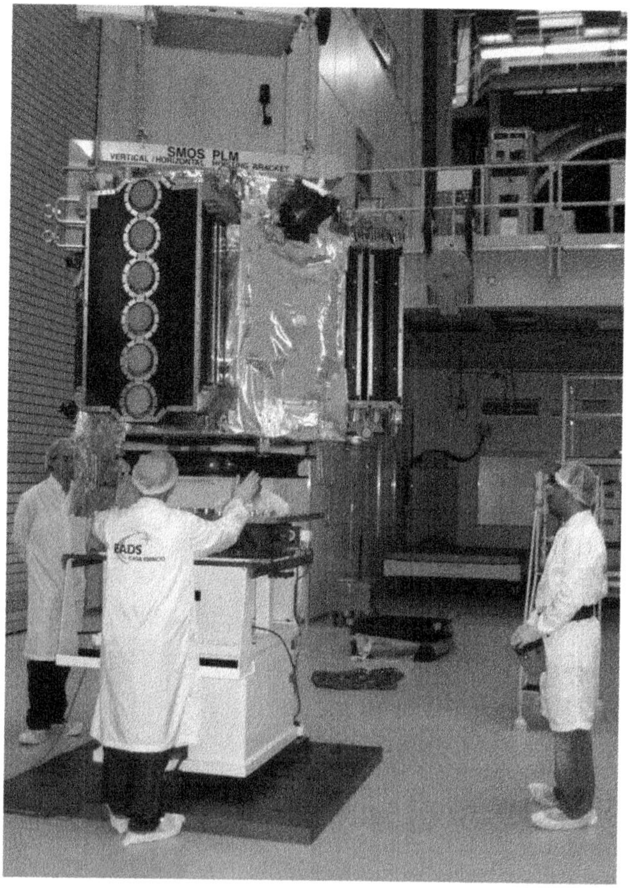

Figure 6.46 *Stowed STM MIRAS model.*

should be mentioned that the DTU was involved in a previous preparatory programme to validate both the facility and the test philosophy for the LICEFs.

Figure 6.47 shows the central hub of MIRAS equipped with the STMs and SMs of the LICEF units during the dry run, ready to be lifted up to the DTU tower mast for antenna testing. The specific MGSE that was designed to allow the transport and pattern testing of the LICEF antenna units was successfully tested.

6.4.2.2 *MIRAS EM*

The EM of MIRAS consisted of a test bench and representative EMs of all the electronic units, or subsets of them (Figure 6.48). The purpose of the EM MIRAS was to verify that the EM is working according to the PLM requirements and that the PLM units are compatible in terms of electrical interfaces in order to have the possibility of reacting to any interface incompatibility between pieces of equipment. The EM build was also checked against the MIRAS EGSE in order to detect any possible communication error.

The EGSE's mission was to establish a connection with the SMOS MIRAS instrument playing the role of the spacecraft, and for that purpose included a platform simulator that supplied both power and communication lines (Figure 6.49). It also included a simulator to acquire science data directly from the CCU and to check the data format was identical as sent by the nominal path through the RF link. It was possible to use this simulator

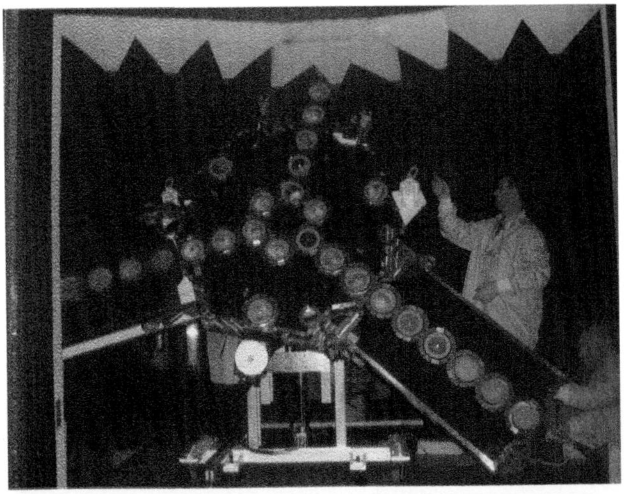

Figure 6.47 Dry-run test of the MGSE and test setup of MIRAS in the DTU Antenna Test Facility.

in the absence of the X-band subsystem or when the assembly, integration and test (AIT) activities did not require specific use of the RF link.

The MIRAS EM (Figure 6.50) included the following units:

- One CCU and SW.
- Two LICEF units equipped with MOHA.
- One NIR equipped with MOHA.
- One CMN.
- Optical harness link between two LICEF units, one NIR unit, one CMN and CCU including:
  - optical splitters (one 1 × 8 and one 2 × 12);
  - the CAS subset including the centralized noise injection network and one segment distributed network;
  - one elegant breadboard of X-band transmitter to check interfaces with the CCU.

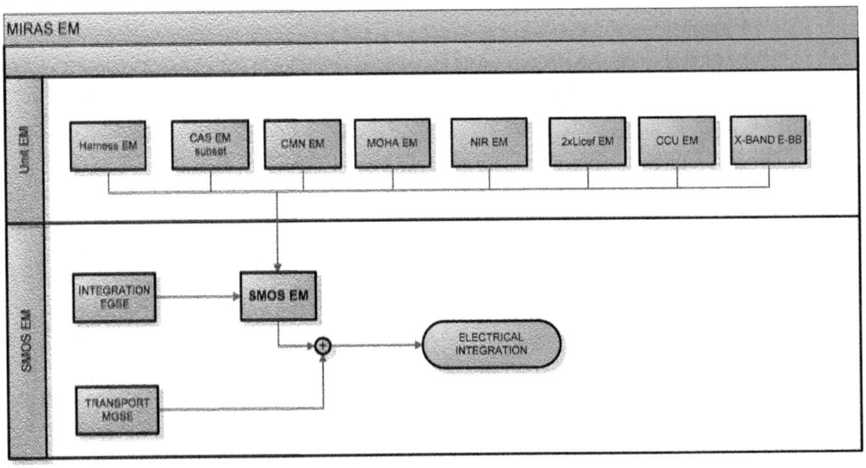

Figure 6.48 SMOS EM model flow.

Figure 6.49 *SMOS payload EGSE.*

The test campaign of the MIRAS EM was intended to verify electronic performance as well as the following functional tests: electrical integration verification, on-board software pre-qualification, functional tests and EMC tests. As mentioned above, compatibility tests between LICEF-CMN, LICEF-CCU, CMN-CCU and the CCU-S/C simulator were of special relevance.

The tests conducted at the SMOS EM stage were used to gain the necessary knowledge and confidence to define and refine the tests at PFM level.

6.4.2.3 MIRAS PFM

The SMOS PFM payload (Figure 6.51) was submitted to a system-level acceptance/qualification test sequence. As mentioned above, each of the subsystems that make up the MIRAS instrument was submitted to acceptance environmental and functional test campaigns before delivery to EADS CASA-Espacio, and after integration the PFM MIRAS test campaign took place.

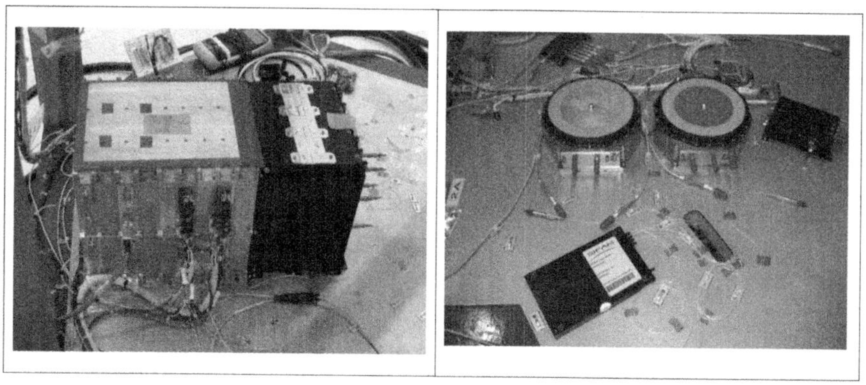

Figure 6.50 *SMOS payload EM.*

Figure 6.51 *MIRAS PFM in the zero-g deployment test.*

Three system development projects supported the SMOS instrument developments: MDPP-1, MDPP-2 and AMIRAS (airborne version of MIRAS) to validate several SMOS concepts and design approaches, including radiation pattern testing at the DTU Spherical Near-Field Antenna Test Facility.

The LICEF antenna radiation pattern test campaign was conducted as an intermediate step before PFM integration. For these tests, a representative model included only RF units and external hardware such as hinges which could influence the individual radiation patterns through scattering and diffraction. This had the significant benefit of reducing the mass of the DUT, alleviating the mechanical loads on the anechoic chamber tower mast.

As part of the test campaign of the integrated SMOS payload, the following tests were conducted:

- Integrated system test (IST)
- Deployment
- Acoustic/sine
- Thermal balance/thermal vacuum (TBTV)
- EMC
- Image validation test (IVT).

A few comments follow on these tests before a discussion of the characterization of the MIRAS LICEF antenna patterns.

IST: After completion of the integration of the electrical subsystems, the objective was to verify the performance and the compatibility of all subsystems with each other in the configurations of the system that are representative of the mission (including redundancies). The verification was done using automatic test sequences built up in the EGSE.

Deployment: This test was performed on the PFM, one arm at a time, given the complexity of having the three arms deployed simultaneously. The zero-*g* test setup developed for STM testing was used.

Acoustic/sine: The PLM was in the stowed configuration and mounted on the vibration test tool at the interface with PROTEUS. The PLM was instrumented with accelerometers at critical points as defined by the structural engineers to monitor the dynamic response and to compare it with predictions.

Figure 6.52 *MIRAS in the Large Solar Simulator at ESTEC.*

TB/TV: This test was performed on the large solar simulator (LSS) capable of simulating the thermal environment expected during the operational phase (Figure 6.52). During the thermal cycling test all the functions and modes of the instrument were verified.

EMC: Tests included conducted and radiated emissions as well as conducted and radiated susceptibilities. No ESD testing was performed on the PFM. The test campaign was conducted in the EMC chamber at the ESA-ESTEC Maxwell Facility and successfully completed in May 2007.

IVT: This test had the objective of performance testing the complete MIRAS instrument, and for that purpose a scene of known brightness temperature within the specified spatial resolution was built in the Maxwell Facility at ESA-ESTEC (Figure 6.53). The reference scene was built inside the anechoic chamber in two different situations, one provided by the anechoic chamber absorbers at ambient temperature and the other by two focal point sources simulated by four probe antennas placed within the useful instrument field of view. The results of this test have been evaluated in [35].

Figure 6.53 *MIRAS at the Maxwell Facility at ESTEC for IVT and EMC testing.*

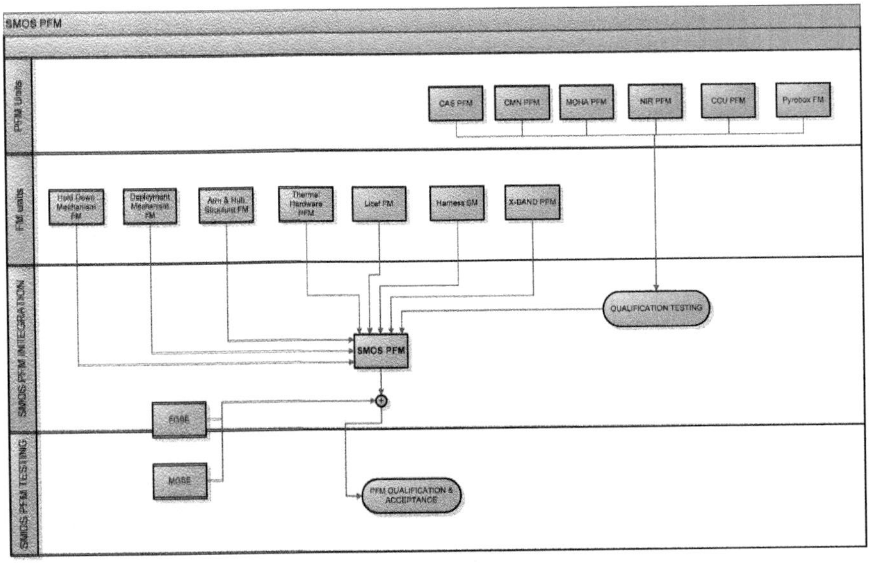

Figure 6.54 *SMOS PFM model flow.*

After the payload test campaign (Figure 6.54), MIRAS was integrated on a PROTEUS platform and an S/C level test campaign was commenced.

6.4.3 Antenna Pattern Test Campaign

In order to retrieve the brightness temperature of the scene, the MIRAS instrument measures the cross-correlations between all possible pairs of LICEF receivers in the three arms to obtain the visibility function. These correlations, performed by the correlator in the CCU, are downloaded to ground and then an inverse Fourier transform of this function permits the required brightness temperature to be obtained. The Corbella equation, a reformulation of the classical visibility function, is the basis of SEPS, the system simulator that governs the MIRAS performance [36], and reads as follows [27]:

$$V_{ij}(u,v) = \iint_{\xi^2+\eta^2\leq 1} \left\{ \frac{\sqrt{D_i D_j}}{4\pi} \frac{T_B(\xi,\eta) - T_r \cdot \delta_{ij}}{\sqrt{1-\xi^2-\eta^2}} F_{in}(\xi,\eta) \cdot F^*_{jn}(\xi,\eta) \right\} \cdot \tilde{r}_{ij} \cdot e^{-j2\pi(u\xi+v\eta)}\, d\xi\, d\eta$$

It links the brightness temperature of the scene (T_B), the receiver temperature (T_r) and the elements' radiation patterns (F_i, F_j). SEPS is also the basis of the error and accuracy budget calculations of MIRAS, and, based on it, the accuracy requirements for the measurement of the MIRAS elements' radiation patterns were established.

Antenna patterns in an aperture synthesis radiometer can only be characterized during the on-ground test campaign, and although studies have been made to assess possible effects on the pattern during flight [37], it is unclear how to perform such pattern calibration during flight, so the usual practice is to perform an on-ground characterization of the antenna patterns of the elements in the array taking into account the mutual coupling between the elements [38,39] and position [40], as well as any other effect related to scattering or the surrounding structure. These effects become error terms only if they are not taken into account in the characterization of the antennas. Thus scattering, mutual coupling or element position in the array become

Figure 6.55 Complete LICEF unit and LICEF antenna.

mere characteristics of the radiating pattern, which are measured with a given measurement uncertainty that depends mainly on the test setup. In summary, antenna patterns must be measured before launch, and they must be assumed to preserve their properties during launch and operation in space [41].

The dual-polarized patch antenna elements of MIRAS were characterized in both co-polar and cross-polar components, taking into account not only the effects previously mentioned, but also the degradation effect that the assembly of the LICEF antenna and the LICEF receiver had on the cross-polarisation of the isolated antenna. LICEF elements in MIRAS (Figure 6.55) had an input 4:1 switch at the interface with the antenna that permitted the selection of H or V polarizations as well as C or U calibration ports. The limited isolation provided by the switch affects the cross-polar level measured at full LICEF level [42]. The NIR units were not affected by this characteristic, since the two polarizations ports of the NIR antennas were connected to fully independent and isolated receivers.

6.4.3.1 *Antenna Measurement and Facility Selection*

The SMOS system simulator was used to define the maximum allowed uncertainty for the characterization of the MIRAS radiation patterns [43–46]. The analysis gives the following uncertainty values for the pattern testing: ± 0.05 dB amplitude and $\pm 0.33°$ phase up to $\pm 35°$ off-boresight 1σ. To avoid any possible confusion, in this subsection it is indicated that all the antenna pattern measurements were conducted on co-polar and cross-polar field components in both amplitude and phase.

On the other hand, it was required to have full patterns in the complete sphere ($\theta \in [0, 180°]$ and $\phi \in [0, 360°]$), so that backlobe effects could be taken into account under any possible situation (e.g. deep space pointing for calibration). Based on these and the accuracy requirement constraints, it became clear that the SMOS MIRAS antennas could only be measured in a spherical near-field facility, and from the possible options it was the Oersted-DTU Facility that was selected. The initial preparatory testing at the DTU started during the MIRAS Demonstrator Pilot Project 2, during which an extensive test campaign was conducted on a MIRAS arm segment demonstrator. The segment was populated with four complete LICEF units. During the test campaign, a number of problems were detected with respect to the EGSE for the polarization switches, and solved.

The initial MDPP-2 test campaign was continued in SMOS phase B, and later on with the preparatory investigations in phase C/D, in which the DUTs were composed of two arm segments populated with eight LICEF units. The test campaign aimed to prepare the testing of the flight model MIRAS instrument, which was expected to be approved for phase C/D. The testing during this phase was focused on an assessment of the accuracy of the testing, the effect of the increasingly larger structure on the radiation pattern of the LICEF

Table 6.7 *MIRAS LICEF antenna characterization uncertainty.*

Error source	Uncertainty	Directivity (dB)	Phase (°)
1. Reflectivity level	< −50 dB	0.004	0.029
2. Multiple reflections	±1.4%	—	—
3. Antenna tower pointing horizontally	±0.01°	0.018	0.260
4. Antenna tower pointing vertically	±0.005°	0.007	0.096
5. Axes intersection	±0.05 mm	0.004	0.039
6. Probe position	±0.3 mm	0.005	0.078
7. Measurement distance	±2 mm	0.001	0.008
8. Amplitude drift	±0.15%	0.006	0.018
9. Amplitude noise	±0.3%	0.005	0.033
10. Amp. nonlinearity	±1%	0.002	0.005
11. Phase drift	±0.27°	0.006	0.082
12. Phase noise	±0.54°	0.007	0.047
13. Phase shift in rotary joint	±0.01°	0.000	0.000
14. Channel balance amp.	±0.6%	0.018	0.034
15. Channel balance phase	±0.2°	0.018	0.034
16. Probe polarization amp.	±0.27%	0.001	0.025
17. Probe polarization phase	±0.07°	0.004	0.006
18. Mode truncation	—	0.004	0.028
Uncertainty (1σ)		**0.036**	**0.313**

under test, the effect of the hub structure on the radiation patterns of the arm elements, the accuracy of the phase retrieval pattern of off-axis antenna units, the initial considerations for automation of the test campaign and the elimination of some stray signals that were detected in the MDPP-2 campaign.

The error budget provided by the DTU for the MIRAS antenna testing, Table 6.7, was generated at the end of the preparatory investigations, with each item either known or estimated during the measurement project, and their influence quantified either by measurement or by simulation.

6.4.3.2 *Logistics and GSE for Antenna Testing*

Logistical issues challenging the SMOS MIRAS campaign were mostly imposed by the large size of the MIRAS instrument (8 m). The DTU-ESA anechoic chamber, which is 12 × 10 × 8 m between absorber tips, could only accommodate objects up to 6 m in diameter; it was therefore not feasible to test the MIRAS instrument in a single run. In summary, this led to an AIT approach which allowed deliveries of the flight model LICEFs and the flight model structure, while not compromising the accuracy of the antenna pattern test campaign.

Automation of the testing through RF EGSE was also covered and implemented in the hardware included inside the structural arms. In the RF test campaign only the RF elements that had contributed to the radiation of the LICEF elements were included, leaving room to include the RF test hardware inside the segment structures. Figure 6.56 shows one of the arms of MIRAS ready for testing. The photograph shows on the right the voltage and current protection circuitry in order to avoid damage to the delicate LICEF electronics, and on the left are the switch matrices that were integrated in order to switch from one LICEF to a neighbouring one. In this way, with a 1:18 switch matrix, hidden in the MIRAS structure, it was possible to perform all LICEF measurements without the need for any reconnection. The V/H switching transistor–transistor logic (TTL) signals were also routed through a protection circuit board and connected to the LICEF DC connector.

Figure 6.56 *Alignment checks at EADS CASA Espacio before test campaign transport.*

The MIRAS arms, as well as the hub structure, were assembled against a supporting structure that was attached to the DTU tower mast. In order to minimize torques in the arm while the pattern testing was taking place in the chamber, the support structure was equipped with adjustable counterweights on the rear side. Needless to say, all the LICEF antenna elements were attached in their final positions, and their locations were measured and registered with laser trackers. A replica of the attachment head of the DTU tower mast was built at EADS CASA in order to verify the positions of the elements and to refer their positions to a common reference coordinate system (Figure 6.57). The AUT was equipped with optical pinballs as well as an optical cube to permit the alignment at the DTU Facility.

The mass of the AUT was carefully analysed in order to avoid erroneous movements caused by an excess in the torque applied to the tower head mast. In the case for the one-arm configuration, the total mass was 105 kg including the arm support, inducing a torque transverse to the roll axis of 0.2 kg m and of 4.2 kg m along the arm. In the hub + three segments case, the transverse torque was 2.7 kg m and 24.5 kg m respectively, being less than a quarter of the allowed values in the worst case.

6.4.3.3 Definition of the AUT

During previous studies conducted in MDPP-2, and investigations in phases B and C/D, apart from the DTU pattern testing that has already been discussed, the coupling terms between the elements of a small hub

Figure 6.57 LICEF block diagram with test port.

demonstrator were measured, giving levels better than 30 dB between elements separated by one baseline distance (0.89λ) and better than 40 dB for elements separated by more than one distance; any other coupling term was at the 50 dB level, whose impact is 0.03 dB, which is very close to the required uncertainty (0.05 dB). This was in excellent agreement with the testing conducted at the DTU SNF Facility, so in summary, when testing one element, it is required to be surrounded by three other elements from the same line array.

The measurement of the LICEF patterns was carried out through a test port connected to a directional coupler included in the RF chain of the LICEF receiver, to sample the main RF path of the LICEF receive chain. The sketch of the receiver in Figure 6.58 shows the location of the TRF (RF test port). The TRF permits the measurement of the directivity patterns of the antennas, without affecting the configuration of the LICEF unit, taking into account the cross-polarization effect induced by the limited isolation of the input switch, as noted previously.

Ideally, the testing of the LICEF antenna patterns should have been made with the fully deployed MIRAS, including all possible effects. However, a number of factors made this unfeasible, the most important being:

- the size of the deployed instrument;
- the delivery schedule of the LICEF;
- the mass of the fully deployed configuration and the maximum load and torque allowed at the facility positioner, as well as by accuracy constraints.

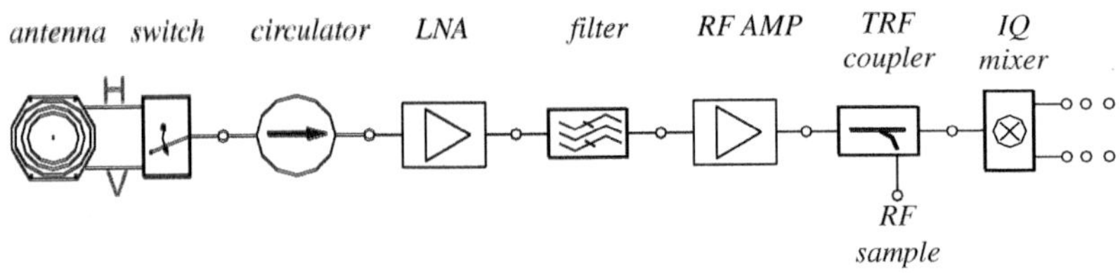

Figure 6.58 LICEF block diagram with test port.

Figure 6.59 Hub + three segments of MIRAS during DTU test campaign.

Each LICEF pattern needed to be measured in fully representative conditions, which meant that each unit had to be surrounded by all the hardware that may affect its radiating performance. In antenna terminology, this means that each unit was measured in *embedded* condition. During the test campaigns conducted in previous phases at the DTU, it was determined that a given antenna element was affected by its three closest neighbours (on either side), and the effect of the hub structure on the elements of the arms was verified; this test campaign provided the necessary information to determine the size of the AUT for measuring a single unit in embedded condition. The minimum setup had to include the AUT plus three more units on each side. In order to minimize the number of transports to the facility and accommodating delivery of the units, it was concluded that MIRAS should have two distinct setups, one setup corresponding to each separate arm and one setup corresponding to the hub. In order to fulfil the constraints found in the exploratory investigations, all units up to the three innermost ones were measured in the arm setup, and all elements except the outermost three elements of each segment were tested in the hub setup (Figure 6.59). In this way the number of transports to the facility was also optimum and this was the agreed AIT sequence.

6.4.3.4 Test Results

Figure 6.60 shows the pattern of one element located in the hub. This pattern can be considered as representative of the whole array, with mean co- and cross-polar levels in the range of 28 dB, for both antenna polarizations and directivity levels in the range of 9.5 dBi. The dispersion between the directivities of all the elements was better than 1 dB over the complete array and both polarizations, which take into account the edge effect on the extreme elements, the effect of the hinge on the elements separating two segments, and the different embedding conditions of the central hub elements with respect to the arm elements. From the point of view of brightness temperature retrieval operation, all these effects are secondary, since the elements are characterized in their real operating conditions, and therefore the pattern data collects all possible effects on the element patterns.

Figure 6.61 shows the measured directivities of all the elements of the complete arm A from the hub to the tip of the extreme element, equivalent to the other arms' LICEF elements and also those of the hub. The summary figure also includes the directivities of the LICEF antenna isolated elements (i.e. not attached to the receiver

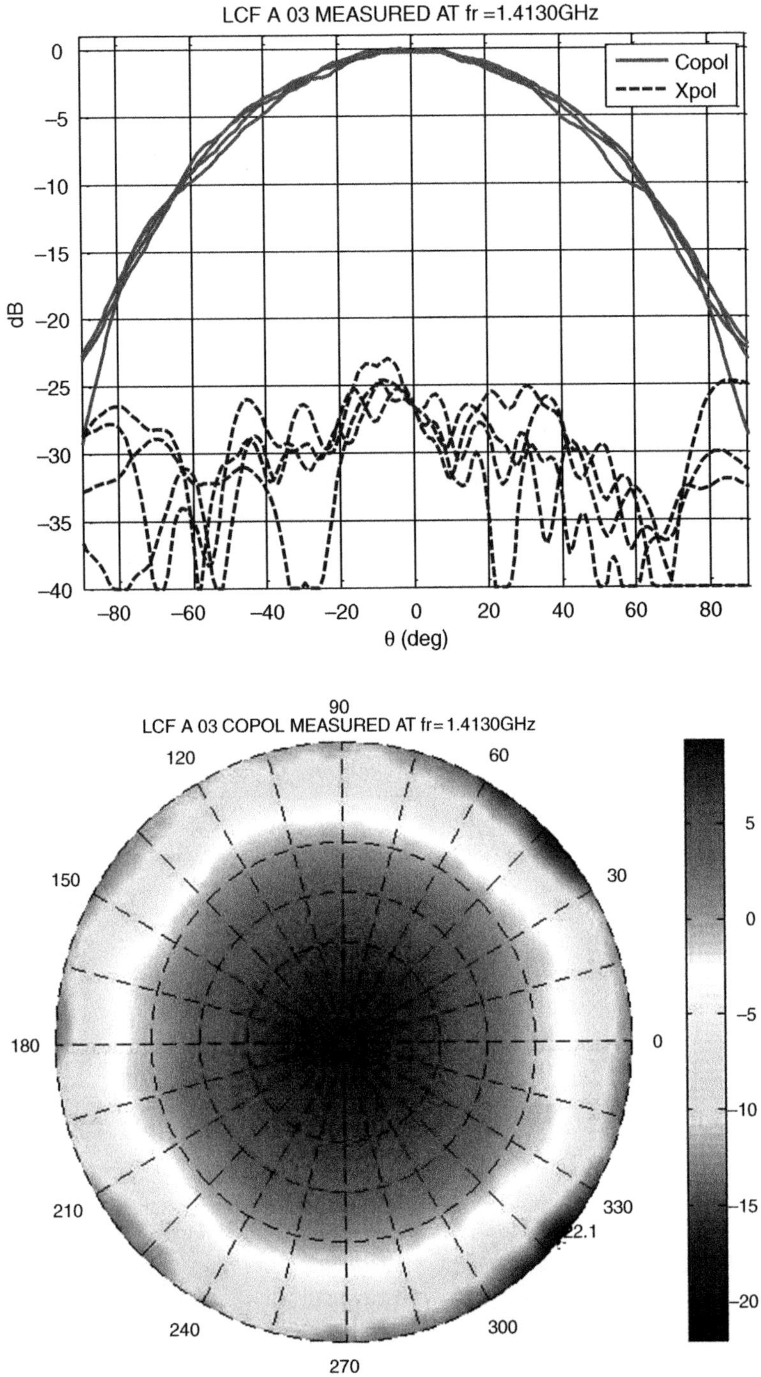

Figure 6.60 Radiation pattern of one LICEF antenna in MIRAS.

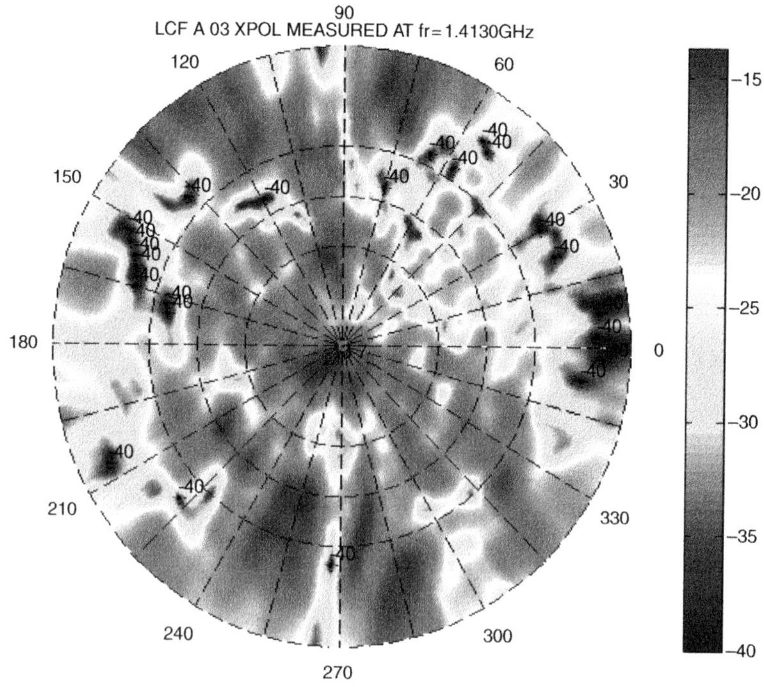

Figure 6.60 (Continued)

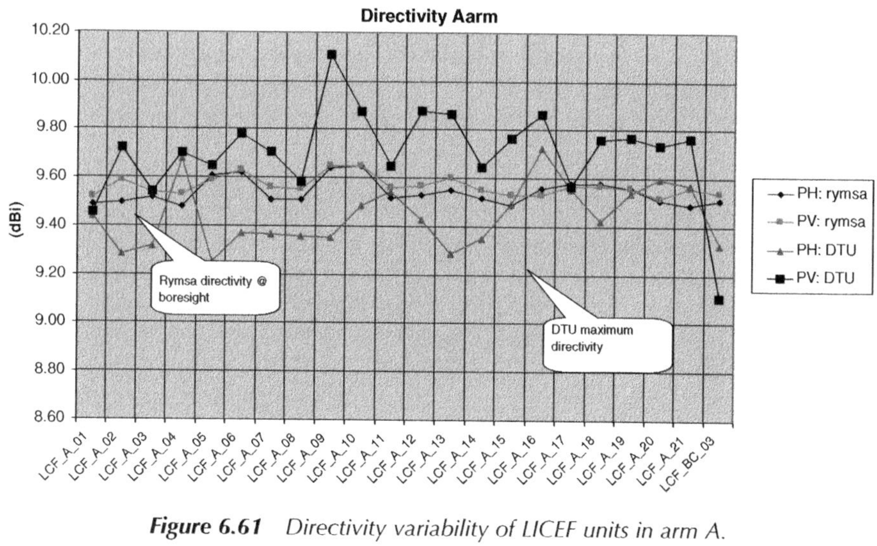

Figure 6.61 Directivity variability of LICEF units in arm A.

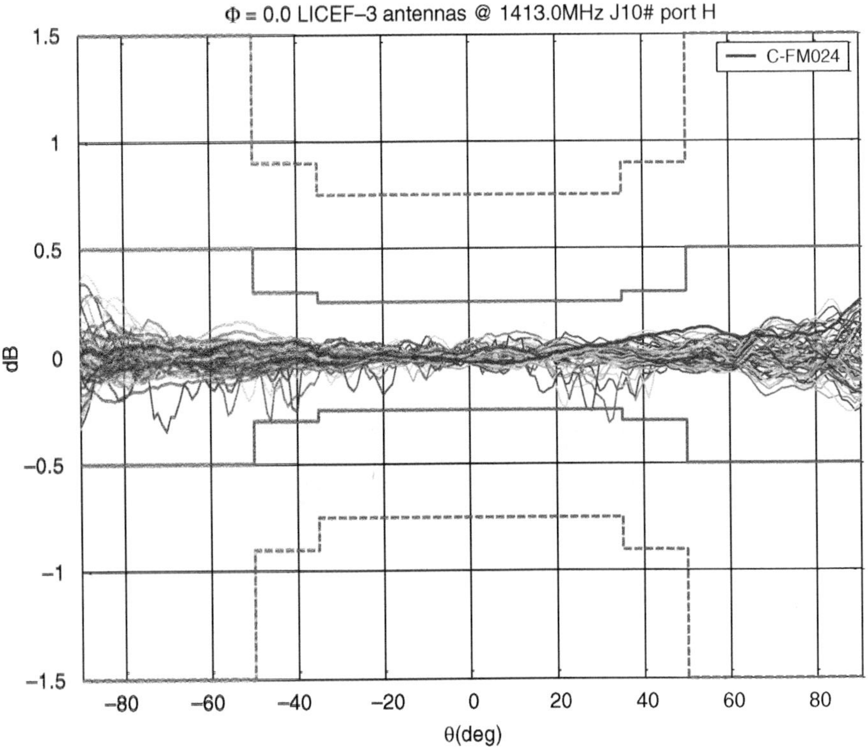

Figure 6.62 *Amplitude deviation mask of manufactured LICEF antennas.*

and measured in isolated conditions). It can be observed how the final locations of the elements affect the isolated element patterns, which were highly reproducible (the standard deviation of the directivity of the complete LICEF antenna population was less than 0.03 dB for both polarizations).

The pattern test data was the main deliverable of the antenna pattern test campaign. The data is entered into the SEPS simulator in order to get the brightness temperature distributions, and has been made available to the scientific community through ESA. It can be concluded that the test campaign conducted at the DTU was successful and the validity and utility of the pattern data have been demonstrated since then.

The excellent pattern symmetry of the LICEF elements in SMOS was unaltered by the scattering environment of the MIRAS structure, or by connection to the LICEF receiver. The symmetry characteristics are inherent to the individual radiating elements of the LICEF, which was due to the electrical architecture of the patch antenna, a balanced dual-polarization patch. One of the specification requirements of the LICEF antenna element was a tight similarity mask for the patterns with respect to the elevation angle, in both amplitude and phase. The compliance of the antennas with the mask was excellent, as shown in Figure 6.62 for the amplitude and in Figure 6.63 for the phase. The mask had the objective of controlling the workmanship apart from guaranteeing the required minimum directivity specification. Only one of the manufactured antennas of the complete batch (71 antennas) was rejected due to non-compliance with this specification. Figure 6.64 shows the four main cuts of the average radiation pattern computed for the complete antenna set. As mentioned above, the average directive gain of the complete population was 9.5 dBi, with a standard deviation of ±0.03 dB and a worst case peak-to-peak dispersion of 0.24 dB.

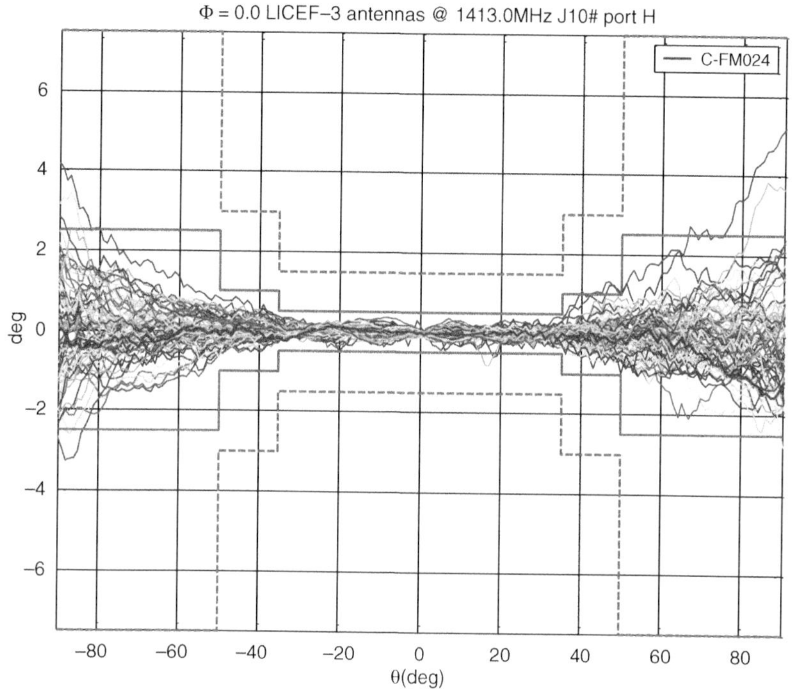

Figure 6.63 *Phase deviation mask of manufactured LICEF antennas.*

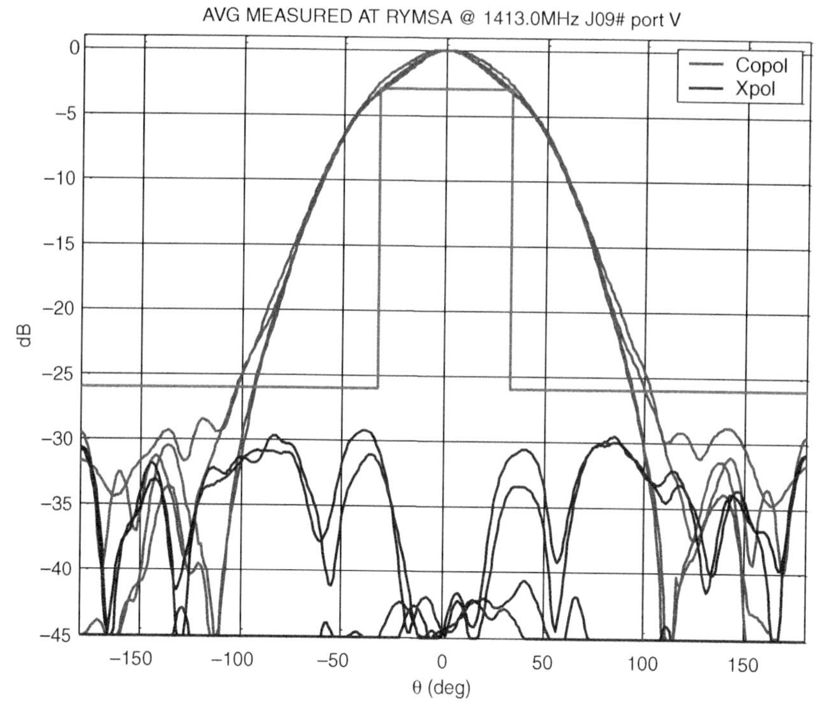

Figure 6.64 *Average pattern of LICEF antennas before integration in LICEF.*

References

1. ECSS (1988) *Verification*, ECSS-E-10-02A.
2. ECSS (2000) *Ground Systems and Operations*, ECSS-E-70 pt 1.
3. Migl, J., Habersack, J., Grim, H. and Paus, S. (2003) Test philosophy and test results of the Intelsat-IX C-Band antennas. Antenna Measurement Techniques Association, AMTA 2003, Irvine, CA, USA.
4. Hollis, J.S., Lyon, T.J. and Clayton, L. (1985) *Microwave Antenna Measurements*, Scientific-Atlanta.
5. Hartmann, J., Habersack, J. and Steiner, H.-J. (2002) A new large compensated compact-range for measurement of future satellite generations. Antenna Measurement Techniques Association, AMTA 2002, Cleveland, OH, USA.
6. Karttunen, A., Ala-Laurinaho, J., Vaaja, M. *et al.* (2009) Antenna tests with a hologram-based CATR at 650GHz. *IEEE Transactions on Antennas and Propagation*, **57**(3), 711–720.
7. Dudok, E. and Fasold, D. (1986) Analysis of compact antenna test range configuration. Journées Internationales de Nice sur les Antennes (JINA), Nice, France.
8. Dudok, E., Steiner, H.-J., Habersack, J. and Fritzel, T. (1989) Design, development and qualification of an advanced, large compact test range. Antenna Measurement Techniques Association, AMTA 1989, Monterey, CA, USA.
9. Dragone, C. (1978) Offset multireflector antennas with perfect pattern symmetry and polarization discrimination. *Bell Systems Technical Journal*, **57**(7), 2663–2384.
10. Habersack, J. and Steiner, H.-J. (2007) High sensitive RF test setup for antenna measurements up to 500GHz. International Joint Conference of the 8th Millimeter-Wave International Symposium (MINT-MIS2007), Seoul, Korea.
11. Migl, J., Lindemer, W. and Wogurek, W. (2005) Low cost satellite payload measurement system. Antenna Measurement Techniques Association, AMTA 2005, Newport, RI, USA.
12. Dudok, E., Steiner, H.-J. and Smith, T. (1990) Scanned quiet zones in a compact antenna test range. Antenna Measurement Techniques Association, AMTA 1990, Philadelphia, USA.
13. Hartmann, J., Habersack, J., Hartmann, F. and Steiner, H.-J. (2005) Validation of the unique field performance of the large CCR 120/100. Antenna Measurement Techniques Association, AMTA 2005, Newport, RI, USA.
14. Kerns, D.M. (1981) Plane-Wave Scattering-Matrix Theory of Antennas and Antenna-Antenna Interactions, in *Theory of Antennas and Antenna-Antenna Interactions*, vol. 162, National Bureau of Standards.
15. Leach, W. and Paris, D. (1973) Probe compensated near-field measurements on a cylinder. *IEEE Transactions on Antennas and Propagation*, **21**(4), 435–445.
16. Brown, J. and Jull, E.V. (1961) The prediction of aerial radiation patterns from near-field measurements. *Proceedings of the IEE – Part B: Electronic and Communication Engineering*, **108**(42), 635–644.
17. Jensen, F. Electromagnetic near-field-far-field correlations, Ph.D. dissertation, Electromagnetics Inst., Technical Univ. Denmark, report LD 15, July1970.
18. Rahmat-Samii, M. and Gatti, Y. Far-Field Patterns of Spaceborne Antennas from Plane-Polar Near-Field Measurements, IEEE Transactions on Antennas and Propagation, Vol. 33, No. 6, pp 638–648, June1985.
19. Brown, J. and Jull, E. The prediction of aerial radiation patterns from near-field measurements. *Proceedings of the IEE*, vol. 108B, pp. 635–644, NOV.1961.
20. Leach, Jr., W.M. and Paris, D.T. Probe-compensated near-field measurements on a cylinder. *IEEE Transactions on Antennas and Propagation*, Vol -21, pp. 435–445, July 1973.
21. Wacker, P. F. Non-planar near-field measurements: Spherical scanning, *Electromagnetics Div.*, Nat. Bureau of Standards, Boulder, CO, Rep. NBSIR 75–809, June 1975.
22. Larsen, F. H. Probe-corrected spherical near-field antenna measurements, Ph.D. dissertation, *Electromagnetics Inst.*, Technical Univ. Denmark, Rep. LD 36, Dec.1980.
23. Hansen, J. E. Spherical near-field antenna measurements, Peter Peregrinus Ltd.,1988.
24. Ruf, C., Swift, C., Tanner, A. and Le Vine, D. (1988) Interferometric synthetic aperture microwave radiometry for the remote sensing of the Earth. *IEEE Transactions on Geoscience and Remote Sensing*, **26**(5), 597–611.
25. Kraft, U. R. Two-dimensional aperture synthesis radiometers in a low earth orbit mission and instrument analysis, in *Geoscience and Remote Sensing Symposium*,1996. IGARSS '96. Remote Sensing for a Sustainable Future., International, 1996, vol. 2, pp. 866–868 vol.2.

26. Camps, A., Bara, J., Sanahuja, I. and Torres, F. (1997) The processing of hexagonally sampled signals with standard rectangular techniques: application to 2-D large aperture synthesis interferometric radiometers. *IEEE Transactions on Geoscience and Remote Sensing*, **35**(1), 183–190.
27. Drinkwater, M., McMullan, K., Marti, J. *et al.* (2009) Star in the sky. The SMOS payload: MIRAS. *ESA Bulletin*, **137**, 16–22.
28. Kudielka, K., Benito-Hernandez, J., Rits, W. and Martin-Neira, M. (2010) Fibre optics in the SMOS mission. International Conference on Space Optics, ICSO 2010, Rhodes, Greece.
29. Colliander, A., Tauriainen, S., Auer, T.I. *et al.* (2005) MIRAS reference radiometer: a fully polarimetric noise injection radiometer. *IEEE Transactions on Geoscience and Remote Sensing*, **43**(5), 1135–1143.
30. Colliander, A., Ruokokoski, L., Suomela, J. *et al.* (2007) Development and calibration of SMOS reference radiometer. *IEEE Transactions on Geoscience and Remote Sensing*, **45**(7), 1967–1977.
31. Brown, M.A., Torres, F., Corbella, I., and Colliander, A. (2008) SMOS calibration. *IEEE Transactions on Geoscience and Remote Sensing*, **46**(3), 646–658.
32. Plaza, M.A., Martinez, L. and Cespedosa, F. (2003) The development of the SMOS-MIRAS deployment system. 10th European Space Mechanisms and Tribology Symposium, San Sebastian, Spain, vol. 524, pp. 231–238.
33. Bueno, J.I., Garcia, I. and Plaza, M.A. (2005) SMOS PLM MIRAS hold-down release and deployment mechanisms. 11th European Space Mechanisms and Tribology Symposium, ESMATS 2005, Lucerne, Switzerland, vol. 591, pp. 235–242.
34. Checa, E., Dolce, S., Rubiales, P. and Lamela, F. (2008) Thermal design and testing of SMOS payload. International Conference on Environmental Systems, June 2008, Portland, OR, USA.
35. Corbella, I., Torres, F., Duffo, N. *et al.* (2009) On-ground characterization of the SMOS payload. *IEEE Transactions on Geoscience and Remote Sensing*, **47**(9), 3123–3133.
36. Corbella, I. (2003) End-to-end simulator of two-dimensional interferometric radiometry. *Radio Science*, **38**(3), 1–8.
37. Le Vine, D. and Weissman, D. (1996) Calibration of synthetic aperture radiometers in space: antenna effects. IEEE International Geoscience and Remote Sensing Symposium, IGARSS'96, vol. 2, pp. 878–880.
38. Straumann, T. (1993) Effects of mutual antenna coupling on SAIR performance. IEEE 8th International Conference on Antennas and Propagation, 1993.
39. Camps, A., Torres, F., Corbella, I. *et al.* (1998) Mutual coupling effects on antenna radiation pattern: an experimental study applied to interferometric radiometers. *Radio Science*, **33**(6), 1543–1552.
40. Camps, A., Bará, J., Torres, F. *et al.* (1997) Impact of antenna errors on the radiometric accuracy of large aperture synthesis radiometers. *Radio Science*, **32**(2), 657–668.
41. Skou, N. (2003) Aspects of the SMOS pre-launch calibration. IEEE International Geoscience and Remote Sensing Symposium, IGARSS'03, vol. 2, no. C, pp. 1222–1225.
42. Garcia-Garcia, Q. (2007) Cross-polarization in dual polarization switching antennas. *International Journal of RF and Microwave Computer-Aided Engineering*, **17**(3), 295–303.
43. Torres, F., Olea, A., Garcia, Q. and Martin-Neira, M. (2008) Antenna error budget in the MIRAS-SMOS instrument. IEEE Antennas and Propagation Society International Symposium, 2008, pp. 1–4.
44. Kerr, Y., Font, J., Waldteufel, P. *et al.* (2000) New radiometers: SMOS-a dual pol L-band 2D aperture synthesis radiometer. Proceedings of the IEEE Aerospace Conference, pp. 119–128.
45. Corbella, I., Torres, F., Camps, A. *et al.* (2000) L-band aperture synthesis radiometry: hardware requirements and system performance. IEEE International Geoscience and Remote Sensing Symposium, IGARSS'00, vol. 7, pp. 2975–2977.
46. Torres, F., Corbella, I., Camps, A. and Duffo, N. (2004) Fundamentals of MIRAS-SMOS error budget. The 5th SMOS Science Workshop, Rome, Italy, pp. 150–150.

7

Historical Overview of the Development of Space Antennas

Antoine G. Roederer

Delft University of Technology – IRCTR, The Netherlands

7.1 Introduction

Based on the application of Maxwell's theory and equations (1873), antennas were first materialized by Hertz in 1887.

When Sputnik I was launched on 4 October 1957, antenna development had been under way for 70 years, driven by communications, broadcasting, radio astronomy and, since the Second World War, also by radar.

In the 1960s, theories and various implementations of the main types of antennas presently used on board spacecraft, such as wire antennas, reflector/lens antennas and array antennas, were already described in textbooks [1–4]. Since then, the specific requirements of evermore challenging space missions, together with the special environment conditions of launchers and space environment, have led to the development for space use of a new branch in the family of antennas. This has driven major innovations and progress in antenna modelling, optimization, beam forming, architectures, technologies and test techniques which have been beneficial to the entire antenna engineering community.

In this chapter, antennas will be essentially classified in three categories:

1. Low-gain TT&C (Telemetry, Tracking & Command) antennas. They were the only antennas on the first generation of low Earth orbit satellites for which simple wire and horn antennas were used to command the satellites, monitor their parameters from the ground and transmit limited science data collected by the spacecraft. TT&C antennas are required on all spacecraft and the scope and complexity of their requirements have grown considerably with the amount of data involved, in particular for science data transfer for deep space and planetary observation missions. For such missions high-gain aperture antennas are required, mostly with mechanically driven beam pointing (reflectors or arrays)

Space Antenna Handbook, First Edition. Edited by William A. Imbriale, Steven (Shichang) Gao and Luigi Boccia.

2. Reflector or lens antennas. They focus waves incoming from restricted angular sector(s) which are then collected by one or a group of small feed antennas. Satellite reflector antennas, initially of modest size with a single beam and feed, can now reach diameters in excess of 20 meters with hundreds of feed elements to generate hundreds of beams formed digitally on board or on the ground.
3. Array antennas. They use a number of identical small antennas which are combined to receive more incoming power or to concentrate the transmitted power in restricted angular sector(s) or beams. Passive arrays of wire antennas were used first, followed by larger arrays of slot elements and later of printed elements for synthetic aperture radars. For space applications, active arrays, with distributed amplifier modules, although much in use for radar, have been slow to appear, mainly because of the poor efficiency of solid state power amplifiers and the complexity of the beam forming.

Figure 7.1 shows the TDRS satellite which uses small TT&C antennas, reflector antennas and array antennas.

For all three categories of space antennas, associated disciplines can be defined:

- Accurate electromagnetic modelling of small antennas in a complex spacecraft environment, of multiple reflector systems at frequencies up to terahertz and of finite arrays with three-dimensional and/or non-periodic features is very challenging. Moreover, in order to implement optimization techniques for such antennas, the accurate analysis must also be fast.
- The spatial and/or temporal processing of signals involved in array and multi-feed antenna systems is often called beam forming. Beam forming can be performed in the frequency band of operation of the antenna, at intermediate frequency or at baseband and can be analogue, digital or optical. Beam forming and its circuits and processors have become very complex for some remote sensing, radio astronomy and telecommunications missions, to the point that they sometimes have to be exported to the ground.

Figure 7.1 *TDRS satellite with TT&C, reflector and array antennas. Courtesy of NASA.*

- Aperture synthesis and signal processing are presently expanding fields with space applications in remote sensing, radio astronomy and wireless communications.
- Environmental conditions during launch and in orbit as well as the growing size and accuracy requirements of space antennas have driven the development of special thermo-mechanical and material designs.
- Antenna and payload electrical and thermo-mechanical testing and verification for large systems with hundreds of reconfigurable beams, or at very high frequencies where phase is inaccessible, can contribute significantly to the cost of satellite payloads. Novel accurate and fast measurement techniques have been developed over time to validate analysis software and to demonstrate the performance of complex payloads.

It is not possible to exhaustively review antennas developed for over 35 000 spacecraft launched in the last 55 years. This chapter will attempt to illustrate, using examples, some of the key developments of space antennas and their modelling and technologies. A good source of material on early antenna developments for space is [5].

7.2 The Early Days

7.2.1 Wire and Slot Antennas on Simple Satellite Bodies

7.2.1.1 Sputnik I, II and III

Unlike today's stable space platforms, early satellites like Sputnik I had no mechanism for stabilization that would keep their antennas in any particular orientation relative to a station on the Earth. Moreover, Sputnik I was tracked passively on its trajectory and orbit (apogee 939 km, perigee 215 km) by several radar and ground telescopes but without on-board response.

Sputnik I consisted of two half spheres of 58 cm diameter hermetically sealed together and pressurized at 1.3 atmospheres. They were 2 mm thick with a 1 mm thermal shield of aluminium–magnesium–titanium alloy polished to reflect the Sun's radiation. The enclosure contained three silver zinc batteries, two 1 W VHF transmitters, a controlled fan, temperature and pressure sensors, and the communication equipment to produce the modulated signals. The battery mass amounted to 61% of the 83.6 kg of the satellite.

Sputnik I's antennas, developed by M. V. Krauyshkin and his team in a laboratory of The Experimental–Design Bureau No. 1, OKB-1, had to radiate in all directions around the spin axis.

Downlink telemetry data included temperature and pressure. Scientific data on the propagation of radio signals in the ionosphere and on the density of the higher atmosphere was also collected.

Sputnik I's telemetry frequencies of 20.005 and 40.002 'megacycles', used one at a time, were selected to take advantage of the satellite's tracking potential by amateur radio operators spread over all of the Soviet Union and beyond. According to Tikhonravov in [6], the telemetry used the frequency, the length of transmitted pulses (0.2–0.6 seconds) and pulse-position modulation (TRAL telemetry).

Figure 7.2 shows a model of Sputnik I with its antennas. According to [6], the antennas consist of two pairs of opposite monopoles, inclined at an angle of 35° from the satellite axis, with lengths of 2.9 and 2.4 m respectively for the low and the high frequencies.

Made of aluminium alloy, these monopoles are maintained closer to the satellite body during launch and released to reach the 35° angle at separation from the launcher.

Not much detail can be found on the electrical design and the feeding of the antennas. One can assume that opposite monopoles are fed with equal amplitudes and phases with, as a result, two radiation nulls along the satellite axis, but only two moderate dips in the plane around it. Use of these dips to monitor changes in the spin rate is mentioned in [6].

Figure 7.2 *Model of Sputnik I showing its telemetry antennas operating at 20.005 and 40.002 MHz. Courtesy of NASA.*

At the low VHF used, the monopoles and the satellite body are small in terms of wavelength. The modelling of the complete satellite and antennas is amenable to the use of integral equation techniques. These, however, as well as computers, were not sufficiently developed at the time to perform an accurate analysis.

According to [6], the two 1 W transmitters were of 'vacuum valve-type'. The transmitter units were mounted adjacent to the antenna connections in the pressurized enclosure.

The testing of the radio system including the antennas and the transmitters was conducted using a helicopter and a ground station [7].

Sputnik I operated until its batteries lost power after 22 days. About a week later, on 3 November, Sputnik II was launched, carrying a much heavier payload, including a dog named Laika.

Sputnik III, also called Objet D, was launched in May 1958. It was originally supposed to fly before Sputnik I. It carried a complete telemetry tracking and command radio system. For telemetry at 66 MHz, it used a turnstile antenna made of four short-circuited wire antennas. One is visible in Figure 7.3.

7.2.1.2 Explorer I

On 1 February 1958, the first US satellite, Explorer I, developed by the Jet Propulsion Laboratory [8] under the auspices of the US Army Ballistic Missile Agency, was launched successfully by a Juno I rocket. Its scientific objectives, configuration and antennas are very different from those of Sputnik I.

The telemetry of Explorer I uses two frequencies of 108.00 and 108.03 MHz that are much higher than those of Sputnik I.

According to [9,10], Warren Hopper and Tom Barr came up with the innovative design of the Explorer I 'fiberglass slot antenna'. Two fibreglass slots/rings were introduced around the metallic enclosures of the spacecraft(S/C) (see Figure 7.5) and could be excited to induce radiating currents on the metallic sections of the S/C producing omnidirectional dipole-like radiation around the spacecraft axis.

This design was also used on Explorers II, III and IV.

Figure 7.3 Sputnik III with element of the 66 MHz turnstile antenna. Reproduced by permission of Sven Grahn.

Used as a backup, a turnstile antenna using four flexible wire monopoles kept normal to the satellite axis by the satellite spinning also provided fairly circular patterns. Because of Explorer's spinning instabilities, this antenna was not used beyond Explorer III.

Figure 7.4 shows a model of Explorer I and Figure 7.5 a sketch of its configuration.

Key characteristics of Explorer I and Sputnik I have been grouped together in Table 7.1.

7.2.1.3 Echo 1

Although it did not actually involve a real antenna on board, NASA's Echo 1 had a historical impact in the early development of space communications since it was the first satellite to relay voice messages from the ground to space and back. It was also very visible. Launched on 12 August 1960, Echo 1 (see Figure 7.6) was a 30.5 m diameter balloon made of metallized polyester film 0.0127 mm thick. Echo 1 was used to reflect 960 and 2390 MHz signals for telephone, radio and television on its elliptical orbit (apogee 1679 km, perigee 1510 km). Launched nine years later, Echo 2 was a 41.1 m diameter Mylar balloon used at 162 MHz.

7.2.2 Antenna Computer Modelling Takes Off

During the 1950s, although the theory of electromagnetism and antennas was established with complex radar antennas already developed and operational, computer modelling was virtually non-existent and calculations were made with slide rules, trigonometric and logarithmic tables or by human computers (see Figure 7.7).

Figure 7.4 *A model of Explorer I, held by JPL's Director William Pickering, scientist James Van Allen and rocket pioneer Wernher von Braun. Courtesy of NASA.*

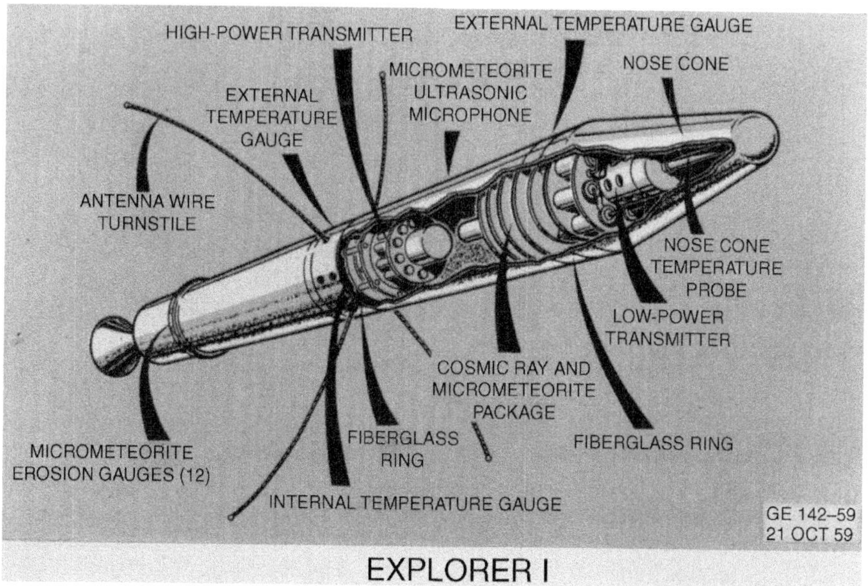

Figure 7.5 *Sketch of Explorer I showing the two fibreglass slots/rings of the dipole-like antennas and the flexible wires of the turnstile backup antenna. Courtesy of US Army SMDC.*

Table 7.1 *Key characteristics of Sputnik I and Explorer I and their antennas.*

	Sputnik I	Explorer I
Agency	Soviet Ministry of Radio Industry	US Army Ballistic Missile Agency
Mission objective(s)	Propagation in ionosphere Density of high atmosphere Micrometeorite detection Temperature sensing	Cosmic rays/particle detection Micrometeorite detection Density of high atmosphere Temperature sensing
Payload/instruments	Temperature sensors Pressure barometric sensor (pressurized enclosure)	Geiger tube detector Wire grid detector and crystal transducer Temperature sensors
Prime Contractor	Experimental Office No. 1, OKB-1	Jet Propulsion Laboratory
Launcher	R7- Sputnik rocket	Juno I
Launch date	4 October 1957	1 February 1958
S/C dimensions	Sphere $\Phi = 58.5$ cm	Cylinder and cone $\Phi = 16.5$ cm, $L = 94.6$ cm + fourth stage (110 cm)
S/C mass	83.6 kg	13.91 kg
Power	Three silver–zinc batteries	Nickel–cadmium chemical batteries
Orbit	Apogee 939 km, perigee 215 km Inclination 65.1°, period 101.5 min	Apogee 2535 km, perigee 360 km Inclination 33.24°, period 114.9 min
Attitude	Non-stabilized spinning	Nominally spinning at 750 rpm
Thermal control	Heat shield of aluminium–magnesium–titanium 1 mm thick Fan and thermal control switches	Thermal paint in dark and light strips
Mission duration	22 d	High power, 31 d Low power, 105 d
Key results	First Earth satellite launched Meteoroid detection Density of high atmosphere Propagation in the ionosphere	First US satellite launched Van Allen radiation belt discovered 45 impacts of cosmic dust detected in 78 750 s
Decay	4 January 1958	31 March 1970
Antennas:		
Coverage	Earth coverage below spinning S/C	Earth coverage below spinning S/C
Tx frequencies	20.005 MHz/40.002 MHz	108.00 MHz/108.03 MHz
Tx power	1 W/1 W (vacuum valves)	10 mW/60 mW (transistors)
Polar	Linear	Essentially linear
Special requirement(s)	Limited S/C attitude control	Limited S/C attitude control
Key antenna description	Whips inclined 35° w.r.t. S/C axis Smaller angle: 22.5° during launch Two whips of 2.9 m/two of 2.4 m	Two fibreglass slots exciting enclosed dipole-like antennas Four flexible whip turnstile backup
Major innovation	First satellite antennas	First slot-driven enclosure antenna
Technology	Aluminium alloy	Flexible whip using S/C spin
Modelling	No data available. Integral equation?	No data available. Integral equation?
Testing	Tested with helicopter and station	No data available
Key reference	http://www.russianspaceweb.com/sputnik.html	W. A. Imbriale, *Spaceborne Antennas for Planetary Exploration* (2006)

Figure 7.6 *Echo 1 (1960) undergoing inflation tests. Courtesy of NASA.*

In the 1960s, large mainframe computers became available and accurate computer analysis of wire antenna current distribution and radiation progressed rapidly. Pioneering work on wire antennas [11] and later on point matching solutions of Hallen's integral equation for straight and curved wires [12], which is now widely known as the method of moments (MOM), was expanded and consolidated in [13]. The development of these *integral equation techniques*, including wire grid modelling [14–16], was mostly driven by defence applications (log periodics, interactions of antennas and vehicles, ships, aircraft, missiles, etc.). Wire segments of carefully chosen lengths, typically smaller than 0.1 wavelength, were used to model the antenna elements, the conducting bodies and their interactions.

At that time, with the small size of the early satellites and the low frequencies used for TT&C at VHF (108, 137, 148 MHz), integral equation techniques were very suitable for the modelling of space antennas and were further developed for space applications on both sides of the Atlantic for ESRO in Europe [17,18] (see Figure 7.8) and for NASA JPL [19].

Figure 7.7 *During the 1940s and 1950s, JPL used the word 'computer' to refer to a person rather than a machine. The all-female computer team, with many of its members recruited right out of high school, were responsible for doing all the maths by hand required to plot satellite trajectories and more. Courtesy of NASA/JPL-Caltech.*

Frederiksborg Amts Avis

ESRO er mere end rumforskning og satellitter

Figure 7.8 Jesper Hansen and Niels Jensen validate MoM models of the VHF TT&C antenna for European ESRO satellites in 1972 [18]. Photo Per Alan Lunding – Courtesy of Frederiksborg Amts Avis.

With the increase in size of the satellites and the introduction of higher frequencies at S-band, spacecraft became too large in terms of wavelengths and, in many cases, in spite of the development of faster computers, the time needed to fully analyse antenna–spacecraft interactions with integral equation techniques became excessive.

Based on the pioneering US work in [20,21] on the geometrical theory of diffraction (GTD), which extended well-established geometrical optics to include *edges, shadow regions, surface slope and so on*, a sizeable development was initiated by ESRO in Europe to apply GTD *ray tracing techniques* to satellite antennas, including a large satellite body with booms and solar panels. Figure 7.9 shows a comparison between a measured and a GTD-calculated radiation pattern for a dipole interacting with the simple model of a spacecraft.

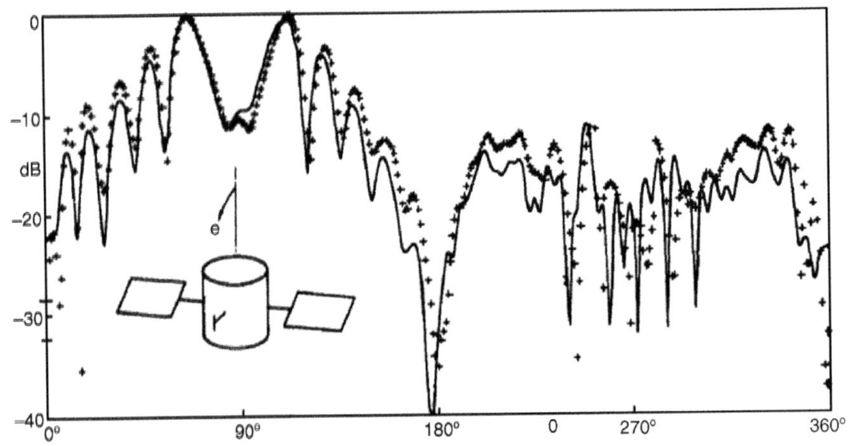

Figure 7.9 Comparison between measured (____) and GTD calculated (+++++) radiation patterns. Courtesy of Aksel Frandsen, TICRA.

Until the late 1960s, although shaped reflectors with cosine-squared patterns were used for some radars, spacecraft *reflector antennas* were mainly symmetrical prime focus fed or Cassegrain configurations. At the time reflector antenna design was essentially based on the techniques described in [1]. In the physical optics (PO) method, the evaluation of induced currents on the reflector surface allows the far-field radiation patterns to be computed by integration. In the aperture field technique, the electric field in a planar aperture is evaluated by geometrical optics and then integrated to provide far-field patterns.

In [22], application of GTD to speed up the computation of reflector antenna radiation patterns outside the main lobe region is also outlined.

This technique, already studied in the USA for subreflectors and some specific reflector geometries [23,24], was perfected further and implemented to replace physical optics for directions away from the main lobe in the early versions of the GRASP programme (General Reflector Antenna Software Package programme), widely used later for space reflector antennas.

7.2.3 Existing/Classical Antenna Designs Adapted for Space

Early satellites with their low-frequency monopoles and slot antennas were soon followed by larger and better stabilized spacecraft with ambitious observation and communication objectives.

These satellites still required quasi-omnidirectional antennas for basic telemetry and command functions just after launch, during manoeuvres and until a stable position and attitude of the spacecraft was reached, and also to cope with any spacecraft malfunctions.

Such low-gain antennas used classical designs such as cross-dipoles, discones or quadrifilar helices. Because of their larger distances from the Earth and the higher data rates to transmit, they also needed more directive antennas with medium to high gains, operating with larger bandwidths at higher frequencies. Such antennas, mostly using reflectors, would often need a pointing system with a rotary joint to point the beam towards the Earth.

7.2.3.1 Ranger

As an example, the telemetry and command of the early NASA JPL Ranger spacecraft (1962–1965), designed to gather data from Venus, used a quasi-omnidirectional discone antenna based on a classical design of the 1950s [2]. A diplexer separated the transmit signals at 960 MHz from the command receiver operating at 890 MHz. This ‘omni’ antenna, integrated in a fibreglass cylinder, is visible in Figure 7.10.

The Ranger directional high-gain antenna, for transmission of telemetry/science data at 960 MHz, also used a classical basic design [2] suitably adapted to the mission space requirements. It was a 1.2 m symmetrical parabolic reflector fed by a dipole at $\lambda/4$ from a small 23 cm diameter disc. The reflector surface was made of a 0.63 cm square mesh supported by six parabolic ribs and three circular rings.

The Ranger spacecraft were among the first to have three-axis stabilization with their longitudinal axis continuously kept pointing at the Sun to ease power supply and thermal control. Pointing of the high-gain antenna towards the Earth could be done in azimuth by appropriately rotating the whole spacecraft around its longitudinal axis and in elevation by activating a deployment/hinge mechanism. This implied use of a rotary joint on the feeding line. A dedicated coaxial rotary joint, with RF chokes which greatly limit metal-to-metal contacts, was specially developed by JPL for this mission. Ranger omnidirectional and high-gain antenna designs were inherited by the NASA JPL Mariner spacecraft.

These antennas, as well as those of most NASA JPL planetary exploration spacecraft, are described in detail in [8], from which the above section and some of the subsequent ones on NASA JPL spacecraft antennas are abstracted.

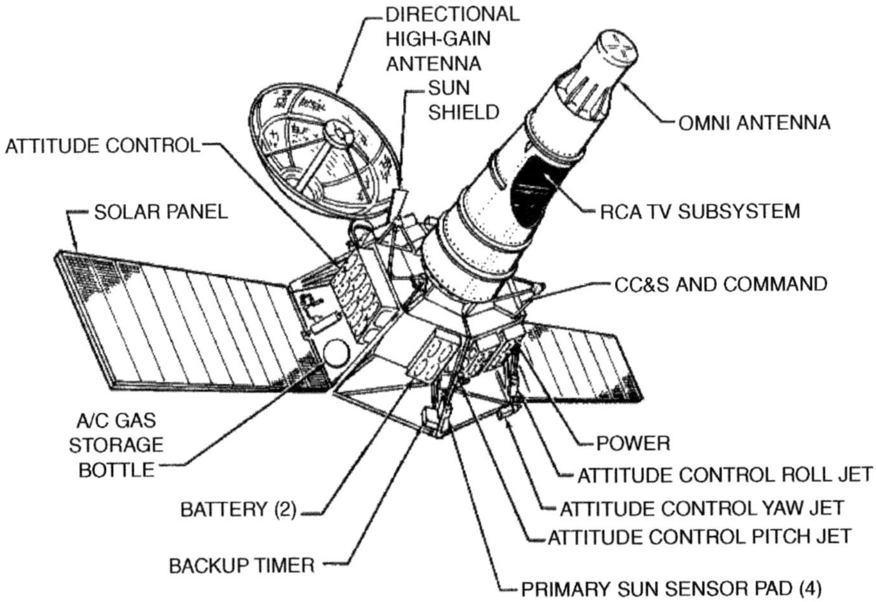

Figure 7.10 *Ranger spacecraft (1962) with omnidirectional and high-gain antennas. Courtesy of NASA.*

7.2.3.2 Venera

Another example where early spacecraft antennas were based on classical designs is that of the first Russian Venus probes Venera (18 were launched between 1961 and 1983), designed to observe and impact or land on the surface of Venus. The Venera spacecraft were evolutions of the Mars probe 1M, the first spacecraft to fly a high-gain parabolic reflector antenna (1960). Just like the Ranger spacecraft, Venera needed low-gain quasi-omnidirectional antennas and a high-gain antenna.

Commands from the Earth were received by the Venera spacecraft at 770 MHz and low-bit-rate telemetry signals were transmitted at 923 MHz.

Typically the same semi-directional circularly polarized antennas were used for both functions. These were initially (Venera 1) crossed dipoles, placed behind the spacecraft's solar panels. They were replaced later by conical spirals which had more bandwidth and more potential for radiation pattern shaping, from pear shape to funnel shape, by changing the angle of the spiral [25]. With the spacecraft rotating slowly around the solar axis, different angles of the antenna's funnel-shaped pattern were needed at different times for the best reception from the Earth's direction.

On the spacecraft bus, there were a few different conical spirals, all aligned with the solar axis but with different funnel angle patterns. The best one was selected via switches for the current relative orientation of the spacecraft with respect to the Earth [26]. The same type of antennas was used for the Mars probes.

Such conical spirals are visible in the photo of Figure 7.11, showing the antennas of Mars 3.

The high-gain data transmission antennas of Russian planetary probes were usually classical centre-fed parabolic reflector antennas. Rigid dishes of 1.7 m diameter were initially used for the Mars probes.

The Venera lander capsules (with typically a diameter of 1 m and a mass of 400 kg and parachuted through the dense Venus atmosphere) also used spirals and Yagis to communicate directly with Earth stations at 922 MHz, at a rate of 1 bit per second. Figure 7.12 shows a spiral and also two Yagi antennas on a model of the Venera 4 lander. The details of a two-arm spiral are also shown.

Figure 7.11 Antennas of Mars 3 with TT&C conical spiral antennas and 1.7 m reflector for science data transmission. Reproduced by permission of Sven Grahn.

For the Venera probes, larger dishes of over 2 m diameter were developed with a rigid central part in front of which a peripheral copper mesh section was folded during launch (see Figure 7.13). Both 922 and 5800 MHz were used for the telemetry sessions for which the high-gain antenna was pointed towards the Earth by rotating the spacecraft around the sun axis, and then locking by a sensor to the Earth [26].

(a) (b)

Figure 7.12 Telemetry and command antennas of early Russian landers. (a) Venera 4 lander with its spiral and Yagi antenna. (b) Blown-up model of a spiral antenna. Reproduced by permission of Sven Grahn.

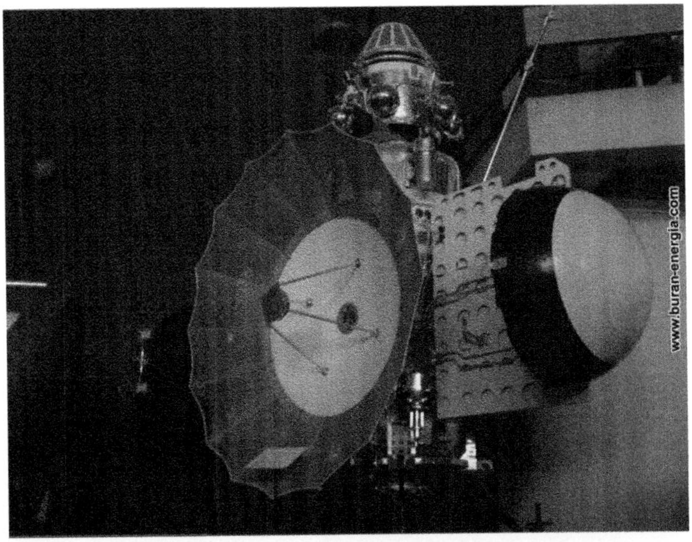

Figure 7.13 High-gain antenna reflector of the Russia Mars–Venus spacecraft series (1961–1983) with peripheral copper mesh section foldable during launch. Courtesy: Kaluga Museum.

7.3 Larger Reflectors with Complex Feeding Systems

7.3.1 Introduction

The most complex active phased array technologies and systems have been mainly developed for non-space military applications and are now preferred configurations for sophisticated radar systems. However, reflectors are better suited for many space missions with a limited field of view and the most advanced multiple beam reflector, lens and feed designs and technologies have been developed by the space sector and are preferred for more sophisticated non-radar space missions.

Table 7.2 gives an overview of different generations of reflector and lens antennas from the 1970s up to 2011.

First, single-spot-beam reflector antennas of increasing size were needed for telemetry and command in one or more frequency bands, for data relay and for communication applications.

Sensors for radio astronomy and for planetary and Earth observation, such as synthetic aperture radars, altimeters, radiometers or sounders, also had challenging requirements with large single-beam and often multi-frequency antennas. Multiband horns and frequency-selective surfaces were developed for these applications.

Until around 1990, mainly single circular and elliptical beams were used to approximate the shape of countries for early TV broadcasting and communications, often with dual polarization.

Single contoured beams were more efficient and were used for Canadian, CONUS (Continental United States), Japanese, European and other coverages.

The multiple contoured beams required by INTELSAT to cover continents at C-band over large bandwidths kept lots of antenna engineers busy for decades.

Multiple spot beams and then contoured beams were also developed at other frequencies: at L- and S-bands for mobile communications with ever-larger mesh reflectors, but also at Ku- and Ka-bands.

Table 7.2 Space reflector antenna evolution.

Period	Beam (s)/coverage/configuration	Application comments
1970–2010	Single circular or elliptical beam from single/ dual shaped mesh or CFRP reflector. Multi-frequency feeds. FSS subreflectors	L-, S-, C-, X-, Ku-, Ka-bands, mm, sub-mm
		Telemetry/data relay. Fixed and mobile communications, broadcasting. SAR, radio astronomy, radiometers.
1972–2004	Shaped beam (s) from parabolic reflector with multiple feeds per beam	C- and Ku-band
		Includes INTELSAT hemispherical and zone beams
1977–2004	Shaped beam from single or dual CFRP or PS shaped reflector. One feed per beam	Ka (CS satellites from 1977) and Ku-bands. Telecom and TV broadcast
1982–2010	Reconfigurable multiple pencil or shaped beams. Lens or reflector optics	X- and Ka-band military
1989–2004	Steerable and rotatable circular or elliptical beam	X- and Ku-band Earth deck – fixed telecom/ TV
1995–2010	Multiple pencil or shaped beams. Large mesh reflector. Matrix amplifiers. RF or digital (remote) beam forming	L- and S-band mobile communications (M-SAT, INMARSAT, ETS VIII). Circular polarization
2002–2010	Regional coverage by multiple pencil or shaped beams. Multiple reflectors	Ku-, Ka- and S-band communications and linguistic TV broadcast
2004–2009	Reconfigurable multiple pencil or shaped beams. One reflector	X-, Ku- and Ka-bands. Telecom + Internet + TV and military

FSS, Frequency-Sensitive Surface; PS, Polarization Sensitive; SAR, Synthetic Aperture Radar; CS, Communication Satellite.

Lenses, with their superior scan properties, were favoured for reconfigurable multibeam military communications.

Multiband, multibeam reflector antennas were also developed and used for various sensing applications (limb sounders, radiometers, etc.) and for radio astronomy telescopes up to IR frequencies.

7.3.2 Multi-frequency Antennas

Early planetary explorer spacecraft such as the Ranger or Venera series used frequencies around 960 MHz for telemetry.

In the mid 1960s, in addition to 960 MHz, NASA introduced S-band. From the mid 1970s onward, more spacecraft used the two-way links at S-band and also transmitted at X-band around 8.422 GHz. These links operated generally in circular polarization.

Deep space probes such as Cassini also use two-way links at Ka-band frequencies around 32 and 34 GHz.

In addition, radio astronomy space telescopes as well as remote sensing instruments such as synthetic aperture radars and radiometers operate in several bands up to submillimetre waves. For these applications, multiband horns or printed feed elements as well as frequency- and polarization-selective surfaces were developed over the years.

(a) (b)

Figure 7.14 *Multi-frequency reflector coaxial feeds. (a) Mariner 10 for S- and X-bands (courtesy of NASA). (b) Quasat feed for 1.6, 5 and 22 GHz. Courtesy of TILAB – Telecom Italia.*

7.3.2.1 Multiband Feeds and Sensors up to Submillimetre Wave Frequencies

The high-gain antenna of Rangers 8 and 9 had, with the 960 GHz telemetry, to accommodate two-way links at 2113 and 2295 MHz. This required the innovative design of multiple band feeds and coaxial rotary joints [8,27]. A broadband backfire conical spiral was used for the feed.

In the early 1970s, NASA added an experimental X-band to the S-band two-way links. The X-band downlink was around 8.45 GHz with a 50 MHz bandwidth.

Mariner 10, launched in 1973 towards Venus and Mercury, was one of the first spacecraft enabled to use the two bands. Its high-gain antenna used a dual-band feed, shown in Figure 7.14a and reproduced from [8]. The inner cylindrical conductor was used as a waveguide feed for the highest frequency band. The lower frequencies were excited by probes in coaxial cavities. Pattern symmetry and cross-polarization were improved by annular chokes around each aperture.

At the time of Mariner 10, a similar design was being studied for Earth-based radio astronomy [28]. Based on this, a three-band version, around 1.6, 5 and 22 GHz, was developed by CSELT in Italy [29] for the ESA projected QUASAT space VLBI mission. It is shown in Figure 7.14b. Patterns are very similar and symmetrical in the three bands.

From the mid 1970s onwards, more spacecraft used the two-way links at S-band and also transmitted at X-band around 8.422 GHz. These links operate in circular polarization.

The requirements of microwave remote sensing of the Earth, planets and their atmosphere in several bands up to submillimetre wavelengths have led to the development of innovative designs for multiband feeds.

The ancestor of the millimetre wave passive sensors seems to be the microwave radiometer flown on board NASA's Mariner 2 in 1962. It observed emissions from the atmosphere of Venus and its surface in two frequency bands at 13.5 and 19 mm. It operated in the standard Dicke mode of chopping between the 48 cm diameter parabolic radiometer antenna, pointing at the target, and a reference horn pointed at cold space. It detected surface temperatures of 500–600 K and a very dense atmosphere.

A good example of a multiband sensor antenna is that of the Scanning Multichannel Microwave Radiometer for Nimbus 7 and Seasat A, flown in the late 1970s. Details on the SMMR can be found in [8,30–33].

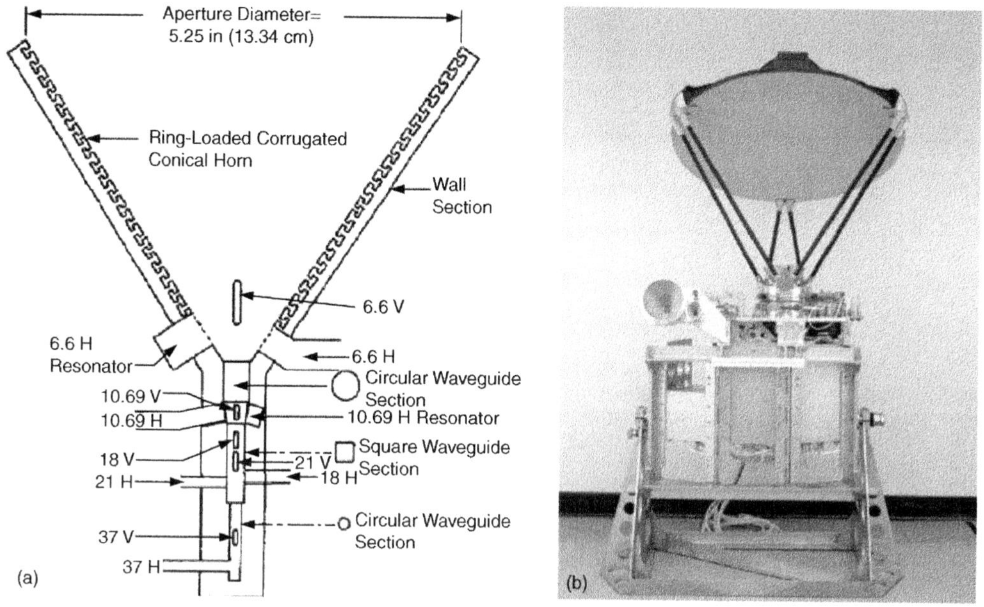

Figure 7.15 Seasat scanning multichannel microwave radiometer. (a) Multiband feed horn. (b) SMMR instrument [8]. Reproduced by permission of John Wiley & Sons.

This instrument is a 10-channel imaging radiometer which measures dual-polarized radiation from the sea and the atmosphere around 6.63, 10.69, 18, 21 and 37 GHz over a 50° swath angle.

Using the retrieved data, sea surface temperature, wind speed, ice cover and also water content in the vapour, cloud and rain in the atmosphere can be mapped.

The single offset reflector of 0.79 m diameter, fabricated from graphite epoxy, is fed by a single multiband corrugated feed to ensure collocation of the beam footprints in the various bands (see Figure 7.15).

Except at 37 GHz, where both polarizations are continuously received, for each band a heterodyne radiometer is switched from one polarization port to the other on alternate scans. Figure 7.16, from [8], shows the antenna configuration and the multiband horn. The bandwidth of the horn is widened by use of ring-loaded corrugations [34]. The various bands were extracted in each polarization by use of slots appropriately positioned along the conical horn, for excitation of the HE_{11} mode at the lower two frequencies, and feeding circular waveguides with a low-pass filter at the higher ones.

For an antenna of this complexity and with closely spaced bands, the overall performance was excellent, in spite of a loss of efficiency due to significant scattering by the reflector support struts.

In the early 1990s, a Ka-band experiment was introduced in the Mars Observer (1992). This was continued with the Mars Global Surveyor (1996) which used 1.5 m Cassegrain optics with a dual-frequency feed shown in Figure 7.16. Described in detail in [8] and [35], the feed uses a corrugated horn for X-band and, along its axis, a disc on rod feed for Ka-band.

Following these pioneering achievements, many multiband feeds and array elements were developed for space antennas, for example the Cassini 3 band (X, Ku, Ka) horn [8], but also for ground stations, for radio astronomy telescopes and for radar arrays. Some examples and many references on the topic can also be found in [36–38].

An alternative configuration for separating multiple frequency bands and polarizations is to use a quasi-optical demultiplexer, including frequency- and polarization-sensitive plates. This was done for the first

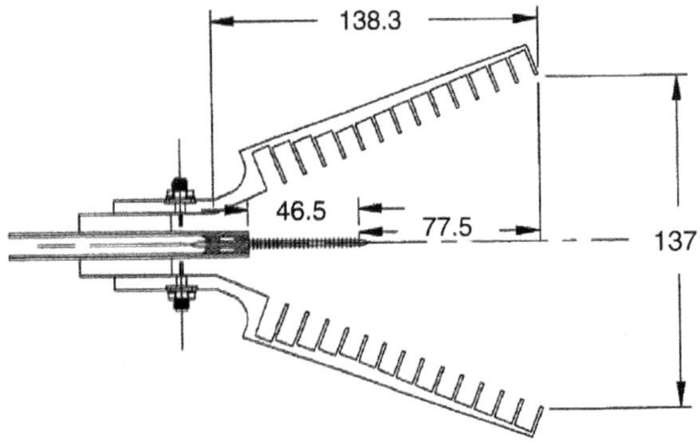

Figure 7.16 Mars Global Surveyor (1996) X-/Ka-band high-gain antenna feed – dimensions are in millimetres [8,35]. Reproduced by permission of John Wiley & Sons.

spaceborne microwave limb sounder (MLS) flown on board the Upper Atmospheric Research Satellite (launched in 1991). This passive instrument, with beam scanning in the vertical plane, measures incoming electromagnetic radiation from 'slices' of the upper atmosphere and routes it to three heterodyne radiometers operating in bands around 63, 205 and 183 GHz and retrieves the concentration of chlorine monoxide, ozone, oxygen, and water, helping to understand the ozone chemistry in the upper atmosphere.

The MLS, described in detail in [39], is shown in Figure 7.17 and uses 1.6×0.8 m aperture Cassegrain optics. A shaped mirror is used to go from the incoming elliptical beam to the circular beams of the horns. Gaussian beam analysis [40] is used as the design technique. Then a dichroic plate, consisting of an array of circular holes in a metal plate, separates the 63 GHz beam. A gridded polarizer is used to separate the 183 GHz and the 205 GHz beams which are directed to the receiver horns by focusing mirrors. A Fabry–Pérot resonator

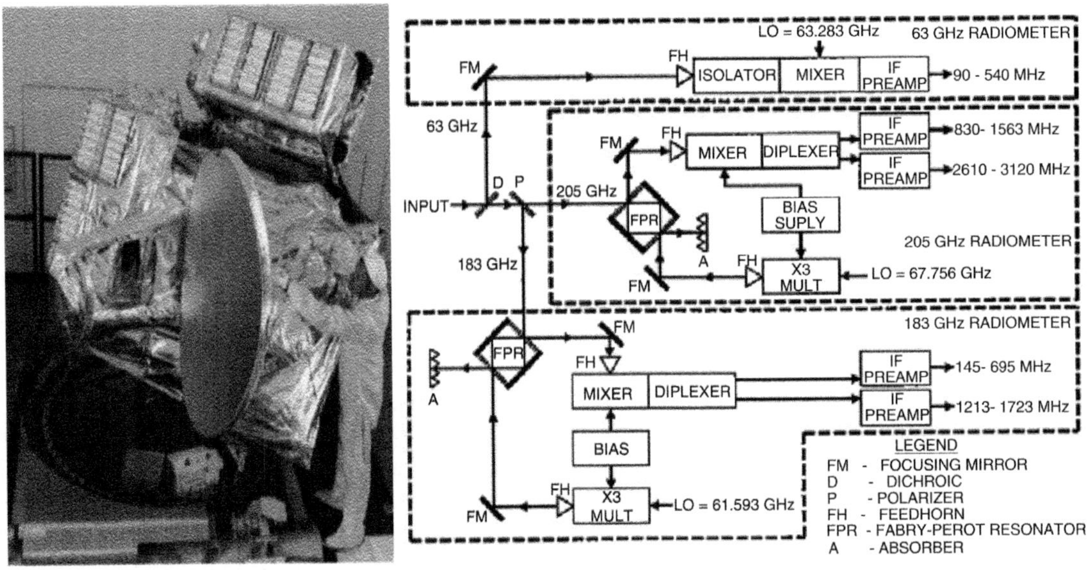

Figure 7.17 Photograph of the first spaceborne microwave limb sounder on NASA's UARS satellite and diagram of its quasi-optics radiometers [39]. Courtesy of NASA/JPL-Caltech.

combines the signal and the local oscillator (LO). A modified Potter horn [41,42] receiving the combined signal and LO signals feeds them to the mixer. The loss of the optical demultiplexer is between 1.0 and 1.5 dB.

Since that time, progress has been made in the field of quasi-optical demultiplexers in particular with the development of improved analysis tools and quasi-optical component technology.

A recent example of a multiband antenna at submillimetre waves is that of ESA's Planck and Herschel Telescopes, launched together on 14 May 2009. Planck will investigate cosmic background radiation from 27 GHz to 1 THz [43]. Herschel, observing from 450 GHz to 5.45 THz, will study the formation of stars and galaxies and investigate the relationship between the two. The 3.5 m Herschel Telescope uses a symmetrical Cassegrain reflector with a wavefront error specification of 6 μm RMS.

The Planck antenna is a modified Gregorian reflector with an off-axis tilted aplanatic design with a 1.5 m circular aperture (1.9×1.5 m reflector with 10 μm RMS accuracy) supporting two scientific instruments:

- The LFI (Low Frequency Instrument) which images the sky at three frequencies between 27 and 77 GHz with a feed array of 22 radio receivers using HEMT transistor mixers. Feeds for the LFI are corrugated horns as described in [44].
- The HFI (High Frequency Instrument) which images the sky at six frequencies between 84 GHz and 1 THz with 52 microwave detectors using bolometers.

The Planck detectors can be cooled to levels down to only a few tenths of a degree above absolute zero.

The Planck Telescope is shown in Figure 7.18. The main and the subreflectors are made of thermally stable carbon fibre reinforced plastic (CFRP) sandwich structures, with a reflecting front face coating (Al + Plasil). A baffle surrounds the telescope to prevent stray light from the Sun and Moon.

The 3.5 m reflector of the Herschel Telescope [45] is a symmetrical Cassegrain reflector. Initially segmented in 12 petals of silicon carbide and following processing of each petal, brazing large pieces

Figure 7.18 Planck Telescope mid-view showing feed array. Courtesy of ESA.

Figure 7.19 *Herschel Telescope mid-view showing feed array. Courtesy of ESA.*

together, milling to 100 μm and polishing to meet a wavefront error specification of 6 μm RMS, it ended up as a monolithic reflector as shown in Figure 7.19 with no gaps between the segments. It was ground tested to 80 K (in-orbit conditions).

7.3.2.2 Dichroics

Dichroic or frequency-selective surfaces are engineered to be reflective in one frequency band and transparent in another. They can be used as an alternative to the multi-frequency horns or elements discussed in the previous subsection.

In the early 1970s such dichroic reflectors, consisting of equispaced circular holes in a metal plate, were used to allow the use of both S-band and X-band transmissions for deep space missions with the 64 m NASA Deep Space Network reflector [46].

Early analysis of such plates can be found in [47] for circular holes. Similarly, plates with equispaced rectangular holes can also be accurately analysed by resolving waveguide modes and Floquet modes at the surface of the grids. This is only accurate for infinite and very large arrays. Finite plates can be modelled by applying Fourier windowing to the infinite plate results as was done in [48] for waveguide arrays. A similar approach can be used for dichroics made of printed elements, like crossed dipoles [8,49], replacing the waveguide modes by the appropriate method of moments unknowns. Curvature remains a problem, although the tangential plane approximation works in most cases.

This was the start of dichroics for space. In the subsequent decades, numerous other patterns such as Jerusalem crosses were analysed and developed.

NASA's Voyager (1977) and Galileo (1990) are notable examples of spacecraft with dichroic subreflectors whose designs, based on [49], are described in detail in [8]. These were successfully used to accommodate S- and X-bands on the same high-gain antenna reflectors. They both use crossed dipoles and exhibit excellent performances with worse case losses around 0.24 dB.

A tradeoff study performed in Europe [50] found that rings are more suitable as elements than crossed dipoles, Jerusalem crosses and square loops. This was verified for a double-gridded subreflector, fanned from a triple Kevlar sandwich dichroic subreflector reflecting from 10.95 to 12.25 GHz and reflective from 18.2 to 20.2 GHz. Good agreement was found between measured and predicted results [51] with losses below 0.4 dB in both the reflection and transmission bands. This approach was later used for the high-gain antenna of Cassini [52].

The US–European Cassini–Huygens orbiter probe to Saturn, launched in 1997, involved one of the most challenging multi-frequency space antenna developments to date. The brief description below is abstracted from [8].

The high-gain antenna (see Figure 7.20) generates on-axis pencil beams at S-band (2035–2303 MHz), X-band (7150–8450 MHz), Ku-band (13 676.5–13 876.5 MHz) and Ka-band (31 928–34 416 MHz). These are all in circular polarization except Ku-band, supporting a planetary synthetic aperture radar (SAR), which operates in linear polarization and must, in addition, generate four fan beams for its imaging mode. The high-gain antenna (HGA) supports a low-gain antenna at X-band.

The selected configuration is a Cassegrain design. The 4 m diameter onset main CFRP reflector uses a thin sandwich (7 mm) with a rib and ring backing structure.

A set of three cascaded frequency-sensitive subreflectors, all transparent for the S-band signals, reflects the X-, Ku- and Ka-band signals. The S-band feed is located at the prime focus of the main reflector and a triple X-, Ku- and Ka-band feed is located at the Cassegrain focus, together with four sub-arrays of waveguide horns generating the SAR fan beams.

The Ku/Ka-band subreflector has a hyperbolic profile and uses double rings resonating at Ku and Ka frequencies. The X-band subreflector uses an array of single rings resonating at X-band frequencies and is shaped to improve efficiency. A shaped screen, with optimized spacing to the X-band reflector, is then added to match the S-band feed. A block diagram of the HGA is shown in Figure 7.20. Each of the screens includes dichroic elements etched on a Kapton substrate supported by a Kevlar honeycomb.

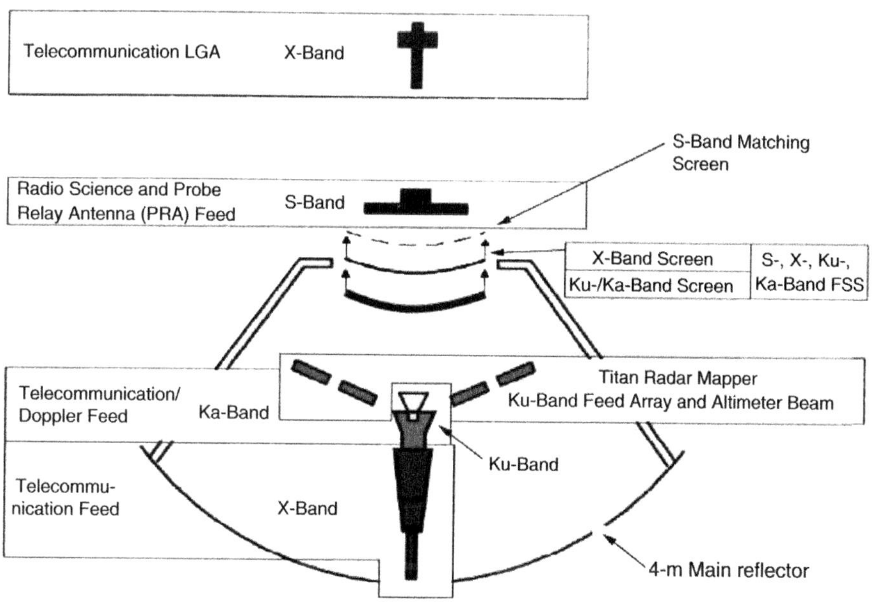

Figure 7.20 Functional block diagram of the Cassini HGA [8]. Reproduced by permission of John Wiley & Sons.

Figure 7.21 *Feed system of Cassini's four-band HGA. Courtesy of Thales Alenia Space Italia.*

Pictures of Cassini's HGA feed system and of the dichroic subreflectors are shown in Figures 7.21 and 7.22.

Numerous dichroic subreflectors have been developed, in particular in Japan for the N-Star Ka-band antenna (launched 1995) and Adeos (1996) for S/Ka-bands.

Figure 7.22 *Dichroic subreflectors of Cassini's HGA. Courtesy of Thales Alenia Space Italia.*

7.3.3 Large Unfurlable Antennas

Since the early 1960s, to accommodate reflectors with diameters larger than that of the launcher shroud, partially deployable reflectors with a rigid central part and a deployable mesh rim were used extensively by the Russians with the Venera reflectors (see Figure 7.13) and later in the USA, for example with the FLSATCOM [53] shown in Figure 7.23.

The reflector antenna of the geostationary FLSATCOM satellite was used to transmit in the UHF band from 240 to 400 MHz. It was made of a rigid central section with a 3.35 m diameter extended to 4.9 m by a folded outer wire mesh screen supported by stainless steel ribs deployed by ground command. The receive antenna, a helix antenna 4.10 m long with a 33 cm diameter, is also folded during launch and deployed by ground command [53].

The next subsections give an overview of some of the many concepts of large unfurlable antennas developed to an advanced stage over the last 40 years. Several detailed reviews of large reflector antenna development can be found in the literature [54–58]. Part of the text in the sections below is abstracted from these references.

7.3.3.1 Wrap-Rib Antenna

Lockheed's wrap-rib antenna principle and its main constituents are shown in Figure 7.24. The reflector includes a central hub and deployment mechanism, CFRP parabolic ribs of lenticular cross-section and a wire mesh of gold-plated lightweight molybdenum [59].

This antenna, with a diameter of 9.12 m, was flown on the first geostationary three-axis stabilized spacecraft, namely the NASA ATS-6 experimental broadcast satellite, operating at in C-, S-, L-, UHF and VHF bands, launched in 1974. It had 48 ribs and weighed 60 kg. Lockheed demonstrated for JPL a sector of a 55 m diameter

Figure 7.23 *US Navy communication FLSATCOM (1978–) with helix and reflector antennas. Courtesy of US Air Force.*

Figure 7.24 *Lockheed wrap-rib antenna for NASA's ATS-6 (1974). Courtesy of NASA.*

offset wrap-rib antenna. A 5.7 × 5.3 m 16-rib wrap-rib offset reflector was developed for the L-band multibeam antennas of the M-SAT mobile communication satellites launched in the mid 1990s.

7.3.3.2 Hoop–Column Antenna

The hoop–column antenna was initiated by the Harris Corporation for NASA in 1975. A 15 m engineering model scaled from a 100 m design was built and tested [60,61]. The antenna is shown in Figure 7.25. The aperture is divided into four quadrant providing four 6 m non-confocal 6 m offset apertures.

The structure includes a central telescopic column deployed from the central hub and a hoop which supports the periphery of the mesh reflecting area, the surface of which is adjusted by a network of cables attached to the top and bottom of the column. The stowed volume is a 2.7 m high cylinder with of diameter of 0.92 m. The mass of the antenna is 241 kg and the RMS surface accuracy is 3.4 mm.

7.3.3.3 Rigid Radial-Rib Antennas

The Harris Corporation radial-rib antenna design uses pivoting tubular ribs to support the gold molybdenum mesh. It was flown from 1983 on the first generation of TDRS satellites in a symmetrical 4.8 m diameter configuration for operation at S- and K-band (Figure 7.26). The same configuration was adopted for NASA's Galileo Jupiter probe (S- and X-band) launched in 1990. The 'double-mesh' design for TDRS and Galileo uses quartz wires connected to catenaries for precision adjustment of the reflecting mesh [8].

An offset edge-mounted configuration with straight hinged ribs with standoffs and integrated boom assembly was developed in the 1990s and is now used on many mobile telecommunication satellites with very large diameters exceeding 20 m. Figure 7.27 shows the 12 m offset antenna of the Asia Cellular Satellite System, ACeS, launched in 2000.

Figure 7.25 Harris hoop and column antenna (1978–1986). Photograph courtesy of Harris Corporation.

Figure 7.26 TDRS radial rib antenna. Photograph courtesy of Harris Corporation.

Figure 7.27 ACeS 12 m Harris hinged-rib antenna. Photograph courtesy of Harris Corporation.

7.3.3.4 Hughes Springback Reflector

Hughes (now Boeing) developed an innovative Springback or 'Taco' reflector for the M-SAT mobile communication satellite (1995) operating at L-band (Figure 7.28). The two elliptical M-SAT reflectors, 6.8×5.25 m, were made of a flexible thin-walled graphite mesh made from material with a triaxial weave with an integral lattice of ribs and connecting elements, with a stiffening hoop along the rim [62,63].

The reflector was folded against the spacecraft during launch in the form of a cone advantageously filling the unused top of the shroud. The reflector had a very low mass of 20 kg.

This concept was adapted for operation at Ka-band and was selected for the antennas of the second generation of TDRS satellites operating in the S-, Ku- and up to Ka-bands.

7.3.3.5 Inflatable Antennas

Inflatable Space Rigidized Reflector This European development started in 1980 aiming at reflectors of 12 m for mobile communications [64,65] and of 15–20 m or more for very large baseline interferometry (VLBI) radio astronomy applications [66]. The inflatable space rigidized reflector is divided into three parts:

- A torus which stabilizes and stretches the structure.
- Reflector/radome membranes, one to support the RF reflecting/transparent surfaces and one to seal the structure.
- A boom to join the reflector to the spacecraft.

Figure 7.28 *Hughes Springback reflector. Courtesy of Canadian Space Agency www.asc-csa.gc.ca.*

The reflector wall consists of a thin fibre-reinforced composite lamina (Kevlar fibres impregnated with a special matrix) with a Kapton foil metallized on one side. After inflation in space, solar radiation hardens the resin in six hours at a temperature of 110 °C and rigidizes the structure, from which the nitrogen is evacuated 48 hours after deployment. Three engineering models of the antenna, with diameters of 3.5, 6 and 12 m, have been built but no flight testing has been performed and the programme was eventually stopped.

An early design study showed that a 20 m diameter reflector would have a mass of 134 kg (0.41 kg/m^2). The 12 m engineering model is shown in Figure 7.29.

Inflatable Antenna Experiment This programme was initiated by JPL to explore the potential of inflatable structures and particularly of inflatable reflector antennas for future space missions.

L'Garde, a company with a long heritage of inflatable space objects, developed a 14 m diameter inflatable antenna with a low-pressure canopy structure, high-pressure torus and three high-pressure struts, which support the feed.

The front canopy was transparent and the back one aluminized to reflect. The torus and the struts were made of Kevlar 0.3 mm thick and the canopies of 6.5 μm Mylar film. The total mass of the inflatable structure was 60 kg.

On 29 May 1996 the Inflatable Antenna Experiment was deployed aboard the free-flying Spartan-207 carrier spacecraft. The struts, the torus and finally the canopy were inflated (see Figure 7.30). Details of the inflation sequence and measurements can be found in [67].

Subsequently a 25 m inflatable reflector was studied for a for radio astronomy applications [68].

Both the 12 m self-rigidizing reflector of the previous subsection and the 14 m IAE reflector suffer from increased surface errors at the reflector periphery.

Figure 7.29 Contraves' self-rigidizing 12 m reflector.

Figure 7.30 The 14 m diameter IAE reflector antenna deployed in space in 1996. Courtesy of NASA.

7.3.3.6 *Tension Truss Antennas of HALCA and ETS VIII in Japan*

A large unfurlable antenna with high surface accuracy was developed in Japan in the 1990s for the radio astronomy satellite HALCA (Muses-B) launched in 1997 and part of the international VLBI Space Observatory Program, VSOP. It is well described in [69,70]. A photo of the antenna is shown in Figure 7.31.

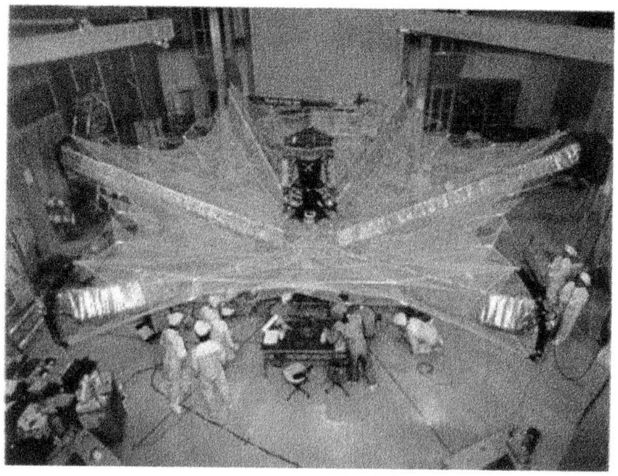

Figure 7.31 *HALCA (Muses-B) antenna. Courtesy of JAXA – Professor T. Takano.*

Unlike most of the larger antennas described above, this 10 m diameter antenna must operate at frequencies as high as 22.15 GHz and also at 4.85 and 1.66 GHz, which implies a surface accuracy of the order of 0.5 mm RMS.

A symmetrical displaced-foci Cassegrain optic was selected and both the main and the subreflector are deployable once in orbit.

The tension truss design of the main reflector uses six extendible masts supporting, via Kevlar and Comex cables, a mesh network itself connected to the finer gold-plated molybdenum reflecting mesh. HALCA was launched in 1997. The antenna deployed and measurements were made until 2005.

Another larger tension truss reflector was developed for the National Space Development Agency of Japan (NASDA, now JAXA) for the Engineering Test Satellite VIII to support S-band communications (2.5/2.6 GHz). This implies a surface accuracy of 2.5 mm RMS [71]. It is based on a modular approach using 14 hexagonal modules of 4.8 m in diameter each with a tension backup truss supporting a metal mesh via quartz cables. The total dimension of the 170 kg antenna is 19.2 × 16.7 m (Figure 7.32). The stowed size is 1 m in diameter and 4 m in height.

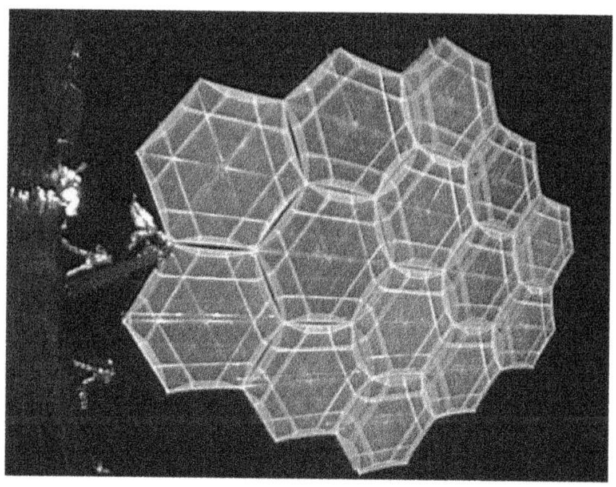

Figure 7.32 *ETS VIII 19 × 23 m antenna. Courtesy of JAXA.*

Figure 7.33 *AstroMesh reflector antenna of INMARSAT-4. Reproduced by permission of Northrop Grumman Corporation.*

7.3.3.7 *AstroMesh Antenna*

The AstroMesh antenna development [72] started in 1990 at AstroAerospace. It uses a tension truss concept. The structure includes a deployable ring truss with two identical parabolic nets with triangular cells attached to it, their mirroring nodes connected by prestressed ties. The reflective mesh is attached to the back of one of the two nets. This reflector has been successfully flown for operation at L-band on several mobile communication satellites. For Thuraya (2000) the reflector had a diameter of 12.25 m and a mass of 57 kg. The reflector of INMARSAT-4 is shown in Figure 7.33.

A more accurate version for Ka-band applications has also been developed [73].

7.3.3.8 *Deployable Truss Reflector of the PRIRODA Module from MIR*

The Priroda module of the Mir Station was equipped with an SAR instrument (Figure 7.34) with a deployable mesh reflector, operating at 1.28/3.28 GHz. Dimensions of the reflector were 6×2.8 m deployed and $0.6 \times 0.4 \times 0.58$ m in stowed states. The back structure of the reflector consists of a rod truss system which creates triangular facets for the mesh attachment. The mass of the reflector was 35 kg and the corresponding surface density 2.1 kg/m^2. Such antennas are produced with dimensions up to 22 m. Several similar reflectors with different diameters were flown in orbit starting from 1985.

7.3.3.9 *Analysis of Mesh Reflectors*

In the analysis of mesh unfurlable antennas two specific areas have been given special treatment: the reflecting mesh with its pattern, contact imperfections and losses; and the gored or faceted reflecting surface. These are reviewed in [56,57] from which this subsection is abstracted.

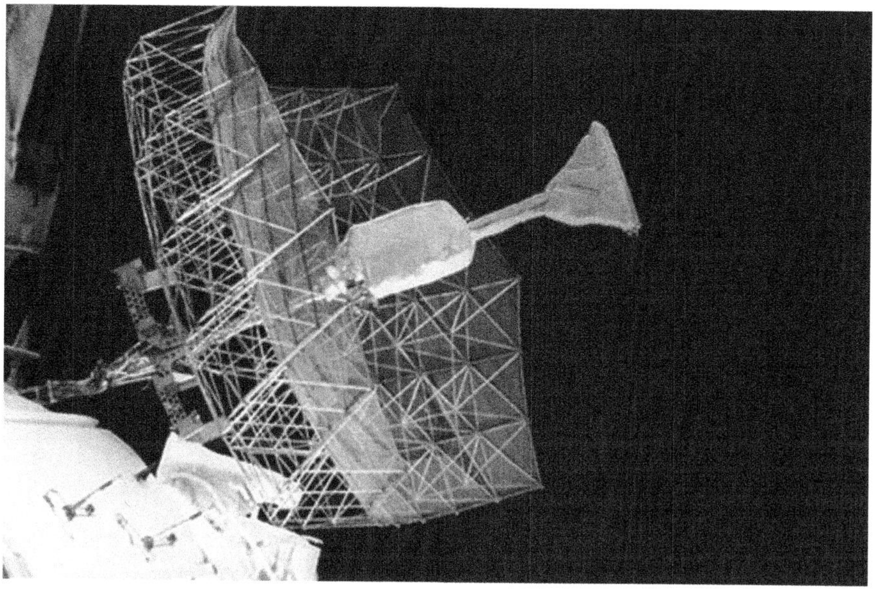

Figure 7.34 *PRIRODA SAR deployable truss antenna. Courtesy of NASA.*

For the mesh two models have been used: the wire grid model [74–76] based on derivation of the relevant Green's dyadics for square lattices in [77], and in [78] for rectangular ones; and the strip-aperture model, introduced in [79], which uses six parameters instead of four to characterize a cell and allows better modelling of certain shapes. Both models have been well validated experimentally. A mesh analysis which includes the complex tricot knit geometry is given in [80].

The main conclusions are that the practical impact of the mesh is on cross-polarization and that a gold-plated mesh with less than $\lambda/50$ openings and with a wire diameter of $\lambda/500$ should give satisfactory performance in most cases.

Passive intermodulation product generation is a sensitive area, particularly if the same reflector is used for the transmitting and receiving functions. The modelling of gores for mesh umbrella reflectors with parabolic ribs was analysed in 1972 [81]. Then, for such antennas, the impact of distortions was covered in [82] by path corrections in the physical optics analysis. For both umbrella and truss reflectors physical optics give adequate results [74,75]. The much faster aperture field technique also gives good results in a more limited field of view [76].

7.3.4 Solid Surface Deployable Reflector Antennas

The surface accuracy of many large mesh and inflatable reflectors has been marginal for use at Ka-band and above which are of interest for radio astronomy, radiometry and, more recently, for high-rate telecommunications.

Several concepts have been proposed, several have been developed and a few have actually flown. Only the key developments will be briefly reviewed here.

7.3.4.1 TRW Sunflower Reflector (1974)

An early (1984) development of a rigid deployable reflector antenna is TRW's Sunflower rigid petal reflector [83,84]. It uses simple joints between the panels.

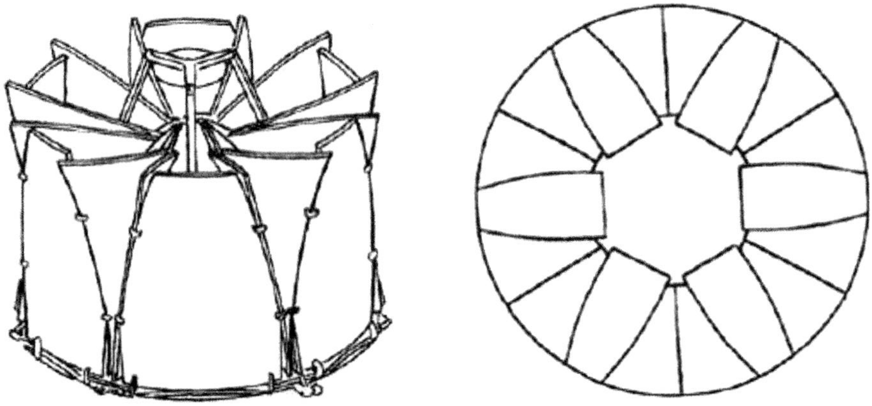

Figure 7.35 *Sunflower design principle.*

A 4.9 m diameter model was built, with a stowed diameter and height of 2.15 and 1.8 m, respectively [85]. Figure 7.35 shows the TRW Sunflower reflector principle.

7.3.4.2 Hinged-Tip Reflectors

In 1978, ESA started the development of a multibeam 20/30 GHz antenna at Thales Alenia Space (at the time called Selenia). This included the development of a 3.7 m projected aperture reflector with hinged tips (see Figure 7.36) and also a complete feed system for the generation of 0.3° spot beams and contoured beams with overlapping feeds [86].

The reflector had a sandwich rib-stiffened structure with an aluminium honeycomb core of 6.35 mm. A reflector density of 3.4 kg/m^2 with a surface tolerance of 0.3 mm RMS was achieved. Accurate spring-driven

Figure 7.36 *The 4 m hinged-tip reflector. Courtesy of Thales Alenia Space Italia.*

hinges were developed by Contraves Italiana. The reflector was used for a mission implementation but the same technology was used (without hinges) for the 4 m Cassini antenna and hinges were sold in the USA.

Some years later a similar reflector was successfully built by Toshiba for the NASDA ETS VI project [87]. The 3.5×4 m projected aperture antenna has a mass of 43.6 kg and an RMS surface accuracy of 0.23 mm in the worst case. It is made of a honeycomb sandwich with CFRP face sheets and is folded by segments for launch and then deployed by a spring. The pointing accuracy of $\pm 0.015°$ is outstanding.

7.3.4.3 Multi-petal Deployable Reflector

Several designs involving a high number of solid petals have been proposed and were developed in the 1980s and later for high-accuracy large reflectors. Driven by radio astronomy and science applications, they have mainly been based on symmetrical configurations and, as a result, were not competitive for offset configurations.

Dornier, in Germany (now Astrium-D), developed in the early 1980s a Deployable Antenna Integral System (DAISY) with radial parabolically curved panels hinged to the central hub. Very high accuracy was achieved: 0.2 mm RMS. A 4.7 m diameter model has a stowed diameter and height of 1.7 and 2.4 m, respectively. The mass was 94 kg [88].

A much lighter evolution of this design was developed further by Astrium and a 3 m thin-shell panel reflector was built. The manufacturing accuracy of a single panel was measured as 0.15 mm. Total mass of the reflector is 10 kg (1.4 kg/m^2). The structure of the reflector can be extended up to 5–6 m diameter.

More recently, a somewhat similar design has been developed in Russia for the RadioAstron satellite [89–91]. The antenna is shown in Figure 7.37.

This space radio telescope has a centre-fed configuration with a diameter of 10 m. It operates in four frequency bands around 0.327 GHz (P-band), 1.665 GHz (L-band), 4.830 GHz (C-band) and 18.392–25.112 GHz (K-band). In order to achieve a high surface accuracy, the telescope has a centre-fed configuration and consists of a solid central mirror 3 m in diameter surrounded by 27 solid petals ($34 \times 115 \times 372$ cm) made of carbon fibre. The maximum deviation for the surface from a paraboloid must not exceed ± 2 mm. The total mass of the dish is about 1340 kg.

Figure 7.37 *RadioAstron 10 m radio telescope antenna (2004). Courtesy of Lebedev Physical Institute of Russian Academy of Sciences Astro Space Center.*

7.3.5 Polarization-Sensitive and Shaped Reflectors

7.3.5.1 Polarization-Sensitive Reflectors

While the antennas of space probes and for mobile communications operated mostly in circular polarization, which is easier to track, linear polarization was used from the mid 1960s for the many Fixed Satellite Service (FSS) geosynchronous communications satellites such as INTELSAT 1 (Early Bird), Syncom 3, Anik 1 and Westar 1 in either the C- or the Ku-band. It is still used nowadays for several FSS and broadcasting applications.

A polarization-sensitive antenna using one gridded reflector in front of a solid one was proposed in 1949, well before satellites existed (see Figure 7.38 from [92]). This approach is still used today in space for parabolic and shaped gridded reflector antennas, generally with offset feeding.

A configuration for frequency reuse in space, and with two separate orthogonally gridded overlapping surfaces, was proposed in 1974 (see Figure 7.38 from [93]). It was used for the two RCA Satcom satellites for US coverage, launched in 1975 and 1976 (see Figure 7.39 from [94]).

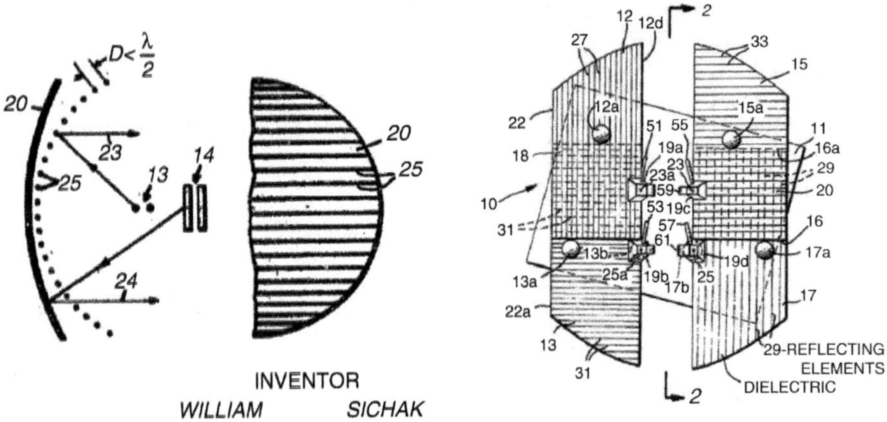

Figure 7.38 First polarization-sensitive reflector for RCA SATCOM (1975) frequency reuse: principle and implementation [92,93].

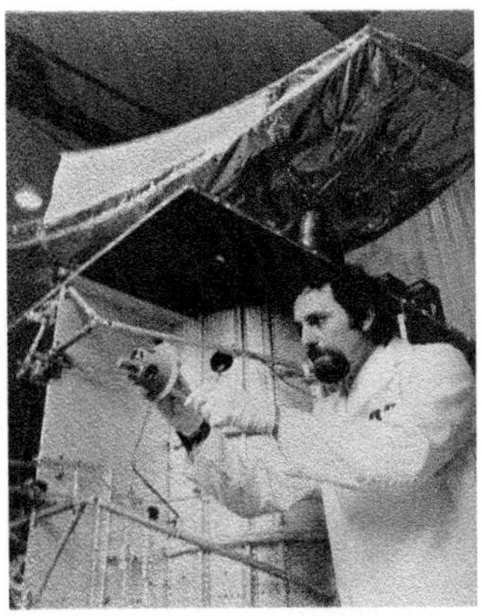

Figure 7.39 First polarization-sensitive reflector [94]. Courtesy of NASA.

In both cases, the wires are seen to be parallel to the incoming polarization by a user in the centre of the field of view. When the coverage footprints in both polarizations coincide or overlap partially, their feeds should not interfere with each other. For a long time, shells with two different shapes (and moulds) were used, as in Figure 7.39. This can generally be avoided in an offset system by properly rotating the offset shells and the feeds [95].

An added advantage of the grids is that the cross-polarization generated by the offset geometry and by the feed can be reflected outside the coverage.

This same filtering property was used by Hughes for COMSTAR 1. Launched in 1976 for communications at C-band, it has two reflectors, each with a polarizing screen placed in the aperture plane in front of each reflector [96]. COMSTAR 1 is shown in Figure 7.40. These polarizers use a parallel grating of conducting strips with 12.7 mm spacing and 25.4 mm depth. The conducting strips are fabricated as a sandwich of aluminized Kapton. The insertion loss for this design is less than 0.1 dB and it provides over 33 dB cross-polarization isolation.

Many parabolic and shaped polarization-sensitive reflectors (and some subreflectors) have been developed and flown in the USA, Canada and in Europe. These involve two reflecting shells with a fan-type set of intercostal beams and a backing structure, typically from a Kevlar sandwich.

The reflecting grids initially were obtained by bonding a gridded film onto the reflector surface. Later, laser etching of an aluminium film through a specific mask was used to produce the required gridding on the parabolic or shaped surface.

In the dual-grid polarization-sensitive reflector, thermal design is all important and involves an RF transparent (loss < 0.1 dB) sunshield, typically using Kapton with vacuum-deposited aluminium and a coat of germanium for electrostatic discharge (ESD) protection, which, in the early days of gridded reflectors, was extremely expensive.

Figure 7.40 COMSTAR 1 antennas with polarization screens. Courtesy of NASA.

Figure 7.41 *An early gridded shaped reflector of 2.3 m for C- and Ku-bands. Courtesy of EADS-CASA.*

A 2.3 m shaped gridded reflector of the 1990s for C- and Ku-bands using a Kevlar front and CFRP backing structure is shown in Figure 7.41. Since then, many parabolic and shaped gridded reflectors up to 3 m in diameter have been flown.

7.3.5.2 *Shaped Reflectors*

Double-curvature, single-reflector shaping using a two-step, one-dimensional procedure based on the conservation of energy and the simple laws of geometrical optics was introduced with success in the 1940s to achieve cosecant squared patterns for air-surveillance radars [97].

Dual-reflector shaping to control the main aperture amplitude and phase distribution, also using geometrical optics principles, was first introduced in the 1960s [98] to maximize the aperture efficiency of high-gain spacecraft antennas. This was used for the X-band antenna of the Voyager spacecraft (launched in 1979).

In the mid 1970s, the same approach was also used to design elliptical beam antennas with low cross-polarization for the ESA OTS satellites by shaping a single circular reflector [99]. The method was extended to design dual shaped offset reflectors [100]. An example of a dual shaped reflector flown on board the ESA ECS satellites is shown in Figure 7.42.

Shaped reflectors for global Earth coverage at L-band with uniform flux were also studied in the mid 1970s in the USA and Europe. The 2.2 m centre-fed shaped reflector antenna of the ESA INMARSAT MARECS satellites [101] provided 19.6 dBi gain at the edge of the Earth, over 3 dB more than a classical horn (Figure 7.43).

In the same period, a related and more general two-dimensional shaping procedure was then developed to design an antenna for the CS satellite with a beam footprint matching the contours of Japan [102].

Later, alternative diffraction synthesis techniques were developed in Europe [103]. These have been applied in a number of studies supported by, among others, ESA and INTELSAT.

Nowadays, physical-optics-based designs have become the norm and software is available to synthesize and analyse large single or dual shaped reflectors [104].

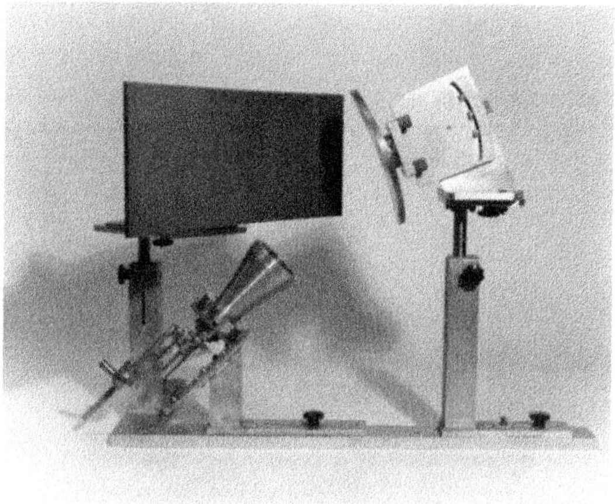

Figure 7.42 Dual shaped reflector model for ESA ECS satellite (1978). Courtesy of Thales Alenia Space Italia.

Figure 7.43 MARECS satellite (1981) with shaped reflector. Courtesy of ESA.

Single reflectors are usually gridded to eliminate the cross-polarization induced by the offset geometry, or to generate two different coverages in the two polarizations.

Gridded reflectors can often be avoided by use of dual shaped reflectors with an initial geometry based on Mizuguch's low cross-polarization condition [105].

One of the problems with shaped reflectors, as opposed to multi-feed parabolic reflectors or lenses, is that only one beam can be generated per reflector profile.

7.3.6 Multi-feed Antennas

Most of the antennas reviewed in the previous sections generate a single circular, elliptical or contoured beam, typically for broadcasting or regional communications. For higher rate communications and also to allow more reuse of frequencies, and more reconfigurability, multiple spot beams and contoured beams provide more capacity.

The single feeds then give way to feed arrays. For reflectors, offset-fed configurations are preferred to avoid feed blockage. Aberrations, when beams are pointed several beamwidths away from boresight, can be a problem. They generate high sidelobes and cause a loss of gain unless a large number of feeds are used to form the off-boresight beams.

Lenses, in spite of their mass and bulk, are attractive because they do not suffer from feed blockage and have fewer scan aberrations.

7.3.6.1 *Multi-feed Lens Antennas*

Centre-fed lenses are well suited for reconfigurable space communication antennas.

Mostly military satellites, such as the DSCS 3 series, use lens antennas and detailed publications on these are scarce. A review of some early developments of lens antennas can be found in [5].

Waveguide lenses for reconfigurable multibeam Earth coverage were studied in the late 1960s for the US Air Force at Lincoln Laboratory, Massachusetts Institute of Technology [106,107]. The X-band square waveguide lens is zoned for minimum thickness. It has 19 feeds and can potentially produce reconfigurable coverage ranging from 3° spot beams to a full Earth coverage beam, due to variable power dividers. The lens studied in [106] is shown in Figure 7.44. Similar lenses but zoned for minimum phase error have also been studied at General Electric Co. [5]. Equal group delay waveguide lenses [108,109] were developed by Hughes.

Designs with 61 feeds were also studied [106] and implemented in the 1980s.This work seems to have been the basis for the development of the multiple waveguide lens antennas of a number of military satellites including the DSCS 3 series (first launched in 1982), one of which is shown in Figure 7.45.

These geostationary satellites provided uplink services in the 7900–8400 MHz band and downlink services in the 7250–7750 MHz band over the Earth, with reconfigurable spot or shaped beams including anti-jamming features due to the use of complex reconfigurable feed networks involving low-loss variable power dividers and phase shifters.

Lenses for multibeam civil communication applications were also studied in the 1970s at Ford Aerospace, which built and measured waveguide and TEM lenses [110,111].

7.3.6.2 *Multi-feed Reflector Antennas*

Fixed Communication at C-band Anik A1 (1972) A multi-feed reflector antenna includes the reflector optics, and just like the lens multibeam antennas, a feed array and a beam-forming network. Amplifiers can be

Figure 7.44 *Multibeam waveguide lens for the US satellite LES 7. Reprinted with permission of MIT Lincoln Laboratory Lexington, Massachusetts.*

Figure 7.45 DSCS 3 satellite with waveguide lens antennas. Reprinted with permission of MIT Lincoln Laboratory Lexington, Massachusetts.

directly at the feed elements, separated from the feed by an output network, at the beam port level or even 'deeper' if multi-port amplifiers are used.

Anik A1, shown in Figure 7.46, was built by Hughes and launched in 1972 for Telesat, Canada. Anik A1 was the first geostationary telecommunication satellite with a multi-feed antenna generating one contoured beam. Anik A1 provided for 12 C-band channels (uplink 5.925–6.425 GHz, downlink 3.7–4.2 GHz) over Canada.

Figure 7.46 Telesat's Anik A1 (1972) with first shaped beam antenna. Source unknown.

The antenna feed assembly includes three transmit/receive horns approximating the image of Canada in the 1.52 m gold mesh reflector [112]. With three feed horns, the transmit feed network can use two waveguide directional couplers to provide the right signal distribution to the horns.

A directional antenna receive feed assembly, receiving a ground signal, is used to control the de-spinning rate of the antenna around 100 rpm.

COMSTAR 1 An evolution of this shaped beam antenna design was implemented in COMSTAR 1 (see Figure 7.40), by using two offset reflectors with added polarization screens (one polarization per reflector) [96]. This provides over 33 dB of isolation between the two polarizations and allows doubling of the capacity to 24 C-band channels.

COMSTAR 1 also uses an innovative, Hughes-patented, dual-mode feed network. Using hybrid couplers, this power divider provides, from the two isolated input ports, two orthogonal excitations to the feeds with the same amplitudes but with conjugate phase laws. As a result, the complex antenna patterns on the Earth from the two input ports are orthogonal, with essentially the same amplitudes, and thus beam footprints, but different phases. By multiplexing the odd channels in one port and the even channels in the other, filtering is much easier with no impact on the coverage.

INTELSAT As indicated in [113], progress in antenna design and technology has been the key to the evolution of the INTELSAT system: Early Bird (1965), with a wide toroidal beam, radiates mostly in deep space; INTELSAT 3 (1968) with a global beam, mostly radiates on the Earth disc. INTELSAT 4A (1975), with two shaped hemispherical beams with 27 dB sidelobe isolation, offers two-fold frequency reuse, INTELSAT 5 (1981) four-fold with in-orbit reconfigurability for two beams and INTELSAT 6 (1989) six-fold frequency reuse with more reconfigurability.

The trend continues with INTELSAT 7 with two hemispherical beams and three zone beams.

While INTELSAT 4A (Figure 7.47) uses a 1.2 × 1.3 m square with 37 feeds, INTELSAT 5 uses a 2.4 m reflector with 89 square feeds and a triplate feed network [114]. INTELSAT 6 uses a 3.2 m transmit antenna

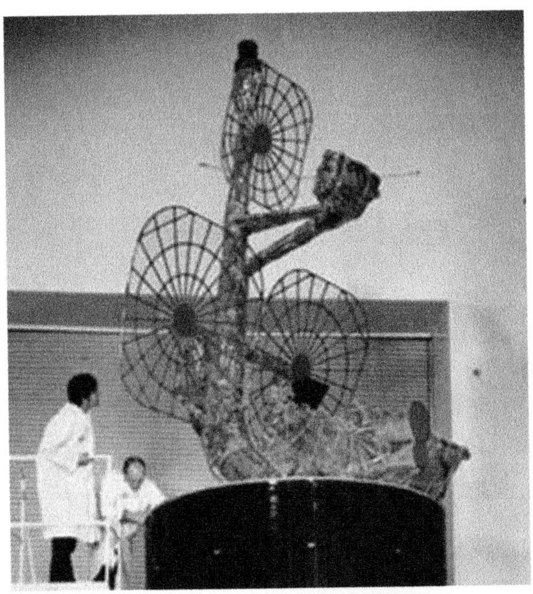

Figure 7.47 INTELSAT IV A (1975). Courtesy of NASA.

Figure 7.48 INTELSAT VIII (1997) antenna. Courtesy of Astrium-D.

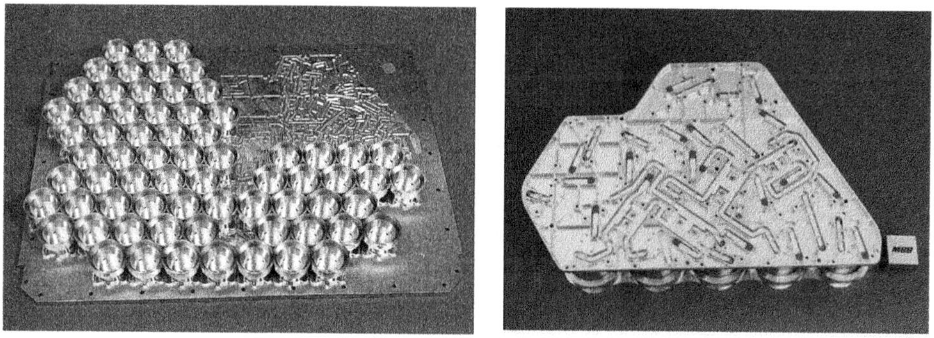

Figure 7.49 INTELSAT VIII 'scrimp horn' and square coaxial feed technology. Courtesy of Astrium-D.

with 145 Potter horn feeds [115]. INTELSAT 7 uses a 2.4 m transmit antenna with over 100 feeds of 1.3λ diameter [116].

The 2.6 m INTELSAT 8 transmit antenna, shown in Figure 7.48, uses compact four-probe-fed dual-mode 'scrimp horns' and square coaxial line feed network technology [117] (Figure 7.49).

INTELSAT 9, with a much increased capacity, uses similar technology with a 3.2 m transmit antenna reflector and 126 'scrimp horn' feeds [117,118]. Major innovations in INTELSAT 9 are the use of digital beam forming and of the flexible semi-active beam-forming scheme [119] for the transmit antenna, also used at L-band in the INMARSAT satellites from INMARSAT-3 onwards.

Mobile Communications at L- and S-bands The evolution of mobile communication satellites at L- and S-bands and their antennas is summarized in Table 7.3.

Marisat did not use reflectors but helices at UHF and L-band.

Table 7.3 Evolution of mobile communications at L- and S-bands.

Period	Coverage/beams	Satellites/antenna design
1976–	Single Earth coverage circular pencil beam	Three Marisat satellites with four helix arrays at L-band (and three UHF helices)
1981–	Shaped Earth coverage beam	Two ESA MARECS satellites with 2.2 m shaped reflector[
1990–	Shaped Earth coverage beam	Four INMARSAT-2 satellites with arrays of cup-type elements
1994	North America	Two M-SAT satellites with two 5.5 m reflectors fed by
	Six shaped beams	23 cup elements, beam-forming networks and two multi-port amplifiers
1996–	Shaped Earth coverage beam and five spot beams	Four INMARSAT-3 satellites; 2.4 m Tx reflector fed by a 22 multimatrix semi-active helix array
2000–	Regional cell phone ±250 beams	Thuraya, ACeS, ETS VIII, ICO G1, Terrestar, SkyTerra: 12 to 22 m reflectors
		>100 feed multimatrix semi-active feed array Digital beam forming on board or ground implemented
2005–	Earth coverage > 200 pencil beams	Three INMARSAT-4 satellites; 9 m mesh reflector fed by a 120 helix multimatrix semi-active array Digital beam forming

INMARSAT, founded in 1979, and regional mobile communication ventures were technology drivers, first for reflector shaping, then for the development of tennis-court-size mesh reflectors, double-bed-size feed arrays, multimatrix power distribution as well as on-board and ground-implemented digital beam forming.

As for the fixed service at C-band, the early generations, producing a global beam, did not use multi-feed reflectors.

The payload of the M-SAT satellites, launched in 1994 and 1995, was developed in Canada and was very advanced at the time with six shaped beams to cover North America [119]. The coverage and an artist's impression of M-SAT are shown in Figure 7.50. M-SAT used two separate transmit and receive 5.5 m

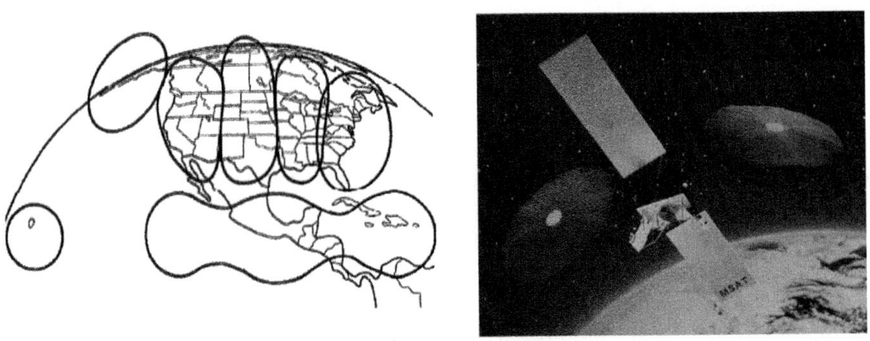

Figure 7.50 M-SAT transmit coverage and spacecraft view. Courtesy of Canadian Space Agency www.asc-csa.gc.ca.

Figure 7.51 *INMARSAT-3 feed array. Courtesy of Astrium-UK.*

reflectors each with 23 feeds. Two multi-port amplifiers distributed the power to beam formers with power to beam flexibility.

The concept and development of multi-port amplifiers had also been consolidated at S-band for the Japanese ETS VI satellite [120]. Further refinements were introduced in [121] and the suppression of the input Butler-like matrices in [122].

INMARSAT-3 had to provide, from various geostationary orbital positions, a global Earth coverage beam and five spot beams.

The INMARSAT-3 transmit L-band antenna is described in detail in [123]. It uses a 2.4 m reflector illuminated by an array of 22 short helices [124] shown in Figure 7.51.

Small Butler-like matrices are introduced between the amplifiers and the feed elements. This multimatrix semi-active design [122] allows a flexible distribution of the amplifier power to fixed or reconfigurable beams. This is done by proper phasing for each beam of the equal amplitude signals before they are combined at the amplifier inputs. As a result all amplifiers can operate around the same nominal level with optimal power efficiency.

In 2000, the regional mobile communications satellites Thuraya [125] and ACeS Garuda were launched with 12 m reflectors and large feed arrays using elements of the type shown in Figure 7.52.

Figure 7.52 *Feed arrays for mobile communications. Courtesy of Ruag Space.*

Then, INMARSAT launched INMARSAT-4 for global coverage. The three INMARSAT-4 satellites each provide a global beam, 19 regional beams and some 200 spot beams at L-band. The basic architecture is described in [126,127]. Heritage from INMARSAT-3 is used here. In addition to the reflector size (9 m) and the number of feeds and amplifiers (120), the main difference is that beam forming uses multi-port amplifiers and a digital signal processor.

An elaborate synthesis process [128] determines signals to be applied for each channel at the multi-port amplifier element beam ports.

Recently, regional communication satellites such as MSV SkyTerra (2010) use reflectors of over 20 m [129]. Others, such as ICO G1 (2008) and Terrestar (2009) resurrect ground-based beam forming [130] with up to 500 beams digitally formed on the ground. This concept was studied in 1975 and used in 1983 for the TDRS satellite for multiple access at S-band.

Communications and Broadcasting at Ku-band The 1971 World Administrative Radio Conference for Space Telecommunications, WARC 71, made allocations for the FSS between 10 and 15 GHz, around 20 and 30 GHz. The WARC conferences in 1977 and 1979 further defined norms for the Broadcast Satellite Services (BSS) with the allocation of elliptical beam coverages for various countries.

The downlink band was within 11.7 to 12.5 GHz with the band of 17.3 to 18.1 GHz for the uplink.

The first satellites to use the Ku-band were Anik B in 1978 and the SBS (Small Business Satellites). The SBS antenna system, described in [131], is a good example of the multi-feed contoured beam antenna at Ku-band. It uses two 1.8 m gridded reflectors made of Kevlar honeycomb with Kevlar sheets and a Kapton sheet with 0.4 mm etched copper grids spaced at 0.8 mm.

The two waveguide feeds have 8 and 13 horns for transmit and receive functions respectively. Some of the horns are corrugated in the E-plane or dielectric loaded in the H-plane to control their aperture distribution.

Biased dual-mode waveguide feed networks are used to ease the multiplexer design.

In Europe, WARC 77 elliptical coverages with stringent requirements were enforced. All reflectors were elliptical but various feeding techniques were used.

The Germans developed for their TV-SAT [132] (launched in 1987) a metallized CFRP elliptical corrugated horn based on a Dutch electrical design [133].

Figure 7.53 shows the elliptical feed horn. The horn includes an RF sensing receive function around 17.3 GHz. Figure 7.54 shows the TV-SAT antennas under test.

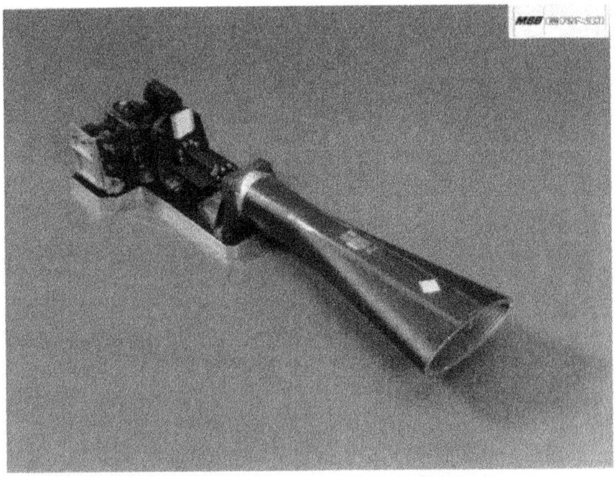

Figure 7.53 TV-SAT feed horn. Courtesy of Astrium-D.

Figure 7.54 TV-SAT antennas. Courtesy of Astrium-D.

The French chose a feed array of nine hexagonal elements with a spacing slightly above one wavelength. The feed network uses WR 75 waveguide with losses below 0.4 dB. Unlike the German antenna, the RF sensing frequency has been chosen at 11.2 GHz. The RF sensing signals are received by four elements producing two sum and two difference lobes. Figure 7.55 shows a photo of the TDF-1 feed.

The US STC satellite antenna [134] generated, with a 2.16 m reflector, an eastern beam shaped using 16 circular horns, a central beam, using six larger horns, and a separate receive spot beam at 17.3 GHz pointed near Las Vegas. A low-loss waveguide feed network fed the horns without feed sharing.

In Europe Ku-band satellite communications were initiated in the 1980s by ESA with its OTS and ECS satellites with circular and elliptical spot beams, which were taken over by EUTELSAT from 1983 onwards. Since the 1990s, the ECS and the WARC 77 elliptical beams have been replaced by shaped reconfigurable beams covering the whole of Europe and higher gain zones.

A good review of this evolution is given in [135] and a description of the antennas of the second generation of EUTELSAT satellites is given in [136].

EUTELSAT 2 can direct channels in a global beam or in a higher gain central zone with 6 dB more gain. It uses two antennas with gridded reflectors, one transmitting and receiving and one transmitting only. The rectangular feed horns are fed by a waveguide feed network. A photo of the antenna is shown in Figure 7.56.

Figure 7.55 TDF-1 feed assembly. Courtesy of Thales Alenia Space Italia.

Figure 7.56 EUTELSAT 2 antenna. Courtesy of Thales Alenia Space Italia.

Multi-feed antennas are still used at Ku-band when reconfigurability is required, but less costly shaped gridded reflectors or Mitzuguch optimized CFRP dual offset reflectors are often preferred nowadays.

Communications at Ka-band Frequency allocations around 20 and 30 GHz for the FSS were made at WARC 71 and consolidated at WARC 79. One problem is the much stronger attenuation by rain than for Ku-band, as was already evidenced by propagation experiments on ATS-6 (1974) and later confirmed with the extensive ESA Olympus campaigns [137,138].

Higher gains and narrower beams are therefore required than at Ku-band. Accordingly, multiple spot beam antenna designs were looked at early on.

Pioneering work was conducted in the USA on the design of multiple-beam reflector antennas to cover the continental USA [139].

In Europe, ESA started a programme in 1978 to develop a multi-feed 3.7 m dual-offset reflector antenna with contoured beam and spot beam coverage capability over Europe at Ka-band [140]. This involved:

- fast analysis of large reflectors, making use of the Whittaker sampling theorem;
- design of wide-scan offset Gregorian optics with feeds disposed on a sphere;
- design and manufacture of a 3.7 m low-mass reflector with hinged tips (Figure 7.36);
- design of 4λ diameter profiled feed horns with dual polarization and RF sensing;
- design of a beam-forming network allowing feed sharing between beams [141].

The beams include low sidelobe city beams (seven feeds per beam) and area beams.

Figure 7.57 shows a photo of the breadboard feed array and network.

Figure 7.57 *The 20 GHz feed array and network. Courtesy of Thales Alenia Space Italia.*

Some of this technology was used in the ITALSAT Ka-band satellite (1991), covering Italy with six 0.3°/0.4° spot beams from two 2 m transmit and receive antennas.

In 1993, NASA launched its Advanced Communications Technology Satellite, ACTS (Figure 7.58), with two large Ka-band multibeam antennas [142]. The 3.3 m transmit antenna uses a polarization-sensitive subreflector and two feed arrays, one for each polarization, to generate five 0.3° spot beams hopping into 51 positions across the USA. The feeds are conical multiflare horns and the beam-forming switch network includes fast (<1 μs) ferrite switches, power dividers and combiners.

Japan was the first country in 1977 to use Ka-band with the CS-1 satellite [102] and subsequently has been very active in the development of Ka-band technology.

One example is the N-Star Ka-band antenna described in [143]. N-Star refers to a series of Japanese domestic communication satellites operated by NTT (Figure 7.59). The 2.2 m array fed offset Gregorian antenna generates three spot beams and a country-coverage-shaped beam on transmit and eight spot beams and

Figure 7.58 *NASA ACTS satellite. Courtesy of NASA.*

Figure 7.59 N-Star Ka-band antenna. Courtesy of NTT.

a shaped beam on receive. A frequency-selective system of two ellipsoidal subreflectors with one common focus is used for space diplexing of the two feeds. One subreflector has double-ring resonant elements and the other is plain. The feed includes 26 pyramidal horns for the uplink and 14 for the downlink. The same horns are used in sub-arrays for both the multiple and the shaped beams, by use of orthogonal waveguide feed networks.

So far, for continuous coverage of large regions, Ka-band satellites have mainly used three or four separate multi-feed reflectors with one feed per beam to produce a set of interleaved multiple spot beams. This is illustrated in Figure 5.60, showing the ESA-developed Alphabus platform in such a configuration.

Figure 7.60 Ka-band satellite with multiple reflectors. Courtesy of ESA – J. Huart.

Generating all the transmit beams from one single aperture, as is done for mobile communication satelites at L-band and S-band, remains a challenge at Ka-band because of aberrations, pointing accuracy and feed network complexity.

7.4 Array Antennas

7.4.1 Conformal Arrays on Spin-Stabilized Satellites

7.4.1.1 Arrays with Toroidal Patterns

Early satellites were spin stabilized. One method to maintain two-way communications with the Earth while spinning is to generate a beam with a symmetry of revolution around the spin axis, for example with a toroidal pattern.

This was the solution used early on by the first communication satellites. A good example is the early experimental communication satellites Telstar 1 and 2 of AT&T, launched in 1962 and 1963 respectively into elliptical orbits (Figure 7.61). They were spin stabilized and their antennas are described in detail in [144].

For the telemetry at 136.05 MHz and for command at 123 MHz, a quadrifilar helix with gold-plated beryllium wires of 2.54 mm diameter was selected. For the communication transmissions, a circular array of 48 rectangular waveguide elements, fed through a hybrid power divider by a 3.3 W travelling wave tube amplifier (TWTA), was used around 4.17 GHz. A similar array of 72 elements, connected to a solid state receiver, was used around 6.39 GHz on receive.

To produce circular polarization, the waveguides are excited diagonally by a probe and the dimensions of the waveguide are selected to create a phase differential of 90° between the two polarizations. Both arrays produced a toroidal pattern of 60° beamwidth around the satellite axis.

Figure 7.61 *Telstar communication satellite (1962) with circular arrays. Courtesy of NASA.*

Figure 7.62 Meteosat L-band EDA antenna. Courtesy of Thales Alenia Space Italia.

7.4.1.2 Electronically Despun Arrays

Mechanically or electronically despun antennas (EDA)s produce a beam spinning at equal and opposite speed to that of the satellite so that a narrower beam, with more gain than a toroidal pattern, can be kept pointing to the Earth.

An early example is the series of European geostationary meteorological satellites Meteosat which have used the same EDA basic design to transmit meteorological data from 1977 until now. Meteosat antennas are described in [145,146].

The antenna (Figure 7.62) is a cylindrical array with 32 columns of four dipoles each. A counter-rotating subgroup of four or five columns facing the Earth is continuously activated with gradual takeover provided by seven intermediate switching states optimized in time for the cylindrical configuration.

GTD models, with 10 different kind of rays, were developed in 1975 to analyse spacecraft–antenna interactions and the antenna near-field measurement setup [147].

7.4.2 Arrays for Remote Sensing

Arrays were used early on for active instruments such as SARs and for scatterometers. They are also used by passive imaging radiometers.

7.4.2.1 Seasat SAR Antenna

A good review of the Seasat, and also of SIR-A, SIR-B and SR-C antennas, can be found in [8] from which some of the following sections are abstracted.

After it was demonstrated from airborne SAR campaigns that an L-band SAR could image ocean waves, it was decided to fly a SAR on Seasat, a near-polar orbital platform at 800 km altitude launched in 1978.

Figure 7.63 *Seasat SAR array [8]. Reproduced by permission of John Wiley & Sons.*

The radar [148], operating at 1.275 GHz, was designed to provide 25 m resolution over a 100 km swath. Figure 7.63 shows a picture of the SAR antenna. The 10.74×2.16 m planar array is subdivided into eight microstrip panels.

Patches are printed on fibreglass sheets bonded to a Nomex honeycomb. Panels are mounted on a composite truss structure that is deployed once in orbit. They are fed by a corporate coaxial feed network from a 1 kW peak power transmitter. The losses introduced by this corporate feed network were less than 1 dB to the individual panels. Key antenna parameters of the antenna, abstracted from [8], are given in Table 7.4.

7.4.2.2 ERS SAR Antenna

The success of Seasat and further studies demonstrated the potential benefits of a C-band SAR instrument, in particular for soil moisture measurements and coastal ocean monitoring. It was decided to fly a SAR on the first European Remote Sensing Satellite, ERS-1. It was launched in 1991 into a Sun synchronous polar orbit at 783 km altitude.

The 10×1 m antenna, shown in Figure 7.64, operates at 5.3 GHz in linear polarization. It consists of 10 electrical subpanels of 1×1 m associated by pair with five mechanical panels folded during launch. Each subpanel includes 24 metallized CFRP waveguide resonant arrays of broadside slots fed by a coupling

Table 7.4 *Key Seasat SAR antenna parameters [8].*

Parameter	Value
Centre frequency	1.275 GHz
Polarization	Horizontal
Radiation gain	37.2 dB
Antenna efficiency	60%
Effective gain at sensor input	35 dB
Beamwidths	$1.7 \times 6.2°$
Polarization isolation	20 dB

Figure 7.64 ESA ERS-1 SAR slot array. Courtesy of Astrium-D.

waveguide on the back of the subpanel. The 10 coupling waveguides are fed from a main feed waveguide network via shunt-series slots.

A special feature of the antenna is the use of metallized CFRP for the high-tolerance slotted waveguides. The ERS-1 SAR operated until 2000, far exceeding its planned lifetime [149,150].

7.4.2.3 Radarsat 2 Antenna

Radarsat, shown in Figure 7.65, is a second-generation Canadian SAR at C-band [151]. Its main applications are in agriculture, disaster management, forestry, geology, hydrology, ice, marine surveillance and mapping.

Developed and utilized by MacDonald Dettwiller and Associates (MDA), it has several new features with respect to the previous generation. Its ground resolution range is 3–100 m. It has a large number of beam modes and the ability to look left and right. It offers fully flexible polarization options.

The 15×1.5 m antenna operates at 5.4 GHz in dual linear polarization and includes 512 sub-arrays, each connected to a transmit/receive (T/R) module. Each sub-array includes 20 dual-polarization microstrip patches [151].

Signals from each T/R module are phased according to the selected beam and directed to the wanted polarization input port of the sub-array. In the receive mode, two polarizations can be received simultaneously.

7.4.2.4 Microwave Imaging Radiometer Using Aperture Synthesis – MIRAS

MIRAS is an L-band (1.4 GHz) 2D passive interferometric radiometer with a Y-shaped three-arm synthetic aperture array. It is the main instrument of the ESA SMOS mission launched in 2009 to map soil moisture and ocean salinity. The instrument, its principle and its antenna are described in [152] and in Chapter 6 of this handbook. A photo of SMOS is shown in Figure 7.66.

The antenna is a Y-shaped array with a total of 69 dual-polarized elements which all receive, from 800 km altitude, radiation in the band from 1.420 to 1.427 GHz from sea or land in their field of view of 70°. Signals

Figure 7.65 Radarsat 2 and its 15 × 1.5 m C-band antenna. Courtesy of MDA – Eric Amyotte.

received by all elements are routed to a central processor and correlated by pairs. The process is repeated as the satellite moves, with an increase in sensitivity and resolution as a result. Since seawater salinity and soil moisture density strongly affect the local radiation intensity between 1 and 2 GHz, this process allows mapping salinity or moisture in a swath below the spacecraft. Calibration is particularly critical but the instrument performance meets expectations.

Figure 7.66 SMOS satellite and its Microwave Imaging Radiometer MIRAS. Courtesy of ESA – P. Carril.

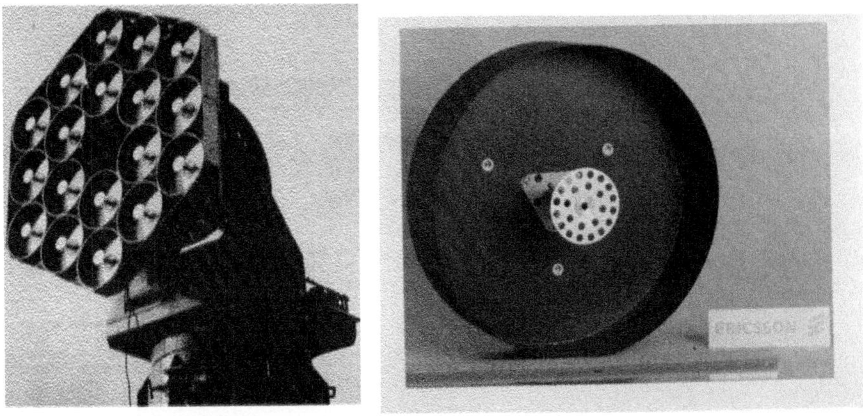

(a) (b)

Figure 7.67 ESA multibeam array model (1976). (a) 18 element array. (b) Qualified short backfire element. Courtesy of Ruag Space.

7.4.3 Arrays for Telecommunications

7.4.3.1 ESA Multibeam Array Model (1976)

Array antennas were considered early on for mobile communications at L- and S-bands. The focus of the ESA Multibeam Array Model programme [153,154] was to study, develop and demonstrate multibeam array designs and technologies, including radiating elements and beam formers, for communications with ships, aircraft and land mobiles. The target was 24 dB of coverage gain using 19 beams from an 11-wavelength array, an improvement of 6 dB from a single shaped beam.

A model of an 18-element array of short backfire elements (Figure 7.67) was designed, built and tested with beam formers at L-band and at also at intermediate frequencies. The elements, with a diameter close to 2.2 wavelengths, were double tuned for transmit and receive. An aluminium and a CFRP model were also qualified. This work has been the seed for the successful development and export of feed elements for many of today's mobile communication satellites.

7.4.3.2 NASA Tracking and Data Relay Satellite Multiple Access Array (1983)

The NASA geostationary Data Relay Satellites relay signals from low orbiting user satellites to the ground.

TDRS 1, launched in 1983, had an impressive antenna farm as shown in Figure 7.1 and in Figure 7.68. It included a multiple access array antenna operating at 2.3 GHz, to relay simultaneously to the ground data from up to 20 low-data-rate user satellites within a field of view of $\pm 13°$. The TDRS can also transmit data to one user satellite at 2.1 GHz. The proposed design, used for seven TDRS satellites, is described in [155–157] and has several innovative features.

The array includes 30 cup helices with 23 turns, of which the top three are tapered. The 4.8 cm diameter helices, with a 12° pitch angle, are made of beryllium–copper wire wound around a Kevlar support tube. Twelve of the helices are diplexed and used for transmission to one single user.

To limit grating lobes, the rings of the helices are rotated so as to break their periodicity [157]. Another innovative feature is the use of the Adaptive Ground Implemented Phased Array (AGIPA) beam-forming system [158].

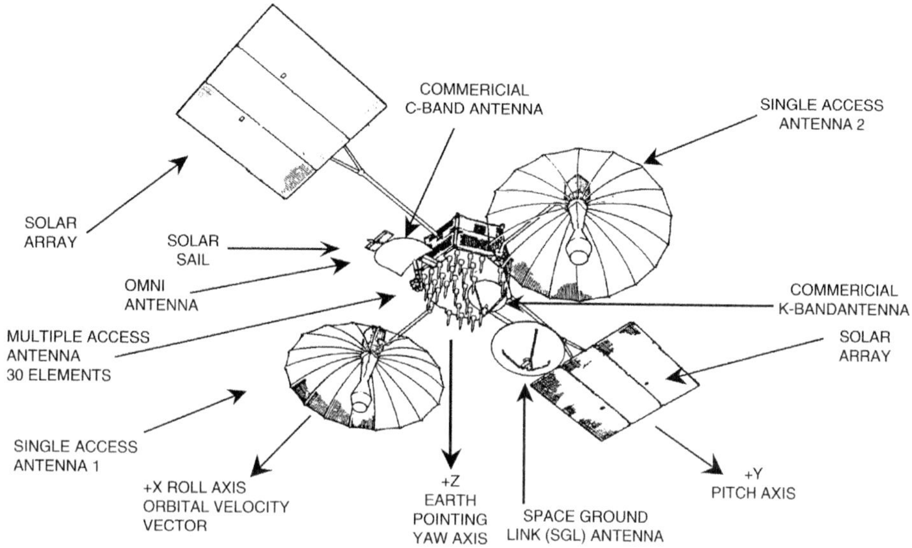

Figure 7.68 *TRDS 1 antenna farm. Courtesy of NASA.*

Signals received at S-band on each helix are multiplexed and sent down separately to a ground station where they are processed adaptively to form one beam per user. This technique, although requiring large frequency spectrum resources and complex beam formers, has now been revived for mobile communication satellites at S-band [159].

7.4.3.3 Japan's Multibeam Array Antenna for Data Relay Payload on ETS VI

In order to support the future data relay needs of Japan, the Radio Research Laboratory undertook, in the mid 1970s, developments in S-band multibeam array technology and a demonstration model described in [160]. During this work, the important concept of sequential rotation in circularly polarized arrays to improve the polarization purity was proposed [161]. Subsequently this array has been developed to space hardware level for the ETS VI satellite [162].

A picture of the 19-element array flight model is shown in Figure 7.69. The elements, with a diameter of 325 mm, are septets of circular microstrip patches fed by two points and with notches which suppress the cross-polarization generated by the feeds. The substrate, honeycomb sandwich plate, has a low dielectric constant and a thickness of 10 mm, allowing operation with the necessary bandwidth. The sub-array gain exceeds 14 dB in the $\pm 10^\circ$ field of view.

The antenna is active in both transmit and receive modes. It receives signals at 2.3 GHz with 19 elements, 16 of which are diplexed to transmit at 2.1 GHz.

Two independent steerable receive beams are formed by the use of 2×19 4-bit phase shifters. In the transmit mode, one steerable beam is generated, using 16 phase shifters.

The array is attached to the tower during launch and deployed once on station.

7.4.3.4 Active Arrays at Ka-band

With the exception of some military missions (such as the X- and Ka-band Wideband Gapfiller Satellite), few arrays have been flown for geostationary communication satellites because of the difficulty in deploying and

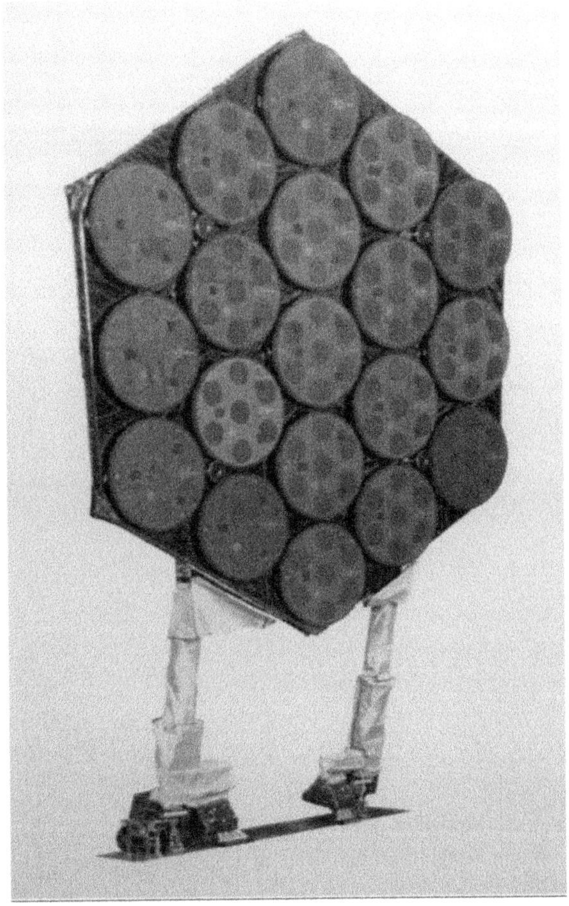

Figure 7.69 Multibeam array antenna for data relay payload on ETS VI. Courtesy of NICT and JAXA.

thermally controlling large active arrays at L-band down to C-band and because of the poor efficiency of solid state amplifiers at Ku- and Ka-bands compared with TWTAs.

There are some exceptions. One is Boeing's SPACEWAY satellites at Ka-band. SPACEWAY (2005–) is the most complex commercial satellite system ever manufactured with approximately five to eight times the capacity of today's satellites. It was initially developed to provide high-speed, two-way communications for multimedia applications in Ka-band, but has evolved to enable DIRECTV to deliver more than 1500 local and national high-definition channels.

The spacecraft includes a flexible payload with a fully steerable and reconfigurable transmit antenna. The antenna front end includes 1500 dual-polarized square horn elements in a square lattice and 3000 solid state power amplifiers (SSPAs).

The fully agile and digitally controlled analogue beam-forming network (BFN) produces 24 high-rate (less than 1 ms) reconfigurable beams (12 in left-handed circular polarization (LHCP) and 12 in right-handed circular polarization (RHCP)) compatible with beam-hopping.

An artist's impression of SPACEWAY can be found on the website of Boeing http://www.boeing.com/defense-space/space/bss/factsheets/702/spaceway/spaceway.html.

7.4.3.5 Arrays for Non-geostationary Satellites

The requirements of medium or low Earth orbit satellites differ significantly from those of geostationary ones with a much wider field of view.

GPS navigation satellites, for example, generate a single beam in circular polarization to cover the Earth from an altitude of 20 200 km. Figure 7.70 shows one of the early GPS 12 satellites produced around 1985. These satellites use an array of helices with longer helices in the centre to generate a more flat-top-like shaped beam.

More recent European navigation satellites use printed technology. Figure 7.71 shows a test satellite for the ESA Galileo programme, GIOVE B, with a dual-frequency array of 42 printed elements. The frequencies are around 1.20 and 1.57 GHz.

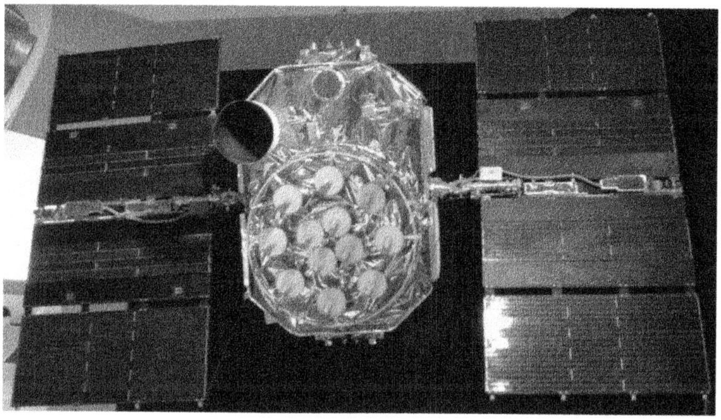

Figure 7.70 GPS 12 satellite with array of 12 cup helices. (Approximately 1985 by Rockwell, now Boeing – Courtesy of Scott Ehardt.

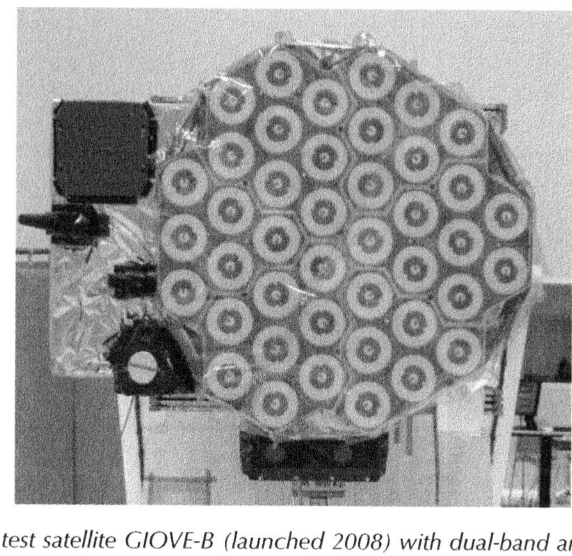

Figure 7.71 ESA Galileo test satellite GIOVE-B (launched 2008) with dual-band array of 42 printed elements. EADS CASA Espacio – Courtesy of ESA – P. Müller.

Figure 7.72 Iridium and its three L-band arrays.Courtesy of Daniel Deak.

Low Earth orbit satellites with multiple beams such as Iridium and Globalstar have moving beam footprints and the necessity to hand over users smoothly from one satellite to another.

The 66 Iridium satellites (1997–), orbiting at 781 km altitude, each have three L-band phased array antennas (see Figure 7.72). Each array has 106 elements over an area of 188 × 86 cm, distributed in eight rows with element spacings of 11.5 cm and row spacings of 9.4 cm. Each antenna generates 16 beams by the use of a 2D Butler matrix. The total service area is therefore covered by 48 beams. The edge of coverage gain is around 24 dB. A special feature of this system is that each satellite operates intersatellite links at 23 GHz with its neighbours.

The 48 Globalstar satellites (1999–), orbiting at 1392 km altitude, each have two active arrays: one receiving at L-band (around 1.62 GHz) and one transmitting at S-band (around 2.49 GHz). The Globalstar satellite antennas are described in [163].

The S-band circularly polarized transmit antenna forms 16 beams optimized to ensure quasi-isoflux conditions across the coverage. It is made of 91 elements in a triangular lattice organized in six sectors of 15 elements each around a central one. Each element is powered by an amplifier module which also includes a filter.

A signal coming from a beam port is split into 91 signals each of which is directed after phasing to the proper 16-to-1 combiner of its destination element chain where it is amplified with the other beam signals. The L-band receive array only has 61 element chains and works along similar principles but with amplitude and phase control.

The launches commencing in 1999 were interrupted after 24 satellites but were restarted in 2010 with 24 improved satellites.

The development of digital beam forming and of more efficient and flexible power amplifiers should guarantee a bright future in space for active and semi-active arrays and feed arrays.

7.5 Conclusions

From the VHF monopoles of the Sputniks and Explorers of the 1950s to the gigantic multiple-beam reflector and active array antennas for communication, navigation and remote sensing spacecraft of today operating up to terahertz frequencies, antennas have grown to become the key to successful space missions. Ground, shipborne and airborne radars have clearly been driving the early development of active phased array theory and technology. But space exploration, communications and Earth observation applications have driven most of the key developments in multiple reflector and lens antenna analysis, in contoured and multibeam antennas with complex feed networks, beam formers and demultiplexers and with multiple shaped, reconfigurable or unfurlable reflectors, some as large as tennis courts.

For the evermore demanding and complex space missions of the future, the antenna system will continue to remain vital and a key to success.

Many areas of antenna engineering are now well under control, but more and more challenging multidisciplinary R&D and co-design encompassing systems, cost analysis, electromagnetics, integrated microwave circuits and signal processing, as well as thermo-mechanical engineering and materials, will be required for these missions.

Acknowledgements

The author wishes to thank the many colleagues (and their institutions) for their support with suggestions, material and permissions for this chapter, including, in particular: Christian Albertsen (TICRA), Eric Amyotte (MDA), Bertram Arbesser-Rastburg (ESA), Susan Barclay (Wiley), Luigi Boccia (University of Calabria), Pasquale Capece (Thales Alenia Space - I), Salvatore Contu (Thales Alenia Space - I), Gerald Crone (ESA), Richard Davies (Wiley), Keith Dickinson (MDA), Aksel Frandsen (TICRA), Steven Gao (University of Surrey), Sven Grahn, Tim Hartsfield (Harris Corporation), Nadia Imbert-Vier (ESA), William Imbriale (NASA/JPL), Per Ingvarson (Ruag), Niels Eilskov Jensen (ESA), Frank Jensen (TICRA), Jake Johnson (NASA Images), Pat Joyce (Northrop Grumman), Takashi Katagi (Mitsubishi Melco), Kees van't Klooster (ESA), Yuji Kobayashi (Mitsubishi Melco), Réjean Lemieux (CSA), Cyril Mangenot (ESA), Jacques Maurel (Thales Alenia Space - F), Roberto Mizzoni (Thales Alenia Space - I), Gerry Nagler (Telesat), Knud Pontoppidan (TICRA), Yahya Rahmat-Samii (UCLA), Michael Schneider (Astrium-D), Alistair Scott (Astrium-UK), Michael Shields (MIT Lincoln Laboratory), Anna Smart (Wiley), Simon Stirland (Astrium-UK), Tadashi Takano (JAXA), Masato Tanaka (NICT), Tasuku Teshirogi (ANRITSU), Hendrik Thielemann (Ruag), Michael Thorburn (SSD Loral), Thomas Tsymbal (Buran Energia) and Helmut Wolf (Astrium-D).

References

1. Silver, S. (1948) *Microwave Antenna Theory and Design*, McGraw-Hill.
2. Kraus, J.D. (1950) *Antennas*, McGraw-Hill.
3. Schelkunoff, S.A. (1952) *Antenna Theory and Practice*, John Wiley & Sons, Inc.
4. Jasik, H. (1961) *Antenna Engineering Handbook*, McGraw-Hill.
5. Rudge, A.W., Milne, K., Olver, A.D. and Knight, P. (eds) (1986) *The Handbook of Antenna Design*, Peter Peregrinus.
6. Tikhonravov, M.K. (1973) The creation of the first artificial Earth satellites: some historical details. 24th International Astronautical Congress, Baku.
7. Siddiqi, A. (1997) Korolev, Sputnik, and The International Geophysical Year, Spacebusiness, http://history.spacebusiness.com/sputnik/files/sputnik52.pdf (accessed 5 December 2011).
8. Imbriale, W.A. (2006) *Spaceborne Antennas for Planetary Exploration*, John Wiley & Sons, Inc.
9. Schantz, H. (2005) Nanoantennas: a concept for efficient electrically small UWB devices. Proceedings of the IEEE 2005 International Conference on Ultra-Wideband.
10. Schantz, H., The iPhone antenna's space age origins. *Aetherczar*, http://www.aetherczar.com/?p=1145 (accessed 5 December 2011).

11. King, R.W.P. (1956) *The Theory of Linear Antennas*, Harvard University Press.
12. Mei, K.K. (1965) On the integral equation of thin wire antennas. *IEEE Transactions on Antennas and Propagation*, **13**, 59–62.
13. Harrington, R.F. (1968) *Field Computations by Moment Methods*, Macmillan.
14. Richmond, J.H. (1966) A wire grid model for scattering by conducting bodies. *IEEE Transactions on Antennas and Propagation*, **AP-14**, 782–786.
15. Harrington, R.F. and Mautz, J.R. (1967) Straight wires with arbitrary excitation and loading. *IEEE Transactions on Antennas and Propagation*, **AP-15**, 502–515.
16. Miller, E.K., Morton, J.B., Pierrou, G.M. and Maxum, B.J. (1969) Numerical analysis of aircraft antennas. Proceedings of the Conference on Environmental Effects on Antenna Performance, pp. 55–58.
17. Albertsen, N.C., Hansen, J.E. and Jensen, N.E. (1971) Numerical prediction of radiation patterns for antennas mounted on spacecraft. *IEE Conference Publication*, **77**, 119–228.
18. Albertsen, N.C., Hansen, J.E. and Jensen, N.E. (1974) Computation of radiation from wire antennas on conducting bodies. *IEEE Transactions on Antennas and Propagation*, **AP-22**, 200–206.
19. Imbriale, W.A. (1975) Applications of the method of moments to thin-wire elements and arrays, in *Topics in Applied Physics, vol. 3, Numerical and Asymptotic Techniques in Electromagnetics* (ed. R. Mittra), Springer.
20. Keller, J.B. (1962) Geometrical theory of diffraction. *Journal of the Optical Society of America*, **52**, 116–130.
21. Albertsen, N.C., Bach, H., Balling, P. *et al.* (1975) A study on radiation pattern prediction for high frequency satellite antennas. Final Report, Technical University of Denmark, Lyngby, Institute of Electromagnetics.
22. Pontoppidan, K. (1976) Applications of GTD to antenna problems. Proceedings of the 6th European Microwave Conference, pp. 71–75.
23. Ratnasiri, A.J., Kouyoumjian, R.G. and Pathak, P.H. (1970) The wide angle side lobes of reflector antennas, ElectroScience Laboratory, Ohio State University, Columbus, Report 2183-1.
24. Rusch, W.V.T. and Sørensen, O. (1975) The geometrical theory of diffraction for axially symmetric reflectors. *IEEE Transactions on Antennas and Propagation*, **AP-23** 414–419.
25. Prigoda, B.A. *et al.* (1972) Circularly polarised antennas with controlled radiation patterns, Equipment for space research, Nauka Press, Moscow, NASA Technical Translation TT F-785, pp. 163–167, http://ntrs.nasa.gov/archive/nasa/casi.ntrs.nasa.gov/19750002916_1975002916.pdf (accessed 5 December 2011).
26. Mitchell, D.P. (2004) Venera: the Soviet exploration of Venus, http://www.mentallandscape.com/V_Venus.htm. (accessed 5 December 2011).
27. Jet Propulsion Laboratory (1962) The Ranger Project Annual Report for 1961, Technical report No. 32-241, JPL, Pasadena, CA, June 15.
28. Jeuken, M., Knoben, M. and Wellington, K. (1972) Dual frequency, dual polarised feed for radio astronomy. *NTZ*, **25**, 374.
29. Savini, D., Figlia, G., Ardenne, A.v. and van't Klooster, K. (1988) A triple frequency peed for the QUASAT antenna. Proceedings of the IEEE AP International Symposium.
30. Gloersen, P. and Barath, F. (1977) A scanning multichannel microwave radiometer for Nimbus-G and Seasat-A. *IEEE Journal of Oceanic Engineering*, **2**(2), 172–178.
31. Njoku, E., Stacey, J. and Barath, F. (1980) The Seasat Scanning Multichannel Microwave Radiometer (SMMR): instrument description and performance. *IEEE Journal of Oceanic Engineering*, **5**(2), 100–115.
32. Green, K. (1975) Final Report on Design Study of NIMBUS-G SMMR Antenna Subsystem, Technical Report, Microwave Research Corporation, North Andover, MA.
33. Green, K. (1981) Multifrequency broadband polarised horn antenna, US patent 4,258,366, March 24.
34. Takeda, F. and Hashimoto, T. (1976) Broadbanding of corrugated conical horns by means of the ring-loaded corrugated waveguide structure. *IEEE Transactions on Antennas and Propagation*, **AP-24**(6), 786–792.
35. Milligan, T. (1995) Compact dual band feed for Mars Global Surveyor. IEEE AP-S International Symposium and USNC/URSI Radio Science Meeting, Newport Beach, CA.
36. Ward, W.W. and Floyd, F.W. (1989) Thirty years of research and development in space communications at Lincoln Laboratory. *MIT Lincoln Laboratory Journal*, **2**, 5–34.
37. Lee, J.E. (1991) The development, variations, and applications of an EHF dual-band feed. *MIT Lincoln Laboratory Journal*, **4**(1), 61–79.

38. Olsson, R., Kildal, P.-S. and Weinreb, S. (2006) The eleven antenna: a compact low-profile decade bandwidth dual polarized feed for reflector antennas. *IEEE Transactions on Antennas and Propagation*, **AP-54**(2), 368–375.
39. Barath, F.T., Chavez, M.C., Cofield, R.E. *et al.* (1993) The upper atmosphere research satellite microwave limb sounder instrument. *Journal of Geophysical Research*, **98**(D6), 10751–10762.
40. Goldsmith, P.F. (1982) Quasi-optical techniques at millimetre and submillimetre wavelengths, in *Infrared and Millmeter Waves*, vol. 6 (ed. K.J. Button), Academic Press, pp. 277–343.
41. Potter, P.D. (1963) A new horn antenna with suppressed sidelobes and equal bandwidths. *Microwave Journal*, 71.
42. Pickett, H.M., Hardy, J.C. and Farhoomand, J. (1984) Characterization of a dual-mode horn for submillimeter wavelengths. *IEEE Transactions on Microwave Theory and Techniques*, **MTT-32**, 936–937.
43. Tauber, J.A., de Chambure, D., Crone, G. *et al.* (2005) Optical design and testing of the Planck satellite. Proceedings, URSI General Assembly.
44. Villa, F., D'Arcangelo, O., Pecora, M. *et al.* (2009) Planck-LFI flight model feed horns. *Journal of Instrumentation*, **4**, T12004.
45. Doyle, D. (2010) ESA: the Herschel space telescope in-flight performance. International Conference on Space Optics.
46. Imbriale, W.A. (2003) *Large Antennas of the Deep Space Network, Deep-Space Communications and Navigation Systems* (ed. J.H. Yuen), John Wiley & Sons, Inc.
47. Chen, C.C. (1973) Transmission of microwaves through perforated flat plates of finite thickness. *IEEE Transactions on Microwave Theory and Techniques*, **MTT-21**(1), 1–6.
48. Roederer, A.G. (1971) Etudes des Réseaux Finis de Guides d' Ondes à Parois Epaisses. *Onde Electrique*.
49. Agrawal, V.D. and Imbriale, W.A. (1979) Design of a dichroic Cassegrain subreflector. *IEEE Transactions on Antennas and Propagation*, **AP-27**, 466–473.
50. Bielli, P., Bresciani, D., Contu, S. and Crone, G. (1985) Dichroic subreflectors for multifrequency antennas. *CSELT Technical Report S 1.6 7*, pp. 443–448.
51. Bresciani, D., Bruno, C. and Crone, G. (1989) Design of a 1m dichroic subreflector for Ku/Ka frequency bands. Proceedings of the IEEE AP International Symposium, pp. 354–357.
52. Mascolo, G., Contu, S., Mizzoni, R. and Borchi, S. (1994) A double dichroic sub-reflector reflective at X, Ku, Ka bands and transparent at S-band. 8èmes Journèes Internationales de Nice sur les Antennes Jina 94, November 8–10, Nice.
53. Krebs, G.D., FLTSATCOM 1, 2, 3, 4, 5 (Block 1), http://space.skyrocket.de/doc_sdat/fltsatcom-1.htm (accessed 5 December 2011).
54. Freeland, R.E. (1983) Survey of deployable antenna concepts. *NASA Langley Research Center Large Space Antenna Systems Technology*, **Pt. 1**, 381–422 (SEE N83-26853 16-15).
55. Rusch, W.V.T. (1984) The current state of the reflector antenna art. *IEEE Transactions on Antennas and Propagation*, **AP-32**(4), 313–328.
56. Roederer, A.G. and Rahmat-Samii, Y. (1989) Unfurlable satellite antennas: a review. *Annales des Telecommunications*, **44**, 475–488.
57. Rahmat-Samii, Y. and Densmore, A. (2009) A history of reflector antenna development: past, present and future. SBMO/IEEE MTT-S International Microwave and Optoelectronics Conference (IMOC), Pará - Belém, November 3–6.
58. Tibert, G. (2002) Deployable Tensegrity Structures for Space Applications, Doctoral Thesis, Royal Institute of Technology, Stockholm.
59. Freeland, R.E. (1979) Final Report for Study of Wrap Rib Antenna Design, LMSC D714613, Contract No. 955345, prepared by Lockheed Missiles and Space Company for Jet Propulsion Laboratory, Pasadena, CA, December 12.
60. Harris Corporation (1986) Development of the 15 meter diameter hoop column antenna, Final Report. NASA Contract Report 4038, NASA Contract NASI-15763.
61. Martin Marietta (1986) Near-field testing of the 15-meter model of the hoop column antenna. NASA Contract Report 178059, Contract NASJ-18016, March.
62. Boeing, MSAT, www.boeing.com/defense-space/space/bss/factsheets/601/msat/msat.html (accessed 5 December 2011).

63. University of Cambridge, Department of Engineering (2001) Elastic Folding of Reflector Antennas, http://www-civ.eng.cam.ac.uk/dsl/ltdish.html (accessed 5 December 2011).
64. Bernasconi, M.C., Reibaldi, G. and Pagana, E. (1985) Inflatable reflectors for satellite mobile communications. *CSELT Technical Reports*, **X1II**(7), 437–441.
65. Reibaldi, G., Hammer, J., Bernasconi, M.C. and Pagana, E. (1986) Inflatable space rigidized reflector development for land mobile missions. AIAA Communication Satellite Systems Conference, Proceedings, pp. 533–538.
66. Reibaldi, G. and Bernasconi, M.C. (1987) Quasat program: the ESA reflector. 36th Congress of the International Astronautical Federation (October 1986), *Acta Astronautica,* **15**(3), 181–187.
67. L'Garde (1997) Spartan 207/Inflatable Antenna Experiment — Preliminary Mission Report, http://www.lgarde.com/papers/207.pdf (accessed 5 December 2011).
68. Chmielewski, A.B. and Noca, M. (2000) ARISE Antenna, http://trs-new.jpl.nasa.gov/dspace/bitstream/2014/18857/1/99-2154.pdf (accessed 5 December 2011).
69. Takano, T., Natori, M., Ohnishi, A. *et al.* (2000) Large deployable antenna with cable net composition for satellite use. *Electronics and Communications in Japan, Part 1*, **83**(8)
70. Takano, T., Miura, K., Natori, M. *et al.* (2004) Deployable antenna with 10-m maximum diameter for space use. *IEEE Transactions on Antennas and Propagation*, **AP-52**(1), 2–11.
71. Shinttake, K., Terada, K., Usui, M. *et al.* (2003) Large deployable reflector (LDR). *Journal of the National Institute of Information and Communications Technology*, **50**(3/4)
72. Thomson, M.W. (1999) The AstroMesh deployable reflector. IEEE Antennas and Propagation Society International Symposium, pp. 1516–1519.
73. Thomson, M. (2002) AstroMesh: deployable reflectors for Ku- and Ka-band commercial satellites. 20th AIAA International Communication Satellite Systems Conference.
74. Pontoppidan, K. (1981) Study offset unfurlable antennas - electrical design aspects, TICRA Report 5-144-01, July.
75. Schaefer, W., Herbig, H., Roederer, A. and Pontoppidan, K. (1982) Unfurlable offset antenna design for L- and C-band applications. AIAA Communication Satellite Systems Conference, Proceedings, pp. 30–36.
76. Dumont, P. (1984) Modélisation radioélectrique des antennes a réflecteur deployable, Thèse Doc. Ing., ENSAE, Toulouse.
77. Kontorovitch, M. (1963) Averaged boundary conditions at the surface of a grating with square mesh. *Radio Engineering and Electronic Physics*, **8**(9), 1446–1454.
78. Astrakhan, M.I. (1964) Averaged boundary conditions on the surface of a lattice with rectangular cells. *Radio Engineering and Electronic Physics*, **8**, 1239–1241.
79. Rahmat-Samii, Y. (1984) Diffraction analysis of mesh deployable reflector antennas. *NASA Langley Research Center Large Space Antenna Systems Technology*, **Pt. 2**, 715–736 (SEE N85-23840 14-15).
80. Imbriale, W.A., Galindo, V. and Rahmat-Samii, Y. (1991) On the reflectivity of complex mesh surfaces. *IEEE Transactions on Antennas and Propagation*, **AP-39**, 1352–1365.
81. Ingerson, P.G. and Wong, W.C., (1972) The analysis of deployable umbrella parabolic reflectors. *IEEE Transactions on Antennas and Propagation*, **AP-20**(4), 409–414.
82. Imbriale, W.A. and Rusch, W.V.T. (1974) Scalar analysis of non- symmetrically distorted umbrella reflector. *IEEE Transactions on Antennas and Propagation*, **AP-22**, 112–114.
83. Archer, J. (1984) Deployable reflector, TRW, US Patent 677259.
84. Archer, J. (1979) Advanced Sunflower antenna concept development. *NASA Langley Research Center Large Space Systems Technology*, 33–58 (SEE N80-19145 10-15).
85. Hachkowski, M.R. and Peterson, L.D.A. (1995) A comparative study of the precision of deployable spacecraft structures, Technical Report CU-CAS-95-22, Center for Aerospace Structures, University of Colorado, Boulder, CO.
86. Badessi, S., Fei, E., Grimaldi, F. *et al.* (1990) 20/30GHz ASTP multibeam antenna, ESA Report CR (X) 3107, Alenia Spacio.
87. Kumazawa, H., Ohtomo, I. and Kawakami, Y. (1989) Fixed/mobile multibeam communication antennas for ETS-VI satellite. IEEE AP-S International Symposium, pp. 472–475.
88. Schmid, M. and Barho, R. (2002) Development status of an unfurlable CFRP skin reflector. 25th ESA Antenna Workshop on Satellite Antenna Technology, 18–20 Sept. 2002, ESTEC, Noordwijk, pp. 289–296.

89. S.A. Lavochkin Association, Roscosmos, RADIOASTRON, The Ground –Space Interferometer:'radio telescope much larger than the Earth, http://www.asc.rssi.ru/radioastron/_files/booklet_en.pdf (accessed 5 December 2011).
90. RadioAstron mission, http://www.asc.rssi.ru/radioastron/description/intro_eng.htm (accessed 5 December 2011).
91. RadioAstron Science Operation Group, RadioAstron User Handbook, http://www.asc.rssi.ru/radioastron/documents/rauh/en/rauh.pdf (accessed 5 December 2011).
92. Sichak, W. (1957) Antenna, US Patent 2,790,169, April 23.
93. Raab, A.R. (1975) Compact frequency re-use antenna, US Patent 3,898,667, August 5.
94. Raab, A.R. (1976) Cross polarization performance of the RCA Satcom Frequency re-use antenna. IEEE Antennas and Propagation Society International Symposium, pp. 100–104.
95. Roederer, A. and Crone, G. (1987) Double grid reflector antenna, US Patent 4,647,938, March 3.
96. Nakatani, D.T. and Kuhn, G.G. (1977) Comstar I antenna system. IEEE Antennas and Propagation Society International Symposium, pp. 337–340.
97. Dunbar, A.S. (1948) Calculation of doubly curved reflectors for shaped beams. *Proceedings of the IRE*, **36**, 1289–1296.
98. Galindo, V. (1964) Design of dual-reflector antennas with arbitrary phase and amplitude distributions. *IEEE Transactions on Antennas and Propagation*, **AP-12**, 403–408.
99. Doro, G. and Saitto, A. (1975) Dual polarization antennas for OTS. International Conference on Antennas for Aircraft and Spacecraft, 3–5 June, London, pp. 76–82.
100. Westcott, B.S., Stevens, F.A. and Brickell, F. (1981) GO synthesis offset dual reflectors. *Proceedings of the IEE*, **128** (pt H), 11–18.
101. Jones, W.L. (1975) The MAROTS antenna system. International Conference on Antennas for Aircraft and Spacecraft, IEE Proceedings (A76-15926 04-04), pp. 107–114.
102. Katagi, T. and Takeichi, Y. (1975) Shaped-beam horn-reflector antennas. *IEEE Transactions on Antennas and Propagation*, **AP-23**, 757–763.
103. Bergmann, J., Brown, R.C., Clarricoats, P.J.B. and Zhou, H. (1988) Synthesis of shaped·beam reflector antenna patterns. *IEE Proceedings*, **135**, 48–53.
104. Jorgensen, R., Frandsen, P.E., Sorensen, S.B. *et al.* (1990) Study of Advanced Methods for Reflector and Array Antenna Analysis, Synthesis and Design, TICRA Report S·345·04.
105. Mizugutch, Y., Akagawa, M. and Yokoi, H. (1976) Offset dual reflector antenna. IEEE Antennas and Propagation Society International Symposium, pp. 2–5.
106. Dion, A.R. (1970) Variable-coverage communications antenna for LES-7. AIAA 3rd Communications Satellite Systems Conference, Paper 70-423.
107. Dion, A.R. and Ricardi, L.J. (1971) A variable-coverage satellite antenna system. *Proceedings of the IEEE*, **59**, 252–262.
108. Ajioka, J.S. and Ramsey, V.W. (1978) An equal group delay waveguide lens. *IEEE Transactions on Antennas and Propagation*, **AP-26**(4), 519–527.
109. Dion, A.R. (1975) Optimization of a communication satellite multiple beam antenna, TN-1975-39, Lincoln Laboratory, MIT, May 27.
110. Lu, H.S., Scott, W.G., Smith, T. and Smoll, A. (1974) A constrained lens antenna for multiple beam satellites. AIAA 5th Communication Satellite Systems Conference, Los Angeles, April.
111. Scott, W.G., Luh, H.S. and Matthews, E.W. (1976) Design trade-off for multibeam antennas in communications satellites. International Communication Conference.
112. Boeing Defense, Space and Security, Anik A: The world's first synchronous orbit communication satellite, http://www.boeing.com/defense-space/space/bss/factsheets/376/anik_a/anik_a.html (accessed 21 December 2011).
113. Neyret, P. (1985) Antenna technology at INTELSAT. *Annales des Télécommunications*, **40**(7–8), 331–377.
114. Han, C.C. (1977) A multifeed offset reflector antenna for the INTELSAT-V communications satellite. 7th European Microwave Conference, Copenhagen.
115. Lane, S.O., Caufield, M.V. and Taormina, F.A. (1984) INTELSAT-VI antenna system overview. AIAA 10th Communication Satellite Systems Conference.
116. Ersoy, L., Schennum, G., Fenner, G. *et al.* (1992) INTELSAT VII hemi/zone antennas (design and results). Antennas and Propagation Society International Symposium.

117. Hartmann, J., Habersack, J., Steiner, H.-J. and Lieke, M. (2002) Advanced communication satellite technologies. Workshop on Space Borne Antenna Technologies and Measurement Techniques, ISRO, Ahmedabad.
118. Paus, S. (1999) The INTELSAT-9 C-band hemi/zone antennas. 29th European Microwave Conference, p. 162.
119. Belanger, R., Moody, H.J., Hatzigeorgiou, S. *et al.* (1993) The communications payload of the MSAT spacecraft. *Acta Astronautica*, **30**, 229–237.
120. Kawai, M., Tanaka, M. and Ohtomo, I. (1989) ETS-VI multibeam satellite communications systems. 40th International Aeronautical Federation, IAF-89-520.
121. Spring, K. and Moody, H. (1990) Divided LLBFN/HMPA Transmitted Architectures, US Patent 4,901,085, February 13.
122. Roederer, A.G. (1992) Multibeam antenna feed device, US Patent 5,115,248, 1992.
123. Perrott, R. and Griffin, J. (1991) L-band antenna systems design. IEE Colloquium on INMARSAT-3, digest, pp. 5/1–5/7.
124. Cox, G. (1991) Development of L-band feed arrays for INMARSAT-3. IEE Colloquium on INMARSAT-3, digest, pp. 7/1–7/12.
125. Boeing, Thuraya-2,3: Complete System for Mobile Communications, http://www.boeing.com/defense-space/space/bss/factsheets/geomobile/thuraya2_3/thuraya2_3.html (accessed 21 December 2011).
126. Mallison, M.J. (2001) Enabling technologies for the Eurostar geomobile satellite. AIAA International Communication Satellite Systems Conference.
127. Stirland, S. and Brain, J. (2006) Mobile antenna developments in EADS Astrium. Proceedings of EuCAP'06, November.
128. Guy, R. *et al.* (2003) Synthesis of the INMARSAT 4 multi-beam mobile antenna. Proceedings of 12th ICAP, pp. 90–93.
129. Boeing, SkyTerra: Next Generation Mobile Communication Platform, http://www.boeing.com/defense-space/space/bss/factsheets/geomobile/msv/msv.html (accessed 21 December 2011).
130. Terrestar, Introducing TerreStar's integrated satellite-terrestrial network, http://www.terrestar.com/technology-solutions/technology-overview/ (accessed 21 December 2011).
131. Bains, P.S. and Taormina, F.A. (1979) SBS antenna system. Antennas and Propagation Society International Symposium.
132. Arnim, R. (1984) The Franco-German DBS program 'TV-SAT/TDF-1'. 10th AIAA Communication Satellite Systems Conference, pp. 75–83.
133. Jeuken, M. and Thurlings, L. (1975) The corrugated elliptical horn antenna. Antennas and Propagation Society International Symposium, pp. 9–12.
134. Rosen, H.J., Macgahan, J., Dumas, J. and Profera, C. (1984) Shaped-beam antenna for direct broadcast satellites. 10th AIAA Communication Satellite Systems Conference, pp. 717–721.
135. Lindley, A. (1995) Spacecraft antennas - a EUTELSAT perspective. Microwave and Optoelectronics Conference, pp. 607–614.
136. Duret, G., Guillemin, T. and Camere, I.T. (1989) The EUTELSAT I1 reconfigurable multibeam antenna subsystem. IEEE AP-S International Symposium, vol. 1, pp. 476–479.
137. Ippolito, L.I. (1975) ATS-6 millimeter wave propagation and communication experiments. *IEEE Transactions on Aerospace and Electronic Systems*, **AES-11**(6), 1067–1083.
138. Stutzman, W.L., Pratt, T., Safaai-Jazi, A. *et al.* (1995) Results from the Virginia Tech propagation experiment using the Olympus satellite 12, 20, and 30GHz beacons. *IEEE Transactions on Antennas and Propagation*, **AP-43**(1), 52–62.
139. Ohm, E.A. (1974) A proposed multiple-beam microwave antenna for earth stations and satellites. *Bell System Technical Journal*, **53**(8), 1657–1665.
140. Doro, G., Cucci, A., Di Fausto, M. and Roederer, A.G. (1982) A 20/30GHz multibeam antenna for European coverage. IEEE Antennas and Propagation Society International Symposium.
141. Roederer, A., Doro, G., and Lisi, M. (1984) Power divider for multibeam antennas with shared feed elements, US Patent 4710776, December 13.
142. NASA, The Advanced Communications Technology Satellite (ACTS), http://www.nasa.gov/centers/glenn/about/fs13grc.html (accessed 5 December 2011).

143. Itanami, T., Ueno, K., Naito, I. and Kobayashi, Y. (1999) N-STAR Ka-Band Antenna. Mitsubishi Electric ADVANCE, June, pp. 11–13.
144. Bangert, T., Engelbrecht, R.S., Harkless, E.T. and Walsh, E.J. (1963) The spacecraft antennas. *Special Telstar Issue, Bell Systems Technical Journal*, **42**, 4. (Reprinted as Telstar I, NASA SP-32, vols 1–3 (July 1963), http://ntrs.nasa.gov/archive/nasa/casi.ntrs.nasa.gov/19640000959_1964000959.pdf (accessed 5 December 2011).
145. Jensen, N.E., Nicolai, C., and Paci, G. (1975) VHF-, UHF- and S-band low gain antennas for Meteosat. Proceedings of the International Conference on Antennas for Aircraft and Spacecraf, IEE, London June 1975 (A76-15926 04-04), pp. 95–100.
146. van't Klooster, K. (2010) Antennas for Meteosat satellites. 20th International Crimean Conference 'Microwave_&_Telecommunication_Technology', pp. 1216–1218.
147. Albertsen, N.C., Balling, P. and Jensen, N.E. (1976) GTD analysis of near-field test equipment for the METEOSAT electrically despun antenna. IEEE Antennas and Propagation Society International Symposium, pp. 265–268.
148. Jordan, R. (1980) The Seasat-A synthetic aperture radar system. *IEEE Journal of Oceanic Engineering*, **2**, 154–164.
149. Wagner, R. and Braun, H.M. (1979) A slotted waveguide array antenna from carbon fibre reinforced plastics for the European space SAR. XXXth Congress of the IAF.
150. Petersson, R., Kallas, E. and van't Klooster, K. (1988) Radiation performance of the ERS-1 SAR EM antenna. IEEE Antennas and Propagation Society International Symposium, pp. 212–215.
151. Riendeau, S. and Grenier, C. (2007) RADARSAT-2 antenna. IEEE Aerospace Conference, pp. 1–9.
152. Martín-Neira, M., Piironen, P., Ribó, S. *et al.* (2002) The MIRAS demonstrator pilot project. *ESA Bulletin*, **111**, 123–131.
153. Andersson, A., Bengtsson, P., Molker, A. and Roederer, A.G. (1977) Short backfire arrays for space communications. IEEE Antennas and Propagation Society International Symposium, pp. 194–197.
154. Roederer, A.G. and Aasted, J. (1978) A multi-beam array system for communications with mobiles. Proceedings of the AIAA Conference.
155. Imbriale, W.A. and Wong, G.G. (1977) An S-band phased array for multiple access communications. Proceedings of the National Telecommunications Conference, pp. 19.3.1–19.3.7.
156. Agrawal, V.D. and Wong, G.G. (1979) A high performance helical element for multiple access array on TDRSS spacecraft. IEEE Antennas and Propagation Society International Symposium, pp. 481–484.
157. Agrawal, V.D. (1978) Grating-lobe suppression in phased arrays by subarray rotation. *Proceedings of the IEEE*, **66**, 347–349.
158. Schwartz, L. and Smith, J.M. (1975) Adaptive ground implemented phased array (AGIPA) for the tracking and data relay satellite system. IEEE Antennas and Propagation Society International Symposium, pp. 213–216.
159. Walker, J.L., Menendez, R., Burr, D. and Dubellay, G. (2010) Ground-based beam forming for satellite communications systems, US Patent 2008051080.
160. Teshirogi, T., Chujo, W., Akaishi, A. and Hirose, H. (1988) Multibeam array antenna for data relay satellite. *Electronics & Communications in Japan, Part 1*, **71**(5) (Translated from Denshi Tsushin Gakkai Ronbunshi, 69-B (11), November 1986).
161. Teshirogi, T., Tanaka, M. and Chujo, W. (1985) Wideband circularly polarized array antenna with sequential rotations and phase shift of elements. Proceedings of ISAP, August, pp. 117–120.
162. Tanaka, M., Kimura, S., Teshirogi, T. *et al.* (1994) Development of multibeam phased array antenna on ETS-VI for S-band inter-satellite communications. *Electronics and Communications in Japan, Part 1*, **77**(1) (Translated from Denshi Joho Tsushin Gakkai Ronbunshi, **J 76-B-11** (5), May 1993).
163. Meben, P.L. (2000) Globalstar satellite phased array antennas. IEEE International Conference on Phased Array Systems and Technology, pp. 207–210.

8

Deployable Mesh Reflector Antennas for Space Applications: RF Characterizations

Paolo Focardi[1], Paula R. Brown[1], and Yahya Rahmat-Samii[2]

[1]*Jet Propulsion Laboratory, California Institute of Technology, USA*
[2]*University of California, Los Angeles, USA*

8.1 Introduction

Deployable antenna structures are required to generate the large apertures necessary for many space applications. Deployable mesh reflectors have been successfully used in telecommunications applications for several decades. The structures are lightweight and can be packaged in a compact form for launch. Once deployed, the antennas are rigid and can be built with surfaces accurate enough for Ka-band operation. Reflector mesh fabrics are available that are fine enough to be highly reflective at Ka-band, yet have relatively low mass.

In this chapter, the history of spaceborne deployable mesh reflectors is briefly reviewed, and then design considerations are discussed. An earlier review paper on deployable antennas was presented by Roederer and Rahmat-Samii [1], and the mechanical designs of several different mesh reflector types were described in detail by Davis and Tanimoto [2] and Tibert [3]. General reflector antenna design guidelines are available in many other antenna engineering handbooks [4]. The remainder of this chapter is devoted to a case study of the antenna design for NASA's Soil Moisture Active and Passive (SMAP) mission[1]. The antenna is a 6 m deployable offset mesh reflector that is part of a remote sensing instrument consisting of an L-band radar and radiometer. This, we believe, is the first very large mesh reflector antenna system with a spinning platform. Specifically, the following unique design issues pertinent for the characterization of this class of antennas will be addressed:

[1]At the time this chapter was written, SMAP was a proposed NASA mission, preparing for its Preliminary Design Review. It would not launch prior to October 2014.

Space Antenna Handbook, First Edition. Edited by William A. Imbriale, Steven (Shichang) Gao and Luigi Boccia.

1. RF modeling of the reflector, truss, and boom with the spacecraft.
2. Electrical performance of the reflector surface materials, with particular consideration in a low-noise radiometer application.
3. Effects of thermal and dynamic distortions of the reflector on the antenna radiation patterns.

8.2 History of Deployable Mesh Reflectors

The development of deployable mesh reflector antennas began in the 1960s. In 1974, a 9.1 m diameter reflector developed by NASA's Jet Propulsion Laboratory and Lockheed Missile and Space Company was successfully launched and deployed on NASA's Applications Technology Satellite 6 (ATS-6) [5]. The wrap-rib design, shown in Figure 8.1, consisted of 48 ribs that functioned similarly to an umbrella. A copper-plated Dacron mesh was used for the reflective surface [6,7]. The antenna operated over multiple frequency ranges from VHF through C-band in a variety of telecommunications experiments [8].

From 1978 to 1989, a series of six 5.3 m deployable mesh reflectors from TRW were put into operation in geosynchronous orbit for the US Navy's Fleet Satellite Communications (FLTSATCOM) system [9]. In 1983, the first of NASA's Tracking and Data Relay Satellites (TDRS) was launched into geosynchronous orbit, with a pair of radial-rib antennas from Harris Corporation. Five additional TDRS spacecraft with pairs

Figure 8.1 Photograph of NASA's ATS-6 satellite. The 9.1 m diameter telecommunications antenna, built by Lockheed Missile and Space Company, operated at multiple frequency bands from VHF to C-band. The satellite operated in a geosynchronous orbit for five years and demonstrated several new technologies, including direct-broadcast television. Courtesy of NASA.

Figure 8.2 *Harris Corporation radial-rib antennas. (a) TDRS 7 spacecraft with a pair of 4.8 m stowed antennas. Six pairs of the same antenna design were successfully deployed on TDRS spacecraft. (b) The Galileo antenna with the antenna deployed during ground testing. The antenna later failed to deploy in space. Courtesy of NASA.*

of the same 4.8 m Harris Corporation antennas were successfully put into orbit between 1988 and 1995. In 1989, the Galileo spacecraft was launched on its way to Jupiter carrying a Harris Corporation radial-rib design similar to the TDRS antennas. The Galileo antenna failed to fully deploy, and investigators concluded that several of the ribs were stuck in the stowed launch position [10]. The mission was forced to rely on its low-gain antennas for data return, but the mission was a great success despite the antenna failure. Figure 8.2 shows the Harris Corporation radial-rib antennas in stowed and deployed configurations.

Harris Corporation continued to play a leading role in deployable mesh reflectors. It developed a folding-rib design that boasted a shorter stowed package than earlier umbrella-style antennas. The folding-rib design has been deployed on several spacecraft, including ACeS Garuda 1 (12 m diameter, 2000), ICO (12 m diameter, 2008), Terrestar (18 m diameter, 2009), and others [11–13]. In conjunction with NASA Langley Research Center, Harris Corporation developed a hoop-column design for the Large Space Systems Technology (LSST) Program in the late 1980s (Figure 8.3), [14,15]. The compact packaging of the hoop–column antenna enabled larger apertures to be realized. In 2010, a Harris Corporation hoop–column antenna was deployed in space on the SkyTerra 1 telecommunications satellite. With a diameter of 22 m, it was the largest antenna ever deployed on a commercial satellite up to that time [16].

In 1997, an 8 m effective diameter reflector employing a tensioned cable truss design was launched on the Japan Aerospace Exploration Agency's (JAXA's) HALCA satellite [17]. The antenna, developed by the Institute of Space and Astronautical Science and shown in Figure 8.4, was designed to operate as a radio telescope for very long baseline interferometery (VLBI) at L-, C-, and Ka-bands [18]. The Ka-band performance was severely degraded for an unknown reason, possibly because of damage to the Ka-band waveguide during launch [17]. The antenna was unique and different from earlier deployable antennas in that the surface was triangularly faceted instead of being tensioned between ribs. The HALCA orbit differed from

Figure 8.3 The NASA/Harris Corporation 15 m hoop–column antenna. The 15 m model was built as a technology demonstration as part of the NASA Large Space Systems Technology Program. Courtesy of NASA.

the orbits of other mesh reflectors flown to date in that it was highly elliptical, with a perigee of 560 km and an apogee of 21 000 km [17].

In 2000, the first AstroMesh tension truss antenna from TRW (now Northrop Grumman) Astro Aerospace was launched into geosynchronous orbit on a Thuraya telecommunications satellite. The 12.25 m diameter

Figure 8.4 A deployment test of the antenna for JAXA's HALCA spacecraft. The largest diameter across the antenna was 10 m, with an effective aperture diameter of 8 m. HALCA, the first space VLBI mission, operated successfully for nearly nine years. Courtesy of JAXA.

Figure 8.5 *Northrop Grumman AstroMesh reflector shown (a) stowed and (b) deployed. This 11 m unit was delivered to Astrium in 2010 for the Alphasat I-XL communications satellite. Courtesy of Northrop Grumman Corporation.*

offset reflector weighed 55 kg and was stowed in a volume approximately 1.1 m in diameter and 3.8 m long [19,20]. Two more 12.25 m AstroMesh reflectors were successfully deployed on Thuraya satellites in 2003 and 2008 [21,22]. A 12 m AstroMesh reflector was deployed on the MBSAT telecommunications satellite in 2004 [23]. Three INMARSAT-4 satellites, launched between 2005 and 2008, utilized 9 m AstroMesh reflectors for geosynchronous telecommunications [24]. An 11 m AstroMesh antenna for Alphasat I-XL is shown in Figure 8.5.

An exploded view of the AstroMesh antenna is given in Figure 8.6. The reflector surface shape is formed by a precisely fabricated front net that holds knitted metallic mesh material in place. The net is made of a stiff dielectric composite material referred to as the web. The front and rear nets attach to the rims of the outer truss structure and then are connected together through tension tie assemblies. The resulting structure is stiff and repeatably deploys [25].

One final noteworthy design was flown on the Japan Aerospace Exploration Agency's Kiku-8, also known as Engineering Test Satellite 8 (ETS VIII), which launched in 2008 (Figure 8.7). Two 13 m diameter aperture reflectors were on the telecommunications spacecraft, one for transmit operation and one for receive. Each antenna comprises 14 modules, which are approximately 5 m in diameter when deployed. Each antenna was stowed in a package of 1 m diameter by 4 m long for launch. The antennas operate at S-band in a geosynchronous orbit [26].

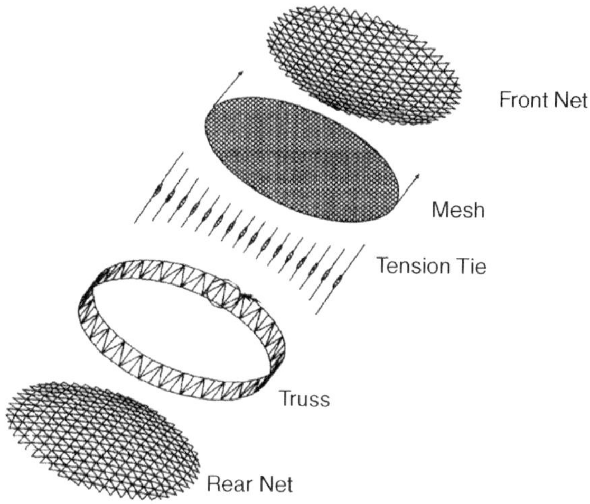

Figure 8.6 *Exploded view of the AstroMesh antenna. The front net lies on top of the mesh to create and maintain the desired surface shape. The front and rear webs are made of strips of dielectric material referred to as webs. Courtesy of Northrop Grumman Corporation.*

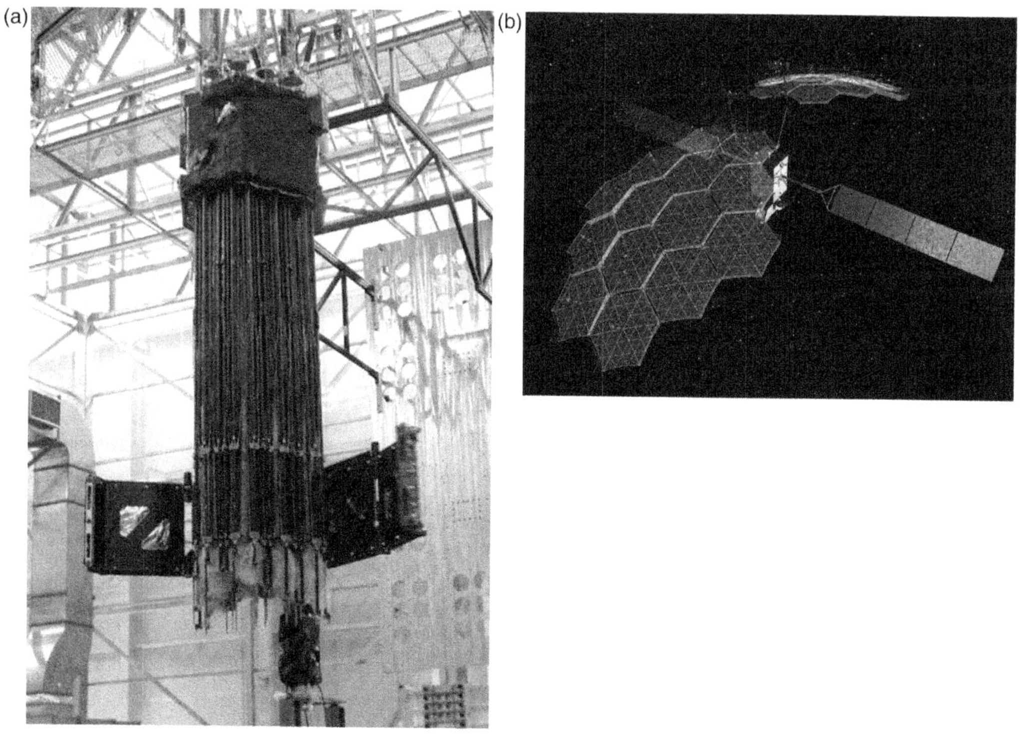

Figure 8.7 *The JAXA Engineering Test Satellite 8 modular antennas. (a) One stowed antenna package is shown. Each antenna stows into a volume 1 m in diameter and 4 m long. (b) An illustration of the spacecraft on orbit. Two antennas are used on the communications satellite, each approximately 19 × 17 m in the deployed state. Courtesy of JAXA.*

8.3 Design Considerations Specific to Mesh Reflectors

Several additional design factors need to be considered when using a mesh reflector instead of a conventional solid reflector. First, the available stow volume may dictate the type of mechanical deployment scheme to be used. For example, the non-folding radial-rib designs require a stowage length approximately equal to the radius of the deployed antenna, whereas the other designs stow in a shorter length, but may be larger in diameter. A second factor is the required surface accuracy. The surface accuracy of most of the antenna designs can be improved by adding more facets or ribs, but at a cost of more mechanical structure, which increases mass and volume. When considering the required surface accuracy, both systematic and random errors should be taken into account. The systematic errors can be a source of grating lobes, which may be undesirable for some applications. The treatment of systematic and random errors has been discussed by Ruze [27], Bahadori and Rahmat-Samii [28], Corkish [29], Rusch and Wanselow [30], and in numerous other articles.

A third design consideration is the fineness of the mesh weave. Meshes are generally classified by their 'openings per inch' or 'OPI.' The more OPIs, the finer the mesh, which entails higher reflectivity; however, the finer meshes have higher mass, are stiffer, and require a larger stow volume. The results of measurements and modeling of different mesh OPIs against various frequencies have been reported [31–34]. The lowest OPI that is adequate for the operating frequency range should be used.

The particular weave of the mesh fabric is the fourth factor to consider because it impacts RF performance characteristics. The weave can increase cross-polarization levels for circularly polarized fields due to the difference in reflection and transmission coefficients for the constituent linear field components [33,35,36]. The weave selection, along with the wire plating and tension of the deployed mesh, also affect the level of passive intermodulation (PIM) products that can be generated at the wire-to-wire junctions of the mesh [37–39].

8.4 The SMAP Mission – A Representative Case Study

8.4.1 Mission Overview

For the remainder of this chapter, a case study of the detailed design of a 6 m deployable offset mesh reflector antenna that is to be used on NASA's SMAP mission will be presented. SMAP's primary objectives are to map soil moisture and its freeze/thaw state over the Earth's land areas. The mission was one of the highest priority missions recommended by the National Research Council's Earth Science and Applications Decadal Survey [40]. The science data applications will include more accurate and longer term weather and climate prediction, earlier warnings for droughts, improved flood and landslide predictions, and improved agricultural productivity predictions. Better knowledge of soil moisture and its freeze/thaw state will also improve understanding of the global carbon cycle [40,41]. The planned launch date for the three-year mission is November 2014 [42].

The SMAP spacecraft will be in a near-polar, Sun-synchronous orbit of 685 km. The 6 m offset reflector antenna will provide a common aperture for a synthetic aperture radar (SAR) operating at a center frequency of 1.26 GHz, and a radiometer operating at a center frequency of 1.41 GHz. The antenna boresight will be pointed at approximately 35.5° off nadir, and the antenna will rotate at a rate of approximately 14 rpm. The resulting conically scanned beam will produce a 1000 km swath on the ground, with a radiometer spatial resolution of 40 km and a SAR spatial resolution of 1–3 km. The revisit time will vary from 3 days at the equator to 2 days at latitudes greater than 45° north [42]. Artist renditions of the SMAP spacecraft, including the 6 m deployable offset mesh reflector antenna and the SMAP swath, are shown in Figures 8.8 and 8.9.

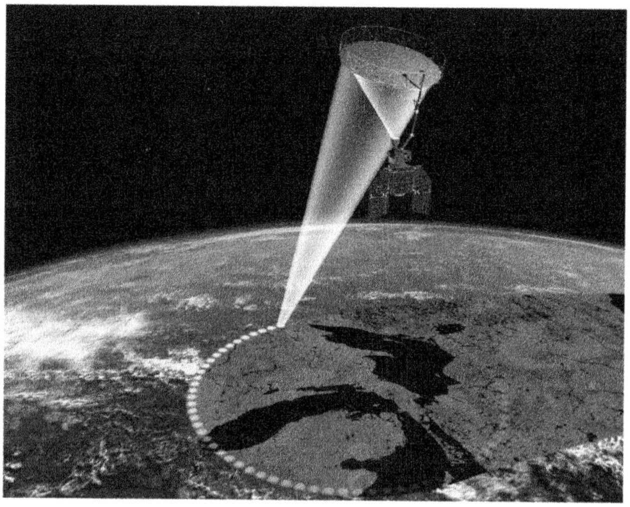

Figure 8.8 An artist's rendition of the SMAP spacecraft. The 6 m deployable mesh reflector is designed to spin at an approximate rate of 14 rpm, creating a 1000 km swath on the ground. The reflector will be shared between the radar, with a center frequency of 1.26 GHz, and the radiometer, operating at 1.41 GHz. Courtesy of NASA.

L-band was selected for the radiometer operation in order to meet the objective of mapping the moisture in the top 5 cm of the Earth's soil [41]. L-band radiometry is sensitive to the emissions from a depth of 5 cm, and it can penetrate through vegetation excluding dense forests. Higher frequency radiometers suffer from more attenuation through vegetation and cannot measure soil moisture to the required depth. Lower frequencies cannot provide the necessary accuracy and resolution given the constraints on antenna size, and the lower frequency bands contain more radio frequency interference (RFI) and galactic noise. The 1.4–1.427 GHz portion of the L-band spectrum is protected worldwide for radio astronomy and passive sensing, which reduces the amount of RFI that the radiometer must contend with.

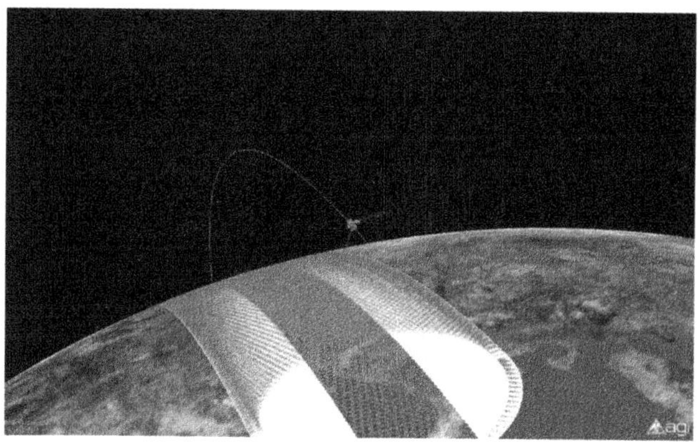

Figure 8.9 An artist's rendition of the SMAP spacecraft swath on the Earth. To create the swath of overlapping beam footprints, the boresight of the antenna is designed to point 35.5° from nadir, and the reflector and feed assembly will rotate. The spacecraft bus will not rotate and its solar panels will remain pointed at the Sun. The 1000 km wide swath enables a maximum revisit time of three days for mapping the Earth. Courtesy of NASA.

The L-band SAR has a much higher spatial resolution than the radiometer, yet provides a less accurate soil moisture measurement because it is degraded more by the Earth's surface roughness and vegetation. The SAR data will be combined with the radiometer data to produce an intermediate spatial resolution for soil moisture. The 3 km resolution requirement for the freeze/thaw state is met with the SAR operating at L-band, given the antenna size and orbit parameters.

The SMAP application of a deployable mesh reflector brings several unique challenges and design complexities:

- The antenna boresight angle from nadir must remain stable while the antenna spins. This was a key factor in the antenna selection and focal length determination, and necessitated accurate RF analyses of dynamic and thermal distortions of the antenna.
- Both the radar and radiometer instruments are sensitive to background signals and noise. With ground-based sources below the SMAP spacecraft, GPS satellites in orbit above the SMAP spacecraft, and satellite radars in many different orbits, accurate antenna pattern knowledge in all directions is critical to the instrument performance predictions. As the antenna spins around the spacecraft, the change in the antenna patterns must also be assessed.
- Knowledge of the RF dissipative losses introduced by scattering off the reflector and the spacecraft is necessary for calibrating the radiometer. The reflector antenna is the biggest contributor to these losses, and very limited assessments have been performed on mesh losses in the past. Except for the HALCA satellite, other deployable mesh reflectors have been used in communications applications which are less sensitive to such losses.
- The low Earth orbit (LEO) of the SMAP spacecraft introduces different environmental concerns than are found in the geosynchronous orbit (GEO) that most previous deployable mesh reflectors have flown in.

8.4.2 Key Antenna Design Drivers and Constraints

8.4.2.1 Baseline Design Selection

The SMAP antenna design is driven by a combination of the SAR and radiometer performance requirements, the orbital environment, and the available accommodation space in the launch vehicle. The key drivers and constraints are detailed in this subsection.

An AstroMesh deployable reflector antenna was selected for the SMAP baseline design because the antenna technology is mature, the design is low mass, the RF losses are low, and a shared SAR/radiometer aperture can be accommodated. The 6 m aperture was determined by the SAR and radiometer resolution requirements. A boresight angle pointing 35.5° off nadir and a spin rate of approximately 14 rpm were chosen in order to create the 1000 km swath size needed to cover the Earth in a revisit period of 2–3 days for the selected orbit. The basic configuration of the spacecraft, showing the spin axis, the boresight direction, and the projected aperture, is illustrated in Figure 8.10.

8.4.2.2 Radar (SAR) Performance Drivers

The primary antenna requirements that flow from the SAR performance requirements are listed in Table 8.1. The SAR operates with dual linear polarizations at a center frequency of approximately 1.26 GHz. A minimum antenna gain of 35.55 dBi is required to meet the SAR data precision requirements. A requirement has also been levied on the gain stability since it affects the SAR calibration accuracy. Requirements on the sidelobes and backlobes are in place to limit ambiguities in the data from the backscattered radar signal, and to limit the

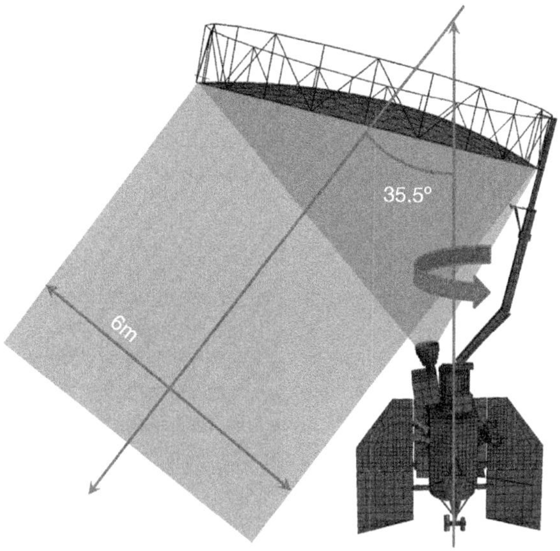

Figure 8.10 Sketch of the SMAP observatory indicating the spin axis, the 35.5° pointing angle off nadir, and the 6 m projected aperture diameter. The reflector, the feed assembly, and a portion of the radiometer electronics spin around the indicated axis while the rest of the spacecraft and the solar panels do not. Courtesy of NASA/California Institute of Technology.

Table 8.1 Requirements driven by SAR performance.

Parameter	Value	Rationale
Polarization	Linear vertical and horizontal ('V-pol' and 'H-pol')	Using both polarizations increases the accuracy of the soil moisture measurement
Radar frequency	V-pol: 1.256 GHz ± 40 MHz	L-band selected for the required resolution Frequencies are in the active Earth exploration satellite band
	H-pol: 1.259 GHz ± 40 MHz	Bandwidth enables frequency hopping to avoid RFI
Power handling	326 W peak operating, 17.5 W average per channel	Needed to meet the SAR precision requirements
Gain	35.55 dBi	
Gain stability	< 0.07 dB	Needed to meet SAR calibration requirements
Beamwidth	< 2.8°	Limits ambiguities in the backscattered signal Needed to meet SAR calibration requirements
Sidelobe levels – peak	< 15 dB below boresight gain	
Sidelobe levels – nadir	< 45 dB below boresight gain	
Off-Earth sidelobe levels	< −6.0 dBi average	Limits amount of interfering signals the SAR will receive from GPS satellites
Return loss at feed input ports	>20 dB	Needed to meet SAR calibration requirements

amount of interfering signal the SAR will receive from other sources, including GPS sources that are in-band with the SAR.

The antenna boresight pointing accuracy, stability, and knowledge are constrained by the SAR accuracy and calibration requirements. The decomposition of the top-level pointing requirements into assembly-level requirements is complicated and depends mostly on the dynamics, control, and structural aspects of the spacecraft and antenna boom. Details of the pointing requirements are omitted here.

8.4.2.3 Radiometer Performance Drivers

Table 8.2 shows the main antenna requirements resulting from the radiometer performance requirements. The radiometer, like the SAR, also operates in a dual linear polarization mode, but at a higher frequency band of 1.4 to 1.427 GHz. The radiometer accuracy requirements, along with the radar requirements, drive the overall antenna boresight pointing accuracy, stability, and knowledge requirements.

The antenna brightness temperature and its knowledge are large contributors in the radiometer calibration budget. The brightness temperature is the product of the RF emissivity and the physical temperature. For a body in thermal equilibrium, emissivity is equal to absorptivity, which is equivalent to the RF dissipative losses at the radiometer frequencies. Accordingly, allocations from the radiometer calibration budget result in loss requirements that are levied on the feed and reflector, as shown in Table 8.2. The reflector mesh emissivity is required to be less than or equal to 0.0035, which means that the maximum dissipative loss in the mesh is 0.35% or −0.015 dB. Similarly, the loss from the front net web material (see Figure 8.6) is required to be less than or equal to 0.00015, or −0.00065 dB, and the feed insertion loss is required to be less than or equal to 0.1 dB. Since the radiometer performance is directly sensitive to the brightness temperature, the mesh loss has the biggest impact on the radiometer performance because the mesh temperatures vary over a much larger range than the web or feed assembly. Consequently, the mesh emissivity is subject to a knowledge requirement to ensure adequate calibration of the mesh brightness temperature error contribution.

Another large factor in the radiometer calibration budget is the antenna pattern correction factor. Requirements on the antenna main beam efficiency (MBE), the sidelobe levels, and the sidelobe-level knowledge flow directly from this factor. The sidelobes introduce unwanted noise into the measurement and can be particularly detrimental in the solar regions if they are too high. A low main beam efficiency reduces the amount of the received power from the desired direction relative to the other error sources, which proportionally scales the weighting of the other error sources in the calibration budget.

8.4.2.4 LEO Environment

The antenna must be able to withstand the LEO environment of a 685 km orbit for a mission life of three years. Of particular concern are UV radiation, atomic oxygen, electrostatic discharge, micrometeoroids, and orbital debris. The gold plating on the reflector mesh wire will protect the mesh from UV radiation and atomic oxygen. Although the mesh is expected to be bombarded with thousands of micrometeoroids and orbital debris particles during its three-year lifetime, it is highly probable that all the particles will be less than 1 mm in diameter, with the vast majority of them being much less than 1 mm. Most of the particles are expected to pass through the mesh without causing any damage. Some particles will create holes in the mesh, but since the mesh fabric does not unravel, the holes will remain approximately the same size as the particles. Since the holes will be much smaller than the 210–250 mm RF wavelength of the radar and radiometer, the performance impact will be negligible.

Degradation of the web material in the front net (Figure 8.6) is potentially more of a concern. A damaging impact by a micrometeoroid or orbital debris is considered a very small risk because of the small size of the particles. If a particle has enough energy to create a small hole in a web, the hole will be very small in diameter

Table 8.2 *Requirements driven by radiometer performance.*

Parameter	Value	Rationale
Polarization	Linear V- and H-pol	Need both polarizations for corrections in the soil moisture measurement
Radiometer frequency	1.413 GHz ± 13.5 MHz	Portion of L-band that is protected for radio astronomy and passive sensing
Beamwidth	2.29° < beamwidth < 2.5°	Needs to be small enough for required resolution, and large enough for footprint overlap
Sidelobe levels – solar region	< 0 dBi peak < 2% of the sidelobes shall be > −5 dBi	Part of the antenna pattern correction factor. Required to meet radiometer calibration requirements
Integrated sidelobe levels – off-Earth	< 10% of the total antenna radiated power Knowledge to within 0.25% of the total antenna radiated power	
Integrated sidelobe levels – on-Earth	< 3% of the total antenna radiated power Knowledge to within 0.25% of the total antenna radiated power	
Main beam efficiency	> 87%	
Return loss at feed input ports	≥ 20 dB	Part of the antenna temperature calibration error. Required to meet radiometer calibration requirements
Feed insertion loss	≤ 0.1 dB overall < 0.0008 dB for radome	
Reflector mesh RF emissivity	≤ 0.0035	
Reflector mesh RF emissivity knowledge	±0.001	
Reflector web RF emissivity	≤ 0.00015	

compared to the width of the web. Analysis of the web materials indicates that a very thin layer of the web may be depleted by atomic oxygen interaction, but it is very small compared to the overall web thickness. The remaining concern is damage caused by UV radiation. A test is under way to verify whether the webs retain enough strength after the expected mission lifetime UV exposure so that the net will maintain the reflector shape within the required surface accuracy.

8.4.2.5 Thermal Environment

The reflector and feed are generally exposed to the Sun, although during some seasons the spacecraft passes through eclipse, and the temperatures drop dramatically. The radiometer calibration requirements drive the need for a stable thermal environment for the feed assembly. As a result, the feed design includes a radome to block direct Sun exposure into the horn, and the assembly is covered in multilayer insulation (MLI). The feed component temperature ranges are listed in Table 8.3.

Table 8.3 *Feed component temperature ranges.*

Feed component	Operational design temperature range
Horn	−90 to +75 °C
Radome	−180 to +75 °C
Titanium thermal isolator	−90 to +75 °C
Orthomode transducer (OMT)	−35 to +60 °C
Waveguide to coaxial adapters (WCAs)	−35 to +60 °C

For design purposes, the operating temperature range for the reflector mesh surface is −115 to +280 °C, and for the web material that holds the mesh in place (Figure 8.6) the design temperature range is −115 to +115 °C. However, the mesh emissivity requirements are defined over a more limited range of −45 to +225 °C, which covers the critical portion of the orbit for radiometer measurements.

8.4.2.6 *Mechanical Drivers and Constraints*

The main mechanical drivers for the reflector antenna are accommodation in the launch vehicle and the ability to maintain the boresight pointing stability while the antenna is spinning at a rate of 14 rpm. Figure 8.11 shows

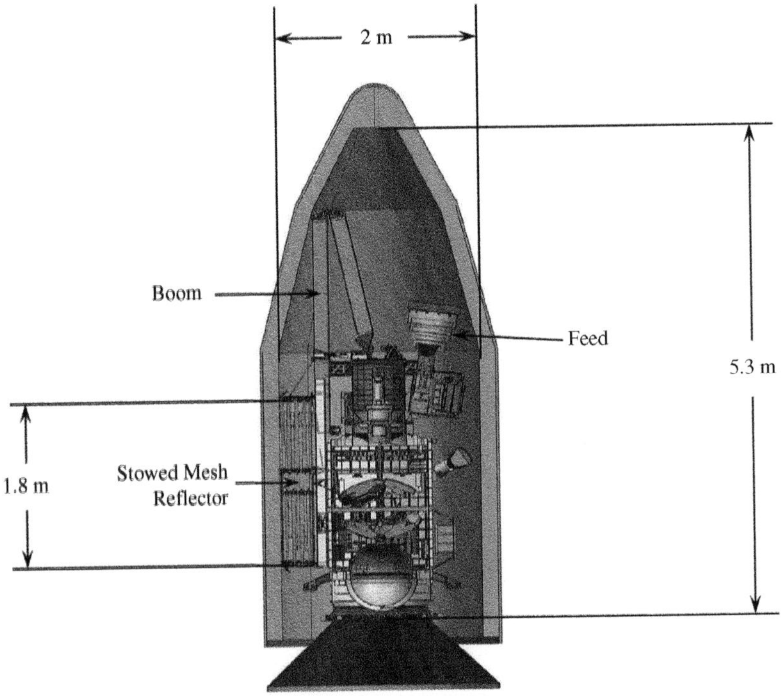

Figure 8.11 *A drawing of the SMAP spacecraft stowed for launch. An envelope of the smallest candidate launch vehicle is shown. Even with the reflector stowed, clearances between the stowed reflector and the hinge point at the boom are small. Courtesy of NASA/California Institute of Technology.*

the reflector and boom in the stowed configuration, inside an envelope representing the launch vehicle. As can be seen, the reflector must be compactly stowed in order to fit into the volume.

In order to maintain the antenna pointing errors within the requirements, the masses and dynamic balancing of the spinning elements of the spacecraft are critical. Since the reflector and boom have large products of inertia, they are the biggest factors in solving the overall balance problem. A trade study was conducted to evaluate the reflector focal length-to-diameter ratio (F/D) and the edge offset height of the reflector against the spin dynamics and the RF performance. A compromise was made in which the selected F/D of the antenna is smaller than optimum for RF performance, and the selected offset height is larger than optimum for RF performance, but the spin dynamics and pointing stability requirements are met.

The mass allocation of the reflector–boom assembly is 65 kg, and the allocation for the feed assembly is 16 kg. Requirements on the center of mass and effective products of inertia have been levied on the reflector–boom assembly.

8.4.3 RF Performance Determination of Reflector Surface Materials

The most common application of mesh reflectors is on telecommunications satellites, where the surface material parameter of most concern is the reflective loss and its contribution to the overall antenna gain budget. For the SMAP radar and radiometer applications, additional reflector surface parameters come into play. Of primary interest for the radar performance is the reflective loss, but secondary concerns include the induced cross-polarization levels, the stability of the reflective loss over temperature and time, and the contribution of the transmission through the mesh to the overall antenna pattern. The mesh transmission is of interest because of potential RF interference from satellites in higher orbits. The radiometer performance does not depend directly on the reflective loss, but it does depend on the dissipative loss, or emissivity, which contributes to the antenna noise temperature. Knowledge of the loss level over the operating temperature range is also important for calibrating the radiometer.

The main reflective surface of the AstroMesh reflector selected for the SMAP mission is a 20 OPI mesh made from gold-plated molybdenum wire. The mesh is formed into a faceted surface approximating an offset paraboloid by a net made of the composite webs, as shown in Figure 8.6 [25]. The net lies on the front side of the mesh reflector surface and covers approximately 5% of the total reflector surface. The effects of both the mesh and the web need to be considered, particularly for the radiometer performance.

For the mesh reflectivity, transmissivity, and emissivity calculation, a periodic finite element method was employed, similar to periodic moment methods reported in [32,33]. An enlarged photograph of the mesh weave was inspected and a pattern to represent a unit cell of the mesh was determined. ANSYS HFSS, a 3D full-wave electromagnetic simulator, was used for the periodic finite element method solution. With HFSS, the mesh wires could be readily modeled as round wires rather than planar strips, which avoided use of the equivalent strip width approximation. The junctions at the wire intersections in the weave were modeled as perfect contacts. The results indicated that the mesh weave is fine enough at L-band such that the mesh performance is independent of the mesh orientation and can be modeled as a flat plate with an effective conductivity (Figure 8.12).

The conductivity change in the mesh wire over the thermal range of -45 to $+225\,^{\circ}$C was investigated in the periodic finite element model to verify whether it had a significant impact on the mesh performance. The mesh wire is gold-plated molybdenum, but the gold plating is only a fraction of the skin depth at L-band. For the mesh HFSS model, published data on the electrical conductivity and the thermal coefficient of molybdenum alone were used. The results confirmed that the emissivity change over the temperature range is on the order of the emissivity knowledge requirement for the radiometer performance, so the change in conductivity over temperature must be accounted for. The emissivity varied as the square root of the resistivity, which is what would be expected for a flat plate model. The reflectivity change is only a fraction of the radar gain stability

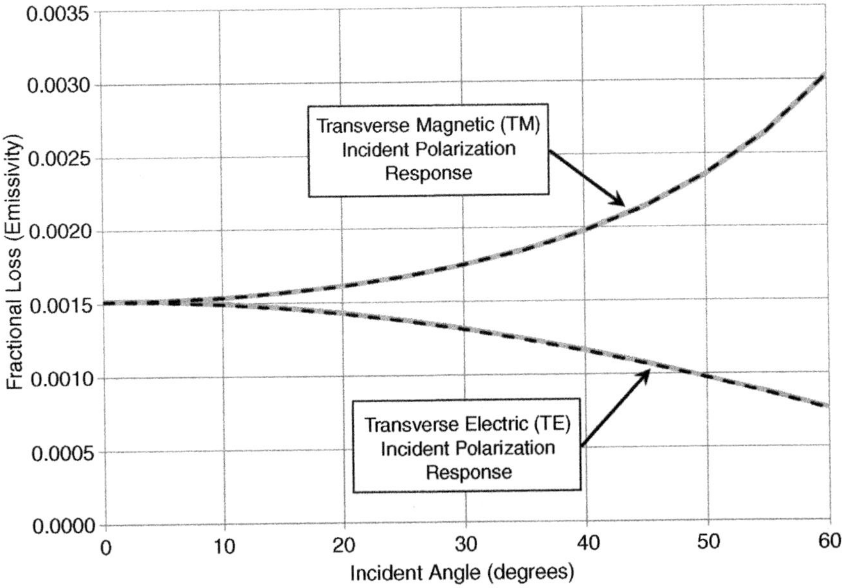

Figure 8.12 *Modeled mesh losses for plane-wave incidence versus incident angle. The light gray curves are the periodic finite element solution and the dashed lines are the flat plate model with an effective conductivity. For the periodic finite element solution, several curves for different mesh orientations are plotted, but they are indistinguishable, indicating that the mesh weave is so fine that the mesh can be modeled as a flat plate. Courtesy of JPL/California Institute of Technology.*

requirement, but large enough that it cannot be neglected in the gain stability budget. Once wire from the SMAP flight production run is available, the DC conductivity of the wire will be measured over temperature using a bridge circuit, but the results are not expected to vary much from the published properties of molybdenum.

The reflection and transmission characteristics are functions of incident angle and incident polarization [32,33,35]. To include these effects in the antenna pattern calculation, the reflection and transmission coefficients will be tabulated and added to the GRASP9 RF model of the reflector [43,44]. Both co- and cross-polarization components can be included in the GRASP9 tabulated coefficient set, so the mesh-induced cross-polarization will be included in the overall antenna pattern. The conduction losses, or emissivity, will be calculated using the tabulated induced currents generated as part of the GRASP model output, combined with the effective conductivity determined by the periodic finite element model and mesh loss measurements.

The validation plan for the RF mesh model includes comparison to radiometric measurements performed by NASA Langley Research Center on 20 OPI mesh [45] and with waveguide *S*-parameter measurements on the SMAP flight mesh. The two techniques are discussed in detail in Njoku *et al.* [45]. A comparison of the techniques is summarized in Table 8.4. In spite of the limitations of the waveguide technique, it was felt that as long as the attainable set of measured data correlated with the RF model, the model results could be trusted within an acceptable level of risk over all the operating conditions of the mesh.

The other reflector surface material of concern is the web composite that makes the AstroMesh front net. Normally, the web is coated with a slightly conductive paint to protect against electrostatic discharge (ESD) events; however, the loss in the paint at L-band may degrade the radiometer performance to undesirable levels. Calculations show that in areas where the web is in contact with the mesh, the losses are negligible because the

Table 8.4 Waveguide measurement versus radiometric measurement for mesh characterization.

Parameter	Waveguide measurement	Radiometric measurement
Incident polarization	Transverse electric (TE) only	Transverse electric (TE) or transverse magnetic (TM)
Range of incident angles	Limited by TE_{11} waveguide mode solution. Angle determined by frequency and waveguide size	If a simple feed is used, a simultaneous range of angle is measured. A single angle can be achieved by collimating the beam with a reflector or lens. The range of angles is limited by the geometry of the reflector optics
Cross-polarization response	Not measured by a simple setup. Could be measured by incorporating a polarizer into the setup	Not measured
Ease of measuring over a thermal range	Possible, but must be careful to separate changes due to the test setup versus the mesh	Difficult to measure over an appreciable thermal range because of the size and complexity of the test bed
Accuracy	Measurements to date have been less accurate than radiometric measurements [45]	Measurements to date have been more accurate than waveguide measurements [45]
Cost	Less expensive	More expensive due to the large complex setup needed to control feed illumination spillover and to control the thermal environment

total electric field is quite small. In some areas of the reflector, the web is not in direct contact with the mesh, which increases the emissivity due to the paint, although the overall contribution is still expected to be less than the mesh emissivity. If the bare web material passes all environmental qualification tests, the web will not be painted and the SMAP reflector may experience ESD events on the order of once per year. The events are not expected to have any observable impact on the antenna long-term performance.

8.4.4 RF Modeling of the Antenna Radiation Pattern

8.4.4.1 RF Model Development for the Antenna and Spacecraft

In order to obtain a highly accurate calculation of the radiation pattern generated by the reflector antenna, an RF model of the entire spacecraft was generated including all the details that were considered important from an electromagnetic point of view. Starting from a simple model, including just the reflector surface and a simulated feed pattern, each part of the spacecraft was gradually introduced. A non-exhaustive list of the most important components includes the reflector truss structure, the scalloped rim of the reflector, the reflector's faceted surface, the boom, the spun platform with all the electronics boxes, the feed horn, the spacecraft bus, the outrigger, and the solar panels. Figures 8.13–8.15 show a comparison between mechanical and RF models and allow the reader to appreciate the complexity with which the RF model evolved to resemble the mechanical design in detail.

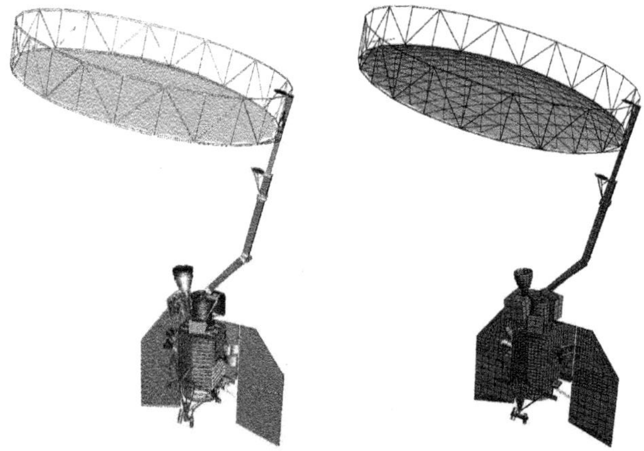

Figure 8.13 *Comparison between mechanical (left) and RF (right) models of the SMAP observatory. As much detail as possible was incorporated into the RF model. Courtesy of JPL/California Institute of Technology.*

The AstroMesh reflector's triangular facets, clearly visible in Figure 8.15, are the result of the front web tensioned on top of the mesh. Very little pillowing is generated on the facets, and they are assumed to be perfectly flat for the RF model. The periodicity of the triangular facets generates grating lobes all around the main beam, clearly visible in the radiation pattern. The boom is made of composite materials and is covered in MLI thermal blankets to reduce the deformations due to thermal variations. The top of the boom, called the prime batten, attaches directly to the truss structure and is visible in Figure 8.15. The motor that deploys the reflector is visible to the side of the prime batten in the same figure. The other two main sections of the boom are more robust and connect the prime batten to the spinning platform. Accurate boom details are included in the RF model since the boom is directly illuminated by the feed radiation. The scattering generated by the boom steers the antenna main beam by a few millidegrees.

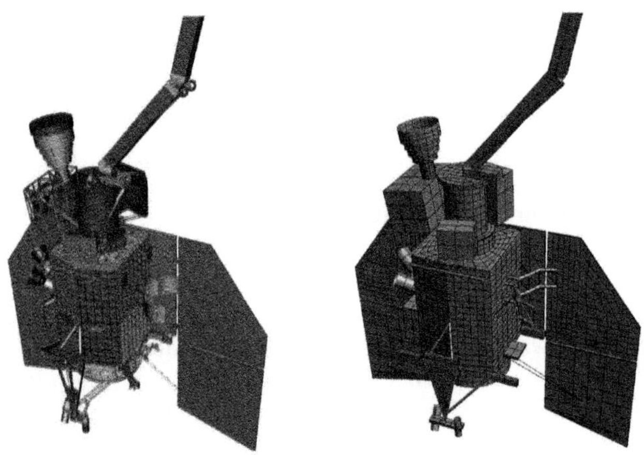

Figure 8.14 *Close-up view of the SMAP spacecraft, and comparison between the mechanical (left) and RF (right) models. Several areas of the mechanical model are shown as open truss structures, but they will be covered in MLI during the spacecraft assembly process. These items were modeled as solid structures in the RF model since MLI is reflective at RF frequencies. Courtesy of JPL/California Institute of Technology.*

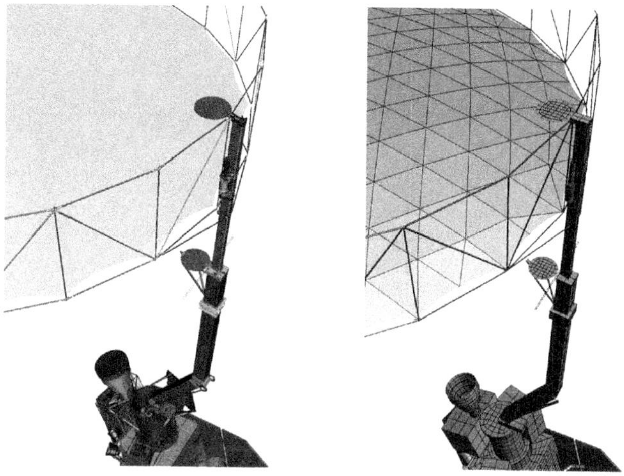

Figure 8.15 Detail of the triangular facets, boom and truss structure, and comparison between mechanical (left) and RF (right) models. In the flight assembly, the boom will be covered in MLI, which is not included in the model. The MLI will conform closely to the boom structure and will have little effect on the radiation patterns. Courtesy of JPL/California Institute of Technology.

While a number of commercial and in-house computer codes were used to simulate the RF performance of the different parts of the complex antenna and spacecraft, the overall RF model was generated using the GRASP9 software package from Ticra. HFSS from ANSYS was used in the design and development of the feed assembly, Ticra's CHAMP software was used to design the corrugated feed horn, and Microwave Innovation Group's WASP-NET was used for the orthomode transducer (OMT). A large number of 'ad hoc' Fortran computer codes were also developed for specific tasks, especially for postprocessing the radiation patterns generated by GRASP9 and for verifying the RF requirements on the different parts of the radiation pattern, such as locating the boresight pointing direction and directivity, calculating the main beam 3 dB beamwidth, calculating the main beam efficiency, identifying peak sidelobe levels in regions of interest, and integrating the sidelobe power in regions of interest.

8.4.4.2 Antenna Radiation Pattern Calculations

The radiation pattern of the entire spacecraft is built step by step by adding to the pattern one major component at a time. The calculation is performed with a hybrid method where the currents on the reflector's surface and the reflector edge diffraction are calculated with physical optics (PO) plus the physical theory of diffraction (PTD), while the currents on other parts of the spacecraft are calculated with a method of moments (MoM) technique. Mutual coupling between the parts in the MoM components are included in the calculation, but since the reflector scattering is calculated with PO, the effects of higher order reflections between the reflector and other objects must be introduced manually by explicitly including all the non-negligible multiple reflections in the calculation. The coordinate system used for the following radiation patterns is depicted in Figure 8.16.

The first component that is considered is the feed horn pattern. As mentioned earlier, the feed horn has been designed with CHAMP, which uses a cylindrical mode matching method to analyze the interior geometry of the horn. The effects of the aperture discontinuity and the exterior geometry of the horn are included in the CHAMP analysis with a MoM calculation that uses the mode matching solution as input [46]. The program

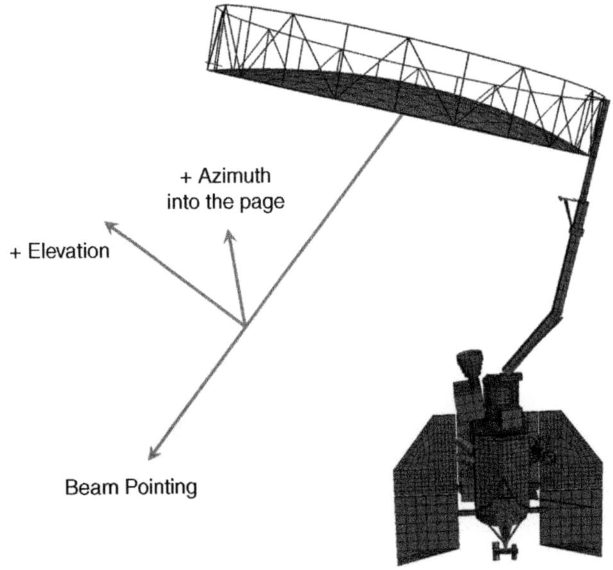

Figure 8.16 Coordinate system for antenna radiation patterns. The antenna beam pointing is nominally at 35.5 degrees from Nadir. Also shown are the positive directions for Elevation and Azimuth from beam pointing.

generates a spherical wave expansion of the radiated field that can be directly imported into GRASP9 as a source. Figure 8.17 shows the pattern of the horn.

The broad main beam has its peak in the location that will be occupied by the center of the reflector and has been designed to produce a controlled taper on the reflector aperture at the different operating frequencies. An optimization procedure was needed to adjust the horn's internal dimensions to obtain the required gain and beamwidths at the different frequencies.

The introduction of the spinning platform assembly (SPA) and the reflector boom into the model generates the pattern shown in Figure 8.18. The scattering of the boom and the SPA are calculated with a MoM procedure and added to the horn pattern. The ripples generated by the boom are clearly visible in the radiation pattern.

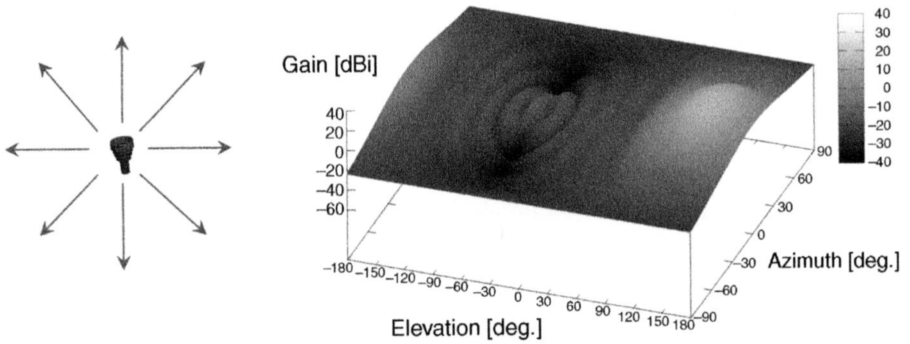

Figure 8.17 Calculated feed horn radiation pattern. The coordinate system is the pattern coordinate system of the 6 m reflector, so that the reflector boresight is aligned to the elevation = 0°, azimuth = 0° point. The feed points to the azimuth = 0°, elevation = 136° point. Courtesy of JPL/California Institute of Technology.

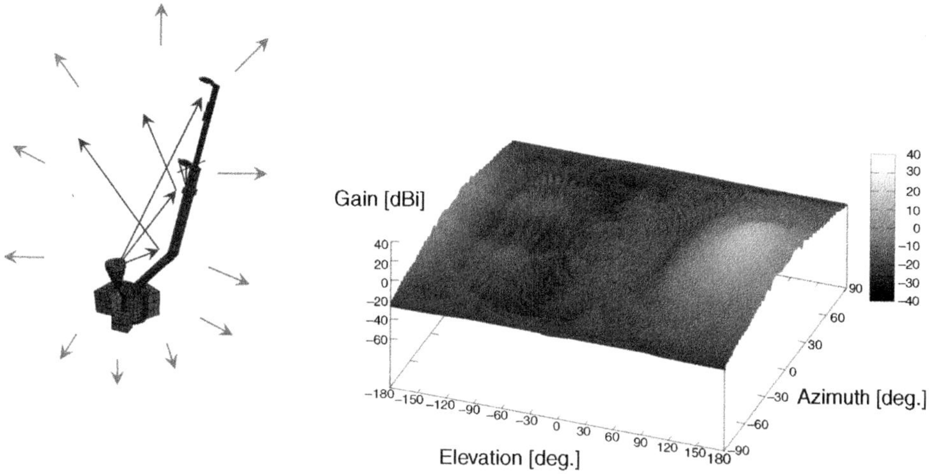

Figure 8.18 Calculated feed horn radiation pattern including the scattering of the boom and SPA. The boom and SPA scattering is calculated using the method-of-moments and then is added to the horn pattern. Courtesy of JPL/California Institute of Technology.

Next, as in Figure 8.19, the lower ring of the truss structure is included in the model. Besides generating more ripples across the entire pattern, the composite ring essentially raises the noise floor level and brings up some low areas of the pattern. Once the reflector is introduced into the model, the pattern starts to look like the final one. The main beam of the reflector is clearly visible in Figure 8.20, with six grating lobes around it generated by the periodicity of the facets on the reflector surface. The area that previously was occupied by the main beam of the horn now contains the reflector's backlobe and its spillover. It is important to note that the currents on the reflector surface are calculated with a PO + PTD procedure and are generated by direct illumination of the feed horn and the scattering coming from the boom and the entire SPA. This accounts for the first reflection of the horn pattern on its surroundings. While this is a secondary effect, it is non-negligible and it affects the pointing of the main beam.

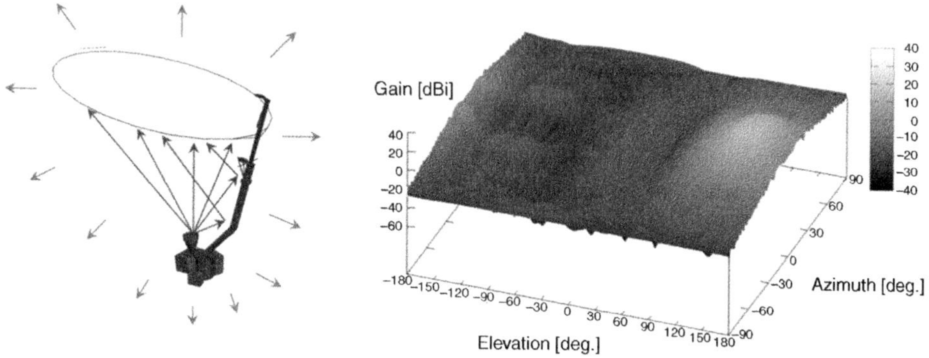

Figure 8.19 Calculated feed horn radiation pattern including the scattering of the boom, SPA, and the lower ring of the truss structure. Addition of the truss structure lower ring raises the overall noise floor on the sidelobes. Courtesy of JPL/California Institute of Technology.

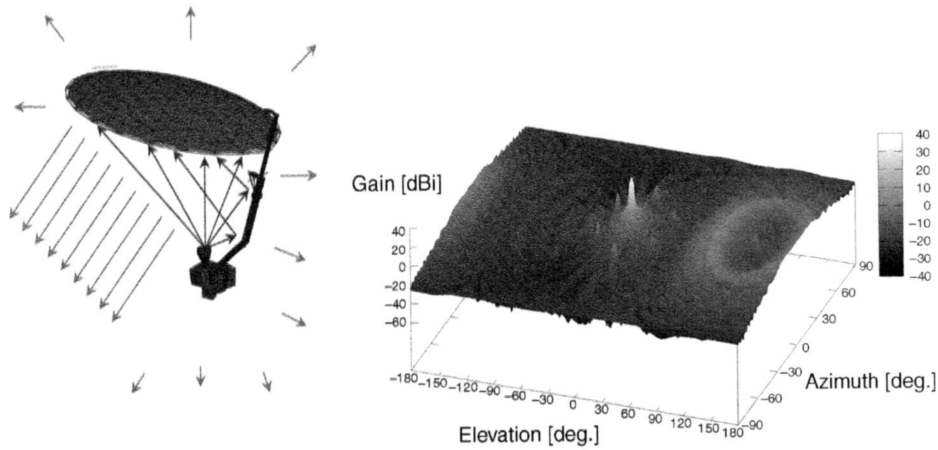

Figure 8.20 *Calculated radiation pattern of the reflector including the scattering of the entire spun side of the spacecraft. Note the six grating lobes all around the main beam generated by the periodicity of the triangular facets on the reflector surface. Courtesy of JPL/California Institute of Technology.*

The last step in this procedure is the introduction of the rest of the spacecraft into the model. The currents generated on the reflector surface at the previous step are now reradiated and the scattering from the spacecraft is calculated. This step, shown in Figure 8.21, is the most time consuming given the large dimension of the entire RF model. Moreover, it is at this stage that the effect of the spinning platform is accounted for and a separate simulation is needed for each position of the spacecraft with respect to the antenna. Figures 8.22 and 8.23 are representative plots of the elevation and azimuth cuts for the final calculation where the results of the complete RF model are compared with a reflector and feed only case.

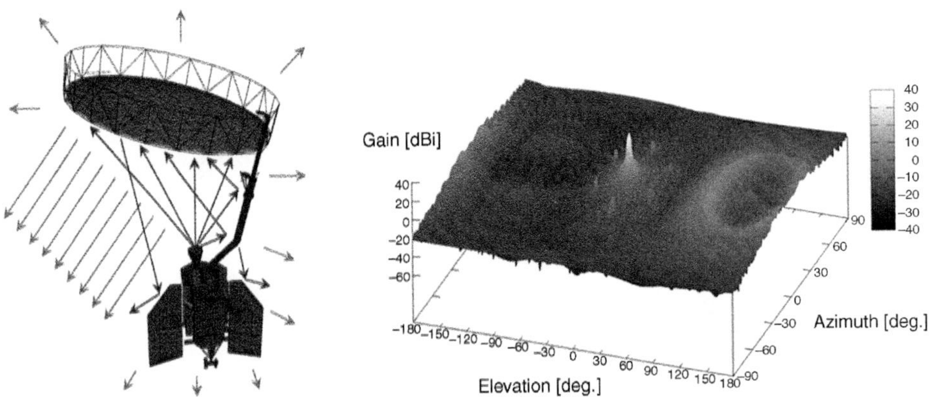

Figure 8.21 *Final calculated radiation pattern including the scattering of the entire spacecraft. The plot shows one representative orientation of the reflector with respect to the spacecraft. Many orientations were simulated to capture the scattering effect of the spacecraft and solar panels. Courtesy of JPL/California Institute of Technology.*

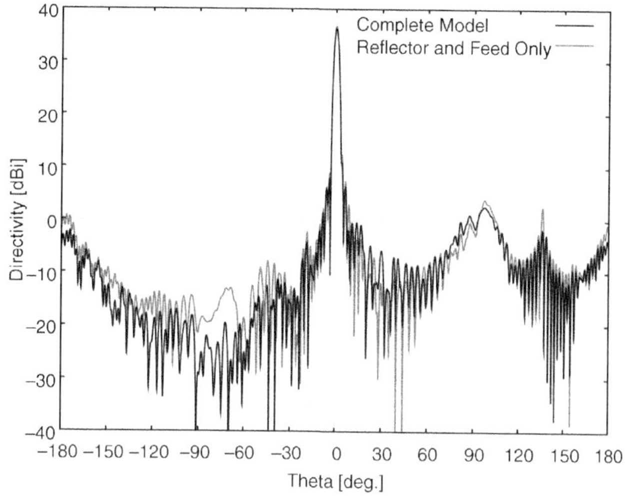

Figure 8.22 *Elevation cut of the radiation pattern at the center frequency of the SAR band. The pattern with all the spacecraft effects included is compared to the pattern of only the feed and the reflector. Courtesy of JPL/California Institute of Technology.*

8.4.4.3 Field of View Study

As the antenna spins above the spacecraft, the antenna patterns change slightly. While the effect of the spacecraft scattering is small in absolute terms (generally about 40 dB below the peak or less), the antenna rotation represents a variable change in the radiation pattern that could compromise the radiometer and radar performance if the variation is large enough and not accurately characterized. The radar is sensitive to changes in the boresight pointing and in the gain stability. The radiometer is concerned with changes in the fractional

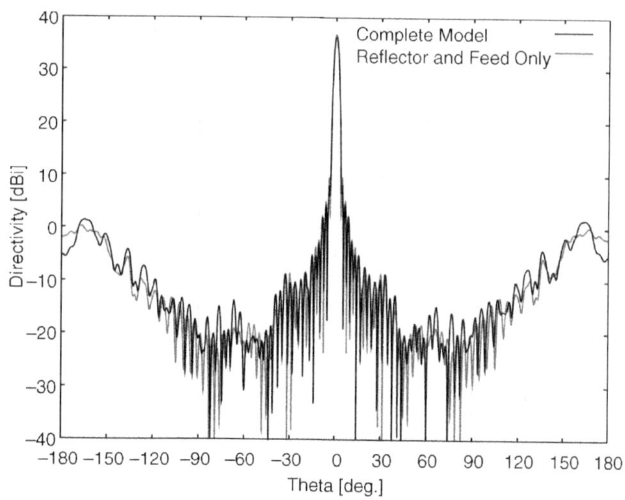

Figure 8.23 *Azimuth cut of the radiation pattern at the center frequency of the SAR band. The pattern with all the spacecraft effects included is compared to the pattern of only the feed and the reflector. Courtesy of JPL/California Institute of Technology.*

power levels in the main beam, the sidelobes looking toward the Sun, the sidelobes looking toward the Earth, and the sidelobes looking into space. Changes in these fractional power levels affect the calibration factors used in the radiometer data processing to account for background levels.

Antenna radiation patterns with the entire spacecraft model, as described in the last step of Section 8.4.4.2, were calculated for every 15° rotation of the antenna with respect to the spacecraft. The largest contributors to the change in the patterns are the solar panels as they go in and out of the antenna's field of view. The patterns are calculated at the low, center, and high frequencies within the SAR and the radiometer operating bands.

Powerful computer resources were needed to accommodate such a complex RF model. Using a 12-core desktop machine with 96 GB of RAM for the calculation, a full set of radiation patterns including the interaction of the antenna with the spinning spacecraft took about three weeks. In the calculation, the maximum size of the MoM patches was one wavelength, although many of the patches resulting from the GRASP9 mesh generator were somewhat smaller. The GRASP9 MoM code utilizes higher order basis functions, which produce accurate solutions for relatively large patches [47].

The conclusion drawn by the SMAP radiometer system engineers from the field of view analysis is that the change in the antenna pattern due to the antenna spinning is on the edge of what the radiometer team can tolerate in their error budget without introducing spin-dependent calibration factors. Changes in the spacecraft design, particularly in the solar panel layout, could force the radiometer team to add the spin dependency to the calibration model.

The spacecraft scattering generates both a bias and a variable component in the main beam pointing of the reflector. The beam pointing bias due to the spacecraft scattering is about 8 millidegrees with respect to the pointing calculated without scattering. The antenna spin creates a ± 1 millidegree pointing variation about that bias. The pointing changes due to the scattering are dependent on the polarization. For instance, the effect on pointing of the scattering of the boom is much more pronounced for V-pol than for H-pol since the V-pol is aligned with the direction of the boom. In the end, the analysis confirmed that the pointing bias and stability components due to scattering are very small factors in the overall pointing error budget.

8.4.4.4 Distortion Modeling

With the calculation of the radiation pattern described above, an accurate prediction of where the antenna is pointed in its nominal configuration is obtained for various positions of the antenna with respect to the spacecraft. Thermal and dynamic distortions of the reflector and boom will introduce more pointing variations.

For antenna pattern distortions, the boom and the reflector surface are the most important components affected by thermal variations. The reflector mesh temperature can vary from -115 to $+280\,°\mathrm{C}$ depending on orbit, season and eclipses. The reflector truss structure and boom temperatures will vary less than the mesh temperature since they have more thermal mass, but the temperature swings will still be on the order of 200 °C. Even though the boom will be covered with thermal blankets, this large temperature range will make it expand and contract and therefore the reflector will move with respect to the feed horn. To assess the effect of these thermal variations on pointing, Northrop Grumman calculated the position of each nodal point in the reflector surface at one-minute intervals along an entire orbit. The extreme cases were then introduced into the electromagnetic model to simulate the antenna pointing in those conditions. The resulting pointing error was less than 10 millidegrees, and changes to the main beam and sidelobe levels were insignificant.

Another distortion effect that was examined was due to dynamic disturbances. While the spacecraft travels on a nominal orbit its unspun portion (the lower part) is kept at the proper attitude with the help of reaction wheels that counterbalance the force induced by the spinning instrument. Vibrations generated by the reaction wheels will propagate across the boom and reflector inducing relatively high-frequency vibrations on the reflector surface. These oscillations can be represented in terms of modes across the reflector surface. Figure 8.24 illustrates the six lowest order modes. The expected mode amplitudes were predicted by the SMAP

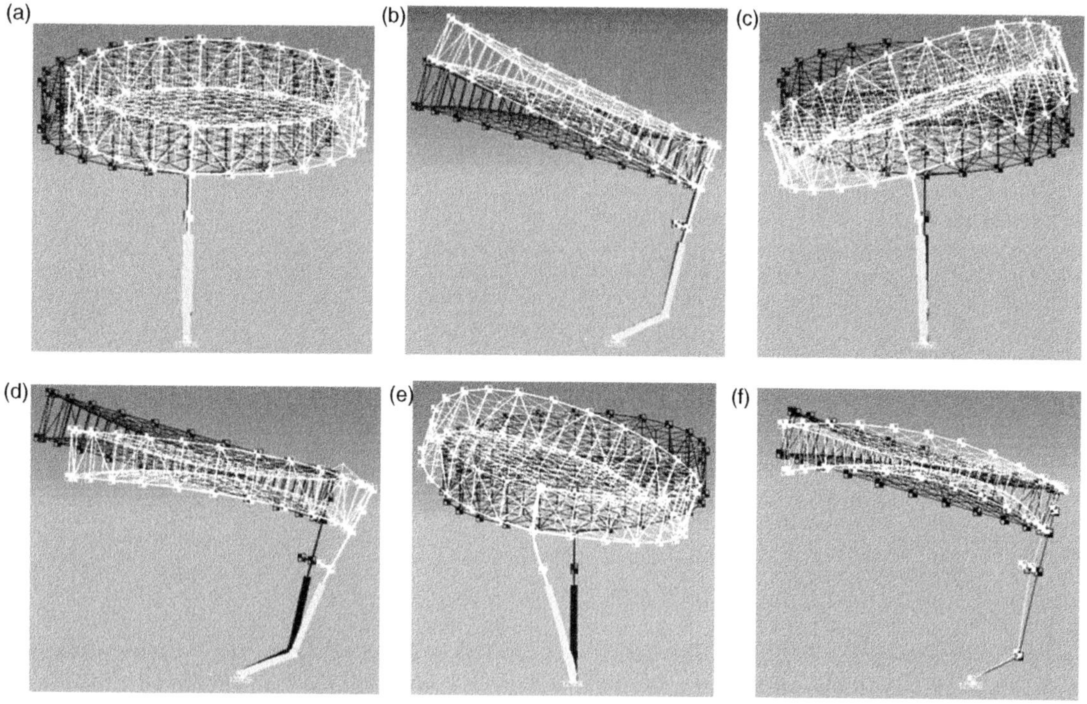

Figure 8.24 *Dynamic modal distortions of the reflector antenna and boom. The amplitudes of the distortions are greatly exaggerated in the figures so the effects can be visualized. The first six modes are shown here; the effects of the 25 lowest frequency modes were analyzed. As can be observed from the figures, some of the modes are simple rigid body translations of the reflector, while others result in reflector surface distortions.*

dynamics engineers and used to generate perturbed variations of the nominal reflector surface. The perturbed reflector surfaces were analyzed in the GRASP9 RF model. The changes were insignificant, resulting in a maximum pointing error of ± 1 millidegree.

Other dynamic forces on the reflector are lower in frequency and generate rigid body movements of the reflector. For instance, a change in the reflector spin rate causes the reflector to pivot on the prime batten. Other effects are mass imbalance caused by asymmetries in the reflector's structure, and oscillations generated by the torque variation of the spin motor. These low-frequency effects do not change the reflector surface shape but they change the reflector's position. In order to map the movement of either the reflector or the feed into a pointing error, a set of rigid body motion simulations was completed by rotating and translating the feed and the reflector along each axis of the pertinent reference systems. This way, once the amplitude of a rigid body movement is known, the resulting pointing error can be quickly determined.

8.4.4.5 Structural Emission Calculation

In order to properly calibrate the amount of power received in orbit, the radiometer team has to accurately calculate sources of noise and RFI that couple into the instrument. One of these is the structural emission of the entire spacecraft, that is, the natural RF emission of the spacecraft structure in L-band. Given a certain part of the spacecraft made of a certain material, its structural emission essentially depends on the specific emissivity of that material, on its physical temperature, and on the fractional amount of the antenna radiation

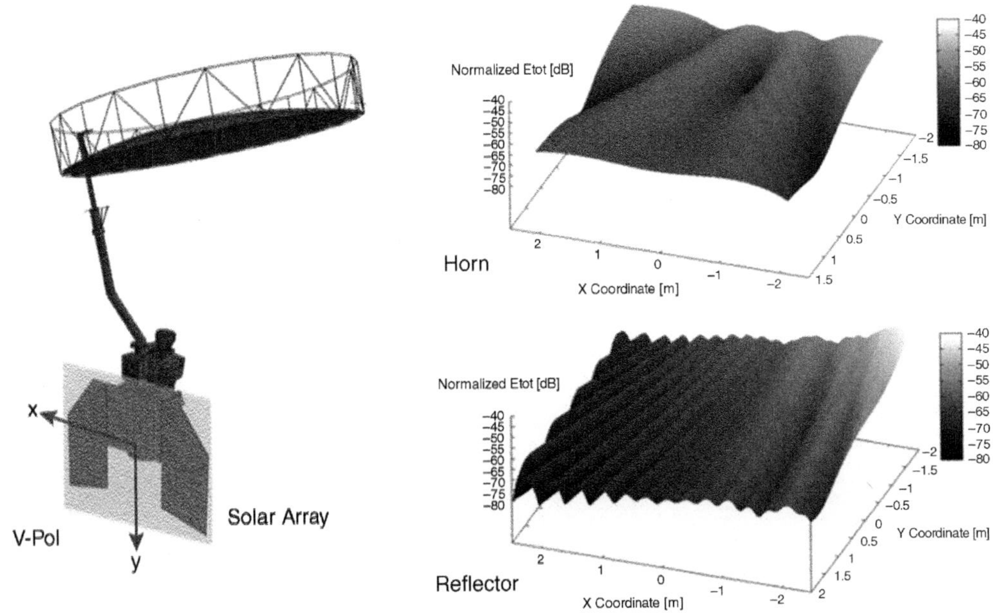

Figure 8.25 Example of calculated total near electric field for the horn (top right) and the reflector (bottom right) in the area roughly occupied by the solar array (bottom left). Note how the reflector's field intensity grows stronger in proximity of the top right corner of the solar panels (coordinates (−2:−2)). Courtesy of JPL/California Institute of Technology.

pattern incident on it. Therefore, the near field incident on every part of the spacecraft needed to be calculated. For each geometric surface of the spacecraft structure the incident near field of both the feed horn and the reflector were calculated. The incident field was then integrated over the pertinent area to give an estimate of the antenna radiation pattern levels impinging on that specific part. The same procedure was then carried out for both polarizations since the orientation of many spacecraft parts made them interact more with one polarization rather than with the other. For instance, the boom interacts much more with the V-pol which is aligned in the same direction rather than the H-pol where the field is essentially orthogonal to the boom structure. Another interesting part is the solar array. Since the antenna spins and the solar array does not, depending on which point in time we look at it, the panels interact more with one polarization than with the other, changing continuously between one and the other while the antenna spins. Especially in the case of the solar panels it is also interesting to look at the difference between the near field generated by the horn and that generated by the reflector.

With reference to Figure 8.25, the relative position of the solar array with respect to the feed horn and to the reflector can be observed. The near field generated directly by the feed horn and by the reflector respectively are plotted on the right side of the figure. Both the effect of the different polarizations and the intrusion of the solar panels into the field of view of the reflector can be observed in these plots.

8.4.5 Feed Assembly Design

The design of the feed assembly is dependent primarily on the focal length and diameter of the reflector, and is not influenced by the fact that it is illuminating a mesh reflector instead of a solid reflector surface. The key drivers for the SMAP feed design are:

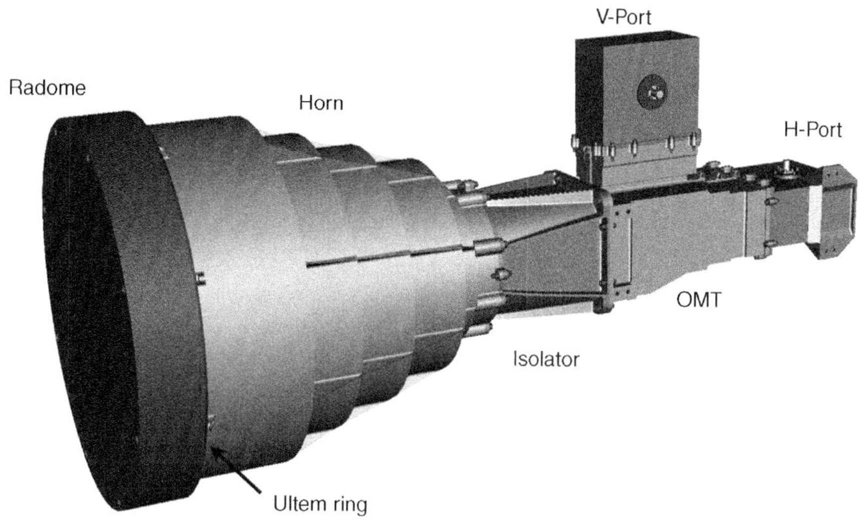

Figure 8.26 Mechanical CAD model of the feed assembly. A circularly corrugated horn is protected from direct sun exposure by a low-loss foam radome mounted on an Ultem ring. The horn connects to a titanium thermal isolator that transitions from the circular horn port to the nearly square port of the orthomode transducer. The orthomode transducer's two linear polarization ports are of standard WR-650 waveguide size.

- A 15.9% bandwidth.
- Dual, simultaneous linear polarization.
- Limited accommodation volume, requiring a compact design.
- A −20 dB return loss requirement.
- A 0.1 dB insertion loss requirement, excluding the return loss.
- A 35.55 dBi gain requirement for the reflector antenna pattern.
- Requirements of a maximum beamwidth of 2.8° for the SAR, a minimum beamwidth of 2.29° for the radiometer, and a maximum beamwidth of 2.5° for the radiometer for the reflector antenna pattern.
- A radiometer main beam efficiency of 87% for the reflector antenna pattern.
- Maintain thermal stability by blocking direct Sun from entering the horn aperture.

The volume available combined with the low insertion loss requirement led to the selection of a feed design concept similar to the feeds used on the Aquarius mission [48]. However, SMAP's larger bandwidth and different illumination requirements needed new designs for all the components. The feed assembly is illustrated in Figure 8.26.

The horn was designed and optimized with Ticra's CHAMP software, a waveguide mode matching code that includes a MoM technique to account for the effect of the horn exterior in the radiation pattern. The optimization parameters used for CHAMP were the return loss, and the desired horn pattern beamwidths in both the E- and H-planes that were needed to get the proper edge taper illumination on the reflector.

The OMT and thermal isolator combination was designed and optimized using WASP-NET. The optimization parameters were the return loss and limiting the excitation of higher order modes. Although functionally a single part from a RF design point of view, the OMT was divided into an aluminum part and a titanium part, referred to as the 'OMT' and the 'thermal isolator,' respectively. The thermal isolator reduces the transfer of heat between the horn and the rest of the feed assembly, which stabilizes the OMT and the waveguide to coaxial adapter temperatures. Since most of the losses of the feed assembly are in those parts, the

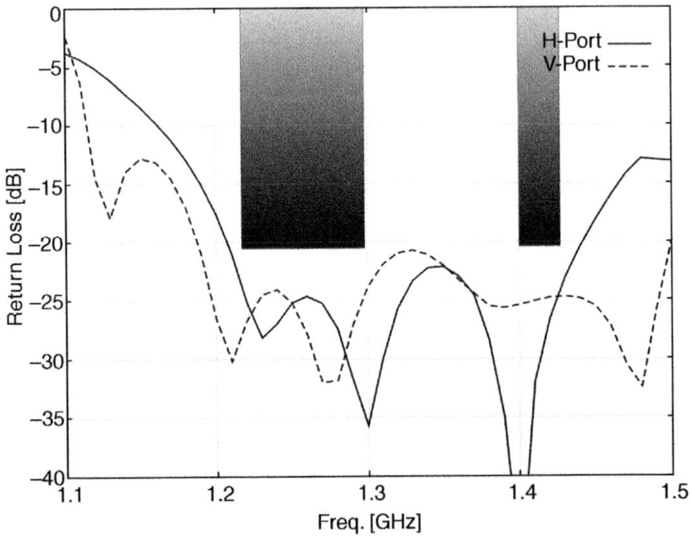

Figure 8.27 *Return loss of the feed assembly at the input coaxial ports. The assembly meets the return loss requirement of −20.5 dB maximum for the radar band and −20.4 dB for the radiometer band.*

thermal stability reduces the radiometer calibration error. The waveguide to coaxial adapters on the OMT ports are of standard WR-650 waveguide size. The adapter design was tuned with the rest of the feed components to achieve the required −20 dB return loss for the entire feed assembly.

All the feed components, including the radome, were analyzed as a single assembly using ANSYS HFSS. The return loss at the two coaxial input ports is plotted in Figure 8.27.

8.4.6 Performance Verification

Antenna pattern measurements of the SMAP reflector antenna will not be done due to the high cost, so the success of the mission greatly depends on the accuracy of the antenna RF models. In order to validate the GRASP9 model, an approximately 10th-scale RF model of the spacecraft will be built and measured on an indoor near-field range. The scale model will not be identical to the SMAP flight design in that the reflector will be solid, and small features that negligibly influence the antenna pattern will be omitted. The RF model will be simplified to match the as-built scale model so that the measured versus calculated data sets can be compared and reconciled.

Feed assembly antenna patterns will be performed, and the resulting measurement data will be used as the feed pattern in the GRASP9 model. The losses of the feed assembly without the horn and radome will be measured over temperature. The horn losses will also be calculated. The complex permittivity of the flight lot radome material will be measured and used to calculate the radome loss.

For the reflector, all distortion effects including thermal, spin dynamics, and gravity offloading, are determined solely by analysis. These effects are backed out from the desired on-orbit antenna geometry to generate the design for manufacturing. Photogrammetry will be performed on the reflector in a 1g environment to ensure that the surface was built to the as-designed values.

Maintaining the integrity of the mesh material during the ground operation phases is critical to the reflector performance. In one of the TDRS mesh reflectors, poor electrical contact at the mesh wire interfaces caused a significant loss in reflectivity [32]. To ensure that SMAP does not encounter similar problems, witness coupons of the flight mesh material will accompany the flight reflector through all ground operations. The coupons will consist of the mesh mounted in waveguide frames, which will facilitate quick *S*-parameter measurements in the field using a vector network analyzer. The coupons will be measured at several predetermined steps right up until launch of the spacecraft.

8.5 Conclusion

In this chapter, a short history and overview of deployable mesh reflectors was provided, followed by a detailed case study of a novel and complex mesh reflector antenna design for NASA's SMAP mission. The design process and technical challenges for the SMAP antenna have been presented to assist others in identifying the pertinent characteristics of deployable mesh reflectors as they apply to their own specific applications.

Acknowledgments

The SMAP development effort described in this chapter was carried out at the Jet Propulsion Laboratory, California Institute of Technology, under a contract with the National Aeronautics and Space Administration. The authors wish to acknowledge and thank several of their colleagues who contributed to the SMAP antenna design and development: Jacqueline C. Chen for mesh performance modeling; the mesh material testing team at Goddard Space Flight Center (GSFC) composed of Bryan Monosmith, Shannon Rodriguez, Manuel Vega, Zhaonan Zhang; Behrouz Khayatian and Daniel J. Hoppe for helping in the design of the feed assembly; Matthew D. Stegman and Frank E. Ramirez for the mechanical design of the feed assembly and generation of countless mechanical CAD models; Michael W. Spencer, Douglas E. Dawson, Eric Slimko, Derek Hudson (GSFC), and Samuel F. Chan for requirements development and system engineering oversight; Lisa A. Sievers and Robert L. Stephenson for generating dynamic distortion models and the antenna modal distortions in Figure 8.24; Ira Katz for performing the ESD analysis of the web materials; Yuki Salinas for managing the reflector antenna contract; Mark W. Thomson for valuable discussions about deployable antennas; Witold M. Sokolowski for materials information and research; Eug-Yun Kwack, A.J. Mastropietro, and Rebecca L. Mikhaylov for thermal engineering support; and Richard E. Hodges, William A. Imbriale, Joseph D. Vacchione, Shannon T. Brown, Edward J. Wollack (GSFC), Edward J. Kim (GSFC), Larry W. Epp, Paul B. Willis, David Staelin (MIT), and several others already mentioned for supporting many design reviews. We would also like to acknowledge and thank the SMAP AstroMesh team at Northrop Grumman Astro Aerospace.

References

1. Roederer, A.G. and Rahmat-Samii, Y. (1989) Unfurlable satellite antennas: a review. *Annals of Telecommunications*, **44**(9–10), 475–488.
2. Davis, G.L. and Tanimoto, R.L. (2006) Mechanical development of antenna systems, in *Spaceborne Antennas for Planetary Exploration* (ed. W.I. Imbriale), John Wiley & Sons, Inc, Hoboken, NJ, pp. 425–454.
3. Tibert, G. (2002) Deployable Tensegrity Structures for Space Applications, Doctoral thesis, Royal Institute of Technology, Stockholm.
4. Rahmat-Samii, Y. (2007) Reflector antennas, in *Antenna Engineering Handbook* (ed. J. Volakis), McGraw-Hill, New York, pp. 15-1–15-63.
5. NASA (2011) National Space Science Data Center. [Online] http://nssdc.gsfc.nasa.gov/nmc/spacecraftDisplay.do?id=1974-039A (accessed 6 December 2011).
6. Love, A.W. (1976) Some highlights in reflector antenna development. *Radio Science*, **11**(8,9), 671–684.

7. Redisch, W.N. (1975) ATS-6 description and performance. *IEEE Transactions on Aerospace and Electronic Systems*, **AES-11**(6), 994–1003.
8. Corrigan, J.P. (1975) ATS-6 experiment summary. *IEEE Transactions on Aerospace and Electronic Systems*, **AES-11** (6), 1004–1014.
9. NASA (1989) NASA Press Release 89–145.
10. NASA (n.d.) Galileo Mission to Jupiter. [Online] http://solarsystem.nasa.gov/galileo/docs/Galileo_Fact_Sheet.pdf (accessed 6 December 2011).
11. Harris Corporation (2000) Harris Corporation Press Releases. [Online] http://www.harris.com/view_pressrelease.asp?act=lookup&pr_id=44 (accessed 6 December 2011).
12. Harris Corporation (2008) Harris Corporation Press Releases. [Online] http://www.harris.com/view_pressrelease.asp?act=lookup&pr_id=2436 (accessed 6 December 2011).
13. Harris Corporation (2009) Harris Corporation Press Releases. [Online] http://www.harris.com/view_pressrelease.asp?act=lookup&pr_id=2809 (accessed 6 December 2011).
14. Campbell, G., Bailey, M.C., and Belvin, W.K. (January 1988) The development of the 15-meter hoop column deployable antenna system with final structural and electromagnetic performance results. *Acta Astronautica*, **17**(1), 69–77.
15. Sullivan, M.R. (1982) LSST (Hoop/Column) Maypole Antenna Development Program, Harris Corporation/Langley Research Center, NASA Contractor Report 3558.
16. de Sekling, P.B. (2010) Space News International. [Online] http://www.spacenews.com/satellite_telecom/101210-skyterra-antenna-deployed.html (accessed 6 December 2011).
17. Japan Aerospace Exploration Agency (2008) Institute of Space and Astronautical Science. [Online] http://www.isas.ac.jp/e/enterp/missions/halca/index.shtml (accessed 6 December 2011).
18. Hanayama, E., Kuroda, S., Takano, T., Kobayashi, H., and Kawaguchi, N. (2004) Characteristics of the large deployable antenna on HALCA satellite in orbit. *IEEE Transactions on Antennas and Propagation*, **52**(7), 1777–1782.
19. Northrop Grumman (2000) Northrop Grumman Astro Aerospace News Releases. [Online] http://www.as.northrop-grumman.com/businessventures/astroaerospace/news_releases/index.html (accessed 6 December 2011).
20. Northrop Grumman (2004) Northrop Grumman. [Online] http://www.as.northropgrumman.com/products/aa_thuraya/assets/DS-409-AstroMeshReflector.pdf (accessed 6 December 2011).
21. Northop Grumman (2003) Northrop Grumman Astro Aerospace News Releases. [Online] http://www.as.northrop grumman.com/businessventures/astroaerospace/news_releases/assets/news2003-0818.pdf (accessed 6 December 2011).
22. NASA (2011) NASA National Space Science Data Center. [Online] http://nssdc.gsfc.nasa.gov/nmc/spacecraft Display.do?id=2008-001A (accessed 6 December 2011).
23. Northrop Grumman (2004) Northrop Grumman Astro Aerospace News Releases. [Online] http://www.as.northrop grumman.com/businessventures/astroaerospace/news_releases/assets/news2004-0329.pdf (accessed 6 December 2011).
24. Northrop Grumman (2008) Northrop Grumman News and Events. [Online] http://investor.northropgrumman.com/phoenix.zhtml?c=112386&p=irol-newsArticle&ID=1208788&highlight= (accessed 6 December 2011).
25. Thomson, M.W. (1999) The AstroMesh deployable reflector. IEEE Antennas and Propagation Society International Symposium, vol. 3, Orlando, FL, pp. 1516–1519.
26. Meguro, A., Shintate, K., Usui, M., and Tsujihata, A. (2009) In-orbit deployment characteristics of large deployable antenna reflector onboard Engineering Test Satellite VIII. *Acta Astronautica*, **65**(9–10), 1306–1316.
27. Ruze, J. (1966) Antenna tolerance theory—a review. *Proceedings of the IEEE*, **54**(4), 633–640.
28. Bahadori, K. and Rahmat-Samii, Y. (2005) Characterization of effects of periodic and aperiodic surface distortions on membrane reflector antennas. *IEEE Transactions on Antennas and Propagation*, **53**(9), 2782–2791.
29. Corkish, R.P. (1990) A survey of the effects of reflector surface distortions on sidelobe levels. *IEEE Antennas and Propagation Magazine*, **32**(6), 6–11.
30. Rusch, W. and Wanselow, R. (1982) Boresight-gain loss and gore-related sidelobes of an umbrella reflector. *IEEE Transactions on Antennas and Propagation*, **30**(1), 153–157.
31. Lawrence, R.W. and Campbell, T.G. (2000) Radiometric characterization of mesh reflector material for deployable real aperture remote sensing applications. IEEE International Geoscience and Remote Sensing Symposium Proceedings, vol. 6, Honolulu, HI, pp. 2724–2726.

32. Imbriale, W.A., Galindo-Israel, V., and Rahmat-Samii, Y. (1991) On the reflectivity of complex mesh surfaces. *IEEE Transactions on Antennas and Propagation*, **39**(9), 1352–1365.
33. Miura, A. and Rahmat-Samii, Y. (2007) Spaceborne mesh reflector antennas with complex weaves: extended PO/periodic-MoM analysis. *IEEE Transactions on Antennas and Propagation*, **55**(4), 1022–1029.
34. Miura, A. and Tanaka, M. (2003) An experimental study of electrical characteristics of mesh reflecting surface for communication satellite antenna. IEEE Topical Conference on Wireless Communication Technology, Honolulu, HI, pp. 218–219.
35. Rahmat-Samii, Y. and Lee, S.-W. (1985) Vector diffraction analysis of reflector antennas with mesh surfaces. *IEEE Transactions on Antennas and Propagation*, **33**(1), 76–90.
36. Miura, A. and Tanaka, M. (2004) A mesh reflecting surface with electrical characteristics independent on direction of electric field of incident wave. IEEE Antennas and Propagation Society International Symposium, vol. 1, Monterey, CA, pp. 33–36.
37. Aspden, P.L. and Anderson, A.P. (1992) Identification of passive intermodulation product generation on microwave reflecting surfaces. *IEE Proceedings H Microwaves, Antennas and Propagation*, **139**(4), 337–342.
38. Lubrano, V., Mizzoni, R., Silvestrucci, F., and Raboso, D. (2003) PIM characteristics of the large deployable reflector antenna mesh. 4th International Workshop on Multipactor, Corona and Passive Intermodulation in Space RF Hardware, Noordwijk, The Netherlands, http://esamultimedia.esa.int/conferences/03C26/index.html (accessed 6 December 2011).
39. Wade, W.D. (1990) Development of low PIM, zero CTE mesh for deployable communications antennas. Military Communications (MILCOM) Conference Record, vol. 3, Montery, CA, pp. 1175–1178.
40. National Research Council (2007) *Earth science and applications from space: National imperatives for the next decade and beyond*, The National Academies Press, Washington, DC.
41. Entekhabi, D., Njoku, E.G., O'Neill, P.E. *et al.* (2010) The soil moisture active passive (SMAP) mission. *Proceedings of the IEEE*, **98**(5), 704–716.
42. Jet Propulsion Laboratory (n.d.) SMAP. [Online] http://smap.jpl.nasa.gov/ (accessed 6 December 2011).
43. Ticra (2011) *GRASP Reference Manual*, Ticra, Copenhagen.
44. Ticra (2005) *GRASP9 Technical Description* (ed. K. Pontoppidan), Ticra, Copenhagen.
45. Njoku, E. *et al.* (2001) Spaceborne Microwave Instrument for High Resolution Remote Sensing of the Earth's Surface Using a Large Aperture Mesh Antenna, Jet Propulsion Laboratory, Pasadena, CA.
46. Ticra (2011) *CHAMP Users Manual*, Ticra, Copenhagen.
47. Ticra (n.d.) GRASP MoM Add-on, Ticra, Copenhagen.
48. Le Vine, D.M. Lagerloef, G.S.E. Colomb, F.R. Yueh, S.H. and Pellerano, F.A. Aquarius: An Instrument to Monitor Sea Surface Salinity From Space. *IEEE Transactions on Geoscience and Remote Sensing*, **45**(7), 2040–2050, July 2007.

9

Microstrip Array Technologies for Space Applications

Antonio Montesano, Luis F. de la Fuente, Fernando Monjas, Vicente García, Luis E. Cuesta, Jennifer Campuzano, Ana Trastoy, Miguel Bustamante, Francisco Casares, Eduardo Alonso, David Álvarez, Silvia Arenas, José Luis Serrano, and Margarita Naranjo

EADS CASA Espacio, Spain

9.1 Introduction

The market for space antennas, in particular in communication applications, is quite conservative and historically dominated by reflector antennas (using travelling wave tubes (TWT)s) whose performance, design methodology, processes and technology have been optimized during the last 24 years to meet really high standards (e.g. shaped multiple feed per beam (MFPB) or single feed per beam (SFPB). However, there are specific driving requirements where antenna arrays are the only viable solution. Active antenna arrays are the dominant solution for newcomer requirements such as pattern flexibility (e.g. power and frequency), electronic beam steering capability, graceful degradation or multibeam capability. Although space antenna arrays are developed for a broad range of applications from communications to Earth observation or scientific missions, there is a common technological background behind such diversified requirements and missions.

Antenna development for space applications involves a wide variety of disciplines due to the difficult thermal and mechanical environment and to the reliability and testing requirements. In the case of array solutions, the development also involves electronics design, components selection and qualification, thermal control of active components, electromagnetic compatibility (EMC), electrical ground support equipment (EGSE), and so on. For these reasons, the development of antenna arrays for space applications requires multidisciplinary expertise along with teaming and coordination skills. The technology described in this chapter embraces the scientific and technical background of the authors, who have worked on several space array projects for the last two decades. Most of the examples proposed in this subject are based on radiating patches photo printed over suspended substrate structures in a single or multilayer configuration.

Space Antenna Handbook, First Edition. Edited by William A. Imbriale, Steven (Shichang) Gao and Luigi Boccia.

Furthermore, some examples describe applications of composite materials, mainly realized using aluminium honeycomb sandwiches.

9.2 Basics of Array Antennas

9.2.1 Functional (Driving) Requirements and Array Design Solutions

9.2.1.1 Passive Arrays

Microstrip array antennas are often used in space applications mainly due to their low profile and low-weight features, flexibility of design and RF performance. They benefit from the well-known and repetitive PCB manufacturing processes, and are very attractive for series production and large array antennas with a large number of constructing units (or sub-arrays) such as antennas for SAR (Synthetic Aperture Radar) applications.

Microstrip patches (see §1.3.5) may come in a great variety of shapes and designs but they are inherently highly resonant elements (the typical bandwidth of a microstrip patch is around 3%), able to operate in dual linear or circular polarization with medium cross-polar performance (around 20–25 dB), with a single-element gain around 8 dBi and medium power handling capabilities (tens of watts up to several hundred watts at sub-array or antenna level). Depending on the mission-specific requirements, several techniques may be used to enhance the typical microstrip antenna RF performance:

1. *Frequency band:* For missions with frequency bands ranging from L-, S-, X- to even Ku-bands, printed circuit technology is well suited. For much lower frequency applications in UHF and VHF, integrated solutions based on sandwich panels are typically selected. In the case of Ka-band, the use of high-precision metallic or ceramic technologies may be required.
2. *Bandwidth:* Wider bandwidth patches can be achieved in different ways. Configurations based on stacked arrangements or with aperture-coupled feeds or fed through electromagnetic coupling, capacitive loading or meandering probes can provide bandwidths up to 20 or 30 %. The array bandwidth is also related to the beam-forming network (BFN) configuration. Simple serial-fed elements can be used for narrow-band applications while corporate feed networks may be required for broadband applications. When size constrains apply, a mixed solution may be required.
3. *Polarization purity:* Depending on the cross-polar (XP) requirements, the typical polarization purity of a patch element can be enhanced at the array level using sequential rotation of the unit elements or sequential feeding techniques. In general, single-feed techniques are normally used for standard polarization purity requirements while configurations based on quadruple feeding may be necessary for highly demanding applications.
4. *Losses:* Losses are mainly determined by the materials used in the radiating element and in the BFN which is typically implemented in microstrip, stripline, suspended substrate stripline (SSSL) or square coaxial technology. Although the losses of these types of transmission lines are typically higher than those of waveguides, designs based on low-loss dielectrics can provide comparable loss performances, at least for low- and medium-size arrays. For large or very large antennas, it is difficult to achieve with microstrip technology same losses of reflector antennas.
5. *Pencil/shaped beams:* Missions requiring antennas with high efficiency allow the use of simple BFNs based on reactive T-shape dividers with fully populated arrays. Typically, such missions provide pencil beams with uniformly illuminated apertures or small taper for SLL (Side Lobe Level) reduction. For this reason, they are robust against errors in the BFN whether they come from internal or external coupling, or feed network inaccuracies. Furthermore, this kind of network can easily handle high RF power. On the other hand, missions requiring shaped beams (including isoflux corrected beams) need sophisticated feed

excitation with amplitude feed coefficient levels of –18 dB and below. This high sensitivity requires BFNs with a tight control of the amplitude and phase errors and a careful study of the radiating element to control external coupling. The design of BFNs with high isolation requires the use of hybrids or of Wilkinson dividers and dissipative devices. To handle high power, sophisticated and complex solutions might be required that employ external loads or heat sinks.

6. *Phase and group delay:* Navigation missions require control of the phase pattern to less than 5° and group delays no higher than 0.11 ns within the whole antenna coverage. Such demanding requirements can be handled by: (i) specific optimization of these parameters during implementation of the BFN; and (ii) careful selection of the radiating elements using the techniques described in the previous paragraph. Sequential rotation may be required to improve the rotational symmetry of the pattern.
7. *Power handling:* Due to high power levels, multipaction, passive intermodulation products (PIMP)s and heat dissipation may impose strong limits on the design and on the technology that can be used.
   Multipaction must be studied considering all the gaps susceptible to provide discharges in the BFN and in the radiator [1]. Local analysis must be performed to compute the field voltages and to verify adequate margins over the susceptibility levels. In case of multicarrier operation, voltage addition of the signals must be considered while more sophisticated techniques should be applied for modulated signals. Specific tests may be done to verify the design and, in some cases, local design modifications can be required.
   In case of multicarrier operation and susceptible Rx chains, *passive intermodulation products* (PIMP)s has to be calculated. If some of the intermodulation products fall in the susceptible Rx frequency band, provision must be made during both the design and it can also involve PIMP sources. This can strongly impact the design and it can also involve PIMP sources in the mechanical structure and in thermal hardware.
   Heat dissipation may influence the material selection, boosting the employment of metallic structures for RF circuitry or for radiating elements. In case of designs including loads (internal or externals), heat sinks or other more sophisticated solutions may be considered (heat pipes, loop heat pipes or fluid loops) to allow heat dissipation.
   To avoid possible *corona* discharges, proper venting of any cavity is required. In some specific applications such as Mars landing missions, antennas operating under low atmospheric pressure are required, thus entailing the study and testing of the corona effect in any susceptible gap.
8. *Thermal requirements:* The thermal environment and the extreme temperatures could induce thermo-elastic distortions and thermal stress. Typically, antenna arrays are quite robust against thermoelastic distortions. This is a great advantage compared with the high sensitivity of reflector pointing. Stress release should be included in the design, which involves isostatic mounting devices, flexible materials, stress-release solder joins or special precautions to avoid force concentration points in the circuits.

9.2.1.2 Active Arrays

Current and future space applications require both transmitting and receiving antennas with high flexibility and robust reconfigurability features [2]. Typically, such requirements are fulfilled by means of active antenna arrays whose main features can be summarized as follows:

1. *Steering capability:* One of the main features of an active array is the ability, by means of phase control, to scan the antenna beam electronically in specific directions, avoiding the use of mechanical controls and actuators. In some missions, such as those based on beam hopping or optical missions (e.g. telescopes or optical Earth observation missions), speed and steering capability are key design drivers.
2. *Reconfiguration capability:* Active arrays also can dynamically or statically reconfigure the defined coverage, allowing, for example, customer reallocation as the market evolves. Reconfiguration capabilities include *tracking* of mobile users, *nulling* to suppress or mitigate interferences, standardization or *coverage redefinition* of payload after launch.

3. *Multibeam capability:* On many occasions, more than a single beam is simultaneously required with a single active antenna. The technology to be employed for the BFN depends on the number of simultaneous beams to be handled by the array. A conventional BFN, based on dividers and digital attenuators and phase shifters can normally handle no more than a dozen beams. In general, the multibeam BFN complexity increases with the number of beams, leading to complex architectures, high numbers of RF connections and digital controls, and large and heavy structures. Therefore, for arrays with a high number of beams, solutions based on lenses, the Butler matrix, optical BFNs (OBFNs) and miniaturized BFNs are required.
4. *Power and bandwidth flexibility:* An emerging feature of the active array used for communication applications is its ability to share power and bandwidth between the different channels regardless of their beam configuration. This can be done dynamically as a function of the traffic in the system, allowing for optimum use of the resources.
5. *EIRP:* Typical EIRP (Equivalent Isotropically Radiated Power) requirements lead to an active array able to provide high directivity and RF power. However, high demanding EIRP requirements are not be easy to realize when size and mass constraints are taken into account. As a consequence, spacecraft EIRP should be increased by maximizing the efficiency of both the antenna aperture and the amplifiers and by reducing the losses of the active chain without increasing the size of the radiating aperture. The advantage of an active array in terms of power handling is that the total power is distributed across the array's unit cells. Therefore, the power handled by each element is small. As an additional benefit, the failure of a few elements is not critical as array performance degrades gracefully.
6. *G/T:* Obtaining high values of the G/T figure of merit for the receiving antenna leads to a design with high directivity and low losses and a minimum noise figure (NF) in the first amplification stage. But typically the G/T requirements have to be evaluated at the interface between the antenna and the spacecraft, including the NF of the receiver chain. In such a common case, extra amplification in the BFN is included to minimize the impact of the NF on the transponder chain. Nevertheless, since there are limitations on the maximum output power of the antenna, some non-negligible degradation of performance is unavoidable.
7. *Optimum aperture efficiency and scanning losses:* In order to maximize the aperture efficiency while minimizing scanning and grating lobe losses, the size of the array cell must be small in terms of wavelengths, being the required size limited by the desired scanning angle. However, as the size of the array cell is reduced, the number of elements required to cover a given aperture increases, as does the complexity of the control system while reducing the available space for the Rx or Tx modules (e.g. filters, amplifiers, dividers, electronics, etc.). Nevertheless, it is desirable to achieve a compact, integrated and modular design for the whole array by pursuing the simplest design for each of these subassemblies.

9.2.2 Materials for Passive Arrays Versus Environmental and Design Requirements

It is desirable to limit the selection of materials for space arrays to space-qualified ones (see §4.3), preferably with demonstrated heritage. In this subsection, we will provide a review of the most common materials employed for lightweight passive antenna arrays:

1. *Circuits:* Implementation of printed circuits is typically done using polytetrafluoroethylene (PTFE)[1] based laminates. There are very low losses when PTFE is used as a resin. Commercially available

[1] PTFE is generally identified by its DuPont brand name, Teflon.

materials include a matrix of fibreglass to reinforce its mechanical performances and stability. However, other supporting materials or structures are typically required to achieve the required mechanical performance and the environment endurance, and to avoid degradation due to radiation. Other materials, such as quartz fibre, have good mechanical and electrical characteristics but there are no commercially available laminates based on such materials, nor is the processing of those materials industrialized.

2. *Ground planes:* Ground planes and spacers can be entirely made of metallic materials (typically aluminium). However, higher flexibility and light weight are achieved using glass fibre reinforced plastic (GFRP-FR4) which shows good mechanical properties. A huge range of materials of this type are available on the market and their processing is well industrialized. FR4 can also be employed for BFN designs in those cases where the losses are not a prime concern.
3. *Dielectric spacers:* Foam or honeycomb dielectric spacers are widely used in space applications. They are used to devise low-weight RF structures and panels. Foam spacers have a homogeneous RF behaviour, with low losses and low dielectric constant. Honeycomb spacers have a low equivalent dielectric constant, but they are less homogeneous and the impact of the adhesive required for the assembly limits their use to low-frequency applications (e.g. under L-band). On the other hand, honeycomb spacers provide excellent mechanical performance for the large panels required for UHF or VHF applications.

In the past few decades, much experience has been gained with the aforementioned materials, which have been tested through several qualification campaigns in order to check their suitability for space applications. As an example, in 2001, a radiation test campaign was undertaken to demonstrate that all these materials were compatible with the INMARSAT-4 irradiation environment. Some samples were manufactured and submitted to irradiation tests representative of the orbit and life of the mission. An example of a patch for the low-band INMARSAT antenna can be seen in Figure 9.1. This probe was composed of three layers of FR4 Nelco-4000-13, two layers of Rohacell (foam spacer), an SMA connector, silver Teflon and aluminium tape with conductive adhesive surrounding the probe.

There are a wide variety of different materials required for the implementation of antenna arrays, such as connectors, electrical pins, conductive (aluminium) tapes, isolation tapes (Kapton) and sunshields.

Figure 9.1 *Irradiation sample from IMARSAT-4 navigation antenna test campaign. Reproduced by permission of EADS CASA Espacio.*

Such additional elements should also be accurately selected to avoid unwanted effects such as galvanic corrosion or sublimation of metallic barriers.

9.2.3 Array Optimization Methods and Criteria

In order fulfil a given a set of antenna requirements, both the array architecture and the aperture illuminations must be optimized [3,4]. In such an optimization, two stages can be clearly distinguished. First, the array architecture should be qualitatively defined by fixing the size of the aperture, the number of elements, the size of the elements and the grid of the aperture. In general, a unique solution does not exist and a tradeoff between element spacing and array size is required. Once the array aperture is defined, a quantitative optimization is done to determine the illumination function which provides the desired beam shaping (sidelobe level, nulling, shaped coverage, etc.) with the minimum impact on the directivity.

In general, there are three different aspects affecting and constraining the optimal array design:

1. *Size of the elementary array module:* Antenna arrays for space applications are generally organized into smaller subassemblies. Typically each subassembly is based on a sub-array, its feeding network (if not a single element), one filter and one amplifier. The most critical aspect for the elementary array module is the space available for the integration of each part of the receive/transmit module. Clearly, larger element distance will ease the integration of the different components.
2. *Topology of the array element:* When the elementary sub-array is based on planar technology it is also necessary to take into account the number of single patches per array element, the polarization and the feed network. Distributing these elements means defining the topology of the sub-array while implicitly allocating the space available for the receive/transmit module and the BFN.
3. *Directivity of the array antenna:* This parameter can be analysed through two different approaches. If the overall size of the antenna array is fixed, the only degree of freedom will be related to the sub-array configuration and design. It is worth noting that, in this scenario, small sub-array sizes and close spacing imply lower pointing losses and higher aperture efficiency. In contrast, this solution requires a high number of small elements, which makes the integration of the receive/transmit modules more difficult. Alternatively, the number of array elements can be fixed so that the cost of the receive/transmit module is the dimensioning factor. In this case, the variable parameter is the size of the sub-array. When the sub-array elements increase, it is necessary to find a tradeoff between the directivity and the losses due to grating lobes.

At an early stage, the impact of the surrounding structure along with interfering signals are also evaluated, allowing the definition of the antenna front-end and the Tx and Rx filtering requirements.

In a second phase, once the aperture of the array and the size of the array elements have been defined, the illumination (amplitude and phase) of the aperture must be optimized to meet the shaping or SLL requirements with minimum impact on the directivity and on the amplifier use efficiency factor. This efficiency factor is affected by amplitude tapering in the aperture illumination that reduces the power contribution from the amplifiers. The efficiency factor is defined as

$$f_{eff} = \frac{1}{N} \cdot \sum_{i=1}^{N} |I_i| \tag{9.1}$$

where N is the number of array elements in the aperture and I_i is the module of the illumination of the ith element normalized to the maximum. Low efficiency factors will result in low efficiency of the antenna, either in receiving or transmitting antennas, being the efficiency of the transmitting antenna more critical.

In parallel with the array element design (see Section 9.1), the optimization of the array illumination is done using numerical techniques where the amplitude and phase coefficients of each array element are optimized against a cost function which typically weights the array directivity, the SLL and the efficiency factor of the illumination. In general, this cost function is minimized using min–max algorithms. As a starting point, the optimization routine can use the aperture illumination function obtained with well-known algorithms, such as Taylor series or Chebyshev polynomials.

As long as the figure of merit is the directivity of the antenna or the SLL, this optimization task could be carried out by defining the proper pattern for each pointing direction. When more specific parameters are optimized (e.g. the group delay) dedicated optimization algorithms may be required.

Once the design of the array element is complete, an accurate model of the element radiation pattern, including coupling effects, is employed to evaluate if the optimized illumination coefficients are still valid to fulfil the antenna requirements. If required, further optimization iterations can be carried out.

Finally, the whole array design, including the BFN, is checked with full-wave 3D antenna modelling software which takes into account both inter-element coupling and edge effects. At this stage, the radiation pattern performance and the optimized array illumination are verified along with the directivity, the SLL or the cross-polar isolation (XPI).

9.3 Passive Arrays

In this section, several examples of passive arrays for different types of space applications will be presented.

9.3.1 Radiating Panels for SAR Antennas

A detailed review of SAR antenna systems is reported on Chapter 13. In this section, we'll briefly focus on the SAR passive radiating panels whose main design drivers are:

- Dual linear operation for multipolarimetric radar measurement with low XPs and two separate BFNs with appropriate sequential feeding.
- Semi-active architecture with large linear sub-array implementation in azimuth with good uniform illumination to reduce the radar uncertainties while scanning.
- Low profile for accommodation in a folded deployable configuration.
- Direct exposure to the space environment.
- Sufficient bandwidth to support chirp modulation.

The BFN topology to be used (serial, parallel or mixed) and its accommodation are critical to harmonize the compliance of all the requirements. Power handling is not a major issue, provided the power handled at panel level after each Tx/Rx module is in the range of 10–50 W in pulsed configuration. However, multipaction analysis need to be addressed in any case.

9.3.1.1 L-band SAR

This section briefly describes an L-band SAR radiating board being developed under ESA contract. The antenna for this SAR instrument is composed by linear arrays of six elements which provide dual linear polarization by means of a parallel division at the input and a 1:3 serial distribution with sequential feeding, allowing implementation (both radiator and BFN) in a single triplate structure. The serial feeding technique is light and compact but it can be only used for narrow band applications. The selected patch design and array

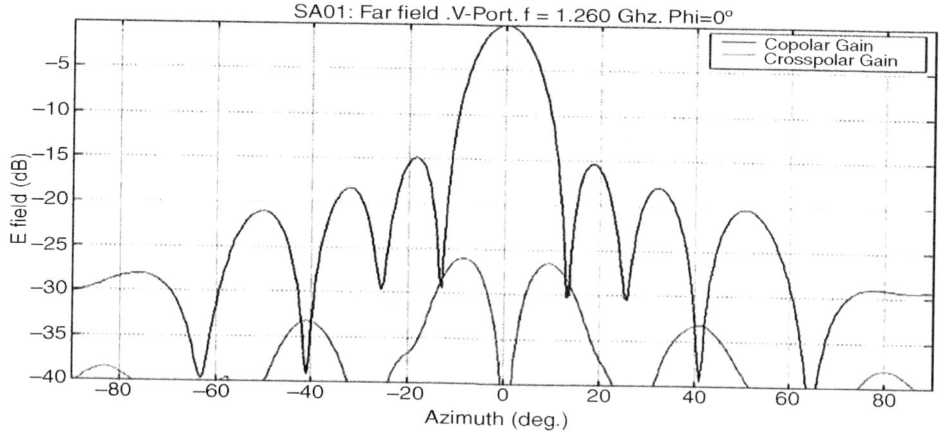

Figure 9.2 *L-band SAR six-element row tested CP/XP azimuth pattern.*

architecture provide excellent embedded configuration and very low coupling level. The array row size is 176.6 × 1096 mm (1 × 6 elements) corresponding to 4.6λ of total length. The row length and element spacing are a compromise between the sub-array losses, BFN complexity, number of Tx/Rx modules, number of panels, panel size, system ambiguities, and so on. The uniform aperture illumination provided 16.2 dBi of directivity, with total ohmic losses of about 0.7 dB (Figure 9.2).

For lower frequency bands, such as L-band, UHF or VHF, the RF structure serves also as a mechanical support. The mechanical interface (I/F) implemented by means of isostatic mounting devices provides extremely good stability, adequate for the large temperature excursion induced by direct solar inputs. The thermal control and ESD grounding is achieved by means of conductive black paint. The mass of the array row is about 750 g including the mechanical I/F and the thermal finish (Figure 9.3).

9.3.1.2 C-band SAR: Envisat ASAR

The development of the C-band SAR radiating board was done for the Envisat ASAR (Advanced Synthetic Aperture Radar) instrument developed within an ESA project. The array row is part of the antenna tile, each tile being composed of 16 array row elements, and each one having 24 radiating elements (19λ total length) mounted over a structural panel. Each row element provides dual linear polarization. The row length and element spacing are a compromise between the sub-array losses, BFN complexity, number of Tx/Rx modules, number of panels, panel size, system ambiguities, and so on. The topology of the BFN is corporate 1:2 at the input, series 1:6 and corporate 1:2, allowing accommodation in a single layer of triplate structure of just 2 mm thickness, with the necessary bandwidth, that is 16 MHz at 5.331 GHz. The selected patch design and row architecture provide excellent embedded configurations and very low mutual coupling. Sequential feeding applied in the BFN and XP of the radiating elements provides very low levels of sub-array XP (< −30 dB at boresight). The uniform feed distribution provides 22.2 dBi of directivity, equivalent to a surface efficiency of 100%, with ohmic losses about 0.9 dB (Figure 9.4).

At this frequency band, the manufacturing tolerances are more critical than at L-band. For this reason, the RF structure cannot serve any mechanical role. The array supporting structure provides support for both the array row and for all the Tx/Rx modules and associated electronics (corporate feed, power supply unit, etc.) including all interfaces with the deployable structure. The mass of the array row is about 100 g including the mechanical I/F and the thermal finish. The mass of the panel is about 5.4 kg including the array rows, thermal

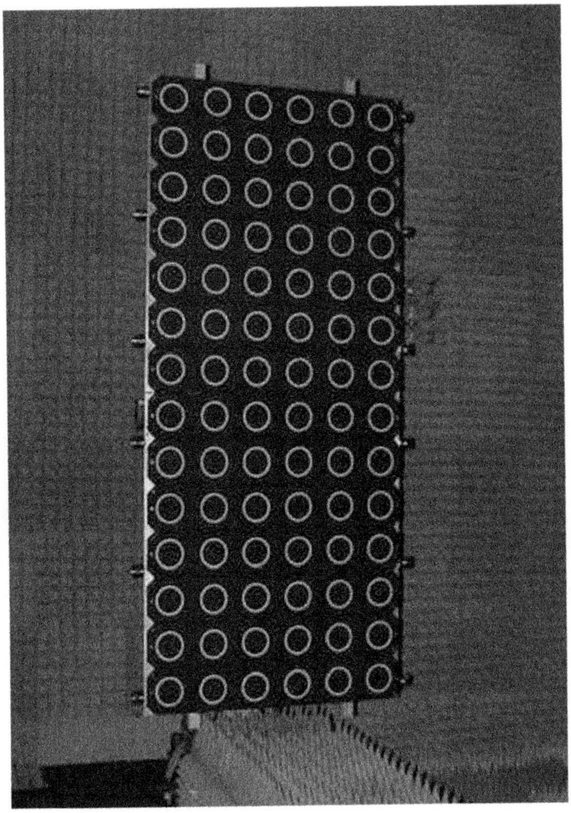

Figure 9.3 L-band complete 6 × 12 SAR panel section during testing. Reproduced by permission of EADS CASA Espacio.

finish and the I/F. The total tile mass supported by the structure is 15.4 kg (Figure 9.5). Although not designed for SAR applications, other interesting examples of L-band arrays are given by those employed on the Japanese multifunctional spacecraft MTSAT for global coverage and as reflector feed. In these designs, mutual coupling was reduced by recessing single or stacked patches into cavities. The same approach was employed also for the deployable array installed on the Russian spacecraft EXPRESS AM33/44. Both designs were developed by Alcatel Space.

9.3.1.3 X-band SAR

In this section it is briefly described the radiating board of the SEOSAR/PAZ spacecraft. SEOSAR/Paz is a Spanish X-band SAR mission based on the TerraSAR-X platform. The antenna, working between 9.5 and

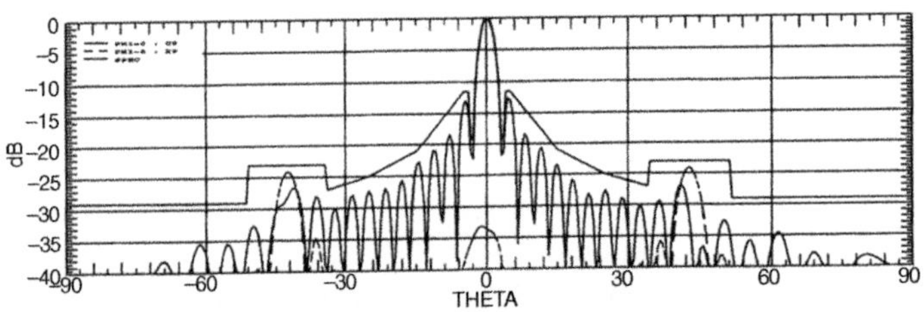

Figure 9.4 Envisat ASAR C-band 24-element row test CP and XP azimuth cuts.

Figure 9.5 Envisat ASAR C-band 16 × 24 panel during testing. Reproduced by permission of EADS CASA Espacio.

9.8 GHz, is composed of 12 panels with 32 sub-array row elements each mounted over a structural panel of 5000 × 700 mm total. Each sub-array element provides dual linear polarization and has 16 microstrip patches (21 × 397 mm, with a length of 13λ). The selected row architecture provides excellent embedded configurations and very low coupling level. The feed law was uniform providing around 20.3 dBi of directivity, with losses around 0.8 dB (including a piece of cable for connection to the Tx/Rx module). A corporate feed has been used from 1:16, implemented in three stacked triplate structures, needed to accommodate the BFN in the two orthogonal linear polarizations (Figure 9.6). Apart from the good linear

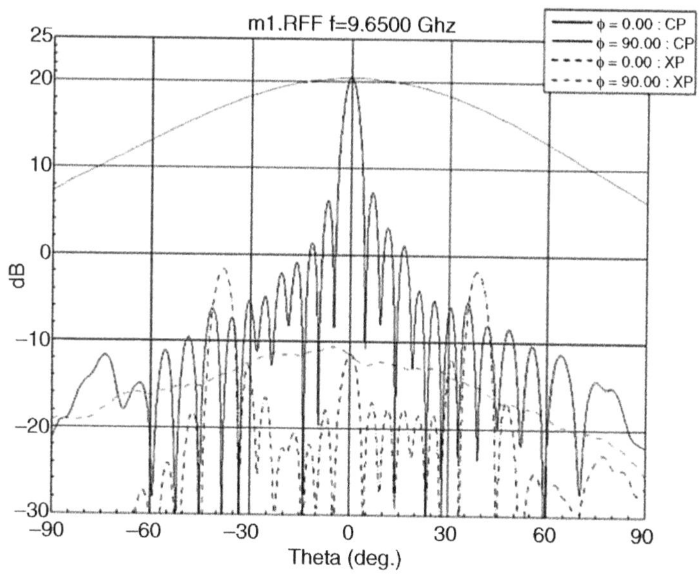

Figure 9.6 X-band SAR directivity CP and XP main patterns.

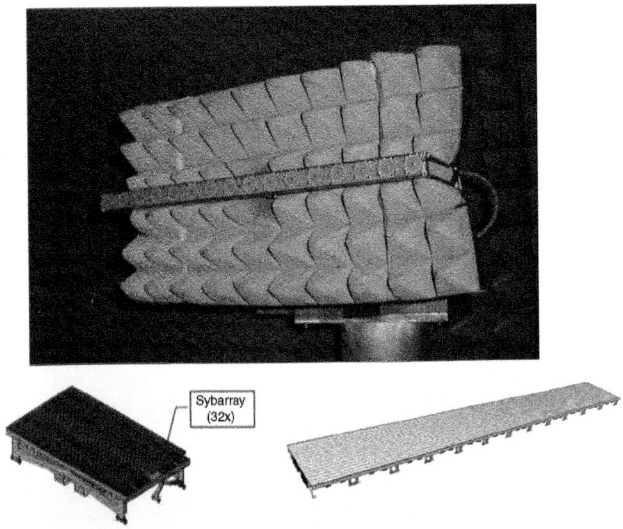

Figure 9.7 X-band SAR radiating row during testing, tile and deployed antenna. Reproduced by permission of EADS CASA Espacio.

performance of the patch radiator, sequential feed between adjacent patches has been implemented in the design to allow low XP (< -30 dB at boresight).

The structure provides support for the array row and for all the Tx/Rx modules and associated electronics (corporate feed, power supply unit, etc.). Thermal control is performed by means of a sunshield single-layer insulator (SLI). The mass of the sub-array is 105 g (Figure 9.7).

9.3.2 Navigation Antennas

Antenna arrays for navigation systems are another typical application of patch arrays in space [5,6]. In this case, low-profile features of printed technology are a key advantage with respect to other solutions such as those based on helix or reflector antennas allowing compact accommodation on the spacecraft and larger multiple launch. For this type of applications the design drivers are:

- Circular polarization operation.
- Global coverage from a geosynchronous or geostationary orbits (i.e. 22 000 km and 36 000 km) with isoflux correction in some cases.
- Low profile, low mass for accommodation on small platforms or as a secondary payload.
- Dual-band or multiple band operation might be required in some cases.

A parallel BFN topology is normally used with sequential rotation of array elements to improve the XP performance. As will be explained in the following sections, power handling, both multipaction and PIM, are major design drivers for these types of applications.

9.3.2.1 INMARSAT-4 Navigation Antennas

This subsection presents the antenna array used for the three units of the INMARSAT-4 satellite that includes the navigation antennas as secondary payload, the primary mission being AstroMesh™, a large unfurlable reflector with multi-feed (single feed per beam) for satellite mobile services. The array operates in right-handed circular

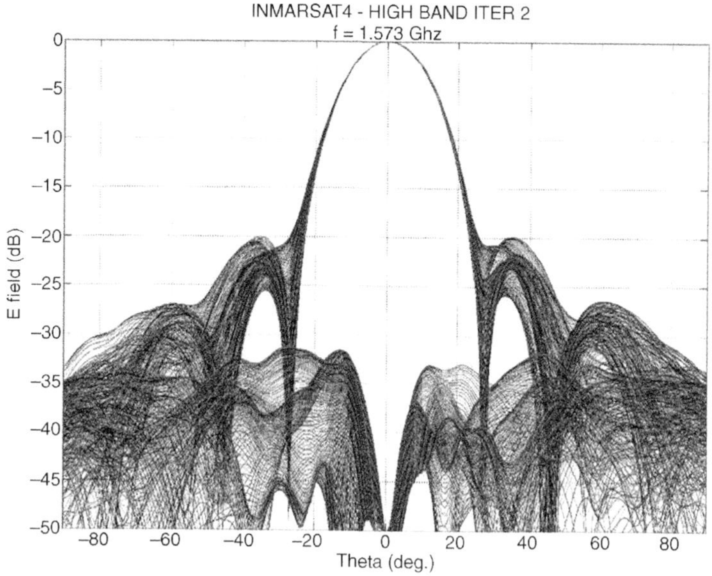

Figure 9.8 *Radiation pattern of the INMARSAT-4 navigation antenna array. Azimuth CP and XP cuts are shown at 1.573 GHz.*

polarization (RHCP) at two frequencies, 1.2 and 1.6 GHz, with a bandwidth of 40 MHz. In order to cover these two frequency bands, two independent scaled arrays were designed for each band. Each array is composed of 12 elements with d/λ equal to about 0.75, in a hexagonal lattice, built in a single piece and a single triplate to minimize the losses, and mounted over a structural panel for mechanical integrity. The selected radiating element is a circular annular slot excited by electromagnetic coupling from a suspended substrate stripline, and 45° notches to generate circular polarization. Although each array element provides good RHCP, sequential rotation of individual elements was applied, significantly improving the XP performance at the array level (< -30 dB in the coverage), as shown in Figure 9.8. In both bands, a coverage cone of 9.1° is achieved with 18.8 dBi peak gain and 16.2 dBi EOC (End Of Coverage). Losses are around 0.4 dB.

The mass of the array, including the supporting structure, the mechanical interface and the thermal finish, is about 3600 g for the low-band antenna and 2270 g for the high-band antenna (Figure 9.9). Thermal control is achieved by means of conductive silver–Teflon over the external surface, which provides excellent optical properties to reject solar heat flux.

9.3.2.2 Galileo Navigation Antennas

Galileo is a navigation system being developed by the European Union and ESA. This initiative was proposed to provide Europe with an autonomous global positioning system from the already existing one, but also to have a certificated system for civil applications in assisting ship and aircraft navigation. The project is structured in three phases: the Galileo System Test Bench Version 2 (GSTB-V2) to demonstrate the system and the key elements launched in 2008; the Galileo In-Orbit Verification (Galileo IOV), composed of four spacecraft to be launched by 2012; and the Full Operative Phase where the complete 28-satellite constellation is deployed. Two antenna designs were developed within the framework of ESA contracts for the first two Galileo phases:

- NAVANT FM antenna developed by TAS-I for the Giove-A satellite
- NAVANT developed by EADS-CASA for GST B-V2 and IOV phases. In the following, the NAVANT antenna type is briefly described. Further details on this topic can be found on Section 3.6 and Chapter 14.

Figure 9.9 INMARSAT-4 navigation antennas during testing. Reproduced by permission of EADS CASA Espacio.

Array design The objective of the NAVANT (NAVigation ANTenna) is the transmission of the encoded navigation signals. The array is composed of 42 (GSTB-V2) or 45 (IOV) array elements in a quasi-hexagonal lattice, mounted over a structural panel. Some of the elements work as dummies to achieve similar embedded performances for all the elements. As can be seen in Figure 9.10, the design is modular and it is formed by: (i) six sectors, independent from the others, with dual-band (1.227 and 1.575 GHz) self-diplexed stacked patches and two superposed BFNs; (ii) a mechanical structure; and (iii) a central power

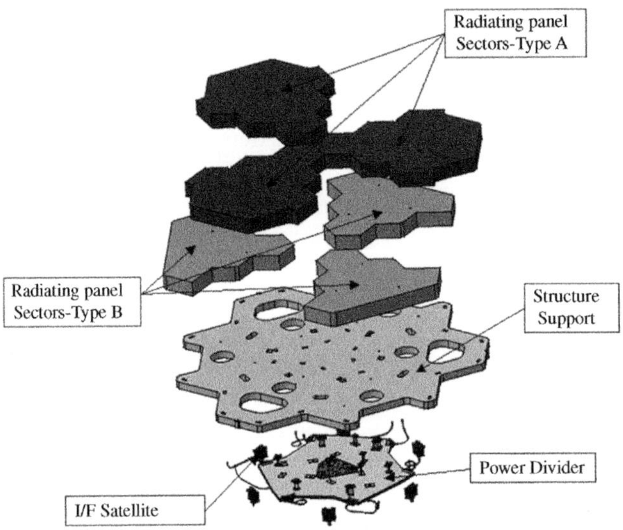

Figure 9.10 Galileo IOV NAVANT modular construction. Reproduced by permission of EADS CASA Espacio.

divider, to feed each sector of the two operating frequency bands in a stacked configuration. Connection between sectors and the central power divider is made by means of a set of 6 + 6 matched coaxial cables.

The key design drivers are:

- dual-band operation;
- isoflux corrected pattern;
- industrialization for mass production;
- low profile, low mass and stiffness;
- phase centre and phase flatness in the coverage area;
- group delay specs in the coverage area;
- PIM, multipaction and power handling.

The feed excitation and the array lattice have been optimized to obtain optimum isoflux corrected gain in the coverage area (12.67° cone) simultaneously in the two bands, while considering the phase pattern and the group delay. The central part of the antenna is excited in phase opposition with respect to the external part in order to get the required isoflux radiation pattern. To optimize axial ratio performance, a feed excitation with real coefficients and sequential rotation between sectors of the array was used. The sectors (Figure 9.11), of two kinds called A and B, the central divider and the structure are interchangeable elements from antenna to antenna.

This design modularity offers important advantages regarding the manufacture of the antenna, industrialization for large quantities and manufacturing yield. Thermal control is performed by a sunshield SLI covering the antenna. The total thickness of the antenna is less than 150 mm and the mass of the array including the structure, the mechanical interface and the thermal finish is around 15 kg (Figure 9.12).

Group Delay By definition, the group delay is calculated as

$$\mathrm{GD}(f) = -\frac{\partial \Phi}{\partial \omega} = -\frac{1}{2\pi}\frac{\partial \Phi}{\partial f} \tag{9.2}$$

The group delay for each channel is defined as the average of GD(*f*). The group delay (GD) is a typical requirement of high-dispersion devices or frequency-selective devices such as filters. In case of antennas, it is

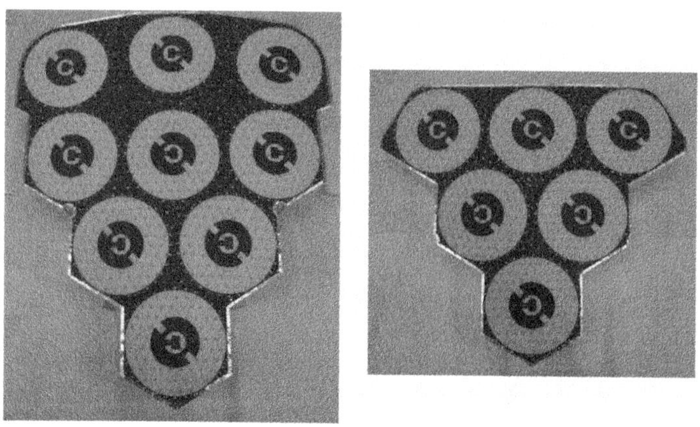

Figure 9.11 *Galileo IOV NAVANT sectors: (a) A-type sector; (b) B-type sector. Reproduced by permission of EADS CASA Espacio.*

Figure 9.12 *Galileo IOV EM antenna during test campaign. Reproduced by permission of EADS CASA Espacio.*

also a function of the coverage direction when the antenna phase pattern is not fully flat. In most antenna applications, the variation of the GD within the coverage is not a mayor issue provided it is below a certain level. Conversely, in a navigation system this is a key requirement and GD variations in all directions must be minimized. The main contributions to GD variation in the coverage are:

- the GD of the radiating element;
- the geometry of the array and sequential rotation;
- the illumination law;
- BFN errors.

Different parameters must be optimized in order to minimize the GD variation; generally, it is required a combination of adequate array geometry and optimized excitation law. Symmetrical array geometries are the most appropriate to reduce the dispersion between phases, which eventually lead to the reduction of the GD variation. This, in addition to an optimized amplitude and phase law, may lead to equalization of the phase radiation patterns for every frequency at each channel. Another key element to be taken into account is the selection of a radiator with low GD. In any case, even if the GD performance of the radiator is not as good as desired, the GD at array level can be reduced by means of sequential rotation of single elements which will also provide good performance in terms of XP. Quite accurate models of the array and of the radiator are required to allow proper optimization of the above-mentioned parameters (Figure 9.13). Nevertheless, it is difficult to optimize the phase response without affecting gain performance. As a consequence, a compromise between these two parameters is required.

Testing the GD is also a quite specific matter. It requires careful calibration of the setup, typically using a known antenna. The far-field (FF) test method is the most straightforward but requires large ranges and provides information only in the tested directions. Further, the near-field (NF) test requires transformation to the far field, gives information on the complete test window and allows filtering of some test range errors (as wall reflectivity) present in the FF method.

Pattern Shaping One of main mission requirements is the isoflux corrected pattern needed to reduce the error between different received signals. In order to transmit the same power to any point of the coverage, the pattern of the NAVANT must be constrained to lie within an 'isoflux' window (Figure 9.14). This mask

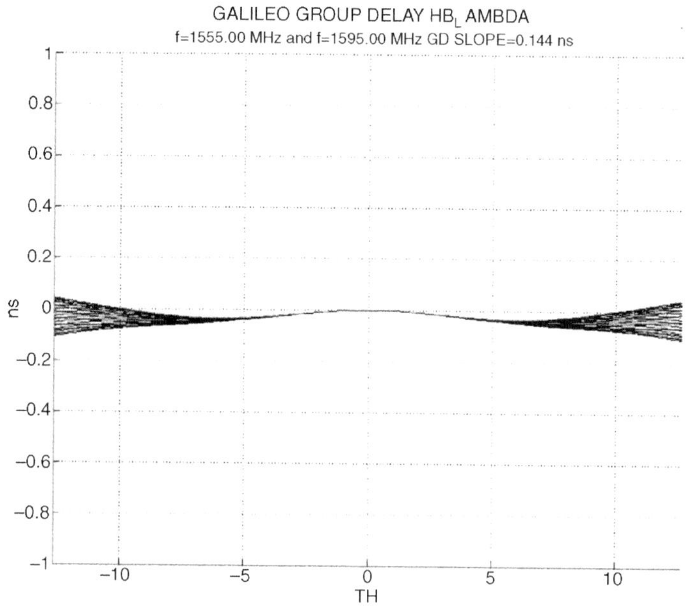

Figure 9.13 Optimized group delay of the NAVANT high band.

determines the minimum and maximum gain levels that the antenna should comply with at boresight and at EOC (12.71° for the Galileo system), being different for both the low and high band.

PIM Requirements and Design Solutions Due to the proximity of the different antennas in the payload, the PIMPs between the transmitted frequencies of the NAVANT may interfere with the reception frequency bands of other antennas. The lower the order of the PIM, the worse the effect. In the Galileo system there are five relevant PIM orders that interfere with the other antennas in the mission: the SAR antenna Rx (second and fifth order), the C-band mission (fourth and fifth) and the telemetry, tracking and command (TT&C) Rx (fourth). A set of requirements of input power and allowed reception power has to be established for each of the possible PIMP situations. In the case of the Galileo system, the NAVANT should not generate PIM levels higher than those presented in Table 9.1.

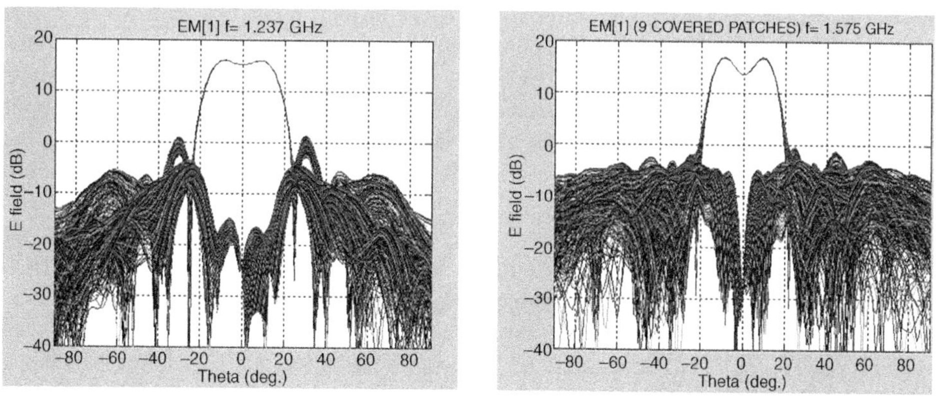

Figure 9.14 Isoflux patterns for low and high bands for IOV mission.

Table 9.1 NAVANT PIM requirements.

Rx band	Order	Input power E5 (W)	Input power E6 (W)	Input power L1 (W)	PIM (dBm)
SAR	2	49.3	—	72.6	−145
	5	49.3	—	72.6	−145
TT&C	4	49.3	—	72.6	−122
C-band	4	49.3	53.5	—	−133
	5	—	53.5	72.6	−127

The details for a typical setup for PIM tests can be seen in Section 9.5.2. One critical setup parameter is to achieve the required noise floor, which is quite challenging due to the extremely low PIM requirement.

Multipaction and Design Solutions The Galileo antenna operates under a single-carrier-modulated signal in the high band and under two carrier-modulated signals in the low band. In the RF path the following areas were identified as multipaction critical areas:

- TNC connector area;
- SMA connector area;
- stripline transmission line area.

The first step is to calculate the margin of the signal peak power with reference to the single-carrier multipaction threshold. The signal peak power of each channel was calculated from the channel-modulated signal in time. The single-carrier multipaction threshold was calculated by the ESA Multipactor Calculator v.1.61. The following results were obtained:

- *High band (single carrier):* In this frequency band, the multipaction margins were higher than 10 dB, which is the required margin [1] for single-carrier and Type 2 components (including dielectrics or other materials with known multipaction properties) for the performance to be demonstrated analytically without requiring test campaigns.
- *Low band (multicarrier):* At low band, under multicarrier operation, the in-phase worst case was considered to calculate the peak power of the band. The results showed that the peak power was lower than the single-carrier multipaction threshold in all the critical areas but with lower margins than those achieved in the high band. As the in-phase worst case criteria can be excessively conservative, the margins were calculated as well with respect to P20 (maximum power for 20 times the crossing time τ_{20}) [1] with improved margins as expected. Due to the lack of reference margins for multicarrier Type 2 components in [1], it was eventually decided to test the three critical areas at low band after a design optimization process which prevented the multipaction effect occurring by avoiding the critical gaps.

Single-carrier multipaction tests were performed at the ESA Multipaction Test Facility TEC-ETM at ESTEC (see Chapter 6). A strontium-90 β-emitting radioactive source was employed to simulate the free electrons in space. Three multipaction detection methods were used continuously during these tests: (i) forward/reverse power nulling (global method); (ii) electron probe detector (local method); and (iii) third harmonic (global method). Very successful results were obtained showing multipaction free performance over the required qualification level, that is up to 2 kW.

9.3.3 Passive Antennas for Deep Space

9.3.3.1 Rosetta MGA

The Rosetta mission is an ESA deep space probe launched to explore the Churyumov–Gerasimenko comet on the periphery of the Solar System. The array installed on Rosetta is working in S-band (Rx 2115 MHz, Tx 2297 MHz) and is composed of six array elements in a hexagonal lattice, constructed mostly in aluminium and mounted over a structural panel [7,8]. The array diameter is 300 mm and the objective is a medium-gain antenna for TT&C and data downlink to Earth. The feed law is uniform providing 14.1 dBi of peak gain for Rx and 14.7 dBi peak gain for Tx, with losses around 0.4 dB.

As can be observed in Figure 9.15 and Figure 9.16, the array elements were sequentially rotated to improve RHCP purity (XP < -21 dB in the coverage, 15° half-cone angle).

Thermal control is implemented on the metallic structures using white paint. The total thickness of the antenna is less than 50 mm and the mass of the array including the structure, the mechanical I/F and the thermal finish is around 700 g.

9.3.3.2 Electromechanical Array: Mars Science Laboratory HGA

One of the most challenging features of an antenna for space applications is the capability to operate under extreme temperature and vibration conditions. In the case of antennas for deep space applications, the operational conditions are especially difficult. This is the case for the Mars Science Laboratory high-gain amplifier (MSL-HGA), where operation is in a Mars environment that includes temperatures between -143 and $+125$ °C (non-operational) and between -135 and $+105$ °C (operational), along with the composition of the planet's atmosphere. Another difference from typical satellite operation is the vibration conditions. In addition to the vibration loads that occur during launch, this mission includes vibration loads due to the descent, landing and also during movement of the rover on the planet's surface. Finally, one important and specific precaution that must be taken into account for this type of mission is disinfection of the specimen before integration and before launch in order to avoid contamination of the Martian environment.

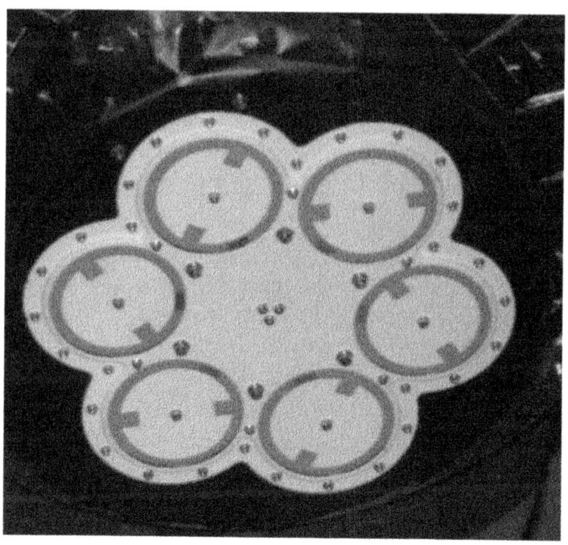

Figure 9.15 Rosetta MGA antenna during testing. Reproduced by permission of EADS CASA Espacio.

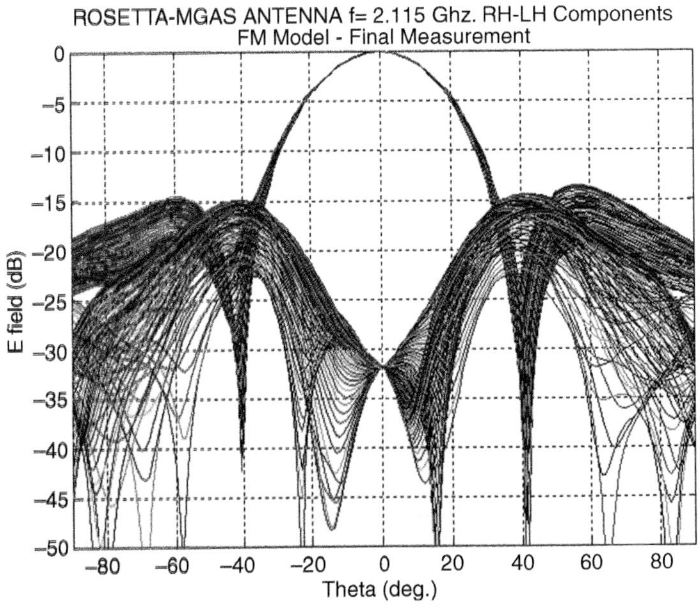

Figure 9.16 *Rosetta MGA normalized directivity azimuth CP and XP cuts.*

In this kind of mission, where the antenna is mounted on the deck of a rover, there exists a very strict restraint on mass and envelope. Typically, printed antennas satisfy both requirements, along with electrical performances. For the case at hand, the high-gain antenna system (HGAS) is composed of two subassemblies, the antenna itself (HGA) and the gimbal (HGAG), which allows mechanical pointing in azimuth and elevation by means of actuators and rotary joints. The total mass of the assembly is less than 1.2 kg (Figure 9.17).

The antenna is a multilayer printed antenna that works in X-band. Its aim is to provide direct communication between the rover and the Earth within two separated working bands for transmission and reception. It is mounted on a sandwich structure which confers robustness to the whole setup. A cover keeps out dust and contamination and provides a thermal and clean enclosure. The HGA consists of an array in a triangular grid composed of 48 photo-printed patches without the central element (Figure 9.18). The sequential rotation between elements enhances the axial ratio performance with no penalty on gain. The radiator array is fed by a suspended stripline BFN corporate feed implemented in three layers of SSSL.

The HGAG provides the mechanical pointing of the antenna in azimuth and elevation to allow full coverage and direct communication with the Earth regardless of the position of the rover on the Martian surface. The gimbal system also allows routing of the RF signal between the antenna and the rover.

In order to test the capability of the antenna to withstand all environmental conditions a full qualification campaign was undertaken. In addition to functional tests, measurements with real operating temperatures (including safety margins), vibration loads and power handling capabilities (including multipaction and corona tests) were performed. In the case of this particular mission, dry heat microbial reduction (DHMR) was required to eliminate any possible way of contamination. Indeed, DHMR is an approved method for reducing the microbial burden on space flight hardware prior to launch to meet flight project planetary protection requirements.

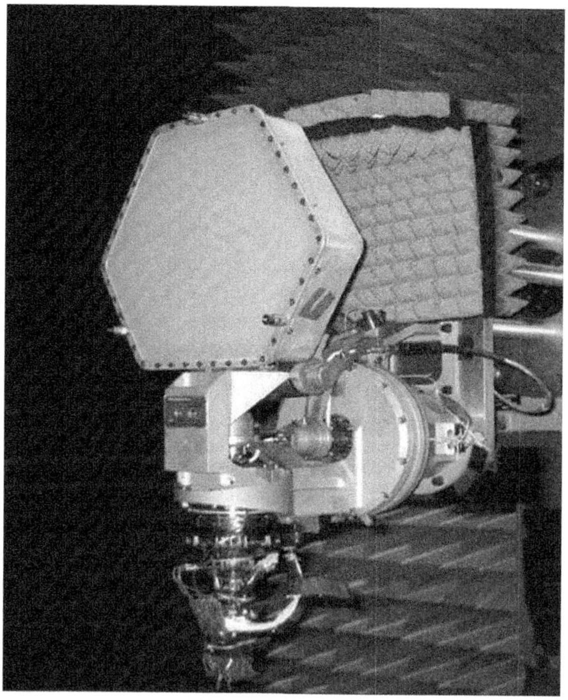

Figure 9.17 Mars rover HGA system during functional testing. Reproduced by permission of EADS CASA Espacio.

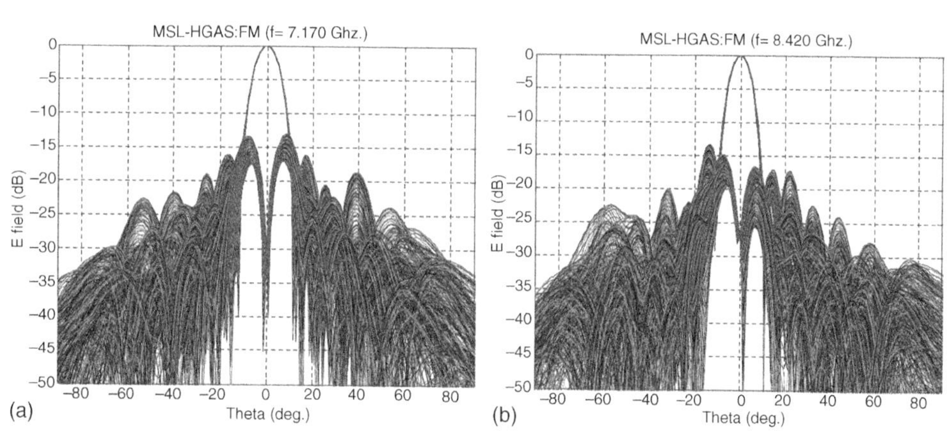

Figure 9.18 Mars rover FM HGA model, Rx (a) and Tx (b) tested patterns.

9.4 Active Arrays

9.4.1 Key Active Elements in Active Antennas: Amplifiers

Signal amplification is one of the essential functions of active arrays. The main specifications affecting amplifiers are EIRP, G/T, linearity (noise power ratio (NPR) or C/I) as well as the power levels handled by the payload. In the following, two different schemes of amplification are presented, namely active and semi-active antenna arrays.

9.4.1.1 Amplification Schemes in Active Arrays

In the most common scheme, each array element is connected to an amplification stage. The use of a distributed amplification helps to optimize the reliability of the array, since a single-point failure yields a graceful degradation of antenna performance. The main complexity of active arrays is related to the high density of active devices which poses critical challenges both for power supply and for thermal control.

Receiving Arrays: LNAs The low-noise amplifier (LNA) should be placed as close as possible to the radiation element in order to optimize *G*/*T* and minimize preamplifier losses. LNAs are normally implemented using GaAs MMIC (Monolithic Microwave Integrated Circuit) devices assembled in hybrid modules. Currently, the most globally accepted high electron mobility transistor (HEMT) technology for space application is pHEMT (Figure 9.19), due to its maturity and proven reliability. Shorter gate lengths are being used in order to optimize the NF (from 0.150 up to 0.050 μm). MHEMT is now seen as a very promising future technology with lower power consumption and better NF. The use of discrete components, even if they offer a certain improvement in terms of the NF, are in most cases not practical due to cost, accommodation, complexity and repeatability.

The dissipated power, the efficiency and the thermal control are less severe problems than in Tx cases. However, thermal stability requires either tight thermal equalization or compensation to achieve fine control. Moreover, the lower the LNA temperature, the better the achieved NF.

In order to avoid degradation of the NF due to combiner/divider losses, after the LNA an appropriate amplification scheme is required to avoid degradation of the signal quality along the antenna Rx chain. Typically, the output power is strongly limited by the transponder input so a careful tradeoff is required.

Transmitting Antennas For the transmitting antenna, the active scheme with distributed amplification significantly reduces the power per element rate, helping to reduce the power requirement in the amplifiers and the temperature of the associated hot spots. However, it increases the number of elements in the antenna and the complexity of the signal distribution. The most significant parameters in this type of configuration are the required output power and the power added efficiency (PAE) of the amplifiers that directly determine the total antenna power consumption and dissipation, and thus are extremely important in space applications. In the most general case, space antennas for communications operate in continuous wave (CW) multicarrier and multichannel modulation schemes. The power amplifiers nonlinear operation

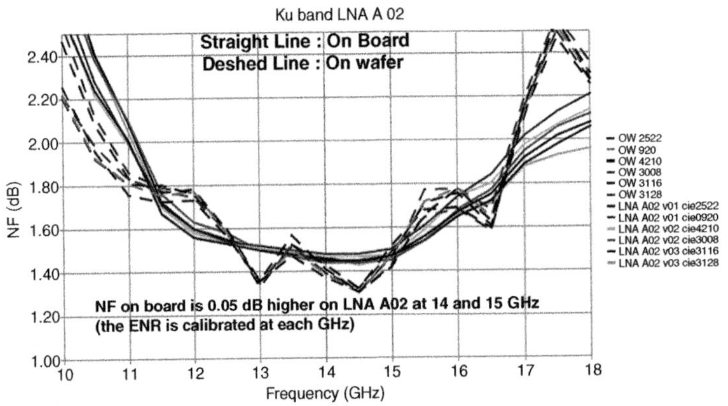

Figure 9.19 *Noise figure in a Ku-band LNA with 0.13 pHEMT process.*

should be carefully analyzed and the selected operation point optimized taking into account different aspects. In particular, the in-band NPR mission requirement should be considered along with the PAE, provided that the IM terms reduce the efficiency and the out-of-band rejections which could increase the complexity of output filters.

Power amplifiers can be implemented based on discrete transistors or in monolithic technology. Normally, MMIC devices are the best approach as relatively low power amplification is typically required and because they are easy to integrate in multi-chip modules GaAs pHEMTs (and HBTs (Heterojunction Bipolar Transistors)) are again the most mature technology for space applications. GaN has still to be confirmed as a mid-term solution for space missions especially because performance requirements have to be evaluated along with the reliability and stability of the foundry process. Similar PAE and NPR for low-power amplifiers are expected, the main advantages of this technology being the higher operational junction temperature and biasing voltages that ease the thermal control.

Antenna configurations with 300 or 500 elements from 1.5 to 0.250 W have been evaluated by the authors in Ku-band while multibeam arrays with 500 to 1500 elements from 1 to 0.1 W in Ka-band have been developed with reasonably competitive performance (see [2]) in terms of power consumption, dissipated power and mass.

9.4.1.2 Amplification Schemes in Transmit Semi-active Arrays

A semi-active array is understood as an array where several elements share the signal though the same distribution network. The signal is distributed to several medium- or high-power amplifiers that deliver enough power to a subset of radiating elements. In this type of configuration, the amplifier stage is far away from the elementary radiating elements and the room available for its accommodation is higher. Therefore DC/DC conversion is less complex, but the hot spot is higher than that obtained in a fully active array. In general, reliability is a critical parameter for semi-active solutions due to the reduced number of amplifiers and due to the fact that each amplifier drives several radiating cells.

As an alternative, although they are challenging to design and implement, schemes based on a combination of several amplifiers through Butler matrices or through hybrids can significantly improve the reliability of the array. Such a network combines the power of the amplifiers at the input and distributes it to the different outputs. This kind of scheme also allows, in some applications, the power of all the input amplifiers to be driven to a subset of the outputs (radiating elements) through phase control only. This solution is useful when a part of the antenna aperture is not fed in some periods of time, optimizing the ratio of radiated power to installed power, and the demanded power per amplifier. The drawback to be carefully addressed is the implementation of this network and the losses introduced in front of the amplifiers.

An example of deep space semi-active array has been developed as part of the ESA GAIA project. The antenna will perform the data downlink to Earth from the telescope on board the spacecraft placed at the second Lagrange point to map the Milky Way. Due to the extremely high stability required by the instrument, electronic steering is required, being performed through phase control only. The array is composed of 28 sub-array elements, in 14 facets of a truncated pyramid, which are connected in a semi-active arrangement to 28 amplifiers. This concept was originally introduced in 1990 by Alcatel Space and it exploits the geometrical properties of the truncated cone allowing to scan a high-gain beam over a large field of view. The amplifier stage of the GAIA antenna is composed of several GaAs amplifiers in cascade with a final high-power stage integrated in a single piece of equipment of 10 W each. The signal is routed using cables to radiating elements through distribution networks. Phase control together with 90° hybrids allows the power of the 28 amplifiers to be driven to 14 sub-arrays as required the Earth following the spacecraft spin. Table 9.2 gives a summary of the amplifiers' main requirements. Of special relevance are the tight stability and repetitive requirements of this particular semi-active scheme.

Table 9.2 GAIA data downlink amplifiers' main requirements for an X-band semi-active Tx antenna.

Requirement	Value
Frequency range	8.46– 8.47 GHz
Input power level	21.7 ± 1 dBm
Input power variation over temperature	0 to +0.5 dB
Input power variation over life	0 to +0.1 dB
Output power level	39.24 dBm
Output level variation with frequency	0.35 dB pk–pk
Output level stability	−0.355 to 0.415 dB
Output level tracking	−0.26 to +0.32 dB
Phase control	< 3° (RMS error)
Phase variation with frequency	0.7° pk–pk
Phase stability over temperature	< 1°/°C
Phase stability over life	±3.1°
Phase tracking	−9.6° to 10.1° (temp. > 0 °C)
Gain compression	< 3 dB
Return loss	21 dB
Harmonic outputs	−60 dBc
DC power consumption	< 33.2 W
DC power dissipation	< 23.9 W
Group delay variation	< 2 ns
Mass	< 1.7 kg

9.4.2 Active Hybrids

Hybrid modules are compact and low-mass devices with a high degree of integration of electrical functions designed to distribute the power and the required BFN controls among all the elements of an array antenna and are mostly employed on direct radiating arrays (DRAs). Hybrids for DRAs are typically designed as multi-chip modules (MCMs) in multilayer ceramic packages. The MCM concept allows a significant reduction of size and mass merely by using components in die format but also by integration of different functionalities. Some typical functionalities of DRA hybrids are:

- amplification stage(s);
- digital phase shifter;
- digital attenuator;
- filtering;
- power splitter/combiner (for multibeam antennas);
- DC and control functions.

To achieve further integration and make better use of available room for the hybrids, MMIC-based solutions are usually preferred for active components such as amplification stages, phase shifters and variable attenuators. For high-frequency applications (e.g. X-, Ku-, Ka-band) the most commonly used MMICs are made from GaAs. Digital control is also preferred in most cases for phase shifters and variable attenuators to simplify the control electronics.

As the number of components increases, to properly set and control the RF active devices, additional digital components are needed to simplify the interface and the functionality of upper level electronics. Different

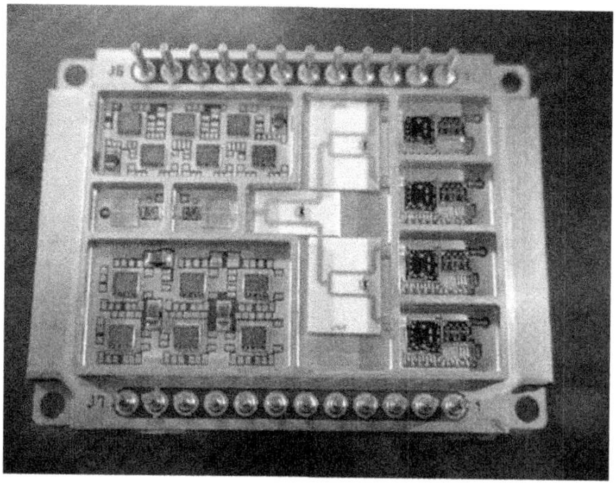

Figure 9.20 Multi-chip module based on HTCC in X-band for Rx in space for a multibeam (four beams) communication application. Reproduced by permission of EADS CASA Espacio.

solutions based on simple devices such as shift registers or more ambitious ones based on ASICs are considered to achieve a higher degree of integration depending on the application.

In the case of pulsed applications, such as for radars, the hybrid modules include both the functionalities, Rx and Tx, through the use of isolators. In the case of communication applications, the huge gain loop in the total transponder between uplink and downlink signals makes the isolation critical and the filtering requirements typically not compatible with an implementation of Tx/Rx chains in a single module.

Hybrid ceramic packages for space applications are made of either high-temperature co-fired ceramics (HTCCs) or, more recently, of low-temperature co-fired ceramics (LTCCs). LTCC hybrids allow the use of embedded components and even RF traces, but the control and characterization of the processes to allow that at microwave frequencies requires a long development cycle.

To mitigate risks and reduce the development schedule, hybrid solutions based on a limited use of buried layers for RF purposes can be adopted, as shown in Figure 9.20, for example, where the use is limited to coaxial biases interfacing with the RF connectors provided. The rest of the RF circuits required are implanted in thin film alumina and brazed to the surface of the hybrid package.

Heat dissipation is a major issue for Tx modules. In such a case, implementation of special techniques in the MCM is required. In general, heat sinks or thermal biasing of the hot areas are employed along with the use of heat pipes as shown in the next subsection.

9.4.3 The Thermal Dissipation Design Solution

Thermal dissipation and control is a major issue in space antennas. Even though array antennas are typically much less sensitive to thermoelastic distortions than reflectors, the presence of electronic circuits and devices in active antenna arrays can be a very critical factor, especially for Tx modules. To reduce the impact of the external heat flux, mainly due to solar rays, thermal blankets (multilayer insulators (MLIs)) are generally employed to cover the antenna body (Figure 9.21). Furthermore, a low electrically conductive sunshield (e.g. single-layer insulator (SLI)) is used in the area of the antenna aperture. The SLI shield is almost RF transparent and allows proper grounding of the space free electrons, avoiding electrostatic charges on the surface.

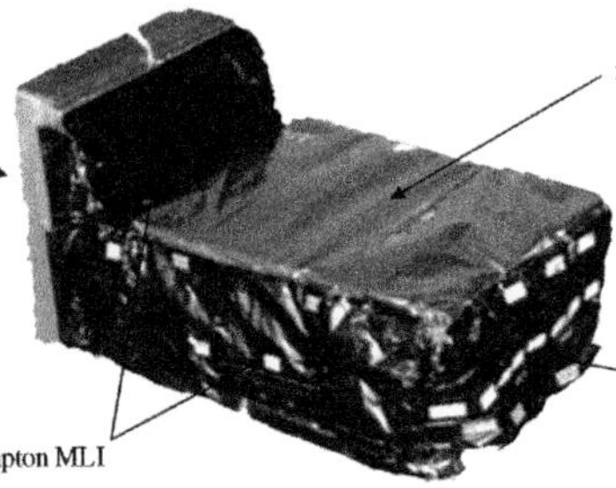

Figure 9.21 *External thermal control of an active antenna (OSR = Optical Solar Reflector). Reproduced by permission of EADS CASA Espacio.*

For ground electronics the main means of heat dissipation are convection and conduction. In space, convection does not occur due to the vacuum conditions so radiation and transmission mechanisms need to be used. In order to extract the heat from the more dissipative elements, good conductivity must be ensured by proper design and careful material selection as described in Chapters 4 and 5. High-conductivity materials are used in manufacturing the hybrid packages and on the body of the electronic boxes. Fillers are placed in the mechanical junctions of equipment to avoid gaps and the lack of conductivity. This also helps to avoid the use of contaminating or outgassing materials. In particular cases, employment of special devices to connect thermally the dissipative components or areas to a cooler area or device is mandatory. In such situations, heat pipes (HPs), loop heat pipes (LHPs) or fluid loops are good solutions for highly dissipative components that require extraction and transportation of the heat to a cold area, as discussed in Chapter 5.

The heat transport capability of these devices varies from a few watts to several hundred watts. Typically the cooler is a thermal radiator panel covered by small, highly effective mirrors called optical solar reflectors (OSRs) facing deep space to reject the heat physically into space. On some occasions, HPs are used not to extract the heat but to equalize the temperature within different elements.

Active control by means of heaters (Figure 9.22) is generally employed and it runs parallel to the electronics and HPs to allow active control and heating during the non-operating cold periods (typically the transfer orbit). Thermistors are used to sense and activate the heater. Accurate thermal models are required to analyse the different operational and non-operational modes. There are specific missions where it is not possible to induce

Figure 9.22 *Internal thermal control of an active antenna. Reproduced by permission of EADS CASA Espacio.*

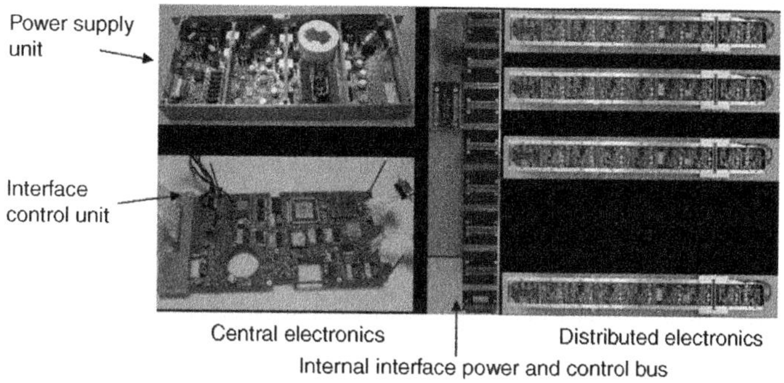

Figure 9.23 Active array electronics. Reproduced by permission of EADS CASA Espacio.

any movement or microvibration, for fluid displacement. Typically, this is the case of satellites including sophisticated optics such as telescopes. In these cases, thermal control relies exclusively on conductive and radiative means that lead to designs based mainly on metallic structures.

In some cases, the different temperature conditions in the RF electronics produces unacceptable drifts in some RF parameters such as gain, phase or NF. In these cases corrections are implemented by complex feedback circuits based on RF parameters or calibration tables as a function of the flight temperatures.

9.4.4 Active Array Control

Active antenna arrays also have some auxiliary electronics (Figure 9.23). Examples are those required to provide power supply to the active components or digitally controlled phase shifters and attenuators that allow modification of the amplitude and phase distribution across the array aperture [9–11]. These functionalities require a digital unit to act as the proper control and for communications with the on-board processor of the spacecraft and for commanding from the ground. In general, such architectures incorporate some software implemented in field-programmable gate arrays (FPGAs) and memories for the necessary lookup tables.

The complexity of this unit is a function of the number of elements, of the required control lines and of the power consumption. However, it normally requires a significant amount of design, testing and budget effort, including an EMC test campaign (Figure 9.24) to make it compatible with the payload and the platform electronics.

Figure 9.24 Active antenna during EMC testing. Reproduced by permission of EADS CASA Espacio.

9.4.5 Active Arrays for Communications and Data Transmission

This subsection presents some examples of active and semi-active arrays mainly related to communications and data transmission applications. The following four test cases are considered: (i) the X-band in-orbit reconfigurable antenna (IRMA) Rx array providing four separately steerable beams and anti-jamming capabilities; (ii) a general case of a semi-active phased-array antenna (PAA) providing a single steering beam using only phase control; (iii) the Ku-band electronically steerable antenna (ELSA) providing four steerable beams; and (iv) an example of an active lens solution for a Ka-band Tx multispot for wide-band applications and with frequency reuse applications.

9.4.5.1 X-band Rx: IRMA, SPAINSAT

SPAINSAT was launched in March 2006 to complete the Advanced Governmental Communications' Programme for the Spanish Ministry of Defence. The SPAINSAT satellite orbits at 30°W and is operated from the satellite control centres in Arganda del Rey (Madrid) and Maspalomas (Canary Islands).

Mission Requirements SPAINSAT is provided with an X-band receiving IRMA devised to fulfil the flexible communication scenarios to be supported by the satellite. The main driving requirements are:

1. *Multibeam reception operation:* The antenna provides four independent beams able to operate in the complete assigned Rx bandwidth of 250 MHz within X-band.
2. *Steering capacity and coverage:* Any of the four beams can be pointed in any direction over the visible Earth from GEO, which means up to 9.8° from boresight. Also, any of the four beams can be shaped to increase the coverage up to full Earth coverage (global coverage). The beams are assigned nominally to a fixed beam centred on the Iberian Peninsula with 4.5° cone beamwidth; two mobile beams with a 4° cone beamwidth; and one designed for shaped global coverage (19.6° coverage width).
3. *Interference mitigation:* The antenna provides interference tracking and mitigation in open loop by nulling in the direction of the interfering signal.
4. *G/T: :* This should be better than −2.5 dB/K for the spot beams and −14.0 dB/K for the global beam with a transponder noise factor NF = 9–9.4 dB, $T_a = 290$ K, a power flux density (PFD) of −75 to −100 dBW/m^2 and a maximum output power (P_o) of −49 dBm.
5. *Polarization and SLL:* The antenna operates in single RHCP with cross-polar discrimination (XPD) better than 25 dB and the SLL is specified to be below −20 dB.
6. Consumption from the S/C power bus should be less than 42 W, and control is through the 1553 Mil-bus.
7. The total mass, including the structure, electronics and thermal control, should be less than 47 kg.

Design Solution The design solution adopted for the IRMA array is a planar array composed of 64 array elements (Figure 9.25) in a rectangular lattice with a total aperture of 600 × 600 mm ($16\lambda \times 16\lambda$).

Each Rx module includes a sub-array of $2\lambda \times 2\lambda$, operating in RHCP, a filter for Tx signal rejection and a ceramic MCM. The latter has an LNA, a driver, a 1:4 Wilkinson divider to generate the four beams, and four attenuation and phase stages for independent control of beam illumination.

The sub-arrays are implemented in low-loss SSSL in printed technology with eight circular slots on a ground plane (Figure 9.26). The circular polarization is achieved by means of notches. Sequential rotation of sub-array elements was used to achieve more than 30 dB of polarization isolation within the pencil beam and 20 dB within the global coverage beam.

The filters provide enough isolation to avoid transmitted signals damaging or saturating the LNA. Extra filters at the output provide extra isolation for the payload.

Figure 9.25 IRMA X-band active multibeam RX antenna. Reproduced by permission of EADS CASA Espacio.

Four independent BFNs, implemented in SSSL, combine the signal (64:1) of each beam. The four beams provide full flexibility in shape and frequency, allowing pencil beams to be directed to any position on the visible Earth, with 33 dBi peak directivity or a shaped global beam with 22 dBi of peak directivity. Full control of the aperture illumination allows the nulling capability in case interference mitigation is required.

Thermal control is done through heat pipes that remove the heat from the units and equalize the temperatures between the different antenna elements. HPs are connected to a dedicated radiating panel covered by mirrors (OSRs) that reflect the heat into space. Heaters are placed over the base plate to inject heat in cold critical periods, such as during the transfer orbit and the non-operating conditions, and to allow proper startup of the electronics. Finally, external MLI and sunshields are used to reject the solar inputs, avoiding sharp hot phases.

The antenna includes control electronics (an interface control unit (ICU)) to communicate with the ground by the 1553 bus I/F platform, and a power supply unit (PSU) to provide secondary voltages for the electronics from the primary S/C bus. In order to comply with the single-point failure requirement, nominal and redundant units of both devices are included. The unit itself is on/off controlled by means of high-level pulse commands.

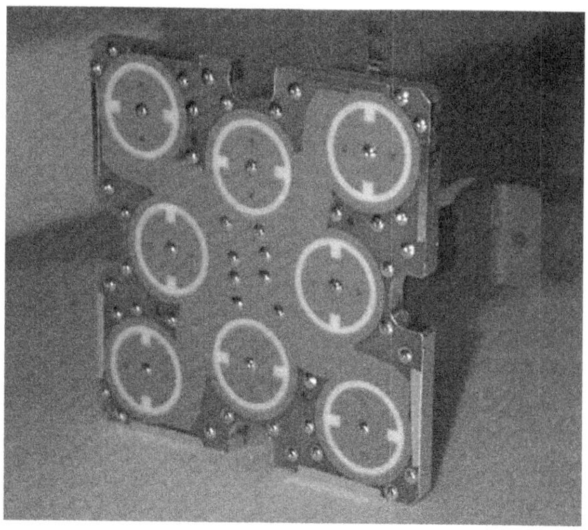

Figure 9.26 IRMA active Rx antenna sub-array. Reproduced by permission of EADS CASA Espacio.

Performances: Flexibility The antenna is defined to operate nominally by providing a fixed, a global and two mobile beams, as shown in Figure 9.27. The electronics allows about 1000 predefined pointing or shaping configurations to be stored and also allows the uploading of new ones throughout the mission life by tele-command. The fast beam switching capability together with the on-board configuration database allows it working in a beam hopping mode.

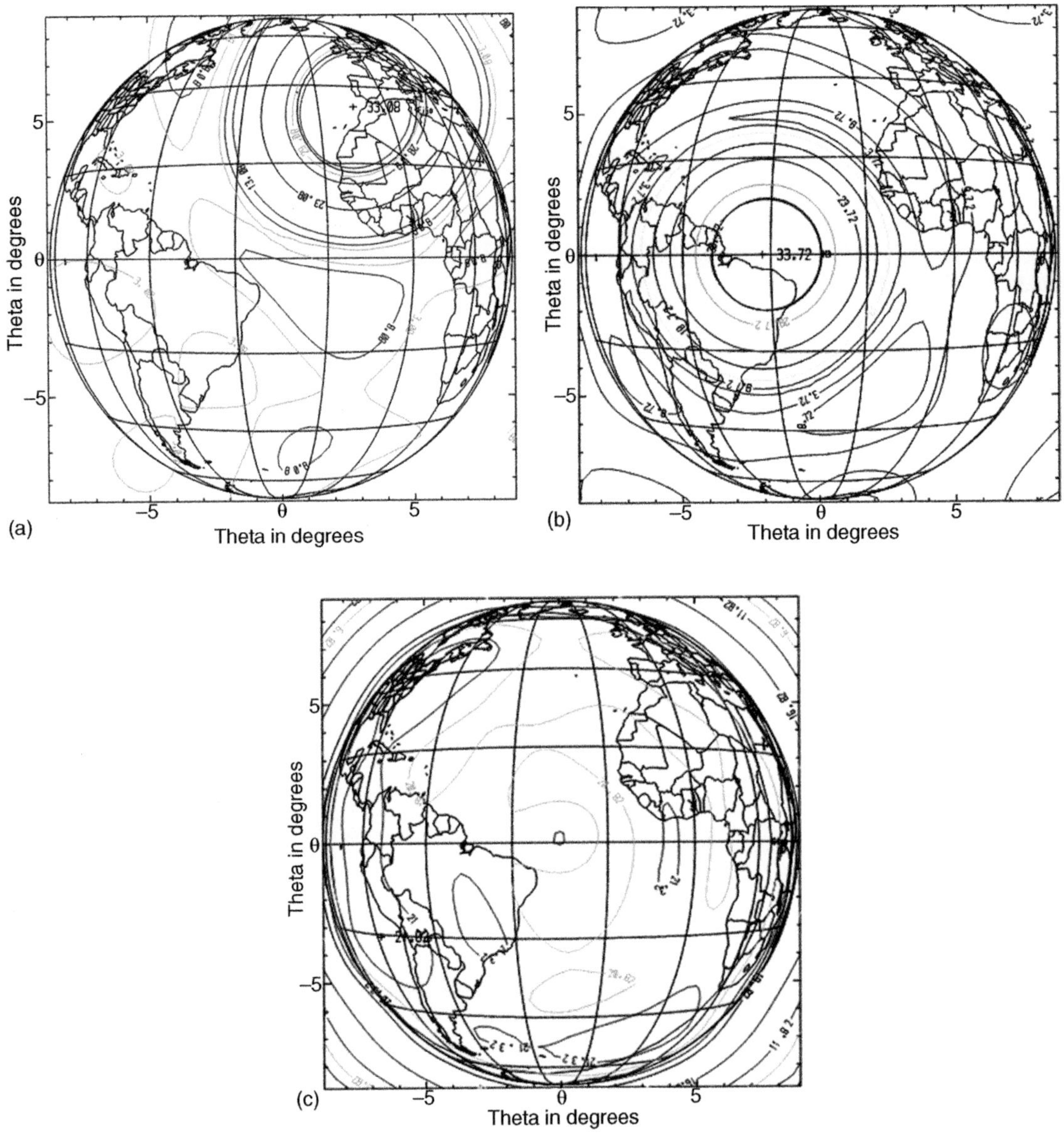

Figure 9.27 *IRMA antenna beams. Directivity coverage diagrams for (a) a fixed beam; (b) an example of a mobile beam; and (c) a global beam.*

9.4.5.2 Semi-active PAA

Earth observation or scientific missions require download of the mission data, typically while the relative position and orientation of the spacecraft change. Use of low-directive quasi-omnidirectional antennas is possible, but use of high-directive steerable solutions provides much better characteristics, while also avoiding dissemination of the information.

Mission Requirements: PAA versus Mechanically Steered Antenna (MSA) In a growing number of cases, the stringent payload requirements require the use of PAAs. Although these requirements typically depend on the mission characteristics, the following general equipment specifications are usually required:

1. *Reliability:* The PAA architecture greatly improves the reliability figure compared with MSAs based on pivoted supports (gimbals), electric-motor-driven mechanisms, RF rotary joints, and so on. In addition, the PAA does not need pyrotechnic release devices and deployment mechanisms, offering a more robust and reliable startup.
2. *Lifetime:* Since the PAA is not based on mechanical elements, which are much more sensitive to the environmental conditions, the operational lifetime of a PAA is significantly longer compared with its mechanically steered counterpart.
3. *Beam pointing capability:* The PAA is based on electronic beam steering, which allows for exceptionally high scanning rates. It is possible to provide a fast (<1 ms) beam pointing capability without compromising precision and accuracy.
4. *Link availability:* Based on the remarkable beam pointing capability, it is possible to use the link while the antenna is moving/rotating, thus increasing the effective data rate.
5. *Hardware fault tolerance:* The PAA architecture offers distributed power amplification in an active or semi-active configuration. This power amplification scheme removes the single-point failures associated with RF amplification. The self-redundant PAA architecture is able to translate a power amplifier chain failure into a slight, sometimes negligible, EIRP degradation.
6. *Power consumption:* Since the PAA beam switching capability is based on the operation of MMIC phase shifters, the PAA power requirements dedicated to the beam pointing are negligible compared with the actuators used in mechanically steered antennas. Therefore, the PAA architecture avoids potential current spikes associated with actuator operation.
7. *Mechanical disturbances:* Scientific payloads may require a minimal level of mechanical disturbances (microvibrations) and electrical variations during operation. Compared with mechanically steered antennas, the PAA architecture offers much higher mechanical and electrical stability. The electrical stability of MSAs is lower than that of their PAA counterpart due to the rotational effect of the rotary joints, which affects both the voltage standing wave ratio (VSWR) and the RF path insertion loss.

PAA Architecture The PAA has a conformal geometry (sphere, half dome, cylinder, truncated pyramid, cone, etc.), adapted to the operational field of view (FOV), which is populated by sub-arrays that are electronically scanned by means of phase shifters. All phase shifters are commanded by the ICUs/PSUs, which perform the required pointing operations and provide all the amplifiers with proper DC power. In order to optimize the transmitted RF power, the PAA architecture generally follows a semi-active or active antenna configuration. Therefore, there are a number of RF power amplifiers distributed over the antenna, with the intent to maximize the RF losses and to enhance the immunity of the antenna system to hardware failures.

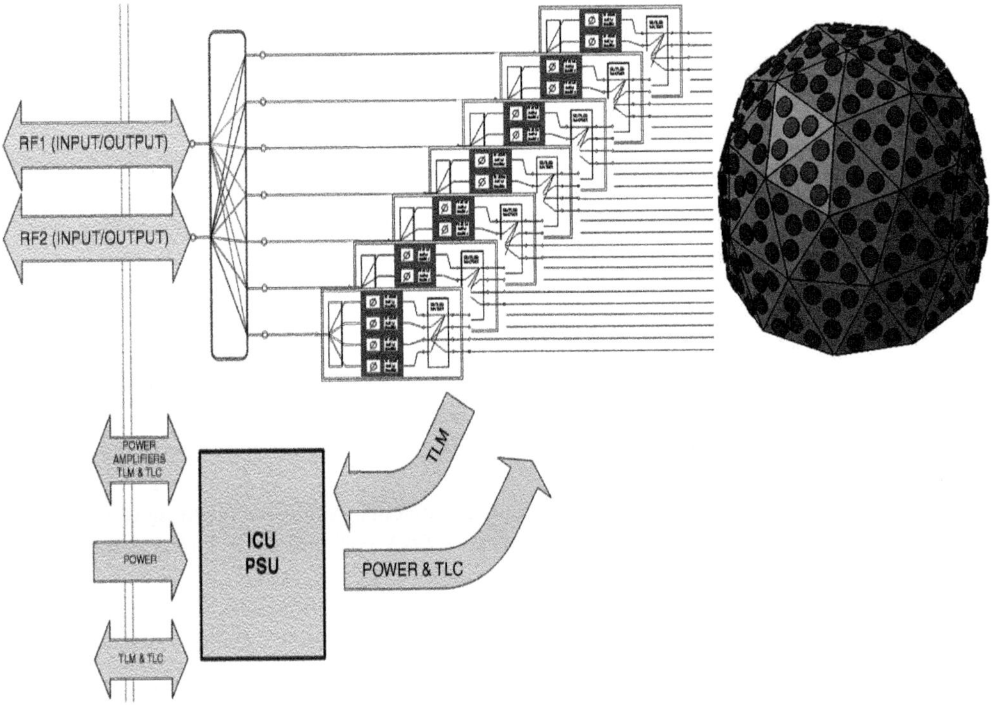

Figure 9.28 PAA block diagram (TLM, Telemetry; TMC, Command). Reproduced by permission of EADS CASA Espacio.

The PAA block diagram, which is shown in Figure 9.28, normally comprises the following elements:

- 2:N-way input splitter, for nominal and redundant inputs/outputs.
- RF harness, from splitter to Tx/Rx module, and from Tx/Rx module to radiators.
- Phase-controlled bidirectional amplifiers (depicted in Figure 9.29), including Tx and Rx active modules, which are shown in Figure 9.30.
- Conformal radiating array.
- ICU/PSU and DC harness.

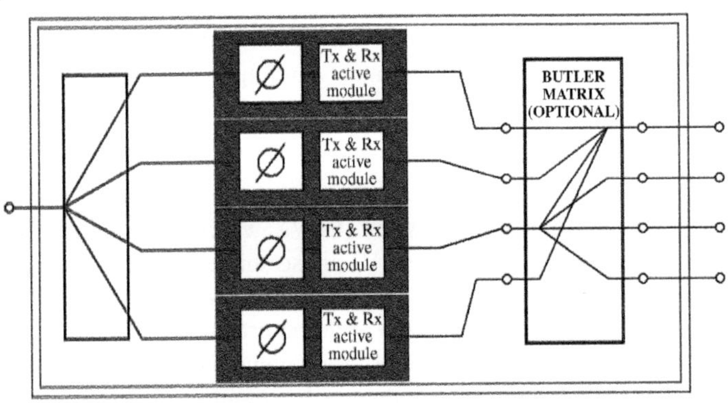

Figure 9.29 Phase-controlled bidirectional amplifier.

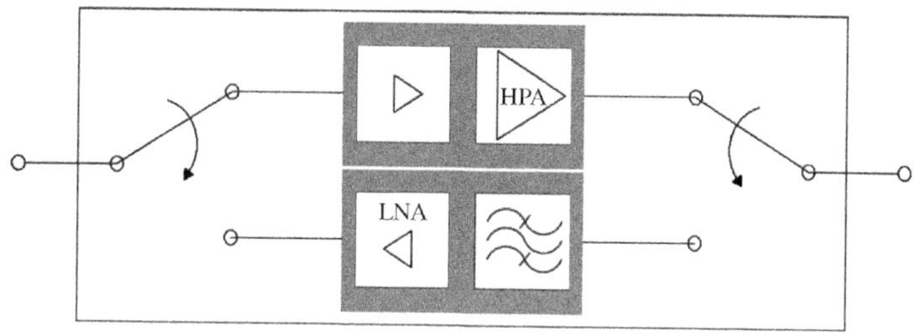

Figure 9.30 Tx and Rx active module.

As shown in Figure 9.28, there are usually two RF inputs from the payload, one from each of the cold redundant transponders that are routed to the 2:N-way splitter inputs. Under normal operating conditions, only one of these two inputs is active and the RF signal is divided into N ways to supply equal RF power at the inputs of all the phase-controlled amplifiers.

The phase-controlled amplifiers, shown in Figure 9.29, provide phase control, power amplification and signal routing to the appropriate sub-arrays within the conformal array by using phase shifters within the individual amplifying chains, allowing accurate steering of the antenna pattern. In order to allow bidirectional operation, each amplifying chain contains a Tx and Rx active module with bidirectional capability (see Figure 9.30). The scheme corresponds to semi-duplex communication. When full-duplex communication is required, frequency discrimination and filtering (diplexer) should be used.

Simple array architectures making use of hybrids allow the power to be driven to only one-half of the antenna radiating elements using all the installed amplifiers. More sophisticated techniques based on Butler matrices allow higher focusing while generating higher losses at the output stage and being more sensitive to circuit tolerances.

Principle of Operation: Pattern Selection Algorithm For proper operation of the PAA, a finite set of beams is selected that together ensure the requirements on EIRP are met everywhere in the coverage (see Figure 9.31)

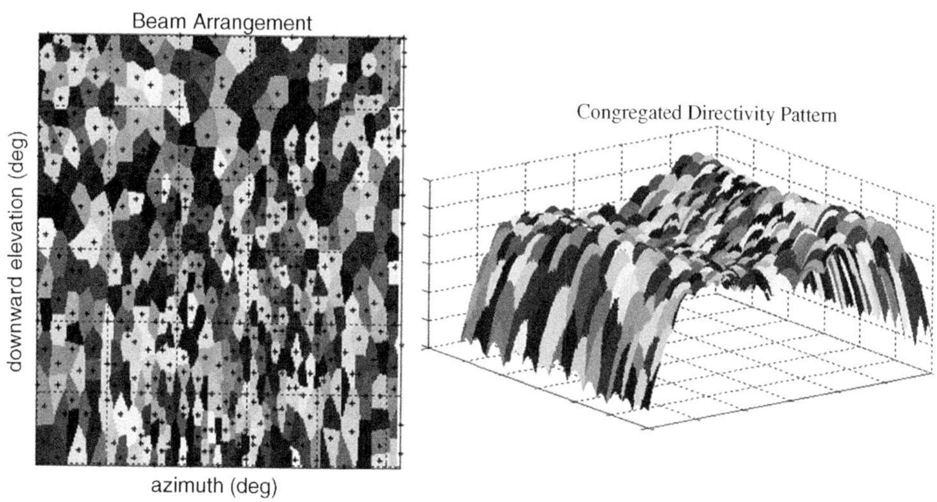

Figure 9.31 Typical congregated directivity pattern of a PAA. Each grey scale represents a different beam and its associated phase status vector.

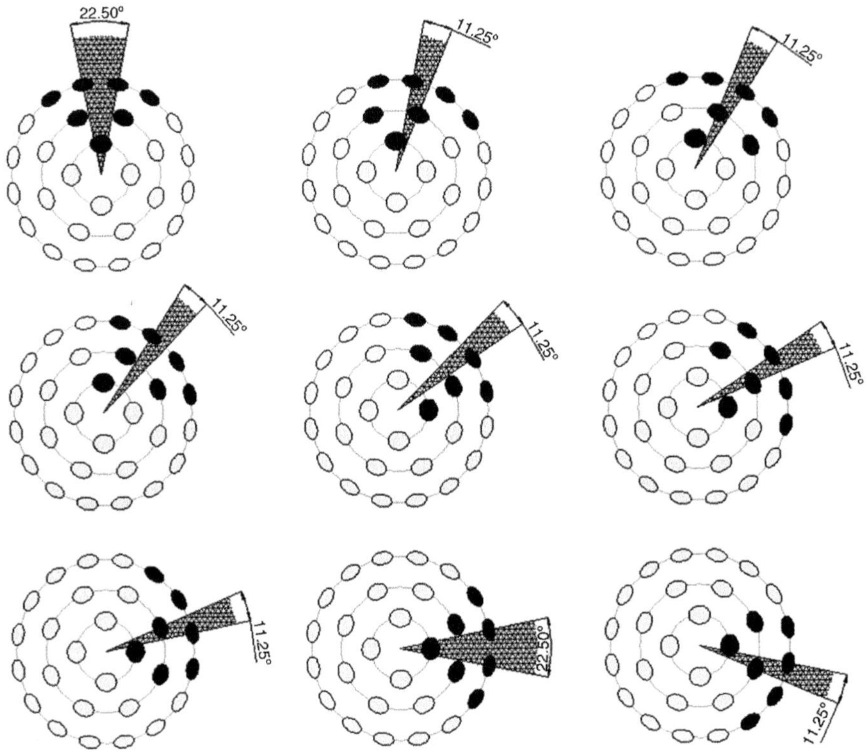

Figure 9.32 Active (dark) and passive (clear) sub-arrays as a function of the operational azimuth angle. Figures show the top views of a hemispherical array.

and the right time switching from one beam to the next, by synchronous control of the phase shifters. Optimum EIRP in every coverage direction is achieved by selecting the correct relative phase difference between the input signals at the Butler matrix or hybrid so that all the RF power is driven towards the most effective sub-arrays. As an example, Figure 9.32 shows the top view of a hemispherical PAA with a four-port Butler matrix, where the dark/clear dots represent the active/passive sub-arrays.

By removing the Butler matrix, and its associated RF harness, it is possible to place a power amplifier just at the input of each sub-array. In this case, selection of the most effective sub-arrays is performed by switching on/off those amplifiers which are/are not efficiently contributing to the operating azimuth–elevation direction. It should be noticed that only the active amplifiers affect the DC power consumption budget.

This solution minimizes the insertion loss after the power amplification and it is not sensible to hybrid or Butler matrix implementation errors. However, only parts of the amplifiers are in use simultaneously.

A second degree of freedom is achieved by introducing different phase offsets to all the inputs of different Butler matrices controlling the array factor value of the selected sub-arrays. Note that the previous sub-array selection is not altered by these offsets.

9.4.5.3 Ku-band-Rx: ELSA

Mission Requirements Under the REDSAT programme of the Spanish CDTI and ESA, an advanced ELSA was designed for a communications mission on a GEO satellite. The main objective was to put in orbit a reconfigurable antenna in Ku-band giving customers the needed flexibility over the typical 15-year mission period [9,10]. The main requirements are as follows:

1. *Flexibility:* The antenna provides four Rx independent beams in the Ku-band (14.25–14.50 GHz) in horizontal polarization. Each beam can be configured as a spot beam (1.5° beamwidth) that can point towards any location on the Earth or configured as a shaped pattern. Furthermore, the antenna can be operated over the beam hopping configuration, where for each beam several spots and/or shaped patterns can be sequentially switched. It is able to store and upload a large number of beam pointing or shaping configurations.
2. *G/T:* The main driving requirement is to get, through a design compromise solution of the internal amplification stages, a *G/T* greater than 5 dB/K for the four beams in any direction from the visible Earth while complying with strong constraints from the payload transponder, such as a receiver noise of 23.2 dB, a maximum receiver input power of −30 dBm and a carrier third-order intermodulation product at the output higher than 52 dBc.
3. *Cross-polarization and SSL:* Another relevant requirement for the radiator polarization purity and pattern synthesis is that the XPI should be higher than 23 dB (in some directions below 28 dB) and the SLL better than 25 dB beyond the first sidelobe.
4. *Filtering:* The antenna is mounted on the Earth deck of a GEO S/C that accommodates Tx antennas of the mission in Ka-band (from 18.1 to 21.2 GHz) and in Ku-band (from 10.7 to 12.75 GHz). The input filter sizing is mainly related to the interfering power flux density (PFD) from the Tx antennas.
5. *Power consumption mass, volume and control:* Other key requirements are a power consumption lower than 82 W, from a power bus line of 50 VDC, a mass below 61 kg, an envelope within 950 × 750 × 400 mm, control through MIL-bus 1553, and analogue and digital telemetries.
6. *Environment:* The environment imposes hard conditions on mechanical and thermal control.

Design Solution As a solution for the ELSA DRA, a self-standing integrated configuration was devised, as shown in Figure 9.33. A base plate structure provides enough stiffness and protection against the environment (first global mode above 90 Hz). Acting as a box for the electronics, the structure provides radiation hardness and EMC/ESD protection and is also the interface and support for most of the equipment. The mechanical interface comprises nine carbon fibre tubular struts and titanium end fittings.

This integrated construction incorporates the functionalities and operational subsystems shown in Figure 9.34.

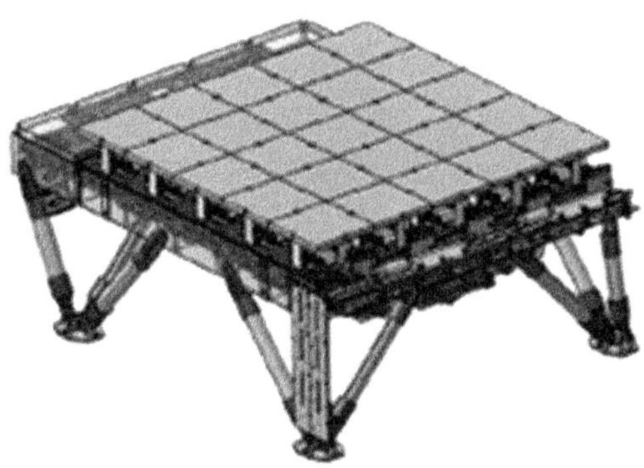

Figure 9.33 *ELSA, Ku-band Rx DRA from the REDSAT mission. Reproduced by permission of EADS CASA Espacio.*

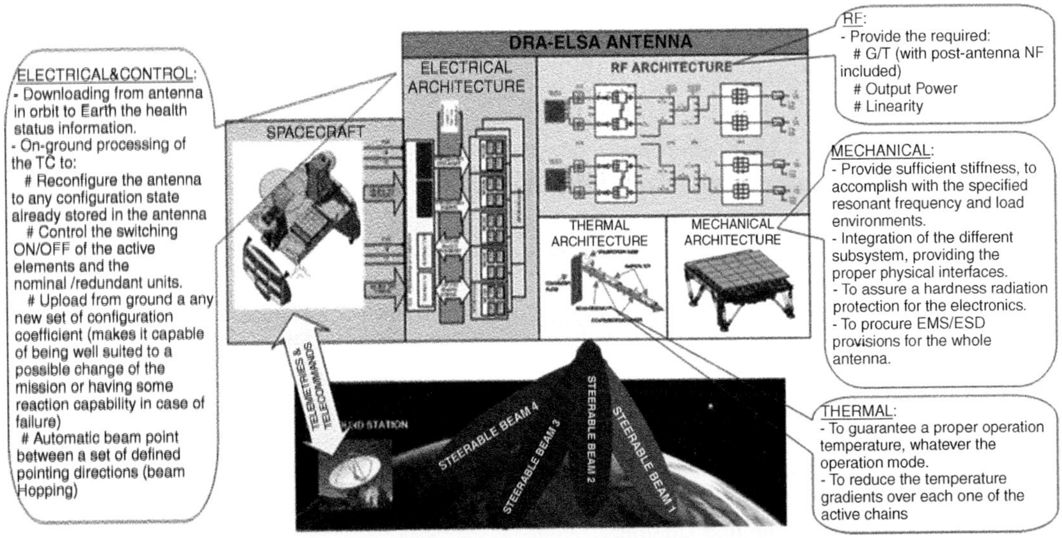

Figure 9.34 *ELSA main functionalities. Reproduced by permission of EADS CASA Espacio.*

The ELSA design is based on the heritage from IRMA (X-band Rx) (see Section 9.4.5.1) adapted to the Ku-band specifications. A configuration with 100 (10×10) sub-arrays of 16 patches each (around 3λ) has been selected (Figure 9.35), over a total aperture of 600×600 mm. In order to reduce sidelobe levels, only 80 sub-arrays are operative in the nominal configuration. The 20 corners of most sub-arrays are switched off and their use is planned only in case of failures.

To avoid the interfering PFD, a filter has been included after each of the sub-arrays (Figure 9.36). In order to simplify the manufacturing and integration, mechanical implementation of the filter is made in groups of four.

Also, as the integration of the electronics for such a large number of elements in a reduced volume is critical, a configuration considering a multi-chip control module (MCCM) that combines the RF and electrical control

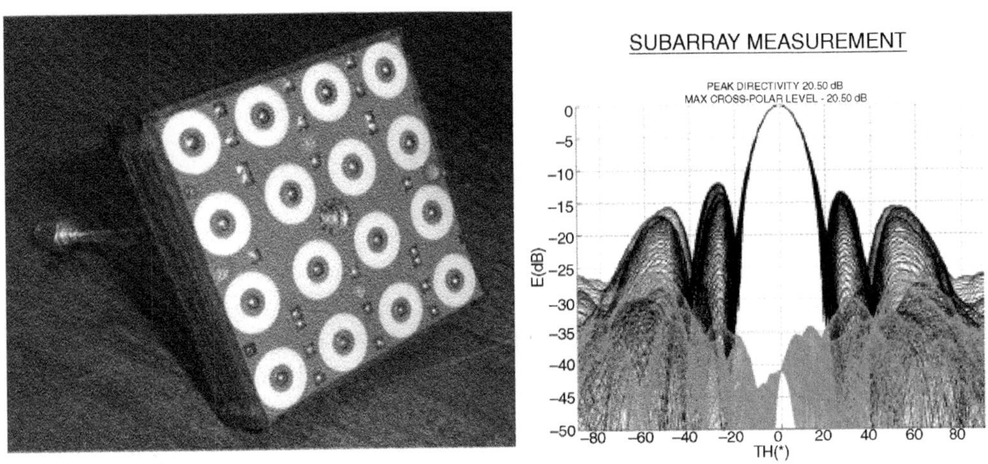

Figure 9.35 *Sub-array of 16 × 16 elements and radiation pattern test performance. Reproduced by permission of EADS CASA Espacio.*

Figure 9.36 Filter block of the ELSA Ku-band Rx antenna. Reproduced by permission of EADS CASA Espacio.

functions for a 2 × 2 sub-array has been developed (Figure 9.37). It has been implemented in multilayer technology over an HTCC package populated on both faces.

From the RF point of view, each MCCM includes a two-stage LNA for each sub-array, a 1:4 divider for each beam, an attenuator and phase shifter MMICs for beam control (per sub-array and beam), a combiner for the 2 × 2 sub-array unit and a final amplification stage per beam to compensate part of the losses of the RF chain.

After that, for each of the four outputs of the 25 MCCMs the signal is combined through four 25:1 combiners implemented in two steps of 5:1 (20 plus 4 BFNs in total). In order to match the output power of the antenna to the rest of the payload, a signal conditioner (MCCM-SC) has been included. The MCCM includes a custom MMIC (LNAs, attenuator and phase shifter) package and ASICs (digital and analogue).

From the thermal point of view, transport heat pipes (THPs) have been included in close contact with each row of the MCCM to extract the dissipated power, and a dual equalization heat pipe (EHP) interconnects all the THPs to two loop heat pipes (LHPs) connected to a thermal radiator panel. Thermal control of the antenna guarantees not only a small gradient over all the 25 MCCMs, but also the maximum and minimum temperature excursion of all the sub-units over the 15-year mission period.

The antenna electronics are implemented in three control levels: (i) the PSU and ICU; (ii) distribution boards; and (iii) MCCMs. Each MCCM includes the electronics needed both to communicate to the ICU of the

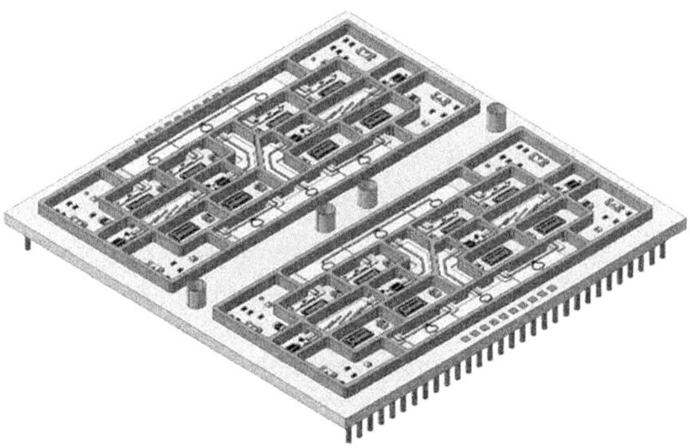

Figure 9.37 MCCM of the ELSA Ku-band Rx antenna. Reproduced by permission of EADS CASA Espacio.

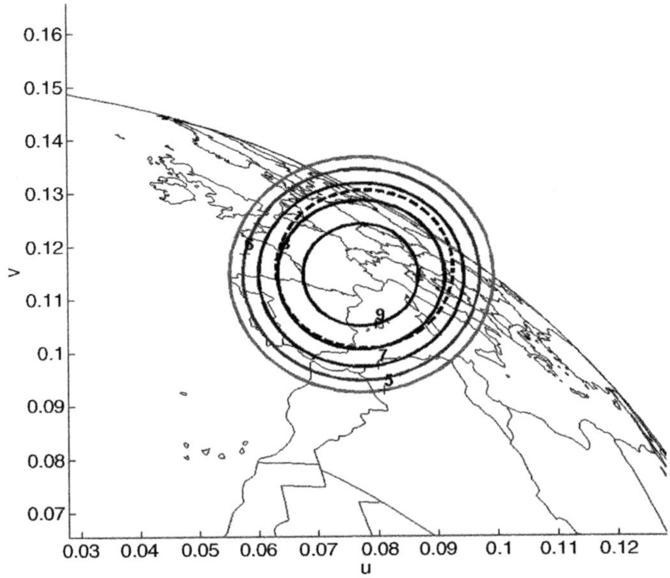

Figure 9.38 *Spot beam G/T over the Earth's edge. Max G/T, 11.20 dB/K; min G/T, 6.65 dB/K; min XPI, 27.5 dB; f = 14.25–14.50 GHz. (Worst case conditions: transponder NF = 23.2 dB; operational temperature of 10 °C.).*

antenna and to command the on/off of each amplification stage, attenuator and phase shifter configuration and telemetry functions by means of a dual-mode digital ASIC (communication/shift register) and analogue ASIC (on/off and telemetries). Each MCCM is connected through a redundant bus to the ICU by means of the distribution boards (DBs) and RF control boards (RFCBs). The architecture, including redundancy of the critical parts and the graceful degradation inherent to this kind of antenna, provides a robust design that avoids any single-point failure (SPF) and good reliability for the mission life.

Performances: Flexibility Apart from the compactness, endurance against the thermo-mechanical radiation environment and compliance with the current functional needs and specifications, this example of an active array provides great flexibility, which is a quite attractive feature when operation over a long period is taken into account. The antenna is able to generate spot beams and shaped beams in any of the four beams and it can be reconfigured just by ground commands. Also, any newly required configuration or scenario can be uploaded to the antenna while in operation. Some examples of beams are included in Figures 9.38 and 9.39 (spot coverage) and in Figure 9.40 (shaped coverage).

9.4.5.4 Ka-band Tx Multispot: Active Lens Solution

One of the challenging problems of active antenna arrays for space applications is the possibility to cope with flexible and multispot capabilities (see Section 9.2.1.2). The flexibility to assign power and bandwidth is inherent in active DRAs but full control is required to achieve beam reconfigurability and steering. This control can be achieved by beam-independent BFNs with amplitude and phase control. However, for antennas with a large number of elements and beams the complexity and number of these BFNs will be a severe limiting factor. The main drawback of alternative solutions based on Butler matrices is related to the network losses and to the high complexity which increases with the number of beams and with the antenna size.

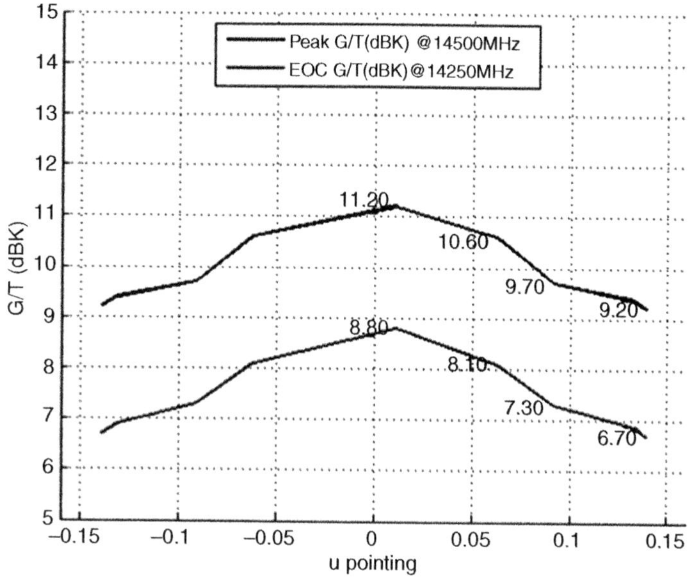

Figure 9.39 *G/T performance of ELSA, peak and EOC for each beam as points away from boresight to the Earth's edge (f = 14.25–14.50 GHz).*

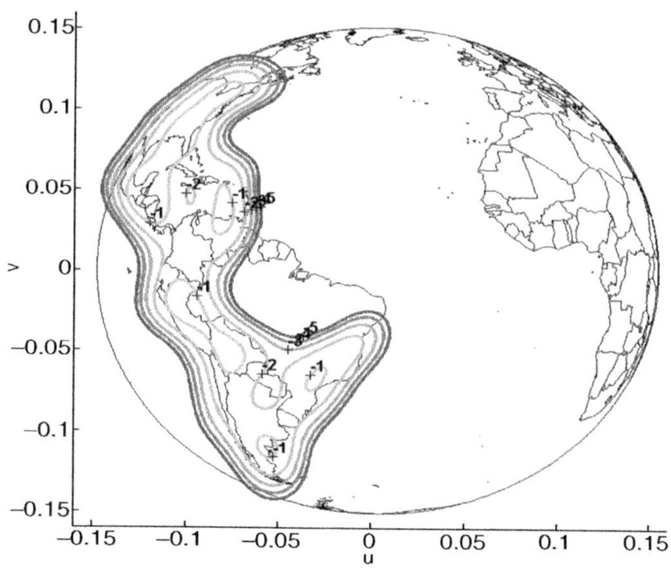

Figure 9.40 *North and South America G/T coverage. In this case the achieved minimum G/T is −2.5 dB/K; that is, 2.5 dB above the classical reflector configuration (−5.0 dB/K) for the same mission, considering the worst case conditions for a transponder NF of 23.2 dB and an operational temperature of 10 °C.*

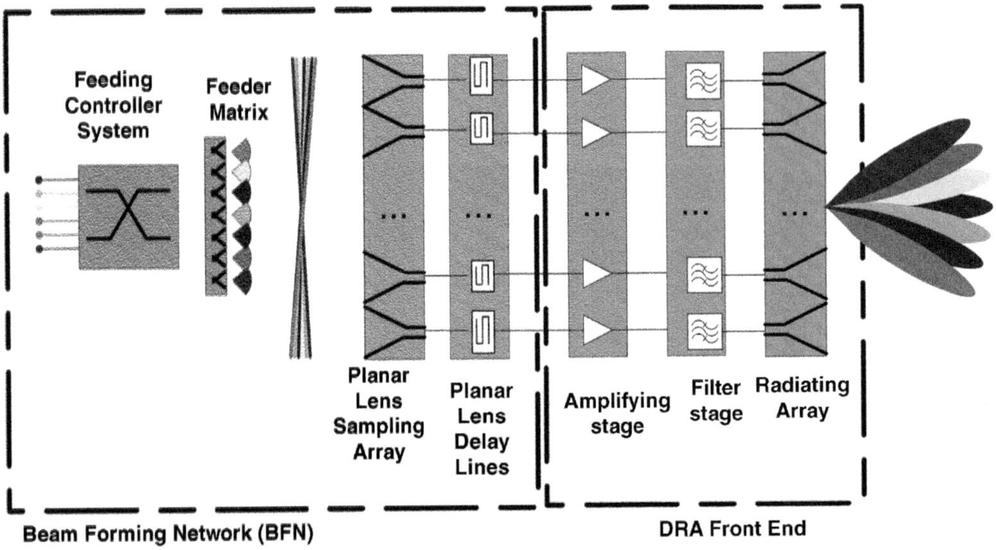

Figure 9.41 *DRA lens-based antenna developed at Ka-band.*

As a possible solution, a DRA lens-based antenna was developed within the framework of the ESA ARTES-5 programme. This architecture provides the possibility to generate a large number of beams and gives additional advantages, such as not needing variable attenuators, phase shifters and dividers, thus greatly reducing the number of components, connections and control lines. The antenna developed is a transmission antenna designed to work in Ka-band frequency (19.7–20.2 GHz) with dual linear polarization whose XP isolation is higher than 28 dB.

The lens–array architecture is composed of two main blocks, the first one being the BFN and the second the DRA front end (see Figure 9.41). The BFN is itself divided into four parts: the feeding control system (FCS), the feeder matrix and the sampling array with the lens compensation delay lines. The FCS provides the antenna beam configuration by means of distributing the signal of the channels among the feeders of the feeder matrix. This feeder matrix consists of a set of low-directivity horns that illuminates the lens and whose different configurations allow the scanning of the beam, null steering and aberration compensation, among other capabilities. At this point, the lens collimates the spherical waves coming from the feeders and conforms planar waves to different pointing directions. In this antenna, a planar lens has been selected to simplify the design, manufacture and accommodation. The DRA front-end works as in a conventional DRA system, with an amplifying stage, a filtering stage and a radiating array.

In the design of the active lens antenna, several tradeoffs have been made while selecting the radiating elements (patches or horns), their sizing and arrangement [3]. The result of these studies led to an antenna composed of radiating and sampling arrays of 454 conical horns of 3.2λ diameter in a hexagonal lattice. The filtering stage in dual polarization, located directly before the radiating array, was subsequently designed in waveguide technology in order to simplify interconnections and minimize losses. The amplifying stages located between the sampling array and the filtering stage have been carefully designed, including a study on the input backoff optimization for solid state power amplitifers (SSPAs) in a DRA [4]. The connection between the amplification stages and the elements in waveguide technology (i.e. horns and filters) is performed through an orthomode transducer (OMT), also designed in waveguide technology. Special care has been taken in the design of the waveguide-based elements in order to cope with the dual linear polarization requirement.

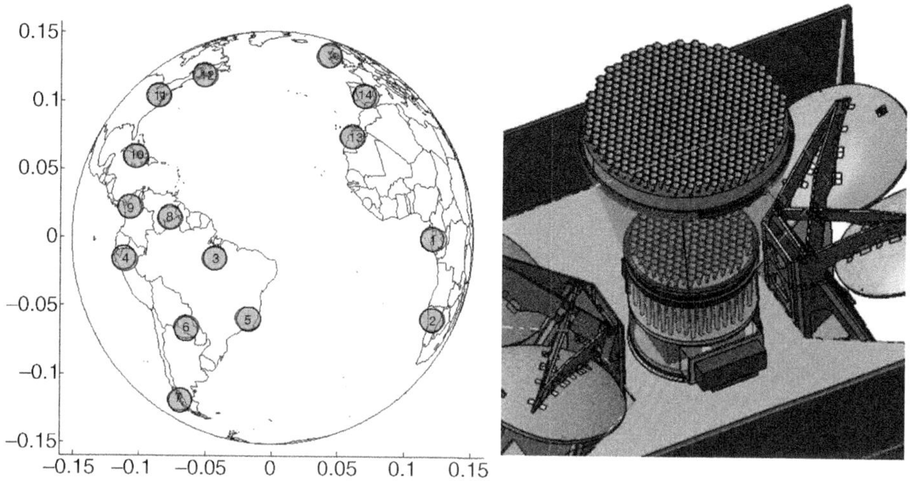

Figure 9.42 *The 15 V + 15 H beam cases and accommodation of a Ka-band flexible DRA. Reproduced by permission of EADS CASA Espacio.*

Additionally, the waveguide elements are used as supporting structures and high integration and compactness efforts have been applied.

At the feed matrix, feeding horns as small as 0.8λ in diameter have to be designed since a cluster of horns was needed to have enough beams to cover the whole Earth seen from the satellite. The total number of feed horns depends on the final number of beams desired and also on the necessity.

Figure 9.42 shows an operational example with 15 beams in one polarization and with full bandwidth reutilization. The other 15 simultaneous beams can be implemented in other selectable directions in the orthogonal polarization. In this case, for 30 channels/beams (15 H + 15 V), with EIRP ranging from 44 to 49 dBW of operating EIRP (51–56 dBW at equivalent saturation), the primary bus DC consumption per channel is 50 W including PSU efficiency. The antenna's dimensions are $1 \times 1 \times 1.4$ m and its aperture is made up of 454 horns. The estimated mass budget of the antenna is 130 kg.

9.5 Summary

Passive and active array applications in space have been presented together with the main driving requirements and design developed for actual space missions, ranging from deep space data transmission to navigation, Earth observation and multibeam communication. The array antenna solutions provide advantages for those solutions based on passive arrays with low profile and low mass, or those based on active arrays with electrical steering or multibeam reconfigurability, and even nulling for mitigating interference.

Acknowledgements

Antonio Montesano would like to thank his parents for their great support, his brother, his teachers, his colleagues at EADS CASA Espacio, his kids and finally his wife for her constant, full and unconditional support.

References

1. ECSS (2003) *Multipaction Design and Test*, ECSS-E-20-01A.
2. Campuzano, J., Olea, A., Montesano, A. *et al.* (2010) Next generation telecommunication payloads based on active array antenna technology. Proceedings of the 32nd ESA Antenna Workshop on Antennas for Space Applications: From technologies to architectures, ESA/ESTEC, Noordwijk, The Netherlands, 5–8 October.
3. Campuzano, J., Montesamo, A., de la Fuente, L.F. *et al.* (2008) Trade-off assessment of sampling distance in a DRA antenna. Proceedings of the 30th ESA Antenna Workshop on Antennas for Earth Observation, Science, Telecommunication and Navigation Space Missions, ESA/ESTEC, Noordwijk, The Netherlands, 18–20 November.
4. Campuzano, J., Montesamo, A., de la Fuente, L.F. *et al.* (2008) Input back-off optimisation for SSPAS in a DRA antenna. Proceedings of the 30th ESA Antenna Workshop on Antennas for Earth Observation, Science, Telecommunication and Navigation Space Missions, ESA/ESTEC, Noordwijk, The Netherlands, 18–20 November.
5. Monjas, F., Montesano, A. and Arenas, S. (2010) Group delay performances of Galileo system navigation antenna for global positioning. Proceedings of the 32nd ESA Antenna Workshop on Antennas for Space Applications: From technologies to architectures, ESA/ESTEC, Noordwijk, The Netherlands, 5–8 October.
6. Arenas, S., Monjas, F., Montesano, A. *et al.* (2011) Performances of GALILEO system navigation antenna for global positioning. Proceedings of the 5th European Conference on Antennas and Propagation, EUCAP 2011, Rome, Italy, 11–15 April, pp. 1018–1022.
7. Olea, A., Montesano, A., Montesano, C. and Arenas, S. (2008) High gain antenna (X-band) for Mars mission. Proceedings of the 30th ESA Antenna Workshop on Antennas for Earth Observation, Science, Telecommunication and Navigation Space Missions, ESA/ESTEC, Noordwijk, The Netherlands, 18–20 November.
8. Olea, A., Arenas, S., Montesano, A. and Montesano, C. (2010) High gain antenna system (X-band) qualified & acceded for Mars atmosphere. Proceeding of the 32nd ESA Antenna Workshop on Antennas for Space Applications: From technologies to architectures, ESA/ESTEC, Noordwijk, The Netherlands, 5–8 October.
9. Monjas, F., Martín, I., Solana, A. *et al.* (2010) X-Band SAR antenna for SEOSAR/PAZ satellite. Proceedings of the 32nd ESA Antenna Workshop on Antennas for Space Applications: From technologies to architectures, ESA/ESTEC, Noordwijk, The Netherlands, 5–8 October.
10. Montesano, A., de la Fuente, L.F., Bustamante, M. *et al.* (2009) EADS CASA Espacio flexible payloads based on Rx DRA: IRMA heritage in X band and REDSAT development in Ku band. Proceedings of the 31st ESA Antenna Workshop on Flexible Payloads, ESA/ESTEC, Noordwijk, The Netherlands, 18–20 May.
11. Campuzano, J., Montesano, A., de la Fuente, L.F. *et al.* (2010) EADS CASA Espacio DRA-ELSA patterns and G/T predicted performances. Proceedings of the 32nd ESA Antenna Workshop on Antennas for Space Applications: From technologies to architectures, ESA/ESTEC, Noordwijk, The Netherlands, 5–8 October.

10

Printed Reflectarray Antennas for Space Applications

Jose A. Encinar

Technical University of Madrid, Spain

This chapter presents an overview of reflectarray antennas including their historical development, state of the art, and recent and future developments. Some examples of reflectarray antennas for space applications are presented.

10.1 Introduction

A reflectarray is made up of an array of reflective elements providing a pre-adjusted phasing to form a focused beam when it is illuminated by a feed, in a similar way to a parabolic antenna. Printed reflectarrays combine certain advantages of reflector antennas and phased arrays. They are manufactured on a planar substrate using printed circuit technology and offer the possibility of beam steering as phased arrays; furthermore, the array elements are fed by an external source as in a reflector antenna, eliminating the complexity and losses of the feeding network used in planar arrays.

The concept of a reflectarray antenna was first proposed in 1963 by Berry *et al.* [1], as an aperture characterized by a surface impedance illuminated by a primary feed, where the surface impedance is synthesized to provide a prescribed radiation pattern. Different implementations of surface impedance were proposed in this reference, such as arrays of dipoles connected to transmission lines or open-ended waveguides terminated in short circuits. The capabilities of the reflectarray antenna were demonstrated by presenting measured radiation patterns for different scan angles, which were implemented by waveguides with different lengths. Furthermore, the electronic scanning or reconfiguration of the beam was also proposed in this pioneering publication by inserting diodes in order to change the distance from each waveguide aperture to its short-circuit termination, and then to synthesize the radiation pattern. Although the results were satisfactory,

Space Antenna Handbook, First Edition. Edited by William A. Imbriale, Steven (Shichang) Gao and Luigi Boccia.

this type of waveguide reflectarray was very heavy and bulky, and for this reason the reflectarray antennas only earned interest in the late 1980s with the development of microstrip antennas.

An alternative concept called spiraphase [2,3] was proposed in the 1970s to provide the phasing in reflectarrays working in circular polarization. The technique consists of switching between different conductive arms, typically arranged in a spiral, resulting in an equivalent rotation of the active spiral arm. The rotation of the reflective element produces a phase shift in the reflected field proportional to twice the rotation angle when the incident field coming from the primary source is circularly polarized. Although this concept provides electronic reconfiguration by switching between the spiral arms using diodes, it did not receive much attention because of several drawbacks, such as the narrow band, despite the use of broadband spiral elements and low efficiency produced by large element spacing associated with the oversized spiral radiators.

The development of microstrip antennas in the 1970s [4] providing low-profile and lightweight implementations was also applied to reflectarrays, see Figure 10.1. The first mention of microstrip elements for reflectarrays was by Malagisi [5,6] in 1977, where a circular printed patch loaded with diodes on the periphery was proposed to provide electronic beam scanning, using the same operating principle as the spiraphase element for circular polarization. The circular disk element was analyzed using the infinite array approach and the results were validated by measuring the reflectarray elements in S-band using a waveguide simulator [7]. More recently, this concept was used to provide the required phasing in circular polarization reflectarrays using identical printed elements, such as square printed patches with attached stubs on the orthogonal sides [8,9], see Figure 10.2a, or loaded ring slot resonators [10] having different angular rotations. Continuous beam scanning was proposed in [8] by inserting miniature or micromachined motors under each microstrip element. The concept of spiraphase was also proposed in [11,12] for reconfigurable circularly polarized reflectarrays, where the reflective element is composed of a circular conductive patch and several radial stubs around the periphery that can be connected or disconnected to the radiating patch by switches. For operation, two opposite stubs are connected to the circular patch and the rest disconnected, working in an equivalent way as rotation of the active dipoles, but operated electronically.

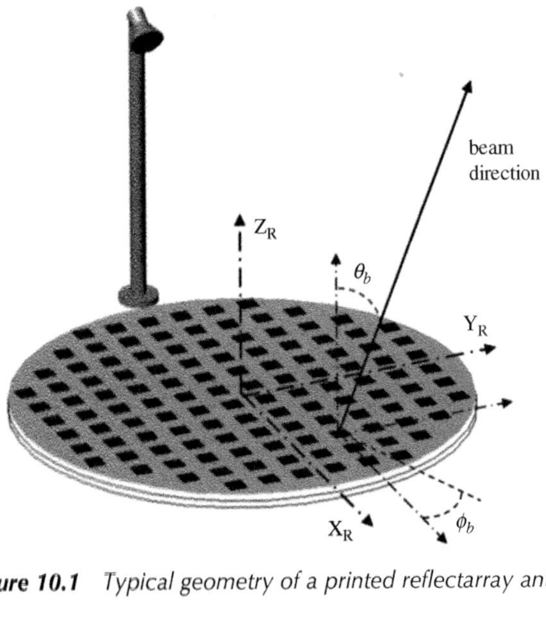

Figure 10.1 *Typical geometry of a printed reflectarray antenna.*

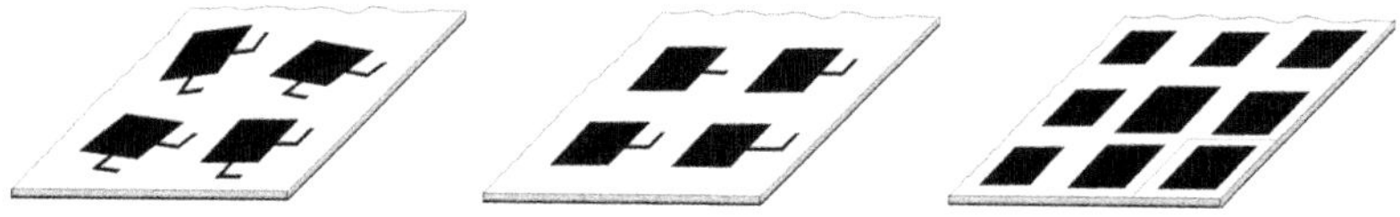

Figure 10.2 Phasing elements in printed reflectarrays: (a) patches with attached stubs and variable angular rotations; (b) rectangular patches with attached stubs; (c) rectangular patches of varying size.

Different concepts were developed in the late 1980s and early 1990s for adjusting the phase shift on printed flat reflectarrays, the most widespread implementations being those shown in Figure 10.2. A classic realization consists of connecting transmission line segments of different lengths to the printed resonant patches [13–16], see Figure 10.2b. The phase delay of the reflected wave is proportional to the length of the stubs, although dissipative losses and spurious radiation produced by the stubs are a disadvantage. Cross-polarization can be reduced by changing the orientation of the stubs, as shown in [16]. In a second implementation, the phase shift of the reflected field is controlled by varying the resonant length of dipoles, crossed dipoles or rectangular patches, as shown in Figure 10.2c [17–21]. These elements produce lower dissipative losses and cross-polarization levels than the printed patches with attached stubs. The main drawback is the inherent narrow-band behavior in microstrip patches, which can be overcome by using two or three stacked array layers [22–24]. Other concepts have also been developed for microstrip reflectarrays, such as aperture-coupled patches with stubs [25], or apertures of different length on a metal plane [26].

In the late 1990s and early 2000s, the development of reflectarray antennas rapidly increased, including steerable beams [27], amplifying reflectarrays [28], reflectarrays in a folded configuration with single and multiple beams [29–31], contoured-beam reflectarrays for space telecommunications antennas [32–36], large apertures for Earth observation [37], inflatable reflectarrays for X- and Ka-bands [38–41] and reflectarrays in dual-reflector configurations [42–44]. Reflectarrays have been proposed in space applications for micro-satellites [45] and also as telecommunications antennas for frequency reuse [46], as an alternative to dual-gridded reflectors. More recently, great effort has been made to develop reflectarrays with beam scanning or beam reconfiguration capabilities [47–59], using different technologies, such as micro-electromechanical systems (MEMS), varactor diodes, ferroelectric materials and liquid crystals.

Reflectarrays have demonstrated their capability to provide high efficiency in large-aperture antennas (as in reflectors) and to produce very accurate contoured beams for direct broadcast satellite (DBS) antennas by using a flat sandwich with printed patches appropriately optimized [34–36]. The manufacturing costs are low due to the elimination of the custom molds, which are required to manufacture the conventional shaped reflectors. A reflectarray with several thousand printed elements can be manufactured by simple and low-cost chemical etching and conventional sandwich processes. For large-aperture antennas, the flat reflectarray allows a much simpler and reliable folding mechanism compared to that required for the doubly curved surface of parabolic reflectors. Similar to the parabolic reflector, multiple-beam capability can also be achieved by placing multiple feeds in the focal area of the antenna. In space antennas, the reflectarray can be flush mounted onto one spacecraft surface, reducing the supporting structure, mass and volume. In addition, the flat reflectarray can be integrated with solar panels, as addressed in [60,61]. The reflectarray technology can be applied throughout the microwave spectrum, millimeter wave [20,30,62], terahertz and even infrared ranges [63,64]. As an additional advantage of reflectarrays when compared to conventional reflector antennas, they can offer some capability to reconfigure the beam if controllable devices, such as switches (based on PIN diodes or MEMS) [12,47–50], varactor diodes [51–54] or other devices, are integrated at each unit element, or at sub-array level [65–67]. The prospect is that reflectarrays can be a moderate cost solution for reconfigurable antennas, as an alternative to phased arrays, which suffer from the drawbacks of high cost, complexity and consumption.

On the other hand, there is one major shortcoming associated with the reflectarray antenna, which is its narrow-bandwidth behavior [68,69], generally lower than 5% and even less for large reflectarrays. Reflectarray bandwidth is mainly limited by two different factors as described in [68]. The first one is the narrow band of the radiating elements, and the second is the differential spatial phase delay resulting from the different lengths from the feed to each point on the wavefront of the radiated beam. Since the major drawback of the reflectarray is its relatively narrow bandwidth, much effort has been made in recent years to improve the bandwidth [23,24], and some broadband techniques have permitted an extension of the antenna bandwidth to 10–15% [33–35] and even have achieved practical design for DBS antennas in both the transmit and received bands [36].

The purpose of this chapter is to present a comprehensive overview of reflectarray antennas, including the operating principles, analysis techniques, design procedures for contoured beams and bandwidth improvement, and some practical developments for space applications.

10.2 Principle of Operation and Reflectarray Element Performance

The operating principle of a reflectarray antenna, see Figure 10.1, can be explained by considering the reflectarray in transmitting mode with a feed horn located in a centered or offset position, and assuming that the reflectarray elements are in the far-field region of the horn. In this case, the electromagnetic field incident on each reflectarray element at a certain angle can be locally considered as a plane wave with a phase proportional to the distance from the phase center of the feed horn to each element. In order to convert the spherical wave radiated by the horn into a focused beam in the direction (θ_b, φ_b), the field must be reflected from each element i providing the following progressive phase, as known from array theory:

$$\Phi(x_i, y_i) = -k_0 \sin\theta_b \cos\varphi_b x_i - k_0 \sin\theta_b \sin\varphi_b y_i \tag{10.1}$$

where k_0 is the propagation constant in vacuum and (x_i, y_i) the coordinates of element i. The phase shift to be introduced on each element is obtained as the difference between (10.1) and the phase of the incident field coming from the feed,

$$\Phi_{Ri} = k_0(d_i - (x_i \cos\varphi_b + y_i \sin\varphi_b)\sin\theta_b) \tag{10.2}$$

where d_i is the distance from the phase center of the feed to the cell $\boldsymbol{i}$. Figure 10.3 shows the required phase shift on a circular reflectarray of 60 × 60 elements with a centered feed to produce a pencil beam in a direction normal to the surface. For the reflectarray design, the phase shift must be adjusted in each element to match these phases. A shaped beam can also be achieved if the phase distribution given by (10.2) is substituted by an appropriate phase distribution obtained by a phase-only synthesis method, as will be discussed in a later section.

The necessary phase shift at each element is obtained by varying one of the geometrical parameters in the reflectarray element. The first implementation of phase adjustment in rectangular microstrip patches consists of connecting transmission line segments of different lengths to the printed elements [13–16], see Figure 10.2b. The signal from the feed is received on each patch and propagated along the transmission line until the end, where it is reflected, propagated back and radiated by the microstrip patch with a phase-shift proportional to twice the line length. For the ideal transmission line model, the phase response is a linear function of the stub length, but in reality there are other effects that produce a nonlinear phase response, such as

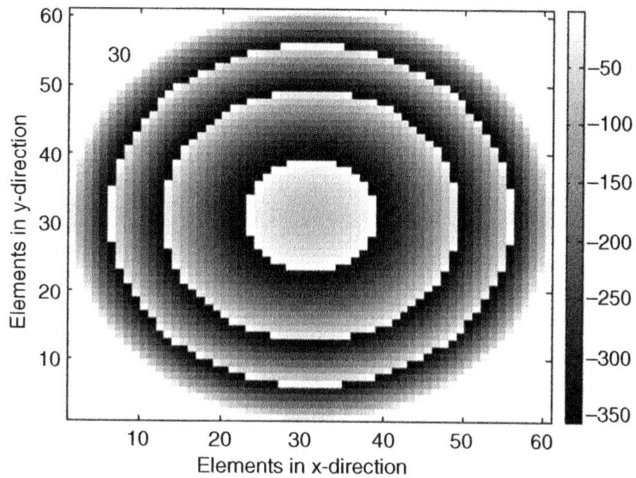

Figure 10.3 *Example of the required phase distribution at 12 GHz for an 84 cm circular reflectarray.*

the specular reflection on the ground plane and the resonances of the stubs. Both effects can be taken into account in a full-wave analysis of the reflective element [70,71], pp. 35–38, as discussed in Section 10.3.1. In addition, the printed line segments generate dissipative losses and spurious radiation that reduces the antenna efficiency and increases the cross-polarization levels.

Another concept for adjusting the phase in microstrip reflectarrays consists of varying the resonant length in dipoles, crossed dipoles or rectangular patches [17–21], as shown in Figure 10.2c. This realization eliminates the drawbacks associated with the attached stubs, such as spurious radiation and cross-polarization. The operating principle of the reflectarrays of variable-size printed elements is based on the fact that the phase of the reflected wave varies with the resonant length of the elements. A microstrip patch is a resonant antenna, so its length should be approximately half a wavelength in the dielectric. If the resonant length is modified in an array of square patches on a grounded dielectric, the phase of the reflected wave will be changed. The total range of phase variation that can be achieved by varying the length of the patches depends on the separation between the patches and ground plane, that is, the thickness of the substrate. For thicknesses smaller than a tenth of wavelength, a 330° range can be achieved as shown in Figure 10.4a for a thickness $t = 1$ mm and frequency $f = 12$ GHz. This range is sufficient for practical designs, but it will diminish for thicker substrates. However, the phase variation versus the length is strongly nonlinear, exhibiting very rapid variations near the resonance, and very slow variations in the extreme values. The rapid phase variation makes the phase distribution very sensitive to manufacturing tolerance errors. Because of the nonlinear behavior, the phase is very sensitive to variations in frequency, significantly reducing the working band of the reflectarray. All these drawbacks are solved by using two or three stacked array layers [23,24].

When the reflectarray cell contains two or more stacked conductive patches, see Figure 10.5a, each of them behaves like a resonator, and the phase of the reflected field varies with the patch size in a similar way to that of one layer, but the phase shift can reach values of several times 360°. Therefore, with several array layers, the separation between them, and the separation between the first array and the metallic plane, can be increased to achieve a smoother and more linear behavior of the phase as a function of the patch size, maintaining a range for phase shift greater than 360°. Figure 10.4b shows the phase curves as a function of the patch size at three frequencies for two stacked arrays above a ground plane, using separators 3 mm thick. For the two-layer structure, the phase variation is very linear over a 360° range and the behavior is similar at the three frequencies. This means a significant improvement in the bandwidth of the element. In addition, the phase

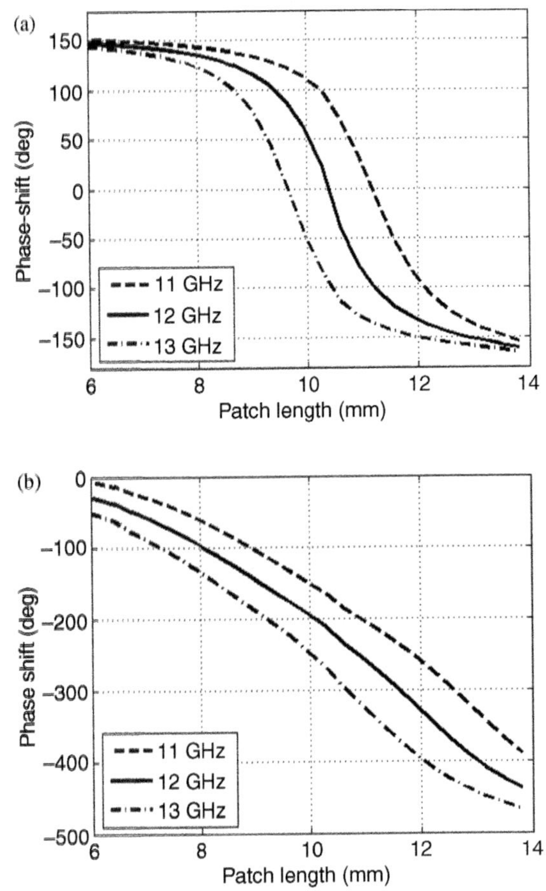

Figure 10.4 *Phase shift at normal incidence for a periodic array ($p_x = p_y = 14$ mm) made of one or two dielectric layers ($\varepsilon_r = 1.05$), with printed square patches versus the patch side. (a) One layer of patches ($t = 1$ mm). (b) Two layers ($t_1 = t_2 = 3$ mm, $a_1 = 0.7a_2$).*

slope versus dimensions (65°/mm) is much lower than in the phase response for a single layer (135°/mm). Then, a tolerance error in patch dimension of 0.1 mm will produce only an error in phase of 6.5°, which means a low sensitivity to manufacturing tolerances. When using three stacked layers of variable-size patches, the phase response shows larger values of phase slope and a wider phase range (more than two times 360°). The adjustment of the dimensions in the three-staked patches provides more degrees of freedom that can be used to increase the bandwidth by using optimization routines as shown in [24,33–36].

Another alternative implementation of the multilayer reflectarray cell that overcomes the drawbacks of attached-stub elements is based on patches aperture-coupled to stubs of different lengths [25], as shown in Figure 10.5b. In this configuration, as in that of the attached stubs, a real phase delay proportional to the stub length is achieved. The aperture-coupled element provides several advantages compared to the attached-stub element, as known from printed arrays. First of all, there is more room for longer lines in order to increase the range of phase delay [72], second, the spurious radiation produced by the stubs is in the opposite direction to the antenna beam and not interfering with the radiation pattern of the antenna, and finally it is very appropriate

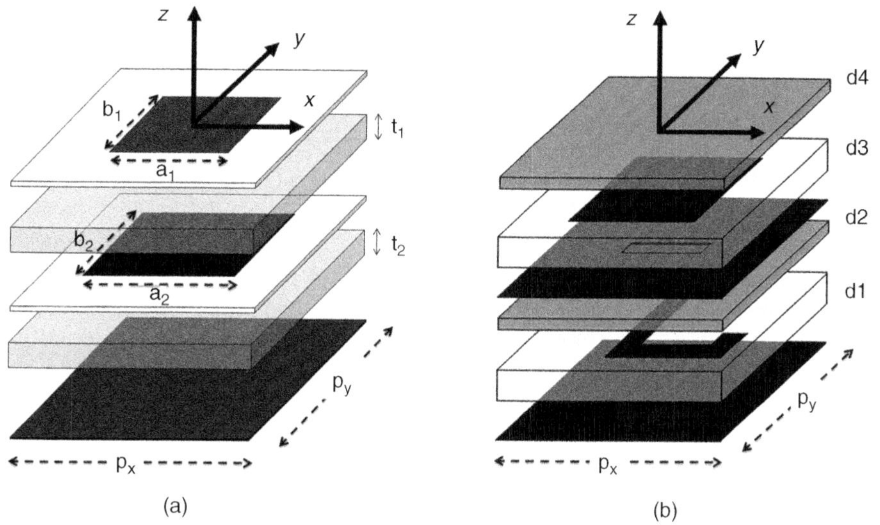

Figure 10.5 *Multilayer reflectarray elements: (a) stacked patches; (b) patch aperture-coupled to a delay line.*

to include active elements or controllable phase shifters in the microstrip lines to reconfigure or scan the beam. The reflectarray element can be designed by adjusting the geometrical dimensions to provide a broadband element and a linear phase response as a function of the stub length [73], by compensating some of the parasitic effects, and also to provide phase delay in a range larger than 360° [74].

10.3 Analysis and Design Techniques

For the design of the reflectarray, any possible value of phase shift must be implemented by varying some geometrical parameters in the unit cell, such as the patch dimensions, stub length or patch rotation angle. The first step in the reflectarray analysis and design is the characterization of the reflective elements, that is, for a given geometry of the reflectarray element, to accurately predict the phase shift and dissipative losses for each polarization of the field. In a second step, some geometrical parameters can be adjusted in the periodic cell to improve the phase shift and bandwidth performance. Once the unit element has been fully defined, the reflectarray antenna must be designed to produce a collimated or a shaped beam by adjusting the phase shift on each element. Finally, the radiation patterns including co- and cross-polar components must be accurately computed to evaluate the antenna performance. These aspects are discussed in more detail in this section.

10.3.1 Analysis and Design of Reflectarray Elements

Although the reflectarray elements can be characterized to a first approximation using simple circuit models [14,15], a full-wave electromagnetic analysis [19,70] is required to accurately predict the phase shift, losses and cross-polarization produced by the element. The reflective elements can be analyzed as isolated elements [15] or in an array environment [19,23]. The analysis in a periodical environment [75,76] is very accurate, to characterize the reflectarray elements [77], because it automatically takes into account the mutual coupling between elements assuming local periodicity. The spectral domain method of moments (SD-MoM) [78,79] is very efficient for a full-wave analysis of periodical structures, assuming planar arrays of patches or apertures in a single-layer or multilayer configuration [79].

Some of the most promising concepts for the phase shifter in reflectarrays, such as variable-size stacked patches [23,24] and aperture-coupled patches [25,72], can be considered as a multilayer structure, with periodic surfaces made of metal patches or apertures in a ground plane, separated by dielectric layers. The multilayer periodic structure can be analyzed using the modular approach proposed in [79], which consists of characterizing each array layer by a generalized (or multimode) matrix, and then analyzing the whole structure by using a cascading process. This approach is very flexible for analyzing diverse geometries, since the matrix that characterizes each layer is computed independently and used as a building block for the analysis of multilayer structures. This modular technique was initially used for the analysis of frequency-selective surfaces (FSS) [79] and was satisfactorily applied to the design of the dichcroic subreflector for the Mars Express and Venus Express missions [80,81]. The technique has been applied to the analysis of two types of reflectarray elements: aperture-coupled stubs and variable-size stacked patches. Both types of phase shifter element allow a significant improvement in the element bandwidth and a range in phase delay of several times 360°.

10.3.1.1 Analysis and Design of Variable-Size Stacked Patches

The phase response and ohmic losses have been studied for several reflectarray elements based on two and three stacked patches of variable size. A two-layer element that provides a linear phase variation in a 16% bandwidth, see Figure 10.4b, has been described in Section 10.2.

The reflectarray element based on three stacked layers of variable-size patches can be designed to provide a broadband phase response over a wide range of phase (more than two times 360°), by appropriately choosing the thickness of the separators (Rohacell or honeycomb), the period and the ratio between the patch dimensions in each layer. Considering a periodic cell of 14 × 14 mm and Rohacell 3 mm thick as separators, a parametric study was carried out by varying the relative sizes of the stacked patches and the following ratios were chosen: $a_2 = b_2 = 0.9b_3, a_1 = b_1 = 0.7b_3, a_3 = b_3$. Figure 10.6 shows the phase of the reflection coefficient as a function of the patch side, at three different frequencies. In this case, normal incidence has been considered and the

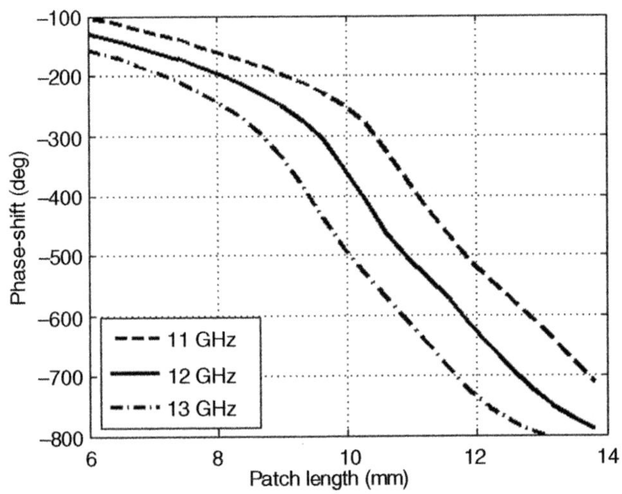

Figure 10.6 *Phase shift at normal incidence for a periodic array ($p_x = p_y = 14$ mm) made of three dielectric layers ($\varepsilon_r = 1.05$), with printed square patches versus the patch side a_3 for different frequencies ($t_1 = t_2 = t_3 = 3$ mm, $a_1 = 0.7a_3$, $a_2 = 0.9a_3$).*

phase in both polarizations is identical. However, the phase shift varies with the angle of incidence differently for each polarization. The differences in phase shift for the two orthogonal polarizations, with the electric field in the x and y direction, can be larger than 60° for an incidence of 40° ($\theta_i = 40°$, $\varphi_i = 0°$). These results show that real angle of incidence and polarization of the field must be taken into account at each reflectarray element for an accurate analysis of the reflectarray antenna.

10.3.1.2 Analysis and Design of Aperture-Coupled Patches to Delay Lines

Aperture-coupled patches, as shown in Figure 10.5b, are well known in microstrip antennas, where the geometrical parameters are adjusted to provide a good matching from the feeding microstrip line. However, when the reflectarray element is designed in the same manner as in radiating patches, that is, to provide low return losses when excited from the microstrip line, the phase response differs from that of an ideal delay line [72] as a result of the resonances of the structure. This effect can be reduced by an appropriate design of the element [73].

For a broadband behavior of the reflectarray element, the phase curves versus stub length should be linear and slightly diverging at different frequencies, as in an ideal delay line ($-2\beta L$, where β is the propagation constant and L the length of the line). The bandwidth of the aperture-coupled element can be improved by adjusting the stub and slot dimensions. Starting from the geometrical parameters obtained for a matched aperture-coupled element, the matching stub and the slot length were tuned slightly to achieve a linear phase curve as a function of the delay line. The resulting reflectarray cell is defined in Table 10.1 and the phase curves for normal incidence obtained by the full-wave method described in [79] are shown in Figure 10.7 at five frequencies. The phase curves are close to ideal phase delay ($-2\beta L$) over a range of 600° for a bandwidth larger than 20% (9.15–11.50 GHz), being practically coincident with the ideal phase shifter from 9.65 to 10.65 GHz. Note that the described aperture-coupled element only provides phasing for one linear polarization, with the electric field in the y direction according to Figure 10.5b. However, the element can be designed for dual linear polarization by adding an orthogonal slot on one side to couple the electromagnetic energy for the orthogonal electric field to another delay line.

10.3.2 Design and Analysis of Reflectarray Antennas

Once the reflectarray element has been designed to provide low losses and a smooth phase response in a certain frequency bandwidth, the next step is to design the reflectarray antenna to produce a collimated or a shaped

Table 10.1 *Definition of aperture-coupled reflectarray element.*

Metal layer	X (mm)	Y (mm)	
Line	1.50	1.50 (stub) + variable	
Slot	7.00	1.00	
Patch	9.30	9.30	
Period	16.5	16.5	
Dielectric layer	**Thickness (mm)**	ε_r	$\tan\delta$
d1	7.700	1.067	0.0002
d2	0.508	3.380	0.0050
d3	2.000	1.067	0.0002
d4	0.508	3.380	0.0050

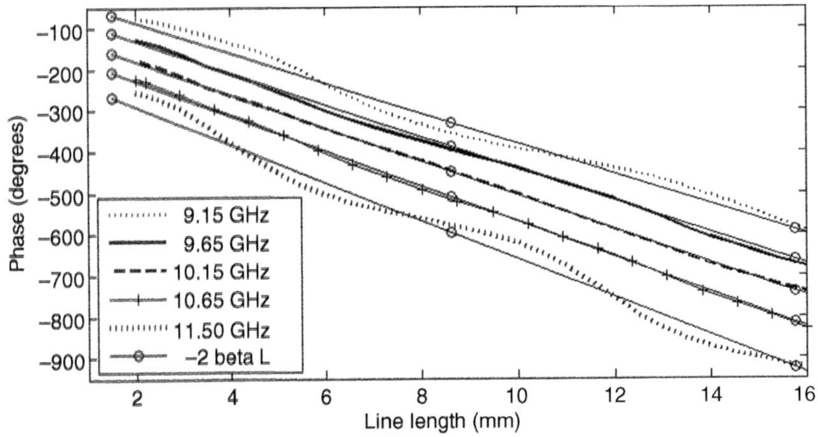

Figure 10.7 *Phase response for the aperture-coupled reflectarray element designed for broadband operation.*

beam, by properly adjusting the geometrical parameters used in the design. Although in a simple approach the design can be carried out by directly using the phase curves obtained for normal incidence, for a more accurate antenna design it is essential to adjust the phase in each element taking into account the real values of the angle of incidence and polarization of the field. The design of the reflectarray antenna can be split into the following steps:

1. *Definition of the phase shift in each element:* Once the working frequency, the position of the feed and the direction of the required collimated beam are determined, the phase shift that should introduce each reflectarray element to achieve a reflected wave with a progressive phase distribution is obtained by (10.2). This phase distribution is defined for each polarization of the incident field.
2. *Adjusting the geometrical parameters on each element at the central frequency:* In this step the dimensions of the patches in variable-size patches, or the line lengths in aperture-coupled elements, are adjusted to provide the phase shift defined in the previous step for each reflectarray cell at the given frequency. An optimization routine can be used for the adjustment of the dimensions of patches (or delay lines) on each cell, which iteratively calls the analysis program, taking into account the angle of incidence and polarization of the field coming from the primary feed, and adjusts the dimensions of each element until the required phase defined in step 1 is achieved. The procedure is repeated for each polarization and for each element in the reflectarray antenna.
3. *Fine adjustment to meet specifications in the working frequency band:* The previous design in a single frequency can be extended to a prescribed frequency band by a further optimization of the dimensions used to control the phase, as will be discussed later in more detail. In this step, the dimensions are adjusted using an optimization routine in order to meet simultaneously the phase specifications at several frequencies within the working band of the reflectarray.

Once the reflectarray has been completely designed and all the dimensions have been determined, an important aspect of the reflectarray analysis is the accurate computation of the radiation patterns, including co- and cross-polar components. Since the reflectarray antenna is made using non-identical elements in order to implement the required phase shift distribution, each element has to be analyzed independently assuming local periodicity, that is, considering a periodic cell with the dimensions, the angle of incidence and the polarization of the incident field associated with each particular element. The local periodicity approach is very accurate for printed patches with aperture-coupled or attached stubs of different lengths, because the patches are identical and only the stub length varies from one element to the next. For elements with variable-size patches, the approach is accurate when the variation in patch dimensions is smooth from one cell to the next, because it takes into account all mutual coupling between patches. The local periodicity approach has been used satisfactorily for the design of several contoured-beam antennas for DBS applications [34,36], providing accurate predictions of the radiation patterns.

For the analysis of the antenna, the feed is typically modeled as a $\cos^q(\theta)$ function, but in some cases the reflectarray elements are not in the far-field region of the feed horn. In those cases, the incident field on each reflectarray element should be computed considering the near field radiated by the feed [82], which can be obtained from full- wave simulations or from measurements. Assuming that the feed can radiate dual linear polarization, with the tangential electric field on the reflectarray surface mainly directed in the X_R or Y_R direction, see Figure 10.1, the tangential electric field impinging on each reflectarray cell (i) is expressed independently for each polarization, denoted as the X/Y, in the Cartesian coordinates of the reflectarray system, as

$$\mathbf{E}_I^{X/Y}(i) = E_{Ix}^{X/Y}(i)\hat{x} + E_{Iy}^{X/Y}(i)\hat{y} \tag{10.3}$$

Note that the incident field for the X/Y polarization contains both components, and although the field is mainly directed along X_R/Y_R, there is also a small component in the orthogonal direction, which is the cross-polarization produced out of the principal planes by projection of the field on the reflectarray surface. The reflected tangential electric field on each reflectarray cell (i) is computed by multiplying the incident field by the reflection matrix $\mathbf{R}^i$ obtained from the analysis of the element in a periodical environment [79]:

$$\mathbf{E}_R^{X/Y}(i) = \mathbf{R}^i \cdot \mathbf{E}_I^{X/Y}(i) \tag{10.4}$$

where matrix $\mathbf{R}^i$ accounts for the cross-polarization produced by printed elements in the Cartesian reflectarray system.

The radiation patterns of the reflectarray antenna are easily computed from the tangential components of the reflected electric field (10.4) on the reflectarray elements, as described in [71], pp. 64–72. The co- and cross-polar components according to the third Ludwig definition [83] are obtained by using fast Fourier transform (FFT) algorithms to improve the numerical efficiency. The antenna gain, which includes the spillover, illumination and dissipative losses, is computed by dividing the squared radiated field by the input power in the feed horn.

This technique is numerically very efficient, because it only requires the analysis of each element, assuming a periodical environment, to compute the amplitude and phase of the reflected field on the reflectarray cells, and then the radiation patterns are computed using FFT and simple algebraic operations. The technique ignores the residual backward field radiated directly by the feed horn and the diffraction at the reflectarray edges; but these effects are not significant in practical cases, because of the low levels of sidelobes for corrugated horns and the low illumination level on the reflectarray edges. In conclusion, this technique provides an accurate prediction of the co- and cross-polar radiation patterns, losses, directivity and antenna gain, as shown in various demonstrators for contoured-beam antennas for communications [82,84,85], and DBS [34,36].

As an alternative to the local periodicity approach, different full-wave techniques have been used to analyze the complete reflectarray antenna [86–88]. These methods take into account all the individual couplings between elements, and then the CPU time is drastically increased. The full-wave techniques are appropriate for an accurate analysis of the reflectarray, but the very high CPU times make unaffordable their use in optimization processes.

10.3.3 Broadband Techniques

Bandwidth limitation is an inherent characteristic of reflectarrays and much effort has been made in recent years to improve the bandwidth. Reflectarray bandwidth is mainly limited by two different factors as described in [68]. The first one is the narrow band of the printed radiating elements, and the second is the differential spatial phase delay resulting from the different lengths from the feed to each point on the wavefront of the radiated beam.

10.3.3.1 Broadband Reflectarray Elements

The bandwidth limitation produced by the printed element is the most stringent for moderate-size reflectarrays [69]. Since microstrip patches printed on a thin dielectric layer are usually used as reflectarray elements, their narrow band (3–5%) limits the bandwidth of the reflectarray. To overcome this problem, different types of reflectarray elements have been proposed in recent years to improve the element bandwidth in printed reflectarrays.

When the phase control is implemented by varying the resonant dimensions of the printed patches, several schemes have been proposed to improve the element bandwidth, consisting of stacking several patches [23,24,89,90], varying the geometry of the conductive patch [91,92], or using multi-resonant elements in a single layer [93–97]. Stacked metallic rectangular patches [23,24] or rings [89,90] have been proposed as broadband reflectarray elements where the resonant frequencies are adjusted to improve the linearity of the phase curves. In a similar manner, reflectarray elements based on several resonant dipoles or crossed loops on the same dielectric layer have been proposed [93–97], where the relative lengths of the parallel dipoles [93–96] or concentric rings [97] are adjusted to improve the bandwidth of the reflectarray element. The reflectarray element based on three dipoles was used to design reflectarray antennas at 300 GHz [95] and at 77 GHz in a folded configuration [96]. Also, a reflectarray based on double crossed loops was designed, manufactured and measured [97].

The reflectarrays based on multi-resonant elements printed on a single dielectric layer are easy to manufacture and can be designed to provide a 10% bandwidth. On the other hand, multilayer reflectarrays based on two or three stacked patches of varying size [23,24], or patches aperture-coupled to delay lines [73], can provide wider frequency bandwidths. For example, a broadband aperture-coupled reflectarray element was described in Section 10.3.1, which provides a 20% bandwidth when compared to the ideal phase delay curves, as shown in Figure 10.7.

For reflectarray elements based on two layers of variable-size patches, see Figure 10.5a, the thickness of separators t_1 and t_2 and the relative size of the patches in each layer can be adjusted to achieve a sufficiently linear phase variation as a function of the patch dimensions for different incidence angles, for different frequencies and over a range greater than 360°. Figure 10.4b shows the phase curves at normal incidence for a two-layer element, where the period is 14×14 mm and the separators are the 3 mm thick Rohacell HF-31. Note that the phase curves are smooth in the 11–13 GHz frequency band (16%). Using this element, a 40 cm reflectarray was designed, built and measured. The results reported in [23] show that a 16.7% bandwidth (11–13 GHz) was achieved for variations in gain less than 1.5 dB, which is a significant bandwidth improvement with respect to other X-band one-layer reflectarrays.

10.3.3.2 Broadband Techniques for Large Reflectarrays

Bandwidth in large reflectarrays, as in the case of space applications, is drastically reduced because of the different path lengths, as discussed in [68]. Parabolic reflector antennas use the physical curvature to equalize signals arriving from the feed and to form a planar wavefront. This is the true time delay compensating method, which in principle is frequency independent. For large apertures, the path length that produces the phase delay in a parabolic reflector is much larger than a wavelength. However, the phase shift in the reflectarray elements is limited to a 360° range, and the phase delay is only exactly compensated at the central frequency. The phase errors at a different frequencies increase for large values of phase delay, and the phase compensation is only valid over a narrow frequency band. This limitation is more significant for large electrical dimensions, as in the case of antennas for space applications.

Several broadband techniques have been proposed to reduce the effect of the differential spatial phase delay. The first one consists of compensating the spatial phase delay, which is proportional to twice the path distance between the reflectarray and an equivalent parabolic surface, in a given frequency band with the phase shift produced by three layers of variable-size patches [24]. A second technique is to implement delay lines in order to compensate on each reflectarray element the real phase delay over the whole range (several times 360°) by using stubs with a length that varies over a range of several wavelengths [74]. A third technique consists of a multi-facet configuration, which approaches a parabolic surface using flat panels [98].

The phase delay $\phi_{di}(f)$ required at element (i) in a flat reflectarray (not limited to a 360° range) is used to define the phase difference at two given frequencies f_1 and f_2. To design a reflectarray in a frequency band defined by the extreme frequencies f_1 and f_2, the dimensions of three stacked patches are optimized for the two linear polarizations, using a Fletcher Powell algorithm, to match simultaneously in each element (i) the phase shift at the central frequency ϕ_{oi} (limited to 360°) and the phase delay difference, defined as $D_{di}(f_1, f_2) = \phi_{di}(f_1) - \phi_{di}(f_2)$, as described in [24]. It was demonstrated that three stacked patches are required to provide sufficient degrees of freedom to match the required phase at several frequencies. Following this optimization process, a 10% bandwidth was achieved for a 1 m reflectarray at 12 GHz by compensating the phase delay in a range of $5 \times 360°$, see [24] for details of the antenna geometry. The radiation patterns on the x–z plane for x polarization at the central and extreme frequencies are shown in Figure 10.8 before and after optimization; the radiation patterns are very similar for the orthogonal polarization. For the non-optimized reflectarray, the radiation patterns deteriorate in the band, with a 6 dB gain reduction, while a very stable gain was achieved for the optimized reflectarray, with only a 0.5 dB gain reduction. These results show that a significant bandwidth improvement can be achieved for large reflectarrays by using optimization routines to accomplish the appropriate phase distribution within a frequency band. For a bandwidth larger that 10%, matching the difference in phase at extreme frequencies is not sufficient to ensure a good performance in the whole frequency band, and the phase delay must be compensated at several frequencies in the band, as shown in [35,36].

The reflectarray element based on printed patches aperture-coupled to delay lines described in Section 10.3.3.1 can be used to implement the true time delay (TTD) over a range of several cycles of 360°. The delay line can be bent to increase the range of phase delay as shown in Figure 10.5b [74]. The phase curves shown in Figure 10.7 can be extended by using a U-shaped delay line, as shown in Figure 10.9a. Then, the range of phase delay increases from 640° to 1080° (three 360° cycles) and the dissipative losses increase slightly with the line length, but they are still less than 0.5 dB.

An 80 cm reflectarray was designed to produce a pencil beam at 9.65 GHz in the direction $\theta_0 = 18°$, $\varphi_0 = 0°$ using the previous aperture-coupled elements, assuming the phase center of the feed placed at coordinates (−288, 0, 793) in mm, with respect to the reflectarray center. The reflectarray was designed in two cases: as customary for the phase distribution truncated to 360° and for the phase delay in three cycles of 360°. The radiation patterns are computed as described in Section 10.3.2 and they are virtually identical at 9.65 GHz for the two phase distributions, but when frequency varies, the pattern is distorted much faster for the reflectarray with the phase limited to 360°. To evaluate the bandwidth improvement produced by the TTD, the antenna gain in the 8.5–11.90 GHz band is compared in Figure 10.9b for the two designed reflectarrays and also for two ideal cases used as reference. In the first ideal case, the phase is limited to a 360° range and the reflectarray elements are considered ideal, providing a frequency-independent phase shift. The second ideal case assumes an ideal

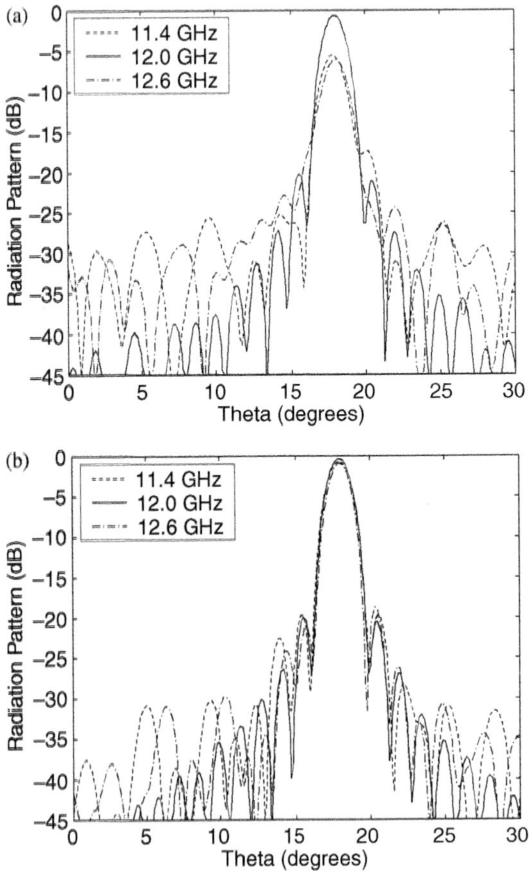

Figure 10.8 *Radiation patterns for a 1 m reflectarray in the 11.4–12.6 GHz band. (a) Non-optimized reflectarray. (b) Optimized reflectarray. From [24], © 2003 IEEE.*

phase distribution using TTD. Both ideal curves coincide at the central frequency and their difference represents the diminution of gain produced by the effect of differential phase delay in reflectarrays. Note that the variation of gain for the reflectarray with truncated phases is similar to that corresponding to the ideal truncated phases, but with a small reduction in gain mainly produced by ohmic losses (0.5 dB) and by some distortion of the element phasing at extreme frequencies, see Figure 10.9a. On the other hand, the gain variation for the TTD reflectarray is similar to the one corresponding to the ideal TTD in the frequency band 9.2–10.8 GHz, but with a reduction in gain smaller than 1 dB, as a result of ohmic losses and small phase errors out of the central frequency. For extreme frequencies, the gain drops more rapidly because the phasing produced by the elements differs considerably from the ideal phase, as shown in Figure 10.9b. These results show that the concept of TTD provides a significant increase in bandwidth. For example, the bandwidth defined for 0.3 dB gain variation increases from 10.1 to 20.0% when TTD is implemented in the reflectarray.

Previous techniques can be used to compensate the phase delay for certain antenna dimensions, defined by a limited number of 360° cycles. The limit for the delay lines is imposed by the room available for the

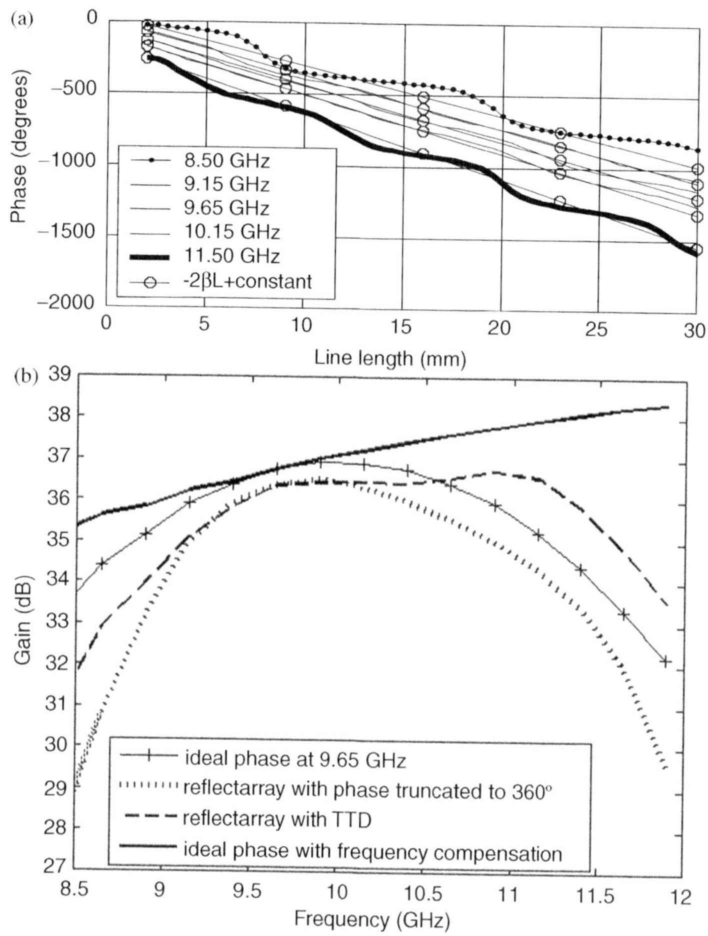

Figure 10.9 *True time delay reflectarray. (a) Phase curve versus the line length. (b) antenna gain versus frequency for different cases. From [74], © 2008 IEEE.*

stub length, while the limit in three-layer reflectarrays is given by the capability of compensating the phase delay by optimizing the three stacked patches. In both techniques the phase delay can be compensated in a 10% bandwidth for reflectarray apertures up to 70 wavelengths, assuming a focal distance similar to the aperture diameter.

For larger reflectarrays, a possible solution to achieve a bandwidth around 10% consists of a facetted configuration that approximates the shape of a parabolic surface using flat reflectarray panels, so that the number of 360° cycles is limited in each flat facet [98]. Figure 10.10 shows three possible configurations from the simplest one-dimensional arrangement to a more complex circumferential structure, similar to that of an umbrella. Each planar facet is a reflectarray that is designed to introduce the required phase shift to simulate the real parabolic surface or to shape the beam. The multi-facet configuration is also compatible with the previous techniques for compensating the phase delay in a limited bandwidth, so the number of panels can be reduced to simplify the manufacture or deployment of the antenna, see Figure 10.10a and b. The multi-panel configurations offer the possibility of being folded and deployed for large antenna apertures.

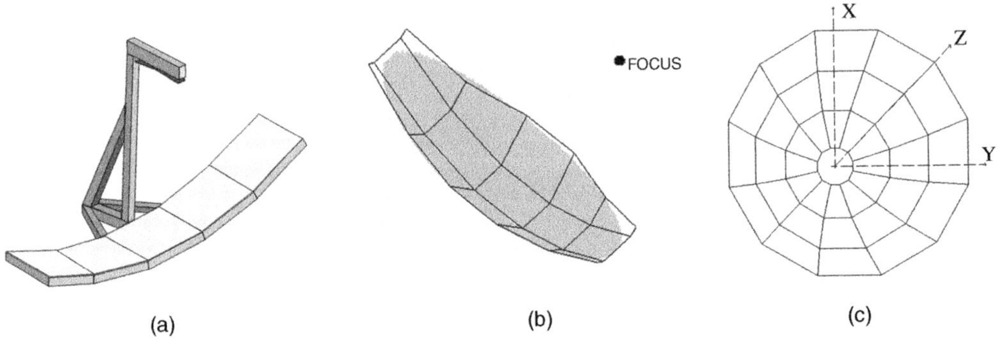

Figure 10.10 Multi-facet configuration approximating a parabolic surface: (a) 1D configuration; (b) rectangular-based arrangement; (c) circumferential arrangement. From [71], Reproduced by permission of John Wiley & Sons, Inc.

10.4 Reflectarray Antennas for Telecommunication and Broadcasting Satellites

The requirements of spacecraft antennas for satellite broadcast and telecommunication missions are very stringent, including shaped contoured beams to efficiently illuminate predefined geographical areas (usually in dual polarization), co-polar isolation in other geographical regions, and in some cases transmit–receive (Tx–Rx) operation. Currently, shaped reflectors are satisfactorily used in many missions to provide all the requirements of coverage, cross-polarization and isolation in both transmit and receive frequency bands. However, the main disadvantage of reflector antennas is the manufacturing of a specific mold for the shaped reflector, which depends on the antenna requirements and therefore cannot be reused for other missions, with an associated impact on the cost of the antenna and the manufacturing time. Printed reflectarrays with contoured beams can be an alternative to the on-board shaped reflectors, because of their lower cost and shorter manufacturing time, since custom molds are eliminated. Note that when a reflectarray antenna is designed for a specified coverage only the dimensions of printed patches or lines are varied and not the structural panel; then the mechanical models and tests can be reused for different missions, leading to a reduction in recursive costs.

Contoured-beam reflectarrays made of one [32] and three [33–36] layers of variable-size patches have been successfully designed for DBS applications. Although the narrow bandwidth is a real limitation in reflectarray antennas, the techniques discussed in previous sections can be used to improve the antenna bandwidth for large reflectarrays. The technique based on optimizing the patch dimensions in a three-layer configuration to compensate the spatial phase delay has been used to design several reflectarray demonstrators for DBS missions [33–36] that are described in this section.

10.4.1 Contoured-Beam Reflectarrays

Contoured beams can easily be generated using reflectarrays by implementing an appropriate phase shift on the reflectarray elements. The amplitude of the incident field on each reflectarray element is fixed by the primary feed and only the phase can be adjusted to shape the beam. A contoured-beam reflectarray providing a cosecant squared pattern [99] was reported for the first time in 1993. Later a reflectarray demonstrator was reported for DBS applications [32]. In that case, the reflectarray was designed using a shaped reflector previously manufactured for a DBS European coverage in Ku-band. The required phase shift at 14 GHz was obtained on each reflectarray element from the distance between the shaped surface and the flat surface where the reflectarray was placed. A significant improvement in the design technique was the implementation of pattern synthesis to obtain directly the phase distribution on the reflectarray without the previous design of a shaped reflector [100]. The pattern synthesis applied to shaped-beam reflectarrays provides some advantages with respect to surface shaping in conventional reflectors. First of all, the phase synthesis is not constrained by geometrical parameters and therefore is more flexible for synthesizing any required radiation pattern. In contoured-beam reflectarrays, the phase can be synthesized independently for each polarization, even to generate a different pattern for each polarization as demonstrated in [34]. Also, the real incident field on each reflectarray element for each polarization can be taken into account in the pattern synthesis, which allows the inclusion of the near field radiated by the feed horn as shown in [82] or to include separate feeds for each polarization [34]. In addition, the reflectarray can be designed to change the polarization, for example, to convert linear into circular polarization just by adding a 90° phase shift in one linear polarization with respect to the orthogonal one.

In reflectarrays, the synthesis of radiation patterns is restricted by the feed since it fixes the amplitude of the incident field on each reflectarray element, and only the phase distribution can be modified. To implement the required phase shift distribution, the dimensions of the corresponding geometrical parameter must be adjusted in the reflectarray elements. A direct optimization process, in which all the element dimensions are simultaneously optimized in an iterative process to achieve the required contoured pattern, is computationally unaffordable, because it would require the analysis of all the elements (several thousand in space applications) to compute the radiation pattern in each iteration of the pattern synthesis. An efficient alternative consists of dividing the design process into two steps as described in [33]. In the first step, assuming a fixed amplitude distribution on the reflectarray surface given by the feed radiation pattern, a phase-only synthesis is applied to compute the phase of the reflected field at each reflectarray element that provides the required contoured pattern. In the second step, the patch dimensions are optimized element by element to achieve the previous phase distribution and the required frequency variations in a given bandwidth, following the same process as the one described in Section 10.3.3.2.

In reflectarray antennas the only variables to be optimized are the phase shifts to be implemented on each reflectarray element; then a technique known as phase-only synthesis must be applied to synthesize the required contoured patterns. Several phase-only techniques have been developed to obtain shaped beams using phased arrays [101–104]. The same techniques can be applied to reflectarrays, but the problem is more challenging because of the very high number of elements in reflectarrays, particularly for space applications. Therefore, only very efficient techniques able to deal with thousands of variables can be applied for the reflectarray's pattern synthesis. The phase-only technique known as the *intersection approach*, which was previously developed for phased arrays [103], has been successfully applied to the design of reflectarray antennas for contoured-beam DBS antennas [33–36,100,105,106] and synthetic aperture radar (SAR) interferometry [107].

In a real mission, the contoured requirements must be fulfilled in a given frequency band, or in two separated bands for transmit/receive antennas. Then, the contoured- beam reflectarray must be designed in the prescribed frequencies. In that case, the required phase shift distribution on the reflectarray must be found at several frequencies in the working band by applying the intersection approach as described in [108,109], and then the reflectarray elements must be optimized to provide those phase distributions. The pattern synthesis together with the optimization of the patch dimensions for broadband operation described in Section 10.3.3.2 have been applied to design a number of shaped-beam reflectarrays for DBS applications [33–36], two representative examples of which are presented here. In all cases, the reflectarray is made of three stacked layers of printed arrays with rectangular patches, in which the dimensions of the stacked patches are optimized in each element to ensure the required coverage at several frequencies in the working band.

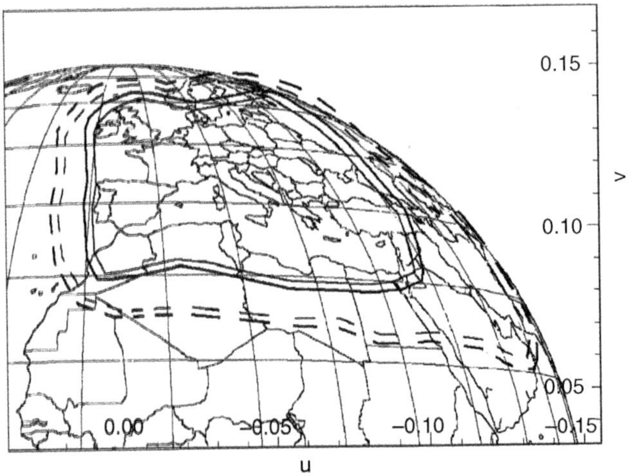

Figure 10.11 Contoured requirements for coverage of Europe. From [34], © 2006 IEEE.

10.4.2 Dual-Coverage Transmit Antenna

The first example is a breadboard of a DBS transmit antenna that provides different coverage in each polarization [34] from a satellite located at 5°W orbital position. In H-polarization, the antenna should provide coverage of Europe in the 11.45–12.75 GHz frequency band, while in V-polarization it should give coverage to several cities on or near the east coast of North America (Washington, DC, New York and Montreal) in a narrower frequency band (11.45–11.7 GHz). The European coverage (Figure 10.11) includes one central region with a requirement of 28.5 dBi (solid line) and a wider region with a requirement of 25.5 dBi (dashed line). Both regions have been enlarged to account for typical pointing errors (0.1° in roll, 0.1° in pitch and 0.5° in yaw). In reflector technology, these requirements were achieved by using a dual-gridded reflector with two superimposed grid reflectors and separate feeds for each polarization, which suffers from high cost and large volume and mass. A dual-fed reflectarray breadboard was designed to provide the required coverage in each polarization using a single reflectarray surface. The reflectarray antenna has been manufactured using space-proven technology and the results were compared to those of the dual-gridded reflector used as reference. The novelty of this breadboard was the demonstration that two independent beams, one for each polarization, can be generated with a single reflectarray, providing a significant reduction in mass and volume with respect to the dual-gridded antenna.

The reflectarray consists of an elliptical flat panel with axes of 1036 × 980 mm, which is the same aperture surface as in the dual-gridded reference antenna. The two feed horns are placed with the phase center at coordinates (in mm) $F_V = (-460, 0, 887)$ and $F_H = (-302, 0, 898)$ for V- and H-polarization respectively, see Figure 10.12. The period for the reflective elements is defined as 14 × 14 mm (0.6λ at 12.75 GHz and 0.53λ at 11.45 GHz). For the contoured European beam, the intersection approach has been applied at the central frequency to obtain the phase distribution on the reflectarray surface. For V-polarization, it was demonstrated that a pencil beam provides the maximum gain and fulfills the gain requirements. The required phases at extreme frequencies were computed by multi-frequency synthesis [108,109] for H-polarization, and by assuming an ideal linear variation with frequency for the pencil beam [24] in V-polarization.

The reflectarray panel is composed of an electrical sandwich and a backside stiffening sandwich made of carbon fiber reinforced plastic (CFRP), as shown in Figure 10.13. The electrical sandwich contains the three arrays of variable-size patches printed on a 25 μm Kapton film, which are bonded to Kevlar composite layers

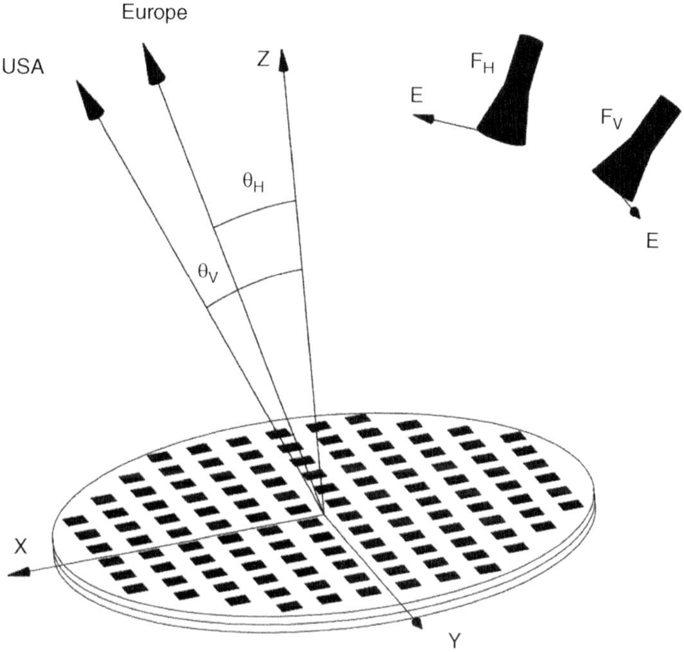

Figure 10.12 Reflectarray configuration to provide different coverage in each polarization. From [34], © 2006 IEEE.

and separated by a Nomex honeycomb 3 mm thick. The details of the materials and layup of the reflectarray panels are given in [34]. To reduce dissipative losses, a low- loss cyanate–ester resin is used for the Kevlar composite layers and also for bonding the array layers to the Nomex honeycomb. A CFRP face sheet (layer A in Figure 10.13) was used as the ground plane, with an increase of dissipative losses around 0.1 dB, which is a typical value in CFRP reflectors.

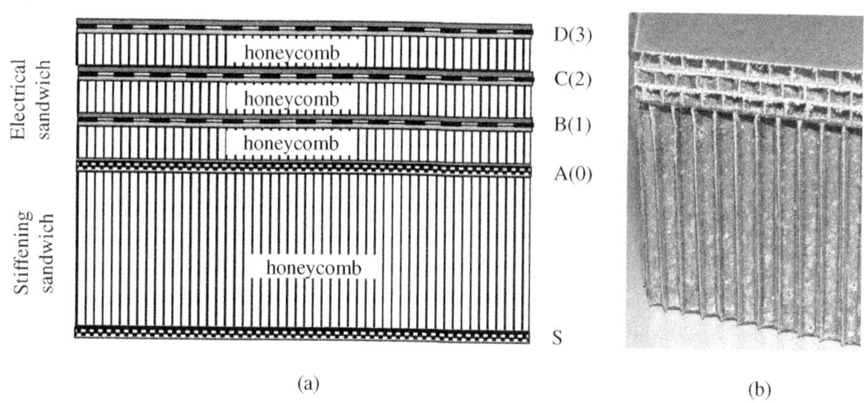

Figure 10.13 Configuration of reflectarray panel with electrical and stiffening layers: (a) layup; (b) section of manufactured sandwich. From [34], © 2006 IEEE.

The reflectarray was designed by optimizing the dimensions of the rectangular patches to achieve the required phase shift at the central and extreme frequencies for each polarization. First, the reflectarray is designed at central frequency by adjusting the x dimensions of the patches to achieve the required phase shift at 12.1 GHz for H-polarization, and the y dimensions to produce the phase distribution at 11.575 GHz corresponding to a pencil beam. Then, the reflectarray is optimized to fulfill the requirements in the prescribed frequency bands, following the process described in Section 10.3.3.2. Note that for each polarization, the required phase shift, the difference in phase delay, the position of the feed (and therefore angle of incidence) and the phase of the reflected field are different. Then, it is more convenient to perform the optimization first for one polarization and then for the other. For H-polarization, with the electric field in the X_R direction, the optimization is carried out by minimizing the following error function:

$$E_{xi}(a_1, a_2, a_3) = C_1\left(\Phi^x_{ci}(f_0) - \Phi^x_{oi}(f_0)\right)^2 + \sum_{l=1}^{2}\left(C_2\left(D^x_{ci}(f_l,f_0) - D^x_{di}(f_l,f_0)\right)^2\right) \tag{10.5}$$

where $\Phi^x_{oi}(f_0)$ and $\Phi^x_{ci}(f_0)$ are respectively the objective and computed phase shift on element i for the reflected x-polarized electric field, $D^x_{di}(f_l,f_0)$ and $D^x_{ci}(f_l,f_0)$ are respectively the objective and computed difference in phase delay defined as $D^x_{di}(f_l,f_0) = \Phi^x_{di}(f_l) - \Phi^x_{di}(f_0)$ and $D^x_{ci}(f_l,f_0) = \Phi^x_{ci}(f_l) - \Phi^x_{ci}(f_0)$, C_1 and C_2 are weighting coefficients and l is a superindex to indicate each extreme frequency in the band. After optimization of the x dimensions of the patches, the process is repeated to adjust the y dimensions by minimizing the associated error function for V-polarization. Note that during the optimization for each polarization the orthogonal dimensions of the patches are maintained unchanged. The process is repeated several times, alternating optimization of the x and y dimensions, to take into account the slight influence of the orthogonal dimensions of the patches. After several alternating optimizations for X- (Europe) and Y- (North America) polarization, the final patch dimensions are obtained. The analysis technique used in the optimization process is a full-wave method of moments in the spectral domain, assuming each element is in a periodic array environment, which takes into account the real angles of incidence at each element, the polarization of the incident field and all the dielectric layers defined in the sandwich configuration. The radiation patterns computed by the analysis routine for the resulting three-layer reflectarray practically fulfill the coverage requirements in the whole frequency band 11.45–12.75 GHz for H-polarization and 11.05–12.1 GHz for V-polarization.

The array layers are manufactured by photo-etching, and the reflectarray panel was manufactured by a multi-step curing process to achieve maximum accuracy and repeatability in the thickness and composition of each electrical layer. The breadboard was assembled, see Figure 10.14, and measurements were made in H- and V-polarization using the same feed in different positions.

For V-polarization, the measured co- and cross-polar contour gain patterns at central frequency are shown in Figure 10.15. The gain requirements of 37 dBi are fulfilled in the whole frequency band used in the design process (11.05–12.1 GHz), which is larger than required. The cross-polarization levels in the coverage of North America are always lower than 3 dBi in the whole frequency band, which corresponds to a cross-polar isolation better than 34 dB.

The measured co-polar patterns for H-polarization are compared to the simulations for three gain levels (28, 25 and 20 dBi) at 12.1 GHz, see Figure 10.16. The gain patterns practically fulfill the mask of requirements in close agreement with the simulated radiation patterns using the analysis routine based on the method of moments and local periodicity. The central European contour is covered with 28 dBi gain in 99% of the enlarged coverage. The antenna gain is reduced by 0.5 dB with respect to the results obtained from the pattern synthesis, because of dissipative losses (0.4 dB) and small errors in phase after the optimizations. The gain contours are close to fulfill the mask of requirements from 11.7 to 13 GHz as shown in Figure 10.17. Note that a

Figure 10.14 Reflectarray breadboard.

small shift in the frequency band was produced in H-polarization, because the breadboard was designed in the 11.45–12.75 GHz band. It was checked that the reason for the frequency shift only in H-polarization is a small anisotropy of the honeycomb produced by its hexagonal structure. In fact the dielectric constant is slightly higher when the electric field is along the ribbon, which in the demonstrator happens for V-polarization. In the breadboard, where $\varepsilon_r = 1.1$ was assumed for the honeycomb in both polarizations, the agreement is good for V-polarization, while a frequency shift is produced in H-polarization as a result of the lower ε_r. In addition, the small error in ε_r for H-polarization can also be the reason for the small deviations in the contour patterns.

The electrical performance of the reflectarray antenna has been compared to those of the reference dual-gridded antenna. For the directive beam in V-polarization, the gain measured in the coverage of North America is comparable to that measured in the dual-gridded reflector. For the contoured beam in H-polarization, the gain is slightly lower than for the dual-gridded reflector and small distortions are observed in the contour pattern, produced by the honeycomb anisotropy and manufacturing tolerances. The losses measured in the reflectarray (0.35 dB) are slightly higher than in the dual-gridded reflector (0.10 dB for the front shell and 0.20 dB for the rear shell). The total mass of the reflectarray panel, including electrical and structural sandwiches, is 2.250 kg, that is, 2.7 kg/m^2, which represents a significant reduction in mass with respect to the dual-gridded antenna (4.3 kg/m^2 for the two-shell configuration). The reflectarray breadboard has demonstrated the capacity to achieve satisfactory cross-polarization isolation, with cross-polar isolation better than −30 dB in the other coverage. For both polarizations, the performances remain stable over a 10% frequency band. In conclusion, this breadboard has demonstrated the ability of reflectarrays to realize two linearly polarized independent beams with a satisfactory cross-polarization level and therefore reflectarrays can be an alternative to dual-gridded reflectors, characterized by a very low cross-polarization but suffering from large volume and mass.

10.4.3 Transmit–Receive Antenna for Coverage of South America

Although, in the previous example, the antenna was operating only in transmission, the use of a single transmit–receive (Tx–Rx) antenna is preferred in telecommunications and DBS applications, because of the

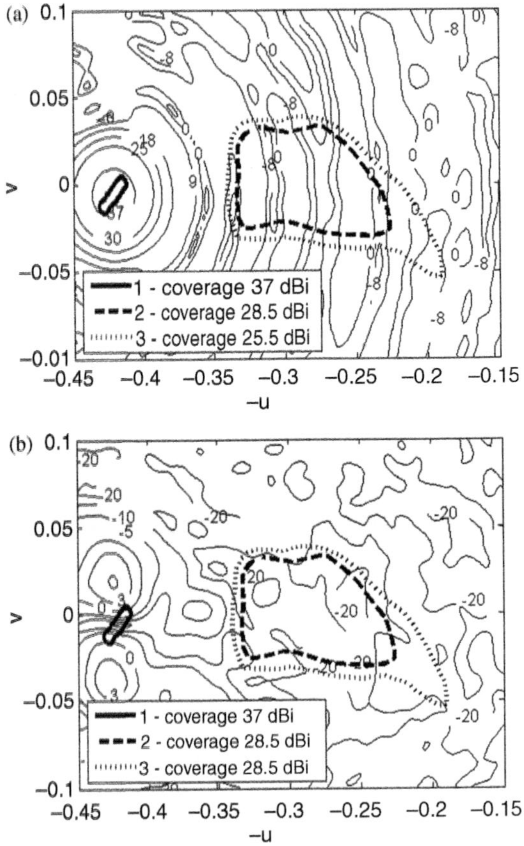

Figure 10.15 *Measured contour patterns for V-polarization at central frequency (11.575 GHz): (a) co-polar and (b) cross-polar patterns.*

reduction in volume, mass and costs. The second example is a 1.2 m reflectarray demonstrator reported in [36] that was designed to accomplish the requirements of a DBS mission in the Tx (11.7–12.2 GHz) and Rx (13.75–14.25 GHz) frequency bands. The requirements of a real DBS spacecraft antenna providing service to South America, the 'South Pan-American mission (PAN-S)' on the Amazonas satellite (located at a longitude of 61°W in the geostationary orbit), have been selected to design the Tx–Rx reflectarray antenna. The PAN-S coverage together with other missions also supported by the same satellite are shown in Figure 10.18. The antenna operates in dual linear polarization, vertical (V) and horizontal (H). The gain and cross-polar requirements, cross-polar discrimination (XPD) for Tx and cross-polar isolation (XPI) for Rx, for each coverage zone, are given in Table 10.2. These requirements should be met, including a typical margin of 0.1° for the antenna pointing errors. In addition, a requirement of 30 dB co-polar isolation with the coverage of Europe shown in Figure 10.18 must be accomplished.

The antenna actually used on board the satellite for the PAN-S mission is a dual-reflector antenna made of a 1.5 m main shaped reflector and a 50 cm subreflector. However, the demonstrator consists of a single-offset elliptical reflectarray of smaller dimensions (with axes of 1248 × 1196 mm) illuminated by a corrugated horn with its phase center located at coordinates (−373, 0, 1482) in mm. The reflectarray elements consist of three layers of variable-size rectangular copper patches sandwiched in a multilayer structure and backed by a

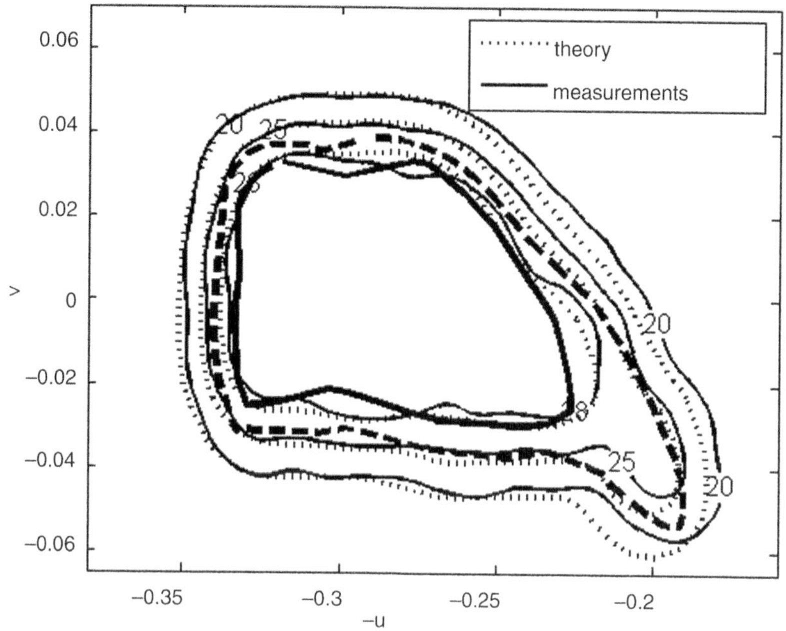

Figure 10.16 *Measured and simulated gain contours for H-polarization at 12.1 GHz.*

stiffener sandwich, as shown in Figure 10.13. The materials and details of the sandwich definition are given in [36]. In order to avoid the errors reported in the previous example, the slight anisotropy of the honeycomb is taken into account in the design, being the X_R-axis oriented along the ribbon ($\varepsilon_r = 1.07$, $\tan\delta = 0.0009$), and the Y_R-axis across the ribbon ($\varepsilon_r = 1.05$, $\tan\delta = 0.0004$).

After carrying out a parametric study, the reflectarray element was defined to provide a linear phase variation in the Tx and Rx bands, small losses and low cross-polarization. The period was defined as 13 × 13 mm and the honeycomb thickness as 3 mm. Figure 10.19 shows an almost linear phase response in both Tx and Rx frequencies for the following ratios: $a_1 = b_1 = 0.7b_3$, $a_2 = b_2 = 0.9b_3$, $a_3 = b_3$. The curves are computed for different angles of incidence by the SD-MoM technique [79] assuming a periodic structure. The effect of the angle of incidence can produce variations in phase up to 100° in both bands, and therefore it should be taken into account in the design process. The dissipative losses are smaller than 0.25 dB and the phase range is around 600° in both frequency bands. Note that the previous size ratio in each layer is only used in the first design step at 11.95 GHz, which is used as the starting point for optimizations. However, in the optimizations all the patch dimensions are optimized independently and the previous size ratio is no longer maintained.

To accomplish the stringent coverage requirements shown in Table 10.2, the pattern synthesis technique based on the intersection approach is applied first at central frequency in Tx (11.95 GHz) to obtain the phase distribution on the reflectarray surface that provides the required shaped pattern. The isolation with Europe is implemented in the pattern synthesis by adding the defined European zones in the mask of requirements and enforcing a level of radiation in these zones below 0 dBi. The near field radiated by the corrugated horn has been computed using full-wave commercial software [110] and used as the incident field on the reflectarray elements for a more accurate design of the reflectarray. After applying the intersection approach at 11.95 GHz with the isolation requirements, the resulting phase distribution for H-polarization is shown in Figure 10.20a. For this phase distribution the radiation patterns fulfill the gain requirements in South America with an excess margin of 1 dB, for possible uncertainties and losses, while the radiation levels in the European regions are

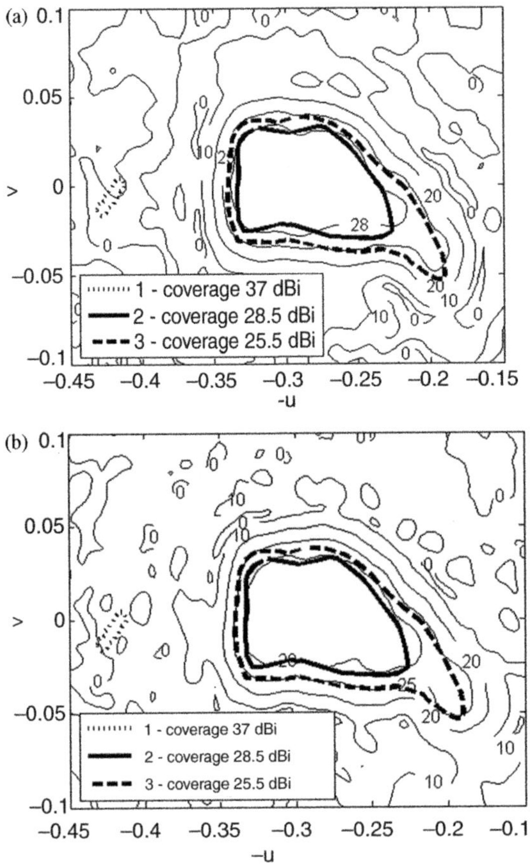

Figure 10.17 Measured co-polar gain patterns for H-polarization at extreme frequencies: (a) at 11.7 GHz; (b) at 13.0 GHz.

below the required level of 0 dBi. A similar performance is obtained for V-polarization (electric field in the X_R direction).

The starting phase distribution for the pattern synthesis at 14 GHz (Rx) is obtained by applying a linear variation with frequency to the phase distribution synthesized at 11.95 GHz, which consists of multiplying the phase delay $\phi_{di}(f)$ distribution (not truncated to a range of 360°) by the frequency ratio. Then, the intersection approach is applied including the near field of the horn at 14 GHz. Since the initial phase distribution already provides a contoured beam close to the prescribed coverage, a small adjustment in the phase should be sufficient to fulfill the requirements; for this reason, the variation of phase in the pattern synthesis is limited to ±45°. This limitation also ensures a smooth variation of the phase distribution required on the reflectarray as a function of the frequency. As a result, it is easier to realize the phases at different frequencies by optimizing the patch dimensions. The resulting phase distribution, shown in Figure 10.20b, provides a radiation pattern at 14 GHz that fulfils the requirements of coverage and isolation with Europe. Then, the required phase distribution for the given contoured beam is obtained at several frequencies in each band (Tx and Rx) by using a multi-frequency pattern synthesis [108,109].

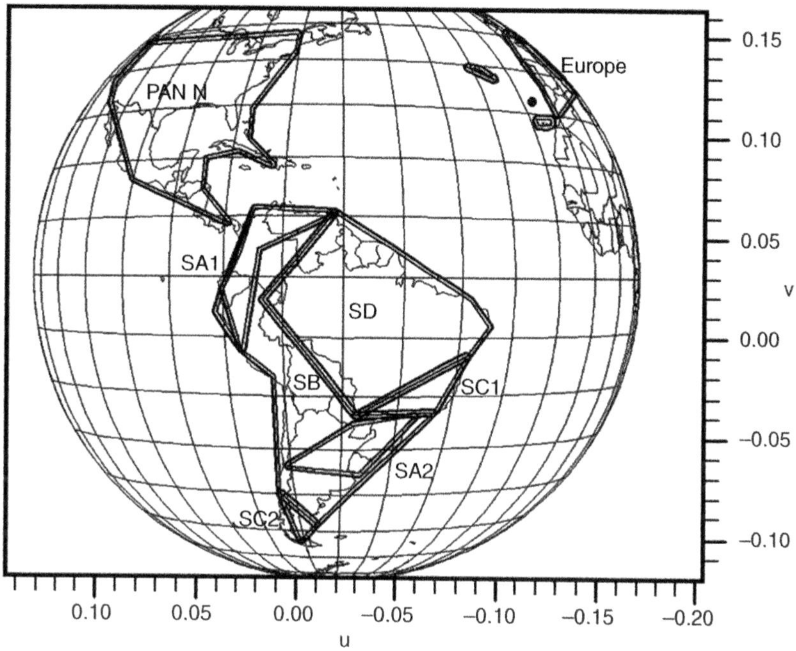

Figure 10.18 *Coverage of the Amazonas satellite. From [36], © 2011 IEEE.*

The patch dimensions are adjusted, element by element, first to match the phase distribution at 11.95 GHz by iteratively calling the analysis routine based on the SD-MoM [79] considering local periodicity and accounting for the angle of incidence and the polarization of the incident filed. In a second step, the patch dimensions are optimized in each element to match simultaneously the phase distribution at central and extreme frequencies in both Tx and Rx bands for each polarization. Since the optimization to minimize the error function involves six variables, three patch dimensions for each polarization, and the phasing performance for the two polarizations is practically uncoupled, as shown in previous works [34,35], the optimization is carried out independently for each polarization as in the previous example, resulting in a significant reduction of CPU time. This process accounts in a simple manner for the anisotropy of the honeycomb ($\varepsilon_r = 1.07$, for X_R polarization and $\varepsilon_r = 1.05$ for Y_R polarization). After a first optimization for H- and V-polarizations, the resulting reflectarray shows a

Table 10.2 *Gain and cross-polar requirements for PAN-S mission.*

	Tx		Rx	
Zone	Gain (dB)	XPD (dB)	Gain (dB)	XPI (dB)
SA1	28.82	31.00	27.32	32.00
SA2	28.81	31.00	27.31	28.00
SB	25.81	30.00	24.31	28.00
SC1	22.81	29.00	22.31	28.00
SC2	20.66	27.00	21.28	28.00
SD	19.81	27.00	18.31	25.00

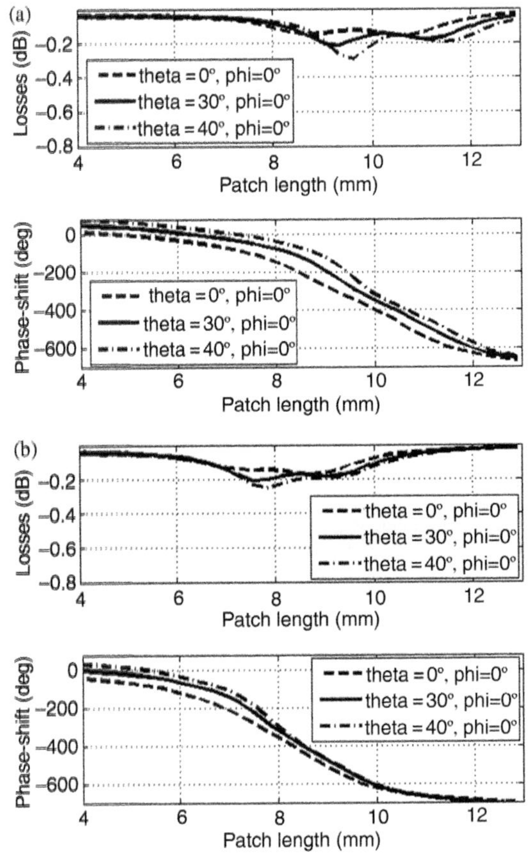

Figure 10.19 Ohmic losses and phase shift computed by SD-MoM at 11.95 GHz (a) and 14.00 GHz (b) for different angles of incidence.

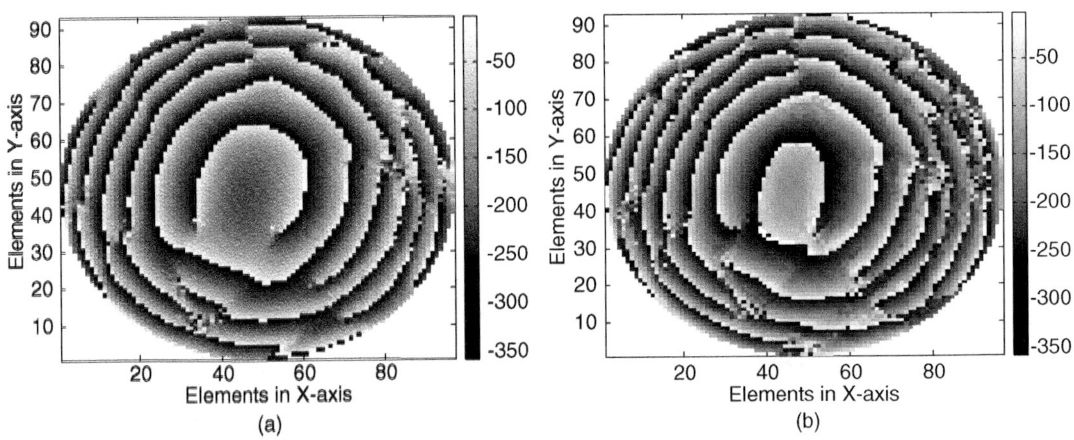

Figure 10.20 Required phase shift to provide the required coverage of South America at 11.95 GHz (a) and 14.00 GHz (b).

significant improvement in the coverage performance for both Tx and Rx frequencies, but the requirements are not fully accomplished. The compliance of the requirements is improved by further optimizations in an iterative process, as follows: (1) the phase distribution produced by the optimized reflectarray is computed by the SD-MoM analysis technique at the defined frequencies in Tx and Rx and used as the starting point in a new pattern synthesis; (2) the intersection approach is applied again at each frequency until the radiation patterns comply with the requirements; and (3) the synthesized phase distributions at the defined frequencies are used in the next iteration for optimizing the patch dimensions to minimize the error functions for each polarization. After four iterations in the optimization of the patches, the improvement was inappreciable and the design was concluded. The computed co-polar radiation patterns are shown in Figure 10.21 at extreme frequencies (11.70 and 14.25 GHz) for H-polarization (electric field in the Y_R direction). The antenna performance is very similar for both polarizations. The levels of gain and cross-polarization in the worst case and the compliance with the requirements at 11.70 and 14.25 GHz are given in Table 10.3 for V-polarization. Note that the gain requirements are always accomplished in more than 90% of the region, in spite of the very stringent coverage of this mission, which includes two separated regions of high gain (28.8 dBi) and several regions of lower gain.

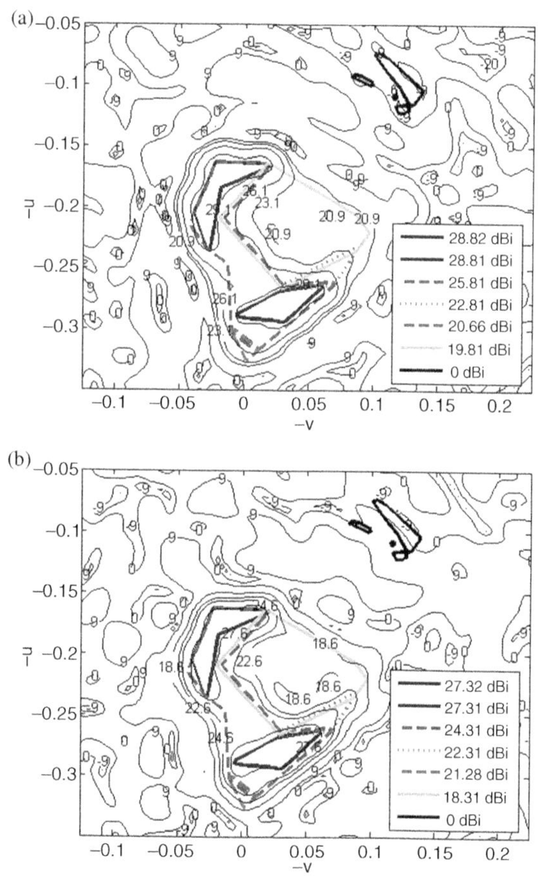

Figure 10.21 *Simulated co-polar radiation patterns (H-polarization) for the optimized reflectarray at 11.70 GHz (a) and 14.25 GHz (b).*

Table 10.3 *Simulated co-polar and cross-polar performance for V-polarization.*

	Tx (11.7 GHz)				Rx (14.25 GHz)		
Zone	Min. co-pol. (dBi)	Compliance (%)	Min. XPD (dB)	Compliance (%)	Min. co-pol. (dBi)	Compliance (%)	XPI (dB)
SA1	29.52	100.0	33.84	100.0	27.79	100.0	26.82
SA2	29.52	100.0	32.12	100.0	27.52	99.0	33.56
SB	26.10	99.8	30.35	100.0	23.70	98.6	22.83
SC1	23.18	100.0	27.74	90.0	23.35	100.0	25.28
SC2	25.34	100.0	35.35	100.0	21.50	95.7	31.28
SD	20.37	100.0	24.83	97.1	17.16	90.7	18.18

The reflectarray panel was manufactured by a co-curing process, in which all the layers in the electrical and structural sandwiches are cured in a single step in order to reduce manufacturing time and costs. The demonstrator is assembled using an aluminum supporting structure to provide the correct position and alignment of the feed horn and the reflectarray panel. The reflectarray demonstrator is shown in Figure 10.22.

Several mechanical, thermal and electrical tests have been carried out on the antenna demonstrator, including modal, thermal cycling, thermoelastic and RF tests, in order to check the antenna performance in the space environment. Co-polar and cross-polar radiation patterns and gain have been measured for both polarizations in a planar near-field system. The measured gain contours for H-polarization corresponding to the levels required in each zone, see Table 10.2, at the extreme frequencies in Tx and Rx, are shown in Figure 10.23. The requirements of minimum gain are practically fulfilled at both frequencies in all the zones except in zone SD, where the minimum measured gain is around 18 dBi at 11.7 GHz and around 17 dBi at 14.25 GHz. Similar results are obtained for V-polarization and for the other frequencies, the gain requirements being better accomplished in the Rx band compared to the Tx band. Co-polar isolation with Europe is well accomplished in the Tx band, but in the Rx band it shows levels up to 5 dBi in some particular locations of the European zones. The reason for this increase of the radiation in some areas out of the prescribed coverage is associated with the remaining phase errors after several iterations of the patch optimizations. The cross-

Figure 10.22 *Tx–Rx reflectarray demonstrator.*

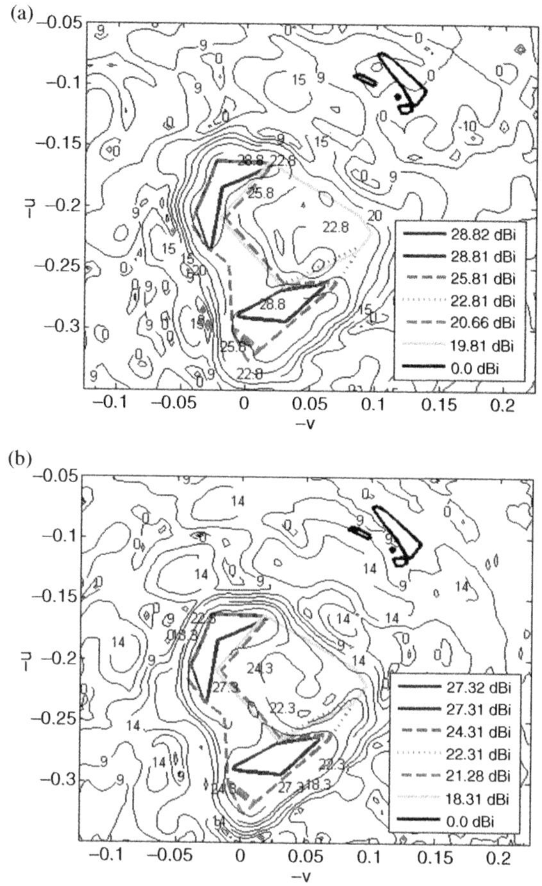

Figure 10.23 *Measured co-polar radiation pattern (H-polarization) for the optimized reflectarray at 11.70 GHz (a) and 14.25 GHz (b).*

polarization is low in all the coverage, with a maximum of −0.4 dBi at 11.7 GHz and 0.6 dBi at 14.25 GHz. The XPD requirements are accomplished in more than 90% of the coverage in the Tx band; however the XPI requirements in the Rx band are not satisfactorily accomplished. The measured ohmic losses of the reflectarray demonstrator vary from 0.26 to 0.3 dB in the Tx band and from 0.36 to 0.44 dB in the Rx band.

The measured gain contours show some distortions particularly in the area of Brazil, which are produced by a deformation of the reflectarray surface. The reflectarray surface has been measured (Z_R coordinate in mm) showing slight deformations within the peak values of −0.96 and + 0.45 mm. It has been checked that these deviations in the surface can produce the distortions observed in the radiation patterns. These results suggest that the mechanical design of the reflectarray should be modified, in order to maintain the surface flatness within a variation of ±0.2 mm, including the mechanical tolerances and the distortions associated with the space environment.

Previous results show that a reflectarray can be designed to fulfill all the requirements of a Tx–Rx DBS antenna provided that the electrical requirements (complexity of the coverage, cross-polarization and co-polar isolation requirements) are slightly relaxed with respect to those of the PAN-S mission. Multilayer printed reflectarrays can be a technological alternative to on-board shaped reflectors, which reduces costs and manufacturing time, as a result of the elimination of the custom molds used in reflector technology.

10.5 Recent and Future Developments for Space Applications

Some recent and ongoing developments in reflectarray antennas for space applications are presented in this section. Examples such as inflatable reflectarrays, large-aperture antennas, multibeam reflectarrays, dual-reflector configurations and beam scanning reflectarrays are presented in some detail.

10.5.1 Large-Aperture Reflectarrays

Large-aperture spacecraft antennas are required in some specific missions for telecommunications, Earth observation and science, such as mobile satellite communications, SAR, high spatial resolution radiometers, orbiting radio observatories, or deep space missions. The development of large space antennas, from 4 to 25 m, requires the fulfillment of stringent requirements on low mass and small stowage volume in order to reduce payload weight and to allow accommodation on the launcher. Different concepts have been proposed for large deployable reflector antennas in the past few decades, including shell solid surfaces, membranes and metallic meshes [111–115]. Mesh reflectors approximate a parabolic surface by a complex system of tensioning cables [114,115]. On the other hand, solid surface reflectors provide better surface accuracy and can be combined with reflectarray technology to correct for the surface deviation as proposed in [98]. Another concept developed for large apertures is based on inflatable reflector antennas [116]. The technology for inflatable antennas has been applied to printed arrays for SAR applications [117] and also to reflectarrays as discussed later.

For large apertures, deployable reflectarray antennas made of multiple flat panels [98] can offer some mechanical advantages compared to deployable reflectors, allowing a reduction in the stowage volume and simplifying the deployment in space. For example, the configuration proposed in [113], which is made of several identical hexagonal panels, can be implemented by using reflectarray panels. The hexagonal panels are stacked along a common axis for storage in a minimum volume, and then the antenna is deployed by using translational hinges and blocking mechanisms. In some particular applications such as SAR, the antenna should be large only in one dimension, requiring multiple panels only in one dimensions, as shown in Figure 10.10a. In this case, the antenna can be deployed to approximate a parabolic cylinder, using mechanisms for deployment similar to those used in solar arrays.

A recent development using the 1D architecture shown in Figure 10.10a is the NASA/JPL Wide Swath Ocean Altimeter (WSOA) in Ku-band [118]. In this application, the required aperture of 2×0.5 m is made of five reflectarray panels, which make up a piecewise planar approximation of a parabolic cylinder considering a 1.125 m focal length. Each reflectarray panel is made of variable-size patches to focus the beam in dual polarizations. The curvature of the multi-panel configuration, on one side, limits the angle of incidence from the feed and, on the other, improves the antenna bandwidth by limiting the differential space phase delay, as discussed in Section 10.3.3.2. An antenna breadboard, shown in Figure 10.24, was manufactured and tested. The results have shown an aperture efficiency of about 50% and good concordance between measured and simulation radiation patterns. Although the aperture in this example is not really large, the same configuration can be employed for large apertures.

Another suitable application for 1D faceted reflectarrays is found in SAR antennas. A three-panel configuration was proposed for an X-band SAR (9.65 GHz with 300 MHz bandwidth) [119]. The aperture was defined as 6×1.6 m and the focal length as 2000 mm. The antenna is made of three panels, where the length of the central and lateral panels is 1.827 and 2.331 m respectively. The effect of differential spatial phase delay in each panel is compensated by using three layers of array patches and optimizing the patch dimensions

Figure 10.24 *Photo of the piecewise flat reflectarray for WSOA in Ku-band. Courtesy of Dr. John Huang, retiree from JPL, Reproduced by permission of John Wiley & Sons, Inc.[71].*

to match the required phase shift for both linear polarizations in the required frequency band (9.5–9.8 GHz), as described in Section 10.3.3.2. Each reflectarray panel was defined considering the patches printed on Kapton, bonded to Kevlar sheets and separated by Nomex honeycomb 3 mm thick. The resulting radiation patterns in the azimuth plane (corresponding to the large-antenna dimension), obtained by an analysis routine based on SD-MoM and local periodicity, are very similar at central and extreme frequencies, as shown in Figure 10.25. These results show that the multi-facet configuration in conjunction with the techniques for bandwidth improvement is a suitable solution for large-aperture antennas.

10.5.2 Inflatable Reflectarrays

The concepts developed for inflatable reflector antennas, based on 'rigidizable' membranes and inflatable support structures [116], have been applied at JPL/NASA to develop a new type of inflatable antenna based on printed arrays [117] and reflectarrays [38–41] on a planar surface, which mitigates the difficulty of maintaining the required accuracy on a large parabolic surface for the lifetime of the space mission. On the other hand, it is significantly simpler to maintain the required surface tolerance for a flat surface by tensioning forces. Two inflatable reflectarrays have been co-developed by JPL and ILC Dover, Inc., the first one of 1 m diameter at X-band [38,39] and the other at Ka-band with a 3 m diameter [39–41] which is briefly described here.

The 3 m inflatable reflectarray at Ka-band is shown in Figure 10.26a. The inflatable structure comprises a horseshoe-shaped tube of 25 cm diameter inflated to 3.0 psi (20.7 kPa) pressure that supports and tensions a 3 m circular reflectarray by 16 catenary points with spring-loaded tension cords. The position of each connecting point can be adjusted to provide the required flatness and position of the membrane. The reflectarray is illuminated by a corrugated horn supported by an inflatable tripod asymmetrically located, shown in Figure 10.26a. The inflatable structure was designed to avoid membrane damage and flatness deviation when the deflated antenna structure is rolled up. In this case, the reflectarray was made using a single polyimide membrane (Uplex) of 5 mil (0.13 mm) thickness with 5 μm copper cladding on both sides.

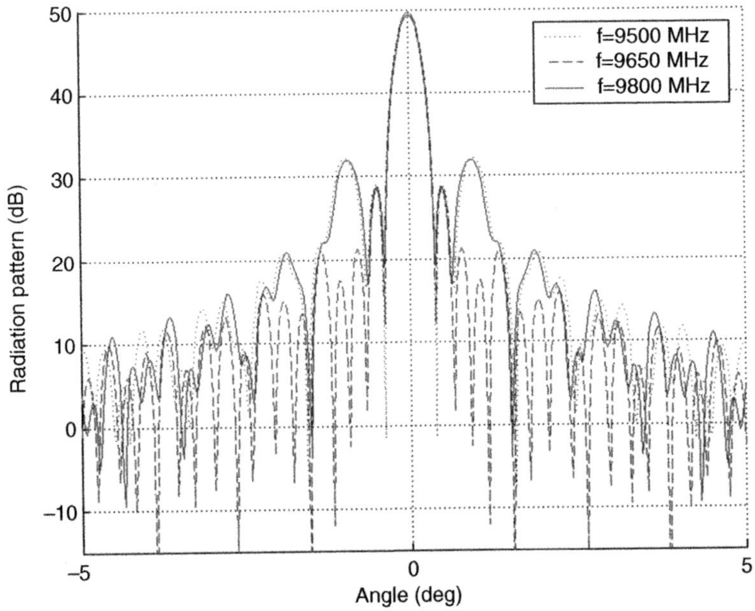

Figure 10.25 *Simulated radiation patterns in azimuth for a three-panel centre-fed reflectarray antenna after optimization in 9.5–9.8 GHz. From [71], Reproduced by permission of John Wiley & Sons, Inc.*

The appropriate phasing in circular polarization is implemented by varying the rotation angle of identical square patches with attached stubs [9] printed on one side of the membrane.

A second model of the 3 m inflatable reflectarray, which includes several electrical and mechanical improvements, was developed [41]. First of all, the reflectarray element was changed to improve the matching of the delay line and consequently the antenna efficiency. The new reflectarray element was a slightly rectangular patch with a matched delay line attached on the corner. Second, the new antenna configuration considers the feed offset located on the spacecraft and the reflectarray surface being independently deployed from the feed. Another improvement proposed in [41] is that the inflatable tubes are made of rigidizable aluminum membrane [120] so that inflation gas is no longer needed once the structure is inflated and the antenna will be more robust for the space environment. The flatness of the improved membrane reflectarray was measured in the old inflatable frame by the device illustrated in Figure 10.26a, showing a surface deviation of 0.2 mm RMS, which is really good for a 3 m aperture. The measured radiation pattern of the improved reflectarray at 32 GHz is given in Figure 10.26b. The beamwidth is 0.22°, the sidelobe level is at least −27 dB with respect to the main beam peak and the level of cross-polarization lower than −40 dB in the main beam. The measured antenna gain is 54.4 dBi, which represents an aperture efficiency of 30%, and the 3 dB gain bandwidth is 550 MHz. The relatively low aperture efficiency results from the dissipative losses in the Kapton membrane and can be improved by using low-loss materials for the membrane, such as liquid crystal polymer (LCP). The results of the 3 m antenna together with the excellent membrane flatness show the viability of inflatable reflectarrays at Ka-band.

10.5.3 High-Gain Antennas for Deep Space Communications

High-gain antennas (HGAs) operating in circular polarization at several frequency bands, X- and Ka-band, are required for communications in deep space missions. The HGA is usually based on a center-fed reflector and a

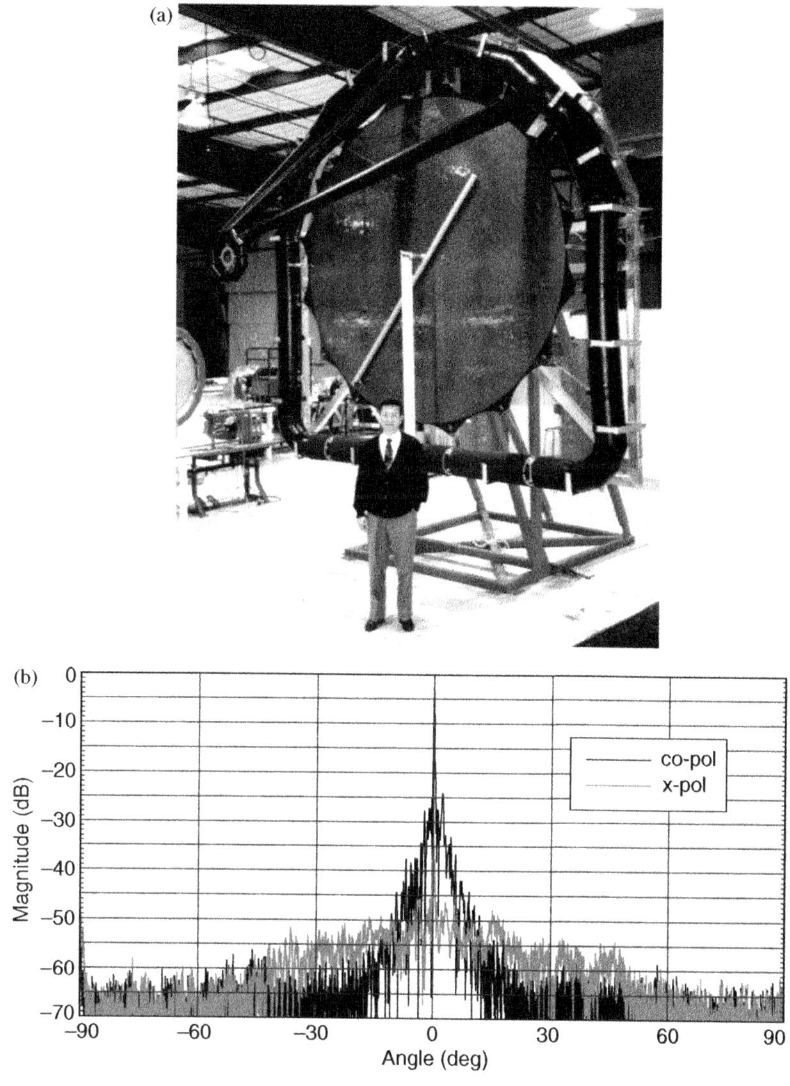

Figure 10.26 *The 3 m inflatable reflectarray antenna in Ka-band: (a) demonstrator; (b) measured radiation patterns (courtesy of Dr. John Huang, retiree from JPL, Reproduced by permission of John Wiley & Sons, Inc. [71]).*

dichroic subreflector to diplex the signals at different frequency bands. Another alternative recently developed by JPL/NASA is a dual-frequency reflectarray [121–124]. This reflectarray should introduce a different phase shift at each frequency band, by stacking or interleaving the printed elements that produce the phase shifting in each band. In the developments reported in [121–124], the low-frequency elements are placed on top of the high-frequency elements. In [121] the X-band elements are crossed dipoles, thin enough to produce a negligible blockage to the Ka-band signal, and the Ka-band elements are square patches, as shown in Figure 10.27a. The phase shift in each band is adjusted by independently varying the size of cross-dipoles for X-band (8.4 GHz) and the patches in Ka-band (32 GHz). A 0.5 m dual-band circularly polarized reflectarray

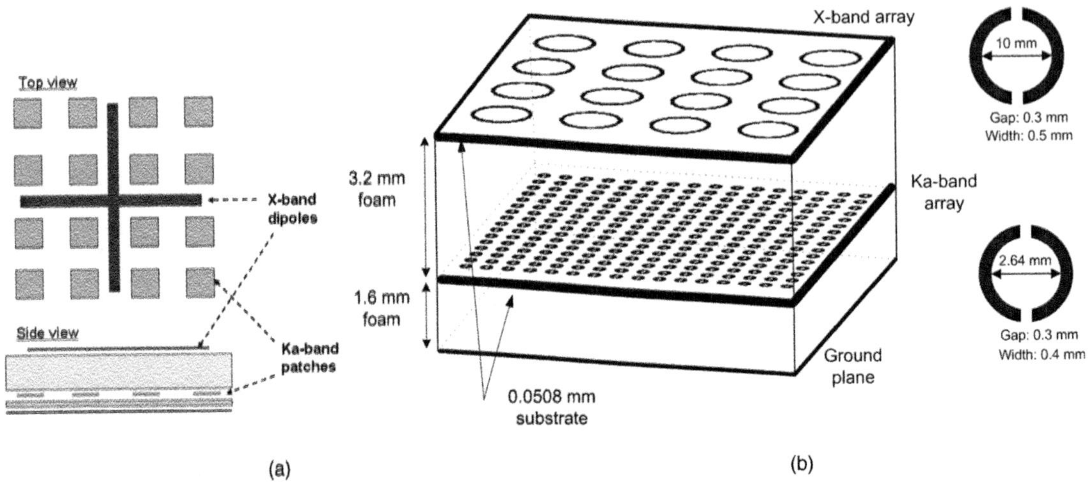

Figure 10.27 X/Ka-band reflectarray with the X-band layer above the Ka-band layer. (a) Using variable-size crosses in X-band and patches in Ka-band, (b) Using ring elements in both bands. Courtesy of JPL, from [71], Reproduced by permission of John Wiley & Sons, Inc.

antenna was manufactured and tested. The measured radiation patterns showed a focused beam with sidelobe levels below −20 dB in both bands. The measurements showed that the Ka-band layer had practically no effect on the X-band performance, while the presence of the X-band layer slightly affects the sidelobe level and gain of the Ka-band patterns.

An X/Ka-band reflectarray using thin membranes was proposed for deep space telecommunications [123], where the phasing in circular polarization was achieved by varying the rotation angle of printed circular rings with two oppositely located small gaps, as shown in Figure 10.27b. The annular ring element with the opposite gaps presents two advantages compared to the cross-dipole elements: first, it reduces the effect on the other frequency band because of the smaller metalized surface; and, second, it provides a broader bandwidth. The two layers of rings are printed on thin membranes to be compatible with future inflatable large antennas. Several reflectarray demonstrators have been developed by JPL/NASA using this type of dual-frequency element: first, a 0.5 m center-fed C/Ka dual-band reflectarray [122]; second, a 0.5 m offset-fed X/Ka dual-band reflectarray [123]; and, third, a 0.75 m Cassegrain offset-fed X/Ka dual-band reflectarray built on thin membranes [124]. The measured radiation patterns of the two X/Ka-band demonstrators showed excellent performance in sidelobe levels, cross-polarization and antenna efficiency (50% in X-band and 48.2% in Ka-band). The measured 3 dB axial ratio bandwidth was 600 MHz (7.1%) in X-band and 2 GHz (6.3%) in Ka-band. The results obtained from the X/Ka reflectarray demonstrators using thin membranes show significant potential for large high-gain inflatable antennas in future space applications.

10.5.4 Multibeam Reflectarrays

Reflectarrays can be designed to generate multiple beams, either with a single feed, or with one or more feeds per beam. Multiple beams can be achieved using passive reflectarrays with a unique feed, as reported in [33], where a reflectarray with two beams separated by 55° was designed to operate in dual polarization at 11.95 GHz. A breadboard made of two layers of variable-size patches was built and measured with satisfactory results. In the case of one feed per beam, the antenna configuration and the design process are similar as in multi-fed reflectors. However, the flexibility provided by reflectarrays to achieve any value of phase shift

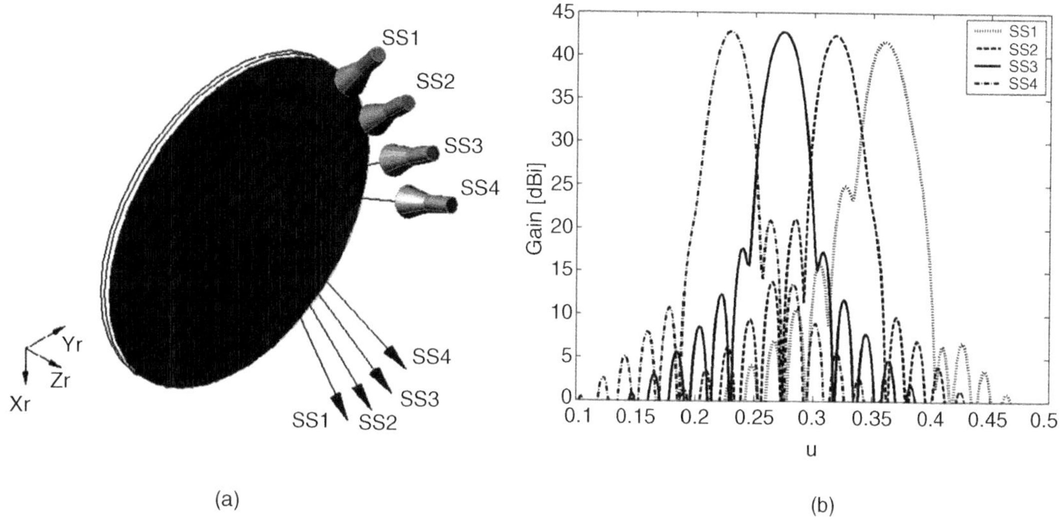

Figure 10.28 Reflectarray antenna with four beams. (a) Sketch of antenna configuration. (b) Simulated radiation patterns in elevation plane. From [71], Reproduced by permission of John Wiley & Sons, Inc.

independently for each linear polarization can be used to generate different beams in each polarization, to improve the antenna performance, or to produce various simultaneous shaped beams.

For multibeam applications requiring several simultaneous beams, such as communication antennas for Ka-band, a reflectarray can be used to generate a beam associated with each feed. In this case, the reflectarray is designed to generate a collimated beam in a given direction (θ_b,φ_b) for a feed located at a focal point. When other feeds are placed in the vicinity of the primary feed, they produce other beams in a different direction, but these beams suffer from aberration effects due to feed defocusing. The optimal feed positions can be found to minimize the aberration using the same techniques as in reflector antennas [125]. For further improvements, not only the feed positions but also the patch dimensions on the reflectarray can be optimized.

The first example is a theoretical study of a reflectarray antenna with four beams for low-cost X-band SAR applications in microsatellites [119]. In this case, the beam scanning can be achieved by switching from one beam to the next. The proposed multibeam configuration consists of a 1.6 m offset reflectarray with four feeds and an f/D around one, as shown in Figure 10.28a, designed for dual linear polarization. The four beams, called SS1, SS2, SS3 and SS4, are radiated in the X–Z plane at an angle θ_b from the Z_r-axis, being $\theta_b = 20.91°$, 18.41°, 15.91°, 13.41°, respectively. The reflectarray is designed at 9.65 GHz for the SS3 beam with the corresponding feed at coordinates $x_{f3} = -540$, $y_{f3} = 0$, $z_{f3} = 1714$ (in mm). The rest of the feeds are positioned on the X–Z plane to radiate the beam in the required directions. The radiation patterns in dBi for the four beams are superimposed in the scan plane in Figure 10.28b for y polarization. Gain and beamwidth are practically the same for the four beams because of the relatively large f/D ratio. It has been checked that the radiation patterns remain without distortion for both polarizations, in both principal planes and in the required frequency band for the X-band SAR, which is 2.3%. The multiple beams can be shaped, for example, to cancel the echo at nadir or to enforce some requirements in the radiation patterns, by implementing an appropriate phase distribution on the reflectarray.

In a second example reported in [84], it was demonstrated that several simultaneous shaped beams can be achieved on the basis of one feed per beam. Although the antenna was designed for a terrestrial point-to-multipoint application in Ka-band (25.5 GHz), the same design techniques can be applied to space antennas. The antenna demonstrator was designed to generate three independent shaped beams, covering adjacent 30° sectors in azimuth with the same squared cosecant pattern in elevation. Each beam is generated by a feed (a standard

20 dB pyramidal horn) that illuminates a square reflectarray made of 30 × 30 periodic cells (5.84 × 5.84 mm). The reflectarray comprises two layers of variable-size patches printed on 0.79 mm CuClad. The central feed is located at coordinates $x_{cf}=-94$, $y_{cf}=0$, $z_{cf}=214$ (in mm), referred to the center of the reflectarray. For the reflectarray design, first the pattern synthesis based on the intersection approach [103] is applied to obtain the phase distribution for the central beam associated with the central feed, and then the patch dimensions are adjusted to produce this synthesized phase distribution for both linear polarizations. Finally, the positions of the lateral feeds are optimized to generate the shaped beams with the appropriate direction and minimum distortion, being the resulting coordinates $x_{lf}=-94$, $y_{lf}=\pm 113$, $z_{lf}=182$ (in mm). The antenna breadboard has been manufactured, including a support structure that allows the feed horn to be placed in the positions corresponding to the central and lateral beams. The measured co-polar and cross-polar radiation patterns shown in [84] agree very well with the simulations considering the near field radiated by the horn [82] in both azimuth and elevation planes. These results demonstrate that several shaped beams can be generated with reflectarray technology.

10.5.5 Dual-Reflector Configurations

Dual-reflector antennas are used to reduce the cross-polarization and the overall antenna volume in many space antennas. The use of a reflectarray as sub- or main reflector in a dual-reflector configuration [42] can provide some advantages in the antenna performance and in the manufacturing process. A parabolic reflector with a reflectarray as subreflector has been proposed for compensating the errors on the surface of very large deployable reflectors [43,126] and for beam scanning in a limited angular range [44]. This antenna configuration, shown in Figure 10.29, combines the broadband of the parabolic reflector with the simplicity

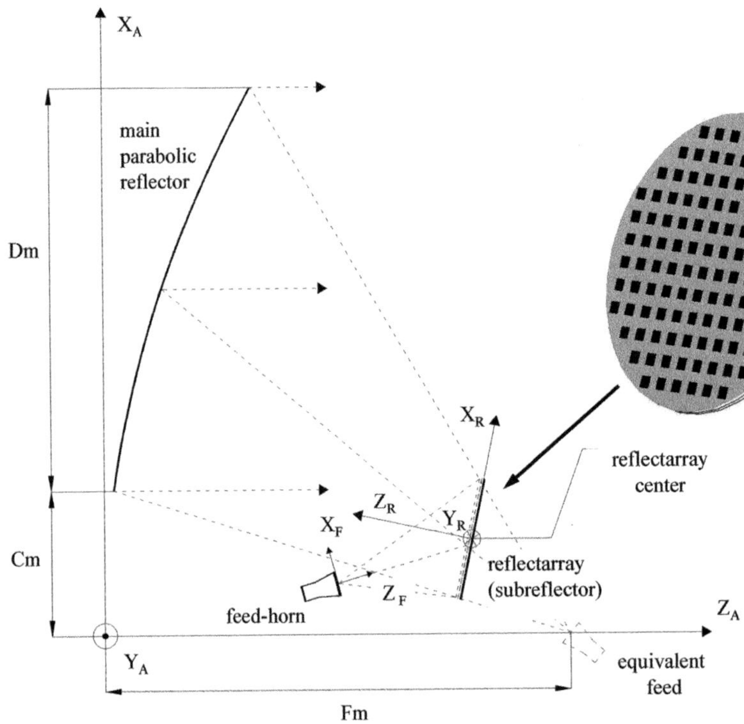

Figure 10.29 Scheme of dual-reflector antenna with a reflectarray as subreflector. From [44], © 2008 IEEE.

of manufacturing a small subreflectarray. In addition, this dual-reflector configuration can be used to scan or reconfigure the beam by electronically controlling the phase shift at the elements of the subreflectarray [44]. The antenna configuration including a controllable reflectarray subreflector can present great potential in several space applications, such as SAR, remote sensing and reconfigurable antennas for DBSs. Although a SAR antenna is usually implemented by active arrays, a reconfigurable reflectarray subreflector can provide beam scanning at a moderate cost and complexity [44]. A second application is in radiometric remote sensing in millimeter and submillimeter frequency bands, which require a beam scanning in a limited range actually implemented by motors [127]. An alternative for electronic beam scanning in this frequency band can be reflectarrays based on liquid crystals [57–59]. Finally, the reflectarray subreflector can be used to shape the beam in contoured-beam DBS antennas. In this configuration, the beam is shaped by adjusting the phase shift distribution on the subreflectarray, while the main parabolic reflector focuses the beam without any bandwidth limitation. If the phase is electronically controlled on the subreflectarray by using MEMS [49,50] or varactor diodes [52], the beam can be reconfigured to change the coverage during the lifetime of the satellite. This aspect is a real need in telecommunications and DBS missions, since the demand in data traffic (TV broadcast, phone, video or data transmission, multimedia direct-to-home, etc.) can change during the 15–20-year lifetime of the satellites.

A first potential application of a reflectarray as subreflector is to compensate the surface errors in very large deployable reflectors, such as mesh reflectors or in inflatable membrane antennas. The surface deformations in large reflectors resulting from thermal distortions or deployment mechanisms produce severe distortions in the radiation patterns that can be corrected by a subreflectarray, as demonstrated in [43,126]. A 1 m subreflectarray was used to compensate the thermal distortion errors in a 20 m offset parabolic reflector antenna operating at X-band (10 GHz) with a focal distance of 20 m and offset height of 15 m. The simulated thermal distortions produced a decrease in the directivity of 3.32 dB (from 65.47 to 62.15 dB). A reflectarray subreflector was designed to recover the original directivity, by using the conjugate field matching method (CFM) previously applied to feed arrays [128], which was adapted to reflectarrays by matching only the phase to correct the surface distortions. The simulated results, considering ideal reflectarray elements, showed an improvement of 2.59 dB in the directivity. The subreflectarray compensation technique has also been applied to correct the aberration in spherical reflector antennas [129]. This technique can be applied to dynamically compensate surface errors in very large deployable reflectors, if the phase correction in the reflectarray is electronically controlled by using MEMS [49,50], PIN diodes or varactors [52].

The dual-reflector antenna described in the previous example was analyzed by a simplified technique considering idealized reflectarray elements. A more realistic analysis technique, which takes into account the angle of incidence and polarization of the field that impinges on the reflectarray elements, was proposed and validated in [44]. The technique is a modular approach that combines the MoM for the analysis of the subreflectarray [23,79] and physical optics (PO) for the main parabolic reflector. Usually, the subreflector is illuminated in the near-field region of the primary feed, and therefore the near field of the horn has to be taken into account for an accurate analysis of the reflectarray antenna [82]. This technique, which is summarized here, provides an accurate prediction of gain, co-polar and cross-polar radiation patterns and is very efficient, so that it can be integrated into optimization routines for practical designs.

The modular analysis technique includes the following steps. First, the incident field on each reflectarray element is computed by using either a conventional far-field model of the feed, or the near field radiated by the feed horn, as described in [82]. Second, the reflectarray is analyzed by computing the total reflected field on each cell by the MoM assuming that each element is in a periodic environment. The calculation of the reflected field includes dissipative losses and the cross-polarization produced by the printed patches and by the antenna geometry. Finally, the analysis of the main reflector is based on the PO technique. The electric current on the reflector surface is computed in terms of the normal vector and the incident magnetic field on the parabolic surface, which is computed by adding the contributions of the field radiated by each reflectarray element.

From the electric current on the main parabolic reflector, an equivalent electric field is computed at the nodes of a regular grid on a projected aperture, which is used to calculate the radiation patterns assuming the first principle of equivalence, and using a time-efficient 2D FFT algorithm, as described in [44].

The proposed configuration can be used as a high-gain beam scanning antenna by adding controllable phase shifters on the subreflectarray. The beam scanning performance of an antenna made of a 1.5 m parabolic reflector and a rectangular subreflectarray (520 × 494 mm) was studied in an angular range of ±2° from the antenna boresight at 11.95 GHz [44]. In this example, the feed horn is modeled as a $\cos^q(\theta)$ function, with the q factor chosen to provide an illumination taper on the subreflectarray around −18 dB, from the center to the edges. The phase center of the horn is placed at coordinates (−294, 0, 326) referred to the subreflectarray coordinate system (X_R, Y_R, Z_R), and the center of the subreflectarray is placed at coordinates (294, 0, 1174) in the antenna reference system (X_A, Y_A, Z_A), all coordinates being in mm. The remaining details for the antenna geometry can be found in [44]. For the particular case of beam scanning in the azimuth plane (X_AY_A), different progressive phase distributions are introduced along the Y_R-axis of the subreflectarray; the radiation patterns are shown in Figure 10.30. In the case of ideal phase shifters considered here, the cross-polarization for the overall antenna is slightly increased when the beam is defocused, being on the order of −30 dB below the co-polar gain. In a real implementation, reflectarray elements with low cross-polarization should be used, in order not to increase the overall cross-polarization of the antenna. In addition, some degrees of freedom in the dual-reflector configuration can be used to improve the cross-polar performance, in a similar way as in conventional reflector antennas.

As another example, a 94 GHz dual-reflector antenna has been designed, manufactured and tested to show that it is possible to steer the beam up to ±5° from boresight by an appropriate phasing on a small reflectarray subreflector [129]. The antenna demonstrator is made of an offset parabolic reflector of 120 mm diameter and a

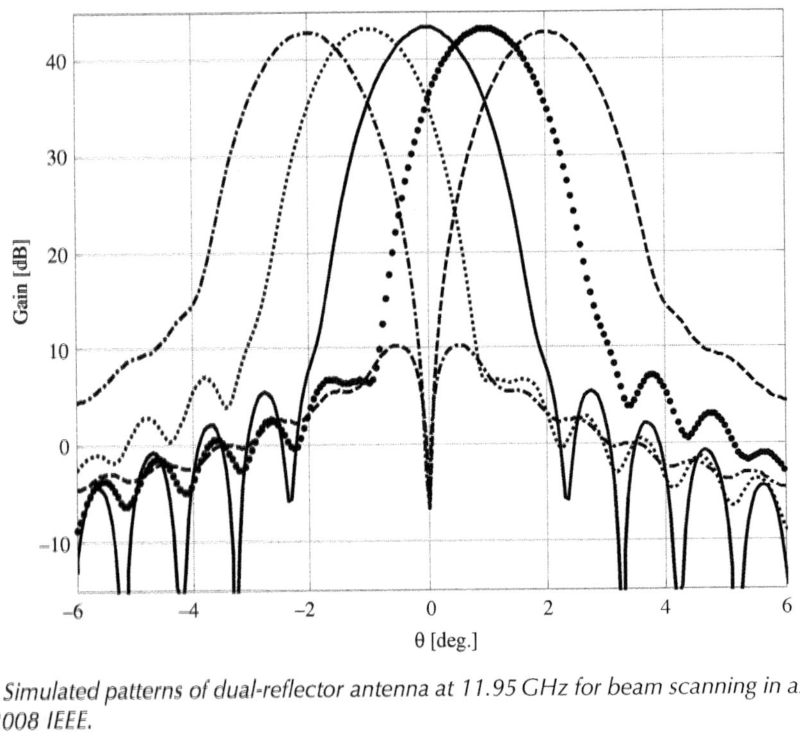

Figure 10.30 Simulated patterns of dual-reflector antenna at 11.95 GHz for beam scanning in azimuth in plane. From [44], © 2008 IEEE.

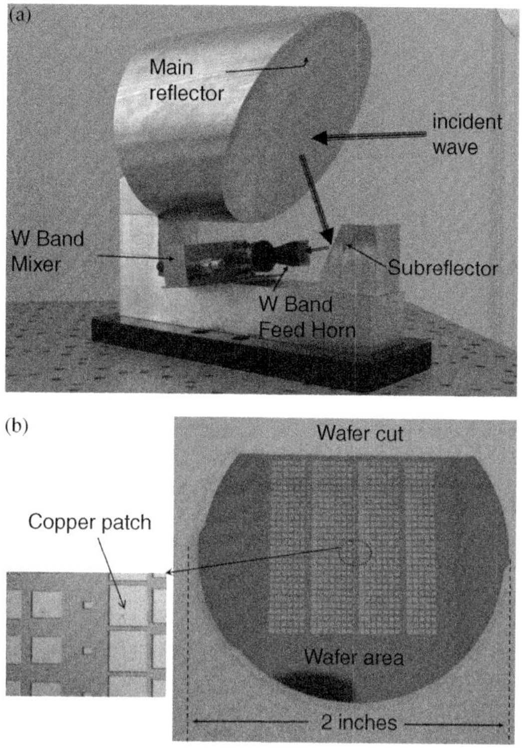

Figure 10.31 Dual reflector antenna at 94 GHz: (a) photograph; and (b) reflectarray subreflector. Courtesy of Dr. R. Cahill from Queen's University Belfast, © 2009 IEEE [130].

flat subreflector printed on a 50 mm diameter quartz wafer, see Figure 10.31. The beam scanning capability of the antenna is demonstrated by replacing a solid metal flat subreflector, which produces a focused beam on the axis of the paraboloid, with a passive reflectarray of 28 × 28 elements printed on a metal-backed quartz wafer 115 μm thick. The reflectarray is made of a single layer of variable-size patches, see Figure 10.31b, designed to produce a progressive phase distribution in the Y_R direction to scan the beam 5° in the azimuth plane. The co-polar far-field radiation patterns of the antenna were measured between −20° and +20° in the azimuth plane at 94 GHz. Figure 10.32 shows the measured radiation pattern using the reflectarray subreflector to tilt the beam 5° and also the results using a flat metallic subreflector to produce a boresight-directed beam. The measured radiation patterns are compared in Figure 10.32 to the simulated patterns obtained through the technique previously described, considering the ideal far-field model (a) and the near field (b) of the horn, see [130] for more details. These results show much better agreement with the measurements when the near-field model is used and confirm the need to use the near-field model to accurately analyze the dual-reflector antenna with reflectarray subreflector. The small distortion of the tilted beam and the deviation of 1.4° in the beam direction are because the antenna was designed by implementing a progressive phase distribution on the reflectarray assuming the far-field model of the feed, which does not provide a realistic estimation of the incident field in amplitude or in phase, as shown in [130]. To eliminate the distortion of the deflected beam, the reflectarray subreflector should be designed considering the near field of the feed horn [82]. To provide electronic beam scanning, dynamic control of the phase distribution on the reflectarray can be implemented by using a tunable liquid crystal layer between the patches and the ground plane.

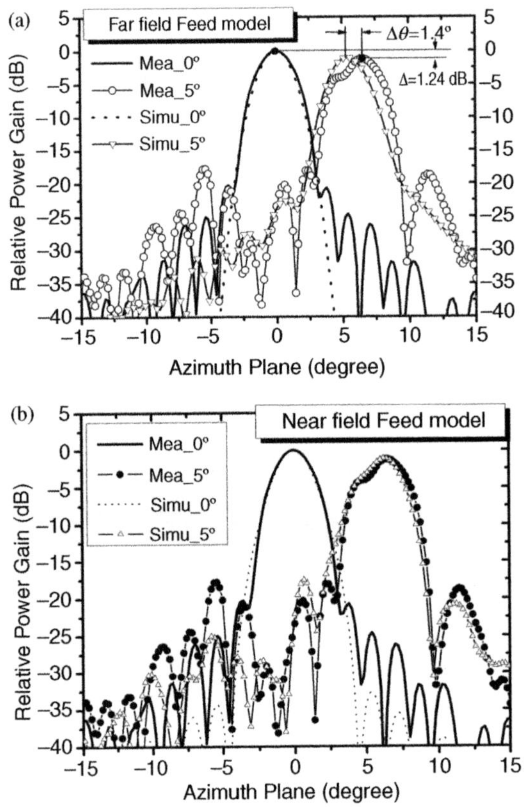

Figure 10.32 *Measured and simulated radiation patterns of the dual-reflector antenna with reflectarray subreflector in the azimuth plane at 94 GHz, using (a) far-field and (b) near-field model of the feed horn. From [130], © 2009 IEEE.*

10.5.6 Reconfigurable and Steerable Beam Reflectarrays

Reflectarray antennas can provide a scanning or reconfiguration of the beam by including electronically controllable phase shifters in the reflectarray elements [47–59] or in small sub-arrays [65–67]. The ohmic losses in reflectarrays can be very low, because the reflectarray is spatially fed, without any feeding network; then the beam can be steered by adding only low-loss phase shifters, instead of the conventional Tx–Rx modules used in active arrays. Several concepts have been developed to implement reflectarray antennas with beam scanning or beam reconfiguration capabilities. Some of them provide discrete phase control by using switches, as PIN diodes or MEMS, and other realizations provide continuous phase control, by using varactor diodes, ferroelectric materials or liquid crystals.

Since a reflectarray generally involves a large number of elements, switched-line phase shifters with low-bit quantization, such as 2 or 3 bits, are sufficient to provide beam scanning. A beam steering reflectarray based on printed patches aperture-coupled to microstrip stubs loaded with MEMS capacitances has been reported in [48]. A prototype made of 12 × 12 elements (6 × 6 cm) was fabricated on a silicon wafer, which contains more than a thousand MEMS to provide 3-bit control to the switched-line phase shifters, see Figure 10.33. The slots are etched on the ground of the wafer, not visible in the figures, and the printed patches are fabricated on another substrate stacked on the back side of the wafer.

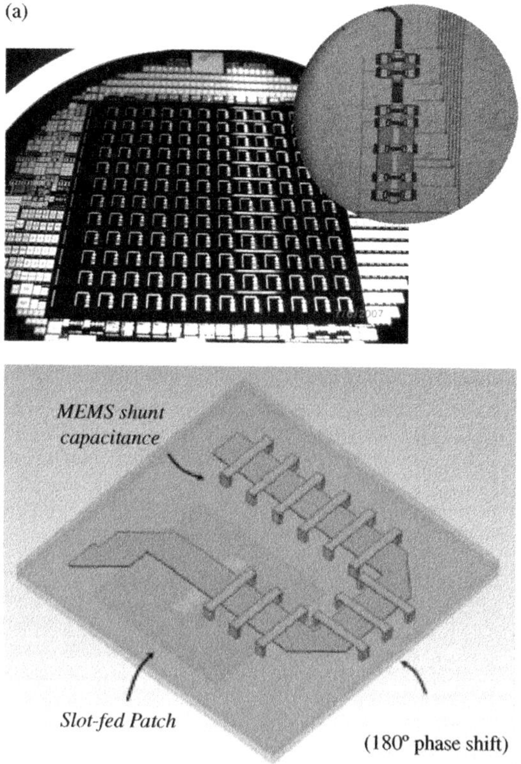

Figure 10.33 MEMS-based reflectarray at 35 GHz on wafer. (a) Bottom side with a detail of the MEMS loaded delay line. (b) Sketch of the aperture-coupled element. Courtesy of Professor Roberto Sorrentino from University of Perugia, © 2007 EuRAAP/THE IET [49].

The number of controls in beam scanning reflectarrays based on switched-line phase shifters can be reduced by grouping several reflectarray elements in a sub-array [65–67], in a similar way as in a feeding network. Figure 10.34 shows the aperture-coupled elements grouped in sub-arrays of two elements. Several groups of two and four elements have been designed at 10.4 GHz to provide a linear phase response as a function of the common delay line [67]. The beam scanning performance of a reflectarray made of 256 individual elements arranged in a circular grid of 18×18 elements (diameter 324 mm) was studied in [67] when the elements were grouped in pairs. The radiation patterns have been obtained considering a feed horn located at $(-100, 0, 330)$, in mm, with respect to the reflectarray center that provides an illumination taper of -11 dB. The main effect of the two-element gathering is the generation of a grating lobe in the plane of the gathering (azimuth), which is around -16 dB below the peak gain for the beam scanned at $11°$. These grating lobes can be significantly reduced by gathering the reflectarray elements in an irregular lattice [131,132]. The simulated radiation patterns for two scanning angles in azimuth obtained by alternating the pairs in the x and y directions are shown in Figure 10.35. In this case the level of the grating lobe in azimuth is reduced to 28 dB below the maximum. For a practical realization, it has been checked that 3 bits is sufficient to provide beam scanning with high resolution and low sidelobe levels.

Several reflectarray demonstrators of scanning reflectarrays have been reported using continuous phase control, implemented by varactor diodes, ferroelectric phase shifters or liquid crystals. A beam scanning

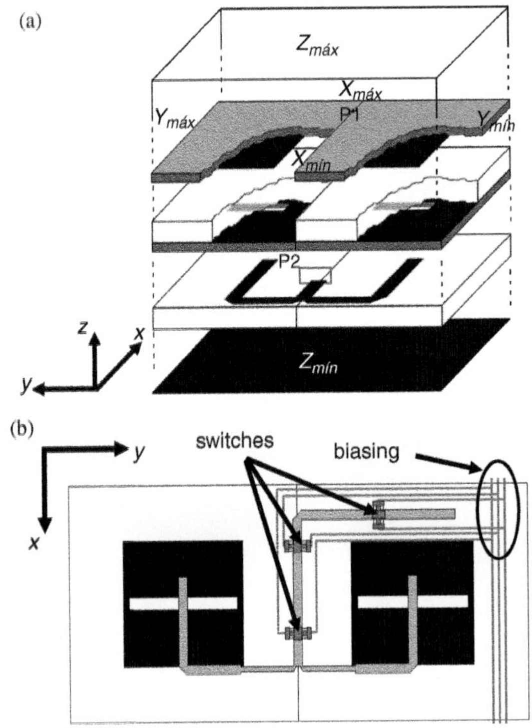

Figure 10.34 *Gathered reflectarray elements based on patches aperture-coupled to delay lines. (a) Expanded view. (b) Top view for two gathered elements. From [67], © 2011 IEEE.*

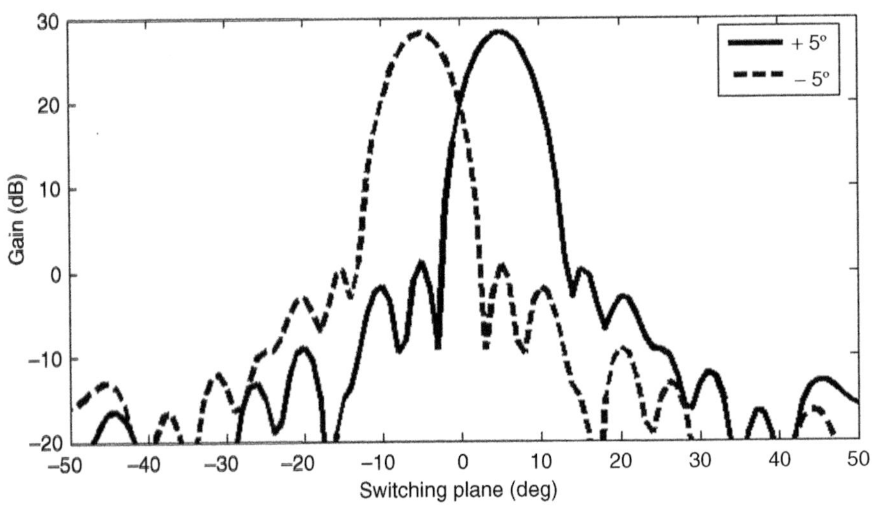

Figure 10.35 *Radiation patterns in azimuth with x and y alternated gathering.*

reflectarray based on printed patches aperture-coupled to a transmission line loaded with two varactor diodes was reported in [51]. The reflectarray element provides a continuous phase variation over a 360° range with a maximum loss of 2.4 dB at 5.4 GHz. A 30-element reflectarray breadboard was manufactured and tested, showing that the beam can be steered up to 40° from broadside by varying the bias voltages on each element.

An electronically tunable reflectarray was proposed in [52] for linear polarization, in which the phase shift is controlled by two surface-mounted varactor diodes that serially connect two halves of a microstrip patch. This concept provides a range in phase shift close to 360°. The same varactor-loaded element was used to demonstrate a reconfigurable reflectarray in real time [53] using a closed-loop control based on genetic algorithms to adjust the control voltages. Initially the algorithm was implemented to produce a retro-directive reflectarray, but the same controllable elements were used to demonstrate a beam scanning reflectarray at 5.8 GHz [54]. The reflectarray breadboard (300×210 mm) was able to scan the beam over a range from $-50°$ to $+50°$. These results demonstrate that a reflectarray antenna can provide beam scanning or reconfiguration in real time by using varactor diodes.

In recent years, new concepts of reconfigurable reflectarrays have been developed based on the properties of tunable dielectric materials, such as ferroelectric thin films or liquid crystals [133,134]. The principle of operation is based on the fact that the dielectric constant of some nonlinear dielectrics strongly depends on the electric field on the material, which is changed be applying an external voltage. These properties have been exploited to develop controllable phase shifters and reconfigurable reflectarrays [55–59].

Beam scanning reflectarray antennas based on ferroelectric phase shifters have been developed by NASA [55,56], and their potential in space missions has been investigated. The phase shifters are based on a series of coupled microstrip lines, patterned over a $Ba_{0.60}Sr_{0.40}TiO_3$ film approximately 400 nm thick. The phase shift is controlled by varying the bias voltage across the coupled lines, which are terminated in open circuit. An inconvenience of this technology is the high dissipative losses associated with the ferroelectric films, which are typically above 4 dB in K-band. However, recent works [56] have reported the development of high-quality ferroelectric phase shifters which allow a reduction in the dissipative losses.

A new class of phase shifter which exploits the dielectric anisotropy property of nematic state liquid crystals (LCs) has been applied to develop reconfigurable reflectarrays [57–59]. The anisotropic properties of liquid crystals and the manufacturing technology are well known from liquid crystal displays (LCDs), but their application to phase shifters at microwave and millimeter wave frequencies is fairly new [62,134]. This principle has been applied to develop reconfigurable reflectarray cells by using a LC layer between a resonant printed patch and the ground plane. When the dielectric constant changes by applying a bias voltage, the resonance frequency changes and so does the phase of the reflected field, in a similar manner as when the resonant length is varied. A reconfigurable LC reflectarray was designed to switch between a sum and a difference pattern [59] at 10 GHz. The prototype was made of a 12×12 array of patch elements illuminated by a 10 dB standard gain horn. The patches were printed on 18×18 cm glass-reinforced PTFE 125 μm thick. A tunable LC layer (BL006 from Merck) 500 μm thick was inserted between the printed array and the ground plane. The reflectarray was designed to generate a focused (sum) pattern at an angle of 19° from the mechanical boresight and to switch to a difference pattern. The size of the patches was adjusted to focus the beam when the permittivity of the LCs in one half of the aperture is $\varepsilon_{\perp} = 2.51$ (unbiased) and $\varepsilon_{\parallel} = 2.66$ (biased) in the other half. When the biasing is reversed, the patches of each half shift the phase by $+90°$ and $-90°$ respectively, thus generating a 180° phase shift between the two halves, then producing a difference pattern. The theoretical and measured patterns demonstrated the pattern switching ability of this reflectarray.

A LC reflectarray made of 16×16 elements has been designed, manufactured and tested to demonstrate beam steering at 35 GHz [57]. First, to simplify the manufacturing and the electronic control, the beam was scanned in only one plane. In this case, all the printed patches in a row are connected by a thin copper line used to supply the bias voltage to each row. A commercially available LC mixture (BL006) with a thickness of 0.127 mm was used, which was kept uniform by separators made of RT-Duroid 5880. This demonstrator provided beam scanning over a range of ±20°. Another demonstrator with the same number of elements (16 × 16) was built for 2D beam scanning [58]. In this case the bias voltage is applied to each reflectarray element individually from the ground plane, which is made of isolated patches separated by small gaps. The measured radiation patterns demonstrate the beam scanning capabilities of this antenna over a range from −35° to +25°. Although there are still some issues to be solved, mainly regarding the high losses of the LCs and their slow switching times, the concept of an electronically reconfigurable reflectarray with LC has some appealing features, such as robustness (once the LC cavity is filled and sealed), low cost and reliable manufacturing processes similar to those used to manufacture LCD optical devices.

For space applications in remote sensing, the LC reflectarray can be used as a subreflector in conjunction with a large parabolic reflector to steer the beam in a limited angular range. This technology can be an alternative to mechanically scanned antennas in instruments operating in the range of millimeter wave frequencies [128], because the losses and switching time of the LC cells decrease for high frequencies. For this type of application, the reflectarray subreflector can be housed in the thermally controlled quasi-optical feed train of the instrument, thereby maintaining the temperature range over which the LCs are in the nematic state [134].

10.5.7 Conclusions and Future Developments

Passive reflectarray antennas have demonstrated their capability to produce very accurate contoured beams for DBS antennas and high gain for deep space communications. The reflectarray can be flush mounted onto one spacecraft surface or integrated with the solar panels in order to reduce the overall mass and volume. For large-aperture applications, a multi-panel configuration can be used to improve the antenna bandwidth and to provide simple and reliable folding and deployment mechanisms.

Dual-reflector antennas including a reflectarray as the sub- or main reflector can offer some capabilities in space applications for both passive and reconfigurable antennas. An accurate and efficient technique has been developed and validated for the analysis of reflectarrays in dual-reflector configuration. The analysis technique can be integrated in design and optimization procedures in order to improve the antenna performance for low cross-polarization, beam scanning or multiple beams. The dual-reflector with an electronically controllable reflectarray subreflector can be used to scan or reconfigure the beam in several space applications, such as SAR, remote sensing and reconfigurable antennas for DBSs. Simple and low-cost phase shifters based on switches (PIN diodes or MEMS), varactor diodes, or other devices, can be integrated at each unit element or sub-array to provide the reconfiguration on a limited-size reflectarray acting as subreflector.

The reflectarray technology can be applied in a very wide range of frequencies from microwaves to infrared. Reflectarrays could potentially exploit the recent advances in tunable reflectarray technology based on LCs to provide electronically controlled beam steering antennas for the next generation of remote sensing instruments in the millimeter wave and terahertz ranges.

Acknowledgments

The work presented in this chapter was supported by the European Space Agency (ESA) and the Spanish Ministry of Science and Innovation under projects TEC2010–17567 and CONSOLIDER–INGENIO CSD2008–00068 (TERASENSE). The author wishes to express his gratitude to E. Carrasco and M. Barba from Universidad Politécnica de Madrid (Spain), to M. Arrebola from Universidad de Oviedo (Spain) and to L. de la Fuente from EADS Astrium CASA–Espacio (Spain) for providing some results presented in this chapter.

References

1. Berry, D.G., Malech, R.G., and Kennedy, W.A. (1963) The reflectarray antenna. *IEEE Transactions on Antennas and Propagation*, **11**(6), 646–651.
2. Phelan, H.R. (1975) Antenna arrays of internally phased elements, US Patent 3925784.
3. Phelan, H.R. (2011) Spiralphase reflectarray for multitarget radar. *Microwave Journal*, **20**(7), 67–73.
4. Howell, J. (1975) Microstrip antennas. *IEEE Transactions on Antennas and Propagation*, **23**(1), 90–93.
5. Malagisi, C.S. (1977) Electronically scanned microstrip antenna array. US Patent 4053895.
6. Malagisi, C.S. (1978) Microstrip disc element reflect array. Proceedings of the Electronics and Aerospace Systems Convention, Arlington, VA, USA, pp. 186–192.
7. Montgomery, J.P. (1978) A microstrip reflectarray antenna element. Proceedings of the Antenna Applications Symposium, Urbana–Champaign, IL, USA.
8. Huang, J. and Pogorzelski, J. (1996) Beam scanning reflectarray antenna with circular polarization. US Patent 6081234.
9. Huang, J. and Pogorzelski, R.J. (1998) A Ka-band microstrip reflectarray with elements having variable rotation angles. *IEEE Transactions on Antennas and Propagation*, **46**(5), 650–656.
10. Martynyuk, A., Martínez, J.I., and Martynyuk, N. (2004) Spiraphase-type reflectarrays based on loaded ring slot resonators. *IEEE Transactions on Antennas and Propagation*, **52**(1), 142–153.
11. Richards, R.J., Dittrich, E.W., Kesler, O.B., and Grimm, J.M. (2000) Microstrip phase shifting reflect array antenna, US Patent 6020853.
12. Richards, R.J. (2001) Integrated microelectromechanical phase shifting reflect array antenna, US Patent 6195047.
13. Munson, R.E., Haddad, H.A., and Hanlen, J.W. (1987) Microstrip reflectarray for satellite communication and radar cross-section enhancement or reduction, US Patent 4684952.
14. Chang, D.C. and Huang, M.C. (1992) Microstrip reflectarray antenna with offset feed. *Electronics Letters*, **28**(16), 1489–1491.
15. Javor, R.D., Wu, X.-D., and Chang, K. (1995) Design and performance of a microstrip reflectarray antenna. *IEEE Transactions on Antennas and Propagation*, **43**(9), 932–939.
16. Chang, D.C. and Huang, M.C. (1995) Multiple polarization microstrip reflectarray antenna with high efficiency and low cross-polarization. *IEEE Transactions on Antennas and Propagation*, **43**(8), 829–834.
17. Gonzalez, D.G., Pollon, G.E., and Walker, J.F. (1990) Microwave phasing structures for electromagnetically emulating reflective surfaces and focusing elements of selected geometry, US Patent 4905014.
18. Kelkar, A. (1991) Flaps: conformal phased reflecting surfaces. Proceedings of the IEEE National Radar Conference, Los Angeles, CA, USA, pp. 58–62.
19. Targonski, S.D. and Pozar, D.M. (1994) Analysis and design of a microstrip reflectarray using patches of variable size. Antennas and Propagation Society International Symposium, AP-S Digest, pp. 1820–1823.
20. Pozar, D.M., Targonski, S.D., and Syrigos, H.D. (1997) Design of millimeter wave microstrip reflectarray. *IEEE Transactions on Antennas and Propagation*, **45**(2), 287–295.
21. Pozar, D.M. and Targonski, S.D. (1998) A microstrip reflectarray using crossed dipoles. IEEE Antennas and Propagation Society International Symposium, pp. 1008–1011.
22. Encinar, J.A. (1999) Printed circuit technology multi-layer planar reflector and method for the design thereof, European Patent 1 120 856 A1.
23. Encinar, J.A. (2001) Design of two-layer printed reflectarrays using patches of variable size. *IEEE Transactions on Antennas and Propagation*, **49**(10), 1403–1410.
24. Encinar, J.A. and Zornoza, J.A. (2003) Broadband design of three-layer printed reflectarrays. *IEEE Transactions on Antennas and Propagation*, **51**(7), 1662–1664.
25. Robinson, A.W., Bialkowski, M.E., and Song, H.J. (1999) An X-band passive reflect-array using dual-feed aperture-coupled patch antennas. Proceedings of the Asia Pacific Microwave Conference, Singapore, pp. 906–909.
26. Chaharmir, M.R., Shaker, J., Cuhaci, M., and Sebak, A. (2003) Reflectarray with variable slots on ground plane. *IEE Proceedings: Microwaves, Antennas and Propagation*, **150**(6), 436–439.
27. Patel, M. and Thraves, J. (1994) Design and development of a low cost, electronically steerable, X-band reflectarray using planar dipoles. Proceedings of Military Microwaves, London, UK, pp. 174–179.

28. Bialkowski, M.E., Robinson, A.W., and Song, H.J. (2002) Design, development, and testing of X-band amplifying reflectarrays. *IEEE Transactions on Antennas and Propagation*, **50**(8), 1065–1076.
29. Pilz, D. and Menzel, W. (1998) Folded reflectarray antenna. *Electronics Letters*, **34**(9), 832–833.
30. Menzel, W., Pilz, D., and Leberer, R. (1999) A 77-GHz FM/CW radar front-end with a low-profile low-loss printed antenna. *IEEE Transactions on Microwave Theory and Techniques*, **47**(12), 2237–2241.
31. Menzel, W., Al-Tikriti, M., and Leberer, R. (2002) A 76GHz multiple-beam planar reflector antenna. Proceedings of the European Microwave Conference, Milan, Italy, vol. III, pp. 977–980.
32. Pozar, D.M., Targonski, S.D., and Pokuls, R. (1999) A shaped-beam microstrip patch reflectarray. *IEEE Transactions on Antennas and Propagation*, **47**(7), 1167–1173.
33. Encinar, J.A. and Zornoza, J.A. (2004) Three-layer printed reflectarrays for contoured beam space applications. *IEEE Transactions on Antennas and Propagation*, **52**(5), 1138–1148.
34. Encinar, J.A., Datashvili, L., Agustín Zornoza, J. *et al.* (2006) Dual-polarization dual-coverage reflectarray for space applications. *IEEE Transactions on Antennas and Propagation*, **54**(10), 2827–2837.
35. Encinar, J.A., Arrebola, M., Dejus, M., and Jouve, C. (2006) Design of a 1-metre reflectarray for DBS application with 15% bandwidth. Proceedings of the First European Conference on Antennas and Propagation (EuCAP 2006), Nice, France.
36. Encinar, J.A., Arrebola, M., de la Fuente, L.F., and Toso, G. (2011) A transmit–receive reflectarray antenna for direct broadcast satellite applications. *IEEE Transactions on Antennas and Propagation*, **59**(9), 3255–3264.
37. Hodges, R. and Zawadzki, M. (2005) Design of a large dual polarized Ku-band reflectarray for spaceborne radar altimeter. Proceedings of IEEE AP-S Symposium, Monterey, CA, USA, pp. 4356–4359.
38. Huang, J. and Feria, A. (1999) A one-meter X-band inflatable reflectarray antenna. *Microwave & Optical Technology Letters*, **20**(2), 97–99.
39. Huang, J. and Feria, A. (1999) Inflatable microstrip reflectarray antennas at X and Ka-band frequencies. Proceedings of the IEEE AP-S Symposium, Orlando, FL, USA, pp. 1670–1673.
40. Huang, J. and Pogorzelski, R.J. (1998) A Ka-band microstrip reflectarray with elements having variable rotation angles. *IEEE Transactions on Antennas and Propagation*, **46**(5), 650–656.
41. Huang, J., Feria, V.A., and Fang, H. (2001) Improvement of the three-meter Ka-band inflatable reflectarray antenna. Proceedings of the IEEE AP-S/URSI Symposium, Boston, MA, USA, pp. 122–125.
42. Khayatian, B., Rahmat-Samii, Y., and Huang, J. (2006) Radiation characteristics of reflectarray antennas: methodology and applications to dual configurations. Proceedings of First European Conference on Antennas and Propagation (EuCAP 2006), Nice, France.
43. Xu, S., Rjagopalan, H., Rahmat-Samii, Y., and Imbriale, W. (2007) A novel reflector surface distortion compensating technique using a sub-reflectarray. Proceedings of the IEEE Antennas Propagation Society International Symposium, Honolulu, HI, USA.
44. Arrebola, M., de Haro, L., and Encinar, J.A. (2008) Analysis of dual-reflector antennas. *IEEE Antennas and Propagation Magazine.*, **50**(6), 39–51.
45. Huang, J. (1995) Analysis of a microstrip reflectarray antenna for microspacecraft applications, TDA progress report, http://ipnpr.jpl.nasa.gov/progress_report/42-120/120H.pdf, pp. 153–173.
46. Profera, C.E. (1996) Reflectarray antenna for communication satellite frequency re-use applications, US Patent 5543809.
47. Gilbert, R. (2001) Dipole tunable reconfigurable reflector array, US Patent 0050650.
48. Mencagli, B., Vincenti Gatti, R., Marcaccioli, L., and Sorrentino, R. (2005) Design of large mm-wave beam-scanning reflectarrays. Proceedings of the 35th EuMC, Paris, France.
49. Sorrentino, R. (2007) MEMS-based reconfigurable reflectarrays. Second European Conference on Antennas and Propagation (EuCAP 2007), Edinburgh, UK.
50. Rajagopalan, H., Rahmat-Samii, Y., and Imbriale, W.A. (2008) RF MEMS actuated reconfigurable reflectarray patch–slot element. *IEEE Transactions on Antennas and Propagation*, **56**(12), 3689–3699.
51. Riel, M. and Laurin, J.J. (2007) Design of an electronically beam scanning reflectarray using aperture-coupled elements. *IEEE Transactions on Antennas and Propagation*, **55**(5), 1260–1266.
52. Hum, S.V., Okoniewski, M., and Davies, R.J. (2005) Realizing an electronically tunable reflectarray using varactor diode-tuned elements. *IEEE Microwave and Wireless Components Letters*, **15**(6), 422–424.

53. Hum, S.V., Okoniewski, M., and Davies, R.J. (2005) An evolvable antenna platform based on reconfigurable reflectarrays. Proceedings of the NASA/DoD Conference on Evolvable Hardware, Washington, DC, USA, pp. 139–146.
54. Hum, S.V., Okoniewski, M., and Davies, R.J. (2007) Modeling and design of electronically tunable reflectarrays. *IEEE Transactions on Antennas and Propagation*, **55**(8), 2200–2210.
55. Romanofsky, R.R., Bernhard, J.T., van Keuls, F.W. *et al.* (2000) K-band phased array antennas based on $Ba_{0.60}Sr_{0.40}TiO_3$ thin-film phase shifters. *IEEE Transactions on Microwave Theory and Techniques*, **48**(12), 2504–2510.
56. Romanofsky, R.R. (2007) Advances in scanning reflectarray antennas based on ferroelectric thin-film phase shifters for deep-space communications. *Proceedings of the IEEE*, **95**(10), 1968–1975.
57. Mössinger, A., Marin, R., Mueller, S., Freese, J., and Jakoby, R. (2006) Electronically reconfigurable reflectarrays with nematic liquid crystals. *IEE Electronics Letters*, **42**, 899–900.
58. Moessinger, A., Marin, R., Eicher, D., Jakoby, R., and Schlaak, H. (2007) Liquid crystal reflectarray with electronic 2D-reconfiguration capability. Proceedings of the 29th ESA Antenna Workshop on Multiple Beams and Reconfigurable Antennas, Noordwijk, The Netherlands, pp. 67–70.
59. Hu, W., Ismail, M.Y., Cahill, R. *et al.* (2007) Liquid-crystal-based reflectarray antenna with electronically switchable monopulse patterns. *Electronic Letters*, **43**(14), 899–900.
60. Gilger, L.D. (2000) Combined photovoltaic array and RF reflector, US Patent 6150995.
61. Zawadzki, M. and Huang, J. (2000) Integrated RF antenna and solar array for spacecraft application. Proceedings of the 2000 IEEE International Conference on Phased Array Systems and Technology, Dana Point, CA, USA, pp. 239–242.
62. Hu, W., Cahill, R., Encinar, J.A. *et al.* (2008) Design and measurement of reconfigurable millimeter wave reflectarray cells with nematic liquid crystal. *IEEE Transactions on Antennas and Propagation*, **56**(10), 3112–3117.
63. Ginn, J.C., Lail, B.A., and Boreman, G.D. (2006) Infrared patch reflectarray. Proceedings of IEEE Antennas and Propagation Society International Symposium, Albuquerque, NM, USA, pp. 4315–4318.
64. Ginn, J.C., Lail, B.A., and Boreman, G.D. (2007) Phase characterization of reflectarray elements at infrared. *IEEE Transactions on Antennas and Propagation*, **55**(11), 2989–2993.
65. Legay, H. and Salome, B. (2006) Low-loss reconfigurable reflector array antenna, US Patent 7142164 B2.
66. Carrasco, E., Barba, M., Reig, B., Encinar, J.A., and Charvet, P.L. (2011) Demonstration of a gathered element for reconfigurable-beam reflectarrays based on ohmic MEMS. Fifth European Conference on Antennas and Propagation (EuCAP 2011), Rome, Italy.
67. Carrasco, E., Barba, M., and Encinar, J.A. (2011) Design and validation of gathered elements for steerable-beam reflectarrays based on patches aperture-coupled to delay lines. *IEEE Transactions on Antennas and Propagation*, **59**(5), 1756–1760.
68. Huang, J. (1995) Bandwidth study of microstrip reflectarray and a novel phased reflectarray concept. Proceedings of IEEE Antennas and Propagation Society International Symposium, pp. 582–585.
69. Pozar, D.M. (2003) Bandwidth of reflectarrays. *Electronic Letters*, **39**(21), 1490–1491.
70. Zhuang, Y., Litva, J., Wu, C., and Wu, K.-L. (1995) Modelling studies of microstrip reflectarrays. *IEE Proceedings: Microwaves, Antennas and Propagation*, **142**(1), 78–80.
71. Huang, J. and Encinar, J.A. (2008) *Reflectarray Antennas*, IEEE Press/John Wiley & Sons, Inc.
72. Carrasco, E., Barba, M., and Encinar, J.A. (2006) Aperture-coupled reflectarray element with wide range of phase delay. *Electronics Letters*, **42**(12), 667–668.
73. Carrasco, E., Barba, M., and Encinar, J.A. (2007) Reflectarray element based on aperture-coupled patches with slots and lines of variable length. *IEEE Transactions on Antennas and Propagation*, **55**(3), 820–825.
74. Carrasco, E., Encinar, J.A., and Barba, M. (2008) Bandwidth improvement in large reflectarrays by using true-time delay. *IEEE Transactions on Antennas and Propagation*, **56**(8), 2496–2503.
75. Pozar, D.M. and Schaubert, D.H. (1984) Analysis of an infinite array of rectangular microstrip patches with idealized probe feeds. *IEEE Transactions on Antennas and Propagation*, **32**(10), 1101–1107.
76. Pozar, D.M. (1989) Analysis of an infinite phased array of aperture coupled microstrip patches. *IEEE Transactions on Antennas and Propagation*, **37**(4), 418–425.
77. Pozar, D.M. (2004) Microstrip reflectarrays myths and realities. Proceedings of the JINA International Symposium on Antennas, Nice, France, pp. 175–179.

78. Mittra, R., Chan, C.H., and Cwik, T. (1988) Techniques for analyzing frequency selective surfaces – a review. *Proceedings of IEEE*, **76**(12), 1593–1615.
79. Wan, C. and Encinar, J.A. (1995) Efficient computation of generalized scattering matrix for analyzing multilayered periodic structures. *IEEE Transactions on Antennas and Propagation*, **43**(10), 233–1242.
80. Encinar, J.A. and Caballero, R. (2001) Design and development of a double-sandwich dichroic subreflector. Proceedings of the 24th Antenna Workshop on Innovative Periodic Antennas: Photonic Bandgap, Fractal and Frequency Selective Surfaces, Noordwijk, The Netherlands, pp. 115–119.
81. Caballero, R., Palacios, C., and Encinar, J.A. (2004) Mars Express and Venus Express high gain antennas. Proceedings of the 27th ESA Antenna Workshop on Innovative Periodic Antennas: Electromagnetic Bandgap, Left-Handed Materials, Fractal and Frequency Selective Surfaces, Santiago de Compostela, Spain, pp. 83–89.
82. Arrebola, M., Álvarez, Y., Encinar, J.A., and Las-Heras, F. (2009) Accurate analysis of printed reflectarrays considering the near field of the primary feed. *IET Microwaves, Antennas and Propagation*, **3**(2), 187–194.
83. Ludwig, A.C. (1973) The definition of cross polarization. *IEEE Transactions on Antennas and Propagation*, **21**(1), 116–119.
84. Arrebola, M., Encinar, J.A., and Barba, M. (2008) Multifed printed reflectarray with three simultaneous shaped beams for LMDS central station antenna. *IEEE Transactions on Antennas and Propagation*, **56**(6), 1518–1527.
85. Carrasco, E., Arrebola, M., Encinar, J.A., and Barba, M. (2008) Demonstration of a shaped beam reflectarray using aperture-coupled delay lines for LMDS central station antenna. *IEEE Transactions on Antennas and Propagation*, **56**(10), 3103–3111.
86. Pilz, D. and Menzel, W. (1997) Full wave analysis of a planar reflector antenna. Proceedings of the Asia Pacific Microwave Conference, Hong Kong, pp. 225–227.
87. De Vita, P., De Vita, F., Di Maria, A., and Freni, A. (2009) An efficient technique for the analysis of large multilayered printed arrays. *IEEE Antennas and Wireless Propagation Letters*, **8**, 104–107.
88. Rius, J.M., Parron, J., Heldring, A. *et al.* (2008) Fast iterative solution of integral equations with method of moments and matrix decomposition algorithm–singular value decomposition. *IEEE Transactions on Antennas and Propagation*, **56**(8), 2314–2324.
89. Misran, N., Cahill, R., and Fusco, V. (2002) Reflection phase response of microstrip stacked ring elements. *Electronics Letters*, **38**(8), 356–357.
90. Misran, N., Cahill, R., and Fusco, V. (2003) Design optimisation of ring elements for broadband reflectarray antennas. *IEE Proceedings: Microwaves, Antennas and Propagation*, **150**(6), 440–444.
91. Bozzi, M., Germani, S., and Perregrini, L. (2003) Performance comparison of different element shapes used in printed reflectarrays. *IEEE Antennas and Wireless Propagation Letters*, **2**(1), 219–222.
92. Cadoret, D., Laisne, A., Gillard, R., Le Coq, L., and Legay, H. (2005) Design and measurement of new reflectarray antenna using microstrip patches loaded with slot. *Electronics Letters*, **41**(11), 623–624.
93. Encinar, J.A. and Pedreira, A. (2004) Flat reflector antenna in printed technology with improved bandwidth and separate polarizations, Spanish Patent P2004 01382.
94. Dieter, S., Fischer, C., and Menzel, W. (2009) Single-layer unit cells with optimized phase angle behaviour. Proceedings of the 3rd European Conference on Antennas and Propagation (EuCAP 2009), Berlin, Germany, pp. 1149–1153.
95. Rossi, F., Encinar, J.A., and Freni, A. (2009) Design of a reflectarray antenna at 300GHz using parallel dipoles of variable size printed on a quartz wafer. Proceedings of the 5th ESA Workshop on Millimetre Wave Technology and Applications & 31st ESA Antenna Workshop, Noordwijk, The Netherlands.
96. Menzel, W., Jiang, L., and Dieter, S. (2009) Folded reflectarray antenna based on a single layer reflector with increased phase angle range. Proceedings of the 3rd European Conference on Antennas and Propagation (EuCAP 2009), Berlin, Germany.
97. Chaharmir, M.R., Shaker, J., Cuhaci, M., and Ittipiboon, A. (2006) Broadband reflectarray antenna with double cross loops. *Electronics Letters*, **42**(2), 65–66.
98. Roederer, A. (2002) Reflector antenna comprising a plurality of panels, US Patent 6411255.
99. Chang, D.C. and Huang, M.C. (1993) Feasibility study of erecting cosecant pattern by planar microstrip reflectarray antenna. Proceedings of Asia Pacific Microwave Conference, Taiwan, vol. 2, pp. 19.20–19.24.
100. Zornoza, J.A. and Encinar, J.A. (2004) Efficient phase-only synthesis of contoured-beam patterns for very large reflectarrays. *International Journal of RF and Microwave Computer-Aided Engineering*, **14**(10), 415–423.

101. Trastoy, A., Ares, F., and Moreno, E. (2001) Phase-only control of antenna sum and shaped patterns through null perturbation. *IEEE Antennas and Propagation Magazine*, **43**(6), 45–54.
102. Kautz, G.M. (1999) Phase-only shaped beam synthesis via technique of approximated beam addition. *IEEE Transactions on Antennas and Propagation*, **47**(5), 887–894.
103. Bucci, O.M., Franceschetti, G., Mazzarella, G., and Panariello, G. (1990) Intersection approach to array pattern synthesis. *IEE Proceedings: Microwaves, Antennas and Propagation*, **137**(6), 349–357.
104. Vaskelainen, L.I. (2000) Phase synthesis of conformal array antennas. *IEEE Transactions on Antennas and Propagation*, **48**(6), 987–991.
105. Zornoza, J.A. and Encinar, J.A. (2001) Multi-layer printed reflectarrays as an alternative to shaped reflectors. Proceedings of the 24th ESTEC Antenna Workshop on Innovative Periodic Antennas, Noordwijk, The Netherlands, pp. 243–247.
106. Zornoza, J.A. and Encinar, J.A. (2002) Design of shaped beam reflectarrays for direct broadcast satellites. Proceedings of the International Symposium on Antennas (JINA 2002), Nice, France, pp. 367–370.
107. Costanzo, S., Venneri, F., Di Massa, G., and Angiulli, G. (2003) Synthesis of microstrip reflectarrays as planar scatterers for SAR interferometry. *Electronics Letters*, **39**(3), 266–267.
108. Zornoza, J.A., Arrebola, M., and Encinar, J.A. (2003) Multi-frequency pattern synthesis for contoured beam reflectarrays. Proceedings of the 26th Antenna Workshop on Satellite Antenna Modeling and Design Tools, ESTEC, Noordwijk, The Netherlands, pp. 337–342.
109. Zornoza, J.A. and Encinar, J.A. (2004) Reflectarray pattern synthesis with phase constraints. 13e Journées Internationales de Nice sur les Antennes (JINA), Nice, France, pp. 200–201.
110. TICRA, CHAMP user's manual for software package for analysis of corrugated and/or smooth wall horn with circular cross section, TICRA, Denmark.
111. Russell, R.A., Campbell, T.G., and Freeland, R.E. (1980) A technology development program for large space antennas. Proceedings of the 31st Congress of the International Astronautical Federation, Tokyo, Japan.
112. Roederer, A.G. and Rahmat-Samii, Y. (1989) Unfurlable satellite antennas: a review. *Annales des Telecommunications*, **44**, 475–488.
113. Kaminskas, R.A. (1989) Stowable reflector, US Patent 811034.
114. You, Z. and Pellegrino, S. (1994) Deployable mesh reflector. Proceedings of the International Symposium on Spatial Lattice and Tension Structures, Atlanta, GA, USA.
115. Thomson, M.W. (1999) The AstroMesh deployable reflector. Proceedings of the IEEE International Symposium on Antennas and Propagation, York, UK, vol. 3, pp. 1516–1519.
116. Thomas, M. (1992) Inflatable space structures. *IEEE Potentials*, **11**(4), 29–32.
117. Huang, J., Lou, M., Feria, A., and Kim, Y. (1998) An inflatable L-band microstrip SAR array. Proceedings of the IEEE Antennas and Propagation Society International Symposium, Atlanta, GA, USA, vol. 4, pp. 2100–2103.
118. Hodges, R. and Zawadzki, M. (2003) Wide swath ocean altimeter antenna electrical subsystem – preliminary design review, JPL internal document.
119. Legay, H., Salome, B., Girard, E. *et al.* (2005) Reflectarray antennas for SAR missions. Proceedings of the 11th International Symposium on Antenna Technology and Applied Electromagnetics (ANTEM 2005), Saint-Malo, France.
120. Fang, H., Lou, M., Huang, J., Hsia, L., and Kerdanyan, G. (2001) An inflatable/self-rigidizable structure for the reflectarray antenna. Proceedings of the 10th European Electromagnetics Structure Conference, Munich, Germany.
121. Zawadzki, M. and Huang, J. (2003) A dual-band reflectarray for X- and Ka-bands. Proceedings of the PIERS Symposium, Honolulu, Hawaii, USA.
122. Han, C., Rodenbeck, C., Huang, J., and Chang, K. (2004) A C/Ka dual-frequency dual-layer circularly polarized reflectarray antenna with microstrip ring elements. *IEEE Transactions on Antennas and Propagation*, **52**(11), 2871–2876.
123. Han, C., Huang, J., and Chang, K. (2005) A high efficiency offset-fed X/Ka dual-band reflectarray using thin membranes. *IEEE Transactions on Antennas and Propagation*, **53**(9), 2792–2798.
124. Han, C., Huang, J., and Chang, K. (2006) Cassegrain offset subreflector-fed X/Ka dual-band reflectarray with thin membranes. *IEEE Transactions on Antennas and Propagation*, **54**(10), 2838–2844.
125. Sletten, C.J. (1988) Multibeam and scanning reflector antennas, in *Reflector and Lens Antennas*, Artech House.

126. Xu, S., Rahmat-Samii, Y., and Imbriale, W.A. (2009) Subreflectarrays for reflector surface distortion compensation. *IEEE Transactions on Antennas and Propagation*, **57**(2), 364–372.
127. Martin, R.J. and Martin, D.H. (1996) Quasi-optical antennas for radiometric remote-sensing. *IEE Electronics & Communication Engineering Journal*, **8**, 37–48.
128. Rahmat-Samii, Y. (1990) Array feeds for reflector surface distortion compensation: concepts and implementation. *IEEE Antennas and Propagation Magazine*, **32**(4), 20–26.
129. Xu, S. and Rahmat-Samii, Y. (2009) A compensated spherical reflector antenna using sub-reflectarrays. *Microwave and Optical Technology Letters*, **51**(2), 577–582.
130. Hu, W., Arrebola, M., Cahill, R. *et al.* (2009) 94GHz dual-reflector antenna with reflectarray sub-reflector. *IEEE Transactions on Antennas and Propagation*, **57**(10), 3043–3050.
131. Carrasco, E., Barba, M., and Encinar, J.A. (2010) Switchable-beam reflectarray with aperiodic-gathered elements based on PIN diodes. Proceedings of the 32nd ESA Antenna Workshop on Antennas for Space Applications, Noordwijk The Netherlands.
132. Carrasco, E., Arrebola, M., Barba, M., and Encinar, J.A. (2011) Shaped-beam reconfigurable reflectarray with gathered elements in a non-periodic layout for LMDS base station. Proceedings of the Fifth European Conference on Antennas and Propagation (EuCAP 2011), Rome, Italy.
133. Weil, C. and Jakoby, R. (2002) Nonlinear dielectrics for microwave applications ferroelectrics and liquid crystals. *IEEE – MTT/AP German Newsletter*, **6**(1)
134. Muller, S., Scheele, P., Weil, C. *et al.* (2004) Tunable passive phase shifter for microwave applications using highly anisotropic liquid crystals. Proceedings of the IEEE MTT-S International Microwave Symposium, Fort Worth, TX, USA, vol. 2, pp. 1153–1156.

11

Emerging Antenna Technologies for Space Applications

Safieddin Safavi-Naeini and Mohammad Fakharzadeh

Center for Intelligent Antenna and Radio Systems (CIARS), Department of Electrical and Computer Engineering, University of Waterloo, Canada

11.1 Introduction

Next-generation space systems require higher data rates, higher precision spacecraft navigation, higher reliability, ultra-broad observation spectrum, and much more cost-effective approaches as compared to today's outdated 'Apollo-era' solutions [1]. X-band telemetry systems are rapidly migrating to Ka-band (32 GHz) systems. Large reflector antennas will be replaced by smaller, less costly and wider bandwidth array systems. On the other hand, space scientists and radio astronomers have been concentrating on the highest range of the radio spectrum. Radio technologies have gone beyond 1 THz and, due to high atmospheric absorption, radio telescopes should be deployed at sufficiently high altitude above the atmosphere [2]. The European Space Agency telescopes, Herschel and Planck, will observe the entire range of millimeter waves up to 5.3 THz. millimeter-wave (mmW) and sub-millimeter-wave (sub-mmW) ranges of frequencies such as windows around 54, 118, 183, 243, 325, 424, and 664 GHz can provide critical information about precipitation and ice clouds [3]. Lower microwave and mmW ranges of frequencies (Qu/Ka-band, V-band intersatellite links) [4] are being extensively used for a wide variety of fixed, land-mobile, marine, and airborne SATCOM systems for broadband information networking with worldwide ubiquitous coverage for a broad scope of scientific, defense, and civilian applications [5]. Although extreme demands on performance required for future missions cannot be easily met with the current state of the art, and this has posed the space technology community huge challenges. New advances in antenna, integrated device/system, and material technologies fuelled by the non-space technology market sectors are offering new opportunities. Rapid progress in microwave, mmW/sub-mmW, and THz integrated technologies, motivated by numerous emerging applications in broadband terrestrial/space radio communication, automotive mmW sensors and radars, mmW

Space Antenna Handbook, First Edition. Edited by William A. Imbriale, Steven (Shichang) Gao and Luigi Boccia.

security imager, sub-mmW/THz spectroscopy for biomedical and pharmaceutical purposes, and land-based and satellite environmental monitoring, offers high-performance cost-effective solutions that can easily be extended to space systems.

In this chapter, recent advances in integrated mmW on-chip/in-package Si-based active antenna technologies, new substrate integrated planar antenna technologies, microwave/mmW micro-electromechanical systems (MEMS) based techniques, and THz integrated antenna systems will be reviewed along with a comprehensive description of low-cost, high-performance phased-array technology for mobile SATCOM, as an increasingly important case study.

11.2 On-Chip/In-Package Antennas for Emerging Millimeter-Wave Systems

Antenna elements/arrays integrated on-chip or in-package using micromachining and semiconductor-compatible technologies offer a number of advantages for space system applications. Antennas integrated with the front-end in highly miniaturized chip-scale packages are being deployed in a number of ground terminal and spaceborne GPS and S/K/W-band systems due to their higher fabrication precision, possibility of direct integration with semiconductor devices (which eliminates lossy and unreliable off-chip interconnects), highly compact package, lighter weight, and low power consumption. Low-temperature co-fired ceramic (LTCC) based integration of miniaturized antennas with a transceiver chip has been proposed and implemented for space wireless sensors, such as those used to measure radiation absorbed by astronauts in extra vehicular activity [6].

Ultra-compact on-chip/in-package integrated antennas and radio front-end have paved the way for a new generation of 'satellite-on-chip' [7] or 'spacecraft-on-chip' [8], tiny combined sensor–transceiver chips that in future will be spread in space to establish an ad hoc wireless sensor network between space nodes (e.g., the International Space Station) and the ground. Prototype structures consisting of antennas, transmitters and sensing devices will be attached to the Materials International Space Station Experiment (MISSE-8) pallet of the International Space Station to collect data on the chemistry and other properties of the solar wind. This experiment will also evaluate the performance of these 'ChipSats' in the harsh space environment. The experiment will pave the way for the application of extremely low-cost/low-complexity antenna-on-chip/in-package devices integrated with radio and sensing devices (lab-on-chip) in chip-scale packaging for a wide range of space-based sensing and data communication tasks.

On-chip antennas are essential for implementing fully integrated radio systems. An on-chip antenna simplifies the matching network and improves the system performance by reducing the front-end loss and noise figure. In addition, for microwave and mmW antenna array systems, where the feed loss of the conventional planar lines is prohibitively large, wafer-scale integration of on-chip antenna elements can enhance the radiation efficiency, minimize feed loss, thus significantly decreasing power consumption, and minimize packaging complexity and cost.

11.2.1 Recent Advances in On-Chip Antenna Technology

Recently, different on-chip antennas have been implemented using low-cost semiconductor technologies, such as silicon–germanium (SiGe) and complementary metal oxide semiconductor (CMOS) due to their maturity, and high integration capabilities [9–14]. With the availability of low-cost devices with f_{max} above 200 GHz and high integration possibilities, advanced CMOS and SiGe technologies are becoming the main technology options for the development of low-cost mmW systems, enabling rapid market growth for low-cost ultra-broadband technologies. This progress will further pave the way for more advanced industrial, space, and defense applications.

So far, most of the reported on-chip antennas have low radiation gains (below −5 dBi) or low radiation efficiency (below 10%) [11]. To radiate the maximum amount of input power, the efficiency of the antenna must be as high as possible, while its size must be as small as possible.

At the mmW frequency range (30–300 GHz), where wavelength is on the order of the device size, implementation of on-chip antennas is more feasible than at microwave frequencies. On the other hand, with rapid scaling of the low-cost silicon technologies, circuits and systems can now operate at the mmW frequency band (even up to 220 GHz) [14]. Since the antenna dimension is the dominant factor in determining the chip area and cost, miniaturizing the antenna while maintaining high radiation efficiency is still a challenge.

11.2.2 Silicon IC Substrate Limitations for On-Chip Antennas

In typical silicon IC technology, the thickness of the silicon could vary between 200 and 700 μm. All the metal layers are placed on top of the silicon and embedded in silicon dioxide. The maximum distance between the lowest metal and top metal varies between 7 and 15 μm. The most significant disadvantage of the silicon substrate for the implementation of passives elements, including the antenna, is its low resistivity, which is the main source of loss in the passive components. Some on-chip antenna configurations can be isolated from the low-resistivity silicon by using the lowest metal as a shield layer. The limited distance between the bottom and top metal is another limitation for antenna implementation. As an example, the radiation efficiencies of a half-wave dipole and half-wave slot versus substrate resistivity and thickness are calculated using ADS Momentum and shown in Figure 11.1.

To increase the efficiency and gain of the on-chip antenna, a DRA on-chip antenna was designed in [12]. Figure 11.2 illustrates the proposed configuration for the integrated on-chip antenna. The thickness of the silicon substrate is about 300 μm after back grinding. The slot aperture is implemented on the top metal layer (MT) of the silicon technology. The lowest metal layer (M1) is connected to the top ground plane through via

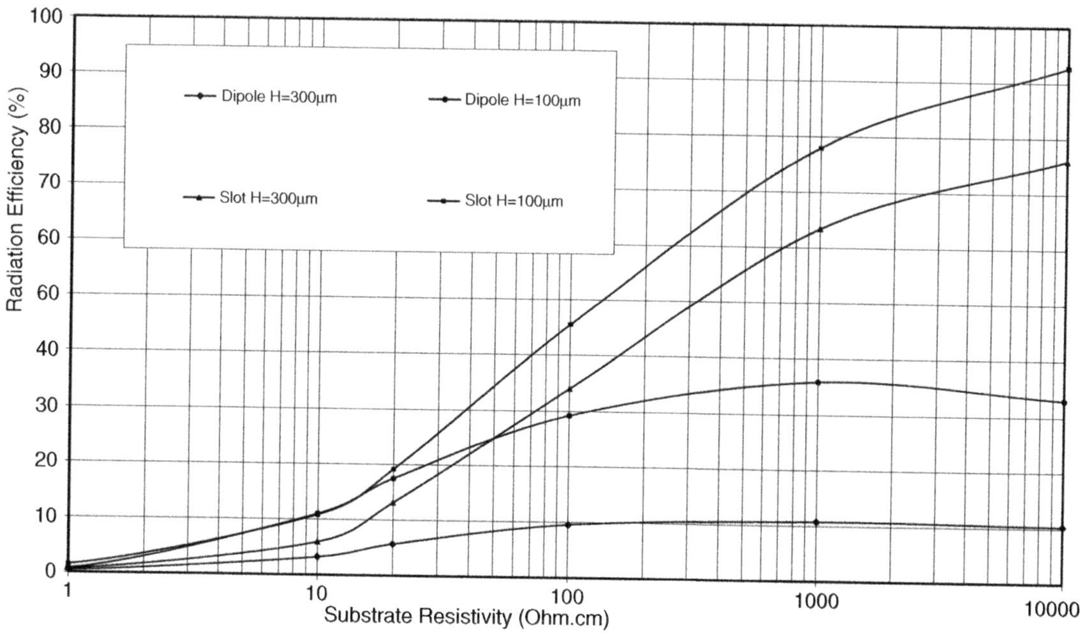

Figure 11.1 *Radiation efficiency of a 30 GHz dipole and slot antenna versus substrate resistivity and thickness on-chip dielectric resonator antennas (DRAs).*

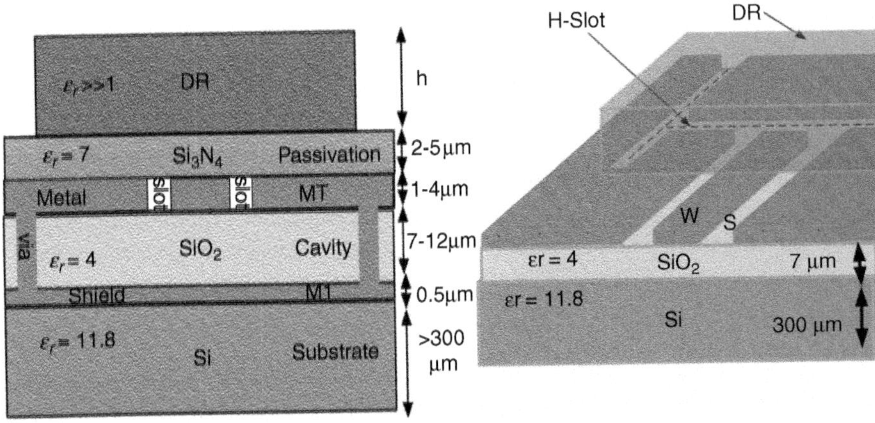

Figure 11.2 *On-chip cavity backed slot antenna with dielectric resonator (DR) on top. Cross-section and 3D structure. Courtesy of [12]. © 2010 IEEE. Reprinted, with permission, from IEEE Transactions on Antennas and Propagation.*

holes to shield the antenna from lossy silicon substrate, thus the antenna substrate is limited to silicon dioxide (SiO_2) and the intermediate dielectric layers between metal layers. Finally, a dielectric layer with a large permittivity is placed on top of the chip to increase the radiation efficiency and improve the antenna matching.

Figure 11.3 shows the die micrograph of the fabricated H-slot antenna, probed by a Cascade Micro-Tech CPW probe. The H-slot antenna was implemented using the IBM SiGe5 a.m. process, described in Figure 11.2. In Figure 11.3b a rectangular high-permittivity layer (with a permittivity of 38 ± 1 and a size of $a = 1.6$ mm, $b = 1.15$ mm, and $h = 0.5$ mm) is placed over the passivation layer to improve the radiation efficiency.

Figure 11.4 shows the measured reflection coefficient (S_{11}) of different samples of on-chip antenna with the passivation layer (PL) over the frequency range of 20–40 GHz. Placing DR reduced the resonance frequency by more than 1.6 GHz and increased the 10 dB impedance bandwidth to 4.2 GHz (12%). The Wheeler method was used to measure the efficiency of this on-chip antenna [12]. The maximum measured radiation efficiency of the on-chip DRA is 59%.

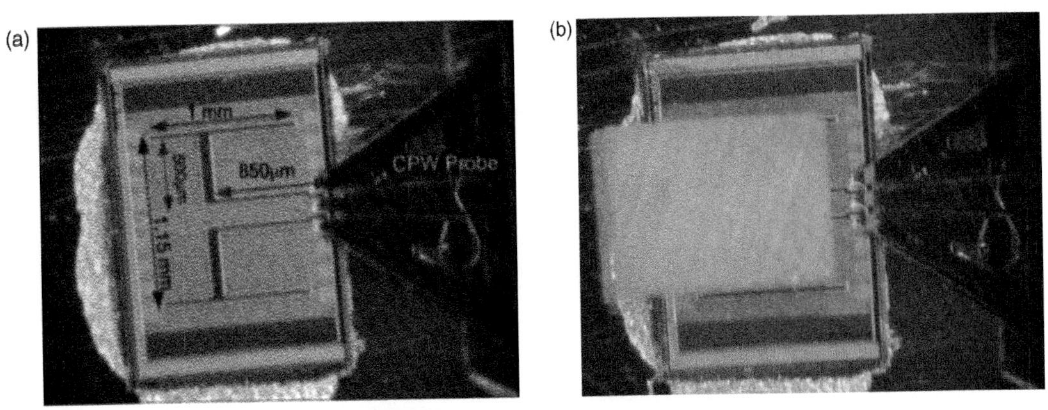

Figure 11.3 *(a) The dimensions of the slots in the metal. (b) Die micrograph of the H-shape slot with DR. Courtesy of [12]. © 2010 IEEE. Reprinted, with permission, from IEEE Transactions on Antennas and Propagation.*

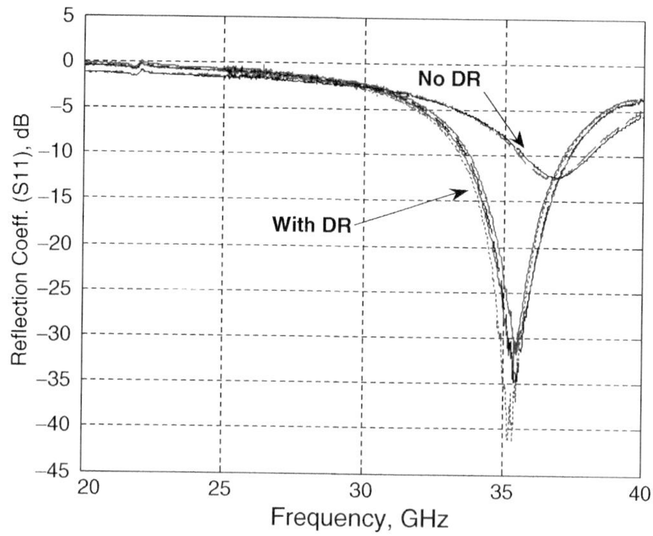

Figure 11.4 Measured reflection coefficient of the on-chip antenna with and without DR for the structure shown in Figure 11.3. Courtesy of [12]. © 2010 IEEE. Reprinted, with permission, from IEEE Transactions on Antennas and Propagation.

11.2.3 On-Chip Antenna on Integrated Passive Silicon Technology

One viable approach to reduce the fabrication cost of on-chip antennas is to implement the antenna structure in passive technologies such as LTCC and IPD (Integrated Passive Device). IPD is a high-resistivity silicon process developed for implementation of high-Q passive components such as resonators and filters. Figure 11.5 shows the cross-section of the metal layers and the silicon of this technology provided by ON Semiconductor Inc., which consists of two thick metal layers (5 μm copper) and a thin metal layer (2 μm copper) on a high-resistivity silicon substrate with $\sigma = 0.1$ S/m and a thickness of 280–700 μm.

A slot radiator with microstrip line excitation in IPD is investigated in [13], shown in Figure 11.6. The thicknesses of the SiO_2 (dielectric of the microstrip line), M_1, and M_3 are 14, 2, and 5 μm respectively.

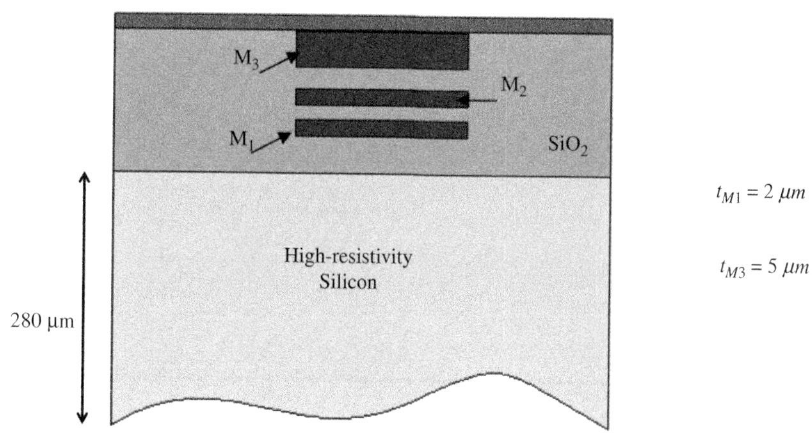

Figure 11.5 Cross-section of the IPD technology with three metal layers. Courtesy of [13]. © 2010 IEEE. Reprinted, with permission.

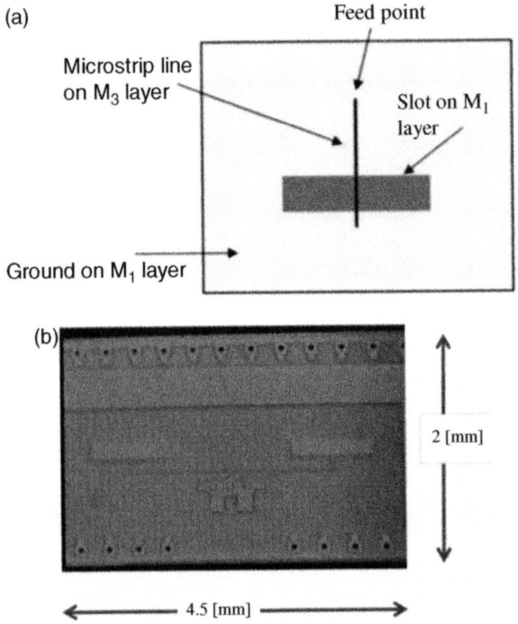

Figure 11.6 (a) Top view of the slot radiator on the M1 (the bottom metal) layer and the microstrip line on M3 (the top metal) layer. (b) Die micrograph of the 2 × 1 array. Courtesy of [13]. © 2010 IEEE. Reprinted, with permission.

The thickness of the substrate is 280 μm, the relative permittivity is $\varepsilon_r = 11.9$, and the conductivity is $\sigma = 0.1$ S/m. The measured gain of a 2 × 1 slot array in IPD technology is shown in Figure 11.7. The plot shows more than 5 dBi radiation gain at 60 GHz with a radiation efficiency of more than 60%. This antenna can be attached to mmW front-end chips or hybrids using flip chip bonding.

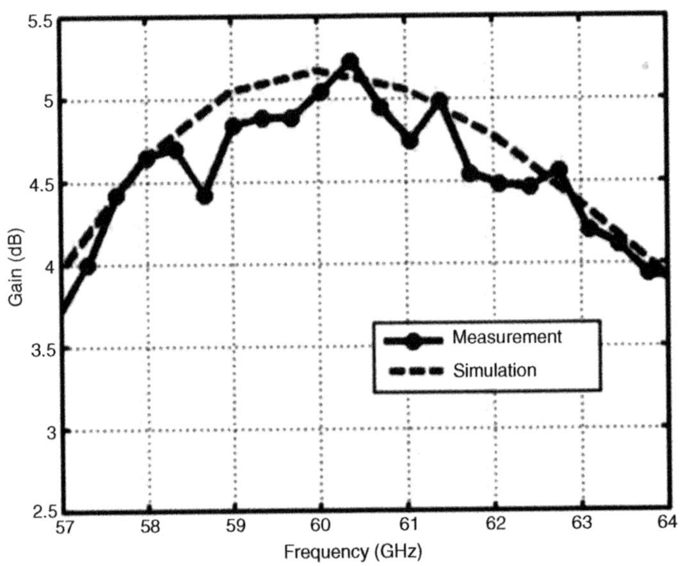

Figure 11.7 The gain versus frequency of 2 × 1 slot array on IPD technology.

11.3 Integrated Planar Waveguide Technologies

Multilayer embedded waveguide technology offers a cost-effective and highly compact solution for high-performance fixed-beam antennas, multibeam antennas [15], and phased arrays for space systems (airplane-to-satellite, land/sea-mobile satellite, etc.) as well as numerous point-to-point and automotive radar applications. Conventional printed lines (microstrip, microslot, co-planar waveguide (CPW), etc.) become prohibitively lossy and dispersive at microwave and mmW ranges of frequencies, particularly for space communication and high-performance terrestrial radio networks.

A planar integrated waveguide, also referred to as a post-wall waveguide [16], substrate integrated waveguide (SIW), or laminated waveguide [17], is a reduced-height rectangular waveguide formed by the top and bottom metal plates of a dielectric slab and very often two sidewalls of metallic via holes. Unlike microwave/mmW components based upon conventional metallic rectangular waveguides (RWGs), which are bulky and have a high fabrication cost, the components based upon SIWs are compatible with low-cost and low-profile fabrication technologies such as printed circuit board (PCB) and LTCCs. Multilayer LTCC implementation of SIW-based components and circuits is a compact and reliable alternative for space applications [18]. A RWG implemented in LTCC, with its simulated modal field distribution, is shown in Figure 11.8. SIW propagation characteristics are similar to those of RWG, especially when operating in the TE_{n0} modes [19]. Almost all the excitation modes are confined within the substrate. Therefore, the SIW is high Q-factor (low loss) compared to other conventional guiding structures, such as microstrip line and CPW, which suffer from severe conduction loss. A SIW-based slot array for 60 GHz applications is shown in Figure 11.9 [20].

For mmW applications, SIW is an attractive guiding (feeding) structure for realizing low-cost and efficient RF circuit components. Moreover, a SIW-based planar antenna technology like a SIW-fed dielectric resonator antenna is an excellent candidate for low-cost and high radiation efficiency mmW transceivers due to the

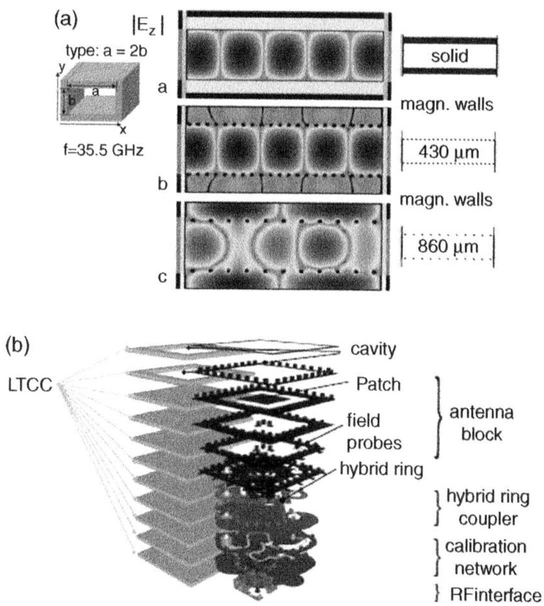

Figure 11.8 Rectangular waveguide in LTCC technology: (a) field distribution; (b) technology layers. Taken from [18]; copyright EuCAP 2009; used with permission.

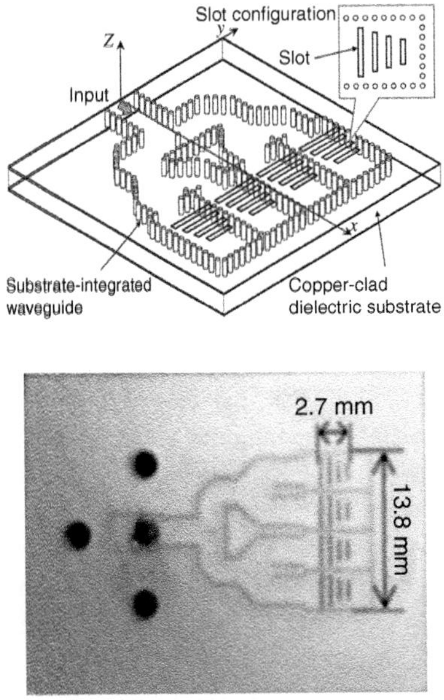

Figure 11.9 *Wideband slot array antenna for 60 GHz band antenna. Courtesy of [20]. © 2010 IEEE. Reprinted, with permission, from IEEE Transactions on Antennas and Propagation.*

absence of conductor loss, which dominates at this frequency band. A rectangular dielectric resonator antenna (RDRA) shown in Figure 11.10 is designed to resonate in its fundamental mode TE_{111} mode for radiation at mmW frequency. A narrow slot cut on the SIW broad wall, as shown in Figure 11.10c, can excite the DRA to radiate two different polarizations (horizontal and vertical). The SIW structure, integrated in a low-loss dielectric substrate with copper-plated top and bottom sides, laterally confines the guided mode by two rows of copper-plated vias of diameter d_{via}. The SIW effective width is W_{SIW}, and the two neighboring vias, are separated by s_{via}. More details about the antenna design and parameter characterization are presented in [21]. A suitable low coupling loss SIW-to-microstrip transition (SIW-MSL) is used to couple the power to the guided fundamental TE_{10} mode. The antennas are fabricated using standard PCB techniques based on low-loss material substrates. The design example shown in Figure 11.10, which is suitable for Ka-band satellite communication and a number of 40 GHz band systems, uses Rogers 5870 ($\tan\delta = 0.0012$) for the SIW substrate and Rogers 6010 ($\tan\delta = 0.0023$) for RDRA, respectively. Fabrication of the RDRA is performed using a customized multi-step, multilayer PCB manufacturing process; more details about the fabrication steps are presented in [21]. The measured radiation pattern (gain) of the antennas at 35 and 37 GHz in two orthogonal planes is plotted in Figure 11.11, with the boresight gains of 5.51 dB (antenna in Figure 11.10a) and 4.75 dB (antenna in Figure 11.10b) respectively.

The optimized single element (SIW-DRA single element) is used as a building block for the N-element SIW-DRA array by using two different slot orientations, a horizontal (transverse slot) and vertical polarization (longitudinal slot), as depicted in Figure 11.12a and b, respectively. The slots are excited by the SIW fundamental mode, TE_{10}, to couple energy to DRA elements. The primary array design parameters are related to DRA position: (1) the distance between the centers of the neighboring elements, D; and (2) the distance between the last element and the SIW short-circuited end, X_S. These parameters define the phase difference between two successive elements, which affects all the radiation characteristics of the antenna. More details on the design criteria and guidelines are presented in [22]. A simple transmission line circuit (T.L.) model has been developed [21,22] to simplify the design and optimization of the SIW-fed DRA array.

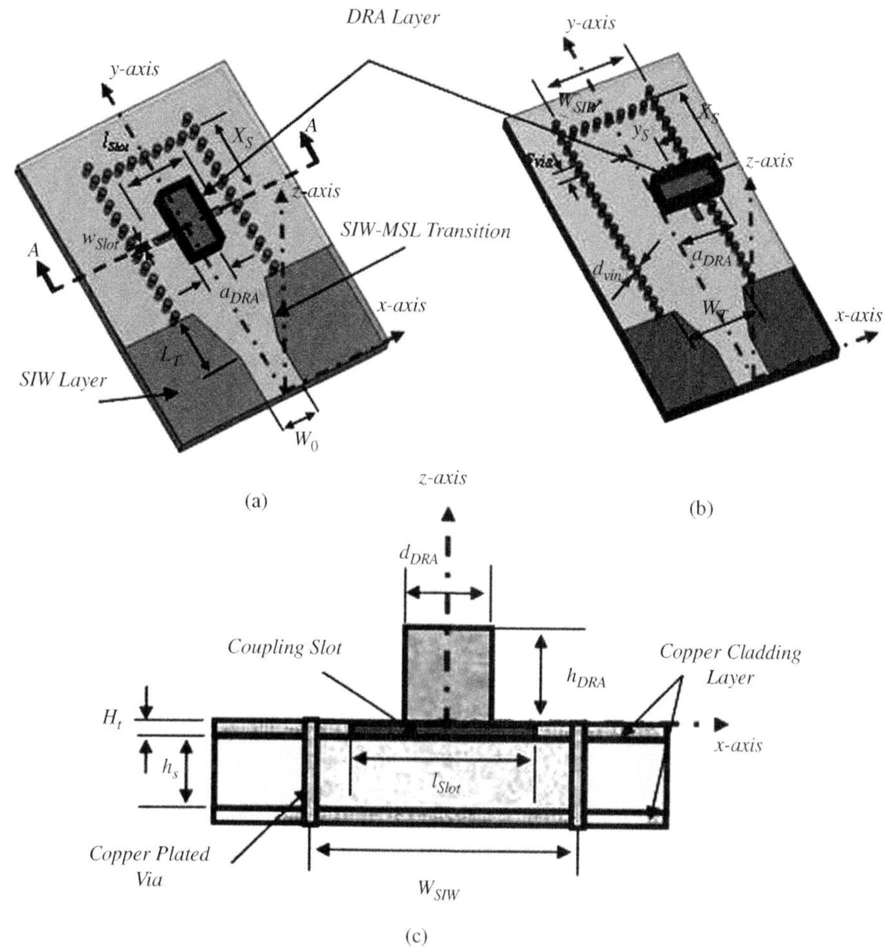

Figure 11.10 DRA-based SIW model [21]. (a) Using horizontal polarized (transverse) slot. (b) Using vertical polarized (longitudinal) slot. (c) Cross-section A–A.

The T.L. model essentially consists of a T.L. model for the SIW TE_{10} mode, which connects SIW-DRA two-port **S**-matrix models. Using the aforementioned model, a number of SIW-fed antenna array structures for Ka-band have been designed, optimized, fabricated, and tested. Two design examples, for the frequency band of 33–40 GHz, are shown in Figure 11.12. The measured reflection coefficients are depicted in Figure 11.13. The impedance bandwidths are 1.60% (around the center frequency 33.87 GHz) and 4.70% (around the center frequency 37.80 GHz), respectively. Good agreement between the measured and the T.L. model results is observed, especially near the frequency at which the reflection coefficient is a minimum.

The simulated and measured *xz* and *yz* plane radiation patterns are shown in Figure 11.13c and d. The horizontally polarized (transverse) and vertically polarized (longitudinal) antenna arrays achieve a maximum broadside gain of 11.70 and 10.60 dB, respectively [22].

Radiation efficiency, particularly the contribution of the feed circuit, is a critical performance parameter for mmW antennas arrays for satellite communications and other space applications. The SIW-fed DRA offers a

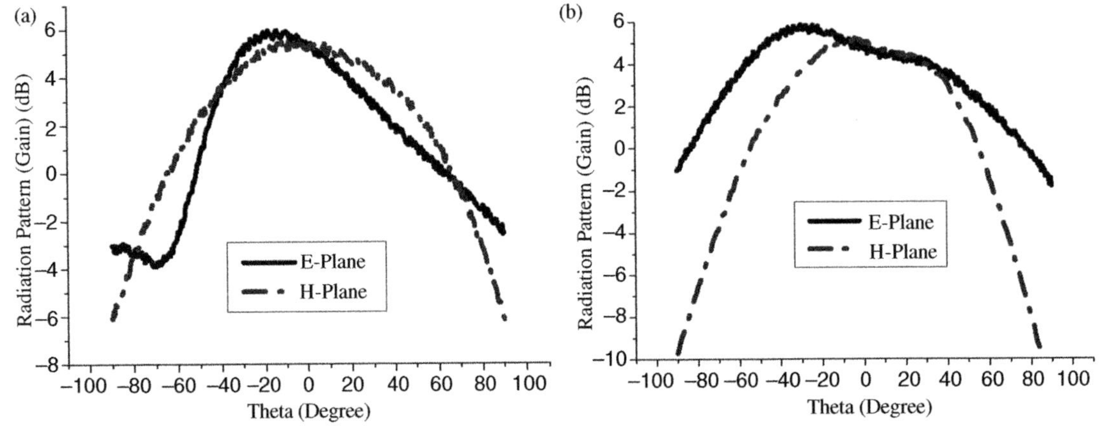

Figure 11.11 *Simulated and measured radiation pattern of DRA-based SIW technology [21]: (a) antenna at* $f=35\,GHz$*, E-plane (yz plane), H-plane (xz plane); (b) antenna at* $f=38\,GHz$*, E-plane (yz plane), H-plane (xz plane), DRA substrate: Roger RT/6010 (*$\varepsilon_{rd}=10.20$*,* $h_{sd}=1.27\,mm$*), DRA dimensions:* $a_{DRA}=3.0\,mm$*,* $d_{DRA}=1.50\,mm$*, SIW substrate: Roger RT/5870 (*$\varepsilon_{rs}=2.33$*,* $h_{sb}=0.7874\,mm$*), SIW dimensions:* $a_{SIW}=4.80\,mm$*,* $s_{via}=0.60\,mm$*,* $d_{via}=0.30\,mm$*, slot dimensions:* $L_{Slot}=3.20\,mm$*,* $W_{Slot}=0.30\,mm$*,* $y_s=0$*.*

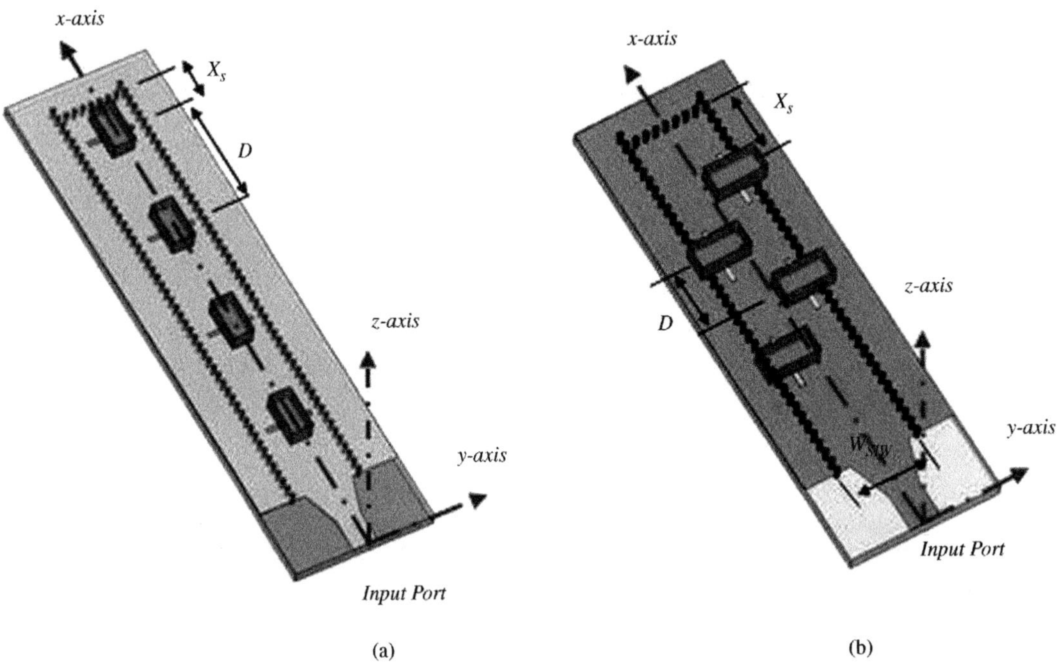

Figure 11.12 *The schematic layout of four-element SIW-DRA array: (a) a horizontally polarized slot; (b) a vertically polarized (longitudinal) slot. Courtesy of [22].*

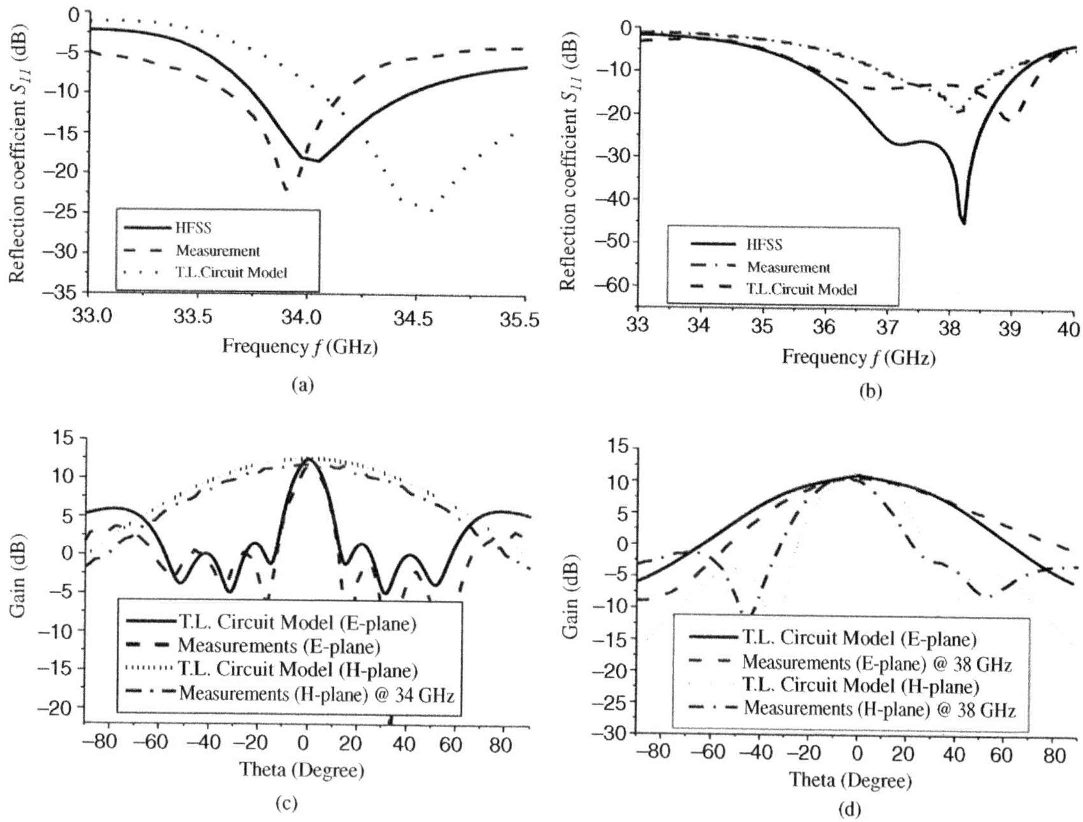

Figure 11.13 *The simulated and measured reflection coefficients (dB) of SIW-DRA array [21]: (a) using horizontally polarized SIW-slot arrangement prototype shown in Figure 11.12(a); (b) using vertically polarized SIW-slot arrangement prototype shown in Figure 11.12b. DRA substrate: Roger RT/6010 ($\varepsilon_{rd} = 10.20$, $h_{sd} = 1.27$ mm), DRA dimensions: $a_{DRA} = 3.0$ mm, $d_{DRA} = 1.50$ mm, SIW substrate: Roger RT/5870 ($\varepsilon_{rs} = 2.33$, $h_{sb} = 0.7874$ mm), SIW dimensions: $a_{SIW} = 4.80$ mm, $s = 0.60$ mm, $d_{via} = 0.30$ mm, slot dimensions: $L_{Slot} = 3.20$ mm, $W_{Slot} = 0.30$ mm, array parameters: $X_s = 2.50$ mm, $D = 7.60$ mm, off-center distance $y_s = 0$. Courtesy of [22]. © 2010 IEEE. Reprinted, with permission, from IEEE Transactions on Antennas and Propagation.*

remarkably high radiation efficiency of more than 90% for small arrays. Typical fabricated structures for the 40 GHz band are shown in Figure 11.14.

11.4 Microwave/mmW MEMS-Based Circuit Technologies for Antenna Applications

New trends in cost-effective and more reliable spaced-based communication systems require flexible microwave/mmW radio circuit and reconfigurable smart antenna architectures, which have been driving demand for low-loss, high-efficiency antenna control and feed circuit technologies. Next-generation multi-functional telecommunication, Earth observation and space exploration satellites and probes, with intelligent on-board bandwidth and radiation beam management capabilities, depend on low-loss, reliable, high-performance RF control devices. Extremely low-loss MEMS switches, varactors, and phase shifters offer unique operational and cost/complexity advantages for multiband/tunable-band filters/multiplexers, smart

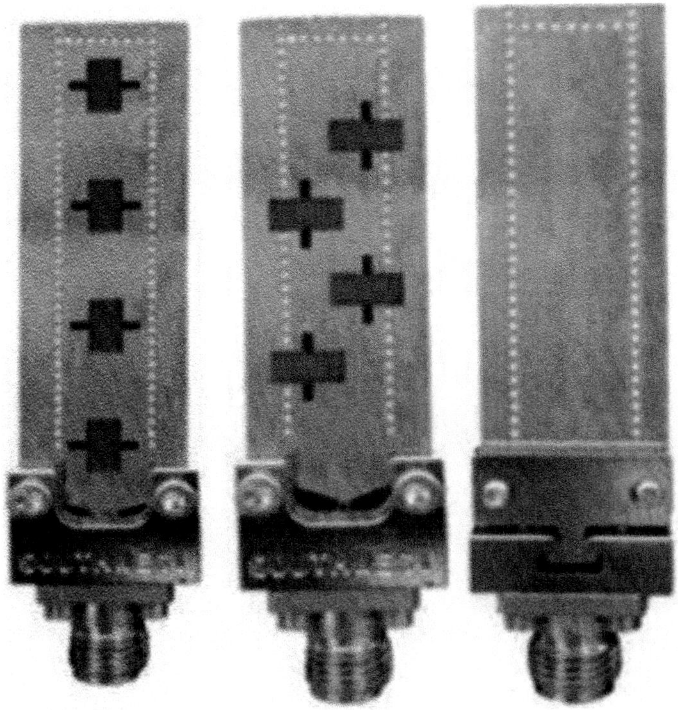

Figure 11.14 SIW-DRA array prototypes using horizontally/vertically polarized SIW-slot arrangement. Courtesy of [22]. © 2010 IEEE. Reprinted, with permission, from IEEE Transactions on Antennas and Propagation.

signal routing, and adaptive/programmable antenna beam shaping/switching and reconfigurability required for such missions [23,24].

Phased-array and reconfigurable antennas are the ultimate solutions for a number of emerging applications for space and other high-performance intelligent radio antenna systems. Moreover, MEMS and surface and bulk micromachining techniques are being widely used for high-efficiency reconfigurable antennas as well as phased arrays at microwave/mmW range [25,26]. For example, low-loss air-filled coaxial line and suspended microstrip antenna elements and have been developed for this range. Shown in Figure 11.15 is a multiple fixed-beam 4 × 4 array [26] with the integrated feed circuit developed by Si-based micromachining.

High-performance phased arrays and multiband dual-polarization reflectarrays require low insertion loss, linear, and wide-band phase shifters [27–29]. Another highly desirable characteristic is minimal insertion loss variation as the phase shift changes from 0° to 360°. As compared to other technologies such as CMOS, SiGe, and GaAs, MEMS devices, particularly when fabricated on low-loss substrates, offer lower insertion loss for switches and phase shifters. Furthermore, MEMS phase shifters/switches have higher power handling capability, making them a suitable candidate for beam-former transmitter array and spatial power combining.

In comparison to mechanical (waveguide/coaxial) and semiconductor (PIN, varactor, and transistor) switches and variable reactance elements, MEMS switches and varactors provide lower insertion loss, higher isolation, better linearity, lower power consumption, and smaller size/weight. However, lower power handling capability, higher driving voltage, medium switching speed, and reliability, particularly for space applications, are the main limitations and challenges that had to be overcome before extensive deployment of microwave/mmW MEMS technology in space systems [30].

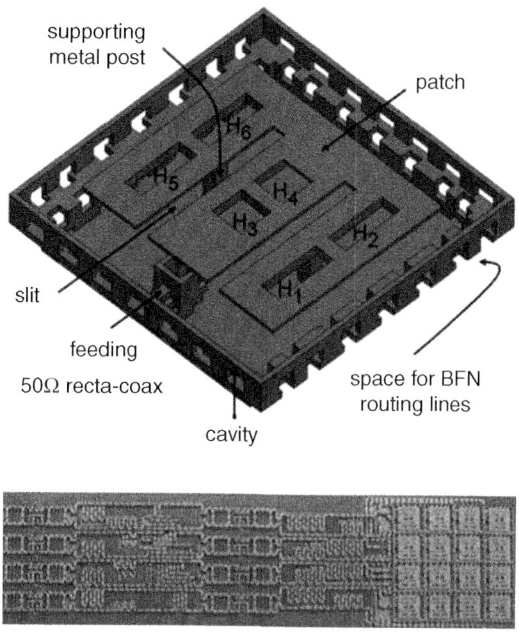

Figure 11.15 A 4 × 4 patch array with integrated feed circuit at 36 GHz (BFN, Beam-Forming Network). Courtesy of [26]. © 2007 IEEE. Reprinted, with permission, from Integrated cavity-backed ka-band phased array antenna.

11.4.1 RF/Microwave MEMS-Based Phase Shifter

One of the first demonstrations of a mmW MEMS-based switched-type phase shifter was reported in [25]. The W-band (70–100 GHz) phase shifter uses CPW capacitive shunt switches on a quartz substrate. The switch has a very low insertion loss (<0.5 dB) and a high isolation (< −30 dB). This switch has been used to develop 1-bit switched-line phase shifters (Figure 11.16) with phase states 0°/90° and 0°/180°, respectively. The measured performance of the phase shifters shows a low insertion loss (−1 to −4 dB) over the frequency range of 75–100 GHz.

Wide-band phase shifting can be realized by changing the propagation constant of a distributed varactor-loaded planar transmission line. A travelling wave varactor-loaded co-planar stripline (CPS) phase shifter for a mmW range of frequency has been reported in [31].

11.4.2 Reflective-Type Phase Shifters for mmW Beam-Forming Applications

The reflective-type phase shifter [32] is a simple and cost-effective structure for a wide range of continuous and digital phased-array applications. The phase shifter uses a 90° hybrid coupler with two reflective loads. The signals reflected by the reflective loads are added in phase at the output port. The same reflected signals are added out of phase at the input port, thereby providing a good input impedance matching. The maximum amount of phase shift depends on the properties of the reflective load. Usually varactors are used as the reflective load. An important advantage of the reflective-type phase shifter (RTPS) is its compact size. RTPS can be implemented using MEMS shunt switches at mmW frequencies. MEMS switches and varactors fabricated on low-loss substrates such as alumina or high-resistivity Si can potentially reduce the phase shifter loss in a significant manner. A 60 GHz 3-bit RTPS with shunt switches using MEMS technology, fabricated on

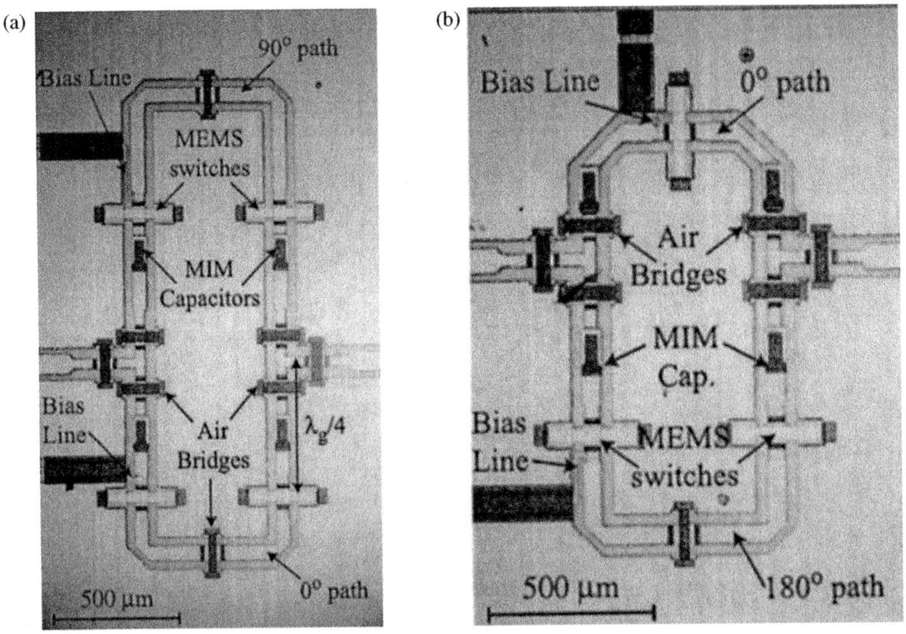

Figure 11.16 Micrographs of CPW-band switched-line phase shifters: (a) 0°/90° phase shifter (1.3 × 2.3 mm) (MIM, Metal–Insulator–Metal); (b) 0°/180° phase shifter (1.3 × 1.85 mm). Courtesy of [25]. © 2007 IEEE. Reprinted, with permission, from IEEE Transactions on Microwave Theory and Techniques IEEE.

an alumina substrate, is shown in Figures 11.17 and 11.18. The hybrid coupler (at the center) uses a suspended broadside coupler, which can provide strong coupling without the need to realize an extremely narrow gap. The reflective loads are short-circuited transmission lines with variable lengths. The short end in these loads is implemented by the metal-to-metal contact shunt switches over a segment of CPW line. The average insertion loss of the phase shifter is 3.9 dB over all phase states at 60 GHz.

The measured performances of the developed 60 GHz MEMS-based RTPS are shown in Figure 11.19 [33].

11.5 Emerging THz Antenna Systems and Integrated Structures

Sitting between light-wave and radio-wave frequencies, the THz range (300 GHz to several terahertz) has some of the best features of both worlds: high spatial resolution of optical wavelength and penetration power and low attenuation of RFs. THz systems and device technologies have been traditionally associated with radio astronomy and remote sensing. Nonetheless, over the last decade, THz techniques have been applied to numerous novel applications from biology and medicine [34] to monitoring and extremely high data rate communications [35]. Despite their unique advantages, the cost and complexity of THz signal sources with sufficient output power, as well as the lack of high-sensitivity room-temperature THz detectors, are the main challenges for commercial applications.

With the current state-of-the-art THz source/detector technologies, high-performance antenna and passive devices can compensate for the low performance of the signal source and detector. The proper antenna technology depends on the physics and structure of the source and detector. THz signals can be generated by an electron beam source, a quantum cascade laser (QCL), semiconductor devices, frequency multiplication of the lower frequency sources, photo-mixing, and parametric interaction in nonlinear crystals.

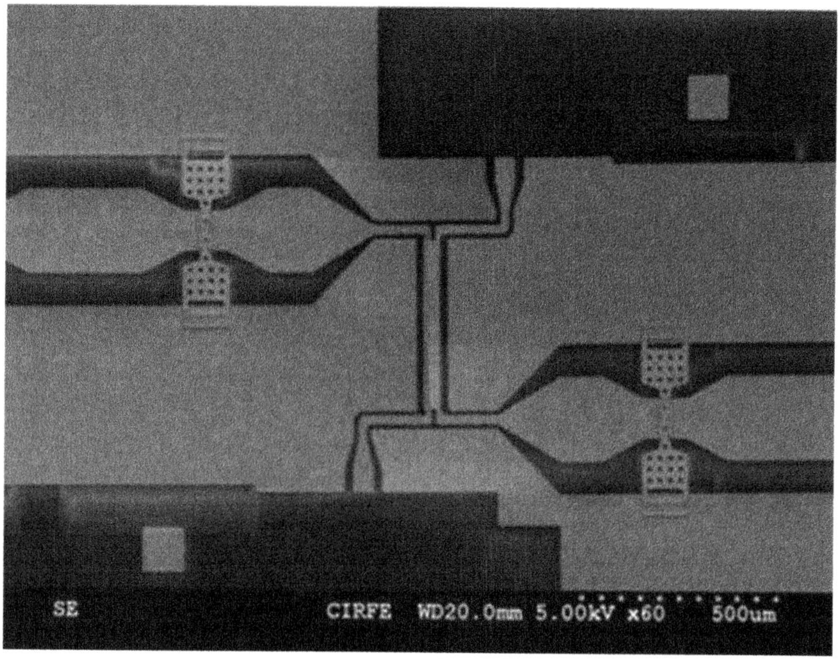

Figure 11.17 SEM image of the fabricated phase shifter, the broadside coupler, and the contact switches of the MEMS-based RTPS for mmW phased-array systems. From [33]; copyright EuMA; used with permission.

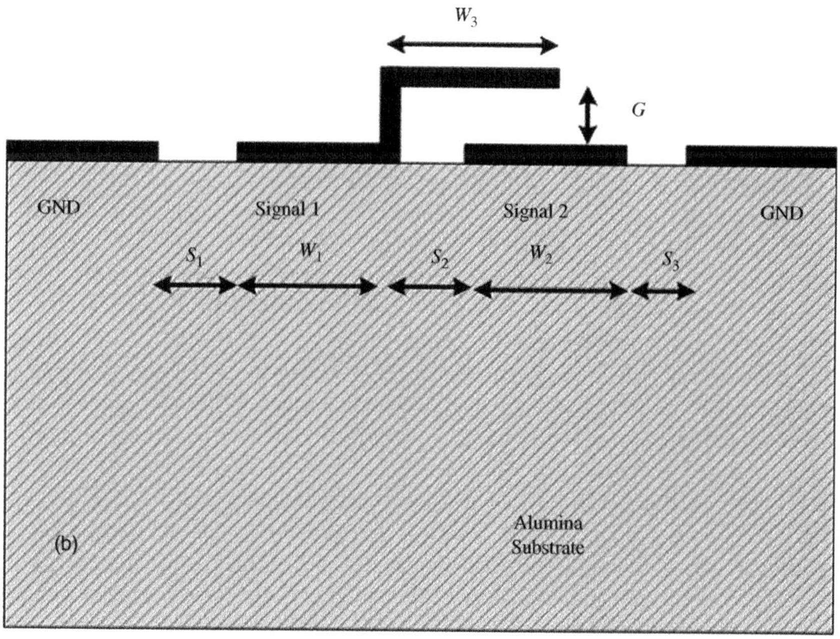

Figure 11.18 Cross-section of the broadside coupler (Figure 11.17) in MEMS technology. The optimized values are $W_1 = 13\,\mu m$, $W_2 = 13\,\mu m$, $W_3 = 22\,\mu m$, $S_1 = 18\,\mu m$, $S_2 = 10\,\mu m$, $S_3 = 10\,\mu m$, and $G = 2.5\,\mu m$. From [33]; copyright EuMA; used with permission.

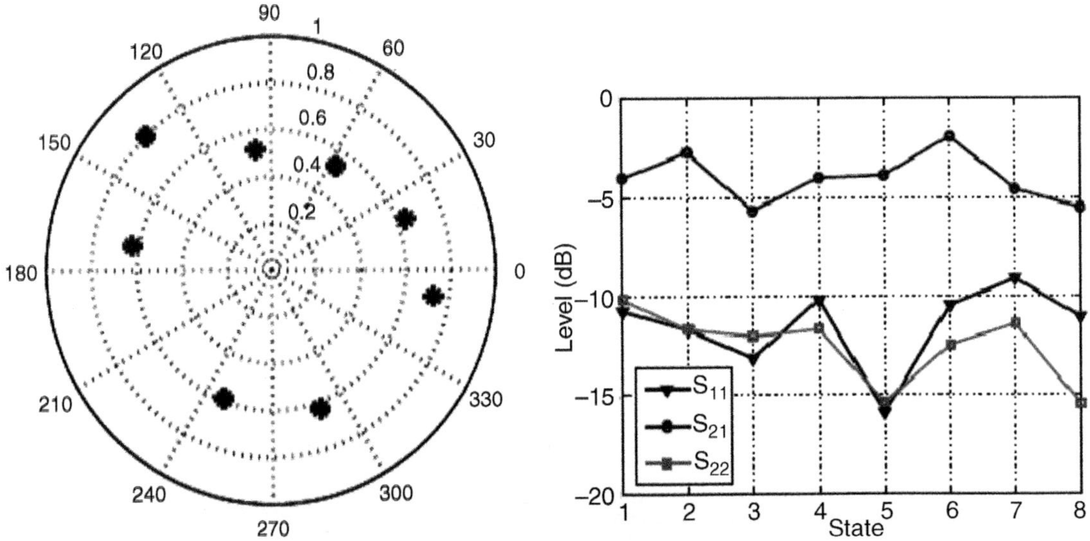

Figure 11.19 Phase shift, insertion loss, and reflection coefficient of the 3-bit MEMS-based RTPS. From [33]; copyright EuMA; used with permission.

The transceiver antenna structure should be compatible with transceiver technology. THz receivers/detectors can be broadly classified into two groups: homodyne (direct detectors) and heterodyne, respectively. Highly sensitive homodyne receivers use a thermal detector to convert the heat generated by the incident power to a change in its physical properties such as volume (Gollay cell), dielectric constant of the material (pyroelectric effect), electrical conductivity (bolometers), or voltage (thermopile). THz detector arrays using Schottky diode arrays have been realized in low-cost commercial CMOS processes [36].

Heterodyne receivers can potentially offer more sensitivity. The incident signal, in such a system, is mixed with the local oscillator (LO) signal in an extremely low insertion loss and high-conversion efficiency mixer to generate an intermediate frequency (IF) signal. Schottky mixers, hot electron bolometer (HEB) mixers, and superconductor–insulator–superconductor (SIS) mixers are among the highest performance types. A typical high-sensitivity heterodyne THz receiver for radio telescopes [37] is shown in Figure 11.20.

The antenna comprising two slot dipoles is integrated with the HEB device and a silicon lens to enhance radiation efficiency and sensitivity. The entire receiver front-end is cooled. Details of the 2.8 THz integrated antenna and HEB are shown in Figure 11.21 [38]. The impedance matching and the filter to the left are designed and optimized to maximize the transfer of signal energy at IF.

Current THz observatories [39,40], high-resolution stand-off imagers [41], and extreme-bandwidth long-distance communication systems [42] use hybrid multi-reflector antenna systems with a single source/detector or focal sub-arrays at the focal point. Furthermore, although metallic waveguides are still a common technology in this range of frequency, as the number of focal array elements increases, conventional waveguide-based technologies and assembly of discrete devices become prohibitively difficult and costly. For emerging applications (space, short-range imaging, and spectroscopy), particularly when lower cost/complexity and compact size are highly desirable, planar multilayer integrated antenna–source and antenna–detector technologies are the most optimal options. Similar technologies are being used in the focal plane of large-reflector and/or lens antenna systems.

Various types of antennas such as dipoles and wide-band planar spirals have been used for mixing and detection. Recent progress in planar integrated antenna–source–detector technologies, with particular emphasis on photo-mixing antennas, will be presented in the next subsections.

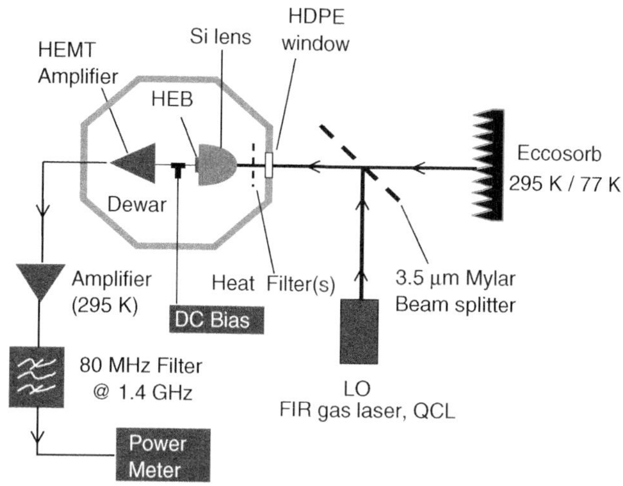

Figure 11.20 A high-sensitivity HEB THz receiver for a radio telescope (HEMT, High Electron Mobility Transistor; HDPE, High-Density Polyethylene; FIR, Far-Infrared; QCL, Quantum Cascade Laser). Courtesy of [37]. Reproduced in an original work entitled, 'Emerging Antenna Technologies for Space Applications', as part of a work entitled, 'Space Antenna Handbook', to be published by John Wiley & Sons, Ltd.

11.5.1 THz Photonics Techniques: THz Generation Using Photo-mixing Antennas

Photo-mixing is a promising approach for realizing compact room-temperature THz systems for a wide range of applications in a cost-effective manner. Photo-mixer source/antennas are compact, low-power-consuming, coherent, and tunable sources. The basic idea of THz photo mixing is illustrated in Figure 11.22, where optical modes generated by two single-mode lasers or a dual-mode laser, with THz difference in frequency, are absorbed and nonlinearly combined (mixed) in an ultra-fast photoconductor [43]. An on-chip/off-chip integrated waveguide or antenna can be used to guide or radiate the generated THz, whose frequency can be adjusted by tuning the wavelengths of the lasers. The metallic bias electrode can be shaped as a half-wave dipole to radiate over some desired range of frequencies. Bow-tie, spiral, and travelling-wave structures have been used for broadband operation.

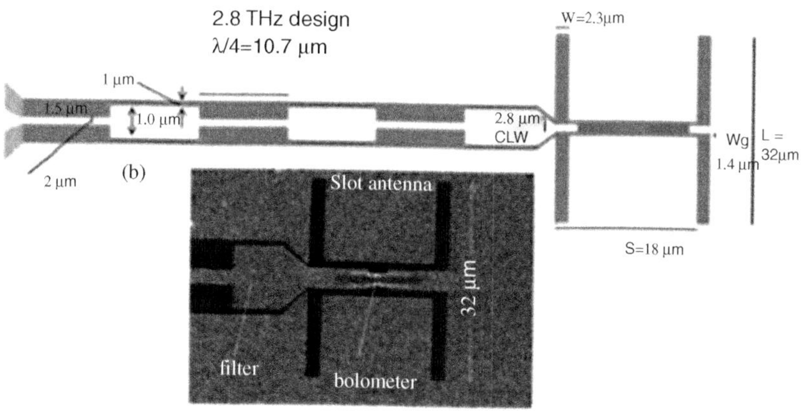

Figure 11.21 Twin-slot antenna integrated with niobium nitride HEB for peak detection at 2.8 THz. Courtesy of [38]. © 2005 IEEE. Reprinted, with permission, from IEEE Proceedings.

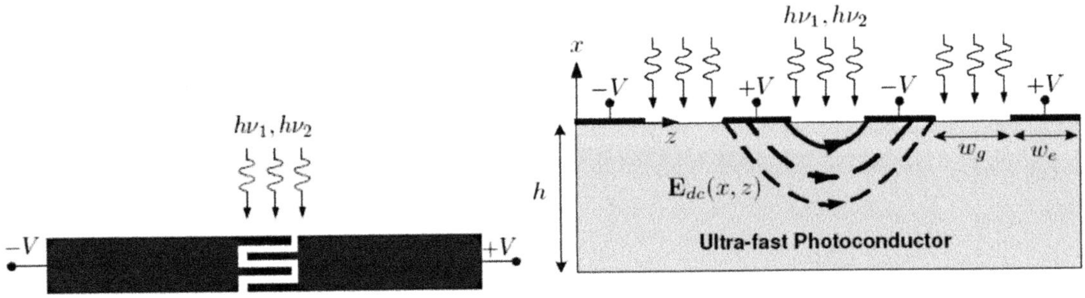

Figure 11.22 *Photo-mixing antenna. Courtesy of [34]. © 2008 IEEE. Reprinted, with permission, from Terahertz Photonics: Optoelectronic Techniques for Generation and Detection of Terahertz Waves.*

The THz power coupled to the antenna can be estimated by [43]

$$P_{THz}(\omega) = \frac{1}{2} R V^2 P_1 P_2 \frac{\tau^2}{1+(\omega\tau)^2} \frac{1}{1+(\omega RC)^2} \tag{11.1}$$

where R is the antenna resistance, V is the DC bias, $P_{1,2}$ are the laser powers absorbed by the photoconductor, C is electrode capacitance, τ is the carrier recombination lifetime, and ω is the THz angular frequency. Edge coupling is a very promising approach for on-chip integration of THz generation, THz signal processing, and antenna for a number of emerging system-on-chip applications. To enhance the radiation efficiency and shape the pattern, the photo-mixing antenna is attached to a lens. Shown in Figure 11.23a are various types of THz antenna elements integrated with photo-mixing developed at CIARS (University of Waterloo). Maximum sustainable optical power and total optical–THz conversion efficiency can be significantly enhanced by distributing the incident optical illumination over a large area. Examples of THz traveling-wave antennas are shown in Figure 11.23b.

(a)

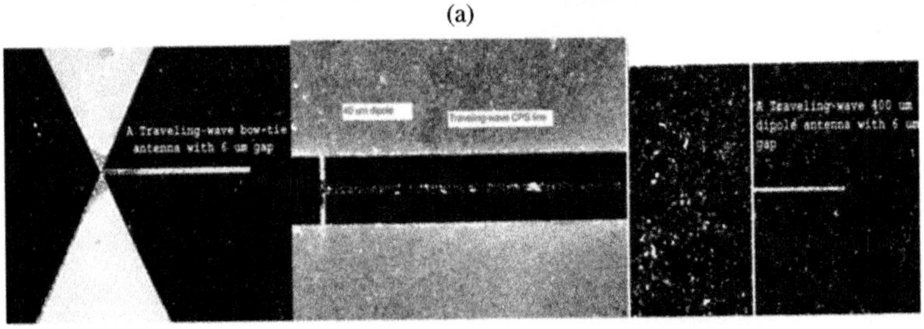

(b)

Figure 11.23 *(a) Broadband spiral photo-mixing antenna. (b) Travelling-wave photo-mixing antennas. Courtesy of THz-Photonics Lab, CIARS, University of Waterloo.*

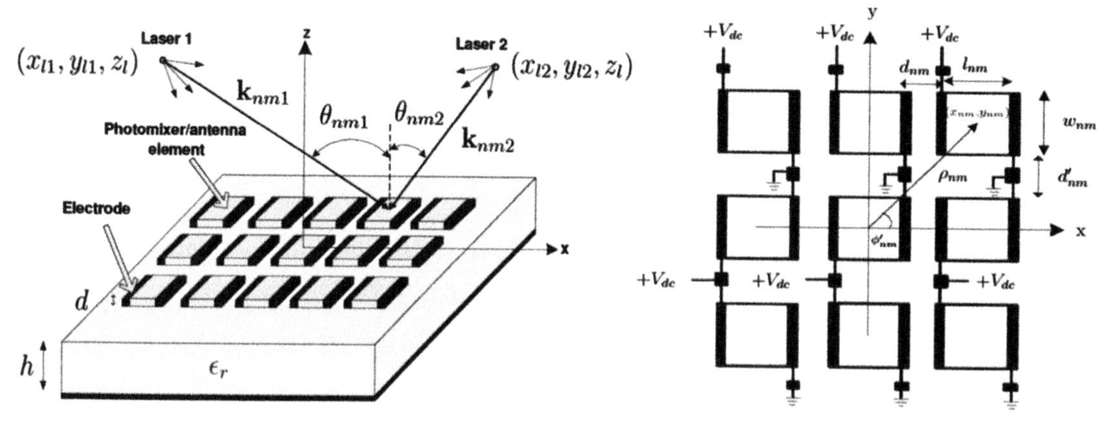

Figure 11.24 Photo-mixing array with scanning capability. Courtesy of [44]. © 2005 IEEE. Reprinted, with permission, from Analysis and design of a continuous-wave terahertz photoconductive photomixer array source.

11.5.2 THz Generation Using a Photo-mixing Antenna Array

An array of photo-mixer source/antennas can withstand more incident optical power and generate more THz signal power. The simplest and the most effective way to excite array elements, shown in Figure 11.24, is to directly illuminate the entire array by two high-power laser beams [44]. No separate feed circuit is needed in this scheme. Furthermore, by changing the angle of one of the laser beams the phase distribution of the photo-mixing antenna elements can be changed and the THz radiation will shift accordingly. This provides a simple and low-cost technique for THz antenna array beam shifting. The locations and lengths of the antenna elements are chosen to maximize the radiation.

The THz radiated beam can be rotated in the $\varphi = 0$ plane by changing the angle between the two laser beams, which results in the variation of the grating vector, $\mathbf{K}_{mn}$, leading to a change in the phase difference between the antenna elements at THz. A complete photo-mixer array with an integrated silicon lens is shown in Figure 11.25.

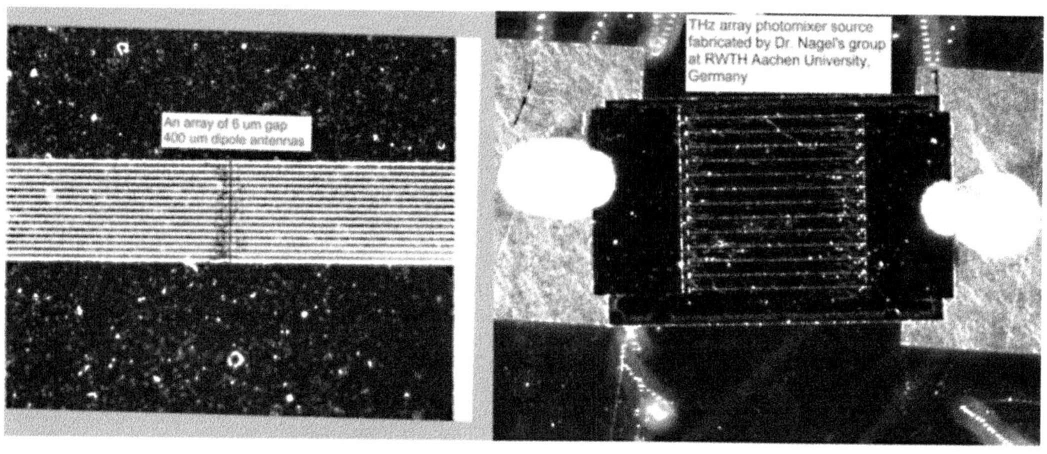

(a) (b)

Figure 11.25 Photo-mixer array antenna structures. (a) THz dipole array (courtesy of THz-Photonics Lab, CIARS, University of Waterloo). (b) Parallel stripline array (courtesy of Dr. M. Nagel, RWTH, Aachen, Germany).

11.6 Case Study: Low-Cost/Complexity Antenna Technologies for Land-Mobile Satellite Communications

11.6.1 System-Level Requirements

Over the past two decades, there has been an increasing demand for mobile satellite broadcasting and wide-band communications to land/sea and airborne mobile terminals [45–55]. A vehicle-mounted antenna is the key component of the satellite communication to moving vehicles. The vehicle-mounted antenna must satisfy the basic link budget requirements, such as high gain, as well as the capability to track satellites during rapid motion. All vehicle motions must be compensated for by an agile beam tracking system to maintain the received satellite signal power above the desired level.

High-gain reflector antennas have been widely used in X-, Ku-, and Ka-bands for a long time; however, such antennas cannot be used on small to medium-size vehicles, due to their large size and air drag issues. Instead, relatively flat and low-profile antennas are desirable for this type of application. Nevertheless, to maintain sufficient antenna gain, the antenna must have a large radiating aperture area. One feasible approach is to use a fan beam, which is very narrow in azimuth and relatively wide in elevation. It is extremely difficult to keep such a beam locked to the satellite direction while the antenna platform is moving. Experiments have shown that medium-size cars may turn as fast as 60°/s with an angular acceleration of up to 90°/s. Considering the narrow beamwidth in azimuth, satellite tracking becomes a serious challenge for any mobile antenna system. Another challenge, especially if the system is required to work in a wide geographic region such as the continental USA and Canada, is to maintain a sufficient gain over a wide range of elevation angles, that is, from 20° to 70°. Obviously, low cost and low complexity are other very important requirements of a mobile antenna system for commercial application. However, such highly demanding performance can be hardly compromised with component costs. The main cost drivers include high-quality antenna and microwave components required to sustain the high-gain, high-precision sensors such as a gyroscope, tilt sensors, and GPS, as well as a robust mechanical platform.

11.6.2 Reconfigurable Very Low-Profile Antenna Array Technologies

Several phased-array antennas for Ku/K-band satellite communication have been developed [45–55]. In [45] the antenna array consists of a planar array structure comprising stacked microstrip antennas fed by a waveguide feed network. The array size is limited to keep the feed loss to a minimum, therefore the achieved gain is not sufficient to maintain the required signal-to-noise ratio in various weather conditions. The elevation coverage is limited as well as the tracking speed of the system. A stair–planar array is utilized in [46] to increase the elevation coverage range of the mobile antenna system. The array scans mechanically in both azimuth and elevation planes. The tracking speed in azimuth is below 30°/s. A stair–planar configuration with one-dimensional electronic beam scanning in elevation and mechanical scanning in azimuth has been reported in [47]. During satellite tracking, the system is operated by the squinted beam tracking with respect to the main beam. Two-level phase shifters are used to form the main beam and a squint beam. In [48], a compact offset reflector for broadband and direct broadcast services in Ka/K-band and Ku-band was developed. This system requires several tracking sensors including a magnetic compass, tilt sensor, gyro, and GPS. Results demonstrate that the system can compensate for a weak disturbance in roll, pitch, and yaw.

Few commercial products have been developed so far, such as the Thinkom Ku-band antenna [49]. This fully integrated mobile satellite antenna system has a height of 4.5 inches (11.25 cm) and a diameter of 60 inches (150 cm), with 2 Mbps bidirectional data communications capability. The power consumption varies from 8 to 40 W. The targeted market is Mobile Internet throughout the continental USA and southern Canada.

Raysat [50] has developed a Ku-band vehicular satellite system for an uplink rate up to 4 Mbps. The satellite acquisition time is less than one minute. The low-profile antenna is capable of mechanical scanning in azimuth and electronic beam scanning in elevation. A comprehensive European project, SANATA, is aimed at a Ka-band (20–30 GHz) phased-array system for applications such as in-flight entertainment or 'office in the air' [51].

A car-to-satellite vehicular terminal must be extremely low profile, particularly for fast-moving regular-size cars. This is an essential requirement for other small- to medium-size fast-moving platforms such as medium-size aircraft. In Section 11.6.5 a low-cost phased-array antenna with electronic beam forming in both azimuth and elevation planes will be described [52–55]. The height of this system is less than 6 cm including the radome, which makes it an ultralow-profile commercial antenna for mobile direct broadcasting satellite (DBS) application. This structure is optimized to have a minimum number of phase shifters and active channels (17 for each polarization) while maintaining the scanning capability and the required gain. The technology can be extended to low-profile structures for two-way phased-array systems.

11.6.3 Beam Steering Techniques

Beam steering is the process of adjusting the array complex weights to rotate the radiation beam. It is often performed by applying the required phase shift or time delay to the different RF signals received from or transmitted to the desired direction to maximize the goal function. Depending on the application, the goal could be to maximize the transmitted power, the received signal-to-noise ratio (SNR) or signal-to-noise and interference ratio (SNIR). An efficient beam-forming algorithm is one which compensates for the hardware inaccuracies, converges quickly, reduces the steady-state error, has low computational complexity, and is easy to implement.

Beam forming for geostationary satellite communication is not an interference-limited problem because, as shown in Figure 11.26, the angular separation of the adjacent co-channel satellites (in azimuth plane) is much

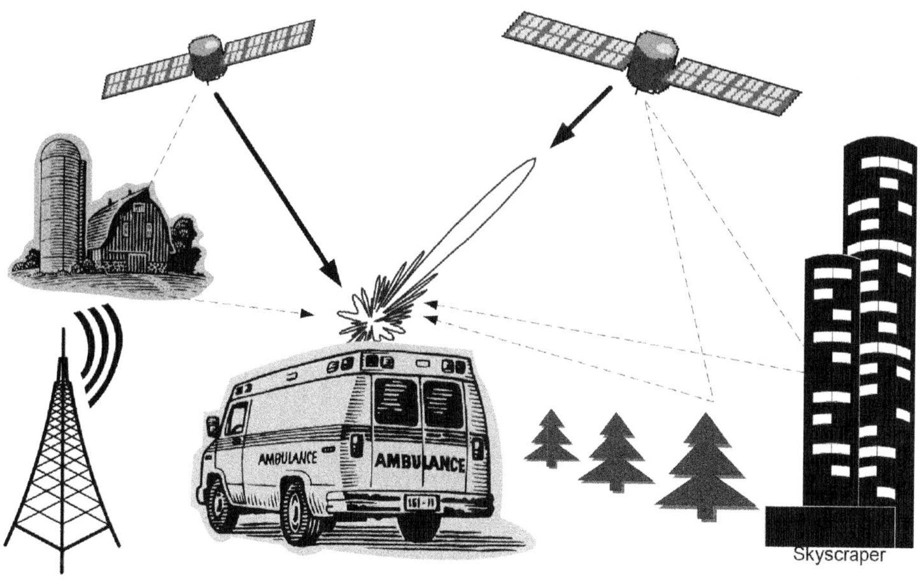

Figure 11.26 A typical propagation environment for mobile satellite communication. Courtesy of [54]. © 2010 IEEE. Reprinted, with permission from IEEE Trans. Broadcast., vol. 56, no. 3, pp. 340–349, Sept. 2010.

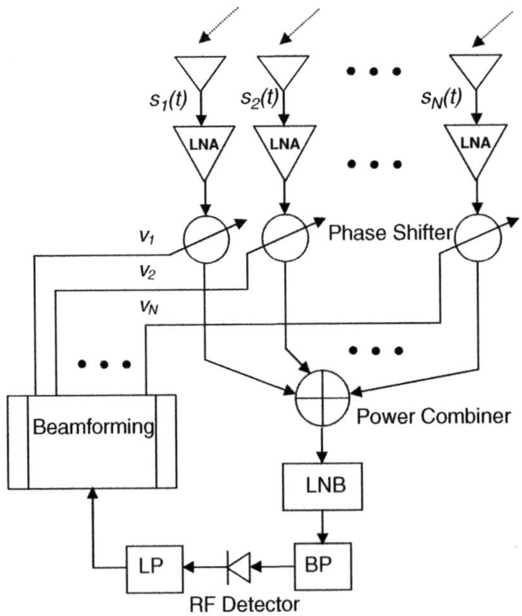

Figure 11.27 *Block diagram of a single-receiver phased-array antenna (LNA, Low-Noise Amplifier; LNB, Low-Noise Block; LP, Low-Pass (filter); BP, Bandpass (filter)). Courtesy of [52]. © 2008 IEEE. Reprinted, with permission from IEEE Trans. Antennas Propag., vol. 56, no. 12, pp. 3667–3679, Dec. 2008.*

wider than the beamwidth of the high-gain antennas (around 2°). Therefore, once the antenna is locked on to the desired direction, the interference from the adjacent satellites is negligible.

When sufficient information from the source (target) and array antenna is available, sophisticated beam steering techniques can be used [53]. Typically, such techniques minimize the least mean square error of a receiver array antenna based on inverting the correlation matrix of the received array signals. To find this matrix an individual receiver chain (mixer, IF, decoder, analog-to-digital converter (ADC), etc.) is required for each signal path in the array as well as a reference signal to estimate the error. If the received array signal and noise are jointly ergodic processes, the array correlation matrix can be estimated by calculating the time average of the received signals, which is a time-consuming process. Thus, adaptive techniques have been used for beam forming to save time [53,54]. Important drawbacks of the optimum and adaptive beam-forming algorithms developed so far, for the application under consideration in this subsection, are: (1) difficulty of accurate measurement of the array correlation matrix; (2) complexity of DOA (Direction Of Arrival) estimation algorithms; (3) variation of the phase–voltage characteristics of the phase shifters with environmental conditions; and (4) the effects of platform motion on the estimated correlation matrix.

When the sum of the signals from all antennas is the only available information (single-receiver phased-array system), as seen in Figure 11.27, the beam forming problem becomes more complicated. There are few reported beam forming methods which maximize the mean received signal power while placing antenna array pattern nulls in the directions of interfering signals [56]. However, beam forming for mobile satellite communication is not an interference-limited problem. In mobile satellite links, beam agility, rapid convergence, robustness, and low-cost implementation are the highest priorities that must be accomplished by a robust algorithm.

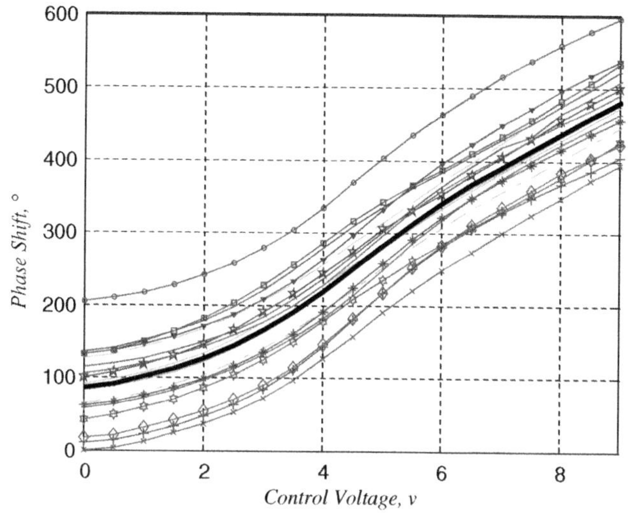

Figure 11.28 The measured phase–voltage characteristics of 17 channels of the phased-array antenna. Each channel includes a microwave integrated circuit (MIC) phase shifter, a cable, and an LNA. The dark solid line shows the average of all curves. Courtesy of [52]. © 2008 IEEE. Reprinted, with permission from IEEE Trans. Antennas Propag., vol. 56, no. 12, pp. 3667–3679, Dec. 2008.

11.6.4 Robust Zero-Knowledge Beam Control Algorithm

'Zero-knowledge' beam forming is defined as a constrained nonlinear optimization problem whose objective is to maximize the received power, without knowledge of the signal DOA. Usually, knowledge of DOA helps to find the proper initial conditions for beam forming, while zero-knowledge beam forming does not depend on a specific initial condition [53]. Figure 11.28 shows the measured phase–voltage characteristics of 17 channels of a single-receiver Ku-band phased array [52]. RTPSs at this frequency band can provide more than 360° phase shift by varying the control voltage. It can be seen that for the same control voltage the achieved phase shift can vary by 200° from one channel to another. The other problem is the imbalanced insertion loss of the phase shifters, which means that the insertion loss varies (up to 4 dB in some cases) when the control voltage is varied [29]. Unlike most of the conventional beam forming algorithms, in the zero-knowledge technique it is assumed that the signal DOA and the phase–voltage characteristics of the phase shifters are unknown. The algorithm adjusts the control voltages of the phase shifters to increase the received power. Using the notations in Figure 11.27, let $\mathbf{s}(n) = [s_1(n), s_2(n), \ldots, s_N(n)]$ and $\mathbf{w}(n) = [w_1(n), w_2(n), \ldots, w_N(n)]$ denote, respectively, the received power from the target by the array elements and the phase shifts applied to each antenna element at time instant n. Then the total signal after the power combiner is given by

$$y(n) = \mathbf{w}^*(n)\,\mathbf{s}^T(n) \tag{11.2}$$

where the asterisk * and superscript T denote the complex conjugate and transpose operations, respectively. The measured RF power at the output of the RF detector is

$$P(n) = E\left[y(n) \cdot y^*(n)\right] \tag{11.3}$$

where $E[\cdot]$ denotes the expectation operation. Note that $P(n)$ is a function of the phase shifts applied to each antenna element.

In general, these phase shifts are controlled by a set of control voltages which can be shown by a $1 \times N$ vector as $\mathbf{v}(n) = [v_1, v_2, \ldots, v_N]$. As discussed earlier, both the phase and amplitude of the array weights depend on the control voltages

$$w_i(n) = f[v_i(n)] \cdot \exp[j\psi(v_i(n))] \tag{11.4}$$

where f and ψ are functions of the control voltage. The zero-knowledge algorithm uses an iterative method to update the control voltages:

$$\mathbf{v}(n+1) = \mathbf{v}(n) + 2\mu\hat{\mathbf{g}}(n) \tag{11.5}$$

where $\hat{\mathbf{g}}(n) = [\hat{g}_1(n), \hat{g}_2(n), \ldots, \hat{g}_N(n)]$ is the estimated gradient vector and μ is the *step size* which controls the convergence rate. Using a two-sided finite difference technique, each element of the estimated gradient vector is calculated as

$$\hat{g}_i(n) \approx \frac{P(v_i(n)+\delta)-P(v_i(n)-\delta)}{2\delta} \tag{11.6}$$

In (11.5), δ denotes the *perturbation* added to/subtracted from the control voltages to find the approximate partial derivatives. The proper values of step size and perturbation can be obtained for the desired convergence rate and the steady-state error as performance measures, through simulation studies [53].

In the following, the results of applying this technique to a Ku-band mobile phased array (described in Section 11.6.5) are presented. A road test including a full 360° turn was performed to examine the beam-forming performance. Rate sensors were installed on the test vehicle (minivan-size car). The sampling rate of the gyro sensors was 100 Hz. For test purposes, the main processor was adjusted such that each iteration of the beam forming was performed in 10 ms. Figure 11.29 shows the measured angular rates of the road test. The yaw rate increases from 0°/s at $t = 3$ s to 30°/s at $t = 5$ s and hovers close to this value until $t = 15$ s. The fluctuations of pitch rate are negligible compared to yaw rate variations. Figure 11.30 shows the received power level by applying the aforementioned zero-knowledge beam-forming technique [54]. Except for a very short period of time (around 12.6 s), the received power is always maintained above the 90% of the maximum level.

11.6.5 A Ku-band Low-Profile, Low-Cost Array System for Vehicular Communication

Figure 11.31 shows the configuration and block diagram of a mobile Ku-band phased-array system with nearly 1000 patch antennas arranged in a stair–planar configuration [54]. The system operates from 12.2 to 12.7 GHz with an antenna gain of 31.5 dBi. Five panels support left-hand circular polarization (LHCP) and the other half support right-hand circular polarization (RHCP). There are 17 active sub-arrays for each polarization. The sub-arrays are mounted on array carriers which can mechanically rotate from 20° to 70° in elevation.

Each sub-array, integrated with its LNA, passes the received signal after preamplification to the phase shifter/power combiner (PS/PC) unit. The phase of the received signal by each sub-array is controlled by an analog phase shifter. The phase-shifted signals from all sub-arrays of the same polarization are

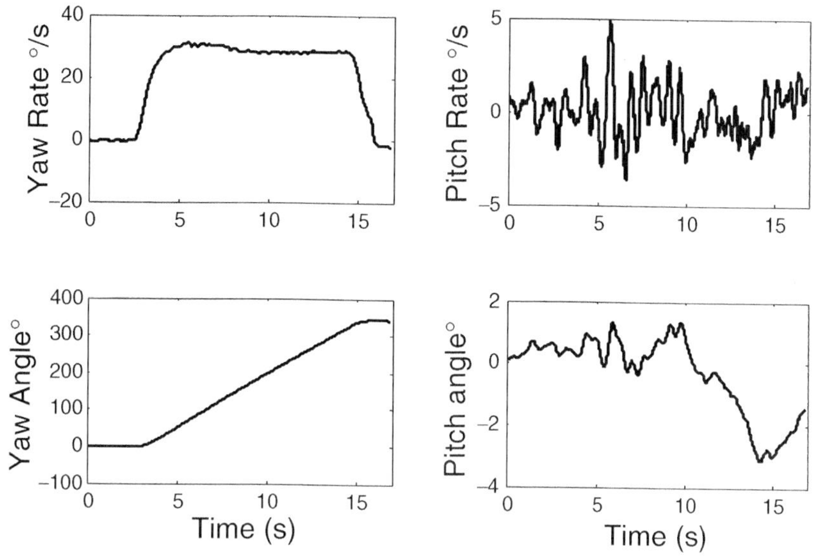

Figure 11.29 Measured angular rates and calculated swept angles for the 360° road test. Courtesy of [54]. © 2010 IEEE. Reprinted, with permission from IEEE Trans. Broadcast., vol. 56, no. 3, pp. 340–349, Sept. 2010.

combined, amplified, and then downconverted by an LNB. The downconverted signal is further processed by the digital video broadcasting (DVB) board for extraction of the satellite ID and is sent to the receiver inside the vehicle through a rotary joint. The RF detector implemented in the DVB board further measures the strength of the signal. The system hardware includes a main control subsystem and several auxiliary units such as a DAC (Digital-to-Analog Converter) to control the voltage of the phase shifters, gyros, motor control, and drivers. A compact, robust, lightweight, and very low-cost mechanical platform has been developed for the phased-array system. The mechanical system is composed of rotating and

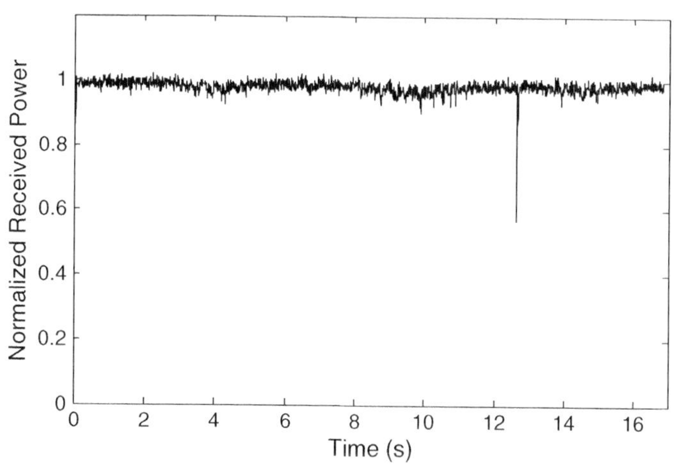

Figure 11.30 Normalized received power by the 17-element phased array for the 360° road test.

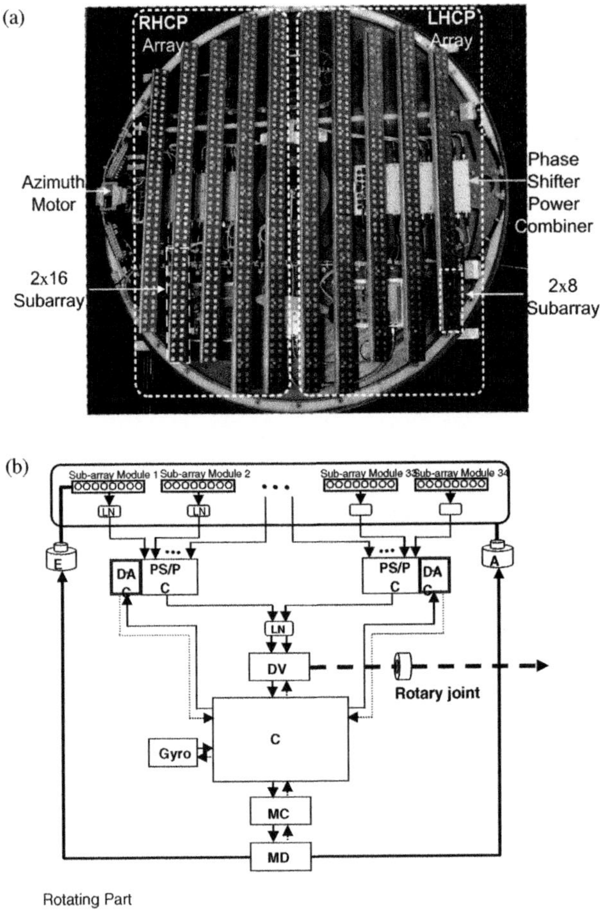

Figure 11.31 *(a) The configuration and (b) block diagram of the 1000-element Ku-band receiver phased-array antenna system (AM, Azimuth Motor; CB, Control Board; DAC, Digital-to-Analog Converter; EM, Elevation Motor; LNA, Low-Noise Amplifier; LNB, Low-Noise Block; MCB, Motor Control Board; MDB, Motor Drive Board; PS/PC, Phase Shifter/Power Combiner). Courtesy of [52,53]. © 2008 IEEE. Reprinted, with permission from IEEE Trans. Antennas Propag., vol. 56, no. 12, pp. 3667–3679, Dec. 2008.*
© 2009 IEEE. Reprinted, with permission from IEEE Trans. Antennas Propag., vol. 57, no. 6, pp. 1645–1654, Jun. 2009.

stationary parts. The stationary part is attached to the roof of the vehicle with roof bars. All above-mentioned electronic parts are integrated in the rotating part. Table 11.1 summarizes the main parameters and dimensions of the phased-array system.

The required average *G/T* for a mobile antenna system in North America is approximately 9.5–10 dB/K with a transmitter effective isotropic radiated power (EIRP) of 50 dBW. This number is basically a compromise between the performance of the antenna under various environmental conditions such as rain and snow and the acceptable profile for vehicular applications. If all RF channel losses (except for different phase shifter insertion losses) are identical, the overall *G/T* of the system can be calculated as

$$\frac{G}{T} = \frac{NG_e\eta}{T_i + T_0\left(L_f F - 1 - \frac{L_f}{g} + \frac{N}{\sum_{k=1}^{N} 1/L_k} \times \frac{L_f}{g} \times L_d\right)} \tag{11.7}$$

Table 11.1 Low-profile system parameters [52].

Parameters	Value
Frequency	12.2–12.7 GHz
Polarization	Dual circular
Gain	31.5 dBi (per polarization)
Tracking speed (azimuth)	60°/s
System height	6 cm
System diameter	86 cm
System weight	12 kg
LNA gain	23 dB
LNA noise figure	0.8 dB
Cable loss	1.5 dB
PS/PC loss	5–7 dB
2 × 16 sub-arrays directivity	21.8 dB
Feed network loss	1.3 dB

where g and F are the gain and noise figure of the LNA, L_f and L_d are the sub-array feed network loss and cable loss respectively, which are assumed to be identical for all channels, L_k is the kth PC/PS individual channel loss, G_e is the effective directivity of each antenna sub-array, and η is the aperture efficiency. The temperatures T_0 and T_i are the room and the sky temperatures, respectively. To meet the G/T requirement, using Table 11.1, the number of 2 × 16 sub-arrays should be more than 15.

A hybrid tracking method composed of two complementary mechanisms, namely, (1) the stabilization loop and (2) electronic beam forming, is used to keep the phased-array beam locked on to the desired

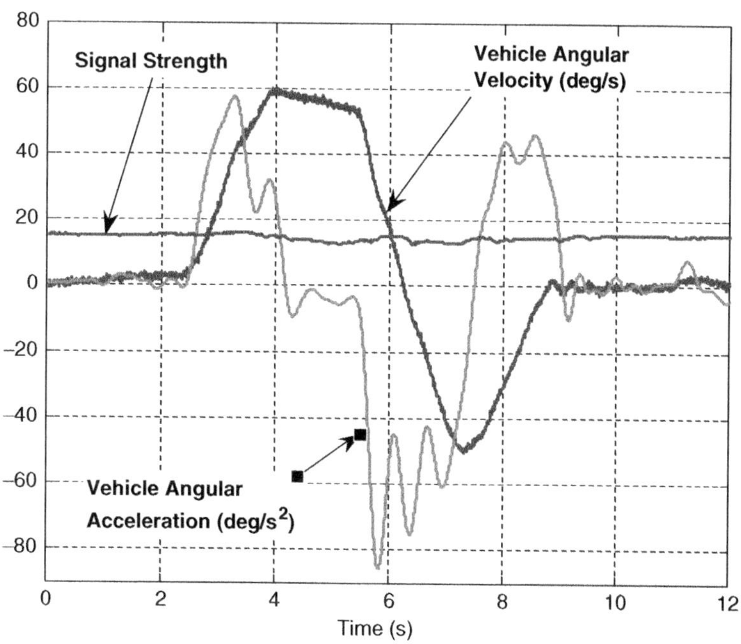

Figure 11.32 The performance of the phased-array antenna in a road test. Courtesy of [52]. © 2008 IEEE. Reprinted, with permission from IEEE Trans. Antennas Propag., vol. 56, no. 12, pp. 3667–3679, Dec. 2008.

satellite. The stabilization loop is mostly responsible for keeping the antenna toward a predetermined attitude during sharp maneuvers of the vehicle. A low-cost MEMS rate gyro, mounted on the antenna platform, provides most of the information required by the loop. However, the low-cost MEMS sensor suffers from high rate drift and strong noise. The electronic beam former compensates for the sensor errors. It assists the stabilization loop and eliminates the residual azimuth angle error. Moreover, integrating the two mechanisms enables the tracking system to nullify slow angular disturbances including those hidden in the noise of the gyro [55].

Tracking system agility was measured by a number of road tests. In order to monitor the test vehicle maneuver an auxiliary gyro was installed on the fixed part of the antenna platform. By manipulating the auxiliary gyro signal the velocity and the acceleration of the vehicle during the road test were calculated. Along with these parameters the scaled output of the RF detector was registered. Figure 11.32 shows a typical result when the vehicle veered away to the left and right. As the results show, the RF signal level remains almost constant even during the vehicle's sharp turns at high angular velocities of more than 60°/s.

11.7 Conclusions

In this chapter a brief review of a number of existing and emerging technologies, which can potentially be applied to current and future space systems, has been presented. The main focus was placed on low-cost technologies that can offer high performance in a compact and reliable package. On-chip antennas were shown to provide an exciting opportunity to develop ultra-compact mmW and sub-mmW systems on-chip and/or in-package with unmatched performance in terms of conformity and integration with other parts of a complex space system, or as a part of a stand-alone tiny 'satellite-on-chip'. Cost/size and performance requirements of several space systems, including radio telescopes, space communication and sensing systems, radar, and satellite communications, can be effectively met using compact planar technologies such as SIW and MEMS-based radio front-end and antenna systems. THz integrated antenna and front-end technologies, particularly those based on THz photonics, have provided immense opportunities for using this increasingly important part of the electromagnetic spectrum for an extremely wide range of R&D activities in space science and technologies, as well as other emerging applications in the life sciences, security, imaging/sensing, and ultra-broadband communications. As an example of an increasingly important class of systems, a low-cost/complexity and low-profile phased-array technology for land-mobile vehicular satellite communication was described in detail. It was shown that by combining low-cost hardware with highly robust and intelligent software, which can compensate for fabrication errors and less than ideal performance of mechanical and electronic devices, the high-performance requirements of several emerging space-based applications can be met in a cost-effective manner. It was obviously not possible to include in this limited space all new developments in antenna technologies of relevance to space systems. The chapter primarily focused on planar integrated technologies, which can potentially lead to highly cost-effective solutions for future space missions, and particularly in those areas where the authors have directly or indirectly been involved.

References

1. Geldzahler, B.J., Rush, J.J., Deutsch, L.J., and Statman, J.I. (2007) Engineering the next generation deep space network. IEEE IMS.
2. IEEE (2009) Special issue on advances in radio telescopes. *IEEE Proceedings*, **97**(8)
3. Charlton, J.E., Buehler, S., Defer, E. *et al.* (2009) A sub-millimeter wave airborne demonstrator for the observation of precipitation and ice clouds. IEEE IGARSS.

4. Nathrath, N., Trümper, M., Purschke, R., Harder, J., and Wolf, H. (2010) Lightweight intersatellite link antenna (LISAMS) operating at Ka-band. EuCAP 2010.
5. Ozbay, C., Teter, W., He, D. *et al.* (2006) Design and implementation challenges in Ka/Ku dual-band Satcom-on-the-move terminals for military applications. MilCOM.
6. Shamim, A., Arsalan, M., Roy, L., Shams, M., and Tarr, G. (2008) Wireless dosimeter: system-on-chip versus system-in-package for biomedical and space applications. *IEEE Transactions on Circuits and Systems—II: Express Briefs*, **55**(7), 643–647.
7. Ouellette, J. (2011) 'Satellite on a Chip' to Launch with Space Shuttle, *Discovery News*, http://www.spacecraftresearch.com/MII/MII_overview.html (accessed May 16, 2011).
8. IEEE (2011) Spacecraft-on-chip. *IEEE Spectrum TechAlert*, 4 August.
9. Natarajan, A., Komijani, A., Guan, X., Babakhani, A., and Hajimiri, A. (2006) A 77-GHz phased-array transceiver with on-chip antennas in silicon: transmitter and local lo-path phase shifting. *IEEE Journal of Solid-State Circuits*, **41**(12), 2807–2819.
10. Chen, I.-S., Chiou, H.-K., and Chen, N.-W. (2009) V-band on-chip dipole-based antenna. *IEEE Transactions on Antennas and Propagation*, **57**(10), 2853–2861.
11. Shamim, A., Roy, L., Fong, N., and Garry Tarr, N. (2008) 24 GHz on-chip antennas and balun on bulk Si for air transmission. *IEEE Transactions on Antennas and Propagation*, **56**(2), 303–311.
12. Nezhad-Ahmadi, M.R., Fakharzadeh, M., Biglarbegian, B., and Safavi-Naeini, S. (2010) High-efficiency on-chip dielectric resonator antenna for mm-wave transceivers. *IEEE Transactions on Antennas and Propagation*, **58**(10), 3388–3392.
13. Biglarbegian, B., Nezhad-Ahmadi, M.-R., Hoggat, C. *et al.* (2010) A 60 GHz on-chip slot antenna in silicon integrated passive device technology. *IEEE International Symposium on Antennas and Propagation.*
14. Gunnarsson, S.E., Wadefalk, N., Svedin, J. *et al.* (2008) A 220 GHz single-chip receiver MMIC with integrated antenna. *IEEE Microwave and Wireless Components Letters*, **18**, 284–286.
15. Ali, A.M., Fonseca, N.J.G., Coccetti, F., and Aubertand, H. (2011) Design and implementation of two-layer compact wideband butler matrices in SIW technology for Ku-band applications. *IEEE Transactions on Antennas and Propagation*, **59**(2), 503–512.
16. Hirokawa, J. and Ando, M. (2000) Efficiency of 76-GHz post-wall waveguide-fed parallel-plate slot arrays. *IEEE Transactions on Antennas and Propagation*, **48**(11), 1742–1745.
17. Uchimura, H., Takenoshita, T., and Fujii, M. (1998) Development of a laminated waveguide. *IEEE Transactions on Microwave Theory and Techniques*, **46**(12), 2438–2443.
18. Wolff, I. (2009) From antennas to microwave systems – LTCC as an integration technology for space applications. EuroCAP.
19. Xu, F. and Wu, K. (2005) Guided-wave and leakage characteristics of substrate integrated waveguide. *IEEE Transactions on Microwave Theory and Techniques*, **53**(1), 66–73.
20. Ohira, M., Miura, A., and Ueba, M. (2010) 60-GHz wideband substrate-integrated-waveguide slot array using closely spaced elements for planar multisector antenna. *IEEE Transactions on Antennas and Propagation*, **58**(3), 993–998.
21. Abdel Wahab, W.M., Busuioc, D., and Safavi-Naeini, S. (2010) Low cost planar waveguide technology-based dielectric resonator antenna (DRA) for millimeter-wave applications: analysis, design, and fabrication. *IEEE Transactions on Antennas and Propagation*, **58**(8), 2499–2507.
22. Abdel Wahab, W.M., Busuioc, D., and Safavi-Naeini, S. (2011) Millimeter-wave high radiation efficiency planar waveguide series-fed dielectric resonator antenna (DRA) array: analysis, design, and measurements. *IEEE Transactions on Antennas and Propagation*, **59**(8), 2834–2843.
23. Daneshmand, M. and Mansour, R.R. (2011) RF MEMS satellite switch matrices. *IEEE Microwave Magazine*, August, 92–109.
24. Yashchyshyn, Y. (2010) Reconfigurable antennas: the state of the art. *International Journal of Electronics & Telecommunication*, **56**(3), 319–326.
25. Rizk, J.B. and Rebeiz, G.M. (2003) W-band CPW RF MEMS circuits on quartz substrates. *IEEE Transactions on Microwave Theory and Techniques*, **51**(7), 1857–1862.
26. Lukic, M.V. and Filipovic, D.S. (2007) Integrated cavity-backed Ka-band phased array antenna. IEEE APS.

27. Ziegler, V., Gautier, W., Stehle, A., Schoenlinner, B., and Prechtel, V. (2010) Challenges and opportunities for RF-MEMS in aeronautics and space – the EADS perspective. IEEE Topical Meeting on Silicon Monolithic Integrated Circuits in RF Systems, pp. 200–203.
28. Legay, H., Bresciani, D., Girard, E. *et al.* (2009) Recent developments on reflectarray antennas at Thales Alenia Space. TAS (EuCap 2009).
29. Fakharzadeh, M., Mousavi, P., Safavi-Naeini, S., and Jamali, S.H. (2008) The effects of imbalanced phase shifters loss on phased array gain. *IEEE Antennas and Wireless Propagation Letters*, **7**, 192–196.
30. Crunteanu, A., Blondy, P., and Vendier, O. (2006) Non-hermetic RF MEMS, Final report, ESA contract no. 17161/03/NL/PA, Alcatel Alenia Space.
31. Kim, H.-T., Lee, S., Kim, J. *et al.* (2003) A V-band CPS distributed analog MEMS phase shifter. IEEE IMS.
32. Biglarbegian, B., Nezhad-Ahmadi, M.R., Fakharzadeh, M., and Safavi-Naeini, S. (2009) Millimeter-wave reflective-type phase shifter in CMOS technology. *IEEE Microwave and Wireless Components Letters*, **19**(9), 560–562.
33. Biglarbegian, B., Bakri-Kassem, M., Mansour, R., and Safavi-Naeini, S. (2010) MEMS-based reflective-type phase-shifter for emerging millimeter-wave communication systems. Proceedings of the 40th European Microwave Conference, pp. 1556–1559.
34. Saeedkia, D. and Safavi-Naeini, S. (2008) Terahertz photonics: optoelectronic techniques for generation and detection of terahertz waves. *IEEE/OSA Journal of Lightwave Technology*, **26**(15), 2409–2423.
35. Koch, M. (2007) Terahertz communications: a 2020 vision, in *Terahertz Frequency Detection and Identification of Materials and Objects* (eds R.E. Miles, X.-C. Zhang, H. Eisele, and A. Krotkus), NATO Security Through Science Series, Springer, pp. 325–338.
36. Sankaran, S. and O, K.K. (2005) Schottky barrier diodes for millimeter wave detection in a foundry CMOS process. *IEEE Electron Device Letters*, **26**(7), 492–494.
37. Gao, J.R., Hovenier, J.N., Yang, Z.Q. *et al.* (2005) Terahertz heterodyne receiver based on a quantum cascade laser and a superconducting bolometer. *Applied Physics Letters*, **86**, 244104.
38. Yang, Z.Q., Hajenius, M., Hovenier, J.N. *et al.* (2005) Compact heterodyne receiver at 2.8 THz based on a quantum cascade laser and a superconducting bolometer. IEEE IMS, pp. 465–466.
39. IEEE (2007) Special issue on optical and THz antenna technology. *IEEE Transactions on Antennas and Propagation*, **55**(11)
40. Rolo, L.F., Paquay, M.H., Daddato, R.J. *et al.* (2010) Terahertz antenna technology and verification: Herschel and Planck—a review. *IEEE Transactions on Microwave Theory and Techniques*, **58**(7), 2046–2063.
41. Llombart, N., Cooper, K.B., Dengler, R.J., Bryllert, T., and Siegel, P.H. (2010) Confocal ellipsoidal reflector system for a mechanically scanned active terahertz imager. *IEEE Transactions on Antennas and Propagation*, **58**(6), 1834–1841.
42. Song, H., Ajito, K., Wakatsuki, A. *et al.* (2010) Terahertz wireless communication link at 300 GHz. IEEE MWP.
43. Brown, E.R., Smith, F.W., and McIntosh, K.A. (1993) Coherent millimeter-wave generation by heterodyne conversion in low-temperature-grown GaAs photoconductors. *Journal of Applied Physics*, **73**(3), 1480–1484.
44. Saeedkia, D., Mansour, R.R., and Safavi-Naeini, S. (2005) Analysis and design of a continuous-wave terahertz photoconductive photomixer array source. *IEEE Transactions on Antennas and Propagation*, **53**(12), 4044–4050.
45. McCarrick, C. (2006) Offset stacked patch antenna and method, US Patent 7102571.
46. Stoyanov, I., Boyanov, V., Marinov, B., Dergachev, Z., and Toshev, A. (2005) Mobile antenna system for satellite communications, US Patent 6999036.
47. Jeon, S., Kim, Y., and Oh, D. (2000) A new active phased array antenna for mobile direct broadcasting satellite reception. *IEEE Transactions on Broadcasting*, **46**(1), 34–40.
48. Eom, S.Y., Son, S.H., Jung, Y.B. *et al.* (2007) Design and test of a mobile antenna system with tri-band operation for broadband satellite communications and DBS reception. *IEEE Transactions on Antennas and Propagation*, **55**(11), 3123–3133.
49. ThinKom Solutions, Inc. (2011) thinAir Ku3020, http://thin-kom.com/news.html.
50. Raysat Antenna Systems (RAS), StealthRay 3000/40 TechSpec, http://www.raysat.com/.

51. Holzwarth, S., Jacob, A.F., Dreher, A. *et al.* (2010) Active antenna arrays at Ka-band: status and outlook of the SANTANA. EuCAP 2010, pp. 1–5.
52. Mousavi, P., Fakharzadeh, M., Jamali, S.H. *et al.* (2008) A low-cost ultra low profile phased array system for mobile satellite reception using zero-knowledge beam-forming algorithm. *IEEE Transactions on Antennas and Propagation*, **56**(12), 3667–3679.
53. Fakharzadeh, M., Jamali, S.H., Mousavi, P., and Safavi-Naeini, S. (2009) Fast beamforming for mobile satellite receiver phased arrays: theory and experiment. *IEEE Transactions on Antennas and Propagation*, **57**(6), 1645–1654.
54. Mousavi, P., Fakharzadeh, M., and Safavi-Naeini, S. (2010) 1 K element intelligent antenna system for mobile direct broadcasting satellite communication. *IEEE Transactions on Broadcasting*, **56**(3), 340–349.
55. Bolandhemmat, H., Fakharzadeh, M., Mousavi, P. *et al.* (2009) Active stabilization of vehicle-mounted phased-array antennas. *IEEE Transactions on Vehicular Technology*, **58**(6), 2638–2650.
56. Godara, L.C. (1997) Application of antenna arrays to mobile communications, part II: Beam-forming and direction of arrival considerations. *Proceedings of the IEEE*, **85**(8), 1195–1254.

12

Antennas for Satellite Communications

Eric Amyotte and Luís Martins Camelo

MDA, Canada

12.1 Introduction and Design Requirements

This chapter focuses on satellite communications antennas. Although it is not possible to do an exhaustive review of all antenna solutions for all types of spaceborne communications missions, the following sections provide an overview of the most common antenna designs used in today's communications satellite systems. An attempt was made to present examples from actual satellite programs for each type of antenna discussed. That was not always possible, however, since approval for the publication of technical details is often difficult to obtain in this highly competitive commercial field, where trade secrets and export regulations are omnipresent. Nevertheless, a lot of hopefully useful information has been included in these pages, supported whenever possible by examples from past programs, and highlighting and discussing the main challenges and constraints facing the antenna engineer for each type of antenna and mission.

The laws of physics cannot be changed (yet), but antenna engineers have learned to use them to their advantage by relentlessly coming up with new and often highly innovative antenna design solutions. These new solutions usually target improvements in various areas including performance, price, mass, volume, power, schedule and reliability. These developments are spurred by the needs of next-generation systems such as higher system capacity and in-orbit flexibility. While most of the developments in the field of satellite antennas are evolutionary in nature, some of them are truly revolutionary and become the new standard for many years before they are replaced by another type of technology in this ever-developing field.

The ultimate purpose of the satellite antenna is to act like a spatial filter, focusing the antenna beam on the desired area on the Earth with the desired polarization, while being isolated from other areas. Satellite-borne communications antennas are located very far from their service area on the Earth, therefore (especially for geostationary satellites) the coverage area represents a relatively small angular region. Moreover, the antenna radiation pattern is usually designed so as to fall off rapidly beyond the edge of coverage, thus maximizing the fraction of the total radiated power that falls within the service area. Consequently, the radiation pattern

Space Antenna Handbook, First Edition. Edited by William A. Imbriale, Steven (Shichang) Gao and Luigi Boccia.

characteristics must be accurately designed and maintained for the mission duration. This precise radiation pattern control must take into account all angular pointing errors, which are the result both of antenna misalignments and distortions and of spacecraft distortions and attitude instabilities. Minimizing this pointing error becomes a critical aspect of the communications system optimization. This is especially important when the nominal coverage areas are relatively small and when the isolation areas are relatively close to the coverage areas (e.g., in modern geostationary high-gain multibeam spacecraft antennas). In this case the pointing error represents a much greater percentage of the coverage area and of the distance between coverage and isolated areas, leading potentially to a much greater deterioration in performance.

12.1.1 Link Budget Considerations

Antennas, both on the satellite and on the ground terminal, are essential components of the communications system link budget. Their performance is critical to the ability of communications systems to generate financial revenue, by enhancing capacity as well as availability and quality of service, while minimizing costs.

On the satellite, antennas are constrained by size and mass, which also act as limits on the maximum achievable antenna gain. For unshaped or slightly shaped spot beams, the peak and edge of coverage (EOC) gain are mostly limited by the size of the antenna aperture and its efficiency, and therefore a useful measure of antenna performance is its effective aperture, A_{Tx} on Tx and A_{Rx} on Rx. The effective aperture is a fraction (typically 60–80% for spot beams) of the physical aperture area of the antenna. As a consequence, limiting the maximum physical aperture size is akin to limiting the maximum effective aperture of the antenna. The effective aperture of an antenna is calculated as the antenna gain multiplied by the square of the wavelength divided by 4π.

For shaped-beam coverage regions, which are defined by geographical areas on the Earth, it is the angular extent of the coverage regions rather than the size of the aperture that tends to limit the EOC gain achievable by the satellite antenna (larger angular coverage areas correspond to lower EOC gain performance). In that case the achievable gain, G_{Tx} or G_{Rx}, which is mainly driven by the angular extent of the coverage region, is a better measure of the limits on performance than the effective aperture. Nevertheless, a larger aperture (in terms of wavelength) allows for a finer beam shaping capability, meaning that the beam shape follows the coverage contour more closely, and thus it results in a higher value of EOC gain. It follows that the allowable aperture size is also a limiting factor for shaped beams, albeit in a less straightforward manner than for spot beams.

For the downlink, the other important parameter defining performance is the RF power transmitted to the antenna, P_{Tx}, which is limited by many factors including the available aggregate output amplifier power and the antenna power handling capability. As a rule, power is a very precious resource aboard a satellite. The downlink performance of the antenna can thus be evaluated as the product $P_{Tx} \times G_{Tx}$, also known as the equivalent isotropic radiated power (EIRP), or by $P_{Tx} \times A_{Tx}$, which highlights the limitations of the satellite antenna when generating spot beams (performance limited by aperture size). For the uplink, the ultimate figure of merit for the received signal tends to be the signal-to-noise ratio (*S/N*). Thus the characteristic of interest of the receive antenna is not only A_{Rx} or G_{Rx} but rather A_{Rx}/kTB or G_{Rx}/kTB, where T is the system noise temperature, B is the Rx signal bandwidth, and k is Boltzmann's constant ($1.380\,622 \times 10^{-23}$ J/K). Since T is the equivalent system noise temperature, and not just the radiation portion of the antenna temperature, a very low-noise receiver front end is a desirable feature of the communications system. For digital communications systems, the bit error rate (BER) is the typical parameter of interest in the link budget, and it is closely related to *S/N*, the exact relationship depending, among other factors, on the modulation and coding scheme used.

At the other end of the communications link, ground terminal antennas are required to form a simple pencil beam focused on the satellite, which is a very narrow beam for large-antenna apertures and a much broader beam for small antennas. Thus, in most cases, the gain performance is greatly determined by antenna size. Therefore, ground terminal antennas are usually adequately characterized by their effective aperture A_{Term}, which most often has very similar values for Tx and Rx (if the same physical aperture is being shared for the two functions and if the aperture illumination efficiency is similar, which is commonly the case). Whereas large fixed ground stations can have very large values of A_{Term}, other applications come with tight restrictions on antenna size, such as for airborne or mobile terminals and especially for handheld and other personal terminals. The Tx power for the uplink, P_{TermTx}, can be limited by the power amplifier capability of the ground terminal or by other factors such as cost limitations in personal terminals. For the downlink, the system noise temperature T_{TermRx}, which is greatly determined by the quality of the low-noise receiver, will degrade the received S/N ratio. That may be a limitation, especially for personal terminals, which are extremely cost sensitive.

For shaped-beam applications, the EIRP (i.e., $P_{Tx} \times G_{Tx}$) is the best characterization of the downlink antenna on the spacecraft and G_{Rx}/T of the uplink antenna, with G_{Tx} and G_{Rx} limited by the angular size of the coverage region. The link budget equations (12.1) and (12.2) below are in a particularly useful form for shaped beams:

$$\left(\frac{S}{N}\right)_{TermRx} = \frac{1}{4\pi R^2} \times P_{Tx} \times G_{Tx} \times \frac{A_{Term}}{kBT_{TermRx}} \times \eta_{ATM} \times \eta_{POL} \tag{12.1}$$

for the downlink, and

$$\left(\frac{S}{N}\right)_{Rx} = \frac{1}{4\pi R^2} \times P_{TermTx} \times A_{Term} \times \frac{G_{Rx}}{kBT} \times \eta_{ATM} \times \eta_{POL} \tag{12.2}$$

for the uplink, where R is the distance between the satellite and the ground terminal, η_{ATM} is a factor between 0 and 1 that accounts for propagation losses through the atmosphere, and η_{POL} is the polarization adaptation factor between the Tx and Rx antennas (also a value between 0 and 1). In a well-designed communications link η_{POL} is usually very close to 1, signifying very low loss. Depolarization due to precipitation may also happen, but it is here considered to be included in η_{ATM}. The factor η_{ATM} is typically between 0.9 and 1 for the frequencies covered in this chapter, in the absence of rain or other precipitation, with the exception of the oxygen absorption band around 60 GHz where it is much less, and the water absorption band around 23 GHz where it is also lower. Note that sometimes these absorption bands can be used to advantage in intersatellite links by shielding them from interference from ground-based antennas. Outside these absorption bands, η_{ATM} generally decreases (increasing loss) as the frequency increases, especially for Ka-band and above. It also decreases as the satellite elevation angle viewed from the ground terminal decreases, approximately as 1/sin (elevation), corresponding to the increase in propagation distance through the atmosphere. The decrease of η_{ATM} with precipitation can be severe and can even lead to loss of communication, especially at Ka-band and higher frequencies. The statistical behavior of η_{ATM} is an essential part of the system design; however, that subject is not within the scope of this chapter and will not be detailed here. The large distances R involved in satellite communications represent large path losses (the first term of (12.1) and (12.2)), of between 125 and 141 dB for low Earth orbit, between 145 and 159 dB for medium Earth orbit, and between 162 and 163.5 dB for geostationary orbit. These losses push the design towards higher gain values, higher Tx power levels, and lower noise receivers, both on the satellite and on the ground terminal.

For unshaped or lightly shaped spot beam applications, an alternative characterization of the downlink and uplink antennas on the spacecraft might be $P_{Tx} \times A_{Tx}$ and A_{Rx}/T respectively, emphasizing the limitation in terms of aperture size. Link budget equations in a particularly useful form for spot beams are thus the expressions (12.3) and (12.4) below:

$$\left(\frac{S}{N}\right)_{TermRx} = \frac{1}{\lambda^2 R^2} \times P_{Tx} \times A_{Tx} \times \frac{A_{Term}}{kBT_{TermRx}} \times \eta_{ATM} \times \eta_{POL} \tag{12.3}$$

for the downlink, and

$$\left(\frac{S}{N}\right)_{Rx} = \frac{1}{\lambda^2 R^2} \times P_{TermTx} \times A_{Term} \times \frac{A_{Rx}}{kBT} \times \eta_{ATM} \times \eta_{POL} \tag{12.4}$$

for the uplink, where λ is the RF wavelength. An inverse relationship with the square of the wavelength is now apparent, implying that for spot beams lower frequencies will require larger apertures or higher Tx power in order to achieve an equivalent level of performance. In fact, at L-band and S-band (which are relatively low frequencies in the satellite communications universe), large-reflector antenna diameters are often realized through the use of deployable mesh reflector technology, achieving diameters of between 6 m and more than 20 m that are impossible to accommodate in a spacecraft and launch vehicle if using more conventional solid reflector technology. Higher frequency spot beam antennas, on the other hand, tend to be realized using smaller antenna apertures, while maintaining the same level of performance.

12.1.2 Types of Satellite Communications Antennas

Communications satellites use a variety of orbits including geostationary orbits (GEOs), elliptical orbits, medium Earth orbits (MEOs) and low Earth orbits (LEOs). Common applications include military communications, digital audio broadcast, television, fixed services, mobile telephony and broadband services such as video-conferencing, and broadband Internet access.

The most popular orbit for satellite communications is the geostationary one, from which the Earth appears as a circular disk with an 8.7° radius. The main attractiveness of this orbit is that, viewed from the satellite, different regions on the Earth remain stationary in the satellite-fixed coordinate system, within pointing error allocations, and viewed from an Earth terminal the satellite remains almost stationary within a small and slow-varying tracking range. There are currently more than 400 satellites around the 360° extent of the geostationary orbital ring. The angular separation between satellites is therefore very limited, calling for stringent spectrum coordination between the different services.

In order to address the needs of today's communications systems, engineers can draw from an extensive set of available antenna technologies and design solutions, including phased arrays, array-fed reflectors, reflectarrays, lenses, shaped reflectors, polarization- and frequency-selective surfaces, printed radiators, and others, as well as advanced technologies for feed elements and feeding structures, including high-performance polarizers, filters, diplexers, orthomode junctions and transducers, and so on. The specific requirements of each type of satellite communications antenna will drive the use of some technologies over others. For example, some applications are characterized by a need for high power handling and will privilege robust metallic implementations over microstrip printed circuits. Other applications will have wide multiple-beam scanning requirements, which will impose the use of arrays rather than reflector technologies.

12.1.3 Materials

The realization of satellite antennas calls upon different engineering skills including mechanical, structural and thermal design, electronics, materials, components and processes, and others. The materials of choice are selected for their properties and in-orbit heritage (see Chapter 4). Carbon fiber reinforced plastics (CFRPs) are used extensively in structures and reflectors for their great stiffness-to-mass ratio and thermal stability. Aluminum is also popular for feed components and even small reflectors, due to its ease of machining into very accurate three-dimensional shapes and its good electrical conductivity. For lower losses, silver plating is typically applied to enhance electrical conductivity even further. In some instances of high-accuracy components such as diplexers, and if especially stringent dimensional accuracy and an ultra-low coefficient of thermal expansion (CTE) are required, special materials such as Invar (a nickel–steel alloy) may be used. Titanium is also used, typically for inserts, brackets and fittings, because of its higher stiffness-to-mass ratio and lower CTE than aluminum.

The use of dielectric materials in space applications is common but needs to be approached with some caution, since the space radiation environment can affect their properties (electrical, structural and thermal) through an often-slow degradation with time. It can also lead to surface and bulk electrical charging which can subsequently lead to electrostatic discharge (ESD), sometimes causing the failure of active electronics nearby. Active (and proactive) ESD management, by appropriate choice of materials as well as by providing conductive paths for discharging the materials before ESD can occur, is a critical and essential part of the design of satellite-borne communications antennas and structures.

12.1.4 The Space Environment and Its Design Implications

The satellite antenna subsystem will be exposed, while in orbit, to the harsh space environment that includes vacuum, temperature extremes, radiation consisting of a variety of particles, static charges, and micro-meteorites. The space environment is described in detail in Chapter 4. The antenna also needs to survive, without structural and dimensional degradation, the usually very severe launch environment. This includes vibration, shock, depressurization and, for larger structures, intense acoustic pressure. This subject is treated in detail by Chapter 5.

Compliance with all requirements has to be achieved with low mass and low power consumption, both of which are preciously scarce resources on a satellite. Reliability is also very important, both to survive the launch and during the in-orbit operation, especially since the required mission duration of communications satellites has been increasing steadily throughout the years and now typically lies between 12 and 15 years. This often demands innovative design solutions. However, the use of innovative designs is often tempered by the customers' preference for past in-orbit heritage, so as to alleviate their level of risk, and consequently their financial exposure and liability.

12.1.5 Designing for Commercial Applications

Telecommunications missions are predominantly of a commercial nature. As a consequence, the financial objectives of a mission are often at least as important as the technical and performance objectives. The antenna designer must take into consideration the influence that the selected fabrication techniques, the required manufacturing tolerances, the ease of assembly, and the integration and test campaigns, will have on the price of the antenna subsystem. Reliability of service is also of critical importance as the satellite ensures a stream of revenues on which the customer and the operators are counting. In practice, the most favoured design from a strictly technical perspective is rarely the optimal satellite communications antenna solution.

For example, sharp corners inside the different components and tight dimensional tolerances are often desired by the antenna engineer so as to optimize performance and simplify the analysis. Interior corners, however, are usually much more efficiently and cheaply implemented as round corners, featuring a finite internal radius consistent with the preferred manufacturing tools, so analysis methods that can account for the presence of such radii are very advantageous. The allowable dimensional tolerances throughout the antenna can usually be relaxed by performing the appropriate dimensional error-sensitivity analysis during the design stage, thus significantly relaxing the demands on the manufacturing processes and containing fabrication costs. A design that minimizes the need for high-precision manufacturing and assembly is generally preferable. Designing generic components, which will cover many sets of requirements, is desirable albeit not always possible. For the program manager, it is a balancing act between the costs incurred in the different phases of the program. When many antennas of the same type are to be produced, it is worthwhile spending

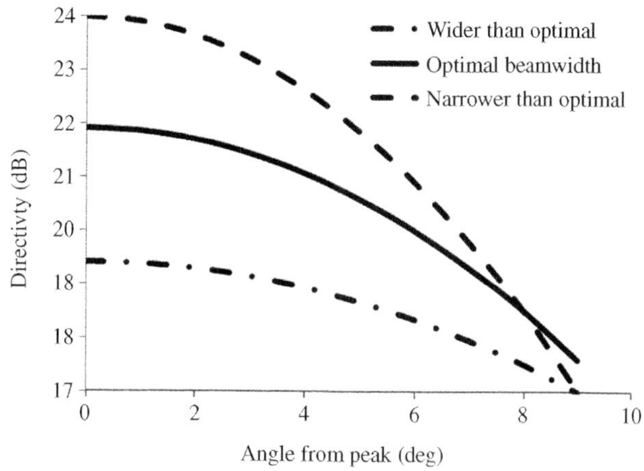

Figure 12.1 *Effect of pencil beam peak directivity on EOC directivity for global coverage.*

more effort during the early stages of the project optimizing the design for high-quantity production, in order to realize the savings due to economies of scale during the later production stages of the program, as compared to the production of a single unit.

12.2 UHF Satellite Communications Antennas

12.2.1 Typical Requirements and Solutions

UHF is a frequency band covering between 300 MHz and 1 GHz according to the IEEE standard definition. In practice, UHF communications satellites often use downlink frequencies around 250 MHz and uplink frequencies around 310 MHz. An advantage of UHF, traditionally used for military communications, is that it easily penetrates deep foliage and other obstacles.

The most common coverage requirement is the full visible portion of the Earth's surface as seen from geostationary orbit (a 17.4° diameter circular coverage region augmented by the pointing error, usually referred to as global coverage). A quick estimate may be made of the maximum possible gain performance over this coverage area. An optimal Gaussian beam pattern shape limits the minimum directivity (commonly known as EOC directivity) to about 17.5 dB (corresponding to approximately 22 dB peak directivity). As shown in Figure 12.1, both narrowing the Gaussian beam shape (larger aperture) to get a higher peak directivity, and broadening the Gaussian beamwidth relative to optimal (smaller aperture), lead to a worse EOC directivity than 17.5 dB. A higher EOC directivity is achievable by beam shaping rather than using a Gaussian beam, requiring an even larger antenna aperture, which is not always practical in the UHF band. Optimal beam shaping generally produces a flat-top beam, thus enhancing EOC directivity. In practice, RF losses reduce the EOC gain to less than the EOC directivity numbers quoted here.

The relatively long wavelengths characteristic of the UHF band (up to 1.2 m) generally mean that the satellite-borne antennas tend to be electrically small, their total size being restricted to only a few wavelengths. The typical size of the spacecraft bus is also only a few wavelengths at UHF. Consequently, the whole satellite structure impacts the way in which the antenna resonates, and it becomes effectively part of the antenna. The satellite can sometimes be used to advantage by the antenna designer in enhancing the basic antenna characteristics, and one common use is to increase the isolation between the Tx and Rx antennas, which is a crucial parameter, especially if the passive inter-modulation (PIM) products requirement is stringent. This can

Figure 12.2 Deployable rigid helices, with separate Tx and Rx. Reproduced by permission of Astrium.

be seen in Figure 12.2, where the spatial separation between the Tx and Rx helix antennas, as well as the presence of the intervening spacecraft, combine to achieve a total Tx/Rx isolation of the order of 60 dB.

Design solutions include single elements and arrays. Achieving the required gain from an antenna with such long wavelengths demands a large effective aperture. End-fire traveling-wave elements, such as helices, have been popular for this reason, since RF propagation along the axis of the helix in effect increases its effective aperture much beyond its physical aperture area. The directivity of a helix is essentially proportional to its length measured in terms of free-space wavelengths. Helices can thus provide a relatively high gain with a relatively small physical aperture diameter.

UHF array antennas may use as radiating elements large patches, cupped dipoles or even helices. In a UHF array the beam-forming network (BFN) is realized using various TEM modes or quasi-TEM technologies, since waveguide technology would have unacceptably large cross-sectional dimensions. The technologies used typically include circular or square-section coaxial lines, suspended stripline, dielectric-filled stripline, and others. Power handling capability will often be a determining factor in the choice of transmission line technology for implementation of the UHF array BFN.

12.2.2 Single-Element Design

The helix is a single-element antenna capable of providing sufficient EOC gain performance for one single hand of circular polarization. Helices can be basically modeled as metallic wire antennas a few wavelengths long, and as such they are readily handled by the method of moments, using one of several widely available commercial software packages. The computer model will also include the metallic cup when it is present, which is usually the case so as to supply an appropriate ground plane and increase forward gain by reducing backward radiation. The presence of the dielectric support structure (usually a thin-walled and mostly hollow cylinder/cone) also needs to be included in the computer model in order to achieve a good agreement with measurements. This dielectric structure changes the equivalent dielectric constant of the helix environment, and this results in a small frequency shift that must be corrected by reducing slightly the diameter of the helix. It may also affect input impedance and thus the design of the input matching structure.

The synthesis of the helix structure for optimal RF performance includes optimization of the pitch between consecutive wire turns, the diameter at the base and the way it tapers along the helix length, the helix length, and other geometric parameters. The optimization usually concentrates on maximizing EOC gain, axial ratio and bandwidth performance. Single-helix antennas are practically limited in peak gain to about 16 dB, which is partly due to the fact that the RF currents usually decrease to virtually zero before reaching the end of the helix,

Figure 12.3 Unfurlable UHF helix, stowed and deployed. Reproduced by permission of MDA.

thus limiting its achievable effective length. The corresponding EOC gain performance for global coverage is then lower than about 14 dB.

The large UHF helix sizes generally imply that they must be implemented as deployable structures, which often are rigid as shown in Figure 12.2. The objective of the deployment is to locate the helix in a precisely aligned position away from the spacecraft bus. Alternative and highly innovative deployment schemes have been implemented, such as the unfurlable helix shown in Figure 12.3 in both stowed and deployed configurations. This antenna can be stowed into a very compact volume for accommodation during launch, and once in orbit it is unfurled to a deployed length of between 3 and 4 m.

12.2.3 Array Design

If the gain of a single element is not sufficient, several elements can be used in an array antenna. Arrays of two to four elements are typically used. Although a two-element array will produce an elliptical beam shape, the EOC gain performance will nevertheless increase relative to a single element, since the 3 dB increase in peak gain is sufficient to more than offset the increased gain roll-off (peak-to-edge) in the plane containing the two helix aperture centers (i.e., the plane of the narrowest beamwidth). A three-helix or four-helix array will increase both the peak gain and the EOC gain further, if that is required, at the expense of increased mass and cost.

Whereas single-helix antennas often use separate helices for Rx and for Tx, achieving a good Tx–Rx isolation through spatial separation alone, an array of helices will generally work for both Rx and Tx frequencies, due in part to the limited volume available on board the spacecraft. Each helix in the array will feature a diplexer to separate the Tx and Rx frequencies, and to provide sufficient isolation between the two signals. The risk of PIM in a common Tx/Rx array antenna is higher than in separate Tx and Rx helices, due to the absence of spatial isolation between Tx and Rx. Any PIM generated on the Tx side of a common Tx/Rx antenna is directly coupled into the Rx port. The design and manufacture of the diplexer at each helix are therefore critical in mitigating the risk of PIM in the array BFN.

12.2.4 Multipactor Threshold

At UHF, multipactor breakdown [1] can be a difficult issue. The power threshold for the occurrence of multipactor breakdown is a function of the internal dimensions of the susceptible RF components, and it is specifically a function of the width of the most critical gap within the component (often, but not always, the smallest gap where a significant field level occurs), expressed in terms of the RF free-space wavelength.

A lower gap/wavelength ratio means that multipactor breakdown will occur at lower operating voltages [1]. This is so since, as the gap width is reduced, a lower voltage is sufficient to generate a given electric field intensity, which in turn governs the acceleration of the electrons. Moreover, an increase in wavelength leads to a longer RF cycle, resulting in a longer time over which the free electrons are accelerated from one surface of the gap towards the opposite surface, and as a result the same electron arrival velocity (which determines the multipactor threshold) can be achieved with a lower voltage level.

As we scale the component dimensions from one frequency band to another, the gap/wavelength ratio would in principle stay approximately the same. However, the problem with UHF, as previously mentioned, is that waveguide-technology-based components are prohibitively large, as their internal dimensions are roughly proportional to the wavelength. Consequently, the different feed components are implemented using TEM or quasi-TEM transmission line technology, such as, for example, coaxial lines, square coaxial lines and striplines. In these types of component designs the dimensions in terms of RF wavelength become much smaller, and as a consequence the multipactor breakdown power threshold is proportionately lower.

A similar problem often occurs at L- and S-bands, where waveguide technology can also be too large to be used. However, the component dimensions tend to be limited by the available physical volume inside the spacecraft bus, thus limiting gap sizes independently of the wavelength. Therefore the gap/wavelength ratio will be approximately five times smaller at UHF than at L-band, since the wavelength is approximately five times longer. This leads to a 5-fold reduction in voltage threshold and therefore a 25-fold reduction in power threshold at UHF.

UHF diplexers are especially prone to multipactor breakdown because of the electric field intensity magnification that occurs inside the resonator cavities. Special techniques have been developed to prevent the occurrence of multipactor breakdown in UHF filters and diplexers, such as, for instance, using pressurized filters, using dielectric-loaded filters (e.g., loaded with quartz) and applying a high DC voltage at critical gap locations to prevent secondary electrons from escaping the surface.

12.3 L/S-band Mobile Satellite Communications Antennas

12.3.1 Introduction

As per the standard IEEE definition, L-band covers the frequency range between 1 and 2 GHz (15–30 cm wavelengths), whereas S-band goes from 2 to 4 GHz (7.5–15 cm wavelengths). Typical mobile satellite communications antennas may use frequencies around 1.52 to 1.56 GHz for L-band downlink, 1.62 to 1.66 GHz for L-band uplink, around 2 GHz for S-band uplink and around 2.2 GHz for S-band downlink.

12.3.2 The Need for Large Unfurlable Reflectors

Since mobile satellite communications systems (MSCS) provide service to (often small, inexpensive and lower performance) handheld terminals on the ground, and since a high link margin is desired to guarantee high availability of service, MSCS require the satellite antenna to generate high-gain spot beams. Since the wavelengths are relatively long, and using (12.3) and (12.4) of Section 12.1.1, these satellites are often implemented using very large-reflector antenna apertures. On transmission, the EIRP could in principle also be increased by boosting the Tx power level, so as to meet the high-capacity requirements of modern MSCS satellite communications. However, power is a very limited and expensive resource on board a satellite. The Tx RF power is also limited by the antenna due to thermal, multipactor and PIM considerations. L-band MSCS, in particular, usually have very stringent PIM requirements, with the typical occurrence of (potentially very strong) fifth-order PIM products with up to 100 W per carrier per feed element.

Figure 12.4 *INMARSAT-4 satellite with unfurlable reflector. Reproduced by permission of Inmarsat.*

Reflector projected aperture diameters can range from as low as 6 m to more than 22 m, and they are typically realized as unfurlable reflectors, made of a reflective metallic mesh surface tensioned to an accurate shape by a system of articulated rigid structural segments. Although there are other available technologies to achieve these aperture sizes, the deployable rib reflector (Harris [2]) and the perimeter-truss reflector (Northrop Grumman [3]) are the most commonly used designs. This type of reflector antenna is addressed in great detail by Chapter 8. An example of a satellite with a 12 m diameter unfurlable reflector antenna is shown in Figure 12.4.

The achievable surface is not exactly parabolic, since the deployed mesh is tensioned into its desired shape at a large, yet finite, number of attachment points, therefore there is usually some amount of faceting or pillowing and a number of attachment dimples which are inherent in the design and construction. These surface errors are periodic and lead to a structure of periodic sidelobes in the far field, which may be significant. Nevertheless, the achievable surface accuracy for these reflector sizes is very good, and it is consistent with meeting the MSCS requirements.

The large parabolic reflector is usually illuminated using an array of feed elements. The elements are designed so as to achieve nearly 100% radiation efficiency in the array environment (including mutual coupling effects). The requirements for these feed elements usually include high power handling capability and, since they are almost always grouped into sub-arrays by beam forming, tight inter-element amplitude and phase tracking over all operating conditions. Previous-generation systems typically used separate antennas for Tx and for Rx, thus mitigating the risk of PIM through a high spatial isolation. The high price of these large reflectors, as well as volume and mass limitations, have led most modern designs into shared Tx/Rx apertures. In these designs the separation and isolation between the Rx and Tx ports is achieved by a high-performance diplexer at each feed element. These designs are far more challenging than those having separate Tx and Rx antennas, especially in terms of PIM avoidance, due to the fact that PIM generated in a feed element can couple directly into its Rx port.

12.3.3 Beam Forming

If each feed array element were used to generate its own spot beam on the Earth's surface, these individual beams would generally have a suboptimal EOC gain, due in part to a relatively high spillover loss (energy radiated by the feed and lost beyond the edge of the reflector aperture). In order to form a large number of spot beams to cover the Earth, many elements are required in the feed array, and each spot beam is formed by exciting a sub-array of those elements [4,5]. The excitation of the elements of each sub-array needs to have appropriate relative amplitudes and phases so as to decrease spillover losses for each beam, and to effect any amount of beam shaping required to optimize performance. The amplitudes and phases of excitation within each sub-array are achieved by means of beam-forming networks (BFNs). Different BFN solutions have been implemented over the years. Typically, for L- and S-band systems, the impact of BFN losses and their power handling requirements are minimized by including a Tx high-power amplifier (HPA) and a Rx low-noise amplifier (LNA) after the diplexer of each feed element, and locating the BFN on the repeater side of the HPAs and LNAs (thus the BFN is a low-level BFN, i.e., a LLBFN). Traditionally, the BFN was realized with a large network of couplers, often implemented in multilayer printed circuit technology. This arrangement of coupler networks is commonly referred to as an analog BFN. In a system implementing analog beam forming, the number of transponder chains is driven by the number of spot beams being formed.

With modern systems like INMARSAT-4 and Thuraya, a dense grid of highly overlapping beams is generated in the digital domain, in a process denoted as digital BFN, or DBFN. An on-board processor performs the complex multiplications and additions required to generate the beam signals from the digitally quantized element signals. In a system employing digital beam forming, the number of transponder chains is driven by the number of feed elements. With this approach, hundreds of different beams can be formed using a relatively low number of elements [4], and coverage reconfigurability is easily implemented.

Recently, ground-based digital beam forming has also been implemented, thus trading lower satellite complexity for higher ground station complexity [6]. With this approach, the element signals are transmitted to, and received from, the ground stations using the backhaul link, and the beam-forming calculations are performed on the ground. Compared to on-board processing, this approach has the benefit of saving payload mass and power. As a disadvantage, a wider backhaul frequency bandwidth is usually required to accommodate the entire frequency spectrum transmitted and received by the feed elements.

Composite beam secondary patterns are formed by adding the constituent single-element secondary beam patterns, multiplied by the corresponding excitation coefficients (amplitudes and phases). It is important to use primary patterns that are calculated (or measured) in the feed array environment, with the other radiating elements appropriately terminated with matched loads, because mutual coupling will typically affect the performance significantly. The performance of each array element depends on its position within the feed array, since its mutual coupling environment is strongly affected by its location. When characterizing element performance by measurements, rather than measuring every single element, it is usually possible (and very advantageous for program cost and schedule) to exploit the symmetries in the array geometry so as to reduce the number of RF measurements required for an accurate characterization of the performance of the different composite beams. The optimization of the element excitation coefficients to form each composite beam is a sensitive and performance-critical process. It is often necessary, for best results, to take into account the actual reflector surface, including all deviations from the nominal parabolic shape such as faceting, pillowing, attachment dimples, errors due to manufacturing and thermal distortions.

If an analog BFN (typically using printed circuit technology) is employed, the antenna designer should run a statistical analysis of the different expected implementation errors present in the feed system and of their impact on the performance of the different composite beams. In this manner the expected degradation of critical performance parameters, as a function of amplitude and phase excitation errors associated with the various BFN implementation issues, can be determined. In a system using digital beam forming these errors can in principle be calibrated out during data processing. There will, however, remain some residual errors which can be significant. This is a rather complex issue that will not be treated in detail here.

12.3.4 Hybrid Matrix Power Amplification

As is well known, each feed element in a focal plane feed array has a one-to-one correspondence with a specific geographic location within the coverage area, and therefore high concentrations of traffic in some areas will result in high concentrations of Tx power in some feed elements and their associated HPAs, known as 'hot spots.' An approach known as 'hybrid matrix power amplification (HMPA)' or 'multi port amplification (MPA)' is generally used in the Tx beam former to address this problem. By using a HMPA to feed the elements of the feed array, the concentration of Tx power into 'hot spots' is spread out into many ports on the repeater side of the output hybrid matrix (similar to a Butler matrix [7]), thus drawing power from many more HPAs than if there were a single HPA at each feed element. With this approach, each HPA will tend to be driven at a much more even power level independently of the geographical distribution of traffic demand. Nevertheless, if a single HMPA is used, many feed elements are typically shared between adjacent spot beams, and this leads to coherent combination of complex voltages at the input ports of the Tx hybrid matrix. This causes significant variation in power levels. A way of circumventing this is to split the output hybrid matrix into interleaved sub-matrices (typically 4×4 or 8×8 matrices), such that the HPAs are effectively grouped into power pools [8,9]. By using similar matrices before the HPA amplification (input hybrid matrices) the input ports of those matrices correspond to the feed array elements and thus to geographic locations on the Earth's surface, where the Tx signals corresponding to each beam can then be input. However, it has been shown that the presence of the input matrices is not necessary, since the BFN excitations can be optimized to include their linear transformations at the inputs of the HPAs directly, and thus only the output HMPA matrices remain. This approach, illustrated in Figure 12.5, allows for power allocation flexibility between beams, and it can be combined with either analog or digital beam-forming architectures [10].

The constituent signals of a Tx composite beam will be spread over many HMPA matrices so as to draw power from the maximum possible number of power amplifiers. The same Tx signal should ideally not be fed to more than one port of any given power pool, since the split signals thus created will be coherent with each other and therefore will not translate into a uniform loading of the amplifiers. If, for instance, 10 elements per beam are used, a minimum of 10 power amplifier pools is recommended. An array using 120 elements, for example, could be connected to ten 12×12 matrices or fifteen 8×8 matrices.

Fewer larger-size HMPA matrices provide more power allocation flexibility than a larger number of smaller HMPA matrices. However, the resulting power handling requirements for the output matrix will also become more challenging, both in terms of average RF power handling (usually a thermal management concern) and of peak RF power handling (a multipactor breakdown issue).

A HMPA matrix needs to be designed such that the relative amplitudes and phases of the different RF signal paths remain close to the nominal design values, even when redundant amplifiers are switched in, over all

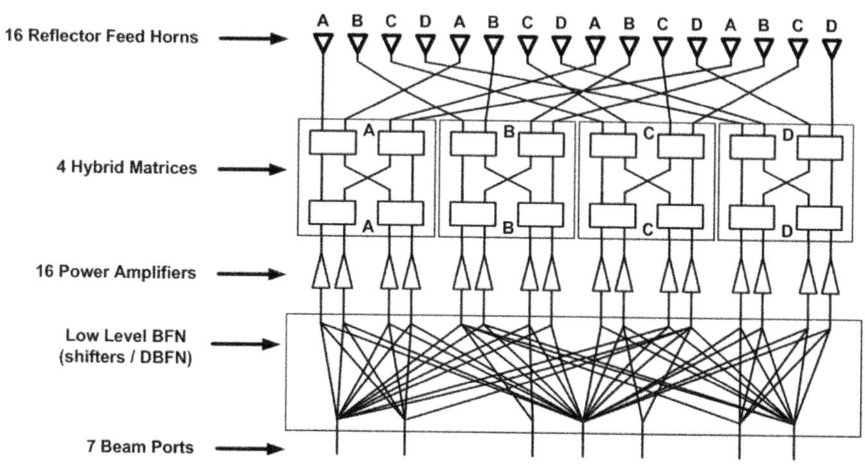

Figure 12.5 Multimatrix feed block diagram. Reproduced by permission of ESA; credit A. Roederer.

operating conditions, and for the lifetime of the mission. Relative insertion amplitude and phase errors between different active amplifier chains, as well as implementation errors in the matrix itself, will lead to degraded isolation between the different paths and to the unplanned excitation of undesired feed radiating elements. This needs to be carefully analyzed, since it will likely impact the interference between beams (measured by the carrier-to-interference ratio, *C/I*) and other performance parameters.

12.3.5 Feed Array Element Design

Usually a single radiating element is designed at first, so as to achieve the desired gain and a good impedance match over the required bandwidth. The radiating elements are then analyzed together in the feed array environment so as to account for mutual coupling effects. Several design parameters, such as cup height and diameter, as well as radiator geometry details, are then optimized. Finite array RF modeling (as opposed to the usually much faster RF modeling of an infinite periodic array) is often necessary, so as to properly consider the variations in performance of the element depending on its location in the feed array. The designer may resort to having slightly different geometries for elements located near the center and near the edge of the array.

Many systems use single circular polarization, in which case a helix is often a good choice of radiator type, achieving high values of aperture efficiency in the feed array grid, thus high element gain. Helices are also capable of sufficiently wide-band performance to support both the Tx and Rx frequency sub-bands. An example of such a feed array is shown in Figure 12.6. More recently, systems appear to be trending towards the use of dual polarization, in which case helices are unsuitable as the radiating element choice. Instead, other radiating element types with the capability of supporting dual circular polarization are needed, such as, among others, cupped patches, cupped crossed dipoles, cup horns [11] and scalar horns.

12.3.6 Diplexers

In shared Tx/Rx antenna designs, which are now increasingly the standard in MSCS, a diplexer must be placed at each radiating element in order to separate the Rx and Tx signals with sufficient isolation between them. The stringent requirements placed on the filtering characteristics of the diplexer and on its level of Rx/Tx isolation make this a highly critical component, whose properly optimized design is one of the most important elements

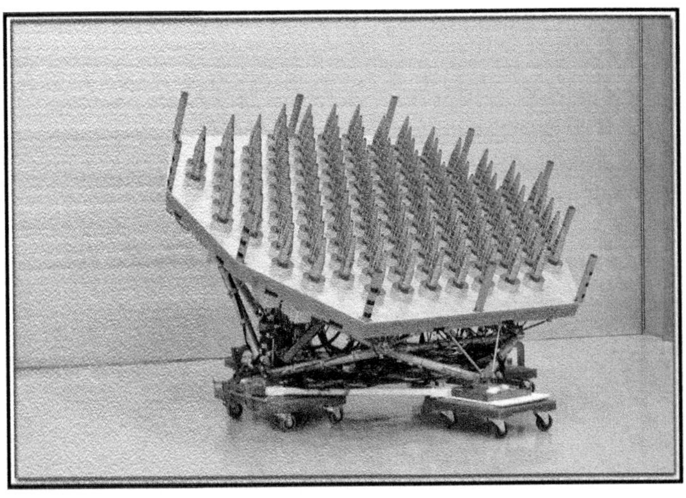

Figure 12.6 *Helix feed array. Reproduced by permission of MDA.*

in the realization of this type of satellite antenna. Temperature compensation is often needed in order to meet the extremely demanding dimensional stability requirements of the diplexer design. PIM mitigation is one of the most important objectives of an optimal diplexer design, because of the close proximity between the Rx and Tx signal paths within the device. Design details (such as either avoiding metal-to-metal contact or using high-pressure metal-to-metal contact) and manufacturing processes that minimize the risk of strong PIM formation in the Tx portion of the diplexer are essential. Dielectric materials are usually avoided in the diplexer design due to the high Tx power levels present, and because their eventual degradation with exposure to the space environment will have an unacceptably adverse impact on the diplexer performance.

12.3.7 Range Measurements

Testing the RF radiating performance of these very large reflector antennas is generally a challenging proposition. This is due to the sheer size of the deployed aperture that exceeds the capabilities of many available RF test facilities. Moreover, these unfurlable reflector structures are designed for optimal surface shape in the zero-*g* space environment and will exhibit non-negligible distortions when deployed in the 1*g* conditions of the test range. Using embedded (i.e., in the array environment) element patterns and BFN excitation coefficients, in conjunction with an accurate reflector surface shape characterization (as is expected to occur in the zero-*g* environment of space), it is possible to rely exclusively on computer predictions for the RF performance characterization of the reflector antenna. The antenna designer should use an embedded element pattern characterization (obtained through analysis or RF test) which includes sufficiently accurate radiated field phase and amplitude information at the location of the reflector surface. This is an important consideration, since using the (asymptotic) far-field characterization of the radiated primary pattern of each embedded feed element will not yield reliable results in most cases. This is due to the fact that the reflector surface is often not sufficiently far to be considered to lie in the asymptotic far-field region of the feed sub-arrays (loosely defined as greater than $2D^2/\lambda$, D being the feed sub-array aperture dimension and λ the wavelength). Linear superposition of the secondary elementary beams that participate in a given composite beam shape, weighted by their respective complex excitation coefficients, usually leads to accurate RF performance predictions for the composite beam. If the elementary beam patterns cannot be measured (because the single-element access ports are not accessible, for instance) a near-field characterization or spherical wave expansion representation of the primary field of each composite beam will generally produce good results.

The prediction model can be validated using an RF range measurement, where the profile of the reflector surface, which is distorted by gravity, is accurately measured (by photogrammetry, for example) and used in the computations instead of the in-orbit surface profile. This serves as a validation of the software prediction model. Since the surface shape is distorted by the 1*g* environment, it will most likely not be an accurate measure of the in-orbit performance, which will then need to be obtained using the now-validated software model and the (predicted or otherwise known) undistorted zero-*g* reflector surface shape [12].

12.4 C-, Ku- and Ka-band FSS/BSS Antennas

12.4.1 Typical Requirements and Solutions

Fixed Satellite Services (FSS) and Broadcasting Satellite Services (BSS) are partly overlapping definitions that cover a variety of communications services between Earth stations and mostly fixed (commercial or personal) terminals on a well-defined geographical region of the Earth. They are normally used for broadcasting to and from television networks and local stations. They can also be used for video-conferencing, broadband Internet and general commercial telecommunications.

C-band is broadly defined (IEEE definition) to cover frequencies from 4 to 8 GHz, Ku-band from 12 to 18 GHz, and K-band and Ka-band together (often referred to simply as Ka-band) from 18 to 40 GHz. For FSS and BSS satellite communications, C-band typically uses frequencies in the 3.5–4.2 GHz range for the downlink and in the 5.5–6.5 GHz range for the uplink, Ku-band usually uses frequencies in the 10.7–11.7 GHz range for the downlink and in the 13–14.5 GHz range for the uplink (and in some cases the uplink uses instead the 17.3–18 GHz range), and Ka-band usually uses frequencies in the 19.5–21.2 GHz range for the downlink and in the 29.5–31.2 GHz range for the uplink. Specific cases may vary from these ranges, within the International Telecommunication Union (ITU) frequency band allocations. Typical bandwidths range from 250 MHz to 1 GHz and, in more recent Ka-band broadband systems, up to 1.5 GHz.

These antennas cover a wide spectrum of frequency bands and applications, but they share many common characteristics that justify assembling them together in one single category. They are by far the largest group among all the commercial satellite communications antennas in operation today, and also among all the missions currently known to be in the planning stages. It is common for FSS/BSS satellites to carry a multitude of these antennas, usually serving different frequency bands. An example of that is shown in Figure 12.7, displaying three deck-mounted Ka-band reflector antennas and two side-mounted Ku-band antennas.

Common reflector antenna configurations for FSS and BSS missions include center-fed reflectors, single offset reflectors, dual-gridded reflectors (DGRs), dual-reflector geometries, steerable reflector antennas, and others.

12.4.2 The Shaped-Reflector Technology

These antennas usually form contoured beams from a geostationary orbit, focusing RF energy onto certain geographical regions while being isolated from other regions as required by the ITU regulations and agreements. Until the late 1980s, this was achieved with multiple feed elements combined with an analog feed network to generate the specified composite contoured beams. The reflector in these cases was simply an offset parabolic dish. Since the 1990s this approach has been replaced in almost every instance by shaped-reflector technology, whereby the energy radiated by a single feed element (usually a high-performance horn) is shaped into a contoured beam that is focused onto the desired coverage area by appropriate reflector shape optimization.

Figure 12.7 *FSS antenna farm. Reproduced by permission of Astrium.*

The reflector shaping process typically starts with a slightly defocused parabolic reflector (beginning with either a divergent-ray or a convergent-ray defocusing) and the surface shape is then finely optimized until the contoured beam closely covers the intended area while suppressing the energy from the isolation regions. The absence of a feed network is a great simplification of the antenna design when compared to the array-fed parabolic reflectors of the past, resulting in great mass and cost savings.

Dual-reflector geometries, such as Cassegrain or Gregorian, are also frequently used, in which case both the main reflector and the subreflector can be shaped during the optimization process.

12.4.3 Power Handling

These satellite systems usually provide FSS or BSS to a relatively large geographical area, thus producing only a moderately high gain. In order to achieve the required high communications capacity the EIRP must be above a minimum threshold, and since the Tx antenna gain is limited by the large coverage area, high Tx RF power levels need to be generated and transmitted. The power handling capability required of these antennas can be a very challenging design constraint, and the usual limiting factors are:

- PIM avoidance (RF power levels per carrier often higher than 150 W and very strong third-order PIM products are common, together with not-to-exceed PIM-level requirements which are often as low as -145 dBm, i.e., 3×10^{-18} W);
- prevention of multipactor breakdown (often a large number of high-power carriers lead to the need for multipactor threshold powers of tens of kilowatts in order to ensure a sufficiently large margin);
- thermal management challenges caused by high total average RF power levels being fed to feed assemblies having non-negligible RF losses.

The analysis of multipactor breakdown [1] for many simultaneous RF carriers has been refined in recent years so as to be less conservative and thus greatly enhance agreement with experience. The peak power was traditionally calculated as N^2P (where N is the number of RF carriers and P is the power in each carrier, thus implying the very pessimistic assumption that the N signals combine by adding their voltages coherently and with the same phases). This estimation method would predict insufficient margin even for a full-height straight rectangular waveguide for many modern payloads, which is in strong disagreement with experience. Nowadays, instead of N^2P, the RF power contributing to multipactor is generally calculated through much more rigorous statistical analysis of the electron avalanche phenomenon. This power level is thus assumed to be the peak RF instantaneous power that is sustained over a pre-specified multiple of the time it takes an electron to cross from one side of a critical gap to the other. The accepted multiples typically vary between 4 and 20, where a multiple of 20 (denoted as the P20 power level) corresponds to a longer time and therefore a lower sustained peak power level than a multiple of 4 (denoted as P4). These stringent multipactor breakdown constraints are critical inside the Tx feed chain components, and in particular in regions where the field amplitude is locally magnified by resonances or other electromagnetic effects, which can be properly calculated by a suitable full-wave software simulator (e.g., using the finite element method or the finite difference time-domain method).

12.4.4 Antenna Structures and Reflectors

Spacecraft and their payload are subjected to a rather extreme environment, both during launch and during the in-orbit mission. Before launch, the spacecraft and the different payload equipment are subjected, at different levels of assembly, to severe qualification and/or acceptance environmental test campaigns, in order to demonstrate that they will survive and perform in accordance with the requirements throughout the mission.

Therefore, all payload subsystems, including the antenna, must be designed to survive without degradation the environments of the test campaign, the launch and the in-orbit operation. The analysis of the different subsystems is performed in isolation and, when required, it is also performed at higher levels of assembly, often including the whole assembled spacecraft in the structural and thermal computer models.

Solid reflector antennas for FSS/BSS can be quite large (up to 3 m diameter) and need to meet very challenging launch and in-orbit environmental requirements. The mechanical design optimization must be accomplished without exceeding the typically very stringent mass allocation, which poses a significant challenge to the mechanical antenna designer. Mass reduction is an ever-present objective when designing a spacecraft antenna, and a high stiffness-to-mass ratio is a very desirable characteristic of reflector designs. Thermoelastic dimensional stability is another characteristic of a good reflector antenna design, since significant RF performance degradation and excessive pointing errors can result from poorly optimized antenna structural and thermal designs. Given these constraints, solid reflector shells are most often constructed from composite materials such as CFRPs, which are the focus of much mechanical and manufacturing research and development effort across the industry. Most often CFRP reflectors do not require metallic coating, since their conductivity is sufficiently high at these frequencies to represent only a relatively small item in the antenna RF loss budget. If especially low loss is needed, as may be the case for some applications such as radiometer antennas for example, then very high-conductivity metallic plating may be applied to the reflecting surface. Another reason for implementing surface plating is to eliminate the surface anisotropy caused by the layup of the CFRP fibers at the reflecting surface, which can generate significant field depolarization, with the resulting increase in cross-polarized levels. A better CFRP layup design, one that manages to orient the surface fibers along a more favorable direction or otherwise reduce the surface anisotropy effects, may be an alternative to metallic plating, and the selection of the preferred solution is often the result of a tradeoff involving both technical and financial considerations.

For small reflector diameters (less than 1 m) a solid aluminum construction is also sometimes used, and its selection usually results from a tradeoff against CFRP-type materials with detailed consideration of mass, thermal distortions, structural stiffness and cost impacts.

For DGRs, also known as dual-shell reflectors, the front shell is often made of Kevlar or similar RF-transparent material, and the rear shell is made of CFRP. Usually, the two shells are kept together at the precise required spacing by intercostal rings, ribs or posts, often made of Kevlar or equivalent. These devices stiffen the two-shell assembly while keeping a low mass. DGR structures are addressed in further detail in Section 12.4.5.5 below.

12.4.5 Reflector Antenna Geometries

12.4.5.1 Introduction

Many different antenna configuration choices are available to the design engineer. On a typical geostationary spacecraft platform, antennas can be side mounted and thus accommodated on the side panels of the spacecraft (usually the east and west panels, but occasionally also the south and north panels), or deck mounted (and thus mounted on the top deck, which is the nadir-facing panel of the spacecraft).

Side-mounted deployable reflector antennas tend to feature the largest reflector diameters and the longest focal lengths. For launch the reflector is stowed, and rigidly supported by hold-down and release mechanisms (HRMs), against the respective spacecraft side panel. Once in orbit, the HRMs are released and the reflector is deployed by means of a deployment mechanism. This mechanism brings the reflecting surface to its proper positioning and alignment relative to the fixed feed horn and assembly, and stably supports the reflector in that position throughout the mission duration.

Deck-mounted reflector antennas usually have smaller diameters and shorter focal lengths. Most often, the feed assembly, main reflector and (when applicable) subreflector are supported and aligned in place by a rigid structural tower, which is often shared by several deck-mounted reflector antennas (different antennas attached

Figure 12.8 Deck-mounted antenna farm on the test range. Reproduced by permission of MDA.

to different sides of the tower). A deck-mounted antenna farm of Ku- and Ka-band antennas is shown (on the test range) in Figure 12.8.

The selection of the final antenna configuration is the result of a multidisciplinary iterative tradeoff, and it is usually based on RF performance, mechanical evaluation, structural and thermal considerations, reliability and risk assessment, program schedule and cost. The RF performance parameters, calculated under the worst predicted distortion, misalignment and pointing error conditions, include gain (minimum gain across the coverage region as well as peak gain), isolation (gain levels on the predefined isolation areas on the ground) and cross-polarization discrimination (known as XPD, and calculated as the co-polarized gain minus the cross-polarized gain in dB). Usually the antenna RF performance also includes gain variation with time and within the frequency band, return loss at the antenna access ports, isolation between the different antenna access ports (between co-polarized ports, between cross-polarized ports and between Rx and Tx ports), group delay and its variation within each channel, and so on. For mechanical evaluation, important parameters are, among others, total mass, mass distribution, volume constraints, and structural and thermal behavior.

12.4.5.2 Cross-polarization Generation in Reflector Antennas

Before addressing the different reflector antenna configurations used in FSS, it is useful to understand some of the mechanisms that lead to cross-polarization in reflector antennas.

Cross-polarization (usually expressed as the cross-polar discrimination (XPD) or the cross-polar isolation (XPI)) in a linearly polarized reflector antenna is caused by several factors [13]: feed horn cross-polarization, reflector surface curvature, feed horn location away from the focal point, and inclination angle of the feed horn relative to the parabolic main axis. Even a perfectly polarized feed horn will result in cross-polarization generation in the reflector secondary pattern due to the other three factors. In fact, a controlled amount of feed horn cross-polarization can be used to counter the other effects and lead to a better XPD performance in the secondary far field (which can be done over a narrow frequency band, for example, by adding other modes, like the TE_{12} mode, to the TE_{11} nominal horn aperture field). This design strategy is not straightforward and is only rarely used. A perfectly polarized feed horn oriented along the main axis would lead to a perfectly polarized secondary field if the reflector were an infinite plane. The reflector curvature will, however, distort the surface current lines (except in the principal planes) and lead to cross-polarization. Nevertheless, for a center-fed reflector, the cross-polarization level is zero in the two principal planes provided that such is also the case for the feed horn primary pattern. That symmetry keeps cross-polarized levels generally low for all angles typical of coverage from geostationary orbit. Moving the feed horn away from the focal point introduces asymmetry, both in the illumination of the reflector by the horn and in the effects of the reflector curvature, such that the perfectly polarized principal planes no longer occur, and that leads to higher cross-polarized field levels.

For linearly polarized single-offset geometries, the XPD performance is further deteriorated by the inclination angle of the feed horn relative to the parabolic main axis, which is necessary in order to point the horn towards the central portion of the reflector aperture (thus minimizing spillover losses). The offset configuration eliminates the symmetry along the plane of offset as well as the occurrence of zero cross-polarization in the principal plane perpendicular to the offset plane, and leads to higher levels of cross-polarization. As shown in Figure 12.9, whereas a center-fed geometry benefits from linear cross-polarization cancellation due to both left–right symmetry and top–bottom symmetry of the currents induced on the reflector, a single-offset geometry preserves the left–right symmetry relative to the plane of offset but it does not benefit from top–bottom symmetry along that plane. The absence of linear cross-polarization cancellation along the offset plane leads to a net resulting cross-polarized component, thus degrading XPD. Since a single-offset geometry preserves the left–right symmetry about the plane of offset, and thus preserves a zero cross-polarization principal plane parallel to the offset plane, the antenna engineer will, whenever possible, try to align the offset plane with the longest dimension of the coverage area in order to maximize XPD performance over the coverage.

It should be noted that XPD performance in a linearly polarized antenna is a rather complex issue. In addition to the factors described above, there is also the fact that the parabolic axis of the antenna is usually oriented towards a point on the Earth selected to optimize performance for a particular coverage region, designated as boresight, and only very rarely does this coincide with the coordinate system used by the customer to define the linear polarization convention. The direction of linear polarization is usually defined as per Ludwig's third definition [14], although sometimes the second definition is also used, with the horizontal polarization direction often defined to be parallel to the equatorial plane. It is crucial that the antenna engineer designs and optimizes the antenna for the specific polarization definition and coordinate system specified by the customer, since failure to do so will introduce an artificial XPD degradation due to the misalignment between the two polarization conventions.

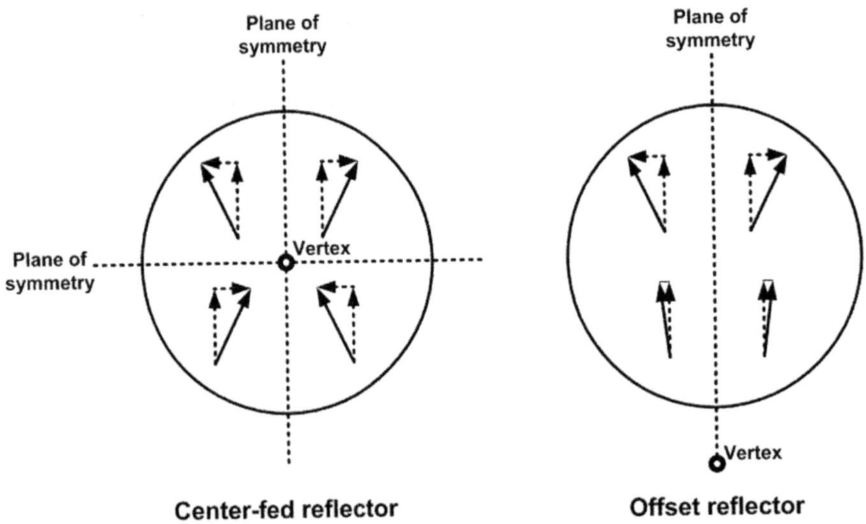

Figure 12.9 *Generation of cross-polarization in reflector antennas.*

In linearly polarized reflector antennas, the cross-polarization cancellation across a plane of symmetry results from the fact that the surface current lines undergo a clockwise rotation relative to perfectly polarized fields on one side of the plane of symmetry, and a counter-clockwise rotation on the other side of that plane (Figure 12.9). For a circularly polarized reflector antenna, however, a rotation of the current lines does not represent the introduction of a cross-polarized component. Instead, it represents a phase shift, which is either positive or negative depending on whether the polarization is right-hand or left-hand, and on the sense of rotation of the current lines. Consequently, rotation of the current lines in opposite senses on the two sides of a plane of symmetry results in a beam squint [13] in a direction perpendicular to the plane of symmetry. For a center-fed circularly polarized geometry the beam squints cancel each other across the four quadrants formed by the two planes of symmetry and thus result in no net beam squint. In a single-offset circularly polarized reflector antenna, however, only one plane of symmetry is preserved (the plane of offset), and therefore there will be a net beam squint effect perpendicular to the plane of offset. This means that, for a single-offset reflector, a right-hand circularly polarized (RHCP) field will undergo, in the far field of the reflector antenna, a small angular beam squint relative to its intended position (in a plane perpendicular to the plane of reflector offset), whereas a left-hand circularly polarized (LHCP) field will suffer a similar angular beam squint, but in the opposite direction. Consequently, for a dual circularly polarized antenna, the two hands of circular polarization (CP) will squint in opposite directions, and the minimum gain performance over a fixed ground coverage region will be slightly decreased in the process. This happens because the antenna optimization will attempt to maximize the worst minimum gain among the two polarizations, in a compromise solution that will always decrease the overall minimum gain.

An additional contributor to cross-polarized levels is the diffraction of the electric currents by the edge of the reflector aperture, where the total current is forced to be predominantly along the edge. This results in reflected currents into the reflector surface that may have a significant cross-polarized content. This is usually more of a concern for large angular coverage areas than for small area spots, and it is an effect that is neglected by the basic physical optics (PO) calculations. It is therefore important, in cases for which this contribution is significant, to account properly for these edge reflections, which can be done, for example, by adding physical theory of diffraction (PTD) terms to the basic PO calculations [15,16]. This edge diffraction affects not only single-offset reflector antennas, but also Gregorian reflector antennas and center-fed reflector antennas.

Other potentially important contributors to cross-polarization are scattering effects, such as reflections of the horn primary pattern and the reflector radiated near fields by neighboring structures including the adjacent spacecraft panels. The impact on XPD performance can be especially strong for circularly polarized antennas, since specular reflections by large surfaces like the spacecraft panels will convert right-hand CP to left-hand CP and vice versa.

Scattering by support struts and feed lines, especially in center-fed reflector antennas, can also add to cross-polarization levels. These effects should be quantified, by analysis or test or both, since they can potentially degrade performance significantly. Some design strategies have evolved to mitigate the problem, such as orienting the struts along the main polarization direction for linear polarization (LP), or laying the struts out in a rotationally symmetric configuration for circularly polarized antennas.

12.4.5.3 Single-Offset Reflectors

These are single-reflector antennas for which the center of the projected aperture is located away from the axis of the parabola. Almost in all cases, the offset of the aperture center relative to the axis of the parabola is sufficiently large that the axis does not intersect the projected aperture at all, and in fact leaves sufficient clearance between the aperture and the axis to avoid blockage of the main radiation beam by the feed assembly. In this case, the feed horn or horn array is pointed towards the central portion of the reflector, at an angle relative to the axis of the parabola. A single-offset reflector antenna is often a relatively low-cost configuration, and

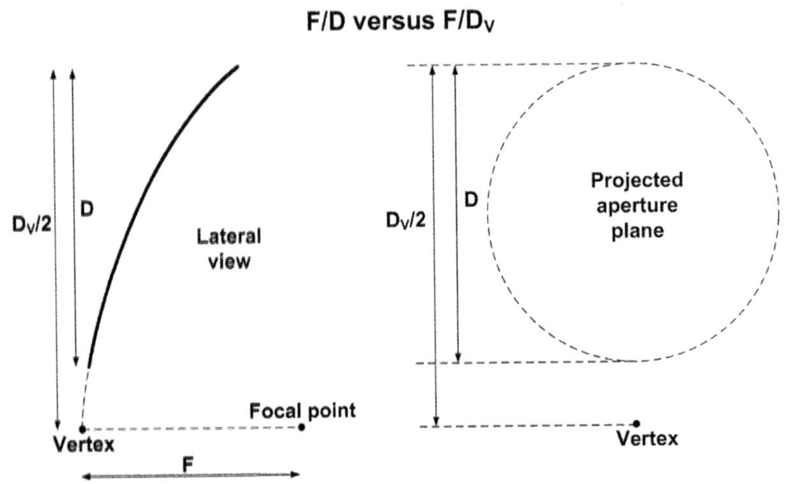

Figure 12.10 *F/D versus F/D_V as a figure of merit for single-offset reflectors.*

tends to be used if at all possible given the performance requirements as well as the mechanical constraints applicable to a particular mission.

Linearly polarized single-offset reflector antennas can exhibit relatively high cross-polarization levels, whereas circularly polarized single-offset reflector antennas will be affected by beam squint, as was described in Section 12.4.5.2. These effects are greater as the inclination angle of the feed and the reflector surface curvature increase, but they can be somewhat decreased by design strategies such as the following:

- Increasing the F/D ratio, that is, the focal length over the aperture diameter. In an offset geometry this parameter would be better defined as F/D_V [13], where D_V is the virtual aperture diameter defined as twice the distance between the axis of the parabola and the reflector point that is the furthest from it, as shown in Figure 12.10. Maximizing F/D (or even better F/D_V) minimizes the cross-polarization levels in linearly polarized antennas and the beam squint in circularly polarized antennas, since in fact it is closely related to minimizing the feed inclination angle and the surface curvature.
- Using a divergent initial step in the reflector surface shape optimization, which results in a final optimized shape that is less concave than the initial parabolic shape. A convergent design would instead yield a final shape that is more concave than the initial parabolic reflector. A divergent design will decrease the severity of the linear cross-polarization generation effects, since it decreases the curvature of the reflector surface. As a disadvantage, it also results in reflected near fields that extend over a significantly greater angular volume than if a convergent design were used. It can therefore cause much stronger fields reflected back onto the feed horn aperture, forcing an increase of the reflector offset in order to control that effect. Scattering by the spacecraft and other nearby instruments could potentially also be greater. In fact, some customers impose a convergent design from the start in the antenna requirements specification.

12.4.5.4 *Gregorian Dual-Reflector Geometry*

Dual-reflector offset antennas have a parabolic main reflector and a hyperbolic (for Cassegrain geometry) or elliptical (for Gregorian geometry) subreflector. Between these two dual-reflector offset geometries, Gregorian designs are by far the most commonly used in spacecraft, due to the fact that they can be designed to greatly suppress cross-polarization and that, for a required projected aperture size, they lend themselves to accommodation within a more compact volume. The basic Gregorian geometry starts with a parabolic main reflector and an elliptical subreflector, such that the focal point of the main reflector coincides with one of the foci of the subreflector, and the feed horn aperture is located on the other focal point of the subreflector. By appropriate selection of the angles between the feed horn axis, the main reflector axis and the elliptical axis (the line connecting its two foci), the Gregorian geometry can be designed to be optically compensated (abiding by the Mitzuguchi condition [17]). In that manner the net electric field rotation that degrades the single-offset design performance (Section 12.4.5.2) can be greatly reduced. As a consequence, a well-designed Gregorian antenna offers better XPD performance than a single-offset antenna for LP applications, and it can be designed to mostly cancel beam squint in CP applications.

These antennas clearly tend to be more expensive than single-offset reflector antennas, because of the additional mechanical assembly comprising the subreflector, which must be properly supported and aligned for performance optimization. This implies more analysis, more support structure design and construction, and more CFRP assemblies, which are all antenna cost drivers. The advantages, besides providing better control of the cross-polarization and/or beam squint effects described previously for a single-offset antenna, include also a slightly enhanced minimum gain performance for a given main reflector aperture size. This can be achieved by shaping the surfaces of both the main reflector and the subreflector, providing more degrees of freedom in controlling the main aperture illumination function and the spillover losses. A difficulty in analyzing the RF performance of Gregorian antennas is that the subreflector is often in, or close to, the near-field region of the feed horn, and thus the far-field pattern of the horn is inadequate in accurately computing the electromagnetic field on the subreflector surface. This introduces some inaccuracy in the calculations, even if the horn radiated field is characterized using a greater number of modes in its description by (as is typically done) a spherical wave expansion. In fact, multiple reflections of the radiated field between the subreflector and the horn itself may occur, and they must be avoided or at least minimized if at all possible through adequate design, since they can degrade performance significantly. This is especially critical if the subreflector is shaped and deviates significantly from its elliptical base. Multiple reflections can also occur between the main reflector and the subreflector, and they sometimes can be contained by increasing the main reflector offset thus decreasing scattering by the subreflector, at the expense of a larger volume occupied by the antenna.

A Gregorian geometry can be rather compact, partly because the fields reflected by the (concave surface of the) subreflector always converge onto the region surrounding the main reflector focal point, and in so doing they create a niche volume with very low field intensity levels that allows for a close-in feed horn placement. By way of contrast, in a Cassegrain geometry the fields reflected by the (convex surface of the) hyperbolic subreflector always diverge to illuminate the main reflector, leaving no volume with low field intensity where the feed horn can be located in a close-in position.

Although Gregorian geometries can be used in a side-mounted configuration, the generally larger dimensions of these antennas lead to relatively large subreflectors, which sometimes also need to be deployable, with the accompanying mass and cost penalty. The most common use of Gregorian geometries is for deck-mounted antennas. In that case an additional advantage over single-offset geometries is that the feed horn and assembly are located closer to the deck, since it is the subreflector, instead of the feed assembly, that is located at the end of the structural support tower. This results in a significant reduction in the waveguide lengths needed to make the RF connection between the feed and the spacecraft, thus minimizing RF losses and enhancing gain.

One concern with Gregorian antennas is that any horn energy spillover around the subreflector rim will potentially hit the Earth and may affect isolation with other regions. This is unlike the energy spillover around the main reflector or in a single-offset reflector antenna, which either radiates into deep space or is scattered by the spacecraft panels over a much wider angle relative to the main beam, and therefore does not often cause isolation problems. Because of this factor, the subreflector of a Gregorian antenna is usually slightly oversized

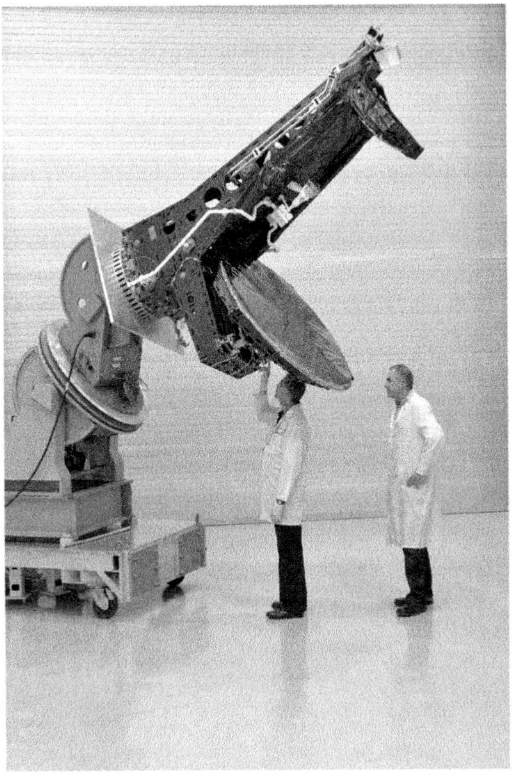

Figure 12.11 Deck-mounted steerable Ku-band Gregorian antenna. Reproduced by permission of MDA.

to minimize spillover, and the designer must then design the antenna so as to minimize subreflector blockage that might arise due to subreflector oversizing.

Figure 12.11 shows a deck-mounted steerable Ku-band Gregorian dual-reflector antenna during integration.

12.4.5.5 Dual-Gridded Reflectors (DGRs)

This type of reflector antenna is used for dual-linear polarization [18]. The DGR assembly comprises two parabolic or shaped-parabolic shells, which are precisely separated, aligned and stiffened by intercostal rings and, when necessary, also by ribs or posts. The front shell, usually constructed from Kevlar (or equivalent), is transparent to one of the LPs and reflects the other LP by way of a metallic grid oriented along the polarization that is to be reflected. The other polarization uses the rear shell (which usually does not have a metallic grid and is often made of CFRP, thus not transparent to RF) and is mostly unaffected by the presence of the transparent front shell. The rear and front shells have separate parabola main axes and separately located focal points, such that the two feed assemblies (one feed chain for each polarization) do not interfere physically with each other.

The DGR antenna consists in effect of two single-offset reflector antennas that share the same volume and aperture, and therefore it uses less spacecraft real estate than would two separate antennas. Being of a single-offset geometry, as was discussed previously, it would in principle suffer from relatively poor XPD performance in either one of the LP polarizations. However, for the polarization that uses the front shell, the metallic grid on the front reflector shell only reflects the co-polarized component of the impinging field, while the cross-polarized field goes through and is reflected away from the coverage area by the rear shell, which is out of focus. For the polarization that uses the rear shell, the field propagates twice through the gridded front shell, and that is the mechanism through which cross-polarized field components are filtered out. The XPD performance of a DGR antenna is usually very good, and in fact it is better than that of a Gregorian-type antenna. The mechanical build of the dual-shell assembly is, however, a rather specialized manufacturing job, and it leads to a relatively high price. In addition, scattering effects as the field corresponding to the rear shell polarization propagates through the intercostal structures can deteriorate performance and make accurate RF analysis more challenging.

The effect of the intercostal structure on the antenna RF performance is sometimes sufficiently large to have a visible impact, and a proper RF analysis of the scattering effects needs to be performed [19]. Mechanical design and construction can sometimes be modified in order to reduce the RF scattering effects of the intercostal structures, by minimizing their quantity (as a tradeoff against the reflector's structural properties), by altering their profile so as to reduce their radar cross-section, by proper selection of materials and surface layer finishes, and even by incorporation of bandgap structures (e.g., periodic metallic patterns designed so as to partly suppress scattering of RF energy onto the main beam region).

Another effect that needs to be properly analyzed in DGR antennas is the spurious reflection of the fields polarized in one LP direction by the reflector shell corresponding to the other polarization. This can be an important contributor to degrading performance, and it must be controlled by proper design. Using the fact that the two focal points are spatially separated, the field radiated by each feed assembly is usually significantly out of focus when reflected by the shell of the other polarization, and therefore the distance between the two foci and the angular separation between the two feed assemblies are design parameters that can be used to mitigate this effect. In addition, it is necessary to ensure that the field radiated by one feed horn is not scattered significantly by the feed horn and feed chain of the other polarization. In each case, the precise geometry of a DGR is the object of an extensive tradeoff, aimed precisely at minimizing these undesirable effects on performance. An important design parameter in a DGR antenna geometry is the rotation of the offset plane of one shell relative to the offset plane of the other shell. This is a parameter also used in optimizing the geometry to further separate the two focal points and to decrease the deleterious effects mentioned above.

12.4.5.6 Center-Fed Reflector

This is a single reflector for which the projected aperture (i.e., projected onto a plane perpendicular to the main axis of the parabola) is centered on the axis of the parabola. In this case, the feed horn or horn array is pointed to the center of the reflector along the axis of the parabola. Center-fed reflector antennas suffer from feed blockage and scattering effects, and as such are only used on spacecraft in special situations for which these effects are acceptable. Besides the feed assembly itself, the struts used to support the feed assembly and the waveguide runs used to make the RF connection between the feed and the spacecraft also contribute to blockage and scattering. These effects tend to decrease gain and increase sidelobe and cross-polarization levels. Center-fed single-reflector antennas would, however, have intrinsically low cross-polarization and sidelobe levels. Therefore the degradation introduced by the feed assembly and its support struts and feed waveguides will subtract from a very good starting performance level and will often result in good overall performance once all degradations are accounted for.

One way to minimize the reflection back into the feed horn and its scattering effects is to cut a hole in the center of the main reflector, thus minimizing the field that would be incident on the horn. Alternatively, a shaped bump or dimple can be used to deflect the reflected field away from the horn aperture. Another means of achieving this is to consider instead a center-fed dual-reflector geometry (usually of the Cassegrain type) for which the subreflector is especially shaped so as to minimize the blockage effects. This dual-reflector geometry also has the advantage of having the feed horn near the spacecraft, thus minimizing feed waveguide length (hence the RF losses) and eliminating blockage and scattering by that waveguide run.

RF modeling of the effect of the struts and feed waveguides on RF performance has been the object of much research effort around the world, and several methods of analysis are now available to the design engineer. The PO method is mostly unsuited to the task, since the cross-sectional dimensions of the struts are usually smaller than one wavelength. One successful analysis approach uses the method of moments. Additionally, R&D continues on methods to minimize the scattering effects of struts, leading to rather innovative strut designs using an optimized cross-sectional profile combined with a periodic metallization pattern aimed at reducing their radar cross-section [20].

In addition to affecting gain, sidelobe levels and XPD performance, the field scattered by the support struts combines with the main reflector field to form an interference pattern that is highly frequency dependent. Therefore, one parameter that is often affected is the gain variation over the frequency bandwidth, and specific analysis is often necessary to assess this performance.

12.4.5.7 Steerable Reflector Antennas

Steerable antennas for FSS/BSS are often needed to generate a steerable coverage on the ground, in response to traffic demand or as part of a planned coverage reconfiguration. Their implementation can take different forms, with varying degrees of difficulty and cost.

One possible realization of a steerable antenna is to mount the complete antenna (reflector plus feed assembly) on a two-axis gimbal mechanism. The beam stays focused and the performance is nominal and constant throughout the required beam steering range. A flexible RF connection is required at the gimbal location, and it may take the form of flexible waveguides, flexible coaxial cables, cable wraps or rotary joints, the selection of which depends on the frequency of the steering operation and the required angular range of motion. Moreover, in order to accommodate rotation of the antenna about two orthogonal axes, two flexible RF connections are usually needed, each one providing flexibility for rotation around one of the axes. Releasable lock mechanisms are also typically implemented to hold down the antenna structure during launch. The two-axis gimbals, the launch locks, as well as the dual flexible RF connections, make this implementation of a steerable reflector antenna the most expensive.

A less expensive and less risky option is to keep the feed assembly fixed and rotate only the reflector shell [18]. This avoids the need for a flexible RF connection to the spacecraft, a well-recognized element of risk and cost in space antenna applications. Additionally, if only the reflector shell is moved, the gimbal mechanism carries less mass and may be a smaller and less expensive model than if the whole antenna is being steered, an additional factor saving cost and mass. In many cases, launch locks can also be avoided because of the much lower mass being supported by the gimbal mechanism. Since a rotation of the reflector shell corresponds approximately to a beam steering by twice the angle of reflector shell rotation, the range of rotation required of the gimbals is about half of what it takes to steer the full antenna; however, this also means that the angular resolution of the steering is coarser by a factor of 2 for a given gimbal mechanism. A drawback of this solution is that the reflector antenna is in fact being defocused by the steering operation, and this can result in a degraded performance (lower gain, higher sidelobes) which is typically worst at the edges of the beam steering range. This degradation is usually minimized by shaping the reflector surface through optimization for a series of beam steering positions spanning the desired steering range. This type of antenna has also been implemented with a steerable dual-gridded shell [18], in which case the two polarizations are steered by slightly different angles due to the different locations of the two shells relative to the common gimbal mechanism's center of rotation. It should be noted that, for this type of implementation with a linearly polarized antenna, the convention for the direction of the main LP should preferably rotate as the beam is steered by more than a few degrees in a diagonal plane, such that it is not accounted for as cross-polarization. This is much less of an issue if the steerable shell is a DGR structure [18], and it is not an issue at all for CP antennas.

Figure 12.12 Deck-mounted steerable single-offset Ka-band antennas in the test range. Reproduced by permission of MDA.

Figure 12.12 shows five identical deck-mounted steerable single-offset Ka-band antennas, where only the reflector shells are rotated by two-axis gimbals.

Intermediate solutions, such that the reflector shell is rotated only about one gimbal axis, and the full antenna including the reflector gimbal are then rotated about a perpendicular axis (which is often an azimuth rotation axis, thus parallel to the main axis of the parabolic reflector), can also be encountered in this type of antenna. In that case, only one-axis rotation flexibility is required of the RF connection to the spacecraft, instead of two axes. Upon detailed tradeoffs, this may be found to be the best solution for a particular application.

12.4.6 Feed Chains

12.4.6.1 Horn

Electromagnetic horns are the preferred feed elements for most FSS/BSS reflector antenna applications, because their size is reasonably small at these frequencies (although C-band horns are still relatively large) and because they have great power handling capability. Additionally, horn design techniques have evolved to the point where very good polarization purity, pattern symmetry, gain performance, reflector aperture illumination control and frequency bandwidth are readily achievable through the synthesis of the internal horn wall profile.

Several horn types are widely used for this type of antenna, including Potter-type horns, corrugated horns and scalar horns, all of which are usually implemented with a circular aperture.

Potter-type horns use internal wall profile steps and taper angle discontinuities to generate, in a controlled fashion, a higher order waveguide mode (usually the TM_{11} mode) that is added to the predominant fundamental mode, TE_{11}, in such a way as to roughly equalize the horn radiation pattern beamwidths in the two principal planes. This horn pattern symmetry enhances reflector illumination efficiency and symmetry, controls spillover losses by minimizing horn pattern sidelobes, and greatly improves polarization purity. Since the two waveguide modes generated have different phase velocities, and since they must arrive at the horn aperture with the correct amplitude and phase relationship for optimal performance, this type of horn is relatively narrow band. Nevertheless, advanced synthesis techniques of rather complex horn wall profiles are able to optimize performance at a number of selected frequencies and therefore significantly widen the frequency band, or achieve a very good dual-band performance covering both the Tx band and the Rx band.

The corrugated horn achieves a similar control of the horn aperture field as the Potter horn, by combining the field distributions of the TE_{11} and TM_{11} waveguide modes. However, the anisotropic surface impedance characteristics provided by the corrugations enable this combined field distribution to propagate as a single hybrid mode, HE_{11}, thus featuring a single phase velocity. As a consequence, the performance of corrugated horns is much more wide band than Potter-type horns, easily maintaining performance for both the Tx and Rx frequency bands. Excellent pattern symmetry and polarization purity are thus achieved over a wide frequency band. A disadvantage of corrugated horns is that they are generally bigger (larger aperture diameter and horn length for a comparable horn gain), heavier and more expensive than Potter horns. Variations on the basic corrugated horn design include profile shaping to minimize length and an impedance matching section near the neck of the horn to minimize reflections over a wide frequency band. New software tools are now becoming available to facilitate and enhance the effectiveness of corrugated horn profile optimization, making it possible to design this type of horn to meet an even wider variety of performance requirements and volume constraints [21].

Scalar horns are used when a very compact size is desired and for relatively low *F/D* ratios, when the reflector subtended angle, as seen from the horn aperture, is relatively large. A series of concentric choke rings around the basic horn circular aperture are used to increase the effective aperture diameter. It is difficult to achieve a large effective aperture size, however, because the outer concentric choke rings are often just barely excited, and therefore greatly limit the horn aperture illumination efficiency.

Other simpler horn types can also be used, and may be appropriate for a particular set of performance requirements. These include, among others, conical horns, rectangular horns and trifurcated horns. Their performance is generally worse than the specialized horn designs covered previously, but they can have a lower cost and in some cases that may be a sufficient advantage.

Modern commercial software packages, using efficient full-wave analysis techniques such as mode matching, are usually employed in horn design. The agreement between predictions and measured results has become excellent, such that horn prototyping is usually no longer required. Other analysis techniques, such as the finite element method and the method of moments, are also occasionally used, but they are not as computationally efficient as the mode matching method of analysis. Nevertheless, and especially for electrically small horns and horns having a significant field level near the rim of the aperture, an approach using the method of moments is often used so as to properly account for the aperture rim and the outer surface of the horn.

12.4.6.2 Diplexer

Diplexers are components used in the feed chain to separate two frequency bands, typically the Tx and Rx bands [22]. They are three-port devices, with a common port, a Tx port and an Rx port. At these frequencies, diplexers are typically implemented using rectangular waveguide technology, and the three ports are then rectangular waveguide ports which may have the same or different cross-sectional dimensions in accordance with the respective frequency bands. The common port is often connected to an orthomode transducer (OMT) or to the feed horn. The Rx signal received by the horn is directed by the diplexer to the Rx port, with minimal reflection and very good isolation from the Tx port, where virtually no Rx signal appears. The Tx signal is fed into the diplexer Tx port, and it is conveyed to the horn with minimal reflection and very good isolation from the Rx port, where no Tx signal is present. Preventing the Tx signal from reaching the Rx port of the diplexer is especially important, since even a small percentage of the Tx power would easily saturate the input amplifiers of the ultra-sensitive low-noise receivers. Rejecting Rx frequencies in the Tx path is also important in attenuating active and PIM products present in the Tx portion of the diplexer that would interfere with the received signals.

Diplexers consist basically of two filters connected by a three-port waveguide junction. The filter on the Rx arm of the diplexer passes the Rx signal frequency band with minimal insertion loss while providing a very high rejection of the Tx band. The filter on the Tx arm of the diplexer passes the Tx signal frequency band with minimal insertion loss while providing a very high rejection of the Rx band. These filters are most often realized as in-line stub filters using a Chebyshev design, but their final behavior differs from a typical Chebyshev response since each filter is accessed through the common port of the three-port waveguide junction, and the other filter appears to it as an additional waveguide stub short-circuited at its end (representing the fact that it is designed to reject its frequency band). Achieving the required low insertion loss, low return loss and high Tx–Rx isolation is usually a challenging objective, and it requires an experienced design engineer equipped with state-of-the-art full-wave software design tools. Most often, and as the catalogue of available past diplexer designs expands with time, it is very efficient to start from an existing design aimed at somewhat similar requirements and modify it for a relatively quick optimization. Although a very first concept can be analyzed using transmission line theory, sophisticated numerical methods such as mode matching, the finite element method and the finite difference time-domain method are required to make an accurate prediction of performance.

Practical component characteristics such as finite metal conductivity, internal surface roughness, achievable minimum radii of curvature at the internal corners, as well as overall dimensional tolerances, are all factors that have special importance in determining diplexer performance. Accounting for these factors in the diplexer RF performance analysis is a challenge. However, experience with many past designs is used as a guide in assessing the expected impacts and in selecting the appropriate modeling assumptions. In that manner it is possible, nowadays, to go directly from the design phase to flight hardware manufacturing, bypassing the need for prototyping.

Diplexers are a critical component of the feed chain in determining its power handling capability, since multipactor is a risk in view of the relatively small gaps inside the filters and the high field intensities that can occur due to field magnification effects in the filter structure. As previously mentioned, it is important not to use an overly conservative approach, which can easily lead to an unrealistically pessimistic result. A P20 or P4 statistical approach to finding the applicable peak power is now widely accepted in the industry, and it has been validated by experience in many programs. This high power handling requirement applies mostly to the Tx arm of the diplexer, the first Rx resonators and the three-port waveguide junction, in which designs one must endeavor to widen the gaps in the interior geometry wherever strong local field intensities occur. Another factor that significantly raises the multipactor voltage threshold is a surface treatment of the machined aluminum component, which in any case is almost never left untreated for space applications. Usually the aluminum surface is anodized, or iridited (also known as alodined), or silver plated, and any of these surface finishes provides additional protection against multipactor (approximately 5.3 dB for the first two and 3.9 dB for silver plating). Silver plating tends to be an expensive feature due to material costs and space-qualified processes, and it is usually only used when a very low insertion loss is required.

On the Rx arm, on the other hand, a main design driver is often to maximize rejection of the Tx signal frequencies. If the desired rejection is not achievable with an in-line waveguide stub filter, or if an unreasonably large number of stubs are needed leading to a higher insertion loss, both the rejection and the insertion loss can be improved by using irises in addition to stubs to create the resonant filter elements. Irises cannot usually be used in the Tx arm because of power handling concerns, but they are suitable for use in the Rx arm of the diplexer.

12.4.6.3 Polarizer

Polarizers are commonly used to convert a linearly polarized signal (at the polarizer LP port) into a circularly polarized signal (at the polarizer CP port), or vice versa. They work by decomposing the incident linearly polarized field at the LP port into two orthogonal linearly polarized fields, and by designing the component such that the two linear components arrive at the CP port with the same amplitude and in phase quadrature (i.e.,

$\pm 90^\circ$ phase difference). The sign of the phase quadrature is generally selected by exciting the LP port in one of two orthogonal polarization directions, and it determines whether a LHCP or a RHCP signal is formed at the CP port of the polarizer. Reciprocally, if a circularly polarized signal is input at the polarizer CP port, a linearly polarized field is formed at the other end (the LP port of the polarizer) in one of two orthogonal directions, depending on whether a LHCP signal or a RHCP signal is present at the CP port. The two orthogonal directions of LP at the LP port of the polarizer can then be conveyed into separate rectangular waveguide ports by inserting an OMT (see next subsection) at the polarizer LP port. This is often realized as a single integrated component incorporating both functions, leading to a three-port device.

Many different polarizer designs exist and are used in the feed chains of BSS/FSS reflector antennas [22]. This includes septum polarizers, pin polarizers, ridged-waveguide polarizers and corrugated polarizers, all of which work along the principles outlined above, thus decomposing the incident LP into two orthogonal LP components and then imparting different insertion phases to these two components in order to achieve the desired sign of phase quadrature. Septum polarizers have a frequency bandwidth (defined as the band where the axial ratio is better than 0.3 dB) that is usually limited to less than about 15% of the center frequency, therefore they are used for single-band (Tx or Rx) applications. The septum polarizer is interesting in that it combines the OMT and the polarizer functions since the septum forms a vein which cuts the waveguide in two halves along its height, providing orthogonal LP ports on each half. Single-band corrugated polarizers have typical percent bandwidths of 20%. Corrugated polarizers can also be designed for dual-band applications (Tx plus Rx) with typical percent bandwidths of 5% in each band. The same can be achieved with ridged-waveguide polarizers.

Turnstile polarizers are also used [23], and they work on a slightly different principle since the two orthogonal LP signals are fed to the turnstile already in phase quadrature, such that the turnstile is only required to combine the two signals into the same CP port. The turnstile is in fact an example of an orthomode junction (OMJ), with typically very good bandwidth and isolation performance. Turnstile polarizers often combine the functions of a polarizer and a diplexer, as described below. They are typically dual-band polarizers (Tx plus Rx) and they can have percent bandwidths of 15% in each sub-band. A worst case axial ratio performance of about 0.3 dB is typical for all types of polarizers and percent bandwidths mentioned above. For typical applications, where the percent bandwidth is smaller, then a better axial ratio, of 0.2 dB or even less, can usually be achieved.

One example of implementation of a turnstile polarizer uses a six-port turnstile OMJ, of which two ports receive two identical signals in one LP on Tx and another two ports receive two identical signals in the orthogonal LP and in phase quadrature on Tx. The fifth port is a CP port used by the Rx band that can then lead to a septum, pin or corrugated polarizer that converts the two hands of CP into the two LP orientations, and the sixth port is the common CP port connecting to the radiating horn and supporting both the Tx and Rx sub-bands. The design described performs the polarizing function but, in addition, it also performs the diplexing function to separate the Rx and Tx sub-bands. Rx band reject filters are usually introduced in the four symmetrical LP arms to prevent the Rx frequencies from reaching the Tx port. Typically, in the Rx portion of the device the waveguide size is selected to be sufficiently below cutoff at Tx frequencies to provide the required rejection of the Tx band, while being sufficiently above cutoff at the Rx frequencies to minimize dispersive effects. Alternatively, filtering to reject the Tx frequencies may also be introduced at this point. Feeding each direction of LP to two ports simultaneously makes for a balanced symmetrical design and it generally improves performance and bandwidth. The two signs of phase quadrature can be obtained, for example, by using a hybrid coupler that introduces phase quadrature between the two output ports (such as a branch-line coupler), and such that one coupler input port generates LHCP and the other coupler input port generates RHCP. The two LP signals in phase quadrature can be transformed into four waveguides (two symmetrical pairs) by means of magic T junctions, for example. A computer model of a turnstile feed assembly is shown in Figure 12.13.

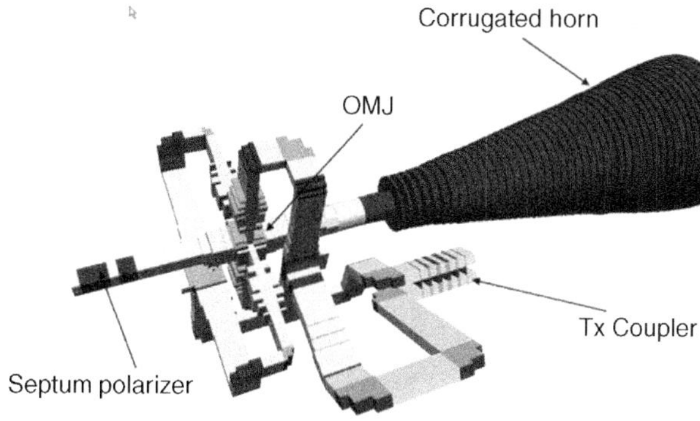

Figure 12.13 *Example of a turnstile feed assembly. Reproduced by permission of MDA.*

The concept described above is an example of a practical feed assembly, but many variations can be constructed, depending on the particular requirements on the antenna and the feed assembly. In particular, the architectures of the Tx and Rx portions of the assembly are sometimes reversed, such that it is in Rx that the turnstile is fed with two pairs of LP signals in phase quadrature, and it is in Tx that the turnstile has a CP port and a polarizer. Such an arrangement is known as a reverse turnstile feed assembly.

12.4.6.4 Orthomode Transducers

OMTs are three-port devices that combine two orthogonal LPs, typically HP and VP, fed into rectangular waveguide ports, into a common (usually square) port comprising both the HP and VP polarizations [22]. This common port may be connected directly to a horn for dual-LP operation, or it may be connected to a polarizer and then to the horn for dual-CP operation. The isolation between the orthogonally polarized ports is generally in the range of 40–60 dB. However, when the OMT is connected to a polarizer and horn as described above for dual-CP operation, any amount of horn return loss causes a reflected CP signal that has the opposite sense of rotation when compared to the incident CP signal, and this reflected signal gets converted into the opposite LP by the polarizer and then significantly degrades the isolation between the OMT orthogonal ports. Depending on the impedance match presented by the repeater at each of those ports, this cross-polar component can be partly reradiated and thus degrade the antenna cross-polarization performance.

Increased OMT isolation can often be achieved by a symmetrical OMT type of design, whereby each LP is fed simultaneously to two symmetrical waveguide branches, thus eliminating field imbalances within the OMT structure. However, structures that rely on the symmetry of the mechanical features in recombining and/or splitting the signal are prone to generating higher order modes, because of the nearly inevitable occurrence of small asymmetries due to manufacturing tolerances in the component. The OMT can be designed so that these higher order modes are below the waveguide cutoff frequency and thus cannot propagate; however, this may result in pockets of stored reactive electromagnetic energy inside the component. These can lead to sharp spikes in the frequency response of the device. If these spurious modes can somehow reach the horn, they may degrade significantly the polarization purity of the radiated field. In recent years, through efforts aimed at fully understanding and characterizing this mechanism, the design of symmetrical OMTs has evolved to the point where this phenomenon can be generally avoided.

Other types of OMTs include the six-port turnstile OMJ described in the previous subsection as part of the turnstile polarizer, as well as other old and newly developed concepts [24] aimed at improving specific aspects of performance such as power handling, isolation and bandwidth.

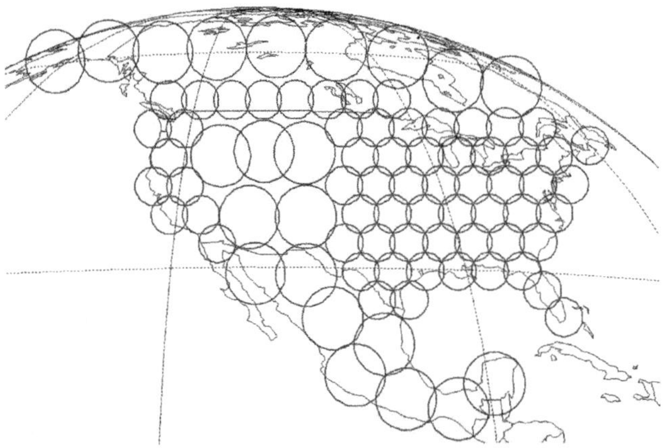

Figure 12.14 Typical multibeam coverage spot grid seen from a geostationary satellite.

12.5 Multibeam Broadband Satellite Communications Antennas

12.5.1 Typical Requirements and Solutions

Applications requiring higher data rates, such as high-speed Internet services for consumers and small businesses, call for much higher antenna gains and wider frequency bandwidths. Large antenna gains are associated with a small coverage region on the ground and with large antenna aperture diameters in terms of wavelength. Realizing wider bandwidths is increasingly easier as the frequency of operation increases, since then the bandwidth as a percentage of the center frequency remains relatively small. For these reasons, this type of service has been typically provided with multiple spot beam antennas (with small circular coverage spots) operating at K/Ka-band. This frequency band is commonly referred to simply as Ka-band in satellite industry circles. The uplink and downlink frequency bands are typically selected within the ranges of 27.5–31.0 GHz and 18.3–20.2 GHz, respectively. The coverage spots are typically arranged in a tight hexagonal grid on the Earth, where either all spots have the same diameter or different diameters are used to address different expected traffic density conditions, as shown in the example of Figure 12.14.

A typical first-generation Ka-band multibeam antenna farm, comprising a total of four Tx reflector antennas, four Rx reflector antennas and two tracking reflectors, is shown in Figure 12.15 (only half of the reflectors are shown).

Several different approaches have been proposed and implemented for Ka-band multibeam antennas [25,26]. They include reflector antennas and direct radiating array (DRA) antennas. Reflector antennas for this application feature multiple feed elements in their focal region, and they are mostly of two types:

- single feed per beam (SFB) antennas, where each spot beam on the ground is generated by using a single feed element, and
- focal array-fed reflector (FAFR) antennas, where each spot beam on the ground is generated by using a subarray of several feed elements excited together through a BFN.

These systems may use different antennas for Tx and Rx or common Tx/Rx antennas. Mass and cost-saving considerations increasingly drive the design into common Tx/Rx antenna solutions. One challenge in this case is to design the antenna for optimal performance in two widely spaced frequency bands (a 1.5:1 ratio) over the same coverage spots on the ground. Designs have evolved to address this challenge. Clearly, the design of

Figure 12.15 Multibeam antenna farm. Reproduced by permission of MDA.

antennas for Tx only or for Rx only is generally much simpler and often leads to slightly better EOC gain performance. The mass and cost penalties are, however, quite high. Currently, common Tx/Rx antennas are mostly used for SFB architectures, where the number of reflectors would otherwise be too large, whereas FAFR and DRA designs mostly use separate Tx and Rx antennas since common Tx/Rx would lead to significant performance penalty and technical risk in those cases.

Additionally, the required carrier-to-interference ratio (C/I) performance (which essentially measures the interference between spot beams using the same frequency channels and polarization) leads to schemes whereby the hexagonal grid of spots is divided into a number of different non-overlapping frequency sub-bands and polarization allocations, also called colors. In this manner, adjacent coverage spots always have different colors, therefore averting the unachievable requirement to maintain isolation between adjacent spots. Generally, four colors are required to ensure that same-color spots are sufficiently separated to minimize interference (thus maximize C/I). Three colors can also work in principle, although this will generally yield significantly worse composite C/I performance.

The different design approaches are addressed in some detail in the following subsections.

12.5.2 SFB Array-Fed Reflector Antennas

In this approach, the feed array is located on the focal region of the reflector, and each spot beam on the ground is generated by exciting a single feed element. Therefore, the number of feed elements corresponds to the number of beams. SFB designs can easily support transmission and reception in the same antenna.

The required EOC gain drives the specified coverage spot size and the diameter of the reflector aperture, although other parameters also affect the EOC gain performance as seen below. The diameter of the feed elements determines their minimum allowed center-to-center separation in the feed array, and therefore, combined with the focal length, it determines the spot beam angular center-to-center separation on the ground. In order to place the feed elements (usually horns) close enough together to cover all the spots on the ground from a single reflector antenna, their aperture dimensions become relatively small, their individual element

Figure 12.16 Ka-band multibeam antennas in the compact RF test range. Reproduced by permission of MDA.

gain is low and their pattern beamwidth is wide, and as a consequence the spillover losses (energy being lost beyond the rim of the reflector surface) are quite high. This high spillover loss precludes achieving an optimal spot beam EOC gain. To constrain spillover losses in an SFB antenna the element pattern beamwidth must be narrowed significantly, thus the aperture of each element must be relatively large (typically between $2F/D$ and $3F/D$ wavelengths, [27–29]). This means that, when illuminating a basic parabolic reflector from a feed array, the spot beams on the ground will not be contiguous, and more than one antenna is required to cover the contiguous grid of coverage spots.

There are two principal types of SFB solutions to overcome these difficulties: the interlaced multi-reflector approach, and the oversized shaped-reflector approach.

12.5.2.1 *Interlaced Multi-reflector SFB Approach*

The interlaced multi-reflector approach uses 3–4 common Tx/Rx parabolic reflector antennas (thus 6–8 reflectors if separate antennas are used for Tx and Rx, as shown in Figure 12.15), each reflector serving non-adjacent spot beams. Since the spots covered by any one reflector are not adjacent, the oversized feed element apertures required to decrease spillover losses can be easily accommodated in the feed array grid. An example of an interlaced multi-reflector approach using common Tx/Rx apertures is shown in Figures 12.16 and 12.17.

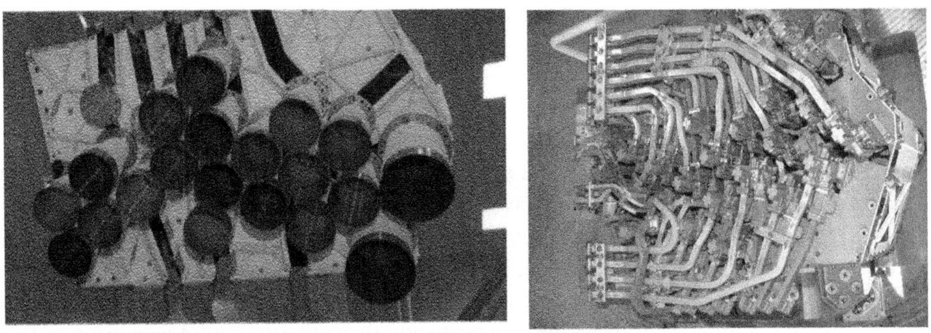

Figure 12.17 Multi-horn feed array and its feed waveguides and diplexer/polarizer assemblies. Reproduced by permission of MDA.

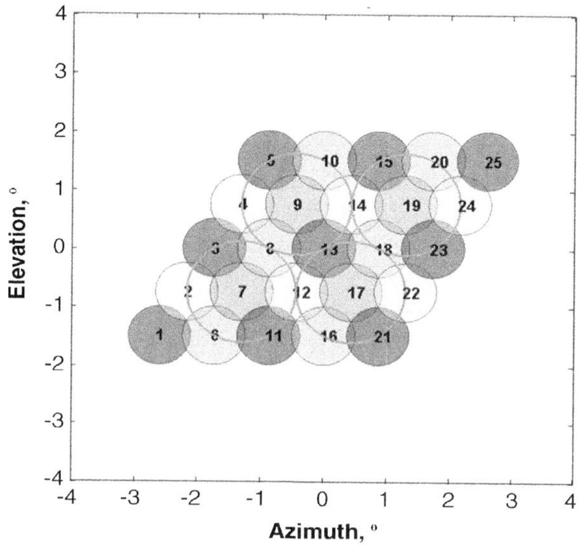

Figure 12.18 *Feed/beam mapping for SFB system using four reflectors.*

In order to optimize the peak-to-edge gain roll-off over the spot coverages, and thus maximize EOC gain, the oversized horns are required to have high aperture efficiency. Many solutions have been used to address this issue, including parasitic discs mounted on the horn axis, dielectric rods and multimode horns. The most elegant solution to this problem is probably the high-aperture-efficiency multimode horn. Multimode horns making use of higher order TE modes have been successfully used in addressing this challenge, achieving aperture efficiencies higher than 90% for separate Tx and Rx applications, and on the order of 80–85% for combined Tx/Rx antennas for which the horns must operate over the two bands (see Figure 12.17). Figure 12.18 shows a four-color coverage spot grid, and superimposed on it is also shown the corresponding horn apertures serving one of the four colours (i.e., spots 9, 19, 7 and 17 in the figure). Interlaced four-reflector SFB antennas are currently the most widely used antenna configuration for multiple spot beam coverage.

While a good performance with separate Tx and Rx antennas is easily achieved with a multimode horn, getting optimal performance for a combined Tx/Rx antenna is more challenging. The different aperture sizes in terms of wavelength for the Tx and Rx bands would tend to produce non-congruent beams. With sufficient aperture efficiency (more than 80% over both bands, and especially Rx) it is, however, possible to underutilize the reflector aperture over the Rx band and thus equalize its beamwidth with that of the Tx band. Because the reflector is oversized for the needs of the Rx band, it is also possible to shape the reflector slightly so as to widen the Rx spot beam and achieve a higher edge of coverage gain without impacting the Tx gain significantly. An alternative implementation of this approach is the stepped-reflector antenna design, which implements a 180° phase reversal on Rx near the reflector edge so as to flatten the beam shape [30].

12.5.2.2 Oversized Shaped-Reflector SFB Approach

The oversized shaped-reflector approach to SFB antennas takes care of the spillover issue by widening the angular width of the reflector as seen from each feed element. The oversized reflector diameter is typically more than twice the diameter of a parabolic reflector required to generate the desired spot sizes. With more than twice the diameter, an oversized parabolic reflector would tend to form spot beams which would be much too narrow for the desired application (thus too large of a gain roll-off between peak and edge of the spot coverage).

To address this issue, the reflector shape is optimized so as to achieve the desired EOC gain performance, by basically flattening the top of the single-element spot beams. This approach achieves generally better EOC gain than the interlaced approach, with its much smaller but multiple reflectors. With the reflector surface shaped for optimum EOC gain, the cross-polarization performance would tend to degrade to unacceptable levels. That must be compensated for by using feed horns with very high polarization purity (e.g., corrugated). These feed horns are also designed so as to achieve simultaneously good spillover and aperture efficiencies.

To first order, the reflector aperture amplitude illumination function is mainly driven by the feed horn design, while the phase distribution is controlled by the reflector shaping. In this case, the reflector is significantly oversized for the spot size requirements so as to mitigate the spillover losses, and therefore its aperture efficiency is relatively low. The ideal feed horn radiation pattern would be mostly uniform over the reflector aperture, but with low spillover losses. When using a modified Potter horn design, rather than the more typically used corrugated horn design, the addition of a TE_{12} mode in phase opposition has been shown to provide advantageous radiation pattern characteristics. The reflector surface shaping needs to avoid sharp discontinuities in the phase slope distribution, so as to restrain sidelobe levels, thus enhancing the carrier-to-interference ratio (C/I) performance.

Based on cost, mass and accommodation considerations, the oversized reflector design can only be a viable SFB solution if one or two reflector antennas are sufficient to cover all the coverage spots. Consequently, the focal length must also be increased compared to each interlaced SFB reflector antenna. This is a feasible but challenging design, and it is easier to optimize it so as to meet all requirements when using separate Tx and Rx antennas. Nevertheless, combined Tx/Rx antennas are feasible and can exhibit good performance, which will depend on the number of beams and on the maximum required scan angle relative to boresight. Since for most available spacecraft and launch vehicles solid composite reflectors are limited to about 3 m in diameter, and taking into account the reflector oversizing factor of this design, the spot beams must be wider than about 1° diameter at Ka-band for this concept to be realizable.

12.5.3 FAFR Antennas

In this concept each spot beam is generated using a sub-array of the complete reflector focal region feed array [5]. Each sub-array is typically formed of a central feed element plus one ring of elements surrounding it (for a total of 7, often denoted as a septet) or two rings of elements around the center element (for a total of 19). Sub-arraying the feed elements requires that they be fed through a BFN. The BFN can be on the repeater side of the amplifiers (low-level BFN, or LLBFN) or on the radiating array side of the amplifiers (high-level BFN, or HLBFN).

The radiating aperture of the sub-array corresponding to one spot beam is roughly of the same size as each oversized horn aperture in the SFB concept described previously, so as to limit spillover losses for each spot beam while maintaining an efficient illumination of the reflector. In order to cover the same grid of coverage spots, the total number of feed array elements is usually 2–8 times higher than the number of spot beams, and the diameter of each feed element is 2–4 times smaller than the diameter of the multimode horns used in the equivalent SFB configuration. The FAFR feed array elements are usually small radiating elements such as simple single-mode horns. The area density of these elements can be 4–16 times higher than for the SFB clusters.

As for all multibeam antennas, the required spot diameter and EOC gain drive the size of the reflector aperture. The reflector is generally parabolic and two reflector antennas, one Tx and one Rx, are usually required. The amplitude and phase distributions within the feed sub-array corresponding to each spot beam are controlled by the sub-array excitation coefficients generated through the appropriate BFN. Many of the feed array elements are shared between adjacent spot beams, which has implications for the subsystem architecture and performance.

12.5.3.1 FAFR Using a LLBFN

When a LLBFN is used on transmission it distributes the Tx signals to the last stage of HPAs that feed the radiating elements. Therefore its RF losses do not affect the final gain performance since the latter is usually referenced to the output ports of the HPAs. When the LLBFN is used on reception it combines the signals received by the Rx radiating elements after they have been amplified by the LNAs. Consequently, the LLBFN RF losses have only a very small effect on the final signal-to-noise ratio of the FAFR system. When using a LLBFN architecture, the radiating elements can be shared between beams without adversely affecting the antenna gain performance since the BFN losses occur on the repeater side of the amplifiers, and therefore the system is rather insensitive to the increases in such losses arising from sharing elements.

Good feed element aperture efficiency is important to mitigate the occurrence and magnitude of grating lobes, thus limiting spillover losses. To avoid increasing the number of elements by an unnecessarily large factor, the element spacing is usually chosen to reflect the beam spacing (slightly more than *F*/*D* wavelengths [27]). A seven-element sub-array (center element surrounded by a ring of six elements) is typically used for each beam. The elements may also be placed on a tighter lattice, although more elements will be required per sub-array to form the beams. Typically 19-element sub-arrays (center element plus two rings of elements) or even more elements may be used.

Since the number of feed elements for a FAFR using LLBFN is 2–8 times higher than for an equivalent SFB system, the number of LNAs and HPAs (one per radiating element) is therefore also 2–8 times higher when compared to an equivalent SFB system. Figure 12.19 shows the schematic of a system where both Tx and Rx use LLBFNs, and it also depicts how a fictitious 25-beam lattice is produced by using groups of 7 elements out of an array of 47 elements, highlighting the element sharing between beams.

One of the major drawbacks of this approach on the Tx side is that, since many elements are shared between beams, the HPAs must be operated in their linear region, thus at a significant backoff from the saturation point, in order to achieve an acceptable noise–power ratio (NPR) performance. Consequently, and compared to HPAs operated near saturation, the DC-to-RF power conversion efficiency is seriously degraded and the overall DC power consumption is significantly higher. Compared to a three- or four-reflector SFB system, this approach has the advantage of using fewer reflectors, but it typically offers lower gain performance and higher power consumption and dissipation. The FAFR with LLBFN is not a solution that is very well adapted to Ka-band systems, especially in transmission.

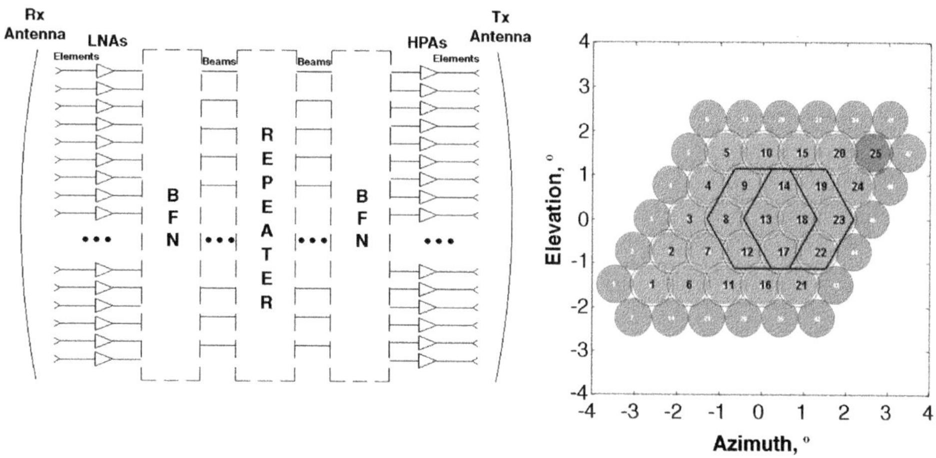

Figure 12.19 *FAFR with LLBFN: left, notional block diagram; right, element-to-beam mapping.*

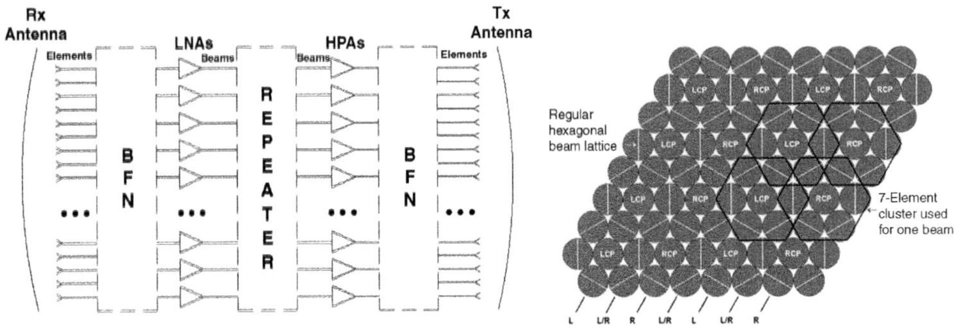

Figure 12.20 *FAFR with HLBFN: left, notional block diagram; right, element-to-beam mapping (in the figure RCP and LCP mean right-hand and left-hand circular polarization).*

12.5.3.2 FAFR Using a HLBFN

In a FAFR using a HLBFN to excite the sub-arrays, the HLBFN is located on the radiator side of the HPAs and LNAs, as shown schematically in Figure 12.20. Since it is located before the LNAs on Rx and after the HPAs on Tx, its RF losses, both ohmic and due to power splitting and combining, detract directly from the gain performance of the antenna and therefore must be minimized. One objective in locating the HLBFN in this fashion is that the elements of each sub-array are mapped onto one single access port, which thus constitutes the spot beam port. The HPAs and LNAs are located at these beam ports, and therefore, as in the SFB approach, their number is the same as the number of beams, that is, 2–8 times fewer than when using a LLBFN. Moreover, the HPAs can now be operated near saturation, which greatly enhances DC-to-RF power conversion efficiency thus decreasing DC power consumption. When using a HLBFN many elements are typically shared between two adjacent spot beams, as can be seen from the beam-to-feed-element mapping concept of Figure 12.20. Since the spot beams sharing feed elements are adjacent, each shared element must be fed with orthogonal excitations (multimode BFN [5]), orthogonal polarization or be somehow differentiated (in frequency, for example), for the two spot beams sharing it, in order to avoid excessive combining or splitting losses. Although this approach ideally seeks to excite a predefined sub-array of elements for each spot beam, a few additional elements will also be excited, albeit at a lower level, due to the element-sharing constraints imposed on the HLBFN architecture. This results in a small performance degradation which is inevitable with this type of HLBFN technology.

The inter-element spacing in the feed array will typically be close to or slightly greater than one wavelength (about $0.5F/D$ wavelengths [27]), and therefore the sub-array might in principle be prone to the occurrence of grating lobes, which if occurring would increase spillover losses. In practice, the aperture efficiency of the feed elements in the sub-array environment is designed to be very high, thus mostly precluding the occurrence of these grating lobes and enhancing the antenna gain performance. High aperture efficiency is thus one of the main objectives of the feed element selection and design.

Since the HLBFN RF losses occur in the high-power portion of the antenna on Tx, and because of the very dense packaging of elements, thermal management of the Tx array and HLBFN can be an issue. Compared to a three- or four-reflector SFB antenna system, this approach has the advantage of requiring only two reflectors (although a third dedicated but usually much smaller antenna will often be required for the gateway beams). However, because of the constrained shared element excitation and the high-level losses, the FAFR solution with a HLBFN has a lower gain performance. A FAFR antenna using a HLBFN is an attractive design solution mainly when required to produce a relatively small number of spot beams and when there is a strong mechanical accommodation constraint, especially if the gateway antenna can be accommodated on the deck with a smaller aperture.

12.5.4 DRA Antennas

In this design solution the array is the antenna and no reflector is present. Each spot beam uses all the array radiators so as to achieve the performance level associated with the full DRA aperture size. The aperture size is driven by the spot diameter and the EOC gain requirement, and it is similar to the reflector aperture size used for the other multibeam antenna solutions. Various schemes have been developed to try to minimize the number of array elements for a given aperture size [31,32]. Even with the most optimized array design, the number of elements required, and the associated active electronics, is at least one order of magnitude higher than for the feed array of the reflector-based solutions. All elements are shared between all spot beams. The HPAs must therefore be operated in the linear bias region, at a significant backoff from saturation, to achieve an acceptable NPR performance. This reduces the DC-to-RF power conversion efficiency of the system when compared to the SFB or the FAFR with HLBFN solution, which is a significant drawback of DRAs. The required BFN is very complex, with the number of ports on the repeater side equal to the number of beams, and on the radiating array side equal to the number of radiators, with all elements being shared by all beams. It can be realized with a very complex network of thousands of couplers, or alternatively as a digital BFN. In the latter type of implementation, the full frequency spectrum needs to be quantized for each element with a sufficient number of significant bits and processed through a complex multiplier/accumulator to generate the spot beams. With the digital BFN implementation the beams can be reconfigured at will, but a very complex and expensive on-board processor is required. The DC-to-RF power conversion efficiency of the system can be somewhat improved by using constant or tapered amplitude with phase-only synthesis of the beams, which is, however, accompanied by some deterioration in performance due to a reduced beam shaping capability [33]. Using time-division multiple access (TDMA) hopping among a series of pre-optimized interleaved 'beam templates' in sequence enables the desired coverage to be achieved with good C/I, EIRP and G/T performance [33]. Optimum utilization of payload resources can in principle be achieved with this TDMA scheme by dwelling longer over beam templates covering high traffic density regions [33].

One approach being considered to minimize the overall complexity and risk is a modular construction, such as building the planar array as an assembly of a large number of active tiles [33], where each tile is built as a highly integrated design incorporating radiating elements, Tx and Rx amplification, electronic power conditioning (EPCs), DC and RF power distribution networks, structural elements, as well as thermal management features enabling the tiles to spread out and radiate the heat so as to keep all semiconductor junction temperatures within their operating ranges. This modular approach, while maintaining the high number of radiating elements and amplifiers, manages nevertheless to contain complexity and thus decrease overall cost and risk.

In summary, the DRA antenna configuration provides the ultimate level of flexibility and reconfigurability, but even with the latest design solutions and technologies it is still the riskiest and most expensive solution for geostationary multiple-spot-beam antennas. Other chapters of this book provide an in-depth view of array antenna design and challenges, therefore the subject will not be further detailed here.

12.5.5 RF Sensing and Tracking

With the very high-gain and narrow-beamwidth requirements of the most advanced multibeam antenna concepts comes a very high sensitivity to beam pointing errors, since they represent a greater percentage of the spot beam diameter as the latter decreases. The 0.10°–0.13° total half-angle pointing error, achieved by most non-tracking antennas on geostationary satellites, can degrade the EOC performance of the spot beams by

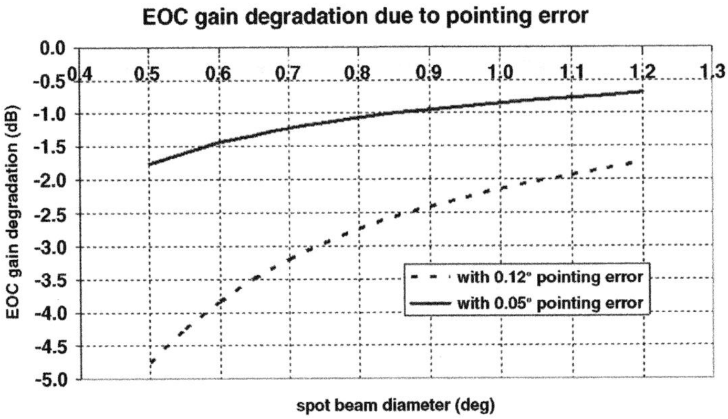

Figure 12.21 *Effect of pointing error on spot beam EOC gain.*

several decibels compared to nominal conditions. This is apparent in Figure 12.21, where the impact of a typical 0.12° pointing error is compared against the impact of a 0.05° pointing error. Smaller cell sizes aggravate the problem, but in all cases reducing the pointing error from 0.12° to 0.05° significantly improves EOC gain. RF sensing systems [34,35] have been used to reduce satellite antenna pointing errors from 0.12° to around 0.05°, and for the best performing systems the error may be reduced all the way down to 0.02°. A beacon with a known location on the ground is detected with a special tracking feed chain located in the focal feed array. This feed chain can be dedicated to RF sensing tracking or it may be shared with the communications function. While some systems have used the traditional four-horn cluster with sum and difference outputs, it is possible to acquire all necessary pointing information by extracting the TM_{01} mode from an impinging CP beacon signal.

12.6 Antennas for Non-geostationary Constellations

12.6.1 Typical Requirements and Solutions

Non-geostationary antennas, particularly for satellites in LEO (with typical altitudes as low as 600 km) and in MEO (with typical altitudes as high as 15 000 km), are characterized by their wide angular coverage requirements, starting at less than 20° half-cone angle for the highest MEO satellites and going up to more than 60° half-cone angle for the lowest LEO satellites. These wide angles result from the large angular size of the Earth's disk when seen from a low or medium altitude. Since non-geostationary satellites move at a large angular speed in their orbits, any ground station will move from one end to the opposite end of the antenna coverage angular range in a relatively short time (typically a fraction of an hour). In order to maintain communication, the satellite antenna must be a global coverage fixed-beam antenna, a multiple fixed-spot-beam antenna, or an antenna supporting steerable spot beams that must accurately track their targets on the Earth's surface. Such a tracking antenna will undergo many end-to-end pointing cycles a day throughout the duration of the mission.

Especially for LEO constellations, for which the number of satellites and the required number of antennas are large, the antennas must be designed so as to benefit from the economies of scale that arise from the manufacture of many identical units. Therefore, the design must address ease of manufacturing, assembly, integration and test. Minimization of the number of piece parts is a common cost-saving objective, as well as minimization of required human intervention in the manufacture, assembly and test of the antennas.

Maximizing the degree of automation of the manufacturing, assembly and testing operations is an essential cost- and time-saving strategy for these types of antennas.

12.6.2 Global Beam Ground Links

Ground links from MEO or LEO use a fixed global beam satellite antenna when only a relatively low gain is required. To maximize the system performance, it is desirable for the antenna to generate an isoflux gain profile, such that a roughly constant power density is radiated towards all Earth locations within view of the satellite. The RF path length, and thus the spatial attenuation, increases as the ground terminal is moved towards the edge of the coverage area. The isoflux antenna gain profile is shaped to compensate for this path loss, which increases from nadir as the square of the distance from the satellite to the Earth's surface. The atmospheric attenuation increases as well from nadir to the edge of coverage, and although it is a relatively small term (in the absence of precipitation) it should also be compensated for by the isoflux gain pattern.

The optimum (ideal) radiation pattern of an isoflux antenna increases from nadir to the edge of the global coverage, beyond which it falls abruptly to zero so as to avoid any gain outside the coverage. Such an antenna pattern is only theoretical and obviously it cannot be achieved in practice; however, it represents a limiting case against which real antenna patterns can be optimized. Globalstar's gateway antennas [36], for example, are implemented as circularly polarized scalar horns achieving a nearly isoflux behavior from their orbit altitude of 1400 km. The optimum isoflux gain patterns for one MEO altitude (10 000 km) and for one LEO altitude (700 km) are depicted in Figure 12.22.

The gain of the medium-orbit isoflux curve is much higher than the gain of the low-orbit isoflux curve since the apparent Earth size is angularly much smaller when seen from 10 000 km than when seen from 700 km. On the other hand, the gain variation from nadir to the edge of coverage is higher as the orbit altitude is decreased, because of the much greater difference between the distance from the satellite to nadir and to the horizon. In the examples shown in the figure, this variation is 3.6 dB for the 10 000 km MEO orbit and 12.8 dB for the 700 km LEO orbit.

12.6.3 High-Gain Ground Links

The wide coverage area inherent in a global coverage, as seen in the previous subsection, limits the gain achievable with a single global beam. Multiple small spot beams are required for high gain, either fixed or steerable. The need to generate these beams over the wide-angle visible Earth calls for a configuration capable

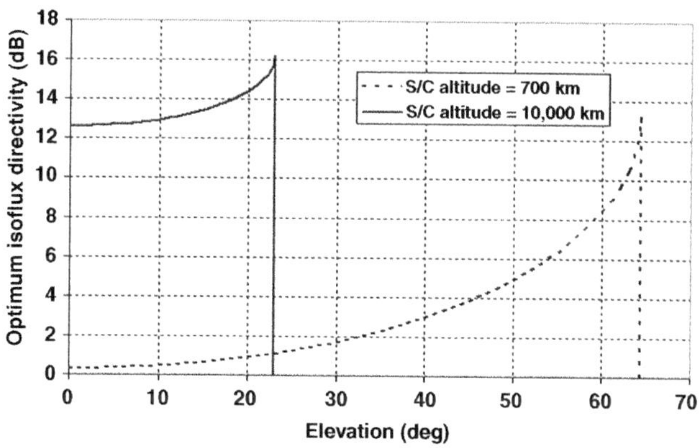

Figure 12.22 Optimum isoflux patterns for a 700 km orbit and for a 10 000 km orbit (S/C, Spacecraft).

of scanning a beam by tens of elementary beamwidths, without significant performance degradation. Some systems rely on a single antenna capable of generating multiple beams from a single aperture, while other systems employ a plurality of antennas, each one generating and steering one single beam. It is sometimes advantageous, in order to optimize the system performance, for the peak gain to also follow an isoflux behavior (as discussed in the previous subsection) as the beam is scanned from nadir to the edge of the visible Earth. Some concepts are based on achieving this type of gain profile, as detailed later. This is not always a practical objective, since the natural behavior of most antenna systems is to have maximum gain at boresight, which is usually near nadir, and then show gain degradation as the beam is steered away from boresight, or otherwise to maintain a constant gain performance in a mechanically steered system.

When there is a need for a few steerable spot beams requiring relatively small aperture sizes, it is possible to use a group of mechanically steerable reflector antennas on each satellite, each antenna generating a single beam.

When larger aperture sizes are needed to meet the spot beam gain requirements, or when the number of simultaneous spot beams is relatively large (usually larger than about 20), mechanical steerable solutions become impractical. Whereas for MEO applications it is sometimes possible to use multiple-feed reflector antennas, for LEO and for many MEO applications, where the range of beam steering angles is large, planar array antennas tend to be the configuration of choice [36,37]. Sometimes, in order to ease the design difficulties inherent in large beam steering angles (e.g., large scan-related losses and even scan blindness effects), the full coverage area on the Earth's surface is segmented into smaller sub-coverage areas [38]. Each sub-coverage area is then served by a separate antenna requiring much lower angles of beam steer.

For each mission, and in view of the specific coverage requirements, link budget objectives and price constraints, a detailed tradeoff must usually be performed to select the most advantageous antenna design alternative. This tradeoff also needs to weigh using separate Tx and Rx antennas versus implementing common Tx/Rx antennas.

12.6.4 Intersatellite Links or Cross-links

In some satellite constellations, such as Iridium, intersatellite links are used between the different satellites to route data through the network of nodes and avoid multiple ground-to-space hops (with their inherent and undesirable time delay). Intersatellite links between satellites in the same orbital plane, which maintain their relative positions essentially constant, use fixed beams to connect to the satellite immediately in front and immediately behind in the orbit. Steerable intersatellite links are used to communicate between satellites located in different orbital planes of the constellation, for which their relative positions change significantly with time. For steerable intersatellite links, the elevation steering angular range tends to be limited, since the altitudes of the satellites are the same, albeit in different orbits, while the azimuth range can be much greater since the satellites are moving at the same speed but in different and intersecting orbital planes. In some applications, steerable intersatellite links are established between each satellite of a constellation and other special-purpose satellites outside the constellation, such as NASA's geostationary Tracking and Data Relay Satellite (TDRS) system or ESA's Artemis satellite.

Steerable high-gain intersatellite link antennas typically require two-axis gimbal mechanisms with limited steering range in the elevation plane. The antennas may be designed to support both Tx and Rx or separate antennas may be used for Tx and Rx. The polarization may be LP or CP, depending on the particular mission.

Although steering the entire antenna using an RF rotary joint is the most straightforward approach to steer the beam over the azimuth range, it is possible to avoid the rotary joint with multi-reflector configurations, whereby a secondary reflector is rotated for azimuth beam steering without rotation of the feed. In these concepts, if using LP, it is important to design the geometry so as to ensure that the polarization alignment is maintained while the secondary reflector is rotated.

There are many design possibilities for intersatellite link antennas, including for example steerable arrays, parabolic reflectors or lenses and reflectors with rotating flat plates.

12.6.5 Feeder Links

Feeder links typically use steerable high-gain ground link antennas. These antennas are usually circularly polarized, require two-axis gimbal mechanisms and need to support both the Tx and Rx sub-bands. The gimbal rotation axes can be arranged in a theta–phi or x–y configuration. The phi (yaw) rotation axis of a theta–phi approach typically intercepts the coverage area, and this generates a kinematic singularity of the gimbal mechanisms, often denoted as the 'keyhole effect.' This may result in extremely high-phi angular rotation velocities in order to maintain uninterrupted tracking of the ground station. This effect occurs when the ground station is momentarily located on, or very near, the phi rotation axis, and it can be quite severe in particular for high-gain spot beams (i.e., narrow beamwidths). This problem has to be assessed whenever a theta–phi gimbal solution is envisaged, since the worst case event would imply loss of communication for up to a few seconds.

The high-gain requirement of feeder link antennas usually implies selecting an aperture antenna design, such as a reflector, an array or a lens antenna. Most often, especially for the higher gain requirements, a parabolic reflector antenna is the preferred solution, since it has the advantages of being very low cost, very broadband (only limited by the feed design) and benefiting from low RF losses.

The most straightforward approach to establishing an RF connection between the spacecraft repeater and the moving antenna feed assembly is to route it through two rotary joints, one for each of the rotation axes. Nevertheless, rotary joints are often a significant risk item, reliability concern and cost element in the antenna, whose total price must usually be relatively low, consistent with the production of a high number of antennas. It is usually desirable to avoid rotary joints, thus reducing cost and increasing reliability. The ground link antenna shown in Figure 12.23, for example, works without RF rotary joints, and with a single steerable reflector shell, which has been shaped to provide good performance over a wide MEO coverage.

A nearly isoflux gain profile can also be implemented using a fixed feed assembly, thus not requiring RF joints, and a single specially configured gimbaled reflector shell to steer the beam [39]. Such a concept is explained by Figure 12.24. As illustrated, this is a single-offset reflector geometry with an offset angle α_H (between the main axis of the parabola and the line connecting the focal point to the center of the reflector aperture). Panel (a) of the figure shows the reflector shell in a position such that the feed horn is located at the focal point, the feed horn axis

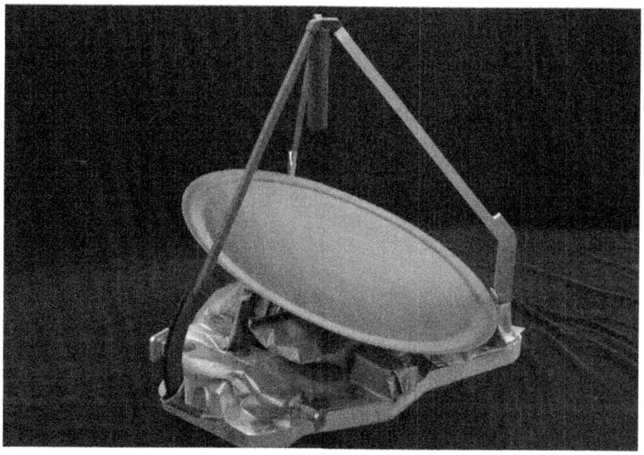

Figure 12.23 Steerable ground link antenna. Reproduced by permission of MDA.

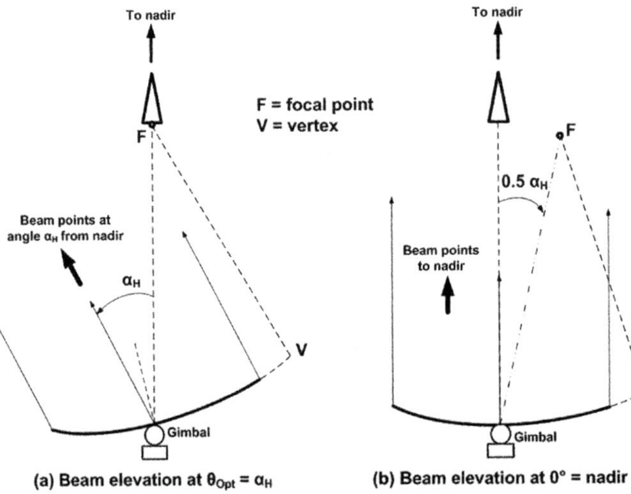

Figure 12.24 *Fixed-feed steerable-beam scanning concept.*

is oriented vertically along the satellite-to-nadir direction, the beam peak is along the axis of the parabola at an angle $\theta_{opt} = \alpha_H$ with respect to nadir (optimum elevation angle), and the gain is maximum since it is a focused geometry. The optimum angle θ_{opt} is usually selected so the maximum gain occurs near the edge of the global coverage area. By rotating the reflector shell in the elevation plane as shown, the focal point moves away from the feed horn (which is fixed, hence its axis always stays vertical), leading to a defocused geometry that degrades the peak gain as the beam is scanned relative to θ_{opt}. The beam is scanned by approximately twice the reflector shell rotation angle. Panel (b) of the figure shows the reflector shell rotated by $0.5\alpha_H$ relative to the focused position, thus the beam peak has been steered by α_H and is now directed at nadir.

Since, when the beam points towards nadir, the horn is no longer at the focal point of the reflector, the peak directivity is lower than at θ_{opt}. This natural behavior of the antenna partly compensates for the increasing path loss when the beam is steered from nadir to the edge of the Earth, and it thus approaches an isoflux beam scanning behavior. With this antenna configuration, in order to steer the beam anywhere within a $\theta_{\max}$ half-angle conical coverage, the angular rotation range of the reflector shell about the azimuth axis is 0°–360°, and about the elevation axis it is about 0° to $0.5\theta_{\max}$. The exact elevation angle range depends on the antenna geometry and the location of the elevation rotation axis.

Acknowledgments

The authors of this chapter would like to acknowledge the contribution of several people, whose solid experience acquired through many past satellite antenna programs was generously shared so as to lead to a better and more solidly based text. Particular thanks must be extended to Jaroslaw Uher, Sylvain Richard, Michel Forest and Joanna Boshouwers, for their invaluable inputs and suggestions.

References

1. ECSS (2003) *Space Engineering: Multipaction Design and Test*, ECSS-E-20-01A, 5 May.
2. Harris Corporation, Unfurlable Antenna Solutions, http://download.harris.com/app/public_download.asp?fid=463 (accessed 21 December 2011).

3. Thomson, M.W. (1999) The AstroMesh deployable reflector. Proceedings of the IEEE Antennas and Propagation Society International Symposium, vol. 3.
4. Brain, J.R. (2002) The INMARSAT 4 mobile antenna system. Proceedings of JINA Internationa Symposium on Antennas, vol. 2, pp. 333–336.
5. Angeletti, P. and Lisi, M. (2010) Multimode beamforming networks. 32nd ESA Antenna Workshop on Antennas for Space Applications, Session 17, ESTEC.
6. Angeletti, P., Alagha, N., and D'Addio, S. (2010) Hybrid space/ground beamforming techniques for satellite telecommunications. 32nd ESA Antenna Workshop on Antennas for Space Applications, Session 21, ESTEC.
7. Delaney, W.P. (1962) An RF multiple beam-forming technique. *IRE Transactions on Military Electronics*, **MIL-6**, 179–186.
8. Roederer, A.G. (1992) Semi-active satellite antennas. JINA'92 Proceedings.
9. Roederer, A.G. (2010) Semi-active satellite antenna front-ends: a successful European innovation. Asia Pacific Microwave Conference.
10. Balling, P., Mangenot, C., and Roederer, A. (2009) Multibeam antenna, US Patent 7,522,116, April 21.
11. Richard, S., Markland, P., Lanciault, F., and Dupessey, V. (2010) High power dual-polarized, combined Tx/Rx feed for L- and S-band mobile satellite service antennas. 32nd ESA Antenna Workshop on Antennas for Space Applications, Session 15, ESTEC.
12. Ilott, P., Patenaude, Y., and Ménard, F. (1994) MSAT antenna measurement program. AMTA 16th Meeting and Symposium, pp. 21–26.
13. Chu, T.-S. and Turrin, R.H. (1973) Depolarization properties of offset reflector Antennas. *IEEE Transactions on Antennas and Propagation*, **AP-21**(3), 339–345.
14. Ludwig, A.C. (1973) The definition of cross polarization. *IEEE Transactions on Antennas and Propagation*, **21**(1), 116–119.
15. Johansen, P.M. (1996) Uniform physical theory of diffraction equivalent edge currents for truncated wedge strip. *IEEE Transactions on Antennas and Propagation*, **44**(7), 989–995.
16. Ufimtsev, P.Ya. (2007) *Fundamentals of the Physical Theory of Diffraction*, John Wiley & Sons, Inc., Hoboken, NJ.
17. Mizuguchi, Y., Akagawa, M., and Yokoi, H. (1978) Offset Gregorian antenna. *Transactions of IECE Japan*, **J61-B**(3), 166–173.
18. Martins Camelo, L. (2010) The express AM4 top-floor steerable antennas: RF design and performance. Proceedings of ANTEM 2010, Session TP9.
19. Yilmaz, E., Yi, J., Yao, H., and Atia, A.E. (2009) Analysis of intercostal effect on dual gridded reflectors. IEEE APS–URSI Antennas and Propagation Society International Symposium.
20. Riel, M., Brand, Y., Cassivi, Y. *et al.* (2010) Design of low scattering struts for center-fed reflector antennas. Proceedings of EUCAP 2010, Session A20.
21. Pressensé, J., Frandsen, P.E., Lumholt, M. *et al.* (2010) Optimizing a corrugated horn for telecommunication and tracking missions using a new flexible horn design software. Proceedings of EUCAP 2010, Session C18P1.
22. Uher, J., Bornemann, J., and Rosenberg, U. (1993) *Waveguide Components for Antenna Feed Systems: Theory and CAD*, Artech House, Boston, MA.
23. Uher, J., Dupessey, V., Hotton, M. *et al.* (2010) High-power compact feed development at C and Ku-band. 32nd ESA Antenna Workshop on Antennas for Space Applications, Session 15, ESTEC.
24. Sarasa, P., Diaz-Martín, M., Angevain, J.-C., and Mangenot, C. (2010) New compact OMT based on a septum solution for telecom applications. 32nd ESA Antenna Workshop on Antennas for Space Applications, Session 6, ESTEC.
25. Amyotte, E., Demers, Y., Hildebrand, L. *et al.* (2010) Recent developments in Ka-band antennas for broadband communications. Proceedings of the 32nd ESA Antenna Workshop on Antennas for Space Applications, Session 2, ESTEC.
26. Rao, S. (2003) Parametric design and analysis of multiple-beam reflector antennas for satellite communications. *IEEE Antennas and Propagation Magazine*, **45**(4), 26–34.
27. Amyotte, E., Gimersky, M., and Donato, M. (2010) High performance Ka-band multibeam antennas. 19th International Communications Satellite Systems Conference, Section 31, #964.
28. Amyotte, E., Gimersky, M., Liang, A. *et al.* (2001) High performance multimode horn, US Patent 6,396,453, April 13.

29. Amyotte, E., Gimersky, M., Liang, A. *et al.* (2010) High performance multimode horn, European Patent 1,152,484, November 11.
30. Rao, S. and Tang, M. (2006) Stepped-reflector antenna for dual-band multiple beam satellite communications payloads. *IEEE Transactions on Antennas and Propagation*, **54**(3), 801–811.
31. Viganò, M.C., Caille, G., Mangenot, C. *et al.* (2010) Sunflower sparse array for space applications: from design to manufacturing. EUCAP 2010 Proceedings, Session C18P2.
32. Catalani, A., Russo, L., Bucci, O.M. *et al.* (2010) Sparse arrays for satellite communications: from optimal design to realization. 32nd ESA Antenna Workshop on Antennas for Space Applications, Session 4, ESTECs.
33. Lier, E., Purdy, D., and Maalouf, K. (2003) Study of deployed and modular active phased-array multibeam satellite antenna. *IEEE Antennas and Propagation Magazine*, **45**(5), 34–45.
34. Demers, Y., Amyotte, E., Apperley, J. *et al.* (2005) RF sensing front end equipment for multi-beam antenna applications. 28th ESA Antenna Workshop on Space Antenna Systems and Technologies, ESTEC.
35. Amyotte, E., Demers, Y., Martins-Camelo, L. *et al.* (2006) High performance communications and tracking multi-beam antennas. Proceedings of EUCAP 2006.
36. Dietrich, F.J., Metzen, P., and Monte, P. (1998) The Globalstar cellular satellite system. *IEEE Transactions on Antennas and Propagation*, **46**(6), 935–942.
37. Hirshfield, E. (1995) The Globalstar system. *Applied Microwave and Wireless*, Summer, pp. 26–41.
38. Schuss, J.J., Upton, J., Myers, B. *et al.* (1999) The IRIDIUM main mission antenna concept. *IEEE Transactions on Antennas and Propagation*, **47**(3), 416–424.
39. Amyotte, E., Gimersky, M., Gaudette, Y. *et al.* (2002) Steerable offset antenna with fixed feed source, US Patent 6,747,604, October 8.

13

SAR Antennas

Pasquale Capece and Andrea Torre

Thales Alenia Space Italia, Rome, Italy

13.1 Introduction to Spaceborne SAR Systems

13.1.1 General Presentation of SAR Systems

Earth observation from moving platforms (aircraft or satellites) has allowed, for many years, monitoring of large areas of the surface of our planet for disparate purposes.

The SAR(Synthetic Aperture Radar) is one of the main instruments for remote sensing from moving platforms. It is an active radar sensor that works in the microwave band. It is widely used because it provides spatial resolutions much better than those achieved by conventional radar techniques. This allows images of the Earth's surface to be produced by satellite with a quality that can be compared to that of optical sensors.

A SAR system is designed to create two-dimensional images of the observed scene (also known as swath) with sufficient spatial resolution in both the directions of azimuth (or along-track motion direction) and range (or cross-track direction normal to the motion); the observed scene is acquired off-nadir. Figure 13.1 shows the acquisition geometry of a SAR. The antenna is assumed to be a rectangular aperture with dimensions L and L_{el} in the azimuth and elevation directions respectively, H represents the flight altitude of the radar and R_0 the distance between the antenna and the center of the observed area. The off-nadir angle is denoted by α_0. The size of the antenna, as well as the frequency being used and the distance R_0, determine the size of the area observed on the ground. To a first approximation, the trajectory of the radar is assumed straight and parallel to the surface, the speed of the platform is constant and the Earth is assumed to be flat.

In Figure 13.1 the two different directions of slant range and ground range can be distinguished. The importance of the slant range is due to the fact that a SAR provides images of the observed surface projected onto this plane, as will become clearer below.

Space Antenna Handbook, First Edition. Edited by William A. Imbriale, Steven (Shichang) Gao and Luigi Boccia.

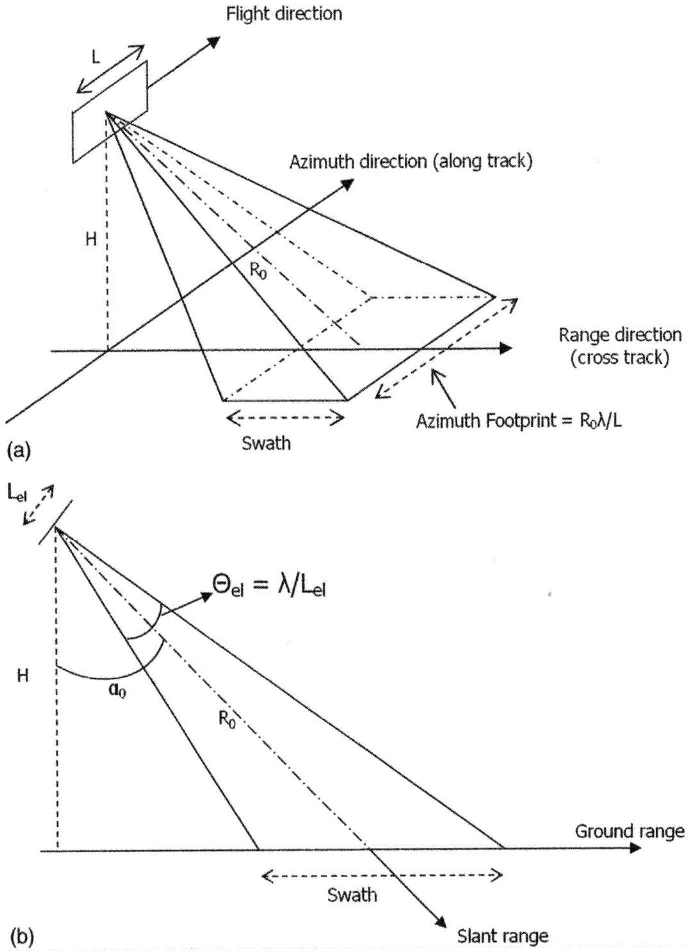

Figure 13.1 *(a) SAR imaging geometry (simplified) and (b) slant and ground range.*

13.1.2 Azimuth Resolution in Conventional Radar and in SAR

Figure 13.2 shows a view of the slant plane. In the absence of suitable processing, the resolving capability of the radar in the azimuth direction would be equal to the footprint and thus dependent on the slant range R_0. This resolution is the usual azimuth resolution of a conventional radar, which corresponds to several kilometers considering a feasible antenna length and a low Earth orbit (LEO) of 600–800 km in height.

SAR systems manage the problem of increasing the size of the antenna length by using the technique of synthetic aperture (hence the name of the class of radar) which consists of synthesizing an antenna much longer than actual size, thus achieving a much better resolution in azimuth. The synthetic aperture principle exploits two features: the relative motion between the antenna and the Earth's surface; and the nature of pulsed radar. In the synthetic aperture technique, the grand opening is not real, but is 'synthesized' by the movement of the platform: the antenna of a SAR, being in motion relative to the ground, can be considered as an array of antennas, that is, as a sequence of many dipoles displaced in the direction of motion and fed gradually. In each

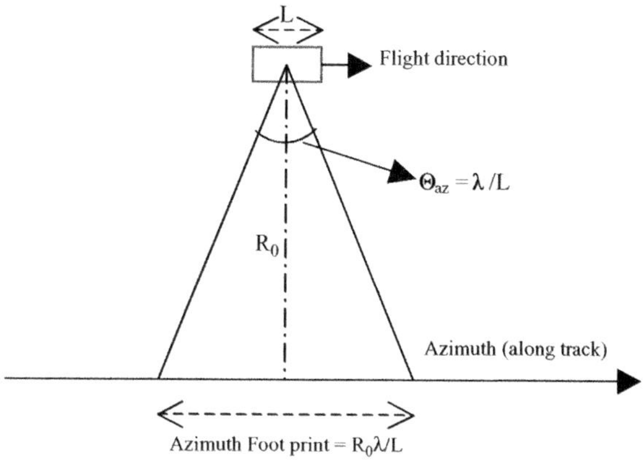

Figure 13.2 Azimuth footprint and azimuth resolution.

position the antenna transmits its pulse and receives the corresponding echo. The echoes received from the virtual dipoles are then processed in a coherent manner. With reference to Figure 13.3 it can be said that, in the absence of antenna beam steering in the azimuth plane, each target (*P*) is observed for a period equal to the azimuth footprint (equal to the distance between the two outer positions in the figure); therefore this size is equal to the maximum length of the synthetic aperture.

As a consequence the achieved azimuth resolution corresponds to an antenna size L_S. Thus the length of the synthetic aperture array, L_S expressed, in radians, is equal to:

$$\vartheta_S = \frac{\lambda}{2L_S} \tag{13.1}$$

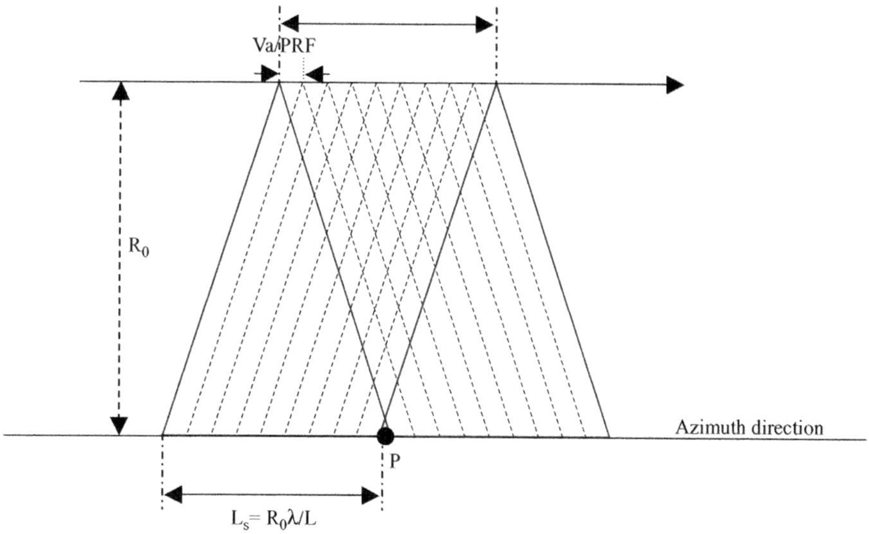

Figure 13.3 Synthetic aperture length (PRF, Pulse Repetition Frequency).

which provides an azimuth resolution R_{AZ} equal to

$$R_{AZ} = \frac{\lambda}{2L_S} R_0 = \frac{\lambda}{2(\lambda/L)R_0} R_0 = \frac{L}{2} \tag{13.2}$$

This result is rather surprising because it shows that the azimuth resolution is not dependent on the flight altitude of the satellite and seems to be improved by simply reducing the physical dimensions of the antenna. However, as will be shown in the next subsection, it is not possible to design antennas with an arbitrarily small azimuth dimension. Hereafter some of the main SAR system parameters will be defined to better understand the antenna dimensioning process.

13.1.3 Antenna Requirements Versus Performance Parameters

The most common operative modes of the SAR are the following:

- stripmap;
- scansar;
- spotlight.

The *stripmap* mode is the standard mode of the SAR: the antenna is pointed at a fixed angle during the data collection along the flight path. This means that the antenna scans a particular area which is parallel to the flight path. The antenna beam can be squinted backward, forward or just perpendicular to the flight path, and its pointing is fixed during data acquisition.

The *scansar* mode is used to increase the size of the swath. This is achieved by a fast switching of the antenna beam in the elevation direction. The swath area is divided into sub-swaths and the antenna beam is periodically pointed to each sub-swath.

In *spotlight* mode the antenna is steered (mechanically or electrically) during the flight path to illuminate a particular area of interest. This mode creates a very long synthetic antenna which increases the azimuth resolution at the expense of smaller illuminated target scene.

All the above radar modes require an antenna with beam steering capabilities. In particular stripmap and scansar modes need only elevation steering to point the beam toward the required swath or sub-swath, while the spotlight mode also requires azimuth steering. The scansar fast switching requirements entail the use of an electronically steerable antenna. In particular SAR antennas operating in scansar mode should possess beam shaping capabilities. Indeed the elevation beamwidth should be adapted to maintain the same swath width when the elevation angle is changed to scan different sub-swaths.

Among the many performance parameters defining a SAR level, the following directly drive the sizing of the SAR antenna:

- access region;
- swath width;
- ambiguity-to-signal ratio;
- noise equivalent sigma zero.

The SAR *access region* is defined as the range of incidence angle coverage where images are acquired.

The *swath* is defined as the on-ground extension of the imaged area (along-track and across-track direction) where the required imaging performance is met. The swath should be measured in km × km. The minimum swath overlap is defined as the minimum on-ground superimposition in the across-track direction between imaged areas related to subsequent (in terms of incidence angle) beams. The minimum swath overlap should be measured as a percentage (on-ground subsequent imaged areas vs. across-track swath extension).

For the *ambiguity ratios*, the distributed target ambiguity ratio (DTAR) is defined as the ratio of the mean intensity of the ambiguous zones in the azimuth and range plane to the intensity of a distributed target located in the unambiguous zone. The point target ambiguity ratio (PTAR) is defined as the ratio of the mean intensity of the ambiguous zones in the azimuth and range plane to the intensity of a point target located in the unambiguous zone.

The *noise equivalent sigma zero (NESZ)* is defined as the backscattering equivalent value assuming a uniform target that produces a signal of intensity equal to the thermal noise. The NESZ should be measured in dB.

In the following the relationships and mutual constraints between these parameters will be examined. In particular the operational constraints related to the basic SAR operating mode (i.e., stripmap) will be reviewed first. Then the criteria for selection of the antenna length and the PRF will be described in order to show their effect on the azimuth resolution and ambiguity. As will be shown, once these parameters are fixed, the analysis can be extended to define the scenarios which guarantee full coverage in the specified access region.

The second step is a description of the bounds on the antenna area driven by the constraint on echo ambiguity and range ambiguity-to-signal ratio. Definition of the antenna area will be linked to the selected antenna length and aspects related to accommodation on the spacecraft.

The analysis of the required power, in order to comply with the needed NESZ, starts with the assumptions on the available power and imaging scenarios.

13.1.3.1 Antenna Dimensioning

The definition of a SAR imaging scenario usually starts from the constraints imposed by the pulsed operation of the instrument. The pulse repetition frequency (PRF) defines the interval (PRI) where the echo signal, related to transmitted pulse n, may be received with no overlap with the echo related to pulse $n-1$ (and $n+1$ if the instrument works in ambiguity of pulses as in spaceborne systems). For a given geometry and PRF, the meaningful extension of each swath is defined taking into account the 'blind ranges' occurring when the antenna is transmitting pulses, and hence not receiving echoes; these can be described in terms of their echo window timing by the inequality

$$\left(\frac{m}{\text{PRF}}+\tau+g_n\right)\frac{c}{2}\leq R\leq\left(\frac{m+1}{\text{PRF}}-\left(\tau+g_f\right)\right)\frac{c}{2} \tag{13.3}$$

where m is an integer (≥ 0), τ is the pulse length, g_n and g_f are guard bands, c is the propagation velocity of light and R is the slant range to the swath.

The guard bands, typically on the order of a few microseconds, allow for switching between transmission and echo reception. The pulse length is normally driven by energy considerations in order to achieve the required performance on NESZ.

For each pulse the echo return begins with the nadir return (at closest range) after a time $2H/c$, where H is the satellite altitude, and spreads out to the horizon. At any particular time the returns from all points on the same range curve will be received. The strength of the return will depend upon the two-way antenna gain, the radar cross-section and the imaging geometry.

The radar cross-section for most targets is much stronger at low incidence angles. Since the SAR transmits a regular stream of pulses, the desired echo return from a particular target location for one pulse will coincide with returns from the other locations for other pulses. These are referred to as ambiguous returns, and are usually suppressed by specifying a null region in the antenna pattern. However, in order to avoid excessive demands on the pattern, it is often desirable to avoid strong returns from the nadir region. This constraint is translated into the equation

$$h+\left(\frac{n}{\text{PRF}}+\tau+g_n\right)\frac{c}{2}\leq R\leq h+\left(\frac{n+1}{\text{PRF}}-(\tau+g_f)\right)\frac{c}{2} \tag{13.4}$$

where n is an integer (≥ 0) and g is a guard band.

13.1.3.2 Selecting the Antenna Length and the PRF

The selection of the antenna length is driven by the required azimuth resolution since it constrains the azimuth beamwidth and the corresponding Doppler bandwidth, B_D, according to the following two equations which represent the same result of Equation 13.2:

$$B_D=\frac{4V_S}{\lambda}\sin\left(\frac{\vartheta_a}{2}\right) \tag{13.5}$$

$$\rho_A=\frac{V_S}{B_D}=\frac{\lambda}{4\sin(\vartheta_a/2)}\cong\frac{\lambda}{2\vartheta_a}=\frac{\lambda}{2\lambda/L}=\frac{L}{2} \tag{13.6}$$

where V_S is the sensor velocity, λ is the carrier wavelength, ϑ_a is the antenna azimuth beamwidth, ρ_A is the azimuth resolution and L is the antenna length.

The required azimuth resolution drives the selection of the antenna length (i.e., $L=2\rho_A$). The same azimuth resolution can be achieved by using shorter antenna lengths but, in this case, higher PRFs will be required. However, this possibility can potentially cause a reduction in the swath extension which must fit within the time between two consecutive pulses, that is, the pulse repetition interval PRI = 1/PRF.

In order to maximize the unambiguous swath coverage, the PRF should be selected in order to satisfy the following equations:

$$\text{PRF}=kc/2h \tag{13.7}$$

where

k is an integer
c is the propagation velocity of light
h is the satellite altitude

with the further constraint of

$$\text{PRF}>K_aB_D \tag{13.8}$$

where k_a is the azimuth oversampling factor (e.g., 1.2). This ensures that the nadir returns coincide with the pulse transmission periods and that the Doppler bandwidth is sampled with no aliasing.

13.1.3.3 Selecting the Antenna Width

The antenna size is closely linked to the resolution and coverage requirements. The minimum width–length product for a SAR antenna is set to avoid overlapping echoes and to sufficiently sample the Doppler bandwidth. Moreover a shaping factor has to be considered to reduce the sidelobe level of the beam in receiving mode. The minimum antenna width is approximately given by

$$\frac{W}{\lambda} = Kw \frac{4hv \sin(\psi)}{cL \cos^2(\psi)} \tag{13.9}$$

where

W is the antenna size in the elevation plane
ψ is the incidence angle
h is the satellite altitude
L is the antenna length
c is the speed of light
Kw is a shaping factor (typically from 1.3 to 1.5).

Equation 13.6 showed the relationship between the azimuth resolution and the antenna length. Typically the same resolution is required also along the elevation plane (range) of the swath. The resolution across the swath depends on the signal bandwidth B and on the incidence angle according to

$$\rho_E = \frac{c}{2B \cos(\psi)} \tag{13.10}$$

where

B is the chirp bandwidth
ψ is the incidence angle
c is the speed of light.

The NESZ can be computed by using the following equation [1]:

$$\text{NESZ} = \frac{4(4\pi)^3 R^3 k_B T_0 B_{tx} N_f V_s \sin\theta_i \cdot L_{tot}}{cP_m D_{TX} D_{RX} \lambda^3} \tag{13.11}$$

where

R	slant range;
k_B	Boltzmann's constant;
T_0	standard temperature (290 K);
B_{tx}	transmit pulse bandwidth;
N_f	noise figure;
V_s	satellite velocity;
θ_i	incidence angle;
L_{tot}	total losses;

c velocity of electromagnetic propagation;
P_m mean transmitter power for each transmit polarization (i.e., for quad polar mode P_m is half the actual mean transmitter power);
D_{TX} transmit directivity calculated for look angle offset by mechanical boresight angle and satellite roll error;
D_{RX} receive directivity calculated for look angle offset by mechanical boresight angle and satellite roll error;
λ wavelength at radar center frequency.

It is now possible to evaluate the total power to be generated by the transmitter or, in the case of active antennas, the RF power which each TR module has to generate.

13.2 Challenges of Antenna Design for SAR

The size of SAR antennas depends on three main factors:

- the required performance of the SAR system;
- the size of the satellite body and the selected launcher;
- the technologies available.

The required beam flexibility is the major antenna requirement in the SAR performance. According to this parameter a choice between reflector antennas and a phased array has to be made. If limited performance can be tolerated, the reflector solution appears to be the most promising and cost effective. If azimuth and elevation steering and beam shaping are considered mandatory, the electronically steerable phased array has to be considered at the expense of high complexity and cost.

13.2.1 Reflector Antennas

13.2.1.1 Reflector Dish

Compared to planar array antennas, reflector-based antennas can achieve higher efficiency with large apertures, and have a lower cost and wider bandwidth. However, they have a limited capability of beam steering and shaping, depending on the flexibility and complexity of the feed system. Another critical aspect of reflector-based antennas is the possibility to stow the antenna on the satellite in the launcher fairing. Two kinds of reflectors can be considered:

- solid reflectors;
- unfurlable meshed reflectors.

Solid reflectors are typically constituted by a sandwich made of dual skins of CFRP (Carbon Fiber Reinforced Plastic) with aluminum honeycomb in the middle: this is a very stable and mature technology, having been used for decades in space applications. It is quite cheap and has very good surface control, but the maximum dimensions allowed at launch are in the range of 3–4 m for the longer dimension. A larger size will require the reflector dish to be divided into a number of sections, but the allocation on the spacecraft still remains quite difficult due to the non-planarity of each section.

An example of a solid reflector with a single horn used for SAR applications is the SAR Lupe mission. A larger reflector has to adopt deployable/unfurlable technologies. This approach has been used to realize reflectors for radar missions from 3.5 m in diameter (TecSAR) to medium-size antennas (6 m in diameter – Kondor, Russia) and up to tens of meters (it seems that a very large unfurlable reflector has been adopted on the US Lacrosse or Onyx spy missions). A number of unfurlable reflectors have been also used in recently for low-frequency communications satellites such as the Thuraya mission (USA) (12 m diameter) or INMARSAT-4 (UK) (9 m diameter).

13.2.1.2 Feed System

The use of multiple feeds can enhance the flexibility of the reflector-based SAR antennas. They are suitable for implementation with some limitations in stripmap, scansar and spotlight modes, but also moving target indication (MTI) functions.

These feeds are consistent with the realization of a limited number of (fixed, switchable or reconfigurable) beams realized by single/multimode sectorial horns in single polarization or by a cluster of dual-polarization beams using high-efficiency small horns for segmentation of each sectorial horn. The number of horns to be used can range from about 8–10 up to several dozen, according to the segmentation and the flexibility required. Dual-polarized beams are achievable by using OMTs (orthomode transducers) at the input of the small horns and two separated beam-forming networks. In the case of the two polarizations are not needed simultaneously, switches can be used to avoid the RF network duplication. Concerning the beam forming, this can include high-power variable devices, mechanically driven by small motors, in order to transfer in continuous mode the power from one horn to another one, realizing in this way a slow beam steering capability. The drawback is that high-power variable passive components are expensive, bulky and require dedicated electronics, increasing the overall complexity and cost. All the above considerations are applicable in the case of a single high-power amplifier (e.g., the TWTA, Traveling Wave Tube Amplifier) of some kilowatts of RF peak power is used to generate the whole transmitted power. Alternatively a cluster of medium-power amplifiers, of some hundreds of watts each, can be used to feed a number of sectorial horns by a multimatrix approach. It consists of using a low-power input matrix, placed before the cluster of TWTAs, which divides the RF input to all the TWTAs, and a high-power output matrix which recombines all the amplified signals into one or a set of radiating horns. A low-power beam former, variable in amplitude and phase, can be used at the input to the first matrix to select to which horn or group of horns the RF power has to be addressed. It includes amplitude and phase controllers to change the excitations to the TWTA inputs. In this case fast beam shaping and steering is possible, acting on the low-power electronic devices, and thus achieving a fast reconfigurability. This is done at the expense of using a high-power output matrix, which can be massive and introduces some additional losses. Considering the high-power handling requirement, the output matrix has to be realized by using waveguide technology, so this approach is limited to the higher frequency bands (C- or X-band) due to the large size for lower bands. Moreover the overall system, based on the cluster of TWTAs, needs to track both the gain and the phase of each amplifier to allow good sidelobe control of the radiated beam.

13.2.2 Active Antennas and Subsystems

The SAR antenna requirements include several RF, electrical, mechanical and thermal parameters. The most important one is the required beam flexibility, such as shaping and steering capabilities, which implies the need to consider an active phased array instead of alternative solutions such as multi-feed reflector antennas, or passive phased arrays. The technologies available for the radiating linear array, transmit/receive (TR) modules and electronics have a strong impact on other fundamental parameters such as mass, RF efficiency, DC power efficiency, and thermal dissipation. Consequently, the antenna dimensioning should be the best compromise among all these aspects.

The radiating aperture of the antenna is typically arranged into a rectangular grid with a number of rows and columns which defines the antenna dimensions taking into account the size of the elementary radiator. This fixes also the number of TR modules to be used inside the antenna and therefore defines the amount of electronics, such as digital and power printed circuit boards, needed to drive the TR modules. Considering that the steering angles required for the antenna are typically in the range of 15–20° for the elevation (range) plane, while low steering (1–2°) is needed for the azimuth plane, the elementary radiator consists of a linear array of small sources such as slots or patches. Typically the SAR antennas are organized in mechanical panels to be deployed on the satellite, each including a number of tiles. A tile consists of a number of TR modules, RF radiators, RF distribution and all the electronics to feed and control them. The tile configuration is therefore the fundamental brick used to build the whole antenna.

13.2.2.1 RF Subsystem

The main RF parameters to be considered for the design of SAR antennas are the following:

- the beamwidth in the azimuth and elevation plane;
- the range of beam steering in the azimuth and elevation plane;
- the central frequency and bandwidth;
- the equivalent isotropic radiated powerEIRP;
- the noise figure;
- the dual-polarization capability;
- Tx and Rx active gains.

Antenna Length vs. Azimuth Beam Size and Minimum Elevation Beamwidth The antenna length is directly related for standard operating modes, such as the stripmap or scansar, to the required radar resolution along the azimuth plane. The simple formula is

$$\text{Az-res} = L/2$$

where Az-res is the azimuth resolution of the SAR and L is the antenna length.

This relation is realized for antennas with a length in the range of, typically, 4–5 to 10–12 m, and this is independent from the working frequency of the radar which can range from P/L-bands (400–1200 MHz) up to X-band (10 GHz) or higher.

An important aspect for space application is that an antenna of such a size has to be fitted on a satellite which should have its total envelope at launch minimized as much as possible to meet the allowed launcher fairing in both diameter and height. Planar antennas allow reduced envelopes at launch by folding into a number of flat panels, each having a length of 2–3 m.

Just in ase the overall antenna length is compatible with the launcher envelope (up to 4–5 m in height for small and medium-size launchers), a deployment system can be avoided as was the case of the German TerraSAR-X, or a very large vehicle is used, as the SIR-C mission in which the large cargo bay of the Space Shuttle was able to carry on three antennas with a total envelope of 12 m × 4 m.

The RF distribution along the azimuth plane of the antenna is uniform in both amplitude and phase in order to guarantee maximum aperture efficiency, minimum antenna length, taking into account that no particular sidelobe control is required. The typical −13 dB for the first sidelobe level is well tolerated by the radar.

Dividing the antenna into panels along the azimuth direction gives rise to grating lobes along this plane. Such grating lobes are caused by:

- mechanical alignment among the panels;
- the average of the RF amplitude and phase of each panel;
- the efficiency achieved inside each panel vs. the theoretical uniform RF distribution.

The first issue is typically the most important and its impact increases with the working frequency of the radar. Deployment errors, misalignment and rotation have to be minimized to avoid the increment in sidelobe level and grating lobes. An antenna planarity of $\lambda/15$ or $\lambda/10$ maximum is therefore required, which is one of the major requirements for the design of deployment mechanisms and also for the overall antenna thermomechanical design, that has to guarantee very low thermal distortion under any operating condition.

Accurate design of the RF distribution is also needed. In this case the most critical aspect is the need to guarantee very good phase stability vs. the antenna thermal condition and in particular the thermal gradients along the antenna azimuth. Finally a uniform distribution along each panel is achieved by subdividing the panel aperture in columns of linear arrays, each of which has to guarantee good aperture efficiency (typically 0.5–0.7 dB or better) over the whole frequency bandwidth to achieve good aperture efficiency for the whole antenna.

Concerning the antenna height, it depends on the required beam sizing and shaping along the elevation (range) plane. For a higher steering angle, which corresponds to a higher access angle, a narrow beam is necessary. This implies narrow beamwidths, typically in the range from 1° to 3° minimum, which corresponds to an antenna height from a minimum of 15 up to 50 wavelengths. Larger beamwidths are then achieved by controlling the aperture field amplitude and phase distribution; in this case the beam is broadened by up to two or three times the minimum required beamwidth.

Azimuth and Elevation Plane Steering Angles The steering angle along the azimuth plane is used to realize spotlight or top SAR radar modes and ranges typically from 3–4 to 6–7 beamwidths. The antenna includes a number of linear arrays along the azimuth plane. The length of this radiator is determined by the acceptable level of grating lobes when the beam is steered to th emaximum angle. In Figure 13.4 the grating lobe level is shown for different steering angles and linear array lengths, where the linear arrays are arranged in a rectangular grid. As an example, grating lobes less than -14 dB along the azimuth plane can be assured for a steering of $\pm 1°$ by a linear array having a length of 9λ. Reducing the length of the linear array reduces the

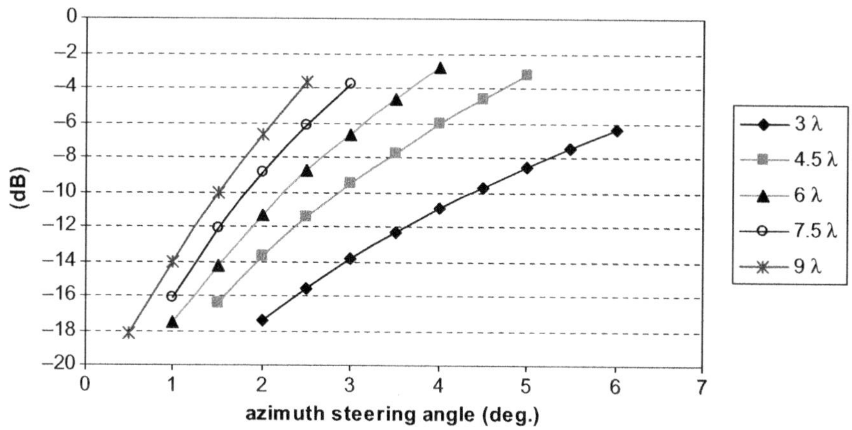

Figure 13.4 *Grating lobe level vs. azimuth steering angle and linear array size.*

grating lobe level as well: it will be better than -17 dB with 6λ linear array length, but this is achieved at the expense of a strong increment (50%) in the number of elementary radiators in the antenna and therefore of the TR modules and the relevant electronics.

The required elevation plane steering angle is typically $\pm 15/20^\circ$, to reach the required access zone up to 60° from the nadir axis, with the aim that the radiator spacing should be less than 0.7–0.75λ, to avoid grating lobes on this plane also.

Central Frequency and Bandwidth SAR systems have been developed at different central frequencies. The ones most used so far are the L-, C- and X-bands. The first SAR system launched by NASA (1978) was embarked on the Seasat satellite. Its radar RF frequency worked in L-band, single HH polarization, and it was operative for some months, demonstrating the validity of the SAR system concept with a microwave sensor. In the following years NASA launched another two missions (SIR-A and SIR-B) using the Shuttle, still working in L-band and in single polarization. In the 1990s, ESA prepared and launched two missions, ERS-1 and ERS-2, working in C-band and single vertical polarization. In this case the antennas were realized in slotted waveguide technology. In parallel NASA realized the SIR-C experiment on the Shuttle where three separate antennas working in L-, C- and X-band were used (the X-band antenna was a passive antenna, vertically polarized, designed and realized in Europe by Italy and Germany). The L- and C-band antennas were dual-polarized active phased arrays designed and manufactured in USA, and they allowed realization of the first quad polarization mission. Several other programs by Canada, the former USSR, Japan, Italy and others confirmed in recent years the interest in these frequency bands. The main difference when working in L-,C- or X-band is that the antenna dimension changes drastically along the elevation plane to keep the beamwidth in the range of 2–3°. An example is the SIR-C mission with three antennas in the L-,C- and X-bands. They had the same length of about 12 m, while their range dimensions varied from 35 cm of the X-band, to less than 1-m for the C-band and up to 2.8 m for the L-band antenna. Different antenna size implies different electronic density inside the antenna. Increasing the frequency decreases the available space for distributed amplification and so a more compact approach has to be followed. In a standard rectangular grid of rows and columns in L-band the sub-array spacing typically allows 160–180 mm of inter-axis distance for the amplifiers, while in C- and X-band this value reduces to about 40 or 20 mm. This impacts in a very significant manner the whole thermal–management design, due to the different amounts of heat generated by each square meter of the antenna.

Another important aspect is the bandwidth: in L-band it is typically in the range of a few tens of megahertz up to 70–80 MHz, while in C- and X-band this range is exceeded, up to some hundreds of megahertz, but still maintaining to within a few percent the bandwidth w.r.t. the central frequency. This is important in order to use relatively simple technology for radiating elements such as printed dipoles, patches or slotted waveguides. The instantaneous bandwidth also affects the beam pointing. Two main parameters have to be considered: the steering angle and the beam size. Since the phase distribution needed to point the beam in a given direction depends on the frequency, this implies that, when changing the frequency in the working bandwidth, the beam moves slightly. This movement can be tolerated if it is small w.r.t. the beamwidth, otherwise countermeasures have to be considered to reduce or eliminate the problem. The beam pointing variation vs. frequency is given by the formula

$$\Delta\vartheta = \vartheta_{st} \cdot \Delta F / F_c$$

where

- $\Delta\vartheta$ is the pointing variation
- ϑ_{st} is the steering angle
- ΔF is the frequency bandwidth
- F_c is the central frequency.

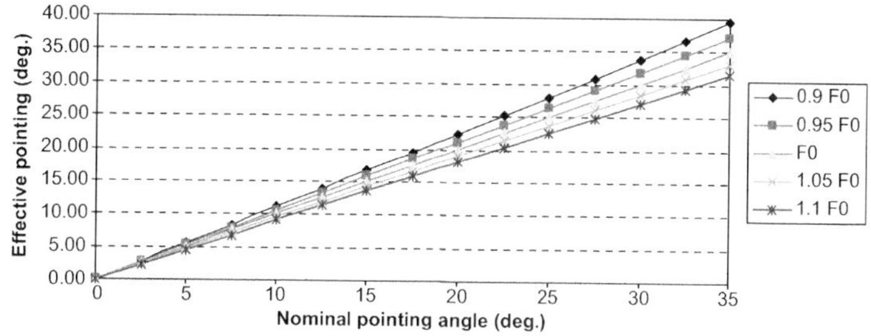

Figure 13.5 *Beam pointing vs. frequency in degrees.*

In Figure 13.5 the beam pointing vs. frequency variation at different steering angles is shown for a generic phased array. For SAR antennas with a typical bandwidth of 1–2% and 15° of steering angle along the elevation plane, the pointing variation expected will be on the order of 0.15–0.3°. This value has to be compared to the antenna beamwidth along that plane. For antennas having limited bandwidth and relatively large beamwidth in the range of 2.5–3°, the above pointing is less than or equal to 10% of the beamwidth; in this case the pointing variation can be tolerated. If the beam pointing variation corresponds to an important fraction of the beamwidth, it is necessary to introduce true time delay lines (TTDL)s into the beam-forming network, each acting over a part of the antenna surface (typically a tile). The TTDL realizes a variable delay, in terms of multiples of wavelength, according to the steering angle and the position of the sub-arrays, implementing a phase slope vs. frequency which allows the phase distribution to be corrected over the antenna surface in the frequency band. Fast switching is necessary in case fast beam change or steering is required. Ferrite circulators could be used as switches with waveguide technology; more integrated solutions based on microwave integrated circuit (MMIC) switches and microstrip lines can be also adopted, but they require additional amplification.

EIRP The radiated power of the antenna in each direction depends on the beam directivity and on the power generated by the distributed amplifiers. Since the antenna length is fixed on the basis of the desired beam size and azimuth resolution, the directivity of the antenna beam will derive mainly from the antenna height. In fact, by increasing the antenna dimension along the elevation plane, it will be possible to improve the beam directivity to some extent, having tighter beam control w.r.t. the required beam masks. On the other hand, the power generated by the amplifiers will be directly proportional to the number of TR modules and the power generated by each one.

Noise Figure The noise figure of the antenna is affected mainly by three things:

- the losses in the radiating element;
- the noise figure of the receiving amplifiers (TR modules);
- the losses in the beam-forming network.

The radiating element consists of a linear array of a number of elementary radiators such as dipoles, patches or slots. This number typically ranges from a minimum of 8–10 up to 22–24 elements; increasing their number makes the design of the two polarization RF dividers more difficult, and also means higher losses. They will also depend on the technology adopted for the RF distribution, the central frequency and the bandwidth. As will be discussed later, slotted waveguides and printed dipoles or patches are typically adopted for SAR antennas, all of which present positive and negative aspects for consideration. Typical values achieved in L-,C- or X-band for ohmic losses range from 0.5 to 1 dB.

Concerning the amplification placed after the radiating arrays, the noise figure will depend on the selected technology used for the low-noise amplifiers (LNAs); in particular better noise figure could be achieved by using discrete active RF circuits, but the need to have greater integration to reduce production cost leads to extensive use of MMIC LNAs, which are simpler to integrate into the TR modules. Finally attention must be paid to the case of intermediate amplification needed in the Rx path; in fact, for shaped beams, the attenuation introduced to taper the antenna aperture can degrade the overall noise measured at the output port of the antenna.

Dual Polarizations and Cross-polar Purity Almost all the antennas designed and realized for SAR so far are required to work in dual linear polarizations: namely, the horizontal (H) and vertical (V). While in receiving mode single, alternated or contemporary dual polarizations are required, in transmission mode a single polarization is typically used (selectable by command), and only in a few cases and for experimental purposes is it required for the antenna to combine, in Tx mode, both the linear polarizations to have a 45° polarized transmitted signal or, by suitable phasing of the two polarizations, to generate a circularly polarized field.

The dual-polarization functionality is one of most demanding requirements for active planar antennas due to the complexity and cost impacts. As for the radiating element, its impact is large: on the one hand, it has to double the beam-forming network inside each radiating array to feed the slots/patches used to radiate the electromagnetic field; on the other hand, the cross-polar requirements for the radiators are increased to improve the isolation between the contemporary received signals in H and V polarization. For the amplification, in order to avoid the ohmic losses caused by RF switches (on the order of 0.5–0.7 dB; PIN diodes are normally used), two Tx and two Rx chains have to be implemented inside the antenna for each radiating element, almost doubling the cost of the RF electronics in the antenna. Cross-purity is affected both by the performance of the radiating element and by the isolation of the H and V channels inside the TR modules. In the radiating element, linear arrays based on slots (typically slotted waveguides) have the advantage of generating a very low cross-level (it depends mainly on the slot orientation) while printed patches suffer from internal coupling among the orthogonal ports, which drastically reduces the cross-performance in the frequency band. Sequential rotation techniques are usually adopted to cancel the cross-field level in the field of view. In all cases 30–35 dB of cross-purity is achievable in a bandwidth of some percentage. The TR modules for dual polarization can be designed and manufactured in a dual-channel assembly, but the single-channel approach represents the simplest way to avoid internal coupling in the TR module cavity between Rx channels when it is required to receive two contemporary polarizations.

RF Chain Main Elements The RF chain of such antennas consists of the following main elements:

- radiating element;
- TR module;
- intermediate amplification and/or TTDL;
- beam-forming network.

The performance of all these devices depends on several parameters such as the central frequency, the bandwidth and the selected technology to realize them. These items are discussed in more detail below.

Radiating Element Central frequency, bandwidth, ohmic losses, cross-polarization purity, but also mass and manufacturing cost, impact the selection of the technology used to realize the linear array. Two technologies have been used up to now for the RF radiator of SAR antennas: slotted waveguides and linear arrays of patch elements.

The slotted waveguides have been proposed for several years to realize large antenna apertures. The radiation mechanism is based on the fact that each slot cuts the current lines on the waveguides walls, radiating a small amount of RF signal. Vertical polarization is achieved by resonant slots cut on the broad wall of the waveguide, alternately placed w.r.t. its center line and spaced about a half wavelength in the waveguide. Horizontal polarization can be obtained by cutting slots on the narrow wall of the waveguide. In this case the slots are slightly inclined to cut the vertical currents which flow on the narrow wall of the waveguide. The waveguides are centrally fed by a launcher and short-circuited at their extremes. The bandwidth reduces, increasing the number of slots and therefore the length of the waveguide. The slotted waveguide presents low losses (in the range of 0.3–0.5 dB for 20–24 slots), and can be used for narrow bandwidth, giving the possibility to realize quite long linear arrays. The main drawback with this radiator is the strong mutual coupling, which affects closer waveguides in particular for vertically polarized slots, and the consequence that there is a strong active return loss with particular sets of excitation. Dual polarization requires a more complex approach to allow two slotted waveguides to be accommodated in the allowed spacing of about 0.7–0.75λ. Recent designs [2] for dual-polarized radiators are based on a ridge slotted waveguide for vertical polarization alternating with a second waveguide with slots cut on the narrow wall. Internal discontinuities such as wires or irises are used to locally modify the waveguide's electric field, in this way causing the slots to radiate. Slotted waveguides are frequently used for medium and high frequencies (C- and X-bands), while lower frequency applications are limited by the dimensions that the waveguide should have to work in L- or S-band.

The patch solution is typically based on the use of one or two stacked square or circular patches, excited by one slot or two slots (for the dual-polarized case) cut on the ground plane. This configuration is adopted to increase the working bandwidth. In this case the RF distribution is achieved by using microstrip or stripline beam formers placed under the ground plane. This design has the advantages of minor cost w.r.t. the waveguide solution, larger bandwidth and good applicability at lower frequencies (P-, L- and S-bands) and up to higher frequencies (C- and X-bands). Two of the major drawbacks of this approach are: the ohmic losses for the high frequencies (which can range from 0.7 to 1.2 dB or more, depending on the frequency and length of the array) due to the losses on the beam former; and the cross-polar level which is affected by the radiation from both the patch and the microstrip lines. In general, good control of the cross-level in the main beam region can be achieved by exciting the patches out of phase and from opposite sides, thus strongly reducing the generated cross-polar levels. Optimizing the microstrip layout under the ground plane, the contribution to the cross-polar levels generated by the discontinuities on the microstrip RF dividers can be minimized.

TR Module Distributed amplification is realized by the so-called TR modules (TRMs). The TR module is typically a hybrid circuit containing a few MMICs which realize the main RF functions. The TR modules can be realized in single-channel or dual- and multichannel configurations. In the first case a hybrid will have one Tx and one Rx chain, while the multipack approach will contain from 2–4 up to 8 Tx channels and the same number of Rx channels. The single-channel approach is simpler to produce and guarantees better isolation among the several channels. A single-channel TR module has typically three RF ports: one input Tx, one output Rx and one antenna port. Moreover it needs digital and power electrical interfaces (I/Fs) to control the setting of the amplitude and phase of the module and to supply all the voltages and currents required by the internal amplifiers. A typical TR module has three main sections:

- A control section, to drive the amplitude and phase of the Tx and Rx RF signal.
- A Tx amplifier section, which amplifies and generates the RF power to be transmitted.
- A Rx amplifier section which amplifies the received signal with the desired noise level and gain.

The control section is implemented by dedicated attenuator and phase shifter chips which have to be driven by parallel digital signals (6 + 6 bits) or by more compact multi-functional chips, as depicted in Figure 13.6.

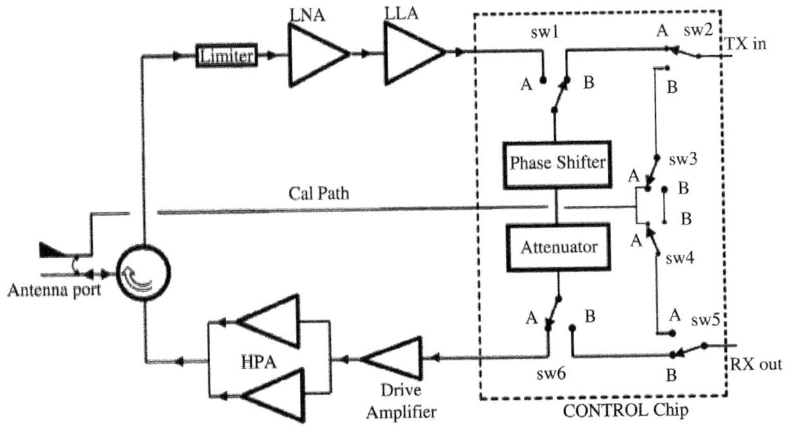

Figure 13.6 *TR module typical schematic layout.*

This chip, also known as the 'core chip,' is used in modern TR modules to allow amplitude and phase digital control, Tx and Rx amplification functions and a set of switches which gives the possibility to select Tx or Rx functionality or the calibration path for the purpose of calibrating the radar amplification chains. These switches have to guarantee high isolation to reduce possible internal coupling and to allow more precise measurement of the gain and phase of the active Tx and Rx chains during calibration. The control chip also includes a shift register to convert the input serial data into parallel format to set the attenuator, phase shifter and switches. The switching time is in the range of tens of nanoseconds. Tx amplification is achieved by a cascade of two chips: a driver chip used to amplify the signal coming from the core chip; and a HPA (High-Power Amplifier) to generate the required power to the output of the TR module. If higher power should be needed, two HPAs can be used to gain 3 dB on the output signal level as shown in the schematic. The RF amplified signal is than routed to the antenna port through a circulator. A proximity coupler, placed between the circulator and the antenna connector, is used to extract or inject RF signals from or to the Tx and Rx ports of the module. The Rx signal is routed by the circulator to the LNA; it can be followed by a low-level amplifier (LLA) to improve the gain of the Rx path and increase the dynamic range of the received RF signal. The typically achieved Tx power per TR module, for space SAR antennas, is reported in Table 13.1 considering mature and available technologies for space such as MMIC GaAs MESFETs (L-band) and PHEMTs (C- and X-bands).

The total RF power generated by the antenna will correspond to the product of the power generated by a single amplifier and the number of amplifiers inside the antenna. The number of TR modules can vary from about 100 elements (as is the case of L-band systems), to 300–500 TRMs (C-band) and up to 1000 modules or more for X-band. This implies that the overall generated RF power is typically in the range of some kilowatts of peak power.

Table 13.1 *TR module, typical peak power level.*

Frequency band	Peak power level (W)
L	35–40
C	16–20
X	4–8

The TRM noise figure depends on the losses introduced by the limiter and by the LNA noise figure. GaAs PHEMTs today allow values from 1 to 1.5 dB of noise figure from L- to X- band to be reached. The limiter's main function is to protect the LNA from the reflected signal coming from the antenna during the Tx pulse. It can be realized by an additional MMIC chip or by distributed networks which allow switching of the reflected power to a RF load. Consequently, the noise figure, at TRM level, will range over 2.5–3.5 dB according to the frequency band, bandwidth, circulator and limiter losses.

Electronic Front End The TRMs are realized by using hybrid technologies: they are typically assembled in a number that can vary from 2 to 8 or more into an electronic unit which is able to drive the TRMs and supply the required DC power and digital interfaces [3]. This circuit is usually called electronic front end (EFE) and includes:

- RF distribution networks to divide/combine the RF power to the TR modules of the EFE; three RF dividers are necessary (one Tx divider and two Rx combiners in case of dual polarization).
- A digital section to drive the TRMs.
- A power section to supply all the power required during the Tx and Rx pulses by the TRMs. In particular the EFE includes the capacitors used to store the energy to sustain Tx or Rx amplification during the Tx/Rx pulse.

TTDL The effect of frequency variation in the band is to move the beam around the pointing at the central frequency. This movement is linear and derives from the fact that all the phase values set on the TRMs are fixed and are computed at the central frequency. By changing the frequency in the operative bandwidth, it is necessary also to change the phase slope along the aperture of the antenna. This can be achieved by using a number of TTDLs distributed over the antenna aperture.

They are included in the beam-forming network of the antenna in order to drive a group of radiators.

Increasing the number of controlled radiators, which corresponds to reducing the number of TTDLs used, degrades the correction introduced over the pointing and grating lobes appear at an angular distance which depends on the size of the group of radiators fed by the single TTDL.

Figure 13.7 shows the elevation plane cut of a beam steered at 20° for three different frequencies (the central one and the extremes of the 5% band), considering a linear antenna having 64 elements. The figure shows three different cases:

(a) No TTDLs are used: the frequency steering in the band is evident.
(b) Four TTDLs are used grouping 16 elements each.
(c) Eight TTDLs are used grouping eight elements each.

Without TTDLs the beam steering is about 1°; this variation reduces by up to 0.16° for the four-TTDL case and to 0.08° if eight TTDLs are considered. Grating lobes appear with a step of 5° for case (b) and 10° for (c), positioned as a function of the TTDL granularity.

RF Power Distribution Networks The RF power distribution/combination networks for Tx and Rx signals inside the antenna are based on coaxial lines and/or waveguide technology. The RF networks interface the radar electronics from one side and several sub-arrays on the other, including a number of RF dividers/combiners. The main requirements for these networks are high phase stability vs. possible thermal gradients along the antenna to avoid impacts on the beam antenna pointing, low ohmic losses compatible with the overall RF budget and very low coupling among them so as not to affect calibration issues. Very high phase stability

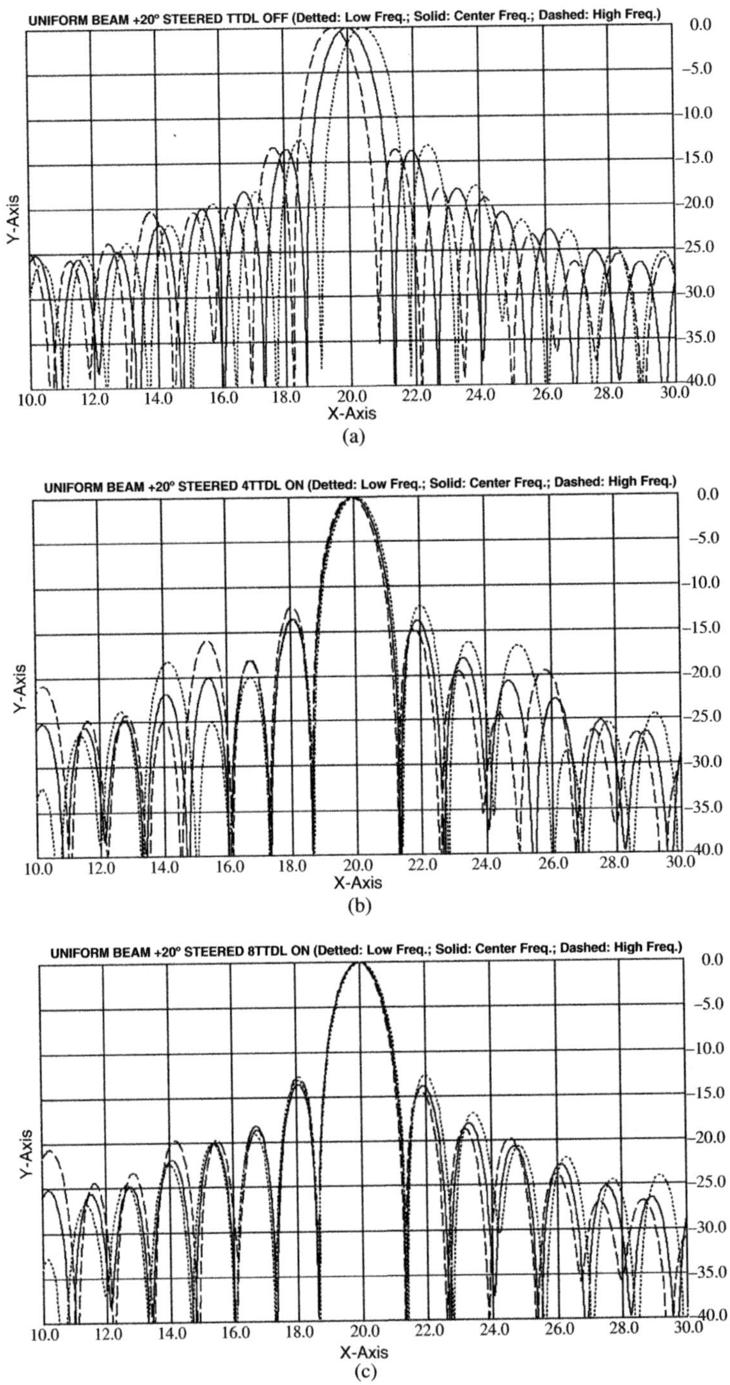

Figure 13.7 *Uniform beam for 64 linear array steered at 20°: (a) no TTDL used; (b) four-TTDL case; (c) eight-TTDL case.*

can be achieved by using carbon fiber waveguides: they are light weight, low loss and very stable vs. temperature. Two main drawbacks are shown by this technology: the high cost and the compatibility with mounting these waveguides when the antenna has an aluminum frame. In this case the different thermal expansion coefficients between the carbon fiber and aluminum could cause mechanical stress in the waveguides if no compensation systems are adopted. Moreover, the RF dividers should also be realized in a very stable technology, still using carbon fibers or materials such as Invar (a nickel–steel alloy) silver plated to improve the ohmic losses. Waveguide technology can be adopted from C- to X-band frequencies: at lower frequency the size of the waveguides become too large, preventing their use.

A coaxial line-based solution is indeed much simpler to adopt due to its reduced envelope; it is based on the use of RF coaxial cables that are stable to temperature changes. These cables can be flexible, and in this case the dielectric material inside is based on Teflon, or is semi-rigid, with a steel jacket and SiO_2 (silicon dioxide) dielectric. To evaluate the beam-forming phase stability required, it is possible, for example, to consider a thermal gradient of about 20 °C between the two wings of a SAR antenna working in C-band, with a total length of 10 m. In this case the beam-forming length is close to 7 m, which corresponds to about 100 wavelengths. The first solution analyzed is the aluminum waveguide: the phase difference between the two beam-forming arms is on the order of 17–20°, which results in a beam pointing error of about 1/10th of the azimuth beamwidth. This is caused by the fact that the aluminum will generate a total variation of about 500 ppm (parts per million) for the 20 °C of thermal gradient between the two wings of the antenna. Using SiO_2 coaxial cables, the expansion with temperature can be about 100 ppm or less. The impact on the beam pointing therefore will be 1/50th of the beamwidth. Moreover, mass, envelope and mounting complexity are all aspects which are favorable to coaxial solution, but at the expense of poorer losses and significantly higher mismatch w.r.t. the waveguide solution.

13.2.2.2 Digital Subsystem

The objective of the digital subsystem of the antenna is to control the setting of the TRMs in order to allow the antenna to generate the wanted beams. It allows each single Tx or Rx chain to be switched on/off and gives to the TRMs the amplitude and phase settings, or to configure the modules in the required calibration condition. It consists of a central beam controller, which sends all the settings for each TRM inside the antenna and for each desired beam, and a number of distributed controllers, usually known as 'tile controllers', which directly act on a group of TRMs of a tile (typically from 16 to 32) to set the amplitude and phase of the modules according to the desired excitations, implementing at the same time the electrical compensation required by the amplitude and phase drifts of the amplifiers when the temperature changes in the operative range. Intermediate amplifiers are also driven by the tile controller to compensate possible drift vs. changes in temperature. The central beam controller is part of the central electronics of the radar, while the tile controllers are distributed on the antenna. A digital serial bus such as the 1553 B or the CAM bus are used to exchange data from/to the tile controllers. TRMs are connected by a standard serial bus (RS 422 or RS 485) to the tile controller to exchange data: in particular the modules send the temperature telemetry and receive from the tile controller the amplitude and phase excitation needed to implement the required field distribution over the antenna surface. Tx and Rx synchronization signals are used to allow the contemporary switch-on/off of all the Tx and Rx chains. The TRM excitations are preloaded inside the tile controller in order to allow fast switching from one beam to the following one. This can be done by suitable broadcast commands sent by the antenna controller to the tile controller. In this way it is possible to change, with a simple command, the antenna Tx or Rx beam from one pulse to the next on the whole antenna. It is important to note that the compensation loop acts almost in real time to minimize the RF drift of the device in the thermal range. The control loop for temperature compensation is repeated at a rate of fractions of a second, and it is based on a 'lookup table,' stored in the tile controller, which contains the compensation data for each channel and each TRM. The temperatures are transmitted in analog or digital form to the tile controllers by the TRMs. Usually the thermal range is sampled by 32 or 64 steps within the overall thermal range from −30 to −20 °C to +50 to +60 °C. An alternate approach based on internal compensation made at a single TRM has been used in the past, but it results in more expense due to the need to check the electrical performance at single TRM level in temperature. Amplitude and phase variation in the thermal range depends on the amplification gain: the drift can reach a value of about 0.1 dB/°C and 1°/°C. After compensation the residual errors, expressed in terms of root mean square values, are close to one-half the amplitude and phase LSB (least significant bit). This means that for TRMs with 6 bits for amplitude and 6 bits for phase control, which will correspond to a minimum step of 0.5 dB and 5.6°, the residual errors computed over the possible 4096 states are about 0,3 dB RMS and 3° RMS.

If intermediate amplifiers are used in the antenna RF network, their gain and phase also have to be compensated vs. temperature. This is achieved by internal amplitude and phase settings capability driven by a dedicated lookup table (LUT) or still acting at the TRM level (the TRM LUT will include a dedicated section to compensate the intermediate amplifiers). The digital controller can be internally or externally redounded, to avoid loosing all of the TRMs driven by the single controller in case the tile controller should fail.

Further, the digital electronics in the EFE circuit can be considered as part of the digital section of the antenna. This circuit has the following functionalities:

- To receive the digital commands on a serial bus from the digital controller and to send them to each TRM (on/off command, amplitude and phase setting, calibration settings, etc.).
- To protect the TRMs from dangerous unwanted settings which could damage the module (i.e., a contemporary switching on of the Tx and Rx chains).
- To collect from the thermal sensors all the temperature data and send it to the tile controller to implement thermal compensation.

13.2.2.3 Power Subsystem

A suitably dimensioned power subsystem is needed to supply the required energy to the distributed amplifiers. The satellite power system should be able to supply this power under all possible conditions and therefore has to be dimensioned taking into account the real scenario of possible usage of the radar, that is, the peak power required by the radar, the maximum number of consecutive images along each orbit, the average power consumption per day, the worst case condition from solar radiation in the different seasons, lifetime degradation of the solar arrays and of battery capacity, and so on.

The antenna power subsystem typically consists of the following items distributed on the antenna surface:

1. A capacitor bank placed close to each TRM: its objective is to supply the wanted energy to the Tx and the Rx sections during the radar pulses. Its sizing depends on the current absorbed by the TRM and on the maximum pulse length. The capacitor banks are mounted on the EFE circuits.
2. A power supply which has the objective of regulating the voltage received from the satellite power bus, and to provide all the secondary voltages and currents to a group of TRMs and intermediate amplifiers if present.
3. A power harness, whose dimensioning depends on the maximum current required from the DC/DC converters.
4. A power control unit for the antenna, which has the objective of supplying all the needed power to the antenna and, at the same time, of protecting the main bus of the satellite from possible overloads inside the antenna.

The power consumption of the antenna is a large part of the total power consumption of the radar. It depends mainly on the following terms:

- power consumption of all the electronics needed to control and drive the TRMs;
- duty cycle and efficiency of the Tx section of the TRMs;
- duty cycle and power consumption of the Rx channels;
- efficiency of the DC/DC converters.

The efficiency of the Tx section depends on the technologies adopted for the HPA chip: typically 40% is achievable for mature technology today, such as PHEMT. Nevertheless, the efficiency of the TRM is reduced by the other chips used on the Tx section, such as the driver and the core chip, and in particular from the section of the core chip which remains permanently switched on during the operative modes of the radar to exchange data with the digital section of the EFE. Another important aspect is that the Tx duty cycle can range from a few percent for stripmap mode to 25–30% for spotlight mode. This has a significant impact on the average transmitted power of the radar and on the relevant power consumption. In some way, increasing the Tx duty cycle increases the overall efficiency due to the increment in the generated RF power, while the power consumption of the rest of the electronics switched on at 100% of duty remains the same. In Rx the duty cycle is larger w.r.t. the Tx duty and ranges from 40 to 80%. In this case the power consumption depends on the technology adopted for the LNA used and on the gain required for the Rx chain of the TRM: the minimum configuration is based on one LNA chip followed by the core chip Rx amplification. If the required gain increases, it will be necessary to place an additional chip on the Rx chain, at the expense of the power consumption of the TRM.

The continuous switching on/off of the Tx chain of the TRMs, with the pulse repetition frequency of the radar, generates a ripple on the power bus that can disturb other units. As a consequence the control loop of the power supply has to be adjusted in order to avoid strong current ripple on the primary bus, but also to limit as much as possible the voltage variations on the TRMs that can affect the HPA performance inside the pulse. The number of power supplies on the antenna depends on the number of TRMs to be fed, their power consumption, the duty cycle, their reliability performance and the total number of TRMs in the antenna. If a large antenna is considered, the number of power supplies can reach several dozen. It is possible to avoid the redundancy of the power supplies, saving at the same time both cost and mass, in case the reliability figure of the single power supply is able to guarantee that only one or two units out of the whole number of converters could be lost at most in the whole operating lifetime of the antenna.

13.2.2.4 Mechanical Aspects

Antennas working in C- or X-band have a large aperture of several square meters and are divided into mechanical panels to allow mounting of the antenna on the satellite in stowed condition at launch. The antenna is therefore deployed after launch when the satellite has reached its final orbit. Antennas having a length of 5–7 m can be organized in two deployable panels or three mechanical panels [4], with the former deployable externally. Longer antennas can reach dimensions of 10–12 m up to 15 m. In this case the number of mechanical panels increases up to four or five, each having a length of about 3–4 m [5]. Then a couple of the panels are arranged into lateral wings, each having a double deployment. One of the most critical aspects of mechanical design is to guarantee the antenna planarity, that is, reduced thermal distortion during functioning of the antenna, to achieve stable performance of the antenna patterns. The amount of acceptable distortion in the antenna plane depends on the frequency wavelength, which has to stay within $\lambda/20$; 1.5–2 mm is still acceptable for X-band, while 3–4 mm can be tolerated in C-band.

The use of a very low CTE (Coefficient of Thermal Expansion) material for the antenna structure, such as CFRP with a CTE on the order of 1–2 ppm per K, improves the dimensional stability at the expense of poorer thermal conductivity due to the low CTE of the resins. Alternatively, aluminum, with its quite good thermal conductivity, can be used to realize the antenna structure, improving, w.r.t. the CFRP, the heat transfer from the TRMs to the main frames and toward to the thermal radiator areas. Nevertheless, higher CTE (24–25 ppm per K) implies that, in the case of a different temperature between the upper surface and bottom side of the antenna, the aperture surface can be warped along the azimuth plane (thermal distortion). This problem can be solved by accurate thermal design of the heat vias from the hot to the cold areas of the antenna. Heat pipes are usually used to minimize all the thermal gradients and consequently the thermal distortion.

After launch, when the satellite has reached its working orbit, the antenna is deployed into its final configuration. This implies the use of two different kind of mechanisms:

- hold-down and release mechanisms, able to hold the antenna in stowed configuration during the launch and to support the mechanical loads in this phase;
- deployment mechanisms, with or without latching systems, to open the antenna panels when in flight.

The hold-down and release mechanisms are based on pyro-cutting devices or non-explosive systems, while the deployment mechanisms can be driven by motors or springs. If motors are used for the deployment mechanism, a driver electronics unit is also necessary to control the deployment, while spring-based mechanisms are mechanically controlled by dumpers and do not need control electronics. Consequently, the spring-based deployment mechanism results in less expense and has to be selected when possible.

Concerning the mechanical panels, two configurations are possible: (1) the electronics are integrated in subassemblies (the tiles), which are mounted on a panel-supporting frame; (2) all the electronics and radiating arrays are integrated directly in large and closed panel structures. Examples of antennas organized in tiles and supporting frames are the Sentinel 1 and the COSMO SkyMed SAR antennas, while Radarsat 2 is an example where the electronics and radiators are directly integrated on the mechanical panel.

13.2.2.5 Thermal Design

The thermal design is particularly important because it has to guarantee an adequate thermal range of functioning for the electronics inside the antenna. Especially for space applications the maximum allowed junction temperature of the active devices is derated with respect to the maximum value granted by the component manufacturers, typically to 60 or 70%. Therefore the junction temperature of power chips, drivers, power switches, and so on, has a maximum acceptable value around 110–120 °C. The junction temperature inside the chips is driven by the technology adopted, the chip thickness and the pulse length, and, in general, can reach a value of 40–50 °C w.r.t. to the back side of the chip. This implies that this side of the chip has to be maintained at the maximum value of 60–70 °C to comply with the derating limits. As a consequence, the TRM packaging design has to minimize the thermal resistance vs. the external environment constituted by the antenna frame.

In cold conditions the minimum temperature is also driven by the electronics, in particular by active and passive components that are limited by their qualification ranges, mounting technologies and different CTEs. All these bring into consideration −20 to −30 °C as the minimum operating temperature for the SAR antenna.

The design should optimize the working thermal range of the antenna to meet the stringent requirements from the electronics, considering functional requirements such as the orbit of the satellite, the number of images per orbit, duration of the continuous acquisition in image modes, the thermal capacitance of the antenna and its parts, single or dual looking operations, and so on, which can determine a rapid increase in the temperature. Another important aspect is the electronic density and therefore the heat dissipation per surface unit, which depends on the working frequency of the radar. Increasing the central frequency increases the power dissipation per square unit significantly.

When the antenna is switched on, the rapid temperature increment is controlled mainly by the thermal capacity of the antenna, in particular by all the parts which are in good thermal contact with the hot spots of the

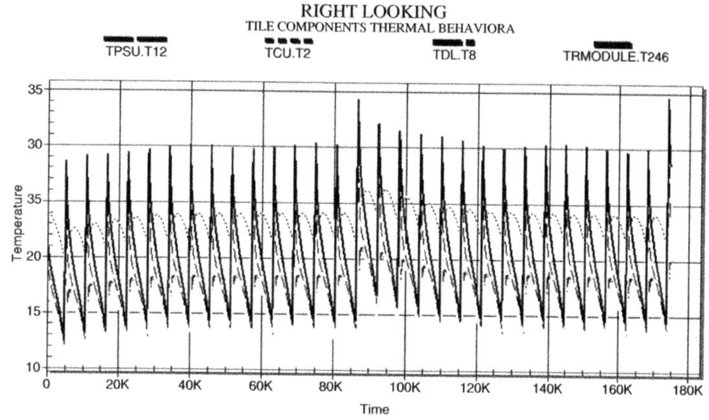

Figure 13.8 Antenna in front-to Sun condition. Heat dissipation vs. time (1 orbit 6000 s).

electronics (the TRMs and power supplies). Typical temperature curves vs. time are shown in Figure 13.8 for the different electronics inside the antenna, such as TRMs, TTDLs, the TPSU (Tile Power Supply Unit) and the TCU (Tile Control Unit). In this case the simulations refer to an antenna made of aluminum panels, which dissipates the internal heat generated by the electronics mainly from its back side, covered by an optical solar reflector (OSR), while the front side is populated by radiating elements characterized by low thermal conductivity and protected by a sunshield. The figure reports the analysis results relevant to the temperature at the interfaces of the electronics inside the antenna for a period of two days (15 orbits/day), with the antenna having its RF radiating surface exposed to the Sun and considering a given number of acquired images for the first 14 consecutive orbits (single-orbit duration is 6000 s) and with about 25% more images for the last orbit of each day. The temperature of TRMs at their interface increases by 15–20 °C during image acquisition (which has a simulated duration of less than 1/10th of the orbit). After the antenna is switched off, the accumulated heat can be dissipated during the rest of the orbit. The same analysis is carried out in Figure 13.9 considering the antenna in back-to-Sun condition.

In this second case the electronic temperature is higher due to the fact that the back side of the antenna is illuminated by the Sun and the heat dissipation through the radiating elements is constrained by the reduced thermal conductivity of the RF radiators. Also, if the average temperature is increased, the maximum

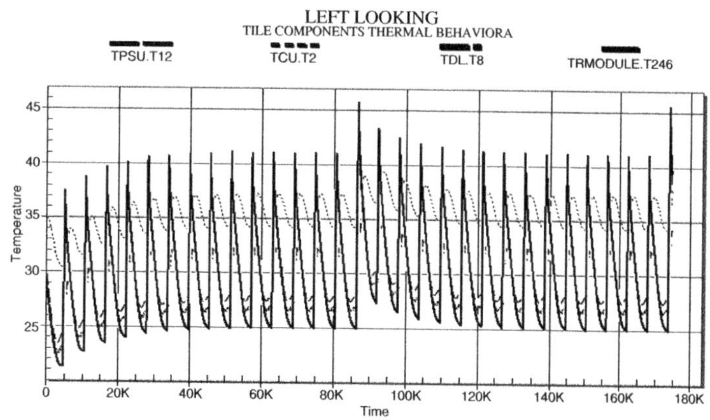

Figure 13.9 Antenna in back-to-Sun condition. Heat dissipation vs. time (1 orbit 6000 s).

temperature is well within the limits for the TRM interfaces, allowing the possibility to work in both front- and back-to-Sun conditions, that is, having a dual-looking capability.

This result can be more easily achieved when metallic panel frames are used instead of more complex carbon fiber structures, due to their better thermal conductivity. On the other hand, a transverse thermal gradient of a few degrees (4–5 °C) is sufficient in metallic structures to generate thermal distortion (warping) of the antenna panels. In conclusion, the main objective of the thermal designer is to reduce as much as possible the whole thermal resistance from the electronics to the antenna frames and the thermal radiators. Finally it is necessary to evaluate the case when the antenna is not used for a long period of time. In this case a heating system has to be dimensioned to prevent the antenna temperature falling below the minimum non operating temperature allowed by the electronic devices (typically in the range of −35 to −45 °C). This is achieved by placing heaters within the antenna, with a maximum dissipation in the range of several hundred watts.

13.3 A Review of the Development of Antennas for Spaceborne SAR

In this chapter the most recently developed projects for SAR antennas are summarized. All of them were launched in the first decade of the new millennium, except for Radarsat 1 which was launched in 1995. Their working frequency ranges are L-band (Palsar), X-band (TerraSAR-X and COSMO SkyMed) and C-band (ASAR, Radarsat 2). All of them are based on active phased arrays except for TecSAR and SAR-Lupe which are reflector-based instruments.

13.3.1 TecSAR

TecSAR is an X-band space radar based on a mini-satellite approach. It was designed as a military mission and launched in 2008. It was funded by Israel's MoD (Ministry of Defense) and designed and developed by IAI/MBT (Israel Aerospace Industries Ltd.). An overview of the TecSAR satellite is given in [6]. The satellite had a launch mass of 300 kg and the radar uses a deployable reflector antenna 3 m in diameter. A lightweight mesh is used for the reflector surface. The reflector is center fed by nine different single-polarization horns as shown in Figure 13.10. The beam has some electronic steering capability and is also mechanically steered by rotating the satellite. The horns are fed by a switching system which can concentrate all of the power in one horn. The power is generated by eight TWTAs working in parallel. Two other TWTAs are used in cold redundancy to increase system reliability.

Figure 13.10 TecSAR reflector antenna. Reproduced by permission of ELTA SYSTEMS LTD, © All rights reserved.

13.3.2 SAR- Lupe

SAR-Lupe is a constellation of five small satellites funded by the German MoD. All of the satellites were launched between 2006 and 2008. It was developed by OHB-Stystem AG, in Germany, with the collaboration of other European companies. The radar works in X-band. The satellite is small with a launch mass of less than 800 kg and average power of 250 W [7].

The SAR antenna employs a solid offset reflector manufactured by Ruag (Sweden) having a diameter of 3 m. A single horn mounted on a deployable boom is used to feed the reflector. The radar electronics were built in France by Thales Alenia Space. A TWTA (in cold redundancy) is used to generate the entire power needed by the radar. The same antenna is also used to download the radar data to the ground.

13.3.3 ASAR (EnviSat)

The ASAR (Advanced SAR) instrument works in C-band. Its payload is mounted on the EnviSat satellite, funded by ESA (European Space Agency) and developed in the 1990s by several European industries. EnviSat is a huge satellite weighting 10 tons and containing nine different instruments devoted to Earth observation. This antenna is the first example of an active phased array developed in Europe for SAR applications [8].

The antenna is 10×1.4 m and made of five mechanical panels organized in two lateral wings (each having two panels) and a fixed central panel. Each panel contains two tiles. A tile includes 32 linear arrays utilizing patch technology, it is dual polarized and employs the relevant dual-channel TRMs. A tile controller and four DC/DC converters complete the tile instrumentation.

13.3.4 Radarsat 1

The Radarsat 1 satellite was jointly funded by the CSA (Canadian Space Agency) and private investors. The satellite was launched in 1995 with the commercial market of Earth images as its target. The radar works in C-band and utilizes a deployable antenna 15×1.5 m. The antenna is a planar phased array with four mechanical panels that are stowed in two lateral wings at launch. The radiating elements are slotted waveguides working in HH polarization. In the elevation plane there are 32 rows of radiating elements each driven by a ferrite phase shifter. This configuration allows beam steering in the range dimension to generate different stripmap beams and implement the scansar mode. Details relevant to the design approach utilized to develop the radiating linear arrays are reported in [9].

13.3.5 Radarsat 2

Radarsat 2 was the follow-on satellite to Radarsat 1. It was funded by the CSA with the dual mission of guaranteeing images to the Canadian government and providing a commercial use. It was launched in 2007. As Radarsat 1, it works in C-band, but has full polarimetric capability. The satellite had a launch mass of 2.2 tons. The antenna, described in [10], has the same dimensions as Radarsat 1 (15×1.4 m) and includes 512 radiating linear arrays arranged in 32 rows and 16 columns. Each sub-array includes 20 radiating dual-polarized patches. The antenna is an active phased array organized in four mechanical panels, grouped in two lateral wings which are deployed after launch. Each panel includes 128 linear arrays, 128 TRMs and the relevant power supplies and digital controllers.

13.3.6 Palsar (ALOS)

Palsar is a fully polarimetric radar developed in Japan by JAXA and the Japan Resources Observation Systems Organization (JAROS). It is flown on ALOS (Advanced Land Observing Satellite) and works in L-band. ALOS has a mass of 4 tons and contains two other main instruments, namely, the Panchromatic

Remote-sensing Instrument for Stereo Mapping (PRISM) and the 10 m resolution Advanced Visible and Near-Infrared Radiometer type 2 (AVNIR-2). ALOS was launched in 2006. Palsar [11] is based on a planar active antenna sensor. It is an active phased array organized in four deployable panels and the total antenna dimensions after deployment are 8.9 × 3.1 m. Each panel includes 20 dual-polarized linear arrays and the relevant TRMs. The transmission peak power is 2 kW.

13.3.7 TerraSAR-X

TerraSAR-X is a program funded by a public–private partnership between the German Aerospace Center (DLR) and EADS Astrium GmbH. The satellite has a launch mass of 1350 kg and a radar working in X-band [12]. The sensor is based on an active phased-array antenna mounted on a lateral wall of the satellite. The antenna dimensions are 4.8 × 0.8 m, and it is divided into three leaves each having four panels. Each panel is made of 32 dual linearly polarized arrays, each made of two slotted waveguides, one for the H and one for the V polarization. The waveguides are made of carbon fiber to improve overall thermal stability. There is a dual-channel TRM for each couple of waveguides. 384 TRMs populate the antenna surface, each generating the transmitted peak power of 6 W with a noise figure of 4.3 dB [13]. The steering capability is ±20° in range and ±0.75° in azimuth.

13.3.8 COSMO (SkyMed)

COSMO-SkyMed (Constellation Of Small Satellites for Mediterranean basin Observation) is a four-satellite constellation funded by ASI (Agenzia Spaziale Italiana) and the Italian MoD [14]. It has the objective to meet the needs of both the military and the civil (institutional, commercial) communities. The four satellites from Thales Alenia Space Italia were launched from June 2007 to December 2010. Each of them is equipped with a SAR operating in X-band with multimode (spotlight, stripmap and scansar) and multi-polarization capabilities. The multimode and multi-polarization capability constrained the architecture and drove the selection of an active phased array for the antenna subsystem.

The antenna consists of an active phased array [15] having an aperture of 5.7 × 1.4 m, with a total mass less than 600 kg. It is divided into three mechanical panels, two deployable lateral panels and a fixed central panel, mounted on the spacecraft's top floor. The phased array includes 1280 linear arrays arranged in 64 rows and 20 columns. The TRMs are grouped in 40 tiles together with the relevant control unit and DC/DC converters. The 40 tiles are organized in eight rows by five columns. Moreover each tile is equipped with a TTDL having the capability to delay the RF signal up to 15 wavelengths to stabilize the beam pointing vs. the frequency band. The SAR antenna mounted on the satellite is shown in Figure 13.11 for both the stowed and deployed configuration. The 16 tiles of each lateral panel are mounted on an aluminum panel frame, which also supports the mechanisms for hold-down and deployment and the antenna harness. The panel frame of the central panel supports the tiles, a heat system based on a number of heat pipes and two lateral thermal radiators.

The eight tile rows are mounted in a staggered configuration on the panel frames. A double periodicity has been adopted to improve control of the grating lobe of the radiated pattern along the azimuth plane, keeping them under −13 dB w.r.t. the main beam at the maximum steering angle. The tile is organized in four columns of eight radiating elements, which corresponds to four columns of eight TRMs each. The tile is internally divided into seven separated rooms, four for the TRMs, two for the power supply units and one for the digital controller and the TTDL, each closed with a proper cover. The external and internal views of the tile are shown in Figure 13.12.

The radiating element consists of a dual-polarized linear array of 12 stacked patches, electromagnetically coupled through slots to the distribution networks. Eight radiating linear arrays are manufactured together and assembled into a single radiating board. The board is shown in Figure 13.13. Two distribution networks are used, one for each linear polarization, to feed the 12 stacked patches. Blind mate connectors allow simple assembly with the active modules.

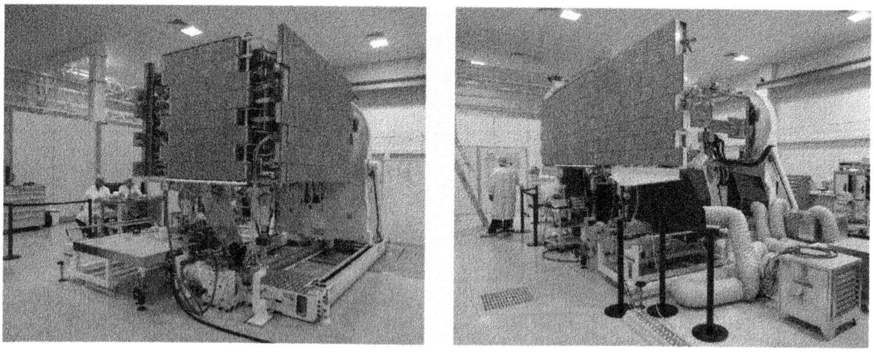

Figure 13.11 COSMO SkyMed active SAR antenna (stowed and deployed). Courtesy of ASI and Thales Alenia Space Italia.

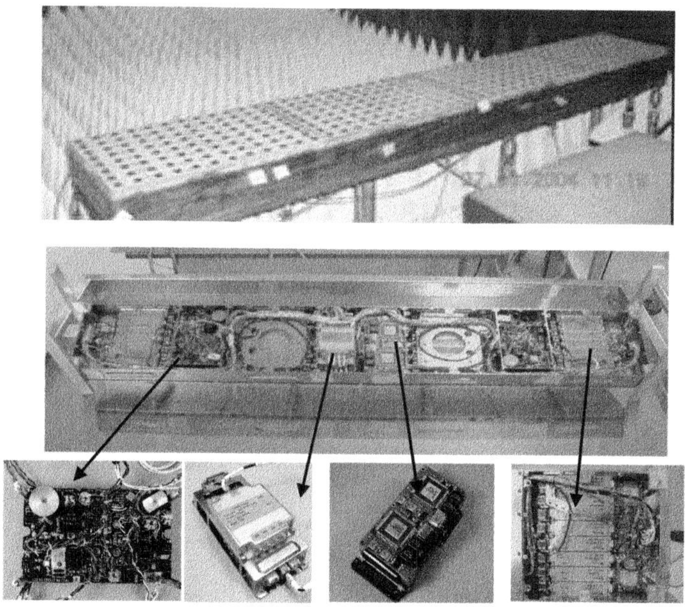

Figure 13.12 External and internal view of the tile (from left to right: TPSU, TDL, digital controller, EFE with eight TRMs). Courtesy of ASI and Thales Alenia Space Italia.

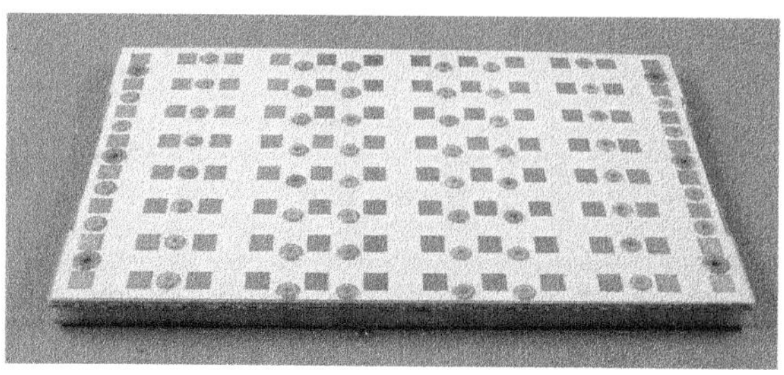

Figure 13.13 Radiating board with eight linear arrays in double polarization.

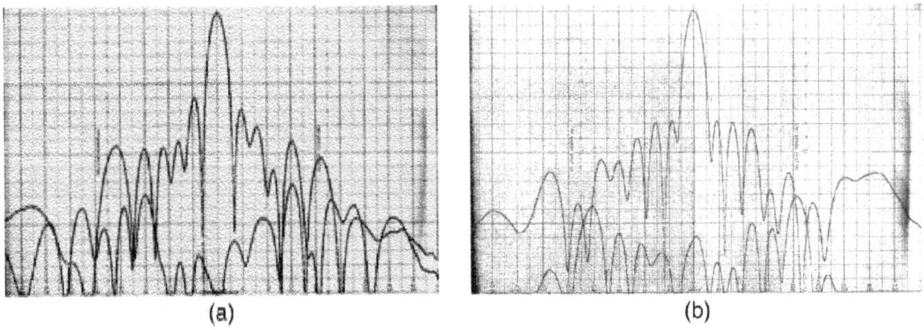

Figure 13.14 Radiating board measured pattern azimuth plane at 9.6 GHz, H pol. (a) and V pol. (b) (plot scale: 40 dB dynamic y-axis and 180° x-axis).

Measured patterns in the H and V polarizations are shown in Figure 13.14. The ohmic losses of the radiating elements have been minimized by supporting the microstrip dividers with dielectric spacers, avoiding glue on the microstrip radiator.

The TRM is a dual-channel hybrid. Its design is based on the use of MMICs in GaAs technology and includes two Tx channels and two Rx channels to reduce ohmic losses and to improve the output power and noise figure. Amplitude and phase control is achieved by a 6-bit attenuator and a 6-bit phase shifter. The schematic layout of the dual-channel TRM is shown in Figure 13.15. The module has one RF input/output port for beam forming and two output ports for the H and V polarization networks of the corresponding linear array. Finally two proximity couplers are used for calibration of the module.

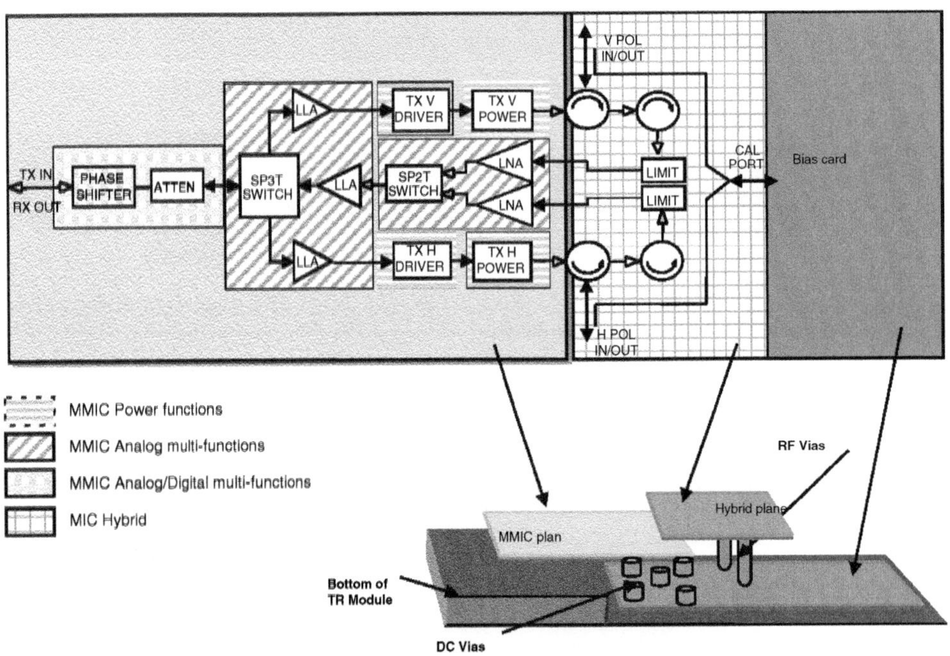

Figure 13.15 TRM, schematic layout.

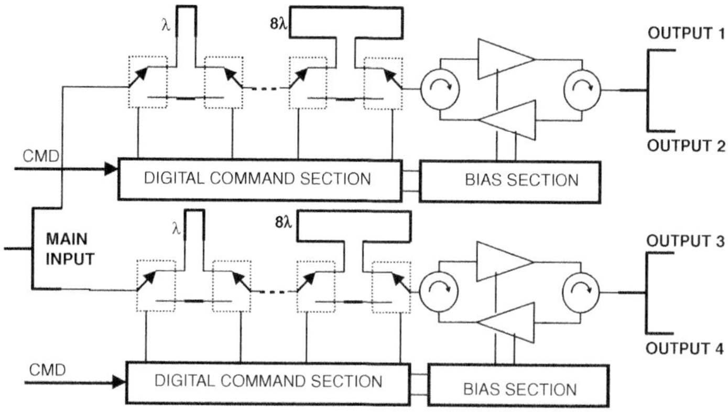

Figure 13.16 *TTDL block diagram.*

The TTDL is a 4-bit device able to reach up to 15 wavelengths of slope compensation. It is based on MMIC switches and microstrip lines. The unit has also the function of amplifying both the Tx and the Rx signals. The TTDL includes RF dividers to feed the four EFEs, each interfacing with eight TRMs. The TTDL block diagram is shown in Figure 13.16. The unit provides amplification and phase control of the RF signals traveling between a common I/O port and four I/O ports.

13.4 Case Studies of Antennas for Spaceborne SAR

A SAR system working in C-band is presented as an example of dimensioning a SAR instrument and active antenna.

13.4.1 Instrument Design

The key requirements considered for dimensioning a SAR in C-band are the following (stripmap mode is considered):

- Incidence angle: 20–45°.
- Elevation steering angle: $\pm 13°$.
- Satellite height: 700 km.
- Satellite velocity: 7 km/s.
- Azimuth and elevation resolution: 5 m (stripmap).
- Swath: 80 km minimum.

13.4.1.1 Antenna Length and PRF

The azimuth resolution required for the stripmap mode (5 m) is used for the sizing of the antenna. Hence, the maximum antenna length to comply with the required resolution is $L = 10$ m.

A shorter length can guarantee the required azimuth resolution, but will force the selection of higher PRFs with the possible reduction of the swath extension.

The effects of the PRF value vs. the azimuth ambiguity for different antenna lengths are shown in Figure 13.17. It is assumed that the azimuth beamwidth is obtained by a uniform illumination function, that is, with a pattern given by the sinc 2(x) function.

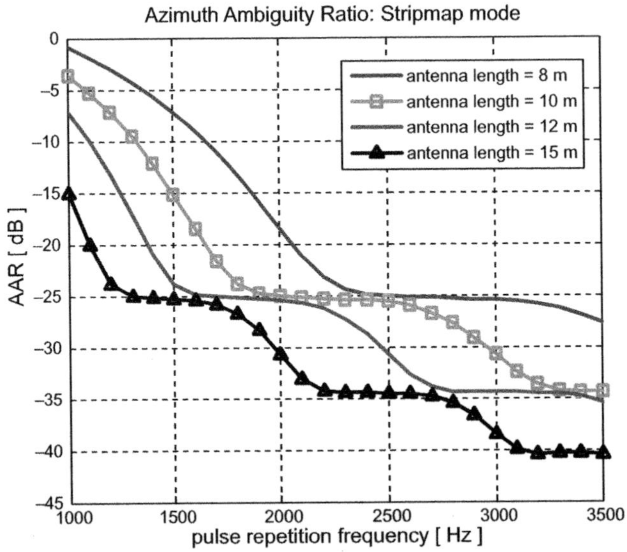

Figure 13.17 *Azimuth ambiguity ratio vs. PRF.*

As expected, the analysis shows that for a constant value of ambiguity, the longer the antenna, the lower the PRF required to achieve that result; moreover, for a fixed value of ambiguity, there exists a certain set of PRF values satisfying that value of ambiguity, since this effect is related to the structure of the sidelobes of the azimuth beam pattern.

For an antenna length $L = 10$ m, and the assumption of an azimuth ambiguity ratio (AAR) of -23 dB, a lower bound PRF ≥ 1800 Hz is derived.

13.4.1.2 Antenna Width

Antenna size is closely linked to the requirements on resolution and coverage. The minimum width–length product for a SAR antenna is set to avoid overlapping echoes and to sufficiently sample the Doppler bandwidth. The minimum antenna width is approximately given by Equation 13.9 and corresponds to about 1.3 m. The antenna width grows rapidly with incidence angles above 40°. When the incidence angle is increased from 40° to 50°, the required antenna width is increased by 70%. In addition, Equation 13.10 shows that a signal bandwidth of 50 MHz is sufficient to guarantee the required resolution of 5 m in the range direction.

Finally, Equation 13.11 allows one to compute the required power generated by the transmitter.

Considering a sigma noise equivalent of about -25 dB, Equation 13.11 gives a value of $P_t \cdot \tau = 0.12\ \mathrm{W \cdot s}$. For a typical pulse length of about 40 μs the transmitted power is 3 kW.

If an active antenna having a dimension of about 10×1.3 m is considered, it can be composed of an array of 320 TRMs each having the capability to generate a minimum of 10 W each.

13.4.2 SAR Antenna

The above SAR antenna dimensioning has led to an antenna of about 10×1.3 m. It has to be a planar active phased array with a regular rectangular grid to meet the steering and beam shaping capabilities required to implement the stripmap and scansar operating modes.

The height of the antenna and the steering requirements along the range plane fix the number of rows. To fill the 1.3 m length with a spacing of 0.75λ, 32 rows have to be provided.

The following aspects have to be analyzed to fix the antenna architecture:

1. The number of columns of the antenna. This depends on the performance achievable at the TRMs, and on the feasibility to realize linear arrays with acceptable losses.
2. The arrangement of the active electronics in tiles to organize both production and testability.
3. The assembly of a number of tiles into mechanical panels to allow the mounting of the antenna in stowed condition at launch.

For the first point, the TRMs in C-band, realized in mature technology and considering all the derating factors imposed for space application, can generate an output pulsed RF power from 10 to 18 W, depending on the use of single or dual chips as the final HPA stage.

A noise figure of about 3 dB is also achievable for large production utilizing LNAs in MMIC technology. The radiating elements can be realized by using patches or long slotted waveguides of up to 1 m and including 22 or 24 patches or slots. This would result in ohmic losses of 0.8–1 dB for the patch cases or 0.5–0.7 dB for the waveguide case. Both solutions yield a bandwidth of about 2/3%. On this basis it is possible to fix the length of the linear array at 1 m, with power generated by a single column of the antenna of about 500 W (if the dual-chip HPA stage is considered).

In this case it is easy to define the tile configuration: it consists of a column of 32 linear arrays with 32 TRMs plus the electronics needed to drive and supply it. The TRM can be single or dual channel. The first solution is preferred in the case of a massive production of TRMs and to guarantee high isolation between the two polarizations. Thus the tile will have 64 single-channel TRMs. They have to be grouped into a number of EFEs. An EFE configuration of eight modules has often been utilized and shows good handling and testability.

The tile controller will have an internal RAM to guarantee a number of preloaded beams (a typical number of beams can lie in the range from several dozen up to a few hundred), while a LUT for thermal compensation of the TRMs is necessary for correct operation over the whole thermal range. The RAM dimensioning for the preloaded beams can reach the value of 512 KB, while 1 MB is typically needed for the LUT implementation of the 64 single-channel TRMs. Another important parameter is the speed needed to transfer the setting data to the TRMs, in order to allow the data transfer within each single PRI to change the beam on a PRI basis. This is done on a serial bus with a typical clock speed of 10–12 MHz. The data to be transferred to the 64 TRMs (one set of Tx amplitude and phase values and two sets of Rx amplitudes and phases for the H and V polarizations) is about 1500 bits (48 bits for each couple of TRMs, considering one Tx channel and two Rx channels). The whole data transmission occurs in less than 150 microseconds to be compatible with a PRF in the range of 2000–2500 Hz.

In principle a single power supply could be used inside the tiles. However, redundancy is needed to take into account the impact on the azimuth plane of the beam of the antenna in case one power supply fails. The more common approach is to use a couple of DC/DC converters, each supplying half of the TRMs of the tile. In this case the impact of one failure can be tolerated. The electrical layout of the tile is shown in Figure 13.18.

The tile includes the two power supplies (TPSU A and B), the TCU internally redounded, eight EFEs and 32 dual-polarized linear arrays.

The simplest way to group the tiles in mechanical panels is to consider a couple of tiles assembled together. In this way the antenna is made of five panels: one central panel fixed on the top floor of the satellite and the remaining four arranged in two lateral wings as shown in Figure 13.19. The tile approach is based on the box concept. All electronic equipment, relevant to the two tiles, is enclosed inside a metal panel structure. This improves thermal control of the tile and also allows more flexible solutions for TRM interconnections to protect the electronics from the external environment. The panel will have a height of 1.3 m and a length of 2 m.

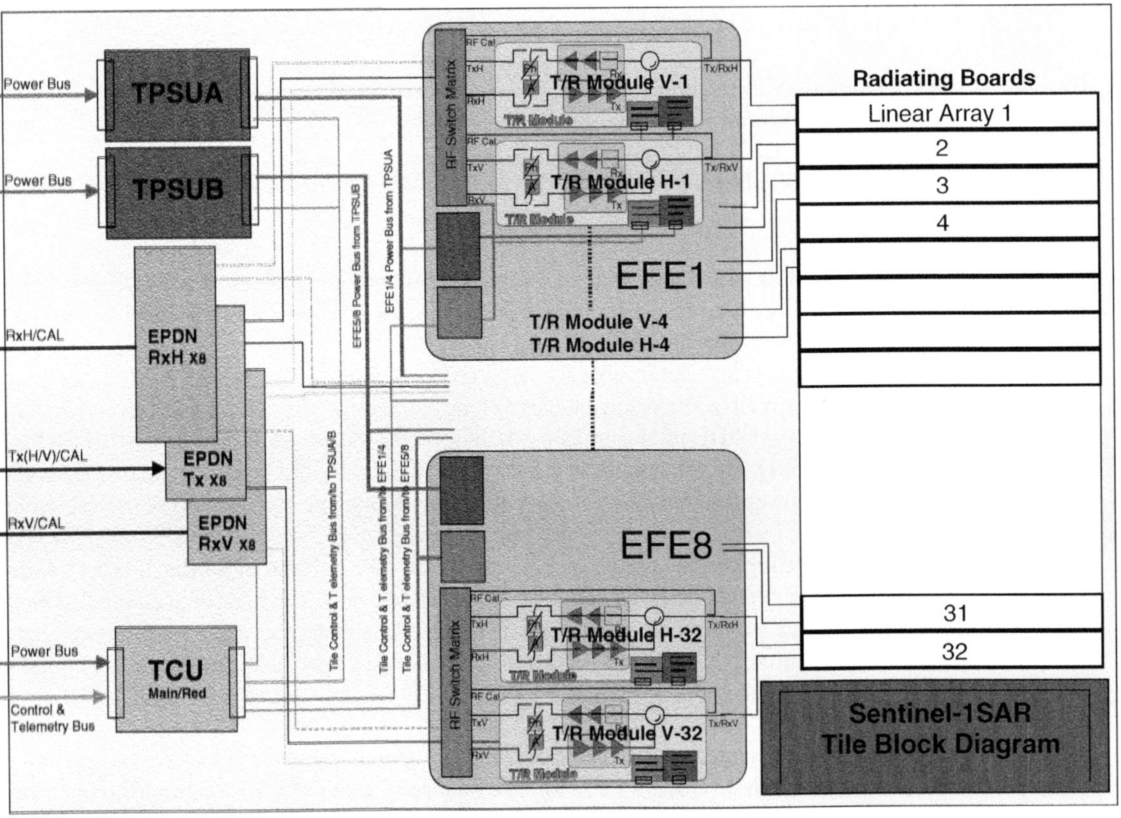

Figure 13.18 *Tile electrical layout.*

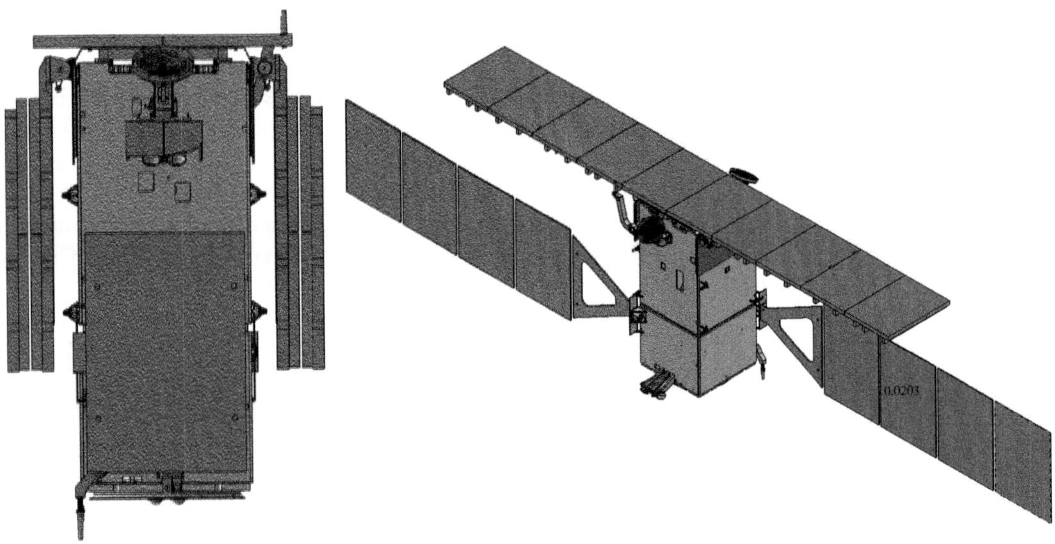

(a) (b)

Figure 13.19 *SAR antenna on the spacecraft: stowed (a) and deployed (b).*

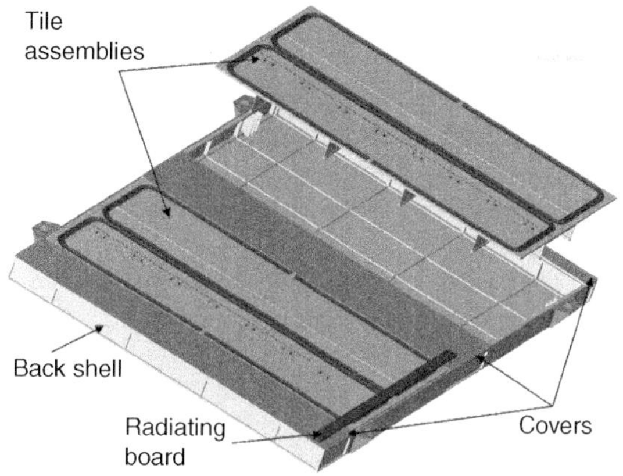

Figure 13.20 *SAR antenna panel: one tile opened.*

Figure 13.20 shows the panel configuration with one tile assembled and the second tile removed. This concept allows all the electronics to be mounted, such as TRMs, the TCU, tile power supply units (TPSUs) and the elevation plane distribution network (EPDN) along a supporting frame, and to perform the electrical characterization in temperature of the tile electronics in a single testing activity. The tile frame with electronics is shown in Figure 13.21, while the EFE with the eight single-channel TRMs is shown in Figure 13.22.

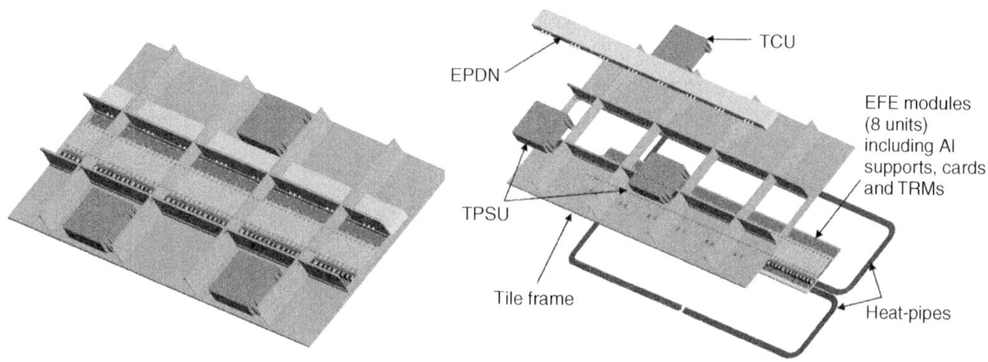

Figure 13.21 *Tile electronics assembled on mechanical frame (a) and exploded view (b).*

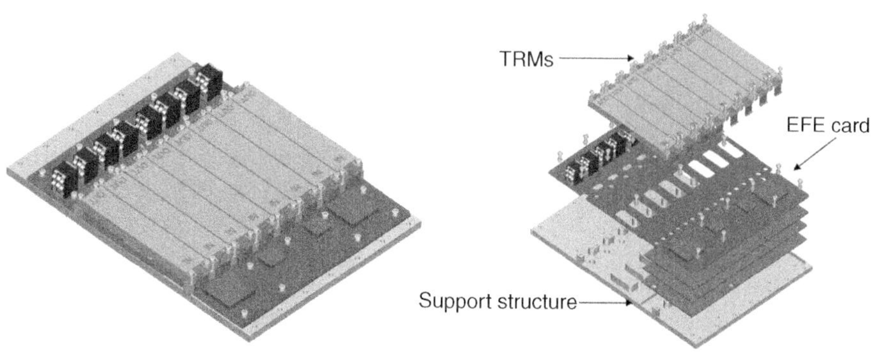

Figure 13.22 *EFE card with eight TRMs.*

13.5 Ongoing Developments in SAR Antennas

In the following subsections the ongoing developments in the field of active planar array antennas related to SAR missions are briefly reported and discussed. Almost all the active phased arrays under development follow on from those discussed in Section 14.3. Next-generation SAR systems will relay massively on full polarimetric features. Some of the future SAR missions under development are:

- Sentinel 1 (Europe), working in C-band, the follow-on of the EnviSat ASAR system.
- Saocom (Argentina), with an L-band active phased array.
- ALOS 2 (Japan), the follow-on of Palsar.
- COSMO Second Generation, the follow-on of COSMO SkyMed.

13.5.1 Sentinel 1

The Sentinel 1 project is funded by ESA. It is the follow-on of the ASAR instrument. The mission includes two satellites: Sentinel 1A and Sentinel 1B. Each satellite, whose mass is about 2.2 tons, is under development by a consortium of several European companies driven by Thales Alenia Space Italia as the prime contractor, while Astrium-GmbH (the EADS Company in Germany) is responsible for the whole SAR payload. The main instrument is a C-band radar based on an active planar antenna [16] consisting of 20 rows and 14 columns for a total of 280 independent radiators, organized in 14 tiles grouped in two lateral wings and a central panel. Each wing includes two panels with three tiles each, while the central panel supports two tiles. Each tile is an active antenna with:

- 20 dual-polarized radiating arrays made of two waveguides each, one for each polarization;
- 10 EFEs, each including four single-channel TRMs [17];
- a tile amplifier (with a second unit in cold redundancy);
- digital controller (with a second unit in cold redundancy);
- two power supplies;
- one elevation plane beam former;
- digital and power harnesses;
- structural elements.

All the supporting frames and the radiating waveguides are made in CFRP. The overall antenna dimensions are 12×0.8 m with a mass of about 900 kg, including the supporting frame to mount the two lateral wings on the lateral walls of the satellite. A pictorial view of the Sentinel 1 satellite is shown in Figure 13.23.

13.5.2 Saocom Mission

The Saocom mission is a two-satellite constellation (Saocom 1A and Saocom 1B) under development in Argentina by CONAE (COmisiòn National de Actividades Espaciales). The objectives of the mission are to provide data relevant to natural disasters, and to support economic activities such as agricultural development, ocean monitoring and monitoring of the Antarctic continent.

Each satellite carries a SAR instrument working in L-band, equipped with an active phased-array antenna [18]. The antenna is about 10 m long and 3.5 m high. It is organized in seven panels, one centrally fixed on the satellite wall and the other six grouped in two deployable lateral wings [19]. Each panel will include 20 RF radiators, based on dual-polarized annular slots arranged in a linear array, connected to a

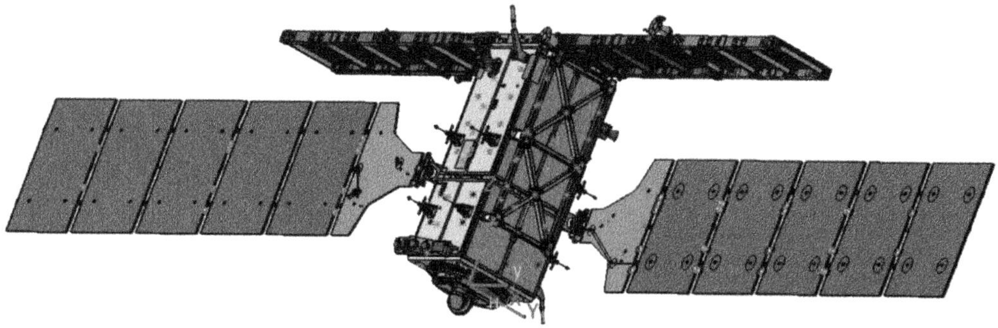

Figure 13.23 Sentinel 1 Satellite. Courtesy of ESA and Thales Alenia Space Italia.

dual-channel TRM. They will be supplied by ASI (Italian Space Agency) and produced by Thales Alenia Space Italia, in the framework of an international agreement named SIASGE (Sistema Italo Argentino de Satelites para la Gestion de Emergencias). This system foresees the use of the two Argentinean Saocom satellites and the four Italian COSMO SkyMed satellites working in parallel.

13.5.3 ALOS 2

Following the Palsar instrument developed by JAXA and operative from 2006, new developments are ongoing in Japan for the ALOS 2 satellite. This instrument will be working in L-band in quad polarization with improved capabilities for stripmap and scansar modes while also including spotlight mode operation. The employment of active phased arrays is confirmed and a new antenna is under study [20] having the possibility to steer the beam in both azimuth and elevation. It will have dimensions of about 10×2.9 m divided into five panels. The antenna will be equipped with 180 TRMs realized with advanced technology able to generate more than 5 kW of RF peak power.

13.5.4 COSMO Second Generation

COSMO SkyMed Second Generation (CSG) [21] is a two-satellite constellation funded by ASI and the Italian MoD. It is the follow-on of the COSMO SkyMed program with the overall objective of assuring operational continuity with the COSMO SkyMed constellation by providing, at the same time, enhanced performance in terms of image quality and operational capability. The SAR of CSG operates in X-band with multimode (spotlight, stripmap and scansar) and multi-polarization capability. The SAR will have improved capabilities mainly in terms of:

- pulse bandwidth that will cover the complete band allocated today by the ITU for spaceborne remote sensing civil applications;
- simultaneous dual-polarization acquisition;
- full quad polarization modes;
- MTI/ATI (Along-Track Interferometry) modes with a multiple sub-aperture antenna.

The electrical design of the antenna is based on the configuration defined by the preliminary tradeoff, which has led to an antenna of 48 rows $\times$ 32 columns, organized into 48 functional tiles. The total number of TRMs will be in the tens of hundreds. Such a high number is due to the need to transmit in H or V polarization and to

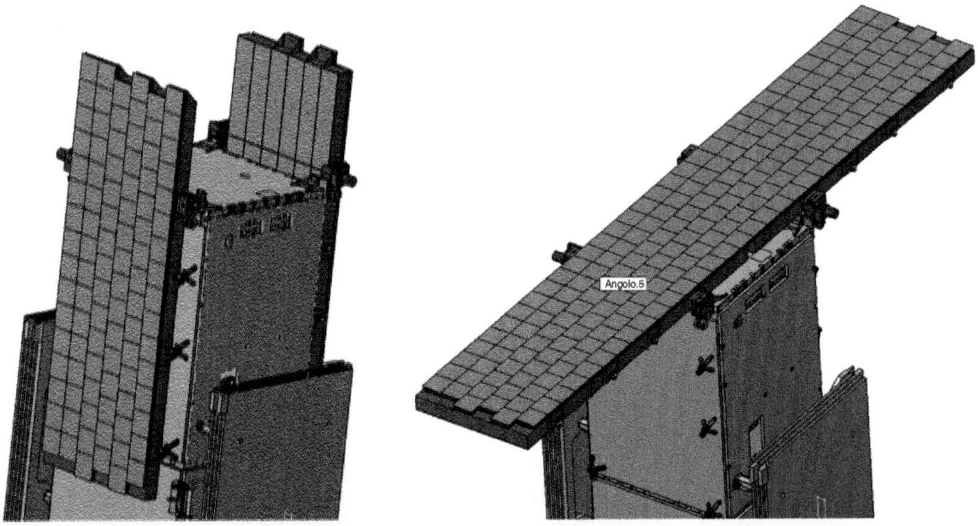

Figure 13.24 *COSMO Second Generation, stowed and deployed configuration.*

receive in one or both the polarizations simultaneously. The TRMs will be grouped in blocks of single-polarization hybrids mounted on a single EFE. Thermal requirements are much more stringent than for the first COSMO Sky instrument. This is mainly due to increased requirements in terms of number of consecutive images and therefore heat dissipation. For this reason, the antenna configuration is now based on two deployable panels with an extensive range of heat pipes. All the electrical parts, such as beam forming, digital and power harnesses will be included in the mechanical frame of the panels, leaving all the back side available for improved heat radiation. A system of six HRMs (Hold-Down and Release Mechanisms) and reinforced LADMs (LArge Deployable Mechanisms) will be used to stow and deploy the two antenna panels, as shown in Figure 13.24. The electrical configuration of the antenna is similar to that of COSMO SkyMed, but with increased electronic functionalities and higher integration. The EFE will include 32 single-channel TRMs, while a single TPSU will be able to feed two EFEs. Finally a single TCU will drive up to eight different EFEs.

13.6 Acknowledgments

This chapter was written in memory of Leonardo Borgarelli, who was a colleague of the authors for many years and the leader of the Radar development in Thales Alenia Space Italia since the nineties.

References

1. Elachi, C. (1988) *Spaceborne Radar Remote Sensing Applications and Techniques*, IEEE Press.
2. Wang, W., Jin, J., Lu, J.-G., and Zhong, S.-S. (2005) Waveguide slotted antenna array with broadband, dual-polarization and low cross-polarization for X-band SAR applications. IEEE International Radar Conference.
3. Imparato, M., Del Marro, M., Fantozzi, M. *et al.* (2009) Sentinel-1 SAR C-band electronic front-end (EFE). ESA Workshop on Advanced RF Sensors and Remote Sensing Instruments.
4. Trento, R., Fabiani, G., Meschini, A., and Carinci, L. (2010) Key thermo-mechanical design aspects of SAR antennas at TAS-I. 32nd ESA Antenna Workshop on Antennas for Space Applications.

5. Herschlein, A., Römer, C., Østergaard, A., and Pitz, W. (2008) Development of Sentinel-1 phased array antenna. 30th Antenna Workshop, ESA-ESTEC.
6. Naftaly, U. and Levy-Nathansohn, R. (2008) Overview of the TECSAR satellite hardware and mosaic mode. *IEEE Geoscience and Remote Sensing Letters*, **5**(3), 423–426.
7. Koebel, D., Tobehn, C., and Penné, B. (2005) OHB platforms for constellation satellites. 5th IAA Symposium on Small Satellites for Earth Observation.
8. Torres, R. (2002) ASAR instrument stability. CEOS WGCV SAR Workshop.
9. Wood, P.J. and Whelpton, J.P. (1991) Characterisation and design of slotted waveguide arrays for Radarsat. Seventh International Conference on Antennas and Propagation, ICAP'91.
10. Riendeau, S. and Grenier, C. (2007) RADARSAT 2 antenna. Proceedings of the 2007 IEEE Aerospace Conference.
11. Rosenqvist, A., Shimada, M., and Watanabe, M. (2004) ALOS PALSAR: technical outline and mission concepts. 4th International Symposium on Retrieval of Bio- and Geophysical Parameters from SAR Data for Land Applications.
12. Pitz, W. and Miller, D. (2010) The Terrasar X satellite. *IEEE Transactions on Geoscience and Remote Sensing*, **48**(2), 615–622.
13. Stangl, M., Werninghaus, R., and Zahn, R. (2003) The TerraSAR-X active phased array antenna. Proceedings of the IEEE Phased Array Conference.
14. Caltagirone, F., Angino, G., Coletta, A. *et al.* (2003) COSMO-SkyMed program: status and perspectives. Proceedings of Third International Workshop on Satellite Constellations and Formation Flying.
15. Capece, P., Borgarelli, L., Di Lazzaro, M. *et al.* (2008) COSMO SkyMed active phased array SAR instrument. IEEE International Radar Conference.
16. Herschlein, A., Römer, C., Østergaard, A., and Pitz, W. (2008) Development of Sentinel 1 phased array antenna. 30th ESA Antenna Workshop, ESTEC.
17. Del Marro, M., Giordani, R., Amici, M. *et al.* (2009) Sentinel-1 SAR C-band electronic front-end (EFE). Advanced RF Sensors and Remote Sensing Instruments.
18. D'Aria, D., Giudice, D., Monti Guarnieri, A. *et al.* (2008) A wide swath, full polarimetric, L band spaceborne SAR. IEEE International Radar Conference.
19. Vega, J., Quiroz, H., and Kulichevsky, R. (2003) A new deployment concept for a space based SAR antenna. ESMATS Conference.
20. Kankaku, Y., Osawa, Y., Suzuki, S., and Watanabe, T. (2009) The overview of the L-band SAR onboard ALOS-2. Progress in Electromagnetics Research Symposium Proceedings.
21. Caltagirone, F., Scorzafava, E., Marano, G. *et al.* (2011) COSMO-SkyMed Second Generation: the SAR instrument. IGARSS 2011.

14

Antennas for Global Navigation Satellite System Receivers

Chi-Chih Chen[1], Steven (Shichang) Gao[2], and Moazam Maqsood[2]

[1]*The Ohio State University, USA*
[2]*Surrey Space Centre, University of Survey, UK*

14.1 Introduction

Table 14.1 lists four major active and planned global navigation satellite systems (GNSS) that allow the determination of the position (longitude, latitude, altitude) of a GNSS receiver. Each receiver receives pseudo-random noise (PRN) code sequences from a constellation of GNSS satellites that are orbiting the Earth at different medium Earth orbits (MEOs) to broadcast satellite signals covering the entire Earth's surface. The total system is separated into three segments: namely, the control, space, and user segments [1]. The control segment includes a network of command centers on the ground and is responsible for maintaining satellite position, adjusting satellite clocks, and uploading navigational data. The space segment includes a constellation of satellites that broadcast their times and positions. The user segment receives satellite signals and derives its position using trilateration methods based on estimated distances to four or more known satellite locations from which the signals are received [2]. Each satellite signal contains the identification, positions, and sending time of the initiating satellite. Although this trilateration positioning concept is mathematically straightforward, in practice, obtaining an accurate range between a receiver in the user segment and a satellite is subject to errors caused by additional delays in the ionosphere, troposphere, receiver antenna, receiver circuitry, as well as interference from multipath signals and other RF emitters transmitting at GNSS frequencies.

Various satellite-based augmentation systems (SBASs) have been developed to broadcast corrections to these errors from geostationary satellites based on errors observed from carefully surveyed ground GNSS receiving stations. For example, the US Wide Area Augmentation System (WAAS) [3] uses a network of ground-based reference stations in North America and Hawaii (soon to be expanded to Mexico and Canada) to

Space Antenna Handbook, First Edition. Edited by William A. Imbriale, Steven (Shichang) Gao and Luigi Boccia.

Table 14.1 *Frequency allocations of GPS, GLONASS, Galileo, and COMPASS signals.*

System	Country	Coding	Frequency (MHz)	Bandwidth (MHz)
GPS	USA	CDMA	L1: 1575.420	C/A ~ 2 × 1.023
			L2: 1227.600	P ~ 2 × 10.23
			L5: 1176.450	I5,Q5 ~ 2 × 10.23
GLONASS	Russia	FDMA	L1: 1602.000 + $k \times 0.5625$[a]	SP ~ 2 × 0.511
		CDMA	L1: 1575.420	HP ~ 2 × 5.11
		FDMA	L2: 1246.000 + $k \times 0.4375$[a]	
		CDMA	L2: 1242.000 (planned)	
		CDMA	L3: 1202.025 (planned)	
		CDMA	L5: 1176.450 (planned)	
Galileo	European Union	CDMA	E1: 1575.420	~24.552
			E6: 1278.750	~40.920
			E5b: 1207.140	~20.460
			E5: 1191.795	~51.150
			E5a: 1176.450	~20.460
Compass	China	CDMA	B1: 1559.052–1591.788	~4.092
			B2: 1162.220–1217.370	~24
			B3: 1250.618–1286.423	~24

[a] *$k = 0$ to 24; each satellite has an individual FDMA channel.*

enable aircraft to use GPS (Global Positioning System) during flight and landing at airports within its coverage area. A WAAS-enabled receiver operated in L1-band (see Table 14.1) can achieve horizontal accuracy better than 3 m and vertical accuracy better than 6 m most of the time. Other similar regional SBASs in operations are the European Geostationary Navigation Overlay Service (EGNOS) [4] by the European Space Agency and the Multi-functional Satellite Augmentation System (MSAS) by Japan. Several more SBASs have been proposed by Japan, India, and China.

Alternatively, the ground-based augmentation system (GBAS) broadcasts position correction information from ground VHF or UHF stations for regional GPS correction service. The US Local Area Augmentation System (LAAS), in support of aircraft landing systems and differential GPSs, is an example of a GBAS.

The successful demonstrations of the US GPS and Russia's Global Navigation Satellite System (GLONASS), in conjunction with low-cost advanced digital technology, have been rapidly reshaping how we conduct our daily lives, especially in developed and developing countries. The European Union's Galileo positioning system and the People's Republic of China's Compass navigation system will soon join the GNSS service. Although the development of a country's own GNSS is probably mainly driven by national security considerations, the results will undoubtedly benefit the general public more.

In the USA, GPS has become an integral part of daily activities in personal travel, military missions, commercial operations, scientific studies, and engineering solutions. For example, satellite operations, surveying and mapping, precision agriculture, trucking and shipping, fishing and boating, disease control, navigation, and power grids are well-known areas that have been increasingly relying on GPS availability. Most of them are actually working without being noticed by the public. It would be shocking to see what would happen if the GPS service is interrupted. The authors, for a start, would probably be lost somewhere and scrambling to find maps!

Table 14.1 lists the countries, coding, and frequency information of four major planned GNSSs. The frequency spectrum covers 1150–1300 MHz and 1559–1611 MHz in two bands. The frequency band from

1300 to 1560 MHz is reserved for other radio users such as military, telemetry, radio astronomy, Earth-to-space operation, and satellite Internet services. The actual bandwidth of individual GNSS signal channels ranges from 2 to 50 MHz, and is determined by the chip clock frequency, code clock frequency, power density spectrum of coding methods, and isolation specification between adjacent channels. Note that GLONASS has an individual frequency channel for each satellite, a different approach from other systems. It is apparent from this table that there are many frequency overlaps among different GNSSs. Within these overlapped bands, channel isolation relies heavily on code modulation. It should be pointed out that there are also other radio systems operate within the GNSS frequency range. For example, amateur radios operate from 1240 to 1300 MHz and air traffic control (ATC) radars operate from 960 to 1215 MHz. These signals can potentially cause interference to GNSS operations.

Antenna design and performance affect GNSS operations in coverage, position accuracy, acquisition time, and interference mitigation. This chapter aims to provide a discussion on various aspects of GNSS receiver antenna design and operations. There are a few military, commercial, and experimental GNSS receiving antennas available to date. These antennas could be categorized according to satellite system, bandwidth, accuracy, and user type as follows:

- *Satellite system:* Most current antennas in low-cost personal navigation products are designed for receiving L1 (1575.42 MHz) signals. As more GNSSs become available, more and more new GNSS receivers will be capable of receiving multiple types of GNSS signals. Most of these will still operate only at 1575.42 MHz, a common frequency to the GPS L1-, GLONASS L1-, and Galileo E1-bands. A truly universal GNSS antenna that can receive signals from all frequency bands of all four GNSSs are more challenging and still in the development stage. The antennas for these receivers need to have good gain response and pattern coverage over different frequency bands.
- *Bandwidth:* Almost all GPS antennas for personal travel navigation are designed only for L1 (1575.42 MHz) with relative narrow bandwidths (2–20 MHz). They are simple single-mode designs that allow for optimal front-end RF circuitry designs such as impedance matching, filters, hybrids, and amplifiers. Some GNSS antennas are designed to operate at several narrow GNSS bands such as L1, L2, L3, L5. These antennas do not work outside these bands and are therefore classified as multiband antennas. In order to cover all possible current and future GNSS bands, an ideal antenna probably should perform well continuously within two GNSS frequency regions from 1150 to 1300 MHz and from 1559 to 1611 MHz. Such an antenna would be classified as a wide-band antenna. Obviously, it is more challenging to design a wide-band GNSS antenna with all the desired impedance, polarization, and pattern properties.
- *Operating mode:* The GNSS antenna designs depend very much on the intended operating modes. Different operating modes could require different position accuracy, traveling speeds, operating locations and altitudes, physical constraints, and operating environments. Thus they demand a very different antenna design strategy. For example, precision GNSS receivers used for geodetic survey or differential mode reference require the antenna design to have exceptional pattern and phase performance for maximum sky coverage, minimum multipath effect, minimum antenna introduced group delay variation, and code phase distortion. Antennas designed for moving terrestrial or aerial vehicles need to take platform motion and interaction into consideration. In military applications where anti-jamming capability is crucial, array configurations are often adopted for enabling digital beam forming and null steering.

Since GNSSs are still being improved or developed among different countries and regions, it is very difficult to find absolute antenna specifications or requirements for future GNSSs. Therefore, most of our discussion on antenna performance requirements and design parameters are focused on electromagnetic considerations for general GNSS receiving operations without direct links to specific systems or missions. A brief review of antennas developed for spaceborne GNSS receivers as well as their specifications will also be provided in Section 14.5.

14.2 RF Requirements of GNSS Receiving Antenna

Although most GNSS antenna designs adopt existing designs [5,6] such as patch antennas[7–15], spiral antennas, helical antennas, conical antennas, dipoles, and loops used in other applications, there are some specific performance requirements that are important to both general and special GNSS operations. Therefore, proper antenna design modification optimizations and additional treatments are often needed to design a good GNSS antenna. Like all other spaceborne antennas, spaceborne GNSS antennas also have to meet mechanical and thermal requirements to ensure reliable operation in space. Section 14.5 will discuss spaceborne GNSS antennas but the reader can learn more about these requirements in Chapter 5. For now, we will focus on the electromagnetic or RF aspects of performance requirements.

14.2.1 General RF Requirements

Table 14.2 shows an list of typical RF requirements for general GNSS receiving antennas operating at L1-band. Although this table is only for the L1 frequency, all the basic requirements on gain, bandwidth, polarization, pattern, and reflection coefficient can be applied to other GNSS frequencies. Some advanced operations may have different requirements for achieving better position accuracy and interference suppression. These advanced requirements with be briefly discussed later in Section 14.2.2.

For the convenience of current discussions, an example of a L1-band GNSS antenna design and its RF performance characteristics are shown in Figure 14.1. This antenna design is based on a simple dielectric-loaded patch antenna design with right-hand circular polarization (RHCP) achieved by using a corner cutout, offset feed, and slightly asymmetrical patch.

14.2.1.1 Gain Requirements

In a satellite receiving antenna, it is important to have a sufficient antenna gain, G, to ensure sufficient sensitivity. The antenna gain affects the signal-to-noise ratio (SNR) of the received signals prior to demodulation and decoding. The noise is mainly dominated by the white noise produced at the first low-noise amplifier (LNA) after the antenna. The typical noise figure (NF) for such a LNA is from 1.5 to 3 dB. A signal with a higher SNR can tolerate more line-of-sight (LOS) blockage, weather conditions, and antenna

Table 14.2 *Typical basic performance of narrow GNSS antenna for L1-band.*

Center frequency	1575.42 ± 1.023 MHz
Frequency bandwidth	9 MHz min.
Gain (zenith, 90° elevation)	+3 dBi (typ.)
Gain (10° elevation)	−5 dBi (typ.)
Polarization	RHCP
Axial ratio (zenith)	< 3 dB
Axial ratio (10° above horizon)	< 20 dB
Reflection coefficient	< −10 dB
Impedance	50 Ω (typ.)

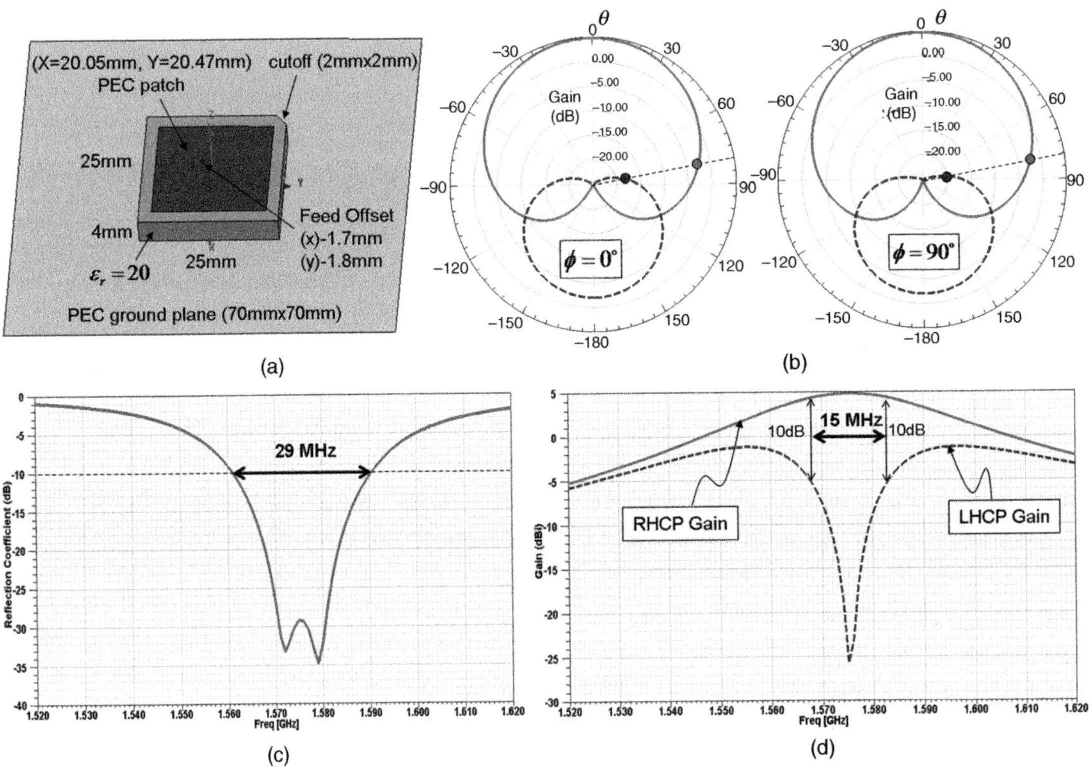

Figure 14.1 *An example of a GNSS receiver antenna design operating at a single L1-band. (a) Square patch antenna geometry (PEC, Perfect Electric Conductor). (b) Elevation patterns in X–Z and Y–Z planes for RHCP (solid line) and LHCP (dashed line). (c) Reflection coefficient as a function of frequency. (d) RHCP and LHCP realized gain at zenith.*

orientation variations. This will shorten satellite acquisition time and prolong reception from the same satellite without the need to acquire new satellites due to signal fading.

Antenna gain is defined as a product of antenna efficiency, e, impedance matching efficiency, $1 - |\Gamma|^2$, and pattern directivity, $D(\theta, \phi)$, defined in Chapter 1. That is,

$$G(\theta, \phi) = e \cdot \left(1 - |\Gamma|^2\right) \cdot D(\theta, \phi) \tag{14.1}$$

The impedance matching efficiency is related to the reflection coefficient, Γ, caused by impedance mismatch between the antenna and the rest of the receiver circuitry. Antenna efficiency is the ratio of radiated power to the total power received by the antenna. The lost power is associated with electromagnetic energy converted into thermal energy and is often attributed to electrical conduction loss or material (dielectric or ferromagnetic) loss. From (14.1), it is obvious that a good GNSS antenna should have minimal loss, good impedance matching between antenna and receiver, and proper antenna pattern to achieve good SNR over a large portion of the sky. Most GNSS antennas preferably have a very compact size and adopt a design approach of using a small electric or magnetic dipole above a ground plane. The theoretical peak gain value for these types of antennas should be from 4.76 to 5.2 dBi on an infinite ground plane under good impedance matching and 100% efficiency conditions. The 4.76 dBi is when the antenna's physical size is much smaller than wavelength at the operational frequency, and 5.2 dBi is achieved when the antenna size is approximately a quarter of the wavelength. At the bottom right corner of Figure 14.1, the computed broadside (zenith) gain curves for both right-hand circular polarization (RHCP) and left-hand circular polarization (LHCP) components are plotted from 1.520 to 1.620 GHz in 1 MHz increments. The figure shows a RHCP gain of 4.76 dBic and a low LHCP gain of less than −25 dBic at zenith.

Although this design example shows excellent gain performance near L1-band (1575 MHz), such performance degrades rapidly as frequency deviates from the center frequency, and thus is only good for receiving satellite signals from GPS L1, GLONASS L1, Galileo E1, and Compass B1 using C/A code.

14.2.1.2 Pattern Requirements

A land GNSS receiving antenna with a higher RHCP gain level produces a higher SNR, desirable for achieving better satellite signal reception sensitivity along the direction of peak gain. However, this could be at the expense of sky coverage. An antenna with a low gain can be caused by impedance mismatch between antenna and receiver as indicated in (14.1). As long as the mismatch loss is minimized, further increase in antenna has to come from a more directive pattern. In most applications where the platform on which the receiving antenna is mounted maintains reasonably stable roll and pitch angles, it is desirable to minimize the antenna pattern below the horizon, often referred to as backlobes. In theory, the gain in the upper hemisphere would be increased by 3 dB if the antenna pattern had no backlobes. This will make the theoretical gain of a hypothetical antenna with a uniform RHCP coverage over the entire upper hemisphere be 3 dBic, as illustrated in Figure 14.2. Therefore, any gain level higher than 3 dBic can only be achieved in some directions at the expense of the other directions. Consider the antenna patterns in Figure 14.1 for example, where the peak gain at zenith is 4.76 dBic but decreases down to approximately −2.5 and −20 dBic at 10° above the horizon for RHCP and LHCP, respectively.

Now let us consider the sketch in Figure 14.3 showing the Earth and a GNSS satellite. The typical height of the GNSS satellite is approximately 20 000 km and the average radius of the Earth is approximately 6370 km. As shown, the distance from the satellite to a receiver on the Earth increases as the receiver moves away from nadir around the Earth's surface. Such an increase in distance causes a reduction in received signal strength due to electromagnetic wave expansion. The range propagation loss can easily be determined from (14.2). This

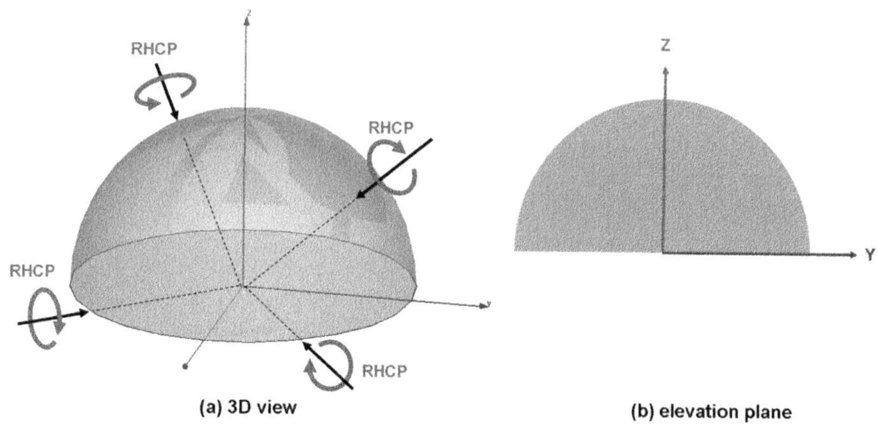

Figure 14.2 Theoretical uniform RHCP coverage over the upper hemisphere.

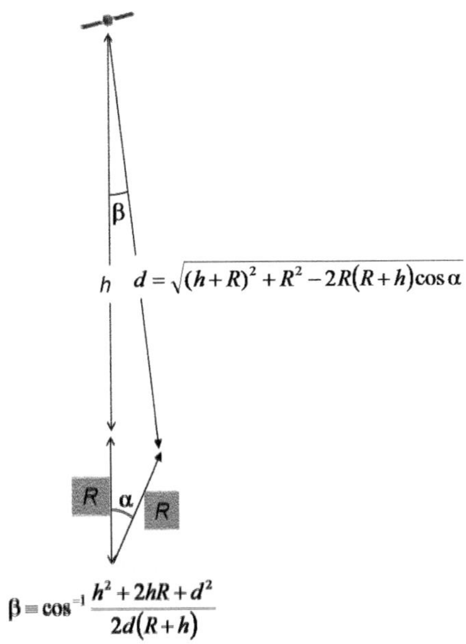

Figure 14.3 Propagation loss related to distance from satellite to a receiver on the Earth.

leads to an approximately 2 dB drop in signals received from a satellite close to the horizon compared to those received from a satellite directly above the receiver.

$$\text{Propagation loss} = 20\log_{10}\left(\frac{h}{d}\right) \tag{14.2}$$

In order to illuminate the Earth's hemisphere with nearly constant signal strength, the transmitting pattern of a GNSS antenna should behave like that in Figure 14.4 which plots the theoretical gain pattern shaping factor as a function of nadir angle from nadir to approximately 14°. This is why most GNSS antennas on board a spacecraft synthesize a shaped pattern similar to that in Figure 14.4 by using two concentric rings of antenna arrays with the inner ring producing a higher power broader beam, and the outer ring producing a weaker power narrow beam [16].

It should be noted that for a GNSS antenna to be operated across different current and future satellite constellations, the stated pattern property should be met across all GNSS frequencies. Also, the above propagation loss does not consider the additional attenuation associated with GNSS signals propagating through the Earth's atmosphere.

14.2.1.3 Polarization Requirements

As discussed in Chapter 1, GNSS adopts the RHCP sense of signals, which is orthogonal (or cross-polarized) to the LHCP sense of signals. Therefore, GNSS receiving antennas should be designed to have high RHCP gain and low LHCP gain. The latter helps reject undesired multipath signals as a result of GNSS signals reflected from a large flat surface. In the design example in Figure 14.1, the RHCP and LHCP gains at zenith around 1575 MHz are 4.65 and −25 dBic, respectively, which corresponds to an axial ratio (AR) of 0.56 dB. Therefore this design exceeds both the gain and axial requirements in Table 14.2 at zenith ($\theta = 0°$). At 10° elevation angle (or $\theta = 80°$) the axial ratio increases to 2.33 dB, which is still much lower than the typical

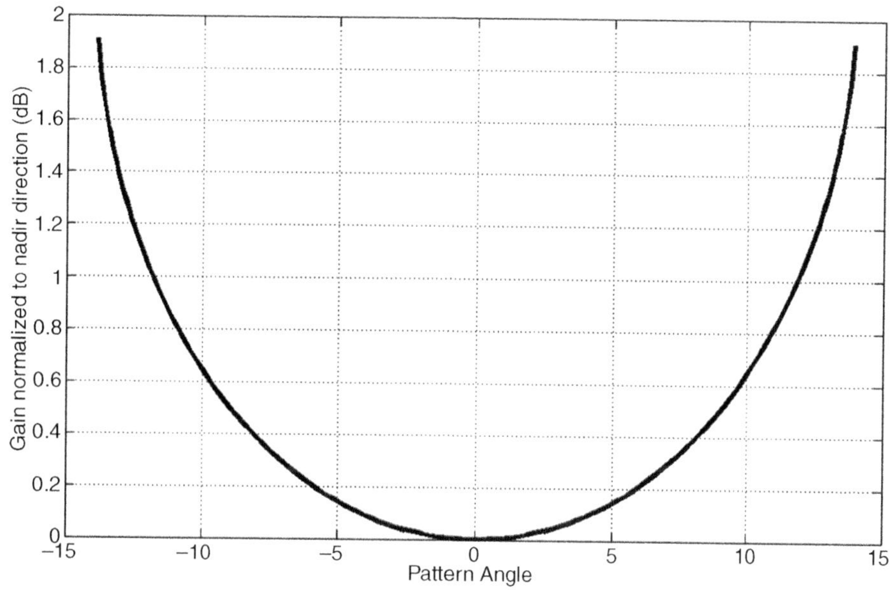

Figure 14.4 *Theoretical satellite transmitting pattern for uniform coverage from nadir to horizon.*

required maximum value of 20 dB. It should be noted, though, that such a good performance can only be achieved within a very narrow frequency bandwidth around 1575 MHz for this kind of narrowband antenna design, and is only useful at L1 using C/A code. A good GNSS receiving antenna should be able to maintain good axial ratio and bandwidths at all GNSS frequencies.

14.2.1.4 Frequency Requirements

Single-frequency GNSS receiving antennas need to maintain good impedance, gain, phase, pattern, and polarization properties over a 2–30 MHz bandwidth, depending on the coding scheme (see Table 14.1). Currently, 1575.42 MHz is the only common frequency band among four major GNSS constellations (GPS L1, GLONASS L1, Galileo E1, and Compass B1). In the near future, 1176.45 MHz will be available from GPS L5, GLONASS L5, Galileo E5a, and Compass B2. Although a GNSS antenna can operate at L1-, L2-, or L5-bands, being able to simultaneously receive more than one band such as L1/L2, L1/L5, L1/L2/L5, and so on, allowing for correction of the delay time error caused by the dispersion effect in the ionosphere. In addition, since many different bands within 1150–1300 MHz and 1559–1611 MHz will soon be added to GNSS services on Galileo, GLONASS, and Compass constellations, future GNSS receiving antennas and receivers need to be designed to be functional at as many frequency bands as possible. This calls for an antenna design that can (a) simultaneously operate at multiple narrow (~20 MHz) GNSS bands, or (b) continuously operate in 1150–1300 MHz (150 MHz bandwidth) and 1559–1611 MHz (52 MHz bandwidth) bands without frequency coverage in between, or (c) continuously operate from two 1150–1611 MHz (460 MHz bandwidth) bands, and meet reflection coefficient, gain, pattern, and polarization requirements at all GNSS frequencies. Approach (a) is most challenging but has the best out-of-band rejection. However, this approach is most susceptible to frequency drifting and in-band dispersion due to its narrowband design. Approach (b) requires 12.2% bandwidth at lower GNSS band (1150–1300 MHz) and 3.2% bandwidth at upper GNSS band (1559–1611 MHz). Approach (c) is probably simplest in terms of antenna design since there are many existing wide-band antennas such as bow-tie dipole, spiral, and conical antennas, and so on. This type of antenna has least in-band dispersion and is immune to frequency drift. However, since it receives all signals from 1150 to 1611 MHz, separating different channels and rejecting interference relies solely on the correlator in the receiver.

14.2.2 Advanced Requirements for Enhanced Position Accuracy and Multipath Signal Suppression

The above basic RF performance requirements of GNSS receiving antennas should be sufficient for most general navigation and tracking operations without requiring very high position accuracy and in the absence of strong interferences. This subsection discusses additional antenna performance requirements for advanced precision positioning and interference mitigating operations. It should be noted that these advanced requirements merely serve as examples and do not represent the requirements of actual GNSS receiving systems, which vary with hardware, algorithms, and intended operating modes.

As mentioned in Section 14.1, a GNSS receiver at the user segment determines its absolute position using the three-dimensional trilateration method based on distances to four known GNSS satellite positions at the time of transmitting. Each distance is derived from the speed of light and time delay between the moment when a segment of a GNSS signal leaves the satellite and the moment when the same segment is received by the receiver. In addition to synchronization error between the clock on the satellite and the clock on the receiver, Figure 14.5 illustrates the main error sources in pseudo-range estimation based on the time delay of a GNSS signal. The first error source arises at the satellite segment and includes satellite position, time synchronization, group delay in the transmitting antennas, and multipath signal from the satellite platform. The second error source arises from dispersion and slower wave velocity within ionosphere layers and produces pseudo-range error ranging from 0 to 45 m. The third error source is related to refraction and slow-wave effects within the troposphere layer and produces pseudo-range error from 2 to 30 m. Both ionosphere and troposphere effects are maximum at the horizon and minimum at zenith. In the user segment, the propagation delay in the receiving antenna and receiver electronics leads to pseudo-range errors that are functions of frequency and

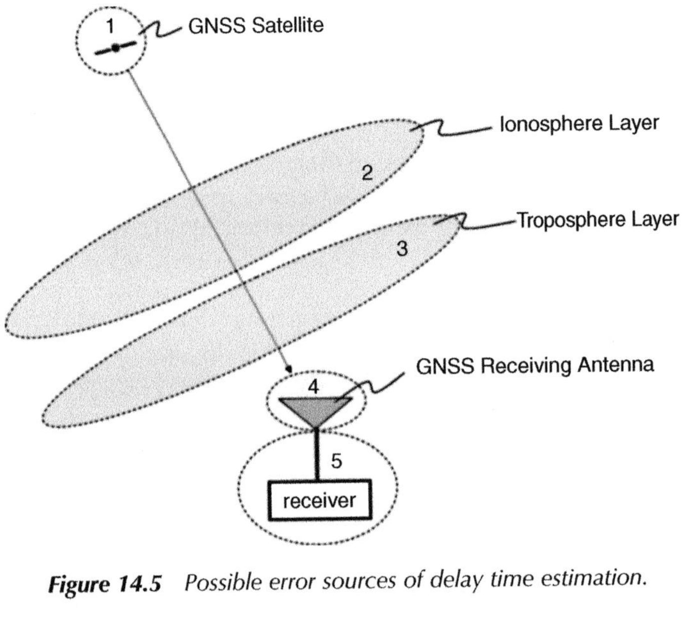

Figure 14.5 Possible error sources of delay time estimation.

angle. The range errors associated with the satellite and receiver can be corrected via factory calibration measurements or real-time monitoring from reference stations. The range error associated with propagation through the ionosphere and troposphere vary with time and place, and thus rely on real-time augmentation corrections. More detailed discussions about these errors can be found in [2]. In this subsection, we will focus on the pseudo-range error associated with the receiving antenna volume.

14.2.2.1 Antenna Delay Considerations

Figure 14.6 defines the antenna volume to be the smallest imaginary sphere that encloses the antenna structures (antenna element, radome, ground plane, etc.) and centered at a manufacturer-defined physical reference point. The radius a corresponds to the true range associated with the antenna volume. However, the pseudo-range is derived from light speed c and delay time $T^{antenna}$ and is defined as

$$R^{antenna} = cT^{antenna} \tag{14.3}$$

In most antenna designs, the satellite signal will experience additional delay within antenna structures such that $R^{antenna} > a$. The antenna range error is then defined as $R^{antenna} - a$,

$$\Delta R^{range} \equiv R^{antenna} - a \equiv c\Delta T^{antenna} \tag{14.4}$$

We will herein refer to all forms of additional signal delays associated with antenna volume as 'antenna delay,' $\Delta T^{antenna}$. The antenna delay can usually be determined experimentally and used for pseudo-range correction at the receiver's processor. In simple antenna designs such as patch antennas or dipole antennas, the antenna range error is around a quarter wavelength, $\lambda_0/4$, where λ_0 is the free-space wavelength at the operational frequency.

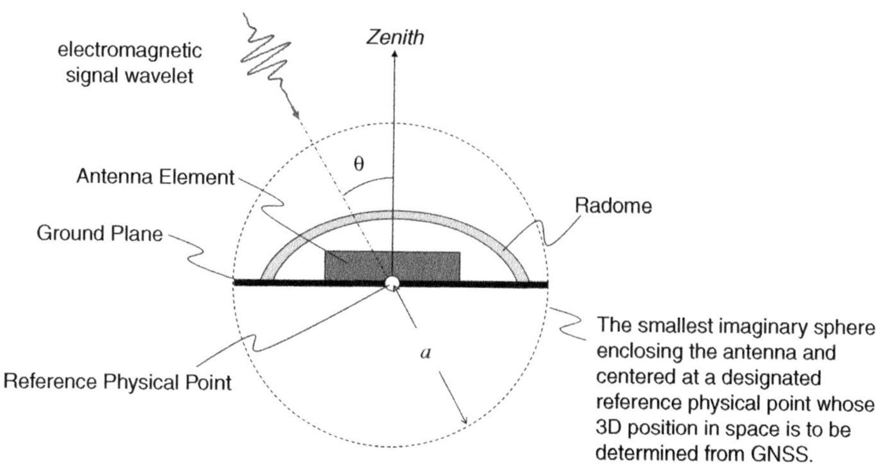

Figure 14.6 *Pseudo-range associated with the GNSS receiving antenna volume that is defined as the smallest imaginary sphere that encloses the antenna structures (antenna element, radome, ground plane, etc.) and centered at a manufacturer-defined physical reference point.*

Let $x(t)$ be the satellite signal arriving at the antenna region defined in Figure 14.6, $y(t)$ be the signal at the antenna output terminal, and $h(t)$ be the antenna's impulse response. Then we have

$$y(t) = h(t) * x(t) \tag{14.5}$$

Considering the received signal to be a sinusoidal signal at the carrier frequency ω_c with an amplitude modulation $C(t)$, then

$$x(t) = C(t)e^{j\omega_c t} \tag{14.6}$$

The output frequency response $Y(\omega)$can then be derived from the Fourier transform of (14.5) and (14.6) as

$$Y(\omega) = |H(\omega)|C(\omega - \omega_c)e^{j\phi(\omega)} \tag{14.7}$$

where $H(\omega) = |H(\omega)|e^{j\phi(\omega)}$and $C(\omega)$ are the Fourier transforms of $h(t)$ and $c(t)$, respectively. If the phase function $\phi(\omega)$ is expanded using Taylor series around ω_c and keeps only the first two dominant terms as shown in (14.8), then (14.7) can be approximated by (14.9):

$$\phi(\omega) \approx \phi(\omega_c) + \left.\frac{\partial \phi}{\partial \omega}\right|_{\omega_c} (\omega - \omega_c) + O\left(|\omega - \omega_c|^2\right) \tag{14.8}$$

$$Y(\omega) \approx |H(\omega)| \cdot C(\omega - \omega_c) \cdot e^{j\left\{\phi(\omega_c) + (\partial\phi/\partial\omega)|_{\omega_c}(\omega - \omega_c)\right\}} \tag{14.9}$$

Depending on antenna design approaches, its bandwidth could be wider or narrower than the signal bandwidth as illustrated in Figure 14.7. If the antenna bandwidth is much wider than the signal bandwidth (i.e., case I in Figure 14.7), the output frequency response (14.9) can be approximated by

$$\begin{aligned} Y(\omega) &\approx |H(\omega_c)| \cdot C(\omega - \omega_c) \cdot e^{j\left\{\phi(\omega_c) + (\partial\phi/\partial\omega)|_{\omega_c}(\omega - \omega_c)\right\}} \\ &\approx |H(\omega_c)| \cdot \left\{C(\omega - \omega_c)e^{j(\omega - \omega_c)\left[(\partial\phi/\partial\omega)|_{\omega_c}\right]}\right\} \cdot e^{j\omega \cdot (\phi(\omega_c)/\omega)} \end{aligned} \tag{14.10}$$

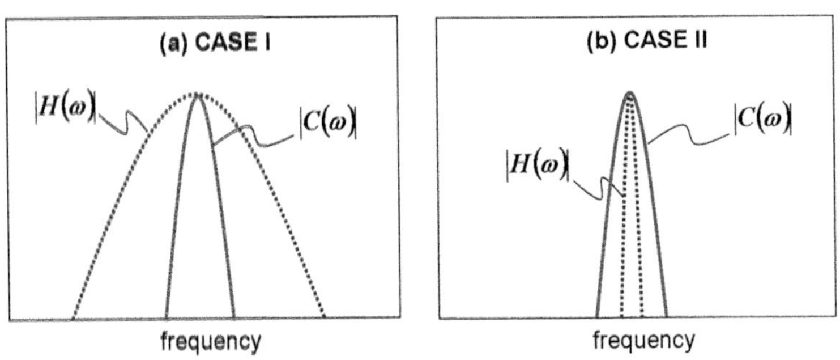

Figure 14.7 *Two possible scenarios of relationship between antenna and signal bandwidths.*

Applying the inverse Fourier transform to (14.10), we can obtain the corresponding time-domain response of the output signal as

$$y(t) \approx |H(\omega_c)|c(t-\tau_g)\cos(\omega(t-\tau_\phi)) \tag{14.11}$$

Comparing (14.11) to the original input signal expression in (14.6), one can immediately see that the receiving antenna scales the signal amplitude by a factor of $|H(\omega_c)|$. It also introduces a group delay of τ_g and a phase velocity of τ_ϕ, which are defined as [17]

$$\tau_g(\omega) = -\frac{\partial\phi(\omega)}{\partial\omega}\Big|_{\omega=\omega_c} \quad \text{and} \quad \tau_\phi(\omega) = -\frac{\phi(\omega_c)}{\omega} \tag{14.12}$$

For GNSS signals, the group delay adds a time delay to code chips, and the phase delay adds a time delay to the carrier signal. If the phase function is a linear function of frequency, then the group delay becomes the same as the phase delay. This leads to same time delay in code chips and carrier signals. This is a desired feature in applications requiring position accuracy of millimeters. For instance, every degree of carrier phase measurement error corresponds to approximately 0.53 mm range error at 1575.42 MHz.

On the other hand, if the antenna bandwidth is much narrower than that of the signal (i.e., case II in Figure 14.7), the code information contained in the signal will be partially lost. Therefore, it is of utmost importance to ensure that the receiving antenna has a bandwidth wider than that of the spectrum of the intended codes. Figure 14.8 illustrates the power spectrum density of planned C/A, Y, and M codes [18,19]. As one can see, 20 MHz bandwidth is needed in order to receive P(Y) codes, and at least 30 MHz bandwidth is needed to receive M code. The code delay from which pseudo-range is derived is determined from the peak location of the cross-correlation $R(\tau)$ between the local signal and the received satellite signal as

$$R(\tau) = \int C(f)H(f,\theta,\phi)e^{j2\pi f\tau}\,df \tag{14.13}$$

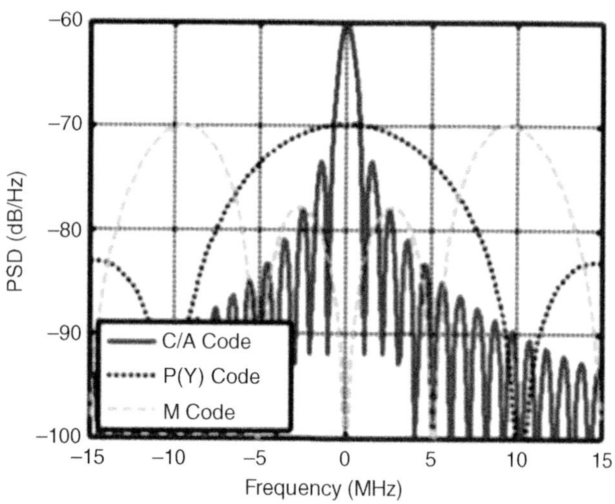

Figure 14.8 *Power spectral densities of baseband C/A code, Y code, and M code signals.*

where $C(f)$ is the code-dependent signal power spectral density (SPD) illustrated in Figure 14.8 and $H(f)$ is the antenna response that often depends on frequency and angle. This code delay depends on signal type and is in general not the same as the group delay defined in (14.12). For a well-behaved antenna, the code and group delay will be very similar. For the GPS landing system reference antenna, the RMS carrier delay variation and code delay variation within the coverage area should be less that 7 mm and 25 mm, respectively [20,21].

14.2.2.2 Advanced Pattern Requirements

One of the major pseudo-range errors is related to multipath signals caused by satellite signals scattered off nearby structures as illustrated in Figure 14.9. The correlator in modern receivers can effectively suppress multipath signals at delays (relative to direct signal) greater than 1.5 chips [1,22]. In GNSS signals, each digital bit '1' or '0' time is transmitted through a fixed number of short pulses called chips of pseudo-noise code sequence designated for each satellite. Many techniques have been developed to reduce the effect of multipath signals with shorter delays [23–27]. For instance, it is reported that using a narrow correlator spacing ($\sim$0.1 chips) can suppress multipath delays greater than 1 chip [22,28]. This corresponds to approximately 293 m and 29.3 m for 1.023 MHz and 10.23 MHz chip clocks, respectively. Clearly, this is still not sufficient for suppressing much closer multipath signals to achieve position accuracy within a few centimeters.

On open fields, most multipath signals come from directions near or below the horizon. Therefore, an ideal GNSS receiving antenna for stationary user segments should have a uniform hemispherical pattern above the horizon and diminish rapidly below the horizon as illustrated in Figure 14.10a. In practice, the truncated edges of the ground plane or host platform produce diffracted fields in the lit region (i.e., above the horizon) and shadow region (i.e., below the horizon) [29] as illustrated by the dotted line of Figure 14.10b. The resultant antenna pattern is similar to the solid line of Figure 14.10c. Such diffractions weaken reception of satellite signals at low elevation angles and increase undesired reception of multipath signals below the horizon. For example, the gain pattern requirements for the GPS landing system reference antenna are -9 to -6 dBic at 5° above the horizon, a gain slope of greater than -2.5 dB/° between 5° and -5° around the horizon, and maximum gain below -5° from the horizon is less than -30 dB below the peak gain at zenith [20]. These requirements ensure that multipath RMS error will be less than 9 cm over a smooth sea surface, that is, the worst case [21].

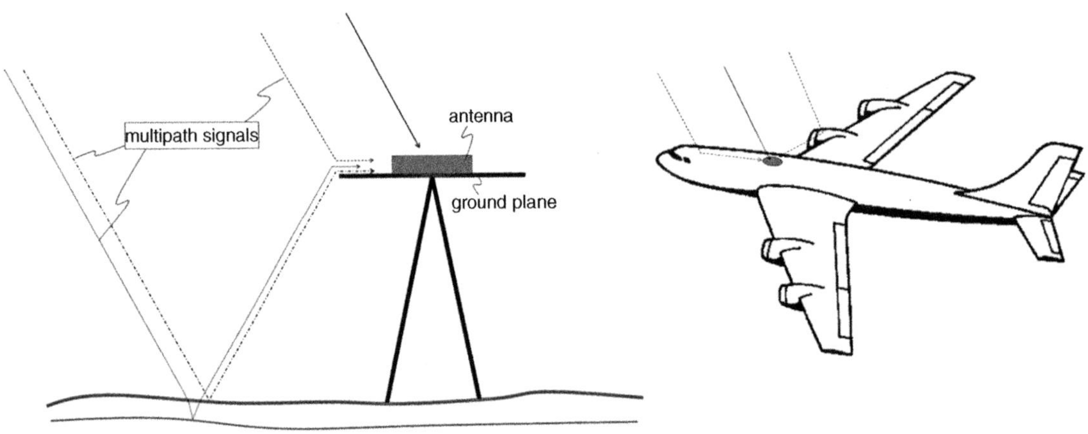

Figure 14.9 *Direct and multipath signals arriving at a GNSS antenna.*

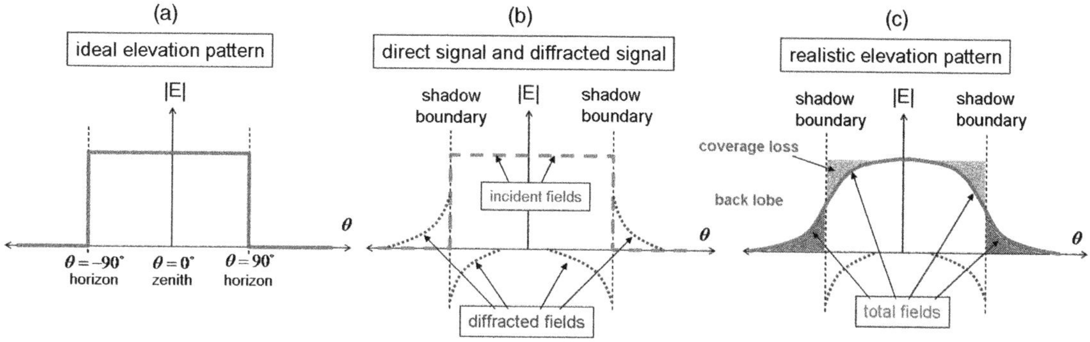

Figure 14.10 *Ideal and practical antenna elevation patterns, assuming a uniform azimuth pattern.*

Designing a high-precision GNSS antenna for stationary position measurement should begin with a good RHCP antenna design and then finish with good diffraction control. To achieve this, some GNSS antennas adopt array architectures capable of digitally shaping the antenna pattern by combining signals received from different array elements with proper amplitude and phase weightings. For example, a large antenna array adaptively forms multiple narrow satellite tracking beams to maximize the carrier-to-noise (C/N) ratio of direct signals. In addition, antenna arrays can also synthesize pattern nulls and steer them toward directions of interference or multipath signals. However, such advanced capabilities increase the size, weight, complexity, and cost of the receiver.

14.2.2.3 Advanced Polarization Requirements

The RHCP signals transmitted from GNSS satellites become LHCP after being reflected once from a large flat surface. Therefore, this kind of multipath signal can be rejected via good antenna design with low LHCP gain at GNSS frequencies in all directions [30]. It should be noted that only in rare cases are multipath signals from more than one bounce strong enough to become a concern. An antenna with a LHCP gain of −20 dB below RHCP gain provides 20 dB suppression of single-bounced multipath signal and 3 dB suppression of linearly polarized interference signal. The design in Figure 14.1 shows a LHCP gain of −30 dB below RHCP gain at zenith at center frequency. However, in this case, the RHCP-to-LHCP isolation decreases as frequency deviates from the center frequency and as the elevation angle lowers to the horizon. A good GNSS antenna for precision positioning should maintain at least 20 dB RHCP-to-LHCP isolation down to −30° below the horizon. Note also that not all single-bounce multipath GNSS signals are LHCP. For instance, incoming satellite signals diffracted off a linear edge of the ground plane or receiver platform are often linearly polarized, which produces equal RHCP and LHCP levels.

14.3 Design Challenges and Solutions for GNSS Antennas

Now that we know the desired RF performance requirements for a good GNSS receiving antenna at the user segment, the question is how to design one. Currently, there are quite a few commercial GNSS antennas operating at L1-band or both L1- and L2-bands [31–34]. These antennas perform adequately for most applications that do not require high position accuracy and, most importantly, are compact and inexpensive. This type of user often accepts occasional large position error caused by multipath or loss of reception. However, for military operations or aviation landing systems, each occurrence of large position error or reception loss could result in dire consequences. This is the perfect reason for many ongoing efforts in developing better waveforms, receiver technology, and antenna designs to take advantage of more frequencies and multiple constellations of GNSSs to achieve better and more reliable position accuracy.

Key GNSS antenna design challenges include: (a) being able to receive all GNSS signals in 1150–1300 MHz and 1559–1611 MHz bands for maximizing future receiver capability and flexibility; (b) minimizing antenna delay variations with frequency and angle; (c) minimizing antenna size and weight for improved mobility and portability; and (d) low manufacturing cost for broader user acceptance. What makes it even more challenging for users and antenna engineers alike is that, to the best of the authors' knowledge, there is no standard set of technical requirements available, and there will probably be none for quite some time, due to the vast varieties of applications. The following subsections discuss the challenges (a)–(c) and possible design solutions.

14.3.1 Wide Frequency Coverage

As illustrated in Figure 14.11, GNSS antenna designs can adopt either dual-band or wide-band approaches to provide good coverage for intended GNSS frequencies in 1150–1300 MHz and 1559–1611 MHz bands. By 'coverage,' it means that the antenna should meet all gain, pattern, polarization, and delay properties at the intended frequencies.

The dual-band approach requires the antenna to have bandwidths of 150 MHz (or 12.24%) and 52 MHz (or 3.28%) at the lower and upper GNSS bands, respectively. This approach is more challenging in order to achieve good delay and polarization performance within the entire sub-bands, but it also offers better rejection of out-of-band interferences.

The wide-band approach requires the antenna to operate continuously from 1150 to 1611 MHz, a bandwidth of 451 MHz (or 33.39%). Such bandwidth is certainly within the reach of traditional wideband antenna designs [6,35–37] such as bow-tie dipoles, spiral antennas, log-periodic dipole antennas, horn antennas, and dielectric antennas. Among them, spiral antennas are probably the most attractive due to their inherent wide bandwidth, circular polarization, and rotational symmetry. The gain and axial ratio of a sufficiently large spiral antenna are fairly independent of azimuth angle and frequency. Figure 14.12 illustrates a couple of commercial GNSS antennas based on spiral antenna designs. Four-arm spiral designs are often used for superior axial ratio and azimuth pattern uniformity over two-arm designs. This will require additional feed line arrangements.

14.3.2 Antenna Delay Variation with Frequency and Angle

Let us now examine a simple square patch antenna above an infinite ground made of a perfect electric conductor (PEC). This example (Figure 14.13) includes three different patch heights of 1, 2, and 3 mm. The

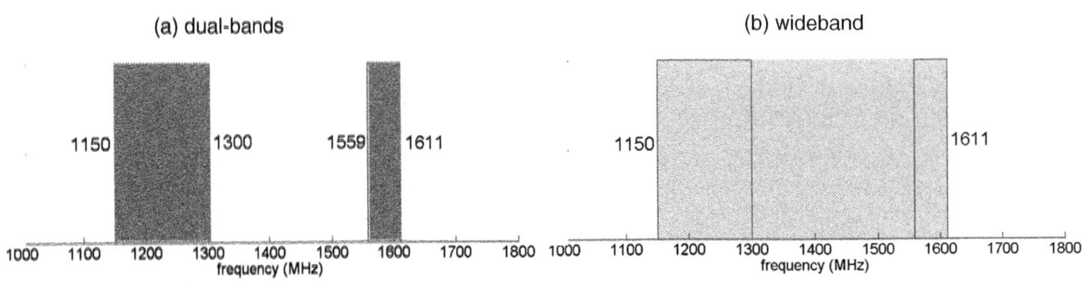

Figure 14.11 Dual-band vs. wide-band GNSS antenna design approaches.

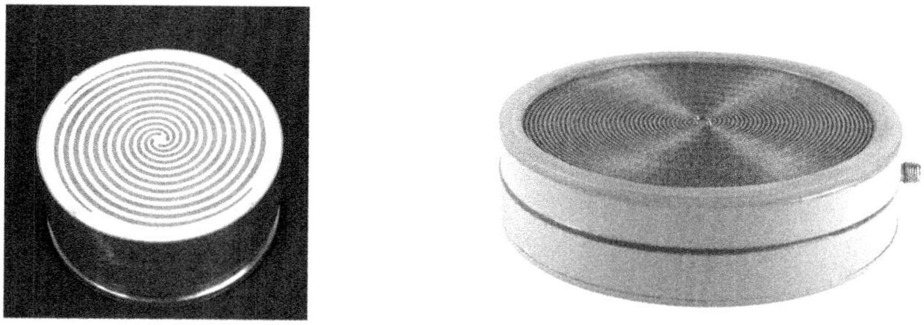

(Image courtesy Roke Manor Research Limited) (Image courtesy Wang Electro-Opto Corporation)

Figure 14.12 *Commercial GNSS antennas based on wide-band spiral antenna designs.*

table inset on the right of the figure shows the optimal design parameters that minimize the reflection coefficient (S_{11}) at 1.575 GHz for each height case. The antenna is connected with a single 50 Ω probe feed that is properly offset from the center. Although this antenna design is linearly polarized for clearer demonstration, the findings and conclusions can be readily applied to its circular polarization versions.

Figure 14.14 plots the calculated phase pattern in the elevation plane (x–z) of the Figure 14.13 design at 1.575 GHz. Each line corresponds to one of the 1, 2, and 3 mm patch height cases. Note that this phase is referenced to that of the direct path from the satellite to the physical reference point. In this case, the reference

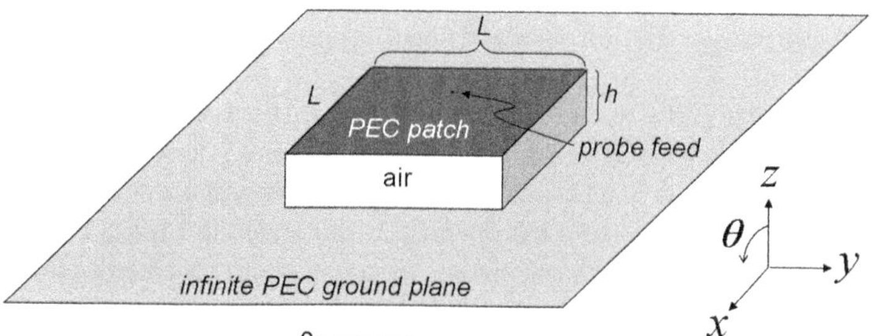

Optimized Parameters for L1 band

H (mm)	L (mm)	Feed Offset (mm)
1	92.74	13.80
2	90.91	14.37
3	89.40	15.00

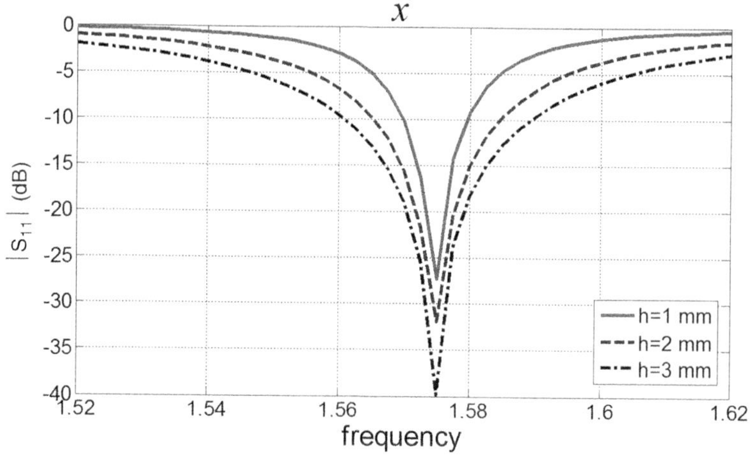

Figure 14.13 *A square patch above an infinite PEC ground plane at different heights. The table on the right shows optimal design parameters for minimizing the reflection coefficient at 1.575 GHz.*

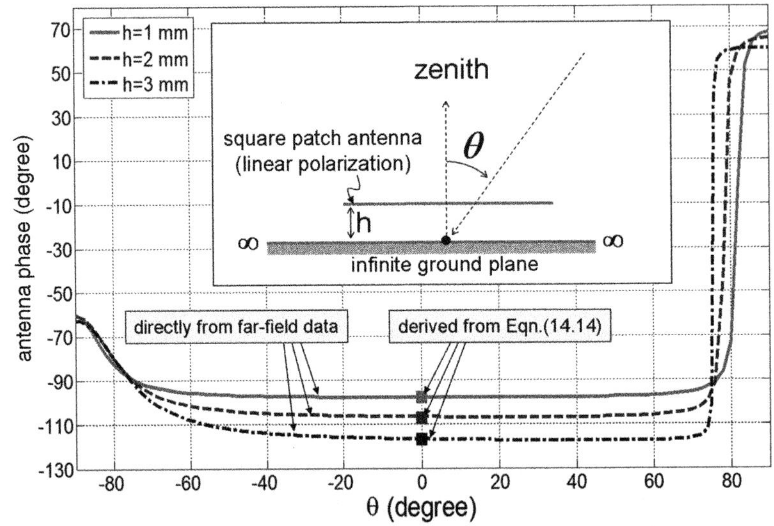

Figure 14.14 *Elevation phase patterns at 1.575 GHz in the x–z plane for Figure 14.13 antenna design.*

point is located at bottom center of the ground plane (i.e., $x = y = z = 0$). Therefore, any non-zero phase value indicates pseudo-range error with negative phase corresponding to positive range error and positive phase corresponding to negative range error. Note that a phase of 10° corresponds to approximately 5.3 mm pseudo-range error at L1 frequency. Therefore the large phase variation observed near the horizon in the E-plane is not desirable.

At zenith direction ($\theta = 0°$), the phase is related to additional propagation delay from the antenna's effective aperture (where most radiation or reception of electromagnetic fields occurs) to the antenna's terminal. Since a thin patch antenna has a patch size of approximately $\lambda_0/2$ in electrical length, in order to establish its resonant condition the electrical distance from the center to either side of the opening as illustrated in Figure 14.15 is approximately $\lambda_0/4$, which should produce a phase of −90° relative to the direct path from the satellite to the antenna's reference position. The additional inductance associated with patch height slightly increases the phase lag from −90° to −98° in the 1 mm height case.

As patch height increases, the series inductance of the waveguide formed by the patch and ground plane increases, causing additional phase lag as observed in Figure 14.14. It was found that the observed phase lag at zenith could be accurately estimated from (14.14) for small heights where the phase delay Φ is obtained from the phase angle of the centroid of reflection coefficient loci on the Smith chart as illustrated in Figure 14.16.

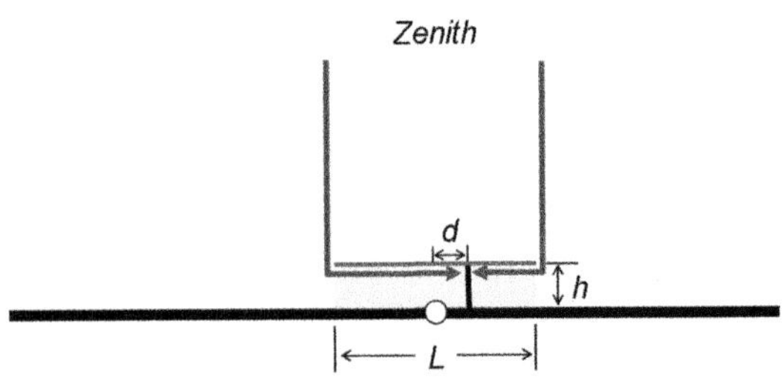

Figure 14.15 *Propagation path in a patch antenna for signals from zenith.*

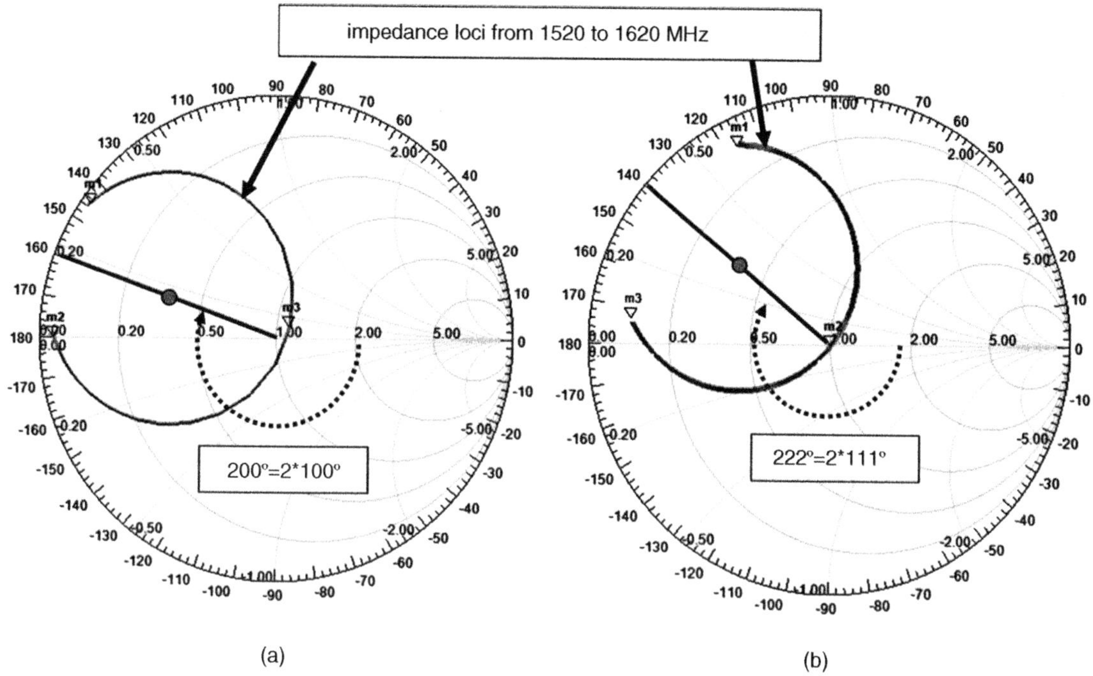

Figure 14.16 *Impedance loci on Smith chart from 1520 to 1620 MHz for Figure 14.13 patch at (a) 1 mm and (b) 2 mm heights.*

Since Φ is related to round-trip delay from the probe to the effective aperture, the one-way phase lag in reception mode is therefore $-\Phi/2$. In the current three cases, $\Phi/2$ values were found to be 100°, 111°, and 123° for h = 1, 2, and 3 mm, respectively. The last term in (14.14) accounts for the phase advance as the patch gets closer to the satellite at zenith as height h increases, as if signal reception occurs at the patch edges. The square markers in Figure 14.14 indicate the resultant antenna phase predicted from (14.14), which agrees very well with the actual phase directly obtained from far-field data from full-wave simulations.

$$-\Psi_{zenith} \approx -\frac{\Phi}{2} + k \cdot h \quad \text{(radians)} \tag{14.14}$$

Let us now examine the phase variation with elevation angle in the E-plane (x–z plane) and H-plane (y–z plane) for the 1 mm height case shown in Figure 14.17. The phase in the E-plane remains stable for most elevation angles because the phase advance and phase delay from the near and far sides of edges cancel out in total fields. At low elevation angles, the contributions from two patch edges cancel each other and result in gain reduction as demonstrated by the gain patterns shown in Figure 14.18. The residual reception near these angle regions is actually from the probe directly. As a consequence, the phase behavior switches from patch mode to probe mode and results in rapid phase change near the horizon. The offset probe arrangement causes different phase lags at either side of the horizon to be approximately +65° and −65°. If the probe is located at the patch center as in the cases of aperture-coupled feed or symmetric probes (see Figure 14.23 below), the net phase lag associated with direct probe reception will be close to zero near the horizon since the signal path becomes the direct path from the satellite. This is also the reason why the phase in the H-plane remains stable over all elevation angles.

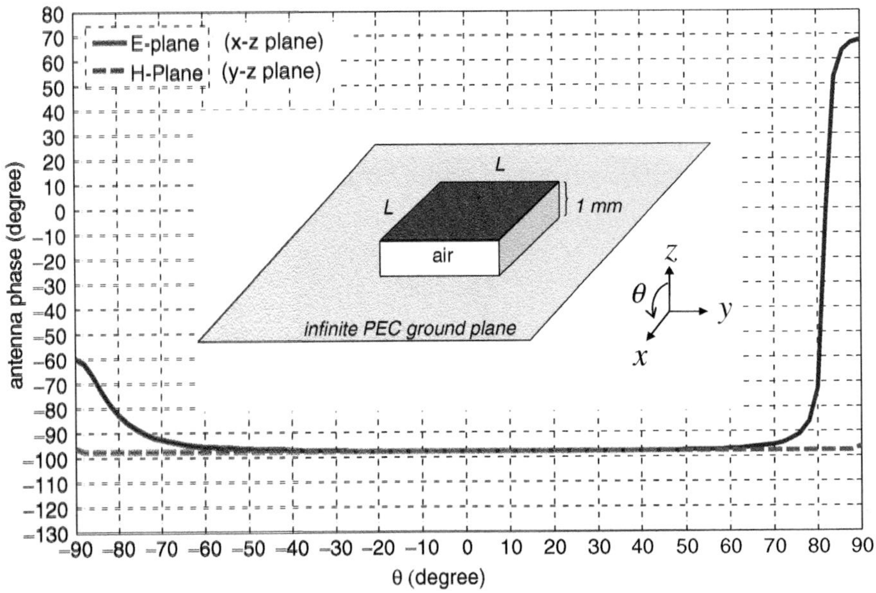

Figure 14.17 *E- and H-plane phase patterns of a linearly polarized patch antenna at 1.575 GHz.*

Most patch antennas have a dielectric substrate for improved mechanical stability or reduced antenna sizes (see next section). Figures 14.18 and 14.19 show the calculated gain and phase patterns in the E-plane at 1.575 GHz for a patch antenna with different dielectric substrates and dielectric constant ranging from 1.01 to 2. Note that the patch size and feed position in each case were slightly different, and were individually

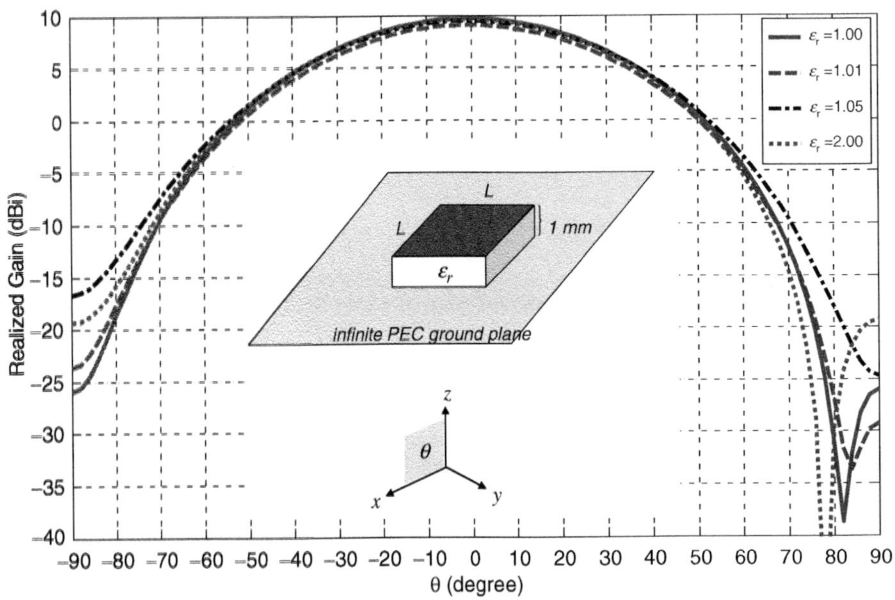

Figure 14.18 *Calculated E-plane gain patterns at 1.575 GHz for a patch antenna with different dielectric substrates. The patch size and feed position in each case were optimized for minimum S_{11}.*

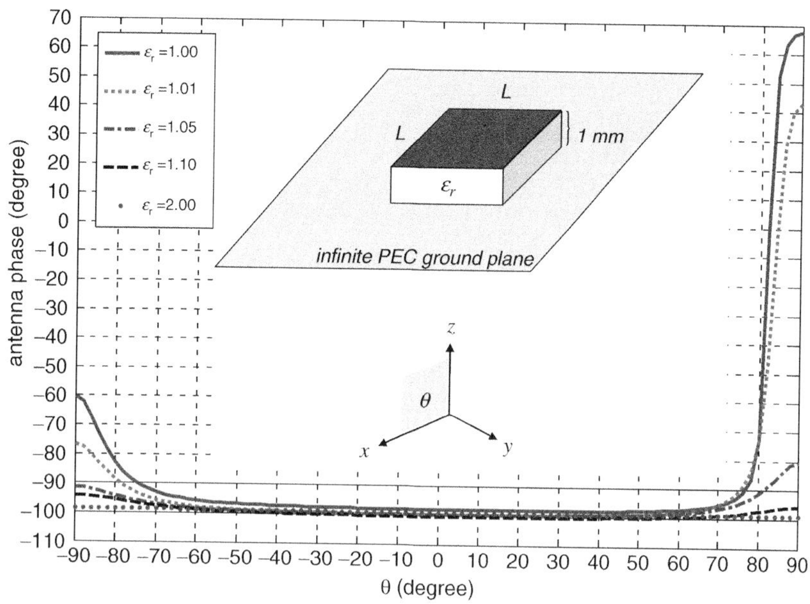

Figure 14.19 *Effect of dielectric substrate on patch antenna's phase pattern in E-plane at 1.575 GHz.*

optimized for achieving minimum S_{11} at 1.575 GHz. The patch height for these cases is kept at 1 mm. It is interesting to observe in Figure 14.19 that even a small amount of dielectric loading significantly reduces the direct probe reception effect near the horizon. As a result, the phase pattern becomes more stable down to low elevation angles. The phase patterns in the H-plane still remains very stable over all elevation angles, and therefore are not shown.

14.3.3 Antenna Size Reduction

Many small antenna designs have been developed by researchers and engineers worldwide and can be found in many references [38]. By 'small,' we mean that the antenna size is much less that $\lambda_0/2$. Most of these designs employ some sort of miniaturization technique to reduce the antenna's physical dimensions while maintaining its electrical dimensions. These small antenna designs often explicitly or implicitly make use of the artificial transmission line (ATL) approach to artificially alter the wave velocity and impedance within the antenna structure [38]. Miniaturization is achieved when the phase velocity of electromagnetic waves propagating within the antenna structure is decreased via increasing equivalent shunt capacitance and/or series inductance. In patch antennas, the former can be implemented using dielectric loading or capacitive ceiling posts as shown in Figure 14.20. The latter can be implemented using slots as shown in Figure 14.21. The ATL miniaturization technique can also be equally effective in wide-band antennas as demonstrated by Kramer *et al.* [39].

It should be noted that, although the physical dimension of the antenna is reduced after being miniaturized, its electrical size remains similar in order to operate at the same frequency. Therefore, the antenna delay should be similar to that of unminiaturized ones. However, miniaturization does help to reduce phase variation as a function of elevation and azimuth angles as illustrated in Figure 14.22. Such variations result from interference among satellite signals arriving at different parts of the antenna and thus have different phase. As the physical

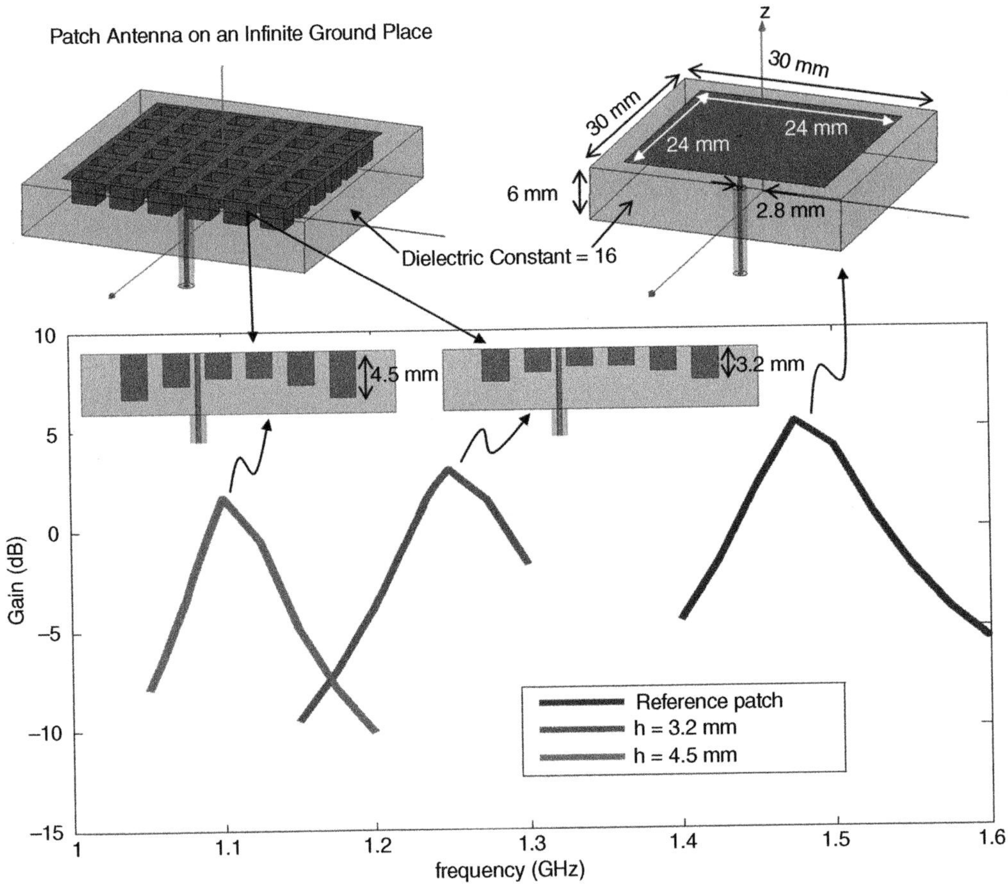

Figure 14.20 *Lowering operational frequency by increasing shunt capacitive posts and dielectric loading [38].*

dimensions of the antenna decrease, the difference in phase among these different reception paths also decreases.

It should be pointed out that the maximum achievable instantaneous bandwidth will also decrease as the antenna's physical volume decreases, especially when such volume is much less than $\lambda_0/4$. This natural consequence dictates the minimum antenna dimensions for GNSS operations which require modest bandwidths in order to operate in the modes shown in Figure 14.11. Of course, this size limitation can be relaxed if one does not need instantaneous coverage of entire GNSS bands and uses a smart receiver that is capable of dynamically tuning its impedance to match that of the antenna at a single GNSS channel at a time. However, the antenna's bandwidth still needs to be wide enough to cover the code spectrum shown in Figure 14.8.

14.3.4 Antenna Platform Scattering Effect

All GNSS antennas are mounted on some kind of platform and have radomes of different sizes, shapes, and compositions. Each GNSS antenna also has it own conducting ground plane for supporting the feed mechanism and receiver electronics. Electromagnetic scattering arising from reflection, refraction, and

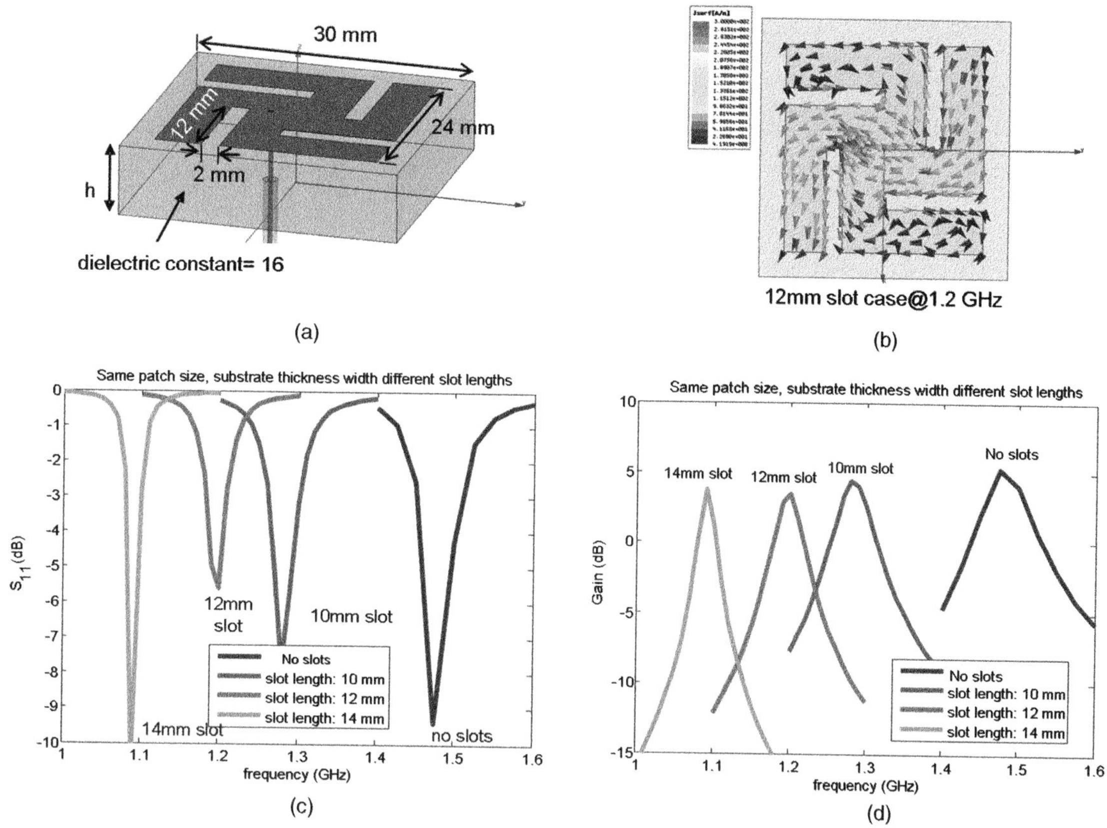

Figure 14.21 *Patch antenna size reduction by increasing shunt capacitance and series inductance implemented using dielectric loading and slots [38]. (a) Geometry of slotted- patch antenna. (b) Induced currents on patch for 12 mm slot case at 1.2 GHz. (c) Reflection coefficient for four different slot widths. (d) Realized gain at zenith for four different slot widths.*

diffraction associated with such a ground plane, radome, or platform structure within one wavelength radius has a significant effect on antenna characteristics. To demonstrate this, we truncate the infinite ground plane in the Figure 14.17 design to three finite circular ground planes with diameters of 15, 25, and 35 cm, respectively. Figure 14.23 shows that the reflection coefficient (S_{11}) and antenna impedance clearly deviate from the infinite

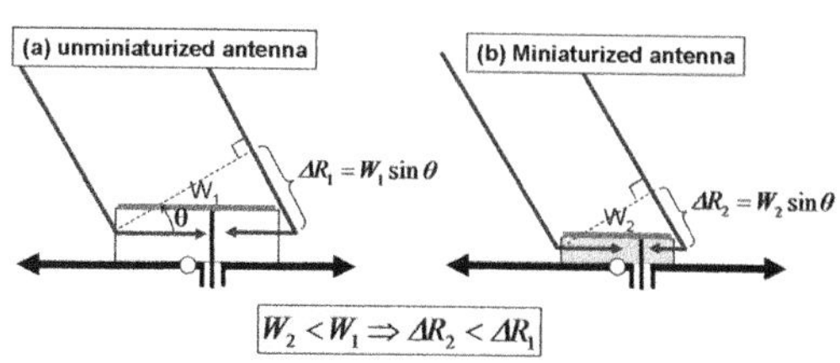

Figure 14.22 *Less delay and phase variation with pattern angle for smaller antennas.*

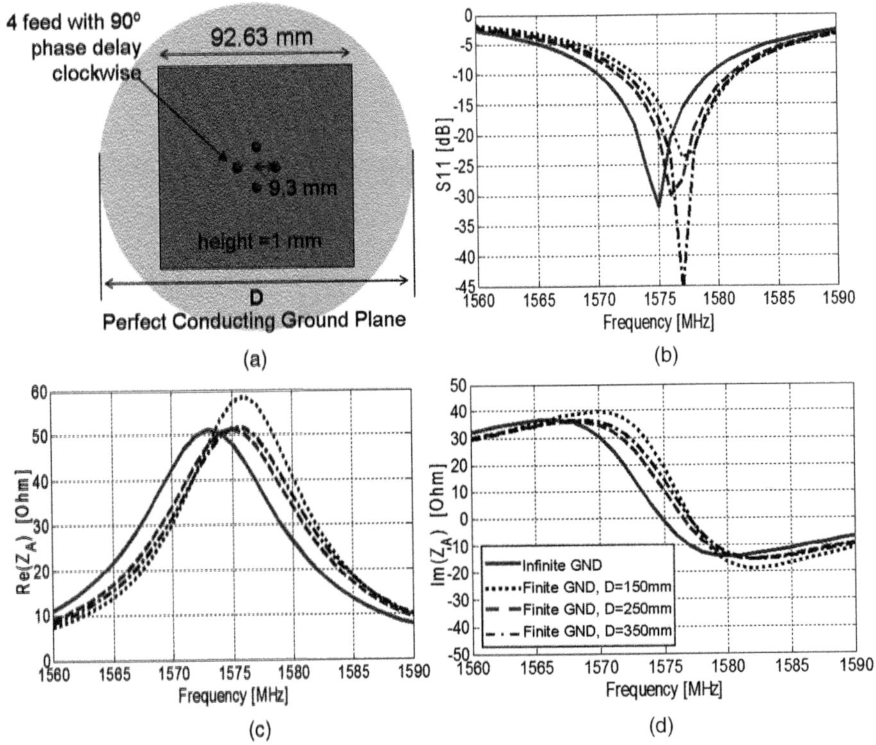

Figure 14.23 *The detuning effect from edge diffractions of finite PEC ground planes. Four probe feeds with progressive phase provide better RHCP and azimuth uniformity. (a) Top view of a square-patch antenna. (b) Reflection coefficients. (c) Realized gain at zenith. (d) Input reactance.*

PEC ground plane case. The edge of the truncated ground plane produces additional diffractions that are superposed upon the direct signal from the satellite to antenna element as illustrated in Figure 14.9. As a result, the antenna appears to be detuned from 1575 MHz. Such a detuning effect will be less of a problem in a wide-band design. Similar multipath effects can occur from any nearby scattering sources such as radome and host platforms like the human body, vehicles, or aircraft.

The impact on antenna characteristics from this ground plane edge diffraction depends on its relative phase and magnitude compared to the direct signal. More specifically, it depends on the size of the ground plane in wavelengths and satellite angles. As a result, it increases gain and phase variation as a function of frequency and angle as demonstrated in Figure 14.24.

The detuning effect observed in Figure 14.23 can be corrected by adjusting the probe position. The retuned S_{11} results for each case are shown in Figure 14.25 which also includes the gain pattern in the elevation plane. It shows that in this particular example the truncation has a minor effect on gain pattern and polarization above 15° from the horizon due to dominant direct signals. Within 15° of the horizon, the direct signal weakens as shown by the solid line in the infinite ground plane case and the edge diffraction becomes dominant. Such diffraction is dominated by the vertically polarized field which produces equal LHCP and RHCP components near the horizon (near $\pm 90°$) as seen in the bottom plots of Figure 14.25.

Several techniques have been developed for reducing edge diffraction. These techniques can be grouped into three categories by how they work: (a) absorbing, (b) choking, and (c) redirecting. Both absorbing and

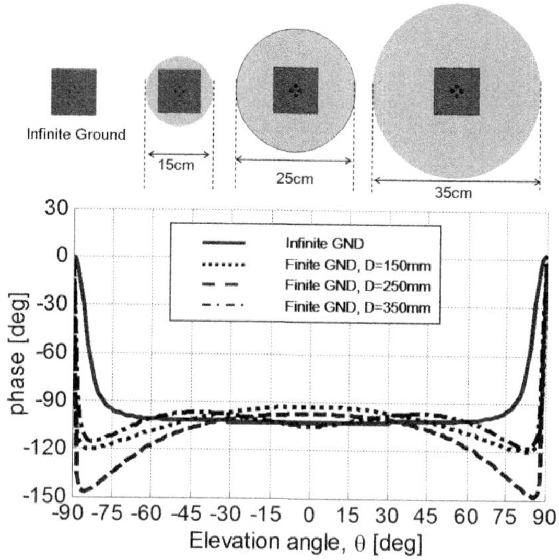

Figure 14.24 *Effect of ground plane edge diffractions on phase pattern of a patch antenna after retuning the patch size and feed location to obtain a reflection coefficient less than −30 dB at 1575 MHz.*

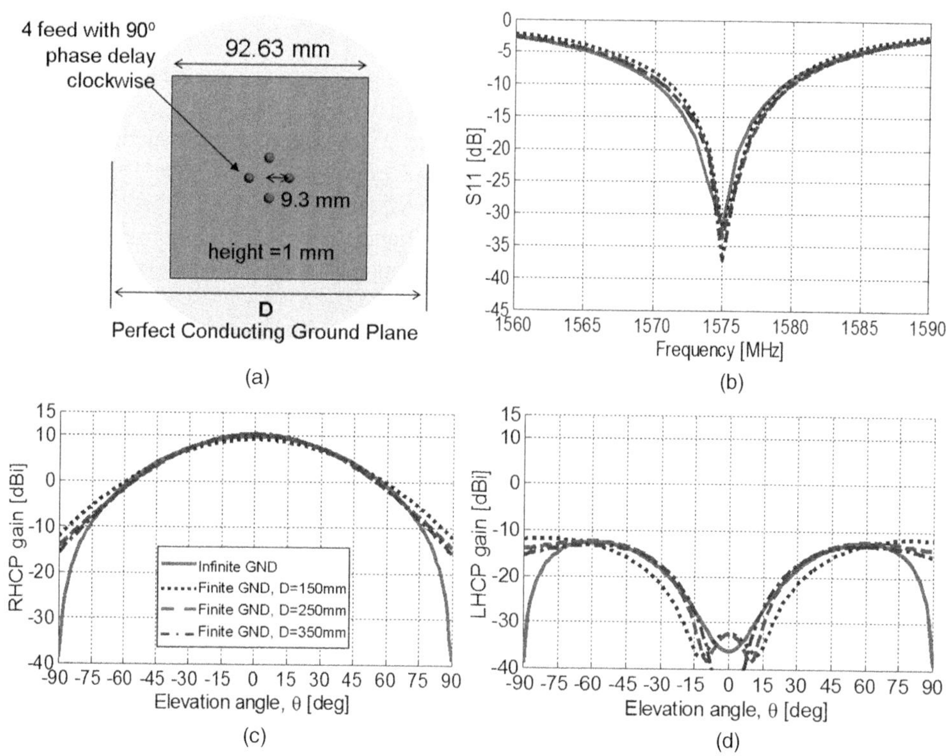

Figure 14.25 *The performance of patch antenna sizes after optimizing feed positions for the finite ground plane cases discussed in Figure 14.23. (a) Top view of a square patch antenna on finite and ground plane. (b) Reflection coefficient. (c) RHCP realized gain pattern in elevation plane. (d) LHCP realized gain pattern in elevation plane.*

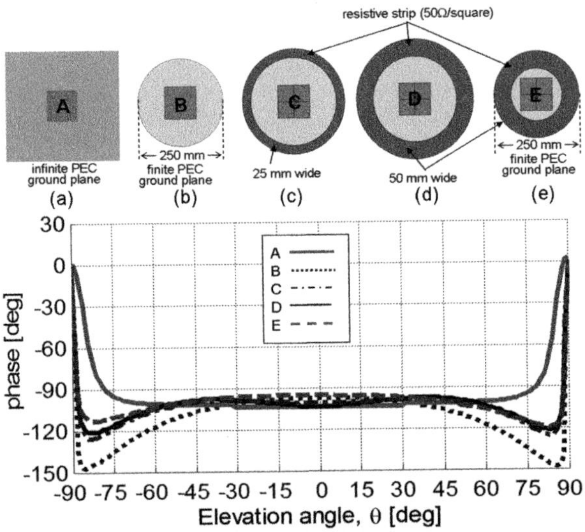

Figure 14.26 Reducing edge effect from a truncated ground plane using resistive loading. (a) Infinite PEC ground plane. (b) The 25 cm PEC ground plane. (c) The 25 cm PEC ground plane terminated with a 2.5 cm wide, 50 Ω/square resistive sheet. (d) The 25 cm PEC ground plane terminated with a 5 cm wide, 50 Ω/square resistive sheet. (e) The 15 cm PEC ground plane terminated with a 5 cm, wide 50 Ω/square resistive sheet.

choking approaches aim at removing wave propagation between the antenna element and diffraction source, that is, the ground plane edge in this case.

The absorbing approach employs RF absorptive materials to attenuate electromagnetic waves propagating along the ground plane between the antenna and truncated edges [40,41]. Figure 14.26 demonstrates the utilization of attaching a resistive strip to the truncated edge to reduce the diffraction effect and thus reduce phase pattern variation. The similarity between the D and E cases also demonstrates that effective resistive loading also makes antenna performance less dependent on the size of the ground plane compared to the untreated cases shown in Figure 14.24.

The choking approach utilizes choking rings [42–49] to produce a high-impedance surface around the antenna element to prevent vertically polarized waves from propagating along the ground plane. Such choking treatment is effective in suppressing signals coming near and below the horizon. This can effectively suppress multipath signals from truncated ground plane diffractions or other ground reflections. The disadvantages of the choking-ring approach include ground plane thickness and reduced reception of low-elevation satellites. The former increases the overall weight. The latter reduce the number of available satellites. One unique variation of choking-ring design for less ground plane height is to operate the choking rings under the cutoff frequency [50].

14.4 Common and Novel GNSS Antennas

14.4.1 Single-Element Antenna

14.4.1.1 Single-Band GNSS Antennas

Dielectric-loaded patch antennas have been commonly used as GPS GNSS receiving antennas for operation at L1-, L1/L2-, or L1/L2/L5-bands due to their simplicity, compactness, and low cost. We have already shown an example of a patch antenna design operating at L1 frequency in Figure 14.1. The circular polarization property of that design was achieved by using two orthogonal modes along the x and y directions via a carefully positioned probe and slightly asymmetric patch. One of the modes was resonating slightly below 1575.42 MHz and the other mode was resonating slightly above 1575.42 MHz, such that the phase difference between the two modes at 1575.42 MHz was approximately 90°. The dielectric substrate was also cut at one of the corners to introduce additional symmetry to help excite the two orthogonal modes more effectively. Other methods for introducing asymmetry or perturbations on patch antennas include the use of slotted, notched, and protruded patches, and so on [5]. In addition to the probe feed approach shown here, patch antennas can also be fed using an aperture coupling method from a slot opening at the ground plane [51,52]. Such a slot is usually fed with a microstrip line on the opposite side of the ground plane. The input resistance and capacitance increase with the width and length of the slot, respectively. The geometry of the coupling aperture can take many different shapes and sizes for achieving different coupling efficiency and impedance matching conditions [53]. Compared to the probe feed approach, the aperture coupling approach offers better E-plane pattern symmetry due to its central location. A wider bandwidth can also be achieved in the aperture coupling approach due to its ability to tune out the large inductance present in patches with large heights. However, the aperture-coupled patch antenna is more difficult to implement in dual-mode designs and requires additional shielding or a cavity beneath the ground to avoid the mounting effect and aperture radiation below the ground plane [54].

14.4.1.2 Dual-band GNSS Antennas

There are many existing antenna designs for dual-band operations. Some designs support two resonant modes via proper combination of two antenna structures. These kinds of antennas include slot-loaded patch antennas [55–59] and stacked-patch antennas [9,60]. Others excite different parts of an antenna structure to produce two resonances. E-patch antennas [61,62], H-slot antennas [63,64], and inverted-F antennas [65] are well-known examples of this kind.

Figure 14.27 shows a stacked-patch antenna design that has two resonant modes for covering L1, L2, and L5 GPS bands. It contains two vertical stacked-patch antenna elements, each with a dielectric layer and conducting patches. Each layer is responsible for resonating at one frequency. A branch-line hybrid located at the bottom of the antenna excites two orthogonal modes with approximately 90° phase difference at each resonant frequency to achieve circular polarization operation. This is a better approach to achieve circular polarization compared to the approach in Figure 14.1 for GNSS operations, due to the wider frequency range. In this particular design, the phase difference at the two orthogonal ports of the hybrid varies linearly from approximately 80° to 100° from 1200 to 1575 MHz. This design also features proximity probes conveniently located on the side [60] and is thus much easier to fabricate compared to the internal probe approach adopted in Figure 14.1. The realized gain versus frequency at zenith is shown in Figure 14.27d, which includes both measured and simulated data for RHCP and LHCP components. This data was obtained on a ground plane 6 × 6 inches in size.

The realized gain and axial ratio patterns at 1200 GHz and 1575 GHz in the entire upper hemisphere are shown in Figure 14.27e. The zenith is located at the plot center. The radial length corresponds to elevation angle θ from 0° at the center to 90° at the outer rim. The rotation angle corresponds to the azimuth angle ϕ in the spherical coordinate system. The average decibel values of the axial ratio at 50°, 60°, 70°, and 80° elevation angles from zenith are indicated in the boxes at the bottom.

Figure 14.28 shows a modified quad-F antenna with two dielectric layers, introduced by Zhou *et al.* [65]. The antenna size is 38 × 38 × 22.3 mm. The electric fields inside the antenna at 1175 and 1550 MHz in the bottom right plots reveal the resonant modes excited by the upper pair and middle pair of fingers of the F-shaped probe. The circular polarization is achieved by combining the four probes with 0°, −90°, 180°, and +90° phase difference. The calculated gain curve at zenith shows good bandwidths at upper and lower bands that almost cover all GNSS frequencies.

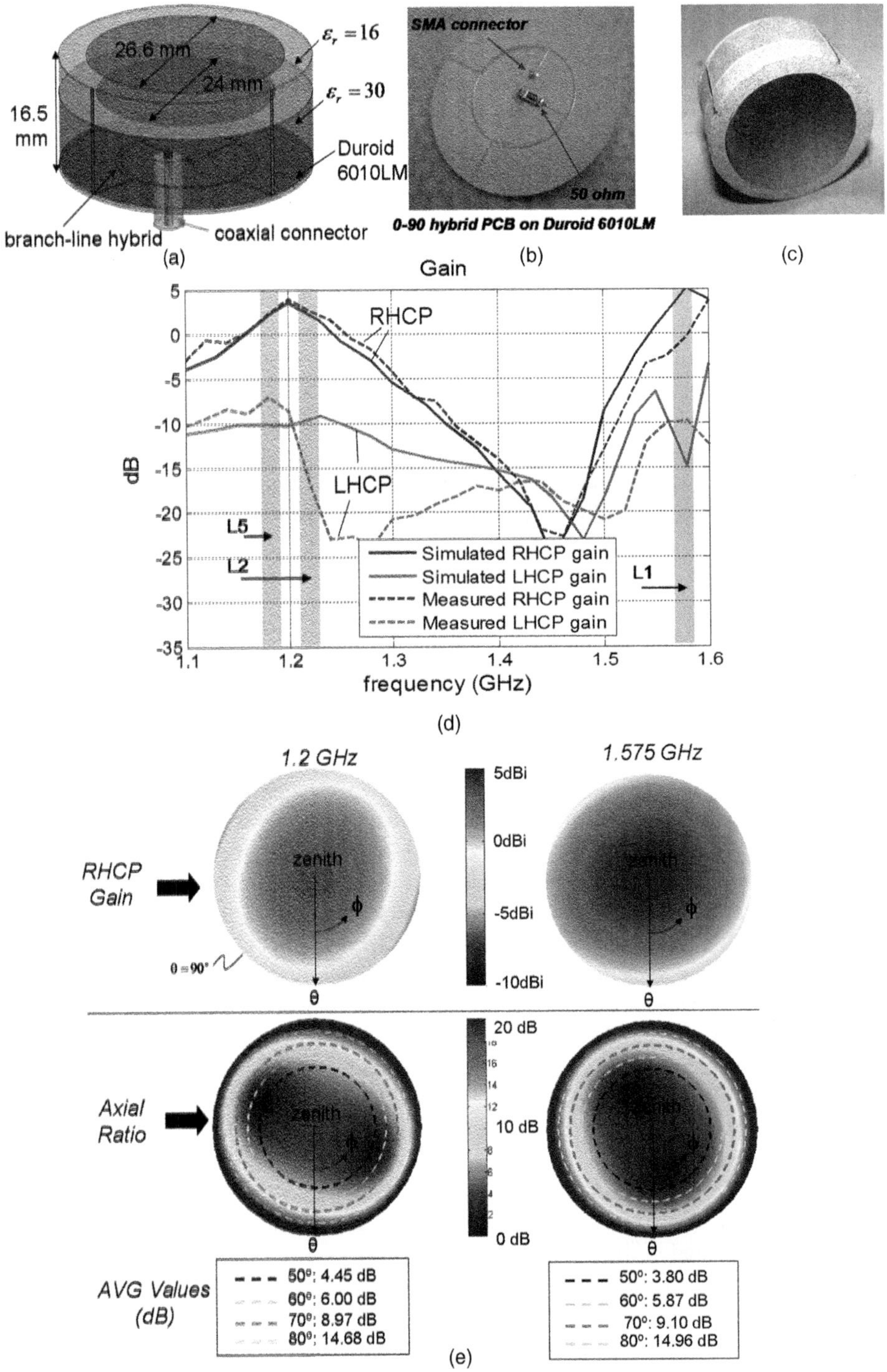

Figure 14.27 *A dual-mode triple-band (L1, L2, and L5) stacked-patch antenna fed with a branch-line hybrid and proximity probes. (a) Physical configuration. (b) Fabricated 0°–90° hybrid. (c) Assembled stacked-patch element without ground plane. (d) Comparison of calculated and measured RHCP and LHCP gain at zenith. (e) RHCP realized gain (top figures) and axial ratio (bottom figures) over the entire sky at 1.2 and 1.575 GHz.*

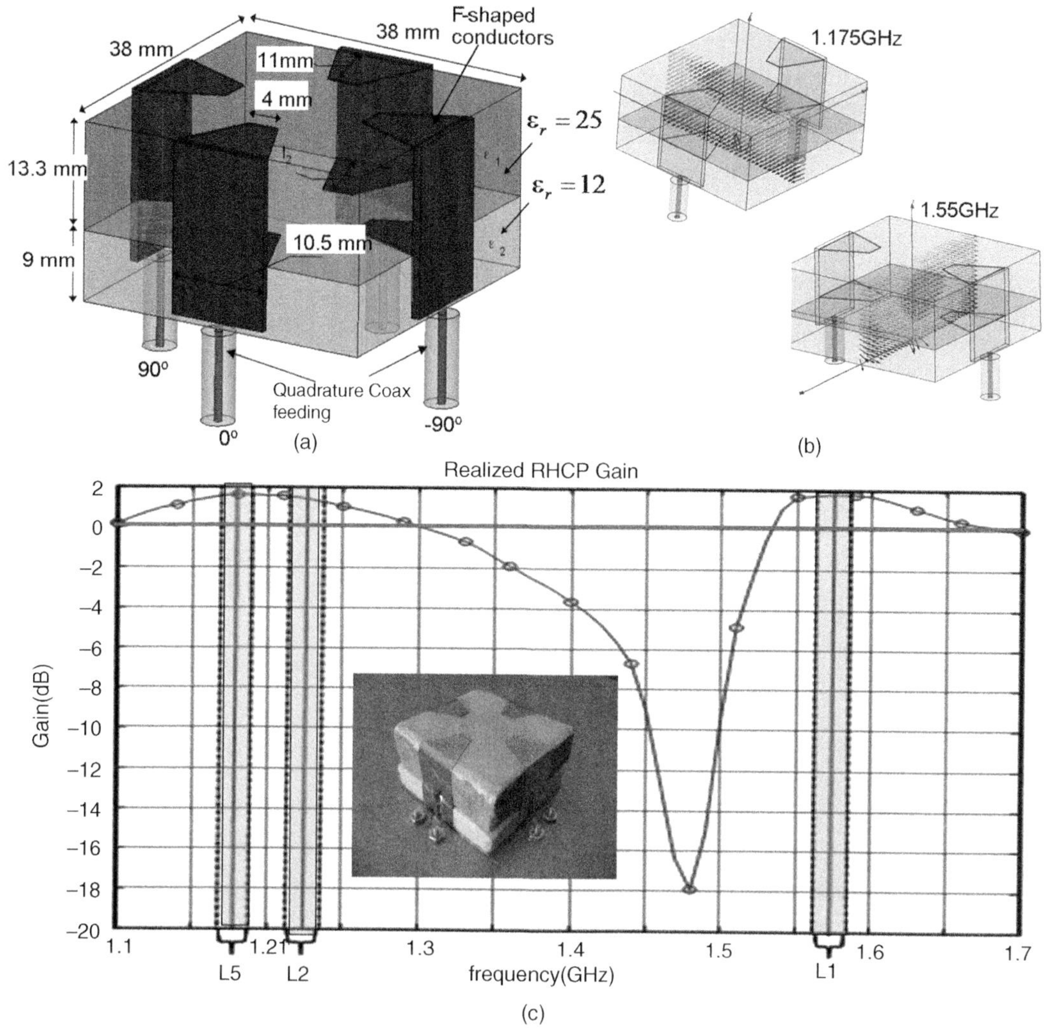

Figure 14.28 *Dual-band quad-F GNSS antenna design developed by Zhou et al. [65].*

14.4.1.3 Wide-band Antennas

Many wide-band antenna designs can also be used as GNSS receiving antennas. Among them, turnstile bow-tie antennas and spiral antennas are commonly used. A bow-tie dipole is known for its wide bandwidth and can routinely operate over a 2:1 bandwidth. Spiral antennas commonly achieve a 9:1 bandwidth.

Figure 14.29 shows an example of an unoptimized cross-bow-tie antenna design. RHCP is achieved by combining signals from the two orthogonal bow-tie dipole elements with quadrature phase. The calculated realized RHCP gain versus frequency at zenith clearly shows sufficient gain level from 1 to 1.8 GHz. These simulation results include three different cases: (A) free-standing cross-bow-tie antenna positioned on the reference plane where the position reference point is located at the center (solid line); (B) same antenna elevated 54.3 mm ($\lambda_0/4$ at 1380 MHz) above the reference plane (dotted line); and (C) same elevated antenna with an additional infinite PEC ground plane placed on the reference plane (dashed line). The signal phase

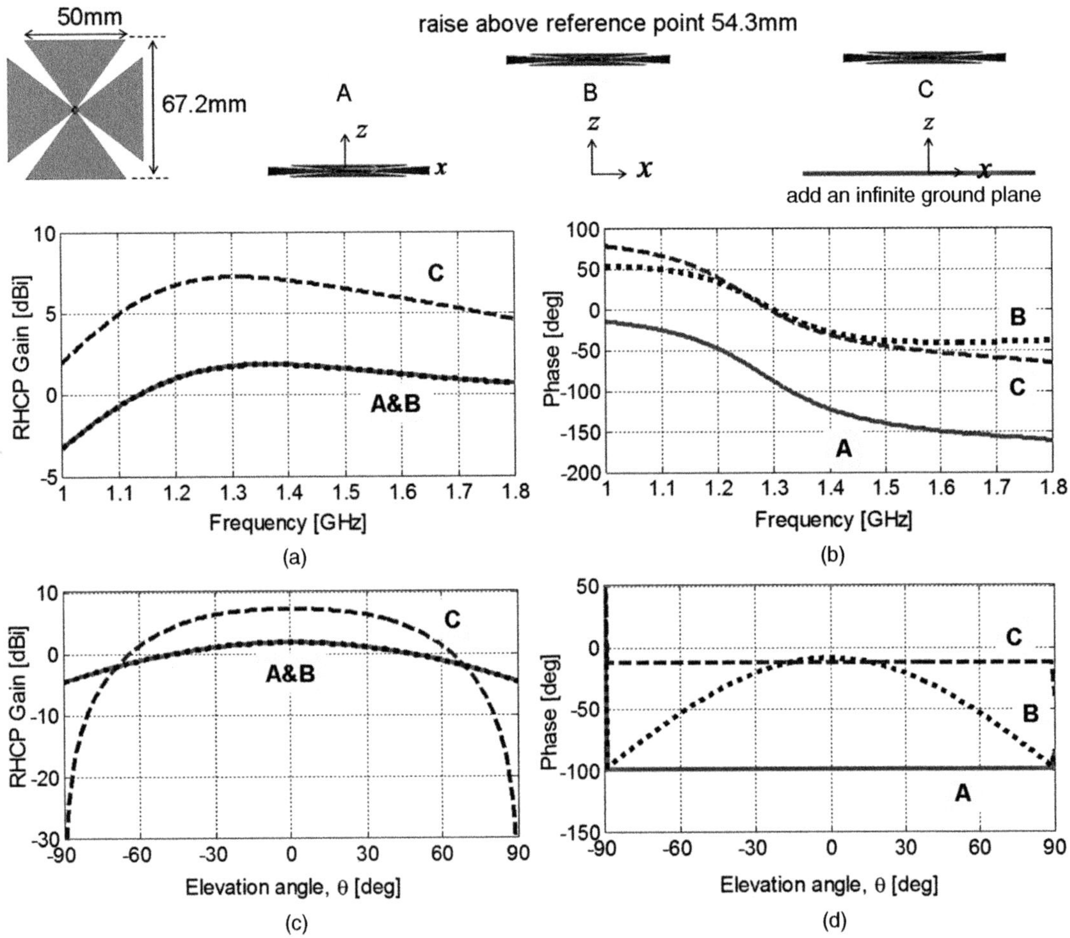

Figure 14.29 *Cross-bow-tie antenna with and without a PEC ground plane backing. (a) RHCP gain response at zenith. (b) Phase response at zenith. (c) RHCP gain elevation pattern at 1380 MHz. (d) Phase elevation pattern at 1380 MHz.*

versus frequency at zenith is shown in Figure 14.29b. The phase variation is related to the impedance matching condition and the location of the antenna's effective aperture (or phase centers). Similar to the previous patch antenna, the bow-tie dipole receives signals from its ends. This should lead to fixed antenna delay (or linear phase) behavior as a function of frequency if the antenna is well matched at all frequencies. The nonlinear phase behavior is caused by finite impedance bandwidth associated with finite bow-tie dipole antennas.

The phase pattern appears to be similar to the previous patch antenna with the phase at zenith for case A (solid curve) to be approximately $-98°$. In fact, this phase remains almost constant for the entire elevation plane, more stable than the previous patch antenna shown in Figure 14.14, due to the absence of undesired probe radiation. In case B, the antenna is elevated (closer to the satellite) which explains the $+90°$ phase shift at zenith ($\theta = 0°$) compared to case A. The resultant phase variation as a function of elevation angle is simply due to the height offset, h, between antenna and reference point, that is, $(2\pi/\lambda_0)h \cdot \cos\theta$. The undesired phase pattern variation can be removed by introducing an infinite PEC ground plane at the reference plane as in case

C (dashed line). The reflection of the ground plane produces the effect of an image antenna positioned at equal distance, h, below the reference plane, effectively forming an array with two out-of-phase elements. This also explains the approximate gain increase at zenith when the height is close to $\lambda_0/4$ at 1380 MHz.

Spiral antennas are also commonly used in GNSS antennas due to their wide bandwidth and circular polarization properties. The electrical circumference of the spiral antenna needs to be greater than the operational wavelength and its arms are wound tightly [38] in order to obtain frequency-independent gain, pattern, phase center, and good axial ratio. Both Archimedean and equiangular spiral designs are commonly seen. They perform similarly if the arms are tightly wound. Otherwise, Archimedean spirals exhibit more undesired frequency variations in impedance, gain, and patterns.

The spiral arm of an equiangular spiral is typically generated using the following geometrical function:

$$\rho = \rho_0 e^{a\phi} \tag{14.15}$$

where ρ is the radius from the center, ϕ is the sweeping angle, a is the growth rate, and ρ_0 is the initial radius. Cheo *et al.* [66] derived an expression for the fundamental mode of the far-zone field of an infinite two-arm equiangular spiral antenna as

$$E_\phi \approx E_0 \beta^{3^0} A(\theta) e^{j[\phi+(\pi/2)-\psi(\theta)]} \frac{e^{-j\beta r}}{r} \tag{14.16}$$

where E_0 is a constant related to the excitation strength, and $A(\theta)$ and $\psi(\theta)$ are the magnitude and phase functions defined in (14.17) and (14.18), respectively, as a function of elevation with zenith at $\theta = 0$. Note that the radiated fields are perfectly circularly polarized with $E_\theta = \pm jE_\phi$. Thus

$$A(\theta) = \frac{\cos\theta \cdot \tan\theta/2 \cdot e^{\tan^{-1}(a\cos\theta)/a}}{\sin\theta\sqrt{1 + a^2\cos^2\theta}} \tag{14.17}$$

$$\psi(\theta) = -\frac{1}{2a}\ln\left|1 + a^2\cos^2\theta\right| - \tan^{-1}(a\cos\theta) \tag{14.18}$$

The phase behavior depends on the wavelength, growth rate, and distance, S, along the spiral arm from the feed to the center of the 'active region' at an operational frequency. From (14.15), the distance S can be found to be

$$S \approx \frac{\lambda}{2\pi a}, \quad r_0 \ll \lambda \tag{14.19}$$

If one neglects the capacitive coupling between adjacent spiral arms as in large growth rate cases ($a \gg 1$), the phase of currents at the active region is approximately equal to

$$\phi = -k_0 S \approx -\frac{1}{a}, \quad r_0 \ll \lambda \quad \text{(radians)} \tag{14.20}$$

Notice that this phase is independent of frequency, a natural consequence of equiangular spiral geometry. This would be a very desirable feature of GNSS antennas. Unfortunately, large growth rates are not desirable in practical spiral antennas with finite sizes and will cause undesirable gain, phase, and axial ratio variations as a function of angle and frequency [38] due to the stronger diffraction from the ends of the spiral arms.

Practical spiral antennas often employ tight windings or more spiral arms for improved gain, phase, and axial ratio variations in the azimuth plane. This inevitably creates strong coupling between adjacent arms near the center region due to the large capacitance associated with smaller gaps. As a result, the phase of currents in the active region becomes a function of frequency.

Figure 14.30 shows a design example of a two-arm equiangular spiral antenna with and without (solid line) the presence of an infinite PEC ground plane backing. The ground plane is placed at the reference plane and the spiral antenna is elevated to $\lambda_0/4$, $\lambda_0/8$, and $\lambda_0/16$ at 1380 MHz, three different heights. A larger antenna diameter of 44 cm is used to reduce the effect of end diffractions. The middle row shows RHCP realized gain at zenith and reflection coefficient (S_{11}) as a function of frequency from 1.0 to 1.8 GHz. The bottom row shows calculated axial ratio and phase at zenith as a function of frequency. Gain, axial ratio, and S_{11} all show excellent performance without the ground plane (solid line). Its presence causes performance degradation due to first- and high-order reflections associated with the ground plane. These results indicate that antenna height above the PEC ground plane should be kept above $\lambda_0/4$. The fluctuations observed in the presence of the ground plane in all plots are caused by non-radiated currents reflected back from the ends of the arms, which also contribute to undesired elevated LHCP gain in the upper space.

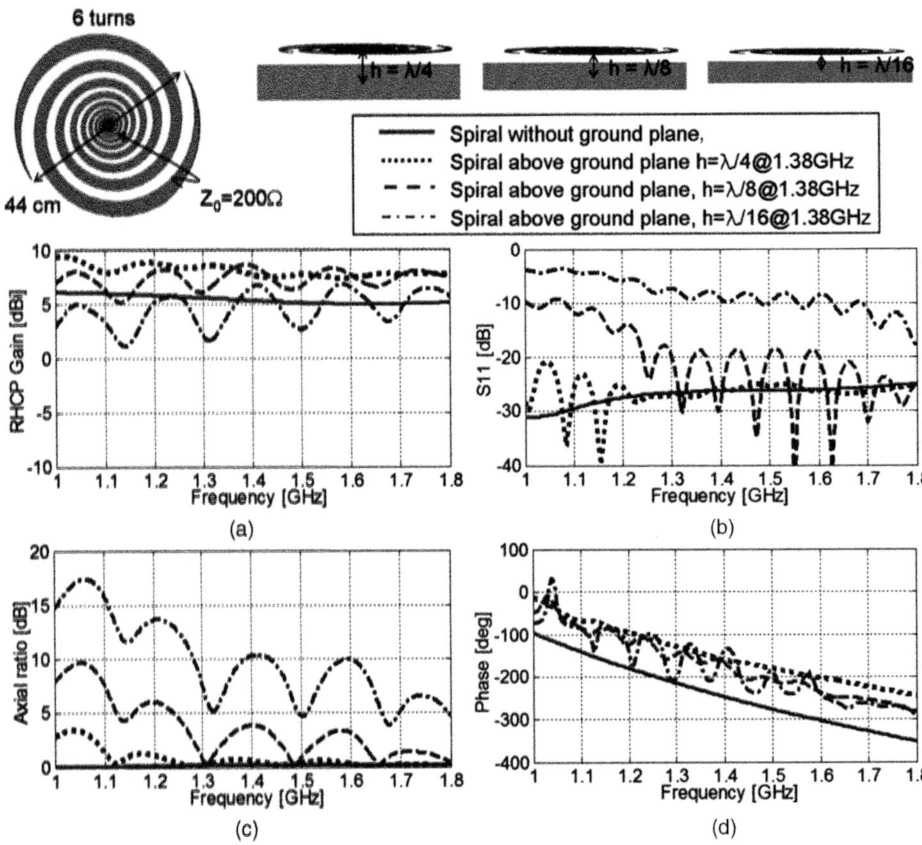

Figure 14.30 Geometry and performance of a two-arm equiangular spiral antenna with and without a ground plane backing. (a) RHCP realized gain response at zenith. (b) Reflection coefficient. (c) Axial ratio response at zenith. (d) Phase response at zenith.

Table 14.3 Group delay and pseudo-range error for the spiral in Figure 14.30 without the ground plane.

Frequency (GHz)	Phase	Group delay	Pseudo-range error
1.15	−160°	0.387 ns	11.6 cm
1.60	−300°	0.521 ns	15.6 cm

Table 14.4 Group delay and pseudo-range error for the spiral in Figure 14.30 elevated $\lambda_0/4$ (at 1380 MHz) above an infinite PEC ground plane (dotted line).

Frequency (GHz)	Phase	Group delay	Pseudo-range error
1.15	−80°	0.193 ns	5.8 cm
1.60	−200°	0.347 ns	10.4 cm

The phase shift from solid line to dotted line is approximately 90° at 1 GHz and is due to the similar image antenna effect discussed in Figure 14.29. The phase also shows a decreasing trend as frequency increases. The corresponding group delay and pseudo-range error can be calculated from these phase using (14.12) and are shown in Tables 14.3 and 14.4.

The pattern expressions in (14.17) and (14.18) clearly indicate that the amplitude and phase patterns of an *infinite* equiangular spiral antenna are functions of growth rate and elevation angle. Figure 14.31 plots the phase pattern using (14.18) for four different growth rates of $a = 0.01, 0.1, 0.5$, and 1. First, it can be seen that tighter windings produce less elevation phase pattern variation. In this case, the antenna phase center is located at $a\lambda/(2\pi)$ directly beneath the center of the spiral [66]. Second, the phase lag at zenith ($\theta = 0°$) increases with growth rate. Note that the previous dipole antenna or patch can be considered as a spiral antenna with an infinite growth rate. In those cases, we found the phase at zenith to be approximately −98°.

Figure 14.32 compares the antenna performance of two similar spiral antennas with different growth rates. Calculated results with and without an infinite PEC ground plane are included. For the cases without a ground plane, the antennas are placed on the reference plane. For the cases with a ground plane, the antennas are

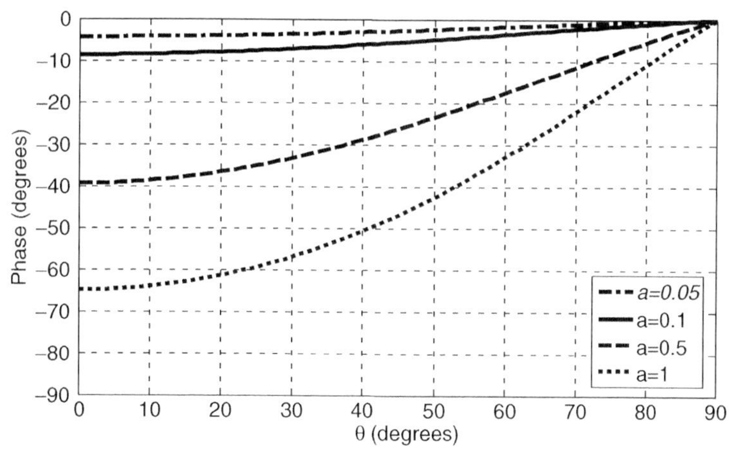

Figure 14.31 *Elevation phase pattern of a free-standing, infinite, equiangular spiral antenna.*

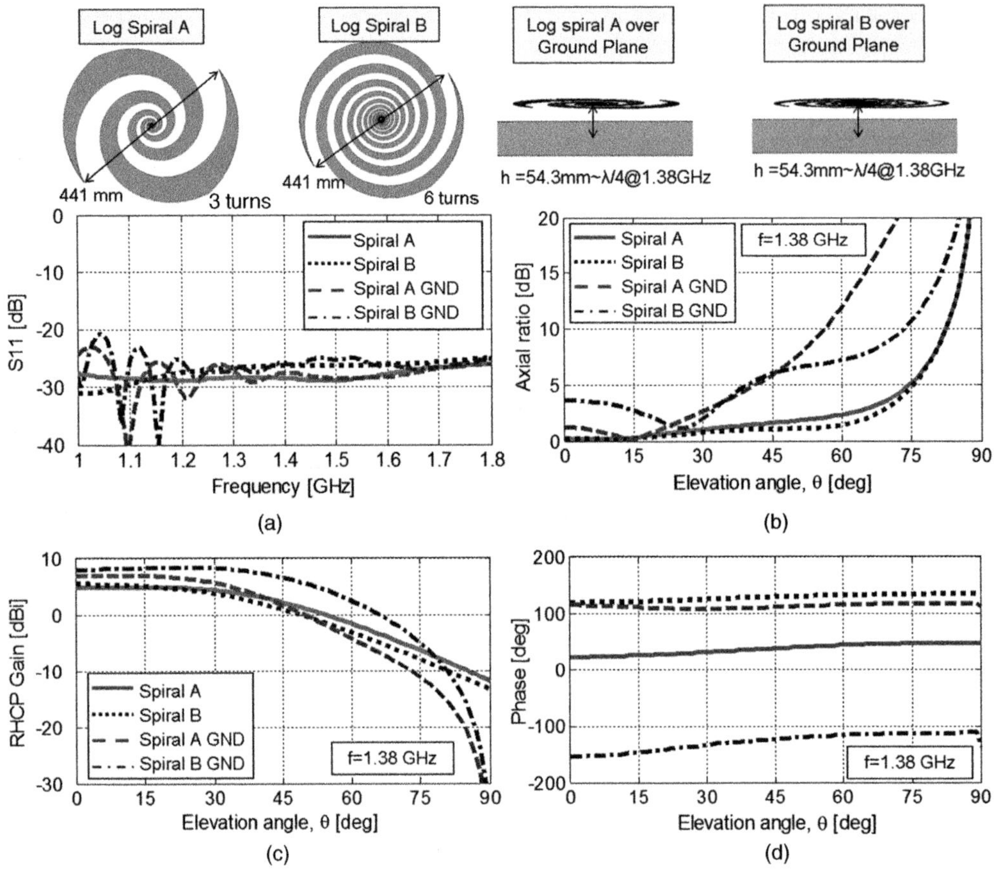

Figure 14.32 Performance of two equiangular spiral antennas of different growth rate with and without an infinite PEC ground plane backing. (a) Reflection coefficient vs. frequency. (b) Axial ratio pattern in elevation plane at 1.38 GHz. (c) RHCP realized gain pattern in elevation plane at 1.38 GHz. (d) Phase pattern in elevation plane at 1.38 GHz.

elevated $\lambda_0/4$ (at 1380 MHz) above an infinite PEC ground plane which is positioned on the reference plane. These results include S_{11} data as a function of frequency, and patterns of axial ratio, gain, and phase taken at 1.38 GHz in the elevation plane. The antenna port impedance is 200 Ω. Again, the presence of the ground plane degrades most of the performance except for higher gain at high elevation angles. The infinite ground plane also causes gain to vanish at low elevation angles. However, this pattern null will be filled by truncation diffraction in practical finite ground planes. It can also be seen that the spiral with tighter windings has a better gain and axial ratio at low elevation angles in the presence of the ground plane. However, the spiral with looser windings shows better phase pattern stability. The pseudo-range error variation over the upper space appears to be less than 3 cm.

14.4.2 Multi-element Antenna Array

Multiple GNSS antenna elements can be operated simultaneously in an array configuration for synthesizing a special coverage pattern by combining signals received from individual elements with special amplitude and phase weightings. Such pattern synthesis could be implemented in a fixed or adaptive manner. Examples of fixed-pattern arrays include GNSS/GPS transmitting antennas on satellites [67,68] and reference antennas for local augmentation systems [20,69]. Adaptive-pattern arrays adaptively form narrower beams directed toward known satellite directions for achieving better carrier-to-noise (*C/N*) ratio, which leads to better position accuracy and longer operation without needing to acquire new satellites [70–75]. An adaptive array can produce pattern nulls directed toward the direction of interferences to suppress their effect on satellite reception [76].

When multiple antenna elements are placed close to each other, strong mutual coupling among different elements may arise from interactions between the antenna's near fields, feed lines, substrates, or radome [77,78]. The common consequences of such mutual coupling include frequency detuning, impedance matching degradation, pattern shape distortion, pattern beam tilting, and increased cross-polarization level. As long as there is no loss of coverage related to deep pattern nulls and reception sensitivity related to the poorer matching condition, the presence of coupling does not necessarily affect beam forming or nulling performance [79,80].

Figure 14.33 demonstrates the coupling effect of two stacked-patch antenna elements (see Figure 14.27) separated by 4 cm (center to center) compared to the results obtained with a single antenna (solid line). The difference between the left and right plots is the different feeding structures. The probes in the left plot are connected to microstrip lines embedded under the antenna whereas the probes in the right plot are connected to coaxial connectors directly beneath the probes. In both cases, the realized RHCP gain at zenith in the two-element case (dashed line) drops as a result of impedance matching degradation. The high-frequency mode in the left configuration suffers from more frequency detuning related to different feeding method since both have the same patch size and dielectric arrangements.

Figure 14.34 shows an example of a six-element array based on the tri-band GPS antenna design [81] shown in Figure 14.27. The RHCP realized gain and axial ratio patterns at 1575 MHz in the upper space from a single active element (top center) in array configuration are plotted in gray scale. Comparing the gain and axial ratio patterns to those from a single element, one can see that the gain peak tilts away from zenith towards the active

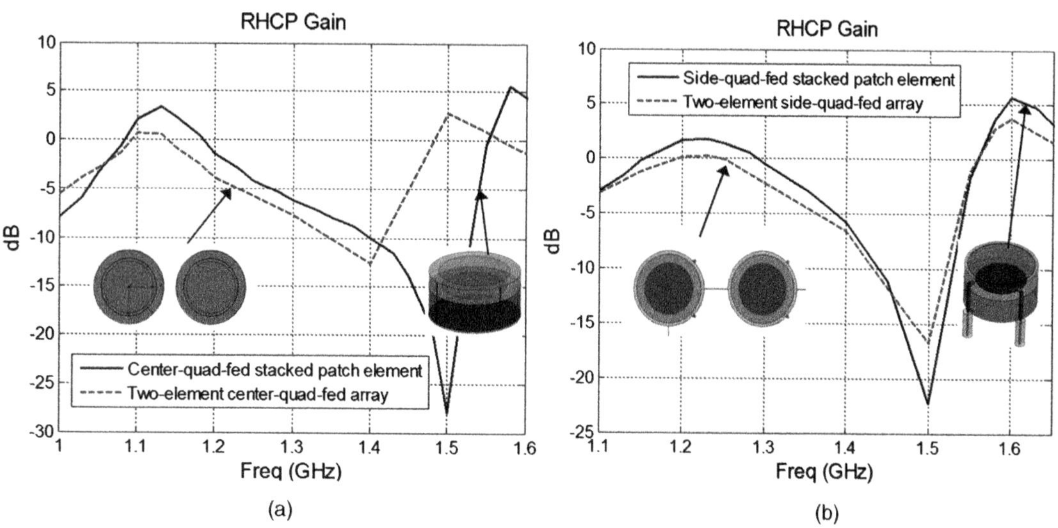

Figure 14.33 Different mutual coupling effect caused by different feeding structures. (a) Feed using outward microstrip lines. (b) Feeding a patch antenna using coaxial probes.

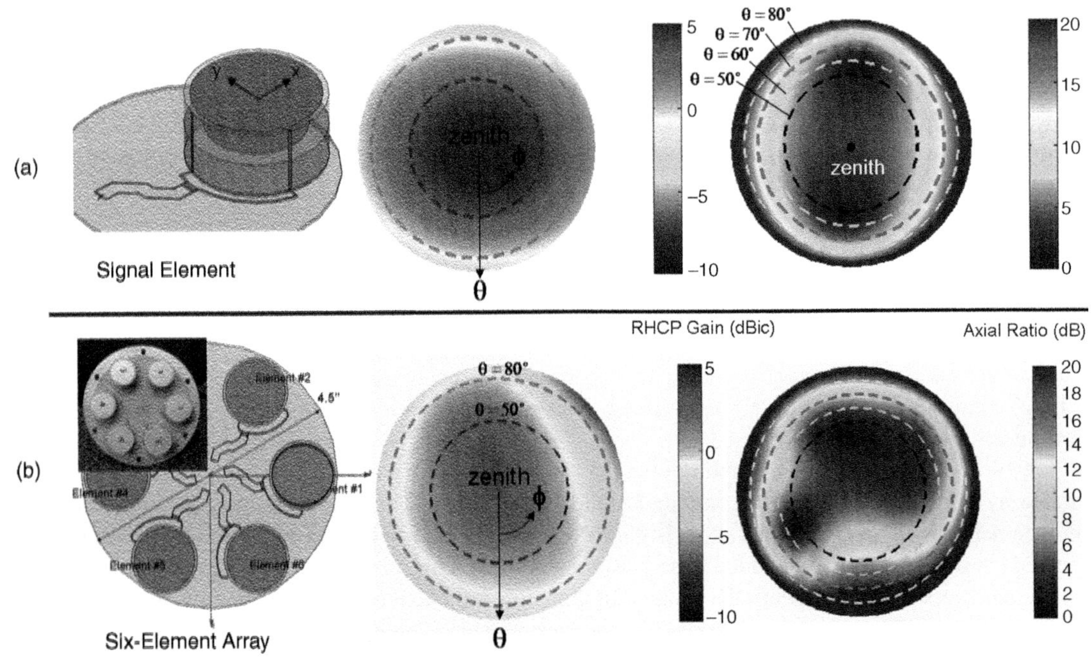

Figure 14.34 Comparison of gain and axial ratio of the same antenna design in (a) stand-alone and (b) six-element array configurations at 1575 MHz. In the array mode, only the highlighted element is active.

element direction in the array configuration and the axial ratio deteriorates at low elevation angles on the far side of the active element. It should be pointed out that the beam tilt could be toward or away from the active element direction, depending on the phase relationship between the signal received directly from the satellite and the signal received via coupling from other array elements.

Figure 14.35 shows a novel compact GNSS array based on four square Archimedean spiral antenna elements [82]. The overall antenna dimensions are 8.9 × 8.9 cm at 2 cm height. Each antenna element is connected through a standard SMA coaxial connector located at four corners. This antenna can provide continuous coverage of all GNSS frequencies. The calculated gain, axial ratio, group delay, and carrier phase in the upper hemisphere at both 1575 (L1) and 1175 MHz (L5) were obtained by combining signals from all four elements to form an RHCP beam toward zenith. The average RHCP gain at 10° above the horizon was found to be −5.5 dBi and −5.7 dBi at L1 and L2, respectively. The average axial ratio at 10° above the horizon was found to be 13 dB and 16 dB at 1575 MHz and 1175 MHz, respectively. This design was found to provide effective suppression of up to three interference signals.

14.5 Spaceborne GNSS Antennas

14.5.1 Requirements for Antennas On Board Spaceborne GNSS Receivers

Spaceborne GNSS receivers are useful for spacecraft precise orbit determination and remote sensing applications. On board the satellite they are able to calculate the position of the spacecraft and can also be used to determine accurate velocity and time. Sections 14.2 and 14.3 discussed the requirements for GNSS receiving antennas, such as multiband operations, hemispherical radiation patterns with a sharp roll-off near

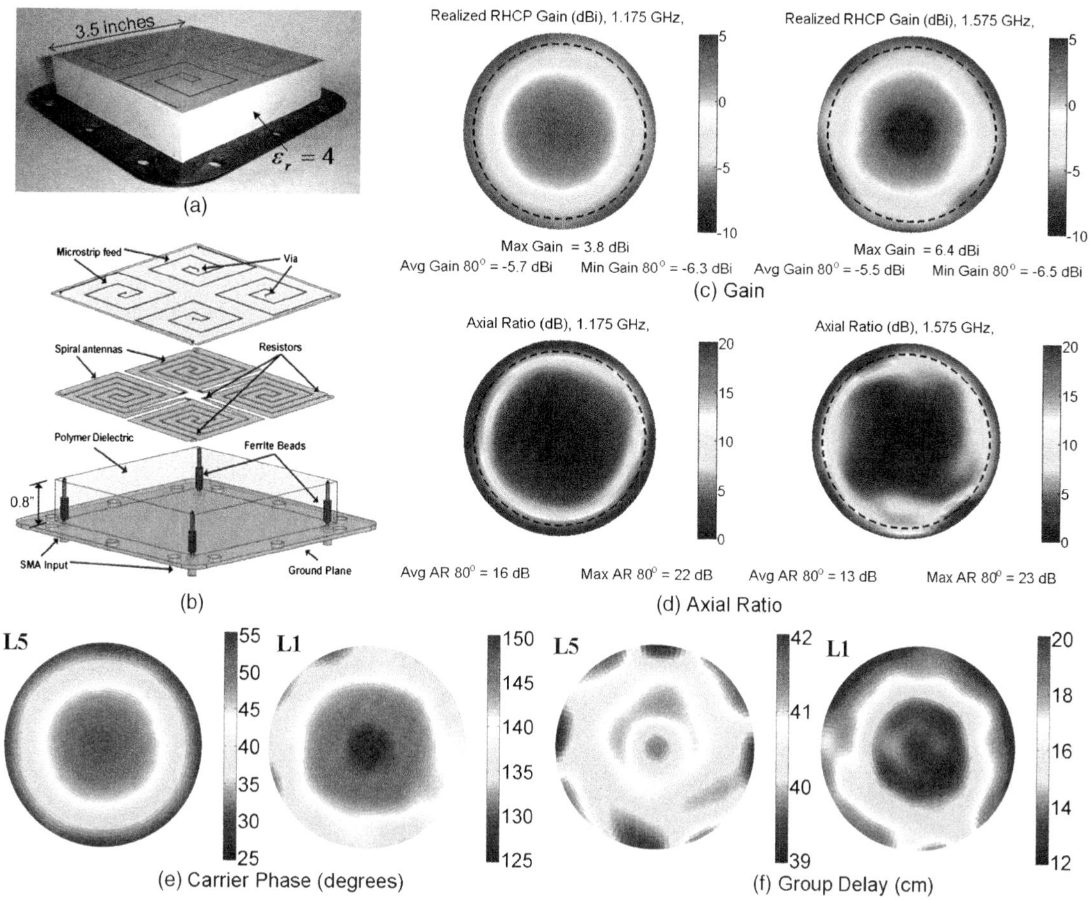

Figure 14.35 A compact GNSS antenna array covering the entire GNSS spectrum. The gain, axial ratio, carrier phase, and group delay shown here are calculated by combining all four elements to form an RHCP beam toward zenith. After Kasemodel et al. [82]. © 2008 IEEE. Reprinted, with permission from IEEE Antennas and Wireless Propagation Letters, 7, 592–595, 2008.

the horizon, high polarization purity, multipath mitigation and phase center stabilities. In addition to these, antennas on board spaceborne GNSS receivers have to consider a few other issues as follows:

Materials: As discussed in Chapter 5, the space environment is rather different from that on the ground. To survive in the space environment, all the materials for fabricating the spaceborne antennas need to be space qualified. Issues such as low outgassing, thermal stability, and so on, should be taken into account when selecting the materials for antennas. The antenna has to undergo a series of tests including electrical, vibration, shock, thermal, and vacuum tests, as discussed in Chapter 6.

Mechanical and thermal issues: The antenna structure must be robust so that it will be able to survive during the launch. The mounting of the antenna on the satellite body also has to be very rigid. Thermal design of the antenna is needed. To protect the antenna from the Sun, a radome is usually employed to cover the antenna. Thus the effects of the radome on antenna performance should be considered during the antenna design process.

Interactions with satellite body, satellite subsystems and payloads: When the antenna is on board the satellite, the mutual coupling between the antenna and the satellite body could affect antenna performance.

Also, the antenna will be coupled to payloads (e.g., cameras) and satellite subsystems (e.g., solar panels, downlink antennas) nearby. It is necessary for the antenna engineer to analyze the interactions between the GNSS antenna and satellite body, satellite subsystems, and payloads, and find an optimum location for the antenna.

Electromagnetic compatibility issues: It is important to avoid any static charge deposition on the metallic structure of the antenna, as this may harm the LNA and other circuits. This problem can be solved by grounding all the metallic parts of the antenna. The other issue is the rejection of other RF signals which may saturate the GNSS receiver. As the signal for the GNSS receiver is weak, the antenna and GNSS receiver front end need to reject RF signals from other RF systems on board the satellite, in particular the RF transmitter for data downlink. This can be achieved through the use of filters. It will be beneficial if the GNSS antenna itself has the function of filtering.

Size, mass, and efficiency: It is important to reduce the size of the antenna, due to the limited space available on board the satellite. A smaller size of antenna may also enable it to be more distant from other instruments on board, thus reducing the coupling between them. The antenna needs to have a low mass to reduce the cost of satellite. It is also important for the antenna to be highly efficient, as the power on board is very limited and high-efficiency RF systems will enable the satellite to have a longer life.

14.5.2 A Review of Antennas Developed for Spaceborne GNSS Receivers

Many antennas have been used for spaceborne GNSS receivers. These mainly include microstrip patch antennas and arrays, patch excited cup antennas, and quadrifilar helix antennas. A brief review of them is given below:

- *Microstrip patch antennas:* Due to its advantages of low profile, low mass, and easy integration with RF circuits, the microstrip patch antenna has become very popular for spaceborne GNSS receiver applications. The ground plane of the patch antenna enables it to have a natural broad-beam radiation pattern with low backlobe. Also the antenna can be placed directly on the metallic surface of the satellite body. Figure 14.36 shows a GPS antenna for a SGR-GEO receiver [83]. The antenna is basically a square microstrip patch, fed by two orthogonal feed lines with a 90° phase delay for achieving circular polarization. A circular patch is put on the top for broadening the impedance bandwidth as well as increasing the gain of the antenna. The antenna operates at L1-band, and has been flying on board the GIOVE-A satellite (launched in 2005). Figure 14.37 shows the microstrip patch antenna on the top of the CHAMP satellite for precise orbit determination [84]. As shown in the figure, the choke-ring ground plane has been used with a circular patch antenna, for achieving a good performance in terms of multipath mitigation and phase center stability. It is important to note that the position of the antenna is symmetric with respect to the satellite structure, for achieving a stable boresight pointing radiating pattern. The microstrip patch antenna with choke-ring ground plane has also been used in different satellite missions such as Topex/Poseidon, Jason-1 and Jason-2 for precise orbit determination and remote sensing applications.
- *Patch excited cup antennas:* RUAG developed these antennas for GNSS precise orbit determinations. An example is shown in Figure 14.38 [85]. It is a dual-band antenna (L1, L2) consisting of two metallic disk elements stacked together with four capacitive feeds and a broadband feed network at the bottom. Instead of a choke-ring ground plane, a cup-shaped ground plane with two layers of choke rings is used for shaping the radiation pattern and reducing the backward radiation and cross-polarization levels. Parameters affecting the radiation patterns mainly include the diameter and height of the cup and layers of corrugations, the dimensions of the patch, and the distance between two patches. The antenna achieved wide coverage, low loss and low interference from the satellite body.

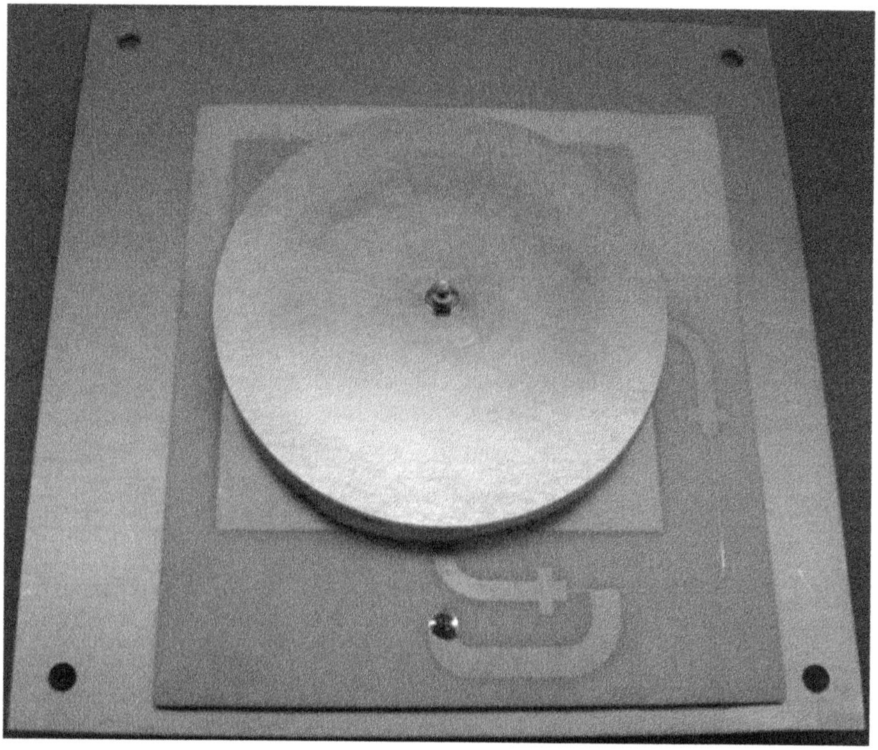

Figure 14.36 *GPS patch antenna. Courtesy of Surrey Satellite Technologies Ltd [83].*

Figure 14.37 *Patch antennas with choke-ring ground plane on CHAMP (POD, Precise Orbit Determination). Courtesy of GFZ German Research Centre for Geosciences [84].*

Figure 14.38 PEC antenna. Courtesy of RUAG Space AB [85].

- *Quadrifilar helix antenna:* The quadrifilar helix antenna (QHA) is able to produce a hemispherical circularly polarized radiation pattern with a sharp roll-off near the horizon. Figure 14.39 shows a miniaturized active QHA antenna which has been flying in space [85]. The radiating structure consists of four printed helical arms which have been significantly reduced in size by the high-permittivity dielectric loading. A balun is integrated with the helix. To improve the noise and gain performance, a low-noise amplifier is also integrated with the antenna, as shown in the figure. The antenna operates at L1-band.

Besides the antennas described above, other antennas have also been developed for spaceborne GNSS applications, such as shorted annular patch antennas, slot antennas, patch arrays, and so on.

14.6 Case Study:[1] Dual-Band Microstrip Patch Antenna for Spacecraft Precise Orbit Determination Applications

14.6.1 Antenna Development

In this case study, the development of a dual-band microstrip patch antenna for spacecraft precise orbit determination will be described [87]. The specifications of the antenna include:

- Frequency bands: L1 (20 MHz) and L2 (20 MHz).
- Gain: 6 dBi at both bands.

[1] The spaceborne GNSS antenna work in this section was in collaboration with SSTL, UK.

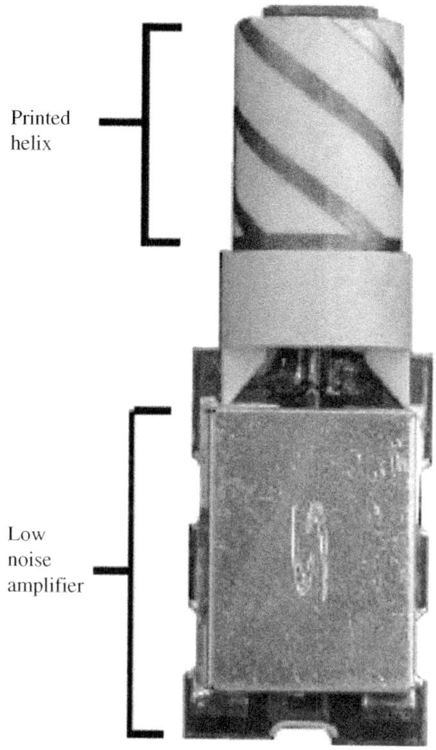

Figure 14.39 Quadrifilar helix antenna. Courtesy of Sarantel Ltd [86].

- Maximum size: 150 × 150 × 20 mm.
- Maximum mass: 200 g.
- Axial ratio: < 3 dB at boresight.
- Voltage standing wave ration (VSWR): < 2.
- Material: low outgassing and thermal stability.
- Environmental and mechanical: to survive thermal, vibration, and vaccum tests.

The first step is to choose a suitable antenna technology. A comparative study was carried out among different antenna technologies such as for microstrip patch antennas, QHAs, miniaturized QHAs with high-permittivity materials, dielectric resonator antennas, and so on. It was concluded that the microstrip patch antenna is the most suitable technology for this application, due to its advantages of low profile and easy integration with the satellite body. The QHA is capable of achieving hemispherical circularly polarized radiation patterns, but its height may lead to aerodynamic problems. A QHA with high-permittivity materials can make the antenna size much smaller, but the gain and efficiency performance of the antenna will deteriorate.

To achieve dual-band operation, a stacked-patch configuration is used. Different materials were considered for fabricating the patch antenna. It was found that Duroid substrate materials ($\varepsilon_r = 2.2$, $h = 1.575$ mm), which are popular for many space and terrestrial applications, are not suitable for this application. This is because two such Duroid substrates together will lead to an antenna mass much larger than the specifications (<200 g). To reduce the mass, it was decided that the patch antenna should employ a combination of Rohacell substrate and Duroid substrate 5880 ($\varepsilon_r = 2.2$, $h = 0.127$ mm). All the layers are joined together by using a 3M VHB 9469 double-adhesive tape. This leads to a very lightweight antenna.

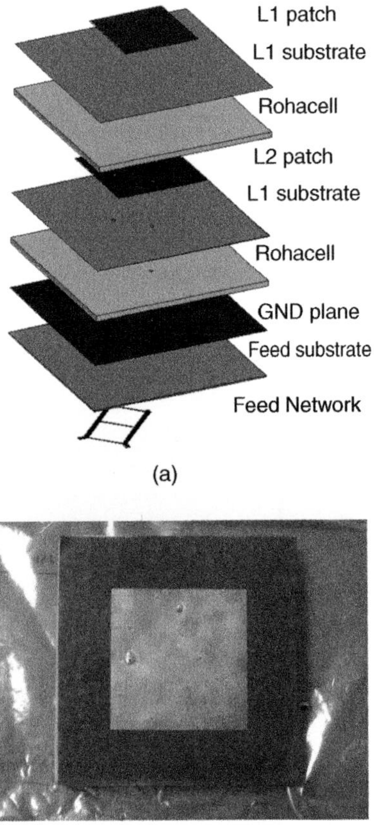

Figure 14.40 *(a) Exploded view of the antenna and (b) top view of the fabricated antenna.*

The antenna design is based on the transmission line model of the patch antenna [88]. Figure 14.40a shows an exploded view of the antenna. A photo of the fabricated antenna is shown in Figure 14.41b. As can be seen, two square patches with different dimensions, denoted as L1 patch and L2 patch, are printed on different layers of substrates, for radiating at both L1- and L2-bands. The initial dimensions of the L1 patch and L2 patch are determined separately, by using the transmission line models, and allowing each patch to resonate at 1.575 and 1.227 GHz, respectively. Further optimizations are then carried out to find the optimum dimensions of the patches. The antenna is fed by two orthogonal feeds. To produce two signals with equal amplitudes and a 90° phase shift between them, a broadband feed network using a three-branch microstrip hybrid coupler is designed to cover both L1- and L2-bands. To connect the antenna to the feed network, two vias are made through the patch on the top to the feed network below the ground plane. The center of the L2 patch is shorted to the ground plane, to avoid any static charge deposition on the metallic structure of the antenna.

14.6.2 Results and Discussions

The final antenna prototype was manufactured with the following dimensions:

- L1 patch = 82 × 82 mm
- L2 patch = 105 × 105 mm

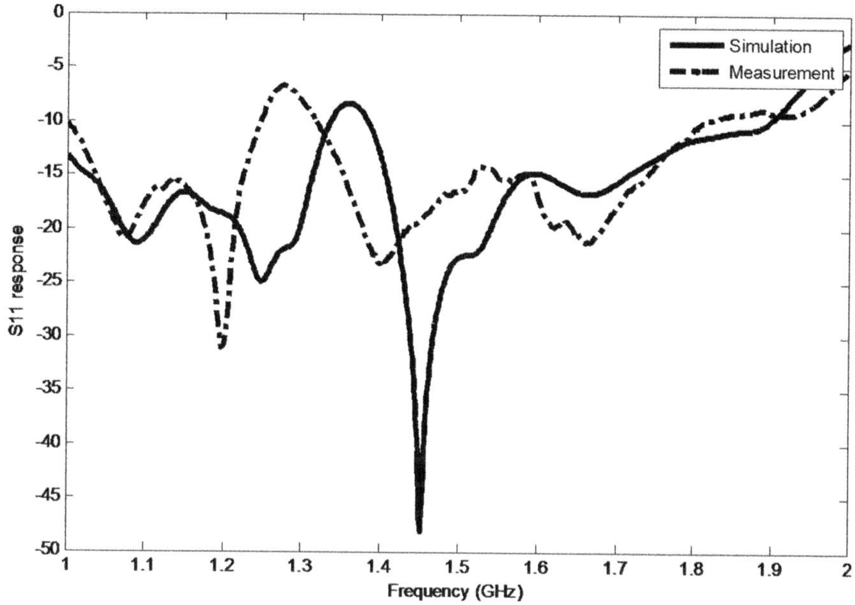

Figure 14.41 *Reflection coefficient (S_{11}) of dual-band patch antenna shown in Figure 14.40.*

- Via (feed) position = 25 mm off-center
- Overall dimensions = 150 × 150 × 15 mm
- Via radius = 0.5 mm
- Mass = 140 g.

The manufactured prototype shown in Figure 14.40b was tested and the results compared against simulations. Figure 14.41 shows the reflection coefficient (S_{11}) of the antenna. It can be seen that the measured and simulated results agree reasonably well and the antenna has achieved a return loss over 10 dB at both L1- and L2-bands.

Table 14.5 gives the measured results of antenna gain. As shown, the antenna achieves a gain over 6 dBi for both L1- and L2-bands. Figure 14.42 shows the measured results of the radiation pattern at L1- and L2-band. It can be seen that the antenna achieves broad beamwidth, low cross-polarization levels below −10 dB, and low backward radiation at both bands. The antenna achieves a 3 dB beamwidth over 65° at both frequency bands.

Table 14.5 *Antenna gain measurement results.*

Freq (GHz)	Gain (dBi)
1.225	6.60
1.227	6.51
1.229	6.43
1.5675	6.26
1.575	6.29
1.5825	6.43

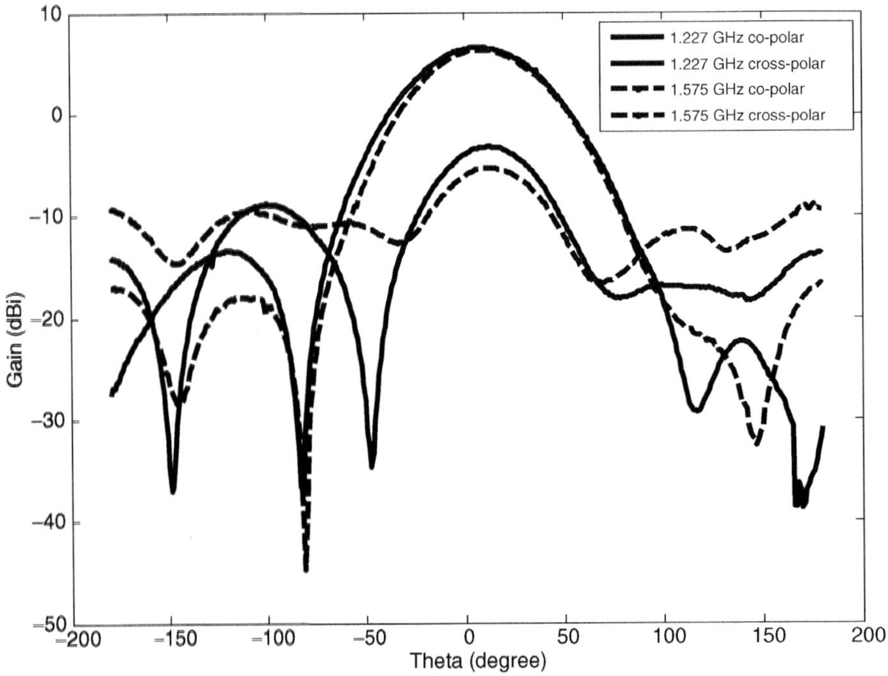

Figure 14.42 *Measured radiation patterns at 1.227 GHz and 1.575 GHz.*

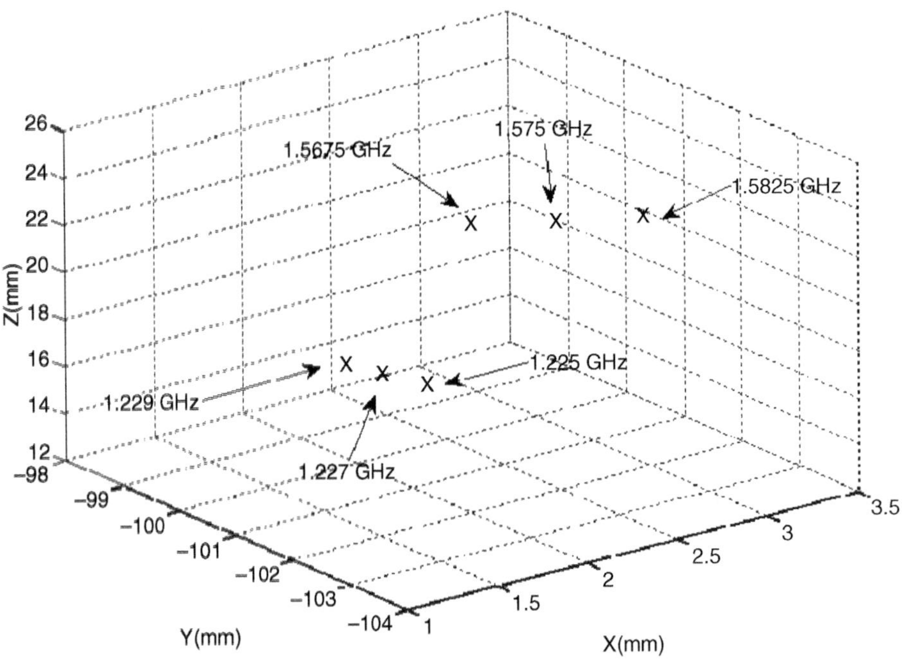

Figure 14.43 *Phase center variation across L2- and L1-bands.*

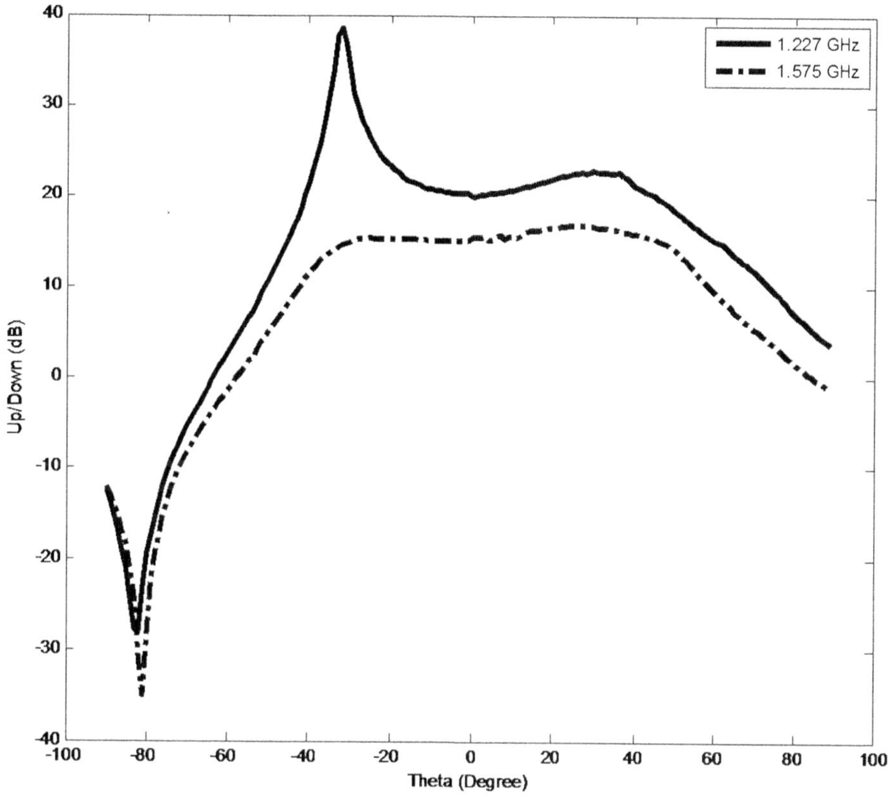

Figure 14.44 *Multipath rejection ratio at L1- and L2-bands.*

The antenna phase center is an important factor for spacecraft precise orbit determination. It is the point in the antenna radiation pattern where the transmitting power is emanating from or, vice versa, all the receiving power is converging. The phase center variation is critical as the distance between the transmitting and receiving GNSS antennas is calculated with reference to antenna phase centers. Figure 14.43 shows the antenna phase center variation at both L1- and L2-bands. It shows that the antenna phase center variation across the L1- and L2-bands is less than 6 mm. Further improvement of the antenna design is still needed to improve the phase center stability across both bands.

Apart from cross-polarization and axial ratio performance, the multipath rejection ratio or up/down ratio is another figure of merit used to characterize the multipath mitigating performance of GNSS antennas. The multipath rejection capability or up/down ratio is defined as the ratio of antenna co-polar gain above the horizon to the antenna cross-polar gain below the horizon. Thus a higher up/down ratio means better multipath rejection capability. Figure 14.44 presents the multipath rejection (up/down) ratio of the prototype antenna at $\Phi = 0°$ cut (E-plane). It can be observed that the antenna achieves better multipath mitigation performance at L2-band than that at L1-band.

14.7 Summary

This chapter described the general spectral requirements of current and planned GNSS operations. The performance requirements of GNSS receiver antennas for general users and for high-precision signal monitoring were also discussed. These requirements involve gain levels, pattern coverage, phase pattern variation, group and phase delays, and polarizations. It should be noted that the requirements discussed in this chapter serve only as examples as these requirements are constantly revised and vary with the applications. Many simulation and measurement results for both theoretical and practical GNSS antenna design examples were also provided for the reader who is experienced in antenna designs. These examples included narrow-band or multiband designs such as patch antennas, qradrifilar helical antennas, and patch excited cup antennas, as well as wide-band designs such as bow-tie dipoles and spiral antennas. Obviously, these are only representative design examples chosen for the convenience of discussion. Although there could be infinite variations of GNSS antenna designs, the best design, in our opinion, is the one that is smallest, cheapest, and, most important of all, works!

References

1. The Global Positioning System, http://www.gps.gov/systems/gps/ (accessed 13 December 2011).
2. Misra, P. and Enge, P. (2006) *Global Positioning System: Signals, Measurements, and Performance*, 2nd edn, Ganga-Jamuna Press.
3. US FAA (2001) *Specification for the Wide Area Augmentation System (WAAS)*, FAA-E-2892b, August 13.
4. ESA (2007) EGNOS – The European Geostationary Navigation Overlay System – A Cornerstone of Galileo, ESA SP-1303.
5. Garg, R., Barthia, P., Bahl, I., and Ittipiboon, A. (2001) *Microstrip Antenna Design Handbook*, Artech House.
6. Volakis, J.L. (ed.) (2007) *Antenna Engineering Handbook*, 4th edn, McGraw-Hill.
7. Huang, C.-Y., Ling, C.-W., and Kuo, J.-S. (2003) Dual-band microstrip antenna using capacitive loading. *IEE Proceedings: Microwaves, Antennas and Propagation*, **150**(6), 401–404.
8. Boccia, L., Amendola, G., and Di Massa, G. (2004) A dual frequency microstrip patch antenna for high-precision GPS applications. *IEEE Antennas and Wireless Propagation Letters*, **3**, 157–160.
9. Pozar, D.M. and Duffy, S.M. (1997) A dual-band circularly polarized aperture coupled stacked microstrip antenna for global positioning satellite. *IEEE Transactions on Antennas and Propagation*, **45**(11), 1618–1625.
10. Su, C.M. and Wong, K.L. (2002) A dual-band GPS microstrip antenna. *Microwave and Optical Technology Letters*, **33**(4), 238–240.
11. Peng, X.F., Zhong, S.S., Xu, S.Q., and Wu, Q. (2005) Compact dual-band GPS microstrip antenna. *Microwave and Optical Technology Letters*, **44**(1), 58–61.
12. Shackelford, A.K., Lee, K.-F., and Luk, K.M. (2003) Design of small-size wide-bandwidth microstrip-patch antennas. *IEEE Antennas and Propagation Magazine*, **45**(1), 75–83.
13. Rao, B.R., Smolinski, M.A., Quach, C.C., and Rosario, E.N. (2003) Tripleband GPS trap-loaded inverted L antenna array. *Microwave and Optical Technology Letters*, **38**(1), 35–37.
14. Luk, K.M., Mak, C.L., Chow, Y.L., and Lee, K.F. (1998) Broadband microstrip patch antenna. *Electronics Letters*, **34**(15), 1442–1443.
15. Peng, X.F., Zhong, S.S., Xu, S.Q., and Wu, Q. (2005) Compact dual-band GPS microstrip antenna. *Microwave and Optical Technology Letters*, **44**(1), 58–61.
16. Czopek, F. (1993) Description and performance of the GPS block I and II L-band antenna and link budget. Proceedings of the ION Conference, pp. 37–43.
17. Papoulis, A. (1977) *Signal Analysis*, McGraw-Hill.
18. Betz, J.W. *et al.* (2000) Overview of the GPS M code signal. Proceedings of the ION National Technical Meeting, pp. 542–549.
19. Betz, J.W. (2001) Binary offset carrier modulations for radio navigation. *ION Journal of Navigation*, **48**(4), 227–246.
20. Lopez, A.R. (2010) GPS landing reference antenna. *IEEE Antennas and Propagation Magazine*, **52**(1), 104–113.
21. Lopez, A.R. (2008) LAAS/GBAS ground reference antenna with enhanced mitigation of ground multipath. Proceedings of the ION National Technical Meeting, pp. 389–393.

22. Braasch, M.S. (1996) Multipath effects, in *Global Positioning System: Theory and Applications* (eds B.W. Parkinson *et al.*), American Institute of Aeronautics and Astronautics (AIAA), vol. 1, pp. 547–568.
23. Braasch, M.S. (1997) Autocorrelation sidelobe considerations in the characterization of multipath errors. *IEEE Transactions on Aerospace and Electronic Systems*, **33**(1), 290–295.
24. Ertin, E., Mitra, U., and Siwamogsatham, S. (2001) Maximum-likelihood-based multipath channel estimation for code-division multiple-access systems. *IEEE Transactions on Communications*, **49**(2), 290–302.
25. Soubielle, J., Fijalkow, I., Duvaut, P., and Bibaut, A. (2002) GPS positioning in a multipath environment. *IEEE Transactions on Signal Processing*, **50**(1), 141–150.
26. van Nee, D.J.R. (1993) Spread spectrum code and carrier synchronization errors caused by multipath and interferences. *IEEE Transactions on Aerospace and Electronic Systems*, **29**(4), 1359–1365.
27. Townsend, B.R. and Fenton, P. (1994) A practical approach to the reduction of pseudorange multipath errors in a L1 GPS receiver. Proceedings of ION GPS, pp. 143–148.
28. Dierendonck, A.J., Fenton, J.P., and Ford, T. (1992) Theory and performance of narrow correlator spacing in a GPS receiver. *ION Journal of Navigation*, **39**(3), 265–283.
29. Kouyoumjian, R.G. and Pathak, P.H. (1974) A uniform geometrical theory of diffraction for an edge in a perfectly conducting surface. *Proceedings of the IEEE*, **62**, 1448–1461.
30. Lopez, A.R. (2003) LAAS reference antennas – circular polarization mitigates multipath effects. Proceedings of ION Annual Meeting, pp. 500–506.
31. GPS Antenna Reviews, http://www.gpsantenna.org/ (accessed 13 December 2011).
32. NavtechGPS, http://navtechgps2.intuitwebsites.com/index.html (accessed 13 December 2011).
33. Nextag, http://www.nextag.com/gps-antenna-lowrance/stores-html (accessed 13 December 2011).
34. Antcom, http://www.antcom.com/index.html (accessed 13 December 2011).
35. Balanis, C.A. (2005) *Antenna Theory and Design*, 3rd edn, John Wiley & Sons, Inc.
36. Kraus, J.D. and Marhefka, R.J. (eds) (2002) *Antennas for All Applications*, 3rd edn, McGraw-Hill.
37. Balanis, C.A. (ed.) (2008) *Modern Antenna Handbook*, John Wiley & Sons, Inc.
38. Volakis, J.L., Chen, C.-C., and Fujimoto, K. (2010) Chapter 5, in *Small Antennas: Miniaturization Techniques and Applications*, McGraw-Hill.
39. Kramer, B., Chen, C.-C., and Volakis, J.L. (2008) Size reduction of a low-profile spiral antenna miniaturization using inductive and dielectric loading. *IEEE Antennas and Wireless Propagation Letters* **7**, 22–25.
40. Rao, B.R., Williams, J.H., Rosario, E.N., and Davis, R.J. (2000) GPS microstrip antenna array on a resistivity tapered ground plane for multipath mitigation. Proceedings of ION GPS, pp. 2468–2476.
41. Westfall, B.G. (1997) Antenna with R-card ground plane, US Patent 5694136, December 2.
42. Tranquilla, J.M., Cam, J.P., and Al-Rizzo, H.M. (1994) Analysis of a choke ring ground plane for multipath control in global positioning system (GPS) applications. *IEEE Transactions on Antennas and Propagation*, **42**(7), 905–911.
43. Basilio, L.I., Williams, J.T., Jackson, D.R., and Khayat, M.A. (2005) A comparative study of a new GPS reduced-surface-wave antenna. *IEEE Antennas Wireless Propagation Letters*, **4**, 233–236.
44. Milligan, T. and Kelly, P.K. (1996) Optimization of ground plane for improved GPS antenna performance. Proceedings of IEEE AP-S International Symposium, vol. 2, pp. 1250–1253.
45. Lee, Y., Ganguly, S., and Mittra, R. (2005) Multiband L5-capable GPS antenna with reduced backlobes. Proceedings of the IEEE AP-S International Symposium, vol. 1A, pp. 438–441.
46. Counselman, C.C. (1999) Multipath rejecting GPS antennas. *Proceedings of the IEEE*, **87**(1), 86–91.
47. Huynh, S.H. and Cheng, G. (2000) Low profile ceramic choke, US Patent 6040805, March 21.
48. Westfall, B.G. and Stephenson, K.B. (1999) Antenna with ground plane having cutouts, US Patent 5986615, November 16.
49. Ashjaee, J., Filippov, V.S., Tatarnikov, D.V., Astakhov, A.V., and Sutjagin, I.V. (2001) Dual-frequency choke-ring ground planes, US Patent 6278407, August 21.
50. Sciré-Scappuzzo, F. and Makarov, S.N. (2009) A low-multipath wideband GPS antenna with cutoff or non-cutoff corrugated ground plane. *IEEE Transactions on Antennas and Propagation*, **57**(1), 33–46.
51. Pozar, D.M. (1985) Microstrip antenna aperture-coupled to a microstrip line. *Electronics Letters*, **21**(2), 49–50.
52. Sullivan, P.L. and Schaubert, D.H. (1986) Analysis of an aperture coupled microstrip antenna. *IEEE Transactions on Antennas and Propagation*, **34**(8), 977–984.

53. Kumar, G. and Ray, K.P. (2003) *Broadband Microstrip Antennas*, Artech House.
54. Waterhouse, R.B. (ed.) (2003) *Microstrip Patch Antennas: A Designer's Guide*, Kluwer Academic.
55. Lee, K.F., Lung, S., Yang, S., Kishk, A.A., and Luk, K.M. (2010) The versatile U-slot patch antenna. *IEEE Antennas and Propagation Magazine*, **52**, 71–78.
56. Chen, Z.N. and Qing, X. (2010) Dual-band circularly polarized S-shaped slotted patch antenna with a small frequency-ratio. *IEEE Transactions on Antennas and Propagation*, **58**, 2112–2115.
57. Row, J.-S. (2004) Dual-frequency circularly polarized annular-ring microstrip antenna. *Electronics Letters*, **40**, 153–154.
58. Maci, S., Biffi Gentili, G., Piazzesi, P., and Salvador, C. (1995) Dual-band slot-loaded patch antenna. *IEE Proceedings: Microwaves, Antennas and Propagation*, **142**, 225–232.
59. Rafi, G. and Shafai, L. (2004) Broadband microstrip patch antenna with V-slot. *IEE Proceedings: Microwaves, Antennas and Propagation*, **151**, 435–440.
60. Zhou, Y., Chen, C.-C., and Volakis, J.L. (2007) Dual band proximity-fed stacked patch antenna for tri-band GPS applications. *IEEE Transactions on Antennas and Propagation*, **55**(1), 220–223.
61. Ge, Y., Esselle, K.P., and Bird, T.S. (2004) E-shaped patch antennas for high-speed wireless networks. *IEEE Transactions on Antennas and Propagation*, **52**(12), 3213–3219.
62. Yang, F., Zhang, X., Ye, X., and Samii, Y.R. (2001) Wide-band E-shaped patch antennas for wireless communications. *IEEE Transactions on Antennas and Propagation*, **49**(7), 1094–1100.
63. Rassokhina, Y.V. and Krizhanovski, V.G. (2010) Analysis of h-slot resonators in microstrip line ground plane. International Kharkov Symposium on Physics and Engineering of Microwaves, Millimeter and Submillimeter Waves (MSMW).
64. Pozar, D.M. and Targonski, S.D. (1991) Improved coupling for aperture coupled microstrip antennas. *Electronics Letters*, **27**, 1129–1131.
65. Zhou, Y., Koulouridis, S., Kiziltas, G., and Volakis, J.L. (2006) A novel 1.5-inch quadruple antenna for tri-band GPS applications. *IEEE Antennas and Wireless Propagation Letters*, **5**, 224 –227.
66. Cheo, B.R.-S., Rumsey, V.H., and Welch, W.J. (1961) A solution to the frequency-independent antenna problem. *IRE Transactions on Antennas and Propagation*, **AP-9**, 527–534.
67. Brumbaugh, C.T., Love, A.W., Randall, G.M. *et al.* (1976) Shaped beam antenna for global positioning satellite system. IEEE Antenna and Propagation Conference Proceedings.
68. Czopek, F. and Schollenberger, S. (1993) Description and performance of the GPS block I and II L-band antenna and link budget. Proceedings of the ION Conference, pp. 37–43.
69. Lopez, A.R. (2000) GPS ground station antenna for local area augmentation system, LAAS. Proceedings of ION National Technical Meeting, pp. 738–742.
70. Griffiths, L.J. and Jim, C.W. (1982) An alternative approach to linearly constrained adaptive beamforming. *IEEE Transactions on Antennas and Propagation*, **30**(1), 27–34.
71. Compton, R.T. Jr. (1979) The power-inversion adaptive array: concept and performance. *IEEE Transactions on Aerospace and Electronic Systems*, **15**, 803–814.
72. O'Brien, A.J., Gupta, I.J., Reddy, C.J., and Werrell, F.S. (2010) Space-time adaptive processing for mitigation of platform generated multipath. Proceedings of ION International Technical Meeting, pp. 646–656.
73. Seco-Granados, G., Fernandez-Rubio, J.A., and Fernandez-Prades, C. (2005) Ml estimator and hybrid beamformer for multipath and interference mitigation in GNSS receivers. *IEEE Transactions on Signal Processing*, **53**, 1194–1208.
74. Ray, J.K., Cannon, M.E., and Fenton, P.C. (1999) Mitigation of static carrier-phase multipath effects using multiple closely spaced antennas. *ION Journal of Navigation*, **46**(3), 193–202.
75. Brown, A.K. and Mathews, B. (2005) GPS multipath mitigation using a three dimensional phased array. Proceedings of ION GNSS 2005, pp. 659–666.
76. Li, R., Wang, Y., and Wan, S. (2003) Research on adapted pattern null widening techniques. *Modern Radar*, **25**(2), 42–45.
77. Rama Rao, B., Williams, J.H., Boschen, C.D. *et al.* (2000) Characterizing the effects of mutual coupling on the performance of a miniaturized GPS adaptive antenna array. Proceedings of ION GPS 2000, pp. 2491–2498.
78. Zhang, Y., Hirasawa, K., and Fujimoto, K. (1987) Signal bandwidth consideration of mutual coupling effects on adaptive array performance. *IEEE Transactions on Antennas and Propagation*, **35**(3), 337–339.

79. Gupta, I.J. and Ksienski, A.A. (1983) Effects of mutual coupling on the performance of adaptive arrays. *IEEE Transactions on Antennas and Propagation*, **31**(5), 785–791.
80. Griffith, K.A. and Gupta, I.J. (2009) Effect of mutual coupling on the performance of GPS AJ antennas. *ION Journal of Navigation*, **56**(3), 161–174.
81. Zhou, Y., Chen, C.-C., and Volakis, J.L. (2008) Single-fed circular polarized antenna element with reduced coupling for GPS arrays. *IEEE Transactions on Antennas and Propagation*, **56**(5), 1469–1472.
82. Kasemodel, J.J., Chen, C.-C., Gupta, I.J., and Volakis, J.L. (2008) Miniature continuous coverage antenna array for GNSS receivers. *IEEE Antennas and Wireless Propagation Letters*, **7**, 592–595.
83. Surrey Satellite Technology Ltd, www.sstl.co.uk (accessed 13 December 2011).
84. GFZ German Research Centre for Geosciences, The CHAMP Mission, http://op.gfz-potsdam.de/champ/index_CHAMP.html (accessed 13 December 2011).
85. Öhgren, M., Bonnedal, M., and Ingvarson, P. (2010) Small and lightweight GNSS antenna for precise orbit determination. Proceedings of ESA Space Antennas Workshop, Section 16, pp. 1–5.
86. Sarantel Ltd, http://www.sarantel.com/ (accessed 13 December 2011).
87. Maqsood, M., Bhandari, B., Gao, S. *et al.* (2010) Development of dual-band circularly polarized antennas for GNSS remote sensing onboard small satellites. Proceedings of ESA Workshop on Antennas for Space Applications.
88. James, J.R. and Hall, P.S. (eds) (1989) *Handbook of Microstrip Antennas*, IEE Electromagnetic Waves Series, IEE.

15

Antennas for Small Satellites

Steven (Shichang) Gao[1], Keith Clark[2], Jan Zackrisson[3], Kevin Maynard[2], Luigi Boccia[4], and Jiadong Xu[5]

[1]Surrey Space Centre, University of Surrey, UK
[2]Surrey Satellite Technology Ltd, UK
[3]RUAG Aerospace, Sweden
[4]University of Calabria, Italy
[5]Northwestern Polytechnical University, Xi'an, China

15.1 Introduction to Small Satellites

15.1.1 Small Satellites and Their Classification

Early satellites were small. However, with the requirements for more and more complicated functions, satellites became increasingly large and heavy. A typical telecommunication satellite nowadays has a mass of several tons. Small satellites here refer to the new generation of satellites which are not only smaller in size, faster to respond and cheaper to build, but also highly capable and reliable. This is in contrast to mainstream military and commercial satellites which are usually very large, heavy and very expensive. Small satellites recently became viable due to rapid technological developments during last decades, such as miniaturization, very large-scale integrated circuits, microprocessors, solid state memories, sophisticated software tools and manufacturing processes.

UoSAT-1 was the first modern small satellite to carry a reprogrammable on-board computer and 2D charge-coupled device (CCD) array camera (256 × 256 pixels). It was developed by university students, staff and radio amateurs and was launched in 1981 [1]. Small satellites provide 'affordable access to space' and can be built at low cost by using commercial off-the-shelf (COTS) devices (developed for terrestrial applications) rather than the more expensive, less technologically current, space-qualified components.

There are several different types of small satellites, including mini-, micro-, nano-, pico-, femto- and cube satellites. Table 15.1 gives a comparison of small, medium and conventional large satellites. As shown in the table,

Space Antenna Handbook, First Edition. Edited by William A. Imbriale, Steven (Shichang) Gao and Luigi Boccia.

small satellites can achieve a significant reduction in mass and cost, and allow 'fast' access to space. Conventional large satellites require five years to a few decades from proposal to launch while small satellites such as microsatellites can require as little as one year from concept to launch. Small satellites are quicker to build and cheaper to launch, and give the option of flying multiple small satellites at the same cost as a single standard unit.

Small and large satellites are complementary to each other. Small satellites are becoming increasingly important as people require more frequent access to space at low cost. Small low-cost satellites are very useful for applications in communications, Earth observation, navigation, science missions, technology verifications and education [2,3]. In the following subsections, several examples of small satellites and applications will be illustrated.

15.1.2 Microsatellites and Constellations of Small Satellites

Small satellites can enable a quick response to natural disasters and provide a unique opportunity for affordable constellations to achieve high-resolution global coverage of the Earth with fast response times. In this aspect, multiple small satellites can achieve tasks which are very difficult for a single large satellite.

Figure 15.1 shows the Disaster Monitoring Constellation, which is a constellation of five microsatellites providing global imaging on a daily basis for the purpose of disaster assessment. It has worldwide daily imaging capability and allows fast delivery of images to the disaster relief community in cases of flooding, fire and earthquakes. Each three-axis momentum bias stabilized microsatellite carries an optical imaging payload to provide a ground resolution of 32 m with a swath width of over 600 km [4].

Figure 15.2 shows one of the microsatellites in the Constellation [4], which has a mass of 100 kg. It consists of various subsystems including RF systems (antennas, uplink, downlink, telemetry command and control), a power system (solar panel, batteries), thermal control, a propulsion system, on-board data handling (OBDH), an attitude determination and control system (ADCS) and a payload module (e.g. Earth observation equipment). Constellation phasing is achieved by using a cost-effective gas propulsion system and the orbit is determined via an on-board GPS receiver. On-board computers and sophisticated attitude and data handling systems enable complex housekeeping and payload operations to be carried out autonomously.

The satellites are in low Earth orbit (LEO) and carry an array of antennas, as shown in Figure 15.2. These include an X-band quadrifilar helix (QFH) for the 80 Mbps payload downlink, an S-band QFH for the 8 Mbps payload downlink, dual redundant S-band patch antennas for reception of tele-commands, and S-band

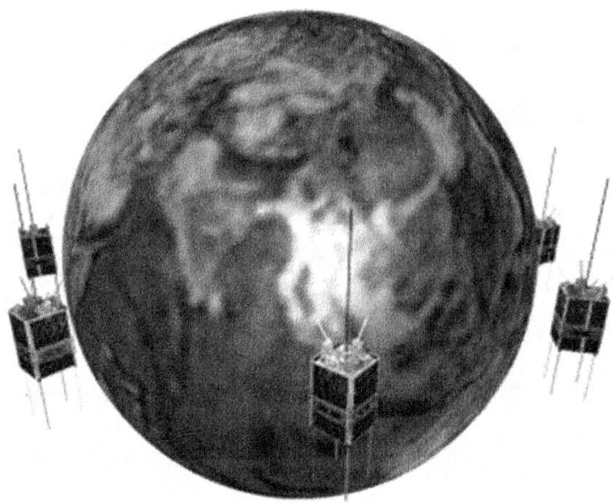

Figure 15.1 The Disaster Monitoring Constellation. Courtesy of Surrey Satellite Technology Ltd, UK [4].

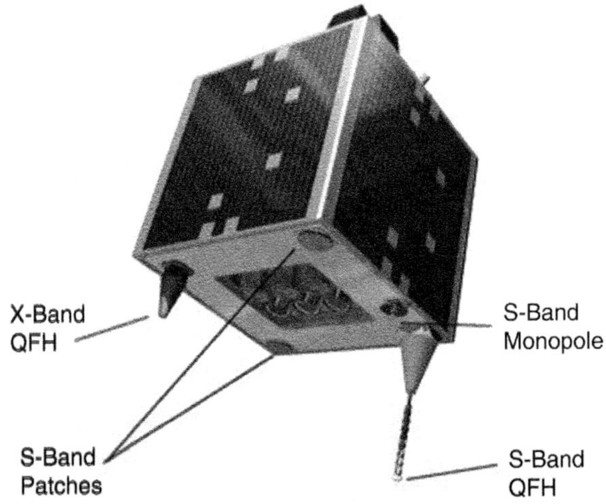

Figure 15.2 *One of the microsatellites used in the Disaster Monitoring Constellation. Courtesy of Surrey Satellite Technology Ltd, UK [4].*

monopoles for the transmission of on-board telemetry. In this case, the requirement for a high-speed downlink was driven by the amount of data generated by the multispectral imager, seen on the Earth-facing facet.

The imagery from the Constellation is used to exploit the full range of applications for large-coverage images with medium spatial resolution and high temporal resolution.

Another example of small-satellite constellations is RapidEye, a commercial multispectral Earth observation constellation comprising five minisatellites [2]. It provides high-resolution multispectral imagery along with an operational GIS (geographic information system) service on a commercial basis. Each satellite has a mass of 150 kg and a size of 1 × 1 × 1 m. The satellites are equally spaced in a single Sun-synchronous orbit at an altitude of 630 km, which ensures consistent imaging conditions and a short revisit time. The RapidEye constellation can image any area of the Earth within 24 hours.

15.1.3 Cube Satellites

Cube satellites, often called CubeSats, are tiny cube-shaped satellites with dimensions of only 10 cm per side and a mass of less than 1 kg. They belong to the category of picosatellites shown in Table 15.1. Figure 15.3 shows a CubeSat [5]. The cubic structure comprises an enclosed aluminium box with solar cells fitted to the outside walls. Antennas, usually flexible monopoles, are deployed perpendicular to the faces, at the corners. The inside of the satellite holds a variety of small circuit-board-sized subsystems typically including power conditioning, a transmitter, receiver, sensors and occasionally a camera. One unique feature of the CubeSat is the use of a standard deployment system, the Poly Picosatellite Orbital Deployer, or P-POD. The functions of the P-POD are to provide a standard interface between the CubeSats and the launch vehicle, and to protect the launch vehicle and primary payload. CubeSats have been mostly used for university projects, giving students practical experience in designing and testing hardware for space. However, to those involved in CubeSats, it is clear that the utility value of CubeSats is rapidly increasing. Reliable, high-performance, commercial CubeSats are now available and there is an increasingly compelling case for CubeSats to be used in many applications, including science, communications, Earth observation and technology demonstration. CubeSats usually employ VHF and UHF communication. It is very challenging to design compact high-performance antennas for CubeSats due to their small physical dimensions. Usually, independent flexible monopoles are

Table 15.1 Types of satellites.

	Types	Mass (kg)	Cost (US $)	Development time: from proposal to launch (years)
	Conventional large satellite	>1000	0.1–2 billion	>5
	Medium satellite	500–1000	50–100 million	4
Small satellites	Minisatellite	100–500	10–50 million	3
	Microsatellite	10–100	2–10 million	~1
	Nanosatellite	1–10	0.2–2 million	~1
	Picosatellite (including cube satellites etc.)	≤1	20 000–200 000	<1
	Femtosatellite	<0.1	100–20 000	<1

mounted on one face of the box to provide up- and downlink capability, as shown in Figure 15.3. The antennas are rolled around the satellite before deployment and held in place with monofilament secured using a short length of Nichrome wire inside the spacecraft. Upon deployment, a current is passed through the Nichrome wire, which heats and melts the monofilament releasing the antennas [5]. The use of monopoles, though simple, severely limits the data transmission rate for CubeSats.

15.1.4 Formation Flying of Multiple Small Satellites

A single small satellite has limited capabilities due to its size and mass. A large satellite, though very capable, is very expensive and vulnerable to single points of failure. Formation flying is a technique which can significantly enhance the capability of small satellites. It involves the use of multiple small satellites employing an active control scheme to maintain the relative position and attitude of several craft. Carrier-phase differential GPS techniques have been used to track and control autonomously the relative position and attitude between the satellites. This technology will enable the development of a 'Virtual Large Satellite' where several small satellites fly in close formation so that they can accomplish a common mission. Many future space missions will benefit from using formation flying to perform distributed observations. Compared with the traditional 'monolithic' approach which uses a single, large and expensive satellite, formation flying uses a distributed array of many simple, low-cost and highly cooperative small satellites, leading to advantages of lower cost, higher reliability, flexibility and reconfigurability.

Figure 15.3 *CubeSat [5]. © 2001 IEEE.*

Figure 15.4 Multiple VHF, UHF, L- and S-band antennas on UoSat-12. Courtesy of Surrey Satellite Technology Ltd, UK [4].

For antenna engineers, this technique will require small, efficient antennas for intersatellite communications and intersubsystem communications, as well as other functions such as uplink and downlink communications, and so on.

15.2 The Challenges of Designing Antennas for Small Satellites

The goal of the satellite design engineer is to maximize the resources available to the payload while providing adequate support for the mission duration at minimal cost. Superficially, the antenna requirements may appear trivial. However, once the entire breadth of mission constraints is examined, the challenges for the antenna design engineer can be considerable.

Figure 15.4 shows multiple antennas mounted on a 400 kg satellite (UoSat-12) [4]. The antennas served a variety of functions. There was a VHF crossed dipole for reception of tele-commands, single-blade antennas for transmission of telemetry, a UHF QFH for a high-power orbital pager experiment, a small patch antenna for a GPS receiver and dual redundant L- and S-band QFH antennas for the digital MERLION transponder payload.

15.2.1 Choice of Operating Frequencies

The frequency spectrum is a valuable and much sought after commodity. To ensure its best use, various bodies including the ITU (International Telecommunication Union, part of the United Nations) legislate its use. Operation within given bands is defined by the use of the proposed link. Specific bands are allocated for such functions as radio location, space operations (space to Earth, Earth to space, space to space), radio navigation, broadcast services, Earth exploration, radio astronomy and amateur use. Furthermore, the frequency use

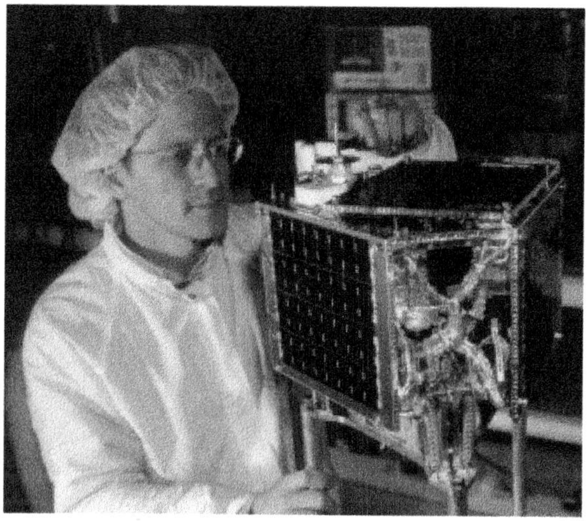

Figure 15.5 Satellite structures may be physically small. Courtesy of Surrey Satellite Technology Ltd, UK [4].

within a band can be defined as either a primary or secondary allocation. This allotment characterizes that secondary frequency allocations shall cause no harmful interference to *or claim protection* from those given primary frequency status. Yet, even when the correct band has been selected, there must be sufficient bandwidth available for the operation of the new satellite. Frequencies must be coordinated through the ITU to ensure existing satellites and even terrestrial services do not suffer from interference as a consequence of the new request being granted.

15.2.2 Small Ground Planes Compared with the Operational Wavelength

It has been shown above that the selection of operational frequency depends on satellite function. There may be a considerable difference in the frequencies needed to provide command and control of the satellite and that of the payload. This can mean that frequencies to be adopted may have wavelengths greater than the longest dimension of the satellite structure. When the responses of the adopted antennas are checked one may find that the pattern is considerably different from the assumed textbook patterns measured over ideal infinite ground planes. Nowadays antenna designers rely heavily upon numerical electromagnetic simulators which can reliably predict the antenna pattern and effect of intrusive obstacles mounted on the satellite structure (see Figure 15.5).

15.2.3 Coupling between Antennas and Structural Elements

The external surfaces of small satellites are seldom flat. Valuable external surfaces are adorned with a variety of solar panels, sensors, baffles, antennas and structural elements (see Figure 15.6). These elements all contend for optimal position according to their importance to mission success. Inevitably compromises have to be made and extraneous elements distort the ideal antenna pattern, upset polarizations and can impact antenna line-of-sight angles. However unfortunate, it is often found that additional obstructions are placed within the influence of the antennas late in the development programme. Designers must take account of this and build in realistic margins of safety and/or backup systems. Prodigious use of electromagnetic simulation will alleviate or confirm suspicious pattern degradation.

Figure 15.6 Patch, helix and monopole antennas on a microsatellite. Courtesy of Surrey Satellite Technology Ltd, UK [4].

15.2.4 Antenna Pattern

Antenna pattern requirements will be dictated by the function of the link, robustness of the modulation coding to errors and stability of the satellite itself coupled with the data rate requirements of that link. Low-data-rate signals from a tumbling satellite providing telemetry data will often require omnidirectional coverage. However, a three-axis stabilized Earth observation satellite in geostationary orbit (GEO) linked to a fixed ground station will need a tightly directional beam. The pattern requirements often dictate the size and complexity of the antenna configuration selected.

15.2.5 Orbital Height

Small satellites are frequently used in LEO where their orbital height dictates the need for tracking ground stations as they describe an arc across the sky as they move through their orbit. In contrast, those in GEO appear to remain stationary at the same point in the sky over time.

15.2.6 Development Cost

Small satellites are often built using limited funds and long costly development programmes to optimize antenna systems are unfeasible. System-level decisions must be made to ensure that there are enough margins within the calculated link budgets to absorb non-optimal antenna performance.

15.2.7 Production Costs

Similarly, budget constraints restrict the complexity and elegance of antenna production. Often cost-effective off-the-shelf solutions may be adopted instead of internally optimized systems.

15.2.8 Testing Costs

Commercial antenna ranges are expensive to hire and unlikely to be available to the antenna designer of a small satellite for significant periods of time. Often the completed satellite with associated antennas will get, at best, a few days in a commercial chamber to confirm correct operation of the entire communications link.

15.2.9 Deployment Systems

Large satellites with their need for high-gain directional beams often require physically bulky structures to support them. To enable them to fit within a launcher fairing, they frequently have to be made deployable. This in turn impacts the complexity, mass and volume requirements [6] which are often too restrictive for small satellites. While elegant solutions have been incorporated for antenna deployment including inflatable structures and STEM (Storable Tubular Extendible Member) systems (a type of antenna which unrolls when the satellite reaches orbit), the mantra of 'keep it simple' continues to provide missions with the highest success rate [7]. As a consequence the majority of small satellites rely on more traditional, non-deployable antennas such as simple sprung monopoles, low-profile patches and compact helices.

15.2.10 Volume

Small satellites often have a low bus system to payload ratio with no room for large-volume antennas. Mechanically complex deployable systems are often excluded from the mission purely on account of the volume required for the deployment mechanism. In a similar manner, highly directional but physically large horn antennas may be replaced with smaller, less capable antennas at the cost of providing lower data rates.

15.2.11 Mass

Small satellites are, by definition, of limited mass. Much of the total weight comprises the satellite structure, solar panels, batteries, computers, transmitters and receivers, attitude determination and control systems, and, most importantly, the payload. Antenna systems are seldom provided with a significant portion of the total mass, hence the drive towards smaller, lighter antenna systems.

15.2.12 Shock and Vibration Loads

A further impact of the launcher is the need to make the structure of the antenna physically robust. While the physical environment is relatively benign once the antenna is placed in orbit, the stresses placed on the antenna mechanics can be considerable during the launch and separation phases of the mission. Shock loads due to the physical separation of the satellite from the launcher coupled with vibration and acoustic stresses need to be considered during the design phase. Small satellites often ride as secondary payloads and must simply accept the conditions that are placed on them. Furthermore, primary payload suppliers and launch agencies will insist that no part of these low-cost secondary missions becomes detached during launch, threatening the very costly rocket or principal satellites.

15.2.13 Material Degradation

The choice of suitable materials is crucial to ensuring the long-term success of small-satellite antenna systems. The space weather environment and vacuum conditions can impose restrictions not present on terrestrial systems and a poor selection of mechanical and electrical materials can produce a product whose performance often degrades with time or can fail catastrophically.

15.2.14 Atomic Oxygen

Satellites travelling in LEO encounter the highly corrosive effects of atomic oxygen. This is molecular oxygen that has been broken down by the effects of ultraviolet light at high altitudes. The effect of this is most predominant at lower altitudes and on the surfaces of antennas. This effect is important as protective coatings and platings can be eroded, changing their surface features and hence their thermal, mechanical or electrical properties. This may be of particular importance to the thermal performance of the antenna.

15.2.15 Material Outgassing

When exposed to the vacuum of space many polymers, plastics and glues outgas condensable volatiles, often leading to mass loss of the material or, more importantly, the possibility of contaminating the optical components. These optical components could be as diverse as solar cells, bus system star sensors and payload cameras. Contamination of these elements could compromise the operation of the satellite by reducing the power available and degrading image quality. Metal coatings such as cadmium have a similar effect and can sublimate, forming conductive paths and potentially shorting out antenna insulators to prevent the transmission or reception of RF signals.

15.2.16 Creep

Creep or cold flow is the increasing deformation of a material under a constant compressive load. The use of materials such as PTFE is not recommended where dimensionally stable insulators are required; for example, where radiating elements are required to be held at precise fractions of a wavelength from one another. Differential creep could cause misalignment of antenna beam patterns potentially causing the loss of communication links.

15.2.17 Material Charging

A common reason for vehicle anomalies in orbit is material charging due to such natural causes as a satellite passing through the LEO plasma environment or direct particle bombardment occurring during geomagnetic storms and proton events. Surface charging due to low-energy electrons accumulates to a point, often several kilovolts, at which an electrostatic discharge is initiated. This discharge can cause the failure of electronic components and subsystems, degradation of thermal coatings, spurious switching events, communication channel degradation and physical damage to the antenna structure. Victims of surface charging on antennas can be such items as protective thermal blankets and radomes. Further problems are caused by high-energy electrons penetrating shielding and causing a build-up of charge deep within dielectric insulators, such as antenna feed cables, up to a point where damaging electrical discharges occur. Charging effects can be mitigated by providing a high-impedance bleed path to satellite ground via conductive components or suitable ESD (electrostatic dissipating) paint. The choice of conductance is critical as too high a value will be insufficient to dissipate the charge while too low a value will adversely affect the antenna performance.

15.2.18 The Interaction between Satellite Antennas and Structure

Small satellites and particularly micro- and nanosatellites are often packed with instrumentation and support systems, leaving scant space for complex antennas and associated deployment mechanisms that could provide precisely defined antenna patterns. The tradeoffs between design simplicity, development programme duration and link budget requirements are taxing to begin with, but are further complicated by the need to co-locate a variety of sensors, cameras and payloads on the very surfaces that, ideally, antenna designers would like to keep specifically for their own requirements (see Figure 15.7). Many sides of a small spacecraft will be covered with solar cells to produce the very energy required to power the satellite. Since shadows cast across these cells by high-profile antennas would cause severe degradation in electrical output power, it is usual

Figure 15.7 Multiple antennas mounted on a 50 kg satellite. Courtesy of Surrey Satellite Technology Ltd, UK [4].

to restrict antenna location to faces free of solar panels, typically the Earth- and space-facing facets. Invariably these are the very faces that experimenters also want to use. Thus the engineer will invariably have to take into account a variety of obstructions fashioned from both conductors and insulators in addition to the limitations of available ground plane and possibly resonant dimensions of the satellite itself.

Broadly speaking there are two categories of antennas that need to be provided: low-gain tele-command, tracking and control antennas that have to be able to transfer commands to the subsystems and link back status information no matter what the orientation of the satellite; and, when the satellite is in its nominal condition, high-gain antennas that are required to provide the bulk of payload data transfer. Ideally the placement of these antennas will keep to a minimum the cross-coupling and pattern disturbance due to both the antennas themselves and the adjacent structures and instrumentation.

Careful choice of frequencies consistent with the ITU requirements and data rates required can help alleviate matters. However, invariably the use of either a suitably complex electromagnetic computer model or a full or representatively scaled physical model will allow the investigation of antenna pattern disturbances. The use of fractional scaled models involves decreasing the model size and scaling up the test frequencies. One advantage of this is that physically small measurement chambers could be utilized. Full-scale models are often tested outdoors in open-air test ranges. However, the effects of parasitic ground reflection have to be subtracted to give realistic results for satellite systems. While scale models were frequently used in the past, they have been largely superseded by the introduction of fast, cost-effective electromagnetic simulators coupled with high-end computers. It is now possible to transfer quickly mechanical data about the satellite structure into a

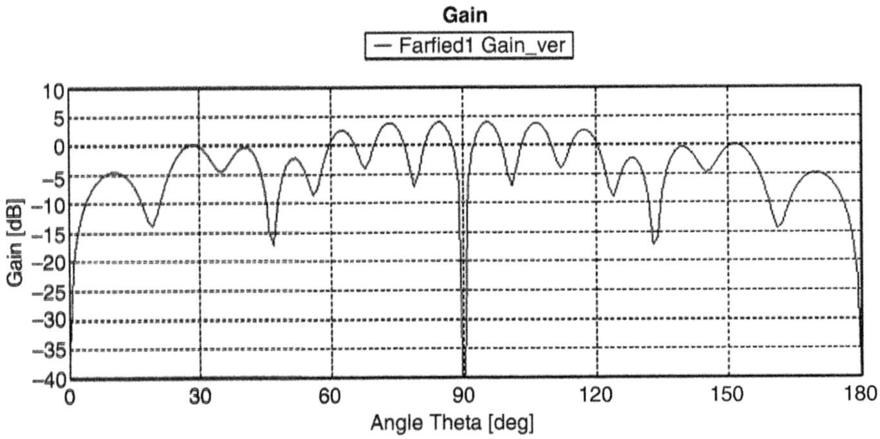

Figure 15.8 Simulated results for twin monopoles on a simple box.

simulator to build a representation of the final configuration and move the antennas about in the design space to optimize their location.

The power of computers has evolved such that simulations that took three days to simulate can now be performed within half an hour. The method of moments (MoM) is in common use and has gained considerable credibility due to the agreement between simulated and measured results. Electromagnetic simulators are supplied with a wide range of solvers, the speed and applicability of which depend upon the nature of the problem to be addressed. Many hybrid methods are being introduced in order to increase the speed of analysis, particularly those that include dielectric bodies. Modern simulators such as the multilevel fast multipole method (MLFMM) allow the calculation of antenna patterns of physically large structures which would be impractical to address with the MoM. This method allows more of the simulation to be held in semiconductor memory rather than continually accessing the hard drive, with a consequent increase in calculation speed to excellent accuracy.

Figure 15.8 shows the simulated results of two monopoles on a simple metallic box; that is, a simplified model of the satellite body. As can be seen, there are lots of ripples in the radiation patterns, which are due to the interactions between the antennas and the satellite body. The resultant radiation patterns of installed antennas rarely match those predicted by their simulation in free space. The effects of the relatively small size of the satellite, small and cluttered ground plane and nearby radiators all conspire to produce bulges and nulls in the response that must be cross-checked with a link budget analysis to ensure that the required data throughput specified can be accomplished. If issues are found, the height and orientation of the antennas, the addition of extra ground planes, movement of the offending structures or in extreme cases the replacement of an antenna with a different type or configuration may be employed.

15.3 Review of Antenna Development for Small Satellites

15.3.1 Antennas for Telemetry, Tracking and Command (TT&C)

15.3.1.1 VHF and UHF Monopoles in the Early Days

The early UoSAT spacecraft used VHF and UHF transmitters with corresponding wavelengths of 2 m and 70 cm on structures typically measuring 58 × 35 × 35 cm. It can be seen (Figure 15.9) that the body of these

Figure 15.9 Co-located VHF and UHF antennas. Courtesy of Surrey Satellite Technology Ltd, UK [4].

spacecraft was shorter than the wavelengths involved. These early spacecraft typically used arrays of monopoles, made from the same spring steel that is still used in measuring tapes, to give a low-cost antenna system with a good approximation towards omnidirectional coverage. The use of this system was crucial to ensuring the continuity of tele-command and telemetry data during the commissioning phases of the mission when the orientation of the satellite was random due to the tumbling caused by separation from the launcher.

At this early stage, small linearly polarized antennas were used on the satellite while the ground station used larger circularly polarized antennas. Although this gave an extra 3 dB of link loss compared with a matched polarization system, it provided a link that was insensitive to the spacecraft's orientation. An added advantage was that antenna arrays could be placed on the space-facing facet leaving the Earth facet free of obstructions impinging on the camera's field of view. The use of monopole arrays with inter-antenna spacing of less than a wavelength was found to give symmetrical far-field coverage with minimal destructive perturbations caused by the spacecraft body. Later on it was found that circular polarization was possible by using appropriate phasing between individual antennas. Indeed, similar effects were possible by shaping the individual antenna blades.

However, over time the need for higher data rates dictated the use of higher frequencies. As the frequency increases or the size of the spacecraft body increases, the modification of the monopole radiation pattern into that more like a dipole becomes less pronounced and the satellite body starts to block the signals. It then became necessary to place antennas at either end of the structure to improve coverage. This, however, had the

Figure 15.10 *Patch antenna with protective radome. Courtesy of Surrey Satellite Technology Ltd, UK [4].*

effect of causing ripple in the pattern due to the interaction between the electromagnetic fields of the two antennas.

The UoSAT satellite group eventually became Surrey Satellite Technology Ltd (SSTL) and the radio amateur frequencies previously employed were no longer suitable for the missions envisaged. There were a number of factors which caused these small satellites to migrate up in frequency. These included a desire for physically smaller satellite antennas, the increasing availability at competitive cost of small, reliable high-frequency and microwave semiconductors, and the availability of high-gain parabolic reflector dishes for use with ground stations. A further factor was that later mission profiles required much more data to be transferred between satellite and ground station, which dictated the use of higher bandwidths and correspondingly higher frequencies. A careful balance of low-power, high-frequency transmitters and receivers, compact satellite antennas and high-gain ground station antennas provided the opportunity to service the higher data rates with acceptable link budgets.

15.3.1.2 Patch Antennas

The monopole antenna had been the preferred workhorse solution due to its simplicity and robustness and still has its place as a TT&C antenna. However, the implementation of patch antennas on small satellites with their low profile, small mass and footprint has made much higher frequency operations feasible. Figure 15.10 shows the top view and bottom view of a circular patch antenna which has been used on board small satellites [4]. Note that a protective radome has been employed in the antenna. Compared with the monopoles, the patch has a lower profile and relaxes issues of stowage aboard the launch vehicle. The patch antenna can easily be made linearly or circularly polarized and its use in phased groups can provide higher gain or electrically steerable arrays. Although it is often possible to multiplex the transmission and reception functions into a single antenna for the sake of reliability and redundancy, these are often kept separate. To provide adequate coverage two antennas are used per receiver, one at either end of the satellite. Since there are usually two independent receivers a total of four antennas is required purely for the TT&C function, making small physical dimensions crucial. While the design of S-band patches is relatively straightforward, the effects of dielectric tolerances, temperature variations and parasitic effects make their manufacture and tuning at higher frequencies troublesome.

15.3.1.3 QFH Antennas for TT&C

Figure 15.11 shows an S-band QFH antenna to be used for the TT&C function with a hemispherical-type pattern [8]. The antenna is built up of four separate metal wires on a dielectric central support in a conical shape. The four helices are fed with a four-point isolated feed network to achieve a broadband antenna working

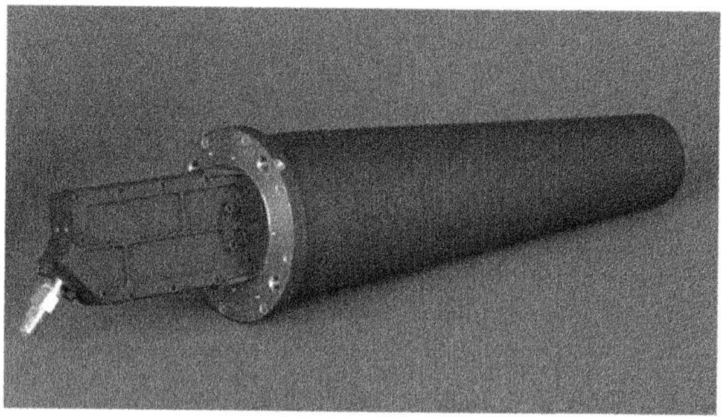

Figure 15.11 *S-band quadrifilar antenna with hemispherical type of radiation pattern. Courtesy of RUAG Space AB [8].*

at receive and transmit S-band TT&C frequencies without the need for tuning. The antenna has low weight and achieves a broad coverage and low cross-polarizations. Another benefit of using the QFH antenna is the possibility to shape the radiation pattern. Measured characteristics of the hemispherical-type antenna are shown in Figure 15.12.

15.3.2 Antennas for High-Rate Data Downlink

15.3.2.1 QFH Antenna for Data Downlink

Helix antennas use individual wires formed into extended coils. Quadrifilar helix antennas (QHAs), shown in Figure 15.13, using four individual volutes (or wires), and monofilar helix antennas (MHAs) using single volutes, become feasible at S-band and X-band where their dimensions become more suitable for small satellites and only low to medium gain is required. Helices have the advantage that they can be quickly

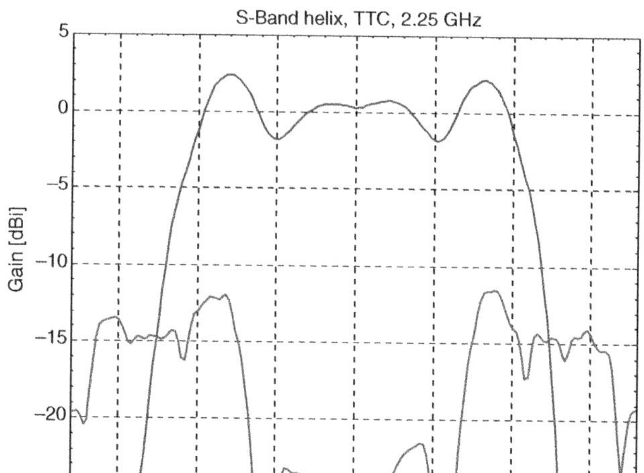

Figure 15.12 *Radiation pattern of hemispherical helix antenna.*

Figure 15.13 A space-qualified quadrifilar helix antenna. Courtesy of Surrey Satellite Technology Ltd, UK [4].

customized by varying the pitch, length and coil diameter to produce different radiation patterns. However, the disadvantage is their rigidity. Unlike bendable monopole antennas, they are not flexible, and their overall physical dimensions are larger. For optimum antenna patterns they are often mounted some distance from the satellite body, which once again raises issues with their accommodation on a launch vehicle.

The SSTL QHA is designed to give equal energy at the ground station from 62° (where the satellite appears on the horizon when viewed from the Earth) to 0° from nadir (directly underneath). To achieve this, the antenna pattern is tailored to take account of the extended path losses when viewed at the horizon compared with that when the satellite is directly above. The variation in antenna gain required for a satellite in an 800 km orbit amounts to approximately 12 dB.

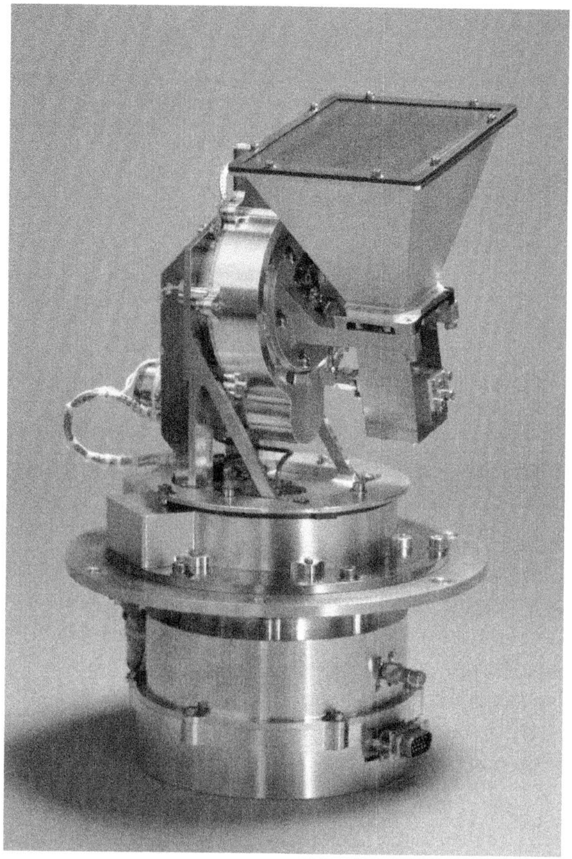

Figure 15.14 *Small-satellite antenna pointing mechanism with horn antenna. Courtesy of Surrey Satellite Technology Ltd, UK [4].*

15.3.2.2 X-band Horn Antennas

As small satellites were used for increasingly complex missions the data rate requirements grew. Typically, early X-band data links worked at 20–40 Mbps using an RF power of 6 W. This power was adequate to provide a good RF link to a ground station with a reasonably sized (5.5 m) tracking dish antenna. These links were supported by low-profile, shaped QHAs on the satellite. More recently, customer demands for higher resolution images have forced the use of even higher data rates extending up to 200 Mbps. Link budgets have been further complicated by the additional requirement for satellite manoeuvres to capture images while maintaining a high-data-rate communications link. Thus for an antenna system that can provide a large bandwidth, a steerable high-gain radiation pattern is required. In order to meet this requirement SSTL developed a mechanical antenna pointing mechanism that could direct either a conventional horn or a small antenna array (see Figure 15.14).

Mechanically steered antennas are applicable to the larger end of the small-satellite range but pose significant design challenges. Typically the satellite travels around the Earth at 7 km/s, completing an orbit in around 90 minutes. The time that communications can be carried out between satellite and ground station depends on latitude but a 10-minute pass is typical. Mechanically steered antennas have to be able to maintain track on the target ground station even when the rate of change of pointing angle is greatest (typically when the

Figure 15.15 The LCROSS antenna. Courtesy of RUAG Space AB [8].

satellite passes overhead). There is a tradeoff between tracking speed and antenna beamwidth. While increasing the beamwidth will reduce the pointing accuracy required, it will also reduce the data rate which can be supported. The pointing mechanism itself is far more complex than simpler fixed antennas, so mechanical designs must ensure the various motors and rotatable RF joints can survive the rigours of the space environment for the total mission duration. The horn antennas currently used on the SSTL antenna pointing mechanism (APM) are circularly polarized to keep the polarization losses in the link budget to a minimum. This means that the polarization purity needs to be controlled over the significant beamwidth over the pass profile.

15.3.2.3 Patch Excited Cup Antennas

For data downlink at S-band when higher gain is needed a medium-gain antenna built as a patch excited cup is a good solution. One example is the antenna to the LCROSS (Lunar Crater Observation and Sensing Satellite) mission [8]. The antenna was used on Moon-crashing probes.

The LCROSS antenna (Figure 15.15) was delivered in 2007 and was launched in 2009; the mission ended in 2009 [8]. The main objective of the mission was to search for ice near a lunar pole. The antenna uses three patches within a shallow cup with a rim height of about a quarter wavelength. The two lower patches form a resonant cavity allowing double or broadband tuning while the upper patch acts as a reflector to enhance the gain. The antenna is circularly polarized and fed from a four-port feed network. The maximum gain is about 12 dBi with a gain better than 9 dBi in a $\pm 20°$ cone (Figure 15.16).

15.3.2.4 X-band Helix Antenna

A shaped type of radiation pattern is often needed to compensate for the difference in path losses between nadir and edge of coverage. One solution is to use a QHA with isoflux coverage. The helix antenna at X-band is a compact and lightweight antenna of less than 400 g (Figure 15.17) that has been developed to respond to the need for cost-effective antennas for downlinking data from small LEO satellites. The antenna has an innovative design of the radiator part which can be adapted to different requirements regarding gain, coverage, polarization and frequency band. The all-metal design of the radiator part, combined with a radome for support and a septum polarizer, creates a modular system.

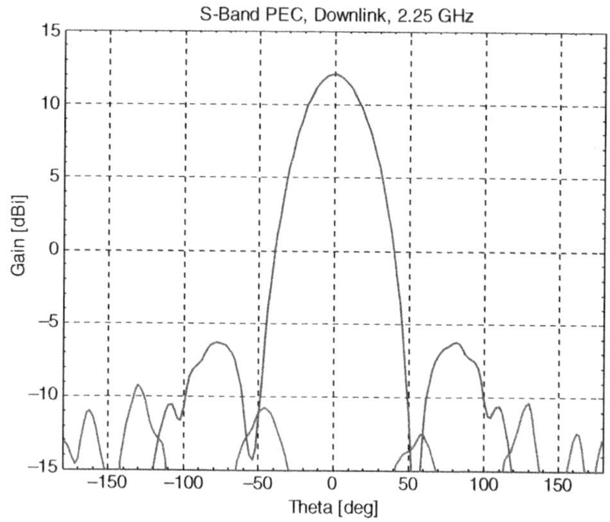

Figure 15.16 Antenna radiation pattern.

While reducing the size and weight compared with the normally used reflector or bicone-type antennas, the antenna retains the characteristics of the traditional helix antenna, namely large bandwidth and good RF performance. The RUAG antenna has edge-of-coverage angles of $\pm 60°$ and $\pm 70°$ depending on the radiator used. The characteristics for an 8.2 GHz $\pm 60°$ antenna are shown in Figure 15.18. The gain at edge of coverage is around 5 dBi for this antenna.

15.3.2.5 X-Band Corrugated Aperture Antenna

Another solution for data downlink in X-band is to use a circular waveguide aperture with corrugations (Figure 15.19) giving a flat-top pattern type of coverage [8]. It was developed as a high-power-capable antenna. It is an all-metal design with only two parts, the radiator part and a septum polarizer. The antenna characteristics are shown in Figure 15.20. The gain within $\pm 60°$ of coverage is around 3.5 dBi for this antenna.

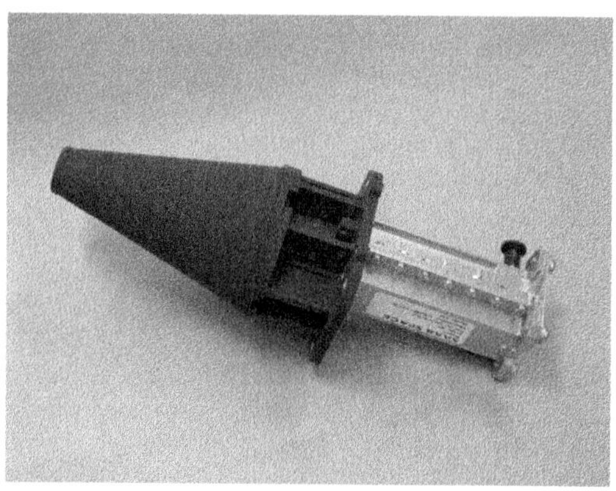

Figure 15.17 Space-qualified X-band helix antenna. Courtesy of RUAG Space AB [8].

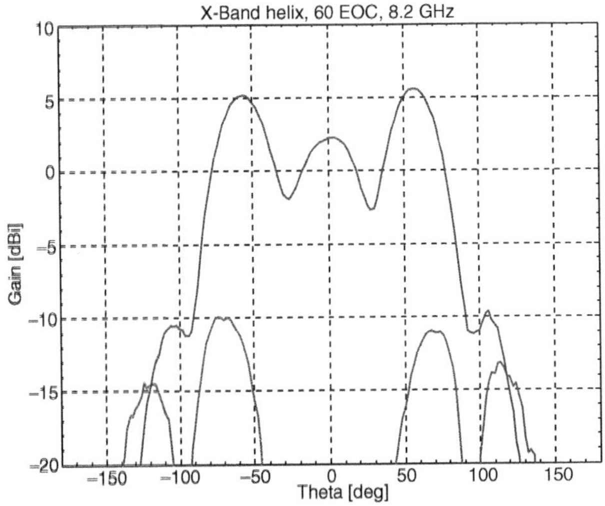

Figure 15.18 X-band helix antenna radiation pattern.

Many other antennas have also been developed for such applications. For reflector antennas, the rigid, single-piece reflectors and deployable mesh reflectors are two types of reflectors useful for achieving very high gain. However, the apertures of rigid, single-piece reflectors are limited by the size for fitting inside a launch vehicle, and deployable mesh reflectors are very expensive because they require complex mechanisms. Presented in [9] is a deployable solid surface reflector antenna which can be packaged into small, low-cost launch vehicles.

Figure 15.19 Space-qualified X-band corrugated aperture antenna. Courtesy of RUAG Space AB [8].

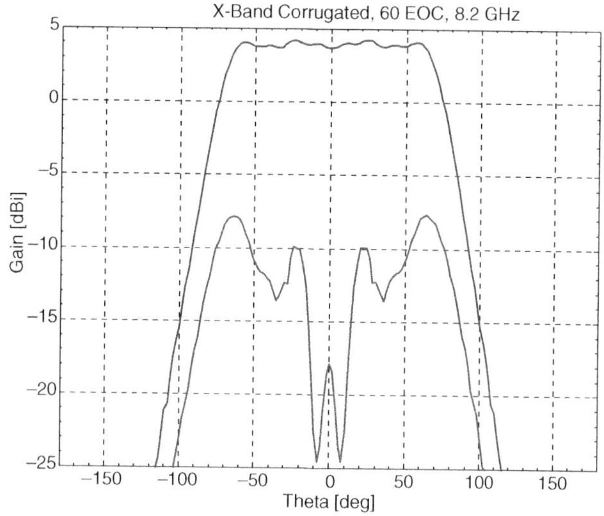

Figure 15.20 *X-band corrugated aperture antenna radiation pattern.*

15.3.3 Antennas for Global Navigation Satellite System (GNSS) Receivers and Reflectometry

15.3.3.1 Antennas for GNSS Receivers for Positioning, Attitude and Orbit Determinations

RUAG has worked with both low-profile patch excited cup (PEC) as well as helix antennas for GNSS applications.

GNSS antennas for small satellites need to be small and lightweight antennas with low back radiation and good coverage. The largest error contribution is usually the local multipath; that is, signals reflected in the satellite structure and received by the antenna. Thus the antenna radiation in the back direction must be suppressed while keeping the coverage to be able to track GNSS satellites at low elevation angles. This is normally achieved by using a standard wide-coverage antenna element placed in a larger choke-ring structure. The disadvantage with this arrangement is that the antenna becomes large and relatively heavy.

To overcome this RUAG developed two PEC antennas: a smaller one (Figure 15.21) suitable for satellites without large flat mounting surfaces; and one where the low-elevation gain was traded against the back

Figure 15.21 *GNSS PEC antenna. Courtesy of RUAG Space AB [8].*

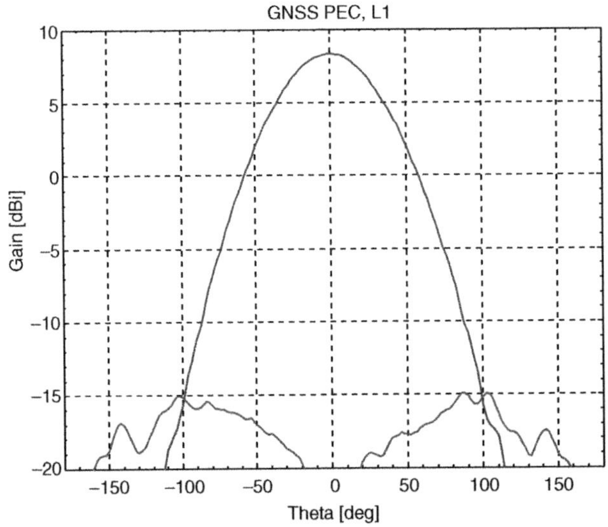

Figure 15.22 GNSS PEC antenna radiation pattern.

radiation and a good compromise was achieved using only two narrow choke rings to allow for mounting on larger flat surfaces with low-multipath interaction [8].

The PEC antenna type consists of two patches, placed in a circular cup. To achieve a stable antenna covering the GNSS frequency bands a four-point feed with capacitive coupling of the bottom patch is used in combination with an isolated feed network. Measured characteristics of the smaller antenna are shown in Figure 15.22. RUAG also developed helix antennas for GNSS application and one example is shown in Figure 15.23 [8].

15.3.3.2 Shorted Annular Patch Antenna

An innovative class of compact high-precision GPS antennas has been introduced in [10] where a shorted annular patch (SAP) element has been demonstrated to reduce significantly the multipath interferences of GPS-based attitude determination sensors.

Figure 15.23 GNSS helix antenna. Courtesy of RUAG Space AB [8].

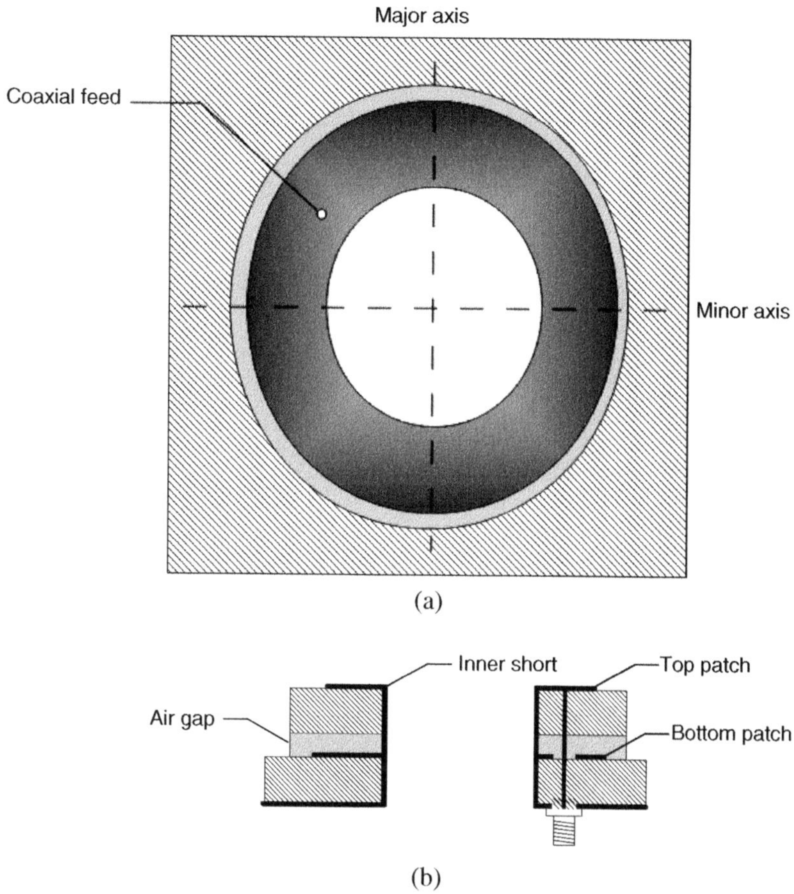

Figure 15.24 Dual-band stacked SAP antenna: (a) top view; (b) side view [11]. © 2004 IEEE. Reprinted, with permission from IEEE Antennas and Wireless Propagation Letters, vol. 3, 2004, pp. 157–160.

Shorted rings possess radiation characteristics similar to those of circular patches; that is, the equivalent magnetic current distribution flowing on the two external boundaries is identical. However, the SAP inner border, being shorted, does not radiate and can be used to tune the antenna resonant frequency, while the external radius can be chosen to achieve narrower radiation patterns and higher directivity. Compared with other concurrent solutions, SAP antennas thus possess the flexibility for controlling the radiation patterns which can be employed to match the low-multipath radiation requirements with a single radiator without employing ground plane extensions or choked rings.

Depending on the application type and on the accuracy requirements, different SAP configurations can be used. Although dual-feed or four-feed SAP antennas are required for higher circular polarization purity and phase centre stability, single-feed solutions can be used in many practical applications. In the latter case, circular polarization can be enforced by designing the SAP antenna to have elliptical borders and by placing a single coaxial feed at 45° with respect to the ellipse axes. In this way, two resonant modes with a 90° phase difference between them are excited to radiate a circularly polarized wave. An example of stacked shorted elliptical antennas which cover both L1 and L2 GPS bands is reported in [11]. As can be observed in Figure 15.24, each of the stacked patches operates in a different band. The resonant frequencies can be controlled by adjusting the outer boundaries, the common inner short and the gap between the two antennas.

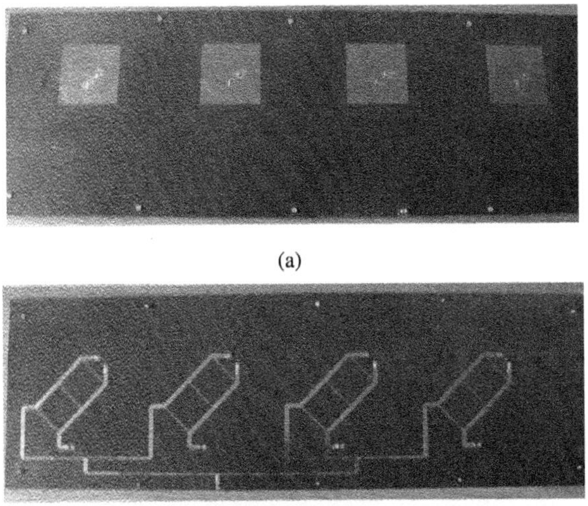

(a)

(b)

Figure 15.25 Dual-band circularly polarized array for GNSS reflectometry: (a) top view showing patch array; (b) bottom view showing the microstrip feed network.

GNSS attitude determination uses GNSS phase measurements from spaced antennas in order to determine the angle of arrival of the signals, and hence derive the attitude of the satellite platform. Multipath from nearby protrusions is a problem and should be minimized, and the antenna needs to be designed to reject multipath. The relative phase anisotropy between antennas is also important; this could be reduced by designing antennas with more symmetric phase as a function of azimuth and elevation. Alternatively, given imperfect antennas on a real platform, the effects can be reduced by ensuring antennas are mounted in the same orientation to subtract common distortions. Furthermore, work has been done numerically to map phase biases and model them either as a lookup table or as a polynomial spherical harmonic model to provide phase difference corrections. Analysis has shown that such a map can improve measurements such that antenna phase distortion and multipath are no longer the dominant errors in GPS attitude determination [12].

15.3.3.3 Antennas for GNSS Reflectometry

GNSS reflectometry measures the signals from the GNSS after they are reflected off the sea, and could revolutionize oceanographic remote sensing. The technique could enable satellites to measure the height of waves and wind speed on the high seas, providing important data to ship owners and operators. It could also enable the measurement of atmospheric parameters. As the reflected signals received on board the satellite are very weak, the receivers for GNSS reflectometry need to be highly sensitive and a compact-size high-gain antenna is required. Figure 15.25 shows a dual-band, circularly polarized printed array antenna for GNSS reflectometry [13]. It is a four-element array. Each element is a stacked microstrip patch antenna operating at both L1- and L2-bands. The patch antenna is fed by two orthogonal feeds, which are connected to a broadband microstrip three-branch hybrid circuit at the bottom, as shown in Figure 15.25.

15.3.4 Antennas for Intersatellite Links

Intersatellite communication is required in the formation flying of multiple small satellites, and in satellite constellations. When nanosatellites or CubeSats are used in these networks, the satellites are too small to

Figure 15.26 Circularly polarized, cross-shaped, microstrip patch retro-directive array mounted on one face of a CubeSat. The resonant frequency is 10.5 GHz [14]. © 2005 IEEE. Reprinted, with permission from Proceedings of the IEEE/ACES International Conference on Wireless Communications and Applied Computational Electromagnetics, Honolulu, HI, April 2005, pp. 606–609.

contain a sophisticated attitude control system on board. An electronic beam steering antenna will enable the satellite to establish and maintain reliable communication links with other satellites. However, it is not possible to realize such an antenna using traditional phased-array technologies or smart antennas, which are too large in size, expensive and power hungry for small satellites.

One promising solution to this problem is to employ the retro-directive antennas. A retro-directive antenna is a multi-element array which can automatically reflect incident radiation in the direction of the source radiator by retransmitting the phase conjugate of received signals at each antenna element. This technique does not require expensive microwave phase shifters as in phased arrays, or the digital signal processing components in smart antennas, thus enabling significant reductions in cost, complexity and power consumption. To achieve the phase conjugation, the corner reflector, the Van Atta array and the heterodyne-based arrays are commonly used. Researchers at the University of Hawaii have done extensive work on retro-directive antennas and their applications for small satellites. Figure 15.26 shows a 10 GHz, circularly polarized, two-dimensional phase-conjugating array [14]. It is based on quadruple subharmonic mixing, which relaxes the local oscillator (LO) frequency requirement. The method uses the fifth-order mixing product, requiring an LO frequency that is half that of the RF.

15.3.5 Other Antennas

15.3.5.1 Distributed Multi-Function Antennas for Micro- and Nanosatellites

Due to their limited size, mass and power budget, small satellites have limited RF capability in terms of number and quality of available wireless links. To overcome this limitation, distributed multi-function antennas for micro- and nanosatellites have been proposed [15]. They are based on the idea of 'structural radiators', borrowed from the handset antenna community. Here the satellite is employed as a key part of the radiator, instead of being considered as a ground plane only. Traditional 'stand-alone antennas', provided they are very small in size, can serve as 'exciters', and a distributed array of 'exciters' is placed all over the satellite structure to stimulate a time-varying current on the satellite platform surface and to achieve versatile control of the radiation patterns from the satellite. Contrary to most space antenna systems, the overall satellite structure plays a major role in all antenna performances leading to the need for a global combined optimization. Figure 15.27 shows an example of an exciter which uses an inverted-F antenna [15]. Each exciter is reconfigurable as it is loaded with reconfigurable lumped elements at different positions and these lumped elements could be varactor diodes, PIN diodes or RF

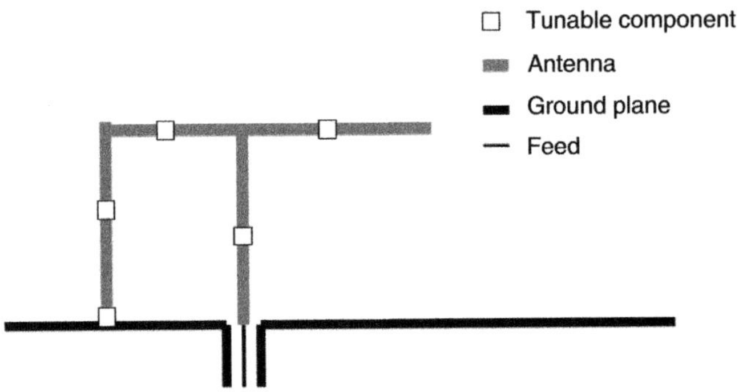

Figure 15.27 *An example of an exciter which uses an inverted-F antenna [15].*

MEMS switches. By controlling the DC biasing of these reconfigurable lumped elements, the exciter can radiate as a monopole, a half loop or a nearly omnidirectional antenna. These exciters can be placed along the edges or in between solar cells. By controlling the exciters, the distributed antenna system can produce reconfigurable patterns and achieve multiple functions such as TT&C, data downlink, RF tracking, and so on.

15.3.5.2 Solar Panel Integrated Antennas

The solar panel occupies the majority of the surface area of a small satellite. As small satellites get smaller in size and mass, it becomes more and more challenging to manage the limited surface area of the satellite body to fit solar panels, space instruments and also many antennas for TT&C, data downlink, GNSS, intersatellite links, and so on. One possible way to solve the problem is to integrate the antennas with solar panels. This can be achieved by either placing the patch antenna under the solar cells, or creating radiating slots on the ground plane under the solar cells. Such designs require, however, custom-made solar cells. Presented in [16] is an optically transparent antenna which can be integrated on after-market solar cells. Figure 15.28 shows the optically transparent antenna, which is a meshed patch antenna having an optical transparency of 93%. The antenna solution is based on two facts: 1. the meshed conductor is optically transparent and can still be an

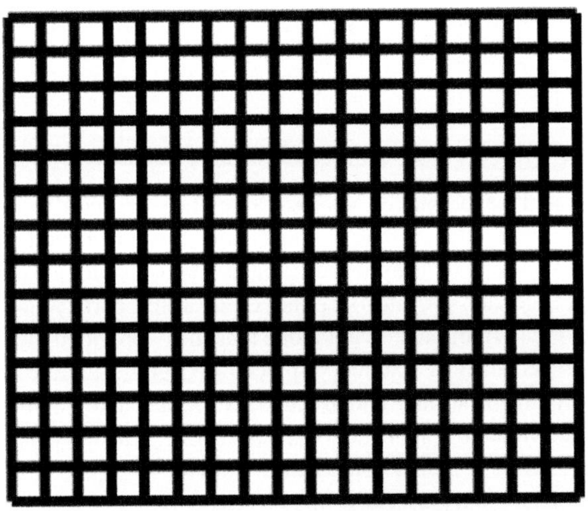

Figure 15.28 *Meshed patch antenna.*

effective radiator; 2. there is usually a cover glass on top of the solar cells and the cover glass can serve as the substrate for the antenna. The prototype in [16] reports an antenna gain of 8.2 dB. Such an integrated antenna could save a valuable amount of satellite surface area, and thus contribute to the size reduction and multi-functionality of solar panels. Other solar cell integrated antennas are also reported.

15.4 Case Studies

15.4.1 Case Study 1: Antenna Pointing Mechanism and Horn Antenna

Increasingly more complex imaging systems and associated data rates forced the adoption of a tracking antenna system aboard a 300 kg SSTL small satellite. The following case study describes the technical requirements and design process that shaped the final successful arrangement comprising a small horn antenna and associated tracking system.

15.4.1.1 Background

LEO spacecraft are, in general, power limited. The power available for translation to RF energy is usually derived from an array of solar cells whose conversion efficiency, while depending upon the technology employed, typically lies between 16 and 27%. As a consequence the RF power available for downlink communications is also limited. In order to assess the maximum RF power required, a satellite link budget must be produced. A link budget can be very complex, comprising a large number of intricate factors, but stated in simple terms the power received by the ground station (P_r) is given by

$$P_r = \frac{P_t G_t G_r \lambda^2}{(4\pi R)^2}$$

Here P_t, G_t, G_r and λ denote the transmitter power, transmitter antenna gain, receiver antenna gain and wavelength, respectively.

In order to make a link possible and provide a margin of safety, let us consider each of these factors in turn. The overall conversion efficiency of a transmitter is typically of the order of 20–40%, thus moderate changes in the output power make considerable demands on the satellite power supply. If possible, the transmitter output power should be kept as small as possible, ensuring power is available to supply the payload and remainder of the satellite bus systems.

The gain of a ground station receiving antenna can be increased without having a direct impact on the satellite. However, to increase the gain, the physical dimensions have to increase. This often incurs a significant increase in weight and cost. Matters are further complicated by the more stringent requirements for housing and driving such an antenna. Higher gain antennas have narrower beamwidths and in order to track fast-moving LEO satellites complex and expensive pointing mechanism are required. If feasible, the ground station antennas for low-cost, small-satellite programmes should be kept as small as possible.

Thus system engineers often look towards the gain of the satellite antennas to make possible a cost-effective link budget. In this case an antenna with a high-gain narrow beam was used in conjunction with a pointing mechanism which kept the centre of the beam focused on the target ground station while the satellite passed overhead. The overall flight system had a considerable number of requirements and constraints, some of which are listed in Table 15.2.

An earlier legacy 20 Mbps baseline downlink system operated with a fixed satellite antenna that had a shaped gain of 3.2 dBi when it viewed the ground station at an angle of 62° from boresight. However, changes

Table 15.2 *Requirements and constraints of antenna design.*

Electrical	Mechanical and environmental	Others
Frequency of operation	Choice of materials	Cost
Polarization	Temperature range	Mass
Gain	Vibration environment	Volume
Gain variation with frequency and temperature	Venting requirements	Power handling
Beamwidth	Atomic oxygen effects	Interfaces
Axial ratio	Surface charging considerations	Mission lifetime
Return loss and return loss bandwidth	Vacuum	Antenna slew rate
Angle of view over which antenna performance had to be maintained	Radiation environment	
Efficiency		

to camera-related data transfer requirements dictated an increased data rate of 200 Mbps. It was desired to keep the heritage 5.3 m ground station dish antennas so an additional 7 dB of link budget gain was required from the new antenna and pointing mechanism. Thus the overall gain needed was at least 10.2 dBi. Table 15.3 shows the SSTL antenna design requirement.

15.4.1.2 SSTL's Horn Antenna

The antenna chosen to meet the derived requirements was a horn. The horn has the advantage that for medium-gain requirements the antenna size is kept small. This is not true for larger gains where parabolic reflectors are a better solution. Calculations of antenna size for the required gain were found to be favourable and the required technology mature. Initial investigations were made to see if off-the-shelf options were viable. However, these showed that all options were either too expensive or physically too large. It was then decided to design a custom module that would integrate a number of features including a coaxial-to-waveguide transition, septum polarizer and horn (see Figure 15.29).

The feed to the antenna was via a coaxial cable terminated with an SMA connector. This in turn fed a launch pin in the waveguide which produced a linearly polarized guided wave. A septum polarizer was used to convert this into a circularly polarized signal. The flared horn itself converted the guided wave into one that could be effectively radiated.

The final design incorporated both left- and right-hand feed sections such that the resulting horn could provide either single or dual polarizations. It was provided with a protective radome which prevented particulate ingress and, due to careful design, helped improve the axial ratio of the antenna.

Table 15.3 *SSTL's antenna design requirement.*

Requirement	Derived requirement	Value achieved
Beamwidth	25°–60°	26°–28°
Gain over 20°	10.2 dBi	15 dBi
Axial ratio	<6 dB	<2 dB over 26°
Mass	<300 g	<150 g
Power handling	>6 W	Operated at 6 W in a partial vacuum
Return loss	>15 dB over 8–8.4 GHz	Achieved
Efficiency	>55%	>85% (worst case)

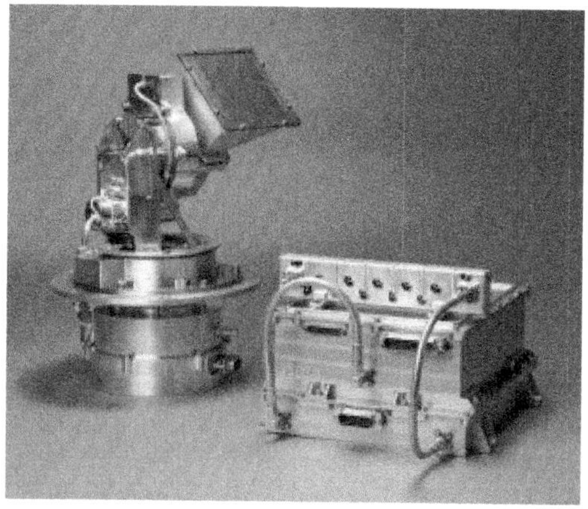

Figure 15.29 A complete tracking X-band transmitter. Courtesy of Surrey Satellite Technology Ltd, UK [4].

The antenna concept was first modelled in a numerical simulator using the MoM and small changes made to optimize the design. A simple prototype was built using double-sided copper-clad PCB material, cut out using a router and hand soldered together. Despite the non-ideal construction method, it was sufficient to prove the concept and it performed well. Once confidence was gained, an engineering model antenna was cut, milled and spark eroded from a block of solid aluminium. In order to optimize the antenna return loss the feeder section was made variable. The SMA connector and launch pin could be slid along a short slot cut into the waveguide section. When the optimum return loss point was discovered the connector was soldered into position and the remaining exposed slot bridged with conductive tape. (The entire engineering horn was nickel plated to allow soldering operations to be carried out.) Often feed points are optimized by making the rear wall of the waveguide adjustable, but in this case the presence of the septum would have made this more difficult. A final, non-adjustable, feed section was then created and integrated, which gave the final desired performance. This section was not nickel plated since a nickel/aluminium interface is known to cause passive intermodulation (PIM) products.

15.4.1.3 Beamwidth and Tracking

The SSTL antenna pointing mechanism (APM) was designed to be mounted on a small LEO satellite and track a small horn antenna at a ground station while maintaining a minimum antenna gain of 10.2 dBiC (see Figure 15.30). This had to be maintained over any specified orbit or off-track pointing activity that might have been required for pointing the Earth observation cameras. The APM was also designed to track the fast rate of change of satellite slew as the satellite passed close to a ground station. The analysis for the tracking requirements was complex but the antenna gain condition of more than 15 dBiC resulted in a 28° beamwidth which relaxed the pointing requirements of the tracking mechanism. The final APM design exceeded all requirements and was able to track at 19°/s and to accelerate at $4°/s^2$.

15.4.2 Case Study 2: X-band Downlink Helix Antenna

15.4.2.1 Background

Small satellites in a low orbit usually need a data downlink antenna. X-band is often used for these applications. For non-steerable satellites, an isoflux coverage antenna can be used to compensate for the change of

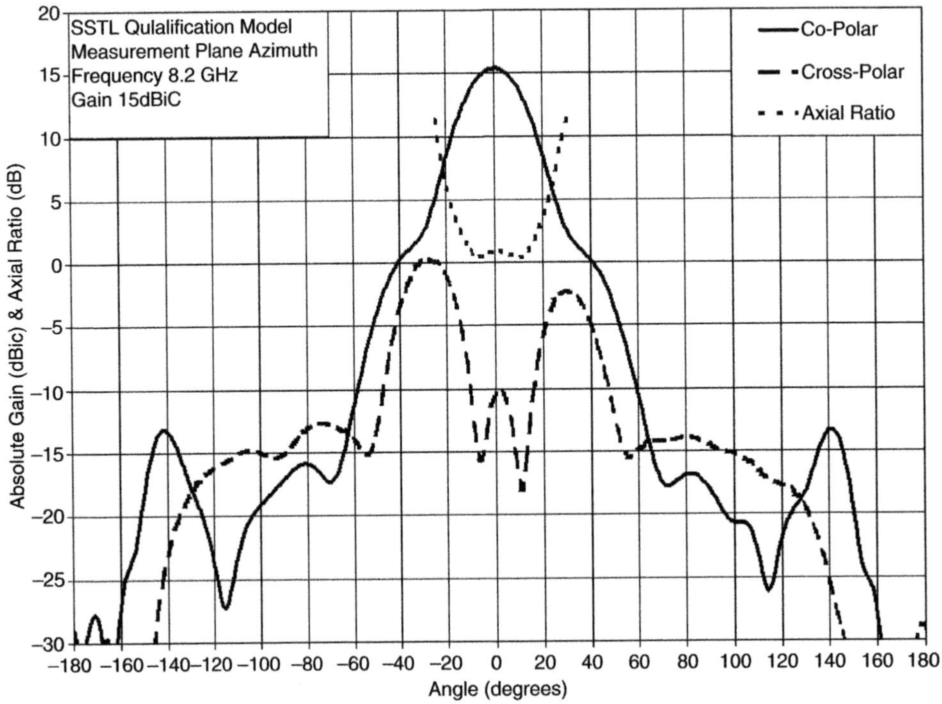

Figure 15.30 *Horn antenna characteristics.*

free-space loss for a user on the ground to see equal power density regardless of whether the satellite is seen just above the horizon or in the zenith position. A suitable antenna radiation pattern will therefore need to have a lower level in the nadir direction than towards the edge of the Earth in order to compensate for the distance difference. Edge of coverage angles between ±60° and ±70° are thus normally to be achieved for such antennas.

The major requirements set up for the antenna design were to have an edge of coverage gain above 4.5 dBi for an edge of coverage ±60° cone variant and above 3 dBi for an edge of coverage ±70° cone variant. Other pertinent requirements were minimized mass of less than 400 g, and size limited to a diameter of less than 100 mm and a height of less than 300 mm.

Concepts considered for this application were small or large circular waveguide apertures with corrugations, bicones, helices, passive arrays and shaped-reflector-type antennas to achieve the needed beam shape. Several of these are normally relatively large and heavy, and thus not an option to be used on small satellites. Gimballed horns or reflectors with pencil beam types of patterns, as used on larger satellites, were also judged too heavy and large.

15.4.2.2 Review of Possible Antenna Concepts

During the design phase, aperture-type antennas (arrays, shaped-reflector antennas, bicones, lenses, corrugated radiators, etc.) as well as line-source-type antennas (helices, slotted waveguide arrays, etc.) to achieve the shaped radiation pattern were considered. There are essentially only two ways to achieve the wanted

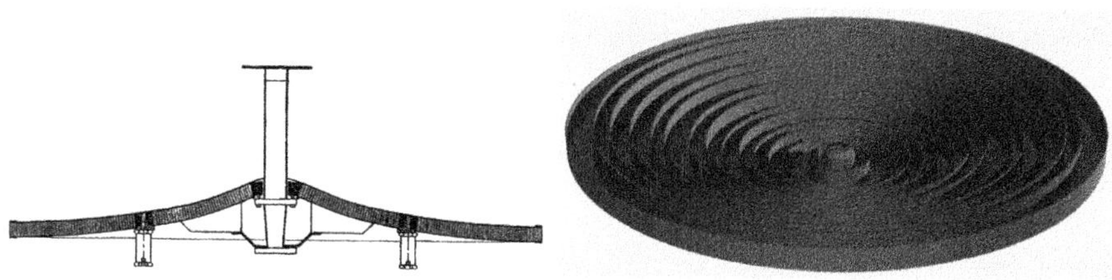

Figure 15.31 Shaped-reflector antenna and corrugated aperture radiator examples.

rotationally symmetric, circularly polarized radiation pattern: either an extended line source, or a relatively large-diameter aperture antenna.

Arrays will have drawbacks such as losses and complexity and were also judged costly for this application. Reflector antennas as well as other aperture-type antennas such as large corrugated radiators or bicones will have drawbacks based on their size and weight. See Figure 15.31 for some examples. Lens antennas will also have drawbacks based on size and weight, as well as the use of dielectric materials, which can pose a problem in the space environment considering charging effects.

For the aperture antennas, under the assumption of a circular aperture, only half of the lateral dimension can be used to generate the wanted radiation pattern due to the requirement of circular symmetry. For the partly focused beam to be achieved in the edge of coverage direction, there is an additional obliquity underutilization factor due to the projected length being smaller, see Figure 15.32. This means that effectively less than a quarter of the aperture diameter can be used to synthesize the shaped pattern required. Taking these effects into account would therefore demand a relatively large-diameter aperture to achieve the wanted radiation patterns. The obliquity effects also lead to polarization purity issues. Apart from this, part of the aperture has to be unfocused to achieve the shaped isoflux characteristic, and the edge illumination kept low to achieve low far-out sidelobes. Synthesis of patterns of this kind is notoriously difficult, and leads often to patterns with significant fast amplitude ripple.

A line source type of antenna is more efficient in that focusing in the edge of coverage direction will have only a small effect on the projected length, see Figure 15.33. Again, a substantial length is needed to achieve the desired pattern with a line source, even if it is about half that of the circular aperture case.

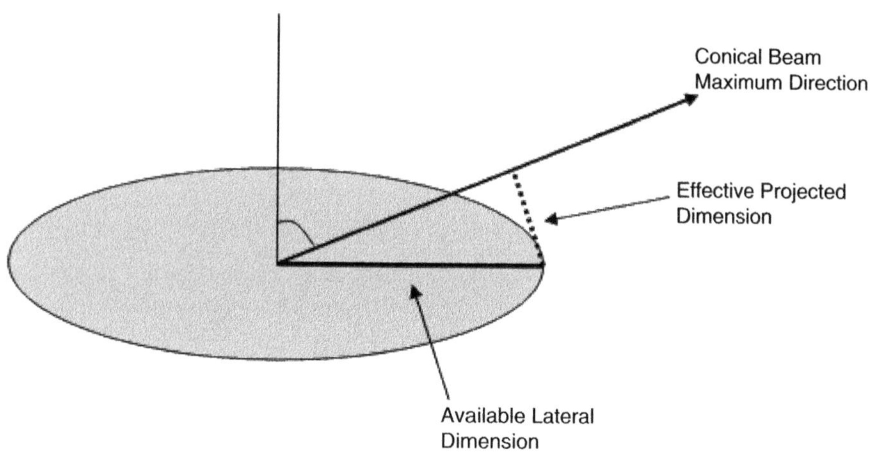

Figure 15.32 Aperture antenna case.

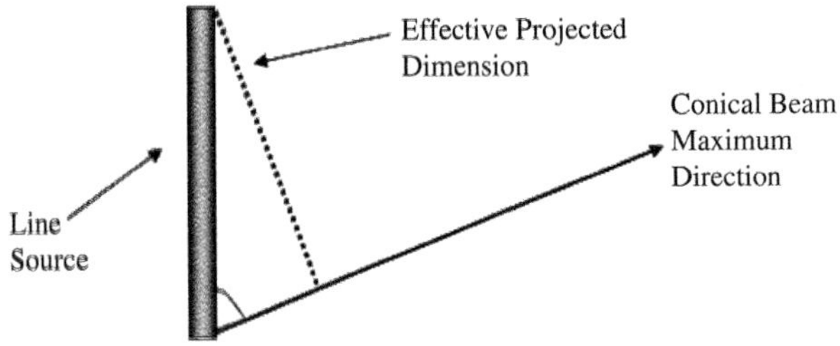

Figure 15.33 *Line source antenna case.*

Based on this, a QHA, being a good approximation to a line source, was likely to be one of the most efficient designs to achieve the required rotationally symmetric and circularly polarized patterns.

Considering the antenna types discussed, a selection was made, and the best antenna type was judged to be the QHA.

15.4.2.3 QHA Design

It was also decided that the QHA concept should be designed as a modular system with the following main parts:

- A helix radiator that can be adapted to the required coverage by changing the helix geometry.
- A polarizer, for all the required X-band frequencies, with a coaxial or waveguide interface.

Traditional helix antennas for space applications consist of helical wires or flat conductors, manufactured to the required shape, supported by a dielectric structure and connected to a conventional stripline-type feeding network. This traditional design was not selected for several reasons, such as that the feed network will have relatively high losses at X-band frequencies and losses can also occur in a dielectric support structure, leading to poor power handling capability and then also low gain for the antenna.

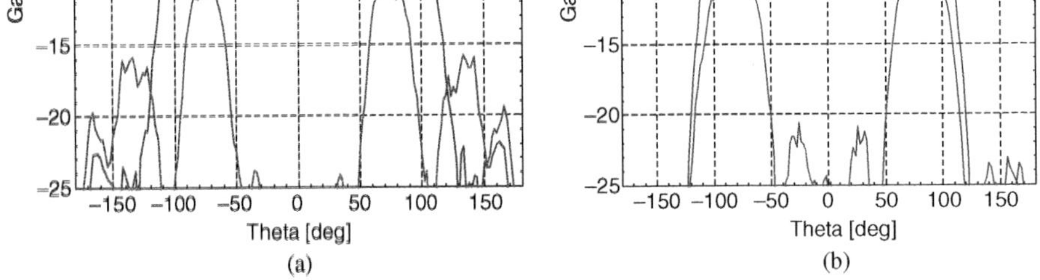

Figure 15.34 *Radiation patterns for 8.1 GHz and ±70° edge of coverage antenna (a) and broadband 8.2 GHz and ±70° edge of coverage antenna (b).*

Figure 15.35 X-Band helix antenna. Courtesy of RUAG Space AB [8].

In the selected design, the helical wires are fed from a waveguide structure instead of a conventional feed network. Via a transition section, the four helices are connected to a circular waveguide. The helix radiator design (diameter, pitch angle and number of turns) was optimized for the desired coverage using a genetic optimization algorithm. Analyses of the optimized helix radiator also gave good agreement with measurements.

The helix radiator can be fed by any waveguide component that generates circular polarization in a circular or square waveguide. For this case a septum polarizer was selected, that is a broadband polarizer covering the frequency range 7.1–8.5 GHz. The polarizer was designed with the option to allow for both a coaxial SMA interface and a waveguide interface within the same design by just adding or removing the SMA connector and a shorting plate.

Measured radiation patterns for a selection of designed antennas are shown in Figure 15.34. They are antennas with edge of coverage angles of $\pm 70^\circ$, and for X-band applications. A photo of the described antenna type is shown in Figure 15.35 [8].

The described antenna has only a few parts compared with traditional QHA designs. Other advantages of the design are good RF performance, low loss, relatively high power handling capability, small size and low mass.

15.5 Conclusions

This chapter presents an overview of the development of various antennas for small-satellite applications. An introduction to small satellites is given at the beginning and different types of small satellites are introduced. Challenges for designing antennas for small satellites are discussed. The chapter then discusses many small-satellite antennas for TT&C, data downlink, GNSS and intersatellite links such as patch antennas, the PEC antenna, QHA, horn antennas, and so on. Finally, two real-world case studies which illustrate the practical design and development of antennas for small satellites are provided. The future development of antennas for small satellites will move towards higher frequencies, electronic beam steering and multiple functions as well as smaller size [17].

References

1. Sweeting, M. (2006) The 'Personal Computer' revolution in space. 20th Annual AIAA/USU Small Satellite Conference, Logan, UT, USA, Paper no. SSC06-I-4, pp. 1–11.
2. Sandau, R. (2010) Status and trends of small satellite missions for Earth observation. *Acta Astronautics*, **66**, 1–12.
3. Gao, S., Clark, K., Unwin, M. *et al.* (2009) Antennas for modern small satellites. *IEEE Antennas and Propagation Magazine*, **51**(4), 40–56.
4. SSTL, http://www.sstl.co.uk/ (accessed 14 December 2011).
5. Suari, J.P., Turner, C. and Ahlgren, W. (2001) Development of the standard Cube-Sat deployer and a Cube-Sat class pico-satellite. Proceedings of the IEEE Aerospace Conference, Big Sky, MT, USA, vol. 1, pp. 347–353.
6. Murphey, T.W., Jeon, S., Biskner, A. and Sanford, G. (2010) Deployable booms and antennas using bi-stable tape springs. 24th Annual AIAA/USU Conference on Small Satellites, Logan, UT, USA, Paper no. SSC19-X-6.
7. Freeman, M.T. (1993) Spacecraft on-orbit deployment anomalies: what can be done? *IEEE Aerospace and Electronic Systems Magazine*, **8**(4), 3–15.
8. RUAG, http://www.ruag.com/Space/Space_Home (accessed 14 December 2011).
9. Barrett, R., Taylor, R., Keller, P. *et al.* (2007) Deployable reflectors for small satellites. 21st Annual AIAA/USU Conference on Small Satellites, Logan, UT, USA, Paper no. SSC07-Xill-4.
10. Boccia, L., Amendola, G., Di Massa, G. and Giulicchi, L. (2001) Shorted annular patch antennas for multipath rejection in GPS-based attitude determination systems. *Microwave and Optical Technology Letters*, **28**(1), 47–51.
11. Boccia, L., Amendola, G. and Di Massa, G. (2004) A dual frequency microstrip patch antenna for high-precision GPS applications. *IEEE Antennas and Wireless Propagation Letters*, **3**, 157–160.
12. Unwin, M., Private communications.
13. Moazam, M., Bandari, B., Gao, S. *et al.* (2010) Development of dual-band circularly polarized antennas and arrays for space-borne satellite remote sensing. 2010 ESA Antenna Workshop for Space Applications, ESTEC, Noordwijk, The Netherlands.
14. Mizuno, J., Roque, J.D., Murakami, B. *et al.* (2005) Antennas for distributed nanosatellite networks. Proceedings of the IEEE/ACES International Conference on Wireless Communications and Applied Computational Electromagnetics, Honolulu, HI, USA, pp. 606–609.
15. Mattioni, L., Bandinelli, M., Milani, F. *et al.* (2010) Distributed multi-function antennas for micro- and nano-satellites. 32th ESA Space Antenna Workshop, ESTEC, Noordwijk, The Netherlands.
16. Turpin, T., Mahmoud, M., Baktur, R. and Furse, C. (2009) Integrated after-market solar panel antennas for small satellites. 23rd AIAA/USU Small Satellites Conference, Logan, UT, USA, Paper no. SSC09-XI-1, pp. 1–4.
17. Barnhart, D., Vladimirova, T. and Sweeting, M.N. (2007) Very small satellite design for distributed space missions. *Journal of Spacecraft and Rockets*, **44**(6), 1294–1299.

16

Space Antennas for Radio Astronomy

Paul F. Goldsmith

Jet Propulsion Laboratory, California Institute of Technology, USA

16.1 Introduction

This chapter discusses antennas used for radio astronomy observations carried out from spacecraft. The considerations for space radio astronomy antennas differ somewhat from those applied to satellite communications and remote sensing of the Earth. Radio astronomy observations from space cover an enormous range in frequency from below 1 MHz to over 1000 GHz. The upper frequency limit is certainly arbitrary as the terminology as well as the technology change gradually as one moves from the submillimeter to the far infrared. The critical antenna parameters also vary considerably depending on the type of observations to be carried out. This necessarily incomplete discussion includes four areas in which space observations have played a major role: cosmic microwave background observations, submillimeter/far-infrared astronomy, low-frequency radio astronomy, and space very long baseline interferometry. For each area, we discuss key aspects of antenna performance and give a summary of the space missions with some detailed information about the antennas that have been employed.

16.2 Overview of Radio Astronomy and the Role of Space Antennas

Radio astronomy deals with the collection and analysis of electromagnetic signals from celestial sources in what is quite imprecisely defined as the radio wavelength or frequency range. The first observations of radiation from the center of the Milky Way by Karl Jansky in the early 1930s were carried out at a frequency of 20.6 MHz. Subsequent observations by Grote Reber were at a frequency of 160 MHz, and he soon moved up to 480 MHz. Technological improvements driven by the development of radar during World War II soon allowed observations of the Sun at centimeter wavelengths, and the first spectral line, the 21 cm hyperfine transition of

Space Antenna Handbook, First Edition. Edited by William A. Imbriale, Steven (Shichang) Gao and Luigi Boccia.

atomic hydrogen at 1420 MHz, was detected in 1951. By the mid 1960s observations of Solar System and extragalactic objects were being carried out at wavelengths of 3.4 mm and even 1 mm.

Astronomical observations from the Earth's surface at centimeter wavelengths are generally possible, but pressure-broadened lines of molecular oxygen and water vapor in the atmosphere result in variable absorption and emission. These conditions are acceptable for typical continuum and spectral line observations, but for observation of very weak signals, and especially of sources having large angular extent, the atmosphere is a real problem. Most notable of sources in this category is the relict radiation from the 'Big Bang' explosion that started the Universe on its expansion some 13.7 thousand million years ago. Following the first detection in 1965, increasingly detailed studies of this cosmic microwave background (CMB) have revealed more and more details about the structure of the very early Universe. To achieve the required sensitivity and limits on pickup of stray radiation, astronomers gradually shifted to using balloon-borne telescopes and satellites with specially designed antenna systems, as described in Section 16.3.

Observations of the lowest rotational transition of carbon monoxide (CO) at a wavelength of 2.6 mm (frequency 115.267 GHz) were first reported in 1970 [1]. This transition of this very abundant molecular species has proven uniquely valuable as a tracer of gas in the Milky Way and other galaxies. The higher rotational transitions are also important probes of warmer regions. With the development of telescopes at high, dry sites, observations at even shorter wavelengths are possible in good weather, but 0.35 mm marks the effective 'short-wavelength limit' for ground-based astronomical observations. At shorter wavelengths, the only possibility is to rise above the atmosphere. This can be achieved to some degree with a telescope in an aircraft or balloon. Airborne platforms including the Kuiper Airborne Observatory (KAO), and now the Stratospheric Observatory for Infrared Astronomy (SOFIA), have made major contributions. But for unhindered access to the submillimeter/far-infrared wavelength range, a spacecraft is necessary, and the systems that have been developed are described in Section 16.4.

At the same time, astronomers were interested in probing longer wavelengths than those used by Jansky. Somewhat lower frequencies could be used under the right conditions, but the Earth's atmosphere again poses a serious obstacle. In this case, it is the ionized component that imposes a cutoff frequency below which radiation is reflected or absorbed. The cutoff frequency varies with time of day and location, but at frequencies below a few megahertz, astronomical observations from the Earth's surface are generally impossible owing to the loss in the ionosphere combined with the strong nose-like signals generated there. These challenges drove astronomers to launch rockets and satellites to carry low-frequency radiometers well above the bulk of the ionosphere. Starting in the early 1960s a variety of increasingly capable satellites were launched to probe from as low as the kilohertz frequency range up to several megahertz. In Section 16.5, we describe some of the antennas and satellites that have been used in this research area.

Radio astronomy has traditionally suffered in terms of angular resolution compared to shorter wavelengths simply because of the diffraction limit for an antenna of diameter D, $\Delta\Theta \simeq \lambda/D$. While single-dish radio telescopes can be made larger than infrared or optical telescopes (with the latter now often being diffraction limited owing to the advent of adaptive optics), the disparity in wavelength still leaves single radio telescopes with angular resolution measured in minutes of arc, although large millimeter and submillimeter telescopes have angular resolutions as high as 10 seconds of arc. Even this is not fine enough to probe many interesting regions, both nearby (such as the disk around a newly formed star) and distant (such as the center of an active galaxy containing a black hole). To overcome this limitation, radio astronomers developed interferometry – the technique of combining signals from two or more antennas, and making use of the phase difference between the signals received by each. In this manner, the effective size of an interferometer can be made as large as the maximum separation of the constituent antennas with a corresponding angular resolution as great as $\Delta\Theta \simeq \lambda/B$, where B is the baseline. This technique gradually evolved from antennas directly connected together to ones spread around a continent or even on several continents, with the signals being recorded (along with a phase reference derived from a stable maser frequency standard) and later brought together and correlated. Thus, very long baseline interferometry (VLBI) evolved to include baselines almost as long as the Earth's diameter. To get even higher resolution requires having at least one antenna beyond the Earth's surface; this is exactly what has been done with space VLBI, and we discuss the missions and antennas used for this research area in Section 16.6.

16.3 Space Antennas for Cosmic Microwave Background Studies

16.3.1 The Microwave Background

The cosmic microwave background (CMB) is the relict radiation produced shortly after the Big Bang, which occurred about 13.7 thousand million years ago and started our Universe on its currently observed expansion. Immediately following the Big Bang, the mean kinetic energy per particle as well as the energy per photon were so large that any atom that might form would immediately be ionized. However, as the expansion continued the average energy per particle diminished, and at a certain time it became possible for atoms to survive. Immediately prior to this transition, the photons were repeatedly absorbed and reemitted and thus acquired the characteristics of a blackbody of about 3000 K peaking in the ultraviolet region of the spectrum. After this transition, a photon would no longer be likely to be absorbed, but instead could scatter and then travel through a largely transparent Universe. However, the Universe continued to expand and, as described by Einstein's theory of general relativity, the wavelengths of all photons increased by a scale factor approximately equal to the redshift of the Universe at the time of this decoupling of photons and matter, which occurred at $z \approx 1000$. Thus, those photons that last interacted with matter about 380 000 years after the Big Bang are now detected as microwave photons. This cosmic microwave background (generally called the CMB) is to first order independent of direction and has a spectrum characteristic of a blackbody at a temperature of approximately 3 K.

Critical characteristics of the Universe were imprinted on the photons that have been traveling since that time. The frequency spectrum averaged over direction gives basic information about the Big Bang starting from the fact that it is, to very high precision, a blackbody. However, the transformation from ionized to neutral gas did not take place simultaneously everywhere in the Universe: density perturbations resulted in variations in the intensity of the photon distribution. These small departures from a uniform (isotropic) radiation field are generally referred to as CMB anisotropies. From the characteristic present-day angular spectrum of these CMB anisotropies (with some complementary information) we can deduce the total density of matter in the Universe. The matter in question includes not only the familiar baryonic matter (protons, neutrons, electrons, and so forth), but also dark matter and so-called dark energy. Variations in the density and temperature at the time and place when photons last interacted with matter produced linear polarization of the photon field. To extract information on the physics of the very early Universe, extremely small fractional polarizations must be measured, and along with determining the polarization itself, the variation of the polarized signal as a function of direction is critically important because different mechanisms that produce polarization result in different characteristic distributions of the polarization vectors.

The CMB was first observed and identified by Penzias and Wilson in 1965 [2]. Since then, there have been increasingly sensitive observations that have used ground-based instruments, systems on aircraft and balloons, and spacecraft. Technology for radiometers throughout the microwave and millimeter wavelength range has improved enormously. This has first allowed extremely accurate confirmation of the basic blackbody nature of the CMB spectrum and precise measurement of its temperature as 2.725 ± 0.001 K [3]. The improved sensitivity has also allowed detection of variations in the temperature of the CMB to a level measured in microkelvins. Extremely high sensitivity as well as low systematics are required for measurement of polarization of the CMB. Studies of the CMB have progressed from centimeter wavelengths to the ability to encompass the blackbody peak, requiring observations to wavelengths as short at 1 mm. Especially at the shorter wavelengths, the atmosphere becomes a significant problem, and for looking for angular variations, ground pickup is a serious limitation. Atmospheric emission and absorption are eliminated, and ground pickup is drastically reduced, by going into space. Many of the most demanding measurements have been made from the COBE, WMAP, and Planck spacecraft, which are discussed individually below. However, there has in addition been a steady stream of other projects that have had a positive synergistic interaction with space missions in terms of proving new equipment and observing techniques, as well as allowing physically larger antennas, particularly interferometric systems.

The CMB is an extended source, and variations of the CMB are limited by fundamental physics to angular scales larger than approximately 1/1000 rad. Consequently, electrically large antennas are not required to extract all of the information it contains. At the 190 GHz peak of the CMB intensity, $\lambda = 1.6$ mm, an antenna of about 1 m diameter is sufficient. What is critical is a clean and very well-understood antenna power pattern. Far sidelobes can be particularly harmful as they can pick up radiation from sources unrelated to the CMB (e.g., from the Milky Way or other bright galaxies) which can be very detrimental to accurate measurements. Polarization measurements require good polarization isolation as well as very good knowledge of the polarized beam shapes. Early measurements did not emphasize high angular resolution and so employed feed horns chosen for low sidelobe level. In systems requiring higher angular resolution, these considerations have led almost universally to the use of off-axis (unblocked) antennas for CMB measurements. And along with this, strong edge tapers to minimize edge diffraction have been employed. Given that antenna gain and angular resolution are not that important, the large edge taper does not represent a significant tradeoff, and is an interesting opposite extreme to considerations that have driven the design of systems for space VLBI, discussed in Section 16.6.

Even with efforts to achieve the highest beam quality and polarization purity, elaborate techniques for scanning the sky (CMB), removing the confusing signals from foreground objects, and dealing with the inevitable imperfect beam patterns and residual polarization issues must be employed to extract the CMB anisotropy and polarization signals. For many experiments, systematic errors rather than statistical uncertainties are the limiting factor on final quality of the data. The feed horns and antennas used for CMB experiments are among the most carefully characterized of any that have been used in astronomy, as is indicated in the discussions of the individual space missions that follow.

16.3.2 Soviet Space Observations of the CMB

Following the detection of the CMB in 1965, measurements were extended to other frequencies to characterize the blackbody spectrum, and studies of angular variations were started using ground-based and airborne systems. The earliest spacecraft devoted to observations of the CMB was the Relikt experiment aboard the Soviet Prognoz 9 satellite [4]. This satellite operated for an eight-month period in 1983–1984. A switched (Dicke) radiometer compared the signal from a feed horn aligned with the spacecraft rotation axis to that from a second horn pointing normal to the rotation axis. The radiometer operated at a frequency of 37 GHz and had RMS output fluctuations of 25 mK in an integration time of 1 s. The feed horns had 3 dB beam widths of 5.5°, and the contribution of the sidelobes was more than 70 dB below that of the main lobe.

An important early result of the Relikt experiment was to measure the dipole anisotropy in the CMB temperature. The effect of the motion of the Sun (and Earth) relative to the CMB had been detected in the early 1970s, but remained a challenging experiment (see [5] for a review of measurements of the CMB dipole). The results reported by Strukov *et al.* [4] gave a dipole amplitude and direction consistent with previous measurements, but having the smallest uncertainties reported until that time.

Subsequent analysis of Relikt data gave conflicting results in terms of detection of a quadrupole term in the CMB radiation, as well as possible detection of an anomalous signal coming from a specific direction [6–8].

The collapse of the Soviet Union apparently precluded the launch of a more capable Relikt-2 mission, which had been planned for launch in mid 1993. By this time, the results from the COBE satellite (launched in 1989) were known to be of exceptionally high quality and went far beyond measurement of the low-order CMB anisotropies, as discussed in Section 16.3.3.

16.3.3 The Cosmic Background Explorer (COBE) Satellite

The Cosmic Background Explorer, known as COBE, was the first NASA mission devoted primarily to the study of the CMB radiation. After a long gestation period and reconfiguration following the disaster with the Space Shuttle *Challenger* (which was to have taken it to orbit), it was finally launched on November 18, 1989. By that point, it was recognized that advancing our understanding of the CMB would require exceptionally well-calibrated measurements. A great deal of the design of COBE thus concentrated on this challenge, and in particular on taking advantage of the space environment.

COBE was placed into a nearly circular orbit of 900.2 km altitude having an inclination of 99.3°. The orbital period was 103 minutes, and the spacecraft rotated at a nominal rate of 0.8 rpm. Boggess *et al.* [9] give an overview of the COBE mission, its instruments, and its performance after two years of operation on orbit.

COBE included three instruments to study different aspects of the CMB, called DIRBE, FIRAS, and DMR.

DIRBE (Diffuse Infrared Background Experiment) was a multiband absolute photometer with 10 bands covering 1.25–240 μm (band center wavelengths) with $\lambda/\Delta\lambda = 1-10$. DIRBE utilized a well-baffled off-axis Gregorian telescope having a 19 cm diameter primary. The very clean optical design was essential to achieve the desired calibration accuracy and well-defined 0.7° field of view. A variety of incoherent detectors were cooled to cryogenic temperatures as low as 1.55 K for the bolometers, which were used at wavelengths of 140 and 240 μm.

The boresight direction of DIRBE was oriented at 30° from the spin axis of the spacecraft. As the spacecraft rotated, a 32 Hz chopper switched the DIRBE detectors from looking at the sky to looking at a cooled internal calibration load. The very good rejection of stray radiation allowed a map of the absolute flux of radiation from the entire sky to be made in a six-month period.

FIRAS (Far InfraRed Absolute Spectrophotometer) was a polarizing Michelson interferometer with two independent spectral channels [10]. The long-wavelength channel ($\lambda = 0.5$–10 mm) was designed to compare the flux from the CMB to that from an internal calibrator. The short-wavelength channel ($\lambda = 0.1$–0.5 mm) was designed to measure the emission from interstellar dust grains in order to remove this confusing signal from the measurements of the CMB. The FIRAS instrument was coupled to the sky through a carefully designed compound parabolic concentrator (Winston cone) described by Mather, Total, and Hemmati [11]. This device produced a very clean beam with no discrete sidelobes to approximately 80 dB below the on-axis response. The highest spectral resolution of FIRAS was 6 GHz.

DMR (Differential Microwave Radiometer) comprised two independent radiometers at each of the three frequencies 31.5, 53, and 90 GHz [12]. These frequencies were chosen to separate the CMB from foreground emission from dust grains and electrons in the Milky Way. The input of each radiometer was switched symmetrically between two corrugated feed horns producing 7° FWHM beams separated by 60° on the sky. Each beam was oriented at 30° relative to the spin axis of the spacecraft. Thus, the rapid Dicke switching (effected by a ferrite switch) took the difference between two regions of the sky separated by 60°, and the two regions were interchanged every 37.5 s (half of the 75 s spacecraft spin period) to remove offsets in the radiometer system.

The 31.5 GHz radiometers shared a single pair of feed horns with two orthogonal circular polarizations separated by an orthomode transducer. The two higher frequency channels had separate feed horns for each sense of linear polarization. Information on the performance of these circular corrugated feed horns can be

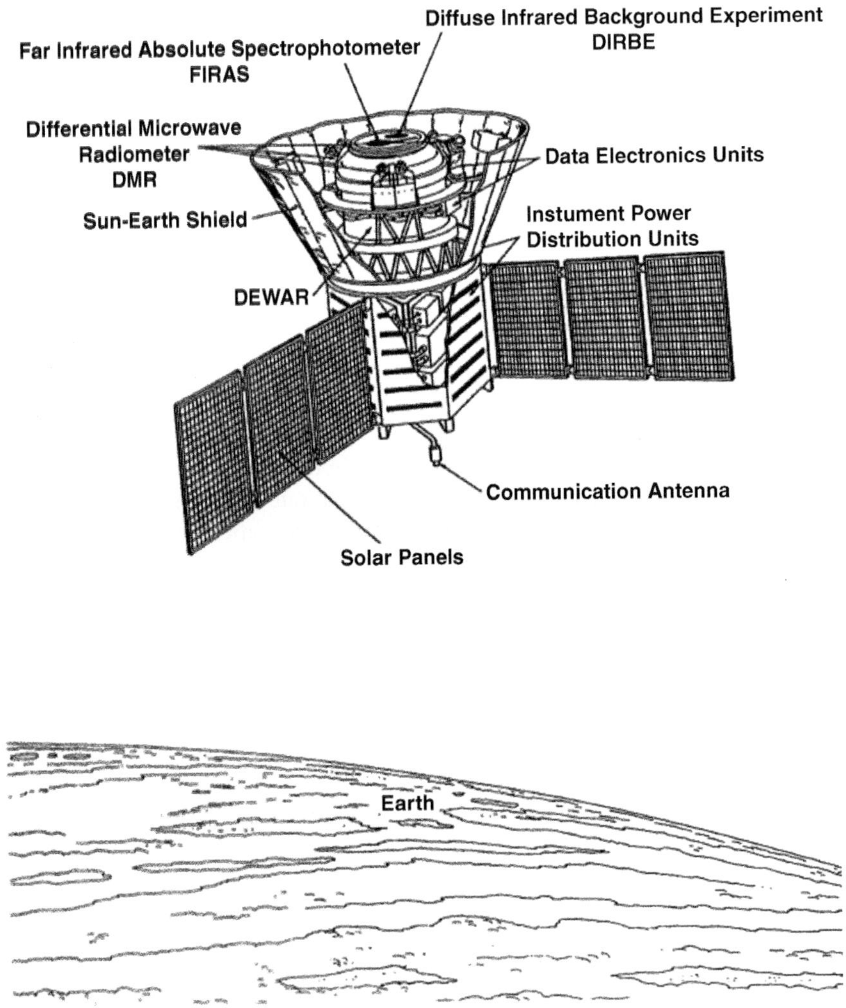

Figure 16.1 *Cutaway drawing of COBE satellite showing the sun- and earth-shield, the three scientific instruments, and other subsystems. From Boggess et al. [9]. Reproduced by permission of the AAS.*

found in Janssen *et al.* [13] and Torall *et al.* [14]. The 31.5 GHz system was operated near 270 K while the higher frequency channels were passively cooled to 140 K.

One noteworthy feature of COBE was the use of a sun- and earth-shield. This was designed to further reduce pickup from those sources below that anticipated from the very clean radiation patterns of the telescope and feed horns used in the instruments described above. The shield and COBE instruments are shown in Figure 16.1. The spacecraft's orbit was designed so that the orbital plane precessed 360° in one year, owing to the gravitational quadrupole moment of the Earth. The orbit and its inclination resulted in the spacecraft spin axis being maintained at approximately 94° to the Sun and almost at the local zenith. The Sun and the Earth could thus be kept below the aperture plane of the shield, maximizing its effectiveness.

COBE was an enormously successful mission. Among its highlights were measurements of the thermal nature of the CMB to unprecedented precision: the CMB has a blackbody spectrum with a temperature of 2.725 ± 0.001 K [3]. COBE confirmed the dipole anisotropy, but made the first measurements of the intrinsic anisotropy of the CMB. For these discoveries, John Mather and George Smoot shared the Nobel Prize in Physics in 2006. The unwanted (in terms of studying the CMB) signals from the Milky Way also proved very rewarding and led to much improved knowledge of key millimeter and submillimeter coolants of the interstellar medium [15,16] as well as the distribution and properties of interstellar dust grains.

16.3.4 The Wilkinson Microwave Anisotropy Probe (WMAP)

In the decade following COBE, study of the CMB intensified, with a wide range of ground-based and suborbital (balloon) instruments revealing increasing detail about this unique probe of the very early Universe. A good many of these are enumerated in the introduction to the paper by Page *et al.* [17]. Cosmologists wanted a finer scale view of the angular variations of the CMB discovered by COBE, which would require a real antenna rather than the feed horns alone employed by the DMR on COBE. In addition, a new source of information about conditions in the Universe immediately after the Big Bang had been recognized. This is polarization of the CMB radiation, which can be produced by variations in the density of material in the expanding Universe before it decouples from the background photon field.

Of particular interest are the quantum fluctuations predicted by inflationary theories of the Universe which successfully explain many of the observed characteristics of the CMB. Measuring the angular spectrum of the fluctuations was a key goal of the Microwave Anisotropy Probe (this was the original name of this mission; it was renamed the Wilkinson Microwave Anisotropy Probe and known as WMAP in honor of David Wilkinson, Professor of Physics at Princeton University, a key member of the COBE and Microwave Anisotropy Probe teams who passed away on September 5, 2002). The importance of measuring CMB polarization was recognized only gradually during development of the WMAP satellite, which was launched on June 30, 2001.

The requirements of WMAP are described by Page *et al.* [17]. These include (1) beam size of less than 0.3° for five frequency bands between 20 and 100 GHz; (2) cross-polarization less than −20 dB; (3) sidelobes less than −55 dBi at the position of the Sun; (4) temperature difference measured as accurately for pixels separated by 0.25° as for those separated by 180°; (5) antenna optics cooled to 70 K or below; (6) systematic error in any mode of the final map of the sky before modeling to be less than 4 μK of the target sensitivity of 20 μK.

The results from the first year of WMAP observations were presented in 2003, and have been updated since as the sensitivity has improved owing to additional integration time. The optics and some of the front-end components are passively cooled, but there is no liquid or solid cryogen to limit mission lifetime. Currently, a number of papers describing the results from seven years of observations are available, which present the most sensitive all-sky measurements of the angular power spectrum and polarization of the CMB yet obtained. Some results from the first year of WMAP observations are given by Peiris *et al.* [18], and results from seven years of observations are presented by Jarosik *et al.* [19].

16.3.4.1 Optics and Antenna

The WMAP optical system, described in detail by Page *et al.* [17], employs a pair of back-to-back offset-shaped Gregorian telescopes 1.4 m in diameter. The baseline antenna design is an offset Gregorian system, but the surfaces were shaped to optimize the quality of the beams and the antenna efficiency. The work of Dragone on minimizing the cross-polarization in offset antennas was used as a guideline for the WMAP design, but implementation was based on Galindo-Israel, Imbriale, and Mittra [20], with special effort to maintain polarization purity and low sidelobe levels. Details of the process can be found in the discussion and references given in Page *et al.* [17]. In addition to careful design of antenna and feed horns, shields were used to minimize pickup from the Sun and Moon. Figure 16.2 shows the shields around the feed horns and secondary reflector, as well as the larger, planar shield at the bottom, which is oriented essentially perpendicular to the direction to the Sun.

The antenna system is designed to be well shielded from the Sun and Earth. In consequence, the thermal radiators allow the primary and secondary reflectors, and the structure supporting them, to cool to 70 K once on

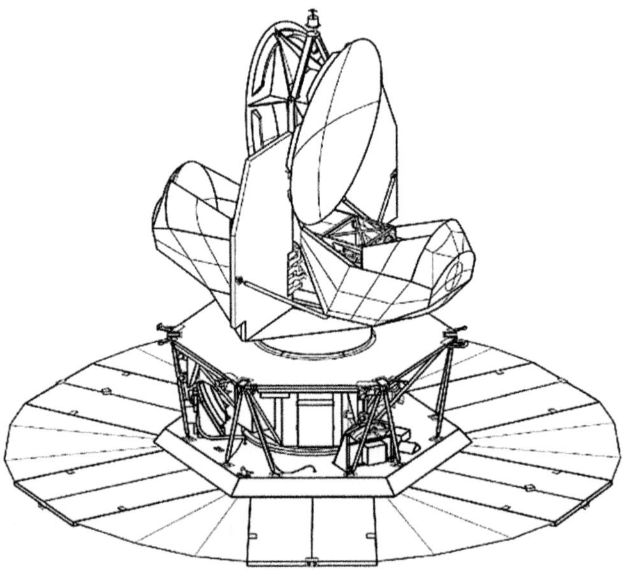

Figure 16.2 Line drawing of WMAP satellite showing the two oppositely oriented Gregorian antennas with their shielding, and below them the spacecraft bus and solar shield. From Barnes et al. [21]. Reproduced by permission of the AAS.

orbit. Together with the requirement of less than 1% emissivity, this guarantees that the noise added by the reflector system is negligible. The reflector surfaces are made of composite material facecloths over a DuPont KOREX honeycomb core 0.635 cm thick. Each primary reflector has a mass of 5 kg, including its backing structure, and the mass of each secondary reflector is 1.54 kg. Table 16.1 gives selected predicted K-band and W-band measures of antenna performance. The relatively high edge tapers mean that there is very low spillover and high main beam efficiency, but the accompanying result is low aperture efficiencies. Information about the beam patterns measured on orbit can be found in Jarosik *et al.* [19].

The WMAP optical system (reflectors and support structure) was optimized to have negligible thermal distortion between 300 and 70 K. The surface RMS deviation at 70 K was specified to be below 0.0076 cm, which corresponds to approximately $\lambda/40$ at the shortest operating wavelength. The beam quality was significantly but not unexpectedly degraded upon cooling. With multiple high signal-to-noise measurements of Jupiter and physical-optics-based modeling of the reflector system, the beam profile could be determined to

Table 16.1 Predicted WMAP antenna main beam characteristics.

Frequency (GHz)	3 dB width 1 (deg)	3 dB width 2 (deg)	Primary edge taper (dB)	A_e (m^2)
20	0.969	0.798	−12.8	0.80
22	0.882	0.721	−13.1	0.76
25	0.787	0.637	−14.7	0.71
82	0.209	0.199	−17.4	0.94
90	0.201	0.190	−21.0	0.83
98	0.198	0.184	−24.8	0.74
100	0.194	0.181	−26.5	0.65

For K-band, beamwidth 1 is vertical cut and beamwidth 2 is horizontal cut. For W-band, beamwidth 1 is +45° cut and beamwidth 2 is −45° cut. The last column is the effective area of the antenna.

Table 16.2 Characteristics of WMAP feed horns.

Band	Freq. range (GHz)	Center freq. (GHz)	Θ_{FWHM}^{feed} (deg)	Feed aperture dia. (cm)	Feed horns per antenna
K	20–25	23	8.8	10.94	1
Ka	28–36	33	8.3	8.99	1
Q	35–46	41	7.0	8.99	2
V	53–69	61	8.0	5.99	2
W	82–106	93	8.4	3.99	4

roughly the −50 dB level. Cooling did affect the beam pointing, with shifts up to about 0.1°, but these shifts could be measured by using celestial sources while on orbit.

Each telescope is fed by 10 feed horns. Table 16.2 gives the allocation of feed horns to different bands along with other information. The number of feeds is greater at higher frequencies to compensate for the increase in amplifier noise temperature with frequency, with the goal of equal sensitivity per unit solid angle to radiation coming from an extended celestial source. Each feed horn is connected to an orthomode transducer (OMT) which separates two orthogonal linear polarizations.

In practice, the output of one polarization from a feed horn feeding one antenna is compared to the output in the same polarization state of a similar feed horn illuminated by the second antenna, in a differential radiometer as described by Jarosik *et al.* [22]. The result is a comparison of the temperature of two beam footprints separated on the sky by 180°. As the spacecraft spins, differences along an arc on the sky are measured, and as the orbit precesses, a map of the entire sky is produced. Since, as mentioned above, the critical data desired from WMAP is the differential temperature and polarization maps, the map of 180° temperature differences is relatively easily transformed into the desired map of variations in the CMB across the sky.

16.3.4.2 Feed Horns

The 10 feed horns illuminating each of the offset Gregorian antennas (Table 16.1) are clustered in the focal region, but because of the relatively large offset angles, the feed horn axes are not parallel to one another. This is illustrated in Figure 16.3 which shows the different feed horns. The smallest feed horns for W-band are located closest to the antenna axis and can be only marginally discerned in the figure. Barnes *et al.* [21] give a detailed discussion of the design of these corrugated feed horns and their measured performance.

16.3.5 The Planck Mission

Development of what became the Planck mission started not long after the first results from COBE became available in 1992. A lengthy maturation process including mergers of different concepts led to the launch of Planck on May 14, 2009, along with the Herschel satellite discussed in Section 16.4.4. Tauber *et al.* [23] give an extensive list of historical references and a guide to the 14 articles in *Astronomy and Astrophysics* volume 520 dealing with Planck. Since this great quantity of detailed information is readily available, we will only highlight here some of the special characteristics of the Planck antenna and optics.

The basic goal for Planck was to measure the intensity and polarization anisotropy of the CMB with exceptional precision. While the frequency range covered by this radiation is defined by its blackbody temperature, Planck included instrumentation to measure the CMB over the maximum frequency extent possible. In addition, an increased recognition of the importance of 'foregrounds,' which to cosmologists means anything between the surface of last scattering of the photons of the CMB and ourselves, led to increased emphasis on accurately measuring the emission from interstellar dust grains. This required

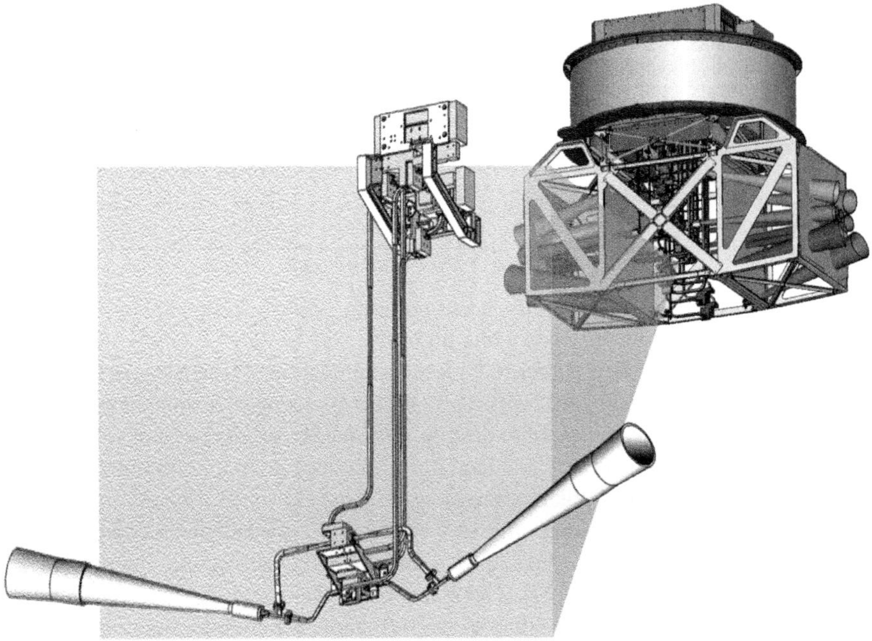

Figure 16.3 Right: schematic of WMAP feed horns showing the feeds for each of the two antennas, each having 10 feed horns. Left: enlarged view of single pair of feed horns, showing the orthomode transitions and waveguides for each polarization. The components located approximately between the feed horns are cooled to 90 K and include hybrid junctions and HEMT amplifiers. The long vertical runs of waveguides transmit the signals to room temperature components for further processing. From http://wmap.gsfc.nasa.gov/media/990180/. Courtesy of NASA/WMAP Science Team.

radiometers with band centers as high as 857 GHz. The result is a very complex focal plane, but one with exceptional capabilities over a 30:1 frequency range.

16.3.5.1 Optics

Following previous CMB experiments, Planck utilizes an unblocked, off-axis telescope design. The design was modified from a Gregorian paraboloid–ellipsoid combination obeying the Dragone–Mizuguchi condition that maximizes polarization purity, to an aplanatic pair of ellipsoidal reflectors [23]. The final surface shapes were determined by numerical optimization using the Code V computer program.

The primary reflector is 1.89 × 1.56 m and thus somewhat larger than the reflectors used in WMAP. The Planck reflectors are made of carbon fiber reinforced plastic (CFRP) honeycomb sandwich material, 80 mm thick for the primary and 65 mm thick for the secondary. Planck was designed for operation at the L2 Lagrangian point for thermal stability and good Earth avoidance. A set of three V-groove thermal shields prevents radiation from the Sun or the Earth from falling on the antenna. The operating temperature is 45 K. Figure 16.4 shows the Planck satellite.

Although the primary reflector is monolithic, the surface accuracy is defined in terms of a number of rings, with accuracy highest toward the center (7.5 μm RMS) and lower at the reflector edge (50 μm RMS). For reasons of sensitivity and sampling, it is not desirable for beam sizes at the shorter wavelengths to decrease as would occur for diffraction-limited operation with the entire surface illuminated. Thus, the feed horns are designed to illuminate a decreasing fraction of the reflector, for which the surface is essentially perfect. For

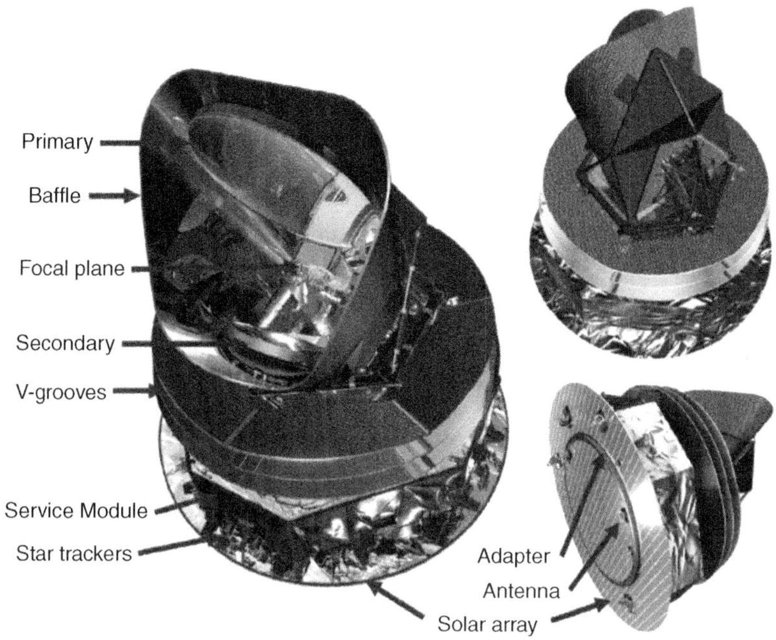

Figure 16.4 *Left: main elements of the Planck satellite. The three V-groove thermal shields are between the optics assembly (above) and the service module (below). Right: rotated views of the spacecraft showing the extensive baffling (above) and the solar array (below). The satellite observes by spinning about an axis passing through the center of the solar array, which is itself pointing directly at the Sun. From Tauber et al. 2010 [24]. Reproduced with permission © ESO.*

example, the highest frequency (857 GHz; 350 μm) band with 4.3′ FWHM beamwidth [25] utilizes essentially only the central ring, for which $\lambda/\sigma = 47$.

In addition to the low temperature, the surface reflectivity is tightly specified to exceed 0.985 (end of life) over the frequency range 25–1000 GHz, in order to limit the thermal emission from the telescope into the radiometers. As can be seen in Figure 16.4, there are additional baffles to prevent far-sidelobe pickup.

One thing is immediately evident from this description and Figure 16.4: unlike previous CMB missions, Planck data taking is *not* based on subtracting measurements from different directions on the sky. The designs of the instruments are described further in the following subsection, but basically each feed horn collects radiation from a given direction, and as the satellite rotates about its spin axis at 1 rpm (and as the spin axis precesses by 1° per day) a map of the intensity as a function of direction is built up.

16.3.5.2 Instruments and Feeds

There are two instruments in the Planck focal plane. The Low-Frequency Instrument (LFI) uses high electron mobility transistor (HEMT) amplifiers in three bands centered at 30, 44, and 70 GHz [26]. The High-Frequency Instrument (HFI) has nine channels centered at 100–857 GHz [25]. The Planck focal plane including all feed horns is shown in Figure 16.5.

Low-Frequency Instrument (LFI) Owing to the differing sensitivity requirements, there are two feed horns for 30 GHz, three for 44 GHz, and six for 70 GHz. Each is fitted with an orthomode transducer that separates

Figure 16.5 Planck focal plane. The 36 feed horns of the HFI instrument are in the central circular structure, which is cooled to 4 K. The 11 feed horns of the LFI instrument surround the HFI structure and are at a temperature of 20 K. From Tauber et al. 2010 [24]. Reproduced with permission © ESO.

two orthogonal linear polarizations. These have better than 40 dB isolation and 20 dB return loss, with less than 0.15 dB insertion loss over the bandwidths of 6, 8.8, and 14 GHz for the three bands, respectively. Each polarization channel is equipped with a correlation (or continuous comparison) radiometer, which uses pairs of hybrid junctions to produce an output which is ideally the difference between the input antenna temperature of the feed viewing the Planck antenna and that of an internal feed horn viewing a blackbody calibration load at 4 K. The effective switching time between the two horns is on the order of the reciprocal of the intermediate frequency (IF) bandwidth, and this very rapid switching rate contributes to the exceptional stability of the system. The polarization properties of the system are discussed in detail by Leahy *et al.* [27] and the antenna illumination by Sandri *et al.* [28].

High-Frequency Instrument (HFI) The HFI uses a very different approach, relying on bolometers, which directly measure the incident power traveling down the waveguide. In consequence, proper filtering at the input RF is essential to have a well-defined bandpass for the measurements. As discussed by Ade *et al.* [29] and Lamarre *et al.* [25], and described in more detail by the references therein, quite elaborate filtering systems employing back-to-back feed horns with quasi-optical filters are employed in preference to waveguide filters which become difficult to fabricate at the shorter wavelengths. The fractional bandwidth of each channel is 33%. The bands are well separated in frequency, and the isolation provided by the quasi-optical filters is as high as 10^{10}.

Of the 36 feed horns in the HFI, 16 feed pairs of polarization-sensitive bolometers, which observe identical patches of the sky in orthogonal linear polarizations. As indicated in Table 16.3, the polarization-sensitive channels are concentrated at the frequencies in which the CMB radiation is significant. These channels are designed primarily for measuring the difference in the input signal between the two polarizations. The higher frequency channels are primarily intended to measure accurately the emission from interstellar dust, which is one of the 'contaminants' of the CMB signal of interest. By measuring the dust's emission at several wavelengths, its spectral behavior can be sufficiently accurately modeled to allow its contribution at lower frequencies to be removed. Maffei *et al.* [30] present data on the HFI feed horn and resulting telescope beam patterns, while Rosset *et al.* [31] discuss polarization behavior and requirements to achieve Planck performance goals.

Table 16.3 Some characteristics of LFI channels; -P denotes polarization-sensitive bolometers and NEP is Noise Equivalent Power.

Channel	Center frequency (GHz)	FWHM beam size (′)	Number of bolometers	Bolometer NEP (10^{-18} W/Hz$^{0.5}$)
100-P	100	9.6	8	10.6
143-P	143	7.0	8	9.7
143	143	7.0	4	14.6
217-P	217	5.0	8	13.4
217	217	5.0	4	18.4
353-P	353	5.0	8	16.4
353	353	5.0	4	22.5
545	545	5.0	4	72.3
857	857	5.0	4	186

In the bolometers used in the HFI, the radiation is coupled to a network of very thin metallic conductors on a dielectric substrate. The current that results heats a resistor, whose temperature rise is sensed by a thermometer. Two different geometries are employed. For the measurement of polarization, a rectangular mesh is used, but with metallization along a single axis, which connects to the resistor located at the edge of the bolometer. For channels that measure total power (unpolarized) two-dimensional *spider web* bolometers are used, in which the metallic mesh geometry is similar to that of a spider web and conducts current to the resistor located at the center of the web. Either geometry offers the advantage that the resistor and thermometer can be much smaller than the area presented by the bolometer to the incident radiation, which results in better sensitivity and reduced susceptibility to cosmic rays [32,33].

16.3.5.3 Current Planck Status

The Planck satellite was successfully launched on May 14, 2009 into a transfer orbit, and eventually into a Lissajous orbit around the second Lagrangian (L2) point, on a line connecting the Sun and the Earth, but about 1.5 million km beyond the Earth. After cooling and testing, scanning the sky began. Following agreement made before launch, the first data to be released was the Early Release Compact Source Catalog, which included both extragalactic and galactic sources. Maps of the Milky Way (with the CMB removed) have already given striking new insight into the highly structured distribution of dust in the interstellar medium, as well as into the properties of the dust itself. The central cosmological data from Planck is scheduled to be released in early 2013.

16.4 Space Radio Observatories for Submillimeter/Far-Infrared Astronomy

16.4.1 Overview of Submillimeter/Far-Infrared Astronomy

The submillimeter region is generally taken to span the wavelength range from 1 mm to 100 μm, corresponding to frequencies between 300 and 3000 GHz. In this wavelength range, the biggest obstacle to astronomy is the Earth's atmosphere, whose opacity is dominated by a combination of individual water lines and the low-frequency tails of even stronger lines at shorter wavelengths. The result is that even from high, dry sites, observations can be carried out only in selected 'windows,' which themselves become opaque in less than optimal conditions. Figure 16.6 shows the transmission at zenith from a site having a precipitable water vapor content (PWV) equal to 0.5 mm, which represents very good conditions on sites such as Mauna Kea, and fairly

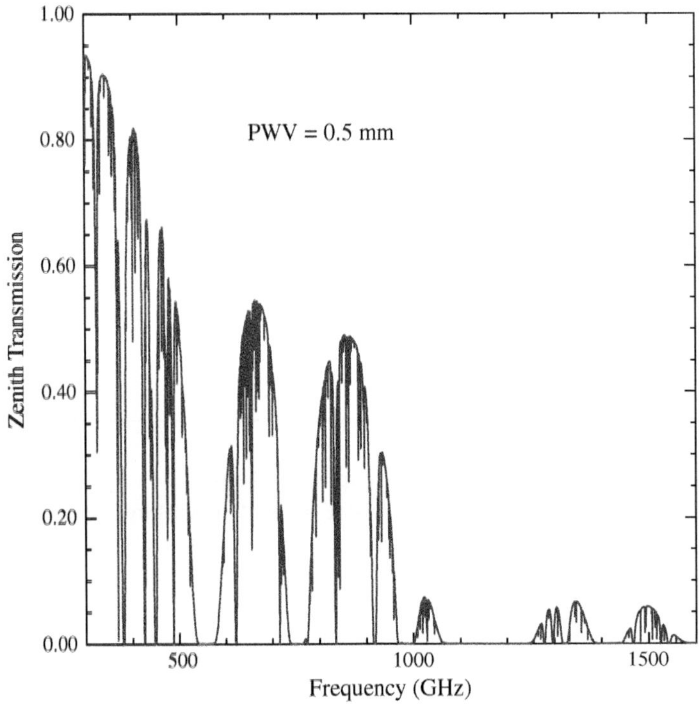

Figure 16.6 *Atmospheric transmission between 300 and 1500 GHz above site with 0.5 mm precipitable water vapor, which corresponds to good conditions on a high, dry mountain site. The transmission is reduced for sources not at the zenith. Transmission for frequencies above 1600 GHz is essentially zero. This data was obtained using the online CSO Atmospheric Transmission Interactive Plotter available at http://www.submm.caltech.edu/cso/weather/atplot.shtml.*

dry conditions from sites such as the Atacama plateau in Chile and the South Pole. The windows at 850 μm (350 GHz), 450 μm (650 GHz), and 350 μm (860 GHz) are clearly visible in Figure 16.6, and are widely used for astronomical observations from good sites.

The windows in which submillimeter atmospheric transmission is above 10% have enabled a great deal of remarkable astronomy. This includes results as varied as the composition of Titan's atmosphere to observations of some of the most distant galaxies in the Universe. However, the transmission significantly attenuates the signal, and its average effective temperature adds to the noise temperature of a coherent receiver and to the average photon flux of interest to an incoherent (bolometric) receiver. Of greater impact are variations in the emission, which can be orders of magnitude greater than the inherent equivalent input noise of the detector system. For broadband observations, this is often the limiting factor for the overall sensitivity and greatly limits the ability to observe weak sources.

Having unrestricted (i.e., not limited to 'windows') access to the entire submillimeter wavelength range has been one strong driver for getting above the Earth's atmosphere, and thus for observations from spaceborne platforms. At aircraft altitudes ($\approx$ 13 km), the situation is much better than shown in Figure 16.6, but there are still frequencies at which the atmosphere is essentially opaque. Most of these are due to lines of water vapor. The situation at balloon altitudes ($\approx$ 30 km) is yet better.

Much of the residual atmospheric absorption is due to transitions of water vapor, which can still be troublesome even from balloon altitudes. This is ironic, because water is one of the key molecules in the interstellar medium, playing an important role in cooling molecular clouds and in permitting them to contract to form new stars [34]. The three space missions discussed below, SWAS, Odin, and Herschel, take advantage of the absence of the Earth's atmosphere to observe at frequencies that have until the last decade been unavailable to astronomers and have yielded their own crop of interesting and, in some cases, very surprising results.

16.4.2 The Submillimeter Wave Astronomy Satellite

The Submillimeter Wave Astronomy Satellite (SWAS) was one of the first missions selected in NASA's Small Explorer Program. Melnick *et al.* [35] give an overview of the mission, and articles in volume 539 of *The Astrophysical Journal Letters*, devoted entirely to SWAS, report the early results from the mission. SWAS was launched on December 5, 1998 by a Pegasus XL rocket into a 70° inclination circular orbit having an initial altitude of 650 km. SWAS was designed to study four different species felt to play an important role in the interstellar medium. For each of these, a single submillimeter transition was targeted, and two separate Schottky barrier diode receivers were developed, with each observing one line in the upper and one line in the lower sideband. Information about the lines and receivers is given in Table 16.4.

The local oscillators (LOs) were tunable to compensate for Doppler shifts due to spacecraft and source motion. The local oscillator for receiver 2 could also be retuned to observe the $H_2{}^{18}O$ 1_{01}–1_{10} transition at 547.676 GHz. Critical components of the front end (mixers, LO multipliers, and first IF amplifiers) were passively cooled by Winston cone radiators to approximately 175 K, and the system noise temperatures were 2500 K (double sideband (DSB)) for receiver 1 and 2200 K (DSB) for receiver 2. Figure 16.7 shows a cutaway view of the SWAS instrument. In flight, the aperture is covered by a flat Goretex membrane to protect the telescope and instrument in the event that the satellite were to inadvertently point toward the Sun.

The nutating hyperbolic Cassegrain secondary reflector could move the beam direction 8.5′ in a direction parallel to the smaller beam dimension, at a rate up to 2 Hz. This was used primarily for continuum observations of pointlike sources such as Jupiter. For spectroscopic observations of molecular clouds, which are generally more extended than the beam throw, the entire spacecraft was repointed (nodded). The pointing system enabled nodding over an offset of up to 3° with combined slew and settling time below15 s. For smaller slews, the loss in observing time was smaller, and generally *on* and *off* observations of 50 s duration were employed, yielding an observing efficiency of over 85%. The star tracker field of view was 8° and stars as faint as 6 mag could be utilized, enabling an absolute pointing accuracy better than 5″ (1σ) over the entire sky.

SWAS employed an offset Cassegrain system, with an elliptical primary reflector 54 × 68 cm in diameter. The choice of this configuration was driven by two factors. First, for best spectral baselines, a system with minimum reflections and RF standing waves was desired, and only an offset system could guarantee this. Second, it was judged that conical feed horns would be used owing to the difficulty in fabricating corrugated feeds at these very high frequencies. The feed horns were electroformed integrally with the mixers to minimize loss, so that a simple and unproblematic design was felt to be essential. The ellipticity of the SWAS antenna was chosen to match that of the main lobe of the radiation pattern of the TE_{11} mode horn electric field distribution. This fortunately was consistent with the constraints imposed by the spacecraft size, Winston

Table 16.4 *SWAS targeted lines and receivers.*

Species	Transition	Frequency (GHz)	Wavelength (μm)	Receiver	Sideband
O_2	3,1–3,2	487.249	615.3	1	Lower
CI	$^3P_1$–$^3P_0$	492.161	609.1	1	Upper
^{13}CO	5–4	550.926	544.2	2	Lower
H_2O	1_{01}–1_{10}	556.936	538.3	2	Upper

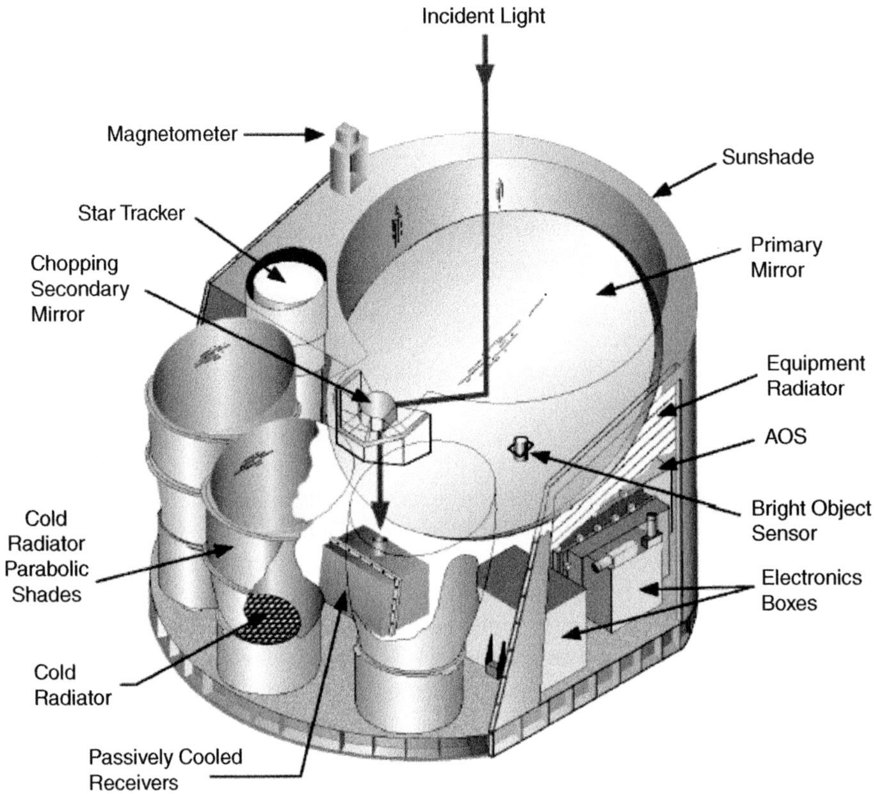

Figure 16.7 *Cutaway view of SWAS instrument. The off-axis parabolic primary reflector with elliptical projected aperture 54 × 68 cm in size is at the center right and the nutating (chopping) hyperbolic secondary reflector, also elliptical in shape, is at the center of the figure. The three Winston cone radiators which cool the submillimeter components and first IF amplifiers are at the center left. (AOS, Acousto-Optical Spectrometer). From Melnick et al. [35], reproduced by permission of the AAS.*

cones, and star tracker, which had to be pointing in essentially the same direction, since SWAS observed while pointing away from the Earth.

The two SWAS reflectors were machined from aluminum by diamond turning to a surface accuracy below $\lambda/50$ at 557 GHz. The beam sizes measured by scans of Jupiter were $3.5' \times 5.0'$ at 490 GHz and $3.3' \times 4.5'$ at 557 GHz, in reasonable agreement with the values expected from the antenna size and illumination. The aperture efficiency was measured to be 0.66 and the main beam efficiency 0.90. The aperture efficiency was somewhat less than the value of 0.80 expected for an unblocked, perfect antenna with Gaussian illumination and a 10 dB edge taper. This may have been due to imperfect phase distribution of the feed horn patterns, a small contribution from the surface error (see below), and slight system misalignments. The main beam efficiency was close to that expected. Figure 16.8 shows far-field beam patterns measured using Jupiter.

The modest size of the SWAS antenna allowed measurement of the complete feed and antenna system using a compact near-field range [36] before launch. This measurement was facilitated by the fact that the Schottky barrier diode submillimeter receivers employed could readily operate at ambient temperature, albeit with reduced sensitivity. The measurements were carried out at frequencies of 490.6 and 551.9 GHz, sampling the ranges covered by the two radiometers. While there were some issues arising from the probe used to sample the

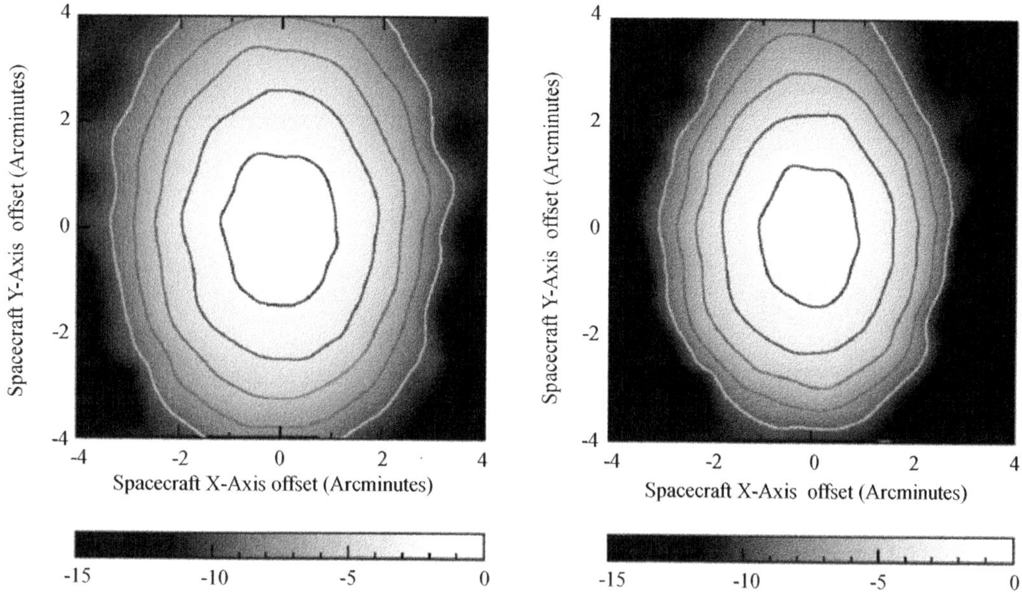

Figure 16.8 SWAS far-field antenna patterns measured using Jupiter at 490 GHz (left) and 553 GHz (right). The contour levels are −1, −3, −5, −7, and −9 dB relative to the boresight response. The ellipticity of the beam pattern resulting from the elliptical aperture with TE_{11} mode illumination is apparent. From Melnick et al. [35], reproduced with permission of the AAS.

field, the illumination pattern was found to be properly located with respect to the antenna, and the edge taper was consistent with that expected from the previously measured feed horn patterns. The surface accuracy could be measured only over the central portion of the aperture owing to the edge taper and the limited signal-to-noise ratio. The total system amplitude-weighted RMS error implied was 12 μm. This value is approximately a factor of 1.5 times greater than found with a Hartman test on the primary and secondary reflectors following fabrication. While still less than $\lambda/40$, these errors (which include feed horn, wire grid diplexer, secondary reflector, and primary reflector) contributed a small amount to the reduction in aperture efficiency relative to that of a perfect antenna. Figure 16.9 shows the aperture errors measured at 491 GHz.

SWAS operated almost flawlessly through more than six years of observations. The performance of the radiometers was particularly good, in terms of both stability and spectral baseline behavior. The on-orbit results were even better than those obtained in an extensive program of laboratory testing before launch, confirming the excellent characteristics of a thermally stable spacecraft for high-resolution spectroscopy. Results of ground and on-orbit performance tests are described in detail by Tolls *et al.* [37].

One of the major findings from SWAS was that the abundances of both water (H_2O) and molecular oxygen (O_2) are several orders of magnitude below those predicted by models of gas-phase chemistry in quiescent dense interstellar clouds. This means that the general cooling rate is lower than expected, making it more difficult for these clouds to radiate energy and contract toward becoming stars. This result produced a major change in the view of interstellar chemistry, forcing the inclusion of grain surfaces into chemical models, with a large fraction of available oxygen being frozen as water ice on these grain surfaces. The inclusion of grains and grain surface chemistry has an influence on the gas-phase abundance of many other species and is being actively followed up at the present time by observations with the Herschel Space Observatory, discussed below.

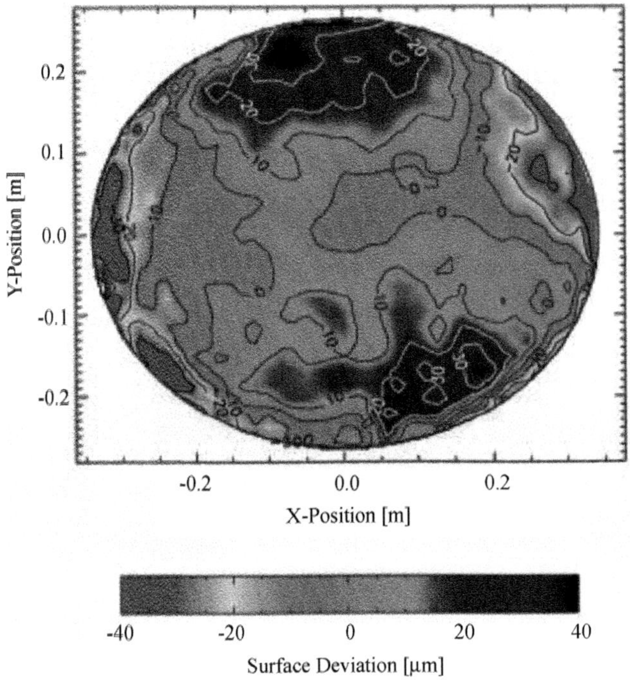

Figure 16.9 *Surface errors of SWAS antenna system measured with near-field compact range at 491 GHz. The errors include those of alignment, feed horn, wire grid diplexer, and the primary and secondary reflectors. From Tolls et al. [37], reproduced with permission of the AAS.*

16.4.3 The Odin Orbital Observatory

The Odin satellite – a collaborative project between space agencies and scientists in Sweden, Canada, Finland, and France – was designed to carry out observations at submillimeter wavelengths, but for two very different purposes. The first was radio astronomy spectroscopy (similar to a certain extent to the objectives of SWAS discussed above). The second was quite different: to study the Earth's atmosphere, measuring the altitude distribution of water, ozone, chlorine monoxide, and a number of other species. Odin was launched on February 20, 2001, by a START-1 rocket from Svobodny, in eastern Russia, into a Sun-synchronous circular orbit of 600 km altitude. The overall mission and early results are reviewed by Nordh *et al.* [38]. A number of papers about Odin and astronomical results from the first year of operation can be found in volume 402 of *Astronomy and Astrophysics*.

Odin employed a 1.1 m diameter offset Gregorian telescope. The reflectors were manufactured from CFRP honeycomb with face sheets of the same material. Reflectivity was assured by vacuum-deposited aluminum. The measured accuracy was 8 μm for the primary and 5 μm for the secondary reflector. Table 16.5 gives the expected characteristics of the telescope. On-orbit measurements at 557 GHz yielded beamwidths very close to those expected and beam efficiencies only a few percent below those anticipated.

The telescope and instrument were held by a CFRP support structure that also included the two star trackers. Figure 16.10 gives an idea of the antenna, support structure, and instrument. The Odin antenna and feed system were tested in two different test setups. A reflective collimator was used to measure beam patterns in a broad bandwidth at a wavelength of 200 μm. This system employed a mercury lamp at the focal point of the collimator, with a 4 K bolometer detector replacing the receiver in the antenna focal plane. This verified

Table 16.5 *Odin telescope performance characteristics.*

Frequency (GHz)	119	480	540	580
Beam size (arc minutes)	9.96	2.42	2.20	2.05
Aperture efficiency	0.72	0.70	0.69	0.69
Main beam efficiency	0.91	0.89	0.89	0.89

beamwidths and pointing direction. A hologram-type compact test range [40] was used at 119 GHz, which verified the antenna patterns of this lower frequency channel.

Odin included five radiometers; four of these operated at submillimeter wavelengths covering 486–504, 541–558, 547–564, and 563–581 GHz. These employed Schottky barrier diode mixers preceded by quasi-optical single-sideband filters. The fifth radiometer covered 118.25–119.25 GHz and employed a HEMT preamplifier. All of these units were processed through a quasi-optical input system. A Dicke switch coupled the beams either to the telescope, or to either of a pair of large solid-angle *sky* beams, or to a calibration load. The sky beams were offset by more than 40° from the telescope boresight and had FWHM beam sizes of 4.7° at 119 GHz and 4.5° at the submillimeter wavelengths. The sky beams were used to obtain reference spectra, which avoids having to repoint the entire spacecraft (as was done for SWAS). This sky beam switching was essential for the aeronomy part of the mission where the satellite was scanning the Earth's limb. To achieve the best possible broadband spectroscopic baselines for astronomy, Odin (as SWAS) nodded the complete telescope pointing between an on-source position and a preselected nearby (signal-free) off-source position.

A Stirling closed-cycle refrigerator cooled the millimeter and submillimeter receivers to 140 K and the IF amplifiers to 160 K. The single-sideband receiver noise temperatures were 600 K at 119 GHz and 3300 K at submillimeter wavelengths. A broadband acousto-optical spectrometer as well as two flexible (low to high spectral resolution) digital autocorrelation spectrometers were available, for observations with any combination

Figure 16.10 *Odin satellite in the laboratory. The CFRP structure holding the reflectors, the radiometers, and the star tracker is dark black. The secondary reflector is just visible at the upper right, and behind it is the instrument package. From Frisk et al. 2003, A&A,* ***402****, L27 [39]. Reproduced with permission © ESO.*

of three (out of the five) receivers mentioned above. In addition to the millimeter and submillimeter instruments, Odin included an optical spectrograph/infrared imaging system (OSIRIS), which was used only for aeronomy.

Odin carried out pioneering submillimeter spectral surveys and was notably successful in studying emission from H_2O and detecting $H_2{}^{18}O$ in comets. Odin pushed the upper limits on the abundance of O_2 in dark clouds to lower levels than reached by SWAS, owing to its ability to observe the $1,1 \rightarrow 1,0$ transition of O_2 molecules at 118.75 GHz. There was one claimed detection of this species, but also at a very low abundance level, confirming the challenge for astrochemical modeling raised by SWAS results. Despite being a comparatively low-cost project with an expected minimum lifetime of two years, it is impressive that after six years of successful shared astronomy/aeronomy operation, Odin has now spent four additional years productively monitoring the terrestrial atmosphere.

16.4.4 The Herschel Space Observatory

The Herschel Space Observatory is a Cornerstone mission of the European Space Agency with major contributions to the science payload from NASA, which is opening up the entire submillimeter wavelength range for continuum and spectral line astronomical observations. Herschel is a much larger and more capable observatory than its precursors, SWAS and Odin, and covers the approximately 670–55 μm wavelength range with three science instruments (discussed below). Herschel was launched on May 14, 2009, together with Planck, into a large-amplitude quasi-halo orbit around the Earth–Sun L2 point. The telescope was launched at ambient temperature and was kept at approximately 170 K by heaters for several weeks to ensure that water outgassing from the entire spacecraft did not stick on the optical surfaces. The telescope was then allowed to cool, and after one month the cryostat cover was opened, and test observations began. The steady-state temperature of the primary reflector is approximately 88 K.

As seen in Figure 16.11, solar radiation is kept off the telescope and cryostat by a large sunshield, which also has the solar panels on its back side. The spacecraft is oriented so that the line to the Earth and Sun is close to

Figure 16.11 *Left: the Herschel spacecraft showing the solar panel/sunshield (solar panels facing away), telescope, cryostat, and service module with the warm electronics. Middle: cutaway of the cryostat showing the three focal plane instruments at the top with the liquid helium reservoir below. Right: Herschel being prepared for acoustic testing. From Pilbratt et al. 2010, A&A,* ***518****, L1 [41]. Reproduced with permission © ESO.*

perpendicular to the sunshield, and so sources in a swathe around the sky perpendicular to the line to the Sun can be observed at any time. The result is that sources are visible during two periods each year.

Herschel employs a classical on-axis Cassegrain telescope. It is the largest monolithic telescope yet launched for astronomy, having a 3.5 m diameter primary reflector, although only 3.29 m of the aperture is illuminated by an individual science instrument at any time. This allows for the off-axis placement of the instruments in the focal plane as well as beam chopping, without having any power spillover past the edge of the primary reflector. The wavefront error requirement of less than 6 μm RMS at 80 K under zero gravity is designed to allow diffraction-limited operation to wavelengths as short at 90 μm. The various requirements on the telescope demanded significant developments in terms of fabrication and testing. The telescope primary is constructed of 12 SiC segments that were sintered together to form a lightweight monolithic mirror which could be polished to the required surface accuracy. The 30.8 cm diameter secondary reflector is also fabricated from SiC and diamond turned, including a central 'scatter cone' to break up multiple reflections that could cause standing waves in the frequency transmission of the telescope.

It was not possible to test the completed Herschel telescope using submillimeter waves because of the combination of atmospheric absorption and the difficulty of making a test facility of sufficient size. The surface smoothness is such that testing was possible using visible wavelength measurement techniques. These were carried out at ambient and cryogenic temperatures, and verified that the telescope surface accuracy met requirements. However, there was an issue with the position of the focal point when the telescope was cold, which was traced to issues with the coefficients of thermal expansion (CTEs) used in the design and modeling of the hexapod secondary support structure. The CTEs turned out not to be sufficiently accurate or representative of the materials actually used. When improved measurement data was incorporated into the models, the measured focal point shift measured was predicted. The issue was resolved by mechanical shimming. Details on the Herschel and Planck telescopes, including useful references to design and test procedures, can be found in Doyle, Pilbratt, and Tauber [42].

Herschel has three instruments in the cryogenically cooled focal plane (shown in Figure 16.11). Two of these are medium-resolution spectrometers and photometers. The Photodetector Array Camera and Spectrometer (PACS) covers wavelengths between 60 and 210 μm in three photometric ($R = \lambda/\delta\lambda \simeq 2$) bands [43]. The detectors are bolometer arrays cooled to 300 mK by a ^{3}He sorption cooler. Each array covers the same $1.75' \times 3.5'$ field of view. The array for the two shorter wavelength bands has 32×64 pixels each subtending $3.2'' \times 3.2''$. The array for the longer wavelength band has 16×32 pixels each subtending $6.4'' \times 6.4''$. PACS also includes a grating spectrometer. In this mode, radiation from 25 spatial pixels on the sky is reimaged by an integral field unit onto a grating. This provides a spectral resolution $R \sim 1500$–4000, generally increasing at shorter wavelengths. Each pixel has 16 stressed Ge:Ga photoconductor detectors to measure the spectral signal. PACS covers 55–210 μm in three contiguous bands and can be used in a number of modes including scanning and chopping in wavelength by tilting the grating, as well as spatial chopping and scanning.

The Spectral and Photometric Imaging Receiver (SPIRE) also combines an imaging photometer and spectrometer [44]. The low-resolution ($R \sim 3$) imaging photometer has three bands, centered at 250, 350, and 500 μm. Arrays of spider web bolometers with Ge temperature sensors are used as detectors covering a $4' \times 8'$ field on the sky. Different numbers of pixels are used for each band, but as all are effectively undersampling the sky, data is generally taken by scanning the telescope pointing direction by repointing the spacecraft. The resulting fully sampled maps of large fields have proven exceptionally valuable for studying gas and dust in interstellar clouds in the Milky Way as well as galaxies throughout the Universe. The SPIRE spectrometer employs a Mach–Zender interferometer, which has two hexagonally close-packed detector arrays each covering a $2.6'$ field of view. There are 37 detectors in the short-wavelength array covering 194–313 μm, and 29 detectors in the long-wavelength array covering 303–671 μm. A spectrum has resolution determined by the maximum path length difference scanned, with the highest resolution being

$0.04\,cm^{-1}$ (1.2 GHz). For this highest resolution R varies from about 370 at the long-wavelength end to about 1300 at the short-wavelength end.

The Heterodyne Instrument for the Far Infrared (HIFI) is a very high spectral resolution heterodyne receiver [45]. There are seven frequency bands covering 480–1250 GHz (using superconducting insulator superconducting (SIS) mixers) and 1410–1910 GHz (using hot electron bolometer (HEB) mixers). Each band observes two orthogonal linear polarizations simultaneously and is equipped with both a wide-band acousto-optical spectrometer with 1.1 MHz resolution over a 4 GHz bandwidth and a narrow-band (high-resolution) digital autocorrelation spectrometer offering spectral resolution as high as 0.14 MHz. Bands up to 1120 GHz employ corrugated feed horns for coupling mixers to free space. Bands at higher frequency employ twin-slot planar antennas and lenses. Jellema *et al.* [46] and Jellema *et al.* [47] give detailed information on near-field measurements and electromagnetic simulations of the HIFI instrument. These include propagation from individual mixer feeds through the complete telescope.

Pilbratt [48,49] and Pilbratt *et al.* [41] give an overview of the mission. Volume 518 of *Astronomy and Astrophysics* contains a number of articles describing Herschel, its instruments, and their performance, together with a selection of the early astronomical results from Herschel. Volume 521 of *Astronomy and Astrophysics* includes a number of additional articles on the early results obtained using the HIFI instrument. At the time of writing, Herschel is carrying out a varied program of observations, with the telescope and all instruments performing essentially according to or exceeding prelaunch expectations. An extensive set of measurements using HIFI confirms that antenna efficiencies and beamwidths are consistent with diffraction-limited operation [50].

16.4.5 The Future: Millimetron, CALISTO, and Beyond

16.4.5.1 Millimetron

Notwithstanding the enormous progress made in utilizing submillimeter wavelengths for astronomy, there is continued interest in yet more capable instruments and space missions. While these are still in various stages of definition, they do serve to give an indication of possibilities in the (hopefully) not too distant future. Considering only submillimeter (rather than infrared) wavelengths, and space (rather than suborbital) missions, there are at least two interesting mission concepts that are especially interesting from the point of view of space antenna technology.

Millimetron is an observatory being developed by a team led by scientists and engineers from Russia, but including a variety of institutions in a number of European countries. Wild *et al.* [51] give an overview, but the concept has evolved significantly since the publication of that article. The basic idea includes a submillimeter telescope with a 12 m diameter deployable reflector. This follows the general design of the antenna for the RadioAstron mission, discussed in Section 16.6.3 below, but with far higher surface accuracy. The stated goal is that the central 3.5 m monolithic central portion will have a surface accuracy of 2 μm RMS, and the overall accuracy will be 10 μm. Figure 16.12 shows conceptual views of the Millimetron antenna in stowed and deployed configurations [52]. While not as accurate as the Herschel telescope, this should allow efficient operation at wavelengths as short as 200 μm, and the very large collecting area and high angular resolution will make this a unique facility up to the short-wavelength limit of approximately 60 μm (or possibly even shorter wavelengths). The current design concept is for the antenna panels to be fabricated from aluminum or (as Herschel) SiC.

A deployable 12 m diameter telescope that must end up with this relatively high accuracy is certainly a technical challenge. Current designs call for active control of the position of the petals making up the panels of the primary reflector, along with a metrology system. In order to keep the photon background to an acceptable level, the optical elements must be cooled. The concept for doing this is currently evolving, but a multilayer

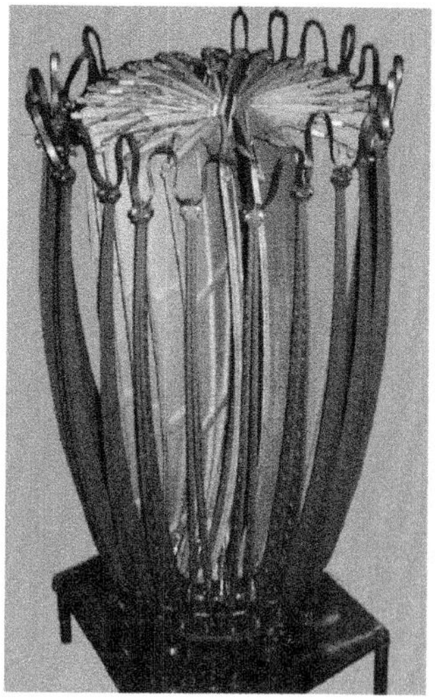

Figure 16.12 *Conceptual views of the Millimetron antenna in stowed and deployed configurations. Reproduced with permission by N. Kardashev.*

thermal screen (similar to that employed for Planck) is part of the design. Modeling indicates that a temperature of 50 K can be achieved purely by passive cooling. A lower temperature of about 4 K is desirable, requiring an active cooling system.

The instrument complement for Millimetron is still being studied, but will likely include a number of high-resolution heterodyne receivers for spectroscopy and an imaging photometer/spectrometer that might be similar to the PACS instrument on Herschel [53]. An unusual aspect of Millimetron's instrument complement derives from the desire to perform Earth–spacecraft VLBI in the 18–26 GHz band and in a number of bands in common with those of the Atacama Large Millimeter Array (ALMA; see, e.g., http://science.nrao.edu/alma/index.shtml), between 31 and 720 GHz. As discussed in Section 16.6, this enables an interferometer with a baseline equal to the distance between a ground-based antenna and the spacecraft, and thus increases the angular resolution available at a given frequency beyond what can be obtained on the Earth's surface. Spacecraft-to-spacecraft interferometry is also envisioned. The issues for Millimetron include having an orbit that allows good VLBI performance along with offering a stable thermal environment with low pickup of thermal radiation from the Earth. For up-to-date information about Millimetron see http://www.asc.rssi.ru/millimetron/default.htm, http://www.sron.rug.nl/millimetron, and http://www.sron.rug.nl/millimetron/MillimetronWorkshopSicily2010.

16.4.5.2 CALISTO

CALISTO is an acronym for Cryogenic Aperture Large Infrared Space Telescope Observatory. This is a concept for a mission to cover the wavelength range 30–300 μm for extremely high-sensitivity photometric and low-resolution spectroscopic observations, primarily of distant galaxies. CALISTO could properly be considered to function at frequencies beyond the upper limit of *radio astronomy*, but much of the science builds on the results coming from Herschel, so it is included here. The goal of CALISTO is to achieve a sensitivity limited by the photon noise of the astronomical background itself. To achieve this, the optical elements must be so cold that their thermal emission is weaker than the astronomical sky at high galactic latitudes, and detector noise must also be lower than the photon fluctuation noise from the background. This is increasingly difficult as the spectral resolution increases. For $R \sim 1000$, as required for good measurement of

Figure 16.13 Conceptual view of the CALISTO observatory after deployment showing the off-axis optical configuration which includes a 4 × 6 m primary reflector. The multilayer V-groove sunshade is below the cooled baffle which acts as a cold stop, as well as the primary and secondary reflectors, are cooled to about 4 K. From Goldsmith et al., CALISTO: The cryogenic aperture large infrared space telescope observatory. Proc. SPIE, 7010, 2008 [57].

redshifts and galaxy line shapes, the current state of the art approaches this limit for individual detectors, and the challenge is to develop arrays of detectors that achieve a noise equivalent power (NEP) of about $10^{-20}\,W/Hz^{0.5}$.

The concept for CALISTO evolved from an earlier study of a mission to accomplish this same goal of extending submillimeter sensitivity to orders of magnitude below that which can be reached by Herschel, which is limited by the approximately 88 K temperature of its telescope. The initial work on the Single-Aperture Far-Infrared Observatory (SAFIR) envisioned an on-axis fairly classical telescope cooled to 4 K [54,55]. Subsequent detailed modeling of system performance revealed that the scattering of the strong radiation emitted by the Milky Way Galaxy into the beam even when pointing in a very different direction would dominate the background and raise the system noise level significantly. The solution was to go to an unblocked off-axis design. In addition to lower noise, this approach offered increased aperture and beam efficiencies [56]. Folding only the secondary and support structure allowed an efficient configuration for launch in an available fairing. The result was a system that offers astronomical background-limited sensitivity and a very clean beam. As shown in Figure 16.13, there are five V-groove sunshades below the 4 × 6 m ellipsoid-shaped parabolic primary. The Gregorian configuration allows there to be a cold stop at the image of the primary, minimizing pickup of stray radiation that would spill past the edge of the secondary reflector [57].

16.5 Low-Frequency Radio Astronomy

16.5.1 Overview of Low-Frequency Radio Astronomy

We can define low-frequency radio astronomy to involve frequencies that are low enough to be affected by plasma effects produced in the Earth's ionosphere. The cutoff frequency varies with sunspot cycle, time of day, and location. It can be as high 15 MHz on the sunlit side of the Earth, about 10 MHz on the night side, and as low as about 2 MHz at favorable spots near the Earth's magnetic poles. Generally speaking, at frequencies of a few megahertz and below, it is generally necessary to be above the Earth's ionosphere to be able to carry out astronomical observations. At frequencies comparable to or even somewhat above the cutoff frequency, variations in the properties of the ionospheric plasma produce both shifts in beam pointing and variable phase delays that impede high-resolution interferometry.

There is also a lower frequency cutoff of about 30 kHz imposed by the plasma of the interplanetary medium (IPM) which blocks extremely low-frequency observations of astronomical sources outside the Solar System. It is thus appropriate to talk about a *low-frequency astronomy window* extending from about 30 kHz to about 30 MHz. Low-frequency observations of Solar System objects, particularly Jupiter and the Sun, are also of great interest, especially the study of the relationship between solar activity such as coronal mass ejections (CMEs) and human activities on the Earth.

Some of the most interesting work in low-frequency radio astronomy took place in the very early period of research in this area when investigators were determining what types of antennas and receivers to use and how observations could best be carried out. Astronomers were interested in determining the spectrum of the radio emission from the Milky Way. It was known from ground-based observations that synchrotron emission was the dominant source of low-frequency radio emission and it was anticipated that its flux would increase at lower frequencies. Observations were expected to inform astronomers about the energy density of cosmic rays in the Galaxy and also the effect of ionized hydrogen on the propagation of low-frequency radio waves. During this period, the first observations in a *truly unexplored* frequency range were being carried out. Some highlights of the equipment used are given in what follows. This is by no means a complete overview of the technology and missions, but gives an idea of the development of this field.

16.5.2 Early Low-Frequency Radio Space Missions

Early work in low-frequency radio astronomy utilized both rocket and spaceborne instruments. One of the questions of interest was how well the ionosphere shielded a radiometer on a spacecraft from terrestrial interference and also from different types of ionospheric noise [58]. Naturally this led to observations over a relatively wide range of frequencies. These early observations were carried out with relatively simple hardware on quite short timescales compared to current space research projects. They did, however, suffer from a range of problems which must have kept this a very unpredictable area of research.

A frequency of 10 MHz corresponds to a wavelength of 30 m, and 1 MHz to a wavelength of 300 m. For any small satellite, the antenna dimensions in terms of wavelengths are very limited, and the angular response will be close to isotropic. Thus, most systems would be sensitive to any signals from 'below' the spacecraft coming through the ionosphere. A description of different types of antennas used in the early years of low-frequency radio astronomy, as well as a general discussion of ionospheric effects on low-frequency radio astronomy, can be found in Huguenin [59].

One of the earliest satellites carrying out low-frequency observations was the Alpha–Beta satellite, which was launched in June 1962 into a 75° inclination orbit with an apogee of 270 km and a perigee of 200 km. These last values were somewhat lower than anticipated, and the satellite decayed after only 15 days [58]. The system had radiometers operating at frequencies of 4.040 and 6.975 MHz. Little information is given about the antenna system used, but the data presented shows a very large variation in the radio noise level, with a maximum level at 7 MHz corresponding to $T_b = 10^{12}$ K. Observations taken at a period of minimum terrestrial noise level are indicated to be representative of the cosmic radio background, which at 7 MHz has a brightness temperature $T_b = 1.5^{+1.5}_{-0.5} \times 10^6$ K. The level of maximum background clearly demonstrates why dynamic range is a critical issue for low-frequency radio astronomical systems.

Another interesting experiment was developed by the Cavendish Laboratory, Cambridge, UK [60], and flown on the UK-2 satellite. This mission was launched on 27 March 1964 into a 51.67° inclined orbit having an apogee of 1360 km and perigee of 290 km. The experiment employed a swept frequency radiometer covering the frequency range 0.65–3.5 MHz, with a bandwidth of 20 kHz. The full frequency range was covered every 25 s. The antenna was a dipole with an overall length of 40 m. The antenna was kept taught by centrifugal force arising from the satellite's spin rate, which was 5.6 rpm after deployment. The wire antenna's deployment from a short boom on the equator of the satellite was controlled by an electric motor, and the change in moment of inertia was utilized to reduce the spin rate of the satellite.

The presence of a plasma, such as is the case here, complicates the behavior of even a simple dipole antenna [61]. For this experiment the preamplifier input was designed to have a very high input impedance so that there is always a large impedance mismatch between it and the antenna. Two ferrite loop antennas operating at 2.3 MHz were also included to provide some data in the event of failure of the dipole antenna to deploy correctly.

A dramatic increase in the ambition of low-frequency satellite design is evident in the Radio Explorer Satellite RAE-1 [62] which was launched on July 4, 1968. Stepped frequency radiometers covering 0.45–9.18 MHz with a bandwidth of 40 kHz were the main data gathering instruments. This experiment included, in addition to antennas for telemetry and plasma experiments, two traveling-wave V antennas 229 m long. The large size reflected the interest in achieving an order of magnitude improvement in angular resolution compared to the dipole antennas previously employed.

The orbit of the RAE-1 satellite was nearly circular with an altitude of 5850 km and an inclination of 121°. This was a compromise between a high altitude to reduce plasma effects of the ionosphere and the requirement that the gravitational forces be sufficient for gravity-gradient stabilization of the long antenna booms. In operation, one V pointed toward the Earth and the other toward the local zenith. During each orbit, the antenna boresight scanned a great circle on the celestial sphere, and this circle precessed by 0.52° per day.

RAE-1 employed two 60° V antennas aligned back to back, hence pointing in opposite directions. Each leg of the V is cut one-quarter of its length inward from the tip and a 600 Ω resistor inserted in series. Thus the antenna operates as a traveling-wave antenna at frequencies for which the length beyond the resistor is equal to a quarter wavelength. This suppresses the backward response and delivers a front-to-back ratio of better than 10 dB. References in Weber, Alexander, and Stone [62] give additional information about the design and testing of scale models of this antenna. The antenna legs were formed from heat-treated beryllium–copper alloy tapes 0.005 cm thick by 5 cm wide, which were stored on a spool. After launch, they were deployed by a motorized mechanism and formed hollow cylindrical tubes 1.3 cm in diameter. Considerable care was taken to minimize thermal gradients across the tubes due to solar heating.

RAE-1 produced data on solar bursts as well as valuable information that allowed separation of galactic and extragalactic contributions to low-frequency radio emission. Noise events from the ionosphere were found to be significant even at the high altitude of RAE-1 and suggested that future missions could benefit from a lunar orbit that would provide shielding from terrestrial emission.

This goal was realized with the RAE-2 satellite which was launched on June 17, 1973 into a lunar orbit [63]. The frequency range was extended to cover 25 kHz to 13 MHz. The antenna system was similar to that employed on RAE-1, but the included angle of the V antennas was 35°. Table 16.6 (from [63]) gives some of the key parameters of the RAE-2 V antenna system.

The lower V antenna was used for a period only partially extended to a length of 183 m, and later was fully extended. The orbit was at an altitude of 1100 km above the lunar surface, and the Moon's gravitational field provided the gravity-gradient stabilization of the antenna pointing direction. Additional information can be found in the article by Alexander *et al.* [63] and in references given therein, including some information on astronomical results.

Table 16.6 Summary of radiation characteristics of 229 m RAE-2 V antenna.

Frequency (MHz)	θ_E (deg FWHM)	θ_H (deg FWHM)	First sidelobe level (dB)	Front-to-back ratio
9.18	37	61	−2	≈10
6.55	27	55	−4	≈15
3.93	80	63	−5	≈15
1.31	180	120	−12	≈15
0.87	220	160	—	≃15

The Imp 6 radio astronomy experiment [64] was launched on March 13, 1972 into a highly eccentric obit with an apogee of 206 000 km and a perigee of 354 km. The orbit evolved to a perigee of 1600 km by the 10th orbit from drag effects. The satellite was spin stabilized with 11.13 s period. The satellite spin axis was aligned with the direction of the ecliptic pole. Two radiometers were employed. The first radiometer sequentially sampled 32 discrete frequencies in the frequency range of 30 kHz to 9.9 MHz. The bandwidth was 10 kHz and the time constant was 6 ms. The second radiometer covered 30 kHz to 4.9 MHz with a 3 kHz bandwidth and 40 ms integration time. The cycle time to cover the full frequency range was 5 s for both radiometers.

The Imp 6 experiment utilized a pair of monopole antennas 45.7 m long oriented parallel to the ecliptic plane (perpendicular to the satellite spin axis) that were combined to form a dipole 91.4 m long. In free space this would be appropriate for a frequency of about 1.6 MHz, but clearly this dipole could not be simply matched over the very large frequency range covered. An antenna of this size must be deployed after launch. No information on how this was done is given by Mosier, Kaiser, and Brown [64] but it could well have been a tape antenna deployed by centrifugal force. The observations in this paper are of noise produced in different regions of the ionosphere; no astronomical results are reported.

16.5.3 The Future

Low-frequency radio astronomy observations from space to date have been restricted to very low angular resolution. The obvious way to overcome this is to use an interferometric array of antennas in space, and plans for various arrays have been developed (see [65–67], and http://rsd-www.nrl.navy.mil/7213/weiler/lfraspce.html, for additional information). An interferometric low-frequency array on the far side of the Moon has also been proposed [68], which, although working at frequencies nominally above the Earth's ionospheric cutoff, would have the very major advantage of being shielded from the RF emission from the Earth.

16.6 Space VLBI

16.6.1 Overview of Space VLBI

Interferometry at radio frequencies involves combining the voltage output from antennas in pairs. The phase difference between the outputs produces an interference pattern, which is a function of the baseline between the two antennas projected onto a plane perpendicular to the direction of the source. Observations having a range of baselines yield an interferogram. Its coordinates can be considered as the components of the baseline expressed in units of wavelength, and are usually denoted u, v. The Fourier transformation of the interferogram in the u–v plane results in an image of the source.

For space VLBI, one has an antenna in orbit around the Earth, which is used either with a single antenna or with many antennas spread across the Earth's surface. Every baseline involving the space antenna has an amplitude proportional to the square root of the gain of the space antenna. Since space antennas are typically much smaller than ground-based radio astronomy antennas, a consequence is that a critical parameter for space VLBI is the antenna gain. Owing in part to the limitations of space antenna collecting areas, the sources that are most readily studied with space VLBI are non-thermal, meaning sources that emit much more intensely than those characteristic of thermal, or blackbody, emission. These include maser and synchrotron emission sources. Both of these are most readily observed at modest frequencies (centimeter to meter wavelengths). Consequently, the requirements on space antenna precision are relatively modest, and deployable antennas have been used for space VLBI missions.

Combining signals from multiple antennas and Fourier transformation of the interferogram significantly reduces the effect of sidelobes on the power pattern of any individual antenna. This applies to antennas utilized for space as well as ground-based interferometry. Thus, the usual tradeoff between sidelobe level and aperture efficiency is weighted toward maximizing the efficiency and thus the effective area. Polarization is in general a significant aspect of interferometric measurements, and dual-polarization systems are used on almost all ground-based interferometric antenna systems, but single-polarization systems were employed for the HALCA space VLBI satellite (Section 16.6.2). Finally, since the purpose of interferometry is to recover information on small angular scales, the information of interest in the image that is obtained usually is restricted to a fraction of the main beam, and often only a very small fraction thereof. Consequently, the imaging capability of antennas for interferometry, including space VLBI, is not a significant consideration, so again the feed system can be optimized to maximize aperture efficiency without consideration of the field of view of the overall system.

16.6.2 HALCA

The HALCA (High Altitude Laboratory for Communications and Astrophysics) satellite [69], which was developed by the Japanese Institute of Space and Astronautical Science (ISAS) and the National Astronomical Observatory (NAO), was launched in February 1997. HALCA was the orbiting element of the VLBI Space Observing Programme (VSOP) and was used in conjunction with arrays of radio telescopes on the Earth to produce a synthetic aperture up to several Earth diameters in size. The spacecraft's orbit was inclined 31° to the Earth's equator and had an apogee of 21 400 km, a perigee of 560 km, and a period of 6.3 h. Thus the maximum baseline is on the order of 20 000 km, as compared to the maximum baseline of approximately 10 000 km available on the Earth's surface. At a wavelength of 6 cm (see below) a maximum angular resolution $\lambda/B \simeq 3 \times 10^{-9} = 0.6 \times 10^{-3}$ arc seconds (0.6 mas) was achieved [70]. A highly elliptical orbit is desirable because it means that a wide range of baselines are available during the course of an observation. This good sampling of the u–v plane means that the resulting Fourier transformation is a relatively clean representation of the source intensity distribution. Five tracking stations spread around the Earth's surface were used to bring data in a 32 MHz bandwidth from the spacecraft to the ground, after which this data could be correlated with signals from multiple ground-based antennas that were observing the same source simultaneously.

HALCA employed a deployable wire mesh antenna approximately hexagonal in form, having a maximum diameter of 10 m and an aperture area of 50.1 m^2 [71], corresponding to an effective diameter of 8 m. Figure 16.14 shows the mesh antenna and the spacecraft behind it. The secondary reflector is also hexagonal, fitting into a 1.1 m diameter circle. The antenna is deployed by extending six approximately radially oriented masts, seen in the figure.

The feed and receiver system covers three frequency bands: L-band (1.60–1.73 GHz), C-band (4.7–5.0 GHz), and Ka-band (22.0–22.3 GHz). Since for interferometry the region of interest is very small, the beams at all frequencies have to be co-aligned, so that a single feed is utilized for this very wide frequency range. One consequence is that only a single circular polarization for each band was implemented. The signals

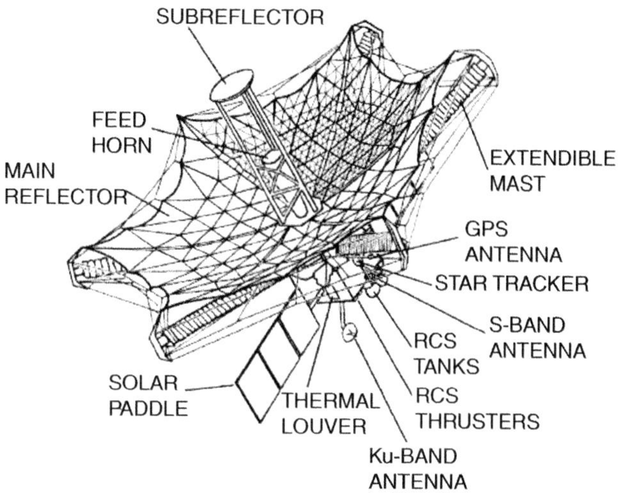

Figure 16.14 *Schematic of HALCA antenna deployed. © 2008 IEEE. Reprinted, with permission from IEEE Transactions on Antennas and Propagation, 52, 1777–1782.*

from the feed are brought to ambient temperature low-noise amplifiers (LNAs) by coaxial cables at the lower two bands, and by waveguide for Ka-band.

A summary of the antenna gain performance before launch is given in Table 16.7 (adapted from [69]).

The design and performance of the HALCA antenna system was necessarily a compromise imposed by launch vehicle constraints and the wide frequency range specified. In particular, the low-frequency spillover was increased by the limited size of the secondary reflector needed to fit within the launch vehicle fairing. The performance for the two lower frequency bands was within about 1 dB of that expected, but, as seen in Table 16.7, the Ka-band performance was dramatically degraded. Damage to the waveguide between feed and LNA resulting from launch vibration is thought to be the most likely explanation.

HALCA was used over a five-year period for a wide range of radio astronomy observing programs, producing images of extragalactic radio sources in combination with ground-based antennas and antenna arrays. In particular, high-redshift quasars were imaged with sub-milliarcsecond angular resolution giving new

Table 16.7 *HALCA gain performance before launch.*

Frequency (GHz)	1.60	4.70	22.15
100% gain (dBi)	42.54	51.90	65.36
Aperture illumination (dB)	−1.07	−1.55	3.62
Spillover (dB)	−2.44	−0.85	−0.40
Blockage	[a]	[a]	[a]
Surface roughness	[a]	[a]	[a]
Pillow effect (dB)	0.00	−0.01	−0.27
Repeatability (dB)	0.00	−0.02	−0.34
Mesh loss (dB)	−0.01	−0.01	−0.06
Dielectric loss (dB)	−0.01	−0.01	−0.08
Feed loss (dB)	−0.42	−0.70	−0.34
Net antenna gain (dBi)	38.59	48.75	60.25
Degradation after launch (dB)	0.94	1.06	24.2

[a]Included in the aperture illumination.

Table 16.8 Parameters of the RadioAstron mission.

Wavelength (cm)	92	18	6.2	1.2–1.7
Frequency (GHz)	0.326	1.67	4.8	25–17.6
Interferometer angular resolution (μas)	540	106	37	7.1–10
Antenna efficiency	0.5	0.5	0.5	0.3

insights into the cores and jets within these remarkable objects. Yet higher angular resolution as well as sensitivity would be desirable, and other space VLBI missions have been developed to push the boundaries by operating at higher frequencies with bigger antennas and in larger orbits.

16.6.3 RadioAstron

The RadioAstron project has been under development for a long period of time in the Soviet Union and now Russia. The mission has also evolved considerably over time, and there has been a lack of published articles giving comprehensive information about antenna and mission parameters. The information presented here has been obtained from Kardashev [72,73]. Updated information can be obtained from the RadioAstron web page http://www.asc.rssi.ru/radioastron/.

RadioAstron is expected to mark a considerable advance over HALCA in terms of sensitivity and angular resolution. The orbit will be inclined by 51° and have a semi-major axis of 189 000 km. The perigee will vary from 10 000 to 70 000 km and the apogee from 310 000 to 390 000 km, giving a maximum baseline of 350 000 km. The mean orbital period will be 9.5 days, but the orbit will evolve with time due to the interaction with the Moon. The minimum mission lifetime anticipated is five years. Some parameters of the RadioAstron mission are given in Table 16.8. The unit μas (10^{-6} arc seconds) corresponds to an angular resolution of 4.9×10^{-12}.

RadioAstron employs a circular antenna 10 m in diameter. The antenna is formed of 27 composite carbon fiber panels and has an overall RMS surface accuracy of 0.5 mm. The $f/D = 0.43$ paraboloid is fed by a multi-frequency prime focus feed, and the system can observe simultaneously at two different frequencies, or in two circular polarizations at a single frequency. A maximum bandwidth of 32 MHz can be processed and sent to a ground station for recording and correlation.

The segments of the RadioAstron antenna are rigid, so the deployment is quite different from that used for HALCA. Here, each petal is supported by a radial tube also made of carbon fiber. The inner end of each tube is on a circle about 3 m in diameter and can rotate to move the petals from a stowed configuration, in which they are nearly parallel but nested together as shown in Figure 16.15, to the deployed configuration shown in Figure 16.16. RadioAstron was successfully launched on July 18, 2011.

For sources that emit over a broad range of frequencies (continuum sources) and which have a structure which is independent of frequency, one can improve the quality of interferometric images by combining data taken at different frequencies. For such sources, this is equivalent to filling in the u–v plane by having a better range of baselines. It is anticipated that frequency-agile broadband receivers will enable RadioAstron to exploit this technique.

16.7 Summary

This review has covered space antennas for radio astronomy that span six orders of magnitude in frequency – from megahertz to above 1000 GHz. For this reason alone it is not surprising that a wide variety of antenna

Figure 16.15 RadioAstron antenna in configuration for launch in which the petals of the primary reflector surrounding a central hub are pivoted to be nearly parallel to the antenna symmetry axis. Reproduced with permission by N. Kardashev.

designs have been created to carry out a range of measurements. The lowest frequency systems, put in space to avoid the cutoff imposed by the Earth's ionosphere, employed deployable wire antennas with relatively low directivity. Interest in the megahertz frequency range for astronomy today is focused on studying the so-called 'dark ages' – the period after the Big Bang, but before the first stars were formed, and during which hydrogen

Figure 16.16 RadioAstron antenna in deployed configuration. The spacecraft bus, solar cells, and telemetry antenna are seen below the 10 m diameter primary reflector. Reproduced with permission by N. Kardashev.

atoms were neutral. A number of space missions have been proposed, and large antenna arrays on the lunar surface are a long-term possibility. At centimeter wavelengths, the primary research area has been investigations of the cosmic background radiation – the relict of the Big Bang redshifted to form the well-known 3 K blackbody radiation. A sequence of space missions has provided increasing detail on its intensity and distribution. This work initially employed feed horns as antennas, with modest angular resolution but well-defined beam patterns. As the demand for higher angular resolution grew, off-axis unblocked antenna systems have become standard, again in order to have very clean antenna patterns so that confusion from extraneous signal sources is minimized.

In the shorter wavelength portion of the RF range, the Earth's atmosphere again becomes problematic, primarily from absorption by water vapor which becomes appreciable at millimeter and at submillimeter wavelengths. Here a variety of antenna configurations, both off-axis and symmetric, have been employed. These increasingly resemble classical optical telescope designs as the wavelength decreases, with focal plane array receivers employed to increase the rate at which extended sources can be imaged. Future space radio astronomy missions envision larger and colder antennas to increase sensitivity.

For studies of distant radio sources, angular resolution is paramount, and to achieve this, interferometers with one antenna in space working with one or more antennas on the ground provide larger baselines and hence higher resolution than can be achieved solely with antennas on the Earth. For this work, antenna gain is the most important parameter, and relatively conventional symmetric antennas operating at centimeter wavelengths have been used. In order to provide the required collecting area, antennas for space VLBI have been and will likely continue to be deployable reflectors.

Space radio astronomy has included a relatively small number of missions and antennas, but which include the full gamut of designs optimized for the varying frequencies and scientific requirements. Future studies of the origin of the Universe, and of the evolution of galaxies and the formation of stars throughout cosmic time, will undoubtedly be evermore demanding in terms of antenna performance and will thus provide an ongoing stimulus to antenna design, analysis, and testing.

Acknowledgments

Thanks are due to Donald Goldsmith, Åke Hjalmarson, Richard Huguenin, Bill Imbriale, John Mather, Gary Melnick, David Murphy, Göran Pilbratt, Lyman Page, Jan Tauber, and Ned Wright for careful reading of the manuscript and helpful suggestions.

This work was carried out in part at the Jet Propulsion Laboratory, California Institute of Technology, under a contract with the National Aeronautics and Space Administration.

References

1. Wilson, R.W., Jefferts, K.B., and Penzias, A.A. (1970) Carbon monoxide in the Orion nebula. *Astrophysical Journal*, **161**, L43–L44.
2. Penzias, A.A. and Wilson, R.W. (1965) A measurement of excess antenna temperature at 4080 Mc/s. *Astrophysical Journal*, **142**, 419–421.
3. Fixsen, D.J. and Mather, J.C. (2002) Spectral results of the far-infrared absolute spectrophotometer instrument on COBE. *Astrophysical Journal*, **581**, 817–822.
4. Strukov, I.A., Skulachev, D.P., Boyarskii, M.N., and Tkachëv, A.N. (1987) A spacecraft determination of the dipole anisotropy in the microwave background. *Soviet Astronomy Letters*, **13**, 65–66.

5. Lineweaver, C. (1996) OT The CMB Dipole: The Most Recent Measurement and Some History. arXiv:astro-ph/9609034v1.
6. Klypin, A.A., Strukov, I.A., and Skulachev, D.P. (1992) The Relikt missions: results and prospects for detection of the microwave background anisotropy. *Monthly Notices of the Royal Astronomical Society*, **258**, 71–81.
7. Strukov, I.A., Bryukhanov, A.A., Skulachev, D.P., and Sazhin, M.V. (1992) Anisotropy of the microwave background radiation. *Soviet Astronomy Letters*, **18**, 153–156.
8. Strukov, I.A., Brukhanov, A.A., Skulachev, D.P., and Sazhin, M.V. (1992) The Relikt-I experiment – new results. *Monthly Notices of the Royal Astronomical Society*, **258**, 37P–40P.
9. Boggess, N.W., Mather, J.C., Weiss, R. *et al.* (1992) The COBE mission: its design and performance two years after launch. *Astrophysical Journal*, **397**, 420–429.
10. Mather, J.C., Fixsen, D.J., and Shafer, R.A. (1993) Design for the COBE Far Infrared Absolute Spectrophotometer (FIRAS), in *Infrared Spaceborne Remote Sensing*, SPIE Proceedings, vol. 2019, pp. 168–179.
11. Mather, J.C., Toral, M., and Hemmati, M. (1986) Heat trap with flare as multimode antenna. *Applied Optics*, **25**, 2826–2830.
12. Janssen, M.A., Gulkis, S., Bennett, C.L., and Kogut, A.J. (1993) Design and results of differential microwave radiometers (DMR) on COBE, in *Infrared Spaceborne Remote Sensing*, SPIE Proceedings, vol. 2019, pp. 211–221.
13. Janssen, M.A., Bednarczyk, S.M., Gulkis, S. *et al.* (1979) Pattern measurements of a low-sidelobe horn antenna. *IEEE Transactions on Antennas and Propagation*, **AP-28**, 759–763.
14. Toral, M.A., Ratliff, R.B., Lecha, M.C. *et al.* (1989) Measurements of very low sidelobe conical horn antennas. *IEEE Transactions on Antennas and Propagation*, **AP-37**, 171–176.
15. Bennett, C.L., Fixsen, D.J., Hinshaw, G.A. *et al.* (1994) Morphology of the interstellar cooling lines detected by COBE. *Astrophysical Journal*, **434**, 587–598.
16. Fixsen, D.J., Bennett, C.L., and Mather, J.C. (1999) COBE far infrared absolute spectrophotometer observations of galactic lines. *Astrophysical Journal*, **526**, 207–214.
17. Page, L., Jackson, C., Barnes, C. *et al.* (2003) The optical design and characterization of the Microwave Anisotropy Probe. *Astrophysical Journal*, **585**, 566–586.
18. Peiris, H.V., Komatsu, E., Verde, L. *et al.* (2003) First-year Wilkinson Microwave Anisotropy Probe (WMAP) observations: implications for inflation. *Astrophysical Journal Supplement*, **148**, 213–231.
19. Jarosik, N., Bennett, C.L., Dunkley, J. *et al.* (2002) Seven-Year Microwave Anisotropy Probe (WMAP) Sky Maps, Systematic Errors, and Basic Results. arXiv 2002.47744v1.
20. Galindo-Israel, V., Imbriale, W.A., and Mittra, R. (1987) On the theory and synthesis of single and dual offset shaped reflectors. *IEEE Transactions on Antennas and Propagation*, **AP-35**, 887–896.
21. Barnes, C., Limon, M., Page, L. *et al.* (2002) The MAP satellite feed horns. *Astrophysical Journal Supplement*, **143**, 565–576.
22. Jarosik, N., Bennett, C.L., Halpern, M. *et al.* (2003) Design, implementation, and testing of the MAP radiometers. *Astrophysical Journal Supplement*, **145**, 413–436.
23. Tauber, J.A., Mandolesi, M., Puget, J.-L. *et al.* (2011) Planck pre-launch status: the Planck mission. *Astronomy & Astrophysics*, **520**, A1.
24. Tauber, J.A., Norgaard-Nielsen, H.U., Ade, P.A.R. *et al.* (2010) Planck pre-launch status: the optical system. *Astronomy & Astrophysics*, **520**, A2.
25. Lamarre, J.-M. *et al.* (2010) Planck pre-launch status: the HFI instrument from specification to actual performance. *Astronomy & Astrophysics*, **520**, A9.
26. Bersanelli, M., Mandolesi, M., Butler, R.C. *et al.* (2010) Planck pre-launch status: design and description of the Low Frequency Instrument. *Astronomy & Astrophysics*, **520**, A4.
27. Leahy, J.P., Bersanelli, M., D'Arcangelo, O. *et al.* (2010) Planck pre-launch status: expected LFI polarisation capability. *Astronomy & Astrophysics*, **520**, A8.
28. Sandri, M., Villa, F., Bersanelli, M. *et al.* (2010) Planck pre-launch status: low frequency instrument optics. *Astronomy & Astrophysics*, **520**, A7.
29. Ade, P.A.R., Savini, G., Sudiwala, R. *et al.* (2010) Planck pre-launch status: the optical architecture of HFI. *Astronomy & Astrophysics*, **520**, A11.

30. Maffei, B., Noviello, F., Murphy, J.A. *et al.* (2010) Planck pre-launch status: HFI beam expectations from the optical optimisation of the focal plane. *Astronomy & Astrophysics*, **520**, A12.
31. Rosset, C., Tristram, M., Ponthieu, N. *et al.* (2010) Planck pre-launch status: high frequency instrument polarization calibration. *Astronomy & Astrophysics*, **520**, A13.
32. Bock, J.J., Chen, D., Mauskopf, P.D., and Lange, A.E. (1995) A novel bolometer for infrared and millimeter-wave astrophysics. *Space Science Reviews*, **74**, 229–235.
33. Holmes, W.A., Bock, J.J., Crill, B.P. *et al.* (2008) Initial test results on bolometers for the Planck high frequency instrument. *Applied Optics*, **47**, 5996–6008.
34. Goldsmith, P.F. and Langer, W.D. (1978) Molecular cooling and thermal balance of dense interstellar clouds. *Astrophysical Journal*, **222**, 881–895.
35. Melnick, G., Stauffer, J.R., Ashby, M.L.N. *et al.* (2000) The Submillimeter Wave Astronomy Satellite: science objectives and instrument description. *Astrophysical Journal*, **539**, L77–L85.
36. Erickson, N.R. and Tolls, V. (1997) Near-field measurements of the SWAS antenna. 20th ESTEC Antenna Workshop on Millimetre Wave Antenna Technology and Antenna Measurements, Noordwijk, The Netherlands, pp. 313–319.
37. Tolls, V., Melnick, G.J., Ashby, M.L.N. *et al.* (2004) Submillimeter Wave Astronomy Satellite performance on the ground and in orbit. *Astrophysical Journal Supplement*, **152**, 137–162.
38. Nordh, H.L., von Schéele, F., Frisk, U. *et al.* (2003) The Odin orbital observatory. *Astronomy & Astrophysics*, **402**, L21–L25.
39. Frisk, U., Hagstrom, M., Alal-Laurinaho, J. *et al.* (2003.) The Odin satellite I. Radiometer design and test. *Astronomy & Astrophysics*, **402**, L27–L34.
40. Ala-Laurinaho, J., Hirvonen, T., Piironen, P. *et al.* (2001) Measurement of the Odin telescope at 119GHz with a hologram-type CATR. *IEEE Transactions on Antennas and Propagation*, **49**, 1264–1270.
41. Pilbratt G.L., Riedinger, J.R., Passvogel, T. *et al.* (2010) Herschel space observatory. *Astronomy & Astrophysics*, **518**, L1.
42. Doyle, D., Pilbratt, G., and Tauber, J. (2009) The Herschel and Planck space telescopes. *Proceedings of the IEEE*, **97**, 1403–1411.
43. Poglitsch, A., Waelkens, C., Geis, N. *et al.* (2010) The photodetector array camera and spectrometer (PACS) on the Herschel space observatory. *Astronomy & Astrophysics*, **518**, L2.
44. Griffin, M.J., Abergel, A., Abreu, A. *et al.* (2010) The Herschel-SPIRE instrument and its in-flight performance. *Astronomy & Astrophysics*, **518**, L3.
45. de Graauw, Th., Whyborn, N., Helmich, F. *et al.* (2010) The Herschel-heterodyne instrument for the far-infrared (HIFI). *Astronomy & Astrophysics*, **518**, L6.
46. Jellema, W., Huisman, R., Candotti, M. *et al.* (2004) Comparison of near-field measurements and electromagnetic simulations of the focal plane unit of the heterodyne instrument for the far-infrared. Proceedings of the 5th International Conference on Space Optics (ICSO 2004), Toulouse, France, pp. 303–322.
47. Jellema, W., Jochemsen, M., Peacocke, T. *et al.* (2008) The HIFI focal plane beam characterization and alignment status. 19th International Symposium on Space Terahertz Technology, Groningen, The Netherlands, pp. 448–455.
48. Pilbratt, G.L. (2008) Herschel mission overview and key programmes. *Proceedings of the SPIE*, **7010**, 701002-1–701002-12.
49. Pilbratt, G.L. (2009) The promise of Herschel, in *Submillimeter Astrophysics and Technology: a Symposium Honoring Thomas G. Phillips*, vol. 417, ASP Conference Series (eds D.C. Lis, J.E. Vaillancourt, P.F. Goldsmith *et al.*), pp. 427–438.
50. Olberg, M. (2009) Beam observations towards Mars, HIFI ICC Technical Note Version 0.7 of 2010-06-09.
51. Wild, W. and Kardashev, N.S. (2009) Millimetron – a large Russian-European submillimeter space observatory. *Experimental Astronomy*, **23**, 221–244.
52. Kardashev, N.S., Khalimanovich, V.I., Shipilov, G.V. *et al.* (2010) A structural design of the space observatory MILLIMETRON and its deployable subsystems. Millimetron Workshop, Sicily, http://www.sron.rug.nl/millimetron/MillimetronWorkshopSicily2010.
53. Wild, W., Baryshev, A., de Graauw, T. *et al.* (2008) Instrumentation for Millimetron – a large space antenna for THz astronomy. 19th International Symposium on Space Terahertz Technology, Groningen, The Netherlands, pp. 186–191.

54. Goldsmith, P.F., Khayatian, B., Bradford, M. *et al.* (2006) Analysis of the optical design for the SAFIR telescope. *Proceedings of the SPIE*, **6265**, 62654A-1–62654A-13.
55. Lester, D., Benford, D., Blain, A. *et al.* (2004) The science case and mission concept for the Single Aperture Far-Infrared (SAFIR) Observatory. *Proceedings of the SPIE*, **5487**, 1507–1521.
56. Goldsmith, P.F., Bradford, C.M., Dragovan, M. *et al.* (2007) CALISTO: a cryogenic far-infrared/submillimeter observatory. *Proceedings of the SPIE*, **6687**, 66870P-1–66870P-13.
57. Goldsmith, P.F., Bradford, M., Dragovan, M. *et al.* (2008) CALISTO: the cryogenic aperture large infrared space telescope observatory. *Proceedings of the SPIE*, **7010**, 701020-1–701020-16.
58. Huguenin, G.R. and Papagiannis, M.D. (1965) Spaceborne observations of radio noise from 0.7 to 7.0 MHz and their dependence on the terrestrial environment. *Annual Review of Astronomy and Astrophysics*, **28**, 239–247.
59. Huguenin, G.R. (1963) Long-Wavelength Radio Astronomy in Space. PhD Thesis, Department of Astronomy, Harvard University. Also published as *Harvard College Observatory, Space Radio Project Publication No. 104*.
60. Harvey, C.C. (1965) Results from the UK-2 satellite. *Annual Review of Astronomy and Astrophysics*, **28**, 248–254.
61. Walsh, D. and Haddock, F.T. (1965) Antenna impedance in a plasma: problems relevant to radio astronomy measurements from space vehicles. *Annual Review of Astronomy and Astrophysics*, **28**, 605–613.
62. Weber, R.R., Alexander, J.K., and Stone, R.G. (1971) The radio astronomy explorer satellite, a low-frequency observatory. *Radio Science*, **6**, 1085–1097.
63. Alexander, J.K., Kaiser, M.L., Novaco, J.C. *et al.* (1975) Scientific instrumentation of the Radio-Astronomy-Explorer-2 satellite. *Astronomy & Astrophysics*, **40**, 365–371.
64. Mosier, S.R., Kaiser, M.L., and Brown, L.W. (1973) Observations of noise bands associated with the upper hybrid resonance by the Imp 6 radio astronomy experiment. *Journal of Geophysical Research*, **78**, 1673–1679.
65. Basart, J.P., Burns, J.O., Dennison, B.K. *et al.* (1997) Directions for space-based low-frequency radio astronomy 1. System considerations. *Radio Science*, **32**, 251–263.
66. Basart, J.P., Burns, J.O., Dennison, B.K. *et al.* (1997) Directions for space-based low-frequency radio astronomy 2. Telescopes. *Radio Science*, **32**, 265–276.
67. Jones, D.L., Weiler, K.W., Allen, R.J. *et al.* (1998) The astronomical low-frequency array (ALFA), in *IAU Colloquium 164: Radio Emission from Extragalactic Compact Sources*, vol. 144 (eds J.A. Zensus, G.B. Taylor, and J.M. Wrobel), ASP Conference Series, pp. 393–394.
68. Lazio, J., Carilli, C., Hewitt, J. *et al.* (2009) The lunar radio array (LRA). *Proceedings of the SPIE*, **7436**, 743601-1–743601-11.
69. Hanayama, E., Kuroda, S., Takano, T. *et al.* (2004) Characteristics of the large deployable antenna of the HALCA satellite in orbit. *IEEE Transactions on Antennas and Propagation*, **52**, 1777–1782.
70. Hirabayashi, H., Hirosawa, H., Kobayashi, H. *et al.* (1998) Overview and initial results of the very long baseline interferometry space observatory programme. *Science*, **281**, 1825–1829.
71. Takano, T., Miura, K., Natori, M. *et al.* (2004) Deployable antenna with 10 m maximum diameter for space use. *IEEE Transactions on Antennas and Propagation*, **52**, 2–11.
72. Kardashev, N.S. (1997) RadioAstron – a radio telescope much greater than the Earth. *Experimental Astronomy*, **7**, 329–343.
73. Kardashev, N.S. (2009) RadioAstron: a radio telescope many times the size of Earth. *Physics – Uspekhi*, **52**, 1127–1137.

17

Antennas for Deep Space Applications

Paula R. Brown, Richard E. Hodges, and Jacqueline C. Chen

Jet Propulsion Laboratory, California Institute of Technology, USA

17.1 Introduction

Deep space exploration began in 1959 when the Soviet Luna 1 spacecraft missed its intended target of the Moon and went into a heliocentric orbit [1]. Over the last half century, spacecraft have explored every planet in the Solar System as well as comets, asteroids, and the Sun. The Voyager 1 spacecraft is now the most distant human-made object, at more than 17 billion kilometers from Earth as of this writing. All these spacecraft have relied on antennas to communicate their discoveries back to Earth.

Antennas for deep space applications comprise a wide variety of types and frequency bands. Applications include telecommunications, radiometers, scatterometers, altimeters, and radars. Antenna types range from simple dipoles to complex, multi-frequency, and multi-application assemblies such as the Cassini reflector antenna [2].

The distinguishing requirements of deep space antennas are often due to the extreme environments in which the antennas must survive and operate. Spacecraft assemblies are exposed to high mechanical vibration, acoustic noise, and acceleration loads during launch and again for entry, descent, and landing if the assembly is part of a lander or rover. Assemblies may experience pyrotechnic shock loads, which could be self-generating if the antenna has a deployable component, or could be induced by a nearby assembly. Depending on the spacecraft's destination, the antenna may experience extreme heat, extreme cold, or high radiation environments. Antenna designs for planetary entry vehicles must consider entry temperatures and operation in a range of atmospheric pressures and gases different from the Earth's atmosphere. Planetary surface antennas must also consider atmospheric pressures and gases, extreme temperature cycling, and potential dust contamination.

As with any spacecraft assembly, mass and volume are key drivers for deep space antennas. An antenna large enough to meet deep space data rate requirements can be one of the largest assemblies on the spacecraft.

Space Antenna Handbook, First Edition. Edited by William A. Imbriale, Steven (Shichang) Gao and Luigi Boccia.

Figure 17.1 *Artist's depiction of Voyager spacecraft.*

Figure 17.1 shows the Voyager spacecraft with its large reflector antenna prominently mounted on top of the spacecraft. Mass and volume requirements can drive the antenna size and corresponding mission data rates, the antenna material and fabrication methods, and the need for deployment mechanisms.

Detailed descriptions of antennas from past NASA missions that were managed by the Jet Propulsion Laboratory are presented in *Spaceborne Antennas for Planetary Exploration* [3]. Rather than reiterate material from [3], two NASA missions that launched in 2011 will be examined in this chapter as case studies. The first, Mars Science Laboratory, is a rover mission that employs several X-band and UHF telecommunications antennas, along with a Ka-band landing radar antenna. The second case study is Juno, which features X-band and Ka-band telecommunications antennas and five antennas with operating frequencies ranging from 600 MHz to 22 GHz that are part of the Microwave Radiometer instrument.

17.2 Telecommunications Antennas

Communications between deep space and Earth require highly efficient antennas with specified gains and beamwidths. Spacecraft typically carry multiple antennas of various gains for different phases and operating

Table 17.1 *Deep space communication frequency bands.*

Band designation	Earth-to-space (uplink) frequencies	Space-to-Earth (downlink) frequencies
S-band	2110–2120 MHz	2290–2300 MHz
X-band	7145–7190 MHz	8400–8450 MHz
Ka-band	34.2–34.7 GHz	31.8–32.3 GHz

Table 17.2 *Mars UHF communication frequency bands.*

Band designation	Surface-to-orbit frequencies	Orbit-to-surface frequencies
UHF	390–405 MHz	435–450 MHz

scenarios of the mission. Immediately after launch, a low-gain broad-beam antenna is required to limit the amount of power received by the ground station and the spacecraft, and to avoid the need to accurately point the antenna. As the spacecraft travels farther from Earth, communications switch to a medium- or high-gain antenna. The low-gain antenna may still be used for spacecraft maneuvers in which the medium- or high-gain antennas are not pointed toward Earth, or for anomalous conditions where the spacecraft direction is faulty or unknown. Frequency bands for deep space telecommunications are constrained by international spectrum allocations. The reserved frequency bands are listed in Table 17.1 [4].

On Mars missions, recent landers and rovers on the surface have been equipped with a UHF link to the orbiting satellites, such as Mars Reconnaissance Orbiter and Mars Express [5]. The UHF Mars communication frequencies are listed in Table 17.2 [4]. The communication between the orbiting satellites and Earth are done over the standard deep space frequencies listed in Table 17.1. This relay scheme enables higher data transfer rates than is possible with direct communications between the surface and Earth. The relatively small antennas that are used on the landers and rovers enable high data rates to the orbiters because of the close separation distance. The orbiters have much larger antennas to provide high data rate communications with Earth [6].

17.3 Case Study I – Mars Science Laboratory

17.3.1 MSL Mission Description

The primary objective of the Mars Science Laboratory (MSL) is to determine whether Mars is, or was in the past, capable of supporting life [7]. MSL will land an approximately 900 kg rover that is the size of a small sports utility vehicle [8]. The rover will analyze soil and rock composition with a complex suite of spectrometer instruments. Cameras on the rover will image both the terrain and close-up views of the rocks and soil. Radiation detectors will be used to assess radiation hazard levels for possible past living organisms, and in preparation for future human exploration. The rover will also carry a sophisticated weather station that will provide daily weather reports from Mars.

The MSL mission will go through four distinct phases en route to Mars and during its one Martian year operating life [9]. The first is the launch phase, which lasts for about one Earth day. An X-band low-gain antenna is used for this phase to accommodate a wide range of observation angles between the spacecraft and ground stations. The second phase is the cruise, where the mission relies on low- and medium-gain X-band

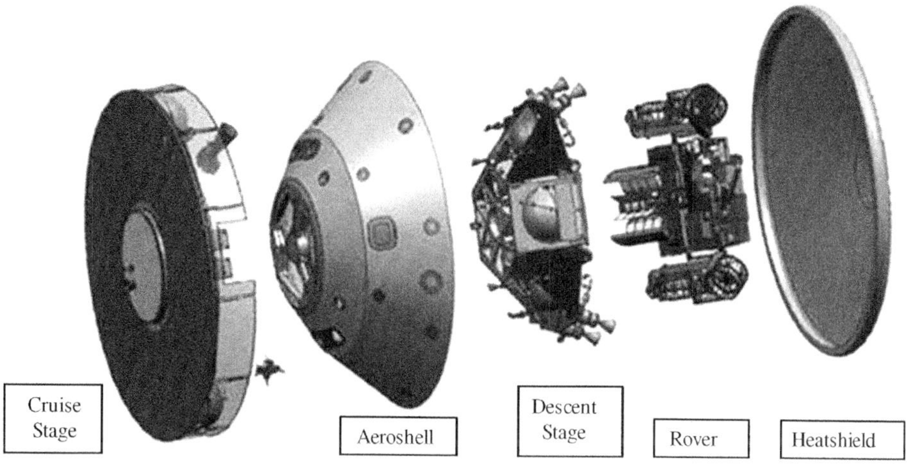

Figure 17.2 *Exploded view of the MSL spacecraft.*

antennas to communicate with Earth. After cruise comes the entry, descent, and landing (EDL) phase, where the spacecraft will use UHF antennas to communicate to Mars orbiters, and low-gain X-band antennas to link directly to Earth. The EDL phase also will engage a landing radar system that includes an array of six Ka-band slot array antennas. The final phase is the rover surface operations, which will use a UHF proximity link to Mars Reconnaissance Orbiter and Mars Odyssey as the main communication path, but will also have low- and high-gain X-band antennas to provide a secondary bidirectional link with Earth.

Figure 17.2 is an exploded view of the spacecraft, displaying the major assemblies when the spacecraft is in the stowed configuration. Figures 17.3–17.6 show the locations of the antennas on each of the major assemblies.

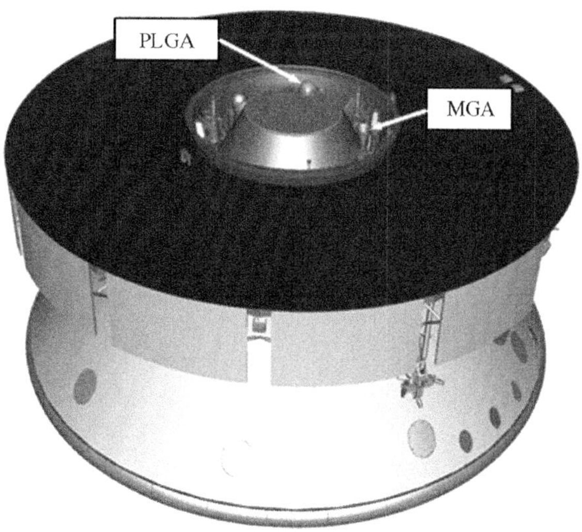

Figure 17.3 *The MSL spacecraft in the cruise configuration. The Parachute Low-Gain Antenna (PLGA) and Medium-Gain Antenna (MGA) are used during the cruise phase.*

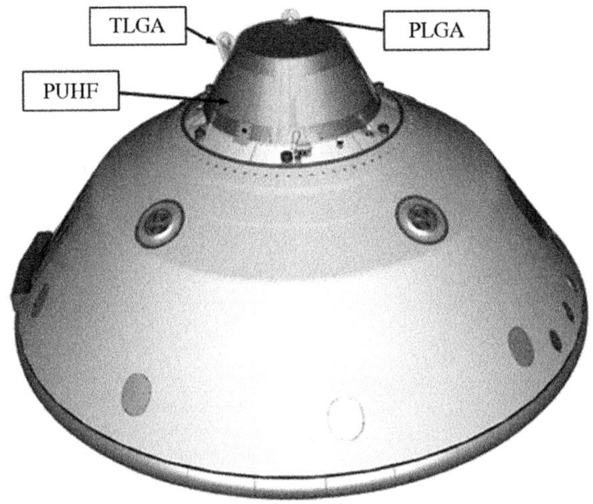

Figure 17.4 The MSL spacecraft in the aeroshell configuration. The Parachute Low-Gain Antenna (PLGA), Tilted Low-Gain Antenna (TLGA), and Parachute Cone UHF (PUHF) antenna are used in this configuration during entry and descent at Mars in this spacecraft configuration.

17.3.2 MSL X-band Antennas

The MSL spacecraft carries six X-band antennas, listed in Table 17.3. The PLGA, TLGA, and RLGA employ the same new antenna design, but each has unique spacecraft accommodation considerations. The MGA is a heritage design that was previously used on NASA's Mars Exploration Rover (MER) mission, with a new separation joint design incorporated. The HGA, developed by EADS-CASA with Sener as the subcontractor

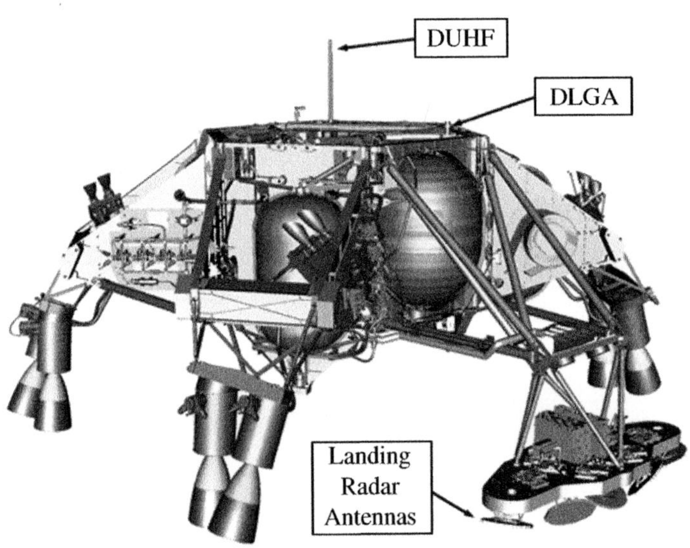

Figure 17.5 The MSL spacecraft in the descent configuration. The Descent Low-Gain Antenna (DLGA), the Descent Stage UHF (DUHF) antenna, and the landing radar antennas are used in this configuration. (The stowed rover is not shown in this figure for clarity.)

Figure 17.6 The MSL rover during integration. The High-Gain Antenna (HGA), Rover Low-Gain Antenna (RLGA), and Rover UHF (RUHF) antenna are used during rover operations.

for the gimbal mechanism [10], was contributed by the Spanish government [11]. All of the X-band antennas are circularly polarized.

17.3.2.1 Medium-Gain Antenna (MGA)

The MGA, shown in Figure 17.7, is a simple conical horn with a pair of irises in the input waveguide for impedance matching. With a gain of 19 dBi, the antenna is directive enough that spacecraft interference effects on the radiation pattern are negligible. The challenge to the MGA design is that the horn is located on the cruise stage of the spacecraft, and the polarizer is located on the entry vehicle. The design accommodates a separation joint in the circular waveguide that connects the horn and the polarizer. To accommodate launch loads, the joint allows a displacement of ±4 mm and an angular misalignment of 0.5° between the two parts of the antenna. The separation joint contacting surfaces are treated with a Teflon-impregnated anodic coating to reduce friction forces during the cruise stage separation event. A plug made from thermal protection system (TPS) material is bonded onto the circular waveguide above the polarizer to prevent hot gases from entering the waveguide during Mars entry.

The measured insertion loss through the septum polarizer, TPS plug, and ball joint is less than 0.3 dB at the nominal position, and varies by less than 0.24 dB over the full ±4 mm range of possible displacement. The

Table 17.3 MSL X-band antennas.

Antenna name	Abbreviation	Antenna type	Mission phase
Medium-Gain Antenna	MGA	Conical horn	Cruise
Parachute Cone Low-Gain Antenna	PLGA	Choked horn with parasitic dipoles	Launch, cruise, EDL
Tilted Low-Gain Antenna	TLGA	Choked horn with parasitic dipoles	EDL
Rover Low-Gain Antenna	RLGA	Choked horn with parasitic dipoles	Surface operations
Descent Low-Gain Antenna	DLGA	Choked horn	EDL
High-Gain Antenna	HGA	Gimbaled microstrip patch array	Surface operations

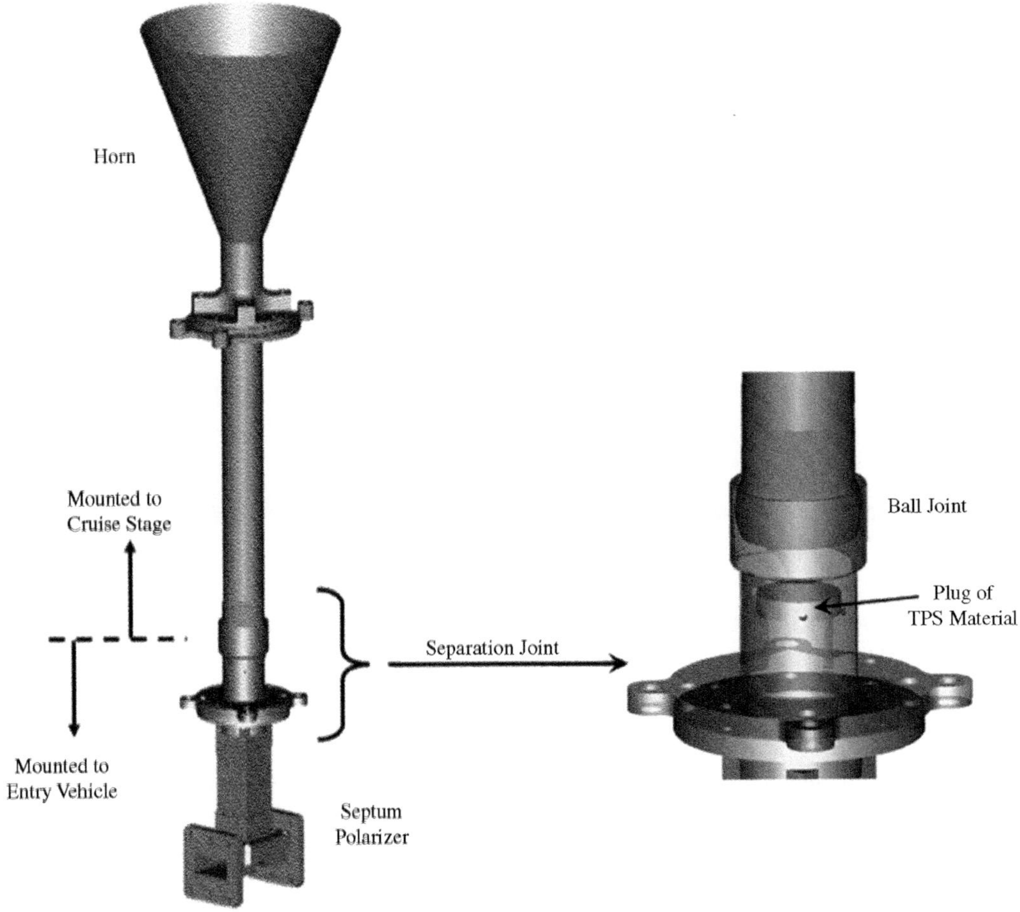

Figure 17.7 *The MSL medium-gain antenna.*

MGA with the separation joint was tested at 340 W peak power in a vacuum environment with no RF breakdown events.

17.3.2.2 The Parachute Cone Low-Gain Antenna (PLGA)

The requirement for a gain greater than 0 dBic in the downlink band and greater than −0.5 dBic in the uplink band over a half-angle beamwidth of 80°, and greater than −1.0 dBic in both bands out to a half-angle beamwidth of 85°, drove the design selection for the PLGA. A choked horn antenna with parasitic dipoles above the aperture [12] was modeled and optimized in Computer Simulation Technology's Microwave Studio to achieve the desired beamwidth, and resulted in an antenna that was inexpensive to manufacture and could be easily accommodated on the spacecraft.

The PLGA is depicted in Figure 17.8, and measured radiation patterns of the antenna by itself are shown in Figure 17.9. The measured gain was slightly lower than the calculated gain at 80° off boresight, but was sufficient to meet the project needs. The antenna successfully passed multipaction and ionization tests at 340 W in vacuum and in simulated Mars gas up to 12 Torr.

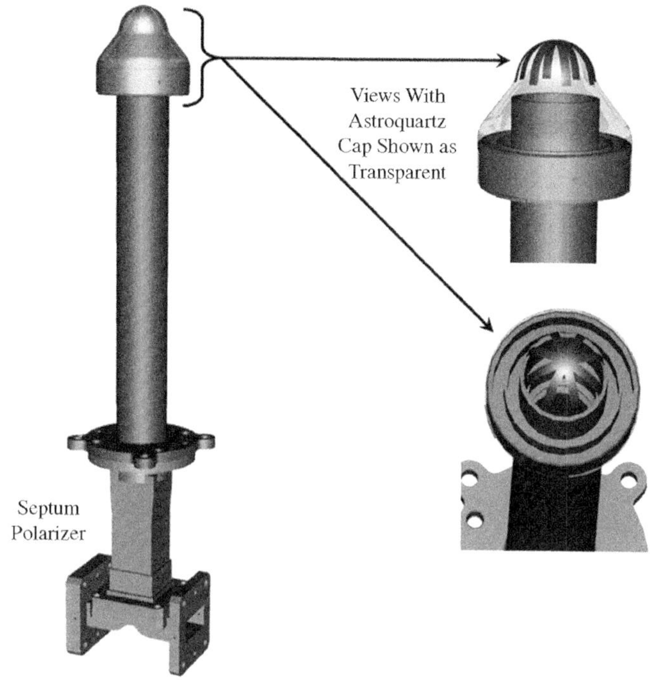

Figure 17.8 The Parachute Low-Gain Antenna (PLGA).

To suspend the dipoles over the choked horn aperture, a thin cap made of a cyanate ester/Astroquartz composite was fabricated, and then the copper dipoles were applied by vacuum deposition. The cap was bonded to the choked horn by injecting adhesive through small holes at the base of the cap.

The PLGA operates in transmit and receive modes during the cruise phase of the mission, but only in transmit mode during EDL. Until the parachute deploys during the EDL sequence, the PLGA is covered with a

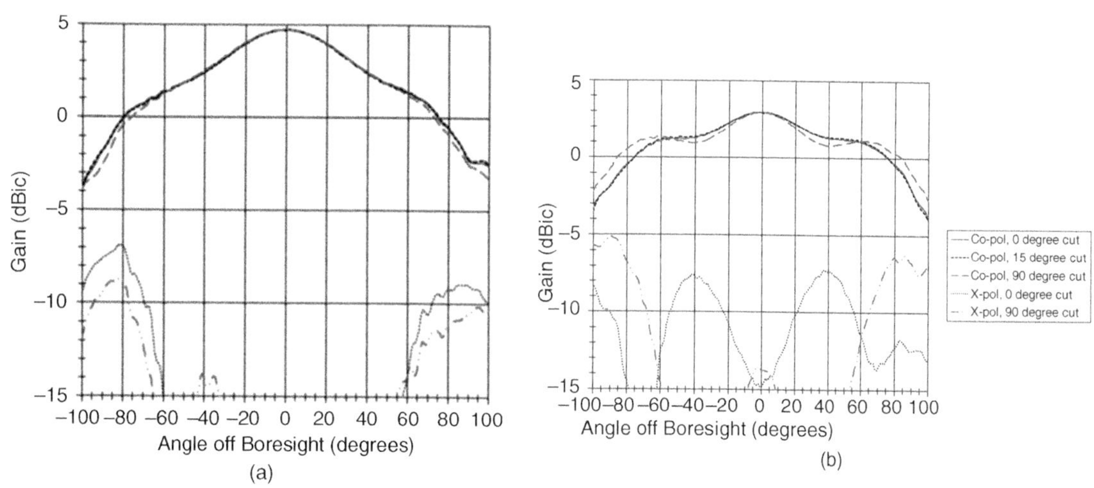

Figure 17.9 The PLGA measured gain: (a) 7.15 GHz; (b) 8.4 GHz.

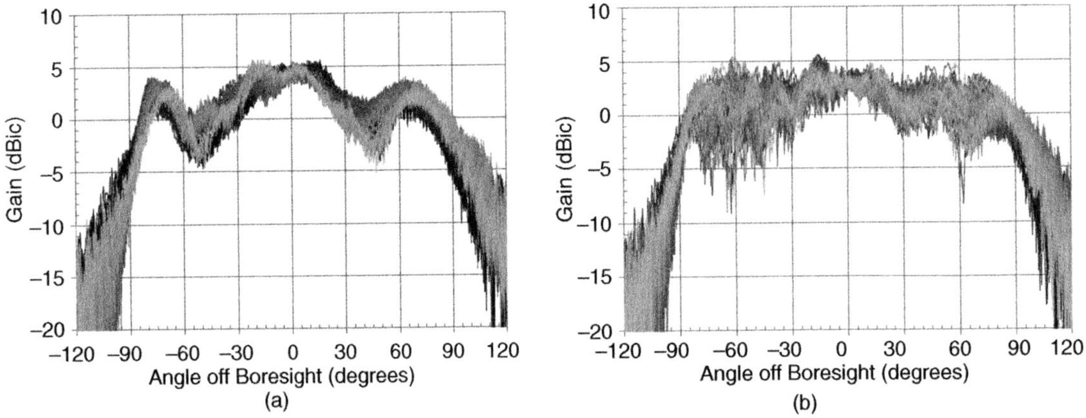

Figure 17.10 The PLGA patterns at 8.4 GHz in the aeroshell configuration (far-field measurements; overlay of multiple phi-cuts): (a) with lid; (b) without lid.

radome that is designed to protect the antenna during Mars entry. The radome consists of a thin base shell made of a cyanate ester/Astroquartz composite and a much thicker layer of Acusil II, a TPS material from ITT. The TPS thickness was sized based on aerothermodynamic entry environment analyses [13]. The dielectric constant and loss tangent of the TPS material was measured and determined to be 1.35 and 0.012, respectively. Antenna pattern measurements showed that the radome has a small but noticeable effect.

Since the PLGA has such a broad beam, scatter off the spacecraft structure was significant. A full-scale mock-up of the top portion of the entry vehicle and the cruise stage was built for antenna pattern testing. Measurements were completed in several configurations, including the entry vehicle and the cruise stage together, just the entry vehicle with the parachute cone lid to simulate EDL before the parachute deployment, and the entry vehicle without the lid to simulate EDL after the parachute deployment. Results of the 8.4 GHz set of measurements are shown in Figures 17.10 and 17.11. The detrimental effects of the spacecraft scattering can be observed in Figures 17.10b and 17.11a. For the aeroshell configuration without the lid, no action was taken to correct the degraded pattern since the data return for EDL will not be mission critical, and the X-band

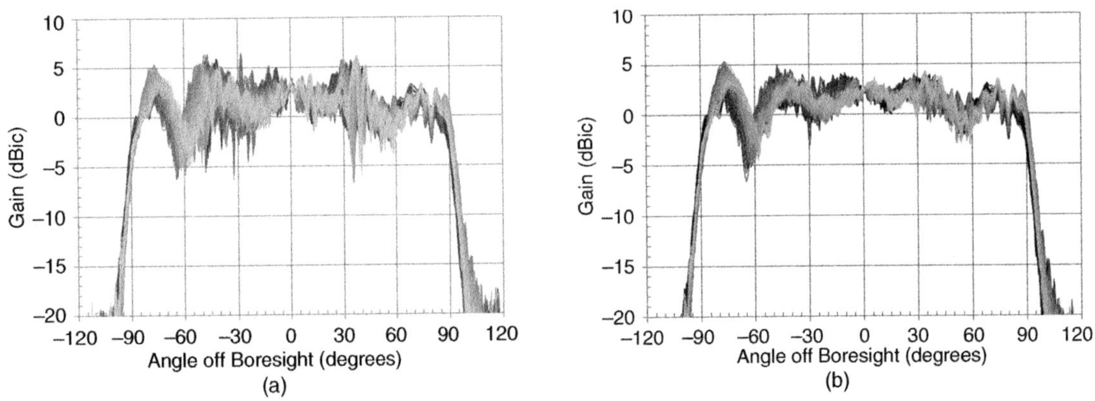

Figure 17.11 The PLGA patterns at 8.4 GHz in the cruise configuration (near-field measurements; overlay of multiple phi-cuts): (a) without RF absorber blankets; (b) RF absorber blankets installed.

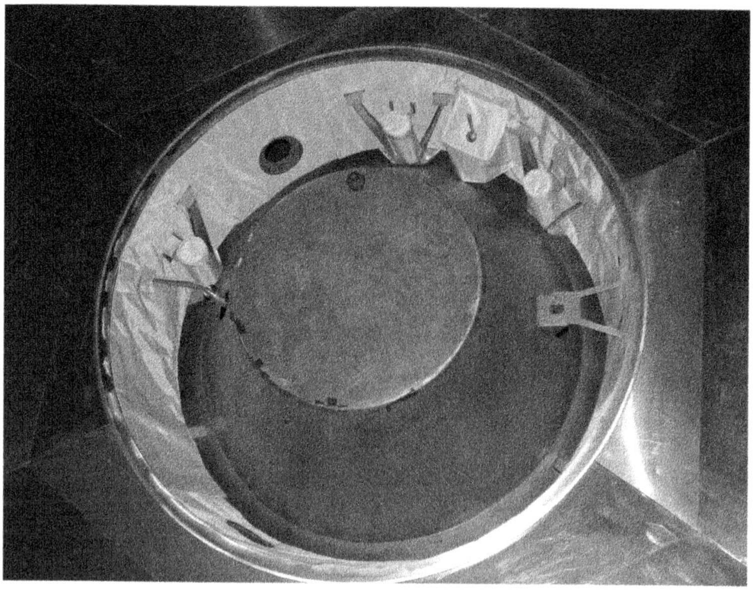

Figure 17.12 *The RF absorber blankets installed on the cruise stage RF mock-up.*

will be a backup to the UHF link for that portion of the mission. For the cruise stage configuration where the X-band will be the prime link and the spacecraft is rotating about the antenna boresight, the deep nulls in the pattern from about 20° to 60° off boresight would cause significant problems and so a solution needed to be implemented. Using a Ticra GRASP model, the cause of the nulls was confirmed to be a double reflection from two conical surfaces on the spacecraft structure. The problem was resolved by installing RF absorber blankets from Millimeter Wave Technology, Inc., onto the launch vehicle adapter. The blankets, shown in Figure 17.12, are a lightweight alternative to other thin conformal RF absorbers and are installed similarly to multilayer insulation (MLI) blankets that are used throughout the spacecraft.

17.3.2.3 The Tilted Low-Gain Antenna (TLGA)

The TLGA uses the same basic design as the PLGA, the only difference being the length of the circular waveguide that was changed for spacecraft accommodation purposes. The TLGA tilts at an angle of 17.5° from the entry vehicle axis of rotation, which aligns it to the nominal anti-velocity vector of the entry vehicle during the guided entry phase of EDL. During guided entry, the velocity vector will vary by ±2.5° about the nominal velocity vector. The view to Earth from the TLGA will remain at a much more constant off-boresight angle than the view from the PLGA, enabling a more stable communications link.

17.3.2.4 The Descent Low-Gain Antenna (DLGA)

The DLGA, shown in Figure 17.13, is a simple choked horn with a septum polarizer. An aperture plug made of TPS material is included in the design to protect the antenna and the waveguide path below it from possible entry of debris during ground assembly and test. An iris with the same width as the aperture plug was included in the design to improve the return loss match. The plug is bonded into the horn by injecting adhesive in small holes through the iris. A GRASP analysis that included the major structural elements of the descent stage was

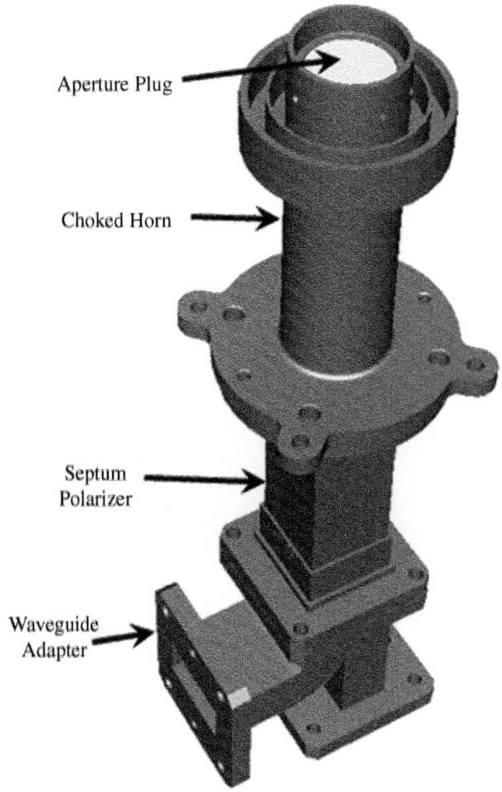

Figure 17.13 *The Descent Low-Gain Antenna (DLGA).*

completed. The analysis concluded that the structural scattering added some ripples to the antenna pattern, but did not significantly degrade the antenna performance.

17.3.2.5 The Rover Low-Gain Antenna (RLGA)

The RLGA, shown in Figure 17.14, is the same basic design as the PLGA and TLGA, but with a shorter circular waveguide portion. A shorting plate covers one waveguide port on the septum polarizer adapter since only right circular polarization (RCP) is required. The driving environmental requirements for the RLGA are a pyroshock overall requirement of 6000 g_{rms}, a qualification temperature range of -135 to $+90\,^{\circ}$C, and a thermal life cycle requirement of 2010 cycles. There is a very small chance that the antenna could be impacted with the electrical umbilical cord from the descent stage during the final stages of EDL. To verify the design against these environmental conditions plus random vibration, quasi-static loads, and power handling, an RLGA qualification unit was fabricated and tested. No changes to the antenna were detected by visual inspection, return loss measurements, or antenna pattern measurements throughout the test campaign.

17.3.2.6 The Rover High-Gain Antenna (HGA)

The HGA, shown in Figure 17.15, is a 48-element microstrip patch array on a two-axis gimbal for pointing directly to Earth. The antenna has a gain of approximately 22 dBi in the uplink frequency band and 23 dBi in

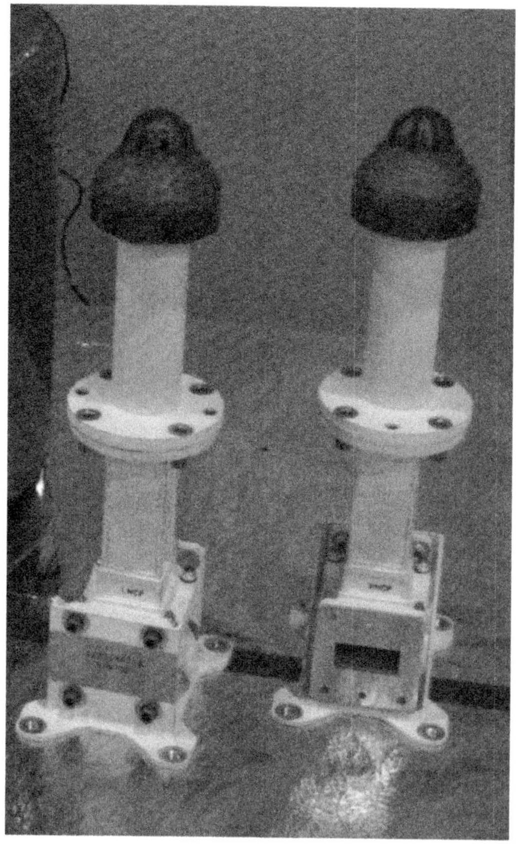

Figure 17.14 *The RLGA flight and flight spare assemblies.*

Figure 17.15 *The HGA assembly.*

Table 17.4 *MSL UHF antennas.*

Antenna name	Abbreviation	Antenna type	Mission phase
Parachute Cone UHF Antenna	PUHF	Conical microstrip array	EDL
Descent UHF Antenna	DUHF	Dipole	EDL
Rover UHF Antenna	RUHF	Quadrifilar helix	EDL and surface operations

the downlink frequency band. Since the antenna is highly directive, the pattern is not degraded from spacecraft structure interactions.

17.3.3 MSL UHF Antennas

The MSL spacecraft carries three UHF antennas, listed in Table 17.4. The Parachute Cone UHF (PUHF) antenna is a modified design from an antenna used on the NASA Phoenix mission. The Descent UHF (DUHF) antenna and Rover UHF (RUHF) antenna are new designs. The details of each antenna are presented in the following subsections.

17.3.3.1 Parachute Cone UHF (PUHF) Antenna

The PUHF antenna will be the first UHF antenna to be used in the EDL sequence, operating from cruise separation to the start of powered descent. The Wraparound antenna was designed and manufactured by Haigh-Farr, Inc., under a subcontract to the Jet Propulsion Laboratory. The antenna is a conical microstrip array with eight patches and is installed on the parachute closeout cone on the entry vehicle, as is shown in Figure 17.4. The antenna provides a close to omnidirectional pattern, which is needed because the location of the Mars orbiters with respect to the MSL spacecraft during EDL will not be known until after the MSL spacecraft launches. The antenna polarization is right circular (RCP), with each of the individual patches radiating RCP. The antenna, like all the EDL antennas, only operates in transmit mode, and the maximum required power handling capability is 15 W.

The driving requirements for antenna selection were an omnidirectional pattern, ability to be packaged compactly or conformally onto the entry vehicle, and ability to withstand high operating temperatures. The antenna operates under a layer of ITT Acusil II TPS material which protects it from extreme entry temperatures, but the temperature at the antenna surface can still reach as high as 200 °C. The Wraparound antenna has an outer thermal protective layer that is integral to the antenna design and enables the antenna to operate at high temperatures. Figure 17.16 shows the parachute closeout cone before and after the TPS application. As can be seen, the PUHF antenna is extremely low profile and conforms to the conical surface.

The costs and potential risks involved with testing the flight antenna on the actual entry vehicle were prohibitive, so MSL pursued a verification program [14] based on the successful Phoenix antenna verification program implemented by Lockheed Martin and Haigh-Farr. During the design phase, the antenna patterns on the aeroshell entry vehicle and on the parachute closeout cone by itself were calculated. Fifth-scale models of the antenna and entry vehicle were fabricated, and antenna patterns were measured on the scale entry vehicle and on the scale parachute closeout cone. For the flight antenna assembly, patterns were measured just on the parachute closeout cone, and then compared to the calculated and fifth-scale patterns on just the cone. The antenna patterns were very similar to the fifth-scale and calculated patterns, so it was concluded that the fifth-scale full aeroshell patterns, shown in Figure 17.17, were accurate.

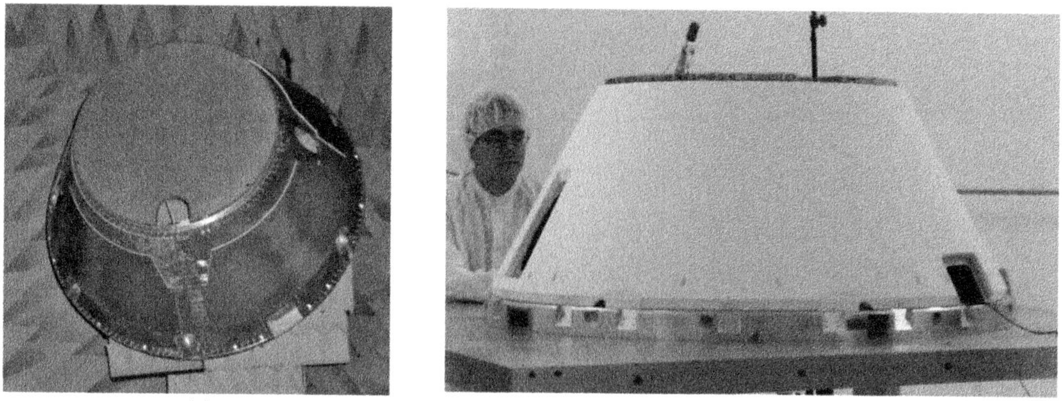

(a)

Figure 17.16 *The parachute closeout cone before and after TPS application.*

(a)

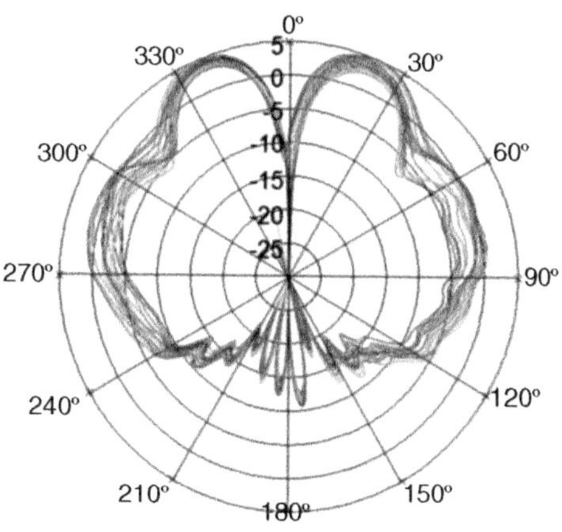

(b)

Figure 17.17 *Measured gain of the fifth-scale model with the full aeroshell (dBi): (a) coordinate system reference; (b) overlay of several phi-cuts.*

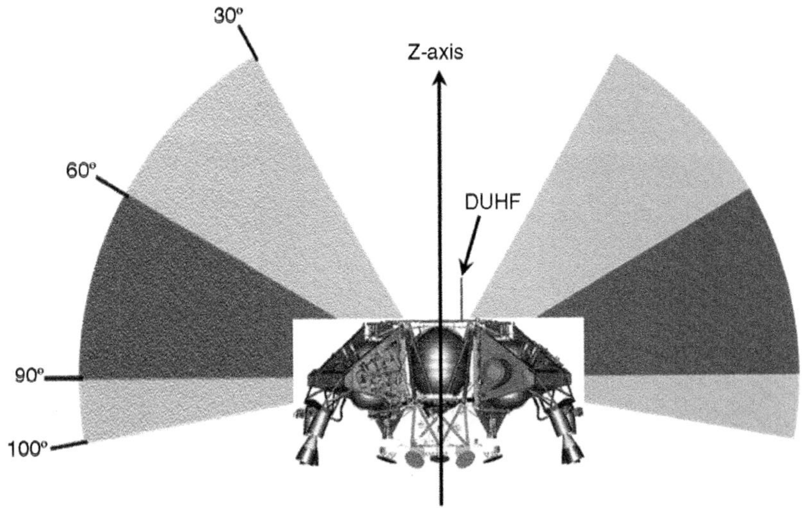

Figure 17.18 *Required angular coverage region for the DUHF antenna. The gray areas indicate the areas of required coverage. The dark areas are the priority regions.*

17.3.3.2 Descent UHF (DUHF) Antenna

For the approximately 40 seconds after the back shell separates from the spacecraft, and before the rover separates from the descent stage, the DUHF antenna is needed to provide coverage in the region of 30° to 100° from the descent stage *z*-axis. Figure 17.18 shows the required coverage areas relative to the descent stage geometry. A patch, a monopole on a ground plane, a deployable dipole, and a non-deployable dipole were considered in the search for an antenna that could be stowed in the available space and provide sufficient gain out to 100° off the descent stage *z*-axis. Although the patch could be the most easily accommodated, it did not provide the necessary coverage at high angles off the descent stage *z*-axis. The major disadvantage of dipoles and monopoles is that they radiate linear polarization while circular polarization was needed, but they best met the packaging constraints. The gain requirements are not very stringent, so the 3 dB polarization mismatch loss was acceptable. Dipole and monopole patterns with the spacecraft scattering contributions were calculated using WIPL-D's moment method analysis software. Both antenna types exhibited pattern nulls in the desired region of coverage, but the dipole nulls were at angles less than 45° off the descent stage *z*-axis, and the monopole nulls were at angles as high as 75° to 100°, depending on the monopole location. The priority region of coverage is 60° to 90°, which is the most likely location of the Mars orbiters, so the dipole was selected as the best choice.

17.3.3.3 Rover UHF (RUHF) Antenna

The RUHF antenna will be the primary telecommunications antenna used during rover operations. The key driving requirements for the RUHF antennas are listed in Table 17.5.

A quarter-turn quadrifilar helix design was selected because it provides the needed coverage, its axial ratio is satisfactory over the specified coverage area, it does not need a ground plane, the front-to-back ratio is good, and it could be accommodated in the available volume. Flat ribbon conductors were used for the radiators to increase the bandwidth over a thin-wire design (Figure 17.19). A 3 dB hybrid was used to phase the coaxial feed lines appropriately. A germanium-coated black polyimide film cover was added to prevent

Table 17.5 RUHF antenna key driving requirements.

Requirement	Value
Transmit frequency band	395–406 MHz
Receive frequency band	434–440 MHz
Polarization	RCP
Transmit gain	>1.0 dBic for angles off boresight < 45° > − 1.0 dBic for angles off boresight < 70° > − 3.0 dBic for angles off boresight < 80°
Receive gain	>0.0 dBic for angles off boresight < 45° > − 2.0 dBic for angles off boresight < 70° > − 4.0 dBic for angles off boresight < 80°
Return loss	< − 10 dB
Power handling	15 W operating; 60 W with test margin
Thermal range	−135 to +90 °C (includes test margin)
Random vibration	Overall 7.9g_{rms} (includes test margin)
Pyrotechnic shock	6000 g
Thermal cycle life test	2010 thermal cycles (includes test margin)

static charge build-up from the exterior and to improve the thermal–optical properties of the exterior for thermal control.

To support the ribbon conductors yet keep the antenna lightweight, a thin Astroquartz composite substrate was fabricated. The radiator pattern was etched from copper-clad polyimide film and then bonded to the Astroquartz substrate. Short copper-foil strips were soldered at the joints of the polyimide film pieces to splice the radiators from the top and bottom to the cylinder side. An Astroquartz dome was bonded to the top of the cylinder to protect the wires at the radiator feed point. The mass of the final antenna assembly, including the aluminum base and hybrid coupler, is 0.55 kg. The design successfully passed all environmental testing, including high-power operation at 60 W in vacuum and in a simulated Mars atmosphere.

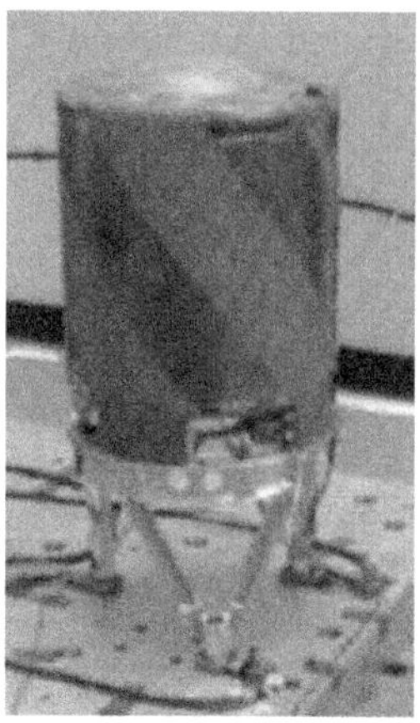

(a)

(b)

Figure 17.19 *Rover UHF antenna (shown with its rover mounting bracket and without its electrical static discharge (ESD)/thermal cover).*

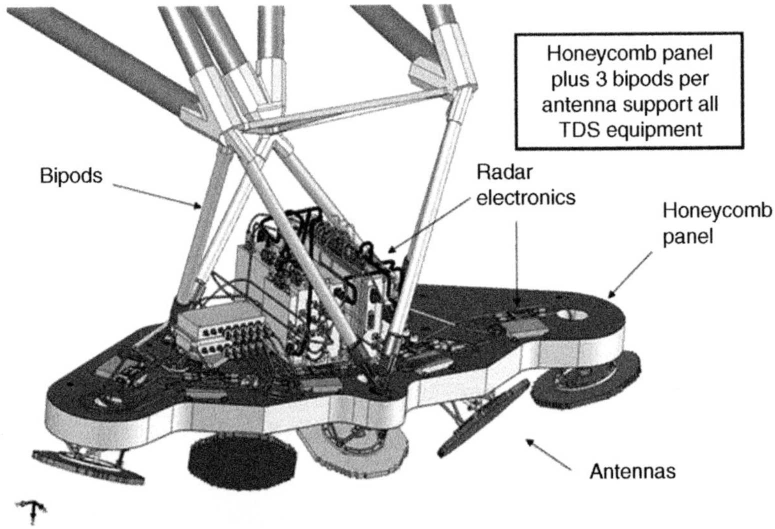

Figure 17.20 *Illustration of MSL Terminal Descent Sensor (TDS) assembly. The nadir antenna is not visible in this view.*

17.3.4 MSL Terminal Descent Sensor (Landing Radar)

One of the most critical aspects of the EDL phase is the final, or terminal, descent that places the MSL rover on the Martian surface. To perform this maneuver successfully, both altitude and ground velocity must be known with high accuracy. The MSL Terminal Descent Sensor (TDS) is a Ka-band radar that provides this critical information during powered descent (see Figure 17.5). The antenna is a key element of this system.

As shown in Figure 17.20, the TDS system uses a set of six individual antennas, pointing in different directions, to perform the altitude and velocity measurements. Table 17.6 lists the key driving requirements for each of these antennas. A brazed aluminum waveguide slot array provided the best technical approach to meet these requirements. However, at the start of this project Ka-band slot arrays had no flight heritage and were an active area of research (e.g., see [15,16]). Nevertheless, EMS Technologies successfully developed slot arrays (Figure 17.21) that met all of the antenna requirements.

Table 17.6 *MSL TDS antenna – key driving requirements.*

Requirement	Value
Center frequency	35.75 GHz
Bandwidth	250 MHz
Polarization	Linear
Peak gain	>34.77 dBi
Peak sidelobe level	< −24 dB (from pattern peak)
Average sidelobe level	< −30 dB (from pattern peak)
Voltage standing wave ratio	<1.5:1
Qualification temperature range	−45 to +70 °C
Mass	344 g
Volume	22 cm diameter, 4 cm depth

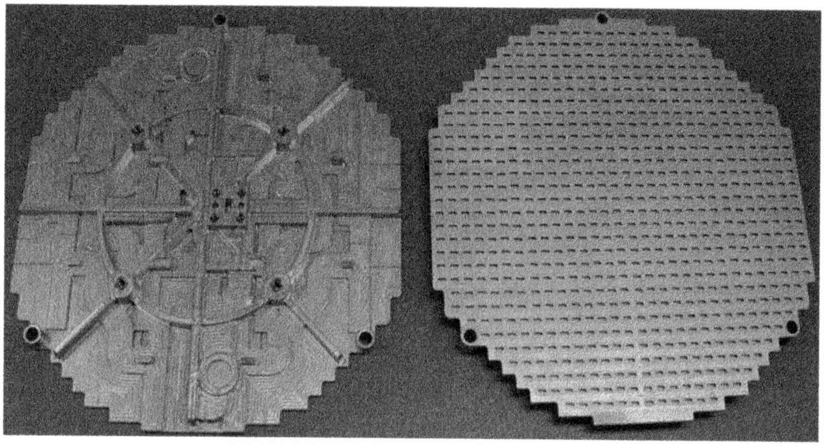

Figure 17.21 MSL TDS waveguide slot array. Courtesy of EMS Technologies.

In addition to the individual antennas, the integration, alignment, and beam pointing metrology was of critical importance for TDS. This was accomplished using the JPL Cylindrical Near-Field Antenna Test Facility. Two orientations (vertical and horizontal) were used to obtain a complete pattern measurement for all observation angles. A laser tracker system provided the very precise mechanical alignment information to characterize electrical vs. mechanical boresight.

17.4 Case Study II – Juno

17.4.1 Juno Mission Description

The primary objective of the Juno mission is to improve our knowledge of Jupiter's interior structure in order to determine how the planet was formed [17]. It is believed that this knowledge will provide a fundamental understanding of the formulation processes and early evolution of our Solar System. The key instrument for this measurement is a six-wavelength microwave radiometer. Juno also includes additional instruments, including a vector magnetometer, plasma and energetic particle detectors, a radio and plasma wave experiment, an ultraviolet spectrometer/imager, and a gravity/radio science system. These instruments will make maps of the gravity, magnetic fields, and atmospheric composition of Jupiter.

The Juno spinning, solar-powered spacecraft was launched on August 5, 2011 (Figure 17.22). Following a five-year cruise phase, including an Earth fly-by for gravity assist, the spacecraft will arrive at Jupiter in 2016 to begin a one-year mission. At Jupiter, the spacecraft will be placed into a highly elliptical orbit that facilitates close atmospheric observation of the giant planet while minimizing exposure to the intense Jovian radiation belts. Juno will sample the entire Jovian surface by executing 32 eleven-day-long orbits.

The Juno instruments and telecom system use a wide variety of antennas (Figure 17.23) and the mission imposes a number of unique requirements on these antennas. First, the Atlas V launch vehicle presents a high launch vibration environment which resulted in the highest vibration requirements of any JPL mission to date. During the six-year mission, the temperature extremes for the Juno antennas cover the range from −150 to 120 °C. Finally, despite efforts to minimize encounters with the Jovian radiation belt, the antennas will be exposed to an intense electron bombardment that largely precludes the use of many common dielectric materials. This subsection describes the engineering approach to meet the required electrical performance subject to the constraints imposed by these environmental conditions.

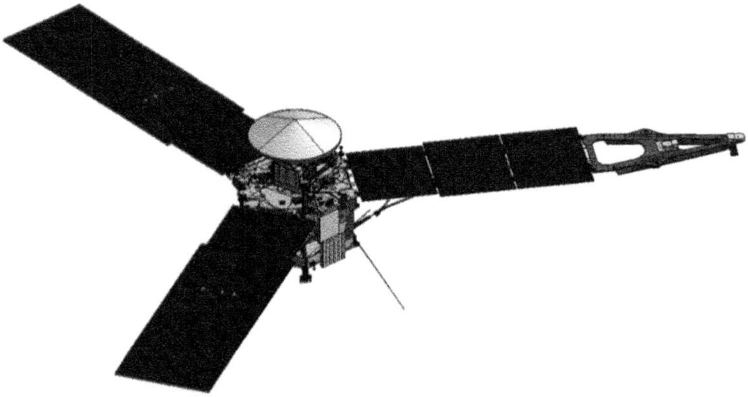

Figure 17.22 Illustration of the Juno spacecraft.

17.4.2 Telecom Antennas

The set of Juno telecom antennas provides the critical link required for spacecraft command and control and sending the science payload back to Earth. These antennas include a high-gain antenna (HGA), low-gain antenna (LGA), medium-gain antenna (MGA), and a unique toroidal low-gain (TLGA) antenna. Since the LGA and MGA are similar to the MSL antennas described above, the subsections below focus on the HGA and TLGA.

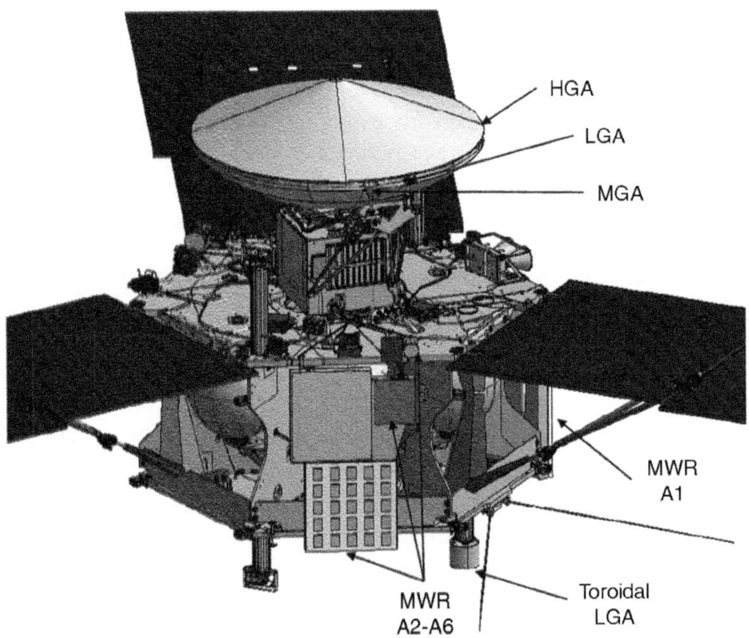

Figure 17.23 Illustration of the telecom and instrument antennas on the Juno spacecraft.

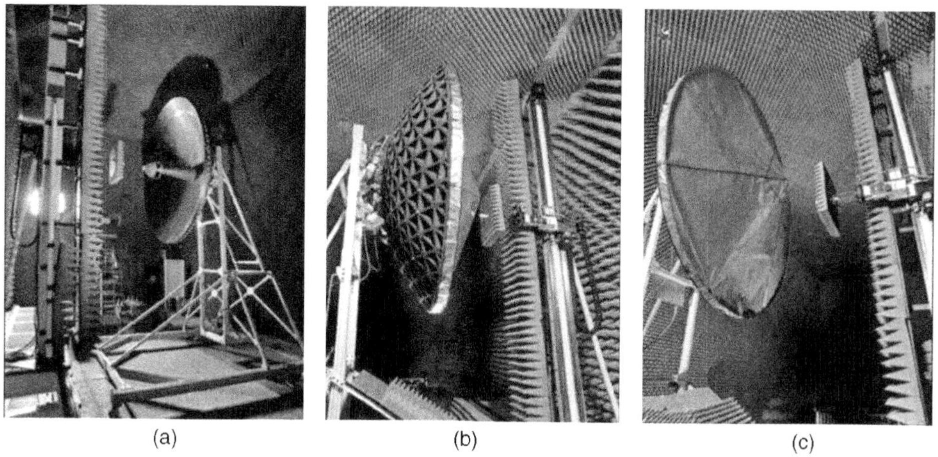

Figure 17.24 Juno 2.5 m HGA in the JPL Planar Near-Field Test Facility. The mechanical design and fabrication of the reflector were performed by ATK, Inc. Custom Microwave, Inc., performed the mechanical design and fabrication of the feed assembly. (a) HGA without thermal radome. (b) Rear view of HGA. (c) HGA with thermal radome.

17.4.2.1 HGA

The Juno HGA shown in Figure 17.24 is a 2.5 m diameter reflector antenna designed to support both X-band and Ka-band operation. The axially symmetric, nominally parabolic reflector is aligned with the spacecraft spin axis to provide continuous, high data rate communications as the spacecraft rotates. A Gregorian configuration was selected over more traditional Cassegrain optics to maximize overall efficiency of the antenna within the available volume. It was necessary to broaden the Ka-band antenna beam to accommodate limitations of the spacecraft attitude control system. This was accomplished through a unique surface shaping technique developed at JPL. The reflector aperture is covered with a thin germanium-coated Kapton radome to minimize direct solar illumination of the composite reflector and thereby improve thermal characteristics of the antennas system. Table 17.7 summarizes the key performance characteristics of the Juno HGA.

Table 17.7 JUNO HGA performance.

Requirement	Value
Transmit frequency bands	X: 8.38–8.43 GHz; Ka: 32.055–32.110 GHz
Receive frequency bands	X: 7.135–7.175 GHz; Ka: 34.335–34.395 GHz
Polarization	Circular
Transmit gain (edge-of-coverage region)	X-band: >43.0 dBic (±0.25° from boresight)
	Ka-band: >41.5 dBic (±0.25° from boresight)
Receive gain (edge-of-coverage region)	X-band: >41.5 dBic (±0.25° from boresight)
	Ka-band: >38.5 dBic (±0.25° from boresight)
Return loss	>20 dB
Power handling	X: 25 W continuous wave (CW) operating; 100 W with test margin
	Ka: 2.5 W CW operating; 10 W with test margin
Thermal range	−170 to +150 °C (tested range)
Random vibration	Overall: 10.4g_{rms} (includes test margin)
	Feed: 34.7g_{rms} (includes test margin)
HGA assembly mass	21.6 kg (includes reflector assembly, feed, and radome)

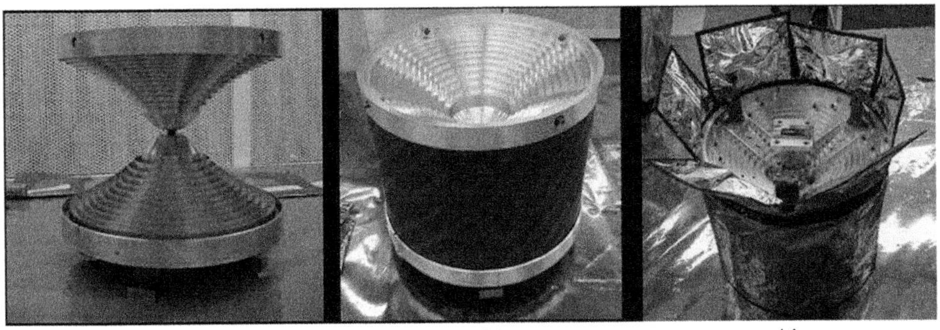

Figure 17.25 Juno Toroidal Low-Gain Antenna (TLGA): (a) TLGA without polarizer; (b) TLGA with polarizer; (c) flight configuration.

17.4.2.2 TLGA

The spinning nature of the Juno spacecraft led to the need for an LGA with a toroidal pattern to maintain continuous communications during cruise maneuvers. This was accomplished with a biconical antenna as shown in Figure 17.25. The antenna is symmetrically fed by a coaxial transmission line and features a profiled and corrugated surface designed to meet the elevation beamwidth requirement. A cylindrical multilayer meanderline polarizer is used to convert from linear to circular polarization. This polarizer uses an Astroquartz/Kapton construction method that is compliant with the aluminum biconical antenna structure. A germanium Kapton thermal blanket provides the required thermal control. Table 17.8 summarizes the key performance characteristics of the Juno TLGA.

17.4.3 Juno Microwave Radiometer Antennas

The purpose of the Juno Microwave Radiometer (MWR) instrument is to measure water and ammonia composition in the Jovian atmosphere. Six wavelengths ranging from 1.3 to 50 cm provide microwave sounding at different depths. The MWR system contains six linearly polarized receive-only antennas. Each

Table 17.8 JUNO TLGA antenna performance.

Requirement	Value
Transmit frequency band	8.38–8.43 GHz
Receive frequency band	7.135–7.175 GHz
Polarization	Circular
Transmit gain, boresight	>6.0 dBic
Transmit gain, edge-of-coverage region	>2.7 dBic min (±10.0° from boresight)
Receive gain, boresight	>4.5 dBic
Receive gain, edge-of-coverage region	>2.3 dBic min (±10.0° from boresight)
Return loss	>20 dB
Power handling	25 W CW operating; 100 W with test margin
Thermal range	−150 to +120 °C (tested range)
Random vibration	Overall: 11.4g_{rms} (includes test margin)
Mass	<2.0 kg

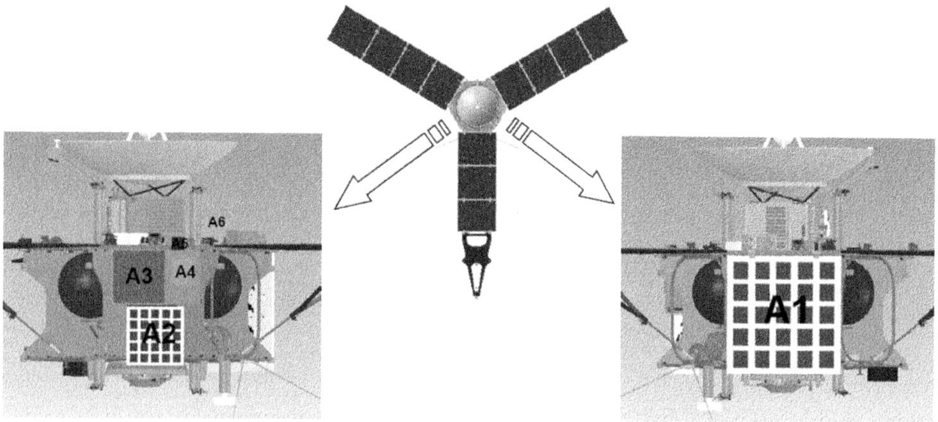

Figure 17.26 *MWR A1–A6 antennas on the Juno spacecraft.*

antenna operates at one of the six frequency bands at 0.6, 1.25, 2.6, 5.2, 10, and 22 GHz, referred to as A1–A6, respectively. The antennas are mounted on two sides of the hexagon-shaped spacecraft as shown in Figure 17.26. The A1 antenna at 600 MHz occupies one entire side of the hexagon and is directly mounted on the spacecraft. The A2–A5 antennas are mounted on another face of the hexagon, with A5 and A6 occupying the upper deck region. Each antenna is connected to the receiver via either coaxial cables (A1–A4) or rectangular waveguides (A5–A6).

All of the Juno MWR antennas require high aperture efficiency, low mass, low average sidelobe levels, high radiation tolerance, and the ability to survive high vibration levels and wide temperature ranges. For example, the MWR antennas are tested at temperatures ranging from -150 to $+120\,°C$ and proto-flight vibration tests up to $40.6\,g_{rms}$. Due to the large frequency range, different types of antennas are used to meet these requirements. These include two patch antennas (600 MHz and 1.25 GHz), three waveguide slot arrays (2.6, 5.2, and 10.0 GHz), and a corrugated horn (22 GHz). The subsections below describe these antenna designs.

17.4.3.1 Patch Antennas

The two lowest Juno frequencies (600 and 1250 MHz) present a unique challenge due to large physical size and low sidelobe requirements [18–20]. From an antenna radiation pattern standpoint, a patch array is ideally suited to this application. However, standard microstrip patch arrays have several drawbacks that needed to be overcome for this application. Patch arrays are typically printed on thin dielectric substrates using photolithography and utilize a stripline feed structure. These dielectric materials introduce unacceptable RF loss (low antenna efficiency) and will not inherently withstand the Jovian radiation environment. To solve this problem, JPL developed an all-metal patch radiating element which is used in conjunction with an air-stripline feed network connected by SiO_2 coaxial cables [21]. This configuration met all electrical requirements and proved to be very efficient, lightweight, and radiation tolerant.

Figure 17.27 shows the 600 MHz Juno patch array. The key requirements for the two low-frequency antennas (denoted A1 and A2) are summarized in Table 17.9. These antennas comprise a 5×5 square grid array with 0.6 wavelength element spacing. A separable aperture distribution is used to simplify the feed network, which uses six identical five-way power dividers in a corporate topology. To meet the sidelobe requirements, a 30 dB Taylor distribution is used in both azimuth and elevation planes. The air stripline power divider network implementation resulted in a total insertion loss of less than 1.2 dB for the overall antenna including patch radiators [22].

(a)

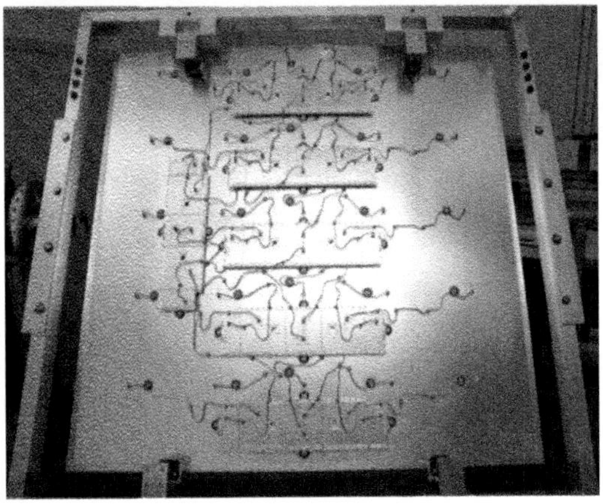

(b)

Figure 17.27 *JUNO MWR A1 microstrip patch array: (a) front view – metal patches; (b) rear view – air stripline power divider.*

Table 17.9 *Juno MWR A1 and A2 patch array antenna requirements.*

Requirement	A1	A2
Center frequency	600 MHz	1200 MHz
Bandwidth	4.5%	4.5%
Polarization	Linear	Linear
Gain (ref. only)	>19.0 dBi	>19.0 dBi
Loss	<1.65 dB	<1.75 dB
3 dB beamwidth	<22°	<22°
Average return loss	>15 dB	>15 dB
Max return loss	10 dB	10 dB
Thermal range	−135 to +120 °C	−135 to +120 °C
Random vibration	$40.6g_{rms}$	$23.5g_{rms}$
Mass (actual)	13.83 kg	4.89 kg
Volume (actual)	$160 \times 160 \times 13.2$ cm^3	$76.8 \times 76.8 \times 9.8$ cm^3

Figure 17.28 Photo of the Juno MWR metal patch radiator.

The radiating elements consist of a 'metal patch' as shown in Figure 17.28. The size of a patch is approximately 0.5 × 0.25 wavelengths with a height of 0.05 wavelengths. The rectangular element is supported by a central column with a diameter of ~0.1 wavelength. The element is fed by a coaxial probe and generates linear polarization. This design achieves an impedance bandwidth of greater than 8% in the array environment (for average return loss of 15 dB), which is possible largely due to elimination of dielectric material. Each patch is machined from a single block of aluminum. The central column is hollowed to reduce mass and provide a convenient location to bolt the patch to the ground plane. The probe connects to the patch through a slip fit connector that is not susceptible to damage by vibration and enables the patch to be removed and replaced if desired.

The Juno patch arrays met all the performance requirements. Figure 17.29 shows measured vs. calculated performance for the 0°, 45°, and 90° pattern cuts of both the Juno A1 and A2 antennas. Note that the metal patch results in a small E-plane pattern squint (0.5° for A1, 0.25° for A2) due to the asymmetrical feed arrangement. Table 17.10 gives a summary of the average sidelobe performance for A1 (A2 is similar).

Table 17.10 Juno A1 antenna average sidelobe performance.

Angular range	Requirement	Calculated (HFSS)	Near-field measurement
25° to 32°	≤−24	−27.3	−27.8
32° to 40°	≤−28	−32.3	−32.6
40° to 70°	≤−35	−37.9	−37.9
70° to 100°	≤−43	−44.9	−45.3
90° to 150°	≤−44	−44.6	−45.3
150° to 180°	≤−40	−41.0	−42.3

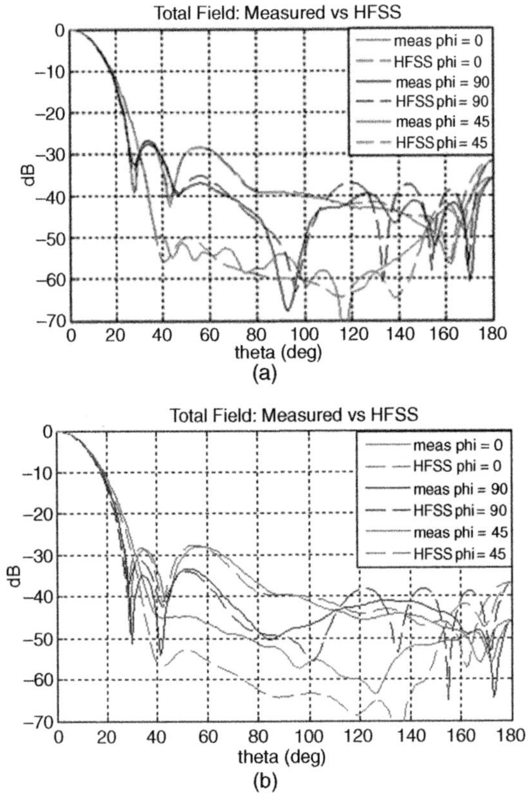

Figure 17.29 *Juno MWR microstrip patch array measured vs. calculated patterns: (a) A1, 600 MHz antenna; (b) A2, 1200 MHz antenna.*

Overall, the measured patterns are in good agreement with predictions. The single patch with the probe feed proved to be extremely rugged and passed vibration testing of 40 g_{rms} in z-axis and 20 g_{rms} in x-axis and y-axis. An acoustic test was performed on the A1 and A2 antenna assemblies successfully. The antennas also passed the thermal vacuum test with temperature ranging from −135 to 120 °C.

17.4.3.2 Waveguide Slot Arrays

The key antenna design challenge for the three intermediate Juno frequencies (2.6, 5.2, and 10.0 GHz) is to achieve low integrated sidelobe levels with an antenna that withstands the extreme vibration, thermal, and radiation environment. A brazed aluminum waveguide slot array is well suited to this application [23,24]. Figure 17.30 illustrates the 10 GHz Juno waveguide slot array (the three antennas are essentially scaled versions of a single design). The nominal performance requirements for these antennas (denoted A3, A4, and A5) are summarized in Table 17.11.

Each antenna is a longitudinal shunt slot array of 8 × 8 elements with ~0.71 wavelength element spacing. A 35 dB, $\bar{n} = 4$ separable Taylor distribution is used to achieve the required sidelobe performance. The shunt slot rows are fed by standard series slot standing wave feeds as illustrated in Figure 17.31, with an H-plane tee used to combine the symmetric array halves. One unique aspect of this antenna is the mirror symmetry offset pattern employed down the centerline of the antenna. This arrangement splits the two-slot periodicity in order to

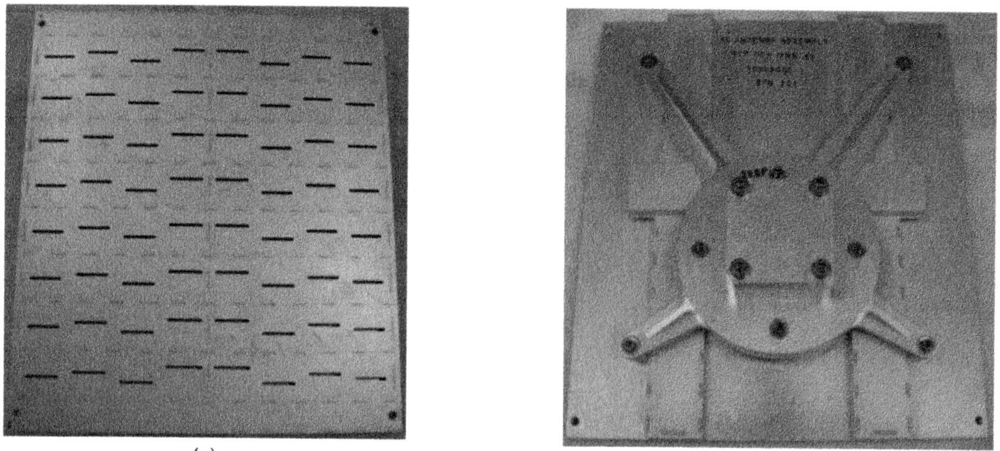

(a) (b)

Figure 17.30 Juno MWR A5 waveguide slot array. Rantec Microwave Systems, Inc., fabricated the A3, A4, and A5 antennas. (a) Front view – shunt radiating slots. (b) Rear view – feed and support structure.

disperse the 'butterfly lobes' that typically appear in the intercardinal plane of a slot array. The feature enabled the otherwise standard slot array architecture to meet the sidelobe requirements.

Well-established methods were employed in the design of these antennas [23]. The Elliott mutual coupling design method was used to tune the resonant frequency of the slots in the array environment. Two separate analysis methods, the moment method and the finite element method (HFSS), were used to analyze the radiation pattern performance. In order to verify that the antennas meet sidelobe-level requirements, Monte Carlo simulations of the array were carried out to account for modeling errors and for systematic and random errors resulting from the manufacturing process. The tolerance and error analysis showed that while there was adequate margin for the average sidelobe-level performance, the return loss bandwidth did not have sufficient margin. To accommodate this, the design includes the option to add a matching iris at the input port of the antenna after the antenna is fabricated. This feature was utilized to tune the A3 antenna to improve the average return loss from 15.7 to 17.7 dB.

Table 17.11 Juno slot array antenna requirements.

Requirement	A3	A4	A5
Center frequency	2.6 GHz	5.2 GHz	10.0 GHz
Bandwidth	4%	4%	4%
Polarization	Linear	Linear	Linear
Gain (ref. only)	>23.0 dBi	>23.0 dBi	>23.0 dBi
Loss	<0.65 dB	<0.75 dB	<0.85 dB
3 dB beamwidth	<22°	<22°	<22°
Average return loss	>15 dB	>15 dB	>15 dB
Max return loss	10 dB	10 dB	10 dB
Thermal range	−150 to +120 °C	−150 to +120 °C	−150 to +120 °C
Random vibration	$21.1g_{rms}$	$21.1g_{rms}$	$11.9g_{rms}$
Mass (actual)	7.25 kg	1.46 kg	0.51 kg
Volume (actual)	$77.1 \times 67.3 \times 8.9\ cm^3$	$38.6 \times 34.0 \times 5.7\ cm^3$	$20.1 \times 17.9 \times 4.4\ cm^3$

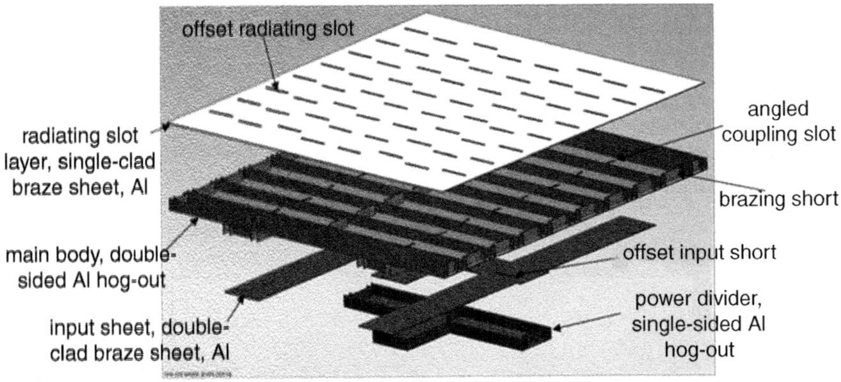

Figure 17.31 *Exploded view showing the Juno waveguide slot array architecture.*

The Juno slot arrays met all the performance requirements. Figure 17.32 shows representative principal plane pattern measurements for the Juno A3 antennas and Table 17.12 gives a summary of the corresponding average sidelobe performance. Overall, the predicted performance is in excellent agreement with measured results. Although not shown, the A4 and A5 antennas have similar performance.

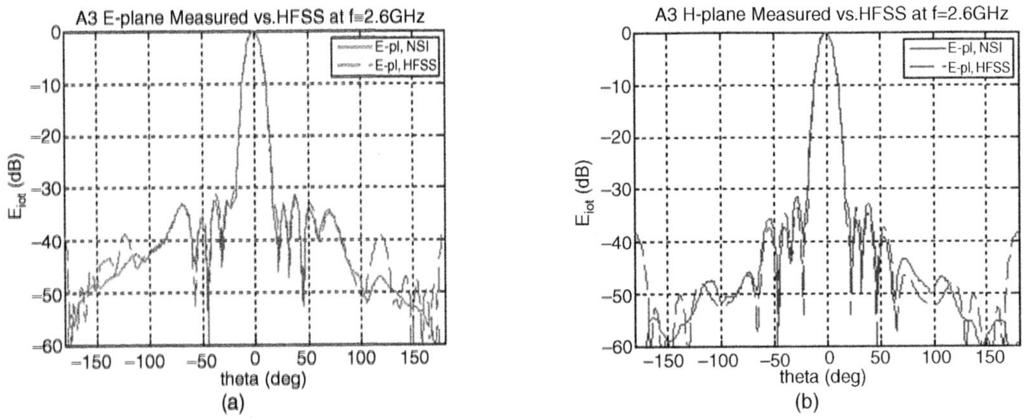

Figure 17.32 *Juno MWR A3 principal plane radiation patterns: (a) E-plane patterns; (b) H-plane patterns.*

Table 17.12 *Juno A3 antenna average sidelobe performance.*

Angular range	Requirement	Calculated (HFSS)	Near-field measurement
20° to 30°	−34	−36.7	−36.4
30° to 40°	−36	−38.4	−38.3
40° to 70°	−38	−43.5	−43.0
70° to 90°	−39	−46.0	−45.8
90° to 150°	−40	−50.8	−52.1

Figure 17.33 *A6 antenna-profiled corrugated horn with a cover.*

17.4.3.3 Corrugated Horn

Corrugated horns are known to have low sidelobe performance. A short corrugated horn at 22 GHz meets the Juno spacecraft mass and volume requirements. Therefore, a profiled corrugated horn a with circular to rectangular transition was selected for the A6 antenna (Figure 17.33). The horn is profiled in order to reduce the overall length while retaining low sidelobes. Machined aluminum construction results in a rugged antenna with low insertion loss and no electrical static discharge issues. The corrugations near the throat of transition are deep and thin, which is challenging to machine as an integral part of the corrugated section. To circumvent this issue, these corrugations are fabricated separately as rings and then press fit into a sleeve.

Table 17.13 lists the key driving requirements of the A6 antenna. The radiation pattern analysis was performed using the mode matching method. Measured and calculated patterns show excellent agreement (Figure 17.34) with the −60 dB level. Sidelobe requirements were met with generous margins (Table 17.14). Indeed, the corrugated horn met all the Juno system performance requirements.

Table 17.13 *Juno corrugated horn antenna requirements.*

Requirement	A6
Center frequency	22 GHz
Bandwidth	4.5%
Polarization	Linear
Gain (ref. only)	>23.0 dBi
Loss	<0.7 dB
3 dB beamwidth	<22°
Average return loss	>15 dB
Max return loss	10 dB
Thermal range	−150 to +120 °C
Random vibration	$11.9g_{rms}$
Mass (actual)	0.75 kg
Volume (actual)	15.3 cm diameter × 34 cm long

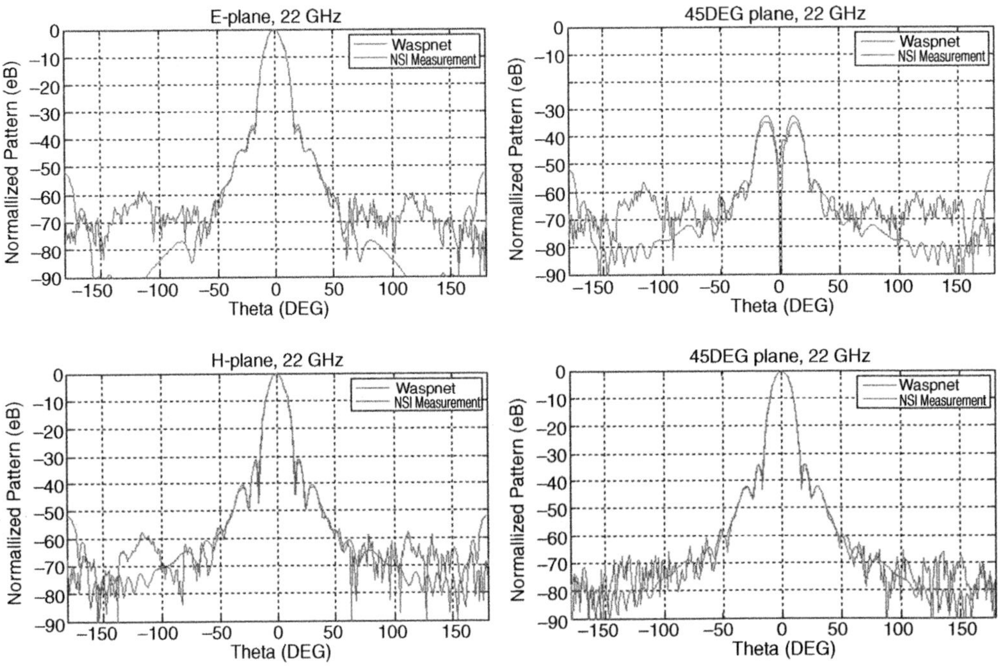

Figure 17.34 Measured and calculated Juno A6 E-plane and H-plane patterns at 22 GHZ.

Table 17.14 Juno A6 antenna average sidelobe performance.

Angular range	Requirement	Calculated (Waspnet)	Near-field measurement
20° to 30°	−31	−38.3	−39.6
30° to 40°	−33	−45.3	−46.0
40° to 70°	−35	−59.6	−59.5
70° to 90°	−36	−70.3	−64.4
90° to 150°	−40	−77.4	−64.0

Acknowledgments

The research on the MSL and Juno projects was carried out at the Jet Propulsion Laboratory, California Institute of Technology, under a contract with the National Aeronautics and Space Administration. The authors wish to acknowledge and thank many of their colleagues who contributed to the MSL and Juno antenna design and development.

The MSL telecommunications antenna team at JPL: Luis Amaro, Nataliya Blyznyuk, John Kent Frankovich, Saina Ghandchi, Jefferson Harrell, Daniel Hoppe, John Huang, Peter Ilott, Behrouz Khayatian, Stephen Pakbaz, Jack Prater, Michael Tsai, and Robert Usiskin. The MSL TDS antenna team: Jefferson Harrell and John Huang from JPL and Michael Guler from EMS. The Juno telecommunications antenna team at JPL: Luis Amaro, Ronald Kruid, Aluizio Prata, and Joseph Vacchione. The Juno radiometer antenna team at JPL: Neil Chamberlain, Jefferson Harrell, Richard Hughes, Julie Jakoboski, Behrouz Khayatian, and Mark Zawadzki.

References

1. National Aeronautics and Space Administration (July 2011) National Space Science Data Center. [Online] http://nssdc.gsfc.nasa.gov/nmc/spacecraftDisplay.do?id=1959-012A.
2. Mizzoni, R. (1994) The Cassini High Gain Antenna (HGA): a survey on electrical requirements, design and performance. IEE/SEE Seminar on Spacecraft Antennas, London, UK, pp. 6/1–6/10.
3. Imbriale, W.A. (2006) *Spaceborne Antennas for Planetary Exploration*, John Wiley & Sons, Inc., Hoboken, NJ.
4. International Telecommunication Union (2009) Factors Affecting the Choice of Frequency Bands for Space Research Service Deep Space (Space-to-Earth) Telecommunication Links, Report ITU-R SA.2167, Geneva.
5. Edwards, C.D. Jr., Arnold, B., DePaula, R. *et al.* (2006) Relay communications strategies for Mars exploration through 2020. *Acta Astronautica*, **59**(1–5), 310–318.
6. Edwards, C.D. (2007) Relay communications for Mars exploration. *International Journal of Satellite Communications and Networking*, **25**(2), 111–145.
7. Jet Propulsion Laboratory, California Institute of Technology (n.d.) Mars Science Laboratory. [Online] http://mars.jpl.nasa.gov/msl/ (accessed 16 December 2011).
8. National Aeronautics and Space Administration (September 2010) Mars Science Laboratory. [Online] http://mars.jpl.nasa.gov/msl/files/msl/MSL_Fact_Sheet-20100916.pdf.
9. Makovsky, A., Ilott, P., and Taylor, J.(November 2009) DESCANSO: Deep Space Communications and Navigation Systems. [Online] http://descanso.jpl.nasa.gov/DPSummary/Descanso14_MSL_Telecom.pdf.
10. Olea, A., Montesano, A., Montesano, C., and Arenas, S. (2010) X-band high gain antenna qualified for Mars atmosphere. Proceedings of the Fourth European Conference on Antennas and Propagation (EuCAP), Barcelona, Spain, pp. 1–5.
11. National Aeronautics and Space Administration (March 2011) Jet Propulsion Laboratory: Mission News. [Online] http://www.nasa.gov/centers/jpl/news/msl20110317.html.
12. Wong, G.G. (1976) A novel hemispherical coverage waveguide antenna. IEEE Antennas and Propagation Society International Symposium, Amherst, MA, USA, pp. 566–569.
13. Edquist, K., Dyakonov, A., Wright, M., and Tang, C. (2009) Aerothermodynamic design of the Mars Science Laboratory backshell and parachute cone. AIAA Thermophysics Conference, San Antonio, TX, USA, pp. 1–14.
14. Brown, P., Farr, D., Demas, J., and Aguilar, J. (2009) Radiation pattern measurements of the NASA Mars Science Laboratory UHF entry antenna using a spherical near-field range. Antenna Measurement Techniques Association 2009 Proceedings, Salt Lake City, UT, USA, pp. 271–276.
15. Rengarajan, S.R., Zawadzki, M.S., and Hodges, R.E. (2009) Design, analysis, and development of a large Ka-band slot array for digital beam-forming application. *IEEE Transactions on Antennas and Propagation*, **57**(10), 3103–3109.
16. Rengarajan, S.R., Zawadzki, M.S., and Hodges, R.E. (2007) Bandwidth enhancement of large planar slot arrays. IEEE Antennas and Propagation Society International Symposium, Honolulu, HI, USA, pp. 4405–4408.
17. Grammier, R.S. (2009) A look inside the Juno mission to Jupiter. IEEE Aerospace Conference, Big Sky, MT, USA, pp. 1–10.
18. Chamberlain, N., Chen, J.C., Harrell, J.A. *et al.* (2008) Patch array antennas for extreme space environments. IEEE International Symposium on Antennas and Propagation and USNC/URSI National Radioscience Meeting, San Diego, CA, USA.
19. Chamberlain, N., Chen, J., Focardi, P. *et al.* (2009) Juno Microwave Radiometer patch array antennas. IEEE Antennas and Propagation Society International Symposium, Charleston, SC, USA, pp. 1–4.
20. Chen, J., Zawadzki, M., Harrell, J. *et al.* (2010) Microwave radiometer antenna and spacecraft interference. IEEE International Symposium on Antennas and Propagation and USNC/URSI National Radioscience Meeting, Toronto, Ontario, Canada.
21. Chamberlain, N., Chen, J., Hodges, R. *et al.* (2010) Juno microwave radiometer all-metal patch array antennas. IEEE Antennas and Propagation Society International Symposium, Toronto, Ontario, Canada, pp. 1–4.
22. Chamberlain, N., Chen, J., Hodges, R., and Demas, J. (2010) Accurate insertion loss measurements of the Juno patch array antennas. IEEE International Symposium on Phased Array Systems and Technology (ARRAY), Waltham, MA, USA, pp. 152–156.

23. Rengarajan, S.R., Zawadzki, M.S., and Hodges, R.E. (2010) Waveguide-slot array antenna designs for low-average-sidelobe specifications. *IEEE Antennas and Propagation Magazine*, **52**(6), 89–98.
24. Zawadzki, M., Rengarajan, S., Hodges, R.E., and Chen, J. (2010) Low-sidelobe slot arrays for the Juno Microwave Radiometer. IEEE Antennas and Propagation Society International Symposium, Toronto, Ontario, Canada, pp. 1–4.

18

Space Antenna Challenges for Future Missions, Key Techniques and Technologies

Cyril Mangenot[1] and William A. Imbriale[2]

[1]*ESTEC-ESA, The Netherlands*
[2]*Jet Propulsion Laboratory, California Institute of Technology, USA*

18.1 Overview of Chapter Contents

This chapter aims to identify the future needs and promising concepts for space antennas. This is quite a challenge. When asked to perform this exercise, the authors considered several ways of presenting their findings. Among them the following classifications were envisaged:

- By products (reflectors, arrays, radiators, etc.): This is a natural choice; nevertheless there is a risk that the product definition and challenges differ significantly with the application. As an example, a direct radiating array will have a high maturity level when used for synthetic aperture radar. However, when considered for a broadcast telecommunication mission, there are several key issues still to be solved, such as power consumption/dissipation, mass, and beam-forming network complexity for a large number of beams. Another consideration is the classification of array-fed reflectors. That is, where to put them? Should they be classified as reflectors or arrays? Finally the same consideration applies for radiators/feeds that are used in both arrays and reflectors.
- By applications (telecommunications, Earth observation, etc.): This is well in line with the actual situation where the antenna design is optimized for a given mission and there are several key differentiating factors. However, several products may be used for more than one application. One may refer to large reflectors but also active antennas, steerable antennas or TT&C antennas. In doing so, there is a large risk of missing the opportunity to identify synergies among developments and possible non-recurring cost sharing and rationalization.

Space Antenna Handbook, First Edition. Edited by William A. Imbriale, Steven (Shichang) Gao and Luigi Boccia.

- By frequency range: This is quite attractive as the sizes of the objects differ so much over the allocated space frequency spectrum that the technologies are also quite different. A mesh reflector will be the natural choice for very large aperture at P-band while a full carbon fibre reflector will be selected at Q/V- band. But there are also some counterexamples such as feed horns that can be used over a very large frequency range up to terahertz level. Antenna modelling and testing techniques are also used over a large frequency range.
- By key space antenna challenges: Considering on one hand the requests from projects and on the other hand the R&D push for technologies/techniques with high potential, it is proposed to identify and focus on some key aims where effort should be concentrated. In doing so, the authors take the risk of not providing an exhaustive list of needed developments. However, it is believed that such organization should be of more interest to the reader and best to stimulate technical exchanges.

Following this last approach, the very diverse perceived needs have been grouped into the following sections. For each key space antenna challenge a section briefly describes the problem area. It is followed by the identification of present and expected future space missions and finally by the anticipated most promising antenna concepts and technologies. It should be understood that this discussion cannot be exhaustive but should provide some trends for the future. A list of acronyms is provided at the end of the chapter.

18.2 General Introduction

The topic of space antenna subsystems is very extensive due to the large variety of applications, including Earth observation, telecommunications, navigation, science, planetary exploration as well as TT&C, manned spacecraft and user terminals. Performance optimization, mandatory for link budgets and spatial resolution (mainly gain), and the need to increase overall system capacity and avoid ambiguity (mainly polarization purity and beam isolation), require the development and maintenance of diverse antenna technologies, concepts and architectures for the different frequencies, bandwidths and radiating aperture diameters.

The antenna subsystem is characterized by strong links with overall system design as well as spacecraft and mission optimization. Satellite and antenna design is an iterative process that needs to consider the RF, mechanical and thermal constraints as well as the specificities of the space environment, such as radiation, vacuum operation, and so on, for the contemplated mission. The limited available volume under launcher fairing requires a careful accommodation exercise in the early development stage and reducing the antenna stowage volume is a major design driver. Despite the tight accommodation the antenna field of view shall be checked to preserve the radiated performance. The antenna design is also constrained by the available spacecraft power, the thermal dissipation allowed, the on-board computing power, the international regulations (flux limitations, radio astronomy band protection, etc.) as well as by the availability and cost of test facilities (e.g. for testing very large antennas). Besides the technical constraints, the end user requires solutions that are affordable.

Therefore the challenge for the space antenna designer is, for the specified mission, to define a competitive compromise between radio electrical, mechanical and thermal performances, considering the technologies available or which can be made available in time, the capacity of the spacecraft and the launcher. This needs to be done in close collaboration with system engineers during all the mission phases. Ensuring overall mission performance at affordable cost most often leads to the elaboration of a rather specific product for the mission. The importance of electromagnetic technologies and techniques has been growing over the years with the increased complexity and power of telecommunication payloads and with the increasing sensitivity of remote sensors.

In this context, it is rather difficult to elaborate a global strategy. In fact, except for science applications, the absence in the last 40 years of long-term plans for space missions together with the scarce resources for innovative R&D funds have prevented the implementation of long-term technology applied research. Substantial antenna R&D funding has mostly been concentrated on near-term R&D needs. The relatively modest and scattered funding allocated to space R&D in Europe and the few European military space applications have resulted in a drastic selection of R&D European developments. This is different than in the USA where developments for military applications have provided a solid foundation for the commercial market. Specific examples are unfurlable antennas, active space phased arrays and aircraft terminals.

Regarding Europe, the globalization of industry and the planned cooperation with the European Commission should help to take advantage of synergies in research between the space sector and other areas.

18.3 General Evolution of Space Antenna Needs and Requirements

Although the L-band to X/Ku-band frequency range is most often used for space applications (telecommunications, Earth observation, etc.), lower and higher frequencies have been implemented in recent missions and are to be considered for future ones:

- Below L-band, there is renewed interest in civil applications at VHF and UHF as used for several early developments. Defence programmes are also using this frequency domain. The global coverage from space and the existence of several low-frequency terrestrial and maritime terminals have helped to develop the space segment. Some missions can be found in the automatic identification system at 165 MHz, search and rescue close to 400 MHz and biomass synthetic aperture radar (SAR) at 435 MHz.
- Above Ku-band, high-frequency antennas for broadband telecommunications are finally being implemented. Even though a full Ka-band satellite was expected to emerge far earlier, it has now become a reality with several commercial satellite orders placed recently: namely, Viasat-1, Ka-Sat, Yahsat 1B, Jupiter as well as INMARSAT-5 (using Ka-band to complement L-band mobile communications offered at a lower data rate). Ka-band is also being used for new missions such as the European data relay satellite or interferometric SAR. It is anticipated that Q/V-band will also be used intensively in the coming future for civil broadband applications.

In both cases, this tendency to extend the frequency domain is pushing the technologies to their limit due to either the very large size of the antenna apertures at low frequencies or the surface accuracy required for high frequencies. This also has a strong impact on all the antenna design stages covering concept elaboration, design, modelling and RF/mechanical/thermal testing and facilities. Another major challenge is related to propagation, since scintillation effects become critical at low frequency and propagation losses are very high at Q/V-band.

An even larger frequency range is required when the mission purpose is to observe the different atmospheric layers for the identification of traces of gas or vapour content. Sounding/imaging instruments operating at high resolution up to terahertz range have been successfully implemented in the last few years. Examples can be found in NASA's developments as well as with the European Herschel and Planck missions launched in 2009.

Another trend is the need for continuity of service for telecommunication, navigation and Earth observation users at lower cost. This is achieved by system, satellite and equipment design, manufacturing and integration choices, and increased operational lifetime of the satellites. The situation can be rather different for science missions, which often target specific needs of the scientific community that can lead to very complex instruments providing 'the best' measurements. For example, the use of cryogenic instruments at very high sensitivity can cause a drastic increase in complexity.

For most applications, reflector antennas are by far the most used types of antennas for space and often remain the best solution for applications with a single high-gain beam or with multiple beams over a narrow field of view. This is expected to remain valid in the future. Reflector antennas provide gain from a simple reflecting surface and can operate at multiple frequencies with single or multiple beams. They can also be shaped to provide one or more contoured beams. On the other hand, arrays and lenses perform better over a wide angular domain and are well adapted to medium-gain requirements. This justifies their use in several LEO missions for Earth observation, navigation and telecommunications. Nevertheless, active arrays are by far more complex to realize, bulky and costly, and developments are needed to tackle these features.

Considering both planned missions and technological status, the following aims have been identified as key for future developments:

- Aim A: Develop large-aperture antennas.
- Aim B: Increase telecommunication satellite throughput.
- Aim C: Enable sharing the same aperture for multiband and multipurpose antennas.
- Aim D: Increase the competitiveness of well-established antenna products.
- Aim E: Enable single-beam in-flight coverage/polarization reconfiguration.
- Aim F: Enable active antennas at affordable cost.
- Aim G: Develop innovative antennas for Earth observation and science instruments.
- Aim H: Evolve towards mass production of satellite and user terminal antennas.

In addition to the above aims, closely related to specific missions, there are two transverse directions where developments should also be performed:

- Aim I: Technology push for enabling new missions.
- Aim J: Develop new approaches for satellite/antenna modelling and testing.

Table 18.1 maps the set of aims considered as key for the future with the planned missions/instruments/antennas. They will be addressed in the following sections.

Table 18.1 *Space missions, instruments and antennas.*

Application	Mission	Frequency band (GHz)	Aim									
			A	B	C	D	E	F	G	H	I	J
Telecommunications	Mobile interactive communication satellite services	L, S	×		×	×		×		×	×	×
Telecommunications	Mobile broadcast satellite services: DARS	S	×		×	×					×	×
Telecommunications	Mobile broadcast satellite services: DMB	S	×		×	×	×			×	×	×
Telecommunications	Fixed and broadcasting satellite services	C, Ku	×		×	×	×	×		×	×	×
Telecommunications	Broadband satellite services	Ka	×	×	×	×	×	×		×	×	×
Telecommunications	Air traffic management satellite services	L	×								×	×
Telecommunications	Tactical Sat Com system	UHF, X	×					×			×	×
Telecommunications	Emergency communication systems	UHF	×		×			×		×	×	×
Telecommunications	Data relay services	S, Ku, Ka	×								×	×
Earth observation	Radiometer	P, L	×						×		×	×
Earth observation	Synthetic aperture radar	P, C, X, Ka	×			×		×	×		×	×
Earth observation	Altimeter	C, Ku, Ka			×	×			×		×	×
Earth observation	Scatterometer	C				×			×		×	×

Table 18.1 (*Continued*)

Application	Mission	Frequency band (GHz)	Aim									
			A	B	C	D	E	F	G	H	I	J
Earth observation	Ice sounding radar	P	×						×		×	×
Earth observation	Data transmission antenna	X/Ka				×		×	×		×	×
Navigation	Global navigation satellite system	L	×						×		×	×
Science	Deep space communications	X/Ka	×						×		×	×
Science	Radio astronomy missions: space very long baseline interferometry	L, C, Ku	×					×	×		×	×
Science	Astronomy											
Submillimetre							×		×	×		
Earth observation	Data transmission antenna	X/Ka				×		×	×		×	×

18.4 Develop Large-Aperture Antennas

18.4.1 Problem Area and Challenges

Even if the trend today is towards miniaturization of all RF equipment, antenna size is expected to grow for higher sensitivity instruments and access to end users with small terminals. To answer this need one may consider:

- Real apertures such as reflectors, reflectarrays, membranes, arrays and lenses.
- Synthetic apertures generated by a set of individual radiators installed on one satellite or by arrays in formation-flying satellites. Main applications are foreseen in the receive mode.

In this chapter, a large aperture is defined as an aperture that must be 'built' in space because its final size is not compatible with launcher accommodation. Considering the current large-satellite platforms, one may say that all apertures above 4 m in diameter fall into this category. Also, when smaller launcher fairings are considered, reflectors of 3 m in diameter may also have to be built in space, so solutions for this range are also of interest. Even if this diameter might appear relatively small at low frequency, it corresponds to a size above 300 wavelengths at Ka-band and therefore perfectly justifies this denomination. One may also note that antennas operating close to the terahertz regime with a reflector size of 1.5 m are definitely very large in terms of wavelength but are not addressed here as they are made in one piece only.

To summarize the present situation, large real space apertures benefit from flight heritage mainly with telecommunication systems based on mesh reflectors and SARs implementing direct radiating arrays. The large-reflector market is dominated by two key US players while the direct radiating array concept has flight heritage all around the world. Lenses have also been used for some military programmes. Reflectarrays have been developed and the multiple flat panel concept allows better RF performance as well as easier satellite accommodation, but these are penalized by higher aerial mass and additional losses compared with reflector solutions. Overall, reflector antennas are used for more than 80% of telecommunication space apertures, and are now being proposed also for SAR with benefits in cost and mass per unit aperture ratio.

As a complement to real apertures, the use of distributed apertures has been considered recently as a way to expand the limit of radiating size that can be generated. When, for the selected frequency, the aperture size imposed by the mission reaches the limits of the technology, the overall aperture can be split into sub-apertures implemented either on a common mechanical structure or in formation flying.

Combining the signals from the different apertures allows very high resolution to be obtained due to the very large effective receiving area. Nevertheless, the sensitivity of the instrument is determined by the real aperture sizes available.

An example of such synthetic apertures is given by the Soil Moisture and Ocean Salinity (SMOS) satellite. In this case we refer to the addition of signals received from individual radiators and the final instrument resolution can be similar to that provided by a real aperture instrument.

Another example is given by space VLBI, mixing Earth-based and satellite-based receiving apertures. Demonstrations up to Ka-band have been made.

A last case is given by the Geostationary Sounder mission where the combination of radiators located on a Y-shaped rotating antenna provides the several baselines requested from the mission.

In all these cases, signal processing plays a major role. Radiating elements are directly connected to low-noise amplifiers and frequency converters and then the signal is processed on board or on the ground. Calibration techniques allow significant reduction of the effects induced by the manufacturing or in-flight distortions.

Real or synthetic large apertures are requested for several space applications over a large frequency domain. These aspects will be further defined and justified next.

18.4.2 Present and Expected Future Space Missions

This subsection identifies several missions requiring apertures with diameters between 4 and 25 m, including well-established telecommunication missions such as fixed and broadcast satellite services, mobile interactive/broadcast services, broadband services, but also new applications such as air traffic management and emergency communications. These missions are spread over the frequency spectrum between L- and Ka-band [1], with surface accuracy requirements becoming, as expected, more stringent as frequency increases. It should be noted that in many cases large reflector become mission enablers and when the product is not available to a specific satellite manufacturer, it may limit the manufacturer's access to a particular market segment. Also, these types of satellites are paving the way for the future in terms of technology and access to some market segments. As such this is a major complement to simple replacement satellites. Several Earth observation and science deep space missions also requiring large-aperture and development synergies among applications should be promoted. These last missions are described in Section 18.10.

18.4.2.1 Telecommunications

Mobile Interactive Communication Satellite Services The system architecture is based on providing connectivity from the user terminal to the satellite in L- or S-band, and from the satellite to a node in the C-, Ku- or Ka-bands. Multibeam systems imply on-board transparent processors which allow reconfiguration of coverage and connectivity between beams. Mobile communications involve mainly voice and low-speed data and implement a global system based on multiple regional coverage. The market is dominated by several players such as INMARSAT, Thuraya, TerreStar Networks and SkyTerra (formerly MSV) implementing GEO satellites. Nevertheless, LEO constellations such as Iridium and Globalstar are operational and next-generation satellites have been recently ordered.

The requests for extended capacity and improved performance imply very large multibeam satellite antennas. This allows smaller user terminals to compete with FSS through the use of transportable/deployable terminals. Key technology enablers are large reflectors (up to 25 m), focal arrays (made of hundreds of Tx/Rx feeds with shared aperture to generate overlapped beams) and digital processors. Today's technology limits are reached through the large number of pencil beams with frequency reuse and the world coverage. Following Terrestar-1 with its 18 m reflector from the Harris Corporation and SkyTerra-1 with its 22 m reflector from AstroMesh, the expected evolution will be reflectors up to 25 m with excellent reflection losses, improved surface accuracy and passive intermodulation (PIM) performance, S-band frequency operation and an increased number of beams. On-ground BFNs are a hot topic for MSS satellites as demonstrated by ICO G and SkyTerra. To enable satellite communications inside buildings, larger autonomous reflectors up to 50 m with the feed system being located and powered on a separate S/C might be considered. Antenna pointing accuracy will be a prime concern with such large antennas.

Mobile Broadcast Satellite Services These services address both the Digital Audio Radio Service (DARS) and Digital Multimedia Broadcasting (DMB).

DARS is well established in the US market with the merged XM Radio and Sirius systems. In Europe the Luxembourg-based DARS initiative is pursued with heritage based on World-Space satellite technology. The European initiative by Solaris Mobile, a joint venture that was formed to sell S-band satellite services to government and commercial users in Europe, has resulted in the very recent launch of the W2A satellite. Solaris, owned by Europe's two biggest satellite operators, SES of Luxembourg and EUTELSAT, is in a competition for the S-band spectrum organized by the European Union's executive commission. This initiative also plans to include interactive services.

Mobile broadcasting missions require very large reflectors in the range of 12 m associated with high-power focal arrays for multilinguistic DMB missions. Some moderate shaping capability is of interest. DARS missions require shaped reflectors in the range from 4 to 7 m. Concerning the primary feed, the main constraints are due to the large radiated power and the feed size.

Broadband Satellite Services This mission requires reflectors operating at Ka-band with diameters up to 5 m. See the section on increase telecommunication satellite throughput for more details.

18.4.2.2 Earth Observation

See the subsections on Earth observation and science missions.

18.4.2.3 Science

See the subsections on Earth observation and science missions.

18.4.2.4 Synthesis of Missions

Based on the mission needs identified above and reported as points in Figure 18.1, three reflector families with similar requirements have been identified covering the different diameters and frequencies. They can be summarized as follows:

- Aperture in the range of 4–7 m with an RMS surface accuracy around $\lambda/50$ (2.5 mm in S-band) for applications mostly in the L- and S-bands.
- Aperture in the range of 4–7 m with an RMS surface accuracy between $\lambda/50$ and $\lambda/100$ (0.5 mm in C-band) for applications from C- to Ka-band.
- Aperture in the range of 9–25 m with an RMS surface accuracy around $\lambda/50$ (2.5 mm in S-band) for applications mostly in UHF and L/S-bands.

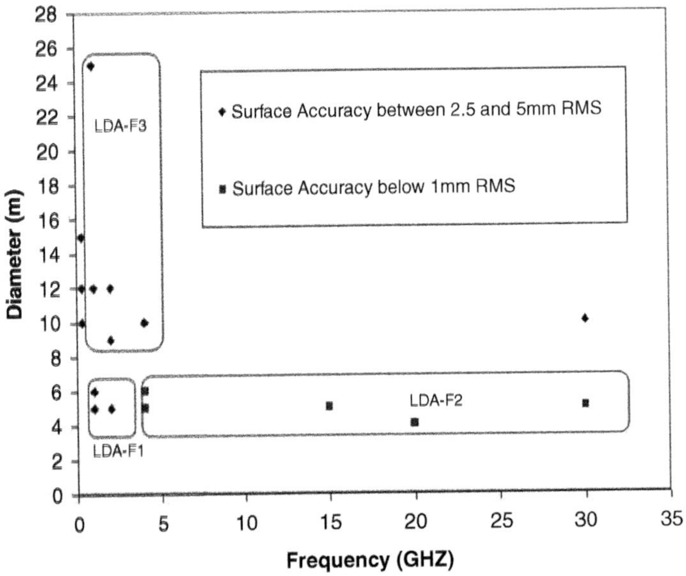

Figure 18.1 *Classification of reflectors by diameter and frequency.*

18.4.3 Promising Antenna Concepts and Technologies

18.4.3.1 Reflector Antennas

To investigate promising antenna concepts, a market survey was performed and perspectives over time examined for large reflectors.

The market survey is illustrated in Figure 18.2. It is based on more than 100 flight reflectors complemented by estimates for the years to come. It confirms that the growth and punctuated developments from the 1980s are now reaching an average of five developments every year. Expectations for the coming 20 years are between 7 and 14 reflectors per year worldwide for the full diameter range above 4 m. Looking at this in more detail, the market perspective for reflectors below 9 m represents approximately two-thirds of the market need, with an average estimate of seven to eight reflector developments per year. This ratio is already apparent in actual developments, as illustrated in Figure 18.3. This figure also sorts the actual reflector developments by provider, confirming the primacy of US companies and the clear market domination of the Harris Corporation due to its very large in-flight heritage and captured markets. Some less well-known developments are related to Russian providers with more than 12 flight items.

This market analysis based on established past data from open literature and provisional figures shows an overall market larger than the one normally perceived.

Some of the advantages of reflector-based antennas are their compatibility with a wider spectrum or multiple frequency bands in comparison with array-based solutions. Combinations of frequency bands are used in SMAP (Soil Moisture Active/Passive) with both passive and active sensing capabilities at L-band, but also in telecommunication missions. Further, reflector-based solutions are compatible with very low stowed volume. Recent developments in sophisticated feed arrays able to provide multiple beam capability for reflector antennas have increased the interest in reflectors versus arrays.

The reflector types considered as having the strongest potential are described below in increasing order of stiffness.

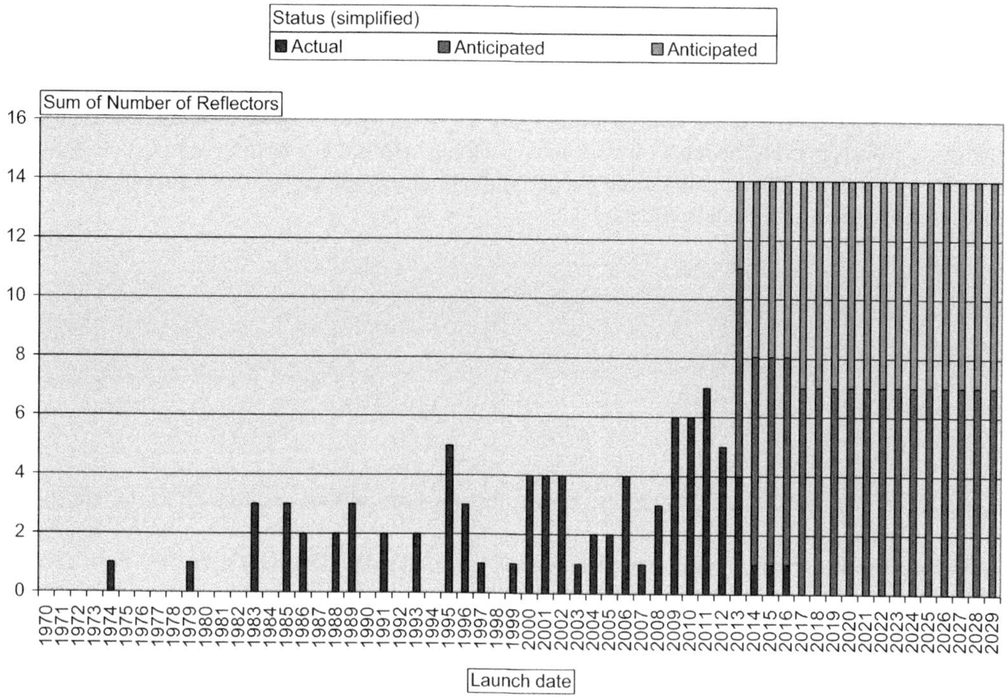

Figure 18.2 *Actual and anticipated large-reflector sales worldwide.*

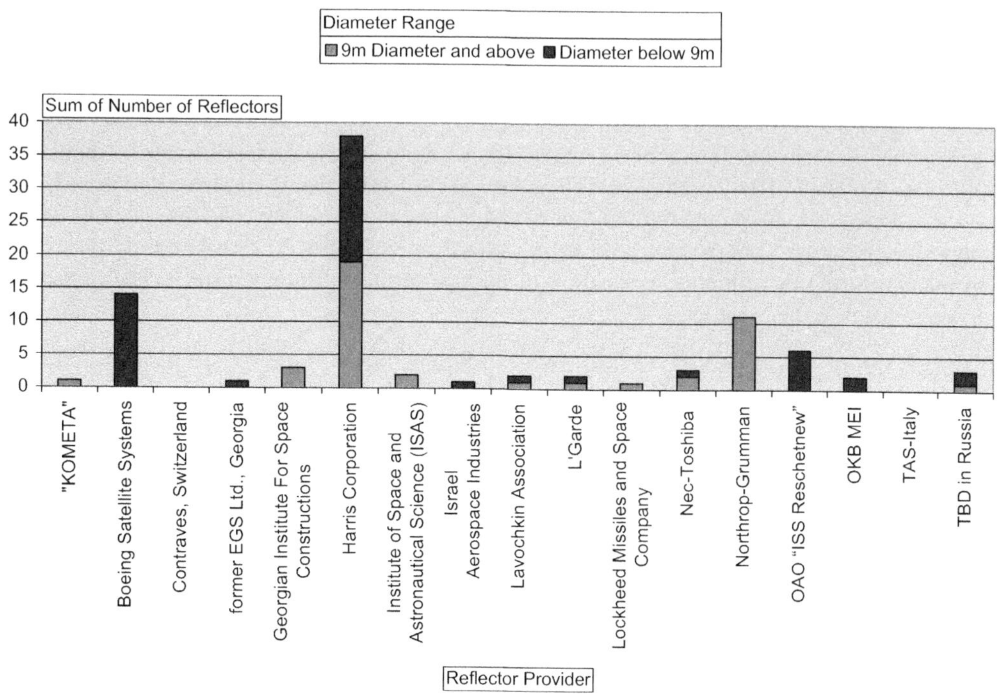

Figure 18.3 *Actual reflector developments sorted by provider and diameter range.*

Metal Mesh Reflectors These are metal-wire-knitted meshes, representing the most numerous group of flown units. They are also called metal tricot meshes. Gold-plated molybdenum and tungsten wires are mostly used. These meshes require tensioning (in the range of 5–10 g/cm) to produce sufficient electrical contact between the wires. The cell size of the knitted meshes is determined by the intended operational frequency. At first sight, mesh reflectors offer the most obvious technological solution due to their heritage, as well as their flight-proven performance. The identification of a particular architecture suitable for all applications requires some compromises among the requirements:

- Size of the aperture
- Surface accuracy
- Mass
- Stiffness
- Complexity (in turn related to cost)
- Package dimensions.

A description of the available architectures is given below, each with its unique characteristics:

- Expandable peripheral ring with symmetrical nets: This concept developed by AstroMesh and recently by the Harris Corporation shows the greatest potential for most of the applications studied. It is suitable for medium- to large-size reflectors. In the AstroMesh reflector, management of the mesh integration is achieved by connecting the mesh in a properly tensioned state to a very rigid (in plane) and thermally stable net. Profiling is obtained by connecting two mirror nets by controlled force tensioners, all probably with the same tension (in principle, simple).
- Modular interconnected elementary reflectors: An example of such a structure is given by the ETS VIII reflector from Toshiba. It is suitable for addressing a large range of diameters from small (one module) to large-size reflectors (as many modules as required). The architecture appears heavier than the others for medium to large reflectors, due to the currently implemented unit cell architecture. This can be improved with other types of cells. The deployment requires a significant number of mechanisms acting in parallel, hence this architecture is of interest in case the elementary cell has to be deployed by simple, very reliable and predictable mechanisms. Interconnecting the RF reflecting surface of the different modules is rather complex and may cause RF degradation. This is the case for the existing implementation in ETS VIII. However, new ways of integrating the front mesh and the backing structure can eliminate this drawback. Decomposition of the backing structure into the modules is anticipated to create a deterministic array effect due to manufacturing or in-flight deformation of the composite reflecting surface. The impact on RF performance needs to be quantified.

Shell-Membrane Reflectors This relatively new type of reflecting surface does not require tensioning. It has a small bending stiffness and requires a slight back support structure. When folding, it acts as a membrane since its bending stiffness becomes negligible. Regarding shell materials, carbon fibre reinforced silicon offers a clear perspective. The concept has been matured in relation to the materials and processes within ESA programmes [2]. The mechanical and thermoelastic stability and RF performance have been measured and show promising results up to C/Ku-band so far. Thus there is a certain knowledge base for the characteristics and potential. In addition, this reflector can be combined with the construction principles of mesh reflectors, that is an expandable peripheral ring or a modular construction based on unit cells. This concept is therefore worth exploring further and is based on the principle of in-plane stiffness, but predetermined surface shapes due to the built-in bending/shear stiffness. Consequently, there is no need for a driving support structure other than a positioning device, which could be a ring or something similar. The resistance to space environment constraints needs to be investigated further.

Solid Surface Reflectors These types of surfaces have the highest bending stiffness, therefore the reflector surface has to be split into many pieces that are joined together to allow folding. This can be envisaged as parabolic-shaped pieces generating a parabolic profile or flat pieces approximating the parabolic profile and complemented by reflectarray technology to achieve the final surface with the required electrical reflective properties (parabolic or shaped).

This type of reflector offers an immediate solution to short-term missions requiring apertures in the range of 5–6 m, both at S-, C- and Ku-bands for fixed communication and broadcast satellite services and at Ka-band for broadband satellite services. The risk involved is smaller compared with other technologies due to the previous heritage such as Ultra Light Reflectors or Stiffened thin shell reflectors. Achieving the error surface as requested for Ka-band still remains a challenge.

The limitation of this reflector type is, however, clear. In principle, and considering spacecraft accommodation, this design will not be capable of growing beyond the 7 m projected aperture. It should be noted that, as a possible evolution, a hybrid solution using conventional CFRP at the reflector centre and non-rigid panels at the periphery (possibly based on a CFRS membrane) might be of strong interest as demonstrated in the past by one of the Russian Venera spacecraft. Another aspect worth mentioning is the easy realization of an elliptical aperture as requested for Earth observation applications (typically 3×7 m). A major limiting factor is related to the overall mass. There is at least a factor of 3 difference for the antenna mass of the solid shell compared with mesh concepts. Also, the large booms used to ensure the large requested F/D (above 1.5) impose severe accommodation constraints.

18.4.3.2 Direct Radiating Array Antennas

Used as SARs for satellites or the Space Shuttle, direct radiating arrays benefit from a long space heritage over the past few decades. The earliest was Seasat in the 1980s. Since then, SARs have been developed from L- to X-band with growing beam agility. They can provide elliptical footprints with low sidelobe/main-lobe ratios which are of interest for SAR. Wide-angle access (inducing coverage, and thus higher revisit time) can be achieved. The on-board power dissipation is facilitated by the large surface and in most cases passive thermal control can be utilized. The flat surfaces of these arrays ease satellite accommodation. Nevertheless they are intrinsically more complex and have larger mass and higher losses when the amplification is not distributed.

Considering the long flight heritage, the key element for the wider use of direct radiating arrays in space is a reduction in the mass and losses. For large apertures the use of a passive array configuration is anticipated. One promising concept makes use of a 1D membrane antenna deployed on a pantograph basis supporting the RF distribution network as well. Some synergies with the technologies used for solar arrays might be beneficial. In both cases, the achievable surface accuracy might limit the applicability to frequencies below C-band.

18.4.3.3 Reflectarray Antennas

Compared with direct radiating arrays, reflectarrays are space fed and do not require a beam-forming network. Passive reflectarrays are made only of pieces of metal etched on the surface (no active devices) and allow compensation of the deterministic surface errors with respect to the ideal reflector by adjusting patch dimensions.

Active reflectarrays have variable phase shifters behind or embedded in the elements to allow single-beam control/reconfiguration. They are very promising for future radar and active sensors, if necessary in a few flat panels to increase the frequency bandwidth.

In both cases, 2D deployment devices (possibly inflatable) may be suitable for generating a very lightweight planar structure or a convex/concave shape even with a medium surface accuracy. This can be obtained by the deployment devices or by the inflation of cavities with major benefits regarding a very low stowage volume. Besides reflectarrays, transmitting arrays or holographic surfaces can be generated without feed blockages.

Even if membranes can be used in several antenna configurations (reflector, array, etc.), it is believed that they have a strong potential for reflectarrays by taking full benefit of their ability to compensate for surface deformation and the absence of a beam-forming network. Membranes are usually made of thin metallized polyimide films requiring tensioning by an external device. They are probably among the most efficient solutions when considering very large apertures (above 20 m) and low operational frequencies. The state of the art of inflatable structures in space is such that rigidized inflatable booms are approaching maturity to the point of providing L-band accuracy within a 5 m length. This, however, has not yet been verified in a representative environment. The reflecting surface, on the contrary, following patents and previous developments, including the Contraves 10 m inflatable antenna and the L'Garde experiment, was already developed to an acceptable degree of performance more than 10 years ago. Some noteworthy progress in rigidization methods has been made since then. The global picture seems to show that, nevertheless, the development of a large inflatable reflector to the level of performance currently met by mesh reflectors could take 8–10 years. The areal mass of such a concept is very attractive but should be balanced by the mass of the tank needed for pressurization.

18.4.3.4 Synthetic Aperture Antennas

In the case of a synthetic aperture, the technological complexity moves from the reflector area to the deployable structures and associated mechanisms to support the large high-gain arrays, as well as the attitude control when radiators are implemented on different bodies and signal processing combines the different received signals. However, the large number of interferometric baselines allows some natural redundancy. These types of structures have the potential for observation over a large field of view as each radiator has a broad radiation pattern. Such distributed apertures are mainly used in receive mode as ITU regulations, poor beam efficiency and power loss in the grating lobes might prevent their use for transmitting even if grating lobes can be spread by the use of non-regular lattices. When transmit/receive mode is required, as for example for SAR, bi-static architecture with a broad beam in transmit can be considered with major benefits. In receive mode, the noise coming from outside the main lobe has to be carefully considered. Nevertheless, these synthetic apertures are considered as key for increasing the instrument resolution when the technological limits of real apertures are reached. This is particularly relevant at high frequency. Development should continue in this domain.

18.4.3.5 Synthesis

Large space apertures are needed in the range of 4–7 m for C- and S-bands with high-power feeds for communications and mobile digital broadcasting. Use of reflectors, artificial lenses or reflectarrays should be considered. Innovative solutions with foldable, unfurlable, inflatable or membrane configurations will be required to achieve the reflector shaping and fine pointing required for the beam(s) and to avoid increasing the requested TWTA output power with risks of breakdowns in the high-power feeds.

Very large space apertures (up to 12 m and above) for S-, L- and P-bands with millimetre surface accuracy are mandatory to allow direct access to mobile end users for telecommunications and also to get the required resolution for low-frequency remote sensing. Lens- and membrane-based solutions might be beneficial.

Associated technologies for very large apertures, such as control and stabilization, fine pointing system, deployable transmit/receive focal arrays, feed circuits and (reconfigurable/digital) beam formers, fast analysis and optimization tools, and so on, will require innovative solutions and early preparatory studies. This includes the study of systems with separate and autonomous reflectors and feeds as in the framework of the ESA science mission XEUS, successor of XMM. XEUS will be a very large free-flying X-ray observatory, consisting of two spacecrafts (one carrying the mirror, the other carrying the detectors) in formation, flying 50 m apart.

Aerial mass is significant for large diameters and mesh reflectors can keep it to around 1 kg/m^2 for diameters up to 13 m, while that of solid reflectors increases with diameter and is close to 3 kg/m^2. On the other hand, solid surface reflectors provide the lowest RMS values. These reflectors provide on average a surface accuracy 5–10 times better than alternative technologies. Inflatable reflectors or lenses appear as promising technology for very large apertures, but can only be considered for long-term developments as there are still major technological challenges to be tackled and lifetime in orbit has still to be improved. Most probably, previous in-flight validation so far has been performed only over short durations.

Several antenna types such as reflectarrays, membranes, active and passive lenses have to be investigated, with major research focused on mass and stowage volume reduction.

18.5 Increase Telecommunication Satellite Throughput

18.5.1 Problem Area and Challenges

Although the broadband satellite market was expected to emerge much earlier (in the 1980s), it is a reality today. For better usage of the scarce frequency resource, the coverage area is split into several pencil beams by frequency/polarization reuse. Several Ka-band multiple pencil beam satellites, namely Eutelsat's Ka-Sat in Europe, Viasat-1 and Jupiter in the USA, and Yahsat 1B providing coverage over the Middle East and Africa, have been recently ordered. There is a continuous evolution of Ka-band broadband multimedia missions towards increased overall capacity by the implementation of payloads with a higher number of beams and a very high level of frequency reuse. Currently flying or planned satellite missions are implementing 50–100 beams with an overall throughput of up to 100 Gbps. Future missions may go well beyond 150 beams, with the objective of reaching a 1 Tbps throughput.

Increased operational frequency and higher bandwidth for users induce new constraints on reflector surface accuracy, stability, pointing system and feed manufacturing tolerances, while multibeam technology requires new high-efficiency feeds and a large number of reflectors.

18.5.2 Present and Expected Future Space Missions

18.5.2.1 Telecommunications

Broadband Satellite Services A large number of initiatives aim to narrow the digital divide in less populated areas. It is expected that the next generation of multimedia networks will see the satellite as an integral part of the global information infrastructure. In this framework, GEO satellites are expected to provide both trunking and access networks, higher throughput and lower delivery cost. The satellite component must be able to cope in a flexible and efficient way with the evolution of terrestrial networks.

To achieve the above targets and respond to future communications requirements by means of space-based solutions, it is necessary to develop truly innovative systems. At antenna level, a new generation of antenna subsystems including antenna tracking systems has been developed and flight heritage is available. The requirement is for larger size reflectors such as a 5 m diameter one to be compatible with a beam size of 0.2°. This will allow either national coverage or European coverage with a very large frequency reuse factor due to the large number of beams. Multiple-beam antennas based on a single feed per beam approach with three or four reflectors, or an array-fed reflector associated with only two reflectors (one in Tx, one in Rx), need to be considered.

18.5.3 Promising Antenna Concepts and Technologies

These are as follows:

1. High-capacity Ka-band broadband multimedia payloads will make use of large reflectors in the 4–6 m range. The classical approach of using four single feed per beam Tx/Rx reflector antennas on board to obtain the full grid coverage of beams is difficult to accommodate on a satellite, expensive and does not offer flexible coverage. This configuration is even less suitable when a limited number of beams are required to cover a linguistic area.

   Thus antennas able to generate all Tx or Rx adjacent beams with only one aperture (multiple feeds per beam concept) will be needed to improve both satellite accommodation and cost. In fact, these same constraints have driven the payload architecture of current L-band mobile payloads (INMARSAT-4, Thuraya, etc.) towards a focal-array-fed reflector architecture and the use of a beam-forming network (either on board or on the ground). To implement the beam-forming networks, two main solutions can be considered: a digital BFN (included on a transparent digital processor) or an RF BFN approach. While at L-band the requested frequency bandwidth is compatible with existing digital BFN technologies, at Ka-band both solutions need to be assessed at payload level considering the high number of beams and feeds, the power consumption, the mass and the cost. The requested bandwidth might require demultiplexing before digital conversion.

   The implementation of an active BFN for multiple feeds per beam antennas can compensate for beam pointing/shaping errors due to satellite attitude variations and thermoelastic reflector distortions. These last aspects are crucial at Ka-band when considering the very large aperture in terms of wavelengths. This should be evaluated in detail to identify and assess the best techniques for on-board implementation. Besides the multiple feeds per beam antennas based on conducted BFNs, recent studies have demonstrated the potential for solutions based on feed sharing by a parasitic structure in front of horns located in the focal plane.
2. When increasing the reflector size, due to the very narrow beamwidth, there will be a high variability and non-uniformity of traffic for many of the beams. This will certainly demand the implementation of on-board flexibility (in terms of bandwidth and power to each beam). Therefore, payloads based on broadband transparent digital processors (based on deep submicrometre technology and high-speed converters) and flexible output sections based on multiport amplifiers on flexible TWTAs and/or mini TWTs will be needed. A hybrid implementation of a mix of standard non-flexible beams for the 'known to be fully loaded beams' and flexible bandwidth/power beams for the 'partially loaded/variable traffic beams' is expected.

   Another key aspect of these future payloads is the implementation of very high-capacity feeder links able to transmit all the bandwidth corresponding to the high level of frequency reuse in the user link. Q/V-band is expected for implementation of the feeder link depending on future regulatory authorization.
3. The implementation in the future of very high-capacity Ka-band broadband multimedia missions will require the development of not only large-reflector antennas, but also of high-efficiency feeds, generic wideband feed systems (user + gateway bands) as well as improved RF performance and analysis/optimization tools for antenna and customized radiating elements. Due to the complexity and size of the focal arrays, the RF routing length to the feeds increases and active feeds with integrated LNAs may be required.
4. The request for beams of $0.2°$ for broadband applications pushes the technology to its limit. For example, a 100 μm RMS surface accuracy on a 5 m aperture has not yet been achieved. This requires the investigation of high-accuracy antenna pointing systems with possibly knowledge of the reflector profile distortions. When a digital or analog in flight reconfigurable BFN is implemented, correction can be applied based on a calibration process.

The current design process for broadband satellite systems usually considers meeting target specifications separately for the antenna and system. However, there surely would be significant benefits from a joint system/payload/antenna design approach, where subsystem configurations are optimized based on overall system performance. This is particularly true in the case of adaptive coding modulation systems, where current specifications for payload design do not maximize actual system performance. Another example is given by the dynamic beam allocation. In this case, a given user is no longer allocated to a given beam but benefits from the signal with the highest level. This adds complexity to the system and requires dual-polarization/dual-frequency user terminals but reduces the constraints on the antenna pointing system.

It is beneficial to devise a joint system, payload and antenna design methodology and simulation tools for maximizing predetermined performance figures of merit. This will allow assessment of the impact at system and payload levels of the utilization of the adaptive DVB physical layer in regenerative systems. System studies are necessary to reduce the difficult *C/I* requirements to avoid oversizing the antennas. Therefore there is a need for system/payload/antenna capacity and power assessment simulation tools for flexible and beam hopping systems. This will permit evaluation of the capacity performance and optimization of the user forward link in a multiple-beam satellite system and determination of the overall architecture in terms of:

- Multicarrier bandwidth management
- Dual polarization at system and beam level
- Number and fixed/flexible TWTAs and MPA architecture
- HPAs to beam routing scheme (multiple number of beams per TWTA or MPA port, post-HPA losses and P_{sat}).

And for the system performance:

- Temporal and spatial availability
- Co channel interference
- Symbol rate (Mbaud)
- Non-uniform power and bandwidth allocation to beams.

18.6 Enable Sharing the Same Aperture for Multiband and Multipurpose Antennas

18.6.1 Problem Area and Challenges

The satellite resource is scarce and the accommodation of several missions or instruments is a real challenge. The last decade has seen the development of both very large GEO platforms with typical payload capacity close to 20 kW and 2 tons and small platforms compatible with payloads of 3 kW and 350 kg. Also, for LEO very large satellites co-exist with small ones. In all cases, the satellite accommodation is always constrained by the launcher fairing and there is a mandatory need for antenna solutions that share one aperture (reflector or array) for several missions. Some examples are multi-frequency antennas implementing multi-frequency feeds or FSS, antennas based on polarization-sensitive surfaces (DGR, polarization grid, etc.) or multiple beams originating from a focal array feed. The increasing number of antennas leads to severe challenges in modelling antenna interactions (coupling, PIM) and methodologies to control interference. These impose new constraints on modelling tools at the antenna/satellite level due to the large number of beams at different frequencies, the very high radiated power and the extreme sensitivity of receivers (PIM, increased noise temperature, radiation pattern degradation).

Using multipurpose antennas with reflector sharing would permit hosted payloads and increase satellite revenue. It is anticipated that hosted payloads on board LEO or GEO satellites could be for military use, Earth observation purposes or emerging commercial markets. This will pave the way to the fusion of applications with, on the same satellite, Earth observation instruments, telecommunication antennas and navigation receivers.

18.6.2 Present and Expected Future Space Missions

18.6.2.1 Telecommunications

Telecommunication missions mainly implement reflector antennas. Due to their wide operating frequency band and large power handling, as compared with arrays, reflectors can be fed by several primary feeds at different frequencies allowing their aperture to be shared. Reuse of the reflector is compelling for large sizes. One may consider combining L- and Ku-band missions or Ku- and Ka-band missions, but also Ka- and Q/V-band ones.

18.6.2.2 Earth Observation

A classical example of multi-frequency instruments is the altimeter. There is also a strong interest in combining altimeters with radiometers. Another example is given by SAR operating at P-band that can be combined with an ice sounder provided that some reconfiguration of the beams can be achieved.

18.6.3 Promising Antenna Concepts and Technologies

It is considered important to address the following domains:

- To increase the number of antennas that can be accommodated on the same spacecraft. This induces accommodation complexity and requires checking all possible interactions between antennas, antennas with structure(s) and optimum co-existence with all other subsystems, such as thermal control, attitude control, propulsion and mechanical structure.
- To find new ways to combine/integrate multiple missions/frequencies/polarizations in single-antenna systems. As an example, radiometer antennas already combine more that five different frequencies due to quasi-optical networks. This will allow the number of service areas covered by a single aperture to be increased.
- To increase the frequency bandwidth and power handling of the different constituents of the antenna. This will allow multiband antennas and more channels.
- To develop a new generation of full CFRP DGRs allowing dual-polarization beams with different shapes and high XPD. Such technology is expected to provide high thermoelastic stability and low losses.
- To develop a circular polarization selective surface for combining beams in circular polarization from a single aperture.
- To standardize interfaces for hosted payloads.
- To make the technologies available for high frequency. This will allow the size of the antenna aperture to be reduced. Nevertheless, this induces new constraints on feeding systems and thermoelastic reflector behaviour, material requirements and new needs such as reflector metallization. Relevant modelling tools and measurement facilities are required to cope with the desired surface accuracy and stability.
- To make booms available for very long focal length and possibly triple reflectors deployed on the east/west sides of satellites.

18.7 Increase the Competitiveness of Well-Established Antenna Products

18.7.1 Problem Area and Challenges

Antennas are a critical and strategic technological domain for current and future space projects. If the time to market is extensive, the technologies may be obsolete by the time they are available. Competition among mature products is based on price and products can be pushed out of markets if standardization is not done quickly. There is a need to cope with the market trend asking for higher power, higher beam isolation, larger bandwidth and a larger number of beams. In all cases, the long qualification programme to reduce risks endangers the timely delivery of new products.

Over the last decade, the satellite market reached a relatively stable level of about 30 launches per year. The vast majority of these satellites are for telecommunication applications. In this domain, operators are requesting reduced costs per transponder and delivery time.

Concerning antenna technologies, reflector antennas are by far the most used for all telecommunication missions as they are more efficient for applications with few beams and limited reconfigurability requirements. Even with the rapid development of solid state devices, the achieved performance and mass/cost budget for direct radiating array active antennas cannot compete with reflector antennas. To benefit from the large aperture at low mass, as provided by reflectors, and generate several overlapped beams, as provided by arrays, complex reflector solutions based on semi-active antennas have been implemented for L/S-band mobile missions.

Over the last 30 years, reflector antennas have undergone several configuration changes to comply with user needs in terms of cost and performance. The move to new frequency allocations such as Ku-band, Ka-band and later on Q/V-band, and the move from linear to circular polarization over Europe, have impacted and will impact reflector antenna architectures and technologies due to an increase in both frequency band and bandwidth.

Different generations of telecommunication reflector antennas have succeeded over the years. For example, C-, X- and Ku-band antennas generating one circular or elliptical beam per aperture, Ku-band antennas radiating shaped beams originating from multiple feeds and later on from a single feed located at the focus of a shaped reflector, Ku/Ka-band Earth deck steerable antennas, Ku/Ka-band lateral face antennas generating multiple pencil beams over a continental area, Ku/Ka-band Earth deck antenna modules and more recently X-, Ku- and Ka-band reconfigurable antennas. Most of these antenna concepts are still in use nowadays and these different products must remain available to the antenna designer for selecting the optimal solution. In addition, innovative antenna concepts must be elaborated to increase the portfolio and respond to new requirements.

Competitiveness is also requested for Earth observation and science missions, mainly for all the generic products with production volume, such as TT&C and data transmission antennas. They should be made available at all frequency bands (UHF, S-, C-, X-, Ku- and Ka-bands) at low cost and with a short delivery time.

Maintaining competence in a large number of companies is difficult when considering the relative small volume of the space market. Highly accurate manufacturing and space qualification require extensive expertise and test facilities and a minimum of production to maintain the know-how and competitiveness. Synergies should be searched for each product considering other applications/instruments and identifying possible shared developments.

18.7.2 Present and Expected Future Space Missions

18.7.2.1 Telecommunications

There is strong interest in the different telecommunication missions as industry competitiveness needs to be improved. The main driver is the most successful space market: the DTH/TV market with consideration of the FSS and BSS markets. Also, government services, the VSAT and mobile markets, and the interactive/broadcast and broadband markets will benefit from reductions in antenna product cost and delivery time.

Despite the continuous progress witnessed in the capabilities of terrestrial fixed communication networks, satellites are an attractive solution for broadcasting, multicasting and point-to-point communications because of their well-known coverage capability and limited ground infrastructure. The fixed telecommunications service represents by far the largest space market.

Nevertheless, satellite networks today often lag behind terrestrial ones and are not flexible enough to support efficiently the current trend towards bursty non-uniformly distributed traffic, in both space and time. However, this is not an intrinsic limitation of satellites; provided the necessary technology is developed with proper timing, they can actually offer advantages because they inherently serve larger areas.

18.7.2.2 Earth Observation

Data Transmission Antennas There is a well-expressed need to increase the data transmission rate for Earth observation and science missions. The move from X-band to Ka-band will have a major impact on antenna products.

18.7.3 Promising Antenna Concepts and Technologies

It is proposed to split the actions to be pursued into two main categories.

First, to *improve existing products and production processes* to reduce the production cycle time and recurring cost. Satellite interfaces and modular configurations on spacecraft (lateral face and Earth deck) should be standardized. This can be done while maintaining or improving today's performance (reflector as-built and thermal distortion surface accuracy, RF losses, XPD).

- Prepare the next generation of high-power feed chains: This covers the feed architectures and the design of building blocks to comply with the operator's needs in terms of increased radiated power (mainly C-, Ku- and Ka-bands) and the evolution towards standardized interfaces and generic wide operating bandwidths.
- Develop generic high-power building blocks (OMT, filters, polarizer, etc.) with relevant multipaction analyses.
- Investigate alternative concepts for high-power combination.
- Extend the material qualification range to make the antenna products more resistant to thermal constraints and allow more powerful missions. This is illustrated by advanced sunshield development, new resins/fibres for CFRP reflector development, and so on.
- Standardize, as much as possible, the I/F and geometries for the C–Ku reflector products to be implemented on spacecraft.
- Ensure that any payload baseline design solution (even with PIM at order 3) can be based on combined transmit/receive antennas. The use of separate transmit and receive antennas has been considered in the past as an alternative to a single Rx/Tx antenna system. This solution has been adopted by some systems (i.e. MSAT, ACeS and ETS VIII) and simplifies certain issues such as asymmetrical mass and moment of inertia distribution on the platform, PIM generated by antenna feed elements and the reflector, diplexer complexity, and so on. Nevertheless, the savings relevant to a single-reflector solution are really considerable and Tx/Rx solution should be the baseline whenever possible. There is a need to consider real signals and not just pure tone when doing PIM analysis in such a way as to relax the constraints on the hardware.
- Make available standard products for TT&C, data transmission and navigation user antennas. As an example, interference has been experienced when using X-band data downlink antennas because many users who use isoflux antennas share the frequency allocation. To overcome this limitation, steerable beam antennas have to be developed to provide spatial isolation. This will allow use to be made of existing X-band on-ground equipment even for users demanding quite large data rates and even more with more advanced modulation schemes. Dual polarization should be considered. Concepts and mechanisms should be transposable to Ka-band for access to higher data rates.

- Develop compact global horns and medium-gain antennas. Telecommunication missions require global coverage from GEO and scientific missions such as BepiColombo or Mars Sample Return have identified the need for medium-gain antennas. For such antennas, with directivities in the range of 16–22 dBi, direct radiating arrays or single elements associated with a multilayer parasitic structure, such as the Fabry–Pérot concept, provide a very good and compact solution when compared with conventional horns or reflector antennas.
- Improve the S–X-band antenna to generate generic products to support data transmission for science missions and prepare the next generation based on the X–Ka-band antenna.
- Make available standard products for TT&C, data transmission, GNSS low-multipath receive antennas and put in place solutions compatible with increased data rate: data relay satellites for real-time communication between LEO Earth observation satellites, ground infrastructure, and so on.

Second, to identify *innovative antenna concepts* using the most recent R&D:

- Use reflectarrays to replace shaped reflectors with benefits in cost and development time.
- Make available low-cost reflector moulds as an alternative for the low-CTE materials of current moulds. Aluminium or steel alloys, when combined with a dedicated and accurate prediction of the distortions arising during the curing process and their impact on the antenna shape, can offer a solution. This may require the use of low-curing-temperature resin systems in the reflector itself. Also, recycle used Invar and graphite moulds, moulds made of a CFRP sandwich or carbon fibre reinforced cement, with a machinable and coated surface, low-CTE foam blocks, such as machined and surface-treated carbon foam, and so on.
- Develop high-gain Ka-band data transmission antennas for Earth observation and science missions. For the former, satellites require high downlink data rates (1.5 Gbps is expected). The present solution based on isoflux X-band antennas with a 50 MHz frequency bandwidth may allow 350 Mbps to be reached. For science missions, typical data rate values are 10 Mbps but environmental constraints can become more stringent (such as the maximum temperature of 400 ° C encountered by BepiColumbo flying near Mercury) or there might be a need to avoid any platform attitude perturbations induced by the antenna's moving parts. To cope with programme needs, several solutions are anticipated. For Earth observation links requiring high data rates, the frequency band 25.5–27 GHz has been allocated, while the 31.8–32.3 GHz and 37–38 GHz bands are reserved for deep space exploration. One may also mention the needs of space VLBI applications which require, as on the VSOP2 mission, data rates up to 4 Gbps. New designs and preliminary critical breadboarding of Ka-band mechanically or electronically steerable antennas are required. They will allow to reduce the need for on-board processing and their added complexity in order to comply with the very large amount of science data generated by the instruments.

18.8 Enable Single-Beam In-Flight Coverage/Polarization Reconfiguration

18.8.1 Problem Area and Challenges

Considering the worldwide heritage of Earth observation SAR with beam agility using both arrays and reflector antennas, this section will focus on telecommunication missions where major challenges still limit the implementation of in-flight reconfigurability.

At present, most of the satellites deployed by operators to provide C- and Ku-band telecommunication services still make use of fixed antenna architectures, which do not allow for any change in the radiation characteristics within the coverage area. Today, flexible coverage is achieved by operators only by means of two- or three-axis steerable antennas and rotating subreflectors, which are implemented on the Earth deck of the platforms.

It is likely that, in the long term, these payloads will be progressively replaced by more advanced systems, which will allow for a higher degree of flexibility and reconfigurability. Future on-board antenna systems should be capable of in-flight changes of radiation performance, such as beam repointing, coverage reshaping and polarization agility. This will allow adaptation to (1) traffic demand changes during the satellite's lifetime, (2) change of orbital location and (3) compensation of the degradation due to thermoelastic distortions and/or rain attenuation.

Having said this, given the high costs associated with such fully reconfigurable active antennas and the cautious evolution of the market, the most likely medium-term scenario is the use of passive antennas with a partial degree of flexibility. It is considered that an antenna allowing for some flexibility to shape the coverage area during flight would be an attractive feature that operators would welcome when ordering their next satellites. It is not evident, though, that operators would accept a much higher cost and mass or a reduced RF performance for these payloads. For this reason, the added flexibility should be achieved by means that do not significantly increase the overall cost, as compared with current antennas. Such an approach might allow tradeoffs among 'generic, modular, scalable' products with low recurring cost and short time to market.

18.8.2 Present and Expected Future Space Missions

18.8.2.1 Telecommunications

This aim is mainly applicable to the broadcast satellite services described in the previous section.

18.8.3 Promising Antenna Concepts and Technologies

To answer the question of partial radiated performance reconfigurability, changes in antenna architectures and technologies are required. If only a single beam is required it opens the door to mechanically reconfigurable antennas, reflectarrays with phase shifters/switches and passive (phased-)array antennas. In addition, hybrid electromechanical antennas may be considered.

Mechanically reconfigurable antennas can only use mechanisms to reconfigure the beam. Besides the existing two- or three-axis steerable antennas based on rotary actuators, and the zoomable antenna with a linear actuator to translate the main reflector, there is a strong interest in mechanisms that allow selection of an antenna aperture to generate a predefined coverage. This aperture can be a (sub)reflector, an array or a reflectarray.

For increased freedom in the radiation pattern shape, reconformable reflectors possibly using several miniaturized linear actuators implemented behind the reflecting surface offer a good alternative. Conventional 2D rotary actuators are contemplated for the scanning while the shaping is obtained from the set of linear actuators.

Passive (phased) arrays can use an aperture split in radiators with central amplification and, possibly, phase and amplitude control in between. In this case, the BFN is an assembly of transmission lines, phase shifters and/or variable power dividers fed in transmission by a high-power amplifier. Passive arrays are used as direct radiating arrays in transmission/reception or array-fed reflectors. Power distribution can also be done in a radiative way with phase shifters included in a discrete lens antenna. The main developments could be undertaken using different building blocks:

- Efficient, flexible and compact TWTAs are key to the viability of array-fed reflectors in Ku- and Ka-bands.
- Ferrite components such as variable power dividers and phase shifters can be used. Reduction of output losses, especially in semi-active array-fed reflector antennas, is a high priority. Connectivity of ONET to feed is complex and lossy even for large L-band antennas, and more difficult for Ku- and Ka-bands.
- Centralization of power generation requires major developments to deal with power, multipaction and PIM issues.
- Antenna RF/mechanical/thermal efficient integration and associated thermal control may be required to handle the large power dissipation.

Active reflectarrays can use an aperture with radiators illuminated by a feed system. By changing the phase reflection characteristics of the radiators, the beam can be steered or shaped. Two in-flight reconfigurable independent beams can be generated in each of the orthogonal polarizations. This concept presents several advantages due to the signal distribution in a radiative way but puts severe constraints on the active devices used to change the phase characteristics. Major efforts should be made on technologies and components such as PIN diode,liquid crystal, RF MEMS, ferromagnetic thin films, ferroelectric, ad so on to enable such configurations considering the space environmental constraints (temperature range, radiation, etc.), the requested low loss/low cross-polarization, the frequency bandwidth and the power handling capability for transmit antennas.

Hybrid electromechanical antennas can use a mechanism to point the aperture towards the desired geographical area and a feed system to shape the beam. Another configuration implemented mainly for mobile user antennas uses mechanical steering for azimuth and electronic steering with a low number of active controls for elevation. Such configurations can provide very efficient antennas with reconfiguration capability at low cost. In the first configuration the requested field distribution in the feed plane can be generated by:

- A set of feeds excited by a BFN.
- A feed or set of feeds illuminating a spatial filter placed in the beam path from the feeds to the main reflector.

In both cases, the reflector illumination can be adjusted to achieve the required performance in terms of shaping and reconfiguration capability (efficiency, polarization purity, etc.) and the best implementation features (manufacturing, assembly, integration and testing, as well as overall cost).

18.9 Enable Active Antennas at Affordable Cost

18.9.1 Problem Area and Challenges

This section addresses active array antennas for multibeam broadband telecommunications (mainly Ka-band), broadcast (mainly Ku-band), mobile communications (mainly L/S-band), navigation, SAR, radiometers, and Earth observation/science high-speed data transmission antennas. These types of array architectures are required when independent agile, reconfigurable and adaptive multiple beams have to be generated (or received) or when the minimization of losses after high-power amplification (or before an LNA) is necessary for the mission.

Active phased-array antennas that radiate or receive the signal directly are called direct radiating arrays (DRAs). They can also be used to illuminate reflectors and, in this case, the antenna systems are called array-fed reflectorss (AFRs). Feed arrays can compensate for the inherent distortions of large reflectors deployed in space and allow relaxation of the surface accuracy requirement, synthesize the desired beam radiation pattern and/or move the complete multiple beam patterns over the coverage area to compensate for seasonal satellite deviations. Both types of arrays, DRAs and AFRs, are addressed in this section.

Two main arrays of active antenna architectures can be distinguished:

- *Semi-active arrays* utilize custom-designed hybrid circuits introduced between the elements and the power amplifiers. This allows operation at nominal levels with optimum efficiency. Beam scanning or reconfiguration is achieved by phase-only signal control at the inputs of the amplifiers. These arrays are often used to feed multibeam reflector antennas.
- *Active (phased) arrays* use, at the element level, distributed high-power generation for transmit and low-noise amplification for receive as well as phase and amplitude control (if required). Active arrays are well

adapted for low earth orbiting communication satellite constellation system when pointing is implemented to maintain the link with the Earth over a large angular range. The most common approach to achieve fast beam scanning is through integration of monolithic microwave integrated circuit (MMIC) phase shifters with the antenna elements.

Currently the RF BFN is often replaced by a digital beam forming in combination with signal processing. In digital beam-forming arrays, only the high-power components, circuits and LNAs located near the antenna aperture use analogue circuits. All the amplitude and phase control functions are performed digitally. With digital beam forming (DBF), very accurate beam control can be obtained and various array signal processing methods can be applied. This includes multiple beams with very low sidelobes, adaptive pattern control and direction of arrival (DOA) estimation. Digital signal processors can discriminate between incoming signals and interference allowing for smaller user antennas.

The scan volume can change drastically with different applications and this has a major impact on selection of the antenna concept. For SAR applications, there is interest in covering about $\pm 13°$ in elevation combined with no or a limited scan in azimuth. For radiometry there is interest in covering about $\pm 45°$ to 60° conical scan volume (SMOS or other push-broom solutions). For telecommunications the field of view is $\pm 9°$ from GEO and for navigation the Earth is seen within a cone of $\pm 12°$.

Array antennas benefit from space heritage mainly for Earth observation missions and military telecommunications. SAR (SIR series, ERS, EnviSat, Radarsat, etc.) are less complex because one single beam with agility is required and the pulsed mode significantly reduces the constraints on power dissipation. It should be noted that future instruments for Earth observation, such as the wide-swath high-resolution radar, will also require very demanding beam agility and control. This implicitly paves a way for digital beam forming and for more consideration of signal processing. Besides Earth observation applications, recent developments such as Globalstar, Iridium NEXT and INMARSAT-4 for mobile communications and SMOS for passive radiometry have demonstrated the major interest in and potential of array antennas. Outside Europe, major recent array antenna developments include SPACEWAY and WGS (Wideband Gapfiller Satellite) in the USA and WINDS in Japan. Arrays with digital BFNs have been implemented in several defence programmes such as Skynet 5, Syracuse 3 or SPAINSAT, but also in most of the mobile interactive telecommunication missions (Thuraya, INMARSAT series).

Since the beginning of the 1980s, although active DRAs were often originally considered for several programmes, after a detailed technical and economical tradeoff performed at the time of final mission implementation, the solutions selected have been almost exclusively based on mechanically steerable reflector antennas. The following examples can be mentioned:

- Mobile user terminals for cars (ERA) and aircraft (Connexion).
- SAR (reflector-based SAR with beam scanning capability is now more often considered).
- Telecommunication multiple-beam antennas (Skybridge, split into several single-beam antennas).

Preliminary selection was often justified by the great advantages of active arrays in terms of flexible coverage, in-flight reconfigurability, multiple-beam generation by only one aperture and avoidance of moving parts for beam steering. Solutions based on active antennas were nevertheless disregarded at that time due to their intrinsic drawbacks in terms of power consumption and dissipation, the limited radiated power capability for broadcasting, the achievable performances, the complexity and maturity of technologies, the mass budget when high gain is requested and the overall cost. This generates the need for new, ultra-light technologies and thermal hardware. With the progress in developments of a more efficient Tx/Rx module, the thermal aspects may be different, but the improvement of this parameter has not been as quick as expected.

Because active phased arrays are complex and expensive antennas, only a few space missions (around 10%) have considered them. This is not the case for ground and airborne radar where important funding mainly from defence programmes has led to large-scale production of Tx/Rx modules, reaching a price of $100 per module as early as 1985. Unfortunately, this background only partially applies for space missions due to the completely different requirements and operational constraints.

Today (active) phased arrays for space missions are mostly considered when no other alternatives with the requested functionalities are available. This is mainly the case for:

- In-flight multiple-beam agility (LEO or MEO telecommunication constellations).
- Multiple-beam reconfigurability or possibly beam hopping to cope with mission requirements during the satellite's lifetime (GEO).
- Inertia-less beam steering for Earth observation or science data transmission antennas.
- Secure transmission (the higher security level, e.g. anti-jamming, required even by commercial applications reinforces the demand for this product).
- L–S-band GEO mobile communication space antennas (spacecraft not suitable for accommodating four passive 12 m reflector antennas).
- Missions with major constraints in terms of volume and requiring planar apertures.

To summarize, array antennas have strong features and advantages and are expected to be more used in several space applications where they can be produced at affordable cost. Innovative array architectures and technologies should be proposed to face in the most efficient way the very large number of constraints associated with array antenna subsystems with an emphasis on missions requiring a medium gain and number of beams.

18.9.2 Present and Expected Future Space Missions

18.9.2.1 Earth Observation

In the past, SARs operating in C-band were mounted on board satellites which also hosted other types of instrument (e.g., ERS, EnviSat). The tendency now is for dedicated satellites, possibly used in constellations, with multiple-beam instruments.

The next generations of SARs will require array antennas with cost-efficient solutions and more flexibility (bistatic operation, receive-only arrays; beam shaping, wide scanning beam agility; polarimetric operation with high polarization purity, interferometric capabilities). It is expected that there will be a need for wider bandwidth (reconnaissance), higher amplitude/phase stability, more sensitivity (science) and simultaneous multibeam reception possibly combined with large apertures in receive only.

Beside reflector based solution with active or passive focal arrays, there is also a need to reduce the cost of SAR based on active phased arrays as such antennas are the main cost driver. For this objective, there will be an evolution towards a new generation of SAR active antennas with Tx/Rx modules using advanced technologies (GaN, SiGe, advanced packaging and interconnect concepts, new materials and innovative technologies enhancing the thermal control etc.) and possibly including frequency conversion or digital conversion.

Special attention should also be given to advanced calibration concepts to obtain significant reductions in cost by possible relaxation of the equipment (and reflector if any) requirements compared with currently used techniques.

Besides C- and X-band instruments, side-looking SAR and nadir-looking ground-penetrating radar operating in P-band (435 MHz) for biomass and ice-sheet thickness measurement impose severe constraints on antennas and very large-aperture (typically 60 m^2) development is required.

Also in the area of passive reception of signals, for example GNSS signals, there is a requirement for antennas operating at L-band to be capable of beam steering and tracking.

18.9.2.2 Telecommunications

Secure Communications Recently, cases have been reported of communications satellite transponders being accessed by unauthorized transmitters. Such transmitters can be located within or outside the nominal coverage area, the latter being possible since satellite antennas using shaped reflectors have relatively high sidelobes. As well as unauthorized access, the communication payloads can also suffer denial-of-service (DOS) attacks through the broadcast of strong disruptive signals which interfere with the wanted signal. Unauthorized ground transmitters are able to cause a serious reduction of the transponder capacity, resulting in a loss of revenue for the satellite operator.

To mitigate the impact of such unauthorized transmitters, auxiliary Rx antennas can be included on the spacecraft and used to reduce the antenna gain, that is to create a null, within the antenna coverage or sidelobe region when, following the proper setting of amplitude and phase, it is combined with the output of the Rx antenna at the most adequate level in the payload receive section. Future flexible coverage payloads employing array-fed reflector antennas in receive will facilitate adaptive nulling, allowing for reconfiguration of the shaped beam by setting the excitation coefficients of the feed elements. This would be applicable to satellites implementing broadcast missions with contoured beam coverage via a shaped reflector and AFR, operating in Ku-band.

The majority of current satellites use a 'bent-pipe' transponder whereby the payload is transparent and broadcasts whatever is received within its frequency band. Consequently, it is necessary to develop and implement the additional payload hardware and processing required by combining the outputs of the main and any auxiliary antennas with the correct relative amplitude and phase in order to produce an effective null towards the unauthorized transmitter or jamming source. It might be expected that high-power jamming will necessarily be narrow band and that nulling can, therefore, be limited to narrow-band signals, and that nulling in a single Rx channel, typically 36 MHz (TBC) located within the complete Rx band, can be considered. It is likely that coverage performance away from the null will be compromised. However, this will be limited to a single channel. In the case of AFR antennas it will be necessary to develop the appropriate processing (algorithm) to determine the DOA of the interfering source and to reconfigure the antenna coverage accordingly.

Mobile Communications See the section on large antennas.

Conventional Fixed Telecommunication Service (Mainly Ku-band) See the section on competitiveness.

High-Capacity Multibeam Systems (Mainly Ka-band) See the section on high throughput.

18.9.3 Promising Antenna Concepts and Technologies

For array antennas, the optimal antenna and payload/instrument architectures are largely interlinked and mission dependent. The final choices strongly depend on:

- The EIRP and *G/T* over the coverage area (having an impact on the number of radiators with their associated amplification which translates into accommodation, thermal and mechanical constraints and functional complexity of the BFN).
- The number of beams to be generated (having an impact on beam-forming complexity).
- The overall frequency band and the allocation per beam/user for telecommunication missions (having an impact on channelization).

- The type of flexibility, that is beam shape, power per beam, frequency, polarization (having an impact on power flexibility, channelization, routing, etc.).
- The on-board available resources, type of platform and satellite orbit (having an impact on mechanical/thermal design and payload/satellite AIT).

The satellite transmit section is the most critical one with regard to the limited on-board resources. The overall power efficiency should be maximized to achieve the targeted EIRP on the desired coverage areas. Considering that the total amount of required RF power generally exceeds the generation capabilities of the single amplifiers available for space applications, the required approach combines a plurality of amplified replicas of the desired signal individually amplified by highly efficient high-power devices such as TWTAs or SSPAs. Different power combining techniques exist (e.g. network combining, polarization combining, spatial combining, etc.), each offering specific advantages as well as disadvantages. They all suffer from combining losses (or cross-pol degradation in polarization combining) caused by non-perfect matching in amplitude and phase between combined signals or the failure of a contributing amplifier, as well as ohmic losses. Combining losses strongly depend on the selected antenna architecture.

The multibeam antenna and its associated RF front end are the key challenging elements of the entire system. To maximize use of the generated RF power, that part of the BFN connecting the high-power amplifiers to the radiating elements should be almost lossless and the antenna should be capable of handling very high RF power levels.

Antenna power reconfigurability among the different beams is made possible through distributed amplification. Major configurations are listed below sorted by increasing number of power amplifiers contributing to a given beam:

- High-power section based on MPAs connected to a self-standing high-power BFN (number of BFN ports equal to the number of beams, no shaping flexibility). This solution is mainly used for AFRs, for example M-SAT.
- Semi-active multimatrix architecture (constituted by a fully interconnected low-signal-level BFN directly interconnected to the TWTA inputs). This solution is mainly used for AFRs, for example INMARSAT-3, Artemis, Thuraya, INMARSAT-4 and now Alphasat.
- Low-power BFN feeding a set of SSPAs or flexible TWTAs directly interconnected to the radiating elements in a redundant-less configuration. This solution is mainly used for AFRs associated with shaped and/or non-focused reflectors, imaging systems and DRAs having a regular spacing.
- Low-power BFN feeding a set of SSPAs or TWTAs directly interconnected to the radiating elements in a redundant-less configuration. This solution is mainly used with DRA and imaging systems. The use of sparse arrays having an irregular spacing allows the requested aperture tapering to be generated while all amplifiers are at the same operational point, and so on.

Block diagrams of these architectures are presented in Figure 18.4 (redundancy elements are not shown). Note that, for the last two cases, the diagram is also valid without a reflector.

Considering the various mission requirements, the following elaborate on the most promising antenna architectures.

High EIRP: Current system trends show a clear need for higher EIRP performance to counteract mobile channel shadowing and to increase quality of service (QoS) and transmission throughput. It can be envisaged that reflector antenna architectures based on MPAs connected to self-standing high-power BFNs will shortly reach their limit in maximum allocable power to a single beam and be confined to non-reconfigurable coverage in shape. High EIRP can nevertheless be obtained when no coverage overlap is required or when non-optimal excitations are accepted to keep them orthogonal. This was implemented for MSAT.

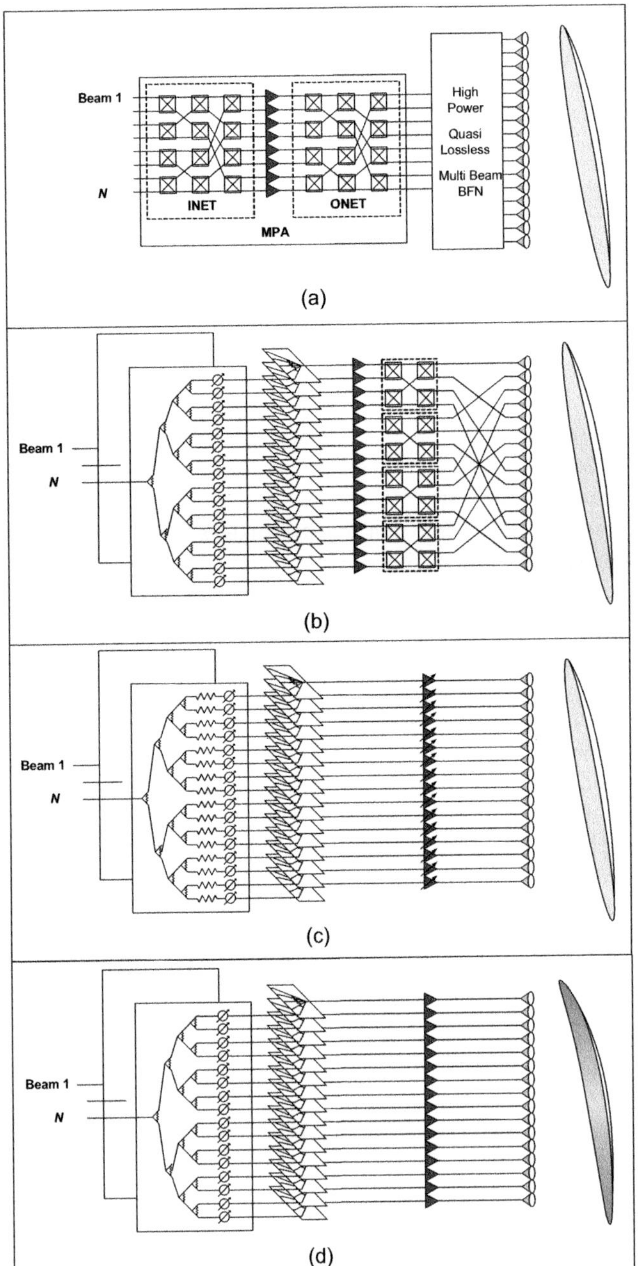

Figure 18.4 *Possible antenna architectures. (a) MPA connected to a self-standing high-power BFN. (b) Semi-active multimatrix. (c) Low-power BFN feeding a set of flexible TWTAs. (d) Active array feeding a shaped reflector.*

The growing request for full power and shape reconfigurability among linguistic beams calls for further improvements in flexible architectures while maintaining highly efficient DC power to EIRP conversion and low mass, volumes and costs. In this respect, the *array-fed shaped (sub)reflector* can offer a good alternative as it includes the high directivity brought by the reflector to the power sharing originating from different feeds in the focal array. Also, *imaging array* antennas can generate high EIRP but are penalized by a large number of active controls and reflector aberrations. The wide dimensions of the feed arrays and reflectors involved lead to severe constraints on the number of antennas that can be carried on board a satellite platform. Integrated Tx/Rx solutions should be preferred but put strong constraints on the single-feed array design with adequate overlapping and spillover efficiency.

Large number of beams: As the number of beams grows, RF BFN complexity drastically increases and greater flexibility both in routing between beams and in frequency planning is often required. The demands of such complex missions are best satisfied with on-board digital processors (OBPs) which can perform adaptive beam forming in addition to channelization and routing. The advantages of a digital implementation include precision, predictability and freedom from factors such as ageing, drift and component value variations. The increasing functionality of OBPs will require continuous advances in digital processor technologies and architectures. Beam bandwidths may be lower than that of the total repeater bandwidth (as in the case of frequency reuse schemes), and this could in principle lead to a complexity saving, first by performing demultiplexing of each wideband feed signal into narrower band signals, and then by performing beam synthesis at a lower sampling rate. This type of payload architecture is also of great interest for the higher security level required by commercial applications. Besides the large number of beams, it is expected that digital BFN tiles for SAR applications will allow implementation of additional radar modes such as MTI, will make available true delay line for high resolution and will make central electronics more generic.

Medium gain: DRAs, either planar or conformal, can be very competitive for modest aperture size, for example for LEO or MEO mobile communication missions or Earth observation instruments as well as for secondary GEO telecommunication payloads. The major advantage of DRAs is to offer flexible/reconfigurable multibeam operation with distributed amplification and graceful degradation. Also, the distributed lower power levels induce a lower risk of multipaction in the antenna. They are nevertheless strongly penalized by the current power-added efficiency (PAE) of SSPAs. A low efficiency impacts not only the achievable EIRP, but also the power dissipated on the satellite. Typical figures of multicarrier PAE at Ku-band of TWTAs (50%) and of SSPAs (10–20%) act most often as a show stopper, yet the situation of SSPA efficiency is not expected to change drastically in the coming years, even considering new technologies such as GaN which will mainly affect the achievable power density due to the higher operational temperature.

Aperiodic arrays with equipower radiators: These allow a drastic reduction in the number of active elements and BFN complexity with expected major impacts on cost as well as power efficiency due to the use of mini TWTAs with higher PAE as compared with the SSPA. Unequally spaced arrays have interesting characteristics and offer several advantages with respect to periodic arrays [3]. Up to now sparse and thinned arrays have been rarely used because they require more complex analysis and synthesis tools. Generic block diagrams of active multibeam DRA transmit antennas based on regular and aperiodic lattices are shown in Figure 18.5. Recently [4], tradeoffs on active DRA architectures based on the two design solutions have been made on the basis of simplified antenna models comparing:

- a periodic array (Figure 18.5 left) with sidelobe control by means of amplitude tapering with an active section using the same HPA design for all the elements (the HPAs are operated according to the required RF power for each radiating element with different amplifiers at different output backoffs);
- an aperiodic array (Figure 18.5 right) with sidelobe control by means of appropriate element positioning with an active section using the same HPA design for all the elements (all the HPAs are operated at the same working point).

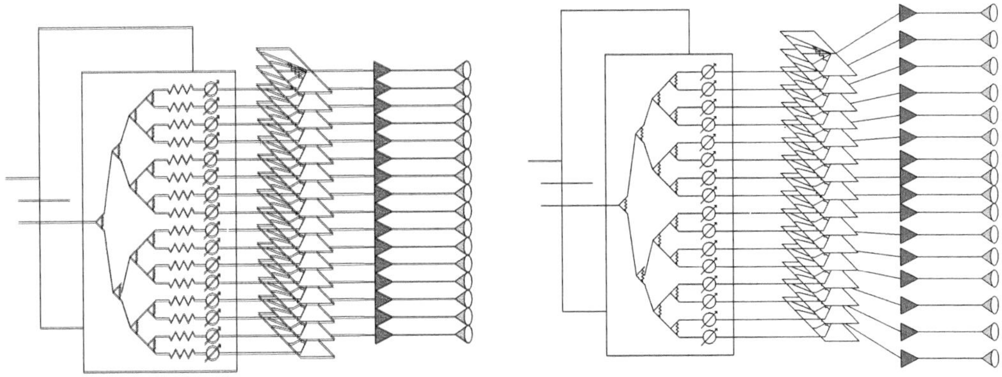

Figure 18.5 *Regular and non-regular radiating aperture spacing of transmit array antenna architectures.*

Recent preliminary investigations clearly show the high potential of active aperiodic arrays in terms of reducing the number of elements, active control elements and amplifiers while improving the overall antenna DC to EIRP conversion efficiency. When the optimization leads to a drastic reduction in the number of power amplifiers, in order to keep the same EIRP, SSPAs can be replaced by either mini or standard TWTAs with additional benefits for the PAE.

Note that breaking the periodicity and reducing the number of elements in sparse arrays allow a significant increase in the spacing between the elements. This is detrimental to the number of grating lobes in the field of view. However, the layout and the tapering of sparse arrays may be designed to control, in addition to the beamwidth and the sidelobe level, the position and shape of the grating lobes. Compared with regular arrays, sparse arrays offer unique freedom to adjust the maximum gain independently (playing on the number of elements active in the antenna aperture) and the beamwidth (playing with the array's overall dimensions for a constant number of active controls). In fact, the beamwidth essentially depends on the aperture of the array and only slightly increases when reducing the number of elements. On the other hand, the antenna gain realized depends strongly on the number of elements and on their characteristics. When the link budget allows, a large array with a very limited number of active controls can generate very narrow beams with the potential to increase frequency reuse over a given coverage.

Another interesting concept using so-called *overlapped sub-arrays* deserves some attention. In this case, the DRA is associated with a flat top pattern at sub-array level. This allows a similar level to be generated from each sub-array over the coverage area while minimizing the subarray quantization lobes outside the coverage. To achieve this sub-array pattern, an amplitude excitation of the $\sin(x)/x$ type is applied at sub-array level with adjacent sub-arrays sharing elements.

Phased-array antennas with a flat-topped element pattern and a limited field of view can be obtained in several ways, as documented in the literature. Examples from the Orchard synthesis procedure for equispaced linear arrays to Elliot and Stern's two-dimensional arrays, or those of Krumar and Branner, and particular procedures or polynomial distributions like the Legendre one, are used to obtain a pattern with a shape similar to a rectangle. Probably the most interesting results have been obtained by Skobelev *et al.* [5] who have been studying, manufacturing and testing limited-size sub-arrays able to generate a pseudo-rectangular shape. The work of Skobelev *et al.* has focused on the case of a single-beam antenna, while for many space applications the case of a multibeam antenna is targeted.

For the array design, the generation of the $\sin(x)/x$ distribution at sub-array level requires a rather large dimension in wavelength, thus generating subarray quantization lobes over the Earth. Eliminating the subarray quantization lobes is done by overlapping the sub-arrays in the aperture plane in such a way as to obtain an inter-amplitude/phase control distance compatible with subarray quantization lobes outside the Earth. This overlap is performed by a set of dividers/combiners after power amplification (in transmit).

The two concepts that have been presented, namely 'sparse arrays' and 'overlapped arrays', allow a significant reduction in the number of controls with an expected decrease in array cost. Another interesting capability, due to the medium directivity of the sub-arrays, is their ability to radiate over the required field of view instead of more broadly as in regular arrays with a small element spacing. This spatial filtering property can be compared with that obtained from reflector antennas.

To enable affordable active antennas and increase their use in space missions, the focus should be on system aspects, innovative architectures and critical technologies such as the following:

- *Develop innovative array (DRA and AFR) architectures* to reduce the number of amplitude/phase controls with major improvements in cost, complexity and power efficiency.
- *Optimize the array amplitude and phase settings* in the command and calibration philosophy, accounting for imperfections, evolution over the temperature domain and dispersion for the active antenna equipment, in such a way as to relax the requirements and achieve cost savings. The very large degree of freedom offered by the large number of RF paths radiating in parallel should be fully utilized.
- Analyse the impacts on both space and user antennas (synchronization, protocol, etc.) when using *beam hopping techniques* to cope with users' demand changes.
- Investigate the benefits of the *generic, modular and scalable* (low recurring cost, short time to market) offered by active concepts and compare performance/costs with highly optimized, highly integrated, low-mass customized solutions.
- Propose solutions for *reducing array mass*. Even if AFR antennas or DRAs allow the generation of several overlapped beams by only one antenna instead of several fully passive reflectors, arrays exhibit areal mass typically 10 times higher than reflector antennas and this greatly reduces their use in highly directive antennas. Major efforts should be made to reach a high integration level of all functions (RF, mechanical and thermal) considering the platform interfaces.
- Improve the bulk and mass of filters in transmit arrays after power amplification. *Filtennas* and self-diplexing antennas, where all or part of the filtering function is integrated in the radiating plane, might be an avenue worth exploring to improve the integration.
- Develop *feeds and radiating elements* offering the best mass/cost/schedule/volume and performance/bandwidth. Guided and printed technologies should be considered for Earth observation and telecommunication missions using DRAs above C-band.
- *Multiband operation* of arrays is difficult when a common array element spacing is used, since this tends to create grating lobes at higher frequencies and oversampling at lower ones. Consider log periodic or pseudo-log periodic configurations, which might have the potential to overcome this limitation.
- For *power amplifier efficiency*, one key problem of active array antennas is the low efficiency of SSPAs operating in multicarrier mode (30% at L- and S- band, 15% at Ka-band), in particular when loading varies. Only limited improvements have been achieved over the last 20 years. Consider that one ideal product would be a power amplifier with *constant efficiency over a large dynamic range*.
- *GaN developments* are vital in this area. Develop GaN SSPAs further, closely linked with packaging issues and integrated thermal control to handle the large array power dissipation.
- Include *efficient, flexible and compact TWTAs* since they are mandatory for Ka-band AFRs, as well as Ku-band, and mini-tubes with a reduced footprint are requested for DRAs.
- Improve *analogue BFN technology* to increase RF/DC/TT&C integration and packaging and to allow coping with mission needs in terms of number of beams and radiating elements.
- Extend *digital beam forming* from mobile telecommunications to other applications, as expected.

- Incorporate *Si/Ge circuits* since they are compatible with large-volume production and are expected to reduce costs and allow the integration of RF and command/control on the same device with a major impact on overall equipment size. It is foreseen that this process can be used for phase shifters and local oscillators with a direct use for SAR tiles implementing downconversion.
- *Active thermal control* is required to cope with the large dissipated power of the arrays. Extend new GaN developments and thermal control tools (heat pipes and two-phase loops) to higher temperatures, in order to explore fully the potential to operate array antennas at higher temperatures and thereby reduce the required radiator area.

18.10 Develop Innovative Antennas for Future Earth Observation and Science Instruments

18.10.1 Problem Area and Challenges

Space provides a unique vantage point for obtaining information about the Earth's land, atmosphere, ice, oceans and biota that is obtainable in no other way. NASA, ESA, JAXA and other space agencies are studying the interactions among these components to advance the new discipline of Earth system science. Ever since the first Earth observing satellite, Seasat, used its radars to study the Earth and its seas, there have been continual missions to provide observation and monitoring of the land, atmosphere, oceans and ice caps. The data collected provides a wealth of information on the workings of the Earth system, including insights into factors contributing to climate change. The vision espoused by the Committee on Earth Science and Applications from Space is the foundation of robust programmes for Earth science research and applications:

Understanding the complex, changing planet on which we live, how it supports life, and how human activities affect its ability to do so in the future is one of the greatest intellectual challenges facing humanity. It is also one of the most important challenges for society as it seeks to achieve prosperity, health, and sustainability.

Earth observation missions require a large diversity of antenna configurations, frequencies of operation and technologies. This leads to very different passive and active instrument types such as radiometers, limb sounders, scatterometers, radars (including SARs) and altimeters. It is expected that these missions will be upgraded to ensure the continuity and improvement of services. In addition, new instruments are under development, opening the door to new scientific data, access to nowcasting and support to numerical weather prediction. One may refer to a better biomass determination due to the new P-band frequency allocation, increased resolution due to the use of real or synthetic large apertures, or access to better knowledge of atmospheric chemistry due to enabling technologies up to the terahertz range. In most cases these missions will call for data transmission antennas having higher data rates to ground and a move from X-band to Ka-band is expected.

Science missions study how the Universe came to be and how it has evolved. The Planck mission, launched in 2009, analyses to the highest accuracy ever achieved the first light that filled the Universe after the Big Bang, the so-called cosmic microwave background (CMB) radiation. Planck carries a 1.5 m telescope to collect the light from the CMB and focus it onto two arrays of radio detectors made of more than 50 feeds operating between 30 and 857 GHz. Since these detectors will be looking at variations of brightness temperature about a million times smaller than one degree, they have to be highly sensitive and protected against unwanted radiation coming from Solar System sources and the Milky Way. This justified in the past the large efforts carried out by ESA to evaluate the impact of stray light on the sensitivity of Planck's instruments. Planetary exploration within the Solar System and missions around L2 will call for larger science data transfer rates and require large and very accurate antennas at X- and Ka-bands. Some of these missions with high pointing accuracy optical instruments will need inertia-less beam pointing, implying electronic beam control of an array or an AFR.

Microwave instruments for space science and Earth observation have several commonalities that justify addressing them together. In both cases:

- Active and passive concepts are required and synergies in architecture and technology can be found.
- Complementarity with the data products from space instruments operating at other frequencies (optical, IR, etc.) and ground-based installations need to be carefully considered.
- Competitive advantages of (sub)millimetre waves need to be assessed as imaging IR spectrometers can achieve similar results with limited resources.

Nevertheless, major differences exist on environmental and satellite resource constraints (mass, power, etc.) which have large impacts on the final antenna configurations and the devices/technologies used.

18.10.2 Present and Expected Future Space Missions

18.10.2.1 Earth Observation

Earth observation missions fall into four categories: observation satellites (specifically designed to observe the Earth from orbit), weather satellites (primarily used to monitor the weather and climate), environment monitoring (satellites designed to detect changes in the Earth's vegetation, sea state, ocean colour and ice fields), and mapping (mapping the Earth's terrain from space). Further needs relate to a better understanding of the long-term climatological effects of the chemistry of the atmosphere and tropospheric/stratospheric exchange processes. It seems likely that future Earth observation missions will include instruments for measuring the various trace gases that have spectral lines at submillimetre wave frequencies. Limb sounders have been developed since 1970s, and the MLS on board EOS Aura includes spectral line receivers up to 2.25 THz. Due to the specificity of the instruments for the different missions, the classification below is proposed by instrument type instead of mission.

Radiometers Resolution and sensitivity in radiometry are of prime importance. Different ways to realize the antenna aperture (continuous or arrayed) as well as different scanning schemes (push-broom and rotating) provide solutions in which compromises are made.

Resolution requirements physically dictate the instrument sensor, and thus the antenna dimensioning, with an impact on the lower frequency bands. Sensors based on arrayed (sparse or dense) configurations allow synthetic aperture formation, exploiting correlation processing. Formation flying is considered but issues remain in view of the realizable quality of the synthesized beam. The physical sizing of the instrument sensor (antenna) is directly related to the best resolution capability. In the case of an antenna as a passive sensor, the pattern quality (the sampling function of an extended microwave noise distribution) is important. Good beam efficiency as well as sidelobe level will be key driving factors. An example of a sensor based on an arrayed configuration is given by ESA's SMOS mission that observes two key variables of the Earth system, namely soil moisture over land and salinity over oceans, to advance the development of climatology, meteorology and hydrological models. It uses an innovative Y-shaped instrument designed as a 2D correlation interferometer acquiring brightness temperatures at L-band (1.4 GHz).

Another radiometer concept with strong features is the one adopting a push-broom scenario. It uses a real aperture and achieves high sensitivity. Nevertheless, it has implicit coverage restrictions, associated with the extent of the focal array. Providing a rotating scheme with a reflector induces strong constraints on the lower frequency band(s) to be observed and the rotation speed. When a rotating scheme is applied, there is a physical upper limit associated with the sensor rotation, and thus a constraint on the resolution. As an example, the proposed NASA/JPL Soil Moisture Active/Passive (SMAP) mission has a suggested 6 meters diameter sizing for the rotating reflector antenna. The SMAP mission has been recommended by the National Research Council Earth Science Decadal Survey Panel for launch in the 2010–2013 time frame. SMAP will enable global measurements of soil moisture with high resolution, sensitivity, coverage area and revisit via an L-band radar and radiometer. The dynamics of the spinning antenna on a small spacecraft were identified as a potential risk but have been revealed as under control.

SAR SARs provide information on Earth's dynamic processes. They also provide valuable information on ocean dynamics, wave and surface wind speed and direction, desertification, deforestation, volcanism and tectonic activities. State-of-the-art high-performance C- and X-band SAR instruments are equipped with active phased-array antennas and show excellent performance. Recently, X-band reflector-based SARs have been developed and multi-swath capability has been demonstrated in R&D contracts. At lower frequencies, for example P-band and L-band, the same tradeoff of 'array versus reflector' is relevant. The need for a large aperture (over 10 m) due to the lower frequency leads to a difficult tradeoff between a large passive DRA and a large deployable reflector.

ESA's BIOMASS mission entered phase A in 2009 and, if finally selected, will estimate forest biomass by means of P-band (435 MHz) SAR observations of global forest cover. Three concepts have been studied: two based on a very large passive DRA antenna, and one based on a large deployable reflector. The latter has been selected and consists of a deployable dual-beam reflector antenna (14.7 × 9.7 m) on a conventional platform.

Another example is given , at L-band, with interferometric SAR systems. It has been proposed assuming a bistatic configuration of two (or more) satellites whereby a master instrument with full SAR capability is accompanied by a slave and a baseline between the satellites (typically of some kilometres) is maintained for a large proportion of the orbit, providing a bistatic interferometric SAR. Large deployable reflector antennas of over 15 m have been identified as having potential for such an L-band mission. As an example, the National Research Council Earth Science Decadal Survey, Earth Science Applications from Space, recommends that NASA launch DESDynI (Deformation, Ecosystem Structure, and Dynamics of Ice), an integrated L-band InSAR and multibeam Lidar mission, in 2017. This mission combines two sensors that, taken together, provide observations important for solid Earth (surface deformation), ecosystems (terrestrial biomass structure) and climate (ice dynamics). The sensors are: (1) an L-band interferometric synthetic aperture radar (InSAR) system with multiple polarizations; and (2) an integrated L-band InSAR system with multiple polarizations for launch in 2017. JPL studied several deployable SAR antenna types for this mission and has settled on two primary candidate configurations: a direct radiating planar phased array and an array-fed offset parabolic reflector. The latter concept requires a 15 m deployable parabolic reflector.

Another example is given by the DESDynI team currently working with DLR in Germany in studying a Tandem-L mission that will minimize the effects of temporal de-correlation due to single-pass measurements of ecosystem structure. Tandem-L is a German mission proposal for an innovative interferometric L-band radar mission that enables the systematic monitoring of dynamic Earth processes using advanced techniques and technologies.

The Tandem-L mission is currently in the pre-phase A study and is being performed in cooperation with NASA/JPL. One of the most challenging tasks in the Tandem-L realization is the development of two identical satellites with a cost-effective implementation approach and at the same time having a high performance in order to fulfil the demanding scientific needs.

Sounding Radar Ice sounding radars require a low frequency such as P-band (435 MHz) to penetrate the medium. Recent campaigns using an airborne instrument show that it is possible to penetrate 3000 m of dry ice to the bedrock and that up to 4000 m (required for Antarctica) should also be feasible. The issue from space is the suppression of ambiguous clutter returns from the ice surface and internal layers, not at the sub-satellite point, and masking the desired echoes. In the along-track direction, these can be largely suppressed by SAR processing, but in the across-track direction, a large aperture of around 18 m is required in order to provide a narrow enough beam with low sidelobes. In addition, multiple phase centres are required to provide additional beams helping with the removal of ambiguous clutter echoes. For such a mission, a large reflector antenna (15–20 m) and a complex feed system are required.

Atmospheric Sounders LEO millimetre-wave atmospheric sounders are in operational use (MSU, AMSU-A/B, MHS) and being considered for future meteorological satellites in polar orbits. Discussions have started on the next generation of meteorological satellites. It seems clear that continuity of the observational data will be required; that is, the new generation of instruments in polar orbit will cover at least the same frequencies (up to 220 GHz). In addition, a millimetre/submillimetre-wave imager is considered as a potential component with channels going up to 664 GHz. The main objective of this instrument is observation of ice clouds, and in particular of cirrus clouds which significantly affect the radiative balance of the Earth's atmosphere. Scientific analysis indicates that channels between 150 and 664 GHz are necessary to achieve the required product accuracy.

Geostationary observations are also contemplated. Unlike those from LEO satellites, the former have the key potential advantage of providing continuous coverage of the same region, which is essential for nowcasting. Nevertheless, this imposes tight constraints on the antenna aperture for achieving the required spatial resolution (a 40-fold increase in the distance to the Earth compared with the LEO) and the necessity for imaging with 2D scanning due to the absence of a relative spacecraft–Earth movement. Furthermore, the wide frequency range is challenging from the point of view of focal plane assembly and receiver technologies. The proposed frequencies range from 54 to 875 GHz.

Several preliminary instrument designs for the GEO case have been investigated around the world and a real antenna aperture with mechanical scan and one with a synthesized (sparse) array associated with interferometry have been considered. The latter has finally been selected for the GEO-SOUNDER project, which is a potential candidate for the Earth Watch programme in coordination with EUMETSAT. Such an instrument is adequate for the observation of rapidly evolving meteorological phenomena such as convective systems, precipitation and cloud patterns.

18.10.2.2 Science

The main drivers for science missions are much improved resolution (even going up to Hubble Space Telescope performance), wide bandwidth performance and low mass/power consumption. The first will provide a challenge towards the maximum possible diameter of novel reflector configurations or the development of novel methods like interferometry. The wide bandwidth coverage, to study for example the atmosphere of Mars or Jupiter, will provide a challenge for the spectrometer and the availability of sufficient LO power. Low mass and power consumption will require novel manufacturing methods and fine tuning of existing solutions.

Radio Astronomy Missions Radio astronomy has been the major driver for the development of technologies at the highest frequencies and requiring the highest sensitivities. This allows exploration of part of the spectrum that cannot be observed well from the ground. The Planck mission is collecting and characterizing radiation from the CMB – the remnants of the radiation that filled the Universe immediately after the Big Bang some 14 billion years ago – using sensitive radio receivers operating at extremely low temperatures grouped in two scientific instruments: the Low-Frequency Instrument (LFI) and the High-Frequency Instrument (HFI).

The Herschel Space Observatory, observing in a different part of the submillimetre spectrum, is the largest ever infrared space observatory launched to date. ESA's Herschel mission has been designed to reveal phenomena previously obscured from view, such as the very earliest galaxies and stars. Due to its ability to detect radiation at far-infrared and submillimetre wavelengths, Herschel will be able to observe dust-obscured and cold objects that are invisible to other telescopes.

It is expected that there will be a follow-on mission to Planck, due to the key role the observations of CMB have in increasing our understanding of the evolution of the Universe. Such a follow-on would require measurement of the polarization properties of CMB radiation. This calls for significant improvements in polarimetric antenna architectures and receivers, due to the extremely low level of the polarized signals.

Radio Astronomy Missions Based on Space Very Long Baseline Interferometry (VLBI) The first VLBI observation with a space-based antenna was in 1986, using a TDRS 4.2 m antenna, locked in S-band to a maser on the ground and observing in combination with Earth-based radio telescopes. It demonstrated the feasibility with a baseline commensurate with the distance of the geostationary orbit projected on the observation plane.

Radio astronomy exploits VLBI reception, in which case antennas form pairs and observe the same source distribution coherently. Provided a sufficient number of such pairs of baselines is available, this permits images or contour plots of the target to be constructed with high resolution as if an antenna of a diameter equivalent to the baseline had been observing. Model-assisted image reconstruction is important here as the number of baselines is in general not fully complete.

Current Earth-based VLBI networks are limited by the finite size of the Earth's diameter and thus the maximum baselines to be achieved, with consequently a maximum resolution at a given frequency to be obtained. Both low-frequency configurations such as LOFAR and high-frequency configurations such as ESPRIT are studied.

The obvious choice is to realize a satellite-based radio telescope, which co-observes with the terrestrial VLBI network, thus providing such a high-resolution capability. Initial studies were undertaken for the space VLBI QUASAT programme in the 1980s by ESA.

It is expected that antennas in space, creating interferometers interacting with Earth-based stations, will have diameters above 8–10 m ($\sim$1/10 of the maximum diameter of an Earth-based telescope). Previous developments have shown the implementation of frequencies up to Ka-band. In the future, sizes of 10–25 m are of interest for the space-based radio telescope for VLBI elements as well as for stand-alone functionality. Resolution and sensitivity are of prime interest to the radio astronomy community.

Planet and Comet Observation Submillimetre-wave sensors are required for planetary and small-body (asteroids, moons and comets) observations. The aim is to understand the atmospheric dynamics of these bodies and their composition with a view to a better understanding of our own atmosphere and a greater insight into the origins and evolution of our Solar System. To date, there has been little heritage in planet and comet observation instruments and space-based observations utilizing (sub)millimetre-wave frequencies, even if one takes into account ESA's Rosetta mission with its MIRO instrument. However, planetary and cometary targets have been identified for which this type of instrument offers the opportunity for significant scientific progress in the medium term. Surface-based (landers) or orbital remote sensing observations of gaseous species in the Venusian, Martian, and Jovian atmospheres, as well as around Europa and Titan, have all been proposed. The instruments should be able to provide information on composition, temperature, pressure and gas velocity (winds) that will provide a wealth of information unaffected by dust and independent of sunlight conditions.

Deep Space Communications Antennas play a key role in orbiting satellite stations for deep space communications. Communications requirements derived from future human and robotic planetary exploration missions ask for extremely wide bandwidth links. Data rates of 1 Gbps have been proposed for the Mars orbiting data relay satellite. When added to the very long distances for missions to outer planets, this leads to the implementation of relatively large space reflectors.

Space exploration is gaining increased attention from the principal space agencies and the communication aspects play a central role in the future success of these missions. In this respect, the large antennas of the Deep Space Network (DSN) Earth stations have seen a significant evolution in performance and size. Nevertheless, the maturity of the developments and implementations leaves little room for performance improvement, especially at the lower frequencies (i.e. S-band).

An Earth orbiting satellite or a more futuristic space vehicle located at the Lagrangian point of the Earth–Sun system could operate as a relay station for the DSN.

Up to now, deep space communications have been implemented using direct links between high-gain TT&C antennas and the Earth station. In some cases, the antenna aperture is shared between the data transmission and the payload itself.

18.10.3 Promising Antenna Concepts and Technologies

The identified technological needs of future missions and the foreseen missions themselves are often extrapolations of the existing situation, and thus represent mainly incremental improvements. It is also important to build on the emergence of new breakthrough technologies, which may allow completely new solutions.

Some candidates for system-level solutions, technologies and novel devices are listed below. They have been classified according to frequency range.

Very large radiating structures for P- and L-band radar: These applications require physically large antenna apertures with dimensions ranging from 9 to 20 m. For these apertures, accurate parabolic reflectors or planar arrays are difficult to implement and accommodation on small-size platforms is critical. This requires a development effort. Also, new architectures and technologies should be proposed for alternative ways to generate large apertures, such as reflectarrays, deployable 1D or 2D membranes, and lenses, as they may allow drastic cost reduction. They could be realized using non-continuous apertures like sparse arrays possibly coupled with DBF techniques.

Reflector-based SARs: These were introduced during the last decade and have demonstrated the feasibility of high-resolution multi-swath SAR-implementing spotlight and scansar modes. Low revisit time is achieved through beam steering combined with spacecraft agility. Focal array architectures and technologies are challenging with respect to power handling. A low-cost mission is achievable with such an antenna concept in combination with launchers such as Dnjepr or Rockot.

To synthesize the different overlapping shaped or pencil beams, high-power, low-loss feed arrays are required. Great care should be given to multipaction issues. Activities are required at feed array level, where switched solutions or orthogonal feed network solutions might be candidates and simplicity considered as a driving factor. SAR instruments in inclined geosynchronous orbits at X- or C-band with very large antennas could also be implemented for continuous observation.

Low-resource (mass, volume and power) instruments: These require development for interplanetary missions. Active and passive remote sensing of planetary surfaces, including ground-penetrating radars, require new technologies. Besides low resources, multi-frequency radars and sounders are considered for Earth observation and space exploration.

Active SAR antenna cost reduction: This is mandatory for SARs using active antennas and can be done as follows:

- By increasing the effort on the hardware acting on the main cost drivers of the antenna, namely Tx/Rx modules and integration and testing. Technological steps are needed at Tx/Rx module level to increase RF/DC/TT&C integration and packaging and to improve MMIC power amplifier efficiency.
- By implementing new command and control approaches using all the potential provided by active antennas. Many hardware imperfections may be compensated, leading to cost savings in design and

manufacture. This requires careful evaluation of the amplitude and phase control philosophy, the calibration scheme and the temperature compensation scenario.

Special attention should also be given to advanced calibration concepts to obtain significant reductions in cost and time compared with currently used techniques.

SAR antenna alternative architectures: Conventional SAR antennas handle transmit and receive signals through the same aperture and only some degrees of freedom in the array illumination allow the transmit and receive patterns and antenna architectures to be optimized separately. This major limitation can be overcome by adopting new concepts such as:

- Bistatic radar based on a separated passive low-gain, high-power transmit antenna associated with an active high-gain receive scanning antenna using a DBF network.
- Other interferometric configurations, possibly based on more platforms. ERS-1 and ERS-2 with their tandem operation in 1996 provide a first example.
- The 'PARIS' concept, which is basically a 'bistatic' altimeter using GNSSs as a signal source.

These concepts led to a major rethink of antenna architectures with an expected decrease in complexity and cost.

Next-generation radar instruments: These require array antennas with wider bandwidth (reconnaissance), higher amplitude/phase stability and more sensitivity (science). The use of Ka-band is considered and allows interferometry with only one satellite. Recently, the CryoSat mission was launched to measure the variation in thickness of floating sea ice and included an SAR/interferometric radar altimeter called SIRAL. Its high spatial resolution is achieved by exploiting the SAR principle in along track and interferometry in across track. The exploitation of the interferometric SIRAL measurement data requires permanent knowledge of the actual interferometric baseline attitude to an accuracy of a few tens of arc seconds in control. To enable new instruments, development of very large and highly stable structures is required. For example, at Ka-band 10 m booms are considered for interferometry.

SAR antenna integrated tiles with DBF: The main idea is to consider the array tile as equipment made of components rather than a subsystem. This integration allows cost savings due to a decrease in:

- manufacturing costs – no individual packaging of function to be provided;
- number of parts or components – harness, RF dividers;
- RF, mechanical/thermal test sequences – only go/no-go tests performed before tile integration;
- integration costs – reduced number of parts to be assembled.

On the other hand, based on large-scale production, this approach may be riskier and requires well-mastered technological processes for the components and an interconnection technology that allows dismounting if necessary. MEMS switches are another field of technological development which is promising in terms of losses and ability to generate true time delay lines for multiple frequency or large bandwidths. Nevertheless, as of today reliable sources of MEMS for space are missing. This new generation of SAR active antennas is expected to use Tx/Rx modules with advanced technologies (GaN, SiGe, advanced packaging and interconnect concepts, etc.) and possibly including frequency conversion or digitization.

Antenna concepts for interferometric/synthetic instruments: These will allow production of high-resolution instruments with limited overall volume and mass. For Earth observation, the first mission using synthetic aperture radiometers is now operational (SMOS). Extension of this technique to millimetre waves would allow imaging and atmospheric sounding with sufficient resolution even from a geostationary orbit, complementing optical/IR meteorological instruments. Geostationary microwave remote sensing, first with atmospheric sounding from 50 to 800 GHz, is expected to develop further.

Polarimetric concepts for submillimetre-wave instruments: The polarization properties of electromagnetic radiation can reveal important information about the emitting sources. For example, the polarized scattering by water ice particles in cirrus clouds allows the size of these particles to be determined and therefore how they affect the Earth's radiation budget. Astrophysics space missions such as B-Pol would require such concepts as well. They are needed to assess the scientific impact in detail, to define the instrument and to identify the critical technologies.

Imaging arrays: These are of great interest for space astronomy and for atmospheric research. In astronomy, most of the spectral line emitting regions are usually spatially extended over many observing beams in the sky and mapping is required to understand the regions under study astrophysically. This poses severe feasibility issues due to the large size of the focal array and the very high integration of the detectors and readout systems. In terms of performance, keeping a good-quality beam over a wide field of view and possibly over multiple channels is very challenging. The requirements for terahertz imagers are fortuitously aligned with imager development in the security field, potentially creating a valuable area of 'dual-use' technology. Prototypes of emergent airport security scanners show strong technical and market potential.

Atmospheric chemistry millimetre-wave limb sounders: Such spaceborne instruments can provide data products for climatology and for operational meteorology for Earth and planetary missions. Multiple beams allow push-broom measurements, for example in limb sounding, and make use of several spectroscopic regions, from 290 GHz up to 3.5 THz. They provide measurements of a wide variety of atmospheric chemicals, from greenhouse gases (mainly at millimetre-wave range) and ozone depleting agents (mainly at submillimetre-wave range) to hydroxyl free radicals (terahertz range).

High-accuracy/stability reflectors and structures: Antennas for future missions will require a high resolution, pointing accuracy and wide scanning range. Moreover, interferometers require very high accuracy for antenna pointing and differential insertion phase stability. The mechanical stability of reflecting surfaces is a major limiting factor in large-antenna performance at very high frequencies. There is a need for a large, high-precision deployable reflector for a Ka- and W-band Earth remote sensing antenna to support the high-frequency Doppler radar antenna envisioned for the Earth Science Decadal Survey's Aerosol/Cloud/Ecosystem (ACE) mission. A concurrent electrical and mechanical design methodology can greatly improve the achievable performances of such antennas by allowing close matching of the mechanical design to the electrical requirements.

Improved multi-frequency quasi-optical feed boxes: Currently, the operational bands and number of channels for Earth observation missions are moving up in frequency. The quasi-optical boxes required must be compatible with demultiplexing, possess scanning/beam rerouting capabilities, and deal with calibration and LO issues while maintaining very high sensitivity. The requirements for the instruments go well beyond those of the related (sub)millimetre-wave instruments existing today. This requires studies of more thermally stable configurations, to increase the selectivity and bandwidth of frequency-selective surfaces and develop new low-loss components.

In synthesis, as seen by this list of instruments proposed for the future, improved monitoring and management of Earth's resources and environment require the deployment of high temporal, spatial and spectral resolution instruments on board spacecraft.

In viewing the projected demands between 2010 and 2020, higher resolution will drive the instrument data rates from 10 Gbps in 2010 to 100 Gbps in 2020. Hence, not only will improved instruments be needed, but also a vast improvement in space-to-Earth communication links will be required. Thus the future communication technology capabilities must include Ka-band and optical frequencies, in addition to the current X-band communications.

18.11 Evolve Towards Mass Production of Satellite and User Terminal Antennas

18.11.1 Problem Area and Challenges

Over the last few decades, antennas have evolved from a one-time antenna to an antenna family and more recently to large production of the same types of antennas. With the orders placed for LEO constellations having a large number of satellites, such as Globalstar, Iridium NEXT and navigation, and the development of swarms of probes for science, relatively large production volumes are becoming a reality for the space domain. In addition, multibeam antennas require close to 100 feed chain assemblies for a single satellite. New manufacturing/testing approaches are mandatory without compromising the antenna performance. As an example, despite the high volume, navigation antennas remain very demanding regarding isoflux coverage, dual frequency, large bandwidth, phase centre stability and high power handling.

Also, one can see from the past that SARs operating at C-band were mounted on board satellites which also hosted other types of instruments (e.g. ERS, EnviSat). The tendency now is for dedicated satellites to be possibly used in constellations together with multiple-beam instruments.

Regarding mobile user terminals, there is currently significant interest in providing high-data-rate communications to moving vehicles. L-band systems are being used to satisfy many current requirements but, due to high air time costs, and as the data requirements increase, Ku-band is now being considered. In order to best exploit Ku-band broadband communications in mobile environments, commercial and military system integrators are looking for cost-effective low-profile, high-gain scanning antennas. Until now, the only solution available has been the hybrid electrical/mechanical array, which remains too bulky for high-data-rate commercial communications applications. A recent survey showed that low-cost fully active arrays are not presently available. As an example, Boeing Satellite Systems has been studying and prototyping one transmit–receive array antenna at Ku-band for in-flight entertainment for about 12 years. In spite of the product finally being available, it has been decided to commercialize a mechanical steerable passive reflector antenna. It is understood that one of the main reasons for this decision was the cost reduction associated with the reflector-based antenna system even when considering the higher drag due to the antenna height.

18.11.2 Present and Expected Future Space Missions

18.11.2.1 Telecommunications

The major need comes from LEO/MEO constellations with a large number (about 50) of recurring satellites. Also, multimedia antennas with a very large number of beams (about 80) call for recurring feed chains.

18.11.2.2 Navigation

The large number of satellites forming the constellation imposes drastic changes on the development process.

18.11.3 Promising Antenna Concepts and Technologies

18.11.3.1 Space Antennas

Several small-satellite platforms have been developed recently around the world at attractive cost. Payload should be proposed to accommodate reduced stowage volume as well as cost. With the overall low volume and mass, access to low-cost launchers is possible with a strong benefit to the overall mission cost. At antenna level, one may anticipate interest in developing:

- multi-frequency shared aperture reflector and array antennas;
- low-mass and low-stowed-volume antennas;
- frequency- and polarization-selective surfaces.

When dealing with large quantities, expensive and lengthy manufacturing processes should be avoided. For example, as far as possible, spark erosion should be replaced by conventional machining. Also, rapid prototyping techniques and screwless assemblies should be considered.

Regarding integration and testing, the large quantities impose drastic changes and transfer of expertise and know-how from ground production.

Interest has been identified in small satellites, microsatellites or even probes that can for example be released in large quantities, Such satellites will be useful for future distributed-satellite space missions, where hundreds or even thousands of small satellites are released simultaneously and form a cooperative network to realize sophisticated functions and monitor the physical parameters of a planetary atmosphere.

Designs of small satellites are strictly constrained by size, mass and power. To make small satellites even smaller, smarter, faster and cheaper, much research work is going on. Recent examples at the University of Surrey include PCB-Sat, where the whole satellite is built on a single PCB, as well as Chip-Sat, where the satellite is built on a single chip [6].

The use of printed antennas also allows integration with solar panels [7]. As the high-gain satellite antenna and the solar panel are the two largest components on board a satellite, their integration into a single component will significantly reduce the size, mass and cost of the satellite while saving the limited space on board. This will also facilitate spacecraft manoeuvring and attitude control, and increase the field of view of scientific instruments when the Earth and the Sun are seen in similar directions. The use of PCBs also opens the door for distributed radiators over the spacecraft body. The combination of their signals can generate several types of beams for the payload, the TT&C or the localization [8].

18.11.3.2 Mobile User Terminal Antennas

Low-cost mobile user terminals with low profiles are needed for cars, trains, boats UAVs and aircraft. The main challenge is to make available steerable pencil beam antennas operating at Ku-band and/or Ka-band with pointing and polarization tracking compatible with the high data rates required for TV reception and two-way interactive communications on the move. Until now, the only solution available has been the hybrid electrical/mechanical array, which remains too bulky for several applications.

Breakthroughs are required to reach this objective with strong efforts being made in the areas of:

- Array architectures ensuring low-profile and low-mass solutions.
- Low-cost production. Use of multi-function MMIC, possibly using Si/Ge, is of interest. Reduce as much as possible the number of active controls.
- Low cost and low volume with horizontal integration. Use of multilayer PCBs combining RF, command and power lines is anticipated.
- Tx/Rx isolation. This requires stringent filtering at element level, which is hard to reconcile with low-mass and low-cost production.
- Low loss front-end. Use of adequate antenna architecture together with high-performance LNAs regarding gain and noise figure.
- Multi-function/multi-frequency user terminals. This will require innovative concepts to maintain the performance even when Ku-band and Ka-band links with different satellites are envisaged.
- Low-multipath antennas for navigation and professional user antennas.
- Low loss and low depolarization multifrequency radomes with strong mechanical properties.
- High pointing accuracy and tracking speed relying on the robust processing of fast algorithms to allow continuous links even with dynamic environments.

Besides mobile users, fixed terminals for emerging systems require flexibility in selecting and/or combining wide-band signals at user terminal level to support multiple Ka + extended Ka, Ka BSS,

Ku- and Q/V-bands for handling both broadcast and interactive services. Frequency and polarization agility is needed to cope with beam pointing instability issues (beam handover) and to switch to the beam having the strongest characteristics. Interest has also been expressed in the optical fibre interface between the outdoor and indoor units and in the potential integration of these units. These terminals need to process two or more channels simultaneously.

At L- and S-bands, the lower requested gain makes possible the use of non-steerable beams and technical solutions are available today. Some improvements are expected with respect to lower sensitivity of performance to the near-antenna environment.

18.12 Technology Push for Enabling New Missions

18.12.1 Problem Area and Challenges

This chapter has identified several developments with well-defined missions. It remains to say that technology push should also be implemented as part of a long-term plan. As an example, the maturity and availability of technologies for (sub)millimetre-wave instruments decrease directly with the wavelength. Due to the commercial applications at lower frequencies, all the technological ingredients are generally available, and cost-effectiveness becomes a driver. In this case the space agencies' role is to adapt existing technologies for space use. This is in contrast to the higher frequencies, where many of the technological elements may be missing or immature, including semiconductor devices, mounting techniques and measurement equipment, but also evaluation/qualification procedures and techniques. It is natural to expect that in such cases the space agencies should assume a more proactive role in technological development.

An important obstacle to the emergence of commercial (sub)millimetre-wave products is the high cost of packaging and the challenge of mounting/assembling submicrometre-size devices as passive components, due to the requirements for high mechanical tolerances and low electrical losses and parasitics. Developments are required in this field. As an example, micromachining techniques have shown very promising results in the field of millimetre-wave imaging. These techniques are well suited for, and often developed in conjunction with, novel structures and artificial materials. Periodic electronic band-gap (EBG) structures and metamaterials may result in significant improvements when realizing integrated (sub)millimetre-wave antennas and receivers. These developments should be continued, together with the development of techniques to integrate semiconductor devices into such structures, for example by using membranes.

As an example, the Herschel mission has been an important driver for the development of submillimetre technologies for over 15 years, both in Europe and in the USA. Significant advances in superconducting device technologies have allowed the development of sensitive heterodyne receivers for frequencies up to 1.9 THz. The initially planned highest channel of 2.4–2.7 THz had to be abandoned, mainly due to the lack of sufficient LO power.

18.12.2 Promising Antenna Concepts and Technologies

Some examples of technology push actions are listed below:

- Large arrays of superconducting bolometers (100×100 pixels) are emerging. A more system-level approach also needs to be taken for integration of large arrays of detectors and readout systems within cryogenically cooled instruments. Readout circuits suited for hybridization with planar monolithic arrays need to be developed.
- Earth observation and science instruments operating at terahertz frequencies are hardly feasible today due to the lack of sufficient LO power. In order to meet future requirements, an increase in LO power by a few orders of magnitude at terahertz frequencies is desirable.

- Novel mixer architectures can satisfy the demands of upcoming missions (e.g. image band rejection mixers at the high submillimetre-wave frequencies).
- Novel materials (EBG and metamaterials, carbon nanofibres and carbon nanotubes) are emerging, and initial studies have shown very promising results. EBG and metamaterial technology offers promising alternatives to overcome the limitations of the current technology (e.g. mutual coupling in arrays, gain enhancement of single elements, etc.), and it is envisaged that many new structures will evolve.
- Fibre reinforced polymer materials and especially CFRP are widely used nowadays in space applications. The incorporation of nanotechnology in the field of composites has opened up new horizons for the development of advanced materials with unique functional properties.
- Microsystem technologies and specifically RF MEMS will allow the development of innovative functional concepts to demonstrate adjustability and tunability. This may represent a fundamental change in the way systems are being designed today when the design itself (and the technology) will allow compensation for design and fabrication process inaccuracies by adding MEMS devices to optimize the structure geometry. To a large extent, using MEMS to enable tunable and therefore multi-functional structures will bring strong benefits to the antenna designer. However, the reliability issue has still to be solved.
- LNAs are currently being developed for frequencies up to 200 GHz, and longer term extension to the submillimetre-wave range seems feasible. This will allow direct-detection receivers, which means significant simplification of the receiver architecture (because no LO is needed) and better sensitivity. Integrated technologies also enable configurations that are not feasible with conventional technology using discrete devices in a waveguide.
- Novel low-loss materials for active devices. Indium antimonide (InSb) transistors for very high frequencies are an option because InSb has the highest electron mobility and saturation velocity of any conventional semiconductor.
- Fast digital signal processing allows completely new architectures for satellite payloads. Advanced memory and data handling architectures are needed at millimetre-wave clock frequencies. Direct interfacing to digital circuits at the (sub)millimetre front end should be considered.
- Novel micromachining manufacturing techniques may play a significant role due to their excellent electrical performance, low weight and repeatable fabrication routes with suitable structural properties. With the extremely compact achievable size, new applications requiring large numbers of RF front ends (like imaging arrays) become feasible. Application to FSSs and millimetre- and submillimetre-wave feed horns with micrometre accuracy is expected.
- Spatial combining of sources, also in the case of receive mixer diodes, can result in better power-level control and, as a consequence, an improved instrument sensitivity considering that, for (sub)millimetre-wave frequencies, the achievable performance is very often constrained by the available LO power.
- Additive manufacturing technology should be developed for applicability to both mechanical parts and RF components. Such technology can enable very complex shapes that are not possible with conventional milling or spark erosion processes and an expected 30% mass reduction. Processes to improve the surface roughness should be implemented in the manufacture of RF components. Such techniques are expected to allow BFNs to be built in one piece with a major reduction in the number of flanges.

18.13 Develop New Approaches for Satellite/Antenna Modelling and Testing

18.13.1 Problem Area and Challenges

Since the 1990s much R&D has led to the availability of antenna modelling tools. This coincided with the increases in computing power. Space agencies supported and are still funding antenna modelling tools to master all the development phases from initial tradeoffs to detailed design and AIT. The theory and algorithms developed during these years have been widely implemented in commercial tools. Today, several well-validated tools are available, allowing the design, analysis and optimization for both conductive and radiative components of antennas. In most cases, the computational technique can be selected depending on the component to be analysed and the requested accuracy at a specific development stage, or the need for optimization. Regarding antenna testing, the evolution from single-beam to multiple-beam antennas and the multi-frequency/wide-band operation call for significant reductions in the testing time. Moreover, the increase in antenna size requires a very large quiet zone and complex zero-*g* devices for large reflectors have pushed the transition from far-field testing to near-field testing. These aspects will be further elaborated in this section.

18.13.2 Promising Antenna Concepts and Technologies

Even though commercial tools are in place and widely used, accurate modelling techniques and tools still need to be improved for:

- Antenna farm performance prediction accounting for interactions between antennas and the spacecraft (radiation patterns, antenna coupling and PIM). Fast and reliable multilevel/hybrid methods are needed for the optimization process and antenna interaction modelling.
- Innovative antenna concepts when the analyses require special features not yet covered by existing tools or when existing tools are not fast enough to perform optimization tasks, for example when the array periodicity is broken or for finite arrays in three dimensions.
- Multiple reflector systems at high frequencies (up to terahertz range) especially for large reflectors (in terms of wavelength).
- Refinement of reflector modelling (non-ideal reflecting surface, near-field effects, etc.).
- Array modelling including coupling between radiators and processes to explore new architectures minimizing sensitivity to distortion, component drift and failures. In particular all existing software are rapidly limited when dealing with 100 or 1000s of elements.
- Integrated space–time antenna simulation and optimization accounting effects of active elements and array space-frequency-dependent response in the end-to-end performances with real signals (active element mismatches/nonlinearities, radiating elements, RF modules and BFN space-frequency-dependent responses).
- End-to-end system simulation better to support (concurrent) satellite engineering throughout the project life cycle.
- Optimization of overall system performance by antenna/system co-design, for example implementing adaptive coding modulation.
- Synthetic parametric antenna models for payload simulations.
- Modelling of thruster plume effects on antenna radiation patterns to account for impact on performance.
- Small antennas in a complex S/C environment.

Moreover, there is the need to develop algorithms fast enough to allow numerical optimization of the design, within affordable time, while achieving the high accuracy necessary to avoid costly breadboarding cycles. To make generic components available, such tools should provide both improved modelling and fast computing to optimize the component design and make it more resistant to manufacturing errors. Tool accuracy should be demonstrated to minimize the verification phases.

Tool interoperability also has to be promoted. An example of such an effort is captured by the Electromagnetic Data Exchange language that has been developed as an easy way to exchange data among modelling tools. More information can be found in [9].

Regarding antenna testing, several key challenges are identified:

- Antennas of very large size (up to 25 m and beyond).
- High-power testing.
- Multiple-beam testing.
- Multiple functions and multi-frequency testing.
- Testing antennas utilizing signal processing.
- Characterization of the RF properties of the materials and processes involved in antenna design.
- Radiated PIM measurements in the near field.
- Multipaction testing in multicarrier operation where the input power and phase of each carrier is accurately monitored.
- Fast detection of electrons during multipaction test.
- Accurate characterization of the radiated phase for interferometric instruments and navigation antennas.
- New VAST (VAlidation STandard) antenna to assess the accuracy of antenna ranges for the emerging frequencies, such as 20/30/38/48 GHz.
- Innovative approaches for efficient antenna/payload/AIT-AIV RF characterization.
- Low-frequency, very large-antenna performance verification methodologies and facilities.
- (Sub)millimetre-wave antenna measurements using phase-less near-field measurement techniques.
- Combined prediction–measurement methodology for antenna validation.

In all cases, the cost of testing and the reduction in time should be improved for RF radiation performance testing. Also, thermal distortion measurements in a thermal vacuum chamber will often be required with the operational frequency increases and cost-effective solutions should be developed.

Regarding in-orbit testing, new approaches considering the potential offered by the multiple beams generated by the antennas should be studied.

18.14 Conclusions

A review of current and future space antenna needs and technologies has been presented through an overview of recent developments and future programmes. As anticipated in the introduction, different missions led to very different constraints and requirements at antenna level.

This chapter has implemented:

- A top-down approach with the identification of all space missions/markets benefiting from the development of antennas.
- A bottom-up approach with a critical review of the state of the art and an investigation of enabling technologies requesting further developments.

For all the identified needs, the chapter concluded with recommendations for making available the needed technologies, considering the synergy between applications for sharing the development costs and expanding the domain of accessible markets.

In trying to find synergies among the development needs and technologies, one may identify some common challenges to be addressed:

- Very large space aperture: It is mandatory to get the required resolution for low-frequency remote sensing and to allow direct access to mobile end users for telecommunications.

- Signal processing antennas: These offer new ways of thinking. The first steps with ongoing implementation for mobile communication and Earth observation instruments will pave the way for space use of this powerful and flexible technique. Technology improvements will then go faster than capacity demands in terms of bandwidth.
- In-flight reconfigurability: The time to market and ability to cope with user demands require power and coverage changes during the satellite's lifetime. Besides fully reconfigurable active DRAs, medium-term solutions with less flexibility have to be investigated.
- Reduced spacecraft size: Earth observation planners are increasingly considering small satellites, possibly in constellations, implementing only one main mission, with a reduced accessibility domain. This allows more flexibility in mission design. This approach may be considered for other applications including formation flying.
- Low-cost solutions: Antenna architectures and techniques leading to reduced costs, such as reflector- or reflectarray-based SAR, need careful investigation for all missions.
- Multi-function antennas: These antennas can either allow several missions to be accommodated on a spacecraft with mass and volume compatible with low-cost launchers, or make accommodation easier on large platforms.
- Antenna phase stability: Interferometric concepts as well as navigation antennas require very stable phase characteristics. This calls for more thermally stable structures and reflectors and new modelling techniques. This also requests new measurement techniques for accurate on-ground characterization and innovative strategies for in-orbit calibration.
- Technology push: Early developments in promising domains such as MEMS, micromachining techniques and submillimetre-wave integrated antennas have to be promoted to enable new missions.
- Fast and reliable multilevel/hybrid methods: For the optimization process and antenna interaction modelling, the set of existing tools needs to be refined and interoperability has to be promoted.

As in the past, antennas will remain a strategic technology for telecommunication, navigation and remote sensing space missions as well as for the user segment.

Increase in size, increase in frequency, increase in number of beams and reconfigurability, aperture synthesis, multi-functionality, integration of solid state devices and circuits, as well as digital beam forming, fast design tools, including one-pass design for many components, fast and accurate measurement techniques, all seem to be key trends for antenna technology in the next 20 years.

Acronyms

AFR: Array-Fed Reflector
AIT: Assembly, Integration and Test
AIT-AIV: Assembly, Integration and Test–Assembly, Integration and Validation
AMSU-A/B: Advanced Microwave Sounding Unit version A/B
BFN: Beam-Forming Network
CFRP: Carbon Fibre Reinforced Plastic
CFRS: Carbon Fibre Reinforced Silicon
CMB: Cosmic Microwave Background
CTE: Coefficient of Thermal Expansion
DARS: Digital Audio Radio Service
DBF: Digital Beam Forming

DGR: Dual-Gridded Reflector
DMB: Digital Multimedia Broadcasting
DOA: Direction Of Arrival
DOS: Denial Of Service
DRA: Direct Radiating Array
DTH: Direct To Home
DVB: Digital Video Broadcasting
EBG: Electronic Band Gap
EIRP: Effective Radiated Isotropic Power
EOS: Earth Observing System
FSS: Frequency-Selective Surface
GEO: Geosynchronous Orbit
GNSS: Global Navigation Satellite System
HPA: High-Power Amplifier
I/F: Interface
INET: Input NETwork
InSAR: Interferometric Synthetic Aperture Radar
ITU: International Telecommunication Union
LDA: Large Deployable Antenna
LEO: Low Earth Orbit
LNA: Low-Noise Amplifier
LO: Local Oscillator
MEMS: Micro-Electromechanical Systems
MEO: Medium Earth Orbit
MHS: Microwave Humidity Sounder
MLS: Microwave Limb Sounder
MMIC: Monolithic Microwave Integrated Circuit
MPA: Multi-Port Amplifier
MSG: Meteosat Second Generation
MSS: Mobile Satellite System
MSU: Microwave Sounding Unit
MTG: Meteosat Third Generation
MTI: Multiple Target Indicator
OBP: On-Board digital Processor
OMT: Orthomode Transducer
ONET: Output NETwork
PAE: Power-Added Efficiency
PCB: Printed Circuit Board
PIM: Passive Intermodulation
QoS: Quality of Service
RF: Radio Frequency
SAR: Synthetic Aperture Radar
S/C: Spacecraft
SSPA: Solid State Power Amplifier
TBC: To Be Confirmed
TT&C: Telemetry, Tracking and Command
TWTA: Travelling Wave Tube Amplifier

Tx/Rx: Transmit/Receive
UHF: Ultra-High Frequency
UMTS: Universal Mobile Telecommunication System
VLBI: Very Long Baseline Interferometry
VSAT: Very Small-Aperture Terminal
XPD: Cross-Polar Discrimination

Acknowledgements

The authors wish to thank several European Space Agency colleagues for their support with suggestions, material and permissions for this chapter, in particular Antoine Roederer, Jean-Christophe Angevain, Julian Santiago Prowald, Peter de Maagt, Kees van't Klooster, Giovanni Toso, Piero Angeletti, Jean-Christophe Angevain and David Huet.

References

1. Takano, T., Natori, M., Miyoshi, K. and Noguchi, T. (1998) A large deployable antenna with tension truss scheme and its electrical performances. IEEE Antennas and Propagation Society International Symposium, Atlanta, GA, USA, vol. 4, pp. 2086–2089.
2. Datashvili, L., Baier, H., Schimitschek, J. *et al.* (2007) High precision large deployable space reflector based on pillow-effect-free technology. 48th AIAA/ASME/ASCE/AHS/ASC Structures, Structural Dynamics, and Materials Conference, Honolulu, HI, USA.
3. Toso, G., Mangenot, C. and Roederer, A.G. (2007) Sparse and thinned arrays for multiple beam satellite applications. 29th ESA Antenna Workshop, Noordwijk, The Netherlands.
4. Toso, G., Angeletti, P. and Mangenot, C., (2008) Direct radiating array architecture based on non-regular lattice. ESA Workshop on Advanced Flexible Telecom Payloads, Noordwijk, The Netherlands.
5. Skobelev, S.P., Eom, S.Y. and Park, H.K. (2003) Shaping of flat-topped element patterns in a planar array of circular waveguides using a multilayered disk structure—Part I: Theory and numerical modelling. *IEEE Transactions on Antennas and Propagation*, **51**, 1040–1047.
6. Barnhart, D., Vladimirova, T. and Sweeting, M.N. (2007) Very small satellite design for distributed space missions. *Journal of Spacecraft and Rockets*, **44**(6), 1294–1299.
7. Zawadzki, M. and Huang, J. (2000) Integrated RF antenna and solar array for spacecraft application. Proceedings of the IEEE Conference on Phased Array Systems and Technology, Dana Point, CA, USA, pp. 239–242.
8. Mattioni, L., Bandinelli, M., Milani, F. *et al.* (2010) Distributed multi-function antennas for micro- and nano-satellites. 32nd ESA Space Antenna Workshop, ESTEC, Noordwijk, The Netherlands.
9. Vandenbosch, G.A.E., Gillard, R. and Sabbadini, M. (2009) The Antenna Software Initiative (ASI): ACE results and EuRAAP continuation. *IEEE Antennas and Propagation Magazine*, **51**(3), 85–92, and http://www.antennasvce.org/Public/EDX.

Index

Space Antenna Handbook, First Edition. Edited by William A. Imbriale, Steven (Shichang) Gao and Luigi Boccia.